LAW AND ECONOMIC GROWTH

THE LEGAL HISTORY OF THE
LUMBER INDUSTRY IN WISCONSIN
1836–1915

LAW AND ECONOMIC GROWTH

THE LEGAL HISTORY OF THE LUMBER INDUSTRY IN WISCONSIN 1836–1915

James Willard Hurst

THE UNIVERSITY OF WISCONSIN PRESS

Published 1984

The University of Wisconsin Press
114 North Murray Street
Madison, Wisconsin 53715

The University of Wisconsin Press, Ltd.
1 Gower Street
London WC1E 6HA, England

Originally published in 1964 by The Belknap Press
of Harvard University Press

Printed in the United States of America

ISBN 0-299-09780-3

LC 83-040287

Library of Congress Cataloging in Publication Data

Hurst, James Willard, 1910–
Law and economic growth.

Originally published: Cambridge, Mass. : Belknap
Press of Harvard University Press, 1964.
Bibliography: pp. 613–625.
Includes index.
1. Lumber trade—Law and legislation—Wisconsin—
History. 2. Lumber trade—Wisconsin—History. I Title.
KFW2649.H8 1984 343.775'078674'09 83-40287
ISBN 0-299-09780-3 347.75037867409

FOR FRANCES

AUTHOR'S NOTE

WHEN *Law and Economic Growth* was first published, in 1964, reviewers dealt generously with it; I appreciate their good words. But they also paid the study the compliment of taking it seriously enough to raise some probing questions about its approach, its theoretical framework, and its conclusions. Recognizing the compliment, I am not ungrateful when I take the opportunity this prefatory note provides to comment on some of their reactions. Appraisal of the differences between us may cast into bolder relief some salient features of the book's subject matter.

In the perspective of some twenty years since I finished writing the book, on the whole I feel warranted in standing my ground on most of the points of principal criticism. I trust that this response has some better foundation than vanity; readers will strike their own judgments. On one point at least the record is plain. Since the book was published there has been no material addition to the facts it presents on the legal history of Wisconsin's lumber industry. Nor did reviewers challenge those facts. What are at issue are questions of marking the limits of the lumber industry story and interpreting its meaning.

Reviewers raised three main types of question. Some queried the breadth of the sources and methods of the study. Some challenged the sufficiency of the theoretical framework. The deepest thrust of criticism went to the book's diagnosis of the dynamics of making public policy, in terms chiefly of the relative influence of inertia, rational choice, and conflict of interests.

SOURCES AND METHODS OF THE STUDY

Not surprisingly, what most often bothered readers was the bulk of the book. Criticisms on this score mingled issues of style and method. In style, some critics found that I offered more on certain matters than the reader wanted to know, and that I was guilty of repetition.[1] One reviewer thought that detail drowned

1. Russell E. Brooks, "The Jurisprudence of Willard Hurst," *Journal of Legal Education* 18:257 (1966); Thomas C. Cochran, Book Review, *The Annals of the American Academy,* 359:230 (1965); David H. Flaherty, "An Approach to American History: Willard Hurst as Legal Historian," *American Journal of Legal History,* 14:222 (1970); John P. Frank, Book Review, *Journal of Legal Education,* 18:395 (1966); James A. Lake, Sr., Book Review, *Maine Law Review,* 17:298 (1965); Harold L. Marquis, Book Review, *American Journal of Legal History,* 9:250 (1965); Earl F. Murphy, Book Review, *New York University Law Review,* 39:900 (1964); John P.

the color he felt must be inherent in the lumber industry era.[2] To a large extent these reactions involve judgments of taste on which it is pointless to quarrel; the book's detail will undoubtedly put off some who pick it up. There are objective justifications for the detailed approach. For one thing, aided by a full analytical table of contents and index, readers can use the book selectively for topics that meet their special interests; in effect the book embodies material which could stand as a number of separate monographs—on public lands policy, on contract law and the market, on legal regulation for non-market goals, on economic and social planning through law. The footnotes are elaborate, but the principal text is written to stand alone. The subject matter inherently involves cumulation of relatively mundane transactions and decisions, the significance of which appears only in light of the massing of detail.[3]

Criticism of the book's density went beyond judgments of style to questions of method. The scale of the inquiry might make this a kind of dead-end effort; for want of resources there could not be many studies of such extent, and if progress in legal history depended on such expenditure, it was said, the future of the field might seem doubtful.[4] This is a sobering matter. However, work in legal history and in jurisprudence has suffered from abstraction and from dependence on selected data without adequate testing for representative quality. There will be few studies involving the detail of this one. But learning more about legal order calls for some effort of this scale, to provide a surer sense of where and how far we can safely take short cuts, especially once inquiry pushes beyond analyzing the doctrine of appellate court opinions. There is desirable assurance in generalizing about the course of contract law bearing on the lumber industry when analysis works from the total body of 571 contract cases before the Wisconsin Supreme Court, and in identifying the hierarchy of values reflected in stream improvement franchises on the basis of examining all of the 802 special statutory charters concerning use of Wisconsin inland waters. That almost all of these legal transactions involved narrowly focused relationships and run-of-the-mill dealings enhances the need to look at the whole flow; a scholar can single out a handful of "great" constitutional decisions, but he needs material in bulk to appraise the meaning of what people perceive as ordinary. And where—as seems true of the lumber industry story—themes of broad reach lie buried in the humdrum, the only way to draw out and validate those themes is by examining particulars.

Some reviewers questioned the effort to weave together details of history, values probed by jurisprudence, and theory of social structure and process. The mix

---

Reid, "Legal History," in 1964 *Annual Survey of American Law* (New York University School of Law), 702; Mark Tushnet, "Lumber and the Legal Process," *Wisconsin Law Review,* 1972: 114.

2. Marquis, note 1 above, 256.

3. Some critics of the book's detail conceded force to this last aspect; see Brooks, note 1 above, 259; Cochran, note 1 above, 231.

4. Frank, note 1 above, 409.

might only confuse one trying to hold onto the thread of discussion.[5] So much detail might be necessary for grasping the richness of history, but a philosopher might doubt whether it was needed to confirm rather large general hypotheses.[6] These are not imagined difficulties. But complexity inheres in the social experience the book studies. Unless the observer brings to the materials some organizing ideas about ends and means and about the ways of social accommodation and conflict one winds up with only a welter of particulars. Moreover, I submit that the analysis relates its particulars to generalizations which are not so indeterminate as to produce little yield for the investment. The mingling of concrete detail and general ideas comes to bear on definable, central issues over uses of law in achieving varying degrees of social order. Especially, this Wisconsin experience provides material for measuring functional strengths and limitations of law which appear in relation to the massive weight of social inertia. These are key elements in tackling the core problem which Roscoe Pound posed to jurisprudence and to sociology, that of understanding the limits of effective legal action.

The original preface took account of limitations of a study directed at one state and one industry in that state, and argued that the approach yielded enough gains to warrant the costs of an enterprise so defined.[7] Reviewers did not take basic issue with these choices, but suggested some additional limiting aspects of the Wisconsin lumber experience as compared with the course of affairs in the East and the South. One commentator observed that Wisconsin had not had the longer history of states which had been colonies. He did not elaborate the point, and it seems open to question on two counts.[8] First, Wisconsin had its own form of colonial experience in the territorial years from the Ordinance of 1787 up to 1848, with a legacy of crude, limited governmental capacity and scant resources. Second, in functional terms communities like mid-nineteenth-century Wisconsin felt the weight of colonial limitations in the form of scarcity of fluid capital and dependence on sources of capital in more settled areas. A more fundamental limitation of the local setting was pointed up by the observation that Wisconsin was perhaps the easiest case for studying interaction between law and society. The relative simplicity of its society allowed the historian to infer social conditions and consequences largely from legal materials, in a way not possible where there was markedly less agreement on values and on means to implement them.[9] In contrast to racial tensions of the South or gross inequalities in access to law's benefits in urban industrial areas, the point is well taken.

The book emphasizes the influence of social inertia in the course of Wisconsin policy toward forest wealth. Reviewers did not challenge the fact, but some felt

5. Tushnet, note 1 above, 115.

6. Julius Stone, Book Review, *Harvard Law Review,* 78:1687 (1965), at 1697.

7. See this volume, xxi–xxii.

8. Flaherty, note 1 above, 231.

9. Robert W. Gordon, "J. Willard Hurst and the Common Law Tradition in American Legal Historiography," *Law and Society Review,* 10:9 (1975), at 54.

that at least by implication I over-generalized this factor as a constituent of United States legal history. By one argument, inertia figured less in the settled East, as in Massachusetts under such a policy maker as Chief Justice Shaw; the diagnosis of drift might be more suited to the mid-nineteenth-century midwestern frontier than to the seaboard.[10] To a degree, the point is well taken. Undoubtedly there have been important regional differences in the country's legal history.[11] The original preface might well have emphasized this fact as it cautioned on the limits of this case history. However, the record for the country as a whole makes it doubtful that social inertia has been less influential in other times and places than it was in this Wisconsin story. Public policy in more settled areas long drifted in poor response to community costs and human suffering attendant on the growth of urban industry and crowding, particularly in matters of public health and individual safety and security. A more pointed objection to the implications of the book's emphasis on inertia was that it ignored the presence of more deliberate use of law to muster resources for public-interest programs in states like New York and Ohio which embarked on ambitious canal projects, and underestimated the amount of partisan or ideological conflict over state law on corporations and banks.[12] Insofar as this book can be read as implying denial of existence of such sectors of calculated policy, my presentation is faulty; I did not intend such an implication. What I offer always comes back to the distinctive character of the Wisconsin forest policy experience, which involves a more fragmented course of events than appears in other fields of public concern. However, in its determining orientation about operations of the private market, that Wisconsin experience seems of broad applicability in other times and places in this country, especially with regard to exploitation of natural resources.[13] Moreover, the book notes, though perhaps without enough emphasis, that Wisconsin had its own experiments in calculated waterways projects—the Fox-Wisconsin rivers improvement, and the Milwaukee & Rock River Canal Company—though these failed to match their bold purposes with effective execution.

More serious than questions raised about the inherent limits of the Wisconsin case history was objection taken to the book's prime reliance on formal legal materials. The criticism regretted the absence of such evidence as diaries, letters, or business records of entrepreneurs or laborers, which might tell more about how people reacted to such inducements or prohibitions as law set out.[14] So far as the criticism is of the scope of the research, I do not think it is well taken. I

10. Flaherty, note 1 above, 231.

11. See James Willard Hurst, "Old and New Dimensions of Research in United States Legal History," *American Journal of Legal History,* 23:1 (1979), at 7; also published as "The State of Legal History," *Reviews in American History,* 10:292 (1982), at 292.

12. Harry N. Scheiber, "At the Borderland of Law and Economic History: The Contributions of Willard Hurst," *American Historical Review,* 75:744 (1970), at 751, 752.

13. Marquis, note 1 above, 254.

14. Tushnet, note 1 above, 129.

share regret that such materials as diaries or letters are not woven more into the history, and in the original preface I take careful note that "We must be modest in the inferences we draw from these [legal] records, recognizing that the informal behavior reflected in law was only part of a great flow of action of which we have little other evidence."[15] Unfortunately, as the quoted caution observes, further informal evidence is not available within any workable limits of search. The list of Sources Cited includes six manuscript files in the Wisconsin State Historical Society containing records of private firms, as well as the revealing, if naïve, autobiography of Roujet Marshall, various doctoral and masters theses on the economic history of the lumber industry drawn largely from newspaper sources, and a number of monographs and biographies dealing with forest area experience on the basis of other than legal materials. The available lumber firm papers had little to show on relations between law and the conduct of business. In light of the limits of these sources it seems unlikely that any feasible additional search would produce more informal records of what went on in the industry. Moreover, granted the limitations of the legal sources, these carry some inherent warrant for taking them seriously, given the setting. "In the activist, opportunistic, improvising temper of nineteenth-century Wisconsin, men did not exert themselves to press and oppose, enact, amend, and repeal hundreds of statutes, and fight hundreds of lawsuits, only to celebrate a ritual."[16] Finally, apart from reasons for taking seriously the effort expended in matters brought formally to law, the book's inventory of relevant legal materials fills out an essential dimension for appraising what went on. Even if we had much more evidence than is available of private activity and private reactions, we would still need to weigh it considering the legal record.

The scope of the research aside, a question remains about how much legal records can prove. To one critic, attention to legal records "leads to a failure of evidence of what was the socially dominant consciousness, congruence with which makes for enduring policy"; we cannot automatically assume that the consciousness embodied in law was dominant in such other areas of social activity as religion, formal literature, or informal writings. "Legal history here converges with intellectual history."[17] The general caution is appropriate for legal historians. It seems in a measure irrelevant to the immediate subject, on which there is no evidence that religion, formal literature, or any like sector had direct bearing. However, the book makes plain that key values evidenced in public policy regarding forest industry were found also in general currents of value and attitude in the environing society, including the individualist spirit engrained in much of contemporary religious belief. For example, the reader may consult headings in the index under Awareness, Commonwealth, Dispersion of power, Entrepreneurial

15. This volume, 226.

16. *Id.*, 225–226.

17. Tushnet, note 1 above, 129.

function, Jacksonian values, Middle class values, and Pragmatism; indicative also is inclusion in the list of Sources Cited of works by such students as Henry Steele Commager, Merle Curti, Robert A. Dahl and Charles E. Lindblom, Oscar and Mary Flug Handlin, Leonard W. Levy, Karl Polanyi, Henry Nash Smith, and Alexis de Tocqueville.

THEORETICAL FRAMEWORK

The original preface notes that before this book I had published three others which examined particular aspects of United States legal history but also undertook to develop general hypotheses to frame further historical inquiry. Some questions raised regarding the theoretical bases of this book indicate that I might have done well to include in it a brief summary of central concepts presented in the earlier studies. Especially relevant is *Law and Social Process in United States History* (1960).[18] However, the important organizing ideas are embodied—and, I submit, adequately defined—in this volume as the reader encounters them in the particular context of the Wisconsin forest story.

Rather oddly, some readers who were troubled by the detail of the book also thought it too highly generalized, pursuing details to what they sometimes deemed commonplace conclusions.[19] The study does deal with familiar, basic elements of nineteenth-century society in the United States—wealth of fixed capital matched against scarcity of fluid capital, centrality of the private market, situational pressures favoring present over future values and placing emphasis only on immediately felt problems and solutions. That such factors were basic does not make them commonplace, especially when the literature of legal history has given them relatively scant attention. We need more, not less study of law's functional relations over time to the constant competition between rationally deliberated policy and sheer drift in affairs, to the working requisites of a private market, to the definition of commonwealth interests in tension with those of private bargainers, and to adjustment between real and monetized calculations of the stakes in dealing with limited resources. Moreover, there is need to check what everyone knows against particular experiences; sometimes what everyone knows is in fact wrong or in need of material amendment. Thus, for example, under close examination of the Wisconsin story, the prime movers in destroying a continuingly productive forest turn out to be not robber barons, but individuals of limited means and limited goals responding to the pressures and invitations of their immediate situation.

With better warrant, readers are cautioned against over-generalizing the Wisconsin experience. The book carries its own warnings on this score, qualifying the respects in which it bears on more general themes.[20] But it would have been

18. See James Willard Hurst, *Law and Social Process in United States History* (Ann Arbor: University of Michigan Law School, 1960), 137–152.

19. Flaherty, note 1 above, 230; Stone, note 6 above, 1689; Tushnet, note 1 above, 128, 129.

20. This volume, xxii, xxvi.

good to have made explicit contrasts with experience outside the Wisconsin forest setting. Thus the point is well taken, that more self-conscious uses of law appear where policy makers mustered capital for public works, as in the canal-building states, or where they confronted fears of concentrated private power in banks and other private corporations.[21]

Further questions appeared on the adequacy of the book's four organizing centers—the ideas of property, contract, police power, and political economy. For one thing, the categories were found to overlap.[22] Indeed, they do.[23] But on the whole each appears to designate a materially distinct center of policy attention—"property" as defining and legitimating access to resources, "contract" as providing machinery for helping constitute and administer private markets, "police power" as embodying concern to hold private dealing in acceptable adjustment to social context, "political economy" as introducing some degree of publicly planned and directed attention to fostering commonwealth values. Moreover, just because in reality there was a mix of division and overlap in the centers of policy, resort to the four principal categories seems serviceable in turning the complexity of experience now one way, now another, to catch its varying emphases. Might narrower categories bring the subject into better focus?[24] The four headings do concentrate attention on functionally distinct components of the story of forest production. And the question seems directly answered in the specialized subdivisions examined under the principal heads—such as the variety of goals explicit or implicit in transfer of land from public to private ownership, or the variety of legal regulations contributing to organization of the market. Do the four categories fulfill the book's ambition to help enlarge ideas of the proper subjects of legal history?[25] The text seeks to broaden perception of relations of the four legal concepts to values and influences active in the general life of the society. If it achieves some success in that endeavor, it takes legal history a good deal beyond those earlier approaches which tended to treat law as a self-contained system.

A useful, more specific question addressed the choice to launch analysis with the Property heading, with its emphasis on land. The suggestion was that it may not have been investment in timberland that gave the greatest impetus to the industry's growth, but rather other factors, notably provision of low-cost bulk transport.[26] Transport was indeed critical.[27] However, lawful access to timberland was a practical as well as legal prerequisite to private forest production; if this factor attracted less prominence in public policy activity than did grants of stream-use franchises, probably this was only because rapid transfer of public timberlands

21. Scheiber, note 12 above, 753, 755.
22. Stone, note 6 above, 1690–1691.
23. This volume, 4–5.
24. Stone, note 6 above, 1697.
25. *Id.*, 1690.
26. *Id.*
27. This volume, 144–145, 270.

to private exploitation fit so readily into familiar patterns of public lands policy in agricultural and urban development.[28] The central importance of disposition of public timberland was reinforced by the importance of relatively cheap raw material.[29] This feature was thrown into sharp relief by frauds perpetrated to obtain public timberland at low cost and in violation of acreage limits set by law.[30]

Among the organizing ideas of the book none emerges to greater importance—enforced by the sheer weight of the record—than that the course of affairs was dominated by social inertia or drift. A probing critic challenged the concept as a mere metaphor, or the product at best of literary grace, which lacked the precision desirable in a research instrument. The related notion of judging the sufficiency of "focus" in public policy was likewise challenged as unclear.[31] A particular facet of this dissatisfaction was the claim that inertia did not explain the favor for the private fee simple title manifest in policy governing disposition of public lands.[32] These are pertinent challenges. I believe the book meets them, though this is a respect in which it might have been desirable to have included a chapter dealing explicitly with the general theoretical structure of the inquiry.[33] The uncritical acceptance of transfer of public lands to private fee simple ownership followed from a complex of elements established in an agricultural setting before the years of exploiting the forest—including the farmer-centered model of the independent yeoman citizen, and acceptance of the norm of clearing timber for farm settlement and the succession of cereal crops to timber.[34] The broader pattern shown in this particular feature of policy was repeated in many other contexts—"a pattern formed not much of overt conflict, but mainly of uncritical acceptance of certain very generally held symbols of value (the fee simple title, the virtue of productivity), fashioned upon the warp and woof of circumstance (the imbalance of production factors, the pace of population growth and technical development)."[35] In repeated examples of such application of older policy to new circumstances without conscious, reasoned examination of the benefit and cost implications of the new circumstances, the book seems to me to make quite precise what the factor of social inertia meant in the uses of law.[36]

In a counterpart challenge to that posed to the inertia theme, there was objection that the book did not offer a precise theory of the political dimension of legal history—the choice of basic values.[37] The objection begs the question at

28. *Id.*, 9–12.

29. *Id.*, 108, 120, 473–474.

30. *Id.*, 16, 33, 60, 80, 97, 112.

31. Tushnet, note 1 above, 116, n. 8.

32. *Id.*, 116.

33. See *Law and Social Process*, 9, 11, 31, 42, 63–75.

34. This volume, 24–34, 49, 51, 95–96, 108, 125.

35. *Id.*, 462–463.

36. *Id.*, 14, 23, 461; see also index headings under Accounting, Market, Pragmatism, Real, compared with money, calculus, and Tempo.

37. Tushnet, note 1 above, 116.

the heart of the Wisconsin forest story, by assuming that lawmakers deliberately determined basic lines of policy. There were many specific contests and specific resolutions on particular matters, for example, by legislators on some dam franchises and some land grant subsidies, and by judges in hundreds of lawsuits growing out of individual lumber industry transactions. The book examines in detail processes of decision in these matters.[38] But the overshadowing reality was that for the most part the brute force of facts rather than calculated action determined uses of law that ended in great consequences. Of course there are ways in which legal processes can help shape basic community choices—by attracting desire, cultivating awareness, economizing effort, and legitimizing exercise of will, all to give people more leverage on their experience.[39] But the essence of the Wisconsin forest policy history is that there was little such conscious use of law to determine the large course of events. The central problem of this history is not to explain how affairs moved in the political dimension of deliberated, major value choices, but to understand why that political dimension was generally not in the picture at all.

DRIFT, CONSENSUS AND CONFLICT

Some questions about the inertia factor took another tack, asking how far we may realistically generalize the influence of this element. The point was properly made that those states which embarked on canal projects showed that drift did not inevitably mark public policy processes, even in the relatively crude setting of the mid-nineteenth-century Middle West.[40] In this light, the book might profitably have explored reasons why inertia played so much more dominant a role in the Wisconsin context than in the canal states. A critical difference may have been in the relatively tangible or abstract quality of the interests at stake in these distinct settings—a difference likely to weigh heavily in the narrowly pragmatic temper of the times. Building canals provided immediately felt and definable goals with technological bases for calculated public action; to keep the forest in being as a productive asset required recognizing and pursuing long-range objectives, inevitably uncertain in future pay-off, as well as developing and applying less obvious forms of knowledge or skill (in soil analysis, in administrative organization to prevent or fight fire, in identifying and weighing the hidden subsidy which headlong exploitation of the forest was giving to the trans-Mississippi states). That law makers would supply such factors as these for their calculations was the less likely, given a climate of opinion which favored private entrepreneurial initiatives and tended to regard sustained attention to public policy as an annoying distraction from the main concerns of the time; the business of the country was business.[41]

38. See, for example, in this volume, pp. 227, 244, 266–270 on river improvement controversies.

39. See *Law and Social Process*, 137–152, 182–215.

40. Scheiber, note 12 above, 751, 752.

41. Cochran, note 1 above, 230.

In this latter perspective, one might query whether inertia in policy was less a cause than a symptom, a symptom of the pervasive drive of private entrepreneurial ambition moved perhaps by individuals' felt need continuously to prove their worth and overcome a frustrating sense of scarcity of resources matched against great opportunities.[42] The book reflects the reality in the forest story of people's insistence on private, willed action, as giving meaning to life.[43] But as we search for the deepest level of cause, the facts of the situation argue for weighing heavily the existential limits of human energy, purpose, and imagination in the face of a hard-paced, turbulent, many-faceted rush of immediate experience.[44] One reader questioned whether the element of drift in policy is relevant to twentieth-century experience of the enlarged roles of science and science-based technology.[45] I doubt that our times have put existential limits that far behind us. Changes in science and technology seem often to have produced additional currents of unplanned, unforeseen drift in events along with gains in command. Cases in point are the health costs of rapid introduction of chemicals into industrial processes, or the impact of mass production and use of the automobile on urban growth and public safety.

One commentator felt that I associated rationally calculated public policy making with economic growth, and absence of such directed effort with inertia or drift. In fact, he suggested, in some historical situations heightened awareness of consequences might impede growth in volume of transactions or production, so that a relative lack of awareness would promote growth so conceived.[46] I do not disagree with this rendering of what happened; the comment indicates that I could have made my meaning clearer. In fact, lack of sophistication of rational policy making did help speed exploitation of the forest for quick volume and quick returns. More rational policy making would have impeded economic growth only in the narrow, market-oriented, money-measured rather than real-returns-measured sense. This perspective on the matter is another way of pointing up the absence through most of the Wisconsin forest history of the greater "political" element, the self-conscious weighing of long-term, large consequences as against short-term, immediate outcomes.[47]

The commentators who found themselves most at odds with the book were those who appraise uses of law in the United States as primarily to enlarge and maintain the power of capitalist entrepreneurs at the expense of the working class or the poor. These critics interpret my approach as based on naïvely optimistic assumptions that this society has been united by a set of values widely shared among the population, and that a properly working legal system can resolve so-

42. Marquis, note 1 above, 258.
43. This volume, 24, 49, 118–125.
44. *Id.*, 23, 117, 126, 172, 194, 206; see *Law and Social Process,* 63–75.
45. Marquis, note 1 above, 259.
46. Gordon, note 9 above, 53.
47. This volume, 84–85, 429–437, 525, 595–596.

cial conflicts and promote creation of a just society. In this view a realistic history would concentrate on analyzing social structure rather than ideas of justice. Such a history would deal centrally with the influence of corruption and conflicts of interest in decision-making processes which consistently aid one interest over another, in denial of the consensus which these interpreters believe me ideologically unwilling to abandon. To these readers my type of account points legal history in directions at least incorrect, at worst morally skewed. Future studies should rather focus on repression through law of more democratic, more radical sets of legal alternatives; not the country's economic system, but rather its value system, most needs scholarly inquiry and appraisal.[48]

Where it is relevant, one proper job for historians is to appraise how and where a society has served or disserved humane values. But it is also the historian's job to deal with such judgments in terms of particular settings of time and place; it is not his business to subordinate the record to pursuit of philosophical absolutes. This study seeks to be a work of history and not of philosophy, though I hope that philosophers may find in history some things useful to their thinking. But if they interpret the book as a brief for an abstract theory of social consensus, they miss the point of the enterprise.

I do not "assume" a social consensus as underpinning the Wisconsin forest story. I find some substantial shared values or attitudes—though these were often as much implicit as explicit—evidenced by the facts of the positive uses people made of law relative to exploiting forest wealth. Nor do I assume that the uses made of law in this context promoted justice; indeed the record shows that the policies pursued in the later nineteenth century were unfair to later generations in Wisconsin and outside it.[49] The record also shows that people used law more or less effectively to adjust a good many conflicts among interests—between competing uses of streams, between contracting parties, between creditors and debtors, and over taxes and land-grant subsidies.[50] What some readers find wanting is an account of social class conflict, and particularly of uses of law for the gain of capitalist entrepreneurs at the expense of laborers and the poor. But through most of the Wisconsin forest story this type of clash of interests found only marginal legal expression, for all its reality in other aspects of life.[51] The bulk of conflicts here brought to law were within a class and not between classes. Almost all contests for and against legislation or between litigants were fought by one set of capitalist-minded contenders against another; the main differences were between somewhat bigger and somewhat smaller capitalists, but they all held much of a common consciousness. That fact of the record reflects the actuality of a great range of shared values and attitudes, or of unquestioning acquiescence by

48. Wythe Holt, "Now and Then: The Uncertain State of Nineteenth-Century American History," *Indiana Law Review,* 7:615 (1974), at 622–625; Tushnet, note 1 above, 115.

49. This volume, 118–125, 461–465.

50. *Id.,* 220–225, 278–281, 333–342, 406–407, 503–518.

51. *Id.,* 481, 486–500.

less active sectors of the society in what active interests obtained from law, or of limits of people's perceptions of what was at stake. These factors take us a long way toward understanding the legal history of Wisconsin's lumber industry before we reach a convincing explanation in terms of capitalist oppression or suppression of interests of labor or the poor.[52] Skeptics may argue that such consensus or acquiescence or ignorance was the product of capitalist success in using formal trappings of law to mystify the rest of the people into believing that the legal order was not geared primarily to advance special interests.[53] But in the Wisconsin forest history the argument is wholly speculative; it lacks any citable evidence, and is intrinsically implausible in the crude simplicity of the social setting.[54]

So far as this study is properly labeled as an exercise in "legal realism," it should not be identified with the narrow focus of the 1930s realism on the doctrinal rationalizations of appellate judges or on concern for abstract justice at expense of attention to social structure.[55] Prime attention here is on broader themes of mid-twentieth-century legal realism in which appraisal of class struggle is only a part. The Wisconsin forest history centers on failure to fulfill commonwealth criteria of rational public policy through adequate accounting for social income and cost, to the detriment of the long-term vitality of the whole society. The Wisconsin story belongs also to a related category of deficiency in political processes which is a central engagement of contemporary political history and political science. This is the continuing, strong trend toward fragmentation of policy making into responses only to narrowly focused organized groups, with gains and costs not compartmented within categories of social classes.[56] In these broader terms, it is premature to pronounce the end of legal realism.

52. Tushnet, note 1 above, 115.

53. *Id.*, 130–131.

54. This volume, 118–125, 462–465.

55. Tushnet, note 1 above, 115, n. 7.

56. *Cf.* Robert A. Dahl and Charles E. Lindblom, *Politics, Economics, and Welfare* (New York: Harper & Brothers, 1953); William A. Kelso, *American Democratic Theory: Pluralism and Its Critics* (Westport, Conn.: Greenwood Press, 1978); Theodore J. Lowi, *The End of Liberalism* (New York: W.W. Norton, 1969).

PREFACE

THIS WORK is a legal history of the lumber industry in Wisconsin through the years (about 1836 to 1915) when that industry grew from small beginnings to major consequence and then rapidly declined to a position of relatively limited importance in the economy of the state and region. Because the industry moved in the main currents of its times, this is also a legal history of some key aspects of economic growth in the nineteenth-century United States. I appraise economic affairs here primarily in the light of their deposit in legal records. This is also, thus, a history of roles of law in society—of the distinctive impress which legal process (legislative, executive, judicial, administrative) made upon the general life, measured in part by the distinctive impress which the general life made upon law.

The present study is part of a course of experiment in studying the social history of law in the United States. I have tried to make my work develop useful theory concerning the significant subject matter of legal history; at the same time I have tried to tell some of the rich substance of the place of law in the turbulent growth of the nation through the nineteenth century. Three previous books of mine have expressed this concern to mingle general and particular study. *The Growth of American Law: The Lawmakers* (1950) sketched the development of the distinctive institutional character of the principal agencies of lawmaking. *Law and the Conditions of Freedom in the Nineteenth-Century United States* (1956) sought to define some key values and attitudes out of which men consciously shaped their uses of law in this society. *Law and Social Process in United States History* (1960) enlarged examination of the conscious uses of law, but also emphasized the massive weight of social inertia and the drift of events in affecting the influence which law had on social growth, and the high significance of legal process for cultivating men's awareness and stimulating and legitimizing their will to exert direction in their affairs. The emphasis of the first book was upon organizing and presenting particular data, that of the second and third mainly upon the development of hypotheses for further inquiry. The present study returns to emphasis upon particulars but seeks to present its matter so as to contribute to a general theory of United States legal history.

This study attempts to serve four principal ends. (1) It investigates relations between law and the processes of economic growth. It addresses itself especially to the shifting balance between the market and the political process as a means of allocating scarce resources, and to the definition and legitimation in law of private and public accounting for private and social income and costs. Such matters present important economic aspects of legal-economic history. (2) The study is concerned with identifying the distinctive impress of law-embodied attitudes and choices, and distinctive roles of the principal agencies of law in defining, determining, and implementing values particularly involved in economic effort. So, also, it seeks to identify distinctive problems or challenges which the course of economic development posed to law, according as such problems or challenges fell upon the legislature or the courts, upon executive or administrative officers, or upon lawyers. From such law-oriented or separation-of-powers points of view, the study emphasizes the legal aspects of legal-economic history. (3) Within the limits of its economic preoccupation, the study is concerned to make available data and experience relevant to general problems of social organization and process as these affect or are affected by law. Thus the study pays attention to those aspects of its materials which bear upon the history of ideas (political and social, as well as economic) in the nineteenth-century United States, upon the influence of inertia and drift in social affairs, upon the creative and destructive tensions generated by interplay of general and special interests, and upon the tendency of means to fashion ends. In this aspect, the study tries to make its primary concern with the interaction of legal and economic institutions yield a product relevant to broader social theory. Such seems a proper purpose of any form of hyphenate legal history, whether it be legal-economic, legal-social, legal-political, or pertaining to any other merger of interests found in men's relations. (4) Inherent in the three purposes already noted is a fourth purpose of the study: to seek more meaningful and more useful concepts of the subject matter of legal history. Historians properly pursue history as men paint pictures or compose music, because these activities promise satisfactions of mind and feeling which are ends in themselves. Historians also try to enlarge understanding, to increase men's control of their affairs. From either standpoint, it is profitable to pass from more naïve to more subtle definitions and tests of the meaningful range of subject matter. Such effort is particularly relevant to a field in as rudimentary a state of development as is study of the social history of law. So viewed, the particular story of law and lumber in Wisconsin is matter only of secondary interest in this book; of prime concern is to learn, from trying to tell this particular story, how better to tell the story of the distinctive parts which law has played in the general course of social experience.

To write any history is a kind of economic enterprise. Data take on meaning only in relation to chosen matters of attention. Connections of events ramify endlessly, and limiting choices must be made among relevant data.

Choices have been made to bring this study to meaningful focus. Inevitably, what the choices gain is purchased at the cost of alternatives foregone.

Several choices of subject and data are basic to the organization of this history.

(1) The framework of the history consists of four general concepts of law in this society: property, contract, police power, and general planning by the political process. I have chosen four law-defined ideas as the organizing base of the data because the primary emphasis of such hyphenate legal history as legal-economic history should be upon law-embodied values and attitudes, and upon legal procedures as these become particular focal points of energy in social affairs. On the other hand, the action validated or guided within each of these legal concepts has importance for economic history, for each of these legal concepts embraces issues of behavior basic to economic functions. The law of property delegates the initiative of decision over the allocation of scarce resources. The law of contract and closely related regulatory law help bring into being the units and procedures of trade necessary to the market. The police power expresses the common concern with the good order of relationships in a society marked by increasing division of labor and long-term commitments of capital. The impetus given through legal process to general community planning recognizes the reality of social capital and social costs (the fact that a society is a commonwealth) and the demands which men finally make on the economy, that it work not as if its operation were an end in itself but as contributor to some larger ideal of a full life.

(2) The breadth of concept inherent in resort to the ideas of property, contract, police power, and public planning is offset here by the choice to examine the affairs of one industry. Obviously this restriction is bought at the price of insights which might come, for example, from examining the functions of contract law not only in log or lumber dealings but also in the traffic in grain, minerals, or metal manufactures. However, we are only at the beginning of the study of legal-economic history in the United States; everything cannot be done at once. The growth of an industry provides at once a focus of attention, criteria of specific functions to be performed, and continuity and accumulation of particular experience. Eventually we should compare the operation of legal concepts and processes across many lines of men's activity. But we would beg a substantial question if we assumed that the operation of legal order did not show factors distinctive to the life course of particular industries. In any case, the present study makes no doctrinaire commitment solely to the affairs of the lumber industry. We cannot properly understand the impact of public lands policy on the lumber industry, and of the lumber industry on public lands policy, without first examining policy taken toward the public domain in reference to Mississippi Valley agriculture. In significant aspects, we observe that lumber-contract decisions of the Wisconsin Supreme Court savor more of contract than of lumber, and

hence must be related to general ideas concerning the ordering of the market. Proper perspective upon the police power and the public planning function, as these appear in relation to the affairs of the forest region, requires attention to broad doctrines of due process of law and the separation of powers.

(3) This is a legal history of the lumber industry in Wisconsin. Again, restriction has its price. We might profitably examine the legal history of the Lake States lumber industry as a whole and might make detailed comparison with the earlier course of lumbering in New England. As with the limitation to lumber-industry affairs, this limitation to the industry in Wisconsin has not been followed to dogmatic extremes; especially have I examined the more readily available materials on the Lake States industry, and I borrow support therefrom. Once more, however, it would beg a substantial question to assume that no significant insight may come from studying intensively the action of a single sovereign. Wisconsin geography (most notably the pattern of river valleys) and the total pattern of the Wisconsin economy (with its own balance of agriculture, mining, and trading enterprise) were factors which tended to give special cast to the lumber-industry problems brought to law in this state. Moreover, in the opportunistic bustle of the nineteenth century, men did not look far away from home for their working ideas; it is quite plain from both legislative and judicial records that though, of course, Wisconsin lawmakers adapted experience bred elsewhere, Wisconsin common law and statute law particularly affecting the lumber industry grew principally out of local experience and local precedent. Once past colonial or territorial origins, the study of legal history in the United States has so far been too little attentive to the full-dimensioned growth of public policy within the framework of state legal systems. Of course, in due time comparative analysis will be in order. But it must wait upon more knowledge of particular units.

(4) Within its chosen area, this study has sought to exhaust the materials produced by the central legal agencies of the state. Digests, indices, citators have been used as cross checks to locate relevant material, but the prime reliance was on paging the session laws and statutory compilations, the legislative journals, the volumes of governors' messages and executive-office reports, and the volumes of reported decisions of the Wisconsin Supreme Court. Some records are opaque or incomplete, and the inevitable mistakes of the researcher add to the number of relevant items missed. But, over-all, I believe that the study rests on substantially all the material relevant to the lumber industry to be found in these witnesses to state policy. Labor has no virtue in itself. Nor is there magic in counting items merely to construct tables of figures. More time cost was invested in this project than would be warranted at a more developed stage of legal history research, when more knowledge should make it safe to cut corners. How-

ever, given the present relative lack of studies in the relation of law to social process, it seems useful that some work should err on the side of overinvestment. Research so conducted may at least provide reasonably assured measures by which to gauge how far we can safely study the growth of public policy by less rigorous search for data. Until we have more such bases of measurement, it seems dubious to expand generalizations on a foundation of a few statutes, court decisions, or legislative committee reports, when we do not in fact know whether the items we cite are sports or lone instances of their kind, or by what comparison they may be called representative. The present study summarizes parts of its search in tables. These tables are presented, and are treated, as summary inventories; they are used simply to suggest trends in events or policy, to help appraise the representative character of particular incidents, and to check impressions of the comparative roles of different legal agencies. The data from which the tables are made are too few, and usually too variant in individual character, to lend themselves to statistical treatment, and no such claim is made for them. Intensive firsthand search went only to the materials produced by the central agencies of state lawmaking. To attempt comparable search for materials generated by local government and trial courts exceeded my resources of time and energy, and so far as I use materials from these levels of lawmaking I draw gratefully on the work of others. Likewise, in the necessary research division of labor, I have undertaken to be my own economic historian only so far as the absence of useful work by others enforced such effort upon me. No small problem in the development of hyphenate legal history is that no single student or research team can muster means for extensive and intensive search of the raw materials of several fields of learning; the present project was possible only because I found competent work had been done before me by students of economic history, especially in the field of public lands disposition and in the specific life course of the Wisconsin lumber industry.

(5) The interplay of the lumber industry and the law in Wisconsin was such, in fact, as to yield no large grist of "great cases" or high debates of constitutional policy. Three closely related features characterize these materials. The legal history of the Wisconsin lumber industry resides more in legislative and executive materials than in the Wisconsin Reports, rests about equally in "public" and in "private" law, and in both "public" and "private" law consists more in policy developed out of rather routinely handled instances than in policy developed in moments of drama. These aspects of this history are not the products of my selection; they are the nature of the history. I do not count them as losses, but rather as assets. "Great cases" and constitutional debate deserve their place in the telling of legal history. But most of life is not melodrama, and most of the reality of legal order as men experience it lies in the kinds of action (and default) and the tone of busy practicality which prevail in the lumber industry story. The writing of

legal history has tended to exaggerate the relative importance of the judicial process and of common law; it is a needed corrective to examine a course of events so much involved in the legislative process as is this one. There has been especially gross neglect of the legislative process in the states, and although federal legislation was important in the public-lands aspect of the lumber industry's affairs, it is a utility of the lumber industry story that most of it turns on state legislation. It likewise serves a more balanced view of legal history that so much of the public policy involving the lumber industry is contained in rather staple matters of the law of real and personal property, of contract, of security instruments used by private financiers, and of tort (trespass to land and personal injury). It is harder to write history out of these everyday staples; yet, the fact that they are everyday staples demands that legal history find organizing ideas and modes of analysis which will make such materials yield up their meaning for the tough-fibered, sustained life of society. Of similar virtue for insight is the fact that, whether the issue be of the law of property or contract, or of the constitutional authority of nation or state to manage public lands or enact police power regulations, most of the public policy with which we deal here was hammered out by accumulation of relatively undramatic instances. The humdrum exterior should deceive no one. Great issues of policy were involved in the dispositions made of the public domain and in the decision-making initiative delegated to private hands through contract and property law, notably through statutory franchises. The issues are no less important because they were not more explicitly debated. Indeed, because this was the characteristic style of so much nineteenth-century policy making, the production of significant policy by accumulation of apparently routine operations is more meaningful for appraising law's roles than many colorful or dramatized episodes.

(6) Finally, this is a history of the general operation of large institutional processes, of the market and of the law especially. I am not forgetful that the lumber industry story was made by busy, striving, and often self-contradictory individuals—greedy and disciplined, shortsighted and farsighted, hopeful and fearful. The very human quality of the story comes through in many details. But individuals' reactions and maneuvers are properly subordinated when the primary goal is to appraise institutional performance; I refer one who wants more flavor of individual action to Richard Current's careful portrait of lumberman-entrepreneur and politico Philetus Sawyer, or to the naïve self-revelations of Roujet Marshall's autobiography. Moreover, I must caution that a reader who looks for colorful misdeeds of lumber barons will find little to his taste in this history. I do not underestimate the self-seeking behind this legal-economic record. But neither the legal record which I have examined at first hand, nor the materials which economic historians make available from newspapers and personal and

company files, provide more than limited episodes of substantial controversy or of corrupt or highly calculated manipulation of legal process. Of course, most of what was done was done under the impetus of private interest, and much of it was shortsighted or wasteful or oppressive, as measured by other values which we can erect beyond the scope of the actors' goals. What stands out, however, as the dominant tone and character of the record is that on the whole contemporary community values supported, acquiesced in, or were indifferent or unseeing toward most of what private interest sought and obtained from law concerning exploitation of the Wisconsin forest. Undoubtedly there would have been more conflict, and more skullduggery, had contemporary attitudes and energies existed to bring more of these matters to explicit issue. No more than *The Federalist* do I rate men as angels. The significant aspect of this history in this regard, however, is that the limitations of men's perception, imagination, and will were more potent than their purposes.

To define key choices made in designing this study tends to emphasize negatives — what was not done, or what was left out. Of course, however, the choices were made in the conviction that, on balance, they represented positive gains. The primary focus on legal concepts I would hope casts some new light on economic functions analyzed in relation to legal order; the attention to one industry, and to the range of relations between that industry and (in the main) the policy of one state sovereign, redeems data from abstraction, revealing the meaning they derive (both in depth and in cross section) as products of particular systems of action; the exhaustive search of the deposits of central state policy making lends assurance that judgments rest on about all that the official record can be made to show, and tempers the judgments that must be made on sketchier evidence about action taken privately or within the domains of trial courts and local governments.

Two other observations are in order concerning the positive scope of this effort.

First, the study presents its subject in two dimensions of depth. The text is written to stand by itself, to tell of the principal areas of interplay between the life course of an industry and the environing legal order. Over substantial parts of the text, the notes are much more detailed than necessary for immediate documentation. This note detail reflects the fact that the development of public policy within the context of large goals or functions is a rich and complex affair. Generalization should rest on firm appreciation of this richness and complexity, to guard against facile statement and to foster more sophisticated appreciation of the institutional weight and force of the law and of the economy. The intensive analysis in the notes offers a general reader opportunity to sample these institutional dimensions according to his interests. Further, the text deals with many focused topics — for example, the uses of statutory franchises, multiple-damages sanctions to implement policy against trespass, or creation of statutory liens for labor — which specialists may pursue into more detail; the notes offer means to do so, with the distinctive insights to be had from observing the reach and the limits of policy within a single sovereignty.

Second, in the pattern made by time and place, and the relevant stages of general economic and legal development, the legal history of the Wisconsin lumber industry embodies issues and materials of broad reach in the general history of the United States. (1) Involved here are both general and special features of the disposition of the public domain, to be studied with the more profit because in this context public lands policy is seen in terms of local federal administration and in terms of state policy making. (2) Both public lands law and the various aspects of property, contract, and police-power law relevant to the lumber story show basic attitudes and operative policy toward the role of the market as an institution of social control in shifting relation to the political process. In the lumber industry story as in other phases of economic development, the growth of public policy concerning the relative roles of the market and the political process took the form of determining when income and outlay might be deemed satisfactorily reflected by bookkeeping in terms of private gain and cost and when social efficiency or humane values required that the law enforce attention to income and outlay as measured by social gain and social cost. Where it involved such issues of social accounting, the legal history of Wisconsin lumbering merged into one of the main currents of the country's general history. (3) These materials contribute, likewise, to give solid content to items important in the history of ideas concerning man's place in his society and his proper expectations and demands upon social order — conceptions of the relative values of mind and will, of individual and group action, of present and future satisfactions, for example. (4) Broad dispersion of decision-making capacity and generous delegation of initiatives of decision in matters of public concern (as in the grant of franchises for providing bulk transport facilities) have been basic elements in the structure of power in this society; both in their benefits and in their costs these structural factors found substantial expression in the lumber industry story. (5) Prevailing middle-class values put a premium on conscious contriving and striving in this society, even as the pace and scale of its growth created subtle and demanding pressures which limited men's perception and constrained their wills to goals or calculations often too narrow or shortsighted. Social inertia and the mindless cumulation and patterning of events thus posed profound challenges to legal order; the legal history of the Wisconsin lumber industry gathers into focus such challenges and the problems of bringing law to bear upon them. (6) Legal history can easily fall into the lay stereotype which equates law with regulation, ignoring the uses which this society made of law to allocate resources and to organize men's relations by persuasion, encouragement, and the proffer of facilities which made one course of behavior more attractive than another. By public lands policy, by generous grant of franchises, by assisting the mobilization of money capital and labor through devices of incorporation, lien, contract, and security finance, and by standardizing various aspects of men's transactions, the law as it affected lumber industry development offers varied examples of such affirmative, situation-shaping uses of political process, which are characteristic of the general development of economic and social organization and productive capacity in the nineteenth-century United States.

This project owes much to the long-term support and sympathetic interest of The Rockefeller Foundation, and especially of Dr. Joseph H. Willits, director of the Foundation's social sciences division when the work was begun. I am also indebted for sustaining interest and help over the years from my colleagues of the University of Wisconsin law faculty and the administration and regents of the University, and most recently from the Trustees of the William F. Vilas Trust Estate under whose auspices I hold a chair as Vilas Professor of Law in the University. My appreciation runs also to the Harvard University Press for the care with which it has brought the study into its present form, and in particular to Mrs. Dorothy W. Whitney for the skill and imagination she gave to her work as editor of the manuscript. Of course the book does not purport to speak for any of these persons or institutions, and I take sole responsibility for what I have written.

Madison, Wisconsin
October 1, 1963

James Willard Hurst

CONTENTS

PART IV: POLITICAL ECONOMY

ILLUSTRATIONS

(all are from the collections of the State Historical Society of Wisconsin, except for the last four, which are from the Wisconsin Department of Natural Resources)

TABLES

INTRODUCTION

LAW AND THE COMMONWEALTH

THE RELATIONS men establish among themselves to make land productive go far to determine the quality, reach, and tempo of their lives. Their relation to the land poses basic issues of social organization because it involves the physical basis of life. For more subtle reasons, also, the relation is one of those which fix the framework of society. The terms of access to the land affect the practical power of decision which some men enjoy over the lives of others; access to the land inevitably becomes a prize of power and an object of ambition. Of course, the state of knowledge, as much as men's relation to land, determines what life is like in a given society. The two elements interplay. There must be a good margin of natural wealth above subsistence to permit the leisure and postponement of satisfactions out of which science and technical skill may grow; on the other hand, what is natural wealth is defined by men's knowledge of how to use what they find about them. This spells the importance of organization as a social asset — the contrivance of living arrangements so that men can win time, can learn, and learn how to apply their learning. Organization implies deciding who may legitimately make decisions to plan and allot the use of resources. Since it is the distinctive function of law to sanction the ultimate distribution of power in a society, the law must be deeply involved in so basic a relation as that of men, land, and organization for the use of land. Against the challenge inherent in the ordering of a relation thus elemental, we can best measure the values embodied in a legal system and the efficiency with which it implements its values.

Wisconsin became a state in 1848, handsomely endowed in riches of soil, minerals, water, and timber. Men emerging from frontier simplicity measured their assets a good deal by rule of thumb, and it is not surprising to find a considerable range in estimates of Wisconsin's natural wealth. But the large outline is enough for our present purpose. This was a heavily timbered area. Probably 30 million of the state's 35 million acres bore significant stands of timber. In the southern part of Wisconsin this was mainly hardwood — oak especially — in areas which experience would prove best suited to agriculture; here men cut the trees mainly to clear land for farming and to obtain lumber for farm buildings and fence posts; the long-run use of the stands that were left was to be for farm woodlots. But over an area roughly the northern half of the state stretched a great forest, of size and quality inviting commercial lumbering. These 18 to 20 million acres of northern Wisconsin were part of the larger forest which ran from eastern Canada and New England across northern Michigan, Wisconsin, and Minnesota, to reach up into western Canada. The first prize of this Wisconsin forest was its white pine, offering wood light in weight and easy to work. The forest contained also a variety of valuable hardwoods—notably oak, basswood, and elm, among twenty-nine types—but their exploitation formed a later chapter of lumbering because they were more difficult of access and less easy to work. The white pine and less plentiful conifers stood in the original forest in a proportion probably about five to one to the hardwoods, in total of saw timber.

Two facts stand out boldly in the record of men's use of the Wisconsin forest: the forest was cut to exhaustion, and the economic and legal decisions critical to this commitment of resources were made in the course of a headlong pace of growth within a span of scarcely thirty years.

The original northern Wisconsin timber stand represented at least well over 100 billion board feet, perhaps as much as 200 billion board feet of merchantable lumber. A careful assessment of somewhat uncertain evidence put the original stand of white pine alone at about 129.4 billion board feet on some 18 to 19 million acres; merchantable hardwoods originally totalled probably at least another 16 billion board feet. In 1950 a tally by the United States Department of Agriculture estimated that there were in Wisconsin 15,200,000 acres of land bearing or capable of bearing merchantable timber and not withdrawn by law from commercial use. But only 300,000 acres of this bore old-growth saw timber, and another 1,500,000 acres second-growth saw timber; of the remainder, there were 2,900,000 acres of pole timber, 6,900,000 acres of seedlings and saplings, and 3,600,000 acres of poorly stocked and deforested land. To these totals in 1950 could be added a modest 200,000 odd acres of public forests, only 60,000 of which bore stands of commercial quality. The great cutover area of northern Wisconsin had not on the whole been transformed into farmland. After more than fifty years from the beginning of agriculture in the northern counties, about one quarter of their area had been put into farms, but only a quarter of this farmland had been improved; thus in this time only about one sixteenth of the former forest had been translated into crop land, and by no means all of this proved capable of supporting its owners.

This transformation of the forest to cutover land or land bearing timber suitable only for secondary uses or slow new growth was accomplished in a relatively short time. Lumbering opened in Wisconsin with small pioneering ventures in the 1830's, began to assume substantial proportions in the 1850's, but did not take on great scale until after 1860. By the best estimate, of an original stand of 129.4 billion board feet of white pine, 20 billion feet were cut from 1840 to 1873 and 66 billion from 1873 to 1898, to leave standing in 1898 but 17.4 billion board feet; over this span some 26 billion board feet of white pine had probably been wasted, largely by fire. The years between 1898 and the mid-1920's saw the cutting of about 16 billion of the remaining 17 billion feet of pine and the reduction of the hardwood stand from some 16 billion to 7 billion board feet. The hard pace at which cutting proceeded in concentrated years was matched by the large increase in working resources devoted to it. Thus in 1860 about two thousand men were employed in the industry, but by 1890 the industry

WISCONSIN, 1900: Principal Streams, Railroads, and Lumber Industry Centers

directly employed more than twenty thousand, and the wood working industries of the state altogether employed about fifty-five thousand persons.[1]

The facts of hard-paced exploitation of the Wisconsin forest are neutral in implication, possibly consistent with wise or with wasteful over-all use of resources. But the combination of quick commitment and lasting result raises sharp issues of social order. These are issues to which the law must answer. They are not issues for law alone — and could not be, in a society constituted on the principle that power should be widely dispersed and placed largely in private hands. But it is law's special function to legitimize the allocation of decision-making power in the society and to hold power in some measure accountable to the ends of human welfare which justify it. Where desire or events or both press men hard to speedy decisions which will bind them for a long future, there is peculiar need to examine how responsibility for decision is fixed and under what procedures and guides it is exercised.

The mutual involvement of the law and Wisconsin's lumber industry was expressed in hundreds of statutes and court decisions and countless transactions, plans, and operations channelled and given form in contract, lease, license, deed, mortgage, and lien. Yet most of this bewildering array of legal action comes to a focus under one of four broad heads of legal function. (1) The law of "property" embodied decisions as to who should hold the initiative in determining use of the forest, who should have access to it, and what other resources should be allotted to promote its exploitation. (2) The law of "contract" provided the procedures and sanctions to allow flexible use of the decision-making initiative which the law of property conferred upon title holders; thus the law lent its force and discipline to help create and operate the institution of the market. (3) The law imposed certain measures of accountability on those to whom it allotted policy initiative. Within doctrines of the separation of powers, law more or less adequately channelled and controlled the use of official power, as of the authority to dispose of public lands or to tax. Also under "the police power" the law exercised some control of the ways in which men wielded the power of private property, and imposed distinctive controls on private activities whose framework importance warranted their regulation as "public utilities." (4) Finally, in concern with the problems of adjustment to inevitable change, the law undertook to survey what had come to pass and to attempt selective planning of the community's continuing response to its experience. Thus we shall examine the law's relation to the Wisconsin lumber industry within the concepts of property, contract, the police power, and the authority to plan.

Such classifications are necessary to make intelligible an unruly mass of detail. But they also in a measure impose fiction on fact. In particular, they may give events a more static appearance than fits the fluid reality and thus incline us to beg the pervading question, which is to assess how men come to grips with change. Thus, when we study the law's role in the allocation of power, we must recognize that "power" is not a fixed quantity nor a finished entity. What men did with this Wisconsin forest resource helped create new men of power and realign the relations of power groups; so it was lumber wealth that bulwarked the political careers of United States Senators Philetus Sawyer and Isaac Stephenson and supplemented party treasuries for battles between stalwart and insurgent Republicans toward the end of the century. Again, power has meaning only in relation to objectives, and men change their purposes as they learn new possibilities of action and are taught by the means they employ. Thus, originally settlement was the prime objective, but the success of settlement bred markets whose demand taught operators a new scale of ambition for the wealth that might be had from the pineries. On the other hand, change has meaning because it contrasts with stability, or at least with unvarying continuity. "Stability" perhaps implies that valuable and calculated continuity is present, yet the absence of change may reflect only inertia. Even in a time of such rapid change as marked Wisconsin in the second half of the nineteenth century, the record suggests that the main pattern of events was set more by inertia and undirected drift put in motion by the cumulative impact of countless narrowly focused actions than by plan or conscious choice of values. The Wisconsin forest was cut rapidly to exhaustion to speed the growth of the prairie economy. But in this the lumbermen simply reacted to stimulus; no legislature nor any council of lumber barons or wheat or railroad magnates decided that the forest should be fully spent as a cost of faster growth for the towns and farms of Illinois, Iowa, and Nebraska; on the whole, the thing just happened. In contrast, it is the nature of law to promote and protect rational decisions for deliberated ends of human welfare. Inherently, therefore, a main challenge to legal order, one measure of its quality, is the extent to which through it men oppose intelligent direction to the faceless, voiceless, but powerful weight of their total situation. Inertia may even give opportunity for determined men of vision and ingenuity, acting against a background of general indifference, to exert leverage disproportionate to their apparent strength. The significance of this point of view for the Wisconsin lumber story is mainly that it helps us see more clearly the elements that were lacking in nineteenth-century legal organization to create effective leverage against drift and passivity. Thus recast, our central theme is the study of the law's role in defining and sanctioning the allocation of decision-making power over the use of the forest, and in contributing to intelligent, willed direction of the course of events. As in all social history of the law, the allocation of power and the channeling of change are here the essence of the story.

PART I

PROPERTY

INTRODUCTION

THE CONSTITUTION OF POWER

BY THE test of social function, as well as logic, the legal history of an economy begins in the law of property. "Property" in law means the legitimate power to initiate decisions on the use of economic assets. The power may be put in the hands of official or non-official persons — the property may be public or private, or some amalgam — but, whatever the arrangement, the pattern of property title is the legal constitution of economic power.

In the legal history of the Wisconsin lumber industry it is peculiarly clear that the law of property in land defines the framework of action. Here there could be no escape into a dim past of custom and practice, to cloud responsibility for determining who should hold the initiative in deciding the uses of the land. The whole 35,000,000 acres which became the state of Wisconsin were originally owned by government. Recognizing no Indian title to the full disposal of the land, but only rights of subsistence use, England had made ill-defined western grants out of its right by discovery and occupation to several of its colonies. Virginia in 1784, Massachusetts in 1785, and Connecticut in 1786 formally ceded to the United States their whole claims to the area, and the United States asserted its sovereign and proprietary rights over the land in the Ordinances of 1785 and 1787. The (Northwest) Ordinance of 1787 made plain that it would be the policy of the United States to continue to hold its lands for disposal according to federal policy even after a given area achieved statehood. True, when the Federal Constitution was adopted, under the 10th Amendment the definition and regulation of private land titles would ordinarily be the proper business of the states alone. But through its initial ownership the federal government enjoyed inherent authority to fix the limits within which private ownership would develop. Indeed, as a condition of attaining statehood, Wisconsin was required to "ordain" by Article II, section 2, of its Constitution "that this state shall never interfere with the primary disposal of the soil within the same, by the United States . . ." Nonetheless, the central government was generous in grants of its lands to new states, to help them finance basic public institutions and works of internal improvement. To few states was the federal government more liberal than it was to Wisconsin, to which it ultimately granted 10,200,000 acres, or about 29 per cent of the whole area of the state.[1]

Thus the growth of the Wisconsin lumber industry rested upon the public lands policy of the United States and the state of Wisconsin. Access to lawful opportunity to use the forest and to enter the industry depended originally on how government would exercise its control of the raw material. The existence of the public domain inescapably cast on the law a function of economic planning; the function might be exercised more by default and drift than by deliberation, but the facts created a role and a responsibility by which in any case the legal order must be measured. The massive fact of the public lands created a challenge to social order most striking and unusual in a society little given to philosophy, but — so far as it had a program — deeply committed to a broad dispersion of power with emphasis on a free flow of decision from countless centers of private decision making.

Economic pressure combined with the state of land title to make the public lands policy the beginning of this history. Quite apart from investment in the fixed capital represented by standing timber, the lumber industry made heavy demands for working capital. Labor cost was a high proportion of production cost, and it took many men to get timber from the stump to the wholesaler; the product was bulky and expensive to transport, requiring either large investment in waterways improvements or in railroads; as technology improved, both woods equipment and sawmills became more and more expensive, and the mills were subject to high risks of loss from fire and flood; time costs were high, for at a minimum the product moved to market only over the long span from fall to spring, with a large period of enforced idleness in between (until the railroads and the donkey engine allowed year-round woods work), and so far as operators relied on streams to move logs to market they often must wait out two years rather than one, if water conditions were not favorable to the drive.[2]

At least until the eighties these capital requirements fell on operators who found cash hard to come by. Scarcity of fluid capital was the dour fact about which developed the most recalcitrant problems of political economy in Wisconsin, as in all the other new states that emerged from the nineteenth-century push westward. Businessmen made no more steady and insistent demand upon law than that it help them overcome their lack of working capital. Yet the same scarcity of cash made it difficult to raise large public revenue from taxes. Wisconsin governors and legislators of the post–Civil War generation, while lumbering boomed on toward its peak, repeatedly showed their sharp awareness that they must pinch pennies and curb ambition for large-scale state action involving money. The federal government had more funds at its command, especially as the tariff filled the treasury; but, if it had more money, also it was subject to more demands from all the expanding west.

In contrast to the limitations or embarrassments surrounding government spending stood the obvious and painless opportunity for grants-in-kind, by disposal of the public lands of the state and the national government. At the least, the lumbermen might hope to get access to the timber of the public lands at prices which would require of them a minimum cash investment for their raw material. In addition, they might hope that government would use its lands to promote development of the expensive transport facilities essential to move their bulky and heavy commodity to market.

Solution of the transport problem made another demand upon the law's power to create rights of "property." The improvement of streams, and especially the building of railroads, required mobilization and discipline of capital on a scale that could not be accomplished by individual entrepreneurs, or partnerships, or joint ventures. For such undertakings, men must ask the help of the state through grants of franchises to act as corporations. And more than the grant of corporate status was involved; transport companies required special licenses or franchises that did not enter the business of ordinary concerns. The waterways clearly belonged to the community; men who would dam a stream or change its flow or channel must obtain the law's permission. Left to ordinary contract processes, railroad builders were vulnerable to the blackmail of the hold-out landowner whose particular tract was essential to complete a stretch of line or avoid natural obstacles; railroad builders needed from the state the grant of the power of eminent domain. Both river improvement companies and railroads held the key to the existence of the lumber industry; they could not expect to be left, as ordinary contractors, to their own devices in setting charges for service; their power to collect tolls involved another special franchise. Thus, in charters for dams, booms, channel improvements, and railroads, the law had to assume further responsibility for defining who might hold the legitimate initiative in decisions over use of economic assets, and so was called on to create further rights of "property" as the legal constitution of the lumber industry.

These facts of law and economics — government's possession of the public domain, industry's sharp-felt lack of fluid capital, and the need of special franchises in the provision of essential transport — created an extraordinary opportunity for law to exercise leverage on the situation. The opportunity was the more significant because government — especially state government — was still relatively lacking in experience of economic regulation, lacked a professional political tradition with which to blunt the importunities of private interests, lacked a civil service or a practiced corps of dependable administrators — and thus might the more gratefully seize on such ready-made instruments of power as were given it by possession of the public domain and the sole right to create essential franchises. Conceivably, through these lands and these franchises it might not only subsidize but also channel or control economic growth; it might, as it helped men overcome their want of capital and meet their requirements of organization, impose conditions of sound accounting, and exact limitations against unproductive, unduly costly, or oppressive ways of doing business. There was a future to make, and here, it might seem, the means by which to make it.[3]

CHAPTER I

GENERAL PUBLIC LANDS POLICY

WE WOULD miss a prime fact in the law's formative relation to Wisconsin lumbering if we considered only how government disposed of its forested lands. Not until 1860 did the Wisconsin lumber industry stand on the threshold of major development. By then the general public lands policy of the United States — and (largely by imitation) that of the twelve-year-old state of Wisconsin — had taken on firm-set substance and form. Public lands policy had been and continued to be a focus of lively political interest. But, save for indifferent care of small forest reserves for naval supplies, there had been little attention to the idea that heavily timbered public land might present any distinctive problems. This was not surprising. Our initial expansion into the central area north of the Ohio River taught us to expect that the public land in the north central states, once cleared, would almost invariably prove first-rate farmland. The northern parts of the Lake States presented the first challenge to the presumption that agriculture was the normal destiny of the public domain. It was not until mid-nineteenth century that economic effort pressed strongly enough into the northern forest to create practical occasion to re-examine policy. Meantime, out of the disposition of public lands south of the Lake States forest, there had emerged a pattern of values and procedures which thereafter worked to limit the imagination and will which men brought to handling the timbered public lands. These limitations carried the peculiar power that attaches to ideas born of practice and response to immediate fact rather than deliberated calculation. Their impact was such that policy derived from disposal of lands primarily valuable for agriculture dominated the disposition of lands that were in fact primarily valuable for timber.[1]

We must also strike a proper balance between the account of federal and of state programs. By the measure of acreage, federal policy dominated; liberal though it was in grants to the state, the United States nevertheless kept for itself the direct disposal of about 70 per cent of the state's area. By priority in time, and hence in weight of practice and precedent, federal policy dominated. The central government began building a public lands program with the Ordinances of 1785 and 1787. Wisconsin had no lands in its own disposal until it became a state in 1848; during the territorial years, 1836–1848, the territorial government perforce acknowledged that it had no independent authority over the public domain but could only, within narrow limits, implement such programs as Congress set forth, or petition the Congress for grants in aid of internal improvements within the territory.[2]

The record shows plainly the weight of these facts of acreage and priority in time in giving the master role to federal policy. With occasional demurrer, but without serious contest or material qualification, the state imitated the laws of the United States in fixing the terms of disposal of state lands.

The state's action made clear that Wisconsin desired, or felt impelled, to dispose of its lands as fast as did the United States. In this situation, the competition of federal acreage plainly limited the state's ability to carry through any disposal program markedly less liberal than that of the United States. Only at the cost of holding its lands until the central government had disposed of most of its acreage of substantially like value, not only in Wisconsin but probably also in other states or territories within economically competitive range, could the state have obtained its own conditions for transfer of its holdings to private ownership. Had the United States elected to fix more stringent terms for disposal of some of its holdings, its much greater possessions and its broader revenue base would probably have enabled it to withstand the pressure of an easier policy by the state. But the Congress showed no sustained will to adopt a long-withholding, high-return policy for federal lands. On its side, while it held substantial acreage, the state displayed only episodic, short-lived interest in holding its lands for substantially better terms. Not until well after most of the state lands had been disposed of was there lament for lost opportunities.[3]

If our interest were only in the ultimate results of policy, we might thus emphasize the activity of the United States, noting that the state trailed in the wake of the larger public landholder. But legal history must attend as much to the processes of law as to their outcomes. Hence what follows gives more attention to the course of policy making within the government of Wisconsin than within the government of the United States. This is not because the processes of the central government did not have their own distinctive function and interest. The division of labor within a federal system must always be a concern of students of its law. Thus one might hope that the central government, out of its greater interests and resources and its relative insulation from urgent parochialism, would form policy by a broader calculus than that which might move lesser sovereigns. But this hope was so far unfulfilled in public lands policy affecting the Wisconsin forest that there is little to say about federal policy making. The United States paid no significant attention to the timber resources of its lands east of the Mississippi so long as it held them; the disposition of public timberlands first won serious attention in Washington after the Wisconsin forest had been committed, when a federal forestry program could have substantial meaning only for the public domain far west of Wisconsin.[4]

If it be a weakness of the states in a federal system to see policy in too limited perspective and to act under too immediate pressures, it may be their strength, out of these same elements, to be more sensitive to new emerging issues and to be favored by a closer view of cause and effect in some relationships. The people of the United States in the second half of the nineteenth century confronted a task of economic development that was at once overwhelming in scale relative to their limited resources and bewildering in detail relative to their limited philosophy. Circumstances made it most likely that their interests and their vision would be narrowly "practical," focused on techniques of operation rather than on understanding of cause. In such a climate of opinion, conceivably the better chance that subtle domestic issues might be defined and brought to deliberated decision lay within the action of a political community small enough to feel the immediate impact of fact. And though a state offers a much more confined and therefore more vulnerable forum in which to play out contests of interest, at least here — because of the lesser scale — we may see perhaps more clearly the challenge, and the capacity and limitations of law, in achieving a healthy balance of power. Within this smaller arena contending forces may be better identified, opposing forces may more readily be mobilized, the ordinary preponderance of social inertia may weigh less heavily against the initiative of minority will and intelligence. Though the response to challenge was uncertain and weak, the Wisconsin record offers a good deal of evidence by which to test these propositions.

THE PATTERN OF DISPOSITION

By 1860 the United States and the state had transferred to private ownership in fee simple most of the land they originally owned in Wisconsin south of the great forest. In its general outline, this part of public lands history is important to our story. It meant that as government and people after 1860 confronted the problem of large-scale disposition of public timberland, their experience provided a ready-to-hand, taken-for-granted pattern of policy for public lands disposal. Moreover, the general pattern stayed the same until almost all the public timber in the state had been transferred to private ownership, about 1890. In a time of haste and bustle, when government lacked experience, tradition, or staff for ambitious programming, when men made a philosophy of crossing bridges only as they came to them, a familiar pattern of policy easily imposed itself on a new situation and set subtle limits on the imagination and energy devoted to a fresh problem.[5]

As of 1860 federal law provided three models for disposal of public lands. All involved the transfer of a fee simple for a consideration. Save for an unsuccessful experiment in leasing or licensing use of mineral-bearing land (an effort which ended in Wisconsin in 1847), during the years critical for the disposition of the Wisconsin forest the federal government had no program of long-term reservation of any substantial part or type of its title in the public domain. The United States held its lands only until they could be surveyed and bargained away.

At the threshold of major exploitation of the Wisconsin forest, the most important procedure for disposal of federal lands was by sale according to terms fixed essentially since the act of April 2, 1820. The United States would sell in tracts as small as eighty acres, for cash only, at an initial minimum price of $1.25 per acre (with some allowance after 1854 for graduation downward of the price of land that remained long unsold). Land must first be offered at public auction; if it went unsold, it was then available at private sale at the minimum price set by statute. In practice the auctions usually yielded prices scarcely above the minimum, partly because of the availability of already offered land for private sale, partly because of collusion by both speculators and settlers to prevent competitive bidding.

The second disposal procedure under federal law was one of sale for a mixed consideration, in cash and in kind. This was sale by pre-emption, firmly established in 1841. The head of a family, a man over twenty-one years old, or a widow — being a citizen of the United States or an alien who had declared the intention to become a citizen — who settled upon public land and improved it, was secured in a prior right to buy up to 160 acres at the government's minimum price. The applicant must maintain residence for about fourteen months and must swear that he did not own more than 320 acres of land other than his pre-emption claim. Wisconsin settlers made substantial use of the federal pre-emption privilege; by fraud, speculators likewise employed pre-emption claims to enlarge their holdings. Since the federal land offices did not report pre-emption sales separately but mingled them with cash sales, the data are not available from which to estimate the relative acreage disposed of for cash and under pre-emption. What evidence there is suggests that the bulk of federal land passed on private cash sales.

In 1862 Congress added another principal procedure by which federal land might pass to private grantees. Under the Homestead Act, a settler who lived on and improved a claim for five years might thereby obtain 160 acres free of all charge except a small filing fee — in effect, another type of sale for a consideration in kind. The Homestead Act came too late to figure materially in the settlement of southern Wisconsin, and it did not form part of the operating experience out of which attitudes were shaped to govern the later timberlands disposal. But the long agitation which lay back of it and the terms on which it made homesteads available offer further evidence of the dominant values of United States land policy. With a considerable mixture of fraud, a sizable acreage was to be entered under the Homestead Act in the northern part of Wisconsin.

The third main method of federal lands disposition, established before the Homestead Act and continued thereafter in substantial subtraction from the acreage available for homesteading, was by grants to the states for their further disposition. We have already noted that Wisconsin profited by this federal bounty ultimately to a total of over 10 million acres. These grants were still another form of transfer of federal lands for consideration in kind — constituted in this instance by the undertaking of the states to use the lands to achieve certain public interest objectives.[6]

The state thus did not enjoy full freedom in disposing of lands granted it by the United States. In every case the federal government gave Wisconsin land to be used in aid of a public purpose stated by Congress — notably to establish schools, reclaim swamplands, and promote the improvement of waterways or the building of canals and railroads. Congress sometimes defined its objectives with ambiguity that bred conflict, as in the swampland grant of 1850. But the central fact was clear enough. If the state honored its obligations, it could not give away these lands without condition nor use their yield for any purpose that pleased the Wisconsin legislature. Moreover, in 1848 the Wisconsin constitution makers filled in whatever gap Congress might leave. Article X, section 2, of the Wisconsin Constitution provided that "all moneys arising from any grant to the state where the purposes of such grant are not specified . . . shall be set apart as a separate fund to be called the 'school fund' . . ." and directed by section 8 that "provision shall be made by law for the sale of all school and university lands after they shall have been appraised . . ."

Politicians and public officers often spoke of the state lands derived by grant from the United States as a "trust." Clearly they were so in a moral sense. In law, the United States never took direct action by statute or lawsuit to enforce these "trusts" against the state. And the state Supreme Court declared that whatever obligations attached to these acres or their proceeds while the state held them did not follow them into the hands of the state's grantees. Nonetheless, enough was done to give some practical content to the notion that the state owed enforceable obligations to the United States regarding these lands. On occasion federal executive officers delayed patents to additional acres, to enforce performance they deemed owing from Wisconsin in connection with transfers already completed. In like fashion Congress used its bargaining power in adjusting claims by the state under ambiguous grants. The Wisconsin legislature, it is true, was guilty of substantial diversion of federal lands or their proceeds from aid of schools or drainage, especially to support of local road and bridge building. On the other hand, the governor several times interposed his veto to balk diversion of the "trust" lands, and ultimately the Supreme Court of Wisconsin set aside some of the offending legislation and enforced an accounting for accomplished diversions.[7]

The foundations were early laid in this fashion for a more or less formally recognized responsibility to treat the state lands as a trust for the state's economic and social development. However much this formal policy was violated, it is a part of the record by which we must measure the quality of the state's performance.

By 1860 the state of Wisconsin, like the United States, had a quite firmly settled pattern of public lands disposal. As with the United States, the state's general program, as it stood on the eve of the major exploitation of the timberlands, included no provision for long-term reservation of any type or portion of the state's title. The state's program deviated from contemporary federal policy mainly in that the state was willing to sell on rather long-term credit (ten to thirty years, with as little as 10 per cent of the price initially paid in, and title meanwhile reserved in the state) at a time when the United States, out of an unhappy experience of default, had for forty years elected to foreswear the role of mortgagee and insist on cash. At times the state seemed more insistent than the United States on favoring the settler; for some years thus it required that every applicant to buy school or university lands must swear that he purchased for his own use and for actual occupancy or cultivation, and that he did not own other such lands in quantity such as, added to his proposed purchase, would give him more than 320 acres. But in practice there was no substantial difference on this score between federal and state policy; by low prices and pre-emption, and after 1862 by the Homestead Act, the United States expressed equally strong official favor toward the settler; on the other hand, the oaths required by the state were backed by no effective enforcement.

These matters apart, the general outline of state lands disposition imitated the federal program. First, the state was prepared to sell land for money consideration. It would sell in fee simple, first at public auction, then at private sale, at a minimum price declared by law (set by appraisal, or else a fixed $1.25 per acre); at auction no bid would be taken for more than 160 acres, but this bid limit did not apply in private sales, though in some years the state limited all buyers to a maximum of 320 acres. Second, the state was prepared to sell for a mixed consideration, in money and in kind—by pre-emption. Though pre-emption was a consistent part of the state's land program, it was offered within a considerable range of variation (now limited to 40, now to 160 acres, now limited to persons having effected improvements before a stated cut-off date, now offered to future settlers), which it is not necessary to our purpose to elaborate. The standard payment exacted of the pre-emptor was the familiar $1.25 per acre, usually required in cash in full upon proving up the claim. The state's program never included any general procedure analogous to the federal homestead law, under which land would be transferred solely upon consideration of occupancy and cultivation; probably such disposition would have violated the plain intent of Congress and the state constitution that the school lands should produce capital funds for education; on the other hand, it was consistent with the intent of the swampland grants that the state should, as it did on some occasions, transfer swampland in consideration solely of the grantee's reclamation activity. The third principal procedure of state lands disposition concerned the lands given by Congress in aid of river improvements and the building of canals and railroads. The state had likewise fixed its policy on this head quite firmly by 1860. Only as the work was done would such lands be passed over to those making the improvement. When title was transferred, it was in fee simple absolute, but accompanied by what amounted to a contract that the grantee would in turn sell with reasonable promptness, likewise in fee simple; these stipulations might be coupled with at least a formally declared preference for ultimate sale to settlers, and perhaps a pre-emptive right for previous settlers. So spoke the seventh section of the 1856 grant in aid of the LaCrosse & Milwaukee Railroad: "Within five years after the title of such lands shall become vested in the said railroad company, the same, if any shall remain unsold, shall be offered for sale in limited quantities at fair prices, preference being given to actual settlers." Such declarations apart, however, in these aid grants ultimate disposition of the land was fully delegated to the assisted enterprise, and no specific means was provided to enforce performance of the statutory limitations. In practice, the aid lands grantees, out of their own need of capital, generally shared the state's interest in fairly speedy sale.[8]

No tabulations are presently available to measure the comparative use of the different procedures for disposal of the state's lands, and it may be doubted that the records will ever permit more than approximation to the precise figures. The available evidence indicates that, as in the case of the federal lands, the great bulk passed by sale rather than by pre-emption, at private sale rather than by public auction, and that formal limitations upon amounts sold to individual buyers did not in fact prevent large accumulations in particular hands. These features proved especially to characterize disposal of the state's lands in the northern half of Wisconsin. But they were features already familiar through experience of dispositions made south of the great forest before 1860.[9]

These, then, were the ways in which the United States and the state had disposed of great areas of the public domain before they encountered the problem of large-scale disposition of the public forest. Certain value judgments were plainly implied in the form and practice of this pattern of disposition. But to understand how profoundly this already established approach conditioned the handling of the forest, we must weigh the ideas and the emotion which found expression in these procedures of sale, preemption, and grant. Back of apparently colorless details of prices, auctions, and proofs of claim were highly charged values and attitudes which entered into the very constitution of nineteenth-century American character.

PREVAILING CONCEPTIONS OF THE USES OF THE PUBLIC LANDS

By 1860 we had made plain by our actions, if not always by our words, that we believed in using law affirmatively to shape certain basic social and economic — as well as political — conditions of our life. We deliberated and chose constitutions, first for the states, then for the nation. We pressed a steady enlargement of the suffrage. We separated church and state. We outlawed feudal land titles and declared for freehold ownership. The states legislated to promote wider markets for their goods through quality inspection, and the Congress enacted tariffs to foster industry. Federal and state taxes and borrowed money went to improve navigation, and build canals and turnpikes, and aid the building of railroads. We used law not so much to hold things steady as to direct change, not to maintain ordered status but to encourage mobility, to give scope to private decision making through private property, but also to make public investment of substantial resources to promote equality of political and social power, and to create facilities that might multiply private production.[10]

Public lands policy both contributed to and was influenced by this readiness of our people to make affirmative use of law. From the start of our national life we saw that prudence dictated that government should have original control of our vast "western" lands. It was our practice that government, and not private dealers, negotiated the cession of Indian lands; we made the transfer to the nation of the western claims of the several states an essential condition of union. Because it held the original disposition of the land, the United States could not escape a positive role in defining the terms on which western states and regions would grow. To the extent that the United States transferred its land to a state, there inhered in the grant proportionate power and responsibility for affecting the general course of social development.

Wisconsin was sharply aware of the power which the United States thus held over its growth, and state opinion insistently reminded the central government that this power carried positive responsibility. Governor Dewey spoke common sentiment when, in 1851, he observed:

> The public domain was vested in the National Government in trust, for the general welfare, and to supply the demands of the agriculturist, created by the

increase of population.—It was not designed that the government should derive a profit from the trust, or occupy the attitude of a great land speculator; or that capital should reap the enhanced value added to it by pioneer labor.

So, in 1853, Governor Farwell urged the legislature to memorialize Congress to favor dispositions of the federal lands that would promote settlement and increased productivity in the state. Estimating that there were still in Wisconsin over 20 million acres "of Congress land," he hammered again at the theme of the planning responsibility which such holdings imported. "It is certainly a question of momentous importance to us, what disposition ought to be made, by the Congress of the United States, of this two-thirds portion of our State"; what Congress decided "must necessarily greatly affect us, favorably or otherwise, however insensibly it may be felt elsewhere." Because improved transport was itself so obviously a key to the over-all development of the state, there was no aspect in which the leverage value of the federal lands was more persistently stressed than in pleas for grants in aid of railroad construction. Typical of scores of such statements was the argument of a memorial addressed to Congress by the Wisconsin legislature in 1867 in behalf of federal aid for a north-central railroad project:

That the Green Bay and Lake Pepin railway . . . is a work of great national importance, and will open to settlement a large extent of country that is now unprovided with roads or railroad facilities; that the country between Green Bay and Lake Pepin embraces some of the richest farming and lumbering lands in the state, capable of sustaining a dense population, and the settlement thereof is almost prohibited for the want of railroad facilities; that it only requires an avenue through it to cause immigration to flow in and settle up this vast region, which otherwise must remain for a long period undeveloped, unless congress grants its aid.[11]

The nineteenth-century state government was a modest enterprise in the whole range of its activity. It is not surprising thus to find opinion less sharply focused on the planning and disposing role of the state regarding its lands. Moreover, though nearly 30 per cent of the state's area was finally put at its disposal by the United States, this was not accomplished at one dramatic step, but by a series of grants over the course of a generation. Thus the state's land control was not displayed to public view in the sharp, high light of the original, predominant, federal ownership.

Nonetheless, so far as men paid attention to the general significance of the state as landholder, their responses were consistent with the prevailing emphasis on the positive use of government assets. Expectation of generous land grants from the United States in aid of education and internal improvements proved one of the persuasive arguments for seeking statehood. Even as they forbade the state to become a party in works of internal improvement, the constitution makers of 1848 took care to except from their ban the use of "grants of land or other property" especially dedicated by the grantor to such works. As federal grants to the state accumulated, a legislative committee acknowledged, in 1858, that "the school and swamp land bureau is now one of the most important departments of the government, and involves greater interests than any subordinate division." We have already noted the recurrent theme in official and political discussion of the state lands which characterized them as a "trust." Inherent in this notion was recognition of the state lands as a means for affecting the pattern of community development. This was the case in spite of the fact that the legislature was guilty of substantial diversion of the lands or their proceeds from their assigned purposes. The fact that legislative action was from time to time called to account in these terms reflected the existence in the background of some sense of planning responsibility in the state for use of its holdings. This attitude came to particularly vivid expression in the Message by which Governor Harvey, in 1862, explained his veto of a bill which would have committed the disposal of the state's swamplands grant wholly to the will of the counties in which the lands lay:

> By the adoption of efficient and judicial measures for the reclaimation [sic] of these lands, and their sale, the prosperity of the State may be widely and largely promoted, and a large surplus be realized above the cost of securing the objects for which the grant was primarily made, which the State may in good faith use in restoring the waste of her school fund, and thus render the grant a source of perpetual blessings to her people. Scatter these lands among the counties, as proposed by this bill, and you condemn many, if not all of them, to pass through a similar experience to that recorded in the past history of land grants to the State. It cannot be that the patriotic and public spirited representatives whom I see so anxious for the passage of this bill, have calculated the peril to every public interest likely to result from exposing these prizes of land for competition in the arena of local politics in their counties.

Similar emphasis on the state's planning responsibility provided the basis of the persistent, if largely unavailing, demands that it not let large acreage fall into the hands of speculators who would hold it off the market and so hinder community growth. Pleas against long-term withholding of its land by the state itself, on like grounds, further attested men's uneasy sense of the extent of directing power implicit in the state's ownership. The Wisconsin Supreme Court emphasized the power of choice and direction inherent in the state's landholding, as it ruled repeatedly that the state held plenary title and might modify its dispositions as it chose until it effected complete transfer of its lands. Such dispute as took place over the use of the state lands dealt not at all with the legitimacy of using the lands as instruments of state economic and social planning, but wholly with the choice of objectives, particularly the relative emphasis to be placed on education, on agriculture, and on improvement of transport.[12]

However, the very fact that we were so ready to use law—including the power implicit in original public ownership of the land—to fashion a society to our liking calls for care in interpreting the record. In this context it may be easy to exaggerate the importance of the law and the extent of deliberated decision that went into making public policy. Logic might say that people who were so confidently disposed to guide their destiny must have spent a great deal of time and energy in making big plans, checking them in execution, and reshaping them by experience. There is enough truth in this to insure that anyone will have a difficult time striking a realistic balance between the proportions of plan and lack of plan, consideration and haste, deliberation and drift, that went into most lines of policy development in the nineteenth-century United States, not least that concerning the public domain. Our legal history was deeply affected by the fact that we had learned to see ourselves as men deliberately making a society out of a wild half-continent. But other factors affected the quality and manner of our decisions, including the part that law played in the whole. The country was big in acres and rapidly getting bigger in people and in the size and range of its economy. So much happened so fast that at best it would have been hard to keep a broad and steady perspective upon public policy. As it was, men confronted large issues with little leisure or experience to season their judgment. The time would come when the North American continent would be very wealthy in yield as well as in potential. But through most of the nineteenth century this was still a hard-pressed society, challenged by the promise of its situation, recurrently frustrated by want of means to realize its promise.

In this setting it is understandable that policy was made with much improvising, haste, and rule of thumb, and with a bias toward choices that would yield the most tangible results within the actors' lifetimes. These aspects of the social setting part, the very constitution of our legal order helped insure that public policy making would not be neat, tidy, or stamped with single-minded comprehensiveness. Through federalism and the separation of powers, and in institutions of private property and private association, we expressed the high value we put on dispersing power broadly. It was a legal order calculated to promote experiment, diversity, and the free play of creative will and imagination; part of the cost of these qualities lay in confused or conflicting purposes and in a lack of focused responsibility for the secondary consequences of decisions.

So, in fashion completely of the times, general policy toward the public domain grew by the 1860's out of the pull and haul of variously conceived objectives, and under various importunities of circumstance. The result was not so much a balancing as an overlapping of purposes; such priorities of values as appeared at a given point of time seemed to depend at least as much on what pressure of events was then most immediately felt as on any weighing of relative worth or cost.

Three pairs of values wove together to make the main pattern of general public lands policy: (1) the weight given the political-social objective of making a middle-class society, compared with the stress on the economic objective of achieving a rising curve of productivity; (2) the relative weight given the physical-productive capacity of land (its "real" worth), as compared with its utility as a means to increase the present productivity of labor or other mobile capital, or to attract additions to the stock of labor or cash as the price of obtaining it (its "monetized" value); (3) the relative weighting of present as against future economic satisfactions, or, more subtly, the relative emphasis upon different kinds of present investment pointing to different styles of future economic growth. These pairings did not represent elements necessarily inconsistent with each other. But there was substantial tension between the members of each pair, so that greater emphasis on one member usually meant a higher cost or sacrifice in terms of the other. These pairings of objectives or attitudes represented the principal range of considerations that entered into conscious policy making about the public lands; we shall hereafter take note of certain other facts of institutional environment which set less consciously realized limits upon this field of policy.

BRING THE LAND TO MARKET

Both political-social and economic objectives highly prized in the first half of the nineteenth century converged to create great pressure to bring the public land to market, fast and in quantity. It was characteristic of the general course of public policy development over the century that the emphasis in public lands programs rested primarily on political and social considerations up to about 1850, then for some forty years focused especially on economic values, and only about the turn of the century — after the Wisconsin forest had been committed to a pattern of use that was for the most part beyond reversing — moved back to a substantial attention again to political values.

From the first major Congressional debate in 1796, when Gallatin urged that the United States offer its lands in family-farm-sized tracts of 160 acres, it was a leading idea in federal policy debate that the lands should be disposed of on terms calculated to create a substantial class of independent-spirited freeholders whose intelligence and integrity would insure a decent, stable politics. Secured by law in their private holdings, provided by their acres with means to win a livelihood by their own effort and prudence, such men were the model of the character needed for the successful life of a republic. To Jefferson, "the small land holders are the most precious part of a state." Even when he overcame his earlier fear of the irresponsible "mobs" of the city and decided that urban workers in this country might, too, be trusted with power, he rested this new faith on the availability of open land: our industrial workers showed themselves "as independent and moral as our agricultural inhabitants, and they will continue so as long as there are vacant lands for them to resort to . . ."

For a while this approach contended for place with the notion that the United States should use land sales principally as a means of public revenue. It was an idea attractive to an untried government, none too sure of its strength to collect taxes. But, a few boom years apart, the lands on average yielded in any year only about 10 per cent of federal revenues; men did not have the money to pay much in cash, and defaults were large on credit sales. More important, however, use of the lands simply for revenue soon appeared to a prevailing opinion to be a low use, of mere grasping expediency, compared with the high purpose of building a political community of sturdy, self-respecting yeomen. The cash-sale, small-tract-sale policy begun in 1820 marked the switch from fiscal to political emphasis, as did the succession of general pre-emption acts beginning in 1830; the permanent general pre-emption act of 1841 signalled the victory of the ideal of the agrarian republic with even more practical significance than did the Homestead Act in 1862, for pre-emption became firm policy when the bulk of richest farmland was still available for the taking.[13]

In its own pre-emption statutes, in its declared intent for a number of years to limit the sale of state land to actual settlers and in quantity such that no buyer should hold more than 320 acres, in the creation of a succession of official "emigrant agencies" to promote movement of settlers into the state, in memorials to Congress to limit large speculative purchases of federal lands and to adopt a homestead act, Wisconsin showed that it shared this ideal of a polity of sturdy freeholders.[14] But a significant ambiguity had entered typical discussion of public lands policy by the 1850's, as Wisconsin stood on the threshold of a major lumber industry. The most creative period of constitution making was now nearly two generations in the past, and the fight for an extended suffrage for freemen had been won, so that even Whigs were professing democrats. Meanwhile a tremendous momentum of economic growth had gathered force, and as markets expanded under the encouragement of cheap bulk transport by water and rail, men gave themselves over to more and more single-minded preoccupation with business. The political-social ideal of the small independent farm owner was plainly basic to the ideas and feelings which Wisconsin people brought to their consideration of public lands disposition in the fifties and sixties. But the striking fact is that this aspect of the matter was more and more left implicit, while the relation of public land disposal to the growth of community productive capacity became the more active criterion of policy.

The shifting balance of emphasis appears in the terms in which two Wisconsin governors put the problem of policy to the legislature toward the beginning and the end of the 1850's. In 1851 Democratic Governor

Dewey was concerned over the failure of federal land policy to curb large private accumulations, because "the concentration of the right of soil in the hands of the minority, is followed by results destructive to the liberty of man; and . . . seriously curtails the aggregate of human happiness, physical, moral and intellectual." In 1857 Republican Governor Bashford urged that in disposing of its lands the state "by every practicable method . . . give preference to actual settlers, and . . . use its influence to induce the General Government to adopt a similar policy," because the consequences otherwise would be unfavorable to the growth of the community's productive capacity: "The facilities afforded speculators at present, for obtaining possession of large tracts of wild land, operates to the prejudice of the best interests of the State. It discourages the immigration of the most desirable order of settlers — farmers of moderate means, who constitute the great productive class of the country — and thus a large extent of territory is often left uncultivated and unproductive for years." From the mid-1850's on, almost the exclusive emphasis of Wisconsin policy spokesmen was that the public lands policy should be directed to promote settlement, because this was the way to increase productivity. Such a pattern of growth spelled creation of the maximum number of responsible, self-starting production units. It would speedily enlarge the tax base. It promised to increase the community's stock of human capital by encouraging immigration. The unceasing flow of petitions and arguments for using public lands to promote transportation improvements followed the same line: railroads, especially, would not merely in a general way provide a framework for economic growth; much more to the point, they would give immediate impetus to settlement, and hence to creation of what was seen as the most powerful leverage for economic expansion.[15]

The result of these currents of belief and feeling was to make "settlement" a potent but somewhat ambiguous symbol. So long as they were promoting the occupation and improvement of the land, men could urge programs and press on to urgent action with reassuring confidence that they rode the ascending curve of a desirable future and spoke for a value which their fellows would accept and applaud. "Settlement" became a ritual word, which made it unnecessary to pursue further the definition of objectives. No matter that, in fact, a subtly important shift might go on from political and social objectives of equal status and a broad dispersion of power to an objective of increased productivity which might or might not imply the same political and social ideal.

Bringing public land to market to advance settlement involved two corollary policies of significant implications for the future handling of the public timberlands. First was the presumption — of strength amounting almost to an absolute rule — that government should transfer land to private hands in fee simple absolute. Neither the United States nor the state was to play the role of landlord or otherwise keep in its hands some reins of control as an owner, whether by lease, license, or title on condition, or right to claim a reversion; government conveyances should pass to private owners the complete initiative in determining the use of the land.

The positive aspect of this policy was the simple pattern we have already traced for the bargain and sale of absolute title for consideration in cash, credit, or performance. In auxiliary provisions the statutes expressed concern that government-derived titles should represent readily tradable (because unquestioned and unqualified) items for market dealing; so the law defined what should constitute sufficient evidence of sound title derived from government and provided means to record such evidence. The state showed its readiness to treat absolute private title as the norm by enacting the otherwise extraordinary concession that, after an initial contract of sale, the state should occupy no more than the position of an ordinary private seller, and upon default by its contract buyer, and foreclosure, should recover only the balance of the contract price due; any excess of value upon foreclosure should go to the state's vendee. If doubt arose in the interpretation of a government grant, the Wisconsin court confidently ruled that it was safe to assume that either the United States or the state meant to transfer an absolute title, and a heavy burden of proof rested on him who asserted the contrary. The court was zealous to acknowledge that one who had fulfilled all the substantial requirements to obtain title from the government should enjoy the legal rights and protection of a full owner, though formal transfer had not yet taken place; thus he might sue to vindicate his title against other private persons, might challenge tax assessments, and might make effectual sale, though a patent had not yet issued to him.[16]

Most significant to this history, however, is the negative proposition always implicit, and sometimes explicit, in adoption of the fee simple absolute as the standard title passed from government to its grantees. This was the rejection of any idea that government would play a useful or desirable role by remaining permanent proprietor of any substantial proportion of the public domain, and simply leasing or licensing its use by private operators or granting titles subject to condition or reversion for breach of stipulated terms of use. The roots of popular revulsion against such a landlord's role for government ran deep. In the proprietary colonies the proprietor had familiarly exacted annual quit-rents from his grantees, never allowing them to buy a completely free title; deep distaste for this institution colored demands for social change in the Revolution, and the new state constitutions typically abolished quit-rents and like "aristocratic" limitations upon freeholds; by the time Wisconsin was ready to frame a constitution, in 1848, it was a matter of course, adopted without debate, that the document should stipulate that "all lands within the state are declared to be allodial . . ."[17]

Popular favor toward freehold titles was so deep-seated from the Revolution on as to encourage little challenge. However, occasions did pose the question whether government might usefully hold some strings on the public domain, and hence kept alive the issue of policy alternatives and provoked reaffirmation of the popular faith in the fee simple. In two aspects federal land policy kept the matter fresh in Wisconsin opinion through mid-century. Under legislation dating from 1807, the United States undertook to lease rather than sell such of its lands as bore valuable deposits of lead. In a measure it pursued this policy in dealing with the lead lands of southwestern Wisconsin, from the first exploitation of these resources about 1822 until leasing was formally abandoned and the lands authorized for sale in 1847. The leasing policy was almost a total failure, measured either by rental return to the government or by effective government control of the extent or manner of mining on its lands. The contemporary lack of technical knowledge or means adequate to reliable identification of mineral lands might well have doomed the program at best. But plainly no such leasing program could be carried out without a sizable, vigorous, and competent field staff. Provision of this was so conspicuously lacking as to suggest how little heart Congress had for the policy. Fraud and large-scale evasion of the program expressed the unmixed opposition of the people on the ground, who also spoke boldly their conviction that it was wrong in principle for the government thus to attempt the landlord's role. The mineral lands presented the most clear-cut issue against long-term retention of land by the United States. But similar popular attitudes found recurrent expression in Wisconsin through the middle sixties in repeated pleas to speed the federal land survey, the extinguishing of Indian titles, and the offering of lands to settlers. Whether the government was slow through inefficiency or in part because it suited eastern employers or speculators not to push the opening of the west at a pace which might create a labor shortage and keep down land prices, the impact was about the same on the impatient westerners, who wanted to see their country grow as fast as it could. In contrast to these two aspects of federal public lands administration, there was no substantial respect in which the state's handling of its own lands occasioned any issue of landlordism. Its own policy was almost invariably to sell as fast as it found buyers, at minimum prices. When it offered aid lands to a railroad, the state stipulated that once the road had earned its acres by building its line it should sell the lands within no more than five years. Again, though legislation recognized that waterpower sites might be a special element of value in the state's holdings, the law provided for their sale simply as a particular marketable item, without attempt to retain any public interest in them.[18]

The second policy corollary to the demand to bring the public lands into the market was that there should be no great private landlords, any more than a government landlord. This was the implication of the persistent demand that the law should curb "speculation" in the public domain. But, significantly — in contrast to the quite clear-cut popular notions against government landlordism and in favor of speedy public surveys and sales — there was ambiguity in the current conception of what constituted the evil of speculation. The typical mid-nineteenth-century settler in the Mississippi Valley was a speculator, if a speculator be deemed a man who looks to make money out of land by holding it for a rise in its capital value, rather than being content simply to enjoy what current income it may yield. Tocqueville observed, in the 1830's:

> It seldom happens that an American farmer settles for good upon the land which he occupies; especially in the districts of the Far West, he brings land into tillage in order to sell it again, and not to farm it; he builds a farmhouse on the speculation that, as the state of the country will soon be changed by the increase of population, a good price may be obtained for it.

By experience and common ambition, thus, popular opinion accepted the legitimacy of private capital gains created substantially by general community growth. Except for the minimum period of occupancy and improvement required under federal and state pre-emption laws, and later under the federal Homestead Act, the law made no effort to tie the settler to the land; indeed, quite to the contrary was the effect of the declarations that all land tenure be allodial. What public opinion condemned as "speculation" was not simply profit contributed to by the over-all increase in land values, but profit thus realized (a) upon the holding of lands for long periods without occupation and "improvement" (i.e., without being put into some kind of production), (b) by capitalists whose means allowed them so to hold idle large tracts of contiguous acres while settlement grew up around them. Crisscrossing these attitudes, however, further to blur the definition of the "speculation" which popular opinion found objectionable, was the high favor toward land grants in aid of railroad construction. Some degree of withholding of land — first by government and then by the railroad grantees which earned it — was implicit in such programs, which invariably called for building railroads in advance of settlement and productive development of an area, to promote settlement and production. Before the United States made its first railroad aid grant to Wisconsin, Governor Farwell in 1857 urged that the only safe course, politically and socially, was to see the public lands put directly in the hands of farmer settlers; railroad aid grants, he feared, would create a new type of great landlord, too powerful for the good of the community, whose holdings would delay opening up the country to production. In the face of an issue so sharply drawn in terms appealing to the general hostility to large-scale speculation, the state nevertheless pressed ahead eagerly in the next generation to get and hold all the railroad aid lands it could persuade Congress to give it. Though the legislature might acknowledge anti-speculative sentiment to the point of stipulating that the railroad grantee sell the land within five years after earning it, there was no assured procedure included to enforce this duty, and in any case it was plain that, adding the years of planning and building before the land was earned, these acres would in the typical instance be long held off the market.[19]

Contemporary opinion achieved no neat synthesis of these elements. Yet it was far from clear how fully consistent were the popular acceptance of railroad grants and the settler's harvest of value accrued from the growth of the community, on the one hand, and on the other the popular rejection of profit seeking by big holders who did not put acres into early production. The feeling against "speculation" was spoken with a frequency, heat, and conviction which seemed to attest that it sustained a strong mandate. But in proportion as there were unresolved conflicts and blurred definitions behind this impressive façade, the policy against speculation might prove a weak guide to the handling of novel problems. Such was to prove the case when government confronted the large-scale disposition of its timberlands.

Back of the ambiguity in public opinion toward "speculation" was in no small part the ambiguity, already noted, in the implications of the shift that tended to occur from emphasis on political-social to economic objectives. For the most part similar attitudes lay back of both the rejection of government landlordism and large private speculation. These attitudes at first emphasized democratic equality. As local reaction reached its peak against continued leasing of lead lands by the federal government in Wisconsin, the territorial legislature in 1846 memorialized Congress to end a condition

> not only greatly detrimental to the settlement and prosperity of the country, but at variance with the best interest of the citizens and contrary to those relations which in our republican government should ever exist between the constituted authorities and the community over which such authorities extend . . . [contrasting] the state of prosperity in which a country may be placed when settled and occupied by the owners of the soil, and its relative position when subject to the feudal condition of the government acting as the landlord, and the citizen occupying and cultivating as a mere tenant.

In 1848 a committee of the first state legislature "heartily" endorsed the principle that public land should be so disposed, and private landholding so regulated, as to limit to modest acreage the amount any one person might own:

> All men are born into and upon the earth through the operation of the same principles — the earth holds precisely the same relation to the existence of all men, and to all men is it equally necessary for life. We hold, then, that all men have an equal natural right to it, and no man or government has a right to monopolize it and sell it to one or many men, at their own discretion. It is the duty of government to secure all her people in possession of sufficient soil for their wants, and to prevent an undue monopoly of it by any.

A legal limit on individual landholding would encourage maximum productive use, but the result which was most emphasized was the social outcome. A landholding limit

> . . . would secure eventually the cultivation of the whole surface, a large portion of which will otherwise lay waste in the hands of speculators for years. It will secure homes for thousands who will otherwise be homeless. Whilst it secures homes for all, it will secure competence for all, and do away with that hired serfdom and consequent poverty and degradation which is now the cause from which we seek to remove, convulsing Europe, and which exists as a reproach upon our republican institutions. Such land limitation will interfere with the rights of no one — for no one has any right to monopolize that soil which is necessary to furnish sustenance to his brother, that he may make a slave of that brother, by controlling the means by which he lives.[20]

But, amid the boisterous material development of the 1850's and 1860's, the note most struck in the multiplying pleas that government hasten its land to market and curb the speculator was the importance of a policy to increase production and the stock of usable resources. Thus the 1852 legislature memorialized Congress to dispose of several thousand acres held, but not needed, for the Fort Howard military reservation, because retention of the excess land was "materially retarding the growth and prosperity of the town and settlement in its vicinity, withholding a large property from taxation, to aid in the support of common schools, and town, county, and state government . . ." As the terms of this plea indicate, there was inevitably an issue of federalism involved in these matters as well as a point of economic policy. But, true to the dominant concern of the times, men showed themselves primarily interested in the balance of political power between the central government and the state as this bore on resources for economic growth; there was only occasional objection to the political power over Wisconsin destiny implicit in large, continued federal holdings in the state; what repeatedly moved men to press the United States for more speedy survey and marketing of its lands was the much narrower, very businesslike interest in turning more acres into tax producers by turning them into production. Indeed, we must not exaggerate how much of a distinctively federal issue there was even on this focused concern for growth of the production and tax base; the same criticisms were made, with like vehemence, of long-term holding by the state of its own lands, especially those held in aid of railroad construction. Undoubtedly there would have been a livelier concern for the political implications of federal landholding, had the Congress demonstrated more interest in using the public lands specifically to plan the course of regional or state growth. But, aside from the isolated episode of the lead land leases, Congress showed little ambition or imagination for the role of master planner; nothing, indeed, underlined this fact more than Congress' inattention to the very distinctive planning opportunity presented by the presence of the great Lake States forest.[21]

In like fashion, the economic tended to submerge the political in the emphasis men gave to the problem posed by private speculative accumulation of land. There was a balance-of-power worry, of political cast, involved in the popular reaction against "speculation"; we have already noted that it was not simply unearned increment that identified objectionable speculation, but unearned increment pursued by large-scale investment. There was a degree of kinship between this rejection of the big speculator, because he was big, and later agrarian revolt against railroad arrogance and the political crusade against the "trusts" which set the original tone of Sherman Act policy. In contemporary pattern, there was much in common between the fear of the big land speculator and Jacksonian rejection of other concentrated financial power operating through the mysteries of banking.[22]

But it was easy to talk political melodrama; more indicative of men's substantial faith and motives was what they did. It is revealing that, despite brave talk, this electorate of farmers and townsmen of modest means never enacted a general limitation by law on concentration of private land ownership. The one general land limitation bill which seemed to have substantial pressure behind it was defeated in 1851 after the Assembly Judiciary Committee tersely recommended its rejection as "both unconstitutional and inexpedient." It was revealing that the state made such feeble effort even to limit concentration in purchases of state lands. In 1855 and 1856 the legislature required of every applicant to buy the state's school, university, or swamplands that he file an affidavit that he bought for his own actual occupancy or cultivation and that his purchase would not leave him owner of more than 320 acres of such land. The statute declared void any purchase in violation of these terms. On the other hand, the legislature made no specific provision for the substantial administrative burden that enforcement of these requirements would entail, and in 1863 repealed the limitations altogether. Revealing, too, is the short history of the state's effort to make productive use of funds from school lands sales by lending them in small amounts at reasonable cost on real estate mortgage. This made sense if anything positive were to be done to curb the power of the speculator, for it was the settler's lack of cash or readily borrowed working capital that typically brought him to deal with the speculator. However, the state's lending venture lasted only from 1849 to the early 1860's, when it was brought to an end by executive action; the state was not willing to provide the machinery to enforce a sound lending system as a positive offset to speculation, and the officials hence turned to alternative investment.[23]

When the whole record is taken into account — actions as well as words — it appears that what this community most valued and insisted on, at bottom, was a rising curve of physical productivity. Insistently, the warnings and protests of the time come back to this; it is when large-scale speculative ownership threatens indefinitely to hold land out of active production that it is most often and surely pronounced contrary to public interest. Characteristic was the criterion by which Governor Bashford advised the legislature how to administer the first great grant of lands to be used in aid of railroad construction:

> The lands donated by said grant should be brought into market as speedily as said act will permit, and sold in limited quantities, at fair prices, to actual settlers. They should not be permitted to fall into the hands of speculators, to be kept out of market, and the lands themselves left unimproved, and the growth of the country thereby retarded. I regard this last recommendation of great importance; which if faithfully adhered to, will remove some serious objections to such grants . . .

The Governor subsequently underlined his point by vetoing a particular disposition which failed to require that the assisted railroad promptly bring the lands to market once they were earned. So when Governor Randall, in 1860, criticized the undue liberality of credit allowed in the sale of school lands, his basis of criticism was largely that in consequence

> the entry of large quantities of the best agricultural lands belonging to that fund by speculators, to be put into the market at speculating prices, has retarded the settlement and cultivation of the lands, and financial reverses have returned them, by thousands of acres, upon the fund.

There was like emphasis in the attitude which federal officials took toward certain aspects of the land program of the United States in the 1860's and 1870's. It was known that fraudulent entries were made on a large scale under the federal pre-emption and homestead laws, to obtain land for speculation. Criticism of the frauds rested mainly on their operation to defeat the ambitions of would-be settlers; there seemed little concern for the implications for the waste or cornering of resources valued primarily as national assets; so long as substantial settlement in fact went on, the over-all official estimate was a rather complacent confidence that public interest was being adequately served. Thus such worry as men had for the impact of public lands programs on social and political equality tended, in practice, to be submerged in their preoccupation with the heady experience of general economic expansion. In such a climate of opinion, it would not be surprising if popular hostility to "speculation" would prove of little effect to limit great private accumulations of timberland.[24]

The weight of public lands policy through the nineteenth century was felt mainly through the consequences of this central drive for rapid large-scale transfer to private hands in fee simple absolute. This kind of disposition responded to a popular faith in productive increase and expansion which tended to dominate men's imagination and desires as the century unfolded, to the practical subordination of considerations addressed more to the balance of social and political power. The times were not altogether lacking in suggestions of other policies which might have led government to retain substantial ownership control of the long-term uses of the land. However, few of these indicated alternatives showed any real chance of adoption in the decades of critical commitment. Thus it seems best not to note them here, where our main concern is to trace the patterns of policy that in fact dominated and set the background for handling the public timberlands. We shall return to consideration of the contemporary suggestions of alternative programs when we consider the more or less explicit attitudes of the mid-nineteenth century toward time preferences in the use of resources.

VALUE LAND FOR ITS CASH OR LABOR YIELD

The nineteenth century United States usually knew more clearly what it wanted to do than why it wanted to do it. For whatever balance of reasons, the rapid transfer of public land to private ownership was so far the dominant criterion of public lands policy that it must stand first in analysis of the pattern of that policy's growth. But one particular reason back of this objective emerged with such clarity and so persuaded men as to become itself a policy criterion, second only to emphasis on private ownership as a warrant for action. This was the proposition that the most rational use of land in our nineteenth-century situation was to obtain the largest present return on the investment of cash or labor. So we would best promote continuing increase in over-all productivity. And this was with us an article of faith for a better life that vied with concern for the social-political balance of power as the ground of devotion to the private fee simple absolute.

This interest in getting the maximum fast return on labor or cash investment involved two closely linked value judgments, about the proportions and about the timing of investment. These were in large part different ways of looking at the same problem; we cannot discuss one in isolation from the other. But different points of emphasis were involved, and we can better see the whole pattern by concentrating first on men's reactions to the problem of achieving an efficient mixture of the available factors of production, and thereafter examining the nature of the time preferences asserted in the law of the public lands.

Reactions to a chronic sharp-felt sense of capital scarcity shaped a great deal of the growth of law in this country from the late eighteenth through the late nineteenth century. Probably this factor had as broad influence — on the law of private association, taxation, public spending and tort liability, for example, as well as on public land law — as any other broadly held attitude toward the experience of opening up the North American continent. To the common view, the facts defined the problem by a contrast so clear that it dramatized the solution. The scarcity was not felt as one of fixed capital; without a dollar's investment by men, the continent offered unbroken soil and untouched minerals, timber, water, waterways, and water-power in an abundance which kept the market price of access to these assets very low even after substantial settlement; indeed, as we moved west, successive generations had the experience of seeing such resources stand along the frontier for years with no market value at all, either because they could be had so easily for the taking, or because they actually represented obstacles to settlement, or because the cost of their development so far exceeded men's resources as to exclude them from practical calculation. What men experienced as scarce was fluid or working capital, not fixed capital; the lacks they keenly felt were of labor supply and of cash to hire labor, buy machinery and supplies to sustain current production, or build facilities for cheap bulk transport. What they most needed was the means to buy time, to start in motion what they were sure would be ever-enlarging cycles of production.[25]

Since labor and cash were scarce (expensive) and land was abundant (cheap), general opinion found obvious sense in using land generously wherever this promised to attract labor and money and to allow them high earnings. This stress on enhancing the immediate returns to cash and labor found a range of contemporary expression rather striking when one considers how little given to theory were the men who were pressing ahead to develop production in mid-nineteenth century.

The most popular statement of the notion that the public domain should be used to enhance labor productivity was in the demand to advance settlement by selling public land for labor ("improvement"), or at least at very low prices; as we have seen, this was treated as of almost self-evident social value, precisely because it would immediately give scope to the energies of those who would work the land. A more oblique approach to essentially the same position was the frequent assertion of the moral equity of the farmer as against the man who held land merely because he had money, or as against the government which held on to land merely to get money. "It was not designed that the government should derive a profit from the trust" represented by the public domain, Wisconsin's Governor Dewey told the 1851 legislature; no more was it proper that government "occupy the attitude of a great land speculator, or that capital should reap the enhanced value added to it by pioneer labor." The extent to which large speculators had been allowed to buy the best school lands "has grown into a fruitful and perhaps a just, source of complaint," warned the State Superintendent of Public Instruction in 1853. The practice was wrong, not only because it caused the lands "to lie unimproved for years, awaiting a rise in their speculative value, instead of becoming at an early day the homes of industrious and valuable citizens," but also because it compelled settlers "if they buy the best lands, to buy of monopolists, at an enormous advance upon their first cost, depriving those whose industry would be of most advantage to the State, of all the advantages which the State can bestow, of a cheap price and a long credit." A less obvious but more fateful expression of these attitudes lay in the kind of land use which followed upon transfer to private ownership. Through the nineteenth century the Mississippi Valley in effect measured the land's productivity by yield per man hour rather than yield per acre. Men practiced extensive rather than intensive farming, mining, and lumbering, and reflected in law their satisfaction with this proportioning of the factors of production by the absence of significant legal check upon spending fixed natural capital to enlarge the output of mobile capital.[26]

Contemporary attitudes about money capital were not so simply defined as those about the moral claims and economic uses of labor. The financier was not a popular figure, and least so when he appeared in the role of money lender to farmers. But though its prevailing political vocabulary was agrarian-radical and egalitarian, the Mississippi Valley most consistently acted on the faith that aggressive and shrewd management of economic means was the society's greatest asset, and that its prime need was to acquire working capital, to give managers leverage on the situation. Borne on the mounting swell of industrial revolution — even in the early 1850's Wisconsin's leaders spoke of agriculture as the state's leading "industry" — this was at base an entrepreneurs' culture. When farmers cried revolt, their language might be that of class conflict, but their practical demands were to maintain a community of middle-class ideals.[27]

The value and prestige popularly accorded economic management readily supported the corollary that government should use its means to increase the opportunities for management. This meant that few questioned the wisdom of converting government assets from land to cash and using public land wherever possible as a substitute for cash or as a means to attract and mobilize money.

These attitudes tended subtly to confine our imagination within the limits of a monetized appraisal of the alternative uses of the public lands. Conversely, they encouraged us to neglect the "real" costs and returns of land, measured by the commodities or intangible satisfactions which the physical content and location of land might allow it to yield. The principal offsetting tendency toward viewing public lands for their "real" utility was in the emphasis on settlement. But the contrast is mainly in appearance. For, as we have seen, the economic point of the settlement policy was to enhance the present productivity of another form of fluid capital, labor, at the cost of an expansively generous use of fixed natural capital. Altogether, the familiar pressure to value land for its use to increase the quantity and productivity of mobile capital, cash or labor, would have profound effect upon the handling of the public timberlands.[28]

The monetizing of public land policy found various expression. Prospective federal land grants figured prominently in the argument for seeking statehood, but invariably on account of the "sums" or the "funds" into which the state could convert the land. The contemporary bias for valuing public land for its present money equivalent was expressed by the correspondent of the Prairie du Chien *Patriot* who argued against holding for a school fund, instead of spending quickly for roads, the lands ("funds") with which the United States would endow the new state:

Congress has been repeatedly petitioned for funds to make a part of [the roads needed in the area] . . . but nothing has been granted, except indirectly in the 500,000 acres and in the five per cent fund . . . Now it is well known that Congress granted these funds for roads . . . And we have settled here as pioneers, subjecting ourselves to all the privations of a frontier life, in the expectation that these funds would some day extend roads through our country and not only enable us to get to market with our produce but enable others to settle by us, so that we could have schools, churches, and other means of mental and moral culture near at hand. But the proposition made by the [proposed] constitution to divert all these funds from roads to schools is only calculated to keep the public lands from selling and the country from settling, and thereby keep us from having schools for want of a sufficient number of scholars to form them . . .

This argument as to the particular use of federal land grants failed, but the Wisconsin Constitution declared the same general fiscal emphasis as it created "a board of commissioners for the sale of the school and university lands, and for the investment of the funds arising therefrom," and directed that "provision shall be made by law for the sale of all school and university lands, after they shall have been appraised . . ."

Sometimes it was bluntly stated that present funds were more useful and more profitable to government itself in fulfilling public functions, the more so because it was hard to raise taxes out of people chronically short of money, and because the scarcity of cash meant that government could lend out its funds at very good rates. So Governor Dewey mingled business and policy in advising the 1849 legislature to press sale of the school lands; it is symbolic that he spoke of the lands as equivalent to the school "fund":

The interest of these funds, it is believed, will be advanced by sale of them. It is estimated that the interest of the money that might be received from such sale at this time, would exceed their rise in value in case their sale should be postponed. Other advantages of an important character would result to the State by their sale. A large quantity of valuable wild land would pass into the hands of industrious settlers, thereby materially increasing the productive resources of the State. This would add to the amount of taxable property, and consequently swell the State revenue.

As the Governor's statement shows, the argument of enlarging the tax base — which we noted earlier as an aspect of the general interest in productive capacity — was itself also another phase of the monetized evaluation of the utility of public land. Further attesting the high premium on cash over land was the fact that, despite the strong sentiment to promote settlement, there was never any serious movement to subsidize settlement with money grants to settlers; state policy did seek to increase the stock of working capital represented in labor through modest appropriations for state agents to promote immigration, but this, again, was consistent with the general preference of mobile over fixed land capital. So, too, through the eighties the state legislature was openhanded in grants of state lands to local governments or — less often — to private promoters to be sold to yield funds to build bridges or roads or improve streams; with like generosity it granted strips of state land for railroad or turnpike rights of way, to relieve promoters of so much added investment, for "though it may be but a small pittance, [it] will be of the utmost importance to the different companies, greatly facilitating their organization and reducing the present heavy tax that is paid for this right over lands of private individuals." With like preference to spend land rather than money, the United States through the nineteenth century consistently discharged its moral debts to veterans of its wars not by cash bonus but by land warrants and the privileges of the Homestead Act. The constraint of cash scarcity made itself felt even in the immediate processes of realizing on the land. Field appraisal of federal and state lands was sometimes delayed, and throughout was limited in care and scope, because Congress and the state legislature could not bring themselves to pay the bills for a businesslike job. Because the Wisconsin government of the early 1850's felt that it could not afford a field survey of its own, the state early bound itself to accept the notoriously inaccurate field notes of the federal survey as the basis for determining the tracts Wisconsin should get under the federal swampland grant; a later generation bitterly regretted the consequent loss of thousands of acres which might have passed under a proper checkup, but this was altogether hindsight wisdom; the early decision truly spoke the bias of its own day to save scarce money at the expense of abundant land.[29]

The most common monetized statement of public lands policy argued that the land's prime utility was that it might substitute for cash by providing the bait or the security that would induce men to pay or lend money to those who would put it to work for community growth. It was easier to get men to pay money as a price or loan than as a tax, in a society whose life was so oriented to contract. Thus the Wisconsin legislature repeatedly asked for federal land grants, not for the sake of the land itself, but to overcome local scarcity of working capital. Such an aid grant for a plank road in a growing section of northeast Wisconsin was necessary, it explained, because "the means of the citizens of said counties for the construction of such a road are limited, and entirely inadequate for the successful completion of such an enterprise." Wisconsin needed federal lands in aid of public institutions for the education of the blind and the deaf and dumb, because the expense of such operations "falls heavily upon a state, which like all new states has no means of revenue but direct taxation, and that revenue derived from one-half of the real estate within our borders, the other half being by the laws of the United States exempt from taxation."[30]

The state asked for federal land grants in aid of railroad construction, so that the land "might be applied to the construction" of the roads. When the grants were made, the state's further dispositions made plain that in the contemporary view the lands had utility either as a source of cash or as a basis of mortgage credit; we have seen that legislation stipulated for prompt sales by the railroads after the lands were earned; the statutes also carefully conferred authority on the roads to mortgage the lands, even in expectancy; moreover, the state often declared the aid lands exempt from taxation for varying terms of years, on the announced ground that the exemption was necessary to make the lands an attractive security to mortgage bondholders. The whole point of such grants of tax-exempt mortgageable aid lands "was an inducement to capitalists to so invest," the Assembly Committee on Railroads explained in 1874, as it ruled that morality and the contract clause of the United States Constitution forbade the state to withdraw an exemption on which investors had relied. In a contemporary petition the promoters of the North Wisconsin Railway bluntly expounded the fiscal importance of aid lands:

> Of course, no railroad company can use this grant in the construction of the road except by the sale of its bonds secured by mortgage or deed of trust of the grant, and capitalists would hardly invest in bonds secured by mortgage upon a grant, the title to which could and might at any time be taken away by an act of congress . . . Any company building the road within the short time which congress will probably fix . . . must raise the money by the sale of bonds secured by mortgage on its property, real, personal and mixed, in possession and in expectancy, including the lands; and in the sale of such bonds will be obliged to compete with the bonds of Minnesota and other roads secured by mortgage on lands exempt for a long period. Capitalist[s] may well prefer the latter, for they then *know* that their security cannot be swept away by tax titles, and that their investment will not require constant watching . . .

The prevailing low prices on public land sold to farmers and speculators had analogous practical import for farm finance; the widening gap between the low original price and later increases in value with improvement of the land and growth of the general community offered the security margin which induced the speculator to sell on purchase money mortgage or the capitalist to lend on bond to the farmer who needed working capital. In practice thus our policy was to use the public land to subsidize private finance for agriculture. Current opinion did not like to acknowledge this, and when it did so was apt to criticize the policy; but the significant point was that we did not abandon the practice.[31]

Nothing more clearly showed how far the preference for mobile capital dominated public lands policy than the preoccupation of contemporary critics with the bearing of that policy on the growth and productivity of the supply of labor and investment funds. True, "settlement" continued to symbolize egalitarian values, and there was recurrent attack upon public lands dispositions that did not promote a broad dispersion of social and political power. But we have seen that as the years wore on, more and more when men invoked the ideal of "settlement" it was to express faith and interest in productivity; criticism ran mostly in terms of land held out of active production, and in the context of the times this meant criticism of failure to use land to obtain the maximum present return from investment of labor or money. Federal officials deprecated frauds on the pre-emption and homestead acts so far as these took acres away from bona fide settlers, but showed little concern for what large-scale fraudulent acquisitions might mean for the type of use, and hence for the physical productivity, of the land. Federal and state officers deprecated trespass on the public lands, but overwhelming opinion viewed the squatter as having a plain equity, because he was making the land give value to his labor; so far as government prosecuted trespass, it took complacent pride in the number of dollars wrung from wrongdoers, and no one asked whether the dollars could or did recompense for damage done to the physical potential of the land. Since by mid-century the United States had given up credit sales of its lands, it did not confront the problem of preserving its "real" security in the land by control of the mode of use by its buyer. The state did offer much of its land on long-term credit, and it did look to the land for security for the unpaid purchase money, retaining a security title or taking a purchase money mortgage; consistent with a monetized concept of land policy, however, it made no stipulations to prevent its farmer buyers from mining the soil, and though we shall see later that it stipulated limits on the buyer's cutting of timber, the stipulations ran only to protect the state's interest in the unpaid part of the purchase price. From the sixties there was mounting criticism that the state had been improvident in disposing of its land, but the criticism was entirely because it had not obtained better prices; lament was only for present dollar loss; it did not yet enter imagination that reckoning should include such "real" factors as the physical potential and location value of public domain. Again, though in policy debate the state lands were often called a "trust," and though various official actions in a measure enforced the trust, all this, too, ran only in money terms; the measure of a trust fulfilled was to obtain a good cash yield from the school lands and to see this fund duly invested and the proceeds spent for authorized purposes.[32]

WHAT WE DO NOW COUNTS MOST

The nineteenth-century United States was so plainly in a hurry to go places that its policy seemed often to value only the action-filled present. As in most other matters, prevailing attitudes toward time were not complex; the century had its headquarters in the saddle, and if ideas served action they were good enough to satisfy these impatient years. Nonetheless, there was more to our nineteenth-century time preferences than a simple desire for present satisfactions. This was so as to public lands policy.

True, men felt pressures to use the public domain to relieve immediate needs of the day. They felt so keenly the lack of fluid capital that they were regularly tempted to turn land into cash to pay current operating expenses. So, in 1846, without questioning the wisdom or morality of the action, Wisconsin's first constitutional convention appropriated the money proceeds of a federal land grant to pay the costs of its meeting. In like temper, advocates of statehood argued that the new state's 5 per cent share in the receipts from sale of federal lands within its borders would have paid the expenses of the state government over the five years past. And from 1848 to 1876 the state economized on direct salary paid its three top executive officials other than the governor — the attorney general, secretary of state, and state treasurer — by allowing them, in their capacity as public land commissioners, to collect quite handsome totals of fees for issuing certificates of sale and patents to buyers of state land. Another argument from present need carried greater emotional appeal than these. Men who did the rough work of first opening the land felt the sharp bite of immediate hardship; they were spending their lives now, and they required to be shown in what respects a bloodless abstraction called the future had better claim than they to what could be gained from the public domain, especially since their optimism told them that this future was bound to be richer than they in any case. Urging speedy sale of federal lands given for a university, a territorial legislative committee in 1842 vigorously denied the superior equity of the future:

> But because the number of youth to be educated is small to what it will be twenty years hence, shall we, therefore, leave the present number to be raised in ignorance, or subjected . . . [to] the expense of *foreign* education; for the sake of aiding immigrants not yet in the Territory, or children yet unborn, to enjoy facilities of which we the pioneers of the country, are deprived of? Shall we adopt the anti-republican policy of oppressing the few and the poor, in order to benefit the many and more wealthy who may follow us? . . . [A policy of rapid sale] is conceived to be more in accordance with our Republican Institutions — with the wants of a newly settled country, and with the designs of Congress in granting those lands, with a view to encourage the settlement and sale of the public domain, than to hoard up the treasure like a miser, with a view to make a princely fortune for a person or persons yet unborn.

Nor did settlers acknowledge generosity in a policy which exacted high prices for public land but allowed a long time to pay. "[W]e look upon the extension of the time as no mitigation," some 1852 petitioners tartly informed the state legislature, ". . . we consider it a remedy which proposes to cure our diseases by bleeding us to death." Those who urged speedy and cheap sales, or easy pre-emption or homestead privileges, could be confident they invoked symbols of broad appeal.[33]

Amid these pressures for present realization of assets, the prevailing opinion nonetheless did not seek so much to enhance present consumer satisfactions as to score present productive advances, the more rapidly to increase the long-run productive power of the country. Particularly in the mid-century Mississippi Valley, the prevailing currents of ambition and criteria of accomplishment among men of modest as well as sizable means, among farmers as among financiers, traders, and industrialists, involved investment and looked for capital gains. Men here defined their roles primarily as producers rather than as consumers. These were attitudes which in some ways and in material degree oriented action to future satisfactions.[34]

The idea that the long-term economic and social progress of the society required the present injection of booster shots of capital such as the public lands could supply was more or less explicit in all major aspects of popular and official discussion of programs. The successive federal pre-emption laws attested Congress' acceptance of popular pressure to open up the continent at as fast an exploitative tempo as the supply of labor and other working capital allowed. On both the federal and state scenes, the objections to any substantial long-run landlord's role for government and to large-scale private speculative holding of land rested largely on the fear that these would slow the pace of "improvement"; low prices for public lands were important not only to promote a generous mixture of cheap (abundant) land with expensive (scarce) labor, but also to speed up the rate at which labor (settlement) was added to land. Chafing under the brakes which scarcity of working capital put upon their ambition, "practical" men of the day saw it as common-sense economy to make liberal use of cheap land where this might in a measure speed the otherwise laggard pace at which marginal increments were made to capital; to their impatience, land was not "wasted" if it was used to increase the present productivity of more scarce factors of production. So, in 1842, a committee of the territorial legislature urged speedy sale of the federal lands granted in aid of a university, for "the settlement of the country, which the sale of these lands would encourage, would be a greater public benefit than the enhanced value of the lands, which delay might afford, would be to future immigrants or to a generation yet unborn. We need the strength of a *present* population to share in the burdens of the government, of schools, of roads, and of the public defence in case of war." The greatest present utility of the public domain was that it be used to get the economy off dead center, argued a Wisconsin editor in 1846: Let there be "immediate disposal of all public lands whatsoever under proper restrictions, so as to encourage the settlement of the country by an agricultural and mineral producing population, which inevitably must result in the establishment of manufactures and the extension of commerce."[35]

The booster investment theory was, of course, most confidently urged in the many efforts to apply public lands to aid the construction of transportation improvements. So in 1855 Wisconsin's Governor Barstow congratulated Congress and the state on the wisdom of the federal land grant in aid of the improvement of navigation on the Fox and Wisconsin Rivers: "Undertakings of a magnitude which would repel private enterprize unless thus aided, are encouraged and assisted to completion, and where, without such aid, if attempted, they would languish and die, the whole country including the General Government, by the more rapid sale of its lands, are benefited to an extent not easily calculated." The legislature recurred to the booster argument over the years so long as Wisconsin sought railroad-aid lands. In 1871, memorializing Congress to renew grants in aid of a railroad for the northwestern part of the state, the legislature reasoned that the renewed aid "would secure the speedy development of a large portion of this state, which must otherwise remain for a long time sparsely settled; . . . and . . . while the renewal of said grant would involve no loss to the national treasury, it would bring a large region into actual cultivation and settlement, and very greatly promote the prosperity of this state and of the whole northwest."[36]

These arguments presented considerations of weight and sense; on the other hand, how much weight and sense should be conceded them depended on a closer appraisal of social costs than the proponents of booster investment ever attempted. It was the easier to ignore costs because contemporary economic theory offered little basis for a critical review of the booster position. Like the "practical" man, the mid-nineteenth-century economist took productive growth to be an obvious good in itself, and his theoretical apparatus, designed to analyze the conditions of achieving a static equilibrium in exchange, paid little attention to the evaluation of differential rates of economic growth or the estimate of their costs.[37]

The extent to which the booster justification for rapid disposal of the public domain dominated decision can be measured also by the narrow limits of such attention as was given to long-term realization of public lands values. The story here parallels that already traced of the monetizing of public lands policy. Almost the only consideration given to the possibility of delayed realization upon the public lands was for the sake of the larger cash returns that this alternative might yield. John Quincy Adams was the most distinguished of a small group of easterners who argued that the United States should move slowly in disposing of its lands, so that over the years they might provide a continuing source of funds to finance national institutions of learning and science. No general program came out of this, and as we have seen, by the Pre-emption Act of 1841 federal policy decisively emphasized "settlement," with all the overtones of using cheap land to increase the present productivity of scarce mobile capital. True, federal and state allotments of public land to education were an outstanding expression of concern for a long-term future. But it was the money yielded by land sales, and not the land, which men envisioned as providing a permanent trust for the schools. This becomes very plain in the discussions that attended the creation, in the Wisconsin constitution of 1848, of a trust for education out of the sale of the principal state lands. Almost invariably, talk ran in terms of the "fund" or the "sums" which the public domain would provide for long-term public purposes. Optimism over the generosity with which the United States might be expected to endow the educational needs of the new state was coupled in the same breath with the argument that there should be "immediate disposal of all public lands." Insistence on putting the land to work was never far removed from desire that the land bring into being a continuing educational trust. Wisconsin should profit by the sad example of Illinois, wrote "A Farmer of Grant" to the Platteville *Independent American*: Illinois "has had a large school fund wasted or destroyed by their speculators for want of a provision in the constitution to guard it, to increase it, and to put it in action." [38]

The exceptional arguments for holding on to the state lands were almost always addressed to obtaining a better price by waiting for the rise in values that would accompany community growth. The dominantly monetized definition of the issues emerges by comparison of this kind of talk with the rare instances in which men urged attention to the "real" stakes of power involved. If government held on to the fee, conceivably it doubled its ability in the long run to use the land as an instrument by which to regulate social and economic relations in the general interest. The land might still yield current cash for public purposes, as rent. But periodically, upon the expiration of whatever leases or licenses it granted, government might use its ownership to redefine the modes of use and incidence of benefits of assets which would bulk large enough in the over-all economy to offer a powerful lever of policy.

Writing to the Madison *Wisconsin Argus* in 1846, "Rough Hewer" urged that government hold onto the school lands to enhance its long-term practical powers of regulation. He would stipulate in the state constitution that

> the school sections donated by Congress to the several townships in this state shall not be sold or alienated from the perpetual ownership of the citizens of such township, and shall not be rented for a term of over thirty-three years, and shall be subject only to the control and management of the townships in which they are situated in such manner as may be provided by law.

In 1848 resolutions of a meeting of citizens of Burlington village criticized the proposed constitution partly because it contemplated sale of the state lands. To these critics the possession of a money fund arising out of the lands was a poor substitute for the power which continued holding of the lands would give the people's government:

. . . the sale of any portion of the school lands is fraught with the greatest evil and ere long will prove detrimental to the best interests of ourselves and our posterity. Facts have almost invariably demonstrated that when any portion of the people's money is placed in the hands of agents with a view to the accumulation of interest thereon, they have not only been disappointed in its accumulation but robbed of its principal.

But the practical importance of these expressions is only to point up by contrast the overwhelming indifference of general discussion toward other than the monetized value of the state's lands. The Southport *Telegraph* spoke the general temper in its easy dismissal of the Burlington critics: "Let honest and judicious men be selected to manage these matters, and no fears need be entertained as to results." The brief debate relevant to the problem in the 1848 convention likewise reflected the prevailing attitudes, as it shifted from what sounded like emphasis on the "real" (control) values of continued state ownership to a simple plea that the state should hold for a rising market. To shape an issue, Delegate Sanders moved to strike the provisions for sale of the school lands:

So far as he had heard any expression on this point among the people at large, it was decidedly opposed to the sale of them. Illinois had sold hers and squandered the proceeds. So it would be with us, if we sold ours.

Kilbourn proposed to amend the provision to leave sales to the discretion of the legislature. Sanders' first reaction was that this would not do; he "would by all means have it prohibited in the constitution." Kilbourn pressed his amendment, and Root then brought the matter to a sharper focus when he asked whether this would mean that the legislature might lease the lands. Kilbourn thought that it might be desirable to add a stipulation forbidding lease. Sanders found this additional suggestion unwise, but now switched his objections wholly to the criterion of money yield. To forbid the state to lease

would force the lands into market, and he thought it would be highly inexpedient to do so. The school lands in Racine were probably worth half a million and were rapidly increasing in value. There were doubtless many who wished to have them sold, for their own interests, but he did not think it was for the interest of the school fund that they should be. The improvements made on these lands were a sufficient answer to the argument that lands would not be improved of which the state was landlord.

On further consideration, Sanders softened his position and accepted the Kilbourn amendment allowing the legislature discretion as to sales. No more was heard of the argument for the superior "real" benefits of continued state ownership of the fee.[39]

Of course, had argument been pressed harder for the long-term "real" advantages of continued public ownership of the fee, it would soon have met the powerful political and social objections to government landlords. The rare proposals for a policy of leasing, rather than selling, public lands highlight the range of power which the public domain put in government to shape the country's growth. But the leasing proposals lost out so easily to the prevailing desire to realize the cash value of land that their story bears strongest witness to the trend to monetize the public lands programs.[40]

Finally, we must reckon with one other limitation inherent in the time perspective in which the mid-nineteenth-century United States viewed public policy making. In ways that materially affected policy, the people did believe in "the future." But in most respects the definition of the span of this future could be, and was, left conveniently vague. To spend much time and energy on ambitious forecasts was felt as a luxury that men could not afford when the immediate challenges of an unopened continent pressed so close upon them. So, when they undertook to define "long-term" action programs, their conception of the long term was not in fact very long. In 1842 the Committee on Schools of the territorial House of Representatives debated the choice between short-term and long-term policy, as encompassed within the alternatives to make "present" sale of university lands or "to leave them for ten or twenty years to rise in value . . ." On the few occasions when a term for holding public lands was mentioned in the discussions surrounding the framing and adoption of the Wisconsin constitution, as few as five and never more than twenty years were taken as measures of a holding policy. The short-lived history of one effort to establish a holding program is revealing of the contemporary pressures that worked against substantial postponement of sales. On recommendation of the University Regents, who were concerned to increase the yield of aid lands held for the institution, the 1850 legislature set a high $10 per acre minimum price for these lands; since common school lands stood under a $1.25 minimum and the best of them were currently appraised at an average of $3.44 per acre, while the average appraisal of the university lands had run at only $2.78, the statutory $10 minimum plainly implied that the university lands should be held for a long-term rise in value. A year later the Governor noted that the university lands were not selling, and, though the Regents expressed optimism that all would find buyers within ten years at $10, the 1851 legislature found it prudent to reduce the minimum price to $7 per acre. In 1852, following the recommendation of the school land commissioners, who reported that the university lands were unsalable at the $7 price in competition with other tracts, the legislature again reduced the minimum, this time to $3 per acre, at which point it stayed so long as any material quantity of university lands was available.[41]

General policy toward the public domain was shaped by men whose hard work and driving ambition were sustained by faith in an expanding future and in the socially desirable results to be expected from steady increase in the production base of the economy. Their faith thus reinforced their necessity to emphasize investment in capital goods rather than in consumer satisfactions. On the other hand, their dominant political tradition was egalitarian and pressed them toward rapid and widespread transfer of public lands to private ownership. Sharp-felt scarcity of mobile capital (labor and cash) molded their rule-of-thumb economic policy, so that they tended to value land for its capacity to mobilize and increase the present yield of labor and money, rather than for its long-run physical productivity, and to put a high premium on generous present investment of land which promised to give immediate impetus to economic growth. In this context of ideas and attitudes they met the special problems of the public timberlands.

THE ENVIRONMENT OF PUBLIC POLICY MAKING

The life of any public policy exists in four dimensions: time, action, awareness (thought and feeling), and relation to a total situation. First for our interest is time (sequence and co-existence, contingency, pace, and span) — because our particular concern is to gain such distinctive understanding of the other dimensions of policy as we can get from regarding them in history. Second, we measure the overt extent of policy by action, by what men declare and practice as legally sanctioned values — as the first section of this chapter seeks to do by outlining the patterns of public lands disposition. Third, we measure the direction and energy of policy by what men desire (both in purpose and in motive), and by what they think they are doing to achieve what they want (both their studied and their naive theories of cause and effect) — as the second section of this chapter seeks to do by tracing the values which were the dynamic of public lands handling. Finally, we measure the creativity and the accomplishment of policy by relating it to its moving environment, the shifting context of circumstance in which it evolves — as we shall here try to do in part.

Desire, thought, and action never set their own terms of being. Policy making is always limited by an environment. This environment includes not only resistant fact outside the policy makers, but also the policy makers' own perceptions, and the subtle limits set to their will and imagination, beyond their awareness, by their whole experience. Perhaps, indeed, the most significant dimension of policy is its relation to the massive weight of surrounding circumstance. So great seem the odds at all times against men's limited store of ideas, resolution, and energy, that there is special and poignant need to ask whether, how, and in what degree men add to the opposing inertia or drift of their total situation by the defects of their own contrivances.

PRESSURES OF THE SITUATION

We can the more briefly note some ways in which environment molded public lands policy because these involve matters we have already discussed in another aspect. The preceding section examined key value judgments by which official and popular opinion defined objectives and tactics in use of the public domain. If they were incomplete or mistaken, these values nonetheless responded partly to the felt pressure of surrounding facts of nature, tradition, and institutions; we need now to take account of these framework factors as they existed apart from men's direct perception of them as elements of public lands decisions.

Always in the background through most of the nineteenth century was drastic imbalance among the factors of production in the economy of the United States and of Wisconsin. We had much unopened land, relatively few people to fill it, and relatively little other mobile capital with which to expand the production base by improving land and transport and building factories and machinery. This state of affairs lasted so long as deeply to condition our social and economic calculus. In hard fact, and not just in men's imaginings, land was relatively cheap and labor and cash relatively expensive, measured by their present marginal yields. High returns encouraged the assembly of mobile capital. The cumulating increase of people and investment created a heady atmosphere of fast spiraling growth, accenting action, venture, and capital gains quickly realized. This economic complex rested on low-cost land, not only for its capacity to yield marketable commodities, but also because the spread between modest acquisition cost and acreage values steadily rising with community development provided security for lenders on mortgage to farmers, railroads, and lumbermen. Wisconsin had already experienced the teaching of this pressure of circumstance in the hectic rise and fall of a great wheat industry from 1850 to 1880, before lumbering reached its full growth. People so conditioned by immediately felt terms of livelihood would not readily adopt policies which would sharply raise the price of public lands, or long withhold them from present production, or closely count the "real" costs of their use against their availability as substitute or booster for scarce cash or labor.[42]

The proportioning of the factors of production provided a setting of subtly powerful bias for the exercise of common sense and wisdom, but only a setting. Such initiative and direction as history shows spring not from the impersonal situation but from what men do with it. What they do with it is much determined by what manner of men they are. Part of the answer may lie in the realm of focused choice; so Wisconsin's constitution makers imposed a "trust" upon policy concerning the state's lands, because they placed certain explicit values on public school education. Part of the answer, probably the greater part in most instances, belongs in the pattern of the total situation within which policy is made. For particular decisions not only express focused choices but give visible form to much broader and deeper values, usually only in part articulated and all the more influential because of that, forming the core of meaning that men of a given culture see in their existence.

Thus the public lands policy of the United States and the state of Wisconsin bears the unmistakable imprint of an individualistic, enterprising, instrumentalist, middle-class philosophy born of the Reformation and the commercial and industrial revolutions. No circumstance more influenced this policy history than the fact that it unfolded in a time when such a point of view supplied the taken-for-granted criteria by which we generally ordered affairs. "The maxims and principles of law, as applied to real property, originating in the policy of the feudal ages, are, in many instances, entirely inapplicable in a country like ours," cautioned the Wisconsin Supreme Court in 1849—"and particularly in the new states and territories, where there is such a vast public domain, where the spirit of emigration is so rife, and where the genius of our institutions, as well as an enlightened public policy, favors the removal of all unnecessary restraints upon the alienation of land." Official leadership and popular opinion alike found it sensible and just that we use the public domain to disperse decision making broadly among private owners in fee simple absolute, consistent with the seventeenth-century English inheritance of constitutional government as we found it defined in Locke; that we use public lands to promote saving and investment of fluid capital, and encouragement of population growth, to give expanded scope to the entrepreneurial skill we valued as a prime social asset; that, without doctrinaire concern over the law's intervention in the economy, we subsidize commercial agriculture and private lending and transport development out of public lands wherever this promised to multiply productive energy; that, consistent with our confidence in the creative possibilities of individual mind and will, we make land a freely marketable commodity as fast as we could, and let the terms of its use be set primarily by the market, according to a money calculus, through deed and contract; and that we trust that the infinite transactions of a numerous class of bustling, mobile, self-centered, saving and risking, calculating and plunging individuals would yield a balance of net profit for them and the society. "I would desire," Governor Farwell told the Wisconsin legislature in 1852, "to see such a state policy adopted as would make available all our resources, and means of improvement; and render us, to as great an extent as possible, independent of all foreign aid, except the benefits of market and commerce."[43]

Our middle-class orientation was fostered by the institution of the market, yet likewise itself operated to make the market more important, encouraging us to assign a major role in social control to market processes. Thus, the market, and popular conceptions of its utility, provided a third principal element of the environment in which mid-nineteenth-century policy making went on.

The positive aspect of this influence was the engrossing attraction which the market exerted through most of the nineteenth century to capture men's primary interest and energies, to the detriment of the conduct of public business. From the 1820's on, apart from moments of crisis in the tension between North and South, men were fascinated by the challenge of unlocking the productive potential of the continent and gave only grudging, secondhand attention to constitutions, statutes, and public administration. The work of appellate courts showed more sustained, professional concern with public policy. Yet, at least as concerns the public domain, the courts generally dealt only with policy of the second order of importance. By their offices, constitution makers and legislators held the sustained initiative here. But as one lives with the record he cannot escape the conclusion that first-rate talent was never expended upon the constitutional or statutory foundations of federal or state policy toward the public domain. It would miss the point to see this as a peculiar defect of this sphere of action. Impatient haste, loose thinking, and shoddy technical performance, piecemeal rather than generalized treatment, were the characteristics of mid-century policy in any field it touched. There were secondary causes for this in the imperfections of the legal process; we must note these problems in a moment. But the second-rate quality of the work is too pervasive and persistent to be explained simply by technical deficiencies; it can only reflect the fact that men saw the market and not the forum as the area in which it was important to invest their first ability and their principal time. Herein was a profound bias of the culture which we must always reckon into the balance as we appraise public lands dispositions.[44]

There were negative aspects to the market as an environing institution, which inevitably affected policy regarding the public domain. One of these negative aspects was inherent in the constitution of the market. It was by function a process for decision making, through contracts, about resource allocations which had exchange value. By function, therefore, it could take account only of matters of sufficiently defined and limited contour to engage the pecuniary interest of the dealers, and only of such costs and returns as had direct incidence upon the dealers. So far, therefore, as men accepted market calculations as a central procedure for achieving social order, and so far as they used the law primarily to foster and support the market, implicitly they restricted the calculation of public policy to those factors that had impact on private contractors, and omitted those factors of income and cost the incidence of which was felt only by groups too large or ill-defined to be dealt with by contracts of exchange. A tangible expression of this in Wisconsin economic development preceding the maturity of the lumber industry was the indifference of public policy toward the mining of the soil that subsidized the wheat boom for a generation after 1850. The intrinsic limitations of market calculus exerted the more narrowing effect upon the frame of reference for social decisions because the limitations were so little explored in contemporary economic theory, which — engrossed in analyzing the achievement of exchange equilibria in a basically static situation — had little to say about processes of economic growth.[45]

Other negative aspects of the market represented defects of operation rather than inherent characteristics. Thus, through most of the nineteenth century, certainly in Wisconsin, the handling of resources was adversely affected by the lack of disciplined or efficient financial machinery. There was not a broad-based mortgage market, nor until the latter part of the century was there a substantial investment banking business to service heavy industry. One consequence was a contribution to cyclical instability, another the lack of reasonably priced, long-term credit, indeed, usually a lack of long-term credit at any price — deficiencies which pressed operators to excesses of optimistic expansion to ride the ascending curves of prosperity, and to chronic over-production to meet their fast-maturing notes. These troubles helped create persistent background pressure to maintain easy access to public lands at low prices for quick production. Imperfections in knowledge are a familiar drag upon the efficiency of the market, and this element also loomed large in the background of public lands handling. The growth of bulk grains production for the market in mid-century far outpaced the growth of scientific or practical knowledge of soils resources. The great premium which our egalitarian values and our middle-class ambitions placed on the development of a community of family farms made us anxious to believe that all land was potentially good farmland and that agriculture was the normal end use of all land, whatever its original condition. "Agriculture . . . is our great and leading interest," Governor Farwell counselled the 1852 Wisconsin legislature, "and any measure or policy, having a tendency to embarrass or restrict the cultivation of the soil already appropriated to the purposes of husbandry, or the improvement of lands yet vacant and unsold lying within our boundaries, I regard as militating against our best interests as a state; and on the other hand, any thing conducive to promote and encourage this branch of productive industry, and to extend its limits, is the true policy of the state . . ." Such views, and the assumptions of fact implicit in them, would profoundly condition our handling of the public timberlands. Here it is relevant simply to note that there were extremely important limitations to what we knew of the mineral, agricultural, and forestry potential of our soils — limitations with which the market could not cope, and which were bound, therefore, to exert the greater distorting influence upon our employment of natural resources in proportion as the law resigned to the market the classification of land uses.[46]

THE LAW AND LEGAL PROCESS AS PART OF THE SITUATION

Save under extraordinary conditions, the creation of any given public policy in any generation makes only a marginal addition to the great body of the law. Existing law — doctrine, implementation, and modes of change — is a large part of the environment which conditions any fresh programming. Consider, thus, the dual significance of the institutions of private property and contract. These responded to needs and demands originating outside the legal order; on the other hand, the existence of a great body of property and contract law, providing familiar means to define, legitimize, and adjust resource allocations, in itself encouraged men to think of market processes as the normal mode of ordering the economy and fulfilling the political and social ends of a system of broadly dispersed power.

The United States succeeded in organizing a federal legal order of continental scale. This was a general aspect of law of great consequence for public lands policy. With wisdom we early decided that new states should be admitted to our union "on an equal footing with the original States in all respects whatever." We could not have built federal union on any other basis. But one cost was to add righteous vigor to the interested demands of fresh-settled areas that they suffer no more brake upon their exploitation of natural resources than had older communities, and yield no more preference to national over local advantage than had been enforced on others. Governor Barstow spoke in this temper as he urged on the 1855 Wisconsin legislature the justice of pressing Congress for all the internal improvements aid the state could get:

> Thus far we have had but little to thank the general government for, save those benefits resulting from democratic institutions, and of which we are all the common recipients, saving those embraced in donations for special purposes. We have paid into the United States treasury the fixed value of every acre of land now or heretofore owned by the state or its citizens, and of this amount thus promptly advanced, the country has received the benefit. These reflections are rendered yet more galling, by the fact that, while the other states have been admitted upon a much more liberal footing, with regard to the public lands we have been denied . . . the necessary appropriations we consider our due . . . The general government still owns nearly or quite one third of the domain embraced within the boundaries of the state, while it does not afford any assistance to our treasury, or lighten, in the smallest degree, the burdens of taxation . . .

With like view of the equities, in 1862 Wisconsin's legislature vigorously protested the bill which became the Morrill Act, granting federal lands to all states to help them set up agricultural colleges:

. . . it is manifestly unjust to the western states, within whose borders nearly all the unsold government lands now lie, to take said lands, or any part thereof, to be appropriated to or held and controlled by corporations without the states in which such unsold lands may lie, and particularly to grant such lands to the old states of the union, that have already realized the full benefits of settlement on nearly if not on every quarter section of land in their respective states . . .

The years showed that these were attitudes not peculiar to the Mississippi Valley, but claims which successive undeveloped areas felt to derive fairly from their equal admission to union; indeed, the time would come when still newer states would cite the easy and unrestricted transfer of land from public to private ownership in Wisconsin to prove the injustice of any more limited disposal of federal lands in the far west. On the other hand, if the federal idea was thus twisted to justify parochial demands, the practical success of federalism nonetheless exposed state economies and state policy making to the sweep of pressures generated in much broader areas. The federal Constitution sanctioned and protected the play of supply and demand in sectional or national markets, including the free movement of labor and investment money wherever they might find profitable employment. Well before the prairie states' insatiable demand for lumber exerted its full attraction on the Wisconsin forest, the state had experienced the shaping power of outside markets for its minerals and grain. Federalism meant that even within its proper local sphere each member state must make policy under the stress of forces greater than would be mustered solely within its own borders. For the nation, federalism meant that the heightened impersonality, the more involved chains of cause and effect, and the greater power of multi-state markets would allow freer play to demand and surging growth, less opportunity for the conservatism taught by operation within more stable and limited ends and means.[47]

The circumstances of our federal union specially conditioned the development of the state's policy regarding public land. Inevitably, as the introduction to this chapter pointed out, the policies and practices of the central government toward the public domain provided models, competing pressures, and subtle limits on will and imagination in the evolution of state policy and practice, when the United States originally owned all the land surface of Wisconsin, itself made direct disposition of 70 per cent of the area, and stipulated the general purposes for which Wisconsin should use the 30 per cent put in the disposition of the state. In substance and procedure, the state's land legislation reflected the direct influence of federal example. Though less openly displayed, its indirect influence moved on the more basic level of molding ideas and feelings as to what was right and reasonable. So, recognized limitations of federal action were cited as moral warrant for courses taken by state law. Why, for example, should state law treat a squatter on federal land as one with no interest the law should recognize as the basis for a contract to finance the purchase price, when the United States took no effective steps to oust such settlers? By 1851, said the Wisconsin Supreme Court, it was "too late in the day to say that the occupant was a mere trespasser when taking possession of and improving public lands, with a view to effecting a purchase of the title at government prices." Government tolerance of wholesale illegal entry upon the public domain in effect recognized the true values of the society:

Such alleged trespassers have peopled and fertilized the great west, until the wilderness was made to bud and blossom as the rose, ere the tardy movement of the government provided a means of securing a title by the hardy pioneer. But congress has recognized the rights of the settler, by passing frequent preëmption laws for his benefit; and it is only in the abstract and upon principles more nice than wise, that the courts can hold occupancy and actual cultivation of the public lands as a trespass; or the transfer of such a possession as without consideration.

Again, though the United States might be making mistakes in its land sales program and in its grants of land to states, so long as it persisted in its errors the prudent line for the state was to take what it could get. With this unabashed plea, Governor Barstow in 1854 countered those who feared that railroad-aid grants spelled dangers of corruption and anti-republican concentrations of power in private hands:

And whether it is true policy of the United States to make such grants in aid of internal improvements, to the several States or not it seems to me that there can be no reasonable doubt as to the policy of the State in receiving them. The United States have not yet adopted the laudable policy of giving the lands, in small quantities, to actual settlers — a measure which has been repeatedly and earnestly pressed upon their attention — but still continue to sell the same, and in quantities to suit the purchaser. While this is the policy, I confess I can see no good reason why the States may not accept and prudently expend the avails of such grants in aid of internal improvements, necessary to their growth and prosperity. The fact that such donations have sometimes been mismanaged, furnishes no argument whatever, in my opinion, against receiving and applying them in aid of important public works. Such a course would surely be prudent for us in our individual capacity. Then why not for us collectively? [48]

Property and contract law, and the federal system, embodied powerful influences upon the handling of the public domain. But, along with such particular aspects of legal organization, we must reckon the influence of two general attitudes concerning public decision making. Nineteenth-century federal and state legislation constantly betrayed the fact that our top policy makers lacked the confidence, experience, or will to make substantial investment in planning what they did. Their impatient haste and opportunistic response to the most immediately felt issue was time and again reflected in enactment of three statutes to do what should have been co-ordinated in one, and in repeated repeals and amendments which within a decade made a crazy quilt of important general legislation. Lack of conviction or grasp of the difficulty of basic policy making was reflected in the readiness with which Congressmen and state legislators alike gave time and energy to hundreds of bills of the narrowest local or special application, while greater matters went by default. Jefferson's America exalted the image of the self-reliant farmer. Jacksonian America broadened the ideal to fervent insistence on the all-round capacities of the common man; God "has conferred the gifts of mind upon every member of the human race without distinction of outward circumstances," preached George Bancroft, and so "true political science does indeed venerate the masses," and we could rest confident that "the Spirit of God breathes through the combined intelligence of the people." The American Whig would roundly damn Bancroft, but in his own way he shared a similar bias; if the politicians would only go away, sound men of business would put public affairs in order as a rather simple, and often useful, incident to getting on with the main matter of increasing production through the market. We definitely did not believe in the need or desirability of the professional hand in government. Properly defined, this attitude provided one of the most valuable of the checks and balances in our constitutional tradition. But we misused a sound insight to justify the ignorant man's suspicion of knowledge, the busy man's disdain of reflection, the technician's indifference to understanding. We misread equality to warrant neglect of quality in public decision making; and the truth that the big questions are too important to leave to experts we perverted to warrant improvisation as the regular mode of public business.[49]

This depreciation of the challenge in public policy making was reinforced by another common sentiment, mingled distrust in government's integrity and doubt of government's ability to recruit the competence and interest to do its work well. Calculating the future worth of federal land grants to the new state, a committee of Wisconsin's territorial legislature limited its optimism because, it explained with resigned acceptance, "it is not to be expected that these lands will be located for the benefit of the State with the shrewdness which a capitalist would exercise for his individual interest." The event considerably justified the committee's prediction, but more relevant to our present concern is the submission to the fact that public business would get second-rate attention. Sober fear was expressed, before Wisconsin accepted federal land grants in aid of railroad construction, that the state government would not have the strength to withstand the temptations to partiality or corruption that would be generated out of contests to win these prizes. A major scandal over the disposition of the first railroad grant, in 1856, amply confirmed these forebodings. The experience did not stop the state from seeking further railroad grants — common insistence on public subsidy of economic development was too strong for that — but the 1856 scandal strengthened popular attitudes which were bound to limit enthusiasm for any sustained, large-scale managerial role for government in the use of public lands. The legislature stood under continuing unjust suspicion in discharging its duties regarding railroad-aid lands, a legislative committee complained in 1867: "The indiscriminate prejudice which, to a considerable extent, seems to regard a state of total depravity as the normal condition of members of the legislature, and all managers of railroads, would aid the many, who are too ready to accept suspicion for proof . . ."[50]

All these factors — the predisposition to dispersed decision and diffused and ill-defined responsibility underwritten by the prominence of the law of property and contract, the scope given at once to the influence of broad markets and to local interest within the federal system, and the prevailing attitudes which limited investment in public planning and confidence in public action — pointed to the critical importance of the law's own provisions for policy initiative. For these general elements of the legal environment clearly favored the odds for drift and default in policy. If the legal system itself made no counterprovisions adequate to focus responsibility and concentrate will, the society indeed had compounded the deficiencies of its leadership.

The lack of adequately specialized, high-level agencies to devise and check top policy was a fact which bore heavily on the course of both federal and state handling of the public domain. This was not for want of a sufficient central deposit of power. Without qualification the Federal Constitution invested Congress with "Power to dispose of and make all needful Rules and Regulations respecting the Territory or other Property belonging to the United States." By familiar doctrine the Wisconsin constitution left plenary power in the state legislature, save as it spelled out particular limitations. As to lands which came to the state charged with some purpose other than aid for education, the state constitution left to the legislature full discretion to decide what modes of disposal would fulfill the state's obligations as grantee. As to state lands charged with the trust which the constitution created for education, it divided authority between the legislature on the one hand ("Provision shall be made by law for the sale of all school and university lands") and, on the other hand, the secretary of state, state treasurer, and attorney general, constituted as "a board of commissioners for the sale of the school and university lands" (with power "to withhold from sale any portion of such lands when they shall deem it expedient"). Such ambiguity as this division of authority carried was resolved by the Wisconsin Supreme Court during the decades of major state lands disposition, in favor of broad legislative power to fix the terms of disposal, subject only to the constitutional trust as to the purposes for which the lands' yield should be put, and to the constitutional authority of the commissioners to have the last word on the "expedient" timing of "sale" and to exercise exclusive authority in the execution of state land law.[51]

Of course, Congress and the state legislature did provide a frame of public lands policy. But it was a frame constructed by criteria so generalized (the self-evident worth of "settlement" or "improvement," men's "natural right" of access to the soil, "the principles of republican government"), as to avoid the hard questions, checked by little attention to concrete choices, values, or costs. So far as legislators concerned themselves with the concrete, it was to spend energy on a hodgepodge of special and local acts disposing of particular tracts, providing for particular surveys, curing particular title defects, or adjusting particular claims. Congress and the Wisconsin legislature created standing and select committees on public land matters, and in a later day good committee work might have furnished some leadership. But the public domain was an urgent problem in years when legislative organization was new and crude, and lacking even the semblance of experienced staff. Neither the federal nor the state legislative record shows one example of committee statesmanship in this field. If the committees were not to lead, they might still have contributed by hardheaded review of what was done, pursuing the classic legislative role of inquiring into the operation of law and the conduct of the executive. Congressional and state legislative committees did from time to time review aspects of public domain administration. But this activity was typically episodic, haphazard and limited in incidence, often marked with partisan animus, more concerned with technical forms or efficiency of day-to-day administration or alleged abuses in particular transactions than with long-term direction or goals. The tradition of slack surveillance of public lands administration, which had become established in the years before large-scale disposition of the public timberlands, was acknowledged by a Wisconsin legislative committee in 1855: "No thorough examination of the condition of these several departments [notably the office of the school land commissioners] has ever been made since the formation of our state government, and the real condition of the affairs of this state is entirely unknown." John Quincy Adams had earlier passed a more bitter judgment on the victory in Congress of immediate expediency and a narrow calculus in disposing of the public domain:

> The thirst of a tiger for blood is the fittest emblem of the rapacity with which the members of all the new states fly at the public lands. The constituents upon whom they depend are all settlers, or tame and careless spectators of the pillage. They are themselves enormous speculators and land-jobbers. It were a vain attempt to resist them here.

We must make due discount for the characteristic Adams conviction of righteousness. Even so, the appraisal deserves weight as a contemporary estimate of the environment of federal policy. For it measures the situation by the criterion of Congress' capacity as an institution to contrive and stand by a deliberated, balanced, continuously productive public lands program.[52]

A twentieth-century observer naturally asks whether the executive might not have furnished policy leadership. The nineteenth-century President and Governor enjoyed some advantages inherent to the chief executive's post — opportunity to take counsel in a small circle, the constitutional functions of recommending matters of general welfare to legislative attention and seeing to the proper execution of the laws, the veto power, and enjoyment of some insulation from parochial pressures conferred by responsibility to a wider electorate than chose any legislator. Again, however, events outpaced institutional growth. The chief executive's policy leadership could not be created by fiat. It would develop only by use, and only with the help of yet unborn means of communicating with a broad public, and only with the building of staff and departmental organization which our scarce means and our inherited distrust of the executive branch would not quickly allow. Moreover, in the nineteenth as in the twentieth century, many problems competed for the limited time and energy of chief executives and legislators. If the absolute measurements of nineteenth-century business seem smaller, yet relative to the country's needs and resources they loomed as big as problems of a later day, and the more so because the apparatus to deal with them was so crude and untried. Public lands policy lost out in competition for the prime attention of Presidents, Governors, and captains of legislation, in comparison with problems of internal improvements, sectional conflict, currency, finance, and creditor-debtor relations — perhaps because in these areas the facts defined conflicting interests with sharper immediacy and with more drama. No President contributed special insight to integration of the public domain into the growth of the Lake States economies. On a few occasions Wisconsin Governors lifted policy discussion to a perspective never achieved in any legislative analysis of the time — as in Farwell's successful plea (1852) that, if the legislature seek federal land grants in aid of railroads, it ask for grants in the state's keeping and not merely sponsor ambitions of particular promoters, or in Harvey's effective veto (1862) of delegation of state lands disposition to the mercies of county politics. In the whole story, however, such were isolated, disconnected incidents, of practical meaning chiefly to emphasize the general lack of sustained exploration of major decisions implicit in public lands disposition.[53]

We lost the possibility of leadership at the second level of executive action by excessive reliance on ex officio functions, characteristic of the nineteenth century. Federal public lands matters were committed to Cabinet officers burdened with much other business, first the Secretary of the Treasury, later the Secretary of the Interior. Given the size of the assets at stake, their potential for affecting the country's growth, and the range and diversity of interests involved, the handling of the public domain should early have been committed to a separate Cabinet office. The Senate Committee on Public Lands paid tardy tribute to the good sense of this in 1882, recommending the creation of a separate Department of Public Lands, the benefits of which it thought to be "so patent that it needs no discussion." There is no stronger circumstantial evidence of the mingled naïveté and opportunism of the century than the fact that the "patent" benefits of a strong initiating center for public lands programming carried no effective appeal before or after the 1882 recommendation. The story in the state was similar, but the continued reliance on ex officio administrators there stands out more sharply because it was more pointedly called in question. In the Wisconsin constitutional convention of 1848, delegate Chase squarely questioned the wisdom of committing administration of the school lands to the secretary of state, state treasurer, and attorney general as ex officio commissioners: "They would be selected on account of other qualities than their fitness for this duty, and they would regard it as merely incidental and of secondary importance." Proponents of the ex officio arrangement had no trouble in satisfying the convention with arguments which remind us again how influential was the felt lack of fluid capital and the related tendency to view the state lands as a potential money "fund" rather than as "real" assets. So delegate Root thought that considerations of expertness favored the proposed system, since keeping the school funds properly pertained to the office of the treasurer, and legal questions to the attorney general; moreover, there would be no added expense to the state for official salaries. When Chase moved to substitute the lieutenant governor and state superintendent of schools for the secretary of state and the treasurer, as commissioners, Estabrook objected that the qualities of a good superintendent of schools might not be suitable for the management of the funds, which was the requirement of the commissioners' job. Later critics thought that experience showed that the ex officio arrangement had been a mistake, though, in fashion characteristic of the limited imagination given to state land policy, the criticisms ran to narrow concerns. In 1854 Governor Barstow held the recordkeeping of the land office inadequate, but found excuse for the commissioners: "The ordinary duties required of these officers are such as to preclude their bestowing the requisite time and labor upon such accounts and records." The report of a legislative committee in 1856, finding inefficiency and improprieties in the conduct of the land office clerical staff, was less charitable toward the staff's superiors: "The testimony . . . goes to show almost an entire abandonment by the commissioners of everything, but salary and fees, to the clerks, and sometimes it seems they complained even that the clerks were taking from them too large a share of fees . . ." Some administrative reforms were accomplished, but, though the commissioners themselves thereafter characterized their charge as "vast and complicated," neither they nor anyone else moved to obtain separate management of the lands.[54]

More thoughtful direction and vigor might yet have been given federal and state public lands policy, had legislators backed up the ex officio administrators with adequate organizations. But, through the years when most of the land was disposed of, the United States General Land Office and the office of Wisconsin's land commissioners were badly undermanned and staffed by ill-paid officials who held clerks' status even when they were in effect heads of division, most of the time without any guaranty of tenure or any special qualification by previous training or experience. There was persistent criticism of administrative inefficiency; qualified observers thought the criticism well justified. There were periodic charges of corrupt favors to big speculators, on which it is harder to strike a balance. Shrewd men who dealt regularly in sizable land office transactions undoubtedly learned at least how to expedite their business through crowded dockets. But there is little clear evidence of outright corruption in the central staffs, however much shady dealing went on in the field. In any case, whatever the deficiencies of current operation, these were clearly secondary to, and a consequence of, the administrators' failure to supply more imagination and will to creating programs for a longer future and on a broader calculus of values. Yet surely the basic responsibility remained that of the legislators who kept the land offices on starvation budgets. In no respect did legislative penury do more damage to the potential contribution that vigorous executive action might have made than in the failure to arm the land offices to make competent classification of public lands according to their particular values of soil, minerals, timber, waterpower, and location. Of its own weight, a sound classification system would undoubtedly have generated or speeded demands for different handling of the public domain than took place; it would be naive to ignore the likelihood that this was one reason why powerful men did not press for it. In fairness, too, we must note that Commissioners of the General Land Office and — less often — Wisconsin school land commissioners complained of inadequate budgets and especially of inadequate field staff, and recommended closer classification of the public lands and repeal or improvement of legislation which promoted fraud or waste. They did so with little result. But, compared with the complacent inertia of those who held the real means of power, the feeble land offices showed enough life to suggest how much useful direction might have been given the situation by specialized, high-level, well-armed executive agencies. The realistic conclusion is that the legislators, and the country, got about what they actually wanted or were ready to pay for in the way of a public lands administration. The weakness of executive organization, the complaints and charges, and the lack of decisive remedial action by Congress continued as the pattern of federal lands history through the nineteenth century, into the disposal of the lands of the far west. The fact tells much about the environmental odds against major change of direction in the earlier years when Wisconsin land was in issue.[55]

CHAPTER II

THE DISPOSAL OF THE PUBLIC TIMBERLANDS IN WISCONSIN

THE STORY of public policy in disposal of the forested northern half of Wisconsin presents a striking combination of great influence and little calculation exerted through law. By their original ownership of the land, the United States and the state of Wisconsin as grantee of the United States held the keys of access to this vast resource. It was inherent that the way in which they exercised their ownership profoundly affected the use made of these forest assets and the consequent influence of their use upon economic growth and patterns of power reaching far beyond the pinelands. History corrects our complacency when it shows that so much can result from little deliberation. For, on the record, the most important decisions in law about disposal of public timberlands in Wisconsin were taken by default. Typically, one goal or procedure was adopted as an almost self-evident imperative without measurement against alternatives. That this could happen was due in large part to the state of legal institutions in relation to environing circumstance and opinion. The processes of public policy making were so organized as to respond to the most immediate pressures, whether of impersonal fact or of prevailing ideas, and to acquiesce in such a fragmenting of both official and private decision as to make it almost impossible to focus responsibility for large consequences. This is not to say that there was not a definable pattern of policy toward use of public timberlands. We shall now see that such a pattern existed, and in a context of identifiable general attitudes toward social values and means to their achievement. But, because we can see a pattern, we must not assume it is the product of design, or entitled to the deference due decisions of balanced deliberation.

THE PATTERN OF DISPOSITION

The United States had disposed of all but about 600,000 acres of the 35 million acres it originally owned in Wisconsin when, in 1892, Congress for the first time made any special provision applicable to disposition of timberland held by the federal government in the state. Through the years in which the United States sold or gave away almost all its forested land in Wisconsin, its public lands policy showed no significant recognition that the great forest of the Lake States was an asset which might have distinctive value or present special problems for its wise use.

Prior to 1878 the only affirmative measure taken to exploit the particular values of public timberland was to create small reservations of live oak and red cedar in the eastern and Gulf states to serve the Navy. Such meaning as this naval reserve program achieved for general administration of public lands came about in an indirect way characteristic of the episodic development of public forest policy throughout the century. By the act of March 2, 1831, entitled simply, to punish wrongful cutting of "timber or trees reserved for naval purposes," Congress capped a series of earlier acts making it a penal offense to steal timber from federal land reserved for naval use; Congress now imposed more stringent penalties for these offenses and also penalized anyone who might cut "any live oak or red cedar tree or trees, or other timber, from any other land of the United States . . . with intent to export, dispose of, use, or employ the same in any manner whatsoever, other than for the use of the Navy of the United States . . ." In 1849, in a prosecution for trespass to federal timberlands in Michigan, the United States Supreme Court interpreted the act of March 2, 1831, to apply to unauthorized taking of any kind of timber from any land of the United States without limitation either to naval reserve stands or to timber peculiarly suited to naval use. The court's opinion in *United States* v. *Briggs* offered no explanation of this surprising readiness to give an extensive construction to a penal statute, other than the terse observation that "the enacting clause is general." Though Congress had demonstrated no previous concern for the matter, it acquiesced in the ruling which thus created the first item of policy distinctive to the general forest assets of the nation. If the backdoor approach was characteristic of most decision making in this field, likewise characteristic was the content of the decision taken, that it was negative rather than affirmative in purpose and result. It was simply a decision to maintain the value of a stock of goods, pending its authorized disposition. This was a practical and important auxiliary to making a policy but in itself hardly constituted a program. There is more to say about the safekeeping of public timber inventory, but we shall postpone this, to return to the thin tale of positive policy toward public forest land.

Before 1892, no law authorizing disposition of federal land in Wisconsin made any provision special to tracts which bore timber. Until 1878 federal law provided no terms or procedures special to the sale either of timberland or standing or down timber, or authorizing the separate sale of standing or down timber or timberland as such, on any part of the public domain. In regard to positive realization upon publicly owned resources (as distinguished from merely safeguarding inventory), before 1876 federal policy recognized no classification of timberlands for any purpose aside from creation of naval reserves. The plain policy of the federal pre-emption and homestead laws, to promote agricultural settlement, implied that they excluded lands unfit for farming and productive only for their minerals or timber. But these statutes made no express reference to excluded classes of land, nor did they provide means to classify tracts by productive type, save as they implicitly relied for their administration upon the data contained in the notoriously uneven field notes from the federal survey. The mineral lands leasing system which stood on the books from 1807 to 1846 and the statutory pattern for transfer of mineral lands which began with the act of July 26, 1866, by their presence emphasize the absence of effective interest in any distinctive disposal of forest assets.

In 1878, in statutes applicable only to states in the Far West, Congress made first (and unfortunate) essays at separate provision for disposal of timber resources. In the Timber Cutting Act it authorized persons in certain far western states and territories to cut timber without charge on publicly owned mineral lands, for mining and domestic purposes; the privilege soon proved the excuse for wholesale abuse, for want of adequate provision for classification of mineral lands. In the Timber and Stone Act, Congress authorized the sale to citizens or to persons who had declared their intention to become citizens of public lands chiefly valuable for timber and stone, in quantities not over 160 acres to each buyer, and subject to affidavit that the timber or stone was for personal use and the purchase not made for speculation. Partly because they were intrinsically impractical, and partly for want of will or means for vigorous field enforcement, these provisions were soon turned to account to create great concentrated private holdings rather than the homestead woodlots which the act formally contemplated. In the face of repeated pleas by responsible administrators to repeal the Timber and Stone Act, Congress not only held to it but in 1892 declared that it should be "applicable to all the public-land States." Thus, at a point when most federal forest land in Wisconsin had already been sold or granted away, for the first time the United States provided a special mode of disposition for its timberlands in that area.

The event had no material effect on Wisconsin lumbering; fewer than eighty thousand acres were entered under the Timber and Stone Act in Wisconsin in the remaining years of substantial woods operations there. But the tardy and limited character of the 1892 measure makes plainer the total lack of a program specifically adapted to disposal of federal forest lands in Wisconsin through the decades that counted. The point is brought into sharper relief if we note that when the United States finally took some affirmative action to realize on the timber potential of Wisconsin land, this could only be by way of rebuilding, through small purchases of private land suitable for growing trees, to create new national forests in the state (under the Weeks Law of 1911).[1]

In the absence of any special program for disposition of federal timberland in Wisconsin, men who acquired it directly from the United States by legitimate means — or at least according to the available forms for legitimate acquisition — did so by buying it at public or private sale, by establishing a pre-emptive or (after 1862) a homesteader's claim to it, or by commuting a pre-emptive or homestead claim for cash. Outright purchase for cash accounted for the bulk of such transfers in Wisconsin, as in the Lake States generally; millions of acres of commercial timberland passed into private ownership by this route. Though the federal sales statutes made no special reference to timberland, they did require that initial offerings be by public auction with sale to the highest bidder. But the statutory pattern was rigid. Land which went unsold upon auction was thereafter on the market at the minimum price set by Congress, with no discretion in the executive to withhold it for future bidding when the market might be more favorable to the seller. We shall note that various factors in the environment conspired against the grant of such executive discretion, or the likelihood that it would have been used with vigor had it been granted. Yet it is part of the pattern of disposal policy, that Congress did not leave the door open to this possible mode of developing a special timberland sales program by executive action. The possibilities in administrative initiative were belatedly demonstrated when, after Congress had shown its indifference to repeated pleas to repeal the Timber and Stone Act, the General Land Office after 1908 exercised discretion the statutory language gave it and curbed some of the abuse under that statute by refusing to dispose of tracts except for prices substantially above the statutory minimum, conforming to specifically appraised values. During the years in which the Wisconsin forest lands were sold, however, there were the required auctions, but most transfers for a price were by private sale. Lumbermen commonly countered auctions by collusive bidding, which the General Land Office proved too weak or timid to combat. In fairness it must be added that this was a point at which the administrators were badly hampered by Congress' failure to provide an adequate system for classifying public lands according to their particular values; when auctions were held, the Land Office must typically depend for its knowledge of the special worth of what it was offering upon the sketchy and often unreliable field notes of the original federal survey. In the end, most federal timberland that was sold was sold at the statutory minimum price, which in practice became the maximum money yield sought by the government.[2]

That land must be cleared of trees plainly did not exclude it from preemption or homesteading; much of the best farmland so taken in the

Mississippi Valley was originally timbered, and, in fact, settlers at first preferred heavily timbered tracts despite the hard cost of clearing because they thought such land would be richer for grain production than would the prairies. On the other hand, it was implicit in the policy of favoring agricultural settlement by land subsidy that the privilege should not extend to land the soil of which was not suited to agriculture. Again, the failure of Congress to provide for adequate land classification meant that the substantial content of policy would be determined according to the presumption we followed in practice. The working presumption was invariably in favor of accepting pre-emption or homestead claims at face value, against the possibility that the land's true worth was only in its trees. Proved fraud would upset a claim, and we shall take further note of this aspect of the matter. Here it is relevant to note that fraud was made very difficult to prove, and all the more so because the temper of the times so clearly favored ready access to whatever marketable values could be got out of the land.

This climate of opinion, which must be reckoned as part of the operative pattern of federal lands policy, may be felt in the disposition the Wisconsin Supreme Court made of *Houlton* v. *Nichol* (1896). Plaintiff sued for an $800 fee claimed by him on a contract under which he had successfully advised and represented defendant in pressing a homestead or pre-emption claim for defendant upon federal land worth $8000 in northern Wisconsin. Plaintiff, the Report notes, was "a woodsman and pine-land explorer of large experience, well versed in respect to the location of valuable timber lands on the government domain in northwestern Wisconsin and the methods to be pursued in order to legally acquire lands under the land laws of the United States." Plaintiff's investigations at the General Land Office had satisfied him that certain lands hitherto kept out of market as subject to grants in aid of railroad construction were in fact free of valid claim under the railroad grants and legally open to settlement under the homestead and pre-emption laws. Having acquired the land with plaintiff's help, defendant resisted payment of the agreed fee, arguing that the contract was illegal because it stipulated a payment contingent on success in obtaining desired action from public officers. So presented, the argument posed little difficulty; the court properly answered that no public policy forbade rendering skilled assistance for a fee in presenting a claim involving simply the application of existing law. More interesting is the fact that the court's opinion, written by a judge of long experience at the bar and on the trial bench in handling problems of the lumber industry, showed no disquiet over the economic character of the land transaction itself, unless by reiterated insistence that all plaintiff had done was to apply to defendant's advantage "the established practice of the general land office." Though the court's recital of the contract circumstances, and the $8000 value of the 160-

acre tract, plainly indicated that this acreage was primarily valuable for its stand of timber, this was not seen as in itself raising any question of the propriety of resort to the settlement laws to obtain the land. Perhaps there was no impropriety; perhaps the defendant held a good faith intention to make a settlement; the practical presumption of policy is nonetheless significant.[3]

Information is unavailable, and probably unobtainable, to measure the extent of north Wisconsin timberland entered or acquired from the federal government by good faith transactions under the pre-emption and homestead laws. Indeed, there is no way to tell how much acreage was entered by pre-emption, honestly or fraudulently, since the General Land Office merged these transfers into the general reports of land sold. Given the obvious character of the land, it seems likely that the overwhelming bulk of pre-emption and homestead entries in the north were not made in good faith. Such, we shall see, was the general opinion of knowledgeable contemporaries, though acknowledgment of the fact did not affect official practice. In any case, the greater part of these timberlands passed by direct sale. However, "sale" in this connection is a term that requires further definition, which we shall attempt after sketching the pattern of disposal of state timberland.[4]

Wisconsin's lawmakers lived too much in the shadow of the great forest to ignore as completely as did Congress the special value problems of public timberland. But what they did was so little that it has less significance as policy than as evidence of the limitations of the policy-making process.

There was no opportunity for local decisions on disposal of timberland assets until Wisconsin became a state; the territory made law wholly within the framework set by Congress, and we have noted that Congress had no positive policy specially adapted to the forested public lands. Though in other respects the territory chafed at what it deemed inadequate provision for its development, it was characteristic that no complaints appear regarding the lack of an affirmative timber disposal program.[5]

Aside from the lively expectation of gifts of land from the United States, no questions of substantive program for disposal of such lands as the new state might own figured materially in public discussions preceding adoption of the Wisconsin constitution in 1848. So far as public lands were discussed, it was almost wholly for their present usefulness in overcoming the community's scarcity of mobile capital. The contemporary estimate was that the north contained an "almost boundless" supply of pine. Hence, it is not surprising that timber received no mention in the constitutional debates as a special component of the value of such lands as the state might own, though it was already taken for granted that the lumber industry would be a material factor in the state's economy. Timbered state lands were thus implicitly included with all others in the two general provisions the constitution makers provided for the state's holdings. In the first place, Section 2 of Article X stipulated that

> the proceeds of all lands that have been or hereafter may be granted by the United States to this state, for educational purposes, (except the lands heretofore granted for the purposes of a university,) . . . and all moneys arising from any grant to the state, where the purposes of such grant are not specified, and the five hundred thousand acres of land to which the state is entitled by the provisions of [the act of Congress of September 4, 1841] . . . and also the five *per centum* of the net proceeds of the public lands to which the state shall become entitled on her admission into the union, (if congress shall consent to such appropriation of the two grants last mentioned,) shall be set apart as a separate fund, to be called the school fund . . .

for support and maintenance of common schools, academies, and normal schools, with separate like provision for devotion of university aid lands to the university. Secondly, Sections 7 and 8 of Article X constituted the secretary of state, state treasurer, and attorney general "a board of commissioners for the sale of the school and university lands, and for the investment of the funds arising therefrom," granted the commissioners "power to withhold from sale any portion of such lands when they shall deem it expedient," but also committed broad power and responsibility to the legislature:

> Provision shall be made by law for the sale of all school and university lands, after they shall have been appraised . . . and [the commissioners] shall invest all money arising from the sale of such lands . . . in such manner as the legislature shall provide . . .[6]

Plainly the constitution makers intended that these be omnibus provisions, covering substantially all the land the state might acquire from the United States. The only exceptions would be lands given by the federal government for other specified purposes, such as the lands which were in fact given in aid of railroads and waterways improvements, and the swamplands so far as the state legislature judged their proceeds necessary to finance drainage. As matters developed, neither the railroad, waterways improvement, or swamplands-reclamation grants added anything to state timberland policy. In none of these instances did Congress make any reference to the timber assets of acreage it gave the state for special purposes, or fix any limitation or guide for positive realization upon the timber value. And in no case did the state legislature see fit to impose any terms of disposition special to the forest value of these properties, save for the modest stipulation that tax exemptions granted the Wisconsin Central and North Wisconsin railroad companies on aid lands so long as held by them should end upon sale of the lands' timber as well as upon sale of the land.

Thus, what the state did regarding the timbered school lands embodied practically the whole measure of such distinctive program as it adopted toward the timber component of any of its landholdings.[7]

On their face the school lands provisions of the Wisconsin constitution contemplated simply the "sale" of those lands and the devotion of their "proceeds" to the support of education. In this the constitution makers reflected the community's urgent sense of capital scarcity and its pervasive desire to use its fixed capital to mobilize or enhance the present productivity of cash and labor. Nothing in the constitution or in the discussions surrounding its adoption excepted timberland from this prevailing emphasis. So, in 1894, in *State* ex rel. *Sweet* v. *Cunningham,* the Wisconsin Supreme Court properly ruled that the legislature lacked power to dedicate and set apart as a state park any of the lands within the description of Article X, Section 2. So far as Laws, 1878, Chapter 324 purported to create a state park system out of school lands, it must be held without effect. The court deemed the reason so plain as to require little but statement: "The school fund is a trust fund, and is placed by the constitution beyond the power of the legislature to divert it to any other use than the support of the schools of the state. It could not set them, or any part of them, apart for a state park." [8]

The state park act of 1878 dealt with too small an area to affect the main course of the Wisconsin lumber industry, and the court decision of 1894 came long after most forested state land had passed to private ownership. These matters are relevant to definition of state timberland policy mainly because of the reliance the Wisconsin court put on the *Sweet* decision in 1915, in *State* ex rel. *Owen* v. *Donald*. In *Donald* it held unconstitutional legislation which sought to create out of the remaining school lands and by purchases with educational trust funds the beginnings of a system of state forest reserves. The majority opinion in *Donald* built on the 1894 decision to interpret the state constitution, in effect, to ban any disposition of "trust" lands other than by sale of the fee for a cash yield, carried through as rapidly as the market would reasonably allow. In this, as in other respects, the *Donald* case was an important part of the pattern of state policy dealing with the aftermath of a declining lumber industry and the place in the state's economy of those cutover lands most suitable for forest production. Though Mr. Justice Marshall's opinion in *Donald* purported to declare what had always been the law, there had not in fact been any such clear definition of doctrine during the boom decades of lumbering. *Donald* stated the law only for the period after 1915. For the years before then, Marshall's analysis amounted only to an historical summary of past attitudes affecting the disposition of state timberlands. There is reason to question whether Marshall here wrote good history.[9]

No evidence from the lumber industry's flourishing years supports the view suggested in the majority opinion in *Donald*. There Mr. Justice Marshall indicated that the constitutional trust for "sale" of school lands might be fulfilled, as to timberland, only by transferring the state's whole title in the land to private owners; the state must press transfer so that it should have no long-term forest holdings, but only a capital fund in money, for support of education.

In the first place, we must be cautious as we read meaning into silence. It is quite plain from the record that when in the years 1846 to 1848 men devised a framework for disposing of state lands, they were not thinking much, if at all, about special problems or opportunities of timberland. Granted, they spoke simply of "selling" the state's land; granted, too, they were much concerned to realize cash, and to increase private ownership, and did not like the idea that government might be a long-time landlord on a large scale. Yet, it is equally clear that in all this they more or less tacitly assumed they were planning the disposition of state lands the natural destiny of which was to be turned into farms. Thus, their generalized attitudes toward the best use of state lands imply little of any particularized attitude toward land of special character.

In the second place — and especially given this absence of focused attention on the peculiar problems of state timberland — "we must never forget that it is a constitution we are expounding." The public land provisions of the Wisconsin constitution sought to create a frame of policy within which might be developed continuing support of public education for an indefinite future. Whatever the limits of their imagination on immediate program, Wisconsin's founders plainly intended that the school lands be used to produce the most fruitful yield that good management could bring to the school trust.

The 1894 opinion in the *Sweet* case recognized that the constitution authorized the school land commissioners to "withhold" land from sale "when they shall deem it expedient," and empowered the legislature to fix in all other respects the terms and procedure of sale — and that in these flexible provisions "the beneficial purpose of the constitution" was to allow the state the "chance of advantage from the rise in value of public lands consequent on the increase of population and the development of the country in districts where the public lands lie." Legislative practice (embodied in special provisions for sale of town sites and waterpower locations, and special stipulations to protect standing timber as security for unpaid balances) treated the constitution as a charge to increase the assured yield of the state lands. The Wisconsin Supreme Court ruled that the constitution and statutes under it should be liberally construed to fulfill this standard. So in *Smith* v. *Mariner* (1856), over a sharp dissent, the majority of the court held constitutional the statute which provided for the "sale" of state land on credit under a certificate reserving to the state full title and rights of absolute forfeiture on default of the buyer. True, authorizing "sales" on credit, the constitution specifically directed that the commissioners take mortgage security for the unpaid price, and in this context, the majority conceded, by "sale" the constitution meant transfer of an absolute fee. Nonetheless, "the main purpose, the great objects of the constitutional provision" should be accomplished; these clearly were "the preservation of the school fund," and since the device of reserved title was a less expensive, equally safe, and more expeditiously applied security measure for the state, the court should not construe the constitution to forbid the legislature thus to contrive its own terms of disposal apart from those spelled out by the constitution makers. As early in the lumber years as 1863, in *State* v. *Weston,* the court accepted without question the legislature's authority to fix terms of disposition peculiar to the sale of timberlands, by reservation of cutting rights as against a buyer who had not yet paid in full. Where the court dealt with lands granted in aid of specific transportation improvements, it regularly cast the balance of doubt in favor of public policies and interpretations of statutes which it felt would promote the fullest realization on the aid lands. So, in the *Sturgeon Bay Ship Canal* case (1874) the court held that, where the statute ambiguously said that the lands should be disbursed in "proportion" to work done on the aided project, the measure of installments must be by market value and not simply by acreage. "The spirit and policy of the acts" governing the case demanded this result,

> because otherwise, upon the facts returned by the commissioners [that market values varied much within the aid grant, largely because of varying timber stands], the object of congress and of the state, in donating any part of the lands, may be ultimately defeated. Other circumstances being equal, or not preventing, that construction is always best which will best subserve the great primary object of the lawgiver.

Of like emphasis were decisions under both federal and state land legislation that the presumption favored intent to require fresh offerings at public sale under competitive bidding before lands might be bought at private sale, and that in the absence of specific directions to the contrary a price named in a statute should be construed as a minimum and not a fixed price when land was offered either at public or private sale. The *Sweet* case (1894) was itself one of these decisions; legislation directing public sale of lands re-offered after having been withdrawn from sale following a prior public offering was "remedial" and "to be construed liberally, to advance the remedy intended by the legislature," and so should be construed to apply to lands withdrawn from sale before as well as after its enactment.[10]

Thus, as established in contemporary discussion and practice, the pattern of state lands disposal set by the Wisconsin constitution was flexible enough to validate special terms for realizing the distinctive values of state-owned timberland, so long as these special measures promised reasonable yield to the educational trust fund. This is to treat the constitution as intending to provide a standard rather than particular rules for action — an approach which both logic and history warrant as befitting the function of constitutional policy. Insofar as it apparently denied this creative potential to the legislature, Marshall's 1915 opinion in *Donald* was bad history and bad law.

On the other hand, we speak here only of the opportunities of choice which the constitution held out to the legislature. This is relevant and meaningful, because it helps define the responsibility that rested on the policy makers and helps also to measure the quality of their work. But we are speaking only of opportunities, not of necessities. The story beyond this point must take account of other very important elements besides the permissiveness of the constitution — must look at the values, hopes, and ambitions which men held, and at the limiting conditions of physical and social environment within which they operated. Within the first year of statehood, Wisconsin's legislature showed promise of making some constructive use of its broad powers to realize on the state's timber assets. Chapter 212 of the Laws of 1849 defined terms and procedures for sales of school and university lands, fixing a frame which was to last in essentials through the years of principal disposition. The statute armed the land commissioners with authority to adjust sales programs for the best yield upon distinctive values of the state's holdings. With the governor's approval, the commissioners might "in their discretion, reserve and withhold from sale, such portions of the school and university lands, as in their opinion it may not be advantageous to sell and dispose of, and for so long a time as in their opinion will be most beneficial to the university and school funds." Whenever the lands included a water-power site, the commissioners might "sell together all the tracts or lots upon which such water power is situated, and such other tracts or lots as are necessary for the use and enjoyment of the same, not exceeding, however, one hundred and sixty acres, or they may sell each such tract and lot separately, as in their opinion will be most beneficial to the interests of the school and university fund." And they might cause any part of the lands to be laid off into small parcels or village lots "whenever, in the opinion of the commissioners, the school or university fund will be improved" thereby. These items did not deal explicitly with timberland exploitation. But they implied a standard of policy capable of growth to include measures specially adapted to any exceptional assets of the state lands. Particularly important was the implication they bore that there should be positive planning to secure the full potential yield that the diversity of the state's resources might hold.

The 1849 act made two kinds of special reference to the timber component of the value of state lands. The first concerned down payments. The statute allowed sales on ten years' credit for as little as a 10 per cent initial payment in cash. But the commissioners might exact a higher cash payment at the time of sale with "reference to the amount of permanent improvements on the land sold not liable to be destroyed, the proportion of timber, mineral or prairie, if any thereon, and the general situation of such land relative to its liability to be injured, or its becoming less valuable by trespass, waste or otherwise." Moreover, the commissioners might require security by bond and mortgage on other unincumbered lands whenever in their opinion "any school or university lands are so situated that additional security should be given for the payment of the balance of the purchase money." Secondly, the 1849 act limited the use rights of the state's buyer on credit, to preserve standing timber as an element of the land's value, supplementing the general security the state held by reserving title to the fee until the full price was paid. The holder of a certificate of sale from the state had thereby the right of possession and might invoke the possessory actions against third persons to protect his interest. However:

No such certificate shall be deemed to confer upon the purchaser, or any person claiming under him, the right to cut down or destroy, or carry off from the land any wood or timber standing or growing thereon, or any mineral therefrom, without the written consent of the commissioners, and only to the extent of such consent; but nothing herein contained shall prevent such purchaser from actually using and applying any wood or timber on the land to the erection of fences or buildings thereon; nor for taking firewood therefrom necessary for the use of his family; nor from actually and fairly improving any such land for the purposes of cultivation.

Upon forfeiture for failure to pay principal or interest due under a certificate of sale, the defaulted buyer of the land was declared "liable to be sued for any waste or unnecessary injury which he may have done to the same, or to the timber or mineral thereon," in an action to be brought by the commissioners "in the name of the state for the benefit of the school or university fund." Pre-emptors were by reference subjected to the same restrictions on cutting.[11]

Between this first act and the Revised Statutes of 1878 the legislature modified provisions regarding timberland sales, but without expanding the scope of its interest in these lands. Acts, 1850, Chapter 236, dealing with sale of the 500,000-acre gift which the federal government made to Wisconsin when it achieved statehood, dramatically enlarged the generosity of credit terms by allowing up to a thirty-year credit on these lands wherever in the land commissioners' opinion the lands "are an adequate security for the purchase price, and will remain so during the time for which a credit shall be given on them." The Revised Statutes of 1858 extended this authorization to cover sales of all school and university lands, and the matter stood so until the Revised Statutes of 1878 returned to the uniform ten-year limit on credit set originally in 1849.

Meanwhile the legislature did other things, the character of which serves mainly to emphasize the piecemeal, uncoordinated approach it took to administering the state's forest assets. In 1852 the legislature relaxed the terms of sale for university lands, authorizing the commissioners to waive any down payment provided that the buyer give additional security "sufficient to protect the fund against loss from any diminution of the value of the said tract by trespass, waste, or otherwise." Chapter 21, Laws of 1855—which launched the short-lived experiment in limiting sales of school or university lands to persons who filed affidavit that they bought only for actual occupation or cultivation—added the terse stipulation "that no person shall purchase any pine lands without paying the whole of the purchase money at the time of the sale." The legislature offered no explanation of this drastic change from the broad discretion previously given the commissioners; perhaps, in its immediate context, the stipulation was intended to help insure that such land would be bought only for clearing, though the hard-pressed finances of the typical farm settler would seem calculated to make the pay-in-full requirement work against this purpose. In any case, the requirement of full cash payment for pinelands survived the repeal of the rest of the 1855 act, and was carried forward in subsequent revisions of the statutes as an independent term of sale. Also in 1855 the legislature declared that no one claiming swamplands by pre-emption should "be allowed to take away from said lands more timber than is necessary for the use of said lands . . ." In 1856, in the first sales act concerning the swamplands granted the state under the federal act of 1850, the legislature in general adopted the terms and procedures set for the school lands. However, it separately stipulated that "at least one half of the purchase money of all of said lands described or designated in the plats and field notes [of the federal survey] . . . as timbered lands, shall be paid at the time of the sale thereof." Chapter 121 of the General Laws of 1866 left the commissioners no discretion to withhold the newly available Morrill Act lands from market; it was "hereby made [their] . . . duty . . . to immediately offer [these lands] for sale at public auction." The 1866 act stipulated a minimum down payment of one fourth of the price on ten years' credit, and authorized the commissioners in their discretion to require a higher portion or all of the purchase money to be paid at the time of purchase; whether the flat requirement of full initial payment for pinelands otherwise applied here was thus left in doubt. In 1867 the legislature at least eliminated the discrepancy in payment terms between pine-valuable school land and swampland by stipulating that all swamplands should thereafter be sold only for cash. The Revised Statutes of 1878 summed up the result of these payment terms in a pattern which lasted through the years of principal state lands disposal: sales of all swamplands and of all pinelands should be for cash only, to be paid at the time of sale; all other sales of state lands should be for at least one fourth of the price paid in cash at the time of sale, with discretion in the commissioners to require a higher down payment, or payment in full, and additional security by bond and mortgage on other unincumbered real estate for any unpaid balance, with a maximum of ten years for payment.[12]

The first thing that stands out as one reviews this pattern is its skimpiness. The 1849 statute had dealt only in limited measure with timber holdings. Its specific references to timber values were altogether negative, directed merely to holding fast such timber worth as the land might have, as security for the unpaid price, and obviously assuming that the whole realizable value of the timber to the state was what a buyer would pay for the absolute control of all of it on the day of a completed sale. Basically, thus, the timber was valued only as security for the land's price, and generally the land's price in practice was the statutory minimum. Despite the limits of its specific terms, however, the 1849 statute paid enough attention to special values problems to warrant hope that the legislature might build on this beginning. Conceivably it might yet create an affirmative, implemented pattern of diverse terms and procedures designed to realize the potential of the different types of value represented in the state's holdings. In the event, however, the legislature moved so little beyond the limited beginnings of 1849 as to add nothing of major significance. Its most striking action — the flat requirement of full cash payment for pine and swamp lands — only underlined the negative, price-security emphasis of 1849. The land commissioners paid somewhat naïve tribute to the substance, as against the form, of the situation when they asked in their 1882 Report whether "our present system of $1.25 per acre as a fixed price [is] wise?" The price was fixed at the statutory minimum only if there were no competing bidders at public or private sale to push it higher. Obviously, the commissioners had given up hope of competition, and the statutes provided that, once unsold at public offering, land not formally withdrawn from sale must be sold at the minimum unless a competing higher bid came forward.[13]

The thin content of developments after 1849 is the more apparent because of the episodic and arbitrary character of what the legislature did do. No good reason appears why, from 1856 to 1867, the commissioners must exact full cash payment for pine-covered school and university lands, but might accept 50 per cent cash for pine tracts classified as swamplands, or, after 1866, as low as a fourth part of the price in cash for timbered acres received under the agricultural college grant. Again, if at least this much solicitude was due the educational trust lands, it was left unexplained why not even this minimum security for price was stipulated to protect the timber value of lands which the state held in trust for waterways and railroad developments. The tardy growth of a market for hardwoods explains why for the first generation of Wisconsin lumbering the statutes made no provision like that for its pinelands, to protect its interest as seller of hardwood tracts, but it is notable that this omission continued into the period when hardwoods achieved distinct value. In its spotty character the legislation relevant to the state's timberlands showed the want of sustained or comprehensive attention to program that characterized the nineteenth-century legislative process. The point is driven home with special force here only because the matter concerned so great a public asset, so central to the state's economy, that if any matter would provoke a more professional treatment, this, it would seem, should be the one.[14]

More revealing than the uneven character of what the legislature did was what it did not do. At no time while the state held important timber acreage did the legislature authorize the commissioners to lease, or to grant cutting licenses, to test these alternative means of obtaining cash yields for the school funds. Though it had been narrowly confined by the want of a vigorous Congressional policy, in 1842 the territorial legislature had at least provided for sale of dead and downed trees on the sixteenth sections reserved for common schools. Not until 1905 did the state legislature go even this far as to marketable state lands, and only in 1925 did the legislature give the land commissioners authority to make separate sales of standing as well as of dead and downed timber. The only substantial measure the nineteenth-century legislature took, other than for outright sale of timberland, was to set apart certain northern tracts as a state park, under Laws, 1878, Chapter 324. But Chapter 324 simply locked up the timber, declaring that the reserved acreage should not be sold, "nor shall any privilege, license or authority be given to any person or persons whomsoever, to cut down or destroy any timber growing on such lands." In 1894 the Wisconsin Supreme Court properly ruled that this sterile lock-up program violated the constitutional trust for educational funds.[15]

There was no more important omission in the pattern of policy relevant to the state's forest holdings than the total failure to provide an adequate procedure to classify land according to detailed and competent field study of its quality. Here, again, there was a brave beginning but no significant development.

The constitution makers had said that the legislature should arrange for sale of school and university lands "after they shall have been appraised." The first legislature provided that three persons be appointed in each county "to describe the quality, location and general advantages" of such lands as the state then held within the county, and "to appraise" their fair value, unimproved, "giving due consideration to other circumstances such as proximity to settlements and credit for purchase money . . ." In prophecy of the future halting course of land classification in the forested north, less than a year later the legislature repealed the appraisal provision as it related to the school and university lands "lying north of the Wisconsin and Fox rivers." The repeal responded to Governor Dewey's recommendation that appraisals be suspended in these "sparsely settled" and as yet largely unsurveyed areas "until the appraisal can be more economically accomplished, and a result obtained more beneficial to the school funds." Appraisal was resumed in some north-central counties in 1852, and the basic system was carried forward into the Revised Statutes of 1858 in a form which left to the governor's discretion the time when the procedure should be applied in any given county, and which declared it the duty of the appraisers to proceed by "a personal inspection . . . describing [the lands] . . . by forty acre divisions, with the value per acre, as estimated by them."

None of the various appraisal statutes provided any means to assure appointment of qualified appraisers, nor any central supervision of how well they did their work. There was telling evidence of the limited utility of the appraisal system in the irrevocable decision made by Governor Dewey in 1851 to allow the "swamplands" tendered by the direct federal grant of 1850 to be identified from the federal survey field notes rather than by inspection by state agents. Obviously Dewey did not regard the state's appraisal machinery as adequate to the job. His "unfortunate" decision (which responsible contemporary critics later estimated cost the state hundreds of thousands of acres not properly noted in the faulty federal notes) thus reflected his estimate that the state would have to create an adequate classification machinery from a standing start, at more expense than he could face. The appraisal system as it stood was no answer. Contemporary opinion found the appraisals quite uneven in quality and reliability and questioned whether much of the land was ever in fact examined. The field work was weakest in proportion as the land in the northern forested sections was hard to reach. So in 1852 the secretary of state criticized the appraisal system because it made clear provision only for pay of the appraisers, though in rough country they could not do a proper job without camp assistants and experienced surveyors. In 1859 a joint legislative committee recommended that the state abolish appraisals and content itself with uniform prices set by statute; in practice the appraisals were not carefully or honestly made in the wooded country, and the land was selling only at the statutory minimum anyway, so that all the money paid for appraisal was wasted: "The whole system of appraising lands your committee regard as worse than useless." The state kept the form of the appraisal system for nearly another twenty years, but obviously there was no confidence either in its merit or its practical significance. The legislature did not extend it to the new agricultural college (Morrill Act) lands, nor to the swamplands, nor provide classification standards or procedures as aids to realizing maximum returns on the large grants from the United States for transport improvements. When in 1872 an Assembly resolution instructed the Committee on State Affairs to consider the propriety of withdrawing from market all school and university lands then unsold until they should be reappraised, the committee answered shortly that "after due investigation, they believe such a course would not be expedient." No more was heard of the suggestion. In 1878 the revisers of the statutes "did away with" any general appraisal procedure. Not until 1917 did the legislature again provide such a system; the 1917 act threw into sharper relief the shortcomings of the nineteenth-century pattern by setting a standard of competence for appraisers, and stipulating that appraisal be by actual view, and that land and timber be valued separately.[16]

Such, then, was the formal pattern of federal and state policy toward disposal of public timberland. Neither the United States nor the state of Wisconsin had made any important effort in declared program to realize the special yield possibilities of forest assets.

But this is not necessarily the whole story. Conceivably practice or custom might shape a distinctive forest management plan where formal law making failed to do so — as in southeastern Wisconsin, in the 1830's, settlers' land claims associations defined and regulated titles in advance of lagging federal survey and pre-emption programs. To look at practice is not to depreciate form. The history of formal decision making is an essential part of the history of substantive legal order. Law has no function more basic than to define by regular processes where power shall legitimately reside in the community. So legal order creates an accepted framework within which responsible reason and feeling rather than simple interest and violence can guide will into action. In these aspects, form is substance. By these criteria, defects in the law's formal processes or gaps in its formal patterns of decision have substantive importance. However, limitations inherent in formal decision making make practice and custom inevitably important elements in the history of any legal policy. Form fulfills its function by its relative insistence on precision and regularity. Given the infinite variety of circumstance and of men's visions and ideas, this means that before formal decisions can be made with any material effect a good deal of give and take must already have gone on, to bring ends and means into conception and to measure competing wills and forces. Its functions require of formal law making an emphasis on deliberate procedure and a front of firmness in declaring values. Yet these qualities are likely to make formal processes deficient in initiating change, or adjusting to new facts, or acknowledging mistakes or misalignment with attitudes or values in the community too weighty to be ignored. What men do may thus determine, amend, supplant, or repeal what they declare.

What men did with the public forest in part negated the limited formal provisions that dealt with timber values, and in part supplanted the fee simple title — which in form was all that government stood ready to convey — by what amounted to *de facto* cutting licenses. Practice thus departed far from formal policy in the disposal of federal and state lands alike.

The lack of adequate federal or state land classification programs, and the related failure to provide sufficiently staffed, qualified, and secure field forces for general administration, were facts of profound effect upon the working pattern of timberlands disposal. For one thing, these deficiencies fostered large-scale fraud upon the single major classification written into federal and state land statutes, by making it easier for men to enter land by pre-emption, or as a homestead, though the land was unsuitable for farming and was entered only for its timber.

Of course, the law gave the government power to penalize the fraudulent entryman and to void the entry, and a scattering of administrative orders and cases fought through the courts attest the reality of the problem. There is no means now by which to measure how much forest acreage was taken and cut by those whose frauds were not brought to book. As early as 1849 the Commissioner of the General Land Office warned that "not three in a hundred" pre-emption entries in heavily timbered areas were followed by completed sales; most entrants aimed only to "despoil" the land and move on. Contemporary official opinion was that the frauds committed on the pre-emption and homestead laws as a cover for obtaining timber were, as Commissioner Burdett observed in 1875, "notorious" and "innumberable" in the trans-Mississippi West, as they had already been in the Lake States. "In all the pine region of Lake Superior and the Upper Mississippi, where vast areas have been settled under the pretense of agriculture under the homestead and pre-emption laws," Commissioner Williamson noted in 1876, "scarcely a vestige of agriculture appears." The law, he complained bitterly, was "educating thousands of men in the crime of perjury." In 1879 the Acting Commissioner calculated that nine-tenths of the homestead entries in the Wisconsin pinery "were made for the purpose of stripping the land of its timber." The chief federal land officer in the Eau Claire area of Wisconsin estimated in 1888 that but 10 per cent of pre-emption claimants proved up their claims; in the same year the chief officer for the Wausau district estimated that 99 per cent of the pre-emption entries there "have been made in order to secure possession of valuable pine timber, the land upon which it stands not entering into consideration . . ." In 1898 United States District Judge Romanzo Bunn, long-term resident of one of Wisconsin's great lumbering sections, characterized the government's experience as at best one of misdirected generosity: ". . . no doubt from a noble and generous sentiment of magnanimity," the United States tendered northern pinelands to those who would improve them for homes. But:

It must be said that these lands were not well adapted to this purpose, and that the great and beneficent designs of the homestead law, so far as the pine lands in the extreme northern portions of the state are concerned, have been much abused and frustrated, and the law very generally used as a means for getting possession of the lands in order to cut the timber for commercial purposes. The lands were valuable for the pine timber, but poorly adapted to farming purposes,—at least in the present generation. It would be a noteworthy and instructive chapter in the history of the land laws if the proceedings under the law in that part of the state could be truthfully written out, so that it could be seen what the proper portion of cultivated farms made under the law would bear to the cases where the pine had been stripped from the lands, and the farm left desolate. We apprehend it would then appear that the bounty of the government has been much abused.

In 1909 the chief of field service for the General Land Office in northern Wisconsin summarized the experience there with homestead entries, as he had checked it at firsthand. The entries had "almost universally" been commuted for cash without intent to make a settlement:

It has been my personal experience to examine solid townships in northern Wisconsin, in which practically all the even-numbered sections had been acquired under the homestead law, quite generally [by] commutation. The timber had been cut off after patent, and yet not a single voter or inhabitant could be found in the township. The Government got $400 a quarter section for lands frequently worth from $10,000 to $20,000 . . .[17]

Though the records give us no over-all estimates of fraud on the state's pre-emption laws, it is clear that this was a continuing problem. A short-lived flurry over swamplands evidenced both the problem and the difficulty of interesting the legislature in any effective dealing with it. An 1855 statute first offered pre-emption privileges in state-owned swampland. Within a year, Governor Barstow recommended its "immediate repeal." Experience already showed that pre-emption entries were being made on timbered swampland solely as a cover for stripping it, so that the privilege "amounts in effect, simply as a license to the pre-emptor to commit such waste upon the land as he may see fit." In 1856 the legislature repealed the swamplands pre-emption privilege, only to restore it in 1857. In an 1884 case before the Wisconsin Supreme Court, the commissioners of public lands noted that for twenty years past a continuing problem had been to check pre-emption papers for fraud.[18]

The state's larger difficulty, though, arose out of its readiness to sell its land on long-term credit. We have noted that the statutes finally stipulated full cash payment at the time of sale for "pine land," and authorized the commissioners to increase down payments on other land, to protect the state's interest in the unpaid balance of the price. The first down payments provisions had been enacted out of a lively enough sense of the problem. As early as 1848 the Assembly's school lands committee warned that "extensive speculations might be made with a direct view to forfeiture," as a cover for cutting off the pine. Governor Dewey in 1849 urged that sales terms should "depend . . . upon the situation of the land as to timber and its liability to waste."

The nineteenth-century legislature characteristically mixed moralistic and narrowly "practical" thinking into a compound as much naïve as hypocritical. So it often satisfied by exhortation its worry over some vaguely perceived general good, while it avoided the pain of denying a more immediately felt interest by failing to implement its general decision. Thus it was here. The state early encountered evasion of its policy of maintaining timber values as security for unpaid prices, and the legislature attested the problem by tightening the sanctions for wrongful cutting by state vendees. But none of the statutory stipulations had substance.

The legislature had provided no effective means to determine which of the school lands were "pine land," or which of its other holdings bore timber stands requiring special payments protection, nor did it provide a field staff sufficient to check on what its buyers were doing. The commissioners often did not exercise the discretion that the statutes gave them to exact substantial down payments on timberland. At first they argued that the state was protected because, as the statute specified, the terms of the certificates of sale forbade the buyer to cut until he had paid up. In 1855 a legislative committee sharply criticized this course of administrative action as wholly lacking in realism:

> The pine lands . . . are in new counties, sparsely settled, and in the vicinity of the pine lands, by lumbermen only, all of whom would be interested in taking off said pine timber instead of informing the government . . . of the same, and most of those who purchase 40 or 80 acres of said pine land for one winter's logging of said large purchasers, would do so for the express purpose, in most cases, of cutting off its timber, and in many cases would not know, but that they had the unquestionable right of so doing.

In 1857 the secretary of state complained that the statutory stipulation of a 50 per cent down payment on all swampland described in the plats or field notes as timbered was "entirely impracticable, for the reason that there is no information in the possession of this office, by which the character of the land, whether timber or not, can be known or designated." In 1860 the land commissioners conceded that most of the state's timberland was still being sold on terms that did not protect the timber values. In effect, the commissioners now refused responsibility, pointing to the legislature's failure to furnish the field procedures which alone could give meaning to the special statutory terms of sale:

> [Timbered] lands ought never to be sold any part on credit. But the practice has been, in the absence of legal evidence of the character of each tract, to consider all lands as farming lands, and to demand only ten per cent of the purchase money down, with a credit of ten years for the balance. The state still retains the larger interest in all pine lands, or other lands chiefly valuable for their timber, sold on such terms of credit, which must be watched with great care, and guarded at great expense, or the value will speedily be removed, and the lands returned, worthless, to the state.

Commenting on this statement, the state schools superintendent observed resignedly that the state should not rely on its buyers to be more careful of the public interest than it was: "It is not to be expected that purchasers will indicate what lands are 'pine lands,' and thus subject themselves to the necessity of paying the full purchase money in advance." At the same time Governor Randall found the state's sales policy "most mischievous," and in no respect more so than in regard to timbered lands, which "have been held until their value has been destroyed by exhaustion of the forests growing upon them, and then forfeited." Because the existing credit sales program had "proved a cheat and a delusion," Governor Harvey in 1862 recommended that it be discontinued, "except perhaps in the sale of really *farming* lands in limited quantities, and in good faith, for immediate settlement and improvement." Nevertheless, though the issue was thus sharply drawn at a time when the state held large quantities of timberland, the legislature did nothing to correct the situation, nor was there further pressure from either the governors or the land commissioners for remedies. Summing up costly shortcomings of the state's land program, the state superintendent of public instruction could still say in 1892, "It is notorious that state lands covered with valuable timber have been sold, a fraction of the purchase price paid, the timber removed and the land then allowed to lapse to the state."[19]

The United States and the state lost, also, because the working pattern of timberland disposal proved to be to sell the bulk of it at the statutory minimum prices. We shall say more of this hereafter as we explore value attitudes underlying the form and practice of land programs. Here it is relevant to observe that the lack of a sufficient land classification procedure or a sufficiently vigorous, qualified, and protected field organization invited slack sales practice. The fact was most sharply demonstrated in the acknowledged failure of the public auction system to produce real competitive bidding for the bulk of the lands. The Wisconsin land commissioners in 1882 summed up a generation of experience on this score, matching that under the federal program: "Public sales of large quantities of new lands, experience has proven to be a very insufficient test of the value of such lands, except in cases of a few tracts well known for their great comparative value, producing competitive bidding." Of course there was more than the government's lack of knowledge of its assets to explain these low yields. Collusion among timber buyers was commonplace at public sales. A classic example was provided in 1869 when the United States put on the block its last large tract of pineland in the Chippewa Valley; after an initial flurry of free bidding, a combination of big buyers by agreement divided the offered land among them without further competition. Another episode in 1873 afforded some measure of the difference between potential and actual yields on public sale; failure of bidders to cooperate pushed prices as high as $19.75 an acre on the first day, but after a night of negotiation the big buyers next day forfeited their bids, and on re-offerings with a disciplined combination in control the land went for about $3 an acre.[20]

If practice largely nullified the limited special provisions made in public lands sales policy concerning forested tracts, practice had still larger implications in the way in which it reshaped the kind of title to timberland that was actually transferred to private owners in northern Wisconsin. In form, the only title the United States or the state of Wisconsin was ready to transfer to timberland was a fee simple absolute; the laws made no provision for leases of timberland or the grant of licenses to cut. As we saw in the previous chapter, popular intent and understanding was that when government transferred the fee to private hands, the purpose and the normal result would be to put the future disposition of the resources so conveyed permanently into the sphere of private decision. Government was not to be a landlord. But, in practice, private timberland owners treated themselves as lessees or licensees of the bulk of the land they held. They did so mostly by the simple procedure of holding the fee only so long as suited their plans for realizing on what they regarded as the valuable timber, and then letting the land be sold for unpaid taxes. In the last and more private-cost-conscious phase of the industry, at the turn of the century, the larger lumber companies sold stripped acres in large quantities at very low prices to speculative promoters of agricultural settlement, but the ultimate risk here also was on the public, to which much of this acreage finally returned through tax delinquency. Thus, over-all, government became in effect a seller or grantor of leases or licenses for times and purposes set wholly at the will of the "lessees" or "licensees." Such *de facto* leases or cutting licenses, in alternation with periodic return of the land to public ownership via tax sale, were often brought into being on the same land two or three times over a span of twenty to forty years, as rising prices and new uses of wood gave profit to harvesting the smaller pines or the hardwoods which had been passed by in an earlier marketing phase of the industry.[21]

What we have just spoken of must be distinguished from the practice of fraud by pre-emptors, homesteaders, or purchasers on credit who cut timber in breach of the terms on which they occupied the land. To strip off the valuable timber, and then "let the land go for taxes," or sell it wholesale to farmland speculators for scrap value, might be inconsistent with social functions which the institution of private property was supposed to fulfill; Locke's stricture that one might legitimately acquire property only in "as much as [he] . . . can make use of to any advantage of life before it spoils" reminds us that title carries some notions of at least moral responsibility for fruitful use. Of this we must take account later. But, when the United States or the state completed a transfer of the fee, typically it imposed no explicit conditions on term or manner of use. It could not complain of fraud or breach of contract when grantees of such unconditioned titles cut the trees and thereafter allowed the land to be sold for unpaid taxes or a speculative residual worth.

The limits of public policy in this respect were recognized by the Wisconsin Supreme Court in *Paine* v. *White* (1867). In an action of replevin for logs, defendant's title traced to Elbridge Smith. Smith had been caught cutting the logs from state land; through their agent, the land commissioners seized the logs and then sold them to Smith under Laws, 1864, Chapter 233, which allowed settlement of such a trespass by sale of the land to the trespasser for its appraised value plus 50 per cent. Since Smith took no patent to the land, however, the adverse claimant in the replevin suit argued that it must be inferred that the commissioners had not required him to buy the land, and hence the sale of the logs to him was invalid. But, the court ruled, the ordinary presumption of the regularity of official action was not rebutted merely because Smith might not have chosen to complete his title; in effect, the court held that it was no public concern whether the man who got the timber was interested in getting the land, providing he made the payment of money which the state asked:

> . . . we must assume, nothing to the contrary appearing, that the commissioners required the proper amount to be paid before releasing the lumber. The commissioners were only authorized to release the logs to Elbridge Smith upon condition that he paid the minimum price at which the land had been appraised, and fifty per cent in addition upon the appraised value. That Elbridge Smith did not see fit to take a patent for the land cannot invalidate the sale made to him by the commissioners.[22]

Though we lack data for close measurement, official comment and complaint from the 1860's on testify that the cycle of formal purchase of the fee from government, logging, and then what amounted to return of the fee to government for defaulted taxes, became a substantial part of the pattern of ownership in the northern forest. As of 1898 over half a million acres in the northern forest were involved in this process; counties then held outright on tax deeds more than a quarter of a million acres, and the State Forestry Commission reported that "considerably more" than this total of acres stood under tax certificates which would probably not be redeemed. In a 1908 case the Wisconsin court pictured the situation inherited in the closing years of the lumber boom, when great sections of the north had become "known as 'cut-over lands,' from which the pine and spruce had been removed, and which had been abandoned by the original owners and sold for the nonpayment of taxes." According to expert testimony, "there was no market for such lands in large bodies in the spring of 1900 . . . [and] there were hundreds of thousands of acres for sale in Ashland, Bayfield, Iron, Price, and adjoining counties, with few purchasers, and . . . no demand for such land except for trading or speculative purposes." Sales of cutover land by lumber companies and speculative middlemen ran into tens of thousands of acres around the turn of the century. But this, too, proved in large part an unproductive end use for the forests; by 1900 only 10 per cent of the land in the more northerly counties had been put to farm use, and defaults and tax delinquency were common. By the 1920's land economist Benjamin Hibbard could point to a full cycle of policy history: "Instead of having no longer a state or public domain, as seemed to be the case a few years ago, there is now coming back into the hands of the state —literally the counties—an amount of land of appalling magnitude." It was likewise part of the pattern of policy, and most characteristic, that though men complained of at least some aspects of the results their behavior produced, they did not see results in relation to causes. Even in the years of lumber boom they worried occasionally over the amount of tax delinquency in the forest sections. But they talked of the phenomenon simply as one of tax load and tax enforcement, failing or not choosing to see that at base the problem derived from the working reality of the kind of title government was content to pass to "owners" of timberland.[23]

Finally, we must take account of formal and practical policy toward timber trespass. The problem was important for public timberlands disposal. One might count the stakes in terms of money values; he might count in terms of the power which control of the public domain gave government to bargain for desired action and to influence the course of economic growth; he might appraise the public lands primarily as an unrealized potential, the holding of which might enlarge the range of public choices in a shadowy future. By whatever calculus, if there were to be substance in the power to decide, it was obviously essential that the law preserve these public assets until decisions were made. On the other hand, though important, this was a limited, because a negative and auxiliary, area of public action: preserving assets was not an end in itself. Precisely because it had meaning only as an auxiliary to larger, positive policies, a program concerning timber trespass was not likely to be firm if there was want of decision in definition of the ends it should serve. But objectives were ill-weighed and ill-defined in the basic policies of the United States and the state toward their forest lands. It is not surprising, then, that even on so apparently elementary a matter as that of holding on to what they had until they decided what to do with it, both governments fumbled, evaded, and surrendered to men who had no doubts about what they wanted in the pineries.

Where a man cut trees on public land without the public's consent, the situation was too simple to allow of much complication of doctrine. This was, at the least, a civil wrong: if there were no previous relation between the cutter and the government, the cutting was a tort (conversion); if the cutting was contrary to terms on which the cutter had entered the land, as pre-emptor, homesteader, or purchaser on credit, it was a breach of contract, or a breach of condition warranting re-entry by the grantor. As a property owner, rather than as sovereign, the United States or the state of Wisconsin might sue at common law for conversion or breach of contract. Indeed, in an action of trover arising out of a knowingly wrongful taking of timber from federal land in Wisconsin, the United States Supreme Court enforced the common law remedy to the full; since the government might have recovered from the intentional wrongdoer not only the original value of the standing timber, but its value as enhanced by his transportation of it to a millsite, the court ruled that the same enhanced damages might be had against a defendant who had purchased the timber in good faith for value from such a trespasser. As was generally their practice in the limited opportunities afforded them under weak and haphazard executive action against trespass, the judges declared and ruled boldly against this kind of invasion of public right. Mr. Justice Miller spoke for the Court with his customary blunt vigor:

> To establish any other principle in such a case as this would be very disastrous to the interest of the public in the immense forest lands of the government. It has long been a matter of complaint that the depredations upon these lands are rapidly destroying the finest forests in the world. Unlike the individual owner, who by fencing and vigilant attention, can protect his valuable trees, the government has no adequate defence against this great evil. Its liberality in allowing trees to be cut on its land for mining, agricultural, and other specified uses has been used to screen the lawless depredator who destroys and sells for profit.
>
> To hold that when the government finds its own property in hands but one remove from these wilful trespassers, and asserts its right to such property by the slow processes of the law, the holder can set up a claim for the value which has been added to the property by the guilty party in the act of cutting down the trees and removing the timber, is to give encouragement and reward to the wrong-doer, by providing a safe market for what he has stolen and compensation for the labor he has been compelled to do to make his theft effectual and profitable.

So, too, where the legislature provided that the successful plaintiff in an action for wrongful cutting might recover enhanced damages, the Wisconsin court ruled liberally in favor of the state's successor in interest: he should have the benefit of the statutory damages for a trespass done before the state conveyed the land, since the general land statutes provided that the state's patentee might recover "like damages as if such injury or trespass had been committed after the patent had issued." With like generosity toward the complainant, and despite the general favor of state policy toward the settler, the Wisconsin court put on the defendant the burden of proving that his cutting had been merely in exercise of his statutory privilege to cut for firewood or to open the land for cultivation.[24]

Government was free to apply criminal as well as civil sanctions to misappropriation of its property. However, the United States was slow to declare it a crime to cut timber without authority on any federal lands; it was not until 1849, as economic activity began to reach into the Lake States forest, that the United States Supreme Court by a liberal interpretation of the act of March 2, 1831, made timber trespass a penal offense when committed anywhere on the public domain. Though the Wisconsin territorial legislature felt that Congress had left it little scope to deal with public lands, in this matter it anticipated the central government. Territorial laws in 1836 and 1844 imposed penalties for unauthorized cutting of timber on land held in trust by the territory for education and internal improvements. The 1844 act allowed rather stiff sentences, up to a year in the county jail or $500, against one who "wilfully" cut public timber without authority. The same definition and substantially similar penalties were carried forward into the new state's Revised Statutes of 1849. Thenceforth state policy always paralleled civil with penal sanctions in the formal apparatus for protecting the state's forests.

The only point of substantive doctrine on which policy wavered concerned the definition of the intention which would make a cutting criminal. In penalizing anyone who cut on federal land without authority, "with intent to . . . dispose of . . . [the timber] in any manner whatsoever, other than for the use of the navy of the United States," the federal statute imported some requirement of intent. But its statement was broad enough to be capable of including any civil conversion — any case where defendant consciously exercised dominion over the property, even though he acted under a good faith and non-negligent mistake of fact. However, the federal courts treated this statute as they typically treated other penal legislation of like ambiguity, by reading it to require a showing of a general wrongful state of mind — that the defendant had acted at least in heedless or careless disregard of public policy. The state's treatment of the mental element in the offense was, in contrast, much more favorable to the accused through most of the years in which the state held substantial timbered property. We have noted that the territorial legislation and the Revised Statutes of 1849 required that the state prove a specific intent on the part of the logger to steal state timber: it must be shown that he "wilfully" took the state's property. The revisers of 1858 continued this provision. In the 1860's the land commissioners became more active in pressing the trespass problem upon the attention of the legislature, particularly to obtain a larger staff of field agents. Probably also at their urging, in 1865 the legislature broadened the scope of the offense; dropping the "wilfully," the 1865 act penalized anyone who "shall cut . . . or . . . remove . . . timber" from state lands, "except when authorized by law." Though this plainly extended the definition of the crime, the language bore the same ambiguity now as the federal legislation; taken at full literal force, it would apply the penal sanction to any conduct amounting to a civil conversion. It is some indication of scant prosecuting zeal that the executive never pressed a criminal prosecution of sufficient consequence or severity to produce an appeal in which the Wisconsin Supreme Court might have resolved the doubt as to the statute's scope. We can be quite sure, however, that Wisconsin authority would have followed the general pattern in such matters and held that the broadened statute imported a requirement of proof of a general criminal intent. In any case, the legislature reversed field again. The Revised Statutes of 1878 repealed the 1865 act and restored the narrower definition of the offense, now penalizing only one who should "willfully, maliciously or wantonly" misappropriate state timber. The record is silent as to the background of the return to the narrower form of the law. The reversal came after several years in which the land commissioners claimed credit for increasingly successful prevention and punishment of timber trespass on the school lands, with prime emphasis apparently on remedies by seizure and sale of illegally cut timber. On the scant record, the matter lends itself about as well to either of two interpretations: that the legislature returned to the narrower definition of the offense to curb the zeal of enforcement, or that it did so simply to return to a more traditional emphasis on the guilty mind, now that the trespass threat seemed well in hand. So the law stood until 1893. In view of the "vital importance" of the remaining timbered lands as part of the school trust, Governor Peck then tersely recommended to the legislature "that the unauthorized cutting of timber upon state lands be made a criminal offense, regardless of the intention of the trespasser." The legislature responded with Laws, 1893, Chapter 64, penalizing "any person who shall cut down, injure or destroy any tree or timber" on state land. The statute law remained thus through the remaining years of major lumbering in Wisconsin. Again, enforcement did not produce a case of sufficient importance or controversy to reach the state supreme court, so that we lack an authoritative interpretation of the 1893 act. Certainly, if its broad terms were read in the light of the governor's recommendation to which the act responded, they must have been construed to make the penal sanction potentially available in every instance amounting to a civil conversion.[25]

In such terms, declared public policy denounced unauthorized taking of timber from public lands as a crime, as well as a civil wrong. Such qualifications or ambiguities of doctrine as there were did not materially affect the central position. On the other hand, however, the pattern of values defined by practice tolerated or acquiesced in unauthorized cutting on a scale that denied substance to declared federal policy and materially qualified the reality of state law.

Reviewing experience in the Lake States as well as in the Far West, the Commissioner of the General Land Office in 1874 despaired of the government's effort to protect its timber inventory by any means short of disposing of it as fast as possible:

. . . the wisest policy the Government can pursue in respect to . . . [its pinelands] is that which will most speedily divest it of title in the same for a fair consideration, for the reason that depredations to an enormous extent are constantly occurring, which existing laws are powerless to prevent and seemingly legally powerless to punish.

There was little effort to bring wrongdoers to account in court, either in civil or criminal litigation. Such prosecuting zeal as was shown lacked continuity or breadth and was quickly daunted by the counterpressures which Congressmen were ready to bring to bear at the urgings of affected local interests. Money settlements with trespassers tended to be the prevailing sanction. These were negotiated on terms very favorable to the wrongdoers; a reliable agent reported to Wisconsin's Governor Fairchild that he had reason to believe that local federal land officers had settled cases in northwestern Wisconsin for as little as 5 per cent of the amount actually cut. In the face of "vast" timber trespass, the United States recovered so little against wrongdoers as to mock enforcement; from 1855 to 1877 it collected about $200,000 from trespassers, or a net of about $150,000 after enforcement costs, equalling the purchase price of about five thousand acres of good pine land. Though there were interludes of complacency in which the Land Office congratulated itself that enforcement was improving, the story remained the same through the nineteenth century. In 1897 a committee of the National Academy of Sciences estimated that in the preceding decade over eleven billion feet of timber had been taken illegally from the public domain; in that time the United States recovered about 4 per cent of what it sought in what was always a very limited enforcement effort.

Through the years of lumber boom, Wisconsin provided an early and sustained example of successful large-scale trespass on federal timberland. The surveyor general for the Wisconsin district reported in 1850 that "a swarm of laborers [were] constantly employed" in illegal lumbering there. Through the 1850's the district survey office urged that it be given means to speed its work in order that the pinelands be available for sale as soon as possible, as the only means to defeat wholesale theft of timber. In the sixties the Commissioner of the General Land Office conceded that "immense areas" had been stripped of valuable timber; in the late seventies the Commissioner cited agents' reports that timber trespass had been "extensively carried on for a number of years past" in Wisconsin, involving instances "of vast magnitude"; and in 1881 the Commissioner looked back on a record of failure to protect the government's holdings:

The yet large areas of public lands in Michigan and Wisconsin have also been the field of perhaps the most extensive trespassing in the country. While most of the large lumber companies have manufactured a portion of their products from timber growing on purchased lands, it is a well-established fact that millions of feet of lumber, thousands of railroad ties, telegraph poles, cords of wood and bark have been taken from the public lands. In many places the land has been stripped of both trees and undergrowth.[26]

The state lived closer to its holdings than did the United States; one might reasonably expect it to show more alertness to the security of its property, and in some measure it did. Nonetheless, for at least the first twenty years of statehood, Wisconsin's policy in practice seems to have been no better than that of the federal government, and there are reasons to look skeptically on official claims to improvement of the situation after 1870.

Until the later 1860's responsible state officers reported their concern over timber trespass, which they found to be "large" and "constant," involving "incalculable" losses. In 1860 the land commissioners warned that illegal operations had "gone on from year to year unmolested" in great areas "screened . . . from detection" by their "remoteness from the settlements." "Of all the causes which have operated to the depletion of the trust fund of the State," they concluded, "none has worked greater loss than 'skinning' the pine and other timbered lands of the northern parts of the State." These lands, they complained again in 1864, "have often been remorselessly stripped" by illegal logging, "so as to become at least for many years, wholly unsaleable." In 1865 they again warned the legislature:

Quietly but actively trespassers have been stripping off this timber, sometimes merely for their own use, sometimes making small sales, and sometimes carrying on the shameful business so extensively as to cover the rivers with stolen logs, and to grow rich upon their illgotten plunder. Others, equally guilty, purchase the stolen property, and thus support and encourage the iniquity. Important corporations and prominent men in some parts of the state have been for years engaged in the traffic.[27]

After the legislature responded to the commissioners' concern by modest additional provision for enforcement, the reports in the middle seventies became very optimistic in estimating the reduction of trespass, and then for a long time fell wholly silent on the matter. However, other contemporary testimony from the governor and from state agents in the field pictured continuing heavy losses by timber theft, and recurrent legislative reorganizations of the enforcement machinery seem inconsistent with the notion that all was well. Further, in these same years serious trespass depleted lands held by the state in aid of railroad and waterways improvements, to the point that the legislature in 1876 provided for recovery by the state against such wrongdoers of "damages to an amount ten times the value of the timber or logs so cut . . . or removed . . ." It is hard to believe that enforcement suddenly became so effective as to exempt the school lands from such losses. Indeed, if timber thievery had been so well taken in hand, it is strange that the governor and legislature as late as 1893 found it wise drastically to extend the scope of the applicable penal law,[28]

In any case, we must read contemporary praise for state executive action against timber theft with realistic appreciation: most of what was done consisted simply in bargaining for small or modest money settlements. There was little effort to take either civil or criminal proceedings to court; there were a few actions in trial court, but at no time did the executive press a trespass matter of sufficient consequence to produce an appeal to the Supreme Court in a case in which the state was the moving party. In the years in which the commissioners reported most complacently on the success of enforcement, they cited only dollar results as a measure of their activity. The dollar recoveries they reported were so small as to indicate that either the trespass threat or the enforcement activity was of minor scale; on the whole evidence, the latter is the more convincing inference. Thus, in 1872 they reported $5584 collected in principal and penalties, offset by $5293 paid to agents; the total reported in 1873 was $3360, offset by $5931 paid to agents. In fact, timber theft from public lands ceased to be a large problem only in proportion as government transferred its timberland to private owners. Whatever the intent, the result accorded substantially with the despairing recommendations of those public officers who early urged government to speed disposal of its forest assets in order to enlist the zeal of private owners against thieves. The problem was undeniably a difficult one. We shall see later that private owners, also, experienced timber trespass as a serious threat to their holdings.[29]

There exists an involved history of tinkering with organization and procedures for combatting timber theft on federal and state lands. So little was done, however, that it is not worthwhile to spin out the story of administrative changes. Some of the administrative history helps reveal underlying values and environmental factors which shaped substantive policy; from this point of view we shall pay attention to it later in this chapter. Let us summarize here what relates immediately to action.

First: budget. Though Congress and the state legislature grudgingly increased provision of money and men to police the public forests, all through the years of Wisconsin lumbering the provision was unreasonably small compared to the value of the assets to be protected. Congress made its first appropriation for special timber agents as late as 1872, providing only $10,000, raised by 1879 to $40,000 and by 1882 to $75,000 — to protect half a billion acres. Although for short intervals there were as many as four federal agents assigned to the Wisconsin area, often only one was available. Not until twelve years after lumbering began to emerge as a major industry in the state did the Wisconsin legislature make regular provision for a force of timber agents under the land commissioners. The commissioners rarely had authority or means to employ more than four to eight agents, and these almost always on a part-time basis, for the four most active lumbering months of the year. The second half of the nineteenth century was not generally a time of fiscal stringency for the federal government; Congress' stingy doles for forest protection were not concessions to needed economy. Lack of will rather than of means was the issue. This was shown best by Congress' refusal to create a revolving fund out of proceeds of forfeitures and fines to sustain and enlarge action against timber theft. The excuse of pinched means for not appropriating readily from general funds was more applicable to the state. Yet, the constitution makers had pointed to the device of the revolving fund when they granted to the support of schools the proceeds of all fines collected for any breach of the penal laws, specifying that the grant was only of "the clear proceeds." Moreover, the legislature did provide that the expenses of protecting lands held in trust in aid of railroad construction should be paid out of funds derived from prosecution of trespassers on these lands. Again, it is a token of lack of will that neither the legislature nor the executive branch ever pressed for creation of a revolving fund for the school lands' protection.[30]

Closely related to budget limits was a second key fact of organizational history. Before Gifford Pinchot became chief of the "microscopic" Forestry Division of the United States Department of Agriculture, on July 1, 1898, no one moved to develop either in the federal or the state executive branch a field force with the selection, training, and morale of a body of forest service specialists. Wisconsin's land commissioners talked bravely of picking skilled woodsmen as their timber agents. But selections were by no ordered rule. State executive letter files show that, as in the pre-Pinchot federal organization, the field agent jobs were regarded primarily as patronage items. The record is particularly clear as to Congress. That body defaulted on forest protection as badly in its refusal to use a decent self-restraint and allow reasonable insulation to the executive branch from pressures generated by the affected local and industry interests, as it defaulted in provision of funds. Of course, the two are part of a common pattern.[31]

The third fact of organizational history was the characteristic nineteenth-century reliance upon delegation of public functions to private hands. As we have seen, to the limited extent to which there was a formulated policy or rationalization in the matter, the working program of the United States and the state was mainly to solve timber theft by transferring the forest as fast as possible into the presumably more alert guardianship of private owners. The state expressed this policy with particular clarity in two ways. By statute it decreed that its patentees should succeed to all uncollected civil claims which the state had for timber trespass occurring before the state transferred title. On notable occasions it delegated to railroad and waterways companies the function of prosecuting both criminal and civil remedies for trespass on aid lands. The state confessed failure in this delegation policy only regarding the railroad-aid lands in the northwest, protection of which it resumed in 1869 after a showing of wholly inadequate effort by the promoters to safeguard their timber assets. Otherwise, delegated protection remained the rule. It is striking evidence and symbol of the direction of state enforcement policy, that in all the years of the lumber industry not one timber trespass case prosecuted by the state came before the Wisconsin Supreme Court. On the other hand, the Court was called on to rule in several important suits prosecuted by patentees of the state for wrongs committed while title was still in the public.[32]

Finally, we should note that despite a considerable volume of legislation and executive orders, this activity was expended wholly upon conventional approaches to the problem of guarding the public timber store. Reliance was placed wholly upon suits or settlements for money damages and fines, plus some resort to seizure (libel) of stolen logs. A later chapter inquires whether alternative, more promising systems of inventory protection might have been devised. The failure of conventional sanctions was clear enough to show that the need was for unconventional arrangements. As at every point in the enforcement story, we confront a basic lack of will and imagination in policy making. The immediate practical problems were great. But they were no greater than others which nineteenth-century law makers found the will and wit to surmount. Thus, as at other points, our outline of timberlands policy leads us to further questions, more subtle and less tangible, concerning the official and popular attitudes and values and the institutional environment within which policy was made.

PREVAILING CONCEPTIONS OF THE USES OF THE PUBLIC TIMBERLANDS

We track an elusive quarry as we try to define the popular and official attitudes which lay back of the ways in which we disposed of public timberland in nineteenth-century Wisconsin. Some central ideas are plain enough in action and in expression. They are the same ideas as those which guided or rationalized the disposition of land best suited for farms and towns. Indeed, the most obvious fact of this story is that government disposed of a distinctive public asset without significant discussion of its special attributes and potentialities. The Commissioner of the General Land Office observed in 1874 that he "fail[ed] to find, from the beginning of the Government to the present time, a single enactment of Congress providing any distinctive method for the disposal of that vastly extensive and proverbially valuable class of lands known as 'pine lands'." With equal accuracy he might have noted that this lack of specialized action was matched by lack of specialized inquiry or debate. The fact stayed so through all the years that determined the fate of the original forest in northern Wisconsin.

Such talk as there was of government's special stake in its forest holdings ran almost entirely to the problem of timber trespass. This only highlights the lack of significant discussion. Concern about timber theft inherently pointed to limited and negative objectives. To protect its full range of choice in using the public domain, government needed to safeguard its assets. But to do so was merely a holding action and no substitute for a positive program. Moreover, it fits the general pattern that the only affirmative policy, to which action against trespass was typically seen as auxiliary, was that of realizing the full market value of the land when government was ready to sell the fee, as it was usually assumed it soon would wish to do. Thus Wisconsin policy makers sometimes acknowledged that standing timber was the most important item in the value of much of the educational trust lands; accordingly the state's long-term responsibility under the trust required particular care to protect these assets against theft. But it is plain that those who spoke so saw this responsibility in no more special terms than a duty to preserve the money value which sale of the fee would soon bring into the trust fund; the long-term trust was discussed as one of dollars, not of timber-productive acres. Of course, current talk recognized that timber was useful and valuable. But, typically, men reckoned it as a present stock of expendable, non-renewable raw material, an extra bonus to be reaped in the course of "improving" land for its normal use for farms or towns, and interesting in itself mainly as a means of acquiring cash. Measured by the currency of common discussion, the northern public forest was simply another part of a stock of fungible public assets known as "land."[33]

However, we cannot let the matter go at this. The gap here between the objective demands of the situation and the content of policy discussion poses questions that are part of the history of this area of law making, and no less so though we may not be able to answer them altogether.

In some degree the failure of public discussion to come to a focus upon the peculiar problems and possibilities of the public timberlands was disingenuous. To treat good pineland as if it were good farmland suited many men's pocketbooks — indeed, suited the immediate economic interest of whole communities — too well to incline them to press for a closer examination. This is plain enough where we confront arrant fraud upon pre-emption and homestead policies. The common complaint that government field agents met only indifference or positive hostility in local communities when they sought to curb trespass points to similar elements in the situation. The people closest to the forest did not want it singled out for special protection, opposed to government's traditional tolerance of those who squatted on public lands suited to agriculture. "Combinations were formed between capital and labor," the Commissioner of the General Land Office complained in 1865:

Community of wants, aided by remoteness and beyond the vigilance of executive officers, rendered prosecutions ineffectual and baffled every effort. Even in comparatively well-settled regions local sympathy sealed up the sources of information, and those personally interested to suppress the mischief would sit inactively, mailing complaints to the department a thousand miles off from the scene of depredation . . .

There was less than a clear show of good faith, too, in contemporaries who excused trespass and questionable land entries on the ground that government had not provided lumbermen with straight-forward procedures to buy standing timber or timberland as such, and who yet made no serious effort to obtain legislation specially adapted to disposal of public forest assets. Pleas for the poor settler carry little conviction when they come from regions obviously committed to a lumber economy.[34]

Yet, a state of public attitude and discussion touching resources and behavior on the scale here involved cannot plausibly be explained just in terms of melodrama. Ideas and points of view were at work, more subtle than mere cunning interest, in some ways more powerful and more resistant to direction than the profit motive.

BRING THE TIMBER TO MARKET

Both before and concurrent with the substantial growth of the Lake States lumber interest, federal and state policy rated timber as less valuable than farms and towns. The plainest testimony to this was the acceptance in law and public opinion of the social utility of clearing land for agriculture on terms that paid no heed to what might be got for the standing timber; indeed, the trees on thousands of acres were cut, piled, and burned, as simply an obstacle to farm development. True throughout the East and in the Mississippi Valley, this was likewise so as agriculture took over the southern half of Wisconsin. The fact is relevant here as evidence of a prevailing policy, without passing judgment on its wisdom; there were good reasons—among them the lack of markets—for much of the practice. In the first 250 years of settlement in the United States, timber on some 150 million acres was cleared off potential farmland by simple destruction, for lack of market or disposition to wait upon a slower pace of growth. So far as the forest was not reckoned simply as a hindrance, official policy saw it only as a quickly realizable aid to agricultural development of the prairie. As early as 1838 Governor Dodge urged the Wisconsin territorial legislature to press the Senate for early ratification of the treaty removing the Chippewas from their living area east of the Mississippi, because the working of the Chippewa pinery "is considered of first importance to the people residing on the borders of the Mississippi, by affording them cheap and abundant supplies of pine lumber." And in 1848 the territorial legislature urged the like theme as a reason for speedy extinguishment of Indian titles north of the Fox River, where were "some of the best pine lands . . . in Wisconsin . . . directly connected with a large and extensive agricultural district . . . totally destitute of this valuable kind of timber." Law bore formal witness to this drive of popular feeling. The Wisconsin legislature expressly qualified prohibitions of timber cutting by pre-emption or homestead claimants or buyers on credit, by excepting cutting done by one who was "actually using and applying any wood or timber on the land to the erection of fences or buildings thereon; [or] . . . taking firewood therefrom necessary for the use of his family; [or] . . . actually and fairly improving any such land for the purposes of cultivation." The general favor for rapid, large-scale transfer of public land to private ownership for settlement carried forward naturally into the first stages of growth in northern Wisconsin. And, in practice, a strong presumption of regularity protected homestead claims even when they were staked out in northern areas committed by many years' experience primarily to the timber industry.[35]

As the economy took shape in northern Wisconsin from the mid-1850's on, however, the invocation of "settlement" becomes less and less convincing as an adequate expression of the ends men were seeking. In large part this is because men were in good faith shifting their direction of emphasis. In part, though, the symbol depreciated because men invoked it rather for its utility in manipulating power than because it described what was going on.

Theft of timber from the federal and state lands proceeded wholesale as markets developed; the scale of theft was limited mainly by the limitations of transport and working capital. In 1852 the Secretary of the Interior put four special agents into the field to protect Lake States timberlands. Limited as this effort was, it was pursued with sufficient honest vigor to provoke immediate and overpowering counterpressures on Congress from lumbermen. A Commissioner of the General Land Office was dismissed, partly in consequence, in 1854, and in 1855 an executive order removed the threat of vigorous enforcement by ending the special agent system and giving the protective task, *ex officio,* to the Registers and Receivers charged with local sale of federal lands. No claim was pressed harder in justification of this sabotage than that a free hand with the public's timber was needed to advance settlement. So Wisconsin's Representative Eastman argued with passion to the House of Representatives in 1852. He found the lumbermen ill-rewarded for their hardihood and courage in opening up the country for agriculture:

> The true policy of the Government would be, for the present, certainly, to leave this enterprising class of citizens to use a portion of the timber of the public lands without stint, as a sort of bounty for the hardships they have undergone in piloting the way to this country. But very far from this is the fact; for the whole power of the country, in the shape of the United States marshals and a whole *posse* of deputies and timber agents, appointed by the President, without the least authority of law, have been let loose upon this devoted class of our citizens, and they have been harassed almost beyond endurance with pretended seizures and suits, prosecutions and indictments, until they have been driven almost to the desperation of an open revolt against their persecutors.

In 1856, the governor reported to the legislature great abuse of the swamplands pre-emption act by claimants seeking only to cut timber and then abandon the land. The legislature repealed the law, but in the face of strong local opposition. In 1857 the legislature restored the pre-emption privilege, despite the record of its misuse. The land commissioners indicated the likely reasons behind the step. The hope of settlement must be the justification when there seemed little hope of protecting the lands in public ownership; otherwise "no other result can be expected than . . . that of waste and destruction being speedily committed on the Swamplands . . . [for] no caution can guard against such results, except the encouragement of a speedy sale of these lands, on such terms as our industrious population can afford to embrace." In 1862, without effect, Governor Harvey criticized long-term-credit sales of state lands not primarily valuable for farming. His ineffectual protest served only to emphasize the momentum with which policies created for southern agricultural Wisconsin carried on as a very different economic pattern developed in the north. The trespass threat alone brought suggestions of a special attitude toward sale of public timberland. Yet, even when the threat was seen as one to the sole value of land, no conclusion was drawn except in favor of the quickest disposal. The point is brought into focus by the estimate made in 1863 by the federal Surveyor General of the marketability of federal holdings in the Wisconsin and Chippewa river valleys. Though he recognized that timber was the unique worth of much of the land, he drew from this no suggestion for a distinctive policy:

> The soil, as a general thing, is valueless for agricultural purposes, and when stripped of its timber, will not find purchasers at any price; whereas, if promptly surveyed and thrown upon the market, large portions of it will command a ready sale.

Limited in view though it was, this was yet a comment of more point than most. With whatever mixture of good faith and guile, the symbol of "settlement" tended otherwise to dominate discussion, with the effect of diverting attention from the peculiar problem and potential of the pinery. So, for example, in 1870 when the bulk of remaining federal land in Wisconsin lay in the forest area, the Commissioner of the General Land Office continued to justify a liberal sales policy by the familiar criterion, even as he acknowledged that some federal land might hold out other possibilities of realization:

> It would be unwise financial economy to administer this trust in such a manner as to raise a million or so more of dollars, if this involved the necessity of impeding the settlement of the public domain by raising the price, or by withdrawing any of the present facilities for their appropriation which the law holds out to the humblest industry.

True, in the same report the Commissioner observed that Wisconsin pine "has become an immense source of wealth, and must continue so to be for a century to come." But he drew from his observation no suggestion for adapting federal land policy to the special possibilities of this asset.[36]

Long-standing popular opinion in this country opposed the notion that government might hold on to its original ownership of land, to play a continuing landlord's role over any great part of its territory. Public policy faithfully reflected this popular reaction. However, it was an attitude which men had developed mainly in contemplation of the familiar use of land to grow annual food or fibre crops with relatively modest capital investment. This was a use favorable to building a republic of small freeholders, according to the classic political aims of the institution of private property as it had been adapted from Harrington and Locke. Land primarily valuable as forest yielded according to a very long "crop" cycle. Its continuing productivity demanded correspondingly large capital commitments and risks, with correspondingly limited opportunities for widely dispersed private ownership. Conceivably these differences might be seen and accepted as warranting a different balance of public and private control of forest resources. In fact, dominant popular and official opinion in the nineteenth-century United States did not see these differences, or, so far as it perceived them, did not draw from them a conclusion favoring more public control. Some of the reasons for this lay in the general institutional environment; no factor was more weighty, for example, than the state of knowledge of soil potentials, and the common misapprehension that most of the northern forested land was best suited to farming as its ultimate destiny. We shall consider such elements of circumstance in the last section of this chapter. Here we are concerned with the matter so far as it entered explicit discussion of policy in the years important to Wisconsin lumbering.

Popular opinion early applied to the forest its hostility to government's claims as proprietor or landlord. In colonial days there was wide refusal to acknowledge the Crown's right to the trees which its foresters marked with the "broad arrow" as especially suitable for Navy ships. The insistent demand that land normally be held by private owners took for granted that this would be ownership in fee simple absolute, under which the holder might dispose of the timber on his acres as he saw fit. The contrast with Old Country limitations caught the shrewd eye of a German doctor with the Hessian mercenaries in the American Revolution. "In America," Dr. Schoep observed, "there is no sovereign right over forests and game, no forest service":

> Whoever holds new land, in whatever way, controls it as his exclusive possession, with everything on it, above it, and under it. It will not easily come about therefore that, as a strict statutory matter, farmers and landowners will be taught how to manage their forests so as to leave for their grandchildren a bit of wood over

which to hang the tea-kettle. Experience and necessity must here take the place of magisterial provision . . .

So unquestioned was the private fee simple as the norm of land title that before the 1870's there was scarcely any action or discussion even to raise for debate a question of continued public ownership of public timberland. Till then the United States held on to forest title only for the narrowest proprietary purpose — to assure a supply of timber for its own ships. Its one other policy — the generally ineffectual concern about timber theft — was seen only as in aid of sales of the fee; even so, the first limited steps against theft met with cries of outrage that government was penalizing activity essential to opening up the land for use and settlement. The program of leasing rather than selling lead-bearing lands was abandoned for the Wisconsin area in 1847, partly under the objection that it exposed the timber on such lands to unchecked theft; thus the only analogy that federal experience offered for a policy of leasing public timberland was dropped just as the Wisconsin lumber industry stood on the threshold of major growth.[37]

In 1875 Commissioner Burdett of the General Land Office posed sharply the question whether the United States should not hold on to its timberland title. It was "notorious" that the soil of much of the public forest was of limited fertility for other than forest products; it was "equally certain" that "innumerable frauds" were being committed to obtain such lands for their timber and not for settlement and agriculture. He speculated on a new policy to come to terms with these facts:

Perhaps the best method . . . would be for the United States to retain the title to the lands, selling only the right to appropriate such proportion of the growing timber upon any given subdivision as should yet leave a sufficient quantity standing to secure the shade and moisture necessary [to maintain cover against erosion].

Land disposal had proceeded so far, east of the Mississippi, that the Commissioner acknowledged that even then his question had practical significance "chiefly" for the public domain in the Far West; by 1875 the time was past for the United States to play a substantial landlord's role over its "comparatively limited" forest holdings in the Lake States. But the Commissioner himself rejected a hold-on policy, even for the yet unclaimed western forests. Such an effort, he thought, would run counter to popular attitudes too deep to be overcome save at a cost unwise to pay:

. . . however desirable such a system might prove, it will at once be perceived that to make it in the smallest degree effective would require the constant presence and intervention of agents of the Government, involving an expenditure for their support, and furnishing opportunities for fraudulent collusion and unjust exactions, which might well be considered as overbalancing the possible good to be attained.[38]

The accuracy with which the Commissioner measured the latent opinion of the time was shortly proved, when in 1876 Congress debated whether to keep title to its pine holdings in the southern states or to open them to "settlement" and sale. The conclusion went decisively against a hold-on policy. In 1877, 1878, and 1880 Secretary of the Interior Schurz urged that henceforth the United States no longer sell its forest lands, and instead sell only regulated cutting licenses. The Secretary's view did not prevail. But, thereafter, retention of federal timberlands title grew to become a major issue, ultimately resolved by creation of a system of national forests. In 1891 Congress first authorized the President to create "public reservations" of timbered land of the United States. Policy developed under this authority subject to the continual pull and haul of ardently opposing forces in the executive branch and in Congress. Drama reached its peak in 1910 when President Taft discharged Chief Forester Gifford Pinchot. A generation later important western interests had not yet given up the fight to turn over the public forests either to the states or to private ownership. The stormy course of the national forest system attested the likely strength of the opposition that would have fast mobilized had anyone seriously pressed the issue in the earlier years when the United States was disposing of its forest holdings in Wisconsin.[39]

So long as the state still owned a substantial acreage of valuable timber, no one ever posed for it the alternative course which Commissioner Burdett at least suggested for federal policy in 1875. In 1867 a select committee of the Assembly produced a remarkably perceptive, far-sighted report, warning that even the vast northern woods "are fast passing from us — faster than those who have given but little thought to the subject believe," and recommending that, since timber was "a common want to the whole people . . . it is the duty of the government to take steps to preserve what remains, and increase its quantity where it is scarce." But this analysis, far ahead of its day in other respects, discussed the problem solely in terms of the education of private timberland owners toward a more enlightened interest, perhaps under some implied threat of regulation of their cutting practice. The limits of contemporary thinking are revealed by the fact that the report said nothing of the state's own holdings, or how the state might use its ownership to husband a resource which the report saw as basic to the general economy.[40]

The thoroughness with which the state pursued a program of sale evidenced its total failure to envisage the possibility of holding on to management of its timberlands, rather than its rejection of such a course. State law provided for retention of timber title only as security for the unpaid price of land sold, and was most inadequately applied even within this narrow limit; logically enough, the law further expressed its favor for private ownership by authorizing the holder of its certificate of sale to sue for timber trespass while his own land title was yet not perfected, and by granting the state's patentee full right to any uncollected claims for trespass done before patent.

Some of the most valuable of the timberland owned by the state was held by it in trust to aid construction of railroads or other transportation improvements. The legislature often provided that, when earned, such aid lands should be tax exempt for a term of years so long as the recipients held unsold the land and its timber. So far as this went, it favored retention of timberland by delegates who held it to realize public purposes. But this was no decision to establish managed forests for their utility to the general economy; the obvious and acknowledged end was simply to prevent forced sales of timber, to allow the best money yield in support of the improvement of transport.[41]

Laws, 1878, Chapter 324, created a "state park" by reservation of timbered tracts owned by the state in several northern townships. The only official explanation offered for this measure was that it aimed to protect forest cover at headwaters of Wisconsin rivers, to check the flood hazard. There was no effort to treat the reservation as the nucleus of a general forest productivity program, nor was the reservation ever extended. The total of reserved lands was too small — a little under sixty thousand acres — to provide the base for any action of major economic effect, had there been any provision of executive means to manage them. In the same year which saw creation of the "state park," the land commissioners suggested the need of further "inquiry into the future of the enterprise," involving as it did only "widely scattered" tracts amounting to "but little more than nine *per cent.* of the territory embraced." Desultory efforts were made to repeal the state park law in 1883, 1887, and 1889. There was no indication that anyone felt that an issue of major policy was involved; rather, as the Assembly Ways and Means Committee saw the matter in 1887, it presented only the correction of an uneconomic use of assets too scattered to be manageable and hence the more open to timber theft. Given the lack of defined purpose or provision for positive management of the reservation, the land commissioners properly questioned whether it was not an unlawful diversion of school trust fund assets, and the state supreme court so ruled in 1894; regarded in this aspect, however, the matter was throughout treated solely as a fiscal problem, and produced no discussion of the possibility of public management of state timberland for long-term income and capital gains.

Discussing disposition of the "state park" lands after the legislature in 1897 had provided for their appraisal and sale, Chief Inspector Mullen of the state's Department of Public Lands summed up accurately the limitations which had throughout governed consideration of this ill-assorted forest reservation. To the governor, disposition of these lands was "a question of business, and not one of politics or sentiment." Agreeing with this characterization, the chief inspector enlarged upon the problem in terms which reveal how far removed was this episode from any venture toward a broad program of state forest land management:

. . . what is known as the "State Park" was in reality many small tracts of land scattered through twenty-four townships. Part of them were classed as pine lands, and the timber thereon was in constant danger of destruction by fire, owing to the choppings left by lumbermen on adjoining lands. This danger was increasing each year as logging operations were extended. The timber on the state lands had reached its maximum value by reason of the fact that logging railways had been extended through the territory, dams had been built, streams cleaned out for driving logs and logging roads cut. It was deemed unwise to keep these lands out of the market until the improvements had fallen into disuse, as there was not enough timber on them to warrant rebuilding or repairing them, and therefore it would have had to be sold for a low price owing to the lack of facilities for getting it to the market.

In sharpest contrast was the import of Governor LaFollette's recommendation in 1903 that the legislature decide "definitely and promptly" whether "in [its] judgment . . . it will be wiser for the State to retain . . . [its remaining unsold] lands for forestry culture, or other uses to which it may appear they are better adapted . . ." This call, and the 1903 and 1905 legislation that provided for managed forest reserves, pointed toward a positive ownership function of the state, suggestion of which was wholly wanting in the nineteenth-century record. These twentieth-century developments do not belong to the legal history of the great Wisconsin lumber industry, but they do throw into sharper relief the limits within which the state previously conceived its role as landowner.[42]

Declared hostility to the private speculator was as much in the tradition of our nineteenth-century public lands policy as opposition to a landlord's role for government. The two complemented each other. A republic of small freeholders wanted no great proprietors of either stamp.

Denunciation of the evils of "speculation" figured somewhat in talk, though very little in action, about disposal of public timberlands. Even the talk was empty. As in other respects, so here there was little will or imagination to analyze the distinctive facts and choices involved in forest production. Most of the criticisms condemned "speculation" in public timberland so far as it was accomplished by fraudulent pre-emption and homestead claims filed as a cover for logging without intent to farm, or because it was thought to hold off the market large areas which would otherwise move into productive agriculture. Criticism not based on these grounds generally ran only to fiscal criteria; profiting from lax price policies of government, or by collusive bidding to defeat public auction, large buyers were depriving the federal treasury or the state educational trust fund of dollars which they should be collecting for public purposes. If the talk had little force, this was not surprising, because it was not directed toward key facts. Fiscal argument did not weigh heavily with the post-Civil War federal government, which generally had ample revenues for its needs. Money came harder to the state, but here we meet another powerful current of attitude. The trust funds represented claims of a future which looked quite remote to the Wisconsin people of the seventies and eighties. Most of the criticism of the low prices at which state timberland was sold came only after the business was done. The limited fertility of soil and the costs of clearing land actually made much expressed concern over frauds or delays irrelevant to achievement of farm settlement in the Wisconsin pinery. If there was little pressure behind talk against "speculation" in state pinelands, this conformed to the fact that there was little pressure to farm the pinelands. "Speculation" in timber presented real and debatable issues in terms of the implications of large-scale private ownership for general balance of power and efficiency. But bigness in industry and finance developed at a pace which left public policy far behind on all fronts in the late nineteenth century. It is hardly surprising that issues were not defined with more sophistication as Lake States lumbermen sought public timber in the first relatively crude stages of improvising a new scale of lumber operations.[43]

Measured by action, neither federal nor state timberlands policy included any significant curb on speculative acquisitions in the years of the Lake States lumber boom. Neither Congress nor the state legislature defined standards specifically adapted to disposal of public forest assets; neither provided central or field executive organization adequate even to police against trespass or enforce the limited settlement programs which did stand on the law books; Congress made a poor record of yielding to local pressures to halt such sporadic initiative as the federal executive branch showed against the elementary threats of theft and fraud. Very large portions of federal lands made available in aid of railroad construction, in Wisconsin and in the country west of the Mississippi, were valuable mainly for their timber. Congress set no limits upon the manner in which these assets should be realized, other than that the proceeds go to build the roads. In Wisconsin, as in western states, lands so derived went far to build some of the largest private concentrations of timber holdings. Congress remained inert in the face of repeated warnings that its 1878 laws providing for timber cutting in aid of mining and for acquisition of forest homesteads were being employed wholesale to block up great private investments. As late as 1891 the United States Supreme Court lent powerful help to this process. *United States* v. *Budd* held that the government did not sustain its burden of proof on a bill to set aside for fraud a patent issued under the Timber and Stone Act, by showing a pattern under which numerous individual entrymen, having formally perfected claims, promptly conveyed them to the same grantee. True, the statute denounced any prior agreement for such an outcome. But:

If when the title passes from the government no one save the purchaser has any claim upon it, or any contract or agreement for it, the act is satisfied. A buyer seeking to accumulate a major stand of timber might rightfully go or send into that vicinity and make known generally, or to individuals, a willingness to buy timber land at a price in excess of that which it would cost to obtain it from the government; and any person knowing of that offer might rightfully go to the land office and make application and purchase a timber tract from the government, and the facts [of pattern, as shown by the government] . . . point as naturally to such a state of affairs as to a violation of the law by definite agreement prior to any purchase from the government — point to it even more naturally, for no man is presumed to do wrong or to violate the law, and every man is presumed to know the law.

The Court was later criticized for this decision. But it simply expressed an attitude long implicit in Congress' acquiescence in development of a lumber industry which depended upon cheap large-scale purchases from the public domain.[44]

Only once — in 1855, when the lumber industry as yet stood only on the threshold of its full growth — did the Wisconsin legislature investigate big buyers of state lands. The investigation had little practical result. It contributed to the enactment in 1855 and 1856 of statutes voiding purchases of state land other than for settlement and in quantity exceeding 320 acres to a purchaser; on the other hand, the legislature made no provision to enforce these limits and repealed them in 1863. The 1855 investigation has meaning for us mainly because it reflected so well the typical failure to focus policy on timberlands values as such. The investigating committee reported that of about 200,000 acres sold in 1854 on thirty years' time without down payment or other security than the land, about 130,000 acres had been purchased by nine buyers, in quantities ranging from 5,000 to 34,000 acres. The value of much of the land, it was well known, "consists almost entirely in their timber." The books showed that acreage had been bid up to high prices but was eventually entered only at its appraised value (which was typically the statutory minimum price). Fraud had been charged. Speculators had first warded off competition by high bids, it was said, and had then forfeited their original bids and bought in the land privately at the minimum. The committee did not make an ultimate finding on this charge, nor was the matter pressed further. The committee was firm in its opinion that "speculators" should not be allowed to "monopolize" the state lands. However, it put no clear content into these large words. The context indicates that the committee was thinking simply of access to the land for farming; it said nothing to show that it was concerned for the relative position of big and small operators in the lumber industry as such. The committee was sharply critical, on the other hand, of the land commissioners' failure to arrange adequate security for the price of land primarily valuable for its timber. But this criticism ran solely to protection of the money return from sale of the lands. The committee had nothing to say of the implications of large-scale purchases for the long-run physical productivity of the forest.[45]

The only other point at which the legislature took action about "speculative" holding of timbered lands concerned the railroad-aid grants. The 1870's brought to a head local resentment over the long withholding of lands from market, as railroad promoters were slow to fulfill the conditions which would put the lands at their disposal. Resentment was compounded because the aid lands were exempt from taxation while title remained in the state, and the legislature typically granted continued exemption for some years after the lands were earned. In familiar fashion, thus, argument went largely in terms of "settlement" or the money return from the land, without coming to grips with the peculiar problems of realizing on forest assets. However, these aid lands all lay in the north, where the forest was too big a fact to be wholly ignored. Critics feared that railroad promoters would strip off the timber before they earned it, or would sell it all while they enjoyed tax exemption and then let the land default for taxes. To these fears the legislature made some response; sometimes it required that a railroad prefer *bona fide* settlers in its sales policy, stipulating also that the road must in any case not long withhold the land from market but sell it within a few years of earning it; moreover, when the tax exemption fight became hotter, the legislature finally stipulated that tax exempt status should end if either the land or its timber were sold. No special provision was made to enforce these various limitations. Certainly, the settler preference provisions did not prevent railroads and lumbermen from coming to terms.[46]

Measured by words, the popular hostility to public or private landlords appears to have figured in disposal of forested, as it did in disposal of agricultural, public lands. Measured by behavior, the sum of the matter is not so simple. In the first instance nineteenth-century policy makers applied to the public timberlands the presumption favoring massive rapid transfer of the fee to private hands. In the years that shaped the Wisconsin lumber industry men did this without serious consideration of alternatives. Yet what they did proved to be an illusion. The grantees did not in fact shoulder the continuing responsibility of fee holders of the forest. They acted as lessees or licensees to cut without limitation of waste, and by tax default or by sale at junk prices to farmland speculators they resigned responsibility for the productive forest within a few years of the time when they ostensibly assumed it. Except by fraud there was not even so much formal compliance with the policy against "speculative" (i.e., large-scale) acquisition. Taken at face value, failure to keep public timberland out of the hands of big operators robbed of meaning the policy which denied a landlord's role to government. In this sense, as also in terms of cash yield, the state land commissioners spoke truly when, looking back from 1902, they observed:

> Efforts to sell state lands at their intrinsic value were resisted on the ground that it would retard the settlement of the state, but the records show that the speculator rather than the actual settler has been the beneficiary of the policy which prevailed prior to 1897.

The traditional policy condemned speculation because it impeded growth of a family-farm society. However, this was a development in fact denied to most of the north country by the limited fertility of its soil. There is little meaning in a judgment that the policy against speculation broke down in application to an area where the reason for it did not exist. The significant fact was that the policy-making processes failed to come to definition and debate of choices, gains, and costs relevant to the actual situation.[47]

Part explanation of this failure was a shift in emphasis between objectives of traditional policy. We had valued settlement as the means to accomplish political and social equality in a republic of small freeholders. Before 1850 this was the dominant theme of public-lands policy, its victory over a sales-for-revenue program symbolized by the federal Pre-emption Act of 1841. Early present, but at first subordinate, was another article of faith. We would fashion a way of life which promised increased individual dignity and satisfaction for all, through cumulating, material productivity based on the division of labor, the market, and the machine. Our experience with the politically oriented policy of free or cheap access to the riches of the public domain soon taught us the excitement of large returns for modest investment of scarce money and labor. More and more the marvels of an ascending production curve captured our imagination. This was the trend of attitudes toward disposal of public lands primarily suitable for farming. The shift was complete, in discussion of policy regarding public timberlands in Wisconsin. Men had demanded that the United States sell and not lease its acres in southern Wisconsin, lest we have a great official absentee landlord dominating a servile tenantry. But they put the case for pressing sale of public pinelands simply in terms of getting the forest into production and increasing the tax base; so far as "settlement" was urged, it was because each "improver" represented an added unit of productive energy.

In this, Wisconsin experience moved in a broad current. Men still invoked the symbols of independent freeholder as against servile tenant in recurrent contests over handling federal lands in the South and in the Far West. Production, however, was the favored goal of policy. When debaters worried over the "small man," it was likely to be — as in the Congressional discussions over sale of southern pinelands in 1876 — on the claim that the initiative of the small operator would best get out the timber, or because — a major theme with late nineteenth-century Grangers, Populists, and Progressives — he was denied his fair share in the realization of wealth and fair entry into the market. We never wholly lost hold of the older political and social values attached to the institution of private property. But these were effectively submerged by the common faith in productivity as the criterion of decision during the critical generation of change after the Civil War. Traditional policies against public and private landlords were not practiced in relation to the Wisconsin forest, because there we preferred to use resources in any way that would visibly increase output.[48]

VALUE TIMBER FOR ITS CASH OR LABOR YIELD

Land was plentiful; labor and cash were scarce. Labor and cash were expensive, naturally enough. Naturally enough, therefore, land ought to be cheap. The gross imbalance among these factors of production blocked the economic growth on which we put much of our hope for achieving a better way of life. We had little enough means to exert leverage on our situation. But we did have more "western" land than we could see the end of, all of it held by government for the public welfare. Let us, then, give it away or sell it cheaply. So we might attract scarce labor and cash. So we might enhance the productivity of the scarce factors, allowing them to reap the most ready fruits of the land without counting the loss of utilities to be had only by spending more work and money upon fewer acres.

This was a pattern of thinking well learned out of the experience of opening up the continent before men began to log the Wisconsin pinery. All the promise and the difficulties of the northern forest made application of the familiar ways of thinking seem simple common sense. Here was valuable raw material, so abundant that for many years men could not imagine exhausting it. Here also was the land in all its massive challenge, posing great risks, requiring hard labor, costly machinery, and even more costly transportation improvements, to make it yield. How put our opportunities to work? Surely the sensible plan was to spend the timber generously. So we could minimize the labor and cash investment needed to get output. So, too, we could limit risk to invested capital by shortening the time span for which capital must be committed. The multiplying farms and the booming towns of the treeless prairies created demand for lumber in a volume which meant rich returns, if the source material could be had cheap enough. But, for the firm which must work on a slender margin, there were heartbreaking difficulties and risks in meeting payrolls and due dates on mortgage bonds and bank loans. It became an axiom that men made big money in the lumber industry, not out of processing or handling but out of favorable investments in standing timber. Immediate experience nurtured these judgments as the Wisconsin industry grew from mid-century to the full scale of the eighties and nineties. They were lessons learned by enterprisers of modest means, like Moses Strong, struggling and failing to set up a profitable mill on the Wisconsin River in the fifties—learned by substantial middle-range firms, like the Daniel Shaw Lumber Company of Eau Claire, which succeeded only by surmounting recurrent crises of working capital—learned by the failure of a concern so large as the Union Lumbering Company of Chippewa Falls, whose finances could not stand the stress of the middle seventies.[49]

It is not hard to understand the industry's estimate of its situation. It is more pertinent to a history of public policy to note that the industry's interest was accepted without substantial question as identical with general interest.

Once again we confront the application of attitudes learned from one pattern of experience to a situation in fact significantly different, while the processes of public decision fail to define the issue. The momentum of official and popular favor toward farm labor ("settlement") carried us a good way toward a preference for spending timber freely to increase the yield of woods labor. Pre-emption and homestead laws encouraged the settler's labor by giving him a free hand with the timber on his claim, to cut and use, destroy, or sell it as part of making a farm. As our emphasis in mid-century tended to be more on economic than on political-social considerations, talk ran to the benefits of using timberland so as to increase the productivity of labor in the forest as a social utility in itself, no longer with exclusive reference to promoting agriculture. Wisconsin's Representative Eastman urged that the United States cease to "harass" the loggers who cut its pine without title, and instead "leave this enterprising class of citizens to use a portion of the timber of the public lands without stint, as a sort of bounty for the hardships they have undergone in piloting the way to this country." This appeal to the superior equity of labor over inert capital ran from years when colonial woodsmen refused to acknowledge Crown claims to trees reserved by the blaze of the "broad arrow." The state should lend the aid of lands at its disposal to increase the extent of its forest industry, urged a committee of the Wisconsin Assembly in 1873; hence it should help build a railroad which would enable Wisconsin lumbermen to deal directly with the midwestern market. This would allow local capital to "reap the benefits" of all stages of production and sale, "while at the same time it would give employment to almost double the number of men within our own state, now employed in handling the lumber." Eager to augment its human capital, Wisconsin created various official agencies to promote immigration into the state; through the generation of lumber boom it became a staple inducement to the newcomer that he could find steady wintertime employment in the lumber camps to help him earn in order to buy a farm.[50]

In practice we were ready freely to spend fixed natural capital to promote the yield, not only of labor, but also of the fruitful combinations of men and materials which managers could make with the help of money. Popular and official attitudes were blurred here, because our agrarian-debtor tradition made us distrust that aspect of managerial skill that involved manipulating money and credit. As with farmlands, so with timberlands, hard words were spoken against ill-defined dealings condemned as "speculation." Plainly such criticisms were not levelled at the man of business as such. This private-property, market-organized society put such high value on the entrepreneur's job that it rarely saw need to make its faith explicit. When it did speak, it spoke firmly; a notable example is the common law doctrine which presumes every private contract to be lawful and lays a heavy burden of persuasion on one who would deny it enforcement as against public policy.

In this context, one may fairly read the criticisms of "speculation" in timberland as aimed at those who would profit merely by holding the land for a rise in value without undertaking to put it to work. Condemnation of forest "speculation" in itself evidenced the preference for valuing fixed natural capital for its utility in fostering the productivity of scarce mobile capital; the point of criticism was that the "speculator" did not acquire land to make it part of a pattern of production.

"Speculation" apart, there was little discussion of the fact that federal or state timberlands were disposed of in ways which subsidized the growth of the Lake States lumber business. One reason surely was that the end result appeared so plainly to fit our general faith in production that it was taken for granted to be a public good. Governor Harvey told the 1862 legislature that the significance of Wisconsin's natural wealth was the opportunity it gave for profitable employment of men and money:

> A comprehensive State policy must take in every judicious measure inviting capital and labor to the development of our resources. The census of 1860 shows 3,746,036 acres of the area of Wisconsin to be improved, while 4,153,134 acres are unimproved. Our State is rich, far beyond all she now enjoys, in capabilities of production — in forests of timber — in unappropriated water power, and in latent mineral wealth. These vast resources need but the touch of labor and capital for the realization of the wildest dreams of prosperity for our State.

By the late sixties Governor Fairchild was encouraged by the fact that "while this is, in the main, an agricultural state, other important interests are growing rapidly in our midst and enlisting the energies of a most active and enterprising class of men" — including "the activity displayed in our pineries." Millions of dollars were invested in "the mining, lumbering and manufacturing interests of the state . . . controlled by a class of our citizens among the most enterprising and industrious." "To them all," Fairchild urged, "the state should extend every possible encouragement." With the lumber industry in full course, the Wisconsin Supreme Court in 1880 felt that it could speak with confidence the policy of the community toward this interest. The industry sustained a traffic which had become "a most important branch of commerce in the state," involving "an enormous sum every year," supporting "the wages of thousands of men who are employed to cut the logs and get them to market." Any doubt in construction of statutes affecting this interest should be resolved in favor of the industry's growth and activity, and the court would so rule, unless the legislature "clearly and unmistakably" spoke otherwise.[51]

The policy of spending timber freely to enhance the productivity of men and money came nearest to expression when we talked about public subsidies for building railroads through the northern forest. The issue naturally became more explicit here. Railroads required uniquely large investments; their subsidy meant correspondingly large commitments of public resources. Where they were built, and how fast, and how they might be run, were matters obviously important for the direction, quality, and speed of growth of the whole economy. When a lumber company founded its prosperity on rich timberland bought at very low prices from government, this was accomplished under general land statutes which gave most such transactions the unremarkable appearance of ordinary buyer-seller dealings in market. But the scale and special import of railroad-aid grants loomed too large to be brought under general procedures. Impatient and careless as they usually were with deliberate, formal processes of making public policy, nineteenth-century leaders yet felt the compulsion of this situation. Here they must spell out more fully than was usual the choices and ideas involved in using up fixed public capital to mobilize and energize fluid private resources. The tangible sign of this was that land grants in aid of railroad construction were typically — and properly — made by "special legislation."[52]

Railroad promoters and public officers alike valued the forested aid lands mainly as they might be used as cash substitutes or inducements to private cash investment. The state legislature treated the lands as a cash substitute when it made grants in kind of public timberlands needed for rights of way. Far more substantial, of course, were the general grants which Congress declared that it made "for the purpose of aiding . . . [and to] be exclusively applied in the construction of" northern Wisconsin railroads, to "be disposed of only as the work progresses." Like Congress, hard-pressed promoters looked to the marketing of these lands to yield or attract funds for current construction. "The value of the land grant keeps pace with the increased values of everything that is required to construct the road," a company president reported optimistically to his stockholders in 1864, "and the high price of lumber renders it certain that large sales of our pine lands can be made for cash." The West Wisconsin Railway's "land commissioner" told the State Railroad Commissioners in 1876 that the company had "instructed [him] to push the sales of land as rapidly as possible," and that this was being done. This was in an area in which some settlement was in progress. Where construction was projected through unbroken forest, the timbered aid lands alone could be looked to for any near hope of financial return. So argued the majority of the Assembly Committee on Railroads in 1875, urging a charter amendment to relieve the Wisconsin Central of a burdensome route requirement which impeded its ability to earn its lands. Since the Department of the Interior would issue no more patents until this difficulty was cleared away, the impediment

> shuts out the company from deriving any advantage from their land-grant by building any more of their road toward Lake Superior. As that line is through a country utterly destitute of inhabitants, there is no inducement to build a road in that direction, unless their lands can be given them.

The timbered aid lands were critically important for financing construction and launching the "Omaha" road through the northwestern pinery, the Wisconsin Supreme Court found in 1889. Hence the lands must be husbanded for their trust purpose. The court ruled illegal a contract by which the Omaha's promoters undertook to transfer a quarter of their aid grant to another road in consideration of the latter's withdrawal of its efforts to solicit the grant from the legislature.

> In bestowing the grant the legislature were executing a trust imposed upon the state by Congress. The legislature had no power to pervert that trust, nor any part of it, even for the benefit of the state, much less for the benefit of a railway corporation upon which no part of the grant had ever been conferred, and which owed no duty in the construction or operation of the road in aid of which it was granted. The object of such grant was not only to aid in such construction, but to insure its continued operation. But to sanction such a contract so perverting one fourth of the grant might . . . leave the constructing company insolvent and without any ability to successfully operate the road.

If the emphasis were not on quick sale of timbered aid lands, they were nonetheless valued for their immediate financial utility, because their forest wealth made them acceptable mortgage security on which the roads might float bond issues to raise construction money.[53]

As with public lands best suited to farming, so with those valuable primarily for timber, these specific preferences for mobile over fixed capital fostered the general habit of measuring all government land problems by what was at hazard in dollars, rather than by what was being spent or risked in physical productivity. With fidelity to the statutory pattern, in 1863 the Wisconsin Supreme Court could define the state's interest in its timberland wholly in terms of collecting a price for goods sold. The state had elected to dispose of its timbered assets in the role of a seller on credit. As such, when it complained that its buyer had cut timber contrary to the terms of his contract, its grievance was solely that of a "creditor having a lien upon the land as security for his debt." "This is all the interest which the state has in these lands after a sale, and that being so . . . an injury not to the land merely, but to the security can alone sustain an action." Thus, the state might recover only to the extent that the wrongful cutting was shown to have impaired its security for the unpaid balance. Officials worried over the money lost to the public purse from timber trespass, measured their success against theft by the number of dollars collected from fines and seizures, and congratulated themselves if these returns paid the timber agents' salaries or even yielded a modest net profit. Such concern as they showed over pricing policy was because private speculators reaped dollars which might have gone to the federal treasury or the state's educational trust fund. Their condemnation and ineffectual pursuit of those who got public timberland by fraud were made in the name of the public "funds" as well as the settlement policy. Both Congress and the state legislature lacked will or imagination to treat the public forest as a self-sustaining productive entity, even to the limited extent of creating revolving funds for its protection out of the meager receipts from trespass proceedings or sales of dead and downed timber. With its monetized calculus of land policy, the state legislature reckoned that Wisconsin could not afford the costs of modest care of its forest wealth. The tenacity of a monetized calculus showed itself further when men began to pass retrospective judgment on the state's land dispositions; for years they still reckoned the costs in dollars lost to the "funds" more than in productive potential lost to the northern land. The story does not lack irony. We prided ourselves that we were a hardheaded, "practical" people. But chronic scarcity of working capital taught us to weigh policy more by the symbols of things than by the things themselves.[54]

WHAT WE DO NOW COUNTS MOST

The original forest was the work of centuries; its maintenance as a renewable resource required commitments counted by generations. Measured against the facts, the use men made of timber wealth was peculiarly revealing of their time preferences and the value they put upon pace and scale in economic growth.

In disposal of public timberlands, federal and state policy manifested the same pattern of preference for present over future yield as in disposition of public land primarily suited to agriculture. Larger individual aggregations of capital were often involved, and there was more investment in heavy machinery and in labor organized into integrated operations in woods and mill and on the rivers and railroad lines. Pushing the rails through the unoccupied forests of northern Wisconsin dramatized the deliberated promotion of economic development more vividly than did the construction of prairie lines. Probably because of such elements in the total situation, the disposal of public timberlands produced better defined statements of the booster theory of public investment than attended promotion of the Mississippi Valley family-farm economy. Such difference as there was, however, was in emphasis and not in substance.

Scarcity of working capital created practical pressures to use forest wealth to meet current operating costs and to relieve the immediately felt hardships of those who were doing the rude work of opening up new country. Congress' lack of will to underwrite strong action against timber trespass in the Lake States derived partly from sympathy for the argument that "bounty" was owed the men who "piloted" the way into the difficult forest section. Both Congress and the federal executive felt the insistent influence of these attitudes from the Lake States forest area. It was "the pressing demand of settlers," along with "the avarice of capitalists," which had "laid waste and spoliated immense areas of timber land," warned the Commissioner of the General Land Office in 1865. "Capitalists" as well as settlers sometimes saw in the public timber a means to save on current operating costs; agents reported that among the boldest timber thieves were those who took the state's wood to fire steamboat boilers. State policy expressly approved some use of wood to meet current costs, for the statutes allowed one who bought on credit to cut timber for his own firewood. No nineteenth-century federal or state law limited large-scale harvesting of wood for sale as fuel, or in any way discouraged spending high quality timber upon uses which inferior lumber would as well satisfy. Casualness toward quality of use was not surprising, given the abundance of material. Nonetheless, unquestioning acceptance of these practices implied that contemporary opinion was willing to see the forest spent in some measure to satisfy current operating costs, rather than to insist that it be treated solely as a capital asset.[55]

However, this was not the dominant emphasis with which official and popular opinion pressed for early rather than late realization of forest wealth. In using forest assets, as in using other wealth, we were investment-minded. Here, too, we were more interested in using law — including the power to direct the allocation of publicly owned resources — to promote the growth of productive capacity than to increase present consumer satisfactions. Thus, for all the talk of easing the settler's life, when the Wisconsin legislature memorialized Congress in 1851 to forego strong action against trespass on federal timberlands, its plea was equally for free access to the forest to speed the growth of productive power:

> . . . the people of the northern portion of the State of Wisconsin are suffering great inconvenience and embarrassment in consequence of the proceedings instituted by the officers of the general government, against persons engaged in the manufacture and sale of pine lumber, from timber obtained on the public lands. It is a well known fact, that from the earliest settlement of the State, a numerous class of our citizens have been engaged in lumbering . . . upon these lands without being regarded or molested as trespassers . . . [T]he occupation has been considered as both laudable and lawful, in view of the necessities of the people of the State, and the prosecution of it has materially aided in the settlement and prosperity of the country . . . [E]ncouraged by this seeming acquiescence of the government . . . as well as the increasing wants of the people, a large amount of capital has been invested in the business, and its continual prosecution is essentially necessary for the better comfort and convenience of that portion of the people of the State, who are rapidly filling up its northern section.

The Congressional debate of 1876 produced the decision to open to sale the southern pinelands previously reserved for homesteading. Again, increase of general productive capacity was a central theme. Those who successively opposed enlargement of national forest reserves by presidential orders under the 1891 act argued confidently that the West had a moral claim on the public domain that it be used to advance their economic growth. Moreover, in the eighties and nineties the early conservationists often presented their case ineptly, appearing to have no program except to lock up the public timberlands. Opponents condemned their efforts to "hoard" public assets, appealing in righteous outrage to the country's traditional faith in the social benefits of rapidly rising productive capacity. Most of these debates concerned developments outside Wisconsin in years after the public timberlands there had already been committed to a particular course of use. Even so, the pattern of discussion in the national forum after 1875 is relevant to our interest. It attests latent attitudes which would soon have come to expression, had anyone seriously tried to postpone realization upon public forest assets in the Lake States. The vigor and tenacity with which men pressed these values well into the twentieth century, when issues were explicitly drawn over using public timberlands to speed regional growth, helps us measure the strength of like attitudes which operated substantially without challenge in the years of the Wisconsin lumber boom.[56]

Some preference for future over present satisfactions inheres in any decision to use assets to build production power. But, though prevailing opinion valued public timberlands mainly for this end, the span of its patience was short. People wanted to see production rise in their generation. The pace of the country's development schooled their ambition to demand visible results. The frustrating scarcity of working capital limited their willingness to pass up any chance to increase the present yield of money and labor by generous spending of timber. If the issue were pressed, contemporary opinion was prepared to deny that unborn generations had a moral claim to any particular present allocation of material resources; the future could not serve the present, and the future's needs and means were unknowable, anyhow, beyond a quite limited forecast. Wisconsin's Timothy Howe put this point of view with candor when, in 1876, the United States Senate debated opening to sale the federal government's southern lands. Successfully opposing the proposal by Boutwell of Massachusetts that the United States hold on to the land that was valuable for its timber, Howe explained,

> I am, as well as my judgment informs me, ready to labor by the side of the Senator from Massachusetts for the welfare of the government today, and of the generation now existing; but, when he calls upon us to embark very heavily in the protection of generations yet unborn, I am very much inclined to reply that they have never done anything for me, and I do not want to sacrifice too much.

Such estimates came readily to men whose lives were caught up in pioneering the economic expansion of freshly opened parts of the United States. The same arguments attend later debates over use of the public domain in the Far West. In 1909 the Senate heard from Colorado's Teller substantially the judgment Howe had delivered a generation earlier, now spoken with a nice mingling of appeals to the values of settlement and of productivity:

> I do not think the sacrifice of trees for useful purposes can be set against the advance of our civilization. I would rather have an American home, with an American family, than to have a forest as big as all out doors. I believe that the natural wealth of this country belongs to the people who go and subdue it. I do not believe that there is either a moral or any other claim upon me to postpone the use of what nature has given me, so that the next generation or generations yet unborn may have an opportunity to get what I myself ought to get.

In 1915 Mr. Justice Marshall of the Wisconsin Supreme Court looked back upon his own experience as a lawyer for great lumber interests in the Chippewa Valley of the seventies and eighties, as he expounded the scope of state police power for forest conservation. The measure of this power was what was reasonable for the general welfare. This would depend upon circumstances. So the state might conceivably do things in its developed, crowded, twentieth-century existence which it would have lacked authority to do in simpler conditions. Measures reasonable today would not have been so

> . . . in the early days of the commonwealth when everything was in a primitive state, burdens of taxation to care for the real necessities of civil government were all that could be borne, the state was wealthy in its forests, lakes, and rivers beyond comprehensible danger of diminution, — [and when] any danger in that regard [was] so remote as not to be appreciated or seen at all and the paramount thought of all . . . [was] agriculture, manufacturing, and development in a way which meant destruction upon the one hand for conservation upon the other . . .

In a country, "more than half its territory unseated lands, in the whole sparsely settled and with no congestion probable for a long term of years," where "wood fuel in abundance . . . served more economically" than any other, where "little or no economy [was] in prospect in the use of local timber supply for building purposes," and surface water supply presented no problem,

> to conserve and increase forest products at public expense solely for the benefit of the people of the future, and under such conditions that the benefits would in the remote time be as likely to go to those outside the taxing jurisdiction as to those within it, would seem to be, at best, a great stretch of the police power.[57]

The booster hypothesis provided the link between nineteenth-century interest in treating public timberland as a capital resource to build production capacity and nineteenth-century insistence on early spending of forest wealth. Generous present use of the public timber was warranted where this would give impetus to general economic growth. Labor and money would yield more richly. This would put in motion a cumulative process of development which, once started, would create steadily compounded increments of productive capacity.[58]

This argument was made most clearly and often to justify spending public timberland to help build railroads through the Wisconsin pinery. So in 1868 the Wisconsin legislature asked that Congress grant fresh land to help build a railroad up the Wisconsin River valley, through country which "embraces many of the most extensive lumbering manufactories . . . as well as large quantities of the richest farming and pine lands," development of which "only requires an avenue through it." Again, in 1871, the state legislature besought Congress to renew the land grants in aid of a northwestern Wisconsin railroad net, to "secure the speedy development of a large portion of this state, which must otherwise remain for a long time sparsely settled." In 1873 the Wisconsin legislature weighed the claims of promoters competing for allocations from the timbered aid lands held in trust by the state. Legislative committees made recommendations according to their estimates of the spur which particular projects might give to overall economic growth. Aid should be given the road whose construction would best "furnish inducements to labor and capital to develop the boundless wealth now awaiting the enterprise of the pioneer and skill of the artisan in the northwestern portion of the state," said an Assembly select committee:

> Labor, capital and manufacturing enterprises follow rapidly upon the track of the locomotive, and the prosperity of all communities is proportionate to the extent of their manufacturing interests . . . [A] wise, statesmanlike and far-seeing policy should provide, so far as practicable, for the encouragement of manufactories and converting into articles of commerce the unlimited materials within our own borders.

These considerations, the committee thought, had particular application to the development which new railroads would mean in the pinery. A strategically located road would expand direct lumber trade with the midwest market. Moreover:

It would do more than this, for manufactories of all kinds, following the great laws of commerce, would inevitably spring up on the Chippewa river, with pine and hard timber in unlimited supply, and water power capable of running half of the machinery of the state.

Favoring a line to open other parts of northern Wisconsin, a second committee urged like confidence in applying aid lands where they would have the greatest multiplier effect:

[The proposed road] will open up all of this vast raw material [of northern timber and mineral] to the manufacturing enterprise and capital of the state. It will distribute the lumber and ore through the central and southern counties of the state, in return take back their products, and cannot fail to enrich both the manufacturing and agricultural counties through which it passes.

In 1874 Wisconsin's first board of railroad commissioners found this merit in the general pattern of railroad construction fostered by the grants of timbered aid lands in the north:

Financially speaking, they are all of them constructions in advance of the paying point . . . On the other hand, it cannot be urged that they were built out of time. They were a necessity to the sections of the state accommodated by them, and are also an advantage to the whole state; some of them opening new channels of traffic for competing companies, and others penetrating entirely new portions of the state, thus ensuring their early settlement, and making their extensive resources available for the more rapid advancement of the whole community.[59]

THE ENVIRONMENT OF MAKING POLICY FOR THE PUBLIC TIMBERLANDS

The context of circumstance pressed in with peculiar weight upon the process of making policy to manage such a problem as disposal of public timberland. The stakes were at once too remote and too close to make decision easy. They were remote in time and space, and hence also in experience. In proportion as the great timbered sections had prime quality as forest, they were unsuited to quick or large settlement; the people who did live in them and who generated the strongest sustained pressures on government concerning their use understandably focused upon uses tangibly related to their own present advancement. Both to them and to the larger populations whom distance separated from close knowledge of the forest, the alternatives to maximum present realization of timber wealth involved time spans which appeared unreal or unreasonable, so far as alternatives entered the imagination of men whose experience taught them to expect accelerating pace and scale of economic growth as normal. The forest as a productive resource was a fact too remote from the lives of most people to teach them a lively concern for maintaining it. In contrast, the availability of cheap, easily worked forest products was an immediately felt reality which eased the time and money costs of constructing farmhouses and barns, fencing lands, providing plank roads, bridges, telegraph poles, railroad ties, and building stores, homes, sidewalks and streets in town. By infinite utility and low price, wood became an important present fact in almost every sector of economic operations for the generations which made critical commitments of the public forest. Decision making here confronted alternatives either so remote from experience that people found it hard to define them or be concerned about them, or so close that people found it hard to forego immediately realized results or to pay immediately felt costs.

PRESSURES OF THE SITUATION

Northern Wisconsin was full of trees and scarce of men and money. Illinois and Iowa were short of trees but booming with growth of towns and farms. No one legislated these particular facts into being. That the factors of forest production were grossly out of balance, that an explosive demand pressed upon the supply of wood, were positive relationships created by the cumulative growth of the whole society. Public policy had contributed materially to this growth, and would continue significantly to affect its force and direction. But public policy could not convert into a tidy, controlled flow the rampaging currents of change in the nineteenth-century United States. Population increase, fast-paced improvements in technology, wide dispersal of decision-making power through private property and the market, invention of business and financial techniques to enlarge the scale of operations generated pressures for change on fronts far wider and in tempo much faster than contemporary law could match. Two aspects of the country's growth had special significance, given the dynamic forces at work. In increasing measure this was, first, a division-of-labor economy, and, second, one tied to sectional, national, and international markets. Hence, it was a society of subtly interdependent processes, in which the impact of change was transmitted sensitively and far in ways hard to foresee or estimate. With everyone in motion, in the midst of "continual striving . . . after fortune," it is not surprising that men tended to accept gratefully as data of commonsense the more massive facts of social environment, the secondary, largely unwilled and unforeseen consequences of past behavior. Americans of the second half of the nineteenth century showed no more capacity than other men to rise, godlike, above circumstance. The generation which made critical commitments of the public forest in Wisconsin exercised its will and imagination within the confines of a supply and demand situation which it inherited and took for granted as supplying postulates for policy.[60]

Into the seventies men saw the Wisconsin forest as "boundless," "almost endless," "vast." In 1870 the Commissioner of the General Land Office congratulated the country on the fact that Wisconsin pine "has become an immense source of wealth," which, he was confident, it "must continue so to be for a century to come." Physical abundance loomed so large that men could not see around it. If government held on to its forests and maintained minimum conditions for their productive renewal, they might prove a substantial source of public revenue and a potent means to stabilize and promote regional economies. But official and popular opinion did not rate the public forest as a "source of wealth" for its real utilities or the enlarged range of real choice that its continued productivity would allow. Save as its lavish use allowed big, fast returns on scarce money and labor, the public forest was little valued, because there seemed so much of it. In 1853 a Wisconsin legislative committee recommended that the state let Menominee Indians keep certain land they occupied; the tract was mostly timbered, with some good pine, but with little area suited to farms; unconscious of irony, the committee observed that the state could the more easily be generous because the land "is of no value to white settlers." Many facets of behavior and attitude which we have noted in other connections implicitly testify how the plenty of public timberland and the scarcity of mobile capital subtly fastened men's whole concern upon overcoming immediate limitations, without thought of leverage on long-term development. So, without significant debate of alternatives, mid-nineteenth-century opinion tolerated or justified wholesale timber trespass upon the public domain, wholesale destruction of trees to "improve" land, wholesale disposal of public timber at minimal prices, without stipulation of minimal responsibility for maintaining the land's timber-productive quality. The contemporary interest in "speculation" and the ready dismissal of a landlord's role for government are likewise significant. Opinion condemned public or private holding of timberland which contemplated waiting upon the long-run growth of the community to give the resource scarcity value. It was natural to treat forest land as the cheap commodity that its abundance obviously made it; it was perverse to administer timberland in ways designed to realize scarcity value from it. On the other hand, nineteenth-century opinion generally admired the shrewd management which produced relatively short-term capital gains, whether by "improving" and then selling farms, or by realizing handsome gains out of combining low-cost timberland investments with the milling and distribution of lumber. Capital gain which accrued from the shrewd mixture of money and managerial skill with cheap land spelled progress, measured in visible additions to production.[61]

The way the industry priced timberland likewise showed the impress that the factors' imbalance made on men's minds. Lumbermen typically figured the money value of timberland or standing timber to be a residual amount — what was left out of the price existing demand would pay for lumber after meeting all the other costs of bringing it to market. Business practice thus estimated the situation in terms essentially the same as those reflected in legislation. The forest represented a static element, of worth only as it provided a setting for the play of the dynamic factors of an ascending demand curve on the one hand, and expensive labor, money, and managerial skill on the other. True, it was important that standing timber be available at favorably low prices; big money was not made out of returns for processing and distribution as such, but out of the spread which volume producers enjoyed when they brought to favorable markets lumber bought cheap on the stump. And the industry's pricing practice depended originally on a public lands policy which offered timberland in quantity at minimum cost. Logically, thus, one might say that the law made the pricing practice. However, the pricing practice fell into pattern with other popular and official attitudes which plainly did not derive from public lands policy but rather shaped it — the image of "boundless" natural wealth, the faith in productivity as a beneficent social force, and hence the social esteem accorded scarce labor and entrepreneurial talent, the impatience with public or private title-holding which locked up physical plenty which shrewd managers might otherwise put to work. In this context, the calculus by which businessmen assigned a residual value to timberland expressed in market terms what contemporary common sense concluded from confronting imperious demand with grossly disproportioned factors of production.[62]

The men who fashioned public policy and business growth within this supply and demand setting were sustained by a confident middle-class philosophy. Their credo was another potent situational factor conditioning will and imagination. These men acted in firm conviction of the social benefits and intrinsic worth of individual will. Individuals should have wide opportunity to show what they could do in managing resources to productive ends. Law should foster and protect creative release of man's energy.

These were such basic postulates of their thinking that men did not often find it necessary to express their assumptions as they determined the use of public timberland. More often they implied these values by what they did — when without question they adopted the fee simple absolute as the natural form for granting access to forest wealth, when they looked with complacence upon widespread trespass or fraud in speeding public timber to market, when they spent these assets lavishly to multiply opportunities for productive venture by promoting waterways and railroads.

When men attested their faith, they did so often with a naïveté which evidenced the gospel force of these attitudes. The Commissioner of the General Land Office found it obviously sound policy, in 1873, that the government should "not . . . indulge in vindictive prosecutions" of timber trespassers, but should, with "due regard . . . to the circumstances of each case," be prepared to compromise with the wrongful cutters, "on their paying any costs incurred and a reasonable stumpage for the timber, which is then released, and prosecution waived." The Commissioner was not wholly unaware that this added up to no very vigorous guardianship, but at least it paid its way:

By this course, although depredations continue, yet they are checked to some extent, and that without cost, it being made a rule that the expenses incurred shall not be permitted to exceed the money realized for the Treasury from the sales of timber seized, and stumpage paid in compromised cases.

This easy program prevailed for years before the federal government decided that it was an overindulgence of free enterprise. Where the issue was sharply drawn whether public or private decision should govern use of timber resources, Senator Timothy Howe of Wisconsin had no doubt where control should reside. In 1876 he advocated that the United States not reserve its southern pinelands:

. . . if a full-grown man with unimpaired intellect really thinks he needs to cut down and cut up a pine tree, and is willing to pay a fair price for it to the owner of the tree, why should he not have that privilege? What right have we to say that he shall not have it? Are we not legislating upon rather a dangerous domain when we undertake to say he shall not have it?

In this culture every presumption favored the moral worth and social contribution of the shrewd manager of men and resources. Astute private management sometimes pushed its gains to a point that overtly threatened the free market and the autonomy of ordinary private property owners. Then opinion could be rallied for legal regulation. So, in the name of middle-class values, Wisconsin farmers and small businessmen sought by law to curb railroad arrogance in the seventies and at the turn of the century. In contrast, there was no effort to use the leverage afforded by original public ownership of the forest to regulate the free flow of entrepreneurial energy. We knew that lumbermen were building substantial economic power out of their opportunities. Granted this, and despite the warning example of the railroads, opinion would not be moved to limit by law what energetic men might accomplish, unless gross abuse of public interest were clearly shown. Lumberman J. G. Thorp was one of the majority of the Wisconsin Senate Committee on Railroads which argued thus in 1867 against the wisdom of regulating any business pricing practices. Why single out the railroads? Successful lumbermen were exacting all they could get, too. But surely we would not limit business growth:

Princely fortunes have been accumulated all around us in the lumber manufacture and trade, and the people in the purchase and consumption of this article have been compelled to pay extravagant profits and prices. The paper manufacturing monopoly levies its enormous gains upon every man in the state who writes a letter or reads a newspaper. And there are other oppressive monopolies, many of them created by and operating under acts of incorporation granted by the legislature. Why then single out railway corporations whose dividends on the most economical basis of operation fall below the legal standard of interest, as especial objects of regulation and restriction, and permit those whose dividends drawn from consumers of their products are ten times ten per cent. or more upon the capital invested, to escape entirely? . . . But corporate and individual rights and interests, as regards the use of capital within prescribed and legitimate spheres are alike to be respected and protected. Unnecessary and unjust interference from whatever pretext or motive of temporary expediency results in injury only . . .[63]

When Senator Timothy Howe urged that wise policy would give free play to "a full-grown man with unimpaired intellect [who] really thinks he needs to cut down and cut up a pine tree, and is willing to pay a fair price for it to the owner of the tree," he paid implicit tribute to a third conditioning factor of the situation: the market. Plainly, he viewed the "fair price" as the expression of an economic, not a political, calculus. Clearly, too, he had no thought that, where government was owner, the calculus should be by government's sole reckoning; what government as owner did with its property should be determined by what the total interplay of market procedures showed the property to be worth.

What men learn by long practice is tough learning. It endures, and it weaves itself subtly into their modes of perceiving reality. By the time we came to dispose of the public forest, some two hundred years of working experience had taught us to rely on the market as a principal institution for deciding how to use and allocate resources. However, the lesser significance of the market was as an institutional setting for business decisions. More important, its familiar use taught us to assume that we could adequately define and measure the total interests in most situations by those definitions and measures which market procedures were capable of providing.

What we assumed from practice, we reinforced from philosophy. We put our middle-class faith in the creative energies of individual will and responsibility. In political translation, this spelled generous delegation of power to private decision makers. It spelled delegation not only regarding specific, but also regarding general, allocations of community resources. Wisconsin's "unimproved" riches in timber, waterpower, and minerals "need but the touch of labor and capital for the realization of the wildest dreams of prosperity," Governor Harvey told the 1862 legislature. "The magic influences of industry and capital" would unlock the state's potential wealth of natural resources, Governor Taylor predicted confidently in 1874. The cumulative impact of private decisions, rather than any directing force concentrated in law, would thus shape the general course of growth. In 1883 a committee of the American Forestry Congress accepted this as a premise dictated by the framework values of the society. Because private fee simple ownership was the normal pattern of power, the market rather than the state would here be the focus of policy making:

. . . the differences that exist between the American states and countries of Europe, as well with respect to the tenure of the land as the structure of the laws . . . would prevent any one of [the] . . . European codes of forestry from being applied in America.[64]

As in other aspects of nineteenth-century American development, so in disposal of public timber wealth, the market exerted greatest effect by privatizing men's interests. The original public ownership of the Lake States forest was a fact of massive significance for the future not only of the regions which would produce but also of those that would consume forest products. This was so, whatever the decisions taken on disposal of this public wealth. But in no public forum, national or state, was there focused, sustained, public examination of the choices to be made, in degree even minimally measuring up to the stakes. No fact stands out more sharply on the legal record. No aspect of the record is more characteristic of the times. Decision making on matters inherently of public concern was atomized into many market decisions. This was done so early and so completely that no public agency ever attempted an overview of the whole. It was done so easily, so simply, out of the drift and cumulation of events as to show that men had no interest in alternative ways of handling the choices which the facts required. No man sought state or national office on a platform concerned with disposal of the Lake States forest; no bill became the focus of debate in the state legislature or in Congress over a program for use of this public wealth; men who had made their mark in Wisconsin's lumber industry became state legislators, Congressmen, and United States Senators, and in the halls of law making concerned themselves with little save local and special bills and party matters, and, significantly, found no need even to stand in the breach for their industry against proposals of broader scope. That public forest was truly affected with a public interest, to the degree that general policy for its use should be decided by law, was an idea which had to wait upon the twentieth century for definition, let alone acceptance, in any significant body of opinion.[65]

Inherent limitations of the market, as well as defects in its operation, helped teach public and trade opinion to take for granted a policy of wholesale transfer of public timber to private hands at minimum prices. The market finds its dynamics in the zeal of private contractors. So far as decision was delegated to the market, it tended to run in terms only of such programs as the imagination and means of private contractors could readily encompass; diffused or long-term interests which might be served by public holding or leasing of timberlands found no natural spokesman in this sharply focused process; at worst, as the governor of Idaho Territory gloomily observed, the "public" or the government which represented it was likely to be regarded by aggressive operators as a "huge, helpless impersonality which any may plunder at will." Moreover, the market calculated in dollars. It was a calculus which carried its own bias. The long-term rise in demand for lumber made for a long-term rise in the market value of timberland. Money-measured optimism thus helped obscure from public view the real depreciation in the productive capacity of a depleted forest. This delayed our recognition that there were values in over-all economic stability and reduced future costs which might be served by greater retention of public forest holdings. The money calculus is quick and flexible. So it facilitated men's reckoning of the immediate situation and helped focus their thought on the nearest opportunities and pressures. Again, the bias was against withholding of public resources and in favor of quicker realization through the familiar market processes. Cost-saving technical innovations like the donkey engine and the logging railroad—readily translated into dollar arguments—persuaded both the industry and the community that social good lay in seizing opportunities for maximum present output, rather than in using the power in public ownership to brake the pace of growth. Again, under the market's dollar calculus, maintenance of considerable liquidity became a condition of survival, not only in the lumber industry, but in all business; delegation of economic decision to many private units worked only if most operators met payrolls, satisfied their general creditors, and paid banks or bondholders their principal and interest. The pressure inherent in money reckoning weighed heavier because of the chronic scarcity of fluid capital. Market discipline taught these lessons to a broad range of middle-class opinion outside the lumber industry. Thus common experience built a readier acquiesence in disposal of public timberland on terms which made minimal inroads on the industry's cash position and increased its return from investment in labor, machinery, and distribution.[66]

These were influences exerted out of inherent characteristics of the market. Others proceeded out of defects in market function. One of these deficiencies had an effect closely related to that implicit in the money calculus. Lack of adequate means to mobilize and administer private long-term investment capital was a persistent handicap in the growth of Wisconsin's lumber industry. A good deal of the industry's financing was done by relatively short-term bank loans, by deferred payments to labor and suppliers, and by constant improvisation of means through hard dunning and advances by distributors. Whatever the financing devices, they all meant that the industry worked on close time rations. Moreover, in the absence of a substantial investment-banker influence, the industry did not feel the pressure toward a lengthened span of planning which such an interest might have fostered. These shortcomings of the supply of long-term investment money lent urgency to the industry's desire to get its raw material cheaply from the public stock. Moreover, these factors were so much part of the common experience of American business as to generate sympathy or acceptance for the notion of using "real" public wealth to relieve the strain.[67]

The defect of market operations which most seriously limited their adequacy as a means to decision upon use of the public forest was their failure to produce the reliable factual information and sophisticated theory essential to wise choice. Thus, the market signally failed to correct one key error of fact. This was the common exaggeration of the extent to which Wisconsin's northern timbered acres would be suitable for farming. The comfortable assurance with which official and lay opinion held this belief helped rationalize rapid transfer of the public timber to private operators for quick exploitation; it was a course which in any case cleared the way for the land's assumed ultimate function. The theoretical defects which the market failed to make good related both to technology and to social cost accounting. We lacked knowledge or interest about forestry practice which would manage timber stands for continuing yield. We did not have cost-accounting concepts which would at least define the issue of competition between present and future yields. Nineteenth-century economic theory centered on analysis of achieving equilibrium with given factors; it offered little help to analyzing comparative gains and costs of different rates of growth in the general economy. These deficiencies in technical and accounting theory left the way so much the easier for all those immediately felt pressures which encouraged us to take it as unquestioned common sense to speed wholesale transfer of public timberland to private control and rapid use. The limitations of contemporary knowledge of soil potentials and of technique and accounting were facts which set hard limits upon contemporary policy making. How far was contemporary government competent to press the extension of knowledge? Skepticism is justified. But the point here relevant is a different one. In proportion as we delegated decision over use of public forest resources to market processes, in such measure we relied on procedures then wholly incompetent to find incentives or means for improving the factual and theoretical context of such far-reaching judgments. Indeed, one aspect of our practice and policy reacted upon another to heighten the problem. The market orientation of our thinking favored rapid disposal of public timberland at very cheap prices. But the fact that the raw materials could thus be had cheaply, in turn reduced market incentives to the improvement of technical or accounting knowledge in the interest of more efficient resource use. This, too, was part of the environment of policy derived from sweeping delegation of social functions to the market.[68]

THE LAW AND LEGAL PROCESS AS PART OF THE SITUATION

As symbols of constitutive values, the prevailing concepts of "property" and "contract" profoundly conditioned decisions on use of public timberland. In part their effect was persuasive. Title "in fee simple" was shorthand for rights of privacy, politically healthy dispersion of power, constitutional limitations upon official authority. "Freedom of contract" — the presumption in law that any private bargain was valid and enforceable, unless reason to the contrary be clearly shown — was shorthand for legal recognition of the intrinsic dignity and creative capacity of individual will and the social worth of its economically productive effort. To create private fee titles in public timberland, and to sanction private contracts about its use, were decisions the more easily taken because they appeared to flow with the main currents of three hundred years of Anglo-American constitutional tradition. However, property and contract concepts probably had less effect through the persuasive force of the arguments of social philosophy they embodied than they had by dissuading men from the need of any argument at all. The most striking aspect of the disposition of the Lake States public forest is the failure of the legal process to bring the issues into focus and to hold them there for sustained examination. The record shows that one reason for this was that these large, abstract, and unimpeachably respectable concepts of title and contract offered too ready a refuge from thought. Busy with the urgent present, nineteenth-century Americans were impatient of "theory." That is, they were impatient in examining the relation of particular choices to larger value patterns and in studying cause and effect in terms not only of visible, primary associations, but also of secondary or even more remote antecedents and results. It is only seeming paradox that, because they were so "practical," these men often acted in disregard of reality. Sometimes we can be fairly sure that they disingenuously refused to look at facts that stood in the way of what they wanted. But the failure to bring all the facts into consideration pervaded both official and popular opinion too widely to be explained as the product of conspiracy. What it did reflect was a defect of will more potent than dishonesty in shaping human affairs. When men most pride themselves on "practical" thinking, they are most likely to act on unexamined theory. The more they taste the anodyne of busy-ness, the readier they are to escape the pain of rational choice by substituting words for examination of facts.[69]

Thus men used conventional property concepts to talk about public timberlands disposal in ways that obscured rather than defined what they were doing. In 1851 Wisconsin's legislature asked Congress to cause the Interior Department to suspend prosecutions for timber trespass on federal land "until those engaged in lumbering can have the opportunity to supply themselves with the material by purchases of the necessary lands," and "to enable them to do so . . . [that] the pine lands on the Wolf, Black, Chippewa, St. Croix, and Wisconsin rivers and their tributary streams may be immediately surveyed and brought into the market for sale." In contemporary debate in the House of Representatives Wisconsin's Congressman Eastman argued that ordinary bargain and sale of the fee would quickly solve all problems of handling federal forest wealth:

If the lands from which this timber is alleged to be cut was in market, and could be purchased by the lumber men, there would be more reason for these prosecutions; but they have not yet been offered for sale, and some of them are not yet surveyed.

Though rarely made so explicit, the same assumption — that normal disposition of federal and state timberland involved the once-for-all sale of the productive land as well as of the existing timber stand — provided the formal frame of doctrine within which both governments applied their general land laws to the transfer to private hands of their forest wealth in the Lake States. Particularly in view of the trespass threat, the government's wisest policy towards its pinelands, the Commissioner of the General Land Office argued in 1874, was "that which will most speedily divest it of title in the same for a fair consideration." [70]

Yet, from the outset, the industry's behavior made plain that lumbermen had no interest in buying the "land." What they wanted was the timber. In practice what they bought was an unconditional license to cut, since after they had cut they either let the land default to government for unpaid taxes or sold it cheaply, not as timber-productive land — which, indeed, it scarcely was, in the condition in which they left it — but as land wishfully assigned to agriculture. Had government dealt explicitly in cutting licenses, the form of the transaction at least would have invited attention to defining the licensee's responsibility for the land's productive future. To frame the transaction as a transfer in fee simple cast it in a form the whole meaning of which was to transfer unqualified initiative of decision to the grantee. This was a kind of grant that the law had defined primarily regarding transfer of agricultural land. Given the extent to which American farmers mined their soil, later generations might question the adequacy of this title form even in its traditional field of application. Over-all, however, the fee simple proved compatible with maintaining a working minimum of farm soil productivity. But this kind of land use involved annual crops and permitted soil protection and renewal techniques capable of relatively quick results. Would the same form of title be equally compatible with realizing the real productive potentiality of the forest? Here was a "crop" with a very long renewal cycle, inviting techniques of management which could show results only over extended periods of time. By any truly pragmatic test, the situation posed questions which could be answered only by attention to its distinctive facts. Unexamined application of the fee simple concept evaded all the real problems of choice inherent in the facts.

In some respects contemporary policy makers did recognize that the lumbermen wanted unconditional licenses rather than owners' responsibilities.

The Wisconsin legislature saw the danger that buyers of its pinelands would cut and run, and accordingly made ineffectual provisions to bar logging and to exact special security for the price of the land so long as an unpaid balance remained. Even these provisions recognized the problem only in a monetized context, however; the legislature's concern ran only to securing payment of dollars due for sale of the fee, without regard to what happened thereafter; the narrow bounds of its concern were underlined by the stipulations under which a trespasser might settle with the state by paying the land's price plus a penalty, and thereupon have full title to what he had cut, whether or not he bothered to take title to the land the state had "sold" him. In one of many fruitless pleas for repeal of the pre-emption and homestead laws in their application to forest areas, the Commissioner of the General Land Office in 1876 defined the situation with rare realism: "Under the provisions of the pre-emption and homestead laws . . . [Congress] is granting a license to destroy millions of acres of pine forests of almost incalculable value, which should be preserved as a national heritage." At most, however, these were scattering and ineffective acknowledgments of reality. They have meaning only as they throw into sharper relief the prevailing practice, which evaded all the difficult questions by invoking the comfortable respectability of the fee simple. Surely the situation presented no really new problems, surely we made wise use of public assets, so long as we did no more than extend the reach of this classic form of ownership.[71]

Popular faith in the social utility of "contract" likewise operated to divert attention from the real stakes in public forest policy. It did this especially by promoting a most casual acceptance of the capital subsidy which cheap sales of public timberland gave to the industry. Consequent incongruities in public policy appear most sharply in relation to the classification of land. On the one hand, government made no adequate provision to classify its holdings according to their distinctive worth; on the other hand, following what was routine good business practice, those who bought from government did so on the basis of field inspection by experienced woodsmen representing the buyer's interest. The familiar presumption favored the legality of private contracts. The courts had no trouble in upholding and enforcing agreements by which woodsmen sold their "minutes" (field notes) to timber speculators, or claimed pay for agreed services in cruising public lands to provide data on which the buyers of their service might make profitable purchases from government. Judicial opinions acknowledged the substantial "profit" which was made available to private buyers by the simple fact of purchase at the government's low prices. Thus in *Schriber* v. *LeClair* (1886) the Wisconsin Supreme Court noted that a tract entered at the government land office on payment of $2451.99 was worth $11,000 at the time of entry, according to the field notes of the timber cruiser who had brought these acres to the buyer's attention as "worth at least three times as much as it would require to purchase them from the government." The point is not that the courts should have interfered with these profitable purchases from government or refused enforcement to the private agreements which facilitated them; it was the legislature's business to fix public lands policy, and nothing could be plainer under the federal and state statutes than that these private activities were wholly consistent with the policy that contented Congress and the state legislature. The point is, rather, that in effect substantial subsidies were thus granted without any public examination of their size or any public debate over their utility in comparison to their cost. The only stage in the legal process where men brought the facts of capital subsidy into focus was when proof of them became incident to lawsuits over private interests. The casualness with which courts then acknowledged the facts of subsidy as commonplace elements of the industry's structure is the aspect relevant to our concern with the quality of public decision making. When government chose the role of a contractor in the market, the form of its action obscured the distinctive fact and responsibility of the use of legal power (the ownership of public assets) to shape economic growth. That public land disposal went on within the familiar procedures of "contract" seemed assurance that all necessary weighing of interests was being effected by bargain; if the public offered its wealth in market, it should abide market decisions; presumably government was in the long run getting satisfactory consideration for what it sold, else like any other seller it would stop offering its goods. Resort to contract thus implicitly reassured us that we were proceeding rationally and with fair balance. Sustained by these comforting symbols, we felt less need to press into what might be harder questions.[72]

That Wisconsin was part of a working federal system brought significant influences to bear on policy toward public timberland, as it did in other matters. Though real, these influences won little overt recognition at the time. This was natural enough. The central government itself had almost no distinctive program regarding its Lake States timber holdings. Wisconsin's sister states made themselves felt only in indirect and impersonal, if weighty, fashion, as their polities helped create insistent markets for lumber to expand their towns, farms, and railroads.

Successive generations of frontier states drew moral confidence from their legal status in the federal partnership. New states entered the partnership in legal equality with the old. This we had declared to be a foundation principle. Surely, successive newcomers argued, this meant equality in substance and not merely in form. Whatever help the nation had lent to the growth of older states it owed equally to the growth of newer ones. In the nineties and after, Rocky Mountain and Pacific Coast states argued that, since the United States had been generous of its timberland in aid of Wisconsin's growth, it was morally bound to give like access to its remaining forest wealth to promote the development of the Far West.

This argument never emerged quite so sharply relative to disposal of federal timberland in Wisconsin, if only because there was no need of it; there was no Gifford Pinchot to press an alternative to the freehanded sales and grants which transferred federal forest to private control from the 1850's through the 1880's. Nonetheless, the confident righteousness with which the "frontier" made claims upon national assets figured in the Wisconsin story sufficiently to show why there was so little impetus to earlier development of a positive federal forest program.

When the Interior Department proceeded against timber trespass in the early 1850's, the Wisconsin legislature protested in terms which restricted the federal government to the limits of its past action, while at the same time stressing the developmental needs of the growing state:

> . . . lumbering . . . upon these lands . . . has been considered as both laudable and lawful, in view of the necessities of the people of the State, and the prosecution of it has materially aided in the settlement and prosperity of the country. In consequence of the long forbearance of the general government to prosecute for trespass on these lands, for the purpose aforesaid . . . and encouraged by this seeming acquiescence of the government . . . as well as the increasing wants of the people, a large amount of capital has been invested in the business, and its continual prosecution is essentially necessary for the better comfort and convenience of that portion of the people of the State, who are rapidly filling up its northern section.

In no respect did federalism have greater effect upon the speed and scale of Lake States forest exploitation than by the federal lands grants which hastened the building of railroads, enormously increasing the impact of the midwest market. Northern Wisconsin interests pressed their case for the use and renewal of these grants as a matter of moral right and not of grace, pursuing the familiar argument that new sections had the same claims on public bounty as the old ones had enjoyed. Congress and the state legislature acceded to the north's position with a readiness which attests how deep ran these attitudes in the moral code of our federalism. One can be sure that an argument to slow up railroad development, in order to brake the pace of forest exploitation, would have achieved only summary dismissal. One episode provides some measure of contemporary standards. Revival of a grant of federal timberland in aid of railroad construction in northwest Wisconsin came in question in 1870. There was specially vigorous reaction to indications that eastern interests might dissuade Congress from renewal. The Board of Trade of the city of Superior addressed the Wisconsin legislature, righteously asserting the proper claims of new as against older sections upon the national bounty:

> It is reserved for a member from Pennsylvania, a state running over with abounding wealth, to arise in Congress and ask the government of the United States to put its hand upon the young, struggling state of Wisconsin, and push down its

springing activities in its own proper field . . . Verily, wherewithal have these pampered capitalists been wont to be indulged, that to make them richer, the humble people of Northwestern Wisconsin must be made poorer! [73]

If localism thus pressed in to help confine federal policy, the current of influence did not run all one way. With its original complete ownership, the United States had potential power to control or at least to guide by example the uses the state made of the forest lands put at its disposal. Congress significantly affected state policy by the simple fact of withholding either control or example. Indeed, the silences and defaults of federal policy—the tolerance or ineffectual gestures regarding appropriation of federal timber wealth by trespass or fraud, the lack of classification of federal lands according to real worth, the failure to enforce competitive bidding, the implicit invitation to illicit practices in the long failure to provide disposal procedures specially adapted to timber resources—all these failures of will and imagination on the part of the older, central government fostered a climate of opinion which made it easier for the weak, untried government of a new state to yield to its weakness, to the drift of events, and to uncritical acceptance of the most immediately felt circumstances.[74]

The influence of the whole upon the part fell most hardly upon federal and state timberland policy in the Lake States through the mediation of the broad Mississippi Valley market, guaranteed and protected by the commerce clause of the Federal Constitution and brought closer to the pineries by railroad construction aided by grants of federal timberland. Contemporary federal and state official opinion was explicitly aware of the existence of the massive, swelling, midwest demand for lumber and other forest products. Men accepted the fact, uncritically, as an opportunity. They raised no questions as to costs to future real productivity or even to future money-measured values, in allowing the surge of this broad, federally created market to set the pace of forest production. Nor was question raised, on the supply side, of the social utility of free movement of mobile capital into forest industry. Probably the commerce clause would have forbidden the state to regulate the tempo and scale of growth by directly limiting imports of labor or money, or exports of lumber. But public ownership of the essential fixed capital, the standing timber, gave the United States and the state clear title to regulate the speed and size of the industry's growth, so far as control of these assets in fact permitted—unhindered by the commerce clause or the 10th Amendment, or, for that matter, by concepts of "due process of law." The state, alone, stood perhaps under a peculiar constitutional restriction, because it held its lands for trusts specified in the Wisconsin constitution. Purposes of economic planning plainly would not be allowed to override these trust purposes. But this was far from demonstrating necessary inconsistency between the two categories of objective. Indeed, to find necessary inconsistency would be to beg the very question which the silences of nineteenth-century public policy did beg: whether future returns to the trusts (and future maximum growth of the over-all economy of the state) should be sacrificed to increase present returns.[75]

The lack of suggestion for regulatory use of government's original ownership of the forest did not stem from lack of actual conflicts of interest to give point to the matter. The northern forest was in fact — though men did not at the time make much effort to examine the facts — a unique resource to the Wisconsin economy. It was likewise in some measure a unique resource to the Mississippi Valley lumber market, by virtue of site value. But it was not the unique resource to the consuming market that it was to Wisconsin as producer. Once the railroad net spread out, buyers of forest products could fill their needs from the South or the Far West. From the standpoint of the sectional or even national economy, the faster development of the prairie states and the stretching of scarce working capital by immediate availability of cheap Wisconsin lumber perhaps represented a good bargain; for sustained production consumers could look later to other forested parts of the country, and if they had later to pay higher prices because of longer hauls, presumably they would by then have prospered so that they could afford to pay more. From the standpoint of the Wisconsin economy the matter was by no means so plain. The productive resource here depleted to subsidize prairie state development loomed relatively much larger and more critically to Wisconsin's long-run productive power than the consumption of the product did in the long-term economies of the consumer areas. It was a consequence of successful federalism that the imagination of Wisconsin's public leaders, its businessmen, and its solid, middle-class citizens could not, in the span 1850 to 1890, see that there were costs to be paid for the tempting opportunities of immediate bustle and growth. The heady development of our federalized economy fostered ebullient optimism, ambition, and preoccupation with the exciting present. The philosophy and reassuring symbolism of fee simple title and free contract provided the dynamics by which to seize the opportunities the federal market offered. In this property-contract-federalist context, it seemed common sense — not only beyond argument, but beyond alternative conceiving — to put the public timberland at the call of the Mississippi Valley market.[76]

Congress was indifferent toward providing an adequate procedure to classify federal lands according to their special worth in timber or other distinctive resources. The Wisconsin legislature felt that the state could not afford the cost of more precise and reliable "appraisals" of state forest land. Both federal and state officials favored speedy transfer of timberland title to private hands as the answer to the trespass problem, partly for economy, partly because they professed greater confidence in the zeal and honesty of private owners than of public agents to guard this wealth.

Thus public forest policy took form subject to pressures created not only by the accepted doctrines and symbols of property and contract and by the structure of federalism, but also by general attitudes toward decision making by law. The nineteenth century felt no urgent conviction of the need to invest much time and many resources in fact finding and deliberation for making public policy. The century tended to be impatient with public affairs, in any case, where these competed for attention with the market. We believed in using the positive power of law to promote economic growth wherever we saw obvious opportunity; but this public effort seemed essentially too simple, and incidental to more engrossing challenges of private business, to deserve our best attention. It was not surprising that, unwilling to give public business first call upon talent and devotion, both official and lay opinion was resigned to expect poor performance by government — ineffective at least, and probably in considerable measure dishonest — and hence was all the less inclined to turn to law for large decisions.

The neglect of the distinctive issues inherent in disposal of public timberland fits the pattern of these attitudes so completely as to suggest the force they had in this as in other areas of policy. Moreover, this bias of thought had general consequences of great effect upon use of public forest wealth. One of these consequences was the almost total failure to bring law to grips with the business cycle in the decades of Lake States lumber boom. Cyclical downswings sharply increased the industry's pressure for liquidity; uncontrolled upswings invited over-optimistic expansion; the total effect was to heighten the industry's need to obtain public timber cheaply and quickly in volume, for rapid marketing. Another area of default in general policy that influenced public forest disposal was the neglect to explore the full range of contribution that government might have made to sounder financing of the extractive industries. Because the market did not supply needed long-term, low-cost credit, the lumber industry felt continuing pressure to turn timber into cash. With limited revenues, and confined by strict constitutional limit on its borrowings, the state probably could not have played investment banker to the lumbermen. But the federal government had broad fiscal authority and enjoyed the bounty of the protective tariff. It might have joined term loans to its public timberlands dispositions in ways that would have enabled lumbermen to work under less constant pressure for liquidity. One may doubt that Congress would have voted such a venture; yet, Congress generously underwrote the financing of railroads. Cheap sale of federal timberland was itself a capital subsidy of a crude kind, ill-defined and little discussed, but the key to the industry's growth. However, this want of definition or discussion was the critical aspect of the matter. The point is, precisely, that Congress never voted this *de facto* subsidy as an affirmative, deliberated decision; it simply allowed the subsidy to occur, by the play of market forces upon generalized procedures for disposal of public lands, having no reference to the special values or problems presented by the possession of forest wealth. Whether Congress would have voted to make the United States the investment banker for the lumber industry is, thus, beside the point. The point is, rather, that prevailing attitudes were content with policy accomplished by drift. In such an atmosphere the idea of using public fiscal power to create a more sound financial framework for resources development did not even enter the area of discussion. Nor did this result exhaust the kinds of influence exerted by this style of opinion. Our impatience in framing public decisions had its effects on forest policy also in limiting the extent to which we exploited the potentialities of federalism. In the twentieth-century, federal grants-in-aid to states, carrying stipulations of program and standards, proved useful means to channel the joint energies of both governments, though even then there was much less co-ordination of effort than was desirable. The nineteenth-century gifts of timberland which the United States made to Wisconsin in aid of railroad construction were grants-in-aid, too, and of massive generosity. Yet, save for the most general description of the ultimate purpose, their crudely simple terms provided no direction or standards effectively to realize the self-renewing potential of the wealth thus put at the state's disposal. Of course, Congress' default might have been the state's creative opportunity. However, without significant exception, the Wisconsin legislature viewed its task as simply to decide which railroad promoters should enjoy the bounty. The state took no advantage of the broad trust powers Congress had left it to guide the manner in which the lands' timber capacity should be developed.[77]

Drift and inertia have normally determined the large outline and content of society. Popular inclination to belittle the worth or integrity of legal processes of decision favored these mindless forces. In proportion as such notions figured in the people's working philosophy, it was the more important that legal process be structured for rational fact finding, intelligent definition of choices presented by the facts, and exercise of responsible, positive judgment.

These demands upon legal order took on special urgency regarding disposition of public forest wealth. Various elements combined here to put a high premium on early diagnosis and exercise of as much directing intelligence as either public or private enterprise could muster.

First, the administration of timber resources posed a peculiarly difficult problem of bringing all interests into focus. Much that was at stake — measured by the long term and by the growth and stability not only of particular industries but of whole regional economies — was remote in time, space, and experience from the ideas of most men. Sharply defined and immediately felt interests produced their advocates for particular decisions on disposal of public timberland. In our early confused state of scarcity and plenty there were no comparably motivated or alert spokesmen to represent more general interests. The defined, immediate interests carried the greater force because their advocates could get what they wanted if the law would only expand the reach of the familiar and respectable fee simple title and contract. Any substantially different approach, in contrast, would require some bold invention. The middle-class philosophy of dispersed decision required a good deal of responsible self-restraint if its system was to work. It was a classic role of contract, property, tort, and trust law to educate and discipline this self-restraint at the initiative of private suitors. However, because it held all original title to the forest, government could not avoid direct responsibility for what happened to forest assets; whatever it did, even its inaction, would fix a framework for their use. Yet, to employ the power inherent in its original monopoly to educate and discipline forest industry — passing the limits set to such an undertaking by the contemporary state of knowledge — government must assume an initiative that nineteenth-century opinion was more accustomed to leave to private suitors. Thereby, at a time when the infinite prospects and excitements of the opening continent invited men to attend only to their own business, the law would ask them to re-examine the constitution of their society in a respect — the delegation of power — which involved its total balance. Both the problems and the climate of opinion in which they arose put exacting demands upon the legal process to develop leadership and mobilize sufficient supporting belief and interest to enable leadership to operate.

A second challenge to the law's capacity to provide direction lay in the special impact which the imbalance of the factors of production was bound to have upon the use of forest wealth. The availability of forest products had critical opportunity value for the economies of both supplier and consumer areas; the presence of the real resource vastly enlarged the range of choices open to economic development. But in the market calculus forest products typically were a minor part of the cost of enterprise, all the more so where relative scarcity of cash and labor put high prices on mobile capital. The resulting pressure of the market calculus upon attitudes toward forest use was symbolized in the residual value which the industry assigned to timberland, after meeting the claims of labor, working capital, and entrepreneurial talent. The community's dominant opinion was middle-class, hence market-oriented. This made it easier to frame policy by monetized (market-price) than by real (facts-of-opportunity) reckoning. Market calculation tends to emphasize achievement of present equilibrium and immediate improvisation. The basic given factors were scarcity of mobile capital and abundance of fixed natural capital. In this context, reliance primarily on the cumulative effect of countless atomized market choices to determine over-all use of forest wealth involved inherent bias against maintaining the forest's long-term real productivity. Conversely, the situation posed the greater need for the law to represent interests which could be expressed only by more generalizing processes of choice than those operative through market exchanges of title or contract.

There were radical contrasts in the pace and span of time involved in alternative ways of realizing forest wealth. This posed a third reason why early diagnosis and positive decision had peculiar importance in this field of policy. Between 1850 and 1890 the central United States experienced an explosive, nearly fivefold increase of population. Lumber industry technology moved fast in ways that favored volume production, notably through the logging railroad, the donkey engine, and improved mill equipment. The pressure of these facts was to speed the exploitation of the forest. Frauds upon the settlement laws became a major problem of public timberland administration with a rapidity which revealed the weight of these accelerating influences, as it also offered some plea in mitigation of government's ineffective response. On the other hand, programs designed to maintain public forest land as substantial, long-term, producing facilities implied policy which had been defined and given continuity over several generations, with accompanying high costs for slow-yielding capital and for bearing unpredictable risks of physical hazard and of technical obsolescence of forest products. The time dimension introduced another element of difficulty. At some point preference for present over future yield brought the Lake States white pine and hardwood stands past the minimum level for continuous renewal; to pass this point meant in practice irreversibly to deplete the forest as a facility for producing prime timber — certainly, at least, to create a situation in which we must bear a very large, new, initial capital investment and the costs of waiting at least two generations if we would try to restore first-grade production. Hence, matters already complex were further complicated by their relations in time. Powerful, yet diffuse and ill-defined, situational factors pressed for haste; resource commitments hastily made would speedily become irrevocable and endure long in effect; alternatives to haste imposed long-term immobilizing of capital, subject to unpredictable risks. The time dimension of choice thus presented drastic alternatives in result, extremes in gain and cost, and a wide range from the predictable to the unpredictable. This immensely increased the need for whatever rationality and broad-based calculation the law could provide to supplement, or if need be, correct, the dispersed and limited-focus decision making of the market.

A fourth element of the situation laid a challenge upon legal process to promote full-factored planning for use of public forest wealth. This was the complexity as well as the size of the overhead cost of pursuing any course materially different from that of cutting public timber for maximum present yield. Conceivably we might have managed the Lake States public forest so that it could indefinitely furnish the Mississippi Valley with a substantial supply of prime lumber and its by-products. But sustained-yield forestry involved overhead costs which had qualitative characteristics that made them almost impossible to accommodate through means available to the fledgling nineteenth-century market. Indeed, these were overhead costs qualitatively beyond the competence of all but the most sophisticated private operators of the much more rationalized lumber industry of the middle twentieth-century. This fact presents another aspect of the three situational difficulties to forest use planning that we have already noted. It was peculiarly hard to come to terms with overhead cost because of the problems of focus, timing, and imbalance of the factors of production. Sustained-yield forestry presented baffling problems of private and social cost accounting because it reckoned with interests not all amenable to neat measurement; given scarcity of mobile capital, pursuit of sustained yield called for great staying power of will and money in order not to press for quick volume when high-cost labor and money raised unit overhead while consuming areas promised to take all that suppliers could furnish; the slow growth cycle of Lake States pine and hardwoods meant that sustained-yield operations must bear not only the interest cost of capital thus committed as against alternative opportunities for quick turnover, but also must support the peculiar risks of physical loss and product obsolescence. These elements posed challenges beyond the theoretical or technical knowledge of the mid-nineteenth-century lumber industry. Thus they would have added an almost certainly insupportable factor of incalculable risk to private operations chronically burdened with scarcity of working capital and hence extremely vulnerable to miscalculations or misadventures in the market. In addition to this, however — and the fourth difficulty of the situation — to overcome either the deficiencies of knowledge and know-how, or to brave the extra degree of risk entailed by learning through experience, called for a quantity of investment in research, experiment, and waiting time (acreage held in growth) such as private industry could not afford. The story of the industry's difficulties and failures in the down movements of the business cycle in the second half of the nineteenth century makes this plain. Relatively well-financed private investors experienced strain in holding Wisconsin pine for only ten to twenty years simply to obtain higher returns when they clear cut the land. So long as substantial quantities of virgin timber were marketed by government at cheap prices without restrictions on the buyers' cutting practices, any private lumberman who pioneered in sustained-yield operations would have foundered under the competition of other sellers who were not shouldering the high overhead costs of such deliberate exploitation. If a sustained-yield alternative for realizing Lake States public forest wealth were conceivable, even on the most modest scale and by the rudest methods, resources which only the law could mobilize and institutions which only the law could provide would be needed for the job.[78]

Legal order was not well armed in organization or experience to meet these challenges, and its inadequacy limited the scope of response to the opportunities offered by original public control of forest wealth. The difficulty was not want of substantive authority. The federal and state constitutions put ample authority in the chief executive to formulate and recommend, and in the legislative branch to shape and adopt, public forest programs within a considerable range of alternatives. The difficulties were want of specialized organization and want of morale.[79]

Those characteristics which made the handling of the public forest a challenge to leadership made it also a problem which would yield only to sustained, specialized attention. The difficulties of bringing all major interests into focus, of acting deliberately to offset the market's impersonal bias in favor of scarce mobile capital, of balancing long-term calculation with short-term improvising under pressure of explosive social and technical change, of investing large capital to meet what amounted to social overhead costs — these resistant aspects of the situation invited escape into empty generalization (glorifying creation of freeholds "purely allodial, with all the incidents pertaining to that title as substantial as in the infancy of Teutonic civilization") or into the fiction that the cumulative impact of countless fragmented decisions was the equivalent of a general policy. The situation encouraged us to accept either generalities without sufficient content or particulars without sufficient relation.[80]

Our mid-nineteenth-century practice — grounded partly in habit, partly in false economy, partly in underestimation of the intrinsic difficulty and importance of public policy decisions — was to commit responsibility for any given public affairs complex to the existing legal agencies of most general jurisdiction. The working presumption assigned not only administration but also top planning as *ex officio* tasks of offices already charged with broad ranges of function. Whether in constitution, legislation, or executive order, we begrudged creating specialized instruments to explore need, generate opinion, and press decision in specific areas of general concern.

This reliance on *ex officio* jacks-of-many-trades severely restricted development of public forest programs through the years that were critical for disposition of the Wisconsin pinery. Congress, itself the holder by constitutional mandate of the widest range of responsibilities *ex officio,* provided itself with no committee sufficiently focused or staffed to take creative initiative regarding federal timberland. It was unwilling to respond to suggestions of even modest provision for specialized executive attention to the forest, such as might have been supported by a revolving fund out of proceeds of trespass prosecutions. It committed care of federal forest wealth, along with the whole massive weight of public lands administration, to departments — first Treasury, then Interior — charged with other urgent duties. Though it might have created an agency of leadership in the General Land Office, it starved the Office for budget, was content to leave as clerical posts jobs which warranted division status, and allowed free rein to local Congressional delegations to exert the full force of parochial interest against such limited executive vigor as was shown, for example, in the timber trespass prosecutions of the early 1850's. Wisconsin also, unfortunately, wrote into its constitution the prevailing bias for *ex officio* duties, committing care of its timberlands to land commissioners who as top executive officers — secretary of state, state treasurer, and attorney general — bore other primary responsibilities. During some years of lumber boom the state Senate and Assembly created standing committees specifically entitled to deal with lumber matters. But no inquiry, planning report, or draft bill ever emerged from these committees or any other state legislative or executive source so long as the state held important timberland, to define for discussion any comprehensive program for use of this resource. As with the federal General Land Office, so in the office of the state's land commissioners, administration was given to clerks, and at best was carried on within the imagination of clerks. There were some random, partisan-tinged legislative inquiries into the handling of the timberlands. But this concern never went deeper than questions of favoritism among individual bidders; no legislative investigator and no important executive officer raised the timely question of the end objects of the state's use of its timber. In justice to federal and state executive offices immediately charged with timberland administration, it must be noted that the legislative branch denied them the most promising means of generating information, interest, and attention to the distinctive problems of forest wealth when it failed to provide adequate procedures and means to classify public lands according to their real timber worth. By the time Secretary of the Interior Schurz began to urge the need of an affirmative program regarding public timberland, in the late seventies, the effort was too late to have meaning for the Lake States. In fairness, too, it must be noted that other factors of timing worked against earlier effective leadership from the operating executive level. For one thing, in the decisive years for the Wisconsin forest, there was as yet no civil service status to foster professional morale among the administrators. For another, knowledge of forestry was still all rule-of-thumb, and there was no body of professional foresters to develop plans or standards as a challenge either to public or private practice.[81]

In some measure the organization of legal process was inadequate to develop a rational and vigorous forest policy because people were content, or selfishly interested, that this should be the case. Prevailing middle-class opinion outside the lumber industry and the forest region was content, since it put a low estimate on the utility of investing in specialized, well equipped legal process in any field. Midwest consuming areas would raise no voice in Congress for ideas that might increase costs of building farms, towns, and railroads; the few who in the fore part of the nineteenth century suggested a more deliberate exploitation of the public domain were, like John Quincy Adams, residents of the settled East. Wisconsin lumbermen sat in the state legislature and manned committees which dealt with bills affecting the industry; substantial mill owners controlled local government in their operating areas as a routine aspect of business; Wisconsin lumbermen represented the state in Washington — Philetus Sawyer and Isaac Stephenson in the Senate, Thaddeus C. Pound and William Price in the House. Whenever the industry faced defined occasions for exerting its influence on public decisions, its weight was always cast for spending public wealth to speed up and subsidize present forest production.[82]

It is naïve to ignore the influence of self-centered bias upon the uses of law. It is even more simplistic to assume that men's selfish calculation is a greater influence than the limits of their intelligence, imagination, and energy. For all that it includes of irrationality and inert custom, the law expresses one of mankind's most massive efforts to direct its affairs. Nonetheless, there is much evidence in the legal record that men's limitations have more effect than their ambitions in shaping society. In this perspective legal history teaches a wry conservatism, though certainly not a complacent or quietist one. In this perspective the enemies to civilized order are at least as much men's sloth, bafflement, and weariness as their wickedness. If we so appraise the hazards of the human enterprise, we shall value the more whatever we can do by law to make our limited stock of talent and energy more effective; likewise, we shall attribute the greater influence to defects in the organization of legal process in determining what we do or suffer to happen.

The overt pattern of federal and state handling of public timberland itself is circumstantial evidence that defects in the policy-making process materially affected the content of policy. The dominant fact was default. Even to so rich and diverse an economy as that of the central United States, the Lake States forest was a fact of immense absolute and relative potential. Yet, disposition of this publicly owned asset proceeded from start to finish without one major discussion or debate, without a single major investigation or controversy, about the ends to be served or the accounting of income and outgo. This default of attention was particularly marked regarding relinquishment of federal control of the public forest, where it is thrown into sharper relief by the one exception — the narrow controversy of 1851 and 1852 in reaction to some short-lived vigor in prosecuting timber trespass. The sum of the state record was not materially different, however. Wisconsin's legislature took a little more special note of the state's forest property — mainly by formal stipulations to preserve timber values to secure unpaid balances on land sold — but just enough to point up the very narrow limits of its attention.

Moreover, what the Wisconsin legislature did carried other defects that reflected the influence of poor processes of decision upon the content of policy. State policy was inconsistent — stipulating special terms of payment for "pineland" without providing means to identify it. State policy was arbitrary in its categories — fixing one set of price terms for pineland, another for timbered swampland, and no special terms at all for other state forest holdings, without reason for distinction. It was partial in a parochial way—appropriating state timberland in particular localities to aid particular projects of internal improvement, but defining no comprehensive criteria for efficiently realizing timber values. It lacked co-ordination, so that default in one aspect fostered trouble in another—as the want of procedure specially adapted to sale of standing timber invited trespass and fraud, while the threat of misappropriation was argued as a reason why the state should unload its holdings as fast as it could. State policy was rigid where it should have been flexible—as in denying the land commissioners discretion to bargain for the best price with individual buyers at private sale—and emptily vague where it should have given direction—as in selling the fee without stipulating any responsibility in the grantee to maintain minimum conditions of continued productivity in the resources committed to his judgment.

Through the years in which the United States disposed of its Wisconsin timber holdings, its policy showed similar defects so far as the situation allowed; if the federal record did not manifest all the shortcomings of the state's, this was no credit to Congress but was simply due to the fact that Congress did not try to do even as much as the state legislature. From the late seventies, when Congress began to enact some distinctive policies for federal forest lands in the Far West, what it did paralleled in defect all the crudities in the state's record. The sheer lack of sensible workmanship in the federal program did not altogether escape contemporary notice. In 1876 the Commissioner of the General Land Office observed with restrained exasperation:

> It is an anomalous fact that the Government is giving away the rich alluvial soil of Iowa, Nebraska, Kansas, and Minnesota to any citizen who will plant a few acres of cottonwood or other inferior timber [under the Timber Culture Act of 1873], while under the provisions of the pre-emption and homestead laws it is granting a license to destroy millions of acres of pine forests of almost incalculable value, which should be preserved as a national heritage.

However, it was typical of the halting course of federal policy development that, despite repeated recommendations for its repeal, the Timber Culture Act survived until 1891.[83]

Shortcomings of this kind—internal inconsistency, arbitrary distinctions, parochial scope, lack of co-ordination, and above all else sheer default in examining (let alone debating) what was done—evidenced the influence of poor organization for decision. To read such a record as the product of skilled maneuver by special interests would romanticize history into melodrama. Had the legal process been structured to promote more realistic study of facts and choices of a high order of generality, its greater effectiveness undoubtedly would have provoked impressive reactions. There would then have been more of the drama of conscious conflict in the story. The striking fact is that neither the lumber industry nor its consumers were ever required to meet a substantial challenge to the course that was followed to their immediate profit in disposal of the Lake States public forest. The lumber industry found access to public timber on terms generally satisfactory to it. However, the lumbermen got what they wanted with so little contest that their success was scant test of their pressure-group prowess, their shrewdness, or their will. What their success more plainly evidences is less dramatic, but more pervasively important. An energetic minority moved by a clear vision of its short-run profit has easy leverage on a situation where there is no legal or other institutional process efficiently organized to bring less obvious interests to definition and a sense of their own involvement, and thus to insistence upon choices by deliberation rather than by default.

CHAPTER III

FRANCHISES TO DEVELOP BULK TRANSPORT AND WATER POWER

"PROPERTY" in land or things or in claims upon land or things means that law defines the individuals or groups who may with law's sanction take the initiative in deciding various uses to be made of these resources. To have property in land or goods or in claims upon land or goods typically means to enjoy the widest range of discretion over their use that public policy allows. The delegation of powers here is general. However, because we believed that a broad dispersion of decision-making power was socially and politically as well as economically healthy, and because scarcity of fluid capital pressed us to use varied means to mobilize talent, labor, and money, the law's delegations of power over resources use were much more complex than can be summed up in the creation of general property titles. We created in law specialized as well as general rights to initiate economic decisions. The specialized rights were inherently rights to impose particular compulsions upon general rights of property of the public or of private persons. In this aspect they bore analogy to the public power of taxation, which in effect removes control of part of the taxpayer's resources from the taxpayer in order to delegate control to other decision makers. The legislature made many specialized delegations of economic decision within the frame of official action, as, for example, to executive officers or municipal corporations for building roads or running public schools. The legislature also made specialized delegations of power to nonofficial persons, which, incidentally, presented distinctive value choices and operating tensions. It is these specialized delegations to private persons of the power to initiate economic decisions that we call franchises.

Law conferred franchises upon individuals and upon groups, for small and for large projects. In lumber history the most significant franchises were conferred on groups (increasingly upon groups organized as corporations) and were granted typically for projects which, if small by some measures, were at least large as measured against the bulk of dealings in the market. These franchises concerned chiefly the provision of means of bulk transport by water and by rail; to a lesser extent during the lumber era they involved the creation of water power for industrial uses.

To provide bulk transport facilities required large capital investment; railroad construction especially called for the largest investments that were made in particular enterprises in Wisconsin's mid-century generation. Men needed law's promotional help to assemble and manage capital on this scale. Promotion came in part by direct gifts of capital resources in government's disposition, by grants of public lands as we have already noted, and by investment of public funds raised by local government taxation to buy bonds or stock in transport enterprises. Promotion came indirectly by creation of franchises under which transport entrepreneurs in effect obtained the loan of law's compulsions to levy upon public or private means.

To supply bulk transport where the commodities being moved were not all owned by the transporter typically required conferring authority on someone to provide works and to control the traffic which used the works. Thus men needed the law's promotional help through the grant of franchises, not only to raise capital, but to discipline the use of capital under more or less limited, legally defined monopolies of location and of traffic management.

Once established, bulk transport facilities by water or rail were critical both for the growth of forest production as a whole and for the existence of particular lumber enterprises. Access to bulk transport spelled life or death for particular lumber ventures. This fact generated legal regulation — early and primitive forms of public utility law — which we shall note in Part Three of this book on police power. Functionally, the creation of franchises located and helped generate initiative of decision in centers of private will; functionally, police power (expressed in elementary public utility regulation) sought to adjust relations among diverse centers of private will once these had come into being. This is a line drawn to serve analysis; the distinction is not an absolute. Some legal scrutiny of the use of franchises was necessary for effective creation of new centers of private initiative, as against oppressive or discriminatory charges or practices; some crude public utility regulation entered as an initial condition into the creation of typical transport franchises. However, though we do not deal with absolute distinctions, there is functional logic in treating the law's relations to development of water and rail transport in part as creating specialized styles of property (franchises) and in part as developing the police power (public utility regulation).

WATER TRANSPORT AND WATER POWER

By a wide-flung pattern of rivers the Wisconsin forest had access in the northeast to Green Bay and Lake Michigan and to the wholly in-state gathering point provided by Lake Winnebago, and in the northwest to

Lake Superior and the Mississippi. Without these natural carriage ways the growth of the Wisconsin lumber industry must have waited thirty years upon the building of railroads. Delay might have promoted the long-run health of the state economy. But the rivers were there. Their availability provided the setting within which men made fresh, complex demands upon law.[1]

The lumbermen dealt in commodities — logs, sawed lumber, poles, posts, shingles — that were bulky, heavy and awkward in handling, low in price relative to their bulk. Apart from the physical difficulty of moving such goods, their transport accounted for a great part (typically 40 to 75 per cent) of the cost of getting them to market.[2] Access to navigable streams, and the terms of access, were thus critical factors affecting the market value of timberland, the scale of woods operations, the location of mills, the growth of towns and cities concerned with processing and distributing forest products. It was far cheaper to blast rocks from a stream than to build a railway roadbed, and far cheaper in labor and equipment (though not in time) to move logs and lumber by water than over a rail network. So the waterways continued to be a material — if diminishing — factor in lumber marketing to the end of the lumber boom. Moreover, as the railroads grew to economic power, lumbermen and public officers valued the waterways for such competitive check as they allowed upon railroad rates and service.[3] Our tradition was that law should be both constitutional (responsible to general welfare) and pragmatic (useful). In such a society we should expect to find that law's services and law's controls tend to develop about key intersections of activity from which because of strategic position a few men can command or affect many lives and large sectors of affairs. For reasons of technology and market function Wisconsin's waterways provided just such key points in the growth of the forest economy. Thus even from territorial years and the first decade of substantial lumber industry development in the 1850's, public policy showed lively concern and activity both to promote and to regulate the use of waters navigable for logs and lumber. The continuity of this interest is striking witness to the extent to which we wove law into the life course of that industry.[4]

A GENERAL INVENTORY OF LEGAL ACTION CONCERNING STREAMS

Most in-state water carriage of forest products moved toward the Great Lakes or the Mississippi. However, the commerce of the Great Lakes or the big river yielded little that distinctively belongs to this legal history. Wisconsin lumber interests were among those which petitioned Congress to spend federal money on lake harbors and Mississippi channel clearing.[5] In charters of lake-shore cities, Wisconsin statutes reflect a limited state concern to foster harbor improvements and authorize harbor policing.[6] Reported decisions of the Wisconsin Supreme Court and the federal courts sitting in Wisconsin show a handful of cases identifiable as arising out of log or lumber transport on the Great Lakes or Mississippi; these suits presented routine issues of general law (most of them growing out of collisions or marine contracts).[7] This scant catalogue covers all the significant items of the law's relation to log or lumber carriage on the Great Lakes or Mississippi. Thus our story is based almost wholly on public policy concerning the state's internal waters. Law responds to the intricacy of men's relations. On the state's inland waters shippers of logs jostled shippers of lumber, both contended with those who would improve a stream for tolls, all of these with men who would develop a stream for power, and all of these with riparian landowners who complained of flowage and trespass; because of their size the broad lakes and the great river were useful — at least in this era of the lumber industry — only for carriage, and could be improved only by government action, so that their use presented a smaller range of issues and hence a more limited response in law.

It follows the logic both of fact and of legal doctrine that our story concerns almost wholly the activity of state legal agencies. Under long-familiar rulings of the United States Supreme Court — given fresh application in Wisconsin-based lawsuits growing out of log transport — the state might freely develop policy concerning its internal navigable waters, so long as Congress had not used its superior authority under the commerce clause to pre-empt the field.[8] Congress granted public lands and appropriated federal funds to assist and later to take over a project for improving the Fox and Wisconsin rivers to connect Lake Michigan and the Mississippi; it made modest appropriations to survey and improve the Chippewa River; it gave public land to help build a ship canal at Sturgeon Bay for a shorter, safer route from Green Bay to Lake Michigan. These actions marked the limits of Congressional attention to Wisconsin's inland waters; plainly these measures in no way barred the state from a wide range of promotional and regulatory policy making, and the court so ruled.[9] This pattern of events is consistent with what we have observed in another important area of policy, namely, that the federal government gave no attention to the special problems which forest wealth presented in the disposition of federal lands in Wisconsin. It would be odd had the United States given more attention to the more specialized problems of in-state log transport.[10]

The lumber industry was interested in Wisconsin's inland waters first for navigation, and second for power. The power use was relatively uncomplicated; the single most common type of issue here turned on the right to flow others' lands and the terms of compensation therefor — matters which had no peculiar relation to the lumber industry; otherwise, the considerable quantity and complexity of legislation and litigation which attended the building of power dams in the forest area reflected more the importance of accommodating power use to navigation use than any issues peculiar to water-power development as such. In contrast, reasons of technology and of business organization created diverse problems of navigation use. (1) In some situations men sought simply to move logs or lumber rafts on streams as they found them. (2) Relatively permanent works or continuing operations were necessary to render some water navigable even for logs, by clearing away rocks and debris, straightening channels, cutting short canals across bends, building dams which from one season to the next could impound spring flood waters to be released to carry down the drives. Such improvements might be built along or affect a whole river or tributary or a large stretch of it; some improvements might be limited to a particular site; in either case an improvement might be for general use or use of the builder only. (3) On some streams it was efficient only to build temporary flood dams and conduct clearing operations according to the requirements of weather and limited logging operations in a given season. (4) Works of improvement apart, there were situations in which there was need or demand for specialized driving management and unified traffic control. (5) Terminal or intermediate handling facilities were necessary. Typically these involved the construction of booms, in effect floating fences made of logs chained together to form holding areas, with points of entrance and exit. Boom operators might perform for themselves or others a variety of services. They might simply provide secure temporary storage. They might furnish, in addition, facilities to sort out logs of different ownership, to be claimed or delivered according to ownership. They might undertake not only to sort logs for their owners but to form them into rafts for further transit.[11]

Measured both by doctrinal development and by operating activity, the story of public policy on log and lumber transport has little to do either with public works or with public money subsidies to private works. This history is chiefly one of navigation improvement delegated by law to private hands, and of public aid given these private delegates principally through grants of franchises for their operations. To facilitate lumber (though not log) transport, and to lower its costs, was one objective urged in territorial years (1838–1848) for a Milwaukee and Rock River canal, and from the fifties through the seventies for a canal and river improvement program to connect Lake Michigan and the Mississippi through the Fox and Wisconsin rivers. The Milwaukee-Rock project involved a subsidy of public lands to a private company; in the lumber years the Fox-Wisconsin project passed through phases of direct state operation, delegation to a private company, and finally transfer to the federal government, but in all these phases public financial support was essential. However, the advantage for lumber transport was urged — quite properly so — as but one phase of programs aimed at the total development of the economy, and on the record the claims for the lumber interest were only occasional and were less emphasized than pleas for the advantage of agriculture and of general trade and industry.[12] Moreover, neither of these grand public-subsidy or public-management programs produced significant results. The Milwaukee-Rock project aborted almost at the outset; in order to render large returns, the Fox-Wisconsin program proved to demand a scale and continuity of capital investment that neither the state nor the federal government was prepared to make.[13] The United States gave lands in aid of the Sturgeon Bay ship canal, and the state incorporated the company which built it.[14] The United States made modest appropriations to survey and improve the Chippewa River, but the efforts and investment of the lumbermen far outweighed the small contribution of the government to making the Chippewa a major logging and rafting stream.[15] As a practical matter, at mid-century only the federal government could command the resources to make substantial money investments in Wisconsin waterways. The state government could raise little money in an economy chronically scarce of fluid capital. Local government in the sparsely settled forest area would later strain its means to tax and borrow to subsidize railroad building. But this was a technique yet to be learned. Probably the device would not have been applied in any case to the development of transport facilities so specialized to one use as were the northern streams to the needs of the lumber industry. These practical difficulties apart, a formidable legal obstacle blocked investment of state funds in waterways promotions. Save as the state might spend means given it for such purposes by the federal government, Article VIII, section 10, of the Wisconsin constitution declared that "the state shall never contract any debt for works of internal improvements, or be a party in carrying on such works." The lumbermen did not lack power in the legislature. But they made no effort to change or avoid this prescription in order to obtain state aid for stream improvements. Thus the history of waterways policy affecting the lumber industry is almost entirely that of the grant of franchises and the enactment of a limited body of regulatory law.

The volume of legislative and judicial action concerning the use of Wisconsin's inland waters in relation to the lumber industry matches the volume of legal action concerning the public lands and the operation of the market. We must try to draw the bewildering detail of these waterways statutes and cases into meaningful generalization. To do so, and to put in proper perspective the important elements of the whole, it will be useful to conclude this introduction with some tables of statutes and Wisconsin Supreme Court decisions.

In timing, in content, and in scope, state policy toward lumber industry use of waters was embodied primarily in statute law. Thus we should begin with summary inventories of this legislative activity.

Wisconsin's waterways legislation was both general and special. The general statutes dealt with rights and duties of all persons using specified classes of waters or affected by their use, or conferred franchises for specified types of water use upon all persons of a class described by the statute. Some general statutes concerned all waters of defined classes anywhere in Wisconsin; some (local, though public) statutes concerned only waters of specified location. Special statutes conferred franchises for specified uses of particular waters upon named individuals, or unincorporated associations, or corporations.

There were four significant lines of public policy declared in general waterways legislation of state-wide scope: (1) provisions stemming from the Northwest Ordinance of 1787, which declared that navigation of waters leading to the Mississippi or St. Lawrence rivers should be free of tax and common to all;[16] (2) provisions continuous from a territorial statute of 1841, requiring legislative authorization for works built in waters meandered and returned as navigable by the federal survey;[17] (3) general statutory authorization, existing first from 1840 to 1850, revived and thenceforth continuously on the books from 1857, allowing the building of milldams on nonnavigable streams;[18] (4) general statutory authorization to any corporation duly organized under general law with the appropriate authority conferred by its charter, having taken prior possession for the purpose, to improve any stream for driving logs—first granted in 1876, enlarged in 1880 to include operation of booms, and so continued on the books until 1927.[19]

Most numerous of the general statutes of local application were those which designated particular streams or portions of streams as navigable and hence to be used within the framework of general law concerning navigable waters; other statutes which bore generally on the use of particular named streams defined standards of construction for dams or booms to assure free navigation, or protected traffic against other obstruction.[20] The four main lines of general legislation apart, this statute law of general application to particular localities was limited in total scope, haphazard in incidence and content, significant mainly in its implicit testimony to the uneven operation of contemporary legislative process.

The great bulk of legislative activity concerning use of the state's inland streams grew out of applications of individuals and corporations for special franchises. The applicants sought statutory warrant to make particular use of streams, free of competing claims which other users might otherwise make as members of the public, or accompanied by authority to impose on others burdens which the others might resist if the statutory sanction were absent.

About 90 per cent of the legislature's activity regarding special franchises for stream use concerned forest-area projects. From the first substantial beginnings of the lumber industry, all special franchises for river improvements and booms were for works in the forest region. Special statutory

TABLE 1. General legislation on use of Wisconsin streams

Years	Applicable throughout the state					Applicable to designated streams				
	Freedom from political restriction	Legislative approval required for works	General statutory authority for improvement	Statutory authority for milldams	Miscellaneous regulations against obstruction	Declarations of navigable status	Navigation guaranties as to dam construction	Navigation guaranties as to boom construction	Navigation protected as to riparian uses	Traffic regulation among navigators
1836–1839	Continuous provisions from Northwest Ordinance (in Wisconsin constitution, after statehood)	Continuous provisions from Laws of Wisconsin Territory 1840–1841, No. 9	Continuous from 1876 to 1927	Continuous, 1840 to 1850, and after 1857	. .	2	. .	. .	. .	. .
1840–1844					1	. .	1	. .	. .	. .
1845–1849					1	1	4	. .	. .	. .
1850–1854					. .	5	3	1	. .	. .
1855–1859					. .	9	6	3	3	1
1860–1864					1	1	. .	. .	. .	1
1865–1869					1	13	. .	5	1	4
1870–1874					1	2	3	. .	1	. .
1875–1879					. .	4	1	1	. .	. .
1880–1885					1	3	2	1	3	1
1886–1890					. .	2	. .	. .	. .	. .
1891–1895					. .	. .	. .	. .	. .	. .
1896–1900					. .	1	. .	. .	. .	. .
1901–1905					1	. .	1	. .	. .	. .
1906–1910					. .	. .	1	. .	. .	. .
1911–1915					. .	. .	1	. .	. .	. .
1916–1920					. .	1	. .	. .	. .	. .
1921–1925					. .	. .	. .	. .	. .	. .
1926–1930					. .	. .	. .	. .	. .	. .
1931–1935					. .	. .	. .	. .	. .	. .
1936–1940					. .	. .	. .	. .	. .	. .
Totals					7	44	23	11	8	7

This table does not distinguish between statutes affecting forest-area and non-forest-area waters, partly because many of the statutes in terms apply to both, partly because it seems desirable to reflect the total amount of this type of general legislation; almost all the statutes of local application fall in the forest areas. Because of the generality of this legislation, it has seemed appropriate to list amendments and repeals as distinct items of legislative activity. Where a statute created distinct regulations affecting more than one head of the table, it was listed as a distinct item under each appropriate heading. Regular legislative sessions became biennial rather than annual after 1883; the grouping of years is arranged in this table, as in other tables, so that through 1885 the five-year brackets encompass five sessions; thereafter the alternate brackets include, respectively, two and three regular legislative sessions.

franchises for private power dams on navigable streams constituted the only significant body of such legislation affecting the southern (nonforest) area of Wisconsin; these totalled only a little over a quarter of all such dam franchises granted. Private power dam franchises were the most numerous single class among these statutory grants, toll improvement franchises next, toll and private boom franchises third. Back of the enacted franchises lies a welter of legislative activity which defies neat tabulation: bills which were introduced, withdrawn, rejected, amended, or made the subject of committee hearings in one or both houses, the focus of supporting and opposing "memorials" or petitions from affected interests, or occasionally the object of the governor's veto. The bulk of franchises applied for were granted, though often with material alterations from the form in which they were introduced; a generous estimate — the veiled terms of bill titles and the crudities of nineteenth-century legislative records make precision impossible — would put the maximum of rejected applications at 15 per cent of the franchises sought. The flow of this legislative business reflects broadly the order of development and the growth in complexity of the economy; there is substantial activity in private power franchises in the nonforest (southern, first developed) part of the state almost a generation before enactment of the bulk of forest-area franchises; with firm settlement of the south, the franchise flow abruptly tapers off after the 1850's. Franchise grants drop with like abruptness for the forested north by 1895, as the lumber industry passes its peak; a spurt in private power franchises, 1901 to 1910, reflects rather the fresh interest in electric power development than the continuing demands of lumbermen.

Recognizing that this evidence reflects only the most reliably identified portion of a broad range of legislative activity, let us consider the picture presented by tabulating 673 enacted forest-area water-use franchises and 129 private power dam franchises for projects in the unforested south.[21] (See Table 2.)

Next to this considerable bulk of legislation, the reported decisions of the Wisconsin Supreme Court provide the largest body of explicit, deliberated, public policy determinations affecting navigation and power uses of the state's inland waters. The federal courts played even a smaller role here than in matters concerning use of the Great Lakes or the Mississippi. Congress had enacted only limited regulations affecting commerce on these inland waters. Thus there could be no lawsuits in federal courts in any quantity, save on issues governed either by the (federally defined) law of admiralty, or by state law of property, contract, tort, or police. Five reported cases in federal courts sitting in Wisconsin arose under the federal Constitution or statutes affecting use of the state's inland waters for log or lumber transport.[22] One or two cases in these federal courts touching lumber industry traffic may have rested on diversity of citizenship; in these

TABLE 2. Special statutory franchises for use of Wisconsin inland surface waters

	Forest area					Nonforest area
	Stream improvements with toll rights		Private dams (Power or flooding: no toll)	Booms		Private dams (All power: none with toll right)
Years	Extended area	Limited location		Public (Toll)	Private (No toll)	
1836–1839	1	..	1	..	..	5
1840–1844	..	..	3	..	..	18
1845–1849	..	2	8	1	..	25
1850–1854	4	4	19	4	2	30
1855–1859	5	2	26	10	14	24
1860–1864	5	3	6	5	7	6
1865–1869	14	11	13	11	10	4
1870–1874	18	13	14	5	7	8
1875–1879	18	17	18	1	1	3
1880–1885	41	20	58	4	9	2
1886–1890	18	6	33	2	2	..
1891–1895	13	11	31	1	..	2
1896–1900	..	1	18	1	2	..
1901–1905	2	3	59	..	..	..
1906–1910	1	..	23	1	2	2
1911–1915	2	..	2	..	..	..
1916–1920	1	..	..	..	1	..
1921–1925	..	..	..	..	1	..
1926–1930	..	..	..	..	..	..
1931–1935	..	..	..	..	..	..
1936–1940	1	..	..	..	..	..
Totals	144	93	332	46	58	129
			Total grants: 802			

This table was made by paging the Wisconsin session laws, and it was cross checked against the tabulation in Majority Report, Interim Committee on Water Powers, Forestry and Drainage (1909), Exhibits 1–3. See note 21 to this chapter. Only original enactments are counted; the subsection on "Tenure" in this chapter discusses the extent and nature of amendments and repeals (which were involved in only a minority of franchises under any of the headings of the table). The table does not include 14 charters of industrial companies which granted general dam-building authority among other corporate powers (e.g., Private & Local Laws, 1866, Chapter 63), where the statutory language was quite ambiguous as to whether the intent was simply to define corporate powers or also to confer a stream franchise; most of these 14 instances involved companies incorporated for operations in the nonforest area; the tabulation does include 15 charters of lumber manufacturing companies, the terms of which seem clearly intended to grant dam-building franchises as well as corporate authority (e.g., Private & Local Laws, 1856, Chapter 504, 1866, Chapters 238 and 284). All cases the federal courts declared and applied law familiar in state court doctrine.[23] The want of resort to diversity-of-citizenship jurisdiction reflects the extent to which Wisconsin's lumber industry was developed almost wholly by Wisconsin residents and Wisconsin-chartered corporations.

Litigation over forest-area uses of inland streams for navigation or power brought 158 cases to decision in the Wisconsin Supreme Court. Of these cases, 69 grew out of log or lumber traffic, or out of operation of improvement facilities (including flooding dams) or booms for handling logs or lumber rafts. Eighty-nine of the 158 cases grew out of the use of private power dams; the record shows that most of the 89 suits concerned dams to furnish power for industrial use, or dams that affected log or lumber transport; a few of the 89 cases concerned grist mills; the records do not

---

instances of dam-building authority conferred simply incident to incorporation occur before 1871, when a constitutional amendment forbade further special incorporation acts for business firms. With trivial exceptions, all items catalogued pertain to works on extended watercourses; three toll improvement franchises involved works including small lakes and canals as well as streams, and two public and two private boom franchises seem to have involved lake operations primarily. Where a franchise authorized booms together with a power dam, it was classed as a dam franchise; this combination occurred in 78 forest-area private power dam franchises, distributed in time substantially according to the time distribution of all such franchises; the dam-boom combination did not appear in nonforest-area private power dam franchises. The improvement franchises exclude all statutes dealing with the Fox-Wisconsin improvement project, partly because this had only secondary relation to log or lumber transport, partly because it was always so specialized a quasi-public or completely public undertaking as not to belong functionally with a catalogue of private improvement franchises. Classification of improvement franchises as for "extended area" or "limited location" works could be only an approximation, according to location descriptions in the statutes; where there was substantial ambiguity, the item was classed as an authorization of "extended area" works, since such a grant would seem more often to fit the working needs of the lumbermen. Some forest-area private dam franchises were silent as to purpose, or simply described their purpose as power production; a few such franchises specify flooding as the sole purpose; some mention flooding or navigation improvement with power production as coordinate purposes; the data was insufficient to permit reliable tabulation according to such sub-classes of purpose. The tabulation in the Majority Report of the 1908 Interim Committee cited above, pages 246–248, attempts a classification of purposes according to the phrasing of the statutes, but this does not permit reliable identification of particular lumber industry uses. The 1908 count lists 244 franchises as to "improve navigation, facilitate log driving" or both, 79 as to "facilitate log driving and [for] hydraulic purposes," 52 for "improvement of navigation and hydraulic purposes," and 227 as solely for "hydraulic" purposes. Each improvement franchise stands for one integrated works operation; dam and boom franchises sometimes authorized erection of more than one separate work, so that the total of dam and boom franchises can by no means be taken to indicate the number of separate physical facilities created. One should remember, also, that the general milldam statute authorized the building of dams on "nonnavigable" streams without further specific authorization, and, as noted later in the text, many dams were obviously built on navigable streams without statutory warrant.

permit a precise count of the dam cases directly related to the lumber industry. Lawsuits connected with private power dam operations provided the only substantial body of waterways policy problems to reach the Wisconsin Supreme Court from the nonforested, southern part of the state. There were 111 of such cases. Consideration of these nonforest-area suits casts some additional light upon the comparable litigation arising out of the lumber industry.

As in other affairs, the flow of these private power dam cases to the Supreme Court was in well-defined relation to the growth of the economy. The tally reflects the simplicity of Wisconsin's beginnings; but four dam cases came to the court from the nonforest area before 1850, and none at all from the forest region. The types and extent of the power use of streams were linked closely to the condition of technology and the specialization of industry. The bulk of the dam cases in both sections of the state concentrated in particular generations of industrial growth when water power was peculiarly important to contemporary patterns of production — in the south, in the years 1860–1880 in the first main development of industrial towns and cities; in the north, in the span 1870–1905, while the lumber industry reached and passed its peak. Among the power dam lawsuits which arose out of nonconsentual relations (that is, on theories of liability in tort or for public or private nuisances), claims for flowage damages to lands were the largest single category — forty-seven (about two thirds) of the seventy-one nonconsentual cases in the nonforest area, twenty-two (about one half) of the forty-five such cases in the forest region. The greater relative number of flowage cases in the south accorded with the more extensive development of urban and agricultural uses of land there, compared with the forested north country. Of similar import seems the fact that of the ten cases in which complaint was of public or private nuisance created by a power dam, other than by interference with navigation, eight arose in the nonforest area; moreover, in the south all of the four publicly prosecuted suits for alleged public nuisance charged creation of conditions harmful to public health, whereas the one such prosecution in the forested north charged interference with a public highway. Complaints that a power dam interfered with navigation made a relatively much larger total among forest-area suits than among those from the unforested south; there were twelve charges of harm to navigation among the northern-origin cases, compared with nine complaints of damage to power uses; among cases from the nonforest region there was but one charging interference with navigation, but fifteen which alleged injury to power uses. Issues of navigation and power shared importance in the lumber country, with navigation of leading concern (especially when we add the suits connected with stream improvements which centered on navigation interests); to the unforested south, water power was clearly the central theme of litigation. The only surprise in these relationships is that land-flowage cases should have bulked relatively as large as they did from the forest region. Most of these northern flowage cases, however, fell in the years 1860–1880, in the initial settlement of the forest area and before the railroads helped disperse settlement. Had full forest production continued longer than it did, it seems likely that in the affairs of what proved to be a thinly settled region the land-flowage claims would have appeared in longer perspective a quite limited phenomenon.

A striking number of the private power dam lawsuits grow out of consentual relationships, turning on matters of contract or title affecting the ownership or style of use of water-power sites and water-power rights, or labor performed in developing such sites. The number of suits of this character symbolized the extent to which men depended on contract and property arrangements to provide frameworks of working relations. There were more consentual cases relative to nonconsentual ones in the lumber industry region than in the unforested south. In the sparsely settled north men operated with more elbow room: in the forest region there was less occasion for controversy out of nonconsentual relations. On the other hand, in such a region of rapid, opportunistic development, there was probably no less dependence on contract and property arrangements, relative to the state of the sectional economy, than in the more settled south.

Log and lumber transport on the state's inland waters brought sixty-nine cases to the Wisconsin Supreme Court. Apart from a handful of suits involving liens, mortgages, or other contract or title issues concerning river craft, these lumber-industry cases were the only suits growing out of navigation on in-state waters that came to the court. The fact dramatizes the peculiar importance which water transport had to the lumber industry. The fact reminds us, also, how much the impetus to legal development depended upon the dynamics of the economy. Thus (in degree relatively greater even than in the power dam cases) these navigation-use lawsuits tightly cluster in the years (1875–1895) in which the lumber industry came to major growth. There was no significant number of such cases before 1875; most grew out of river improvement and boom operations which characterized the more developed state of the lumber industry and presented relational issues likely to arise in quantity only with the more intensive exploitation of particular forest areas. Property-title and contract cases stand large here, as in power dam litigation, constituting almost 40 per cent of these navigation-use lawsuits; dependence on the organizational frame which contract and property provided runs through all aspects of the industry. However, the terms of access to navigation bulk largest among these cases. Complaints of physical interference with navigation generated 28 per cent of these suits. If we add those cases which centered on claims for tolls (about 25 per cent of the whole), approximately 53 per cent of these suits focused on the terms of access to navigation use. Only three cases (about 4 per cent

TABLE 3. Private power dam cases before the Wisconsin Supreme Court (Gravamen of suit)

	Arising in forest area							Arising in nonforest area						
Years	Flowage of land	Navigation interference	Interference with power use	Other public nuisance: public action	Other public nuisance: private action	Contract and title issues	Miscellaneous tort	Flowage of land	Navigation interference	Interference with power use	Other public nuisance: public action	Other public nuisance: private action	Contract and title issues	Miscellaneous tort
1836–1839	..	..	..	..	..	..	..	..	..	..	..	..	..	..
1840–1844	..	..	..	..	..	..	..	..	..	..	..	..	1	..
1845–1849	..	..	..	..	..	..	..	1	..	1	1	..	..	..
1850–1854	..	2	..	..	..	..	..	5	..	..	1	1	5	..
1855–1859	1	..	..	..	..	2	..	1	..	2	1	..	2	..
1860–1864	2	..	1	..	..	3	..	10	..	1	..	..	7	..
1865–1869	2	2	..	..	..	1	..	5	..	..	..	..	10	..
1870–1874	2	..	1	..	..	2	..	7	..	4	..	..	2	..
1875–1879	3	2	..	..	..	3	..	2	..	..	..	1	2	..
1880–1885	1	1	..	..	1	6	1	7	..	..	..	..	4	..
1886–1890	1	..	3	..	..	6	..	2	..	3	..	..	1	..
1891–1895	..	1	2	..	..	3	..	2	..	..	..	..	1	..
1896–1900	..	..	..	..	..	1	..	1	..	1	..	..	2	..
1901–1905	1	..	..	..	..	7	..	2	..	1	..	..	3	..
1906–1910	2	..	1	..	..	3	..	..	1	2	1	1	..	..
1911–1915	..	..	..	1	..	3	1	1	..	..	..	1	..	..
1916–1920	2	..	..	..	..	..	..	1	..	..	..	..	..	..
1921–1925	2	..	..	..	..	1	..	..	..	..	..	..	..	..
1926–1930	3	2	..	..	..	..	..	..	..	..	..	..	..	..
1931–1935	..	2	1	..	..	..	..	..	..	..	..	..	..	..
1936–1940	..	..	..	..	..	1	..	..	..	..	..	..	..	..
Totals	22	12	9	1	1	42	2	47	1	15	4	4	40	0
	Total forest-area cases: 89							Total nonforest-area cases: 111						

Cases tabulated in this and in the following table were located by paging the Wisconsin Reports and cross checking these results against Shepherd's Citator and published digests. See Hurst, ed., *A Digest of Regional Sources for the Study of the Economic and Political History of the Law* (1941). Each case is counted but once, under one heading; the count reflects the realities of the subject matter of litigation, which with no more than three or four exceptions fell clearly into one heading of the table. Title and contract cases are merged, because this merger reflects functional reality; if one distinguishes the sub-categories, there were among the forest-area cases twenty-eight title and fourteen contract issues, and among the nonforest area cases thirty title and ten contract issues. Because they grew out of quasi-public projects which at most had only secondary relation to lumber industry venturing, cases directly involved with the Milwaukee and Rock River Canal and Fox-Wisconsin improvement programs are omitted from this and the following table.

of the whole) turned primarily on complaints over transport interference with use of waters for power. This percentage stands in sharp contrast to the substantial number of such complaints arising from private power dam operation in both the forested and unforested sections of Wisconsin. Stream improvements were typically made in open country; boom operations typically were auxiliary rather than competitive to the mills which were the prime users of north-country water power; on both counts there was less likelihood of relational clash than in situations where water-power development was the main industrial activity involved.

This inventory of general and special legislation and of Wisconsin Supreme Court cases provides a frame of fact and furnishes some rough measures of the proportions of various problems and interests, within which we can better define policy concerning the lumber industry's use of Wisconsin waters. The inventory is limited. It is limited to those materials that lend themselves to relatively firm classification and hence to reasonably reliable tabulation. Thus the inventory omits the instances in which Wisconsin governors dealt with waterways matters, because these occasions were sporadic and uneven in incidence; account will be taken of what little the governors did as we consider the specific content of policy. The inventory does not catalogue the massive detail of legislative procedure and maneuver in the introduction, consideration, rejection, or approval of bills. Much of the legislative record is cryptic or opaque. In proportion as the legislative record is revealing, it is so mostly in terms of particular matters, and can be treated most usefully in particular context. The inventory also tallies only material which is accessible without impossible costs of gathering. On this count the inventory omits the over-all activity of trial courts. The cases that reach appellate reports by no means always mirror the quantity and proportions of activity in the trial courts. However, the interests at stake in the important relations involved in stream use were substantial and well-focused enough so that at least the main types of litigated issues seem to have been fought through to the appellate stage. An inventory of the business of the circuit court for Chippewa County — sitting in the heart of one of the principal lumber industry areas — shows no material difference from the over-all pattern of forest-area lawsuits that reached the Supreme Court.[24] The absence of trial court figures is the only omission of first-line or field-agency data which deserves emphasis; through the lumber industry years there was no administrative apparatus either in state or local government to license or otherwise scrutinize works or operations on the state's inland waters. In the years important to this history Wisconsin public policy on use of the inland waters found definition primarily in the statutes and, in secondary measure, in the appellate decisions which we have tallied.

GENERAL FORMULATIONS OF PUBLIC POLICY AFFECTING STREAMS

The prime significance of the public policy that developed about statutory franchises for stream use was the early and (for the times) remarkably strong assertion of the police power, including some forecast of that specialized branch of police power we call the law of public utilities. These develop-

TABLE 4. Forest-area navigation-use cases in the Wisconsin Supreme Court (Gravamen of suit)

	Stream-flow improvements									Booms										
	Toll projects					Private projects				Toll					Private		Navigators		Bridges	
Years	Flowage of land	Navigation interference	Interference with power use	Contract and property	Service: rates and quality	Flowage of land	Navigation interference	Interference with power use	Contract and property	Flowage of land	Navigation interference	Interference with power use	Contract and property	Service: rates and quality	Navigation interference	Contract and property	Navigation interference	Other	Navigation interference	Interference with power use
1836–1839	..	..	..	..	..	..	..	..	..	..	..	..	..	..	..	..	..	..	..	..
1840–1844	..	..	..	..	..	..	..	..	..	..	..	..	..	..	..	..	..	..	..	..
1845–1849	..	..	..	..	..	..	..	..	..	..	..	..	..	..	..	..	..	..	..	..
1850–1854	..	..	..	..	..	..	..	..	..	..	..	..	..	..	..	..	..	..	..	..
1855–1859	..	..	..	..	..	..	..	..	..	..	..	..	..	..	..	..	..	..	..	..
1860–1864	..	..	..	..	..	..	..	..	..	..	..	..	..	..	..	..	1	..	..	..
1865–1869	..	..	..	..	..	..	..	..	..	..	1	..	..	..	..	..	..	..	..	..
1870–1874	..	1	..	..	..	..	..	..	..	..	..	..	..	..	..	..	..	..	..	1
1875–1879	3	..	..	..	4	..	1	..	..	..	1	..	1	..	1	..	..	..	..	..
1880–1885	4	1	..	..	2	1	1	..	..	1	2	..	2	2	1	..	1	..	2	..
1886–1890	3	..	..	1	2	..	..	1	..	..	2	..	..	3	1	1	2	1	..	..
1891–1895	..	..	1	1	1	2	..	..	..	..	..	..	1	1	..	1	..	1	..	..
1896–1900	1	..	..	..	1	..	..	..	..	..	..	..	..	..	..	..	..	..	..	..
1901–1905	..	..	..	..	..	1	..	..	..	..	..	..	..	..	..	..	..	1	..	..
1906–1910	1	..	..	..	..	..	..	..	..	..	..	..	..	..	..	..	..	..	..	..
1911–1915	..	..	..	..	..	1	..	..	..	..	..	..	1	1	..	..	..	..	..	..
1916–1920	..	..	..	..	..	..	..	..	..	..	..	..	..	..	..	..	..	..	..	..
1921–1925	..	..	..	..	..	..	..	..	..	..	..	..	..	..	..	..	..	..	..	..
1926–1930	..	..	..	..	..	..	..	..	..	..	..	..	..	..	..	..	..	..	..	..
1931–1935	..	..	..	..	..	..	..	..	..	..	..	..	..	..	..	..	..	..	..	..
1936–1940	..	..	..	..	..	..	..	..	..	..	..	..	..	..	..	..	..	..	..	..
Totals	12	2	1	2	10	5	2	1	0	1	6	0	5	7	3	2	4	3	2	1

ments are the more striking because they were incident to creating special kinds of private property.

We need recall the functions which "private property" and "police power" stood for in the circumstances of nineteenth-century economic growth. Both concepts, of course, refer to values besides those related to economic productivity; the law of private property expressed values put on individuality and privacy, for example, and the police power expressed values grounded in demands for equality and humane concern for the worth of life. But as a means for enriching individual life and promoting humane regard for its worth, the nineteenth century trusted in nothing more confidently than in the increase of material productivity. To this end it put great trust in delegating initiative of decision over use of scarce resources by creating private property titles and private contracting capacity. A fee simple title was a delegation of power to make productive use of designated land. In the surge of nineteenth-century economic growth this dynamic function of title was treated as more important than protection of acquired expectations; the personal and social significance of title was not so much that law assured a vested right as that law delegated power to decide the use of the resource.[25] True, within our tradition that all power should be responsible (constitutional), the law reserved a veto upon styles of title or agreement which the legislature or the courts found to disserve the values for which property or contract was created; the law might deny legal effect to forms of title or terms of contract which it found to be contrary to public policy. However, the presumption was that any given private property arrangement or any given contract was valid until strong reason was shown to the contrary. This presumption was very strong both in doctrine and in practice because it expressed a faith deeply held. Prevailing opinion thus tended to forget that property and contract existed as creations of law, qualified by the purposes for which they were created, and tended to treat them as declaring absolute rights.[26]

The nineteenth-century United States was a society organized by broad dispersion of decision making and more and more intricate division of labor. The federal system provided an assured framework for the growth of sectional and national markets. The corporation provided procedures for mobilizing and disciplining larger combinations of capital. The developments of technology and the increasing returns open to investment of fluid capital and labor on a base of cheap land and increasing population pressed us to enlarge the scale and complexity of economic operations. Thus the most dynamic factors of our situation tended toward increasing stress upon organization as a social asset. This emphasis generated demands upon law to assure that relations be effectively ordered, especially against abuse of power by men who stood at key points of intersection. Such was the background for the growth of the police power — the use of law to define and enforce at least minimum standards of design and operation, that men's relations might effectively serve the values they sought.[27]

It is plain enough — as quickly appears from their specific content — that statutory franchises for stream use represented the law's creation of "private property."[28] The terms of franchises showed that they were granted as means of releasing productive energy by delegation, fulfilling the central function of the property institution as the nineteenth century viewed it. By their terms the franchises also performed the function of creating assured expectations (vested rights), that those to whom initiative of decision was delegated might know that government would not arbitrarily despoil them. To implement these purposes, in fashion familiar in the general institution of property, the franchises constituted legitimated forms which validated in the grantees the acquisition of delegated initiative of decision over particular resources; a franchise was a title. Like general property, the franchises stood under the law's protection against invasion by private persons, and the law stood ready to enforce agreements made or claims for services rendered pursuant to the franchise. To grant such franchises was to create rights of private property, albeit property rights of special character.

Their special character lay in the unusual extent to which recognition of commonwealth interests was woven into the fabric of these delegations of power. Enforcement of franchise limits was uncertain and sporadic, often nonexistent; the contemporary executive establishment was too simple and too pinched to check what was done under the legislature's grants; remedies lay in court for breaches of franchise, but these remedies required the initiative either of the weak executive or of private litigants (likewise perhaps weak, or possessing interest too slight to spur action). Even so, the assertion of commonwealth interest was — for the times, certainly — an impressive aspect of the legislative response to pressures for productive use of streams.

Constitutional texts, general legislation, and common law embodied only a few principles of policy on stream use, of limited impact on the working realities of the lumber industry; in style characteristic of the nineteenth century, most of the content of policy in this realm must be derived from the particulars of hundreds of special statutes. We shall examine the policy implications of this body of special legislation. But we should first mark out such frame of policy as general doctrine provided.

The Northwest Ordinance limited the authority of territorial legislatures over the principal waters: "The navigable waters leading into the Mississippi and Saint Lawrence, and the carrying places between the same, shall be common highways, and forever free, as well to the inhabitants of the said territory as to the citizens of the United States, and those of any other states that may be admitted into the confederacy, without any tax, impost or duty therefor."[29] In substantially the same terms, Article IX, section 1, of the Wisconsin constitution (1848) limited the authority of the state.[30] This earliest declaration of general policy concerning the state's waterways guaranteed private persons access to navigation, free of some kinds of official controls. Its core meaning seems clear enough, though it was never put to direct test in Wisconsin: the legislature might not directly tax the use of these waters.[31] Perhaps the indicated policy would bar other forms of state regulation which might limit navigation. However, the development of commerce clause doctrine by the United States Supreme Court provided a broader doctrinal base for controlling state policy so far as it directly burdened or discriminated against interstate traffic. This was in effect recognized by federal and state court decisions which interpreted the navigation guaranties of the Northwest Ordinance and the Wisconsin constitution so that their protection to interstate commerce differed in no way from the law under the commerce clause.[32] The practical question thus was not of the possible full reach of the Northwest Ordinance guaranty and its state analogue, but whether this guaranty created a particular limitation on the state, apart from limits imposed by the commerce clause of the federal Constitution.

Legal agencies in the United States have — quite properly — tended to treat constitutional language as declaring standards for exercise of reasonable discretion by government, rather than as creating specific rules of limitation. This was the course of doctrine early implicit in legislative practice, and made more explicit in later judicial decisions, under the navigation guaranty stemming from the Northwest Ordinance.

Probably the guaranty meant that no man need ask leave of the territory or the state to navigate these waters in their natural condition. In large measure this was a moot point; no statute or executive order ever asserted so broad a licensing control.[33] In the absence of legislation, the governing doctrine on navigation of Wisconsin's inland waters was judge-made (common) law. The navigation guaranty derived from the Northwest Ordinance was undoubtedly a limitation on policy making by the courts as well as by the legislature or executive. However, nineteenth-century professional opinion took common law for granted as a part of our jurisprudence as basic as the Constitution; the Wisconsin court developed and applied common law to these waters without ever explicitly acknowledging that this judicial law making was subject to the guaranty in Article IX, section 1. Since the court declared no common law which limited access to navigation of waters in their natural condition, the question of judicial authority was not sharply posed.[34] Wisconsin common law held that any person might use the waters for transport, as he found them, and ruled only that there must be reasonable accommodation among different users. This mutual accommodation included the right to inflict on another without compensation such detriment as was a necessary incident of ordinary transport use of the stream; thus a log driver might detain logs long enough to sort out his logs with diligence from those of other owners, and he was not liable to the landowner whose shore suffered damage from a log jam such as was normal in the ordinary course of a drive.[35]

In practice the important issues arose because men wanted to change the natural condition of waters, to make them more productive. This was especially so for the lumber industry's use of the streams. Even on the larger rivers, to handle the awkward bulk and weight of logs required that men alter channels and manage the flow; this need was the greater as lumbermen had to contrive maximum use of small feeder streams to bring in their raw material from scattered logging points. The striking early acceptance of the police power here lay in the confident assertion by legislature and courts of the state's authority to set terms of use to protect commonwealth interests.

The Wisconsin Supreme Court assumed without question that it held familiar common law authority to give damages for, and enjoin as a public nuisance, the maintenance of works which in fact materially obstructed navigation.[36] The state might sue to abate a public nuisance. Or a private person might sue to enjoin maintenance of such a nuisance or to collect damages for harm caused by it. The private suitor carried a heavier burden than the state, however. He must establish not only that a public nuisance existed, but also that he suffered particular detriment thereby, apart from the general public inconvenience or the infringement of the public's right to be free of works built without authority.[37] It might not always be necessary to show material obstruction of general navigation in order to prove that a public nuisance existed. Where the state showed that works had been built without valid statutory authorization, on a stream suitable for general navigation by boats or rafts, the court indicated that the absence of an effective statutory franchise was enough in itself to make out a public nuisance, without proof of material obstruction in fact.[38] On the other hand, where a stream was suitable only for moving logs in the spring freshets or with the aid of flooding dams, the basic element of public nuisance would not be shown without proof that the offending works caused serious obstruction in fact, even though they existed without statutory warrant.[39] But where works had been built on any stream under a statutory franchise, there was sufficient showing of a public nuisance by proof that the grantee had violated navigation guaranties stipulated by the statute.[40] Such judicial controls over nuisance applied as well to dams or booms built purportedly in aid of navigation as to dams built to create water power.[41] To such extent, then, Wisconsin common law asserted the judges' authority to regulate stream improvements. The cases raised no question whether this assertion infringed the navigation guaranty derived from the Northwest Ordinance; the court sensibly took for granted that the guaranteed rights could exist only within a framework of relations ordered by law. The striking fact, however, was the attention which the judges gave to the presence or absence of a statutory franchise, or to compliance with a statutory franchise, in deciding when and on what terms they would apply their common law remedies.[42] This judicial attitude focused upon the true center of official activity affecting stream improvements, for this center plainly was in the legislative process.

Continuously from Territorial Act No. 9, of February 9, 1841 (carried into Wisconsin Revised Statutes, 1849, Chapter 34), the legislature asserted its ultimate authority to license all works projected on waters meandered and returned as navigable by the United States Survey, "to such an extent, that no dam, bridge, or other obstruction may be made in, or over the same, without the permission of the Legislature." [43] Likewise from territorial years on, in hundreds of special statutory franchises for dams, booms, improvements, and systems of traffic control, the legislature implicitly asserted its authority not only to license works and organizations for the more productive use of navigable streams, but also to exercise a broad discretion in fixing the terms of such grants.[44] In over forty statutes it asserted authority to define particular waters as navigable, implying a mandate that the courts should apply to such waters the protections they extended to navigation on streams held navigable by the measures of the common law.[45] In a few acts of state-wide reach and in nearly fifty statutes applicable to particular streams, the legislature asserted its ultimate authority to make traffic rules and to regulate against pollution or obstruction of waters, albeit it did so, characteristically, in fragmentary fashion and without provision for effective enforcement.[46] From the year 1840 (with a break only from 1850 to 1857), in a general statute setting terms on which men might erect milldams, the legislature asserted its prerogative over nonnavigable streams.[47] The legislature left the bulk of its policy to be collected from the details of special acts. Its general laws were uneven in coverage and lacked effective sanctions. Nonetheless, the record of its over-all activity left no doubt that the legislature claimed and exercised the general control of policy over use of the streams.[48]

Some twenty years of legislative practice had already asserted the broad reach of legislative authority over improvements affecting navigation, before the Wisconsin Supreme Court had occasion to speak to the point. However, the court in effect forecast the approach it would take when, in 1849, it sustained the constitutionality of the milldam act governing the use of nonnavigable waters; the benefits to the general economy which the legislature might reasonably anticipate from encouraging the development of water power justified delegating the power of eminent domain to private dam builders.[49] Thus it was not surprising that in 1856 the court ruled that the legislature infringed no public rights, but acted within its authority to promote the general welfare, when it licensed works on navigable waters.

Significantly, the opinion gave weight to legislative practice in defining legislative authority: "It is a very common exercise of legislative power to authorize such obstructions [as power dams] to be placed in rivers, the legislature taking care to annex such conditions as it may deem essential, to protect the public from injury." [50] Not until 1872, however, did the court rule on the scope of legislative authority regarding stream improvements to serve the lumber industry. Then, in *Wisconsin River Improvement Co. v. Lyons,*[51] the court validated the general requirement of legislative license declared in the Revised Statutes, in terms of the broadest hospitality to legislative prerogative. The opinion was not well-defined in doctrine. But it was clear in effect. (1) The court saw no inconsistency between broad legislative regulation of navigable waters and the navigation guaranty which stemmed from the Northwest Ordinance; the court treated the guaranty and the regulatory statute as parts of a common pattern, in which — the opinion implied — the legislature's licensing control implemented the constitutional assurance that the Wisconsin River should be an effective "common highway." [52] (2) Judge-made law must yield to statute law in determining the legality of works built in navigable waters. The trial court had upheld the claim of the defendant to build his dam without statutory franchise, on the basis of common law riparian right. Plaintiff stood on a specific legislative warrant for his works, with which he alleged defendant's construction would interfere. Since the Supreme Court found that an effective statutory franchise had been granted plaintiff, and that defendant's dam stood contrary to the legislature's requirement of a statutory license, there was no room for defendant's claim of common law right.[53] (3) Defendant had no basis — as individual, as property-owner, as entrepreneur — to insist that his initiative of decision on the use of these waters should prevail over the legislative judgment. If there were debatable questions of fact and judgment as to what works would best improve the use of the river here, the legislature had authority to make reasonable resolution of those questions: ". . . no man can predict or anticipate with any certainty, what the effect of the dam when completed may be upon the navigation, and it is enough for the courts to know, and absolute law unto them, that the legislature, acting within the scope of its powers, has said that the party is a trespasser upon the rights of the public and shall not erect his dam without public permission, even though the same may prove, as he claims, to be a benefit instead of an injury to the navigation. We know of no way of getting along with the positive provisions of such a law, except that of being governed by and giving full effect to its requirements." [54] Thus the legislature held clear title from the court to pre-empt the initiative in regulation, free of any absolute limits set either by the constitutional guaranty of public navigation, the separation of powers, or the vested rights of individuals.

Through the lumber years the legislature wielded this policy initiative with little rebuff. The court never invalidated a general legislative policy concerning stream improvements.[55] The only area of general policy in which tension arose concerned delegation of the power of eminent domain in aid of private improvers. The court recurrently indicated disquiet over its 1849 ruling in favor of the milldam act.[56] But it stood by that precedent, and expressed its concern only by sharp insistence that compensation be paid where works encroached beyond the natural channel, and that there be fair and secure procedures for defining and paying claims.[57]

In *Stoughton v. State* (1856) the court had intimated that the legislature's authority to license works in navigable streams might be conditioned upon its "taking care to annex such conditions as it may deem essential to protect the public from injury." [58] Moreover, the court often emphasized that the legislature might regulate stream improvements only to advance public, rather than merely private, interest.[59] If such statements implied an affirmative duty on the legislature to insert positive navigation safeguards in its franchises, the implication received no judicial implementation; the court never invalidated a franchise for the mere absence of a navigation guaranty which the judges thought it should contain.[60]

In their negative aspect, such judicial pronouncements only applied to navigation law the familiar premise of our constitutional polity that public power must be wielded for a public purpose. In such doctrine there was no emphasis peculiar to regulation of navigation. On the other hand, this was judicially enforceable doctrine; waterways issues apart, there was Wisconsin precedent for invalidating legislation for want of a public purpose.[61] However, in but one instance did the court upset a river improvement franchise because it had been granted on terms which made it available to serve only private interest. This ruling involved a franchise for a dam and boom at Eau Claire on the Chippewa River, granted in 1875 only after a hot contest among competing mill towns and rival groups, colored by charges and countercharges of corruption and abuse of official position for selfish purposes.[62]

Conceivably, it might have been possible for the governor to use his veto with such range and vigor as to constitute an important influence on the content of statutory policy. There was early promise of such a role for the chief executive. In 1853 Governor Farwell vetoed two river improvement franchises because they conferred undue monopolies of water-power development and traffic control. In 1860 Governor Randall vetoed a dam franchise for want of protection to navigation.[63] These vetoes did not make clear whether they expressed only differences of judgment or purported to enforce some ill-defined constitutional limits on legislative power. In either aspect, however, these incidents failed to launch a tradition of vigorous participation by the governor in developing legislative policy. As we shall see, the legislature went on to create more traffic monopolies, and it met no more vetoes.[64] For a generation before Randall's veto in 1860, the well-defined legislative practice was to include navigation guaranties in special franchises for stream improvements.[65] It would attribute more effect than the record warrants, to credit Randall's single intervention with the continuance of this pattern. In 1882 and 1883 Governor Rusk insisted that all statutory dam franchises should include a reservation of legislative power to amend or repeal.[66] Rusk's intervention did set a lasting pattern, important to the sustained vitality of the police power. But this achievement stood alone in its impress of the governor's office upon the development of navigation policy.[67]

The definition of "navigable" waters provided the framework of public policy concerning the transport uses of streams and their adjustment to other public and private interests. The concept of navigability measured the reach of the constitutional guaranty derived from the Northwest Ordinance; it marked the bounds of common law rights and remedies; it identified values and furnished criteria of value preference, to justify legislation. The development of the concept of navigability thus gave form and dimension to the growth of general policy affecting streams.

We can find no chapter of doctrinal history more steeped in this society's instrumental attitude toward law than the story of changing definitions of navigable waters. By the time important issues arose in Wisconsin, the United States Supreme Court had already given the lead in abandoning the English precedent which identified navigable waters with tidewaters; by the Supreme Court's new criterion, those waters should be deemed navigable in law which were navigable in fact for the useful commerce of the country.[68] It was a criterion suited to the expanding market needs of a big country generously endowed with inland waterways, and Mr. Chief Justice Taney said bluntly that this was the basis on which the Court adopted the new definition. The Taney Court acted specifically to define the scope of federal admiralty jurisdiction. But Wisconsin judges saw reason in adopting the new definition for all purposes of state policy.[69] In effect the territorial and state legislatures anticipated the judges by a generation. From as early as the year 1839, by both their range and their terms statutory franchises for stream use recognized as navigable any waters which would bear to market the ordinary produce of the neighborhood, including logs and rafts of lumber.[70]

Though they used the term, neither the Northwest Ordinance nor the Wisconsin constitution defined navigability. Presumably, thus, it fell within the general policy authority of the legislature to put such content into the idea as would reasonably serve public interest.[71] The legislature early made implicit assertion of this authority when it inserted in special dam franchises stipulations to assure free passage of vessels, rafts, and logs; it asserted the authority implicitly when by granting improvement franchises it subordinated the power uses of waters and the agricultural or industrial uses of riparian lands to the development of transport.[72] It asserted the authority explicitly in the territorial act of 1841 which borrowed the designations of the Federal Survey to mark the waters on which no works might be built without legislative license.[73] The court never challenged the ultimate authority of the legislature to make reasonable definitions of navigable waters to promote the general interest in transportation.[74] However, there was no occasion to draw a sharp issue. For the most part, the legislature adopted for its ends the definitions of navigable waters made by the judges. In any case, the legislature never attempted a broader definition than could be found in the Wisconsin decisions.[75]

Though courts in the United States were hospitable to a definition of navigable waters broader than the English precedents, there was doctrine which would limit the concept to waters navigable by vessels or boats. But such a definition was too restricted to serve an economy in which a lumber industry bulked large. Courts of eastern lumbering states had already ruled that a stream was navigable which would float logs. When the question was first presented to it in 1868, the Wisconsin court had no trouble in adopting this broader test, for "it would be exceedingly detrimental to the public interests, especially in the pine-growing regions of the state, if this were not so." [76] However, this ruling did not exhaust the judges' readiness to adapt their definition to the needs of the industry. The 1868 decision had given the protection of navigable status to a stream which would float logs "at an ordinary stage of water." In 1877 the court applied a considerably enlarged version of the log-transport criterion of navigability: (1) The stream need not offer continuous facility for moving logs. It was sufficient that ordinarily in season there was enough water to be useful; the spring freshets gave a watercourse navigable status for purposes of log transport.[77] (2) The potential rather than the natural condition of the stream was decisive; the logger was entitled to claim a creek as a public highway if he could make it usable by clearing away stumps and sunken logs, alder bushes and sand bars, and making hard use of a driving crew.[78] (3) The record showed that the claimant logger had managed to move his drive because his crew trespassed on the shores in order to maneuver the logs. To the court this fact did not deny the navigability of the stream. The court would not imply that operations from the banks were essential; the judges would assume that if the stream would float a log, it would float boats which could be used to direct the drive, even if at considerable risk. Obviously these rulings went the limit in extending the law's protection to log transport. The court conceded that precedent in other logging states would hold such a stream not navigable. It justified its contrary determination because the decision was "essential to the public interest in the pine-growing regions of this state." Even in its refusal to acknowledge that this stream was in fact navigable only by unlawful invasion of riparian lands, the court believed "that we are supported in principle and by authority [not cited], in adopting a rule in this state which appears to us to be important to the great lumbering interests."[79]

In 1889 the court in effect conceded that it had pressed its formulation too far, and qualified some of the broader implications of its extended definition of navigability. "A distinction," it now felt, "may well be made between those streams which are capable of floating logs and timber only at certain periods, and then for a few days in times of freshet, and streams which are capable of more extended and constant navigation." The distinction concerned the strictness of the protection which the law afforded transporters on the one or the other type of water. Works might not be built on a principal stream without license under a statute, but on a stream navigable only by the extended log-transport test a riparian landowner might build a dam without specific statutory approval, so long as he did not in fact obstruct such navigation as the stream allowed.[80] The court justified its distinction as a reading of the total pattern of legislative policy; the legislature had expressly required statutory license only for works built on streams meandered and returned as navigable by the Federal Survey, and its general statute against nuisance penalized only obstructions in fact. This effort to find legislative sanction tacitly acknowledged that the legislature had taken the leading role in adjusting competing interests involved in stream improvements. But the exposition lacks full conviction. The legislature enacted scores of special franchises for works on minor streams. The policy implications of this legislative activity were ambiguous, but at least they suggest that the legislative process did not draw the sharp line of the court's decision between streams suitable for limited and "more extended" navigation. This line was in fact drawn by the court, and thereafter acquiesced in by the legislature. The line was drawn for reasons of policy which the court, and not the legislature, here initially determined. The convincing heart of the decision lay not in the attempted rationalization from legislative policy, but in the court's own crisp declaration: "It seems to us that in reason and common justice a distinction should be made in view of riparian rights." The point was to achieve maximum realization of economic benefits from all the resources involved, land as well as water, in relation to these streams of limited transport utility: "For if the right of floatage is paramount, so that no bridge or dam or other obstruction can be placed in or over the stream by the riparian owner, his use and enjoyment of his property are unnecessarily abridged and restricted."[81]

Even with this qualification, Wisconsin's lumber-era definition of navigable waters represented a very broad legal protection of transport interests. The extended definition not only gave far-reaching scope to common law rights of navigation; it also implicitly validated the broad area within which the legislature ranged in issuing franchises and stipulating for protection of transport. It is not surprising that a court which held to this extended definition of the public interest in navigation found little occasion to curb the jurisdiction which the legislature asserted in these matters. In 1894, as logging neared its peak, a trial court in the forest country passed realistic appraisal upon the result: "In view of the vast interests depending upon the use of small streams for driving logs, it seems to have been the policy of this state to extend the rule of navigability to its extreme limit, and we find that tests adopted by other courts are rejected by ours, and streams are held navigable in this state that would not be regarded as such in others."[82]

If the definition of navigable waters provided the frame of policy, the substance was built largely on attitudes taken toward the relative roles of public and private improvement of streams. We can bring these attitudes into focus the more quickly because they have already been indicated, incident to other aspects of policy.

In retrospect — though contemporary lawmakers never paused for an overview of their operations — we can see that the relations of public and private decision here tended to fall into four heads of policy.

(1) There was no substantial development of lumber waterways by direct public action. The Wisconsin constitution forbade the state to contract any debt for, or be a party in carrying on, any "works of internal improvement." Local governments were unready and lacked means. Congress had authority to appropriate in aid of waterways development, but it showed little concern for the lumber industry's needs. Thus the only major type of governmental promotion of inland stream improvements in Wisconsin was by the courts' declaration of protective common law doctrine and by the legislature's grant of franchises. To say that the absence of substantial public investment was an expression of principle would overstate the case. The internal improvements ban of the state constitution certainly declared a limiting principle. But the values implicit in the limitation are not altogether clear. Since the provision allowed the state to spend subsidies obtained from the federal government, and since its ban did not extend to local governments, evidently the intent was not to forbid all forms of public subsidy to the supply of social overhead capital. The limits of federal aid and the want of local government investments in stream improvements were plainly products of economic and political fact. The same tone of simply practical consideration sounds in the limits which the internal improvements clause set upon its own declaration of principle.[83]

(2) Given these limits of law and practice upon direct public investment, and given, also, the necessities of the lumber industry and the general faith of the community in the beneficence of economic growth, stream improvement naturally proceeded by delegation of powers to private investors and decision makers. From territorial years, unbroken legislative practice assumed the legislature's authority to make such delegations. The practice could claim warrant in the plain implication of Article VIII, section 10, of

the Wisconsin constitution; when the constitution forbade the state to "be a party in carrying on . . . works . . . of internal improvement," clearly it assumed that the law would provide a framework within which others might be enabled to "carry on" such important works.[84] Finding that the legislature could reasonably believe that the general interest in economic growth would so be served, the court sustained a broad legislative discretion to delegate the supply of such facilities.[85] There must be reasonably clear definition of the terms of the delegation, but this requirement worked no serious limit upon legislative action; in only one case did the court upset a statutory franchise on this basis — for want of sufficiently defined grantees.[86] The delegation must be for a public purpose; in one case we have noted (involving works at Eau Claire on the Chippewa River), the court invalidated a franchise the terms of which appeared to warrant serious impairment of navigation on an important stream solely for private advantage.[87] But delegation to private hands of functions of public concern was in itself clearly held a permissible technique of public policy. About this principle the mid-nineteenth-century United States developed its highways, canals, railroads, hospitals, colleges, aids to dependent persons, and provisions for scientific research and adult education.[88] The central place which such delegation had in the development of Wisconsin waterways policy simply demonstrated one of the organizing principles of the times.

(3) If delegation was permissible as a technique of government, it was equally clear that the private improver required government's positive warrant, if his efforts might materially affect other persons' use of resources. So far as common law recognized a landowner's right to build a dam or boom, it was only with reference to streams of the most limited navigable use or only on the assumption that any hindrance to general navigation was trivial or a temporary incident to the builder's own proper transport use of the waters.[89] As we have seen, on streams suitable for navigation over extended periods of time by boats or rafts as well as logs, there was indication that any dam would be held a public nuisance if it were not built under statutory authority.[90] In any event the landowner's rights at common law stood always subject to the legislature's authority to intervene with the requirement that a license be obtained on the legislature's terms.[91] Thus the private decision maker had little claim for a wholly independent sphere of initiative in stream improvement. If the improver's project called for more than the simple construction of works, if he sought to raise his capital or meet his operating costs by levying toll on others, or to realize the stream's maximum carrying capacity by managing the whole flow of traffic, clearly he must obtain the legislature's permission.[92] As a natural corollary of these propositions, legislative practice asserted, and the courts sustained, the authority of the legislature to adjudicate among competing private interests in the use of streams and to fix reasonable orders of preference in the public

interest among them. This authority included the legislature's right to be wrong; if its determinations did not plainly exceed bounds of reasonable judgment, the legislature might resolve uncertainties of fact and prediction in favor of one interest rather than another, and no less effectively though its determination cast risks of damage upon one party to the benefit of another.[93] The range allowed to legislative judgment in this field showed how strongly men felt that basic commonwealth interests were at stake in the dispositions made of the streams. This was a striking aspect of public policy in a time which put so much reliance on private decision making through property and contract. The attitude stands out the more sharply because of the judicial reaction about the turn of the century toward severe limits on legislative judgment in the name of due process of law.[94]

(4) If a private enterpriser needed the authority of a statute, that he might lawfully proceed with a stream improvement which touched substantial concerns of the public or other individuals, the requirement was not without offset. For the legislature held power to grant its licensee valuable privileges beyond its mere leave to occupy particular waters with his works. Five privileges in its gift might be especially helpful; each might be enjoyed only if the legislature gave it. (a) A statute might confer on the private improver the right of eminent domain, so that a stubborn landowner might not deny him access to land necessary to the improvement, or drain his capital by exacting whatever price the improver's necessity might require him to pay.[95] (b) Only by statute might the improver charge toll to those who used the waters he had made more useful; as grantee of a toll right, the improver might levy for capital as well as operating cost upon the users in order to provide the public benefit.[96] (c) The toll right could be the more valuable if the legislature gave it as a monopoly.[97] (d) Quite apart from tolls, effective use of a stream for driving or booming logs might require that someone hold exclusive authority to make improvements or manage the total traffic, allocating water among alternative channels, deciding the order of passage and delivery, or directing operations; the legislature had power to confer such a monopoly of management.[98] (e) Finally, here as for other enterprises, the legislature might confer corporate existence upon its grantees, to help them muster and discipline capital and assure continuity of business life.[99]

Obviously, in the background of such vitality of legal dispositions — back of legislative prerogative, the court's extended definition of navigability, the bold delegations of powers, and the variety of valuable franchise privileges — lay some dynamic, confidently accepted, validating principle. Of course, selfish interest was at work. Men sought these franchises to profit by them. And most who sought franchises got them. But, as we shall note later in more detail, the legislative record shows surprisingly close and continuous attention to the specifications on which franchises were given. We cannot fairly read this record to show that the legal process here did no more than routinely put its formal stamp upon whatever special interests asked for. We should not romanticize the legislative process; the legislature was quite capable of letting a special interest write its own ticket. But it does not appear that business was done this way in the matter of stream improvements—if for no loftier reason than that the affected interests were at once sufficiently focused, diverse, and influential to enforce some real adjudication in the grant of improvement franchises. If this be the fact, it emphasizes the more that there was a dynamic consensus of value behind the bold use of law in this area. This moving principle was the common belief—the article of faith which made almost all men in the country (save Henry Thoreau) Hamiltonians or Whigs—that it was common sense, and it was good, to use law to multiply the productive power of the economy. Productivity was the central test and validating canon. By this criterion the legislature and the judges confidently wielded authority over the waterways.

In 1876 the legislature enacted a second franchise for a major dam and boom to manage the great traffic of logs and rafts at Eau Claire on the Chippewa River. The franchise was drawn to meet the objection which had led the Wisconsin Supreme Court to invalidate a prior grant.[100] The court had upset the earlier grant because its provisions might allow use of the franchise solely for private profit, without adequate stipulations to serve general interest. In passing the second franchise the issue was sharply defined as one of finding justification in terms of public welfare. With obvious confidence that it stated a ground to which none could take exception, the Assembly Committee on Lumber and Manufactures recommended the new grant because it "would seem that considerations of public policy—that policy which fosters manufacturers and seeks to develop to the utmost the natural resources of the State—point with . . . clearness to the course which we should pursue." [101]

Its narrowly practical temper gave the legislature little inclination to expound its decisions. Moreover, in ordinary circumstances—removed from the unusual controversy that surrounded the Eau Claire franchise—this particular policy was so natural to the country's surging growth that men took it for granted. Thus, basic though it was, the value put on using law to promote material productivity must typically be read from what the legislature did, rather than from what it said. The implied judgment and justification came closest to exposition in three aspects of the flow of statutory franchises. (1) This community put great store in releasing men's energies of mind and will through private property and contract. Normally, if an enterprise needed the use of another's land it must obtain the land in market, by bargain. When the legislature delegated to stream improvers the sovereign power to take land by eminent domain, it allowed the improver to insist not on a bargained but on a law-determined price. In doing so, the legislature attested how deeply it felt that law was properly used to foster investments which would multiply productive activity.[102] (2) We shall note later that a rudimentary law of public utility obligations developed, incident to granting stream improvement franchises. Herein the legislature recognized that its franchises provided an enlarged framework within which general production could expand, but on which production might also become so dependent that the law must subject the franchise holders to special measures of public responsibility.[103] (3) Where the legislature granted a toll right, the franchise terms stipulated for a *quid pro quo* — terms which often implied that the legislature found warrant for its action in obtaining an increase in general productive capacity. Toll rights were always conditioned upon achieving some actual improvement in the navigability of the stream.[104] This was a minimum bargain, which might be interpreted as no more than an effort to secure a fair return to the user for the fee he paid. However, a number of franchises carried terms which looked to enhancing over-all productive capacity as a condition of the toll right. Sometimes the grantee must make a stipulated, substantial, minimum dollar investment in improvements.[105] Sometimes he must complete his works within a stated period.[106] An occasional franchise spelled out that more was sought than merely service to the immediate users. The object was to raise over-all operations to a more efficient plane: toll might not be collected "unless the expense of driving said logs or lumber shall be reduced by the use of said dams"; the improvements should be such as "shall seem most likely to cheapen the cost of running or driving and manufacturing said logs."[107]

Like the legislature, the judges had no trouble in deciding that it was proper to use law to increase the productive capacity of the economy. The Wisconsin Supreme Court took this ground in the first major issue which came to it involving legislative regulation of the use of streams. In 1849 the court sustained the constitutionality of delegating the power of eminent domain under the milldam act. The court could see no reason "why the legislature should not favor . . . [the] construction [of milldams], especially in a new country, among a scattered population and where capital is limited." "Such a law . . . by inviting capital into the interior of the state, by encouraging enterprise and diffusing the conveniences of social life, enhances the value of land, advances its settlement, and promotes general civilization."[108] In 1872 the court ruled that the legislature might validly require its license for works built in navigable waters, because such regulation was "in the interest of commerce and navigation." Moreover, to promote general use of the waterways the legislature had authority "to do anything within the banks of the stream itself which may be considered for the benefit and improvement of commerce and navigation" — an authority which "will be found to be a most extensive and absolute one."[109] When the industry moved into its full tide of production in the seventies, the judges had no hesitation in sanctioning the legislature's grant of large-scale river improvement franchises carrying toll rights. Such legislation was enacted "no doubt because there was a necessity for increased facilities to move the products of our forests to market." [110] "Has not the state power to promote, by artificial means, the navigability of streams, and render them more useful to the public?" [111] The legislature might validly authorize the building of flooding dams to furnish water for navigation, as well as dams to create water power: "The practice of building these flooding dams on all the rivers and smaller streams in the lumbering portions of the state has become of late universal, and their beneficial use to the logging interest is immense. It is expressly authorized by statute [Revised Statutes, 1878, section 1777, for example], and would seem to be almost a public necessity, it being the only way in which much of the pine timber can be utilized or made available . . ." [112] That a stream improvement altered the natural flow of the water, at some detriment to riparians, created no right of compensation in the affected landowners. The stream flow was a commonwealth resource; there was a legitimate exercise of commonwealth authority in attending to its maximum use. A forest-area trial court noted that the legislature authorized navigation improvements without compensating riparian owners, while it provided compensation for lands flowed by power dams. The statutory pattern "show[ed] that the legislature have always considered such streams as navigable highways, and the use of flooding dams as a lawful exercise of the right of navigation in them." The heart of the matter was to multiply productive use: "Because in its natural state only 100,000 feet could be floated on [the stream] . . . during a season, the state is not precluded from so using its waters as to make them carry 50,000,000 feet if it can." [113]

These justifications of legislative authority had their counterpart in the extended definition of navigable waters developed by the judges themselves, avowedly to serve the productive potential of the pinery, "in view of the vast interests depending upon the use of small streams for driving logs." [114] The public policy indicated in the grant of stream improvement franchises was essentially the same as that which impelled the broader definitions of navigability; in both instances the emphasis was upon increasing productive capacity. The dynamic of policy was brought to sharp definition in *Volk v. Eldred* (1868). Defendant had built a dam on a navigable stream under statutory franchise. Plaintiff sued for damage to his rafts from the alleged improper condition of defendant's works. Defendant argued that no legal harm had been done, because rafts like those of plaintiff could not have navigated the stream at all in its natural condition at the same stage of water as when these rafts were run. The court ruled that this was no defense: "It would be wanton, arbitrary and unreasonable to say that, in such a case, those navigating the river had not a right to avail themselves of the benefit of any improvement in the navigation occasioned by the dam. The legislature expressly required those maintaining the dam, to keep 'a good and sufficient slide, that will admit the passage of all such rafts, crafts and boats as may navigate the said river.' And there can be no doubt that this act had reference to the navigable condition of the river after the erection of the dam, and intended to require it to be kept in such condition as to admit the passage of rafts properly constructed in view of that condition."[115] Nor could there be doubt — so ran the implication of the opinion — that it was within legislative authority to lay such enlarged duties upon one who built works in the stream. This was a society avid for economic growth. Its legislators and its judges concurred that there was no more assured justification for using law than to enhance economic productivity.

THE PARTICULAR CONTENT OF STREAM FRANCHISE POLICY

Most of what we have just examined has concerned development of definitions of official power, including the power to delegate decision making to private persons. But a history of public policy must move from power to the uses of power. We need to consider both (1) the particular formal decisions which gave content to policy, and (2) the functional meaning of power and the uses of power, taken together in their impact upon the community. We shall postpone to the next section the matter of functional significance. Let us here take inventory of the specific formal uses which the law made of its authority over the streams. The basic operating policy was to delegate almost all detailed planning and execution to private actors. Thus it fits reality to organize our inventory mainly about the key items which defined the terms of this delegation. We should consider grants of title to act (including designations of delegates, location, and duration), aids to financing investment (notably grants of eminent domain power and of toll rights), authorizations to command the will of other persons (in toll rights, and rights of traffic control), the grant of corporate capacity. These four heads present the items which had most urgency with the men who sought franchises. To these heads we need to add two which reflect primarily the concerns of those on whom the franchise holders' activities impinged: the consideration exacted for the grant of franchises, and the extent of vested right assured the grantees. These various aspects of stream-use franchises we shall group and examine under the headings "title," "prerogative," "incorporation," and "tenure."

Title

The most conspicuous feature of the statutory franchises for stream use was that they were instruments of title. They established who might do what, and where, and for how long. These declarations came first in order among the provisions of such grants. This was their proper functional position. For they defined the framework which gave necessary form to all other powers delegated to franchise holders.

This primary character — as grants of title — inhered alike in special and in general legislation authorizing private development of stream uses. The obvious distinction of the general laws — those which gave improvement or booming monopolies to corporations which first took effective possession of waters for the stated objects — was that their authorization ran to persons of a broadly described class.[116] Even in this respect, however, the general franchise laws differed only in degree from special statutory franchises. The special franchises typically conferred title on a class, albeit a more narrowly defined class; a few special franchises ran simply to named individuals, but typically they ran to named grantees plus some category of successors in title — most often "associates or assigns." The Wisconsin Supreme Court spoke of such statutory franchises in terms which cast doubt on the need to designate successors. The court indicated that in the absence of contrary statute provision such a franchise was not limited in duration to the life of any named grantee or his associates. Thus it would appear that every franchise was inherently a grant to a class — to the designated grantee and those whom he might by conveyance or other regular form of transfer designate as his legitimate successors. In this light, explicit designations of "associates or assigns" were perhaps surplusage, inserted out of abundance of caution.[117]

That a primary function of the franchises was to serve as instruments of title was underlined by the court's ruling that a statutory franchise was invalid if it did not designate a class of grantees with at least that minimum of definiteness necessary to orderly development of the use of waters. Thus a grant was ineffective which purported to confer improvement and toll rights without limit of number and without stated order of preference upon anyone who in fact improved a stream.[118] This decision halted at its inception a new and undesirably loose trend in legislation. Measured against the total record, however, the case dealt with a unique instance; the 802 special statutory franchises all ran to named individuals or firms, and the durable general laws limited their grants to those who established prior occupation.[119]

The basic definitions of "what," "where," and "for how long" were so interwoven that they cannot be far separated; they constituted the operating titles conferred by statutory franchises. Treatment of these elements rested on common ground: The legislature considered that authority to alter or manage the natural flow of a stream was of such importance that there must typically be clear and separate grant of it, whether under a general or a special statute. A grant of such authority would not be implied merely because a corporation held a charter which authorized it to engage in stream improvements as part of its business. Prevailing legislative practice indicated that such language in a corporate charter merely defined corporate powers for the internal governance of the corporation. Where an operating franchise was intended, as against the public or third parties, the legislature typically made a distinct grant of it.[120] As we have seen, the interplay of judge-made law and legislative acquiescence created a qualification upon this general requirement of a distinct statutory title. The court was loath to rule that familiar common law principle was displaced by mere implication from legislative practice. At common law, incident to his land title, a riparian owner might build a boom or a dam if he did not materially obstruct public navigation. The legislature enacted that its license must be had for building works in any of the principal waters protected by the navigation guarantee in the Wisconsin constitution. Also, it provided statutory remedies against unlawful obstruction of navigable streams. In the milldam act it set terms under which riparian owners might build power dams on nonnavigable streams. Within this context, the court decided that legislative policy accepted continuance of the common law riparian rights, so far as these allowed building booms which merely gave access to the channel of navigation without impeding transport, or dams which stood in minor waters (navigable only by logs in freshet time) and which were so built as to allow reasonable passage of the limited traffic the water would bear.[121]

Location was quite specifically identified in special dam franchises. Franchises which authorized "improvement" of navigability — to which purpose dam-building authority was typically a named incident — conferred varying degrees of discretion in location of works. Almost always such a grant left the grantee some range of locational choice. Of 237 improvement franchises, 144 (60 per cent) authorized works wherever the grantee found them serviceable to improve a named stream or an extended stretch of stream; 93 grants were for improvement in areas of rather limited definition, but even here typically there was some room for choice.[122] Obviously the difference in specificity between franchises which simply authorized dams and those which authorized over-all improvement responded to practical needs. Likewise, however, the evident bias toward specifying location where this could be done emphasized the title character of the franchises.

The prevailing tendency was to state construction requirements in terms of standards rather than specifications. Rarely a franchise would spell out required dimensions for a slide or chute; typically it stipulated only for a "suitable" slide, or a dam constructed "so as to enable the driving of logs through the same at any stage of water on which logs could be run to the mouth of the river below said dam," or such construction as "shall not obstruct or hinder the safe navigation of said river for the running of logs or lumber." [123] A franchise authorizing general improvement of channel or flow might specify types of authorized alterations in the natural condition of the stream. But it did so always in the context of a broadly defined objective. Thus Private & Local Laws, 1866, Chapter 352, authorized the Keshena Improvement Company to improve the headwaters of the Wolf River in

Shawano County "by deepening the channel, blasting rocks, closing up side-cuts, building dams and wing dams, and side booms, and such other means as shall be necessary to the end proposed, so that loose saw-logs may be run the whole distance from one of said points to the other, in said waters, during the ordinary spring freshets."[124] The Wisconsin Supreme Court took note that the statutes typically gave general improvement powers in liberal terms. Hence, in case of doubt the court should give an extensive rather than a restrictive interpretation to the grant of such authority. So, authority to build "dams" included authority to build flooding dams.[125] And authority to improve navigation by "deepening, widening and straightening the channel, closing up chutes and side-cuts" included authority to close up a navigable slough, to increase flow through the main channel.[126] In significant contrast to the generally flexible handling of construction details, there was a marked tendency for statutes to stipulate a maximum allowed height, in feet, for a particular dam. A maximum height was stated in about two thirds of the forest-area dam franchises, and in about 54 per cent of those granted outside the forest region.[127] The height of a dam directly affected how much land the dam would flow; height thus entered closely into determining the "location" of the works; again we sense the title function of the franchise in the tendency to be specific as to authorized area of operations, so far as the nature of the works allowed.

Time is an important dimension of title. Choosing common law analogies to interests in land rather than to mere licenses, the Wisconsin court indicated that unless a stream franchise expressed a limit it was of indefinite duration.[128]

If the franchise was an inseparable part of a charter of incorporation, contemporary opinion regarded it as subject to the legislative power to amend or repeal declared in the Wisconsin constitution. Had anyone probed deeper, there might well have been doubt of this proposition. The Wisconsin constitution reserved to the legislature authority to alter or repeal any general law or special act by which a corporation was "formed" or "created." Where a corporate charter defined rights of stream use, the definition did double duty. So far as the definition pertained to the internal governance of the corporation — for example, defining business into which the management might venture without exceeding the powers delegated by the stockholders — clearly it was matter within the constitutionally reserved power of the legislature. But so far as the definition had effect to confer privileges in derogation of what would otherwise be public rights of navigation, its effect was other than to "form" or "create" corporate existence. If the legislature took away the life of a corporation by repealing its charter, or amended the charter to strike from the corporation's internal powers its authority to engage in the franchised activity, a stream use franchise contained in the charter might fall for want of a grantee. But the grant of a stream use right or privilege as such would not, it seemed, be subject to alteration or repeal by virtue of the constitution's reservation of legislative power over corporate existence. Doubt on this score may explain why the legislature occasionally included a reserved right to amend or repeal, both in special corporate charters containing stream use franchises and in separate statutes granting stream use franchises to corporations which already existed by virtue of other legislation.[129]

As a distinct item of property, whether held by a corporation or an individual, a franchise enjoyed the protection of the contract clause of the federal Constitution; under this shelter a franchise would hold good indefinitely, in the absence of a stated time limit or reservation of legislative power to amend or repeal contained in the franchise itself.[130] After Governor Rusk's intervention in 1882, prevailing practice was to insert a reserved power to amend or repeal in all franchises not deemed clearly subject to the constitutional reservation.[131] This aspect apart, the legislature wrote a term-of-years limit into 38 (about 14 per cent) of 237 general improvement franchises, and included such a limit in 15 (about 32 per cent) of 46 toll boom grants. So far as it went, the fixing of a term of years was associated with public-utility-type ventures; a time limit was stated in only 10 (about 3 per cent) of 332 forest-area private dam franchises (and in none of the private dam grants outside the forest region), and in only 2 (4 per cent) of the 58 private (nontoll) boom franchises. Fifteen years was the term most often set. Most of the time limits appear in franchises granted in the seventies and early eighties; their absence before then perhaps reflects the relative crudity of operations; their absence thereafter reflects probably the broader inclusion of the legislature's reserved power to amend or repeal, and perhaps also more familiarity with the need of stream improvements.[132]

The prime function of these title aspects of a franchise was to give authoritative dimensions to a delegation of power. To do so served interests of the public and third parties immediately affected by a grant. That the franchise must be cast into reasonably definite form — to say who might do what, where, and for how long — fostered legislative responsibility. This formality required defining at least the more tangible scope of the legislative decision.[133] Further, this formality provided a necessary frame of reference within which affected third parties could measure their own rights and duties and seek orderly adjustment of relations with the grantees.[134] The title aspects of a franchise also served needs of the grantees. Grantees wanted a sufficiently assured basis of expectations to encourage them to take the risks of investing in improvements and of committing themselves in business operations.[135] A franchise conferred a right to use a stream in the ways it defined, undisturbed by public or private prosecution. So long as the grantee held and conformed to a statutory franchise, he had a full defense either to an action to abate his works or to collect damages for their maintenance; freedom from disturbing lawsuits was an important title, beyond whatever right conventionally attached to his ownership of land.[136] By designating grantees and vesting power to transfer or to succeed to rights, as well as by specifying or at least allocating an area of operation, a statutory franchise promoted regularity in claims of power. Thus it helped men muster associates and capital and adjust to changing situations, with confidence that law sanctioned their arrangements.[137] In such varied respects the title aspects of franchises showed that by providing form the law could help create substance.

Prerogative

As a title, a franchise gave the holder the law's protection in the undisturbed use of a stream location for those purposes which the franchise stated. So long as he worked within the franchise terms, the holder was privileged as against other persons on whom his use weighed; a court would not order that his works be abated, nor enter judgment for damages, on account of the ordinary delays or ordinary physical harms which the presence of his dam or boom or other operations might cause to other men's use of the stream itself.

But in large measure men wanted more than such negative protection from a franchise. They sought the law's affirmative help in executing their enterprises. They wanted law to help them muster capital to build their works. They wanted law to empower them to charge for their services, and to enforce the charges. They wanted law to help them discipline the total traffic on a stretch of water, to realize the carrying capacity of the stream. They wanted, thus, (1) the power of eminent domain, (2) the authority to collect tolls, and (3) the authority to manage traffic. A degree of monopoly position inhered in each of these grants. Beyond this, sometimes the grantees wanted, also, (4) an explicit monopoly on tolls or traffic control.

These elements had a point in common, for each authorized the grantee to impose his will on others in a respect beyond what was incident to a simple title to use a stream. They were delegations of power to make rules binding other men's conduct, and as such they shared a quality of law. Backed by law's enforcement, they amounted to delegations of public power. They delegated public power to private hands, to fulfill functions deemed of substantial public interest. Such grants created a hybrid. The franchise holder might not shelter himself behind the state's immunity from suit; though the legislature delegated a public power for a public purpose, it did not intend that its franchise should give the project the status of an operation of government.[138] Yet, the legislature must have reasonable ground to find that a public interest would be served, else the delegation would be invalid.[139] Moreover, though the franchise holder was not a state officer, he might be deemed the agent of a state function, to the extent that in some matters he was answerable only to the state. Thus, as against private parties, the franchise holder might be sued for a public nuisance only by the public authority, unless a private complainant could prove not only the trespass to public rights but also some damage therefrom which was special to him. And if the complaint were, not that defendant was obstructing public waters with no color of right, but that he was violating the terms of a statutory franchise, the franchise would probably have effect to deny a private suitor an order for abatement and limit him to recovery of money damages.[140] So, also, the court held that the franchise of an extended-area river improvement company implied freedom from local tax assessments; the company should be deemed subject only to state taxes on its integrated operations, which otherwise might be fragmented by local levies.[141]

The nineteenth-century Wisconsin legislature was liberal in delegating eminent domain power to private enterprisers to develop transport and energy uses of streams. By this delegation the milldam act offered positive help to men desiring to build mills on nonnavigable streams.[142] The grant of eminent domain power stood equally with grants of toll rights and traffic control among the inducements which the general river improvement statute held out, to encourage corporations to improve navigation.[143] The legislature conferred the power of eminent domain in 354 (about 44 per cent) of the 795 special franchises it granted for stream use in the lumber era (1836–1910). Commonly the legislature incorporated by reference in its special franchises the eminent domain procedure under the milldam act, or that under the river improvement corporation act, or both.[144] It was liberal, too, in the range of uses for which it would give the eminent domain power — most often to obtain flowage rights incident to a power dam, but also to obtain land or rights in land needed in connection with building flooding dams, canals, booms, or generally improving a channel.[145] Its liberality extended alike to individuals and to corporations as delegates of the power. On the other hand, we must recall that, important as it was, a grant of eminent domain power was nevertheless of limited significance, for it related simply to the compensated use of private property. So far as a franchise applicant sought authority to affect the natural flow, or the uses of the natural flow, within the banks of a stream, the rights he would modify were those of the public rather than merely those of private owners. A grant of eminent domain power did not confer rights against the public use of commonwealth resources. Thus, to close a navigable slough in order to increase the flow in the main channel of a river, an applicant must obtain distinct statutory sanction apart from any right of eminent domain he held.[146] Conversely, however, the legislature need not require that its grantee compensate riparian owners when under statutory sanction he altered the natural flow of the stream without otherwise affecting riparian lands; there was no private property in the stream flow itself, hence no "taking," and no occasion to exercise eminent domain.[147]

There is no evidence in legislative records that the legislature refused the power of eminent domain to any substantial number of applicants who sought it. The reason still more special franchises did not give the right was probably that some applicants saw no need for it. This seems especially true of franchises for projects in the wild, thinly settled, forest region. At any rate, the grant of eminent domain power always appeared in a substantial percentage of franchises for dams in the more settled southern half of Wisconsin; the percentage of eminent domain grants was relatively lower among franchises for forest-area dams in the earlier years, and rose as the north country developed. So, too, the relative number of eminent domain grants in river improvement franchises was high in the fifties and sixties, fell sharply in the seventies and eighties, and rose high again after 1890; the lower percentage of such grants occurred in years when the lumber industry had moved its operations from the main rivers into the reaches of smaller streams, but before the industry had fully occupied the forest.[148] Though some applicants apparently did not feel they needed the eminent domain power, and though the legislature gave the power liberally to those who did ask, we must not think that men took it lightly. The power existed only where it was expressly conferred; this was regarded as too weighty a privilege to exist by mere implication from other terms of a franchise.[149] Once made, most grants of the eminent domain power stood unchanged through the years. However, occasional amendments redefined locations or projects in which the power might be used; if a question arose, it became evident that men took seriously the particular terms on which the power was given.[150]

The readiness with which the legislature thus bestowed the eminent domain power upon private enterprisers stands in sharp contrast to later policy. A more densely populated, more intricately organized twentieth-century Wisconsin generally confined the power of eminent domain to public agencies or to public utilities.[151] However, we must not exaggerate the range of nineteenth-century generosity. The more important nineteenth-century delegations were made to public utilities, such as the river improvement corporations authorized to charge toll and control traffic. Public utilities apart, the nineteenth-century legislature made significant delegation of the power of eminent domain only with regard to transport and energy uses of streams. In the twentieth century it would seem odd to confer this great sovereign power upon hundreds of individuals, to help them build local grist mills and sawmills; indeed, as we have noted, on second thought the practice struck the nineteenth-century Wisconsin judges as a dubious extension of "public" interest concepts, though they stood by their original validation of the milldam act and its counterparts.[152] Perhaps the practice seems peculiar only so long as we forget how much the nagging consciousness of scarcity of fluid capital figured in mid-nineteenth-century public and private economic decision making. Like the public lands, the streams were a kind of natural capital which men had in abundance. It seemed only common sense to make as extensive use as possible of this abundant (hence cheap) natural capital, to obtain maximum returns from investing more scarce (hence more costly) resources of manpower and money. The nineteenth-century community felt that there was a public interest in using law to put natural capital to work as much and as fast as the limits of money and manpower allowed. As it found confident expression in public land policy, so also this attitude took shape in nineteenth-century delegations of eminent domain power to increase the use of the streams.

The common desire for economic growth made the legislature liberally responsive to requests for toll rights as consideration for private improvement of navigation. The general statute authorizing works by river improvement corporations granted the right also to "charge and collect reasonable and uniform tolls upon all logs, lumber and timber driven or floated on" the improved waters.[153] Among the 802 special franchises for stream use granted between 1836 and 1940, 237 conferred toll rights or rights to exact shares of common expenses incident to improvements of flow by flooding dams and channel clearing or incident to general drives, and 46 allowed tolls for the operation of booms. Toll grants were made thus in a little more than 35 per cent of all special franchises. The bulk of these grants came in the years of greatest growth in the industry; 74 per cent of the improvement toll grants and 50 per cent of the boom toll grants were made between 1865 and 1890; 25 per cent of the improvement toll grants were made in the short span from 1880 to 1885 alone.[154]

A right to collect toll for navigating improved waters existed only by express grant of the legislature. The court would not imply the existence of such a right from other terms of a statutory franchise. Nor would the court acknowledge a claim in quasi contract, for benefit conferred by stream improvements; if the legislature had not declared that users should pay, a toll would not be created thus by indirection.[155] Given this insistence, toll provisions naturally reflected the applicants' awareness that they must establish a title. Franchises typically described with some generality the works or the channel alterations which the grantees might make. Toll provisions showed more striving for specific, limited statement. The grant always specified the commodities and services regarding which the grantee might charge toll; occasional amendments underlined this emphasis on the specific, as when the legislature added "timber or shingle bolts" to the "logs" which the original grant had alone made subject to charges.[156] Most grants set a specific money maximum upon the allowed charges, often with differing schedules according to the type of commodity, service, or area involved.[157] Toll grants were typically phrased to show that they were made in consideration of benefit in fact provided to those using the stream; some went beyond this, conditioning toll not only upon a general requirement that the grantee's works or operations should benefit navigation, but more specifically upon fulfillment of a stated minimum dollar investment in works of improvement.[158] Both general and special legislation on river improvement enterprises took some care to distinguish between authority to collect a toll — which might include a reasonable profit — and to collect merely a due proportion of "expenses" in situations where the parties were viewed as in a shared or co-operative venture rather than in the relation of public utility and customer.[159]

Many who moved logs to market were shoestring operators of little financial responsibility. The hazards of the business could press hard even on the working capital of more substantial operators.[160] These facts found reflection in the common provision which backed up a toll grant by giving the improver a lien. If the improver could not collect his toll otherwise, at least he might get it out of the logs. Unlike the toll right, a lien for some kinds of toll probably might be had in Wisconsin without statutory grant; where the improver took possession incident to his service, Wisconsin common law would probably recognize such a claim as falling within the familiar category of liens afforded those whose work on others' property enhanced its value.[161] Nonetheless, both general and special legislation commonly provided for a lien; the practice was so general as to indicate that to most improvers the assurance of a lien was of equal practical importance to the grant of a toll right.[162] Yet, many franchises were ambiguous as to the scope of the lien created. A common law specific lien would obtain only against goods which the lien claimant had in his possession, and probably only against the particular articles on which charges were owing.[163] Thus there might be distinctive benefits to be had from a carefully drawn statutory lien which reached beyond the common law remedy. Some lien provisions clearly showed such intention. However, many lien clauses were silent or ambiguous as to whether the improver might assert his lien against logs or lumber which had passed out of his possession, or against one lot of logs or lumber for charges due on account of another lot which belonged to the same owner.[164] Given the pointed interest which led men to include lien provisions at all, such uncertainty in statement is a commentary on the loose drafting which characterized much nineteenth-century legislation.

Rights of traffic management were an even more pervasive type of franchise prerogative than the grant of toll rights. To help another man manage the movement of his logs or lumber by water for a fee — whether by driving, or booming, or rafting — did not of necessity present questions of franchise. True, to "improve" a navigable stream by altering its natural condition or building works in it, a man must have a statutory franchise. But by ordinary private contract men might agree to drive, or boom, or raft other men's logs or lumber. By ordinary private contract owners of logs or lumber might arrange to have other men render such services. So long as they represented no more than dealings which the parties were in fact free to enter upon or not, as they chose, such agreements — as among the parties — presented only questions of contract law.[165] So long as the operations made only reasonable navigation use of the natural flow, affected third parties had no legal ground of complaint; if by design or by negligence an operator exceeded these limits, to the harm of a third person, an action would lie in tort as for any other intentional or negligent harm.[166] But no man needed a franchise from the state to enter such a transaction; a prime function of the law of private property, contract, and tort was precisely this, to validate a wide range of private initiative in economic decisions.

Questions of legal regulation might arise, and still not involve questions of franchise. Men entered some relations characterized by a high degree of practical dependence on one side and a marked superiority of practical power on the other. Here other values than those of ordinary contract, property, or tort law might, indeed, come in question. Here the law might impose its own definitions of acceptable qualifications and standards of performance upon the party who otherwise enjoyed widely disproportionate power in the relationship. This was a domain of "police power" — of the authority of law to adjust relations in order to enforce a fair accounting for the social income from, and the social costs of, private activity. A later chapter on police power examines these aspects of law relating to log or lumber transport.

However, we confront other distinctive issues of public policy where private enterprisers seek their advantage by relationships created not simply from private means and negotiations but dependent upon using public resources or social opportunities in ways which exclude or limit other men's access to these resources or opportunities, save on terms acceptable to the enterprisers. Because some preclusive use of public resources or social opportunities is involved, this style of activity does require that the enterpriser obtain a license from the government. This, then, is a domain of franchise.

In most uses which lumbermen made of streams, whether for transport or power, of necessity one man's use to some extent excluded another's, if only for a limited time or in a limited area. The most widely felt exclusion effect was by activity which limited other men's practical freedom in navigating. One who built a dam or boom necessarily precluded another man from building a dam or boom at the same spot or within the area of normal flowage or maneuver. But this effect was unlikely in itself to impinge on the affairs of many other men. And to the degree that a franchise authorized a dam or boom only at a particular location, as many franchises did, it limited monopoly by leaving other sites open for other grants. On the other hand, by its effects on flow a dam affected all who sought to use the stream for navigation both above and below it for some distance and so long as the dam stood. So, too, under certain circumstances one operator's extensive driving or booming operations might work for an indefinite time to limit all use which any other navigators could make of the stream. Thus a significant prerogative aspect of stream franchises was the delegation to particular grantees of some continuing authority to control traffic within a given area.[167] The delegation might be explicit; most often it was implicit. To delegate rights of traffic control meant inherently to confer some extent of monopoly power, consisting in the grantee's prior or exclusive right, for some time and within some range of affairs, to impose his terms of operation upon all who would use the stream for transport. The presence of a monopoly element affecting use of a common resource or the exercise of common market opportunities is enough in itself to require that we weigh the implicit or explicit grants of traffic control powers as an important aspect of franchise policy.

Wisconsin legislative and judicial practice accepted the legitimacy of several degrees of private control over stream traffic:

(1) Presumably of narrowest scope was the right of traffic control which existed only as a necessary corollary of grants for developing or using water power — under franchises for power dams, or for private or toll booms auxiliary to a particular sawmill. The legislature insisted that anyone who would build a dam on regularly navigable waters must have a statutory license; the Wisconsin court showed itself ready in finding constitutional, common law, and statutory protections for navigation; altogether, thus, if there was conflict between a substantial general navigation use and even a considerable power interest, nineteenth-century Wisconsin policy put navigation in a preferred position.[168] Most forest-region dam franchises expressed this concern for transport by requiring that the dam be built with a suitable slide or chute to permit passing logs or lumber "freely" or without "obstruction"; the minority of such franchises which omitted this guaranty were scattered and evidenced no calculated exception to the general rule.[169] Water transport was critical to forest industry; water transport was much less important than water power to the economy of southern Wisconsin. Over-all, 81 per cent of private dam franchises in the north country contained a slide requirement, compared with 48 per cent in the south.[170] The contrast underlines the navigation preference implicit in the pattern of north-country dam franchises. Similarly, boom franchises (all in the forest area) typically required that the boom be built leaving an open channel, or with a convenient exit, freely available to all who did not wish that the boom service their logs or lumber.[171] In a few general statutes the legislature also imposed slide or exit requirements for all dams or booms built on particular rivers.[172]

Even with such navigation guaranties, however, to authorize a dam or a boom meant, inherently, to authorize the grantee to exercise some control over the quantity and timing of the natural flow available for transport, or to impose some limits on other men's freedom of maneuver in navigating.[173] Most power dam franchises contained no navigation guaranty except the slide or chute requirement; most private or toll boom franchises stipulated only for providing a free exit point; conspicuously lacking in most grants was any declared restriction upon the dam owner's manipulation of the flow, or any requirement that the boom owner at his own expense move out of his boom logs belonging to another who did not want them held there. The implication that a dam grant inherently gave the mill man substantial rights of managing flow is thrown into sharper relief by the contrasting policy stated in some dam franchises. About 20 per cent of the private dam grants added to a slide or chute requirement the stipulation that the dam be provided with flood gates, and that, in season, when logs were ready for passage, the dam owner must open his gates and, without charge, provide sufficient flow to carry a drive or rafts safely through his works and well downstream.[174] A few dam franchises went further, requiring the dam owner to supply at his expense such labor as might be necessary to assure safe, prompt passage of logs or lumber.[175] Franchises for private or toll booms auxiliary to particular sawmills sometimes laid analogous obligations on the grantees. A common formula stipulated that the grantees "shall not detain the logs or lumber of other persons [in grantees' boom] longer than a sufficient and reasonable time to sort out their own logs or lumber from the remainder of the drive," usually adding as a sanction that if the grantees failed so to act, the log owners might move out their logs at the grantees' expense.[176] Such a provision acknowledged public rights of navigation, but at the same time bore implicit recognition that the boom franchise allowed the grantee some substantial right to manage all traffic in reasonable pursuit of his own interests.

(2) Many forest-region dam franchises showed by their phrasing that they were granted primarily or in substantial part to promote navigation. Some boom franchises showed by their terms that they authorized a public-utility-style facility in aid of general transport, not merely auxiliary to a particular mill. Franchises which embodied this positive purpose of promoting navigation thereby carried their own special implications for the management of traffic. The legislature implicitly conferred substantial rights of traffic control where it authorized building or operating flooding dams, clearing or straightening channels, or building and operating booms, in terms broad enough to allow the grantee to occupy or alter the whole breadth of the stream suitable for navigation.[177] There was particularly strong implication that a franchise granted exclusive right to make improvements, and the right to make the controlling decisions in managing stream flow, where the franchise authorized the grantee to build flooding dams, or otherwise improve the channel, at as many locations as he found necessary over a whole river or over an extended stretch of stream.[178] A fran-

chise conferred by implication a particularly valuable right to set terms for all traffic when it authorized a toll for benefit conferred by works which might occupy or affect the whole usable channel.[179] Such a franchise in effect said that all who would pass must pay, because in such a situation all who would pass must of necessity take the benefit of the improved waters.[180]

The implied scope of all such grants is clearer by comparison with franchises in which the legislature expressly stipulated that the grantee's improvements should not pre-empt all usable channels, but that the grantee should maintain some natural channel undisturbed, or supply an alternative channel, for use without charge.[181] However, if a franchise authorized improvements which might occupy the whole stream, the mere fact that it expressly stated some limits on the grantee's rights to control traffic did not imply that the grantee might not still affect all traffic by such management of flow as the approved works in fact made possible. Such franchises, for example, often included a general stipulation that the works should not in any manner "obstruct," "impede," or "hinder" navigation.[182] In context, such qualifications seem only to make clear that the grantee might not wholly deny the stream to anyone, and that his "improvements" must in fact reasonably improve navigation. This limited scope of such general navigation guaranties became clearer where the stipulation was that the grantee should not "unreasonably or materially" impede navigation, or where a franchise which contained such a general navigation guaranty also authorized the grantee to charge toll of all who had the benefit of the grantee's improvements, where the fact was that all who used the stream must of necessity thereby accept the benefit.[183] Some franchises declared that no toll might be charged for benefits derived from the authorized improvements.[184] Some franchises, we shall see, granted specifically limited rights to manage the movement of other men's logs, whether the owners consented or not — for example, to drive and charge for driving logs which became mingled with the grantee's general drive.[185] Such provisions set plain enough limits: the grantee might not charge toll generally, or he might not take over the driving of logs which did not mingle with his. But the highly specific character of such limited provisions left little basis for extending their effect by implication. Moreover, there was no necessary link of function between charging tolls or driving another's logs, and managing stream flow; that a franchise expressly limited the grantee in charging toll or in driving another man's logs thus carried no necessary inference from function that it also limited him in exercising the control of stream flow which his works of improvement made possible.[186]

Usually the facts did not require that all who passed must accept the safekeeping, the sorting, or the raft-assembly benefits of a boom, whether they wished to or not.[187] Even if a boom occupied the whole channel, at the cost of some delay and extra handling a log owner was usually able as a matter of fact to drive his logs through or past the boom if he chose.[188] Franchises for booms authorized as facilities of general log transport (not merely auxiliary to a particular mill) sometimes contained a navigation guaranty analogous to the slide requirement in power dam franchises; the boom must be built or managed to allow free exit for logs or rafts of those who did not desire the boom's services.[189] However, the inclusion of this guaranty implicitly acknowledged that the boom grant inherently conferred on the boom owner some right to manage traffic as a reasonable incident of his boom operation.[190] Some public (toll) boom franchises stipulated that log owners who wished to pass their logs through must do so only after due prior notice and only through the regular exit point provided, or at least required that the log owner pass out his logs without damaging the boom.[191] Thus the franchise allowed the boom owner to impose on all passing traffic the costs of delay and labor entailed in dealing with the conditions the boom created. There was another analogy to the pattern of power dam franchises; a few public (general toll) boom grants expressly required that the boom owner not only provide an exit, but at his cost supply labor to move out of the boom logs of those not seeking its services.[192]

Over-all, these navigation improvement franchises indicated a pattern of policy. Legislation developed some well-defined styles of limited franchises. Though in the minority, there were a substantial number of these limited grants. The limiting language fast settled into standardized formulas. In this context the silences in navigation improvement grants regarding public rights of free passage took on particular meaning where the authorized improvements might occupy the whole usable span of the stream at some point or over some area. The judges accepted the implications of this legislative pattern. True, the Wisconsin court had adopted the doctrine which Taney announced in the Charles River Bridge case; the court should interpret ambiguous terms of a special statutory franchise so as not to enlarge what the grantees took in derogation of public right.[193] But legislative silence did not necessarily create ambiguity. The Wisconsin court thought that here the facts supplied clarity which might seem lacking in the statutory words, if taken alone. If the legislature said nothing of public rights of free passage when it authorized works which might occupy the whole span of a stream, then by necessary implication of fact it gave the grantee authority to impose such terms as the franchise otherwise authorized upon everyone who might pass. In such a case, all who would use the stream for transport must use it subject to such management of the flow as the grantee reasonably determined appropriate to proper operation of the works; if toll was authorized, it fell on all who in fact used the improvement, and no less so though the situation left them no option. Such, the court ruled, was the legislature's intent. Moreover, such a grant did not infringe upon constitutional guaranties of public navigation. The purpose of navigation improvement validated the franchise. The legislature had power to make reasonable choices of means to achieve improvement. The court could not say that, merely because a franchise gave the grantee some control or claim upon all traffic that passed, the legislature's choice of means was thereby shown to be unreasonable.[194] Once again we feel the imperative pressure exerted on public policy by the value which contemporary opinion put upon increasing total productivity.

(3) The legislature granted some franchises which expressly conferred exclusive (i.e., monopoly) rights to provide navigation improvements, or to render booming or driving service, over entire streams or substantial parts of streams, and to charge therefor. Broadest of these grants was section 1777 of the Revised Statutes of 1878.[195] Section 1777 offered its benefits to any corporation, formed under the general business corporations act for improving any stream and driving logs, "which shall have taken prior possession of such stream for that purpose." The Wisconsin Supreme Court construed this language to mean that such a pre-emptor would acquire statutory rights which it would hold as its exclusive prerogative on the whole stream.[196] So read, section 1777 gave the pre-empting corporation three kinds of monopoly: (a) The corporation had "power to improve such stream and its tributaries, by clearing and straightening the channels thereof, closing sloughs, erecting sluiceways, booms of all kinds, side, rolling and flooding dams, or otherwise, if necessary"—and, inferentially, to employ such of these works as appropriate to make reasonable management of the flow. The monopoly here was over works construction and management but carried no absolute right to exclude traffic, for the statute added that the grantee "shall in no case, in any manner, materially obstruct or impede navigation upon such stream, or erect any dam or other obstruction below the head of steamboat navigation, or obstruct any navigable slough, except with the written consent of the owners of the entire shores on both sides thereof."[197] (b) For improving a stream and maintaining and operating its works so "as to render driving logs to the mouth thereof reasonably practicable and certain," such a corporation might charge reasonable tolls "upon all logs, lumber and timber driven or floated on the same . . ." Thus, though the corporation might not insist on taking all transport operations into its own hands, it might yet collect a fee from all who transported logs, lumber, or timber on the waters it had improved.[198] (c) Section 1777 gave the corporation a limited right to take over transport operations. The statute authorized the corporation to "take possession of all logs put into such stream or upon rollways so as to impede the drive, when the owners thereof, or their agents, shall not have come upon the stream adequately provided with men, teams and tools for breaking rollways and driving

such logs in season for making a thorough drive down such stream, without hindering the main drives; and . . . [to] drive the same down and out of such stream . . ." For its services in handling such ill-attended logs, the statute authorized the corporation to charge "the actual costs and expenses of so driving the same." Section 1777 thus distinguished between "tolls" (authorized against "all logs . . . driven or floated on" the improved waters) and "costs and expenses" (chargeable when the grantee itself drove untended logs). The distinction made clear that the legislature intended to give a monopoly on making and charging for improvements of channel or flow, but not on driving logs or navigating rafts.[199]

By Laws, 1880, Chapter 279, the legislature added provisions which both clarified and somewhat enlarged its general monopoly policy. It now made explicit its intent that a qualifying corporation should have "the exclusive power to improve." The new provision provided, apparently, for limited-area as well as whole-stream improvement monopolies; it conferred the exclusive right on "such stream or portion thereof" as the grantee occupied.[200] Like the original statute, the 1880 act continued to list, among the works the grantee might provide, "booms of all kinds . . . necessary or suitable" for the improvement. Unlike the original statute, the 1880 act authorized the grantee to make "boom charges."[201] The new act enlarged the original guaranty of navigation; continuing the stipulation that no grantee should "in any manner materially obstruct or impede navigation," it added that the grantee must "leave a free, open and unobstructed passage way, at least eighty feet in width, on at least one side of such stream or navigable slough."

Obviously all was not clear about the relation of the 1880 additions to the original section 1777, which continued unchanged on the books. No case afforded the Wisconsin Supreme Court opportunity to define the whole pattern made by the two general river improvement corporation statutes. In addition to authorizing limited-area improvement monopolies, the 1880 statute authorized, at the least, one kind of boom monopoly. The continued and enlarged navigation guaranty in the 1880 act did not mean that a corporation qualifying under that statute thereby obtained the right to require that everyone driving logs on the river must use its booms and pay for that use. But the "exclusive power to improve . . . by constructing all such booms of all kinds as may be necessary or suitable for the purposes of" the grantee's improvement scheme, coupled with the new authorization to make "boom charges," gave the grantee the right to assert a monopoly on rendering boom service for a fee to all who wanted boom service. To this extent the legislature made general provision for private monopolies in one phase of log handling (booming for hire), as well as in improvements of channel or flow.[202]

In a few special statutory franchises the legislature created monopolies like those given by this general legislation, and in some cases it created broader ones. The greater generosity of some special grants lay in requiring that all who would use a stream to move logs or lumber must commit to the grantee the actual handling of their goods in some or all of the phases of transport, and pay him therefor. In style analogous to the general legislation, some special grants empowered their grantees to assert exclusive rights to improve the channel and the flow, and to levy toll therefor upon all log or lumber traffic that passed.[203] Some granted the exclusive right to offer boom service for a fee in a given area, though not requiring anyone to use the service.[204] Beyond these grants, however, a few special franchises gave their grantees the exclusive right to drive or boom all logs that moved on a given stream, and to charge tolls for their services.[205] Some franchises conferred a more qualified, yet still drastic, driving monopoly by providing that any person or group which first came equipped to make an effective general drive should be entitled to take possession of all logs put into the stream, to manage them as part of the general drive, and to collect reasonable fees for doing so.[206] In the early years of the industry the legislature granted a number of co-operative driving monopolies; some such grants vested in a particular association or joint stock company the exclusive right to drive all logs, or at least all logs of members, on a particular stream, stipulated that anyone interested in logs or lumber on the stream might join the association on due application, and provided for fair sharing of expenses rather than for tolls.[207] Some special franchises, both of the co-operative and the private stock type, gave a limited monopoly to drive all logs which were not properly provided for or whose owners allowed them to mingle with the grantee's general drive, and to collect a share of expenses or a toll therefor.[208] Boom franchises — generally those carrying toll rights — often allowed the grantee to hold unclaimed logs, saw them into lumber, and sell enough to realize boomage or sawing charges, or both.[209]

Taking all these types together, the legislature did not grant a great number of special franchises which conferred explicit monopolies of improvement, operations, or tolls. However, it granted such franchises on some of the major rivers, and some of its grantees developed important operations; the Wisconsin River Improvement Company, the Black River Improvement and Log Driving Company, and the Wolf River Boom Company were notable examples.[210] Thus the special franchises have significance for the whole pattern of public policy.

General and special legislation on waterways monopolies was sufficiently limited in quantity and cautious in terms to indicate that contemporary opinion would not accept the granting of monopoly powers over many types of business. So, too, the Wisconsin Supreme Court showed a restrictive attitude toward recognizing basic monopoly claims. It did not refuse to recognize any but an express grant, but the court would recognize an implied monopoly right only if the claimant showed that the right must exist, by sheer compulsion of fact, within the framework of what the legislature had clearly allowed him. The court would not read a boom franchise as a monopoly where the statute did not say it was so and where others could in fact maintain their booms without physical interference to the claimant's boom.[211] However, where the legislature had made express grant of a monopoly right, the court was ungrudging in protecting it. Indeed, so long as a statute established a clear-cut core of explicit monopoly right, the court showed a generosity especially striking in view of its announced regard for the strict construction policy of the Charles River Bridge case, in recognizing by implication added privileges or protections it deemed necessary to effectuate the explicit monopoly.[212] But the claimant of statutory rights or protections must show clearly that he satisfied such conditions as the statute put upon using the monopoly. So, for example, one who sought fees for managing another's allegedly untended logs labored under a strict burden to prove that the owner had made no adequate provision for their transport.[213]

On the whole, the legislature and the court bore witness to the fact that contemporary opinion did not unalterably oppose transport monopolies created by law. Obviously there was prevailing acceptance of the use of law to foster significant centers of private economic power. A grant of monopoly rights did not provoke immediate controversy, provided the grant might promote private investment of money and skillful organization in key activities calculated to enlarge the general volume of business. This attitude was part of the prevailing enthusiasm for productivity; express grants of monopolies on the streams fall into pattern with contemporary readiness to give law's favors to railroad promoters. However, as with the railroads, though in lesser degree with river improvements, distrust and controversy tended to succeed the first confident response to promoters' petitions. But this aspect of the subject will be treated later in our discussion of police power.

Incorporation

Existence as a corporation lay in the gift of the legislature, and, indeed, could be enjoyed only by authority of statute.[214] Before a constitutional amendment in 1871 prohibited the legislature "from enacting any special . . . laws . . . for granting corporate powers or privileges, except to cities," in sixty-one instances the legislature combined in the same special statute (1) a charter of incorporation for a particular boom or navigation improvement company, and (2) a grant of authority to affect public rights of navigation by works of improvement or by conducting defined types of driving or booming operations.[215] After 1871 the legislature made available to corporations which existed under the general business incorporation law the river improvement monopolies authorized by Revised Statutes, 1878, section 1777, and the booming monopolies authorized after 1880 by sections 1777a or 1777e. To particular corporations organized for manufacturing purposes under special acts before 1871, and under the general law thereafter for manufacturing or for navigation purposes, the legislature also made fifty-four special-act grants conferring stream-use franchises (mostly for dams).[216]

Neither in law nor in fact was a charter of incorporation required for the effective grant of special rights of using navigable waters. That this was generally understood and taken for granted by contemporary opinion in the legislature and in the industry is dramatically clear in the record of special legislation. Of 802 special statutory franchises for stream use, only 115 (about 14 per cent) ran in the first instance to corporations. Even in the categories of franchises in which special grants to corporations were most numerous, the bulk of grants ran in the first instance to individuals; about 44 per cent (20 out of 46) of the toll boom franchises given by special statutes ran to corporations, while corporations received about 22 per cent (51 out of 237) of those given for toll improvement projects.[217]

Though it was clear that the grantee need not be a corporation, there was some confusion manifest in nineteenth-century doctrine as to whether a franchise for stream use did not itself make the recipients a corporation. Though they contained no grant of corporate being as such, and conferred none of the conventional powers or privileges of corporate bodies, some franchises spoke of the individuals to whom they ran as "the corporation." [218] Confusion derived probably from the fact that both corporate existence and special rights of stream use were forms of franchise; from this common factor some men apparently drew the illogical conclusion that any franchise represented a grant of corporate status. The confusion might have bred considerable trouble in defining the legal responsibilities of holders of stream franchises. However, there is no evidence that the mistake had much practical consequence. Probably this outcome reflects the fact that most works created under stream franchises were small and local, as well as that most such operations went on in wild or sparsely settled country and in an atmosphere of improvisation where men's relations were not so tightly knit as to require close definitions of legal position. The one respect in which the confusion produced practical consequences concerned the reach of the 1871 constitutional amendment, which prohibited the legislature from granting "corporate powers or privileges" by special act. Trouble here originated in the governor's office. From failure to realize that stream use rights and corporate existence were distinct types of franchise, Governor Ludington in 1877 vetoed the grant of a dam franchise to named individuals and their "successors" because "the use of the word 'successors,' makes of the bill such an act of incorporation as to bring it within the constitutional prohibitions against granting corporate powers and privileges by special or private law."

In 1882 Governor Rusk vetoed a dam franchise given by special act to an existing corporation incorporated since 1871 under a general act, arguing that any power or privilege conferred upon such a corporation fell within the ban on special laws.[219] In the seventies the Wisconsin Supreme Court showed some confusion on the matter.[220] Finally, at the end of the lumber era (1909), the court explicitly ruled that a special dam franchise granted to an existing corporation did not violate the constitutional limitation, which, the court said, forbade only special laws giving rights "essential to . . . corporate existence."[221] In any event, though they are part of the story of stream use franchises, these questions pertained rather to Wisconsin policy toward the use of the corporate device than to the state's policy on use of navigable waters.

Likewise, if we ask why some men wanted to use the corporate form for enterprises involving stream use, the answers derive more from the utility of the corporate device in general than from purposes or techniques peculiar to use of waterways. Some franchises conferred corporate existence in terms so terse and limited, sometimes almost casual, as to suggest that the grantees put no great weight upon corporate status, but in matter-of-fact fashion threw it in as a handy instrument. Thus Private & Local Laws, 1855, Chapter 120, spent only sixty words in creating a corporation, declaring simply "that David Jones and his associates and successors, be and they hereby are constituted a body corporate by the name of the 'Oconto Boom Company,' and may so continue for the space of twenty years, and by that name may sue and be sued, make by-laws not inconsistent with the laws of this state for the management of their corporate concerns, and have and enjoy all the rights of similar corporations." Chapter 405 of the Private & Local Laws, 1856, declared no more than that "John H. Knapp, Oliver Gilbert, Alexander O'Neil, George Warren, William Carson, H. T. Allen, Jas. Reid, George Randall, Adin Randall, W. J. Gibson, C. C. Washburn, their associates and assigns, are hereby constituted a body corporate, by and under the name of the 'Chippewa River Improvement Company,' and under such name may have full power to dam all bayous and sloughs that make out of the Chippewa river, and to confine said river to its proper channel."

These short-spoken grants are concentrated in the fifties and sixties, and they perhaps reflect only the simple conditions of a frontier business community.[222] However, the more complex corporate charters do not reveal much that seems special to the situation of the lumbermen's stream improvements. A few river improvement programs involved substantial investments; thus, by the middle 1880's the Chippewa River Improvement and Log Driving Company claimed total improvement investments of $337,000; the Daniel Shaw Lumber Company, a firm of substantial yet middling size, spent about $14,000 over several years on the Thornapple River, and in association with others invested about $35,000 in the Half Moon Lake Canal Company, a Chippewa River booming operation.[223] Where large investment was required, the corporate form provided a useful means to mobilize capital. However, navigation improvement altogether involved only a minor part of the total investment in the lumber industry, and most particular projects were small and local in reach. Taken over-all, the finance of navigation improvement was not calculated to create a major motive for incorporation.[224] As with corporate charters in other fields, so charters for stream improvements showed a mixed pattern of attention to the liability of stockholders for corporation debts. Limited liability was not the invariable rule; a few stream improvement charters stipulated some liability of stockholders for the firm's debts. In any case, the record is almost wholly wanting in evidence that limited liability was a point of lively concern to stream improvers.[225] Continuity of the enterprise seems to have been a consideration of weight, reflected in marked attention to terms of membership and to stipulations of the number of years for which the firm might exist.[226] The creation of a legitimated means of making and enforcing decisions, to order relations among associated or interacting interests, was probably the value of incorporation that was most distinctive to waterways operations—though, of course, this was a consideration familiar wherever the corporation figured in large and complex affairs. It was not by chance that the greatest relative use of the corporate form was for toll booms (44 per cent of the whole number, as we have already noted), and for navigation improvement enterprises carrying toll rights (22 per cent of the whole number). These operations by nature required some management of the general flow of traffic. They were operations which of necessity involved power to fix timing of flow and priorities of movement, and to enforce association of effort or combination of logs or lumber of different owners for driving or rafting. A large toll boom was a kind of freight yard or terminal warehouse, which could function only under a definite, centralized authority.[227] Especially where an improvement franchise authorized altering channels and operating flood dams over an extended stretch of water (perhaps over the whole length of a river of large carrying potential), the operation called for focused and summary decision-making capacity which the corporate form best provided.[228] Both by their reach and by their limits, the navigation guaranty clauses written into stream use franchises of these types reflect the practical problems of traffic management which would invite use of the corporate frame for operations.[229] Controversy swirled for years between Chippewa Valley millmen and independent log dealers, on the one hand, and the Mississippi River Logging Company, on the other, over the latter's operation of the great Beef Slough boom for rafting logs at the Chippewa's mouth for movement to Mississippi River mills. The Mississippi mill interests regimented their own forces and mobilized great business power through the Mississippi River Logging Company. Their success through that device in compelling the Chippewa millmen to negotiate peace dramatized the utility of the corporate form for imposing order on the whole traffic of a busy waterway.[230]

Tenure

Security of tenure entered also into the calculus of policy concerning the particular content of stream use franchises. Grantees, the public, immediately affected third parties (especially customers or competing users) all had reason to be concerned how much security of tenure attached to a franchise once vested, as against later efforts by law to take it away altogether or to limit or regulate its use.

The tenure that attached to a franchise had functional meaning both (1) for the private venture built upon the franchise and (2) for the social gains and costs of that venture in its unfolding relations to other activities. Thus, if we ask what social functions were served, this problem of franchise tenure becomes part of the histories both of title and of police power. (1) Consider, first, title. Implicitly, if not explicitly, tenure of some degree must be given, else a franchise conferred no title in any meaningful sense. Title to use resources or social opportunities means that law recognizes and protects rights in a private person, as titleholder, to initiate some range of decisions over those resources or opportunities — free of arbitrary official or private interference with this initiative, and free of private or official reprisal for, or appropriation of, the legitimated fruits of the sanctioned initiative. The law's assurance may range all the way from conferring a revocable license to vesting a fee simple absolute. Whatever the reach of its assurance, a prime function performed by law which enters into making a "title" is that of helping create a sense of reasonably assured expectations, encouraging commitment of private energies and means. True, title encourages private commitments by other means of assurance — by legitimating forms for defining, vesting, and transferring rights, and by identifying grantees, locations, powers, and privileges. But this commitment function of title also inherently involves some legally sanctioned tenure of rights. This element is present even if the assurance be no more than the minimum pledge that the franchise holder may safely act in certain ways without incurring legal liability, so long as his franchise has not expired or has not been revoked.[231] Because it inheres in the idea of title, the extent of tenure security conferred necessarily enters as a factor (whether realized or not) into decisions upon what types of title shall be deemed consistent with public policy. For tenure has costs; the tenure secured for one right may — as determined often by circumstances other than the law's own force — thwart or unsettle the practical effectiveness of other rights. This is the regulatory impact of creating title, as the promotion of private commitments is its affirmative, social-subsidy element.[232] Though this regulatory factor enters into making titles, it directs attention also to other problems of social values and to social functions other than those which focus mainly on issues of title. (2) Consider, then, franchise tenure as it poses an issue for the police power. We use law to foster a decent, workable society, because a healthy social environment is necessary to healthy individualism. Social well-being involves humane and functional adjustment of relations in time — both (a) in current cross section and (b) in the flux of events experienced in sequence. (a) At a given moment society depends upon effective interplay of a complex of existing relations. (b) Over a given span of years, society depends upon effective adaptation of men's relations to the implacable flow of change which the sequence of affairs brings. Law's competence to deal with the health of social relations either in current cross section or in the dimension of sequence is what we mean by police power. Both the total pattern at a given moment and the pattern as it shifts and blurs over a succession of events present phenomena too far-reaching and too varied for men to keep constantly in full view and comprehension. The title aspects of franchise deal with direct and more readily foreseeable consequences of the law's action in conferring powers or privileges. But the needs of a healthy society and the limits of men's understanding and energies require that law be able to adjust title to the broader pattern of relations as that pattern exists in time — in the context of a given period or in shifting design through a sequence of affairs. Implicitly or explicitly, thus, law must qualify such security of tenure as it attaches to franchises by reserving reasonable authority to limit, or regulate, or on occasion even take away altogether, rights claimed under franchise, in order to serve more extensive interests of social order or continuity.[233]

Wisconsin judge-made and statute law built some provisions for tenure into the inception of stream use franchises. If a statutory franchise stated no time limit, the Wisconsin Supreme Court indicated that the franchise probably should be interpreted to hold good indefinitely, or at least until the legislature took some positive action that withdrew it. We cannot be sure whether the court might have set limits to such doctrine, had cases arisen to provide concrete issues; such cases did not present themselves.[234] The legislature set a franchise life of a specified number of years in about 14 per cent of the river improvement franchises it granted by special act, and in about 32 per cent of the special franchises for toll booms; so far as it went, this policy was applied almost exclusively to public-utility-type grants, and then mostly in a few years of the seventies and eighties, before it had become settled legislative practice to include in each special franchise a reservation to the legislature of authority to amend or repeal what it had given.[235] No time limit was set upon the improvement or boom franchises which sections 1777 and 1777a of the Revised Statutes held out, from 1878 and 1880 on, to any corporation which took prior possession of a stream for the permitted purpose; absence of a time limit here probably reflects contemporary opinion that the constitutionally reserved power of the legislature over acts creating corporate powers or privileges applied to any separate grant of a franchise to a corporation existing by virtue of other law.[236] A franchise granted to an individual was protected by the contract clause of the federal Constitution against subsequent change or repeal, within the scope of what the grant contained.[237] A franchise to a corporation was regarded as subject in its inception to the power of repeal or amendment reserved to the legislature by the Wisconsin constitution. Following Governor Rusk's pressure in the 1880's, it became standard practice to include an analogous reservation in all franchises to individuals. The issues of franchise tenure thus became, for all franchise grantees, first, the definition of the formal extent of such reservations of legislative power, and second, the definition of the operating policies of official agencies in using the formal power.

Conceivably the state might ground its regulatory authority elsewhere than on these reservation clauses. No legislature had authority to give away fundamental police power; hence each franchise should be construed to include the implicit limitation that its terms be subject to reasonable exercise of police power.[238] Conceivably, too, issues might arise — whether under police power implicitly acknowledged in the grant, or under the reserved powers to amend or repeal — either out of need to make better adjustment of a franchise to the total pattern of social relations existing but not adequately appreciated when the grant was made, or out of need to adjust to changed circumstance or fresh experience in the passage of time after the grant. In practice the issues that did arise were dealt with almost entirely as questions of the scope and proper use of the reserved powers to amend or repeal, and they concerned values or tensions brought for decision because of pressures of time-born change rather than because of fresh perceptions upon existing relations. That these were the focuses of attention to franchise tenure seems not accidental. The fact reflects deep-running, prevailing attitudes on social values and organization. During most of the lumber industry years there was relatively little use of the police power to make better adjustment of existing patterns of social relations. There was common acceptance of what were taken to be the self-evident values of releasing creative energies through the market and through the law's related instruments of contract and property. This bias so predominated that there was little impetus to assessing social gains or costs that did not fall within a market calculus.[239] Only the spur of change produced by the sequence of events sufficed to bring into legal focus a few relational issues under waterways franchises. Since time was the spur, such issues as time developed were more readily defined in terms of the reserved powers clauses than of the general police power; the reservations of legislative power to amend or repeal offered the more obvious warrant of legislative capacity to respond to pressures generated out of the sequence of events.[240] There operated, too, the influence of timing in doctrinal growth. The contract clause and reserved powers had been woven into lawyers' thinking on franchises since the cases of Dartmouth College (1819) and the Charles River Bridge (1837). The bulk of police power doctrine waited for development upon the full flow of litigated challenges to state legislative power under the due process clause of the 14th Amendment, which did not set in until the main lines of Wisconsin lumber history had been fixed.[241]

The legislature and the governors, rather than the Wisconsin Supreme Court, made most of the policy concerning change of the legal rights and duties of holders of existing franchises. The judicial role was inherently limited in this realm. Subject to the constitutional guaranties of navigation — which the courts interpreted as general standards which left broad discretion to the legislature — the legislative branch was conceded the superior authority to license and regulate stream use. Such common law as developed might always be ousted by statute.[242] The legislature was relatively active in this domain of policy, mainly by granting franchises; thus the judges had limited scope for influence except as litigation gave them occasion to react to legislative initiative.

Little litigation reached the Wisconsin Supreme Court in which an issue was made over the impact of later legislation upon outstanding franchises. Through all the lumber era no case came to the court involving an issue under the contract clause concerning amendment or repeal of a forest-area stream-use franchise.[243] The scant record of litigation is consistent with the generally conservative attitude which the legislature and the governors took toward altering outstanding franchises. Most special franchises stood without legislative change through the lumber years, and the legislature made no restrictive alterations in the general authorizations for improvement or boom franchises under sections 1777 or 1777a of the Revised Statutes.

Nonetheless, there were explicit repeals or amendments of special franchises. These changes were relatively few, compared with the whole number of franchises, yet they were not trivial in absolute number — 45 repeals and 108 amending acts concerning the total of 673 special statutory franchises given for stream use in the forest region (about 7 per cent repealed, about 16 per cent amended).[244] There was enough repealing and amending activity to throw into sharp relief the relative absence of reported litigation over amendments or repeals. Out of 158 Wisconsin Supreme Court cases involving forest-area log and lumber transport or dams, two cases (growing out of the same legislation) presented an issue of implied repeal; two cases involved interpretation and application of amendments which put new burdens on existing franchises; and 15 cases involved amendments on which franchise holders relied to create or enforce their own claims. Thus out of 158 cases only four (less than 3 per cent) turned on issues of change in the law adverse to holders of existing franchises.[245] If we had an inventory of all trial court business, it is unlikely that it would show any material addition to this scant record of litigation.[246] Repeal or alteration of a franchise against the will of the holder would likely spell a clash of interests of sufficient consequence to cause either the holder or one who benefited from the change to appeal a lower court judgment upon the matter.[247] The inference which can be drawn from the scant amount of reported litigation is that most repeals or amendments of waterways franchises had the full assent of the franchise holders (often, as is apparent on the face of some provisions, being designed to adjust matters to the holders' better convenience or advantage), or at least represented negotiated settlements.[248] However, this state of affairs was not the product simply of the good will or compatibility of all affected parties. Though they did not speak on many occasions, the legislature, the governors, and the court concurred in surrounding the legislature's broad reserved powers of amendment or repeal with operating limitations against change imposed without the assent of the franchise holder.

Since the initiative of formal change lay in the legislative process, the working attitudes there displayed had primary importance for the practical security of franchise tenure. Two basic principles of practice stand out on the record. (1) By its liberality to one grantee the legislature did not typically intend to take away a specific privilege it had already given someone else. Legislative practice showed that, unless the contrary clearly appeared, a new grant was intended to fit into pattern with existing grants. Franchises sometimes expressly disclaimed intent to repeal prior grants on the same stream.[249] Quite often they expressly stipulated that the privileges now given might not be used in such fashion as to interfere with the exercise of privileges already conferred on others.[250] The legislature was generous in giving stream-use franchises. But its record shows that it intended typically that its generosity be at the expense of public rights as they would otherwise stand, and not at the expense of particular rights already bestowed upon particular persons. Legislative practice in this field of policy fully warranted the familiar rule of the courts that repeals by implication were not favored. In two cases the Wisconsin Supreme Court confronted claims that the franchise held out to qualifying river improvement corporations by section 1777 of the Revised Statutes impliedly repealed the franchise which an earlier special statute had given the Black River Improvement Company and which the Company had acted upon. The court found no difficulty in holding that the legislature intended no such repeal of the prior special grant by implication of the later general tender.[251] (2) Where a statutory stream-use franchise had formally vested in a particular holder, and the holder had made substantial money investment or had adjusted his continuing business affairs in reliance on the franchise, the considered doctrine of legislative committees and governors was that (a) the franchise should not be altered or divested save for cause,[252] (b) proved in a fair legislative or judicial hearing after the holder had notice and opportunity to be heard,[253] (c) with the remedy or penalty for misconduct tempered to the gravity of the misconduct, so that the franchise should not be wholly divested except for gross abuse.[254] These were limitations declared as a matter of wise legislative self-restraint. They were not treated as requirements of the contract clause or any other constitutional command. They were operating principles announced on the initiative of legislative committees and governors, without pressure or suggestion from judicial opinions. They were applied where the franchise was subject to some form of explicitly reserved legislative power of amendment or repeal; the question was not deemed one of authority, but of wisdom in using authority. These legislative and executive attitudes illuminate the fact that never through the lumber years did the Wisconsin court face an issue under the contract clause concerning a waterways franchise. So high was legislative regard, not perhaps for vested rights as such, but for vested rights whereon capital or business venture stood committed, as against particular challenges, that the potential issues were disposed of before they could come to court.

The Wisconsin Supreme Court had little occasion to declare policy affecting the security of tenure of franchises for stream use. What the court did showed value preferences similar to those displayed in the legislative process. The court refused to find an implied repeal of a prior special franchise in the later general improvements authorization given by section 1777. On the other hand, where the legislature enacted a revised special franchise which made no reference to the fifteen-year term which the original set upon the granted privileges, the court found an implied repeal of the original time limit and held that the revised franchise continued indefinitely.[255] The seeming difference among these cases in attitudes toward implying change in vested franchise rights resolves into a common concern to protect the beneficial position of the original grantee. In various other respects, under the milldam act and under special franchises authorizing power dams, the court favored the security of an established commitment of capital as against later particular challenges to the right of the particular claimant. These rulings were made mostly in situations that did not arise out of the lumber industry. However, they showed a temper of policy which helps us see the significance of the small amount of reported litigation over security of forest-area waterways grants.[256]

We would distort the record if we read from it nothing but zeal to secure vested franchises against any kind of legal change. Where men had committed capital or venture, the legislative, executive, and judicial policies we have just examined all favored security of vested franchises against later efforts at change directed at a particular franchise with the consequence of focused benefit to some particular adverse claimant. These were policies designed to protect an existing franchise against the impact of an adjudicatory (determinate or particularized) rather than a legislative (generalized) use of the legislature's authority. We must calculate within a different hierarchy of values when we consider the declaration of general policy, under the authority of law makers to provide for the integrity of social structure and the productivity of social processes.

Nineteenth-century Wisconsin policy was strongly Hamiltonian; this was a community keenly conscious of the problems of promoting over-all growth in a new economy, scarce of mobile capital; men were prepared to use law vigorously to obtain multiplier returns from the fixed (natural) capital which they had in abundance. So, we should not forget the constitutional guaranties of public interest in realizing the navigable capacity of streams. We should recall the bold use which the legislature made of its authority to pre-empt the further definition of public policy concerning terms of stream use franchises and general regulation of stream use to foster development of water power and traffic.[257] This framework of police power included authority to adjust men's relationships to growth in experience and change in circumstance.

Exercise of authority to make general adjustments to time's impress might come to bear upon particular existing franchise rights, as upon other products of men's past contriving. The legislature enacted various stream traffic regulations to control the activity of holders of existing franchises as well as of any other users of streams.[258] Governors sanctioned such legislative determinations, and indeed Governor Rusk objected to omitting reserved powers clauses in franchises given to individuals, because unqualified grants might interfere with future proper exercise of general police power.[259] The legislative process responded promptly, consistently, and without contest to the policy dramatized by Rusk's intervention. By their response, lobbyists and legislators alike acknowledged that the governor spoke a community consensus, the strength of which practical men must heed.

Circumstance gave the Wisconsin Supreme Court little opportunity to speak of these matters. We have noted that the court was jealous of particular legislative or executive alteration or repeal of a particular outstanding franchise for the benefit of a particular adverse interest. On the other hand, the court spoke firmly to maintain continuous vigor in the general police power. Thus the court repeatedly ruled that, by reasonable regulations to enhance general traffic or water-power development, the legislature might —without providing compensation—take away established common law riparian rights of using stream flow, or modify such public rights of navigation as obtained on a stream in its natural condition.[260] These rulings are particularly relevant to our present concern, because in every case the statute involved represented legislative adjustment to changed economic conditions — to a new scale of traffic impinging upon the navigation and power uses appropriate to a simpler state of industry.[261] True, none of these cases presented a general statute operating upon particular private rights derived from earlier legislation; no such case came to the court arising out of lumber industry operations. However, after some preliminary fumbling, the court had early made plain that it would dispose of such a clash of statutes in a fashion which would maintain the continuing vigor of police power. The milldam act set no term of years for the rights it conferred to flow land, nor on the other hand did it explicitly grant indefinite tenure. Within this framework, the court held that it would not construe the act as intended to confer a right to flow another's lands for a time beyond the existence of the statute; if the flowed landowner had not relinquished full title by accepting payment (or a contract for payment) for full value of the flowed acres, repeal of the milldam act terminated the dam owner's flowage rights. Delegation of the eminent domain power under the milldam act had legal justification only because it represented a reasonable legislative determination that such flowage of land helped achieve public benefit. Continuance of the statute must be regarded as "the continuous exercise of the sovereign power," determining that such flowage rights continued to serve public interest. Thus it inhered in the milldam act as an exercise of police power that the act "was repealable in its nature" upon legislative determination that the statute no longer served the general welfare.[262] This society sought constitutional order, which meant order continuously responsible. Continuing responsibility was opposed to irrevocable commitments which might prevent reasonable adjustment of values and procedures to growth of experience and change of circumstance. Such a position, we felt, was demanded by due regard for private as well as for public interest. Manifestly, our society wanted to keep its hands on its destiny.

THE OPERATIVE MEANING OF STREAM FRANCHISE POLICY

By nature legal history looks for meaning first in formal actions and in formal processes for producing authoritative action. This priority accords with a prime function of law, which is to help men make their experience significant in terms of their choosing; through law men impose form upon the unruly substance of life. Yet, if we would know the history of law in the life of a society, search cannot stop with formal decisions and formal procedures. Especially when we study the history of policy made within a constitutional legal order, we must attempt to measure the impact of law's formal operations relative to men's informal behavior (both within and outside of legal institutions), and also relative to the formal and informal works of other institutions with which law interacts. So far as the issue is the shaping of overt action — which is typically the purpose of using law — form has significance only as it affects substance, however else the situation may be in art, where the issue is to shape perception and feeling as an end in itself. Within a legal system which aspires after an ideal of constitutional order, law's formal operations must always be measured against law's substantial effects. For it is the constitutional ideal that no form of secular power, including law, should be treated as an end in itself; power has warrant only as it serves to enlarge the humane content of life in terms of a consensus richer in time and broader in reach than the views of any particular power holders.

Thus, as elsewhere in our lumber industry story, so here we should ask about the working significance of the large body of formally declared policy on stream-use franchises. Of course, it is never possible completely to separate definition of public policy from estimates of private and official behavior which environs, complements, or contradicts values and procedures formally declared in law. Courts recognize this fact when they receive evidence of trade custom to put content into the law's standards for construing contracts (what a reasonable man would understand as the intent, given the relations of the parties), or for fixing liability for personal injury unintentionally inflicted (whether defendant's conduct was that of a reasonable, prudent man), or when in interpreting a statute they give weight to the long-continued, uniform practice of the responsible administrators of the statute. Responding to the same pressures of reality, this chapter has already noted various respects in which unspoken official attitudes or the impress of general economic trends and specific lumber industry problems and practices entered into defining stream franchise policy. This reduces the detail in which we need examine other matters. In part by bringing earlier discussion into new focus and in part by fresh analysis, we should now consider the balance of economic and social interests struck by the law in this area, the practical impact of stream franchise law upon the life of the lumber industry, and the location of actual decision making in these stream-use matters, as between private interests and legal agencies, and as among legal agencies. Matched against so ambitious a catalogue of subject matter, the resources available for this study are more often than not quite unsatisfactory. Even so, the effort to treat such a range of matter can add some dimension in depth to what we read in statute books, governors' messages, and opinions of judges.

Adjustment of Interests

Inherent in the enactment of hundreds of statutory franchises and scores of more general acts — as also in the flow of court decisions at common law and under legislation — was a ranking of interests involved in men's use of the forest streams. This ranking was not accomplished in one stroke of creation. It was the product of experience in time, and hence it constituted an aspect of legal order which is peculiarly the business of legal history. The ranking never took form in a comprehensive constitutional or statutory code or in a masterful judicial synthesis. It was a product of fragmented expression and official practice, its content implicit in the cumulation of specific terms of statutes and judgments more than set down in ordered exposition — hence an aspect of legal order partly within but extending beyond the formal but incomplete pattern of public policy which this chapter has previously examined.

Hard problems, both of present judgment and of forecast, arose in creating a broad pattern of stream-use policy. The problems were no easier because they must be met by a lusty but crude new community, which confronted novel issues of growth without benefit of specifically relevant traditions for relating public and private action, which lacked experienced lawmakers, and for whose guidance there existed no professional learning in political economy. But there was sound common sense in the pragmatic approach to policy making, and there was good reason why only time and the experience of action could make a pattern of policy. Sound pragmatism, however, includes reflection upon time and particulars, to make men more aware of what goes on and of what consequences follow, in order thus to enlarge their reach of purpose and control. Over some sixty years of state legal activity concerning the forest streams there was so little generalization of policy that one suspects a bastard strain in the pragmatism that contented itself with churning out a flow of specific responses to specific pressures.

Nineteenth-century lawmaking was narrowly operational by choice, moved by compulsions of feeling and will derived from men's reaction to the experience of opening a new continent. Prevailing opinion was impatient with law making; men wanted to put their prime energies into the challenges they saw in the market. Both in market and in public policy, prevailing opinion was impatient with plans or decisions which looked very far ahead in time or very far abroad in current calculations; when there was obviously so much to do, and such frustrating scarcity of mobile capital with which to realize opportunities, men felt overwhelming impulsion toward improvisation, opportunism, and the quick return. The same temper which set the tone of dealings with public lands moved through dealings with public waters. Lawmakers surrendered to this temper of the times with enough awareness to be uneasy, yet in response to such deep undercurrents of attitude that their imagination could not or would not break free into bolder generalization of policy.

The situation finds a symbol in Governor Rusk's troubled objection to granting statutory dam franchises to individuals without reserving legislative power to amend or repeal. Rusk worried over the fragmentation of policy: "A large number of the lumbering streams in the state are being given over by this means to the control of individuals. This evasion of the constitutional amendment against special legislation may be in most instances necessary. The improvement may be for the public good, the tolls fixed may be fair and reasonable, but nearly all such measures affect the interests of many who know nothing of their pendency and are unheard as to their effect. In the nature of things it is impracticable for the legislature to make thorough and exhaustive investigation in each case, and to know the precise effect upon all interests of the measures asked for. Moreover, what may be an improvement in the situation of affairs to-day, may be very far from an improvement a few years hence; and what may be fair compensation for maintaining dams and other public improvements to-day, may not be fair or reasonable after the lapse of time." In effect, the governor was arguing the need of bolder legislative generalization of stream-use policy, implemented by a licensing procedure under administrative rule and surveillance. But his message showed no awareness of its implications. Of course, we should not expect that in 1882 Rusk would produce a full-blown twentieth-century model of legal control. Nonetheless, the limits of the governor's reach suggest how much we are prisoners of the familiar, and of unspoken assumptions. From his analysis Rusk did not even draw the moral that the time was overdue for legislative generalization of policy; his sole prescription was that the legislature include in each *ad hoc* disposition a reservation of authority to make *ad hoc* changes.[263]

Because the times did not favor exposition of policy, much of the law's ranking of interests lay implicit either in technical details of defining and implementing values or in the practical service of certain social functions or objectives. For example, without formal declaration the law nonetheless made a value choice between competing interests when it assigned to one claimant rather than another the burden of proving facts which must be established as a basis for an adjustment. Again, without formal declaration the law made value choices when, to promote creative release of men's wills, it treated some activity as lawful in the absence of positive regulation or prohibition, but required the law's explicit license as a condition for other action; a man might at his own discretion clear-cut timberland which he held in fee simple, but he must have a statutory franchise to build a dam on a stream that was suited to general navigation. Thus an account of interest-ranking must include: first, some catalogue of technical procedures by which interests were realized in law; second, some catalogue of general social functions or objectives served by specific legal devices; third — illuminated by these two modes of analysis — an inventory of the particular valuations made among interests involved in stream use. These are three ways of looking at the same result. However, we had better begin by cataloguing devices for realizing value judgments and by examining some practical social functions or objectives served by particular legal rules; often we can recognize the existence of value judgments only by the presence of specific devices or specific preferences among modes of operation which give body to values by conferring particular benefits or imposing particular burdens.

Devices for Realizing Value Judgments. A catalogue of specific means by which law defined or implemented policy has relevance beyond an inventory of value judgments. The history of public policy under a constitutional legal order must attempt to assess not only objectives declared in law, but also effects realized through law's use, for constitutionalism means that law finds warrant only as it serves life. Realization of policy depends upon effective means for defining and executing policy. Thus the array of devices employed in realizing interests involved in stream use in itself represents a significant product, as well as component, of the process of adjusting interests.

To say that law ranked interests by its specific definitions of public policy or by its specific techniques for implementing policy, means that law either, out of its own resources, created benefit or detriment for persons affected, or accepted and protected the fact of benefit or detriment generated directly by causes other than law. The distinction is not between completely distinct categories of law-made and fact-made benefit and detriment. The distinction applies to instances where law made its help or its force available to effect some change in the situation otherwise existing, compared with instances where the law made its force available to require men to accept the situation otherwise existing.[264] The distinction is not between benefit and detriment wholly law-made, or wholly fact-made; fact-made benefit and detriment were always involved. Wisconsin's law did not operate in a social vacuum, nor did it spontaneously generate the purposes or motives for which men used it. Use of law derived from experience of scarcity in the human satisfactions available out of men's relations to the world and to each other. Physical and psychological phenomena, and the operation of other-than-legal institutions, did not yield men free and unlimited fulfillment; life in society yielded benefits, but not to everyone in like measure, and only when someone paid costs. Because of scarcity in life experience, the incidence of fact-made benefit and detriment became a concern of legal order; this is why there is a wider range of contact, and a greater similarity of concepts, between law and economics than between law and any other branch of ordered learning. Men brought to law problems of maximizing life satisfactions and preventing, resolving, or repairing overt conflict, where there were not in fact limitless opportunities for exercising will, or for using material resources or developing productive relations, without allocation of opportunities, or without particular loss or limitation of access to opportunity for advantageous action or experience.[265]

One way to inventory specific devices by which law ranked stream-use interests is thus to catalogue (1) law-made benefits or detriments added to, or realigning, benefits and detriments inherent in facts of scarcity, and (2) legal doctrines or procedures which accepted or protected the incidence of benefits or detriments as these arose out of men's relations or activity apart from specific legal intervention.

(1) Law-made benefits and burdens we may profitably classify according to type of impact.

(a) The legislative process — broadly conceived — made its distinctive contributions to the ranking of stream-use interests partly by subsidies out of public resources (notably state lands), and mostly by defining standards and rules of conduct. Identified by function, this definition process went on at the hands of all officers authorized to declare general policy — not legislators alone, but also the governor, and the state supreme court.[266] The technique which gave greatest flexibility to definition as an instrument of policy was the manipulation of classifications or concepts. By enlarging or restricting the scope of such concepts as "property" or "navigability," lawmakers could favor one interest and subordinate another, in a fashion so quick and quiet, so economical of analysis, seeming so routinely logical in its application of accepted values, that unless the observer watched carefully the fact that a ranking of interests was being made escaped his notice.[267] It was a technique to which the rough-hewn pragmatism of nineteenth-century law making was especially hospitable, since it minimized the need of an explicit philosophy of decision.

(b) Quantitatively, most ranking of interests went on by executive and judicial action within the frame of defined general policy. If we identify a process by the function performed, we must recognize that the nineteenth-century state legislature shared with public prosecutors and courts a good deal of the executive and adjudicatory action that was taken. Likewise by the criterion of function, we must recognize that in large measure the execution of public policy was committed to the initiative of private suitors. Devices for particularizing public policy sought either (i) to prevent the incidence of certain fact-made benefits or burdens, or by advance arrangement to organize relationships to select, alter, or limit the incidence of such benefits or burdens, or (ii) to repair or readjust the accomplished incidence of such benefits or burdens. (i) The most sharply defined preventive or organizing device for translating public policy into particular expression was the license (including the situation-shaping effects of terms or conditions set in licenses, like the navigation guaranties in a dam franchise). Licenses were created in largest quantity by general or special statutes, but court injunctions (both in their mandatory and their negative aspects) might serve the same function.[268] Less obvious in their situation-structuring character, but of great practical significance to this end, were law's modes of enforcing payment for services performed: for general benefits, by levy and collection of taxes;

for particular benefits, by authorization and enforcement of tolls, creation and enforcement of rights in contract (and of contractual features of property titles, including creation and enforcement of liens), and creation and enforcement of rights of eminent domain.[269] (ii) The familiar reparative or readjustment devices were the judgment for civil damages, the penal judgment for a fine or a jail term, and the injunction to abate an existing condition.[270]

Preventive or situation-shaping devices and reparative devices together represent a category of law's operations distinct from processes of policy definition. True, in a sense the definitional and application processes represented simply successive phases of a common process of achieving legal order; devices of application never came into play except to implement some definitions of policy; policy often found further definition in patterns of application. But definition and application each may have its separate effect on behavior. Hence each may properly count as a distinct type of legal operation for ranking interests. Definition is entitled to separate status, because the processes of bringing interests over the threshold of awareness, and of identifying their relations — as congruent, in conflict, or indifferent — influence behavior at least by affecting how men perceive their situation, if not by appealing to whatever motives they may have for conforming their conduct to legal norms. Application is also entitled to separate status, because the use of licenses, judgments, liens, penalties, or injunctions depends not only on prior definitions of substantive policy but more directly upon the initiative and vigor of private claimants or public prosecutors, and the firm employment of power by those who administer or adjudicate under law. Definition typically reached further than application in impressing policy on behavior. This fact sometimes reflected deficiencies of legal order; implementation was often timid, weak, and uneven. But it was also true that in a healthy legal order only a portion of defined policy could or should find expression in application of sanctions. At most, law could command only a limited part of a society's resources. Social order was imperiled if there must be more than marginal use of law's force to give reality to defined policy. Measured both by the qualities and the defects of legal order, definition and application constitute separate categories of law's operation.

(c) Most subtle of the law-made benefits and detriments by which interests were ranked were allocations of the risks of law's own operations. This was an impact of law felt through all its branches. In all its branches, law was man-made and man-activated. Always, someone must bear the costs (in time and energy if not in money) of making law and putting it in motion; always, those who found law moving against their interests must submit, or lay out their own costs to satisfy it or turn it aside. Because the operations of law involved many wills and turned on infinite variations of often inscrutable fact and trend, there were always risks of failure to be added to the costs of proposing or opposing a bill in the legislature, or pressing or defending a lawsuit. (i) Risks inhered in the definition (legislative) process. The principal means by which law enlarged or reduced the risks men ran in calculating their substantive positions under law was by the choice it made between expressing public policy through standards or through rules. When the Wisconsin court declared that one committed an actionable nuisance who built a dam on any navigable waters without legislative authorization, it weighted the scales heavily in favor of transport over water-power development; the court indicated that any water capable of moving the produce of the country to market was navigable within this doctrine, and under this broad standard the heavier risk of illegal action thus fell on the dam builder, who must hazard his judgment whether a given stream was capable of bearing any traffic. Conversely, at a later time the water-power interest achieved a higher rank, by reduction of risks inherent in its subjection to such a broad, adverse doctrine; reflecting more complex economic growth in the north country, the court later ruled that a dam with reasonable passage facilities was not in itself an actionable nuisance, though it was built without legislative license, when it was situated on a stream navigable only for logs in spring freshet season. So law gave the security of a specific rule to one category of dam builders.[271] (ii) Risks inhered in the (executive and adjudicative) processes of applying public policy. Law might favor or disfavor a given interest by the procedures provided to bring policy to particular focus — notably by allocating burdens of proof, fixing rules of evidence, and devising presumptions for interpreting legally operative language. First, to allocate the burden of proof in establishing a legal position offered a ready device for distinguishing degrees or phases of favor among competing interests. The court favored transport over water power by ruling that any work built on a major navigable stream without legislative license was in itself a nuisance. But the court tempered this doctrine by ruling also that one whose legal position depended on the navigable character of a stream had the burden of proving the fact of navigability; there was no presumption that all streams were navigable.[272] Again, when the legislature favored a particular water-power development by granting a dam franchise, typically it qualified its favor by stipulating that the dam be built with suitable slides or flood gates, and be so managed as to allow logs or rafts to pass; yet the law qualified this protection to transport by putting on a complaining navigator the burden of proving that the dam builder had not fulfilled the passage guaranties.[273] Second, what law treated as competent and relevant evidence went far to determine in practice how readily, or with how much difficulty, a given interest could establish its position compared with that of another. For example, log drivers must use reasonable care that their drives did not cause unnecessary damage to shore lines; however, the law went a good way toward preferring the log transport to the riparian interest when it accepted evidence of loggers' practice to spell out what constituted due care.[274] Third, by various (often unspoken) presumptions applied in interpreting legally operative language, the courts assigned different weights to affected interests. The court favored navigation improvement by presuming that the navigation guaranties in the Northwest Ordinance and the Wisconsin constitution were not intended as rules which forbade any interference with the natural navigable capacity of waters, but as standards which allowed reasonable legislative discretion to promote increase of navigable capacity.[275] The court would not, by implication only, find a legislative intent that a franchise to develop water power should permit total exclusion of transport uses.[276] The court would not, by implication only, construe a private deed of a water-power site as intended to authorize such development as would reduce the total power potential of the area to the disadvantage of other sites in the grantor's disposal.[277]

(2) Law also in effect ranked interests when it accepted or protected the event as it occurred without law's direct intervention. This meant, in the first instance at least, letting benefit accrue where it would, or letting loss lie where it fell, as the product of causes the immediate impetus of which lay outside the force of law.[278] The deliberate withholding of law's force was not without relation to the positive use of law's force to create benefit or detriment. The potential of law's positive force was always in the background; if the law said that a man was privileged to maintain a dam, though its maintenance caused particular detriment to another, implicit in this ruling was the promise that law would protect the privileged actor if the man burdened reacted with private force or contrivance to negate the practical worth of the privilege.[279] The important privileges affecting use of the forest streams were to be free of legal liability either (a) for causing tangible detriment to another, or (b) for subjecting another to risk of detriment.

(a) Freedom from legal liability for completed tangible detriment was expressed in negative complements to affirmative rights of access to resources or to human relations, defined by the law of private property, the law of contract, and the law of common rights. Thus, by common law or under a statutory franchise an upstream riparian owner might build and operate a flood dam as an incident of his right to make reasonable use of the stream, without liability to a downstream power dam operator whose millsite was rendered the less valuable because of the consequent interruptions to the natural flow; there was no "taking" of the downstream man's "property," because the latter had no "ownership" in the flow.[280] Thus, by statute a log driver was privileged to enter a riparian's land to use reasonable effort to recover stranded logs; there was no "taking" of the landowner's "property," because all property was held subject to reasonable exercise of the police power, which was reasonably used to prevent waste of good timber.[281] So, if a milldam owner contracted for good consideration with flowed landowners to drain his millpond after a stated time, with the agreed right otherwise in the landowners to tear down his dam "without claiming any damages of them whatever," the flowed landowners might have law's help to assert their contract right; an agreement to forego potential water power was not contrary to public policy; by its eminent domain procedure the milldam act offered affirmative help to dam builders, but that statute only permitted or promoted, it did not command, power-site development.[282] So, one who built a dam, subject to legal obligation to provide means of passage, must accept the negative as well as the positive consequences of the law's conditions. For not providing proper passage facilities, he might become liable in damages to a log driver or raftsman; this was the positive, law-made burden which we discussed earlier. But, also, he thereby was guilty of such contributory negligence that he might not himself recover compensation but must submit, without legal remedy, to damage done his works by a log drive which the required slides would have accommodated.[283]

(b) Law in effect preferred some interests and subordinated others, by holding its hand while private actors initiated activity which might eventuate in detriment to other persons. Even if law-made penalties or burdens were thereafter laid on the actor where detriment followed his action, in tolerating interim risks to other interests law gave a measure of protection, if not encouragement, to that action which it would not stay at the outset. This was the negative complement of procedures in applying law-made benefits or burdens, the positive aspects of which we noted earlier. Where reparative sanctions were in question, the law thus ranked interests by allocating the burden of proof in tort. The rule that normally one who suffered detriment might not recover in tort unless he showed also that the defendant was by some measure at fault, was a legal means of protecting and encouraging initiative for action; so, as between several navigators, each might in the first instance use his own ingenuity to move his goods as best he could, becoming liable to another navigator only if the aggrieved man showed that the defendant had not acted within bounds of ordinary prudence.[284] Where preventive sanctions were in question, the law thus ranked interests by fixing its own measures of the timeliness of remedies. The court would not interpret a statutory power-dam franchise as an absolute right intended to authorize works which would bar navigation; on the other hand, the court would presume from such a grant a legislative determination that the authorized works might be built consistent with maintaining reasonable navigation opportunity, and hence would not enjoin their construction or maintenance until sufficient time had passed to obtain reliable experience of the actual effect which the works would have upon transport.[285]

So far the discussion has examined particular legal forms by which law ranked stream-use interests — licenses, judgments for damages, common law or statutory definitions of rights or privileges, for example. Because contemporary lawmakers left so many of their value judgments to be read between the lines, it is hard to perceive by any one scheme of analysis the full range of methods or occasions by which in effect they ranked values. Thus, though on the whole we look at the same substance, we may gain insight by aligning our materials according to still another criterion. Consider, then, how the ranking of interests is revealed when we inquire what principal social functions or objectives were served by particular legal devices.

Social Functions Served by Legal Devices. Given prevailing wants and working philosophy in nineteenth-century Wisconsin, it is not surprising to discover that legal devices used in defining or implementing stream-use policy operated either (1) by setting terms of legitimacy and opportunity for exercising will, or (2) by responding to the community's sharp-felt sense of its scarcity of mobile capital.

(1) To enlarge and enrich the content of individual life experience, to support the ideal of constitutional (responsible) power, to foster more intensive use of scarce mobile capital (management skills, labor, and money), nineteenth-century public policy showed no concern greater than to use law to release the creative energies of men's wills. This concern expressed itself in an intricate interplay of public and private decision making, which became the most characteristic structural feature of legal order in the United States. Law affected the exercise of will by two types of intervention in affairs — (a) by regulating what men might choose as purposes of their overt action, and (b) by regulating the manner in which they pursued their purposes. Of course, the difference between "what" and "how" was not a difference in kind, but only in range of definition; every decision on what to do may, in a larger frame of reference, be viewed as a decision on how to do something else. But the difference helps identify and appraise the impact of various forms by which law ranked interests.

(a) Law most drastically affected the exercise of will by public or private decision makers when it forbade, or insisted on licensing, access of will to defined purposes or areas of conduct. Superior legal agencies might, thus, limit the range of decision open to subordinate agencies and, in so doing, express a ranking of values among objects of private concern. The Northwest Ordinance and the Wisconsin constitution forbade the legislature to lay direct taxes on use of navigable waters; these provisions not only controlled legislative will, but did so to the end of assigning navigation a preferred status over other economic activities, so far as concerned contributions to the overhead costs of social order.[286] By general or special acts the Wisconsin legislature set terms on which dams might be built in navigable waters; such provisions not only limited the policy role which the courts might otherwise play under common law doctrines of nuisance, but in so doing also implicitly preferred the more rationalized and sustained investment and production likely to be encouraged by statutory titles, compared with rights definable only through the hazards of litigation.[287] The law's broadest device for limiting the access of private will to defined realms of behavior was this one of the statutory license. To require a license at all accomplished a sharp ranking of interests; to require a general or special statutory license as prerequisite to lawful improvement of navigable waters at the least meant to rank ahead of water-power development and large-scale organized transport such navigation uses as the natural condition of streams allowed.[288] Moreover, the grant of a license preferred the licensed activity over all interests to which it might cause detriment; a dam franchise implicitly preferred the uses which the dam served, over alternative power or transport uses, to the extent that other men must use extra labor or care, or alter the timing of their operations, or suffer such delays as inhered in the licensed activity, to drive their logs over the slides provided or to run their mills in accommodation to the altered flow of the stream.[289] A license might rank interests more sharply than this by reaching beyond permission to the grantee and excluding action by others within the sphere of the grantee's operations. So Section 1777 of the Revised Statutes favored integrated river improvement enterprises by conferring a traffic-control and toll monopoly upon that corporation which first exercised the privilege the statute held out.[290]

(b) This society had faith in the beneficent effects of releasing men's creative energies. Though to some extent constitutions limited the purposes which should be legitimate objects of public decision making, on the whole our constitutional law sanctioned broad discretion in positive use of official power; characteristic of this bent of policy was the tendency of the courts to construe constitutional language as setting standards rather than prescribing rules, as under the navigation guaranties of the Northwest Ordinance or the Wisconsin constitution.[291] So, too, law insisted on licensing private decision making only in restricted areas of purpose. The general presumption of policy was that initiative of decision should rest with private individuals and associations.[292] Accordingly, the most pervasive types of law's effect on will took the form of defining legitimate methods rather than purposes, and applying reparative more often than preventive sanctions. When law defined legitimate methods of action, the choice made between stating a standard or a rule carried value implications. Where law defined legitimate behavior simply in terms of a standard of conduct — for example, that a log driver must exercise reasonable care against damaging other navigators or dam owners or riparian farmers — in effect it favored enterprising initiative by the actors subject to its standard. The general movement of logs to market was more important than the possibility of particular incidental detriment, so that the burden of persuading law to intervene (and the inherent, law-made risks of success in this effort) was put on him who made particular complaint of the movement; the complainant's scope of will was constrained, the transporter's was left free or encouraged toward expansion.[293] Where law defined specific rules for conduct, typically it did so because it was concerned to qualify the basic favor it showed some predominant interest. The milldam act showed general favor toward developing water power, by giving a blanket license to riparian owners to build dams on nonnavigable streams. But it qualified this favor, and protected the scope of will of the landowner, by absolute requirement that fair compensation be paid for land flowed by such a dam.[294] Save as it stated otherwise, a dam franchise gave the dam owner broad discretion to control stream flow to his purposes. But many special franchises qualified this approved scope of will by stipulating, in limited protection of riparian farmers, that the flood gates must stand open between specified summer dates (as from each June 15 to September 30, or July 20 to October 15); so by its detail law created constraints of will that otherwise would not have existed.[295]

(2) Scarcity of mobile capital was a constant challenge and frustration to the enterprising men of Wisconsin's lumber era. In large measure contemporary public policy ranked stream-use interests in such order as promised maximum leverage for limited investments of mobile capital. Public policy affecting capital commitments represented a special case of legal allocations of decision-making power, but a special case of such substance as to deserve its own catalogue.[296]

The law in effect preferred or subordinated interests by provisions which affected either (a) the amount of fixed capital, or (b) the amount of working capital which must be mustered by those who would use the streams, or (c) the shares of benefits or costs of social capital which should accrue to stream users in comparison with other members of the society.

(a) Law affected investment of fixed capital in particular stream-use enterprises either by law-made subsidy or by mandates implicit in franchise terms. (i) Out of lands given the state by the federal government in aid of specified navigation improvements (like the Sturgeon Bay canal), the state might favor transport development by subsidies in kind which allowed entrepreneurs to use their scarce dollars for other needs than to buy right of way. The state constitution forbade money subsidies, but there was indication that the legislature and the court would interpret liberally the exception from the constitutional ban on state participation in internal improvements, which allowed use of donations made to the state for specified projects. However, we have noted that—in part because Congress showed only limited interest, in part perhaps because capital demands were relatively so much less in stream improvement efforts than in railroads—this style of subsidy did not bulk large in developing the forest streams.[297] A more important, if largely unacknowledged, form of capital subsidy to navigation improvements was by the grant of toll rights. Even when limited to reasonable returns on investment, tolls should provide return of capital committed, hence presumably should encourage initial commitment. Beyond this, since there was no effective procedure to review the reasonableness of charges, grantees favorably situated could enjoy practical freedom to push charges to the point of capital levy on those who had no legal or practical choice but to pay in order to move their logs or lumber to market.[298] (ii) The legislature ranked stream-use interests when in effect it set minimum or maximum amounts of fixed capital investment as conditions of stream-use franchises. Statutory dam franchises usually stipulated that grantees build suitable slides or flood gates, so that logs or rafts might pass safely without delay. To the extent that this condition required of dam builders capital outlays which they otherwise might have elected to avoid, the stipulation preferred the general transport interest over the specific uses of the licensed dams and, more generally, over the water-power interest.[299] Conversely, the law might favor an interest by minimizing capital demands on it. So the legislature preferred water-power development over land-surface (typically, farm) uses when it delegated the power of eminent domain to dam builders, that they might limit the dollars they must pay for flowage rights.[300]

(b) Law might operate to reduce or increase current operating costs of stream users, hence to reduce or increase the extent of working capital they must command, and thereby put a different practical evaluation upon different activities. (i) Public subsidies to meet operating expenses could usually be made only by spending public money. Quite clearly the Wisconsin constitution forbade this procedure to the state when it said that the state should "never . . . be a party in carrying on" works of internal improvement.[301] But the federal government and Wisconsin municipalities (which were not subject to the internal improvements ban of the state constitution) both subsidized the operating costs of water transport, so far as they provided not only lighthouses and harbor improvements (fixed capital subsidies) but also the salaries of light tenders and harbor masters.[302] Given the crude state of public executive functions, however, this type of public subsidy was bound to be small. Again, the pervasive type of subsidy was indirect, by grant of toll rights, which lent the sanction and force of law to private enterprisers to enable them to mobilize and discipline private means to support the current costs of stream improvements.[303] (ii) As with fixed capital requirements, so the terms imposed in stream-use franchises might work to limit either the minimum or maximum working capital which stream operators must have. It inhered in the relation of competing interests that conditions which law put upon franchises typically worked to require one affected interest to incur more, another to incur less, operating cost in consequence. So when a dam or boom franchise required the grantee not only to construct suitable passage ways (raising the minimum necessary fixed capital investment), but also to supply at his expense the labor needed to pass another's logs or rafts safely through the licensed works, the stipulation relieved the navigator of labor cost at the same time as it added to the licensee's labor costs.[304] On the other hand, to license the erection of works in a navigable stream — even with stipulated navigation guaranties — probably meant of necessity to impose on navigators some delay and some physical damage to goods; such imposed time and handling costs proportionately subordinated general transport to particular power developments or particular improvements of navigation or distribution (boom) facilities.[305]

(c) Like everyone else, stream users enjoyed benefits (enhanced income) and paid costs (taxes, and limitations upon practical freedom of action) derived from social structure and processes. In most respects they shared social income and social costs in ways no different from the experience of others engaged in economic activity: like others, they benefited because there was law to enforce contracts and redress torts; like others, they suffered because the market gave play to extremes of dealing which neither private nor public institutions yet knew how to mitigate. To attempt deliberate adjustment of social income and cost factors made the most exacting demands upon law — demands for experience, ordered learning, flexible response, and mature mingling of sophistication and responsibility in confronting contending power groups. There is no cause for surprise, therefore, that the relatively crude processes of nineteenth-century law made little use of social income and cost adjustments as devices for ranking interests.

The particular issues of nineteenth-century political economy derived mainly from the interplay between pragmatic readiness to use law positively to mobilize and allocate economic resources, and reliance upon institutions of private property and contract to release creative energies of private mind and will. Quite naturally, definition of the relative domains of commonwealth on the one hand and private property on the other embodied the most sustained attention which stream-use policy gave to questions of social accounting.

Private property in land surface meant, basically, private right to initiate decisions for use within the range of productive possibilities which the natural situation allowed. Where a dam flowed land belonging to someone other than the dam builder, the flowage almost certainly precluded any use of the land surface by its owner. This total preclusion of surface use the Wisconsin court regarded as so plainly a taking of private property that it never felt need to explain why the dam builder must pay compensation under the constitutional command that "the property of no person shall be taken for public use without just compensation therefor."[306] True, within its reasonable discretion to judge the public welfare, the legislature might decide that flood dams to improve navigation, or power dams to develop the over-all productive capacity of the economy, represented public uses for which private property might be taken.[307] But the cost of these additions to general economic potential should not fall on particular individuals who lacked means readily to shift it; hence law would lay the cost on the improver or mill operator, who might be expected to spread it over his customers.[308] To this core of constitutional doctrine the courts added elements of their own making. So, where the eminent domain statute provided that value added to flowed land might be set off against compensation due for flowage, the Wisconsin court would not construe the set-off as intended to include benefit derived from the over-all economic growth of the neighborhood, stimulated by the presence of the new mill; this benefit was one shared by the whole community, but the damage suffered by the flowed landowner was his loss alone, for which he should be paid.[309] So, too, the United States Supreme Court ruled that it would not construe the intent of Congress to be that riparian titles derived from the United States came burdened with easements of flowage incident to navigation improvement projects.[310]

On the other hand, the Wisconsin court held that a downstream riparian landowner had no right to compensation, nor an action in tort, where a flood dam built upstream according to statutory franchise operated of necessity to alter the stream flow so as to diminish the utility of the downstream millsite. The stream flow was a commonwealth asset; no riparian landowner had private property in the water or its running; hence there was here no taking of property, nor any tort.[311] Of course these propositions merely state the conclusion; they do nothing to uncover the bases of the public policy so declared. The court never clearly explained why a licensed dam builder must pay for land flowed but not for a millsite destroyed or impaired. Utility as a millsite was as real a use value as utility for agriculture or other surface use.[312] Explanation did not lie in terms of the breadth of purpose or reach of consequences of the upstream dam; whether the dam was built to develop a particular power site or to aid navigation on a whole length of stream, if it caused flowage, the flowage must be paid for, but not the impairment of other men's opportunities to use the stream flow as such to develop water power or improve navigation.[313] The closest the judges came to a rationale of their social cost accounting was to indicate that law should not create potential financial burdens of such weight as to discourage all productive development. Flowage might be expected typically to involve only rather limited land areas, but a given improvement might affect a large and uncertain number of potential power sites or opportunities for profitable navigation works; if any particular improver must pay, or risk paying, all these opportunity costs, he might not be willing or able to take the burden, and the community might find itself with so broad a concept of protected private property rights as to stultify its general economic development.[314] This analysis, implicit in the flowage cases, came closer to expression when the court ruled that log drivers must exercise reasonable care to avoid damaging riparians' shore lines, but that they had no greater responsibility: ". . . it was not [the driver's] . . . duty to build booms or other structures along the [riparian owner's] . . . shore to protect it from wearing or washing away. If the enjoyment of navigation on such a stream as the evidence shows Apple river to be was subject to that burden, the right itself would be of little value." [315]

Some attention to tax policy also showed that the lawmakers were not wholly unaware that they could manipulate social accounting to express valuations of activities relative to social context. The legislature after 1891 subjected river improvement companies to a special "license fee" of 2 per cent on gross receipts, with a credit for taxes paid under proper levy by local governments; the residual state levy implicitly recognized that such enterprises had their significance as elements of the general economic framework and so, if not otherwise charged, might properly be required to contribute to the overhead cost of state legal order.[316] The Wisconsin Supreme Court supplied important complementary doctrine when it interpreted the general legislative intent under such stream franchises to preclude local governments from assessing and collecting taxes on separate items of real property (such as a dam site) which were necessary elements in conducting the franchised enterprise. The integrated licensed activity was the taxable subject, to be protected against the dismemberment that would be accomplished by enforcing taxes on separate elements of the enterprise.[317] The court further protected the functional integrity of the franchised enterprise by ruling that only the local government at the principal place of business might lay such tax as was permissible upon the enterprise considered as a whole.[318]

We have examined the principal legal forms (definitions of values, and implementation by licenses or reparative sanctions) and the principal functional concerns (effects on will and on capital commitment) through which law ranked stream-use interests. It remains to sketch the principal hierarchy of interests so established.

Principal Hierarchy of Interests. (1) The base line of stream franchise policy was attention to values, and preference of values, which helped men realize increased output of goods and services from using the streams. This meant preference or, indeed, exclusive attention, for economic compared with noneconomic values. Thus — understandably, given the formative stage of the society — stream franchise policy showed no concern for aesthetic values, and little concern for recreational utilities, affected by developing the power or transport potentials of streams.[319] In another focus, the distinction was like one that was prominent in public lands policy. The law did not give blanket preference to economic values. Rather, it attended almost exclusively to those economic values which the market could translate into a dollar calculus. So stream franchises reflected lively concern that more logs be moved or sawed for market, as well as concern for recovery of the fair money value of lands flowed. But — again quite understandably in view of the stage of community development — until about the turn of the century there was no reflection of substantial concern for public health or safety as these might be affected by dam construction.[320] It was consistent both with preference for activity which fostered the direct productive contributions of streams, and preference for activity which yielded values readily measurable by a money calculus rather than merely by a real calculus, that the Wisconsin court adjusted the assessment of social overhead costs (by taxation) to the functional imperatives of efficient stream use (by construing tax legislation to require assessing stream improvements as productive entities, subject, as such, to the jurisdiction of but one taxing body).[321] Over-all, the basic elements in stream use policy reflected the common faith in (a) rising material productivity as the society's prime dynamic, and (b) the release of creative energies of mind and will primarily according to a market (money) calculus, through the related legal institutions of property (title) and contract, as the society's most pervasive style of operations.[322]

(2) Growth-minded and constantly concerned to obtain maximum productive leverage from limited mobile capital, Wisconsin public policy preferred stream uses in the order of their likely capacity to produce multiplier returns from law's promotion or protection of investment.[323] (a) Transport and water-power development enjoyed high preference, accordingly, over alternative uses of riparian land surface; transport provided a frame for enlarged markets, hence for enlarged industrial activity; water power provided a base for multiple increase of output for given commitments of scarce cash and labor.[324] Clearest symbol of this preference was the delegation of the eminent domain power to builders of flood or power dams.[325] Of like import were rulings that those who moved logs were not liable for erosion of shore lines incident to flood dam operations or drives conducted with reasonable care.[326] So also the legislature conferred on log owners the privilege to make reasonable entry on riparian lands to recover stranded logs.[327] (b) Transport enjoyed well-marked, though qualified, preference over water power and over fisheries; as a transport facility, a stream was the backbone for the entire region accessible to it, in comparison with which any given power site could serve the growth only of its immediate neighborhood (at least so long as rail nets were little developed, and mill machinery was yet of relatively limited capacity); both in scale of output and in dollar value, fisheries represented a minor item in the state's economy during the surging years of lumber industry growth.[328] The basic preference for navigation over power found manifold expression: in the constitutional guaranties of free navigation; in the expanded, judge-made definition of navigable waters to include those capable only of moving logs; in the legislature's insistence that its license be had for works on principal navigable streams, and in its consistent care to condition dam and boom franchises with stipulations to assure passage; in the readiness of the Wisconsin Supreme Court to uphold broad

legislative discretion to authorize flood dams, main-channel improvements, and exclusive public-utility-type boom franchises at the cost of uncompensated detriment to riparian landowners' millsites and mill facilities; and in the court's enforcement of the navigation guaranties contained in statutory dam franchises, through actions for damages or injunction. This marked preference for transport was not modified until after 1910, when the legislature provided a new framework of policy for licensing dams which reflected the rise of interest in hydroelectric power and the sharp decline in the log transport importance of the northern waters.[329] Similar to the nineteenth-century weighing of power interests was the treatment of fisheries. Though fishermen shared the common rights of access to public waters for their trade, they must submit to losses or burdens reasonably incident to transport uses. Not until late nineteenth century did the legislature regularly require fishways in the dams it authorized, and at no time in the lumber era did it show sufficient concern to make adequate provision to enforce such fishways conditions as it imposed. Navigators owed the fishermen the same duty as, but no more than, they owed anyone else to exercise reasonable care in their operations; a tugman towing lumber was liable where he broke nets by careless navigation, but in the first instance the fishermen must bear the risk of damage and make out a case for relief.[330] (c) Improved (technically augmented) navigation enjoyed preference over navigation in the natural condition of waters, as another expression of favor for the multiplier principle. The Northwest Ordinance and the Wisconsin constitution proclaimed the state's principal navigable waters to "be common highways, and forever free" for use "without any tax, impost, or duty therefor." The legislature and the court joined in construing these guaranties as setting no barrier to statutory franchises which empowered men to increase natural navigable capacity by building and operating flood dams and booms, or clearing channels, charging reasonable tolls on all who passed, and managing the movement of traffic.[331] Pursuit of the multiplier principle usually spelled some concentration of economic operations and decision making, and required the discipline of otherwise dispersed economic resources.[332] These were features so analogous to exercise of the taxing power that the Wisconsin court quite naturally held that preference for the multiplier benefits of improved waters over natural waters must be sanctioned by the legislature; there was thus a separation-of-powers value, as well as economic policy, implicit in the requirement that one must have statutory warrant to improve a stream or charge tolls.

(3) Where they confronted competing claims to similar uses of streams, the lawmakers responded by ranking interests according to three criteria which hinted at a policy pattern that never took clear-cut form. (a) Legislative policy favored the claimant who was first in time with effective means to realize the productive potential of a situation which invited concentrated

control. So Revised Statutes, section 1777, gave a monopoly of stream improvement to that corporation "which shall have taken prior possession of such stream for that purpose." [333] And the legislature licensed some general-service, public-utility booms with exclusive rights in waters the location of which favored a large-scale distribution facility; such grants, the Wisconsin court ruled, lawfully displaced the common law rights of log drivers to make their individual arrangements or of riparian landowners to service their mills with their individual booms.[334] (b) If someone had an interest sufficiently focused to prompt him to make an issue, the law was likely to rule against a man who sought to take benefit from a particular public or private investment without sharing the cost. True, this dislike of free riders found expression only within the limits of prevailing nineteenth-century attitudes. This unphilosophic time was not inclined to probe to uncover the concealed subsidies to private venture which lay in unacknowledged social overhead costs; as contemporary stream policy fostered rapid maximum use of the streams for log transport, it was as little attentive as was contemporary public lands policy to the long-term social costs of promoting irreversible depletion of the pinery. But where lawmakers were conscious of particular commitments of scarce cash capital, it offended not only their sense of justice, but also their sense of economy and their concern to promote effective multiplier use of fluid capital, to allow another to reap where he had not sowed. Thus legislative policy — readily sustained by the court — was to authorize private improvers to levy reasonable toll on all who used the improved waters, even though the physical situation left the users no alternative water route.[335] Where public subsidy developed new water-power opportunities incident to a navigation improvement, government might control and reap all benefits from disposing of the fresh-made power, without compensation to riparian owners.[336] Where under a proper statutory franchise a private enterpriser developed an artificial water course, no power site rights accrued to others whose lands were "riparian" to the man-made channel.[337] A municipality might not ground authority to tax land at a high rate, as "improved," because it bore a flood dam which had special value only as part of an integrated stream improvement enterprise which stretched over several counties.[338] One who put his logs into a stream without adequate provision to drive them, and suffered them to become mingled with logs of another operator, became properly subject to a statutory charge for reasonable compensation to the man who supplied the necessary labor and means to bring the whole drive through to market.[339] (c) Where the law gave no monopoly and the physical situation invited efficient use of waters by many users, taking their separate initiatives of decision, the law forbade or penalized styles of individual operations which needlessly precluded or burdened others' uses. Either as a matter of common law or to implement statutory assurances of free navigation, the Wisconsin court allowed a navigator to recover damages against one who let his rafts block passage for an unreasonably long time, or against a duly licensed dam owner whose violation of navigation guaranties stipulated in his franchise caused loss or delay to logs or rafts seeking passage.[340] Legislative policy favored multiple stream use wherever this was practicable, requiring that burdened persons accept the temporary limitation or even preclusion of other uses which the policy entailed.[341] Such was the implicit judgment of hundreds of special and general statutes which sanctioned defined uses of waters without conferring general monopolies; had not the problem of scarce use opportunities lain always in the background (with its inevitable corollary of burdens on some, offsetting benefits to others), there would have been little practical significance to seeking or passing this complicated bulk of legislation.[342] So, too, legislative and judicial policy joined in rejecting any absolute preference for land or water transport. One who would bridge a navigable stream must have statutory license to do so. The legislature had authority to approve bridges — especially for railroads — though these might cause some detriment to navigation; on the other hand, the consistent policy both of Congress and of the state legislature was to condition bridge franchises on suitable guaranties for navigation, and the Wisconsin court ruled that it would not presume legislative intent to authorize such a style of bridge as would in fact materially hinder log or raft navigation at a given point.[343] Apart from such specifications of statutory policy, remedies by lawsuit were made available on terms which favored multiple use of waters. The common law embodied a general preference for the initiator of economically productive action, by casting the burden of persuasion on an aggrieved person to show cause why law should intervene to shift a loss from where it fell as a consequence of the initiative taken.[344] Statutory licenses for action sharpened the preference; merely to plead the holding of a statutory franchise was a sufficient defense against one who complained of detriment suffered from the licensed activity, so that the complainant might establish fault in the actor only by showing violation of terms set explicitly or by clear implication in the franchise.[345]

Despite their diverse aspects, these three specific trends of policy suggested a unifying frame of value. Whether by making limited grants of explicit monopoly, by discouraging free riders, or by protecting opportunities for multiple use, law ranked claimants to substantially similar stream uses by arrangements calculated to promote economy of will. These specific trends of policy all worked to prevent conflicts of purpose or decision which would spend entrepreneurial time and energy in dispute or would increase the risks of frustration which discourage productive initiative. Regarded in a more positive aspect, these policies worked to organize stream-use relationships so as to match the situation with the most efficient quantum of will. Again, in impetus if not in expression, law amounted to applied political economy.

We have catalogued the ranking of stream use interests (1) in terms of legal devices for expressing an order of values, (2) in terms of principal social functions served, and (3) in terms of types of behavior preferred or subordinated. We need thus to classify in order to see better the purposes, processes, and effects involved in the bewildering mass of detail which embodied Wisconsin policy toward the forest streams. A hazard of such analysis is that it may seem to present policy as a set of static concepts. We need recall that the purpose of the analysis is to obtain a more nearly complete view of events in time. The events which concern us moved rapidly to their climax. Here, as in other aspects of the lumber industry's legal history, we find that the bulk of significant public policy remained about the same throughout our period, once law had responded across the range of problems most urgently presented to it. In this sense our classifications represent reality in time, as well as serving their function as instruments for seeing better the details of what went on. But even within this stability contributed by the speed of events there was enough important change to remind us that we are observing history. Thus, as pressures for economic growth impelled law to foster stream improvements, legislative practice and judicial doctrine qualified the seemingly absolute navigation guaranties of the Northwest Ordinance and the Wisconsin constitution. Again, the concept of navigable waters underwent change. The court first enlarged the concept to protect expanding log transport; later, when the north country felt more need of water power, the court qualified its earlier insistence on statutory licenses for damming navigable streams by ruling that without such a license men might build power dams on minor streams if they allowed passage of logs in the time of spring freshet. Finally, toward the end of the lumber story, new interest in hydroelectric power marked the continuing shift of events. Through the principal lumber years the legislature clearly preferred transport over power uses. But from about 1901 to 1915, as the lumber industry sank in relative importance and the electric power industry rose, stream-use franchise measures veered toward fostering the use of waters to produce electric power. Time, thus, was patently an important dimension of policy.

Policy in Practice

Did law materially affect men's behavior in using the streams? If it did, in what respects, and to what extent? We can answer with some assurance so long as we ask only the more general questions.

The law did materially affect stream use. Even within the law's formal records there is ample—if mainly circumstantial—evidence of this fact, in the volume and diversity of matters brought to law concerning the streams. In the activist, opportunistic, improvising temper of nineteenth-century Wisconsin, men did not exert themselves to press and oppose, enact, amend, and repeal hundreds of statutes, and fight hundreds of lawsuits, only to celebrate a ritual.[346]

There are two important types of specific questions which the materials do not let us answer, and there is one particular caution which the shape of the materials invites. (1) First comes the problem of identifying types or norms of informal behavior. Where legal records evidence some particular adjustment, or deviation, or tension between law and practice in stream use, ordinarily we can make some objective determination of the representative character of the incident only if the legal records themselves supply analogies or like incidents. Of course, men did many things which left no legal record of their conformance or nonconformance to law. Law lacked resources to operate formally upon more than a marginal amount of the total behavior to which it addressed itself. Moreover, in a constitutional legal order which emphasized dispersion of power and privacy, this marginal operation of law was more than a fact; it was a value. In any case, most informal behavior in stream use left no durable evidence. Save for simple tallies of quantities of goods handled, men made little private record of the flood dams or booms they built or of the drives they ran; they simply acted. In the archives are papers of a few lumber industry firms, but these limited records deal mostly with phases of the business other than water transport or water power. Evidence of stream-use practice relevant to stream-use law lies most abundantly in the legal records. We must be modest in the inferences we draw from these records, recognizing that the informal behavior reflected in law was only part of a great flow of action of which we have little other evidence.[347] (2) For like reason, we can make only the most general quantitative appraisals of the operative significance of particular types of informal behavior or particular stipulations of law. There is no way to tell how many dollars men invested in building dams or booms or altering channels which they would not have invested had the law not set the terms it did on stream improvement. We find statutes which validate dams built without prior legal license; we have no means to calculate the percentage these instances are of the whole number of unrecorded, unlicensed construction. On the other hand, substantial legal attention to given behavior argues that the point was probably important in the total flow of action. Thus we find not only a high original prevalence of stipulations on height and slides and exit facilities in dam and boom franchises, but also considerable activity in amending such specifications. We may infer that these franchise stipulations were not empty forms but involved matters of lively practical concern, and that there were interests active on both sides of the relation of works builder and navigator in attending to what was done or not done under such terms set by law. We cannot, however, make more finely drawn estimates than these.[348] (3) We should not exaggerate the representative significance of particular controversies which left uncommonly full legal records. At a few river locations there was recurrent but relatively low-key conflict, as over the Kilbourn dam on the Wisconsin River. There were two highly charged interest-group battles on the Chippewa River—competition between Eau Claire and Chippewa Falls millmen over a franchise for a traffic-management boom at Eau Claire, and contests over control of the Chippewa's outlet to the Mississippi at Beef Slough. Disputes which put law under unusual pressure can be especially revealing of structural or functional strength or weakness in legal process. But we must not overlook the uncommon features of such contentious histories, especially the unusual strength, intensity, and stubbornness of the competing interests, and hence the unusually sharp focus upon issues. Most stream-use practice proceeded by a succession of commonplaces. Usually issues were much less in focus, or there was less competition of interests, or competition was episodic. Usually, thus, greater play was given to custom, improvisation, and the simple cumulation of events.[349] Because we have unusually full records of a few hard-pressed contentions, we must not take combat—let alone melodrama—as the norm. Within the common objective of promoting economic growth, the norm was, rather, a matter-of-fact concentration upon operations. Men were "practical." They concentrated on bringing things to pass.[350]

Law made its basic claim upon stream-use behavior where it undertook to define the most elementary terms on which private will might be exercised over use of the streams. Federal and state constitutional guaranties protected such navigation as the natural condition of waters permitted. However, large-scale transport, and any development of water power, required altering natural conditions. Primarily by statute—and, with less emphasis, by common law—Wisconsin public policy required that men have or obtain a statutory franchise before they changed the natural condition of waters in any manner which might materially affect the interests of navigators or riparian landowners.[351] Some eight hundred special statutory franchises, and scores of lawsuits before the Wisconsin Supreme Court presenting claims under general franchise legislation, attest that men established their navigation and power use of forest-area streams under such legislative licenses. The indication is that where the law clearly said that men must act under the law's license they generally undertook to regularize their conduct accordingly.

As we might expect of this energetic, self-willed community, there is evidence that men built some dams and booms without permission of law. In twenty-seven statutes the legislature "hereby legalized" dams "heretofore built," or "now kept up," or authorized a dam at the same location "as the unauthorized dam built and maintained at the said point."[352] Eight statutes apparently legalized booms built without authority.[353] In eight cases which reached the Wisconsin Supreme Court the record showed erection of dams or related works without prior legislative sanction.[354] Several dam statutes legalized structures which had stood without authorization for more than a generation. Thus—authorizing a dam at a specified location on the Big Plover River in Portage County—Laws, 1901, Chapter 261, declared: "The dam heretofore built and maintained at the place mentioned . . . built in or about the year 1853, and the building of said dam and the maintenance of the same to the present time is hereby validated and legalized." [355]

Despite such evidence, the whole record shows a striking concern by private enterprisers to legitimize their use of the streams. The franchises which validate unauthorized works represent between 5 and 6 per cent of the whole number of special franchises; measured by what they say, the rest authorize fresh construction. Except for the first boom works at Beef Slough, the unauthorized works thus validated seem to have been of local and limited importance; the Beef Slough controversy apart, the record shows no case in which men undertook to improve a large collecting area or an extensive stretch of a major stream without positive permission of law.[356] Moreover, we should recall that in 1889 the Wisconsin Supreme Court found that it was not the intention of the legislature to require licenses for building dams on streams navigable only for logs at freshet time, provided the dams were so built as not wholly to obstruct the passing of logs in those limited periods of navigability.[357] There was no legislative reaction to this interpretation, which we can be sure only gave formal expression to established lumber industry practice.[358] There is little evidence of unauthorized local works construction on the major transport rivers such as the Wisconsin, the Chippewa, or the Wolf.[359] Most of the statutes validating particular unauthorized structures concerned secondary streams (which, nevertheless, moved substantial log traffic). Enterprisers gave themselves the benefit of the doubt in not waiting upon clear legal sanction to dam streams which obviously had a very limited navigation potential. But they did not care to commit themselves to extensive dependence upon using improved waters on which they did not know they had an assured right. Of the thirty-five statutes validating originally unauthorized works, all but one were enacted after 1865, and twenty-one were enacted after 1879; as it moved into the full tide of growth, the industry showed more care for the legal basis of its stream improvements.[360] As titles and as grants of prerogative, the law's franchises apparently taught even a rough and tumble industry a substantial respect for the practical values of legitimacy.

To discuss license requirements tends to emphasize the regulatory relation of law to stream-use behavior. It fits reality to begin here: licensing provided the base line both for formal policy and for industry practice in stream improvement. But it would violate reality not to turn attention promptly to the promotional impact of law upon transport and power uses of water. Men had to give some initial attention to regulation because there was at least a minimum of legally sanctioned order that was necessary for market operations. But their hearts were in promotion, not regulation.[361] In any case, there was not a sharp division of function between these two aspects of law. Regulatory order in stream-use policy was not primarily restrictive. It served to facilitate a more effective release of productive energies than would have been possible out of random collision of wills. Formal policy was clear on this point. Hence, for example, Revised Statutes, section 1777, tendered a river improvement franchise only to that corporation "which shall have taken prior possession of such stream." [362] Hence the milldam act qualified its proffer of eminent domain procedure in aid of power development by stipulating that "no such dam shall be erected to the injury of any mill lawfully existing, either above or below it on the same stream, nor to the injury of any mill site on the same stream, on which a mill or mill dam shall have been lawfully erected and used, or is in the process of erection . . ." [363] Entrepreneurs in effect acknowledged that the regulatory features of water law fostered productivity. The legislation which constituted the bulk of public policy on forest stream use was shaped almost wholly by the play of interests within the lumber industry; the record shows little counterpart to the wider play of interests (farmer compared with timberland investor, farmers and local businessmen relative to big city promoters, for example) which marked the growth of policy regarding public lands or railroads. Yet, with the forces in play typically drawn from within the single industry, we find a volume of legal regulation which matches that concerning public lands or railroads. There would not have been such recourse to legislation had not millmen and transporters found productive utility in so employing law. We are unlikely to draw wrong inferences when our interpretation of legal affairs is consistent with the operations-minded devotion of these men to enlarging output for market.[364]

The promotional demand upon law was that it foster investment of money, labor, and management skills in realizing the power and transport potentials of streams. The most direct ways to this end were the furnishing of capital (out of public borrowings, or from taxes, or from public lands) and the furnishing of management (through public enterprises). Only limited recourse was made to these direct methods, and their use was of secondary importance in exploiting the waters. Federal lands and money, and at different periods some state and federal management, went into efforts to create a continuous channel between Lake Michigan and the Mississippi River via the Fox and Wisconsin rivers. The Fox-Wisconsin project aborted, particularly because of the inadequate flow and shifting sands of the Wisconsin. In any event, neither in plan nor in effect was the project of substantial importance to log or lumber transport, though as a by-product it fostered development of water power in the Fox River valley. The United States made a substantial land grant to aid construction of a canal (important to lumber shipping) from Green Bay to Lake Michigan. The federal government also invested about $200,000 in surveys and limited improvements of the Chippewa River. This activity was pressed particularly by lumber interests, but the scale of public investment was too small to be rated a major contribution to the navigation uses of the Chippewa. The United States invested more heavily and continuously to improve Mississippi navigation, but with the prime objective of aiding general steamboat traffic. The Wisconsin constitution plainly and strictly forbade investment of state general funds or management in works of internal improvement. This legal barrier may help explain why the lumber industry made no attempt to obtain such help. However, the state constitution did not forbid local government units to subsidize transport improvements. Municipalities went into considerable debt to assist railroad construction. But municipal aids to navigation were few and small; municipalities could expect little if any direct benefit from facilitating shipment of logs out of their boundaries, and apparently there were few physical situations in which local mills felt need of public help to organize common sorting or storage works.[365]

The minor role of direct public subsidy in aid of log and lumber water transport (especially in contrast to large public aids to railroads) reflected factors of industrial and legal organization and practice, compared with which a doctrinal element like the internal improvements clause of the state constitution was of secondary account. Though it included a few sizable firms, the lumber industry of the nineteenth century developed no giants; more fragmented and commanding fewer dollars or men than did the principal railroad promoters, the lumber industry could not muster like political force to extract large favors from Congress. Lumbermen had practical power to get a great deal of what they wanted from the state legislature or local governments. However, these agencies operated on a very limited tax base, and otherwise — should legal barriers be removed — could help only out of borrowed money or out of lands given by Congress; had lumber interests seriously pressed for such aids, they would have met the overwhelming competition of the railroads.[366] Moreover, not only the dispersed state of the lumber industry, but also the peculiarly local character of stream improvement problems, would have required that effective public subsidies be made with close attention to particular situations. Between 1840 and 1890 there was no federal or state executive establishment competent for such tasks. The time-costly and clumsy processes of contemporary legislative investigation demonstrated the practical incapacity of the legislative branch to meet such a problem; the legislature managed to grind out eight hundred special franchises for stream use only because it could — implicitly, if not explicitly—generalize most of the policy choices presented. Through public lands disposition, the federal government and the state subsidized great numbers and diversity of private operators in timber speculation and logging. They accomplished these subsidies despite the ineptness of the federal and state land offices. But here the problem was seen as a simple one (merely to hand over full initiative of decision upon the use of given resources), and in the familiar form of the fee simple title the governments had a single, well-established device by which to operate. The complete breakdown of the experiment in leasing rather than selling mineral lands attested that contemporary executive apparatus was too simple and crude to carry out any program that called for detailed operations in the field.[367] There is no evidence that scruples of principle restrained those interested in log or lumber water transport from seeking direct public aids. What seemed practically desirable and obtainable, they sought to get. But the pattern of contemporary legal and economic institutions and practice was unfavorable to a substantial direct subsidy effort in this field.

Thus, it was mainly by indirection that law promoted investment of money, labor, and management in exploiting the power and transport potential of streams. Again, we encounter a principal theme of nineteenth-century policy—the pursuit of objectives of public concern by legitimating and supporting the exercise of private will.

As grants of title and prerogative, stream-use franchises fostered investment by creating frameworks of reasonable expectation within which men could act, knowing in what respects law allowed them to impose their will on others and in what respects law protected them against detrimental claims or actions by others.[368] The contract clause of the federal Constitution and the policies of self-restraint which the legislature and the governors observed in applying reserved powers to amend or repeal franchises under the state constitution and statutes offered additional assurance to encourage capital commitments.[369]

Speaking from experience of their community and its ways, contemporary lawmakers time and again asserted their conviction that the grant of franchises in fact promoted investment of private resources in stream improvements. This conviction found firm expression at the beginning of Wisconsin policy. In 1842 the Committee on Internal Improvements of the Territorial House of Representatives observed that the Territory's credit was not such as to allow it to raise money for internal improvements, and that Congress had only limited generosity for such local ventures. Thus the committee put its reliance on the stimulating effects of a liberal franchise policy: "The only remaining resource, therefore, for the accomplishment of these great works of improvement, is private enterprise. It is believed, that if the Legislature would grant to the individuals owning the land on these streams, the right to construct dams, with suitable locks and slides, wherever rapids occur thereon, their navigation would be speedily improved, and the agricultural, manufacturing, and commercial interests of our territory, thereby greatly improved . . . Let then the people immediately interested and who own the soil bordering on these streams, have the right — give them the privilege of bringing into operation the extensive water power which will be created, in connection with the improvement of these rivers — and your committee believe that these important channels of trade will be speedily opened, without the aid of the General Government, and without involving the faith or the responsibility of the Territorial Government."[370] The hundreds of special stream-use franchises granted in the years that followed, as well as the general river improvement statutes, were stated almost invariably as bargains — certain privileges given, for certain performance stipulated. The bargain form attested that these practical men saw franchises as inducements to investment in stream improvements. The persistence of this style of grant, and the care taken in quite a few instances to amend details of the required consideration, attest that men found that the franchise inducement in fact operated and was taken seriously.[371] Referring to the transport needs of the lumber industry after a generation of growth, the Wisconsin Supreme Court in 1877 thought that common experience showed that the state could obtain river improvements by granting toll right franchises, and, indeed, could get the work done on no other terms. "The state itself cannot remove such [natural] obstructions [to navigation], because it cannot be a party to carrying on any work of internal improvement, and, if such obstructions are removed, it must be done by other agencies. It is certain that no corporation or individual would be to the expense of making such improvements in these streams without some reward. The [Wisconsin River Improvement Company, operating under Private & Local Laws, 1853, Chapter 30] . . . alleges that it has expended over $30,000 in erecting dams for slack water, and in removing rocks from the channel of the Wisconsin river at the rapids; and how is it to be remunerated for its outlay, except by tolls upon lumber and timber which pass through the improvement?"[372] When there was unusual controversy, and issues hence were more sharply drawn than in the ordinary course — as in repeated battles over improvement franchises on the Black River and at Eau Claire on the Chippewa — contestants and legislators alike made plain that the stakes were high, precisely because to grant or withhold, or to revoke, a franchise meant the practical difference between action or no action by particular enterprisers.[373]

That men regarded the grant of franchises as stimulating the will to act appears also from the amount of amendment of the detailed terms of franchises. It was exceptional for a stream-use franchise to be adopted just as it was introduced. By a conservative appraisal, 70 per cent of all franchise bills enacted suffered some amendment in course of passage. There was much less amendment of franchises after their enactment, but the amount of such amending activity was nonetheless substantial; about 16 per cent of all forest-area special franchises were so amended.[374] We can better examine some details of this amendment experience in the following section of this chapter. It suffices here to note that the most numerous types of amendments dealt with matters inherently important to the title or the prerogative aspects of franchises, especially the definition of locations of franchised action and of toll rights.[375] This activity in amendment — especially the more pervasive attention to amendments preceding original enactment — offers impressive testimony on men's attitudes. As evidence which is implicit in overt action directed to focused details of projected behavior, it can properly be taken more at face value than can the explicit rationalization men make of their decisions. By its extent, it is evidence generated from commonplace more often than from peculiarly controversial stream-use franchise situations; therefore, it is evidence likely to be the more representative of regular patterns of behavior. What this evidence suggests is that men did not regard the obtaining of a franchise as an empty form, or as of little account in affecting future behavior. Those who sought franchises — and if not they, then those who saw themselves affected by others' franchises — spent an impressive amount of energy attending to details of franchise terms. Men do not so invest energy without cause — least of all these operations-minded nineteenth-century movers and doers. The inference is that they regarded franchises as material influences upon action.

To help develop and maintain operating organization and discipline in using the streams was a promotional objective equally important in economic effect to promoting investment in works of improvement. However, organization and discipline did not present as pervasive and important problems concerning water power as they did in relation to transport. Forest enterprise was considerably dispersed over thinly settled country. Some mill towns grew to consequence, notably Green Bay, Oshkosh, and Fond du Lac in northeastern Wisconsin, and LaCrosse, Chippewa Falls, and Eau Claire in the northwest. Even within these more closely exploited water-power areas such conflicts as arose among competing power users were handled ultimately in law through the same forms of litigation and under the same common law doctrines of reasonable use as served to adjust complaints among more widely separated dam users. The milldam act restricted the benefits of the eminent domain privilege to new dams which did not work "to the injury of any mill lawfully existing." Except for this provision — and an occasional like stipulation in special franchises — the legislature was not impelled to adjust relations among competing millsites. Indeed, from as early as 1853 it expressly resigned such adjustments to the judges, declaring that "the boundaries of lands adjoining waters and the rights of the state and of individuals in respect to all such lands and waters shall be determined in conformity to the common law so far as applicable."[376] Of the relatively few cases which came to the Wisconsin Supreme Court involving competition for flow among contending millsite operators, most turned on questions of title — the interpretation or enforcement of terms of deeds, leases, or licenses of water-power rights — rather than of tort. This distribution among the appealed cases suggests that areas in which water-power development opportunities were physically concentrated were likely also to be areas in which the early private land titles were concentrated. In such situations the common grantors created by private legislation (in the terms of their deeds) the criteria by which conflicts among their grantees might be adjusted; controversies among millsite operators not sharing common sources of title were apparently not numerous enough to generate pressures which could not be dealt with within the ordinary course of litigation.[377]

In water transport of logs or rafts there was likely to be crowding, since the bulk of product moved during short driving seasons. Whether as among competing navigators or as between navigator and millman, navigation more commonly posed factually inconsistent use claims upon a common resource than arose among different millsites. Moreover, overhead cost pressed harder on transport than on power uses of water; good business thus enforced pooling or at least co-ordination of log driving and booming activities earlier and more pervasively than it fostered concentration in woods or mill operations.[378] As the industry developed, the statute books reflected increasing demand for law to support frameworks of organization for disciplining water traffic. The demand was natural. Traffic management meant that some men would impose their will on others, regardless of agreement; this was a relation which invited resort to law, the most distinctive function of which is to superintend the total distribution of power in a community. Too, the situation called for specially contrived legal arrangements of power — that is, for franchises. Traditional forms of private property might accommodate relations among those whose claims rested on titles to riparian lands derived from common grantors. But stream flow was a commonwealth resource; the familiar law of private property would not serve to adjust flow uses as such.[379]

Such economic and organizational pressures help explain why there was demand for franchises to support stream management. These factors have further significance. They were pressures generated by the functional needs of efficient business and power relationships. Because of their functional origins, the growth and persistence of demand for management franchises are facts which in themselves evidence that such franchises were material elements in enlarging the scale of log and lumber transport. In 1867 a legislative committee recognized the force of such pressures of function in explaining why lumbermen on the Black River sought and obtained a franchise for the Black River Improvement Company, and continued to rely on it to legitimize the traffic management which joint use of the river required: "It had been fully demonstrated by the experience of ten years, that individual effort would do but little for the improvement of the navigation of the river, and hence the necessity of some organized corporate effort became apparent, both to the loggers and manufacturers of lumber on the river." [380] In 1879, with forest production mounting toward its peak, the Wisconsin Supreme Court found that a franchise to manage driving on an improved stream was likely to be a material factor in persuading men to the effort necessary to make the stream an efficient carrier; hence amendment of an improvement company's charter to add traffic management authority did not violate the constitutional requirement that no local bill embrace more than one subject. "We are very clear that the driving, running, sacking and sorting of logs on the Yellow River is legitimately connected with the improvement of such river, for the purpose of aiding and facilitating such work, and that such business has a natural and legitimate connection with the improvement of the navigability of such river . . . [I]t might well be the case, that the right to run and drive logs and timber on the river, when the improvements were made, would be the main inducement to the corporators to expend their money in making the necessary improvements." [381]

The behavior of those who used the streams implicitly testified that the availability of a franchise was a material attraction to investment in more effective traffic organization. As their scale of operations grew, the lumbermen attached increasing value to efficient driving and booming organization as such. In the sixties and seventies men organized important joint driving arrangements on a cooperative basis; there was in these agreements no provision for tolls or dividends, but only for sharing expenses; on their face the arrangements indicated that the parties looked for no direct profit from the joint transport effort, but valued it simply as an organizational asset useful to increase eventual output.[382] Later enterprisers commonly stipulated for tolls for their traffic management. Yet this development appears to have been of secondary significance. Most water traffic managers were involved mainly in moving their own logs or rafts; none made a great amount of money from carriage tolls; some who had the formal right to charge tolls did not exercise it.[383] What really counted was better coordination of movement of goods. Compared with the attention given to devising better procedures of driving and booming, the absence of significant direct profit from this activity, or even of much effort to make a profit, points up the importance which lumbermen attached to achieving more effective stream-use organization as such. Men who thus estimated their interest were bound to put great store in obtaining franchises. They wanted organization to facilitate enlarged operations, but they could not safely organize as they wished without law's validation. The capital stakes were always high — not so much the investment required for stream improvements and management as the commitment of a winter's woods output which must reach market. Enterprisers needed assured definition of the terms on which they might have access to public waters, and of the terms on which they might expect to deal with those whose activities would impinge on their own in the rush and hazard of the drive.[384] Discipline would be necessary, but even the large operator hesitated to discipline others without knowing he had support of law.[385] In a showdown those who would manage traffic wanted law's help to enforce tolls and liens and to set bounds to liability for alleged faulty performance or civil wrong. Even when large operators flexed their muscles, they sought color of law for the traffic controls they would impose on others.[386] Though dealings were typically within a small group of operators sharing a watershed, the pressure for organizational regularity showed itself in increasing resort to incorporation; it was not coincidence that the greatest relative use of the corporate form was for toll booms and toll improvement ventures.[387] Of course, stream traffic organization developed as it did for business reasons, but in this instance it was inherent in the nature of the business objectives that the availability of a franchise should be a material inducement to action.

Having paid our respects to the promotional aspects of stream-use franchises, we must return to the fact that they also bore restrictive features — both the burdens they allowed the franchise holders to impose on other persons, and the burdens they laid on the holders for the benefit of other persons. Here especially, when we ask what were the practical effects of franchise terms, we must be content with inferences from scattered evidence. For the practical meaning of these franchise features lay wholly in the cumulative impact of operating incidents which left no regular record.

Grantees enforced their rights to toll for use of the waters they improved, though — as we might expect — sometimes the users contested the legitimacy of the claim or were poor risks for the money they owed. We have noted that toll returns were not a major item of income for improvers, and that some grantees did not assert their right. Nonetheless, there were enough occasions of explicit conflict in the legislature and in the courts, and sufficient activity in amending the toll terms of franchises before and after enactment, to show that men treated these claims with respect for their practical effect. We shall take further note of this matter in a later chapter dealing with the police power in its bearing on franchises.[388] Of sharper significance were the rights of traffic management implicitly or explicitly conferred by improvement franchises. It was over the physical control of the passage of logs or rafts that men fought the most bitter battles in legislature and in court.[389] The episodes of forcible self-help which dot the record concern command of physical movement.[390] Most log and lumber transport flowed without acute controversy. As the volume of traffic grew, more and more of it moved by coordinated effort under single management, and the want of conflict was the most striking evidence that men in fact used their traffic control rights, and used them with at least tolerable restraint and effect. Such an outcome matched the logic of the total situation. Workable traffic management rested on franchise rights which reflected imperatives of fact. So, in 1879 the Wisconsin court observed of such a delegated power that "the utility, indeed the necessity, of booms at convenient points for receiving, assorting and distributing logs . . . is so universal on such rivers [as the Wisconsin] that it is judicially recognized as entering into the law governing their use." [391]

So far as transport was involved, the important regulatory or restrictive aspects of franchises were those that affected practical freedom of passage. Location of works was of basic significance to navigation. On this point the enterprisers' acceptance of the business values of legitimacy of title apparently carried with it general observance of the area limits set by franchises; neither in legislative maneuver nor in litigation was there a significant amount of trouble over infringements of location limits.[392] Some violations of limits set on the height of dams found reflection in legislation or lawsuits arising out of navigators' complaints, but the evidence does not suggest that this was often a major item of practical concern.[393] Most interest centered on franchise terms that dealt directly with passage facilities — provision of slides in dams and exits in booms, and the exercise of care by works operators against unnecessary damage or delay in passing or sorting logs or rafts headed downstream.[394] In the nature of the facts no statutory fiat could foreclose the many disputes and adjustments which such stipulations might involve. Care in operations could be measured only against the infinite variations of the days. What was a "suitable" slide for a given dam depended upon the peculiarities of the location, and — as we can see in recurrent disputes about the Kilbourn dam on the Wisconsin River — might have to be worked out by successive rebuildings.[395] There was the greater likelihood that navigators would be dissatisfied with the observance of such navigation guaranties because, though their application called for flexible adjustment to particular situations, the only means that nineteenth-century law offered for resolving disputes were the relatively slow, costly, and clumsy remedies of litigation. The considerable amendment of such franchise terms preceding and following enactment indicates the reality of the adjustment problem, as well as the fact that men took these passage provisions seriously. But legislative licensing could be effective only in quite general terms, even when the focus was upon particular situations. Again, the demonstration case was the legislature's repeated concern with complaints over the slide at the Kilbourn dam; as he reads the troubled, fumbling efforts of legislative committees to resolve conflicting testimony and bring dispute to some sensible solution, a twentieth-century observer feels the implicit pressure of the problem toward an administrative handling which the times were not yet ready to provide.[396] The inference is that the practical function of franchise navigation guaranties was mainly to provide legitimated criteria by which private parties stood warned and could bargain and argue out their differences. It is impossible now to tell how much informal controversy there was, or how far the franchise holders in fact fulfilled their obligations toward navigation. There is not a great quantity of recorded legislative or litigious trouble, relative to the volume of stream franchise activity.[397] This seems an area of behavior in which we should be especially careful not to generalize from trouble cases which left unusually detailed records. One fact above all makes it reasonable to believe that franchised works were built and used, on the whole, with fair attention to the navigation guaranties. Most works of navigation improvement were built and used by men who themselves were large transporters of logs or rafts; more often than not, the builders were the largest users.[398] Forest-region power dams and mill town developments offered more occasion for conflicts of interest. But millmen, too, needed the streams to move their raw materials and finished products. Moreover, the sharpest conflicts between transport and power — as on the Black River, or on the Chippewa at Eau Claire — did not arise out of simple heedlessness or lack of care of millmen toward passing traffic, but reflected a more complex phenomenon: competition between communities or segments of the lumber industry over the total profits of processing and distribution.[399] It would exaggerate the problems of enforcing franchise navigation guaranties as such to chalk up under this head the cases of total competition between local or regional economies.

Franchises to develop water power carried their own restrictive features. The inclusion of navigation guaranties in a dam franchise acknowledged that the grant in itself allowed the dam builder to impose on passing traffic such control of flow as was otherwise reasonable to accomplish the power use. Most licensed power dams were built, along with other works, under general statutes. Each constructed work, by its very existence, constituted an exercise of privileges to affect traffic or power uses of the waters.[400] Beyond this feature, many works licenses (for developing power, or for navigation improvement) delegated the right of eminent domain. The delegation had two aspects. Employed at the initiative of a works builder, the eminent domain power enabled him to restrict the ordinary right of a landowner to bargain in market for the value of his property. Employed at the initiative of a flowed landowner, the eminent domain power was a limitation upon the works builder, requiring that he pay fair compensation for benefits which he obtained at the cost of inflicting certain kinds of detriment on another.[401] We have no practical means to inventory the total use of these delegations of eminent domain authority. Two things are clear. Draftsmen of franchises put high value on including the eminent domain power; whether grantees used it or not, they wanted it in reserve.[402] On the other hand, prevailing community values were equally clear that only through eminent domain might men be deprived of the use opportunities that went with standard titles to land surface.[403] There was substantial legislative attention in the lumber era to drawing and redrawing the terms and procedures by which eminent domain power was made available for building dams and improving navigation. Both from its timing in the state's economic history and from the face of the statutes, it is clear that this activity had reference primarily to forest-area developments. On the other hand, there is evidence that dams were built in the north country which flowed lands that did not belong to the dam builders, and that no proceedings were taken to pay compensation and validate flowage rights.[404] A good deal of dam building and stream improvement went on in the forest region in wild and unsettled areas. Land titles would often be in owners who held only for the value of the standing timber, or in local governments which had succeeded by tax titles to land then regarded as of little or no market value. In the circumstances, it is reasonable to infer that the operative significance of the eminent domain provisions for much of the forest country was as a reservation against exceptional difficulty or complaint.

Over-all, and despite the limitations of the evidence, we can believe that the practical impact of the stream franchises was substantial. In some aspects the emphasis was on law-fostered values of legitimacy and order. In others it was on the industry's business needs and on still broader functional requirements of the economy. The lines of influence tended to weave together, and through them the stream-use franchises became integral to the whole pattern of the industry.

Location and Quality of Decision Making

We have examined some relations between law and the lumber industry's use of the streams, as factors affecting production and distribution of forest products. This approach emphasizes economic values and results, defined by input and output. Such is the more distinctively economic aspect of legal-economic history. But we miss important dimensions of reality if we deal only with product and not also with process. The dynamics of defining values and achieving results derive from the ways and means of allocating scarce resources. Economic order and legal order are both responses to pressures of scarcity — to the fact that, since life does not offer men unlimited fulfillment, they must make choices. So the organization of decision making is an element basic both to men's economic and to their legal experience; as the economist studies the market, the lawman studies the separation of powers. Herein lies the more distinctively legal aspect of legal-economic history. Even in its relatively crude nineteenth-century condition, Wisconsin law was an institution with its own internal economy of ends and means, generating distinctive challenges and tensions out of its existence and its growth in relation to other institutions, notably the market. Because it held the legitimate monopoly of violence, and as a corollary claimed a veto upon all other forms of secular power, law derived its most distinctive influence from the qualities and defects of its organization of decision making. Thus we should reappraise our materials as they reflect the special contributions of legislative, executive, and judicial processes to the growth of stream-use franchise policy.

This appraisal concerns only the operation of agencies within the state organization. Federal subsidy and regulation were too limited in scope to make federal processes of major significance to use of Wisconsin's forest streams. This outcome was itself in part the reflection of organizational aspects of the national government — the want of legislative or executive staff to promote the formulation of broad, long-term policy on natural resources, the prudent refusal of the United States Supreme Court to construe the commerce clause to give Congress exclusive jurisdiction over waters primarily of state concern, the political pressures upon Congress to let policy drift in ways most satisfying and least disturbing to local constituencies. The sum of these features of federal process was the practically complete freedom of the state to deal as it would with the forest streams. Thus it fits reality to focus upon decision making within state legal order.

It is appropriate, also, to make this appraisal without waiting upon examination of law's relations to railroad development. The rail net became important to log and lumber transport in Wisconsin. But the railroads were a mighty force in their own right, which brought problems and pressures to law much broader than those simply of the lumber industry. The growth of stream-use franchise policy is, in comparison, a relatively self-contained story. Moreover, because of the striking bulk of special legislation in this field — far greater in quantity and variety even than the special laws concerning railroads — the stream franchise history has its own value as a setting within which to study nineteenth-century legislative process.

We shall consider: first, formal and operative definitions of decision-making roles among agencies; second, principal elements in decision making; and third, relative expressions of public and private interest, or of general and particular values, through official decision-making processes.

Definition of Agency Roles in Decision Making. Decision making proceeded, in varying proportions, through (1) processes for enacting constitutional provisions, and through the regular operations of (2) the legislative, (3) the executive, and (4) the courts.

(1) Formally, the law's basic decision-making processes were those which

produced a constitution or constitutional amendments. Obedient to the terms set by Congress for admission to the Union, its makers wrote into the Wisconsin constitution the navigation guaranties of the Northwest Ordinance. So far as this provision included a specific rule — the ban on taxing navigation — it proved to declare a definitive boundary for state policy; at least, the legislature never challenged the limitation. Legislative practice and judicial interpretation restricted the ban closely, as applying to the taxing power alone, and to taxing transport use as such; under the police power the legislature might grant toll rights to one who improved navigable capacity, and might levy a gross receipts tax on the improver. This approach represented no want of respect to the constitution. By familiar doctrine, the state constitution was not a grant of specific legislative powers; the vesting of legislative authority carried full discretion to the legislature to shape the content of public policy, subject only to such limits as the constitution declared. Moreover, the classic tradition in this country valued a constitution as a warrant for orderly growth. Hence we presumed that constitutional language which was not specific should be treated as setting standards and not rules for official action, inviting development of particular content by responsible action of the operating agencies as they dealt with unfolding experience.[405] Reinforcing these doctrinal attitudes was the powerful emotional temper of late nineteenth-century Wisconsin, which ran against ready resort to decisions by the constitution-making process. Constitution making or amendment invited a range and depth of policy consideration which these bustling, market-bent years begrudged to public affairs. Oppressed by an unwieldy bulk of particular business, the governor and legislature put the constitution-making process in motion to obtain the 1871 amendment limiting special and local legislation. It was characteristic that, even so, they dealt with general issues only in those particular forms which chanced to gain focused attention. Thus they forbade further special charters of incorporation, but not special franchises for stream use; there was no magic in the constitution-making process to produce fruitful generalization of policy. On the other hand, again the record showed that specific constitutional commands affected legal process. The 1871 amendment stopped the flow of special corporate charters for stream-use enterprises. So far as deviations threatened, the judiciary committees of the legislature and the governors, as well as the court, showed zeal to enforce the new constitutional limitation.[406] Nor was constitutional language without effect on political behavior simply because it was treated as setting standards rather than precise rules for official action. Opponents of franchise bills appealed to the navigation guaranties, as well as to the constitutional injunction that private property might be taken by eminent domain only for public purposes, to enforce arguments that particular measures lacked warrant of public interest, and legislative committees as well as the governor and the court paid heed to these reminders as they justified or rejected (mainly, it is true, justified) uses of official licensing power.[407]

(2) Within the range of legal action, from the beginnings through the full flow and ebb of the lumber industry, the legislative process produced the greatest volume of basic official decisions on stream use. This is not to say that the legislature as a body was the source or even the producer of most of these decisions. Source implies initiative; the initiative — whether in proposal or opposition to creation of stream-use franchises — was typically with immediately involved private interests. Production implies control; effective control in shaping legislation on forest stream use typically lay with the forest-region legislators and entrepreneurs; wanting knowledge, interest, or means to muster force on matters remote from their constituencies, legislators from the southern half of the state rarely made a record of distinctive participation in this business. Granted these important qualifications, the fact remains that the bulk of significant stream-use franchise policy obtained form and content by what happened in the legislature. The primacy of legislative process was expressed in statutes which required legislative license for works on navigable streams, conferred navigable status in law on defined waters, validated originally unauthorized works, and set terms on which milldams might be built on nonnavigable streams. The most conspicuous mark of legislative primacy, of course, was the accumulation of 802 special statutory franchises for dams, channel improvements, booms, and driving or traffic management enterprises.[408]

The primacy of legislative decisions was confirmed in the behavior of the governors and the Wisconsin Supreme Court. The governors participated in the legislative process and on the whole tacitly accepted it for what it was, as they signed the bulk of what the legislature adopted. The occasional vetoes — objecting to inadequate navigation guaranties or to unfair treatment of particular grantees or affected persons — by their focus upon specifics implicitly conceded that generally policy in this field should be defined by what the legislators would approve. The one intervention by a governor with broad effect on legislative behavior was Rusk's successful insistence that special franchises to individuals should always include reservation of legislative power to amend or repeal. It accords with the whole pattern of agency relationships that the governor used his position here to insist on maintaining the potential vigor of the legislative role.[409] The Supreme Court consistently deferred to the primacy of legislative decision in stream franchise policy. The judges did not elaborate explicitly upon a presumption of the constitutionality of legislative action. But in practice they applied a strong presumption in favor of the validity of the fact findings and the value judgments implicit in stream franchise statutes. The court could the more readily take this line because most franchise legislation was promotional rather than regulatory in its prime objectives, and thus it accorded with the deepest prevailing values of the society, which the judges shared. Nonetheless, what promoted some interests almost always produced a regulatory impact upon some others; had this not been so, men would not have brought the lawsuits which gave the court occasion to rule on legislative authority. Thus it is the more significant that the court supported all major aspects of the legislature's franchise activity; the judges could not bestow their approval lightly, for they always confronted an aggrieved litigant to remind them of the consequences of legislative action.[410]

(3) It fits the general course of nineteenth-century state government that the executive branch played a minor role in waterways policy affecting the lumber industry. Save for urging federal grants in aid of navigation improvements, no chief executive ever expressed an affirmative generalization of public policy in this field during the lumber years.[411] The negative interventions by veto were episodic in concept and in effect, except for Rusk's insistence on the reserved power clause. Even in that instance the governor's energy was bent to a negative rather than a positive end, to retain legislative capacity to correct a mistake or withdraw what time might make unduly burdensome. Governors used the veto power for such limited objectives that they made it more an instrument of case adjudication or administration than an expression of the chief executive's role in shaping legislation.[412] So viewed, the use of the veto might fairly be said to have accorded with the legislature's own working conception of its job. Nothing stands out more sharply on the record than the contrast between 802 special stream-use franchises and the thin output of general legislation on stream use. Despite brave assertion of its general licensing prerogative, in practice the legislature was itself content with a role more administrative or adjudicative than legislative.

Neither legislature nor court had direct impact on the governor's role — which is not surprising, since the governors offered so little to which the others could react. The legislature in each instance deferred to the veto; the rare efforts to override vetoes fell short by such definite margins as to invite respect for the practical power which the governor held — where he chose to use it. Of course, had governors wielded the veto with more boldness or in the interest of greater generalization of policy, their vigor might have stimulated greater force in response. However, as the record stands, it raises question whether imagination and energy might not have made the nineteenth-century chief executive more of a leader and less of a reviewing judge over legislative activity.[413]

The governor apart, there was so small a state executive establishment through the lumber years that there should be no surprise that no other significant executive influence came to bear on stream franchise policy. The Attorney General was preoccupied with general business of the state, and his connections with waterways matters were even more episodic than the governor's interventions by veto. There was no specialized state administrative office for waterways affairs, not even an agency like the ineffective office of state land commissioners. Detailed implementation of state policy was usually at the mercy of the uncoordinated efforts and uncertain valor, energy, and ability of local public prosecutors; the meager record of public nuisance prosecutions reflects how little any development of policy could be expected from this source.[414]

Conceivably, the legislature might have delegated to local government the issue of stream-use licenses and supervision of licensed action. A few municipal charters provided for harbor controls, and in a handful of cases the legislature authorized local governments to subsidize area navigation improvements. However, contemporary thinking did not include economic planning as a municipal role. Local government was crude, and in the forest area it often existed primarily to create petty patronage. Apart from a few mill centers, most lumber industry exploitation of the streams for transport or power was scattered, and much of the important activity went on in wild, open country; the county, rather than villages or cities, would have been the appropriate local agency to deal with such franchise matters, but county government in the forest region was the least developed of all types of municipal organization, conventionally confined to a narrow round of public housekeeping chores centering on the courthouse, with some attention to roads.[415] There was only one occasion when men involved municipal government in a major franchise issue. Chippewa Falls millmen opposed efforts of downstream Eau Claire mills to obtain a statutory license for a great dam and boom near Eau Claire on the Chippewa River. Chippewa Falls attacked the Eau Claire application as a project purely for private advantage. To meet the charge, the proponents recast their proposal. They now asked for a franchise to allow the city of Eau Claire to build a dam, with authority to the city to use the dam to develop a water supply or to exploit the water-power and booming opportunities in the authorized area. The Wisconsin Supreme Court invalidated the first franchise granted in this form because, as drawn, it did not require that the franchise be used primarily to develop a public water supply, but allowed the city to create power and booming facilities primarily for private benefit. The Eau Claire interests obtained a second franchise which required unequivocally that the authorized dam be used primarily to develop a public water supply — with authority in the city to develop by itself or through lessees such water-power or booming facilities as might be reasonably incident to the primary public purpose. The court ruled that the second statute created an effective franchise. The exceptional character of the invocation here made of local government is revealing. Municipal administration was not thought of as a normal means of executing policy of state or regional significance. The city was brought into an important franchise matter simply as a maneuver in a tense and involved legislative battle. Circumstances thus dramatized the limits of contemporary concepts of municipal roles.[416]

The legislative record shows some groping toward definition of a dimly felt need for administrative apparatus to implement statutory policy. The milldam act committed to a jury the determination of the fair value of land flowed, and also empowered the jury by its verdict to regulate "the height to which water may be raised, and the length or period of time for which it may be kept up in each year." The eminent domain procedure made available for river improvement ventures after 1880 provided for court appointment of "three disinterested and competent freeholders," residents of the county, to appraise compensation for flowage.[417] In a short-lived flurry in 1856 and 1857 several special dam franchises provided for appointment of commissioners *ad hoc* to deal with the more basic issue — truly a matter of delegated legislative function — "as to the necessity of authorizing for the benefit and convenience of the public, and for the public good, the erection and maintaining . . . a dam" at a specified location.[418] Episodically, developing no general policy in the matter, the legislature continued to make occasional resort to *ad hoc* commissioners to provide some follow-up on franchise stipulations. So Laws, 1873, Chapter 126, declared that the Stevens Point Boom Company must maintain a free passage in the main channel of the Wisconsin River until three commissioners named in the franchise, or a majority of them, certified to the circuit court that the Company had improved a specified slough in order to afford like ready passage. So in the eighties a number of dam franchises stipulated that decisions upon the times of opening or closing flood gates, and how long the gates should be open or closed, should be made by three arbitrators; the arbitrators should be named annually, one by the grantees, one by the governor, and the third by these two, or, in default of such selection, by the governor "on application of any person or corporation interested."[419] Following adverse recommendations of its committees, the 1895 legislature rejected two bills which would have authorized the governor to appoint commissioners to regulate water flow over all Wisconsin River dams in order to assure most effective use of the stream for log and raft transport.[420]

Such occasional enactments and proposals have significance only as by contrast they make plainer the operating implications of the bulk of stream franchise terms, which made no provision for independent administrative surveillance but, in effect, delegated implementation to the initial discretion of grantees, checked in the first instance by the bargaining and complaints (and occasional self-help) of others immediately affected, and subject to specific official intervention only by uncertain and costly litigation. Of course, franchise grantees had no interest in promoting development of licensing or supervisory administrative apparatus. That a statutory grant should be self-executing in law, requiring no further official validation, was obviously to the grantee's advantage, limiting his burdens of persuasion and proof and vesting his title and prerogatives forthwith.[421] Those who feared that their interests would be adversely affected were quite active in pressing for, and often obtained, amendments in course of passage on matters of location, height, passage rights, and toll terms. But beyond the limited, generally isolated incidents we have noted, the record shows no efforts to obtain administrative devices to discipline the use of franchises. Men's concepts of legal sanctions and procedures had not yet grown to include flexible use of administrative techniques; such was the weight of inertia that current attitudes and ideas confined the imagination even of those to whom a break-through would have been an advantage.

(4) The content and tone of the Reports show that Wisconsin judges fully shared the confidence of their nineteenth-century colleagues elsewhere in the idea that the bench had both title and competence to play a substantial role in making general public policy. It was confidence developed primarily out of the experience of fashioning a great body of common law in areas of conduct for which legislatures showed little concern. It was confidence developed also from the necessity of exercising discretion and responsibility to fill gaps and correct incongruities left by the loosely framed product of the nineteenth-century legislative process. The Wisconsin Supreme Court demonstrated both aspects of the judicial policy role in matters concerning use of forest streams. Building on judge-made doctrine from older lumber states, it extended the common law concept of navigability to protect all feasible log movement by water.[422] Making good the shortcomings of statutory language, it gave liberal interpretation to particular powers of river improvement companies, and insured that improver and flowed landowner alike might initiate eminent domain proceedings.[423]

Though the Wisconsin court showed such vigor, the natural counterpart of the primacy of legislative decision in stream-use matters was a relatively subordinate, supporting role for the judges. The decisions the court made which had the greatest working significance for the lumber industry were decisions which liberally sustained the franchise-granting powers of the legislature. Even where it defined the common law of waters, the court paid marked deference to the legislative process; at Wisconsin common law any works erected in the channel of a regularly navigable stream without statutory warrant constituted a public nuisance in itself, without proof of obstruction in fact.[424] This judicial deference was not the fountainhead of Wisconsin policy; it followed upon vigorous assertion of legislative primacy early in the territorial years, first by the milldam act in 1840 as to nonnavigable waters, then in 1841 by the general requirement of legislative permission for works on navigable waters.[425] Indeed, as is not uncommon in the relation between legislative and judicial action, cases did not present themselves to afford the court opportunity to speak on most of the important constitutional issues affecting stream franchise policy until years of legislative activity had established the legislature's practical construction of its powers.[426] The pattern of the court's responses consistently supported policy as it was declared in legislation; the court construed constitutional navigation guaranties to allow broad legislative discretion in promoting enlarged use of waters, adopted in practice a strong presumption in favor of the validity of legislative findings and judgments, and gave liberal interpretation to the powers for stream improvement conferred by statutory franchises.[427] There was little specific legislative reaction to the court's performance in stream franchise affairs. This was natural enough, since the court broadly upheld almost all that the legislature did; in the rare instances in which the court held legislative action ineffective (notably in the first Eau Claire dam case), there was the less occasion for serious conflict because the court's opinions did not bar all legislative action but simply found defects in the style of particular grants, which the legislature corrected.[428] Had the legislature been more active in general policy innovations, there might have been strain; it is significant that there were sharp conflicts of court and legislature at the end of the lumber era, when legislation introduced new concepts of public interest and created administrative procedures to regulate hydroelectric developments on the waterways.[429] In the late nineteenth-century legislative policy was primarily promotional and relied primarily upon delegation of stream-use development to private entrepreneurs; this was a bias of policy which found deep community consensus, in which judges shared. This consensus made it easier for the court to practice a strong presumption in favor of legislative action, which in itself minimized occasions for legislative-judicial tension. Finally, most waterways interest relationships fell within relatively local or restricted sectional bounds; there is no evidence, for example, that any of the most controversial affairs on the Chippewa River had repercussions within the Wisconsin River valley, let alone in the more distant northeastern watersheds of the state; these forest-area water-use issues did not have the potential for generating large interest-group response such as fired the railroad farm mortgage battles of the late fifties and the sixties.[430]

Legislative primacy in the growth of stream-use policy was built deeply into the total situation. This was the heart of the matter: circumstance contributed more to the central role of the legislative process than did the uneven energy or the narrowly operational viewpoint of the characteristically inept nineteenth-century legislature. Whether used for power or for transport, surface water courses were basic assets for the general operation and development of the economy. The fact was so plain that there was never serious question but that these waters must be treated as a commonwealth resource. It was equally obvious to contemporaries that this did not mean the waters were to be hoarded for some indefinite future, or that their exploitation should wait upon the uncertain and slender fiscal resources of state or local governments. They should be made available in ways calculated to multiply productivity. In contemporary context this meant that they should be made available to enterprises privately devised and privately administered, for uses of public interest. Thus a major aspect of the organization of power in this field was the familiar nineteenth-century delegation to nonofficial actors of the initiative of decision and implementation in matters of general concern. Even so, thinking started and returned to the facts of economic function and community dependence, which men found so fundamental to the situation that they never considered fragmenting water-use development into the seeming simplicity of fee simple titles which they so readily adopted for disposing of the public lands. There would be broad dispersion of authority over detailed exploitation of the streams. But this must be effected within legally formulated objectives, subject to continuing possibility of official scrutiny in operation, and subject to recall in the public interest.[431] The legal tradition which put the control of the public purse firmly in the legislature required that such broad, continuing supervision of community resources likewise rest in the legislature.[432] Moreover, such resource control invited a range of fact finding, an extent of new interest creation, and leeway for innovation in forms and procedures of licensing and sanctions which far exceeded the traditional limits of either "the executive power" or "the judicial power" which the Wisconsin constitution vested in the other principal agencies of government.[433] For all its frailties the legislative process was bound to be the principal center of decision-making activity over use of the streams. Only the legislature had the institutional endowment or the public consensus out of which to make the necessary responses to the pressures of the state's development.

Elements in Decision Making. Making public policy is a complex business. It is nonetheless so, when policy makers do their job crudely or without full awareness of what the job inherently involves. To make official policy men must (1) perceive problems, (2) find facts relevant to workable comprehension and resolution of the problems (whether the finding be by fresh investigation, by draft upon present knowledge, or simply by assumption), and (3) choose among limited means and ends, expressing their choices through formal or informal operations of legal process. The qualities and defects of performance on these counts form the history of policy-making process.

Process is elusive. Much, perhaps the greater part, of the action which expresses process leaves no record save what may be implied from success or failure of the end product. Inventory and appraisal are no easier because often the most significant aspects of the story are the problems not perceived or ill-perceived, the facts not found or ill-found, the choices made in effect but not consciously. We should be modest, therefore, in claiming to tell the history of ways of decision; we can be sure that we can capture only part of what might be told. Yet, if we do not try, we omit a dimension of legal history. Much of what should be said has been indicated already, in cataloguing types of official action and outlining general waterways doctrine and the particular content of stream-use franchises. Here we shall use the same material to a different end by focusing upon process rather than upon product.[434]

(1) Policy begins, for better or worse, with perception. The first contribution that official process can make to men's rational adjustments to life's scarcities — first in importance, and (ideally) first in time — is to advance their awareness of their situation.

Use of the forest streams, like use of the public lands, implied multifactored judgments of political economy, reaching beyond the affairs of any particular venture. At issue was the disposition of resources which by common consent were public property, hence ultimately under the legislature's control. Such a situation offered only limited opportunities for courts to help make general policy. They had no authority to grant franchises or otherwise to make general dispositions of commonwealth resources. They could work only within the frame of reference provided by the accidents of litigation, limited to such factors of judgment as were relevant to the issues shaped by a particular lawsuit. The broadest opportunities for judges to contribute to policy making in such a realm lay in clarifying and rationalizing the initiatives taken in legislation, the better to define what was at stake and what gains and costs must be reckoned.

Within the frame of common law litigation judges were usually impelled to define only factors brought into awareness by the interests of the suitors; even if the court's imagination ranged more widely, the limits of tradition and propriety implicit in the constitutional vesting of "judicial" power held judges to considerations relevant to disposing of specific disputes. Given the scant development of the executive branch, suitors were almost all private litigants, moved to examine only their private concerns. Where men dealt with such a commonwealth resource as the streams, the claims which private suitors might press in lawsuits were quite limited to start with, unless legislation enlarged the rights or privileges of particular operators. Not surprisingly, hence, the common law of stream use represented a quite limited perception of problems. The most distinctive common law waterways doctrine was the standard of reasonable use; a private enterpriser must employ a stream in ways reasonably accommodating the uses of others. It was a standard of common-sense, but limited, operational emphasis — a natural product of building policy out of accretion of particular instances centered upon the concerns of particular operators. Similar — though overlapping the fields of common law and judge-made constitutional law — were the judicially developed criteria for deciding when there was a "taking" of "private" property, so that compensation must be paid under eminent domain proceedings. No root generalization united these rulings; the resulting doctrine was the product of weighing relative deprivation, case by case. Where policy was left to be formed thus by the impress of particular disputes incident to limited enterprises, there was the less stimulus to generalize the grounds or ends of policy.[435]

When the legislature stepped in, it usually made little record beyond what it enacted, to enrich the definition of the situation. Yet, since it alone could make positive disposition of commonwealth resources, legislative intervention expanded the opportunities of judges to expound and rationalize community policy — as they ruled upon the validity of what the legislature had done, or interpreted its meaning. The opinions of the Wisconsin Supreme Court dealt with waterways legislation in pedestrian fashion. But — perhaps because the traditions and the more focused nature of its job put the court under sharper pressure than the legislature to integrate doctrine and explain to the losing party — the court did therein add substantially to eliciting the reasons and the objectives of public policy. Thus it was the court that most explicitly spelled out that legislative license must be had for works built on navigable streams, but only on streams which had sustained use for navigation; that the legislature might grant exclusive or limited-monopoly privileges where such grants might reasonably be deemed to promise multiplier returns in over-all productivity; that the legislature might make reasonable judgments preferring one category of stream use over another, even to the point of precluding certain competing uses altogether; that navigation should generally be preferred over other uses of water; that grants of franchise powers auxiliary to increase of productive capacity should be liberally construed; that in all respects the public interest lay in such ordering of stream use as would more speedily increase forest production.[436] Having paid the judges their due for what they did to illuminate the grounds and nature of legislative activity, we must note, however, that the initiative in all these matters was with the legislature. It was legislative activity which first carried these issues over the threshold of official awareness. Without legislative activity, there would have been no opportunities for judicial exposition, and usually the legislature had had its say for some years before lawsuits presented the court with opportunity to add its gloss to what legislative practice had already established.[437] Legislative initiative might, indeed, even lie in the background of what appeared to be primary perception and evaluation of a problem by judges. So it was with the definition of "navigable" streams. In 1868 the Wisconsin court first declared navigable a stream capable of floating saw logs. It justified its ruling as one established "by the common law"; counsel had furnished the court with relevant citations from Maine, Michigan, New Hampshire, and New York. There is no great importance here in assigning credit for originality. But it is relevant to estimating relative roles in bringing issues to awareness to observe that for about thirty years before the Wisconsin decision Wisconsin legislatures had in practice recognized the log-transport test of navigability by the terms of navigation guaranties in special dam franchises.[438]

Clumsy though it was in expression, the legislative process did contribute to men's perception of problems involved in using the streams to exploit the forest. Its representative mandate and its awareness of its role as holder of the public purse (including control of commonwealth assets) prodded the legislature into some perception of the waterways as social overhead capital. Legislative product reflected this awareness — in the requirement of statutory licenses for stream improvements, the imposition of terms to protect general transport, and the tender of franchises as leverage to induce private investment. However, the nineteenth-century legislative process made its most pervasive contribution to men's awareness of their situation simply by its availability. It held out an ordered procedure through which men might obtain valuable privileges and protections in their access to transport and industrial power. Because the legislature held such gifts, and offered a legitimate means of obtaining them, it constituted an attraction to men's ideas (and schemes), ambition (and greed), invention (and maneuver), stimulating them to sharpen their own perception of the possibilities of their situations. The reality of the stimulus stood most vividly attested in 802 special stream-use franchises.[439]

There were significant limits upon the perception of the situation achieved in the legislative process. Foremost was the extent of reliance upon delegating to private actors the initiative in defining values, opportunities, costs, and hazards. Even a well-organized legislative chamber is ill-suited to original research or invention. But through effective committees and staff it can add substantially to defining interests and problems pressed on it. The Wisconsin legislature of the lumber years had no staff beyond the minimum necessary for clerical business. It created no standing committees on waterways policy as such. The standing committees which handled the bulk of stream franchise matters — chiefly committees on the Judiciary or on Lumber and Manufactures — divided their time among a miscellany of other affairs, and in any case never did more than react to specific issues referred to them. The legislative record contains not one committee report seeking to define facts and choices common to the main flow of stream franchise business. Thus the most fundamental limit of perception arose out of the legislative role itself, as legislative practice shaped that role. In action the legislative role was seen as more akin to a judicial function (hearing and deciding upon specific cases presented by parties moved by particular interests), but lacking in its operation an element analogous to the appellate opinion, which played so important a part in judges' contributions to bringing facts and choices into sharper awareness.

Corollary to this basic limitation in the working concept of the legislative role were two limitations in defining the scope of substantive policy. These were limits on perception of the horizontal (contextual) and vertical (durational) dimensions of choice.

Stream franchise policy developed mainly in response to activities conceived as local ventures — to provide a power dam at a particular place for a particular mill, improvements of the channel of a given tributary or a designated part of a principal river, a boom at a point especially suited to large sorting or rafting operations. Symbolic of the localistic approach to these matters was the occasional action by which a franchise bill was referred for consideration to a select committee consisting simply of the sponsor, or of legislators from the immediate area or from the county in which the licensed action would occur.[440] There is no evidence to deny that many of these enterprises were only of local importance, and in this aspect were correctly viewed in local context. Yet the substantial similarity of terms in hundreds of special franchises bear witness that they carried general policy implications. Where general problems exist, they should be matched by generalization of relevant facts and values. Such is, indeed, the distinctive function of "legislative" treatment of matters of public concern. To fail to generalize where the facts invite it is to fail to perceive one dimension of reality. This failure caused inefficient legislative operation; the legislature burdened itself with a volume of detail it should not have handled, as it recognized in a limited way when it sponsored the constitutional amendment of 1871 to end special chartering of corporations. Failure to generalize perception also raises questions of the substantive justification of limited decisions. If it was fair to specify width, height, graduation of descent, and key construction details for slides required on all dams on the Wisconsin or the Black River, why did the legislature never make similar stipulations for all dams on other great forest-area streams? [441] Why should a statute establish standard guaranties for safe passage of logs and rafts past bridges on the Wisconsin River, while elsewhere the legislature dealt with the matter *ad hoc*? [442] In 1863 a statute dealing with the Wolf River and its tributaries provided that whenever logs of two or more persons were so intermixed that they could not be conveniently separated for driving, and either owner failed adequately to provide for driving his logs, the other might drive all the logs and have reasonable compensation for doing so. If this was a sensible adjustment — as experience bore witness that it was — why did the legislature wait until 1881 to translate it into a traffic regulation applying to all streams? [443] Hundreds of special franchises were built around a common core of provisions as to location, height, navigation guaranties, terms of operation, compensation for flowage. That draftsmen's practice could so far standardize the general content of franchises acknowledged a common core of problems and values. Why, then, for sixty years did legislation spawn variations in secondary details (two general procedures for exercising eminent domain powers, for example, with other departures in statement sprinkled through special statutes), instead of achieving codification of standard formulas?[444] Of course, one may choose to interpret local variations as products either of selfish maneuvers for special advantage or of pragmatic adaptation to different facts. However, the relevant place to consider such appraisals is at the end of this discussion, when we must try to strike some judgments on how the legislative process served the public interest in handling stream franchises. Here our concern is with the quality of perception of issues. On this count, the bulk of variation in secondary details, on the one hand, and the extent of failure to generalize important policy, on the other, defy explanation in terms either Machiavellian or pragmatic. The legislative process was simply not geared to generalize perception out of particular pressures.

The excessive localism and particularity with which legislative process perceived issues delayed the creation of a sensible administrative process capable of coming to come to terms with the real problems of flexible application of general standards. More was involved than mere operational efficiency. Narrow perception of issues carried great danger of putting the legislative process under intolerable strains and encouraging resolution in private terms alone of problems of public concern. The only reason that the gravity of this hazard did not become more apparent was because most lumber industry operations were of relatively modest scale, compared with the maneuvers of the empire builders in railroads, oil, steel, or finance. But the long, stubborn, and bitter battle between Eau Claire and Chippewa Falls mill interests over a franchise for a great dam and boom works near Eau Claire demonstrated the point in a lumber industry context. In the estimation of the protagonists the issue was life or death for the firms and communities immediately involved. With forces closely matched, and their interests defined in terms which seemed mutually irreconcilable, the competitors were impelled to extremes in legislative maneuver. There were the greater temptations to over-reaching because, when the issue was seen as no more than a competition for advantage between two mill towns, there was little in the contest to engage the interest and responsibility of the legislature at large, least of all of the members outside the forest area. The efforts of the contenders were climaxed in charges and countercharges of bribery, which two legislative investigations did not satisfactorily put to rest, and which so troubled Governor Fairchild that in 1871 he vetoed the first successful Eau Claire bill. The Eau Claire millmen finally had their way, in the disingenuous guise of a franchise for an Eau Claire city waterworks; even in success, apparently free of corrupt influence, they subverted legislative process by obtaining a decision upon a false issue. The irony and the waste of this troubled history appeared to the full when in 1880 and 1881 Frederick Weyerhaeuser brought into negotiation not only the Chippewa Valley millmen but also the important logging interests concerned with driving logs down and out of the Chippewa to Mississippi River mill points. Weyerhaeuser persuaded the parties to define the problem not as a narrowly focused conflict between two mill towns, but as a challenge to rationalize the productive use of the Chippewa River system as an entirety. Within this perception of the matter, he obtained a private treaty which apportioned operating rights and spheres of interest and created lasting peace among the principal producers. Of course, making the treaty ultimately involved bringing wills into accord and obtaining decisions. But, especially in contrast to the parochial battles of the two cities in the legislature, it was the definition of the situation within a wider reach of interests and objectives which made these acts of will possible. Because contemporary legislative process was not equipped to facilitate such an imaginative perception of issues, not only did the legislature fail of sound accomplishment, but it brought its own integrity into question.[445]

If localism was the mark of foreshortened horizontal perception, which failed to generalize problems to see their broad context, casually quick response to pressure was the mark of foreshortened vertical perception, which failed to weigh problems by reckoning of long-term consequences. Stream-use petitioners did not always get from the legislature just what they first asked for. But almost always they got some response quickly, within the first legislative session in which they made application. The short span between plea and response was symptomatic. As with the later railroad development, the rapid, broad-scale exploitation of the streams materially accelerated exploitation of the forest, contributing to its irreversible depletion within sixty years. As with the policy adopted toward public lands, probably no other outcome was politically feasible. Yet, another aspect of these events is relevant to legal history. To employ the streams to accelerate forest production was — so far as concerned any calculus of time utilities — less the product of decision than of default or of opportunism responding simply to circumstance. A rational case can of course be made for applying a large discount to values that can be realized only in an uncertain, far-distant future. Most often, however, other factors (laziness, fatigue, habit, impatience, greed, bafflement) enter to incline men to weigh only the gains and costs they see close at hand. Against such odds, a function of law should be at least to stimulate early definition and deliberation, if not decision, on the longer-term consequences of present action, for it is the business of law to help order (enlarge the meaning of) experience. The only concession which Wisconsin policy made to a longer time perspective in stream franchise policy was to provide in the constitution (as to grants of corporate powers) and by statute (under the prodding of Governor Rusk, as to grants to individuals) that there should be reserved an authority of the legislature to change or repeal franchises.[446] The Wisconsin community had to mature a good deal more before a longer-term calculus of values generated substantial issues in the legislative process — as occurred after the lumber era, for example, in debate over policy toward hydroelectric power development on the principal rivers.[447]

(2) The second element in the policy-making process is to find facts relevant to marking the dimensions and understanding the content of choices. Perception and fact-finding play upon each other; we find only what we perceive, but ordered investigation also teaches us where to look.[448] However, the history of legal process shows that the two are significantly distinct operations of decision making, warranting separate notice.

By inheritance the legislature held the most extensive authority of any of the regular agencies of government to inquire into matters of public concern.[449] In practice, good investigation required more patience and skill than the untutored, short-session, scantly staffed nineteenth-century Wisconsin legislature could supply. The legislative journals present only skeleton records of the handling of most special or general stream-use bills. Only in a few unusually contested matters — notably, the disputes over the Kilbourn dam on the Wisconsin River, complaints over the traffic control assigned the Black River Improvement Company, and the battle over the works authorized at Eau Claire on the Chippewa — do we have reflection of substantial taking of evidence by committees.[450] The typical franchise bill underwent some amendment in course of consideration; so far as amendments clarified or limited the impact of the proposed franchise upon those who might be adversely affected, their presence implied attention to facts and arguments tendered by those who sought the amendments, perhaps countered by opponents.[451] However, the legislature sat in relatively short sessions, and stream franchise bills were only part of a bulky burden of business; it is unlikely that the committees did more than hear *ex parte* statements, relying otherwise upon the knowledge of local situations and industry problems which forest-area legislators brought with them to Madison.[452] This inference is the stronger in view of the light cast on current attitudes by the practice of referring proposed franchises for consideration by select committees of legislators from the immediately affected locality.[453] In the few more elaborate inquiries which left records, the level of accomplishment was unimpressive. Lacking staff and time, even on these occasions committees did no more than listen to conflicting presentations of partisans, and — in the outcome — use some verbal ingenuity to minimize the need to make strong or clear-cut resolution of issues. Suggestive of the committees' unhappiness with their investigating role was their readiness to treat the ultimate issue as one of allocating the risks of an uncertain outcome, usually giving the benefit of the doubt to that course which seemed to promise the greatest short-run increase in productivity, and inviting those aggrieved to return to the attack with new evidence at some conveniently distant time.[454]

Within the closer confines of issues defined in litigation, the courts were in practice better equipped than the legislature to elicit facts relevant to making policy; judges had the regular aid of counsel, working within detailed procedures for mustering evidence. However, it exceeded the sophistication of nineteenth-century technique for counsel to initiate, or the court to request, formal presentation of evidence relevant to formulating, as distinguished from applying, general policy; lumber industry waterways litigation includes no forerunner of the "Brandeis brief." What we do find in abundance is sweeping and generally unacknowledged resort to judicial notice, where judicial opinions take for granted as parts of common experience various propositions as to economic behavior or values — for example, that the eminent domain privilege was needed, lest hard bargaining in market discourage milldam construction; that enterprisers would not invest in navigation improvements without the inducements of toll rights; that a sheer boom could give adequate assurance that logs and rafts could pass the chute of a dam without undue delay or damage.[455] Occasionally the court made glancing reference to what it was doing on this score (". . . we take notice of the fact that the capacity of many small navigable streams in this state to float logs and lumber into the larger streams below and to market, has been greatly increased by the erection of dams across them to hold or discharge the water as circumstances may require"),[456] but usually the judges simply wrote into their reasoning their assumed sure knowledge of forest-area conditions, without feeling need to legitimize their practice. There is no case in the Wisconsin Supreme Court in which counsel objected to this tacit reliance upon judicial notice, probably because the judges stayed within the elementary facts of the industry situation.[457] However, that judicial practice stayed within propriety did not alter the result that, since the judges grounded their value judgments on such taken-for-granted premises, they contributed little to ordered investigation of the factual context of stream franchise policy.

(3) The third element in making public policy is that official power plays a significant part in producing decisions, that is, in effecting choices among scarce satisfactions and among alternative ways of using resources to achieve chosen ends. Decisions may be for action or for nonaction, for definitive resolution or for delay or avoidance. The heart of the matter is that what is done is done with awareness, else there is no decision in any proper sense. So long as there is choice with awareness, decision may have formal embodiment (as in a statute or a judgment), or may find expression through informal behavior (as by a calculated failure to act). We may estimate the decision factor in public policy making in two main respects. First, we may ask whether there is decision at all (as opposed to mere reaction to pressures of a situation or to the functional logic of a going institution). Second, we may ask how decisions measure up to some criteria of quality. Depth and range in perception, and skill in investigation, enter intimately into the quality of decisions; hence we may better postpone the problem of quality

to the conclusion of the present discussion, after cataloguing more separable elements. Here we need deal with the more elementary point — whether official process operated as a material factor in determining stream uses.

We have already considered this question from the point of view of the law's influence upon industry practice. We saw that the industry significantly adapted its behavior to requirements and inducements of law, especially in the obvious concern shown to legitimize access to transport and power uses of streams.[458] But this approach emphasizes industry reaction rather than the law's own impress. Did industry practice perhaps respond only as particular interests in the industry chose to make response for their own prudential reasons, or did the intervention of legal process substantially affect the content of lay behavior?

Clearly the legislative process had significant effect on behavior, beyond simply conferring assurance of legitimacy, in the prerogative aspects of stream franchises. Without positive legislative sanction, private enterprisers might impose their will on others involved in stream use only within the restricted discretion allowed by the common law of reasonable use; for rights of eminent domain, toll, or traffic management, there must be legislative license, and once the license was had, others must in turn adjust their conduct to it. Where men found it advantageous to set up a hierarchy of private wills, the intervention of legislative process was of material, indeed essential, effect in determining stream use.[459]

This, however, deals with matters only in the large. We can assess the influence of legislative process from another viewpoint, if we look at the typical record of proceedings in the handling of stream-use bills. Here we must look mainly at what happened concerning special and local laws. There was not much general legislation on stream-use privileges, and it dealt with matters on which there was such apparent consensus as to produce little record of legislative maneuver.[460] The great volume of special franchise applications shows more variety in treatment. Though there are too many gaps and obvious inaccuracies in the records to allow of precise tallies or fine calculations, there is enough definition of trend to let us make some general estimates.

If a stream franchise bill were introduced, there was high likelihood that it would pass, in some form. However, petitioners showed respect for the hurdles they must surmount; lumbermen came to Madison to shepherd their bills, and retained able counsel to help them.[461] The legislative record showed that a man had better attend to his business if he would obtain a franchise. Not every bill passed; over-all through the lumber years the legislature rejected between 10 and 12 per cent of the stream franchise bills introduced. More significant than rejection was amendment. A minority of bills passed in the same form in which they were introduced; at least two thirds of all stream franchises enacted underwent some amendment in course of passage.

Some changes, it seems clear, were made at the wish and for the benefit of the proponents of the franchises, to correct errors or oversights in the original drafts.[462] Most changes dealt with matters likely to affect the interests of third parties — the height of a dam, inclusion of suitable navigation guaranties, the terms of toll — and although at our distance in time we cannot confidently appraise the real balance of interests in all cases, it is not plausible to believe that this much effort was spent on points of no substantial concern to others.[463] Most amendment was done in course of passage. However, about 16 per cent of forest-area stream franchises were amended after enactment; the figure is substantial enough to indicate that the continuing possibility of legislative intervention was a factor with which men must reckon as they used the waters.[464]

Decisions, we observed, might find formal or informal expression. The formal style of decision through legislative process — the enactment of statutes — requires appraisal of a subsidiary but not unimportant aspect of policy making, which is the craftsmanship of the formal expression. Form should serve order, predictability, understanding. On these counts we cannot give good marks to nineteenth-century legislative process. Drafting was clumsy, careless, and erratic, and action was obviously sometimes ill-considered. The legislature was not infrequently moved soon after enactment to correct errors or inequities in franchises.[465] Points which should have been dealt with by clear-cut standard formulas were treated with needless diversity and ambiguity. Most dam franchises, for example, stipulated a maximum height to which the dam or the water might be raised, but draftsmen alternated without apparent reason among different ways of stating the limitation, and often imposed a height limit without defining a base line from which it should be measured.[466] Draftsmen of special franchises often sought to shorten their labors through incorporating by reference eminent domain or lien procedures, or statements of toll rights or service obligations, drawn from other statutes. They often did so, however, with such incomplete or ambiguous description of, or qualifications upon, the general sources on which they drew, as to complicate rather than to simplify their product.[467] Scores of franchises present idiosyncracies of statement in secondary details. Franchises enacted at the same session for similar purposes differed in terms, without evident justification.[468] An examiner bent upon finding covert meanings may interpret these erratic variations to reflect special interests. But the very bulk and inconsequential character of so much variation render this an implausible interpretation; melodrama does not come in job lots. What is indicated, rather, is the effect of headlong economic growth which outstripped the slender technical capacity of contemporary lawmaking — and the effect, too, of the closely related, narrowly operational and impatient temper of the times, which had scant indulgence toward developing a professional approach to public business.

The courts, however, thanks to a more highly organized procedure, employed under insistent pressure of parties and counsel, typically feel under greater pressure than do legislatures to bring each controversy to some formal decision. This does not mean that judges must answer every question litigants would like them to answer. Judges have scope for making their own informal dispositions; the court may find that a party lacks standing to obtain a decision on a given point, or that a rule of constitutional or statutory construction disposes of an issue short of probing its depths. However, there was little maneuver in the Wisconsin Supreme Court to avoid the more basic issues of stream-use policy involved in litigation. In practice the court applied a strong presumption in favor of the constitutionality of statutory franchises calculated to promote general economic growth.[469] In the Eau Claire dam litigation the court stoutly refused to look behind the formal face of a public waterworks franchise, to consider whether the legislature's motive had in fact been only to favor private advantage.[470] In these respects the judges followed doctrine firmly established since John Marshall, and did no more than observe proper limits of judicial review. The judicial behavior which involved substantial informal decision consisted of the broad, unchallenged, and usually unexamined resort to judicial notice to establish the factual underpinnings of value judgments. However, the absence of controversy on this count indicated that the court held itself within the range of what the industry and the community accepted as common truth.[471] Altogether, the court met its responsibilities of decision well, if not with creative excitement. Since stream franchise policy was a realm primarily of legislation, the judicial record here foreshadowed the typical twentieth-century role of state appellate courts, in contrast to the buoyant, confident lawmaking in which judges indulged in affairs assigned to common law.[472]

The Quality of Decisions on Stream Use. The ultimate relevance of considering how decision-making roles were allocated among agencies, and what component operations entered into decision making, is to help appraise the quality of the decisions made. Where official decisions are made subject to constitutional values, the prime criteria of quality are that the decisions should serve "public" rather than merely "private" interest and that they should produce the greatest amount of durable and humanly significant satisfactions at least cost. Obviously such criteria involve a good deal of ambiguity and subjective judgment, and may readily cloak reality or beg more questions than they answer. However, these are difficulties in large part inherent in an undertaking which must nevertheless be attempted if we are to examine all the dimensions of any body of public policy. One function which study of legal history may perform is to introduce more concrete and particularized data into such criteria of the quality of official decision making.

Appraising past stream franchise policy from the vantage point of 1905,

Governor LaFollette showed uncommon mildness as he left criticisms to be inferred from his recommendations suggesting that the legislature pay more positive heed to public interest when it dealt with the new problems of hydroelectric power development. It was now, he thought, "quite apparent that these waterpowers are no longer to be regarded simply as of local importance." (Inference: that past policy had been framed in terms of interests seen chiefly as local.) "A legislative policy which grants franchises without substantial conditions amply protecting the public, and securing to it reasonable benefits in return, is no longer right and just, and ought no longer to be tolerated." (Inference: that past franchise policy did grant franchises with minimum attention to "protective" terms, and that, whether or not this had been right for the times, it had at least then been "tolerated.") "Reserving the right to amend or repeal is not enough." (Inference: that this reserved power had been in the past the prime provision protective of general interest.) For: "Economic conditions are rapidly changing in this state and in the country . . . The capital already invested, industries already established, may in a few years find themselves quite at the mercy of power companies in combined control of the waterpower of the state." (Inference: that past policy had not on the whole subjected to private oppression or arbitrary control the industry then dependent on stream uses, in contrast to the new situation emerging.)[473] The governor's implied analysis fairly weighed some key elements in the record: the legislative process had operated on a highly localistic basis, public policy had emphasized promotion rather than regulation of production, insistence on the reserved power to repeal or amend grants had represented a relatively exceptional effort to establish a matter of generalized, long-term, regulatory policy, and on the whole there were few episodes of acute controversy relative to the volume of franchise business.

Some later official critics were less restrained. In 1909 Governor Davidson thought that nineteenth-century stream franchises had "contained no provision tending to protect the public, except the power to amend or repeal the act."[474] In 1916, the Attorney General argued that new general legislation which empowered an administrative commission to regulate stream flow in the public interest should be construed and held effectively to supersede rights acquired by assignment of a franchise for a dam formerly used in aid of log transport. The Attorney General backed his arguments from authority with an appeal against the moral entitlement of the franchise here assigned, which "was one of many similar grants by complaisant legislators to the pine barons 'to do with these waters what he or they may deem advisable or necessary to carry out the purposes of the act.' There was no thought then of uniform flow, of fisheries, of public health or highways, of hydroelectric power, of navigation, or of 'performing the public duty imposed.' "[475]

Prairie and oak grove country in southern Wisconsin (the Thomas Week log cabin, built about 1860, one mile south of Hartland, Wisconsin). This was the type of land for which the normal public-lands disposal policy was framed.

The north country (virgin timber on Torch Creek, 1900): The type of land for which general public-lands disposition policy was ill-adapted.

Running the big river: Lumber raft on the Wisconsin River, 1887, near the Kilbourn Dam.

Small streams were highways, too: A log drive on the Clam River, Wisconsin, 1902.

These summaries were not accurate, realistic, or fair. The reserved-power clause was not the only item in nineteenth-century stream franchise policy which purported to regard general interest. The state constitution guaranteed the principal navigable waters as "common highways . . . forever free," and legislative and judicial policy faithfully preferred general transport over other uses of navigable streams.[476] Common opinion through the lumber era considered that the public interest had no greater concern than the increase of the productive capacity of the general economy. If Governor Davidson identified as measures "tending to protect the public" only those restrictive or regulatory upon private will, his estimate is correct, in the sense that in economic affairs the dominant attention of nineteenth-century policy was upon promotional rather than regulative use of law. But to join the Attorney General in attributing this bias of policy to the yielding of "complaisant legislators to the pine barons" would vastly underestimate the forces at work. It was not only the legislature, but prevailing community values, which were "complaisant." This complaisance did not reflect special regard to the ambitions of "pine barons." It reflected the general conviction, the roots of which ran deep below rational calculation into naive faith, that no purpose could be more public or more beneficent than to enlarge the output of immediately useful goods and services.[477] Stream franchise policy was not inattentive to the particular content of public interest, so conceived; for the titles and prerogatives it gave, the legislature required a *quid pro quo* for public gain (industrial power developed in return for eminent domain privileges, channel improvements or facility of commodity handling in return for toll or traffic control rights, guaranties of safe and unhindered passage for general transport in return for particular privileges to dam or boom).[478] It is more nearly true that "there was no thought then" of fisheries, public health, highways, or hydroelectric power. However, if the innuendo is that nineteenth-century lawmakers heedlessly put aside such interests in order to serve the advantage or protect the profits of the lumbermen, there is no evidence in the legislative or judicial record to support such a generalization. Through the years in which the lumber industry flourished, there was no strong state policy in these respects touching affairs anywhere in Wisconsin; the want of such policy development was not peculiar to the forest region. It is true, in fact, that stream franchise policy paid no significant attention to interests not closely involved with power or transport use of waters. The Attorney General's comment reflected a new sensitivity to social accounting, characteristic of Wisconsin's entry into the closer-knit living of the twentieth century. In this light his observations are pertinent to help us identify significant shifts in facts and in attitudes, reshaping perspectives upon public policy.[479] But by the same measure his comments are on the whole irrelevant for evaluating the quality of stream policy making during the lumber years. Most franchises in which the lumber industry was interested concerned operations in thinly settled, generally wild country in which there was yet no urban market for fish, little development of roads, no evidence of pressure of population upon supply of water for domestic purposes, and no evidence of dangers to public health other than those to be expected wherever towns grow.[480] Hydroelectric power—and, for that matter, recreational use of streams—created issues of public policy under pressure of changes in technology and economic conditions which began to be felt only at the end of the lumber era.[481]

The conclusion is not, of course, that stream franchise policy faultlessly expressed the public interest, but only that criticism should be relevant to what actually went on and to the context of fact and attitude within which processes of decision operated in their time. Thus, even if—unlike Governor Davidson and the Attorney General—we judge performance in light of the times, we encounter two notable shortcomings in nineteenth-century legislative process. First, after crediting Wisconsin policy with its full array of declared concern for various aspects of public interest in stream use, we are bound to ask how far these declarations were implemented. Circumstantial evidence suggests that most of the declared policy had substance in action.[482] But conspicuous on the record is the want of official apparatus specifically charged to check the extent or quality of fulfillment of franchise terms imposed for the general interest; implementation was left almost without exception to the zeal and means of private persons affected, to whom the law offered no more effective support ultimately than could be had from the processes of time-costly and money-costly litigation.[483] Second, though there is no evidence that substantial public interests in fisheries, highways, or public health or safety in fact suffered material detriment during the lumber years because of forest-stream franchises, we cannot avoid disquiet regarding the competence of the contemporary legislative process to promote a broad calculus of interests.

Franchise policy centered almost exclusively on transport and power uses of water. It was not coincidence that these were the interests in waterways then most readily translated into the money calculations of private enterprisers acting under the spur and discipline of operations in the market. As with the public lands, so here the interests dealt with in the legislative process tended to be limited to those recognized in the market. Yet, experience demonstrated that—as was inherent in its own sharply focused ends, functions, and energies—the market took account of only a part even of the economic values and processes relevant to a vital society, to say nothing of social and political factors which found no representation in market bargaining and could not be defined in the market's money language. In this sense, political process failed to show independent energy, apart from initiative generated in the market, in perceiving and weighing the full range of elements potentially relevant to stream franchise policy.[484] Fisheries, highways, public health or safety probably did not, in fact, from 1836 to 1900, present problems which were neglected at the cost of substantial public harm. However, this should have been determined deliberately and explicitly by an efficient legislative process. But there is little evidence that legislative proceedings even brought these matters over the threshold of perception.[485]

More serious because of deeper import to the integrity of legal order, or because of more lasting impress on the commonwealth, were the analogous failures in perception and decision which were marked by excessively localistic definition of problems and total blindness to the fact that the terms of stream use materially affected the speed and extent of depletion of the forest. These were respects in which public interest was poorly served by current decision-making processes, not out of special-interest villainy or the venality or timidity of legislators, but from causes more serious and durable because they sprang from the limitations of men's minds and wills and the baffling complexity and daunting mass of their total experience.[486]

The "public interest" in an event or course of behavior is no simple single element. Nor does it consist only in the relations of the event or behavior to the concerns of all or even the greater part of the persons who live within a given polity. An event or a course of behavior concerns public interest in those respects in which it has substantial relevance to values or processes by which men achieve community and a workable division of labor. (These terms must be taken in no narrowly political or economic sense, but as referring to creative sharing of secular purpose or power in any province of social experience.)[487] Because public interest consists in effective interlock or interplay of diverse centers of thinking, feeling, will, and activity, the particular content of public interest in a given situation typically lies in the state of relations within and among groups of persons or institutions which, however large in themselves, are nonetheless only small elements of the whole society. In this sense there are many "publics." So viewed, the general public interest has relatively abstract content, consisting most often in a concern that adjustment of relations within and among these smaller publics be achieved with good order and without violence. To this elementary prescription a constitutional legal order adds the requirements that adjustments be made by rational criteria, under fair and humane procedures, to the ultimate end of serving to enrich individual life.[488]

It is a permissible abstraction to consider the lumber industry as one of the smaller publics within the Wisconsin community. We have considered key aspects of the quality of decision making that went into relating the stream-use interests of this lumber public to publics outside it, and especially to the general public concerned with commonwealth resources and the income and costs involved in creating or using social overhead capital. But, of course, the lumber industry was itself no simple entity. It was a complex of diverse interests and functions, the reasonable and fair adjustment of which — in view of the importance of the industry to the economic growth and social and political structure of Wisconsin — was a distinct matter of public interest.

We speak now of more focused human concerns, deriving their self-defined purposes and their energies from self-interest driven under market pressure. Here is the area in which we are most likely to find "public" interest defeated by "private" interest. This is the frame of reference within which we may more plausibly look for the devious maneuvers of "pine barons." If human nature and faceless circumstance offer more convincing (though less engrossing) explanations for some of the larger deficiencies of public policy making, colorful and understandable greed, tinged even with satisfying villainy against legislative virtue, may explain how the millmen of Eau Claire undid the millmen of Chippewa Falls.

Nine out of ten stream franchise matters appear on the legislative record as run-of-the-mill, both in content and in treatment. We have noted that, so far as they were related to transport and power uses, the bulk of franchises paid heed to the more general public interest in the growth of productive capacity and the assurance of navigation.[489] There was also a general public interest in fair and rational adjustment among more immediately affected parties. How shall we appraise the record on this score, in this 90 per cent of run-of-the-mill dispositions? It would be surprising if men who held financial and political power and knew how to maneuver did not sometimes manipulate legislative process to obtain privileges without fair hearing or fair weight given to adversely affected interests. If we read the routine legislative records closely, we can indeed pick up suspicious circumstances. Prominent lumbermen held seats in the legislature, sat on key committees, were elected as presiding officers.[490] Instances occur where a franchise bill passed with remarkable smoothness, as introduced, under the sponsorship of a prominent lumber legislator. Consider bill 88S, of 1875, to authorize to Dan F. Smith "a toll-dam at Clam Falls, on the south fork of Clam River, in Polk county . . . to be used for the purposes of driving logs over Clam Falls . . ." Introduced by veteran lumberman-legislator Henry D. Barron on February 3, the bill was referred to a select committee which consisted of Barron alone; a week later, February 10, the select committee recommended that the bill be passed as it was introduced, and thereupon, on Barron's motion, the rules were suspended and the bill passed its third reading in the Senate; two days later (February 12) the Assembly Committee on Lumber and Manufactures (reporting, as was customary, without printed opinion) unanimously recommended passage, and on February 15 (twelve days after introduction) the bill passed the Assembly and became Laws, 1875, Chapter 45.[491] Chapter 45 contained nothing beyond the conventional provisions for such a grant. Even so, simply to obtain warrant to maintain a dam at a given location might be a decisive prize in itself, in a stiff competition for economic advantage between owners of standing timber, each of whom saw his profit in controlling the means of transport.[492] Thus there still might be reason to suspect that such hurried passage, on what amounted to *ex parte* presentation, reflected the use of legislative process unfairly to prefer a special interest.

That such was the character of any but a marginal number of these run-of-the-mill dispositions, however, seems unlikely. First, there are few instances in which even close and suspicious reading of the record provides evidence of abuse of legislative process to favor one private advantage to the detriment of another.[493] Second, most franchise bills were amended in course of passage in ways calculated to protect other immediately affected parties than the grantees, suggesting that in one way or another adverse interests had a hearing.[494] Third, so many franchise bills were so poorly drawn from the standpoint of the grantee's advantage as to caution against reading great guile into these proceedings. For example, grantees of some flood dam franchises carrying toll rights found it necessary to return to the legislature to amend the lien conferred on them to enforce their tolls; originally given only on logs running over the dam, the lien had to be amended to cover logs driven by aid of the dam's waters, though not passing over it.[495] Again, some improvement and driving franchises failed to authorize the grantees to take control of logs which were impeding the general drive because their owners had not provided an adequate crew to handle them.[496] Some grants left highly ambiguous the important matter of eminent domain power, as they stipulated merely that "said dam shall be subject to all provisions of law relating to the flowing of lands," or that the grantee "shall pay in full for all damages occurring by reason of the erection and maintenance of said dam, said damages to be determined as provided by law."[497] Before we erect large suspicions of adroit, covert efforts for private advantage, we ought first to be able to find that grantees were clever enough to write their grants in terms calculated to yield them full benefit; yet, scores of franchises are studded with omissions or ambiguities which could work only to the detriment of the grantees. Thus, even in the favorably expedited "toll dam" franchise which became Laws, 1875, Chapter 45, there was no provision for succession in title beyond Dan F. Smith, no provision of a lien for tolls, and no authorization of tolls for water furnished to move logs below the dam, and though Chapter 45 said that its grantees should "have at all times control of said dam, its gates and sluices," it did not safeguard them against complaints of navigators by the additional stipulation (which many like franchises contained) that they might refuse passage water to upstream navigators when it was needed to drive logs already below the dam.[498]

These general estimates of the legislative process at work on stream franchise matters find corroboration, if we inventory the particular performance of a given session. No one session can fully represent all, but that of 1880 presents the legislature in the full tide of lumber industry affairs.[499] In 1880 the legislature enacted thirty-seven special stream franchises (fifteen forest-area private dams, eighteen toll improvement enterprises, one toll boom, and three private booms).[500] Ten (27 per cent) of the thirty-seven bills were enacted as introduced; twenty-seven (73 per cent) underwent some amendment in course of passage.[501] Of the twenty-five amended bills for which records remain, fifteen were amended in ways which lessened the advantages or increased the obligations of the grantees, compared with the terms of the bills as introduced. In one boom franchise the authorized location was narrowed and restricted to the stream area in front of the grantee's own land; in eight toll improvement franchises tolls were sharply reduced from those originally proposed; in six private dam franchises navigation guaranties were added or tightened.[502] Where the only amendments were favorable to the grantees, the favor was in most cases probably of a corrective sort; thus in eleven private dam franchises amendments added "assigns" to the description of the grantees or dropped words of succession ("heirs" or "successors"), which might raise doubt whether the franchise offended the constitutional ban on special grants of corporate powers.[503] In addition to these enactments, the 1880 session summarily rejected or buried in committee nine toll dam bills and one private dam bill, without entertaining any of them to a stage which produced amendments, and also rejected two toll dam bills and one private dam bill despite substantial expenditure of committee effort on amendments.[504] Altogether, this 1880 inventory does not lend itself readily to the notion that the legislative process automatically ground out any special bill that was put into it, or that critics did not find substantial opportunity to tighten up the terms of what was given the grantees.

Such is the pattern of the bulk of decision making by legislative process in this realm of stream franchise policy. There are left six episodes which stand out from the pattern because they involved much sharper controversy than any others of which we have evidence and, sometimes, involved explicit charges against the integrity of the legislative process. Four of these episodes (those over the Kilbourn dam on the Wisconsin River,[505] and improvement or driving franchises on the Black[506] and the two Yellow Rivers)[507] we can put aside; despite half-stated complaints of rough handling from some of the losers, on the evidence available to us we can only say that the legislature acted within the scope of reasonable discretion that legislative power carries, and that, if it erred, the legislature has power to err.[508] The other two matters concerned the control of Beef Slough at the outlet of the Chippewa River into the Mississippi, and the conflicting ambitions of the Eau Claire and Chippewa Falls millmen farther up that river.

Beef Slough, one of the mouths of the Chippewa River, offered the only adequate natural harbor for holding and sorting logs moving from the Chippewa Valley pineries to mills on the Mississippi.[509] Concerned to hold for themselves the processing of the region's raw materials, in 1866 a group of Eau Claire millmen banded together and obtained a franchise to themselves as individuals for a boom at Beef Slough. Passage was routine. The franchise was not exclusive in terms, but it was so in practical effect, since the grantees' license covered the only feasible booming area. It contained no reservation of legislative power to amend or repeal it. The licensees obtained this franchise without intent to use it, but simply to preclude operations at the Slough by anyone else.[510] In 1867 a number of investors in Wisconsin timberlands who wanted to move logs to out-of-state mills on the Mississippi formed the "Beef Slough Manufacturing, Booming, Log Driving and Transportation Company," under Wisconsin's general incorporation statute, for the purpose of building and operating works clearly inconsistent with those licensed under the special act of 1866. Also in 1867 the Beef Slough Company sought its own boom franchise, which the Assembly passed (57–26), but which the Senate rejected (20–4), after hard lobbying on both sides.[511] In 1868 the Eau Claire millmen succeeded in defeating (47–36) in the house of origin a renewed effort by the Beef Slough promoters, though the proponents had won a favorable report from the majority of the Assembly Committee on Lumber and Manufactures.[512] However, the Beef Slough men won covertly what they had failed to obtain in open combat. The 1868 legislature also routinely enacted Chapter 331, Private & Local Laws, 1868, incorporating the Portage City Gas Light Company, section 9 of which inconspicuously enacted a piece of general legislation providing that "in all cases where any franchise or privilege has been or shall be granted by law to several persons, the grant shall be deemed several as well as joint, so that one or more may accept and exercise the franchise as though the same were granted to him or them alone." One of the individuals named in the preclusive franchise which the Eau Claire millmen had obtained in 1866 had switched allegiance; assuming the right to act individually under section 9 of Chapter 331, he assigned his right to his associates in the Beef Slough Company, and the Company built its works.[513] The contenders continued to test their strength, by self-help (when Beef Slough crews cut Eau Claire booms because the Eau Claire men held back the logs) and by resort to the criminal law (when the Eau Claire interests assembled the sheriff's posse against boom cutters and put the Beef Slough crews under bond to keep the peace.)[514] In 1870 the Beef Slough interest mustered sufficient force to obtain smooth passage of an act which authorized it to maintain the works already built, subject to specified guaranties of the general navigation of the Chippewa River.[515] The continued battles in and out of the legislature had by now left the Beef Slough Company bankrupt, but threat of its dissolution brought into the picture a much stronger force. Frederick Weyerhaeuser led other Mississippi millmen into organization of the Mississippi River Logging Company, an Iowa corporation, which took over the works at the Slough. In 1872 the Weyerhaeuser group consolidated its legal position by obtaining a statute which was in terms general, but was recognized in the legislative proceedings as designed particularly for the Beef Slough situation. General Laws, 1872, Chapter 105, said that no one authorized to keep a boom on a river should be obliged to maintain a passage for navigation through any outlet of the river "other than the main outlet," provided a free passage was left "in the main channel . . . or channel which has been a usually followed route" for traffic, and that past failure of any such franchise holder to fulfill any previous requirement of law to maintain a passage through an "outlet (other than the main outlet)" should "in nowise . . . forfeit, or work a forfeiture of any right . . . [or] franchise" held by the offender. The practical power of the new Beef Slough interest was attested by the fact that, while a stiff legislative battle was waged over the exoneration feature of the 1872 act, there is no record of a substantial fight over the prospective aspect of the statute.[516] The 1872 act ends the significant legislative battles over the Slough. The Eau Claire millmen now turned, unsuccessfully, to litigation in a federal court to oppose the Mississippi River Logging Company's traffic facilities. Since Congress had not undertaken to regulate the terms of navigation on the Chippewa, and Wisconsin's legislature had licensed the Beef Slough operations, this line of attack, realistically measured, was one of desperation, and it is of secondary importance in this history of contention; it is not surprising that it failed completely and in short order.[517] The long contest finally was put to rest in 1880 and 1881, when Weyerhaeuser used his power and his quiet insistence on rationalizing business arrangements to obtain a private treaty which apportioned shares of Chippewa Valley logs among the Chippewa Valley and the Mississippi mills.[518]

The Weyerhaeuser pact of 1881 also settled the competition between Eau Claire and Chippewa Falls millmen which had kept the legislature under severe pressure from 1867 through 1876, while Eau Claire interests sought a franchise for a great dam and boom works near their city. The episodes of this parochial contest we have already described;[519] here let us recall two features of its handling. First, the pressures on the legislature erupted in 1871 in charges and countercharges of bribery of legislators, which in Governor Fairchild's estimate showed such substance as to taint the bill the Eau Claire millmen had won, which he therefore vetoed. Two unsatisfactory legislative investigations indicated that bribes were offered, that there was want of evidence that any were accepted, but that in any event the legislators in general were negligent toward their good name by tolerating styles of lobbying calculated to cast suspicion on the integrity of special legislation.[520] Second, the legislative process here proved not sufficiently robust or competent to produce under pressure a decision on the true issue. The real problem was to establish grounds for allocating stream uses to promote rational development of a regional economy. Instead, the legislature assented to an artificial disposition (by pretending that the objective was a public water supply for the city of Eau Claire), which shirked the legislative function of defining real issues and making responsible choices upon them.[521]

The handling of the Beef Slough and Eau Claire dam controversies thus show significant failures to fulfill the public interest in good legal process. Lax standards of lobbying allowed a heavy cloud of suspicion to settle over legislative operations in 1871. Only dishonest maneuver could obtain disposition of the Beef Slough matter in 1868 by a terse piece of general legislation buried in an unrelated special charter, and only loose legislative practice could fail to reveal the dishonesty. There was want of toughmindedness in analysis and debate to insist that the Eau Claire dam franchise find a proper basis in legitimate economic planning (if such basis could be found), instead of concealing decision behind a contrivance irrelevant to the real situation.

These, however, are findings of loss to the public interest in the good working of legal process. What of the public interest in the substantive problems whether there should be a harbor at Beef Slough or a dam and boom at Eau Claire? That the outcome of each of these controversies spelled private gain to some self-seeking interests, private loss to others, in itself proves neither that public interest was served or disserved; particular private gain or loss attends most public action.[522] There is no evidence on which to hold that general Wisconsin interest would have been better served by the greater growth of Chippewa Falls, the lesser growth of Eau Claire. Over-all economic and social productivity probably suffered by the intense struggle between interests too parochially conceived; certainly the public interest in sound legal process suffered.[523] The policy represented by the commerce clause of the federal Constitution at least casts the burden of proof on one who would contend that the Chippewa Valley millmen promoted a legitimate public interest in trying to prevent the movement of logs for milling outside the state.[524] Weyerhaeuser's private treaty of 1881 achieved a striking accommodation of interests, in contrast to the energy (and integrity) lost in years of legislative combat and fruitless litigation. The reason the private treaty succeeded was that it did the legislative job of generalization better than the legislative process proved able to do it. The 1881 agreement defined the problem as that of rationalizing the Chippewa Valley productive potential as a whole. Within this broader frame of reference more parochial interests lost their desperate urgency, and with a more assured basis for calculating available resources the parties found confidence to make mutual accommodations.[525] Smoother production and a reduction of effort spent in battle were presumably in the public interest. The unresolved element in this analysis is that the public had no representation in the making or administration of the 1881 treaty. Just in proportion as the treaty effected a workable economic plan for a major producing region of the state, it would seem that public interest was not well served without public representation. However, this point—though well taken from a twentieth-century view—is probably unrealistic in terms of nineteenth-century state policy-making process, which (as we have seen in various contexts) was simply not geared to provide a broad-gauged public-interest scrutiny of such a program as Weyerhaeuser's treaty, even if means of public representation had existed. Nineteenth-century public policy making was concerned with public control of private rationalization of large areas of economic activity, as witness the Jacksonian attack on the Bank, the Grangers' drive for railroad regulation, and the Populist demands for curbs on the "trusts." But the limits of contemporary imagination and experience are indicated by the fact that all these movements were essentially negative.[526] As with the basic defects of perspective upon the contextual breadth and the time reach of public policy, so here we confront the conclusion that the most serious failures of legislative process in serving the public interest were less in the content of particular decisions than in the limitations of the policy-making process itself, which in turn reflected limitations of knowledge, patience, experience, and will in the general community which the law served.

RAIL TRANSPORT

Timberland owners had reason to welcome the building of railroads into the forest. Much valuable timber was located too far from driving streams to be moved to mill and market by water. Hardwoods, which would not float, could be counted useful only after cheap means of land transport were available. On both counts the extension of rails could confer value on land otherwise worthless by market measures. For like reasons loggers needed the railroad, which in addition offered them freedom from seasonal limitations on woods operations; with rail transport—and the donkey engine—woods production need not depend on winter snow to allow moving logs to loading places, nor on spring freshets to carry them to processor or distributor. Wisconsin millmen saw in the railroad the means of access to new markets east and south and the means of direct access to older markets in the treeless prairies west of the Mississippi, free of dependence upon the river port wholesalers through whose hands all water-borne timber and lumber must pass.[527]

However, the main impetus to building Wisconsin's rail network, even in the forested north, did not come from the lumbermen. The railroads offered such promise of profits from financing, construction, and land speculation, quite apart from earnings of traffic, that they provided ample base for the rise of an independent railroad interest—represented, for example, by Byron Kilbourn of the LaCrosse and Milwaukee and by those more solid entrepreneurs, Alexander Mitchell of the Milwaukee and St. Paul, William B. Ogden of the Chicago and North Western, and Gardner Colby of the Wisconsin Central, none of whom was a figure in the lumber industry.[528] The principal railroads represented capital investments which dwarfed the largest Wisconsin lumber enterprises; while railroad development proceeded in calculations of millions of dollars and through ambitious programs of consolidation, logging and milling in the Wisconsin forest industry remained in the pattern of many separate firms, most of them relatively small or of modest reach; the one major consolidated lumbering enterprise, brought together under the leadership of Frederick Weyerhaeuser, did not venture into railroading.[529] Even in the forest region the local appeal of railroad promotion rested largely on the railroad's promise as a frame for over-all growth of towns and sections; in the seventies and eighties northwest Wisconsin held grandiose visions of its future as a country of farms and mines and trade and industry, so that — once settlement was launched — other ambitions than those of lumbermen came to bear on the state legislature to advance the rails.[530] Local ambitions for general economic growth fitted the dominant temper of the times. Thus typically the governors and legislative committees who set forth reasons why the state should use its means to promote railroad building in the north did so in the name of agricultural, mining, mercantile, and industrial development generally, as well as of lumber. Indeed, it is rare that these statements of state policy put special emphasis on forest industry.[531] There may have been some pretense here, some readiness to divert attention from the extent to which policy responded to lumber pressures. But the pattern of emphasis on general regional development is so consistent through a generation of railroad promotion, fits so well with the north's own hopeful illusions, and links so logically to the speculative faith of the railroad industry itself, that it strains credulity to read the record as a thirty-year sustained effort to conceal the power of a forest lobby.[532]

In this perspective the men with stakes in forest wealth and forest industry appear as an influential source of pressure on state government for railroad aids, but only as one important source of pressure, and that marked by limits in time and area and by divisions of interest within their own ranks. Time and area limits stand out most boldly in the record.

Water transport easily satisfied the industry so long as demand could be met with the timber accessible to streams. The tie to the seasons was a handicap and there were substantial costs in sunken and stranded logs and water-damaged lumber. But capital investment for stream improvement was relatively small, and driving and rafting costs were relatively low, compared with rail transport.[533]

Railroad construction in the fifties reached from east to west across the southern half of the state only. Congress conditioned the first grant of federal lands in aid of Wisconsin railroad construction (1856) upon undertakings of the ultimate recipients to build roads through the state's northern forest both to the northeast and to the northwest. However, the rails did not reach far into the northwest until the mid sixties, and construction in that part of the state encountered misdeeds, delays, and disappointments which postponed substantial accomplishment until well into the seventies. Hopeful predictions of the general economic growth which the railroads would stimulate included early forecasts of northern forest production and traffic. However, these predictions were cast in the most general terms, and neither timberland owners or lumbermen appear as movers in this first phase of railroad building.

Men prominent in exploiting forest wealth in northwestern Wisconsin began to interest themselves in local railroad projects in the late sixties. But water transport clearly dominated until about 1870, and it was only after that point that substantial lumber interest showed itself in railroad promotion.[534] This timing was in itself of effect in limiting the prominence of the forest interest as a pressure upon the law for railroad aids, since the most important railroad industry power had already been formed by the time men of forest wealth became much involved.

There was also a marked area difference in the forest industry's involvement in the railroads. Fortune produced better leadership and financing first for the extension of rails into the northeastern timbered area, so that construction was done and the federal aid lands earned there with much less maneuver than in the less favored western side of the state. Moreover, one great railroad — the Chicago and North Western — early predominated there. Thus this section of northern Wisconsin saw much less quantity and variety of railroad promotional activity through legislation than did the northwest. Geography played a part, too. The northeast had relatively more ready access to water transport than the northwest, whether to Lake Winnebago as a collecting point for rail shipment or to Lake Michigan as a direct water route to the Chicago wholesalers; thus the lumbermen of the northeast showed a smaller range of concern with the legal framework for railroad promotion than did those across the state.[535] This was the story of sectional-focused rather than of personal interests. Some of those who made money by timberland and lumber operations in northeastern Wisconsin invested in the northwest area and became active in railroad promotion on the other side of the state; Philetus Sawyer of Oshkosh was a notable example.[536] That circumstance, however, does not affect the sectional distribution of legal activity concerning railroad development. This sectional difference, with the concentration of pressures upon the legal process in the northwestern area, has relevance for legal history as another demonstration of the typically *ad hoc,* opportunistic response in law to the problems of economic growth. Legal activity showed up, primarily, not according to deliberated general policy but in response to the impact of chance, geography, and the uncalculated cumulative effect of separate private ventures.

Most of the promotional efforts and related problems brought to law with the participation of substantial northwest Wisconsin forest industry interests concerned three endeavors — (1) to build a line from the St. Croix River on the state's northwestern boundary north to the western end of Lake Superior, with a branch to Bayfield (the northerly portion of the construction contemplated for western Wisconsin under the federal aid grant of 1856), (2) to build a line up the valley of the Chippewa River, and (3) to build connections across north central Wisconsin. The St. Croix line was finally brought to completion after a tortuous history stretching from about 1866 to 1883, involving successively the West Wisconsin Railway Company (successor to the earlier Tomah and St. Croix), the North Wisconsin, and the Chicago, St. Paul, Minneapolis and Omaha Railway Company (the Omaha), eventually to be brought into the Chicago and North Western system.[537] The dream of a separate Chippewa River valley road was as distinctly the creation of local lumbermen as any rail project of the times, but it came to nothing; the rail facilities of the area were supplied by lesser undertakings which eventually were taken into larger systems. The failure to create this separate Chippewa road is in itself telling evidence of the limits of the lumbermen's purposes or means to become substantial figures in railroad promotion.[538] The north-central connections were provided finally by the construction and assembly of the Omaha and the building of the Wisconsin Central. Parts of the Omaha system contributed by the West Wisconsin and North Wisconsin, and the Wisconsin Central road as a whole, were the railroads which to the greatest extent were built on the immediate prospects of log and lumber traffic. In this sense they were forest industry roads, but not in the sense that they were the product of men whose primary interests were in timberland speculation or forest production.[539]

Had timberland owners or lumbermen played more leading roles, it would be pertinent to this history to follow the details of these three main fields of northwest Wisconsin rail development. Since the forest industry men did not on the whole occupy a central place in these ventures — at least in the successful ones — it seems more pertinent merely to note the nature of the franchises which they, in common with others interested in railroad building, sought from the state.

The basic franchise which railroad promoters needed and wanted from the state was the grant of corporate existence. This was practically necessary to assemble and discipline the capital required for such sizeable enterprises, and to provide for the orderly management of continuing operations. Thus without exception we find that railroad promoters sought and with little evident trouble obtained corporate franchises, typically by special act of the legislature.[540] In their need of a legal framework whose organization and procedures would encourage and facilitate the assembly and management of capital, the railroad builders presented demands upon law no different in kind from those which led entrepreneurs generally to seek incorporation for ventures of substantial size. In these aspects we may pass over the railroad charters here and merely refer to the discussion in Part II of the reasons which led to increasing use of the corporate form in the lumber industry itself.[541]

To build railroads, promoters wanted special franchises also. These special franchises were invariably linked to the grant of corporate status. No constitutional doctrine required this; both the early milldam statute and the special acts authorizing dams and stream improvements showed that the legislature might confer upon individuals franchises similar or analogous to those sought by the railroads.[542] But the same scale and complexity of railroad operations which required these special privileges insured as a practical matter that they would be tied to corporate charters.

The consistent pattern of railroad charters shows that four franchises were treated as of critical importance by those who sought the law's help. Examples of these franchises may be found readily in railroad charters of which prominent lumbermen were sponsors. For example, we may look at the 1866 charter of the Chippewa Falls and Mississippi River Railroad Company. This was a corporate blueprint for the abortive effort at an independent Chippewa valley road, counting among its incorporators such notable lumber industry names as Albert E. Pound, Daniel Shaw, L. C. Stanley, Andrew Taintor, J. G. Thorp, and Thomas Wilson. Or we may turn to the 1869 charter of the North Wisconsin Railway Company. The North Wisconsin was a subsidiary of the West Wisconsin, whose promoters, including leaders in the northwest lumber industry, were concerned to fulfill the terms of construction to earn the long-dangling federal grant for a road from the St. Croix to Lake Superior.

First, railroad charters designated in general terms the route on which, or at least the termini between which, the corporation was "hereby authorized and empowered to survey, locate and construct, and perpetually to have, use, enjoy, maintain and operate a railroad." Second, usually the charter made auxiliary grant of the right to build the railroad "upon and along, across, under or over any public highway, or any private way or road, street, plankroad or railroad, if the same shall be necessary," and to bridge any stream "for the use of its railroad," subject to restoring any land way so as not to impair its use, consistent with the crossing right given the railroad, and subject to building bridges so as not to obstruct navigation. Third, apart from this privilege to cross other avenues of transport, railroad charters and general laws conferred the right of eminent domain, so that the railroad might enter any land to lay out its route and build its line, taking a right of way of specified width together with land needed for auxiliary facilities. This grant was coupled with elaborate procedures for determining the value of private property so taken to build the road. The legislature left much ambiguity as to whether the railroad must pay the landowner before it might use the land taken, but the Wisconsin Supreme Court exerted itself in constitutional rulings and statutory interpretation to insist that payment must precede use. Fourth, charters conferred the key operating privilege to transport persons and property and "to demand and receive such sum or sums of money for the transportation of persons and property . . . as . . . [the grantee] shall from time to time deem reasonable."[543]

These franchises were essential in law and in fact to the objectives of the railroad promoters — and, if not in all respects in equal degree, to the desires of the community and its industries to obtain rail service. In the first place, private individuals or associations might not lawfully undertake to lay out highways for public transport at their sole discretion. There must be public assent, if not public direction, concerning measures which not only provided a framework within which the general economy might develop, but also inherently involved significant power to shape the course of economic growth. Railroads, Mr. Chief Justice Ryan observed in 1874, were too important to the common life to be allowed to behave as "chartered vagrants." That a charter allowed a railroad some freedom to change its route for convenience did not imply authority in the corporation to change the termini stated in the charter, "and so to leave railroad corporations free, by a little management, to change specific charters into roving commissions throughout the state. The legislation of the state is liberal enough of franchise to these corporations, in all conscience, to leave them without excuse for licentious construction of their charters."[544] Ryan's proposition came only after twenty years of railroad building in Wisconsin, but it articulated a principle of policy clear and consistent in legislative practice from the chartering, first of plank roads and turnpikes, then of river improvement companies, and finally of railroads. Without formal declaration, promoters and legislators alike had from the first accepted without question that private design of public transport facilities must be licensed by statute at least as to termini.[545] Grounded thus in legislative practice, this doctrine was striking testimony to a sense of commonwealth values and the proper role of the law in governing terms of dealing which would shape the commonwealth, in a time when so much policy of general importance was fashioned by default. Of like meaning was the equally consistent legislative practice that legislative consent must be had if a new public transport facility were to cross or limit the effectiveness of an older one. With ample analogy in common law doctrines which provided relief against obstruction of public ways, this second form of franchise or license perhaps represented less of an innovation than the first. Even so, insistence on it showed readiness to use law to adjust relations in an economy of growing complexity; the "police power" was not only a formal concept of jurists but a working reality of the legislative process.[546] The police power had the more vigorous reality because regulation which limited some private interests almost invariably benefited others, so that there were private reasons to press for law's licensing and adjustment action. The franchises for routes and public crossings in Wisconsin railroad charters did not contain express grants of monopoly, nor would monopoly be implied. Nonetheless, the fact that promoters of new roads must seek the legislature's license for their routes and their crossings or interferences with existing ways inherently spelled considerable protection against fresh competition to transport facilities already established. Awareness of this appeared in contemporary pleas to either establish competition as a discipline upon existing carriers or to deny competition lest it destroy investment already made.[547]

The ability to appropriate land for a railroad right of way without bargaining for it in market was indispensable to assembling a line and to doing so with economy of scarce capital and freedom from the extortion of a few hold-out landowners.[548] There was urgent need of cheap bulk transport facilities for economic growth in this country of great distances and untamed land. The felt need created general acceptance of the idea that the market should not have its free play in this matter. Again we encounter public policy fashioned initially not in explicit doctrine but in legislative practice. It was largely in the development of the railroads that state legislatures in the second quarter of the nineteenth century, for purposes deemed of public concern, delegated to private persons the authority to obtain the use or title of other people's land at prices fixed by other than market procedures. Thus the concept of delegated eminent domain powers was substantially the product of legislative response to railroad promotion.[549] True to the value which this society assigned to increase of economic productivity and the release of the creative energies of men's minds and wills, the Wisconsin court by 1854 found incontestable basis for legislative authority to make such delegations. "However powerful may heretofore have been the adverse array of logical argument, the suggestions of cautious expediency, or the jealousy of personal rights, it has long since yielded to the irresistible tide of progressive improvement, and been lost in the wake of judicial authority."[550] This prized privilege could be had only by grant of the legislature, in which lay the power of the public purse and hence the authority to define the terms on which the law could sanction levies on private means. This franchise must rest on public grant, for "the only principle upon which such delegation of power can be justified is, that the property taken by these companies is taken for the public use . . . It is not the company as an association of individuals who take the property for their own use, but it is the sovereign power that takes it, through the agency of the corporation, for the public use." [551]

Given the values which mid-nineteenth-century Wisconsin policy put upon private property and contract, one conducting a lawful business ordinarily need seek no franchise to set the asking price upon his goods or services or to conclude deals upon such terms as might suit him.[552] On the other hand, the egalitarian beliefs which provided much of the moral legitimacy of private property and contract freedom likewise promoted distrust of developments which might create severe imbalance of social, economic, or political power. Before and contemporaneous with the growth of the railroads, this factor in community temper showed itself in the recurrent, if largely fruitless, moves to curb large-scale speculation in land.[553] Another element relevant to the terms of railroad charters showed itself in contemporary attitudes and policy. Despite the relative crudity and simplicity of mid-nineteenth-century Wisconsin, because this was a community in obvious process of rapid development men grasped the idea that there were key points in social structure where power might best be wielded to broad effect. They showed this in their eagerness to use the public lands to promote railroad construction. Latent in this viewpoint was readiness to see the need of public consent and scrutiny regarding power delegated and used at such key points — a readiness which came to expression, for example, in the Granger legislation to regulate the roads. Our English legal inheritance included a somewhat ill-defined or misunderstood, but nonetheless operative, notion that those who pursued "common" callings were subject to legal controls over the reasonableness of their charges and the fairness of their service. But, however we interpret the scope of these obligations, they were not enforced by license but by lawsuits after the event.[554] Thus, as with the delegated eminent domain power, it was a development primarily of state legislative practice more than common law inheritance that brought to first definition the idea that public bulk transport facilities must have statutory license to fix their rates. The oldest Wisconsin legislative precedent, borrowing from Massachusetts, had since territorial days provided direct statutory regulation of the rates of grist mills.[555] We have seen that, in years before railroad construction had reached far north, the legislature's liberality in conferring franchises for stream improvement was quite consistently coupled with care to treat the power to exact tolls as a subject of statutory license. So there was foundation for including in railroad charters explicit authorization to collect charges for service. The charters were typically drawn in terms chosen by the promoters. Thus there is testimony that promoters realized their dependence in this respect upon a public franchise, in the care which they typically took to frame their license most favorably to their discretion — that the railroad be entitled to make such charges "as it shall from time to time deem reasonable." [556]

These, then, were the franchises which railroad promoters — including those leading lumbermen who interested themselves in railroad development — regarded as essential additions to the grant of corporate existence, to create an effective working organization for building a road. These grants had in common, as franchises, that they conferred special rights to impose upon the more general rights of the public or of other private persons decisions initiated by the grantees for purposes which the legislature stamped as in the public interest.

So defined, there was a franchise aspect to grants of public lands in aid of railroad building through the forest, as also to state grants of tax exemption on such forested aid lands and to provision of a special system of taxes on railroads (by a state-levied gross receipts tax in lieu of locally levied property taxes). Indeed, there was a like franchise aspect to grants (or low-priced sales) of public lands generally. There was likewise a franchise aspect to the large grants of money (sometimes by purchase of railroad stock) and the large lending of public credit (based on issue of municipal bonds) which local governments made to help railroad building. The previous chapter viewed these policies primarily as forms of direct or indirect capital subsidy to the lumber industry and to the transport facilities which served it. This defines the matter in the terms most meaningful to the beneficiaries of this public bounty. Viewed from the standpoint of others than the immediate beneficiaries, public land grants, low-priced sales of public land, state tax favors, and local government grants of money and pledges of public credit implicitly conferred franchises on the beneficiaries to impose their decisions as to use of these public lands upon all other potential beneficiaries of these public assets, and to press their own economic promotions relieved of the same burden of contribution to social overhead costs which the general taxpayers must bear.

What is most significant about the franchise aspect of these grants of public lands, money, and tax privileges is the relative inattention given to it through the years of decision. This is to repeat from another angle the appraisal that emerged as basic in the story of public lands policy considered as a program of capital use — that the basic decisions were taken mostly in silence and largely by default. The franchise aspect of the public land grants to the railroads came into public debate only by implication, in recurrent but futile complaints that settlement of the north country was delayed by withholding from market the sections earmarked for railroad aid.[557] State tax exemption of railroad aid lands and local government assistance stirred sharper, more focused controversy only in the later seventies, and in connection particularly with roads especially important in opening up the forest for lumber exploitation — the West Wisconsin, the North Wisconsin, and the Wisconsin Central. Issues were drawn much more sharply sometimes in these battles than was normally true in the opaque workings of contemporary legislative process. Critics attacked the tax exemptions and local government aids quite precisely, on the ground not merely that they were favors to privileged corporations, but that they were implicit levies upon the complainants' properties and enterprises, which must make up the tax load the exempted or assisted property did not carry. Under pressure of attack, the railroads brought their side of the issue to a definition more precise than ordinary in contemporary policy making. Extension of the rails was necessary to general economic development; but earnings prospects of roads through the unsettled forest were so slim and uncertain in timing that annual tax bills would eat up the capital value of the aid lands while the railroads held them for reasonably advantageous sale; to sell bonds secured by the earned aid lands, the railroads must be able to assure investors that the lands would not be called on to pay part of the community overhead costs while the road held them for favorable disposition. So, too, the railroad must have local government financial help, to make up for the fact that the rails were laid not on the basis of existing traffic but on the hope of future business.[558]

At first granted or authorized with little evidence of dissent, tax exemption of railroad aid lands and subsidies by local government came into sharp and continuing controversy in the seventies in the forested north. This conflict marked a pronounced development of attitudes from the late sixties, when legislators from northern districts twice voted heavily—without ultimate success—for proposals to amend the state constitution to remove blocks on direct state financial aid to railroad building.[559] The differences are instructive on the processes of legislative policy making. The focus of burdens was important; state money grants for northern railroads would fall on taxpayers the state over, but tax exempt aid lands and subsidies by local governments spelled heavier taxes close to home, in the counties where the particular railroad development lay. The timing of decision was also important; northern settlement was little advanced in the years just after the Civil War, but by the later seventies enough settlement had proceeded to create more diverse interests among local farmers and businessmen. In the first stages of local development, the clash was typically sharp and simple—between the scanty resident population, which wanted the rails, and absentee speculators, who fought the higher tax loads which endangered their capacity to hold for the long-term rise in timberland values. At a later stage, the timberland owners began to gain some local recruits in battles against local government generosity to the railroads, as more of the resident population acquired property stakes which made them more tax conscious.[560]

However, amid these battles over tax exemption franchises the timberland owners and millmen played an opportunistic role which defies neat recital or reduction to principle. Where railroad construction promised value to hitherto inaccessible timber, its owners supported state and local aids. Where construction would mainly benefit lands of competitors, or the grant of tax exemption would spell heavier taxes on timberlands already enjoying access to mill and market, owners opposed the favor. According to the particular situation, the same timberland interests might be found here supporting, there opposing, similar public action.[561] These reactions are understandable, and they followed the logic of individual firm behavior in the market. The point such a record makes is that there was a function to be performed here by processes designed to express other than individual market interest. More general economic and social interest, conveyed in votes of southern as well as northern Wisconsin legislators, created tax exemptions and special tax systems in aid of railroad building in the north during the generation of pioneer construction. As this phase passed, more focused local interest played more heavily on the legislature, always especially sensitive to local pressure. Conservative general viewpoints prevailed to protect outstanding favors from repeal, mainly on grounds of moral obligation to investors, partly on the threat that repeal might be invalid under the contract clause of the federal Constitution. The zeal of local interests, including those of lumbermen, worked together with the general faith in promoting productivity to bring about enactment of one last tax exemption for the Wisconsin Central, in 1877. But from that point on it was clear that legislative policy in practice would permit no further creation of this type of franchise. Likewise, local government aids had passed their peak of significance by the early eighties.[562]

On the whole, the creation of various franchises affecting the development of railroads in the Wisconsin forest area was a concern simply of state law. To charter roads for operation wholly within Wisconsin was naturally the business of the state legislature under the tradition of our federal system, and Wisconsin forest development was off the main track of Congress' attention to transcontinental railroad construction. Conceivably Congress might have used the leverage of federal aid land grants to affect railroad building as a means of influencing the pace and direction of exploitation of the Lake States forest. But, as we have noted in discussing federal land policy generally, such an approach was foreign to contemporary conceptions of federal land policy. In one respect the handling of the federal lands, as of the state's own federally derived holdings, represented an important franchise peculiar to the lands. Though practice fell far short of profession, there was an enduring policy against allowing public or private lands to be engrossed by large-scale, long-term speculators. Before the railroad fever caught on, the Wisconsin territorial legislature declared that this policy should apply to railroads; a ban on land speculation was made a condition of incorporation.[563] However, the point of granting public lands in aid of railroad construction was to allow the roads to sell them advantageously, to raise investment capital. There was early opposition to the grants precisely because some feared the roads would hold their lands off the market for long years, awaiting the rise in values. Concessions were made to this feeling in some instances by stipulations that grantee roads must dispose of their lands within limited periods after earning them, and on terms offering some preferences to settlers.[564] Nonetheless, implicit in the aid grants was recognition that the roads might hold the aid lands for some time, and hold them to realize capital.[565] As matters worked out, the railroads typically were so hard pressed for capital that they marketed their lands rather soon after earning them, under compulsion of their own need more than of law. Even so, the extent to which public policy accorded them an exception from the general disfavor toward land speculation represented a potentially significant franchise, and one which may be viewed as being as much the creation of federal as of state policy.

The lumber industry left its distinctive mark much less upon the development of north Wisconsin railroads than of north Wisconsin waterways. The difference was that the greater opportunity for economic planning in creating a rail network allowed the railroads to take on significance for a much wider range of economic and social interests than those concerned with the use of streams in the forest country. The law's roles in railroad building are part of the lumber story, but the activity of timberland owners and lumbermen on the whole forms just one more part, if sometimes an influential part, of the general promotion of railroads. The pervasive effects of railroad activities upon the general economy point to the reasons why controversy and lawmaking over railroad promotion and railroad regulation tend to make their own chapter of legal history, rather than appearing merely as a facet of the legal history of the lumber industry.

PART II

CONTRACT

INTRODUCTION

THE CONSTITUTION OF THE MARKET

NINETEENTH-CENTURY lawmaking in the United States gave its energy more continuously and more devotedly to building, extending, and implementing the market than to any other institution of the society. This was true of the two primary centers of nineteenth-century lawmaking: the legislatures and the courts. The executive was not as important a contributor to shaping policy; thus it does not break the general pattern that the executive sometimes displayed skepticism about the capacity of the market to express the full needs of the community. Nor did constitution-making processes depart far from the main pattern. In the creative generation of constitution making that ended with the First Congress of the United States, events compelled men to think broadly of the general organization of power, beyond the confines of market theory. In comparison, except for the Civil War amendments, nineteenth-century constitution making went on in a narrower frame of reference. As in Wisconsin of 1846–1848, communications, the supply of money and credit, and public subsidy of economic growth were likely to be most prominent among the subjects of ardent concern in state constitutional conventions; it was significant that these were matters in which public policy bore directly upon the range and effectiveness of market activity.

The market took on legal definition mainly in the law of contract, and quite naturally in the temper of the time the law of contract dominated the nineteenth-century legal order. Because marketing cannot go on save in a context of reasonably assured expectations, the legal order as a whole was, of course, indispensable to the existence of a market. But it was the law of contract which supplied the assurances and the procedures and tools necessary for the immediate operation and steadily expanding energy of the institution. The relevant law was not only that of simple contracts for the sale of land or goods or personal services, but likewise the law of more complex arrangements — of negotiable instruments, of secured transactions (mortgage, pledge, reserved title, lien), of business association (joint venture, partnership, corporation), and of insurance.

The market means an organization of behavior for decisions upon allocation and use of economic resources, by men not holding official power, through private bargain and plan. A classic definition would require that these decisions be made by the wholly impersonal interaction of numerous actors, none of whom alone or in combination is able to exert material effect upon the outcome. But the degree of practical control held by particular actors was always a socially important fact. Concern over such special private power produced response in legal regulation. However, this regulation — quite limited in scope and even more limited in effectiveness — more properly falls within consideration of the police power than of the system of contract. Hence we examine it in Part Three, and in this Part confine ourselves to the system of contract as the law supported and extended it.

Discussion of market power takes on sharper outline by comparison with official power. Official power manifested itself in two principal forms, to help shape the role of the market in the exploitation of Wisconsin's forest wealth. To some extent there was direct allocation of economic resources by official action, most notably by disposition of public lands and the grant of transport franchises. But official decision making also affected the allocation and use of resources at second remove, by official determination of standards of conduct or of the range of available, legally defined and recognized procedures and instruments, within the framework of which market operations might go on. Such secondary legal controls upon resources allocation included definitions of permissible forms of land tenure, formal requisites for the making of contracts, officially standardized terms of deeds and agreements (most commonly declared in the name of "interpretation" of private documents), and officially imposed standards of what should constitute effective operation of private deeds or grants or of what should be deemed performance or breach of private agreements.

The relative roles of official and unofficial decision making regarding resources allocation may be brought into sharper focus by a scale of incidents involving a tract of timberland. (1) If one having no connection with the landowner cuts timber and subsequently is compelled to pay damages for doing so, the law alone has governed the situation. In acting to protect private property, law seeks partly to preserve elementary order and to prevent violence; it seeks partly to bulwark a system of private property dealing, too, but even in this aspect such a case presents not an instance of operation of market processes but rather of legal provision of a framework within which market processes are possible. (2) If one who has a life interest in land is held to do legal wrong to the holder of the remainder by excessive cutting of timber, where both interests trace to a third-party grantor whose grant had not defined the grantees' rights in timber, again it is the law which decides, though the effect may be to define and protect an interest which may be bargained for in the market. (3) Where a deed either "reserves" or "excepts" standing timber for benefit of the grantor, obviously we move closer to a situation where the parties' legal relation to the timber is fixed by their own agreement; yet, if the grantor's right is defined—in the absence of further terms in the deed—by the fact that previous judicial decisions have assigned specific, different meanings to the quoted words, clearly official decision is still playing a major part in the allocation of the resources in question. (4) Where the extent of a grant of standing timber is determined by a court in the light of evidence (derived from the total context of the contract of sale) that the grantee sought assurance of a ten-year supply of timber of certain quality for a sawmill of an indicated capacity, we have moved over a line of rather clear-cut demarcation, to a situation in which the decisions of the contracting parties are the dominant, active factor, and the law plays the background role of lending its force to the fulfillment of their private plan.

In regard to at least four major aspects it is important to inquire concerning the relative roles of official and nonofficial decision making. (1) We need to define who is in fact exercising will upon a given situation. To make this determination has meaning in itself, for it helps us to understand what in fact is going on. Moreover, this determination is a necessary preliminary finding for other inquiries. (2) The range of considerations or interests likely to be taken into account in a decision will probably be affected by who makes the decision. In its nature contract focuses primarily upon the interests and predictions of the immediate contracting parties. This intensity of focus creates the peculiar drive and energy of contract relations; obviously it also may lead to ignoring or underestimating elements which the contractors do not see as vitally affecting their interests. Of course, official action will not necessarily rest on a full-sweep comprehension of all involved interests or factors; indeed, vested interest in official position may create its own distortions of appraisal. But at the least, insofar as official decision proceeds among men who are not contracting for their own interests, it is likely to proceed from an appraisal different from that produced by contract. (3) The kind of power brought to bear upon a given situation will vary, according as the decision makers are armed with the direct power of the state or are unofficial persons who at most may ask the loan of state authority to back them up. By definition, the law possesses the legitimate monopoly of physical compulsion; presumably in all save conditions of extreme social tension or breakdown, this factor will remain in the background. Closer to everyday transactions, the law possesses extraordinary power to mobilize and direct economic resources, through ownership of public property, and by the powers of taxation and eminent domain; likewise the law possesses broad potential authority to require its license or leave for the conduct of activity of public concern, and to ask questions about what goes on. How far nonofficial persons possess either practical or legally recognized powers of physical compulsion, or of control over allocation and use of resources, will depend ultimately upon the legal order of their society. Unofficial persons may in fact exert great power over others. But in the nineteenth-century United States such power was held by delegation under law, or (more often) by default of law. Moreover, private power might be more limited in potentiality than public power; often it was only by indirection or by stealth that men could exert the influence they enjoyed by virtue of their private position. (4) The central legal principle of our society was constitutionalism — the idea that all legitimate power, including that of the state, was limited by the requirement that it serve to fulfill individual life. Private property stood as a symbol of the high value our tradition put on dispersion of decision-making power, and on men's practical freedom to lead private lives and to exercise their wills for other ends than that of political power. Constitutional government expressed, on the other hand, the conviction that a meaningful human life was only to be found in community, and that there were proper limits to private will. Tension between these values was a creative force in our life; our successes and our failures in relating the one to the other marked critical stages in our growth.

CHAPTER IV

CONTRIBUTIONS OF CONTRACT TO THE ORGANIZATION OF THE MARKET

IN ALL phases of producing and marketing lumber and its products, a substantial role of law was to provide an assured framework of contract doctrine and administration within which private arrangements might work themselves out. This was especially true of the law's relation to logging and milling operations and to trade in logs and lumber.

Throughout the industry's legal history in Wisconsin, the contract law which played such an important part in industry operations was mostly judge-made law (common law). The broadest and most explicit representation we have of this body of contract common law is in the approximately seven hundred Wisconsin Supreme Court decisions dealing with lumber-industry contract issues, and in the hundred-odd tort cases in the Supreme Court which, in the course of dealing with unauthorized handling of property, lend especially close support to contract doctrine.[1] We do not know what proportion these Supreme Court cases are of the total amount of lumber-industry contract-type litigation brought before all the trial courts of the state, and it would be impracticably costly to find out. As a measure of cross check on the Supreme Court data, however, this study does include findings derived from a detailed study of the flow of business in the major trial court of one of the two greatest lumber producing areas of Wisconsin — the Chippewa Valley — and we shall supplement our view of lumber-industry contract matters before the Supreme Court with what we find of such business before the circuit court for Chippewa County.

THE GENERALITY AND STABILITY OF CONTRACT LAW

Two general characteristics stand out boldly in the seven hundred or so lumber-industry contract cases. (1) They are almost all contract-law cases first, and lumber-industry cases second. (2) They show the application of an already well-defined and stable body of legal doctrine, with little change over the period of the industry's lifetime.

The legal history of Wisconsin's lumber business makes plain that by the second half of the nineteenth century contract was a firm body of doctrine, of such generality that it could embrace the problems of a new industry growing at headlong pace, without felt need to mold itself much to the peculiarities of the industry. Taken together, these seven hundred cases show comparatively little significant variation in the shape of the contract law they declare for lumber, as compared to the law that applied to other dealings in the market. The generality in contract concepts which made this possible was the source of both strength and weakness. It was a source of strength, so far as it meant that the legal order could efficiently and smoothly adapt itself to varied circumstances. But there was weakness, so far as contract law achieved this generality by intense devotion to a quite limited range of policies, abstracted from the living context in which they arose. Thus there was strength in doctrine which readily treated timberland or standing timber as marketable goods in a community in which they were major assets. But there was weakness in doctrine too abstract to acknowledge that there might be good reasons in public policy why trade in timberland or standing timber should not be treated as if it were trade in grain or dry goods. The weakness was a defect inherent in a quality. The prime quality or function of contract law was to serve the market's need of an assured framework of dealing; certainty might be disturbed by intrusion of a wide range of policy variables. The forms of contract served insistent demands of immediate dealings. But they did nothing to urge men's attention toward the broader or deeper context of transactions.

The body of lumber-contract case law made its most distinctive adaptations to the peculiarities of the industry insofar as it admitted evidence of trade custom in interpreting agreements, and evidence of trade facts in applying contract terms, to decide whether there was adequate performance or a material breach. Consideration of trade fact or custom, however, simply underlined the generality or abstractness of the contract concept and the significance of contract as a framework for delegating power to private hands. Neither mode of adaptation of the law to the industry involved asserting community authority to set the pattern of dealing; to the contrary, in proportion as the judges gave weight to such evidence, they demonstrated that the prime role of law here was to implement the operation of forces outside the law.

Moreover, the decisions limited the use of evidence of trade fact or custom in ways which carefully preserved the parties' basic freedom of contract. Thus, sufficiently specific terms in an agreement would always override evidence of contrary custom, however well-established. And the decisions would hold relevant only the custom of the particular trade area, or custom otherwise established (by a measure of high probability) to have been within the bargainers' awareness. The pattern of the decisions dealing with title to timberland or standing timber presented one qualification on use of evidence of custom, which in effect, however, was probably another contribution to the law's stress on serving the market. The court took in stride the resort to custom in interpreting and applying contracts which dealt with logging, driving, milling, and buying and selling logs and lumber. In striking contrast, no case presented an issue of use of trade custom to define any interest in the title to timberland or standing timber. On the other hand, the law's own "custom" figured prominently in title matters. The effect of words "excepting" or "reserving" standing timber in a deed of land was, for example, declared firmly fixed by the meaning given to the terms in judicial precedent, beyond the judges' power to alter; given certain relations of the parties, an equitable mortgage would be declared to exist, almost as a matter of law; on the basis of established case law, a deed would be ruled to convey the trees on a tract of land, unless they were expressly excepted. Probably the absence of resort to trade custom in this one area of dealing reflected the emphasis our law traditionally put on security of land title; the intrusion of lay practices in determining the content of title might blur definitions of title in a measure which would impair marketability. Land title was thus left to professional — and primarily to judicial — definition.[2]

There were only four striking qualifications to this contentment with the common law of contract in the boom years of Wisconsin lumber. These qualifications were all by legislation; all had to do with matters auxiliary to contract. The legislature passed laws (1) in some measure to standardize trading in logs, by special recording and measurement provisions; (2) creating liens for labor and supplies in connection with logging and milling; (3) providing increased damages for timber trespass, to protect investments in standing timber; and (4) making the privilege of incorporation available to lumbermen. The record shows that these were significant additions to the industry's legal setting. But they represented no large exception to the general contract pattern of leaving policy making to private decision. No aspect of this legislation modified basic contract doctrine. The initiative for the laws plainly came from elements in the industry itself, though, it is true, not without some internal differences. Their common effect was to promote private transactions.

Why the only major additions to the law of private dealing in timber came by legislation, rather than by growth in judge-made law, poses a different problem. Rigorous nineteenth-century jurisprudence would have answered readily enough — because courts do not make law. But, of course, judges had in fact made the basic law of market dealing to which these Wisconsin statutes were supplements. And, indeed, these statutes were not without analogy in past judge-made doctrine protecting bona fide purchasers, creating common law liens for services, imposing damages for timber trespass. The contemporary Wisconsin Supreme Court on occasion showed that it was not averse to rather bold policy making; we shall see that, in the name of interpretation, it used a strong hand upon the legislature's timber trespass laws, the better to adapt them to competing equities in the industry, as the judges saw those equities.

The record gives no clear answer, but probably the prominence of statutory rather than common law growth in these areas reflects some general characteristics of the two processes of lawmaking. The legislature was likely to pre-empt the field where there was (1) sharp clash of interests among groups of too broad reach readily to be represented within the confines of particular lawsuits, as in the case of timber trespass; or (2) demand for facilities and doctrine tailored to the alleged special needs of particular localities, as with problems of lumber inspection and scaling and loggers' supply liens; or (3) a felt need for provision of new, positive machinery (recording or scaling officials, their offices and records), such as could be met only by a body holding the public purse strings or possessing the authority to create offices supported by enforceable fees. Demands of such type highlighted the limitations involved in the more negative and passive aspects of the judicial process.

The second general characteristic of the law expressed in these seven hundred lumber-contract cases was the comparative absence of change in doctrine or administration, so far as contract law came into appellate litigation. There is nothing in the Chippewa circuit court history to suggest that there was any less stability at the trial court level. As one might expect, the data suggest that there was some tendency for increase in the scale and money value of transactions brought to court in the later years, as the industry developed; but larger transactions were accommodated within essentially the same frame of law as the earlier, smaller ones. The largest transactions generally did not come into court; we must not forget that, outside the frame of litigation, great operations were conceived and executed through contract around business bargaining tables and in the give and take of business dealing, as in the building of the Weyerhaeuser empire. Change in scale was dramatic in this realm of dealing outside the courtroom, and carried implications for public policy that added up to issues different in quality rather than merely in size from the problems of the early years of the business. Even so, essentially the same contract law provided the legal setting for enlarged systems of dealing; indeed, the new policy problems arose largely because the generality of contract adapted itself with so little question to radically new directions of growth.

The comparative lack of change in this body of lumber-contract law was a natural consequence not only of the abstract generality of contract concepts, but also of the concentration of the contract activity within a relatively short span of forty years during which the industry underwent a hurried but uniform experience of headlong development. This concentration of activity marked all phases of contract law in relation to the industry. (See Table 5.)

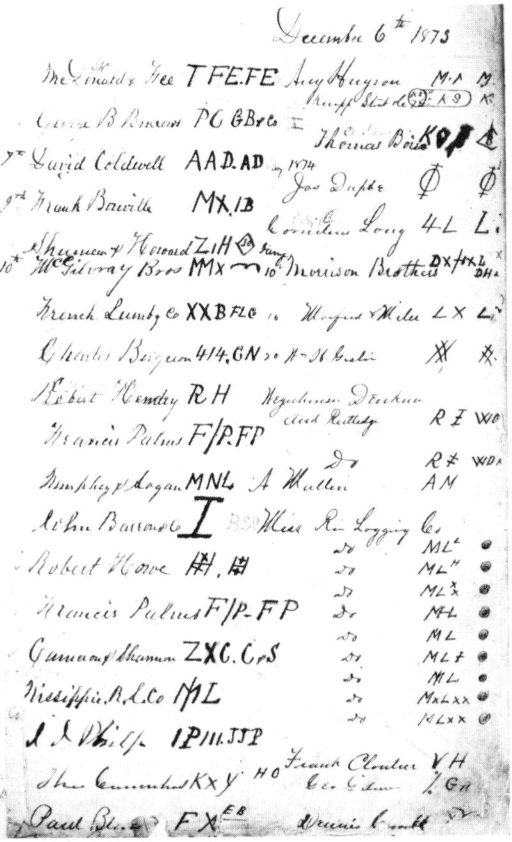

A page from the log mark record, Lumber Inspector, District No. 6, 1873.

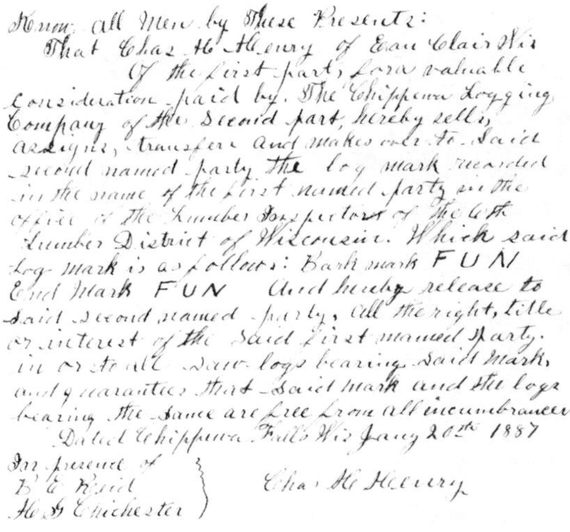

Know all Men by These Presents:
That Chas H. Henry of Eau Claire Wis
Of the first part, for a valuable
Consideration paid by The Chippewa Logging
Company of the second part, hereby sells,
assigns, transfers and makes over to said
second named party the log mark recorded
in the name of the first named party in the
office of the Lumber Inspector of the 6th
Lumber District of Wisconsin. Which said
log mark is as follows: Bark mark FUN
End Mark FUN And hereby release to
said second named party, all the right, title
or interest of the said first named party
in or to all saw logs bearing said mark,
and guarantees that said mark and the logs
bearing the same are free from all incumbrances
Dated Chippewa Falls Wis Jany 20th 1887
In presence of
B. E. Reid
H. G. Chichester
Chas H. Henry

Recorded bill of sale of logs, District No. 6, 1887.

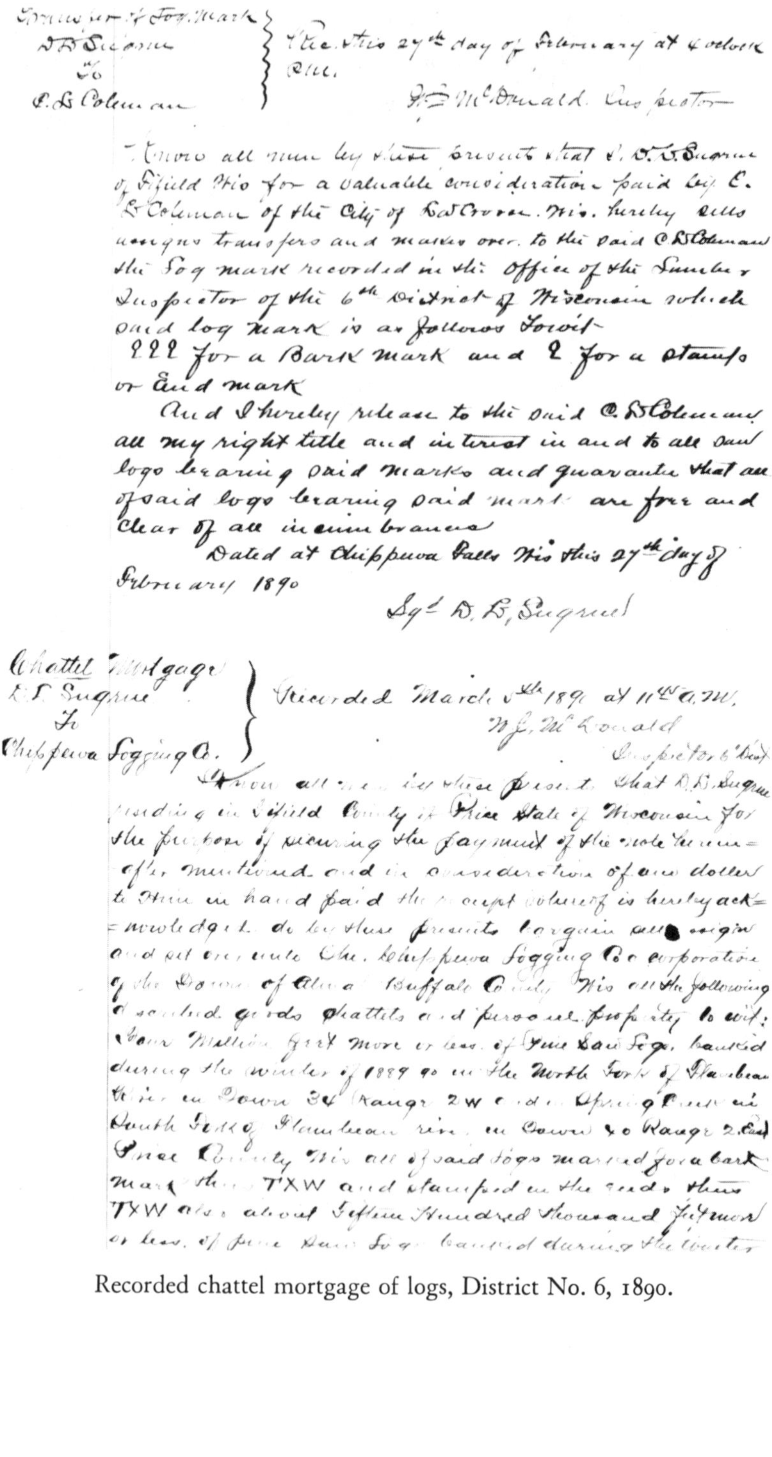

Transfer of Log Mark
D. B. Sugrue
To
C. L. Coleman

Rec. this 27th day of February at 4 o'clock P.M.

Wm McDonald Inspector

Know all men by these presents that I, D. B. Sugrue of Fifield Wis for a valuable consideration paid by C. L. Coleman of the City of La Crosse Wis. hereby sells assigns transfers and makes over to the said C. L. Coleman the Log mark recorded in the Office of the Lumber Inspector of the 6th District of Wisconsin which said log mark is as follows to wit

222 for a Bark mark and 2 for a Stamp or End mark

And I hereby release to the said C. L. Coleman all my right title and interest in and to all saw logs bearing said marks and guarantee that all of said logs bearing said mark are free and clear of all incumbrances

Dated at Chippewa Falls Wis this 27th day of February 1890

Sgd D. B. Sugrue

Chattel Mortgage
D. B. Sugrue
To
Chippewa Logging Co.

Recorded March 5th 1890 at 11th A.M.
W. J. McDonald
Inspector 6th Dist

Know all men by these presents that D. B. Sugrue residing in Fifield County of Price State of Wisconsin for the purpose of securing the payment of the note hereinafter mentioned and in consideration of one dollar to him in hand paid the receipt whereof is hereby acknowledged do by these presents bargain sell assign and set over unto The Chippewa Logging Co a corporation of the Town of Alma Buffalo County Wis all the following described goods chattels and personal property to wit: Four Million feet more or less of Pine Saw Logs banked during the winter of 1889 90 on the North Fork of Flambeau River in Town 34 Range 2 W and on Spring Creek in South Fork of Flambeau river in Town 40 Range 2 East Price County Wis all of said logs marked for a bark mark thus TXW and stamped on the ends thus TXW also about Fifteen Hundred thousand feet more or less of pine Saw logs banked during the winter

Recorded chattel mortgage of logs, District No. 6, 1890.

TABLE 5. Lumber industry contract cases in the Wisconsin Supreme Court

Years	Timber-land sales	Standing timber sales	Joint ventures	Logging	Milling	Labor and supply liens	Transport: River	Transport: Lake	Transport: Rail	Log sales	Lumber sales
1840–1849	..	1	..	..	1	..	..	..	..	..	2
1850–1859	3	..	..	2	10	..	..	2	..	2	3
1860–1869	5	3	..	8	13	4	..	..	..	7	10
1870–1879	7	8	2	11	15	17	4	..	3	17	14
1880–1889	17	8	6	16	29	17	13	1	..	11	19
1890–1899	14	11	3	24	38	24	5	1	1	22	36
1900–1909	15	15	10	27	12	9	..	..	3	7	21
1910–1919	4	2	4	3	6	1	2	1	3	4	13
1920–1929	1	4	1	5	3	1	..	..	6	1	9
1930–1939	..	..	..	3	3	1	..	..	3	..	6
1940–1949	2	1	..	3	..	..	..	..	..	..	..
1950–1956	..	1	..	1	..	..	..	..	..	..	1
Totals	68	54	26	103	130	74	24	5	19	71	134

About 70 per cent of this Supreme Court lumber-contract litigation fell in the years 1870–1909, about 80 per cent in the span 1860–1909; these were the years of economic boom in Wisconsin lumber, and contract, being in nature adapted to serve the market, naturally followed the course of the market.

The changes that did appear in the law, within and outside of the contract frame of reference, were of a character to highlight the functional nature of contract and the limitations of judicial lawmaking. We noted that the striking adaptations of the law to the special conditions of the industry occurred through legislation, and were in their nature auxiliary to contract. We should now observe that this legislation pertaining to records, measures, lien, and trespass came mainly in the late seventies and eighties, as the industry's growth accelerated beyond all precedent. Evidently the sheer pace of events contributed to make case law inadequate to the development of auxiliary devices needed by the industry. The one area in which case law showed marked change fits into the same pattern. Rail transport was critical to the industry's expansion from the seventies on, but few railroad-lumber suits came to the Supreme Court. The absence of litigation testified more to the limited and unsatisfactory nature of common law remedies for aggrieved shippers than to the existence of a state of perfect harmony between roads and clients. The few cases that appeared carried symbolic importance beyond their number; of nineteen lawsuits, eleven turned on the application of regulatory legislation or administrative orders which primarily fixed the terms of the parties' dealings. Moreover, all but three of these nineteen railroad lawsuits came in the twentieth century.[3] This, too, fits into a pattern: When the state's general policy toward timber resources finally changed drastically beyond a contract frame of reference, it was by legislation (as in the matters of conservation and land use control), and in the twentieth century.

THE FRAMEWORK FUNCTION OF CONTRACT LAW

CREATING THE MARKET

The market was a social institution of central importance in shaping the course of the Wisconsin lumber industry. Law did not bring the market into being. But law provided essential conditions for its existence. The tradition of constitutional limits upon official power expressed the fact that a high social value was put upon leaving men alone. Judicial enforcement of the contract and due process clauses of the federal Constitution, and judicial insistence that state power be used only for public purposes, provided dramatic examples of the broad rights of privacy accepted as key values in this society. Such doctrine spelled the broad dispersion of decision making—the protection of private choice and venture—which were the essence of the market. To this policy the law of contract made its contributions.

Contract law did so in part by raising a presumption of the legality of private agreements, and giving to this presumption such force that the issue became material in only two of the approximately seven hundred lumber-contract cases to reach the Wisconsin Supreme Court. This was not necessarily the whole story, however. Without the overt display of authority involved in distinguishing between those agreements which would be held enforceable as consistent with public policy and those which would not, the law might more subtly confine the discretion of contracting parties. The legislature might impose standard terms or forms of agreement. The courts might find and enforce implied warranties or other terms undreamed of by the parties, and in passing upon the effect, separability, or materiality of particular provisions alleged to be in default, might substantially reshape to the judges' ideas of justice the pattern of behavior which the contractors had sought to fix for themselves.

It is thus important to note how far, by these indirect methods, Wisconsin law intruded upon the practical freedom of lumber contractors. We observed that legislation established the only striking adaptations of contract to the peculiar circumstances of the industry, through recording and measurement procedures and the creation of labor and supply liens. Such statutes provided at least a presumptive standardization for certain aspects of transactions, and in this sense the law imposed its terms upon the parties' dealings. But the function of this legislation was plainly to aid the expansion of contract energies, not to confine it. Moreover, we shall see that parties might substitute their own agreements for the official measurements and the statutory liens, though they might not set aside the effect of recording acts.

More difficult to appraise was the extent to which the Supreme Court imposed law-made terms upon contractors. Not infrequently there was room to disagree over how far any given decision simply implemented the parties' intent and how far it stamped their transaction with the court's own terms. Too, there might be argument over a point of definition: Perhaps all decisions which turned on the formal existence of contracts (focusing on offer-acceptance-consideration, or on the Statute of Frauds or the Parol Evidence Rule, or on issues of an agent's authority to make an agreement to bind his principal) should fairly be classed as instances in which the law was fixing for the parties the basic terms on which they might deal. On the other hand, determination whether the parties had so acted as to create a legally binding agreement might be said to go to the form and not the substance of their relations, and hence should not be counted as legal intrusion upon the freedom of private decision. In any case, the record showed a pattern, if one compared the number of cases turning on the legal existence of a contract with those turning on issues of construction or performance, and if—among the cases focusing on construction and performance issues—one listed separately those in which it seemed reasonably clear that judge-made terms were being imposed upon the parties' own course of dealing.

Of 708 lumber-industry contract cases in the Wisconsin Supreme Court, 122 belong to categories (liens and transport) so framed in statute law as to blur assessment of the judges' contributions to parties' contracts; thus we may better confine our attention in this respect to the 586 cases in which the court confronted issues derived in the first instance from private agreements alone. If we count for this purpose issues concerning the existence of contracts, 244 of 586 cases (about 41 per cent) present instances in which the court shaped the parties' relations to the measure of doctrine which was primarily the product of the law rather than of the parties' agreement. If we accept this general total, it represents a considerable qualification upon the extent of discretion allowed to private decision making. But suppose we put aside the issues about the existence of contracts, distinguishing them (for a different focus of analysis) as concerned with the form rather than the substance of the parties' dealings. Then the picture is more like that which we would expect from the concept of contract as a realm of discretion delegated to private persons. Of 434 cases involving questions of interpretation, performance, or analogous issues, 92 (about 21 per cent) seem instances in which judge-made law imposed terms on the parties. Moreover, by this latter tabulation, the number of court-imposed terms was well below the over-all average in four categories (joint ventures, logging, milling, and lumber sales). The categories in which instances of court-imposed terms ran substantially above the general average were such as to point up by contrast the delegation of power which was the expectable norm in the true realms of contract. These were the groups of cases involving transactions in land or standing timber, where the value placed on secure definition of titles put a premium on terms fixed by the law.[4]

TABLE 6. Issues and role of judges in lumber-contract cases before the Wisconsin Supreme Court (Transport and lien cases excluded)

Type of issue and percentage of total[a]					Court-imposed terms[b] (Other than regarding existence of contract)	
Contract type	Total cases	Contract existence	Contract interpre-tation	Contract perform-ance	Regarding perform-ance	Other
Timberland sale	41	15 (.365)	7 (.170)	19 (.463)	5	..
Timberland sale finance	27	12 (.444)	..	15 (.555)	9	..
Standing timber sale	38	14 (.368)	16 (.421)	8 (.210)	11	..
Standing timber finance	16	2 (.125)	..	14 (.875)	1	..
Joint venture (including agency)	26	7 (.269)	9 (.346)	10 (.384)	3	..
Logging	103	32 (.310)	21 (.203)	45 (.436)	7	5
Milling	130	36 (.276)	26 (.200)	68 (.523)	9	..
Log sales	71	12 (.169)	19 (.267)	40 (.563)	14	..
Lumber sales	134	22 (.164)	37 (.276)	65 (.485)	18	10

[a] Definitions: "Contract Existence" issues include questions regarding offer, acceptance, and consideration, agents' authority to contract, and fraud or mistake sufficiently basic to allow rescission, as well as questions (such as those under the Statute of Frauds or the anti-trust laws) touching the formal or substantial validity of agreements. "Contract Performance" cases include those in which the main question was whether one or the other party had fulfilled his obligations; incidental to some such cases might be issues of interpretation of the agreement, but only where an issue of interpretation was the center point of controversy was the case classified as one of "Contract Interpretation."

[b] Figures in first of these two columns duplicate cases counted under "Contract Interpretation" or "Contract Performance"; figures in second column represent cases not otherwise counted, and are included to make the count of "Total Cases" given in the first column.

If we paid attention only to this tabulation of Supreme Court cases, we should probably exaggerate the extent to which the law imposed its own terms on the parties. It was inherent in the situation that "hard" and debatable issues should go to litigation, particularly all the way to the appellate court; we know from contemporary records and from reminiscences of the trade that here, as in business generally, the day-to-day flow of activity saw a vast multitude of contract dealings which went through to consummation as the participants arranged that they should.[5] And in Roujet Marshall's tales of the building of the Weyerhaeuser organization in the northwest, or Isaac Stephenson's recollections of the growth of the industry in northeastern Wisconsin, we are reminded again that nothing in this record is more striking than the absence of legislation or litigation challenging the basic contractual framework within which the big firms were created. On either of these counts, we are in a measure dealing with marginal phenomena when we focus on lawsuits in the Supreme Court. In this perspective, the Supreme Court record does not require that we make any major change in the picture of contract as a frame of doctrine within which sweeping power was conceded to private operators in the market to fix the allocation of timber resources and the directions of timber industry growth.

EXPANDING THE VOLUME OF THE MARKET

For the timber industry as for other business, contract law provided a framework of reasonably assured expectations within which men might plan and venture. The availability of the forms and procedures of contract thus helped the expansion of the market. But simple recognition of the legal validity of agreements and provision for their enforcement were limited and relatively passive roles for law. As in other business, in the growth of the timber industry contract law could be put to more dynamic use than this; various aspects of lumber-contract doctrine, as we see it in the Wisconsin Supreme Court cases, in effect gave a positive thrust, or lent affirmative support, to the expansion of market activity. Contract law could promote the extension of the market both (1) by fostering an increasing volume of transactions, and (2) by facilitating a larger scale for private planning and operation, both in quantity and term. The law in fact did more to extend the volume and size than the time reach of lumber transactions. This division in impact of law marked no limitations in contract doctrine. Rather, it reflected the fact that even in its most dynamic aspects, contract simply provided opportunities and means of action, and responded hence to what men wanted, rather than shaping their wants.

Definition of "Property"

The law basically affected the volume of market activity by the range of marketable values which it recognized and protected. We have seen that the fundamental decision of this character affecting the timber industry was taken outside the realm of contract, when by constitution and statute the people expressed their determination to bring the public lands into market as fast as possible. In another area outside contract the law created a kind of marketable value which was important to lumber history: The legislature provided that title to land might be obtained out of proceedings to collect delinquent real estate taxes, and tax-title speculation was to loom large in the timberland market. But this is tangential to our present concern, for the law created tax titles not as part of a policy of creating and expanding the market, but as incident to the business of government; indeed, we shall note shortly that in some respects contract law policy was at odds with, and prevailed over, the existence of tax titles.

In the law's direct concern with the market, the decisions taken regarding the public domain overshadowed all others affecting the timber industry. Yet, there were specific contract law developments which contributed significantly to expanding the range of legally valid marketing conduct.

The interesting legal activity here centered about the recognition of marketable interests in timberland or its exploitation. The decision of most distinctive import, especially as affecting the net product that would be realized from the Wisconsin forest, was the recognition that standing timber was a salable item, apart from the title to the land on which it grew. As formal doctrine, this was old law — derived from England, and refurbished in the experience of the lumber states of the Atlantic coast. In no respect did the institutional force or inertia of the legal system weigh with greater consequences on the Wisconsin economy than in the unquestioning acceptance of standing timber as an ordinary commodity. Following an accepted pattern, the Wisconsin court construed the state constitution to adopt the common law only "so far as applicable to our condition." But, despite the unparalleled extent and richness of the state's timber resource, it occurred to no one even to question the suitability of treating standing timber as an item open to ordinary trade. Thus, for example, contracts for the sale of standing timber were ruled to be contracts for the sale of a legally recognized interest in land, and hence within the Statute of Frauds requirement of a writing to transfer realty. Repeatedly, Supreme Court opinions recognized that the value of standing timber was the significant component of the value of land in litigation, and determined the suitors' rights in the land in the light of this paramount fact. The right to cut standing timber, stipulated in a contract for the sale of timberland, was ruled to be assignable. It was never seriously questioned that private agreements might create effective security interests in standing timber; the complete acceptance of this practice was underlined by the fact that the serious controversies that arose concerned not the timber interest, but the operation of the market upon the timber interest, when bona fide purchasers claimed freedom from reserved titles in trees or their cut product.[6]

A particular case will make more vivid this ready acceptance of standing timber as an item of lively traffic. *Edminster v. Sturges* (1886) reflects a bustle of dealing which was typical.[7] In January 1880 for $600 the Wisconsin Valley Railroad by written contract granted C. A. and S. A. Sherman the right to cut and remove the pine timber on certain land by April 1, 1881; subsequent endorsements on the instrument extended the removal time to April 1, 1884. On February 9, 1883, the Shermans assigned their interest in this contract to Benjamin Edminster. On July 3, 1883, Benjamin conveyed all his interest in this timber by quitclaim deed to Mary Edminster, plaintiff in the lawsuit. Despite his quitclaim deed to Mary, in September 1883 Benjamin, without Mary's knowledge or consent, assigned the Sherman contract to defendant, as collateral security for a debt owed by Benjamin to defendant. Benjamin later paid this debt. Nevertheless, after the debt was paid, defendant made a quitclaim deed of all his interest in the timber to Whitney, for $500, without the knowledge or consent of Mary and at a time when Mary still had a right to cut the timber under the extensions of the time for its removal. Mary thereafter did cut and take the timber, as the court held she had a right to do; even if Benjamin's assignment to defendant might have been given any legal effect, it was of a security interest only, and was ended when the secured debt was paid. On the other hand, since Mary's title was thus secure, and she had in fact obtained her timber, the court ruled that she had no basis on which to recover from defendant the money he had received for the worthless quitclaim he had sold Whitney. Such a case presents nothing remarkable in the matters with which it explicitly deals. But its tacit approval of a fluid way of dealing gives it meaning as a symbol of value judgments so deep-seated that they were not felt to require explanation or justification.

Where the court did not draw upon precedent so old and firm as that which sanctioned traffic in standing timber, it was led to make its policies more explicit. Thus, though it is of secondary importance in the industry, a line of cases enforcing land locators' claims for services rendered timberland investors illumines the growth of policy. In the leading case of *Bell v. Thomas* (1884), the defendant had agreed to pay plaintiff fifty cents an acre for "minutes" of certain state and federal lands made by plaintiff; the minutes "included the description of the land, the amount and quality of the pine found on each forty acres, the number of thousand feet of timber found, the quality of timber, how many saw-logs there are in a tree, how many logs it will take to make a thousand feet of lumber, the nature of the soil whether rough or smooth, and the nearness of the land to a stream for driving logs." Defendant admitted the contract, but insisted that its proper interpretation was that he need not pay plaintiff for any minutes which were not of use to him, and that he had obtained minutes of the same land before that time from others and had used only those minutes in making purchases. The Supreme Court ruled for plaintiff in terms which might have come from

John Locke: Plaintiff's minutes must be recognized and protected in law as property, for "they had been made by the labor and skill of the plaintiff, and they had value in market. He might have sold them to others had he not sold them to defendant." Public policy must foster and safeguard the marketable quality of the knowledge which plaintiff's energy had gathered, and the contract must be construed to prevent trifling with these values:

The fact that the defendant had minutes of the same lands, made by others, did not by any means render the plaintiff's minutes valueless. They might have been fuller or more reliable. At all events, by his obtaining them, he prevented the plaintiff from disposing of them to others . . . The condition of payment was fulfilled when the defendant purchased or caused to be purchased, the lands described in them . . . Any other construction would be trifling with the plaintiff's rights of property and render the contract futile . . .[8]

On similar grounds, in *Schriber v. LeClair* (1886), the court found sufficient consideration to sustain an agreement amounting to an equitable mortgage; though the "mortgagor" had no title at the time when he arranged for the "mortgagee" to finance a land purchase and take title from a third-party owner, yet the mortgagor "had hunted up the lands, made minutes of their descriptions, and estimated the amount and value of the timber upon them," and "this certainly was of value in securing the lands."[9]

One configuration of cases suggested a qualification upon expanding the categories of marketable property. Where the financier of a land purchase took title, the court would find his title to be simply an equitable mortgage, upon suit of one who showed himself the intended recipient of the beneficial interest in the land; but the opinions stressed that proof must be clear and convincing to make out such an equitable mortgage. Upon a sale of land containing standing timber, no reservation of the timber to the vendor was implied; on the other hand, where the contract price remained unpaid, there was no implication of a right in the vendee to cut before payment. Time limits for cutting stipulated in sales of standing timber were strictly enforced, and were construed as conditions rather than covenants, so that the contract passed no title to uncut timber; this ruling, several times reaffirmed, stood out because the court acknowledged a strong line of authority to the contrary in other states. Where a deed must be reformed to acknowledge the grantor's prior sale of the right to cut timber within a specified time, there was material error in a decree which omitted the stipulated time limit; without the limitation, the holder of the cutting right might let the trees stand indefinitely, "to the great prejudice" of the later grantee's title. Certain rulings defining interests in land also fitted into this picture. A license to cut timber was construed as revocable, in the absence of evidence to the contrary, and passed no title to uncut timber. In the rough conditions of unsettled, unfenced, often ill-surveyed timber country, there must be strong evidence of continued and fully exploitative use to establish title by adverse possession; entry for sporadic timber trespass did not suffice; on the other hand, entry suited to the nature of the land, for purposes of systematic logging, made out a case.[10]

In the bulk of these opinions, the judges were content to ground their rulings on precedent, and did little to express policy. In result, however, the cases suggested a presumption in favor of holding together in law the components making up the total value of timberland tracts. The presumption was rebuttable; the parties' agreements might deal separately with standing timber, create security interests in it, and allow liberal cutting terms. So far as the presumption emerged, however, it braked the creation of a diversity of claims upon a given timber tract. Probably this result was not inconsistent with the law's general bent to foster the market; if the presumption somewhat curbed the variety of dealings, yet it tended also to keep titles simpler and hence more easily traded.[11]

Favoring Venture

Public policy in the nineteenth-century United States favored nothing more highly than putting economic resources in use. Though it was often too narrowly defined and too impatiently pursued, productivity was our key value; the enterprising venturer was our most prized member. Policy toward the public domain, and the promotion of transport development, gave the most powerful thrust to this exploitative temper. But contract law contributed important items to the pattern.

Contract expressed energetic self-interest. Contract law expressed the nature of contract by insisting that men assert their interests, push them, and fight for them, if they were to have the help of the state. Rules of waiver, estoppel, and laches were familiarly invoked in lumber-industry matters to penalize the laggard. Policy called for combative vigor, as well as mere zeal. Thus the court would not help one who, having purchased standing timber under a valid contract containing a time limit for cutting, allowed the limit to go by while he stood inactive because a subsequent buyer of the land asserted a conflicting title. Defendant had made

> a mere claim of title, and a notification that, if the plaintiff cut timber, it would cut the same at the risk of an action for damages. This, however, is a warning frequently given, and rather commendable than otherwise, because it notifies the adverse party of what may be expected to ensue from its acts. People frequently are obliged to choose between two courses of conduct, knowing that the choice of one course will result in a lawsuit and that the choice of the other may forfeit valuable rights; but we have not supposed that this knowledge would relieve from the necessity of making a choice, or excuse one from sleeping on his contract rights.[12]

Public policy favoring venture and productive use of resources implied the law's support only for those who made substantial commitments. The law wanted men to hazard a stake; only such ventures would bring the commitments of responsible will and effort which promised rising production. At least, the outline of policy was clear to this effect, so far as the law dealt with private transactions in the market. In public land policy, and in legal handling of the social cost problems of the lumber industry, this lively concern that there be a substantial private commitment in return for public support was markedly absent. But within the context of private dealings, lumber-contract law in Wisconsin promoted the expansion of true and substantial market venturing (1) by discouraging free riders, and (2) by devoting painstaking attention to defining the protection of the good faith purchaser for value.

Something for nothing always looked attractive, especially in a capital-scarce economy. We shall see that contract law made important contributions to men's efforts to maneuver with scant resources. It was the more striking, therefore, when the judges drew some lines beyond which they would not lend the law's help to the shoestring operator, or to one who would have his gain without paying for it.

The broad and flexible principle against unjust enrichment was the classic expression of policy protecting the exchange basis of the market. This found familiar application in lumber-contract cases. There might be rescission for fraud or mistake going to the essence of the bargained-for exchange, and likewise for failure of consideration, as where a land title proved fundamentally defective. Rescission, on the other hand, was conditioned on the rescinder's return of any value received which he could restore to the other party; reciprocity was the rule. So, too, where plaintiff, having ratified an unauthorized timberland sale made by his agent to defendant, sued for and obtained strict foreclosure under the land contract, this was held a binding election to pursue a contract remedy, barring an action for conversion of timber stripped from the land by defendant in excess of the contract's security stipulation. Similar doctrine lay back of the recognition and enforcement of equitable mortgages in land. Though the proof must be clear, if it was established that a legal title had been taken only as security, it would be treated as a mortgage; the strength of the policy involved was indicated by the strictness with which the court would so rule as a matter of law, once the facts of the parties' relations were established.[13]

Concern to protect the exchange basis of the market was further manifest in the care with which the decisions sought to mark out the legal consequences of the cutting-time-limit clause familiar in sales of standing timber. We have seen that the cases in effect favored holding the value components of timberland together for marketability by giving strict enforcement to such time limitations, ruling that no title passed save to such timber as was cut within the stipulated period. On their face, such sales contracts typically expressed the dominant position of the seller of standing timber, by insisting that the timber sold be cut "and removed" within the time limit set by the agreement. In the face of such language, the court made strict use of traditional distinctions between real and personal property, to insure that the logger would lose none of the fruit of his labor. Thus, in *Golden v. Glock* (1883), where the buyer of standing timber had cut down trees and manufactured the logs into stave bolts on the land, but had not carried away the manufactured product by the date within which the contract stipulated that the "timber" should be "removed," the court ruled that

. . . the stave bolts were cut before the time limited in the deed. The trees from which the bolts were manufactured having thus been severed from the soil prior to the expiration of the time limited, and their character essentially changed by such manufacture, so that the product became personal property . . . they were, in effect, thereby removed from the premises within the meaning of the conditions in the deed, and hence . . . the plaintiff, even after the expiration of the two years [stipulated in the sales contract] had an implied right or license to go upon the premises and take therefrom the stave bolts so manufactured.[14]

The underlying bias in favor of the active exploiter as against the passive timberland investor came nearer to expression when the court pushed the ruling of the *Golden* case a measure further in *Hicks v. Smith* (1890). Again, a contract of sale of standing timber had stipulated that "the said pine timber . . . be cut and removed . . . on or before" a set date. Before that date the buyer cut the trees, and there sawed some into logs, marked the rest into log lengths and caused his mark to be placed on each log and log length, but carried none of them off the land by the stipulated date. The court reaffirmed the ruling of *Golden*; if the timber had been so altered as to be transformed in law from real to personal property, it had been "removed" within the requirement of the sales contract. But the court was obviously troubled, because here the timber had not been so far processed as in the earlier case. Thus it was impelled to argue that sufficient processing had taken place to give the timber the character of personalty, and then to go further and deny that *Golden* required more than severance of the trees from the soil to convert them into personalty. In the face of the plain intimation to the contrary in *Golden,* the court now claimed:

It is not made essential in that case that the trees should have been manufactured into anything after they were severed from the soil to make them personal property. It is a statement of the facts in that case merely . . . Manufacturing the timber, after it is cut down, into any form is no part of the act of its *severance* from the land. It is personal property because it is "severed from the soil". The sap can no longer go up into the tree from the soil, as some writers say.

To demonstrate that timber had been transformed into personal property by severance would not seem necessarily to prove that it had been "removed" within the meaning of the parties' contract. The invocation of a mechanical criterion of decision to apply formal legal concepts ("realty" and "personalty"), and the strong handling of the language of the prior opinion, suggest that some stronger substantive policy was pressing in the background. This policy was evidently to protect the active logger-developer of the property from loss of his effort to one who could not claim to be fulfilling so productive a role. "The plaintiff had bought and paid for [the trees] . . . and cut them down when he had the right to do so, and treated them as his *movable* property." Defendant had bought the timberland from the standing-timber seller after the sale of the timber, with knowledge of plaintiff's contract, and only about five weeks before the expiration of the cutting time limit. The court found nothing in the defendant's situation to induce the positive help or sympathy of the law:

It seems that the value of the land consisted almost entirely of the pine timber on it, and when the defendant contracted to purchase it, he evidently supposed that the plaintiff would be unable to remove said timber so cut and marked by him before . . . January 1, 1888, the day of forfeiture, then near at hand, and that he would probably obtain both the timber so cut and the benefit of the plaintiff's labor in cutting it. Whatever, therefore, the defendants' right may be at law, they have very little in equity.[15]

Another strand of this emerging pattern of doctrine was woven by decisions which refused the law's aid to one who sought a large return for nominal risk or outlay. Contract law in essence meant broad delegation of power to the contractors; consistent with this policy, the law generally would not weigh the adequacy of their exchange as a criterion of whether consideration supported the agreement. But there were ready principles of estoppel or laches to bar gross overreaching by one who sought to gain much for little.

Lacy v. Johnson (1883) invoked familiar estoppel doctrine to preserve the integrity of a contract for the sale of standing timber. Having admittedly cut and not paid for timber the contract price of which was about $2500, the buyer under the contract denied liability on the ground that after he made his contract with plaintiff, and after he had cut the timber, he had bought tax certificates from the county for delinquent taxes on the land, and had later obtained a tax deed of the land based on these certificates. He paid the county $36.12 for this tax title. But the judges were not disposed to allow him thus to obtain for $36 timber he had contracted to buy for $2500. Because the relation of seller and buyer existed between the parties at the time, the buyer was incompetent to acquire a tax title which would cut off the seller's rights to collect the price of the timber:

The injustice of allowing the defendant to avail himself of the tax deed as a defense is manifest. The plaintiff's right of action against him for the value of most if not all the logs converted had accrued before any tax deed was issued, and while the tax certificates were in the hands of the county. If the county had any claim to these logs before a deed was issued, it was . . . only for the amount due on the certificates, and if the defendant had any reason to expect the county would take measures to enforce such lien, he could have defeated it by redeeming from the tax sale . . . When, therefore, he purchased the certificates of sale of the county, as between him and the plaintiff, it must be treated as a redemption of the same, and the deed issued thereon cannot defeat the plaintiff's title to the logs which had been long before converted by the defendant.[16]

The implicit policy bias of *Lacy v. Johnson* came to more open expression in decisions which vigorously enforced the legislature's declaration of policy that the benefits of the recording acts should go only to a "subsequent purchaser in good faith and for a valuable consideration . . . whose conveyance shall first be duly recorded." An originally unrecorded deed for which substantial consideration had been given would thus prevail over the intervening claim of a shoestring operator. The question, said the court in one case, was "as to the effect of actual knowledge of the fact that other persons not in the apparent chain of title have been deeding the land as if they were owners for substantial considerations, in connection with the fact that the apparent title is based on a quitclaim deed executed but yesterday for a nominal consideration." Given this opposition of interests, one whose title rested on quitclaim deeds purchased for about $5.70 per forty-acre tract lost as against an unrecorded deed, bought at about $237 per forty acres, the holders of which were on record as having paid the taxes on the lands for six years.[17] Similar doctrine was applied in another case to defeat one who for a total of $5 had obtained his quitclaim deed to a forty-acre tract which he admitted he valued at between $1.50 and $2 per acre. Defendant, a land cruiser or woodsman, in partnership with a lawyer operated a firm dealing in Lincoln County lands, making a practice of searching the land records for titles which they could buy for nominal outlays. Plaintiff had purchased this forty along with others, by a deed duly recorded; but the register of deeds had by mistake omitted to copy this one description into the record. Defendant admitted that his search showed no delinquent taxes against the land, and that he had not searched to see who had been paying the taxes, nor had he inquired of his quitclaim grantor as to the basis of the latter's title. Granting plaintiff's plea for a decree removing the cloud on plaintiff's title created by defendant's deed, the court observed crisply that the recording act

was not enacted to protect one whose ignorance of the title is deliberate and intentional, nor does a mere nominal consideration satisfy the requirement that a valuable consideration must be paid. Its purpose is to protect the man who honestly believes he is acquiring a good title and who invests some substantial sum in reliance on that belief.[18]

Timberland speculation posed issues the resolution of which was revealing of values held in nineteenth-century Wisconsin. The most serious problems, and the least satisfactory resolutions, concerned the disposition of public lands. But there was need to refine policy also within the framework of purely private dealings. Here, for example, market logic seemed to look two ways in the matter of something-for-nothing. As an institution for production, the market demanded that men put something in, if they were to take something out; and we have seen that the law showed concern for this principle. But venture energized the market, and the nature of speculation was to hope for great returns from small commitment. Wisconsin policy took speculation for granted, as the legislature showed in its eagerness to put the public lands on sale at low prices, and as the court showed in its unquestioning acceptance of the legality of the general run of speculative contracts in timberland.[19] However, the law insisted that there be risk taken in fact, and refused its help to one who first advanced his claim to profit only long after the event had showed success.

On this basis, for example, the court denied a decree of specific performance in favor of the buyer under a timberland sale contract, in *Combs v. Scott* (1890). The contract was dated May 1, 1882, to be performed July 1 of that year; the contract seller died October 7, 1886, and the suit for specific performance was begun against his trustee in April 1888. In some situations this might not be a great lapse of time, but, true to the exploitative bustle of late-nineteenth-century Wisconsin, the court in effect insisted that delay must be measured not in absolute terms, but relative to the pace and scale of events. The court noted that the contract buyer's own testimony showed that at the contract date for performance (1882) the 29 forty-acre tracts found to be in issue were worth only $10 apiece, but that at the time of trial (1888) they were worth twenty to fifty times as much, or from $200 to $500 for each forty acres:

> The timber on these lands has become much more valuable by the long delay, and a railroad has been built and is in operation through these lands, and the country generally has been greatly improved since July 1, 1882 . . . The enforcement of the contract at maturity would have been of merely nominal expense and damages to [seller] . . . but will now impose an enormous claim upon his estate of many thousands of dollars . . . It would be difficult to find a case in the books of greater change in the situation and value of the lands and the circumstances material to the relief, occasioned by the delay, or in which specific performance has ever been granted under such circumstances . . .

Plaintiff could not "lay by for years while these changes were going on," and then cash in on the changed value through a degree of specific performance, grant of which lay in the equitable discretion of the court. Hence he was left to his remedy by an action for damages, which the court carefully noted would be limited to the value of the lands at the date of performance set by the contract.[20]

The court made even more plain its demand that there be good faith assumption of risk, when in *Russell v. Fish* (1912) it denied a laggard plaintiff a decree that the holder of legal title to a timber tract held it as security only for money advanced to buy the land for plaintiff's benefit. After ten years' delay in suit to assert this claim, plaintiff might not succeed:

It is universally recognized that the speculative character of the property involved in a transaction is an important element in determining the effect of delay in claiming a right thereto and that a higher degree of diligence is required to avoid the charge of laches in such cases . . . The facts of this case are clear that the property was highly speculative in character, that this was well known to the plaintiff, and that he delayed enforcement of his alleged right for many years to await the outcome of the investment and enterprise.[21]

Speculation in timberland tax titles particularly tested this judicial reaction against the nominal venturer. The condition of the law and the nature of the situation made for ill-defined policy here. An elaborate, though ill-drawn, statutory scheme provided for the creation of tax titles as a useful instrument of public finance. The legislature thus put these titles into market, but with little evident consideration of how they would fit the values of a market economy. In scores of decisions the court tacitly accepted timberland tax titles as routine items of traffic; more specifically, many decisions implicitly accepted the fact that large-scale speculation in tax titles to timberland went on, in the hands both of resident and nonresident dealers. Though void, a tax deed was enough to give color of title to one in possession of land, so that his relinquishment was consideration sufficient to support an equitable mortgage taken by one purchasing the land for his benefit. And "after much doubt and hesitation" the court ruled that a tax deed issued under the statutes and fair on its face was "*prima facie* a marketable title," which the buyer under a land contract was bound to accept, unless he could make out specific objection to it.[22]

Yet, the court seemed in effect to draw some lines in its acceptance of tax titles. Its acceptance was most fully given where the question arose between tax-title claimants or persons who were deliberately dealing in such titles as a business practice. Thus the judges declared their acceptance of the tax title as a marketable title in a case where the court found that the contract buyers knew that their seller's rights in the large bundle of tracts in question consisted in general of tax titles, in which the seller regularly dealt. Moreover, the court found it not surprising that the contract embraced a waiver of any seller's lien for the price of the lands, for "the lands were purchased evidently for speculative purposes, and not because of their particular adaptation to any immediate use to which the vendees desired to devote them."[23] So, also, it was in litigation between rival tax-title claimants that the court most firmly applied the presumption of the regularity of official action, and the protection of the statute of limitations, to sustain particular deeds; on the other hand, the court showed marked lack of sympathy toward interpreting timber trespass legislation to give the benefits of enhanced damages to tax-title holders for wrongful cutting by the original owners.[24]

The decisions returned to their familiar insistence upon protection of the substantial exchange basis of market dealings, when they refused to accord the tax deed holder the standing of a good faith purchaser for value. Thus, as against the beneficial owner of timberland, a tax-title claimant might not assert the benefit of the statute of limitations to overcome so basic a mistake as the failure of the town treasurer to accept payment of taxes when tendered by the landowner's agent.

As to all such defects or want of authority to sell and convey, the doctrine of *caveat emptor* applies, and the purchaser who takes a title depending upon the exercise of special statutory authority cannot be considered a *bona fide* purchaser without notice. He must see to it, at his peril, that the title he takes is a valid one. Besides, he never pays a full or fair price, but gets, rather, 'acres for cents' . . . This is no hardship, as against the purchaser at the tax sale, who purchases for such a small consideration, and who stands in privity with the tax officers who are alone at fault in the premises, and is, in law, affected with notice of that fact.[25]

Other decisions, however, showed that the court was conscious that it was moving here in an area of policy in which public revenue needs must be recognized, as well as the conditions of market dealing. So, the original owner did not prevail against a tax-title holder where the owner had offered to pay taxes to an official not authorized to receive payment; such an officer's mistakes in tax collection would not prejudice the county or its grantee. And where a failure to redeem land from delinquent taxes was due as much to the negligence of the redemptioner as to that of the county clerk, the tax title stood.[26]

Perhaps for reasons inherent in the situation, it was only regarding timberland or standing-timber transactions that the court confronted a broad array of issues raised by dealers who would reap large gains on a nominal risk or commitment. There was no long-term market in logs or lumber, in which one might hold such goods for a speculative rise in value over time. Few cases came to court regarding sales of chattels for taxes, and where taxes were so collected, the officials could readily sell only so much as would yield the tax due, so that there was less likelihood than in the case of land that valuable property would be bid off for a trifle of its worth. The court did have repeated occasion to apply familiar doctrine to test whether particular transfers of logs or lumber, or particular logging- or milling-operations contracts, should be set aside as in fraud of creditors. But these cases provided the only type of situation involving logs or lumber where the judges found consistent need to voice the law's policy against violation of the exchange basis of the market.[27] Oddly, in view of the fluid market in timberland and the familiar use of timberland title as a basis of financing operations, almost no challenges of land transactions as in fraud of creditors were brought to the Supreme Court.

The particular lines of policy we have examined convey special insight into the law's role in extending the volume of market dealing. For one reason or another, in these instances values were more explicitly declared than in the ordinary course of legislative and judicial decision. To keep a realistic perspective, we must not forget that law's most pervasive influence in expanding market activity in timber was through the routine validation, interpretation, and enforcement of ordinary agreements. Legally guaranteed security of transactions was the foundation of market venture. Men wanted legal assurance that they would get the control of economic resources or goods for which they had bargained and made commitments. They got what they wanted from law largely (1) through application of familiar and pedestrian doctrine (as in the rules deciding what constituted effective delivery to make a completed sale of logs or lumber), (2) through the law's tendency to standardize transactions in a broadening and more impersonal market (by recording acts or legislation defining regular measures of quantity, and by judge-made rules giving binding effect to trade custom and practice, applying "objective" measures of the existence and meaning of agreements, and setting the terms of protection of good faith purchasers for value), and finally (3) by the orderly determination of disputed questions of fact and orderly enforcement of obligations. This is to say that law contributed most to the expansion of the market by providing doctrines and procedures for orderly administration of a regime of contract — a matter which a later chapter will explore further.

Bootstrap Finance

What most astonished Tocqueville in the bustling economy of the early-nineteenth-century United States was, he said, "not so much the marvelous grandeur of some undertakings as the innumerable multitude of small ones."[28] Contract afforded the formal means of expression for this expansion of energies, as characteristic of Wisconsin's growing lumber industry as it had been of the eastern states which Tocqueville had observed a generation earlier. But contract provided more than form; in significant measure it helped create the actual means of releasing upon Wisconsin's forest a torrent of individual and small-firm exploitative projects and ambitions. It contributed this dynamic force to the Wisconsin scene primarily by furnishing the legal means by which men might make maximum use of limited capital.

Once it hit its stride in the late fifties, Wisconsin lumbering involved large capital investment — from about $6 million in the Wisconsin River Valley alone by 1857, to over $45 million for the state by 1897. By the end of the century, the industry was dominated by relatively large, integrated firms, and the big lumbermen had accumulated capital surpluses to invest outside the state. But in the first generation capital was scarce, and statutes and court cases reflect a period when capital scarcity pressed hard on men's ingenuity and put a premium on developing the business through small, independent land speculators, loggers, and mill operators.[29] Even in the later years, while the large firms flourished — and, on the whole, stayed out of court — the flow of decisions showed that the financing of smaller operators depended on the procedures made available by contract law.[30]

Concrete cases alone can convey a full sense of the rich variety with which contract played this role. Consider, for example, *Beckwith v. Philleo* (1862), a quite typical situation in which contract helped a buyer of timberland lift himself by his bootstraps. Plaintiff agreed to sell a timber tract to Palfrey; Palfrey was to pay for the land in lumber at the sawmill on the premises, part upon the execution of the contract, the rest in equal annual installments, plus payment of taxes on the real estate. Plaintiff continued to hold title to the land, to be conveyed upon completion of payment by the buyer, but the contract declared that the buyer should "have the possession and use of said premises without impeachment of waste or claim of damage against him, the said Palfrey, as long as he performs the conditions herein contained." Palfrey assigned the contract to defendants, who cut timber from the tract and made it into lumber. When defendants failed to pay an installment of lumber due under the contract, plaintiff sought to seize the lumber by a writ of replevin. But, by familiar doctrine, replevin would lie only if at the beginning of suit plaintiff had either a general or a special title in the lumber plus the right to possession. The Supreme Court held that by the terms of the contract plaintiff had no such right, since defendants

> were rightfully in possession of the premises; the equitable estate under the contract was in them, and they would be entitled to a conveyance in fee simple when the contract was executed on their part. The cutting of the timber was not a wrongful act, for it was the obvious intention of the parties that the purchaser should have the right to cut off the entire timber upon the lands sold, and manufacture it into lumber . . . Regard being had to the nature, character and use of the property, it is most obvious that it was expected and intended that the timber on the lands sold should be cut off and manufactured into lumber . . . The property in this case was to be used for manufacturing lumber. It was to be paid for in lumber obtained from the lands. And it is idle to say that the purchaser under the contract had not the right to cut timber to any extent and sell the lumber manufactured from it, because the legal title to the lands remained in the vendor . . . Suppose the contract had required the payments to be made in money, would a default have vested the absolute property in the lumber in the respondents? . . . It appears to us not. The appellants [defendants] are in possession of the mill and lands under a contract of sale, and are making the precise use of the timber growing upon them that the parties contemplated . . .

Thus the seller of the land was remitted to his ordinary remedy, by suit to foreclose the land contract, or an action for the price, and the buyer otherwise was left free, as the contract contemplated, to make his way by working the land.[31]

Upham v. Hewitt (1877) shows us problems involved in a common kind of relation between men with cash and loggers without it. On October 17, 1871, Hedges, Garfield, and Warwick, loggers, made a written contract with Henry Hewitt, Jr., and Welcome Hyde, financiers, by which the loggers agreed

> that they will commence immediately and work faithfully and constantly for said Hyde and Hewitt until the logs hereinafter mentioned shall be marketed in Oshkosh, in the summer of 1872, as follows, to wit: They are to take charge of the several logging camps of the said Hyde and Hewitt, to be started at such points as Hyde and Hewitt shall direct, in the county of Shawano; hire men to run the same; cut, haul, bank and run to Oshkosh, all the pine logs they can get out during the coming fall and winter; and do all said work in a good and suitable manner, and for the best interests of all the parties hereto.

The loggers also agreed that the financiers should have the right to sell the logs. On the other side, the partnership putting up the working capital agreed

> that they will pay for all stumpage upon any and all pine logs so to be cut by the [loggers] . . . pay for all hired help and for all teams necessary for getting out said logs and running the same, and all supplies and expenses necessary in cutting, hauling, banking, rafting, running, moving, and towing the said logs . . .

Finally, the agreement stipulated for the return which the two interests should have from the venture:

> . . . when the said logs are sold at Oshkosh, or at the Wood river boom, by said Hyde and Hewitt, all the moneys paid out and expended by the said [financiers] . . . shall be first deducted therefrom, and the balance, and all teams and supplies remaining after the sale of said logs, shall be divided between the parties hereto in equal portions . . .

The present action was for the price of goods which plaintiffs had supplied the loggers. Admitting that the logging contract was not without some ambiguity on this score, the court ruled that their agreement was such as to make the financiers and the loggers partners and not merely joint venturers; hence, though the logging had yielded some $700 short of the amount which would have repaid the financiers their advances, Hewitt and Hyde were liable to plaintiffs for the goods furnished to defendants' logger partners:

> It was a case where labor and skill were contributed to the business on one side, and capital on the other. There was necessarily a communion of profit and loss

. . . But the profits of the business . . . were to be divided between them. The parties were therefore interested in the profits themselves, as profits. It is true, for their security for advances made, Hyde and Hewitt reserved the right to themselves to sell the logs . . . But this stipulation might well be made, and still the parties intend to create, and in fact create, a partnership in the business. So, while the stipulation was that the logs were to be sold only by Hyde and Hewitt, it is plain . . . that the title to the property was not in them alone, but in all the partners . . .[32]

Here had been a characteristic exchange of assurances: the men with some money to put out had obtained a season's commitment from skilled woodsmen; the men who had primarily their woods knowledge to sell, had obtained a line of credit to sustain a continued operation; and in manner reflecting the typical stronger bargaining position of the man with cash, the financiers had also got the security of a contract right to control the sale of the output. The parties' financial situation was substantially like that described by the court in a typical case of advances by a timberland owner to a logger, where there was no partnership relation:

. . . advances for supplies are sometimes made to loggers early in the logging season, and before any logs are banked, to the amount of seven eighths of the contract price for the work . . . [I]n a large proportion of the numerous logging cases which reach this court, it appears that the loggers have little capital, and would be unable to execute their contracts were not liberal advances made to them early in their operations, when they must procure teams, tools, and equipage for the business, involving their heaviest expenditures . . .[33]

To these instances—involving men with fixed capital (landowners, or, sometimes, millowners), men with fluid capital (advancing cash or supplies), men whose capital was primarily their skill in conducting woods operations (logging contractors)—we should add a case which puts in the picture the man whose capital was simply his labor and skill with axe or team. In *Kelley v. Schupp* (1884), plaintiff sued as assignee of John Hennessy to recover an unpaid balance for work done by Hennessy in the winter of 1875–76 as a member of a six-man logging crew working under McDonald, a logging contractor. McDonald had located a "logging chance" (a location where timber was available for cutting), but evidently could not obtain a cutting permit on his own credit. Accordingly, he arranged that a permit be obtained by defendants, a partnership engaged in a general merchandizing and lumbering business; defendants regularly supplied woodsmen, as well as doing some lumbering themselves. Defendants contracted with McDonald that they would furnish him goods from their store amounting to $750, and cash up to $250, which McDonald agreed should be used to log this tract. McDonald agreed:

That soon as said season shall commence, he enter upon those certain [described] lands . . . with a sufficient force of men and teams, and cut and haul therefrom all the pine timber suitable for saw logs. That he will mark all said logs so cut by him with the log mark J*MX, and as soon as the stage of water will admit in the spring or summer of 1876, he will drive all said logs into the St. Croix boom and deliver the same to [defendants] . . . free from all labor liens. That in cutting, hauling, marking and driving said logs he will in all things conform to and comply with the terms of that certain permit dated October 9, 1875, executed by Alfred J. Lindley, agent of Lubelle L. Bruce, to David Tozer, and shall perform all and singular the agreements on the part of said Tozer, made in said contract . . .

For this logging opportunity, McDonald agreed to pay defendants the amount of their advances in goods with 12 per cent interest, the amount of their cash advances with 15 per cent interest, a 2 per cent commission upon the gross receipts of sales of the logs, plus all expenses of preparing the logs for market, and all boomage, stumpage, sluiceage, and other charges or liens against the logs which might be paid or assumed by defendants. Defendants, for their part, agreed to receive the logs through the St. Croix boom, prepare them for market, and sell them on the best terms obtainable, and pay to McDonald the net proceeds after the agreed charges and deductions, or turn over to McDonald any remaining logs unsold.

Like many another farmer, John Hennessy was ready to work in the woods when winter enforced idleness in the fields. But he was not certain, he testified, that he wanted to work for McDonald. He talked the matter over with one of the defendants:

. . . I saw Mr. Tozer in the store at Stillwater. I said I wasn't acquainted with McDonald, and that I would like to know how it would be about my pay. Mr. Tozer asked me what I was getting a month and I told him and he said that I must have a good team, and I told him the team was pretty good, and I says I want to know about my pay, and he says, what time has McDonald promised to pay you? Well, I told him that he promised to pay me in June or July, and he says McDonald can do it. Says I, that is not what I am after. I want to know whether I am going to get my pay or not, and he says, well, I will pay you when the logs are sold. Well, says I, that is all right . . .

A statute gave Hennessy a lien on the logs for his labor. But, like most men, he shied away from the mysteries of formal legal proceedings. He did not press his lien, though he was well aware of it:

Would you have gone and performed that labor if it were not for Mr. Tozer's promise to pay you?

Yes, I don't know but I would. I would have put a lien on the logs only I depended on Mr. Tozer to pay me.

Tozer did not pay, denying that he had promised to, and explaining that the venture had not worked out well: ". . . the state of the market that summer [1876] was very dull . . . [W]e knew by looking at the bills that there was no money in it, after we commenced to sell"; in the outcome, Hennessy's claim apart, defendants paid out more than they realized. Still shy of lawsuits, Hennessy did not choose even to sue on his claim, quite apart from enforcing his lien, but assigned the claim to plaintiff:

I traded this bill to Kelly for a horse. It is hard to say whether he was a pretty good horse. Some might consider him a good horse, and some wouldn't. I would just call him a middling horse . . . I sold a claim against Mr. Tozer, who is perfectly good, amounting to $229.30, and the interest on it from the 27th of March 1876, for a horse that might possibly be worth $150, because I didn't want to get into any law.

The evidence was sufficient to go to the jury on the question whether defendant Tozer had orally promised Hennessy, before the work was done, to pay him for his work under McDonald. But the Statute of Frauds voided a promise not in writing to answer for the debt of another. The Supreme Court cleared this hurdle with a ruling which reflected the realities of shoestring finance. Tozer's promise, if made, was not primarily for McDonald's benefit, but to protect the direct interest of defendants as financial underwriters of the whole project:

. . . [McDonald's] interest was contingent upon there being any surplus, and it turned out that the proceeds of all the logs were not sufficient to reimburse the defendants. There being no surplus, the interest of McDonald in the logs proved to be merely nominal.

Thus holding the legal title of the logs, which the result of the enterprise showed was the only substantial interest in them, and having bound themselves unconditionally to make advances to McDonald to the amount of $1,000, (to say nothing of the commissions and liberal interest they were entitled to, and the cost of stumpage and other expenses they agreed to pay), the defendants had a direct interest in having the logs cut, hauled and run into the St. Croix boom, and in avoiding the filing of liens upon them by McDonald's employees. Hence the labor of Hennessy upon the logs was performed upon the property of the defendants, and inured directly to their benefit, as did also his neglect to assert a lien thereon for such labor. Under these circumstances, the alleged promise of the defendants to pay Hennessy for his labor (if made) was in furtherance of their own interests, and the consideration therefor was a benefit accruing directly to them. This takes the promise in question out of the statute of frauds . . .[34]

A common problem lay back of the varied financial arrangements made possible by the flexibility of contract. The problem was to get production of logs or lumber with a minimum of working capital in cash or supplies —by contract, to introduce movement into an inert situation by giving fluidity to fixed capital. Some men held land, the principal if not sole value of which was in its standing timber. Some men held skills as woodsmen, or —like the winter-bound farmer—had time on their hands, which in their situations could be profitably used only in lumbering. Measured by available alternative uses, the woodsman's skill or muscle represented fixed capital no less than the timberland titles. Mobile capital, cash to pay labor and buy supplies, was not easy to find; either it must be attracted by provision of security for its repayment, or procedures must be devised which would in effect substitute for cash or at least reduce to a minimum the immediate needs of cash.

The simplest substitute for cash was to let a man work out his debt or share. The record showed a considerable variety of these arrangements, in which contract law's decisive contribution was to recognize or protect as binding consideration the effort put out in exploiting timber. Thus a land locator's knowledge was protected as his "property," and his contribution of it made a good consideration for his share in a land venture in which others put up cash. A land locator conveyed the tone of Yankee improvisation in these deals, as he testified to the terms of a typical arrangement he and his partner had made with defendant financier:

> [Defendant] agreed to buy these lands and pay for them and carry a half interest for McArthur and myself,—one-quarter for myself and one-quarter for McArthur. We were to pay for them by profits derived from the lumber. He said he was an old lumberman, and I told him I was, and I would put in my time to conduct the business—the lumber business—on these lands. And McArthur had some teams, cattle, horses, and sleds, and he was to put in such proportion of his stuff as was required, and to log them off, and we were to cut them, or sell them in the market, as the case might be. And after he was to get his money and interest back, we were to divide the profits derived from the logging operations of the timber. The profits were to be divided after paying the purchase money of the land, which was $36,000.

Obviously, though the parties might do business on a shoestring, they were often dealing in quite substantial values. As this testimony suggests, a familiar kind of deal was one in which the buyer of timberland or standing timber might pay for it by logging the land and turning back logs to the seller to settle the account; in similar fashion the buyer of a timberland tract and sawmill, or a sawmill owner buying logs to process, might pay with part of the lumber produced. The general-store owner who advanced supplies to an independent logger might take notes, or set up accounts, payable in logs or lumber. On the other hand, the initiative for such payments-in-kind transactions might come from the owner of timberland; with no cash outlay, he might arrange to have his tract logged or his logs turned into lumber, by paying for the work in a portion of the tangible produce.

Where one party's contribution lay simply in his productive effort, it was natural that there should be little talk of security put up by the operator for advances of money or supplies. In these cases the tangible property (timberland, standing timber, mill, supplies, cash) was already in the hands of the financing member of the deal. Of course the operator might have some separate property to mortgage; a logging contractor would sometimes give a chattel mortgage on his sleds, teams, and other woods equipment, to secure advances. But typically the operator's main stake in the transaction was his own effort, so that the only "security" he could offer for advances of goods or cash was his agreement that the financier might withhold payment for work done, if the operator were in default in repaying advances. In such fashion contract, pure and simple, functioned to provide a framework of mutual obligation which encouraged men to venture on production.[35]

Where someone held an interest in timberland, or standing timber, or logs, his title gave him added leverage to start production going. The variety of arrangements by which a title might be used to secure ventures attested the importance of contract to the bootstrap finance of Wisconsin lumber. This became clearest when security arrangements were pyramided. So in *Hoile v. Bailey* (1883) a timberland seller stipulated that enough pine be left standing to take care of his own still-owing debt for the land, though otherwise the buyer should be free to cut in order to earn the purchase price; the court held that the seller was, accordingly, entitled to an accounting for timber cut by his vendee, but, on the other hand, must subordinate his claim on the timber proceeds to that of a banker who had subsequently made a loan to stave off foreclosure by the original seller of the land. In *Schierl v. Baumel* (1889) defendant, a logging contractor, contracted with Darwin to get in logs from a tract owned by Darwin and his partner; defendant relied on his statutory lien to secure his claim for his services. Plaintiff furnished supplies to defendant, relying initially on defendant's own credit. When plaintiff asked payment, defendant had no money, but instead gave him a draft on Darwin, explaining that he held a lien-secured claim against Darwin. The court held that plaintiff's unexcused delay in presenting the draft for payment within a reasonable time, or in notifying defendant that it had not been paid, gave defendant an effective defense in an action for the price of the supplies furnished. In *Hyde v. German National Bank of Oshkosh* (1897) plaintiff bought certain standing timber, paying one third in cash and the rest in his notes. Thereafter, he sold a part of this timber to Paulding Lumber Company, on six- to eighteen-month notes, with a lien for the purchase price. Paulding later borrowed from the bank to obtain money for its cutting operations, and assigned its timber-purchase contract as security. Thereafter plaintiff also borrowed from the bank funds to make his second payment due on his own timber-purchase contract, and assigned to the bank as security both his own purchase contract and that which he had made with Paulding. In the course of subsequent logging operations, Paulding and plaintiff each made advances to the other, after Paulding agreed to log plaintiff's remaining timber as well as that which plaintiff had sold to Paulding. It is not surprising that the court held that an action for accounting was proper, when plaintiff finally claimed for his share of the proceeds of the lumber ultimately realized from the combined operation.[36]

The owner of timberland or standing timber might create a security interest in the land or timber or its log product, to induce another to risk his effort in its exploitation. Thus he might by contract agree that an independent logging contractor should have a lien on the logs cut, for his services, with the right accordingly to hold possession and sell to repay his claim. However, the most striking use of a security interest to this end was accomplished by legislation, in the statutory lien on logs given to persons who furnished labor or supplies in getting out the timber. We can more conveniently discuss the statutory log liens at a later point.[37]

One who put up sizable cash was not likely to be content with a contract lien as his sole security; some kind of transfer of security title was almost invariably in the picture. The owner of timberland might mortgage it, and might make the mortgage more attractive by an added option in the mortgagee to purchase desirable tracts outright. One who held a contract right to cut standing timber might pledge it or assign it, as security for advances. A sawmill owner might mortgage the mill to finance its acquisition. The most common type of shoestring-finance lumbering — certainly, at least, the single most litigated type before the Supreme Court — was that in which the man with cash bought and took title to timberland or standing timber for the benefit of the logger; equity would treat the financier's legal title as a security interest only, and protect the logger as the beneficial owner, with a mortgagor's right to redeem. This kind of financier's mortgage was undoubtedly common because it neatly fitted the practical need of bringing together men with money and men with the skill and knowledge to locate and cut good timber. The pattern also fitted a special need of sawmill owners. Facing the need to cut unit overhead cost by building a large enough flow of milling business to make for a profitable level of operation, the millman might finance the purchase of a tract by a logger, under an agreement that the financier should have the milling of the logs and meanwhile should hold title to the tract itself as security.[38]

The availability of various security devices also encouraged owners of timberland to venture its exploitation in combination with loggers of scant capital. The seller of the land or standing timber might take back a conventional purchase money mortgage to secure the price, but so far as the court cases indicate, the more common practices were less formal. The type of transaction which appeared most often in the Reports was that of the land contract, in which the seller explicitly retained title to the land or its standing timber until he was paid; the law supplied by implication what the parties' written agreement might not spell out — that the buyer thereby acquired upon execution of the instrument an equitable title, and that the seller's reserved legal title was held by him as a security interest only. Since in the standing timber lay the prime, if not often the only, value of the land in most of these deals, the heart of the transaction was in the provision made about the buyer's right to cut. Usually the buyer could hope to pay only if he might log; on the other hand, the seller needed assurance that the logging proceeds would in fact go to pay him. At this point, the development of land-contract security provisions tended to merge with that of contract liens on logs. Contract liens on logs for the purchase price of standing timber became, as Mr. Justice Lyon observed in 1881, an "important class of contracts" in the headlong growth of the industry. The seller might stipulate flatly that no timber be cut until he was paid in full; in this case he held an unqualified security title in the timber as part of the real estate. The seller might stipulate that the buyer should cut, haul, and bank the logs on stream side — or perhaps, further, drive them to a designated delivery point — but that sales should be only by the landowner, who would deduct his due from the proceeds before paying the balance to the logger. Or he might authorize the logger to sell from time to time, in proportion to the amount of a down payment or installment payments. If the landowner trusted the buyer beyond this point, he might authorize the logger to sell, but with a reserved right in the landowner to take possession and control the further sale of logs if the logger fell in default. Where the logger's credit was good, the landowner might make an outright sale and transfer of title to standing timber, stipulating for himself a contract lien pure and simple, empowering him to take possession of and sell the logs on default. But, generally, it appears that contract liens on logs were associated with a reserved title in the timberland or standing timber. Whether alone or in connection with other security, the contract lien on logs for the price of standing timber threw into sharper relief the fact that the availability of contract law helped put a static situation into motion.[39]

So luxuriant was the growth of timberland and logging security arrangements that it stirred the court to an unusual attentiveness to the issue of contract legality. True, the judges applied the general presumption favoring the legality of private agreements; the striking fact was that they should have been moved to declare explicitly that the creation of a present lien on logs was valid as between the immediate parties, when ordinarily the validity of a contract was taken for granted. However, the court imposed two limits on contractors' discretion in this field. Following Wisconsin precedent laid down regarding security interests in growing crops, the court indicated that an attempt to give a chattel mortgage on logs acquired by the mortgagor after the execution of the instrument would be ineffective, at least as against good faith creditors or subsequent buyers from the mortgagor. Likewise following familiar precedent from general sales law, the court ruled that if a financier or seller of standing timber by contract or practice created a misleading impression of unqualified title in the possessor of logs, any effort to assert a reserved security interest would fail against a purchaser in good faith for value without notice of the mortgagee's claim.[40] We shall examine the pattern of values here in more detail later, when we consider the recording acts affecting timber. At this point it suffices to note that these rulings against an after-acquired property clause and against a concealed security interest bore a common meaning. Both sought to fulfill what the judges deemed to be basic functional requirements of the market for security of transactions in a fluid situation. Thus, if the court here displayed unwonted attention to issues of the legality of contracts, it did not depart from the prevailing concern of nineteenth-century policy to sustain market operations.

The court dutifully enforced the security interest of mortgagee or lien holder where this became necessary against a recalcitrant or evasive debtor. The decisions realistically recognized that the standing timber and its log product were usually the sole security of value in these situations. Accordingly, the mortgagee might have relief against waste, if the mortgagor began to cut beyond contract limits or in such measure as to imperil the security. Here a shift in declared policy marked the care which the judges took to protect the secured interests that made land more marketable. In 1862, the court indicated that it would relieve against waste only on a showing that the mortgagor was insolvent. In 1873, with timberland speculation and exploitation in full flow, the judges disavowed this limitation, ruling that the mortgagee might enjoin timber cutting by a solvent mortgagor, where the value of the security scantly covered the debt and the timber value was critical to the value of the security.[41] But the decisions also showed the court alert to preserve the full measure of maneuver which the parties' relationship contemplated for the debtor; if these security arrangements were to promote trade and production, the debtor must be given his fair chance to work out his obligations. That a contract of sale of standing timber reserved a security title in the seller, did not prevent the buyer from assigning his interest to another; and if the contract allowed the buyer to cut and sell, he might do so, free of the creditor's interference, until the creditor foreclosed on due showing of the buyer's default. If the contract gave the buyer a right to cut and sell, no action would lie against him for waste, but only for an accounting, though the parties had mistakenly overvalued the timber security. The security holder would be bound by conduct amounting to waiver of limitations on the right to cut or sell.[42]

Bootstrap financing — the substitution of effort for cash, the exploitation of timberland through a combination of security interests in standing timber and cutting rights in buyers lacking ready money, the trading in logs and lumber primarily on the security of the traded commodities — obviously was a substantial factor in Wisconsin lumbering. How large a factor it was, we cannot determine just from legal materials; especially because these rather involved and often ill-defined relations tended to generate lawsuits, the law's records may create an exaggerated impression of the economic importance of such dealings. The transactions were not necessarily small because they were managed on a financial shoestring; lands priced by the parties in tens of thousands of dollars were sometimes in litigation. Though the opinions are characteristically sparing in exposition of policy, their total effect is to indicate that the judges handled these cases with a sober sense that at stake was not only justice among the immediate parties, but also the functional efficiency of procedures important to the economy.

In any case, bootstrap finance plainly did present a significant element in the flow of lumber industry business before the Wisconsin Supreme Court. It is impossible to measure exactly how many of our some seven hundred lumber contract cases indirectly involved bootstrap financing, for the issues were often such that it was not relevant to bring these facts into evidence. We can make a more definite tally in those cases directly involving issues of law pertaining to the financing of transactions. Here it is plain that in some areas almost the whole business of the court consisted in untangling problems involved in these efforts to trade in timberland, or get into timber production, with a minimum of cash.

Though the figures in Table 7 are rough, they show contrasts bold enough to indicate that the pinch of cash scarcity was felt the more sharply the closer operations were to the woods. This pattern probably reflected in part

TABLE 7. Bootstrap finance issues before the Wisconsin Supreme Court

	Cases involving non-financial issues		Cases involving financial issues	
Contract type	Evidence of bootstrap finance	No evidence of bootstrap finance	Bootstrap finance situations	Other than bootstrap finance situations
Timberland sales	7	34	25	2
Standing timber sales	7	30	16	1
Joint venture	15	11	0	0
Logging	24	61	12	6
Milling	11	76	22	21
Labor and supply liens	74	..	..	0
Transport	2	46	0	0
Log sales	8	55	3	5
Lumber sales	6	105	4	19

the fact that the nearer one came to the wholesale lumber market, the more he entered the domain of large-scale firms which had working capital to carry both their own dealings and those of their customers. In part, too, the contrast would be a natural consequence of the fact that processed forest products, being the more marketable because they had been processed, offered a readier base for conventional cash financing. In the woods operations of the timber industry, however, the record shows that a significant function of contract law was not merely to supply the parties with a formal frame for dealing, but to contribute mobility to fixed capital and give leverage for productive effort.

EXTENDING THE SCALE OF MARKET OPERATIONS

In the life pattern typical of late-nineteenth-century heavy industry in the United States, Wisconsin lumbering grew to be relatively big business. As early as the 1850's, Isaac Stephenson had begun to build large timberland holdings. The Civil War decade saw the emergence of such giants-to-be as Knapp, Stout & Company and the Weyerhaeuser interests. But it was in the 1880's that large ambitions and investment came to dominate the course of the industry. The generality and flexibility of contract provided ready accommodation to this trend, as contract law helped men enlarge the size and lengthen the time span of capital commitments.

However, the most conspicuous body of data relating the law to the industry's market operations seemed to run counter to the economic trend. For the bulk of the lumber-contract lawsuits that reached the Wisconsin Supreme Court concerned small or modest ventures (the execution of which would typically stretch over but one to three or four years), or limited, short-term aspects of larger transactions. With some variation, this was consistently the pattern of such appellate litigation over the whole span of years from 1840 to the 1950's. For the larger-size, longer-term enterprises, the significance of contract law lay not in litigation, but in providing a legitimated framework for private planning tailored to particular situations; it was the more standardized contract relations which figured in the bulk of lawsuits arising out of lumber industry transactions.

In 177 cases the Reports tell us the valuations which the parties themselves placed on transactions litigated — in the contracted price for a timberland tract, a quantity of logs, or a logging or milling operation. The tables on the following pages show that in 20 per cent of these instances the contracted value of the transaction was less than $500; in 40 per cent, it was less than $2,000; in nearly 70 per cent it was less than $10,000; transactions in the substantial range of $10,000 to $20,000 valuation made up a little short of 16 per cent of the 177 cases; those in the top bracket, involving agreed values over $20,000, were a little short of 15 per cent of the whole. In another — and distinct — group of 160 lumber-contract cases before the Supreme Court, though the Reports do not show a contract value set by the parties on their deal, we do have money judgments or jury verdicts rendered in the trial court. These figures make a misleadingly bold contrast with those of the 177 agreed-values instances, unless we remind ourselves that these two sets of dollar figures measure different aspects of the parties' relations. The verdicts or judgments put a valuation only upon that part of the parties' dealings which came into dispute, or measure only net loss or damage rather than gross commitments. But after one makes this allowance, the 160 verdict-judgment figures, on the whole, confirm that the bulk of this litigation concerned small or modest affairs. Of the 160 verdicts or judgments, something over 40 per cent were for less than $500; a little over 70 per cent were for less than $2,000; about 92 per cent were for less than $10,000; 5 per cent involved verdicts or judgments in the range of $10,000 to 20,000; and 2½ per cent fell in the bracket over $20,000.[43]

These two sets of figures offer some evidence of a tendency for the larger-scale transactions to grow in relative importance in the years of the industry's maximum development, 1880–1919. Thus, though the agreed-value cases below $500 were about 20 per cent of the whole 177 instances, this lowest value bracket represented about 40 per cent of these litigated transactions in the years 1840–1879, fell to about 11 per cent in the great expansion years, 1880–1899, stood at a little short of 20 per cent, 1900–1919, and rose again to about 36 per cent in the declining years of the industry's cycle after 1920. The agreed values below $2,000 followed a similar curve: about 70 per cent of the transactions sued on were in this bracket, 1840–1859; about 60 per cent from 1860–1879; about 34 per cent in 1880–1899, 28 per cent from 1900 to 1919, and then the figure rose again, to over 60 per cent for the years after 1920. The modest ventures in the range from $2,000 to just under $10,000 were a little more stable in number, but definitely rose over all the years to 1920, from 20 per cent to 39 per cent. Too few instances fell in the categories of $10,000 to 20,000 or over $20,000 to warrant great weight on precise percentages, but again, the relative number of these transactions rose steadily over the years to 1920, from an early range around 10 per cent of the whole to really substantial proportions in the years of the industry's peak — about 38 per cent in 1880–1899, and 32 per cent in 1900–1919. The figures on verdicts and judgments show somewhat less change, but still reflect the relative rise of larger-scale affairs. Thus the categories under $2,000 predominate, in declining proportion, from 1840 to 1919; the number of cases in these brackets was swelled considerably by wage claims. On the other hand, suits yielding verdicts or judgments for over $10,000 — practically nonexistent before 1880 — add up to nearly a third of the instances in the boom years, 1880–1899, fall away to about 8 per cent in the span from 1900 to 1919, and, strikingly, represent again a third of the small number of cases brought to judgment after 1920.

TABLE 8. Dollar values of lumber industry matters before the Wisconsin Supreme Court: value of transaction as fixed by the parties

Years — Sums	Timber-land	Standing timber	Joint ventures	Logging	Milling	Log sales	Lumber sales	Totals
1840–1859								
Below $500	1	1	..	..	1	..	1	4
$500–1,999	..	..	..	..	..	..	3	3
$2,000–9,999	1	..	..	..	1	..	..	2
$10,000–20,000	..	..	..	..	..	..	..	0
Over $20,000	1	..	..	..	..	..	..	1
1860–1879								
Below $500	..	3	..	3	..	..	3	9
$500–1,999	..	..	..	..	..	1	4	5
$2,000–9,999	1	1	..	..	1	3	..	6
$10,000–20,000	2	..	..	..	1	..	..	3
Over $20,000	..	..	..	..	..	..	..	0
1880–1899								
Below $500	1	2	..	3	3	..	1	10
$500–1,999	7	4	3	2	..	3	1	20
$2,000–9,999	3	4	..	3	3	1	10	24
$10,000–20,000	4	2	..	1	3	5	2	17
Over $20,000	5	1	..	4	1	..	5	16
1900–1919								
Below $500	2	1	1	1	1	..	3	9
$500–1,999	..	1	1	..	..	2	..	4
$2,000–9,999	9	1	1	1	1	2	3	18
$10,000–20,000	..	1	1	..	1	..	3	6
Over $20,000	2	2	..	1	..	1	3	9
1920–1940								
Below $500	..	2	..	1	1	..	..	4
$500–1,999	1	1	..	..	..	..	1	3
$2,000–9,999	..	..	..	..	2	..	..	2
$10,000–20,000	..	..	..	2	..	..	..	2
Over $20,000	..	..	..	..	..	..	..	0

Summary

	Below $500	$500–1,999	$2,000–9,999	$10,000–20,000	Over $20,000	Totals
1840–1859	4	3	2	0	1	10
1860–1879	9	5	6	3	0	23
1880–1899	10	20	24	17	16	87
1900–1919	9	4	18	6	9	46
1920–1940	4	3	2	2	0	11
	—	—	—	—	—	—
Totals	36	35	52	28	26	177

TABLE 9. Dollar values of lumber industry matters before the Wisconsin Supreme Court: amount of verdicts or judgments obtained in trial courts

Years — Sums	Timber-land	Stand-ing timber	Joint ven-tures	Log-ging	Mill-ing	Log sales	Lumber sales	Totals
1840–1859								
Below $500	..	..	..	1	2	1	..	4
$500–1,999	..	..	..	..	..	..	..	..
$2,000–9,999	..	..	..	..	..	..	..	..
$10,000–20,000	..	..	..	..	..	..	..	..
Over $20,000	..	..	..	..	..	..	..	..
1860–1879								
Below $500	..	..	..	2	5	4	3	14
$500–1,999	..	1	..	4	4	3	2	14
$2,000–9,999	1	1	..	1	1	..	..	4
$10,000–20,000	..	..	..	..	..	..	..	0
Over $20,000	1	..	..	..	..	..	..	1
1880–1899								
Below $500	..	1	1	8	8	5	4	27
$500–1,999	2	1	2	4	7	4	4	24
$2,000–9,999	1	..	1	2	1	3	5	13
$10,000–20,000	1	..	..	..	..	..	..	1
Over $20,000	..	..	..	..	1	..	1	2
1900–1919								
Below $500	..	..	4	2	6	3	5	20
$500–1,999	..	..	..	1	4	3	..	8
$2,000–9,999	7	2	..	2	2	1	1	15
$10,000–20,000	3	..	..	..	1	..	..	4
Over $20,000	..	..	..	..	..	..	..	0
1920–1940								
Below $500	..	..	..	1	..	1	..	2
$500–1,999	..	..	..	..	..	..	1	1
$2,000–9,999	..	..	..	2	..	..	..	2
$10,000–20,000	1	..	1	1	..	..	..	3
Over $20,000	..	..	..	..	..	..	1	1

Summary

	Below $500	$500–1,999	$2,000–9,999	$10,000–20,000	Over $20,000	Totals
1840–1859	4	..	..	..	..	4
1860–1879	14	14	4	..	1	33
1880–1899	27	24	13	1	2	67
1900–1919	20	8	15	4	..	47
1920–1940	2	1	2	3	1	9
Totals	67	47	34	8	4	160

TABLE 10. Dollar values of lumber industry matters before the Wisconsin Supreme Court: percentage of instances falling within dollar brackets

	Below $500	$500–1,999	$2,000–9,999	$10,000–20,000	Over $20,000
	Value of transaction as fixed by the parties				
1840–1859	40	30	20	0	10
1860–1879	39	21.7	26	13	0
1880–1899	11.5	23	27.5	19.5	18.3
1900–1919	19.5	08.7	39	13	19.5
1920–1940	36.3	27.2	18.1	18.1	0
Of total of 177 instances, 1840–1940	20.3	19.7	29.3	15.8	14.6
	Amount of verdicts or judgments obtained in trial courts				
1840–1859	100	0	0	0	0
1860–1879	42.4	42.4	12.1	0	03
1880–1899	40.3	35.8	19.4	01.5	02.9
1900–1919	42.5	17	31.9	08.5	0
1920–1940	22	11	22	33	11
Of total of 160 instances	41.8	29.4	21.2	05	02.5

Neither the data on agreed valuations or on verdicts and judgments show significant trends or patterns when analyzed by type of transaction (land, logging and milling, trading) in relation to time periods. A priori, one might expect dealings in timberland or standing timber to provide a markedly higher percentage of instances in the higher dollar brackets, especially as the industry moved to its peak; but the figures offer only some suggestions of this. Again, one might expect that transactions in the logging and milling phases of the business — where the shoestring operator especially flourished — would provide the predominant percentage of instances in the lower dollar categories; the figures for verdicts and judgments tend to bear this out, but not those for agreed valuations; on the other hand, if one combines the instances involving land and trading, the data on the whole reflect the fact that these were the phases of the industry in which the larger operator was markedly prominent, in contrast to the woods-operations and first-processing stages of the business.

Whatever their specific limitations, the data on lumber-contract cases before the Wisconsin Supreme Court showed that litigation mainly served functional needs of small or modest ventures; data from the survey of business before the circuit court for Chippewa County indicate that the picture was no different within the frame of trial court operation. The big firms

TABLE 11. Dollar values of lumber industry matters before the Wisconsin Supreme Court: percentages of instances classified by subject matter, values, and periods

	Up to $1,999			$2,000–9,999			$10,000 and over		
Years	Timberland, standing timber, joint ventures	Logging and milling	Log and lumber sales	Timberland, standing timber, joint ventures	Logging and milling	Log and lumber sales	Timberland, standing timber, joint ventures	Logging and milling	Log and lumber sales
				Value of transaction as fixed by the parties					
1840–1859	28	14	56	50	50	. .	100	. .	. .
1860–1879	21	21	56	33	17	50	66	33	. .
1880–1899	56	26	16	29	25	46	36	27	36
1900–1919	46	15	38	61	11	27	40	13	46
1920–1940	56	28	14	. .	100	. .	. .	100	. .
				Amount of verdicts or judgments obtained in trial courts					
1840–1859	. .	75	25	. .	. .	. .	. .	. .	. .
1860–1879	03	53	42	50	. .	50	. .	100	. .
1880–1899	13	52	33	15	23	61	33	33	33
1900–1919	14	46	39	60	26	13	75	25	. .
1920–1940	. .	33	66	. .	100	. .	50	25	25

of the industry were not parties to any substantial number of these cases before the appellate court; it is not always possible to be certain who were the real parties in interest in many matters, but even by a liberal count, large firms were parties in not more than 10 per cent of the instances included in our tables.

The preponderance of small-scale, relatively short-term projects in these appellate cases sufficiently fits the temper of the times to suggest that litigation here responded to prevailing directions of men's energies and imagination. Contemporary policy regarding public lands disposal, and the tardy development of either public or private conservation policies, fit into a consistent pattern with the emphasis on bustle, contrivance, and immediacy in these lumber-contract lawsuits. Moreover, there were reasons inherent in business fact why men engaged in the more modest ventures should be most involved in litigation. That they typically must operate with a minimum of fluid capital, dependent upon advances of cash and supplies, and bound by various security devices; that they had a minimum of reserve resources to fall back upon if they met unexpected difficulty; that their own bargaining power or means of compulsion was small — such factors were likely either to bring them into litigation or require them to seek the law's help to enforce claims essential to their survival. The large firm had its own staying power, possessed a bargaining weight which would often substitute for lawsuits, and gained experience in negotiation and administration which kept it out of court. Isaac Stephenson expressed a distaste for formal involvement in the law, which he shared with many businessmen of much less means. Lawyers, he grudgingly conceded, "have their place in the scheme of civilization as well as physicians," but he looked back nostalgically to the days of early, simple settlement around Green Bay when "we did very well without them," in contrast to a later age in which "every community is overcrowded with them" and their activities "complicated" life, while "litigation over trivial things clogs the courts." These were common attitudes. But smaller operators could rarely boast the success in avoiding formal entanglement with the law which Mr. Stephenson recalled of his own large operations:

I have so tried to regulate my own affairs as to avoid lawsuits or legal entanglements and have succeeded, I think, very well. I have never had a personal lawsuit, never gave my note and endorsed only one, never borrowed a dollar. The suits entered against the companies of which I have had charge have been very few and unimportant. I have managed the affairs of the N. Ludington Company at Marinette, which has done an extensive business, for more than fifty-seven years. For the first fifty years of this time less than fifty dollars was paid in legal fees nor was it involved in a lawsuit. All deeds, of which there were thousands, were prepared in the office of the company and there was never occasion to call for legal counsel or aid.[44]

However, to observe that the bulk of litigation served the functional needs of the smaller-scale operations does not tell the whole story, or even the most important part of it. Very large transactions, too, did come into litigation, and in not trivial proportion, as our tables show. For the large as well as the small businessman, thus, the lawsuit played its role in warning of the background force by which the legal order might lend vigor to private planning and compel sober responsibility in its fulfillment.[45] Of much more pervasive practical importance than this, however, was the fact that contract law provided the forms and procedures, and embodied the decisions, within the frame of which large firms committed their capital, contrived and executed what in that day appeared to be long-term programs, and sought to impose their standards of order upon the market. An 1891 opinion reminds us that contract thus provided the framework for a vast body of dealing that never got into court. Having resolved a relatively minor dispute over one phase of a large lumber-sale contract, the court observed that it was

highly commendable to both parties that in this large lumber transaction, involving nearly $150,000, they have had so little cause of complaint against each other, and have treated each other with mutual forbearance, without seeking unworthy advantage. In so large a transaction, their disagreement has been comparatively very slight.[46]

Such unlitigated dealings were the substance of the industry's history. For them contract provided indispensable definition and arrangement, and the scale and sweep of what was done attested how real and important was the delegation of power over economic resources thus accomplished.

The corporation was the single most important instrument of private agreement which law made available to expand the scale and time reach of lumber industry transactions. Through incorporation an enterprise might achieve indefinite life in law, or at least life for a stipulated term of years not dependent on the survival of particular men; incorporation provided ready procedures for gathering in capital in large amounts by stock subscription, ordering the affairs of the business under the discipline of broadly empowered officers and directors, and at the same time perhaps limiting the risks of the investors. The corporate form grew in relative importance as the framework for lumber business from the late sixties on, and was the dominant form of large-scale enterprise by the time the industry mounted to its peak in the 1880's and 1890's.[47] Incorporation presents something of a mixed case, however, for unlike the forms of private agreement generally, it was available only so far as, and on the terms, authorized by the legislature. Hence, we shall reserve details of this aspect of contract for separate examination. But except for provisions affecting transport, there is little evidence that the legislature saw anything peculiar to lumbering which should limit lumbermen's access to the privilege of incorporation. Since incorporators were thus substantially free to write their own terms, it fits reality to include at this point the use of the corporate form as a type of private agreement by which the range of market operations was greatly enlarged.

Incorporation was the law's most efficient instrument for integration of business activity under common direction. Even so, greater reach could also be given men's ambitions and plans by the use of long-term contracts between independent bargainers. Until the economic historians tell us more of the life of particular firms, we lack the knowledge fully to appraise this role of contract as an alternative to the tighter integration achievable by incorporation. Our difficulty is the greater, because most of the story has no expression in court records. Recognizing the limitations of the presently available sources, however, we may at least note some important ways in which men used contract to provide the frame for large-scale plans and operations.

The assurance of continued control over the use of timberland or standing timber was an absolute prerequisite for large, long-term commitments of capital. The capitalist who invested in standing timber, to hold for a rise in values over a substantial time, typically could not manage such a venture without representatives on the ground to take care of taxes, boundary disputes, trespass claims, and sales. Compounded partly of the parties' agreements and partly of fiduciary obligations imposed by the law out of the relationship of the parties, the law of agency contracts thus provided indispensable procedures for timber investment. The generation-long services of Henry C. Putnam to the pinelands investment of Cornell University provided only the most dramatic illustration of a kind of relationship which profoundly affected the exploitation of the Wisconsin forest. Likewise, the availability of legal devices to secure the advance of money to buy land or standing timber — by mortgage, or lesser contract liens — made possible large-scale transactions in timberland. On the other hand, these security devices seem to have affected more the size than the duration of land dealings; the reported cases give little indication of the use of security to finance timberland holding as such; typically, when land was used as the basis of security, it was to finance its immediate exploitation. Somewhat analogous to a security relation, however, in promoting land investment, was the familiar agreement by which a land locator pooled his knowledge and skill with another's cash, to make a joint venture in which an undivided share in the title or eventual sales profits was the locator's return.[48]

In various rulings, the Wisconsin Supreme Court recognized the special character of the timberland investments made possible by the support of contract law. Though the law thus in a measure imposed its own requirements on the parties, it was characteristic of the prevailing temper of policy that this was done to implement and protect reliance built up on private agreement. So, an agent would be estopped to deal in tax titles in his principal's land, especially where he allowed the land to become delinquent while he held funds derived from its current produce with which he might pay taxes. Again, in proportion as large holdings were at stake the court would require clearer proof of authority in an agent to sell and fix sales terms. And in a suit to partition timberland, recognizing that full realization of its value depended on holding sizable tracts together for efficient logging, the court would order a sale of the whole rather than its division into small units.[49]

Long-term contracts might also be used to assure a continuing supply of raw material, either for large sawmill operations or to sustain sizable lumber sales or manufacturing efforts. The Wisconsin court showed itself alert to recognize and enforce guaranties of output or supply in such transactions. In *McMillen v. Pratt* (1895) the Wisconsin court granted plaintiffs foreclosure of a mortgage on an undivided half of about 4,700 acres of pinelands in Lincoln and Langlade counties, given by defendants to secure performance of their agreement to cut and deliver from the land 25,000,000 feet of timber of specified quality; in addition the court affirmed judgment for the plaintiffs for $74,918.67 (of which about $56,000 was damages, and the remainder interest at 7 per cent from the contract termination date to the date of the suit), for damages for defendants' failure to deliver the full amount which the court interpreted the contract to guarantee.

Key terms of the complicated agreement in *McMillen* indicate the elaboration of private economic planning which a regime of contract could support. The sellers had owned these lands for over twenty years. The buyers agreed to pay in a form and on a scale which in effect recognized the sellers' role in thus holding as long-term investors; apart from a stipulated price of $6 per 1,000 feet for delivered logs, the buyers agreed to purchase an undivided half of the lands for $100,000, giving their notes for the price. As woods operators, the sellers bound themselves to more complex undertakings than in their role as land speculators. Defendants agreed to cut and deliver, over the span of the contract, 25,000,000 feet of pine logs and timber, which "shall average not more than four and one-half to the thousand feet, and in quality shall be a fair average of all the pine standing on the lands,— not more than five per cent to be Norway pine." Defendants would build, equip, and operate for seven years from the contract date a logging railroad, to bring the logs to water; the railroad must be ready, so that no less than 5,000,000 feet of the logs should be carried over it in the first year of the agreement. Sellers would deliver the logs and timber, rafted out at the Wolf River boom, free of all liens and incumbrances, with 5,000,000 feet to be cut and delivered by the early drive of 1887 and the full amount within three years and ten months after September 1, 1885. To secure performance, the sellers executed and delivered two mortgages on the land, the first to be discharged when the first 5,000,000 feet of logs was delivered free of liens and incumbrances, the second (the mortgage here in suit) to be released on full delivery of the 25,000,000 feet, free of all costs and charges; further, defendants agreed that until the second mortgage was released, not less than 9,000,000 feet of pine should be left standing on the lands involved in the contract.

The sellers delivered only about 18,000,000 feet of timber and logs of the agreed quality, and argued that since this represented all of such timber available on an undivided half of the lands, it fulfilled the agreement; by sellers' reading, the contract was only that they would get out for the buyers so much of the defined quality of timber as half of the land would yield, and the 25,000,000-foot figure represented only an estimate of that amount. The court found this analysis unconvincing, when the objective of the contract was defined in light of the buyers' business situation and needs. Buyers needed an assured raw-materials flow for their large door, sash, and blind factory at Oshkosh. They did not know the special capacities of the lands; sellers did, having held them for twenty years. The contract excluded buyers from any of the timber on the land other than that of the specified quality, pointing to their concern for a particular type of raw material. The explicit terms of the contract so differed from the rights which the parties would have enjoyed as ordinary tenants in common after the sale of an undivided half of the lands, as to show that the intent was to guarantee buyers a full 25,000,000 feet of specified quality timber. As tenants in common, after buying an undivided half of the lands, the Oshkosh manufacturers would have been entitled only to so much of the specified quality timber as an undivided half of the land would yield, but the contract's express terms "mentioned . . . the lands, in their entirety . . . as the field from which the logs and timber were to be taken to perform the contract," and "the plain stipulation" was that the timber sold should be "a fair average of all the pine standing on the lands." These terms tallied with the proved representations of the sellers in making the contract as to the special utility of this particular timber, if realized on through delivery for manufacture at Oshkosh in lieu of purchases in the open market. Defendants had represented

> that the timber could be taken out on certain tributaries of the Wisconsin river; that it would have to be hauled or transported a distance of ten miles to put it into Wolf river, but if taken down that river to Oshkosh . . . it would be worth much more than if taken down the Wisconsin river . . .

Thus, "the contract was not a mere lumbering contract. It secured to [the buyers] . . . the right to have the stipulated stumpage, the logs and timber, if [such were] to be found on the lands." The heart of the matter was that the contract be interpreted and enforced to assure buyers the raw-materials supply critical to their own long-term production planning.[50]

In like fashion, the court insisted in *Brittingham & Hixon Lumber Co. v. Manson* (1900) that a lumber-sale contract must be construed as binding seller to deliver his entire mill output to buyers whose need, as extensive wholesalers of lumber, was obviously for an assured flow of supplies. Holding that defendant-seller's demurrer to the plaintiff's complaint should be overruled, the court observed,

> . . . it becomes proper to consider the allegations of the complaint as to the situation of the parties at the time of making the contract. From those it appears that the plaintiff, a lumber dealer, was under the necessity of having engagements reasonably definite for considerable amounts of lumber in advance, for the purposes of its business; that the defendant was the owner of this tract of timber and of a mill; that he was unwilling to sell the timber alone, but desired to dispose of it in such a manner as to utilize his mill property; and that the parties met and negotiated upon mutual understanding of those circumstances. From this foundation . . . it is well-nigh inconceivable that the plaintiff would have entered into a contract whereby, entirely at the caprice of the defendant, it might have forced upon it a wholly unknown quantity of lumber, varying from a mere trifle to the entire capacity of defendant's mill, which . . . appears to have equaled or exceeded 6,000,000 feet per season, without any notice beforehand to enable

it to graduate its contracts for sales according to the lumber to be received. Further, the knowledge of this situation gives especial cogency and significance to the provisions of the contract declaring that the timber in contemplation amounted to about 17,000,000 feet, and that the same should be cut and manufactured within three years, and that approximately one-third thereof should be so manufactured and delivered each year . . . [I]n weighing the relative importance of conflicting provisions, the situation of the parties is specially significant, and in the present case leaves no room to doubt that the stipulations as to time and rate of delivery were most carefully considered and deemed highly important by both parties. Upon them the plaintiff must regulate its commercial business, and defendant the operation of his mill and the size of his logging outfit.[51]

The more obvious uses of long-term contracts were to assure continuity in land management, the flow of raw materials, or provision of markets. Less apparent, but no less important, was the fact that either in intention or in effect the long-term contract might work to fix the incidence of risks from price fluctuations. This was likely also to be the point of strain in such a long-term relation; thus plaintiff in the *Brittingham & Hixon Co.* case undoubtedly focused on the critical fact in his seller's default when he alleged that, since the beginning of the first sawing season under the agreement, the price of lumber had proved to be continuously higher than that stipulated in the contract. In another case, granting a decree for specific performance in favor of the buyer under a long-term contract for the supply of pulpwood, on a finding that the contract represented a unique asset for plaintiff in its assurance of the raw materials he needed for long-term mill operations, the court observed,

The defendant entered into a fair contract and for many years it proved to be a very profitable one. Because the market price of the material covered by the contract has advanced, defendant seeks to be relieved of its solemn obligation formally entered into. If the contract were not one capable of specific performance the defendant would have the power to terminate it, but being capable of specific performance it must perform the contract or submit to the legal consequences.[52]

Of course, integration under one ownership and direction was the form of organization which would give the greatest legal assurance of continuity in supply, production flow, and cost control, and perhaps of stability in markets. After 1870 the prevailing ambition in the lumber industry was to promote the growth of the individual firm. Contract was the primary legal instrument of this power politics. As he reviewed his career as leading counsel for the Weyerhaeuser interests in the Chippewa Valley, Roujet Marshall told proudly of hard-fought, shrewdly maneuvered lawsuits. But, significantly, he measured the high points of his professional success and service by the occasions in which his draftsmanship, counsel, and negotiating ability helped integrate and administer the Weyerhaeuser empire—the negotiation of what Marshall himself termed "treaties" between interests warring over use of the Chippewa River, the foreclosure of the $650,000 bond issue of the Union Lumbering Company, the financial reorganization of that firm into the Chippewa Lumber and Boom Company, and purchase of the latter and of other properties building up the Weyerhaeuser holdings.[53] Contract gave form and definition to the building of great firms, but, we have seen, it was only in contests to control waterways that litigation touched any of the vital processes of this empire building. Large firms were occasionally in court over a contract or trespass or title matter. But, though there might be a good deal of money at stake, these suits concerned the administration rather than the existence of the big litigant. Large interests realized their vital ambitions chiefly around private bargaining tables, and occasionally in the legislature; only the small men fought life-and-death battles in the courts.

Beyond the point to which men could achieve integration under one roof, the economics of lumber provided powerful incentives to create more loose-knit arrangements to regulate production and price. The large-scale operator must hazard big stakes in investment in timberland and plant; as production—and overhead cost—rose, demand fluctuated and proved acutely vulnerable to general business depression. Elsewhere we examine the industry's collisions with antitrust policy. It suffices here to observe that experience showed baffling and troublesome tendencies in which the use of contract to enlarge market dealings generated ambitions and interests which in turn threatened the existence of the market. It is significant of the primacy placed on the market as an institution of social regulation that this was the only area in which the courts were brought to a sharp consideration of issues of illegal contract.[54]

THE ADMINISTRATIVE FUNCTION OF CONTRACT LAW

The law's role in the market could not end when, through contract, it provided a framework of delegated power within which private decisions might operate. The proper function of this delegation was to release men's creative energies. But unchannelled energy moves toward destruction or fruitless dispersion. Market conduct meant at least a workable minimum of rationally planned and controlled behavior, based on reasonable calculations of what the actor might expect of people with whom he dealt. The nature of contract was to disperse decision making widely, to encourage ventures all the terms and outcomes of which no one could define at the start, to sanction the expression of self-centered ambitions; out of the accompanying swirl and bustle of action there were bound to be disagreements over definitions of intention, misunderstanding and faulty interpretation of terms, rebellion or evasion before disappointed plans, some trickery and fraud, and —whether the actors were honest men, fools, or knaves—the need to fix the burden of unforeseen, perhaps unforeseeable, loss.

The infinite variety of market activity generated its own imperative demand for means of administration. The law's most obvious response was in the several hundreds of lumber-contract lawsuits which came before the Wisconsin Supreme Court. Some of these cases we have already discussed because they declared important general policy; the great bulk of them, taken singly, bear a more humdrum and limited aspect; altogether, whether "great" cases or little, they represented an indispensable administrative contribution to the market development of Wisconsin's forest. Our Chippewa County circuit court data remind us that trial courts in the timber country carried a load of such business far bulkier than that which went to appeal. Lumber company papers, and histories of business dealings set out in the records of litigated contracts, reflect the infinitely greater number of matters negotiated, disputed, and adjusted across business bargaining tables and in lawyers' offices, within the frame of contract law.[55]

The lumber-contract cases in the Supreme Court can be readily divided into three principal types, according as the main point at issue involved (1) the formation or formal or substantial validity of an alleged bargain, (2) its interpretation, or (3) determination whether one party or the other had duly performed or had justification for not performing. No marked change appeared over the years in the relative distribution of these types of issues; the indication was that, both absolutely and in relation to each other, they represented quite stable problems of administering a system of contract.

The Supreme Court cases fell into a pattern suggesting some division in relative emphasis among these types of issues, according to the size of the transaction. One might expect that the larger the stakes, the more care and skill would be likely to go into the formation and definition of a deal. Of course, this was not necessarily so. In 1886 Roujet Marshall was called on to extricate the great Mississippi River Logging Company from an embarrassingly vague log-driving contract, the ill-defined terms of which might entail claims running to many thousands of dollars; significantly, however, this agreement had been entered in haste and without proper counsel, because it was deemed at the time to involve only a small matter.[56] Ordinarily, the large operator might be expected to act with care and under competent advice. The limited data afforded by the Supreme Court cases tended to bear out this hypothesis, for the suits involving the larger transactions (those involving agreed consideration of $10,000 or above) showed less than their due proportion of issues turning on the interpretation of the contract, and a concentration on issues of performance.

Thus 63 per cent of the cases involving reported agreed values of $10,000 or over turned on issues of performance, compared with 48 per cent among all other cases; on the other hand, only 13 per cent of these upper bracket cases centered on issues of interpretation of the contract, contrasted with 25 per cent among all other cases. Over-all, the cases turning on existence-of-contract issues were almost the same percentage of each main group, though

TABLE 12. Comparison of main issues in lumber-contract cases before the Wisconsin Supreme Court, according to value of transaction as fixed by parties

Contract type	Value unstated in report plus cases of agreed values under $10,000				Agreed values $10,000 or over			
	Total cases	Existence	Interpretation	Performance	Total cases	Existence	Interpretation	Performance
Timberland sales and finance	54	20 (.37)	5 (.09)	29 (.53)	14	7 (.50)	2 (.14)	5 (.35)
Standing timber sales and finance	48	16 (.33)	15 (.31)	17 (.35)	6	0	1 (.17)	5 (.83)
Joint ventures (and agency)	25	7 (.28)	9 (.36)	9 (.36)	1	0	0	1 (1.00)
Logging	90[a]	31 (.34)	21 (.23)	38 (.42)	8	1 (.13)	0	7 (.83)
Milling	124	33 (.27)	25 (.20)	66 (.53)	6	3 (.50)	1 (.17)	2 (.33)
Log sales	65	11 (.17)	17 (.26)	37 (.56)	6	1 (.17)	2 (.33)	3 (.50)
Lumber sales	111[a]	21 (.19)	36 (.32)	54 (.48)	13	1 (.08)	1 (.08)	11 (.84)
Totals	517	139 (.26)	128 (.25)	250 (.48)	54	13 (.24)	7 (.13)	34 (.63)

[a] Totals reduced by five cases under Logging and ten cases under Lumber Sales, as compared with Table 6, to restrict present tabulation to situations of clear-cut contract, excluding some auxiliary matters deemed relevant for inclusion in other tabulations for broader purposes. Transport cases are also excluded in the present table, because many involve issues centering on public utility obligations of a kind not typical of simple bargaining relationships. Lien cases are excluded, because, though ancillary to contract relations, in these lumber situations they were almost invariably creations of law and not of private agreement. The left-hand column includes 123 cases in which an agreed value appears in the Reports. (Cf. Table 8.) It is unlikely that the Reports would not indicate where a suit involved a transaction of agreed value of $10,000 or more; probably, thus, all cases in the left-hand column involved transactions which the parties valued at low figures.

if one excluded the timberland-sales and milling categories, in which an unusual number of such issues appeared, comparison of the remaining categories showed a substantially higher percentage of existence-of-contract issues in the cases of unstated value or involving deals valued under $10,000 (.25), than among those involving transactions valued at $10,000 or more (.09).

Determinations of whether the parties had struck a bargain (with an offer and acceptance, grounded on consideration), or had made a bargain valid in law as to form and substance, added up to a subordinate, but still important, part of the whole function of administering a system of contract. In about 26 per cent of all the cases in Table 12 the main issue was of this kind. There was some marked variation from this general average, as among the different types of transaction.

TABLE 13. Percentage of types of main issues by types of lumber-contract cases before the Wisconsin Supreme Court

Type of case	Approximate per cent of type of issue		
	Existence	Interpretation	Performance
Timberland sales and finance	40	10	50
Standing timber sales, finance	30	30	40
Joint ventures and agency	28	34	38
Foregoing categories together	33	22	45
Logging	34	21	45
Land and logging categories	33	21	46
Milling	27	20	53
Log sales	17	27	56
Lumber sales	17	29	54

The land-dealing, or, more broadly, the woods-operations (land and logging) phases of the business presented the court with a somewhat higher proportion of issues regarding contract regularity. In a community in which land was for a long time the most valuable single article of trade, the courts might be expected to place special stress on clear definition and proper form in dealings; the strictness with which the Wisconsin court enforced the Statute of Frauds, to require that parties reduce to writing their dealings in standing timber and their arrangements as joint venturers, underlines the point suggested by the tabulation. Further, land speculation and shoestring logging operations in their nature invited men more given to plunging activity than to conveyancers' caution; it is not surprising that disputes over offer and acceptance and charges of fraud or mistake bulk up especially in the land and logging suits.[57]

Issues of contract interpretation provided the same proportion of business for the court as did the questions of contract formation and regularity—about 25 per cent of the whole. At first appearance, the percentages do not show such difference between land-transactions and woods-operations cases, and lawsuits involving the other phases of the industry, as was noted regarding existence-of-contract questions. But it was probably not accidental that the highest proportion of interpretation questions arose concerning the generally speculative joint-venture or land-agency contracts. Where issues of interpretation arose, they were generally resolved out of the evidence of intention found in the parties' own documents and actions; of a total of 135 cases, in 24 (about 18 per cent) the court relied upon a rule of law to find an "implied" intent of the bargainers. It was suggestive of the law's emphasis on regularity in land transactions, that 15 of these 24 instances concerned timberland, standing-timber, or joint-venture transactions. These, plus the logging cases, accounted for 18 of the 24 occasions in which a rule of law resolved the issue of interpretation; again, there is the indication that the more loosely organized and shoestring-financed woods operations tended to generate more than their share of problems involving definition of bargaining relations. The point thus suggested by the tabulated data is supported and given vivid illustration in the many difficulties the court confronted in deciding whether various bootstraps-finance arrangements created partnerships, joint ventures, or relations of principal and agent.[58] On the other hand, the general predominance of cases in which issues of interpretation were settled on the basis of the parties' own course of dealing emphasized the reality of the delegation of power which contract law made to private decision makers.

Plainly, the determination of issues of performance was the dominant single aspect of the administrative function of contract law in these lumber industry cases. There is some shadowland in classification, as between issues of interpretation and performance, but in Table 13 doubts were resolved in favor of placing a case under the "Interpretation" column, and correction of such margin of error as remains would probably not decrease any of the "Performance" percentages by more than one or two points. Issues of performance were the focus of almost 50 per cent of all the lumber-contract cases studied, and in all but one category of lawsuit the proportion of this type of issue ranged from 40 per cent up to 56 per cent.

It requires some rather arbitrary generalization to reduce a description of the 135 interpretation-issue cases and of the 284 performance-issue cases even to the many heads of Table 15. The summary description of these issues does, however, convey a sense of the teeming variety of projects, hazards, and human involvements with which contract law must cope, and which it must hold within some workable minimum of order.

The inherent difficulties of this administrative task were compounded by certain facts apparent in the timber business. Its actors were often men of little formal education. In any case, they did much of their work in an atmosphere of improvised means, opportunistic scrambling for chances, and restless ambition. The situation encouraged rough-and-ready ways of dealing, and judges voiced impatience at the slipshod transactions which produced the snarled relations they must untangle. The parties' relations in *Clinton v. Webster* (1886), the court acidly observed, were "from beginning to end . . . conducted on both sides with almost criminal negligence and in the most slattern and slovenly manner . . . [with] no reliable system of bookkeeping . . . and the entries made here and there, on divers scattered books . . . many of them erroneous . . . and . . . a dearth of written memoranda in respect to many of the most important matters."[59] We do not have the comparative studies which might tell us how far such looseness was peculiar to the timber trade, how much it was a general characteristic

TABLE 14. Summary of issues regarding existence of lumber contracts before Wisconsin Supreme Court

Contract type	Offer-acceptance-consideration	Statute of Frauds; Parol Evidence Rule	Agent's authority to contract	Fraud or mistake	Fraud on creditors	Against public policy	Compliance with statutory forms	Mortgageable interest	Other
Timberland sales	4	4	. .	7	. .	. .	. .	. .	. .
Timberland sales finance	9	1	. .	. .	. .	. .	. .	2	. .
Standing timber sales	4	6	2	2	. .	. .	. .	. .	. .
Standing timber sales finance	1	. .	. .	. .	. .	. .	. .	1	. .
Joint ventures	3	3	. .	. .	. .	1	. .	. .	. .
Logging	(8)	(6)	(4)	. .	(4)	. .	(5)	. .	(5)
Contract to log	2	1	1	. .	2	. .	. .	. .	. .
Labor	2	2	2	. .	. .	. .	. .	. .	3
Partnership	2	. .	. .	. .	. .	. .	. .	. .	. .
Ancillary	. .	1	. .	. .	. .	. .	. .	. .	. .
Finance	2	2	1	. .	2	. .	5	. .	2
Milling	(10)	(3)	(7)	(2)	(10)	. .	(3)	(1)	. .
Acquisition, etc.	2	. .	2	1	. .	. .	. .	. .	. .
Partnership	2	. .	. .	. .	. .	. .	. .	. .	. .
Labor	2	. .	2	. .	. .	. .	. .	. .	. .
Sawing	. .	2	1	. .	. .	. .	. .	. .	. .
Mortgage	2	. .	. .	1	1	. .	. .	1	. .
Mechanics lien	. .	. .	. .	. .	. .	. .	3	. .	. .
General debts	2	1	2	. .	9	. .	. .	. .	. .
Log sales	3	3	2	. .	. .	. .	. .	. .	. .
Log sales finance	. .	. .	. .	. .	2	. .	1	. .	1
Lumber sales	3	8	5	. .	. .	1	. .	. .	. .
Lumber sales finance	2	. .	. .	. .	2	. .	1	. .	. .
Totals	47	34	20	11	18	2	10	4	6

of nineteenth-century turbulence, or how far it is natural to the variety and pace of market dealings in any time. Whatever its relative significance, however, the fact was plain in the law's relation to this industry: the single biggest job of administration thrust upon contract law was to impose order upon disorderly or out-of-order bargains, and, incidentally, to remind all traders that the law viewed contract obligations seriously, and would enforce a like attitude upon those involved in them. So, contract might contribute to the greater rationalization of society.

But the record of these performance-issue cases showed that all would not be well if men were only careful. Contract was the instrument of will. Willfulness would present problems for adjustment, if no other cause did. Men sought to get out of bad bargains, or to shrug off a less profitable deal for a better one; they covenanted to give good title, and then sought to avoid responsibility when a forgotten or hidden adverse claim was advanced; they committed themselves in fact — as by a delivery of logs — and then tried to deny that they had committed themselves in law, and sought to treat the logs still as their own; they did not foresee all their costs or risks, and then strove to avoid paying for what they had not anticipated.[60] The special sign of the law's interposition before the arbitrary willfulness of the contractor was its sharp insistence that his rights be measured by the objective situation he created, and not by his own interpretation of it:

> . . . It is entirely immaterial what the *intentions* of the McDonalds were in relation to the delivery. The question is not, in this case, what did they *intend* to do? but what *did* they do? The court will look only to their acts, and not to their mental processes . . . Having placed . . . the logs at the designated place, while they were the owners thereof . . . neither the McDonalds or any other person claiming under them can be heard to aver that they placed the logs there with the mental reservation that they did not intend by the act to fulfill such contract, but to violate it.[61]

The most common task of the court in these lawsuits over performance was, however, to determine the facts of what the parties had done under their agreement. Did the seller of land deliver the promised number of acres, or acres bearing the promised quantity or quality of standing timber? Did the financier make all the promised advances to the logger? Were the logs driven so that they arrived at the stipulated boom at the stipulated time? Of course, the determination of the "facts" might often require resolving issues of judgment or opinion. Did the logger cease cutting too soon, or did he correctly judge that the spring break-up was at hand? Did the land locator use all the skill which his client might expect of him, in selecting the acres now found disappointingly sparse in pine? Whether simple or complex, such questions did not call for the definition of rules of law, but for the determination of facts of conduct and of trade practice or standards.[62] It is impossible to make a precise calculation, but a conservative estimate would be that, in approximately 70 per cent of 284 performance-issue cases, the prime question appeared to be the determination of the facts of the parties' behavior under their agreement, apart from any problem created by slipshod ways of business, or bad faith, or recalcitrance. The sheer volume and variety of men's dealings in a lively field of production and trade generated fact disputes sufficient to create a significant problem in market administration.

The 708 Wisconsin Supreme Court cases dealing with the existence, in-

TABLE 15.[a] Principal points at issue regarding interpretation and performance of lumber contracts brought before the Wisconsin Supreme Court

Timberland sales (26)		Timberland sales finance (15)	
Whether contracted-for title was effectively passed	14	Vendee's right to cut: contract construed	5
Interpretation of scope of title contracted for	6	Vendee's right to cut: general waste doctrine	3
Whether land held agreed quantity of timber	2	Determination whether mortgage satisfied	2
Whether land held agreed quality of timber	1	Security asserted against adverse title	3
Action for price of land	1	Mortgagee's election of remedy	1
Vendee's waste	1	Vendor's lien enforced	1
Laches of vendee	1		
Standing timber sales (24)		Standing timber sales finance (14)	
Construction of cutting time limit in timber grant	8	Security asserted against adverse title	12
Construction of vendor's reservation of timber	6	Replevin a proper procedure to foreclose	1
Construction of other terms	4	Release of security	1
Whether contracted-for title was effectively passed	4		
Vendee estopped to assert tax title against vendor	1		
Action for price of timber	1		
Joint ventures (including agency) (19)			
Determination whether agent's or locator's fee or share was earned	13		
Agent estopped to assert tax title against principal	3		
Accounting among joint ventures	2		
Payment for locator's minutes	1		
Logging: Contracts to log (32)		Logging: Labor contracts (11)	
Fee claimed for logging	16	Wage collections	11
Damages claimed for logger's failure to perform	12		
Damages claimed for not furnishing agreed advance	2		
Accounting among joint venturers	1		
Contract lien construed	1		
Logging: Partnerships (8)		Logging: Ancillary issues (7)	
Accounting among partners	4	Enforcement of rights of way, privately created	3
Debts to third parties	2	Price, or rent of equipment	2
Breach of contract to form	1	Bailment of team	1
Enforcement of mutual obligations	1	Inspector's bond enforced	1
Logging: Finance (8)			
Collection for unsecured advances	5		
Enforcing security for advances	3		

TABLE 15. (continued)

Milling: Acquisition, upkeep, and insurance of mills (34)	
Enforcing contract to sell mill or site	11
Enforcing mill lease	7
For price of mill or equipment	5
Breach of covenant to build mill	1
Remedies of vendee of mill machinery	3
Insured's suits on fire insurance policies	5
Value of mill upon taking	1
Fraud on insurer (crime)	1
Milling: Partnerships (8)	
Accounting	6
Enforcement of mutual obligations	2
Milling: Labor contracts (12)	
Construction of terms of employment: rate, time	5
Collection of wages	3
Employer's breach of term or job contract	2
Enforcing agreed liquor bans	2
Milling: Contracts to saw (18)	
For price of sawing	7
Damages for failure to deliver for sawing	4
Construction of sawing contract: when money due	3
Construction of sawing contract: services due	3
When delivery of product is made	1
Milling: Finance: Mortgages (9)	
Foreclosure suits	3
Construction of coverage	2
Accounting between parties to mortgage	2
Security asserted against adverse title	2
Milling: Finance: Mechanics lien (9)	
Enforcement between the parties	5
Assertion against adverse title	4
Milling: Finance: General debts (4)	
Collection of debt	4
Log sales (55)	
To collect price	20
Dispute over quantity	14
Dispute over quality	5
Effectiveness of title transfer	11
Title transfer and risk of loss	5
Log sales finance (4)	
Enforcing security	4
Lumber sales (86)	
To collect price	38
Dispute over quality	18
Dispute over quantity	14
Effectiveness of title transfer	4
Title transfer and risk of loss	5
Sale of business	1
Miscellaneous	6
Lumber sales finance (16)	
Enforcing security	16

[a] Cases here tabulated are those summarized in Table 12 columns headed "Interpretation" and "Performance."

terpretation, or performance of lumber industry contracts, we should again remind ourselves, represented only a small part of the activity that went on within the frame provided by contract law, in trial courts, in lawyers' and businessmen's offices, in logging camps, and in the woods. Most of this activity left records too fugitive for recall; if we had the entire record before us, we might see that peculiarities of appellate litigation caused the Supreme Court cases in some ways to present a distorted reflection of the whole. That the largest body of accessible evidence comes from litigation would introduce an obvious element of bias if it persuaded us that harsh dispute was the norm of contract dealing. Lawsuits typically show serious or even total breakdown in the parties' relationships. The swelling curve of timber production and trade testifies that such breakdowns were marginal; within their own sometimes too limited frame of reference, the market and its contract instruments obviously did not break down, but functioned effectively. These lawsuits, then, are more important for what they indicate of a pattern of doctrine, procedure, and reserved compulsion by which affairs might be ordered and adjusted, than for their resolution of the particular failures of contract with which they dealt. Checked against the other scattered sources available — our data from the Chippewa County circuit court, the business papers of lumber companies, and the reminiscences of lumbermen — the suits before the Supreme Court seem fairly to represent the probable types and distribution of administrative problems in the general working of contract. Viewed in their double aspect, these scores of cases — most of them routine and enmeshed in peculiar detail, of interest only to the immediate parties — create a cumulative picture of flexible and pragmatic administration without which the turbulent energies released by the market could only spend themselves in chaos.

CHAPTER V

CONTRIBUTIONS OF LEGAL REGULATION TO THE ORGANIZATION OF THE MARKET

FREEDOM OF contract was the rule, and restraint the exception, in the legal history of Wisconsin's timber industry. But there were uses of legal regulation and compulsion consistent with this general bias of policy. Exerted at selected points, the positive force of law might enlarge men's practical range of choice and promote release of their creative energies. So, by penalties and controls the law imposed its will in limited measure to shape Wisconsin timber dealing, to supplement or protect the contributions which contract made to developing the market.

Such legal intervention in the market was mainly by statute. Earlier we noted that both doctrinal and practical reasons made the legislature rather than the courts the likely arena for developing new forms of positive regulation. Possessing both the ultimate policy power and control of the public purse, the legislature might define new rights and create new remedies with a freedom not enjoyed by judges. And, on the other hand, with allegiances close to their home districts, legislators were more likely agents than the governor to press distinctive claims of particular sections, industries, or individuals.[1]

A striking aspect of the development of legislation auxiliary to timber marketing is the almost complete absence of evidence of contributing activity by the Governor or other executive officers. Reactions to the problem of timber theft are suggestive. Successive Governors repeatedly, if rather ineffectually, urged strong measures against timber trespass on the public lands; but the record shows not a single executive reference to thefts of standing timber from private land, though in various statutes the legislature reacted to the problem.

In this contrast, timber trespass exemplified a general pattern in Wisconsin lawmaking. Through the years from 1848 to 1948 there was little initiative taken either by the Governor or other executive officers to improve the staple law of crimes, tort, contract, or property, though the legislature regularly tinkered, and occasionally imposed broad-scale change, in all these areas of policy. It was a pattern of affairs consistent with attitudes which led us to organize so much of our life around the market as a central social institution, and around contract law as its legal embodiment — out of preference for dispersion of decision and delegation of policy making to those most immediately interested, and out of a general impatience with formal, deliberate processes of lawmaking which would require us to take more time away from our central preoccupation with economic growth.

On the other hand, a close view of the development of policy in these areas of regulation auxiliary to timber marketing shows that the impact of the statutes might be materially shaped by the treatment the legislation received in court. The more sharply a statute affected parties' dealings, the more likely that it would stimulate the judges to their own reflections upon the policies at stake, as they weighed arguable points of interpretation. If the initiative lay with the legislature, the outcome rested often in the interplay between it and the court.

CREATING THE MARKET: TIMBER TRESPASS AND THE PROTECTION OF INVESTMENT VALUES

Timber thieves threatened private as well as public holdings. Since public policy promoted private investment in timberland and the marketing of standing timber, the practical corollary was that law must protect such investments and dealings. The problem was the more pressing as the public lands were disposed of, and as values rose under the pressure of millmen and wholesalers to assure themselves of sources of supply. Timber owners had to contend mainly with small-scale trespassers, but there was enough overreaching by bigger operators to show how great this elementary threat to the integrity of the market might be, were it not kept in check. Private owners found that they must employ local agents to police their lands, and that the vigor with which they pressed action against trespassers might materially affect how much they ultimately realized on their investments. The problem was acute for owners who were not residents or who did not carry on substantial production operations in the immediate area of their timber holdings, for local sentiment opposed absentee ownership, especially where it spelled the withholding of land from market or production to await a long-term rise in values. Indeed, if an absentee owner could not properly police his holdings, he might find himself under pressure to sell while the trees still stood. Thus in January 1849 Wisconsin agents advised an eastern investor to sell his Grant County lands to a local buyer who was "an honest . . . Pennsylvania Dutchman" and a "warm believer," even though the deal must be on fairly liberal credit terms. "We think you had better sell it, as it lies in a part of the country where the timber can only be protected by standing on the land with a rifle and shooting the trespassers. The tract has already been seriously injured."[2]

Since theft presented a classic threat to property, it is not surprising to find that Wisconsin law began with a ready array of weapons against it. The common law recognized no larceny of standing timber. But from the time of a Wisconsin Territorial act of January 27, 1844, statutes laid criminal penalties on those who "wilfully" destroyed, or cut down and carried off, trees from another's private lands. Chapter 17 of the General Laws of 1866, and successor statutes, explicitly and broadly analogized the offense to the theft of other kinds of property, declaring that one who severed from another's soil and converted to his own use any growing tree "shall be deemed guilty of larceny in the same manner and of the same degree, as if the articles so taken had been severed at some previous and different time." [3]

On the civil side, Wisconsin common law and statutes offered a range of general remedies, as capable of adaptation to protecting timberland investment as any other property. An aggrieved landowner might seek to protect primarily his interest in the land, by an action of ejectment, or trespass *quare clausum fregit.* These familiar remedies received some statutory modification, the effect of which was better to adapt them to guarding standing-timber values; thus the Revisers of 1878 modified the statutory action of ejectment to allow the previously denied auxiliary remedy of a judgment for waste, and the Laws of 1905, Chapter 264, declared that double damages might be recovered in trespass for willful invasion of the plaintiff's close involving the cutting of timber. Even before the enlargement of the remedy, each party in ejectment might have a temporary injunction against logging by the other, pending determination of the title. Actions to quiet title were used to resolve disputes over timberland, independent of other proceedings. Equity would enjoin waste by timber cutting, at suit of one who could show that he held a present interest in land the value of which might be irreparably damaged by defendant's threatened wrong.[4]

All these were remedies which focused directly on the title to land and the protection of its value in its natural state. The familiar general remedies of trover and replevin were also available for the timberland owner who chose to define his grievance as the taking of logs or lumber derived from wrongful cutting on his property. Trover was an action simply for money damages; replevin might lie for specific repossession of the claimed goods as well as for damages for their taking, though since the plaintiff must post bond to recover possession, the action might always be translated into money stakes. Of course, trover and replevin were also available to protect those who held title to logs or lumber simply as personal property.[5]

The information is not available by which we might satisfactorily measure the whole impact of these various property-protective remedies upon the life of the lumber industry. In the nature of the case, the facts of most trespasses or thefts were such as would leave little lasting record. Our only general measures are in the attention which the legislature gave to the problems of invasion of private property in timberland, logs, and lumber, and in the cases that reached the Wisconsin Supreme Court. The legislature reacted with a vigor that attested the concern with which important timber industry interests viewed the trespass threat. The strength of the legislative reaction is also indicated in the fact that statutory trover provided the most numerous and controverted class of trespass-type cases in the Supreme Court. The civil cases in the Supreme Court added up to a substantial number in themselves, but a number which also bore the duly subordinate relation to the volume of contract cases that one would expect in a secure system of private property. Only one case came to the Supreme Court under the body of penal legislation protecting private titles in standing timber and forest products. Our Chippewa County circuit court data show scant use of criminal sanctions to protect private interest at the trial court level. Contemporary complaints were that local prosecutors, juries, and judges were as little likely to take action against timber thieves on behalf of absentee private landlords as they were when state lands were plundered. Civil remedies bore the load.[6]

A catalogue of principal issues in the suits by owners of timberland or standing timber (see Table 16) suggests the functional role of this range of remedies in protecting titles. Of the forty-five suits (excluding statutory trover actions for the moment), the plaintiff's title was a main issue in eighteen instances, the defendant's in nine; five of the eighteen plaintiff's-title issues turned on tax titles, as did five of the nine defendant's-title questions. Of the other eighteen of these forty-five cases, eight turned on definition of rules of damages, and ten on a miscellany of pleading points. The statutory trover cases presented distinctive problems in interpreting the scope of the special rule of damage created there. Even so, of the thirty-five statutory trover actions before the Supreme Court, nine presented an important issue of plaintiff's title (two of these were tax-title questions), and nine others turned partly at least on an issue of defendant's title (one of which was a tax-title matter); of the remaining seventeen cases, twelve turned on questions of interpretation of the statutory measure of damages, and five presented miscellaneous pleading issues. Thus, of the whole eighty cases in which plaintiff's standing rested on title to timberland or standing timber, forty-five involved important issues of title; in most of these the title question was paramount. The pattern was about the same in the twenty-one cases in which the plaintiff claimed simply as the owner of logs or lumber; of these, five turned mainly on plaintiff's title, and seven on defendant's; nine involved miscellaneous issues of pleading or evidence. It is likely that title questions would come to the Supreme Court in higher proportion than they would bulk up either in trial courts or in disputes settled by negotiation out of court, where questions turning simply on what defendant had in fact done were likely to have their final resolution.[7]

TABLE 16. Cases of protection of real or personal property titles incident to the lumber industry before the Wisconsin Supreme Court

	Suits by owners of timberland or standing timber						Suits by owners of logs or lumber			
Years	Ejectment	Quiet title	Injunction against waste	Trespass, quare clausum fregit	Replevin and trover	Statutory trover	Replevin and trover for logs	Replevin and trover for lumber	Criminal prosecutions	All contract cases[a]
1840–1849	..	..	..	1	..	..	1	1	..	4
1850–1859	..	..	..	..	..	..	..	2	..	20
1860–1869	1	..	1	1	2	..	1	3	..	46
1870–1879	4[b]	..	..	5	6[c]	1	3	..	..	74
1880–1889	1[d]	..	1	7	2	16	2	..	..	106
1890–1899	2[e]	1	..	1	1	6	4	..	1	148
1900–1909	1[e]	..	..	2	..	8	2	1	..	107
1910–1919	..	2	..	..	..	3	1	..	..	36
1920–1929	..	..	..	..	..	1	..	..	..	24
1930–1939	..	..	..	..	..	..	..	..	..	12
1940–1949	..	1	..	2[f]	..	..	..	..	..	6
1950–1956	..	..	..	..	..	..	..	..	..	3
Totals	9	4	2	19	11	35	14	7	1	586

[a] Excludes lien and transport cases from cases tabulated in Chapter IV, Table 5.
[b] Temporary injunction issued in one case.
[c] Includes two cases of common law relief, in which measure of damages is redefined to conform to policy of newly enacted statute, prior to its effective date.
[d] A taking by eminent domain, noted here by analogy.
[e] In each instance, one case involving allowance of damages for waste, ancillary to ejectment.
[f] Includes one allowance of double damages under statute.

It accords with this prominence of title issues, to find that the doctrine of these general remedies was markedly favorable to the security of regularly established claims. Peaceful possession was a sufficient showing of interest to ground relief against a mere wrongdoer. One who held a contract to buy land, with an agreed right of possession, might recover in trespass for his vendor's wrongful cutting of timber. Though defendant entered plaintiff's land and carried off logs in the good-faith but mistaken notion that he had a right to do so, he was nonetheless liable therefor in trespass. An aggrieved title holder might elect to treat the trespasser as his agent and sue for the proceeds of cutting in an action in contract, rather than seeking damages in tort, if he found this course to his advantage. Defendant would not be allowed to claim adverse possession, unseating plaintiff's title, where defendant's possession rested only on the acts necessary to periodic timber trespass. Though defendant had paid taxes on land during his wrongful holding of it, he might not counterclaim for the amounts so paid, when he was sued in trespass; true, the legislature had provided that a defendant in ejectment might recover or offset the value of improvements made during his tenure, but the court ruled that this grace to wrongdoers should be limited strictly to ejectment, the only instance for which the statute had expressly provided. Plainly, in protecting title by civil remedy Wisconsin common law, as declared and slightly modified by statute, was stern against one who wrongly invaded another's dominion, even though the wrong was without evil intent.[8]

However, this stern policy was on the whole not the result of any special concern for the problems of the timber industry. These were doctrines generally applicable to the protection of title, anywhere, anytime. Of the forty-five Supreme Court cases invoking general remedies, thirty-four found no occasion to note that the lawsuit arose out of lumber operations; by a liberal count, in only eleven cases might it be said that the court's analysis was affected by considerations peculiar to timber values. So far as they turned on the interpretation of the legislation fixing a special measure of damages for timber trespass, the statutory trover cases, of course, reflected concerns peculiar to the lumber industry; but the eighteen statutory trover suits which involved important title issues dealt with these wholly within the frame of general doctrine. The same prevalence of general title law, rather than any policies peculiar to the log or lumber trade, marked the cases brought for conversion of logs or lumber as personal property.[9]

Even a mature legal order may find it more difficult to enforce a policy than to define it. The challenge of enforcement is the greater when a legal tradition is yet to be built, and the means of executive and judicial action are limited and simple, as in the Wisconsin forest area during the years of its greatest exploitation. The general doctrine available to protect titles in Wisconsin evidently met the need for definition of substantive rights, since the legislature showed little concern over the formal definition of interests. The one narrow exception to this pattern was the felt need to break the rigid mold of a common law concept and to declare that the theft of standing timber should be punished as larceny. Response here was alert to the special problem of the timber industry—indeed, it came from as early a date (1844) as there was any substantial logging in the state.[10]

It was concern for the enforcement, rather than the definition, of rights that spurred legislative activity regarding timber trespass, affecting both criminal and civil remedies. The statutory changes had a clear general direction—of increasing severity toward the wrongful cutter or dealer—which testified both to the growing worry of certain interests in the face of the trespass threat, and to their awareness that practical enforcement did not keep abreast of the problem. But cross currents in judicial decisions applying the statutes showed that there was a more complex interest alignment here than simply that between honest men and thieves, and that the lumber industry by no means presented a united front to the problems growing out of trespass.

The naïve reaction to continuing law violation is to announce stronger penalties; it requires harder thought and effort, and more money, to take the more effective course of tightening up field enforcement, or—most efficient, but probably most immediately expensive—devising means to prevent trouble. In passing through these phases, Wisconsin timber trespass law conformed to a pattern which stamped most controversial areas of nineteenth-century legal regulation.

The legislature toughened the formal penal sanctions on timber trespass in two ways. Carrying forward the territorial act of 1844, the Revised Statutes of 1849 declared that the "wilfull" cutting of another's timber was a misdemeanor, punishable by imprisonment in the county jail not more than one year, or by fine not less than $5 nor more than $500. In 1854 the legislature increased the minimum penalties from five to ten days or from $5 to $10, and then dramatized the concern of the protected interests over the evident failure to curb timber theft, by providing double minimum and maximum penalties (twenty days to not more than two years, or $20 to not more than $1000 plus costs of suit) for a second offense committed in the daytime, or a first offense done at night. Upon all convictions above the second, both imprisonment and fine with costs were made mandatory, for either daytime or nighttime trespass. Carried into the Revised Statutes of 1858, the device of an increased scale of penalty for successive offenses was dropped in the 1878 revision, and was never restored; evidently it had made no significant difference in enforcement. Under the 1878 revision, timber trespass for a short interval was merged in a general misdemeanor of willful injury to realty or fixtures, with much reduced penalties (imprisonment not exceeding six months in the county jail, or fine not exceeding $100). Evidently this alteration reflected an oversight in revision, rather than a change of policy, for the Revised Statutes continued more severe penalties for willful taking of logs or lumber in transit, and in 1881 the legislature again enacted a distinctive ban on theft of standing timber, with strong penalties. Adopting the general penalty scale set by the Revised Statutes for larceny, the 1881 act set the sole penalty of imprisonment in the state prison for not less than one or more than five years, if the value of the stolen timber exceeded $100 (so far making the offense a felony, where before it had never been more than a misdemeanor); fixed the sole penalty of imprisonment in the county jail from six months to not more than one year, if the value ranged from $20–$100; and provided alternative penalties, of imprisonment in the county jail not over six months or fine not over $100, only if the value of the stolen timber did not exceed $20. For trespass involving intentional violation of another's right, this remained thereafter the pattern of penalties, qualified by eventual restoration of the alternatives of jail sentence or fine. An 1889 amendment enlarging the offense of larceny of standing timber or its cut product to include a principal who knowingly accepted the fruits of a willful trespass by his agents strengthened the form of the law, but at the same time bore implicit testimony to the practical difficulties of its enforcement.[11]

A second way in which the legislature belatedly increased the formal penal sanctions on timber trespass was to expand the reach of the criminal law, so that its penalties substantially paralleled the scope of civil actions for damages. Ever since the late seventeenth century the norm of Anglo-American legal policy had been to punish criminally only breaches of order committed with a guilty mind. Thus the Wisconsin legislature conformed to familiar policy when through most of the years it fixed criminal penalties only for "willful" timber trespass, whether it treated the offense as a misdemeanor or a felony. Of course, this definition marked a sharp distinction from the scope of civil actions; by long-established doctrine, one who exercised wrongful dominion over another's property was civilly liable for damages, though he acted in good faith and under a reasonable, but mistaken, notion that he had title or right of possession. There was tribute to the strength of the traditional limits on criminal law in the slowness with which legislators extended criminal sanctions to ordinary trespass, despite the manifest concern over continued timber theft. Indeed, the legislature did not make the reach of penal sanctions co-extensive with that of civil remedies until long after the timber industry had passed its peak. Thus, Chapter 252 of the Laws of 1949 is relevant to our story mainly as it illuminates by contrast the limits of use of criminal penalties in the years of full-scale exploitation of the forest. Its enactment, perhaps, reflects less tolerance toward invasions of property in a community which had grown more crowded and more cost-conscious than the Wisconsin of the rough-and-tumble nineteenth century.

The 1949 act declared guilty of a misdemeanor "any person who unlawfully cuts or directs or contracts for the cutting of forest products on the lands of another." The language was broad enough to include any conduct that would constitute a civil trespass, though committed without guilty intention, and the limited reach of penalties provided (a $10–$50 fine, or ten to thirty days in the county jail, or both) underlined this as the legislature's meaning. This intent emerged the more clearly from the contrasting penalties assigned for trespass committed in circumstances which the legislature in effect declared to show the existence of intent to steal. For the statute applied the full, severe scale of penalties for larceny to

> any person who unlawfully cuts or directs or contracts for the cutting of forest products on the lands of another . . . and who does not own or control adjoining land; or who, though owning adjoining land bearing merchantable forest products, cuts on an acreage substantially in excess of his adjoining land; or who, as part of an unlawful cutting operation, removes or destroys any survey monument or bearing tree . . .[12]

All the penal legislation so far examined fell within conventional nineteenth-century ideas of the way in which to enforce policy. We must go to 1931 to note the application to a timber-trespass problem of the twentieth century's characteristic approach, through preventive administrative procedure. Creating a pattern of licensing the cutting of Christmas trees, Chapter 404 of the Laws of 1931 dealt with a forest-products trade which the 1880's would have found too trivial for notice. Nonetheless, by its contrasts, Chapter 404 usefully points up limitations inherent in the typical nineteenth-century handling of timber theft. The 1931 statute sought to regularize cutting in such fashion as to reduce mistake, and from the outset eliminate groundless claims. Thus, it forbade anyone to "cut for sale in its natural condition and untrimmed, with or without roots, any evergreen or coniferous tree" from another's lands without the written consent of the owner, specified the contents of such an effective license, and required on pain of penalty that every person in charge of any cutting carry the original writing or a certified copy to be shown on demand to any law officer. More formal licensing was imposed upon persons shipping or transporting such trees outside the county where they were cut; such an operator must first obtain from the Conservation Commission for an annual fee a license as a Christmas tree dealer; each dealer must obtain written evidence that any given lot of trees was cut with due permission, and must keep this evidence for six months, to be open to the Commission's inspection; every rail or truck shipment must bear a tag showing the dealer's name, address, and license number, and no common carrier might lawfully receive a shipment not bearing the required tag. Violation of the general terms of the act was a misdemeanor, punishable by a fine ($10–$100), a county jail sentence (ten days to three months), or both; unauthorized signing or transfer of required consent papers carried heavier penalties (a fine of $100–$500, or a county jail sentence from three to six months, or both). Ruling that the dealer's license requirement was intended as a distinct regulation, apart from that concerning the cutting of trees, and that hence a dealer must have a license even for trees cut from his own land, the Attorney General emphasized that the preventive character of this legislation must be given liberal recognition: the clear purpose of the act was "to license the transportation of all trees in order to enforce the provisions against illegal cutting."[13]

Standard criminal prosecution was the primary mode of official initiative against violations of public policy. True, the informer's fee was not unknown in Wisconsin penal legislation as a stimulus to private persons to set in motion the machinery of the criminal law; it was provided for a time as an aid to policing the public lands against timber theft. However, the legislature never invoked the informer's fee to spur private initiative against timber trespass on private lands — probably because it recognized that the sympathy of local residents in the pineries was so enlisted with the cutters that informers would not come forward, or if they did, local juries would not convict on their testimony. Thus the prime field for private initiative in enforcing public policy against timber theft from private lands was in civil actions for damages brought by the aggrieved owners.

Statutory change in civil remedies paralleled that in criminal sanctions — the declaration of more severe penalties — but in a fashion somewhat more subtle, and involving more acute conflicts of interest among segments of the industry. The simpler changes merely added special items to, or multiplied, conventional damages. As we have noted, the Revisers of 1878 added to the remedy in ejectment a provision for recovery of damages for waste; there is nothing in the legislative record to confirm that this was done primarily in recognition of timber industry problems, but in any case the provision was soon put to use in cases involving timberlands. In 1880 the legislature amended the general statutory provision for interest on money judgments, to provide that in cases of trespass, trover, or replevin arising out of the unlawful cutting of timber, when the value claimed and recovered exceeded $50 the plaintiff might have additional recovery for full costs — including the actual, reasonable cost of one survey of the land, if this were proved to have been a necessary disbursement. Even in this simple change there was recognition of cross currents of policy, however, for the statute provided that if plaintiff was not entitled to recover costs, defendant might recover like costs in the same manner in such actions. Treble damages were provided for conversion of floating or stranded logs or lumber; however, this seems functionally a development in protection of transport rather than of title. The legislature did not apply to timber trespass the simple device of multiple damages until by Chapter 264, Laws of 1905, it declared that a private owner upon whose lands a "willful" trespass was done might recover in a civil action "double the amount of damages suffered." Probably this simpler, multiple damages sanction came late on the books because timberland investors had chosen rather to rely on an 1873 statute — next to be discussed — in which the measure of damages in trover was so defined as to promise recovery not merely double, but four to ten times the amounts traditionally awarded.[14]

The tightening of civil remedies for timber trespass which provoked the most controversy and the most complex pattern of doctrine, in reflection of the diversity of interests involved, was Chapter 263, Laws of 1873, which became section 4269 of the Revised Statutes (1878). The central provision of this act was the declaration:

> In all actions hereafter commenced in any of the courts of this state to recover the possession of logs, timber or lumber wrongfully cut from the lands of the plaintiff, or to recover the value thereof, or damages for such cutting, if the plaintiff shall succeed, he shall, unless possession of such property be restored to him, recover as damages the highest market value of such logs, timber or lumber, in whatsoever place, shape or condition, manufactured or unmanufactured, the same may be between the time of such cutting and the time of the trial of the action, and while it remained in the possession of the party so cutting the same, or any purchaser thereof with knowledge of such wrongful cutting . . .

The statute allowed defendant a chance to mitigate damages. Within ten days of the service of the complaint, defendant might serve on plaintiff an affidavit "that such cutting was done by mistake," and tender judgment for the full value of the timber when cut, plus interest from the time of cutting to the date of the tender, and 10 per cent on the whole amount as damages, together with the cost of the action to the time of tender, to be taken in full satisfaction for the cutting. If plaintiff duly notified defendant that he refused the tender, he must deny the allegation of the affidavit that the cutting was by mistake; if a trial established the fact of mistake and the value of the timber as tendered by defendant, plaintiff would have judgment only for this value without any additional damages, and must pay the costs of the action since the tender; if mistake was proved, but the jury found the value of the timber greater than the amount tendered, plaintiff should recover the value found, with 10 per cent damages and full costs.

In itself, the basic provision of section 4269 did not change the substance of familiar common law actions of replevin or trover in cases arising out of illegal cutting of timber. The unjustified exercise of dominion over another's land and its timber wealth was the essence of the civil offense; as the court's decisions under the original statute made clear, this was no less true though defendant cut with no evil intention, but under a good-faith, if mistaken, belief that he had title. The statute's change was in the measure of damages. Adopting what proved to be a minority view in the general pattern of United States doctrine, the Wisconsin court had ruled that for either intentional or unintentional conversion of personal property plaintiff might recover only the fair market value of the goods at the time of the taking, excluding any value thereafter added by defendant's efforts (or, perhaps, by market fluctuations). The statute now allowed plaintiff to recover not merely the value of converted timber at the time of trial, but its highest value at any time between taking and trial, though such a recovery might include value added by a rise in the market, or by defendant's labor in getting the timber out of the woods and to the mill, or by his effort in causing it to be sawed, rafted, and delivered to an advantageous market.[15]

The dollars-and-cents force of the statute could be severe. Under it, a successful plaintiff would commonly more than double his possible recovery under the common law measure of damages, and the recovery might well be five to ten times greater. Though the stakes in particular suits might be rather small, the potential threat was large. Thus, where the common law recovery was about $1650, that under the statute was ruled to be $3162; in other cases the contrast was between judgments for $121 or $363, for $926 or $3704, for $537 or $2465, for $105 or $1050.[16]

Though in its primary provision section 4269 laid down no new substantive doctrine, but merely changed the remedy for wrongs redressed at common law by actions of trover or replevin, the court in effect paid tribute to the drastic nature of the new measure of damages when it characterized a suit governed by section 4269 as "virtually a statutory action, and not at common law." If the substance of the law declared in the statute incorporated familiar common law doctrine protective of title, yet its particular focus gave the act a distinctive character not quite described by any common law label. It was not simple trover or replevin, for it applied only where converted logs or lumber derived from a wrongful cutting of standing timber on plaintiff's land. On the other hand, it was not an action like trespass *quare clausum fregit,* for the statutory damages were not for injury to the land, but for the wrongful taking of cut timber or its product. The new remedy, the court concluded, was "perhaps nearer an action of trover than any other, with specific damages."[17]

In an area of policy—like that under section 4269—where both legislature and court do their lawmaking with unusually explicit attention to business and economic fact and moral ideas, it is useful to be reminded that affairs may also be shaped by concepts and administrative policies growing out of the legal order itself. Thus, before we consider the economic and moral interests reflected in the development of section 4269, we should note that an array of technical consequences followed from the court's definition of the legal focus of the new remedy. For the court's characterization of the remedy brought into play various considerations that had little to do with timber trespass or any other facet of the lumber industry. Since a suit under section 4269 rested on a showing of wrongful cutting, though the action was not trespass *quare clausum fregit* it partook enough of an action for trespass to land so that the court would not interpret the statute as intended to apply where the land from which the cutting was done was in another state; to this extent, the action was "local." And since in terms the statutory action was for timber wrongfully cut "upon the land of the plaintiff," it had the character of an action regarding an interest in realty to the extent that one holding merely a license to cut might not recover thereunder, for want of an interest in the land. Yet, where plaintiffs brought a bill in equity to set aside defendants' tax deed to land of which plaintiffs claimed to be original owners, and asked also for an injunction and damages for waste by timber cutting, the relief sought went to the title and value of the land, and not to the conversion of timber, and hence section 4269 did not fix the measure of damages for waste. On the other hand, where plaintiff joined in one action the wrongful cutter of timber and one who bought from the cutter with knowledge of the trespass, plaintiff's complaint was not vulnerable to demurrer for misjoinder of causes of action in trover and trespass to land, where the relief sought fitted section 4269, since plaintiff's complaint against both parties derived from the wrongful cutting, as the statute contemplated. So, too, where plaintiff's complaint fitted the scope of section 4269, plaintiff might properly offer proof of his title to the land from which the timber was cut, to establish his title to the converted timber, and since this was a matter of evidence and not of pleading, his complaint was not subject to demurrer for his failure to allege his land title.[18]

But it was not niceties of legal concept, pleading, or procedure which made actions under section 4269 the most numerous and hotly contested kind of timber-trespass suit before the Wisconsin Supreme Court. The spring of controversy was the evaluation made by the legislation and the decisions under it regarding the various economic interests brought into focus by trespasses to timberland.

In effect, section 4269 assigned a high priority in public interest to the protection of long-term, private investment in timberland or standing timber. Contemporaries ascribed its enactment to the reaction of timberland investors to the court's ruling in *Single v. Schneider* (1872) that a timberland owner might recover only the value of timber at the time of cutting, even against one who had knowingly invaded plaintiff's interest. The legislature enacted section 4269 at the session next following *Single v. Schneider*. It was probably not accident that the legislature acted against the theft of logs in transit (on streams or in booms) before it dealt with theft of standing timber from private lands; the earlier legislation had come in years when government was still the principal owner of pinelands and the dominant private interest was in cut timber; the 1873 act came when large firms were beginning to acquire sizable timberland holdings, was introduced by an assemblyman from a lumber area (Dunn and Pepin counties), and was shaped to its final form by a committee including prominent spokesmen for timber interests. Speaking out of his experience in the 1870's and 1880's as lawyer for great timberland interests, Justice Roujet Marshall later observed that the 1872 decision had created "real mischief," and that "as might have been expected, at the first opportunity the legislature was successfully invoked to provide some remedy." [19] But, having been moved to action, the proponents of statutory change did not confine themselves to overriding the objectionable case; the enhanced measure of damages applied, in the absence of special defense, to any timber trespass, willful or not, and the departure from common law was underlined by allowing the plaintiff not merely any enhanced value of the goods at the time of trial, but the highest value of the converted property at any time between the taking and trial.[20]

When hard cases made the judges sensitive to the weight which section 4269 might lay on the defendant, they sometimes described the section as giving plaintiff recovery "over and above the actual value of the property taken," perhaps "several times the amount of the damage actually sustained," indeed even "sometimes unconscionable damages."[21] Such characterizations almost always occurred in cases where the plaintiff showed a shaky title, or defendant had some argument in mitigation of his wrong. Viewing the matter from the standpoint of the aggrieved landowner, it begged the question to appraise section 4269 as giving him more than "actual" damages. This was clearly so, if one assumed that his whole loss consisted in the present value of the wrongfully cut timber. But the purpose of timberland investment was often to profit by the long-term rise in values. The trespasser deprived the landowner of his command of the situation, and forced him to realize on his investment at a time fixed by the trespasser. The resultant loss might perhaps fairly be labeled speculative, but it might nonetheless be "actual." So the more realistic judicial appraisals estimated the intention and justification of the statute. Thus in an opinion contemporary with the legislation (1874) the court observed:

> As an original question, there is much reason for saying that this rule only furnishes a just and proper indemnity to the owner for the injury done him in cases of this description. The timber may be cut by the wrongdoer when it is low, and when the owner wishes it to stand in order to give him the benefit of a rise in the market. The value of the timber at the time of the trespass, and interest thereon to the time of trial, would not in that case give him compensation for the injury. And if one chooses to invade the rights and property of another, he cannot complain if the owner waits for a rise in the market and delays bringing his action.

The statute, the court said in another instance, was enacted because the legislature undoubtedly felt that the common law rule of damages "was likely to be a source of injustice towards those who owned pine lands, and had invested their money in such lands as permanent investments." And the Supreme Court approved a trial court charge which had justified the statutory measure of damages, because

> it is evident that the mere value of the stumpage in a great many cases does not compensate. Many men who have choice timber would be reluctant to have it cut for the price of the stumpage.[22]

Section 4269 implicitly acknowledged another claim of the timberland owner to the law's special protection. Most owners of tangible property were in position to keep a wary eye on their possessions. The man who put his money into standing timber typically was barred by distance and rough country from close supervision of his holdings. The common law measure of damages did nothing to offset the risks inherent in the facts of his situation. Under *Single v. Schneider,* the trespasser hazarded little; if he were caught, he had merely to pay the fair cost of his raw materials and interest for the use of them. The result offended against basic principles of the market, because in effect the law thus allowed the trespasser to gain the immediate benefits of exploitation without having shouldered any of the risks of a fixed capital investment. Justice Marshall came close to this view of the statute:

> That *Single* v. *Schneider* . . . created an exceedingly embarrassing situation for owners of land in the great timber region of the state, — where in the very nature of things to prevent stealing of timber, which was common and profitable, was impracticable —, must be evident. It operated as a great stimulus to very annoying and damaging invasions of property rights. The novel rule was very beneficial to lawbreakers and very disadvantageous to owners of the wild and unoccupied land of the state.[23]

That the statute in effect sought not only to repair the conversion of timber, but also to prevent unjust enrichment out of another's capital ventures, was perhaps implicitly acknowledged in decisions which showed marked reluctance to give the statute's benefits to a tax-title holder against an original-owner cutter, unless plaintiff made very clear proof of a sound tax title.[24]

The legislature's basic judgment on the social value of timberland investment stood firm through the years of boom and decline in Wisconsin's timber industry. The 1878 revision of the Wisconsin Statutes broadened, or at least clarified, the 1873 act, making plain that it covered the taking of down as well as standing timber — an important point, in view of the valuable amount of timber cast down by wind. The 1878 revision further tightened the act against the defendant by striking the clause which had given defendant the chance to avoid the higher damages by returning possession of the property to plaintiff. Bill 127S in the 1879 session would have repealed the act, to substitute merely a provision that defendant in an action for wrongful timber cutting

> shall not . . . be permitted to set up in his defense that such cutting, removing or taking was done by mistake of the boundary lines or locality, or description of such land, unless he has before such cutting procured a survey of such land to be made by some competent surveyor, and the lines of such survey to be distinctly marked.

This change would, indeed, have been damaging to the investor interest, for not only would it return them to the unsatisfactory common law measure of damages, but it acknowledged a defense of mistake which the common law itself did not afford; that the defense was limited did not alter the fact that even less than the common law remedy was offered in exchange for section 4269. On recommendation of the Senate Judiciary Committee, the bill was indefinitely postponed by a recorded vote of 25 to 5.[25]

Nineteenth-century public policy in the United States generally favored action and the venture of capital in production. The stiff protection which section 4269 offered to timberland investors was thus all the more a tribute to the focused political power of some of the large investing firms. Even so, the legislature showed strong sympathy for the problems of the active exploiter of timber. We noted that from its first form section 4269 allowed a defense of mistake, which would reduce damages close to the limits of the common law rule. True, by providing that mistake might be matter of defense, rather than defining the plaintiff's cause of action as one simply for willful trespass, the legislature cast on defendant the burden both of pleading and proving his mistake.[26] Nonetheless, it was remarkable to allow the defense at all in an act so vigorous in protecting the investor interest, especially when the common law, for all its lack of care toward the plaintiff, did not offer this escape to the trespasser. Continuation of the mistake defense did not go without challenge. In the 1876 session, bill 322A, which would have stricken the defense entirely, was favorably reported by committee in each chamber, and passed the Assembly, only to be indefinitely postponed in the end-of-session rush in the Senate. From this point on, however, the defense was accepted without contest, as section 4269 was carried forward into successive revisions of the Wisconsin Statutes.[27]

Inclusion of the defense of mistake recognized the practical difficulties of strictly observing titles and boundaries in a rude country, where record offices were remote from logging operations and survey lines were ill-marked. If there was a public interest in protecting timberland investment, there was also a public interest in current production; we saw how sharply this attitude was expressed in public land policies emphasizing speedy sale, and in the recurrent criticisms of policy which allowed large tracts to be withheld from market for private speculation. It was consistent with such attitudes that, as a contemporary judicial opinion saw it, the legislature should be concerned that its new statute not unduly hamper active production, "appreciating that such a rule would be oppressive and unjust if enforced against a person who had been acting in good faith." And the court undoubtedly was right in sensing that, since the mistake defense recognized the rude conditions of woods operations, it should not be governed by the nice precision of common law pleading. Confronted with "papers . . . very informal, and so badly written that it was very difficult to read them," once the court had determined that it was "evidently the substance" of the writer's intent to make tender of judgment accompanying an affidavit of mistake, the court would recognize what "appears to be . . . a substantial compliance with the statute" and hold the defense adequately pleaded.[28]

The legislature again attested its sympathy to the active operator, in the only other major change it made to section 4269 during its life. Chapter 239 of the Laws of 1882 amended the section to limit plaintiff to "only the actual damages" from a wrongful cutting, "when the defendant shall have in good faith acquired a title to and entered upon the land under the same, believing such title to be valid and shall have cut the timber therefrom under such circumstances." The amendment expressly placed on defendant the burden of pleading and proving this defense. In contrast to the original mistake defense, which was held to embrace only mistakes of fact, the 1882 act was plainly intended to allow the defendant to plead some ill-defined area of mistake of law; mistake of fact — as of the physical location of a boundary line — must be pleaded under the original defense provision. The background of the 1882 addition is not clear. It may have been inserted in part from concern that section 4269 might be unconstitutional if its heavy damages prevented one from acting in good-faith reliance on an apparent title; but if this was the ground, it was later repudiated by the court, which eventually ruled that it was within reasonable legislative discretion to define the allowable defenses to an action of trespass.[29]

The Wisconsin Supreme Court developed section 4269 in a fashion substantially faithful to the legislature's pattern of values — with primary and strong emphasis upon protection of timberland investment, but with qualifications reflecting a sharp sense of the community's desire that production go forward. Chapter 263, Laws of 1873, came before the days when the courts had realized the scope of review they might exercise over the legislature under the due process clause of the new 14th Amendment. Hence, it is not surprising that the court never expressly ruled on the constitutionality of section 4269 during the boom years of the timber industry. But *Schweitzer v. Connor* (1883) plainly indicated that the judges would have found no substantial question under the due process clause. Conceding that the statutory rule "may operate harshly in some cases," the court said, shortly, that "if the rule is an unjust or oppressive one, the remedy is with the legislature." And when the due process challenge was finally made, years after timber had declined to a subordinate place in the state's economy, the court made equally short work of it. Did section 4269 impose penalties which practically forbade the accused party to defend himself, save at unreasonable risk? Not so:

> . . . we do not think section 4269 imposes a penalty at all. It merely prescribes the measure of damages for the wrongful cutting of timber upon the lands of another, the power of the legislature to prescribe which has never been questioned.

Moreover, the reasonableness of the statute was deemed established by the fact that "the rule prescribed [by it] . . . obtains by judicial decision in many states." Nor did it violate the equal protection of the laws, to except from section 4269 those who made out the defense of mistake; anyone who could prove himself within the statute's proviso might have the benefit of the defense, and, the court indicated, provision of the defense represented a weighing of competing values well within the reasonable discretion of the legislature.[30]

Questions of validity apart, the court often declared that section 4269 should receive a liberal and sympathetic interpretation, broadly to effect its beneficent purpose of correcting the injustice which the common law rule of damages did to the landowner. The court early dramatized this attitude when it drew on the statute as authoritative evidence of the public policy of the state to overrule the common law doctrine and apply the new, enhanced measure of damages in timber-trespass cases which had arisen before the effective date of the act:

> Surely the maxim *stare decisis* cannot be invoked to control the rule of damages in an action for tort; and there are very cogent reasons why the statutory rule should have prevailed always. At any rate, after the legislature has settled the policy of the state in that behalf, no good reason is perceived why the courts should not adopt and enforce that policy in all cases, without regard to the date of the trespass.[31]

". . . the statute . . . is salutary in its provisions, and should be fairly enforced," the Supreme Court told trial judges; and Justice Marshall added his special word in favor of a liberal construction, for "only by giving [section 4269] that broad construction could the real mischief of the condition created by *Single v. Schneider* . . . be removed."[32]

Measured by the great preponderance of decisions, the court in fact interpreted and applied the statute as sympathetically in protection of timberland investments as it said that it would. Any unauthorized cutting was "wrongful" within the act's meaning; the statute's severely enhanced measure of damages should apply to the full range of conduct amounting to trespass or conversion under the broad test of the substantive common law. In the absence of special statutory provision, mistake of law was no defense, nor would mistake of fact, however reasonable, avail defendant, if he did not properly plead and prove it under the statute's proviso. Thus construed according to the full thrust of its language, the statute's favor towards the landowner and its rigor against the cutter were made clear; the cutter faced a hard choice, either to stand on a claim to the timber and suffer the weight of the enhanced damages if he lost, or at the outset to forego all claim of property in the cut timber by filing an affidavit of mistake to mitigate damages. Again, one who had a contract right to cut timber not less than ten inches at the top was held liable for section 4269 damages, if the landowner chose to sue thereunder for trees cut in violation of the contract stipulation, for as to such timber the contract cutter was no more than a trespasser. The statute on its face allowed plaintiff to recover the highest market value of the timber while in the trespasser's hands, and this provision must be taken to allow recovery of its value at the most distant market to which defendant's labor had removed it, however much this calculation increased the recovery. The act must be construed to allow recovery for timber suitable for manufacture into any marketable product, and not as limited to timber suitable for lumber only.[33]

No decisions more emphasized the court's desire that section 4269 fully protect the timberland investor than those in which interpretation was guided by the judges' expressed concern to prevent evasion. Thus a partner was liable for the statutory damages, though he was ignorant of wrongdoing, where the trespass was done with the guilty knowledge of a copartner; and likewise a principal was liable for wrongful cutting by his agents, though he had no personal connection with the wrong. Otherwise, "this act . . . would very soon become a very feeble instrument in effecting the purpose for which it was enacted." Despite the heavy weight of the statute, it must be so applied as to have practical meaning in view of the difficulties of enforcement under the rude conditions of woods operations:

Unless this rule be adopted, it would be an easy thing to avoid the statute as to enhanced damages. All a party would have to do would be to employ a dishonest or reckless party to cut and haul logs for him from some tract of land owned by himself, and his employee could cut and haul logs from the land of others in his vicinity, putting on them the marks of his employer, and the employer could insist on holding them, notwithstanding he was duly notified of the trespass before action was commenced to recover the possession of them by the true owner, and then on such trial, when defeated in his claim of ownership, insist that the plaintiff should be limited to a recovery for the value of the stumpage.

Out of the same concern to bar evasion, the court showed that it would require very clear proof to establish that a wrongful cutter was an independent contractor, rather than an agent of defendant. Indeed, the court went to the edge of declaring as a rule of law that no defense was available under the sweeping terms of section 4269 on the ground that the wrongful cutting had been done for defendant by an independent contractor rather than by employees:

It is true the cutting, banking and rafting of the timber was done by Webster & Co., who had charge of the work, and who had agreed to do it for so much a thousand feet, but still, in fact, Webster & Co. were employed by the defendant to cut its timber from its lands . . . They were employed for this precise service, and if in the execution of its employment they committed a trespass upon the land of another, and the defendant received the logs, and manufactured them into lumber, it would be held responsible for the statutory damages, as much so as it would have been had it hired Webster & Co. to work for it in cutting and rafting timber on its lands at so much a day or week . . . The case is within the policy and intent of the statute . . . [T]o hold otherwise would defeat the plain object of the statute. We shall not dwell upon the cases which consider

> the law as to purely independent employment, for we think they have no application here. Under the statute, the maxim *respondeat superior* should and does apply where the persons, in fulfilling their contract, commit a trespass on the lands of another through their negligent acts . . . [T]he defendant, by receiving the timber and using it, must be deemed to have ratified their acts. "Unless this rule is adopted it would be an easy thing to avoid the statute as to enhanced damages . . ."

Similar alertness to thwart evasion led the court to significantly restrictive interpretations of the statutory defenses of mistake. Thus the mistake defense under the original act would not include a case where the cutter had been careless in failing to observe boundary lines; the trial court erred in charging the jury that so long as defendant or his servants did not knowingly cut on plaintiff's land they might have the benefit of the statutory defense, though they invaded plaintiff's bounds out of negligence. The trial court's construction

> would . . . defeat the very purpose of the statute. If a trespasser can avoid the larger damages given by the statute by simply neglecting to take any precaution to ascertain whether he is cutting upon his own land or upon the land of his neighbor, and avoid the effect of the statute by alleging such want of precaution on his part, it will open a wide door for those who are disposed to despoil their neighbor's land of the growing timber to do so and avoid the damages given by the statute . . . He cannot neglect a duty which he owes to his neighbor, and which neglect is in itself an act of bad faith, and still claim the benefit of a mistake under the statute.

A like construction was given to limit the scope of the defense added by Chapter 239, Laws of 1882, for the defendant who entered and cut in good-faith reliance on title. Despite the broad terms of the new defense, the court held in effect that it was available only where defendant's mistake was compounded of fact and law; if defendant knew all the facts which as a matter of law made his title invalid, he might not successfully invoke this bar to plaintiff's action. For,

> if a defendant, with a full knowledge of all the facts which the law says renders his title invalid, can insist that he believed it to be a good title and acquired it in good faith, because he was so advised by counsel, there would be but a slight chance for the owner of standing timber to protect himself against the trespasser.[34]

These were common-sense rulings to advance the policy of the statute, given such familiar facts of the pine country as the frequent lack of well-defined boundaries and the practice of unscrupulous loggers to cut "a big forty" by liberal expansion of their logging operations on all sides beyond the lines of their own land.

Consistent with its generally liberal (i.e., extensive) construction of the basic provision of section 4269 was the court's carefully restrictive interpretation of the scope of the mistake-of-fact and mistake-of-title defenses allowed thereunder. We have already noted the rulings that the first defense did not cover mistakes of law, nor the second a mistake of law unalloyed with some mistake of fact, that a mistake committed through negligence offered no defense, and that in any case the burden of pleading and proving these defenses rested strictly on defendant.[35] So, also, the court would treat statements in defendant's affidavit of mistake as admissions binding on him, and as good ground for refusing to nonsuit plaintiff. Where defendant's agent had done the cutting, the agent's own conduct must be such as to allow him to plead the mistake defense, if his principal were to have the benefit of it. And where a mistake had been made in good faith, defendant must have continued in this good-faith error, successfully to raise the defense; if at some point he became aware of his blunder, and yet continued to deal with plaintiff's property as if it were his own, he would be fully liable under the statute; the same was ruled under the mistake-of-title defense created by the 1882 amendment.[36]

Though liberal construction was the dominant note in the court's response to section 4269, the severity of the statute was enough to induce some reservations in this approach. The court usually announced its decision to limit the extension of the act by calling it a "penal" statute, which because of this character should not be extended by broad implication. Most of the limitations imposed were marginal, and did not affect the court's general readiness to see the full measure of protection given the landowner. Thus, given the severity of the statute, it was not unreasonable to insist that the plaintiff make clear proof of his title, at least where he was not in possession, and especially where, holding only a tax title without possession, he sought damages for timber cutting by the original owner. So, too, plaintiff must make clear proof of the other elements of his cause of action. Moreover, though the court was ready to give the plaintiff the full measure of the new remedy within the legislature's clearly indicated policy, it would not by implication add burdensome incidents; especially it would not imply procedural incidents which might make it impossible for defendant to prove his case in mitigation of damages. Thus, because the act was "penal," the court would not allow plaintiff interest on his judgment thereunder when the legislature had not expressly provided it. And where the general statute providing for survival against an executor or administrator of actions for damage to realty limited recovery to "damages actually sustained, without any vindictive or exemplary damages," this was interpreted to bar recovery from an administrator of the enhanced damages under section 4269 for the wrongful cutting done by his decedent. The court grounded its ruling partly on the policy of construing a "penal" statute "with reasonable strictness." But the decision does not necessarily show lack of sympathy towards the policy of section 4269, for

—as the court itself pointed out—the legislature had provided the defense of mistake, to rest on defendant's affidavit:

This is a matter within the personal knowledge of the wrongdoer alone, and presumably an executor or administrator could not make such an affidavit. Had the legislature intended that the section should apply to a case like this, some provision would have been inserted to meet this difficulty. It scarcely requires a resort to the rule of strict construction to enable the court to hold that this case is not within the statute.

So the court in a later case sustained a section 4269 action against the receiver of a partnership for a timber trespass done by the partners, where, though one partner had died, it did not appear that the deceased was the only person capable of making an affidavit of defense.[37]

However, in one area restrictive interpretation of section 4269 added a major element to the total pattern of policy. The statute gave plaintiff "the highest market value of such logs, timber or lumber, in whatsoever place, shape or condition, manufactured or unmanufactured, the same shall have been, at any time before the trial, while in possession of the trespasser, or any purchaser from him, with notice." In *Wright v. E. E. Bolles Wooden Ware Company* (1880), the court ruled that plaintiff might not recover the statutory damages against one who had bought logs as a bona fide purchaser for value without notice, from the man who had wrongfully cut them on plaintiff's land; against a bona fide purchaser the landowner might have only the common law damages, limited to the stumpage value of the timber. The difference here was between a recovery of 50 cents per 1000 feet board measure (for the value of the standing timber) or $5 per 1000 feet (the highest market value of the timber or logs while in possession of the wrongful cutter).[38]

"The statute is in the nature of a penal one," said the court, for "its application to this case would increase the damages tenfold." The weight of the potential judgment against defendant obviously moved the court. But this alone could not justify a restrictive construction, for a severe judgment might be amply justified by the interests at stake. To characterize the act as "penal" was thus only to fix on it a label implying more detailed policy. This implication was made clear only two years later in *Tucker v. Cole* (1882), when the court rejected the argument that, since in criminal law a partner was not criminally liable for the acts of a copartner, he should not be liable under section 4269 for wrongful cutting done wholly by his copartner. Holding that defendant was liable under the statute for his partner's tort, the court declared that section 4269 "is not in its nature a criminal or penal statute."

It simply prescribes a rule of assessing damages in certain cases . . . If the principal is not liable for the acts or knowledge of his agent, or the partner for the acts or knowledge of his co-partner, under this act, it would very soon become a very feeble instrument in effecting the purpose for which it was enacted.[39]

The same statute could, within two years, be designated to be and not to be "in its nature a penal stature," only because different interests were involved in different applications. *Tucker v. Cole* presented squarely the central policy of the act, to protect the timberland investor against direct aggression; defendant there was in close business association with the wrongful cutter. Moreover, the working ties of defendant and the immediate wrongdoer posed the danger of evasion of the statutory policy, a threat to which the court was obviously sensitive in this field where difficulties of law enforcement were notorious. The *Wright* case, in contrast, presented a defendant who had dealt at arm's length with the immediate offender, and who had himself not been involved in the actual invasion of plaintiff's lands.

Confronted in *Wright* with a defendant thus not directly linked to the wrongful cutting, the court was made the more sensitive to the fact that section 4269 was legislation enacted for the extraordinary protection of a particular interest:

It is confined in its operation to a single class of trespassers, and establishes a rule of damages therefor which may, and often does, give the injured party much more than actual compensation for the injury he has suffered, and much more than he could recover for a trespass upon other property, equally injurious to him and committed under the same circumstances. By all sound rules of construction such a statute should be interpreted with reasonable strictness, and its penalties should be inflicted only in cases clearly within it.

As a bona fide purchaser, the *Wright* case defendant represented a social interest fully as legitimate as that of the timberland investor, and, moreover, an interest peculiarly close to the thrust and ambition of a society in growth. For this bona fide purchaser symbolized the active market, and the dynamic production which it made possible. This symbolism made him of key importance to nineteenth-century Americans who pinned their faith on progressive growth, defined largely in terms of rising productivity. Effective markets demanded a legal framework within which men could deal impersonally and at a distance, in reliance that they would get value for substantial consideration ventured. The court must be careful that it did not extend the broad terms of this specialized statute in a way that might impair market operation:

The purchasing of logs and timber which have been floated down the streams of the state to market, frequently for long distances, is a most important branch of commerce in this state. These purchases amount to an enormous sum every year, and the money paid on account of them, or a large percentage of it, goes to pay the wages of thousands of men who are employed to cut the logs and get them to market. In the nature of things, it is impossible that such purchasers can always, or usually, ascertain where the timber or logs they buy in the market were cut, or whether those cutting them had lawful authority to do so. If such dealers are held responsible for every trespass committed by those who cut the logs purchased by them in good faith, and are subjected to the statutory rule of damages, no one could purchase logs, timber or lumber in the market with any safety. The result of such responsibility would necessarily be to embarrass and check an important business, and greatly injure an important industry of the state. We are slow to believe that by enacting Section 4269 the legislature intended any such result; and it ought not to be held that it so intended, unless the language of the statute clearly and unmistakably indicates such intention.[40]

The court stood fast by its interpretation in the *Wright* case, and tightened it by later making plain that the burden of proof lay on plaintiff to prove that a defendant who had bought logs or lumber from the wrongful cutter was not a bona fide purchaser.[41] But of course one who bought with notice of facts showing a wrongful cutting did not enjoy the benefit of the *Wright* doctrine. Moreover, the court's concern to protect market operations did not blind it to the strength of the statutory policy protective of timberland investment, for it also ruled that a defendant who bought in good faith, but who learned that the logs were wrongfully cut before he entered plaintiff's land and removed them, was liable under section 4269.[42]

There was tribute to the court's perceptive evaluation of the relative interests at stake, in the fact that the legislative record shows no attempt by timberland investors to overturn or modify the *Wright* decision. Nor did the legislature or successive revisers ever tamper with that ruling during the lifetime of section 4269. It was consistent with the main currents of our nineteenth-century public policy that, in a clash between more static, speculative, investment capital, and dynamic capital at work in the market, the latter should prevail. Moreover, this outcome fitted into the pattern of policy drawn by the statutory defenses of mistake of fact and mistake of title, for these, too, reflected the century's favor towards productive action, and its concern that legal liability not add too much to the inherent economic risks of action. The vigor and importance of these qualifications upon the strong, special protection which section 4269 afforded the timberland investor clarified the value judgment at the heart of the central provision of that statute. This was legislation not to protect vested interest as such, but to foster the expansion of the market by encouraging men to bring land into market even at the risk of holding it for the long term. Market-oriented thinking provided the full frame of reference for timber trespass law.

EXPANDING THE VOLUME OF THE MARKET

FAVORING VENTURE: DEFINITION OF UNITS OF TRADE

Rationality is an ideal of trading behavior. But, under the often crude conditions of pinelands business and the bustle of growth in a community lacking settled tradition, the conditions of timber traffic evolved more from practice than from the logic of market objectives. Certainly this was so of the law's role regarding the standardization of timber transactions. Reliable standards for defining and measuring units of exchange are means indispensable to broad extension of the market for most goods. Law helps build the framework of the market when it gives authoritative form to such standards. Logically, thus, an early task in creating a regime of contract should have been the comprehensive legal definition of standard units of trade. In fact, efforts to this end often followed rather than preceded the growth of the industry, and Wisconsin law never brought into one logically complete whole an authoritative definition of units of timber trade. Neither lawmen nor lumbermen took initiative to accomplish this elementary planning for the more efficient conduct of the business. True, the enactment of many statutes bore witness to the reality of the functional needs of the market for such standardized units of dealing. But this lawmaking was tardy and piecemeal. Most of it applied only to particular river valleys or sections, and on its face responded to the desires of local enterprisers, with no apparent effort by the legislature to determine whether the resulting rules might properly be generalized.

The Recording of Timberland or Log Titles

Lawful dealing in timber began with the buying and selling of timberland —or of the standing timber alone, the title to which the court decisions always ruled to be an interest in the land. Here we encounter the one area in which the law made broad provisions to standardize terms of dealing. The rectilinear survey system created by federal law provided the standard units in which all land in Wisconsin was traded. The everyday sign of the pervasive influence of this standardizing measure was the use of "the forty" as the common unit for trafficking in Wisconsin pinelands. Second only to the rectilinear survey as a legal procedure regularizing land transactions was a statutory scheme for public recording of documents representing or affecting interests in land. From the time of the Michigan territorial statutes of 1833, Wisconsin never lacked some kind of recording act. Complementing the survey and recording systems, to give form to land deals, was the general requirement of the Statute of Frauds that in order to be effective even between the immediate parties a transfer of an interest in land must be evidenced by a writing of at least minimum prescribed formality.[43]

These measures sufficed to give regular form to all dealings in timberland or standing timber involving the creation or transfer of conventional interests, such as fee simple titles or mortgages. The rectilinear survey showed its influence even where it broke down in administration. At best it was hard to run accurate lines in rough country, and much of the work was poorly done. When boundary disputes arose, as they did in considerable number, the guiding formula, both in state legislation and court decisions, was to bring boundaries into as close alignment with the federal survey pattern as could equitably be accomplished.[44] Application of the Statute of Frauds presented no special problems in Wisconsin timberland cases. The prevalence of trade in standing timber, separate from the land, was the distinctive feature. But this presented no novel issue. The Wisconsin court was familiar with long-established English and American doctrine that rights in the natural growth of the soil, other than in annual crops, represented rights in land; and it had itself already ruled that a claim to standing timber fell within the Wisconsin statute requiring summary removal from before a justice of the peace of cases involving the title to real property. Moreover, to bring such transactions under the Statute of Frauds, as concerning interests in land, simply conformed law to economic reality in a state whose legislation and judge-made law repeatedly acknowledged that standing timber constituted the principal value of much of its land.[45] Application of the recording acts to the ordinary transfer of timberland was routine. Of the several score of Wisconsin Supreme Court cases concerning contract rights to timberland or standing timber, only two presented important questions under the recording acts, and these turned simply on determinations on the facts whether particular grantees could support their claims to be bona fide purchasers for value without notice. On its face, this record was some tribute to the efficacy of the recording system in ordinary title transfers. True, the original design of the recording system — relying on a cumbersome grantor-grantee index — detracted much from its potential contribution to marketability; development of the simpler tract-index procedure, administered through private abstractors, came in slow, costly, unplanned response to the pressure of circumstance, as the cumulated complexity of realty titles outgrew the capacities of the original indexing. However, this was a general defect of the recording acts, not peculiarly affecting timberland trade; indeed, it was a defect which might have been less burdensome for many years in the pinelands than in farming or urban sections, had it not been for the lush growth of tax titles in the timber country.[46]

The general recording acts were an adequate framework for ordinary dealings in timberland. But when the capital needs of the growing industry produced a distinctive type of land transaction, the lack of policy leadership to meet the legal needs of the business showed itself in confusion of doctrine and by dependence upon cumbersome lawsuits to settle issues which might have been avoided by a specialized recording act.

Sellers of timberland or standing timber often reserved to themselves in the deed or contract a security title in the standing timber or its product until the price was paid; similarly, one who financed the purchase of such land or timber for another's exploitation commonly took the title in himself, to secure his advance; in either type of situation the interest of the seller or financier was recognized as a security interest only, and the equitable title was in the vendee or person for whom the advance was made. The conflicts which concern us here arose between the security holder and purchasers from the equitable owner.[47]

No serious difficulty with third parties seems to have arisen at any time out of such dealings, so far as concerned the title to the land or the standing timber; the numerous cases that reached the Supreme Court all concerned competing claims to logs or lumber derived from land or trees so dealt in. The absence of controversy over title to the land or standing timber probably reflected, in part, simple business fact; these situations of reserved title or security for advances were unlikely to arise except where the equitable owner desired quickly to log the tract; men who already owned timberland or had money to put into long-term holdings of timberland would naturally want to hold for their own account, if the objective was speculation in the rise of stumpage values. Moreover, familiar legal doctrine barred serious controversy over the equitable owner's dealings in the underlying title to the land or standing timber; his possession of a tract did not enable him to convey to another more title than he actually held, even if the deed or contract under which he took possession was unrecorded, for nothing in the recording acts relieved a buyer of the prudent man's obligation to inquire as to his own grantor's source of title.[48]

The trouble arose when the timber was cut, and the cutter, as equitable owner, undertook to sell the logs or lumber to a third party who claimed to take free of the original owner's reserved title or the financier's security interest. There was never serious question that the law recognized that the reserved title or security title in the standing timber continued to be enforceable as to logs, lumber, or money derived therefrom, between the parties to the original transaction — and none the less so, though the deed or contract was unrecorded.[49] However, the temper of the time was alert to the development of the market, and distrustful of arrangements that might impair marketability. The Wisconsin court bluntly declared its awareness that competing policies were at stake when the question was one of pursuing a security title in standing timber into its derivative logs or lumber, not simply between the original parties, but as against buyers in the market. The fact that logs and lumber, as movables, were assets more easily traded, and hence closer to equivalence with cash, underlined the business importance of preserving their transferability in law. After considerable experience with contests arising out of reserved security titles in standing timber and its products, the Supreme Court observed in *Wing v. Thompson* (1890):

It may be premised that a contract of this kind is not favored in the law, and the right to enforce the reservations as against a *bona fide* purchaser without notice must be based upon evidence which shows that the plaintiff has not done anything in regard to such property while in the hands of his vendee which would amount to a waiver of his right or estop him from asserting his title against a purchaser from his vendee.[50]

The doctrine thus formulated in *Wing v. Thompson* had the not inconsiderable virtues of steadiness and consistency; the court had taken this line in the first case in which a third-party buyer contested the claim of a holder of a security title to follow his security from trees to logs. *Andrews v. Jenkins* (1876) presented, not a contest over a reserved title as in the *Wing* case, but over the "mortgage" interest of the financier whose money had bought the land from which the logs in suit were cut. Though stipulating for his security in the standing timber and its product, the financier had contracted with the cutter that the latter should log the land and drive the logs to market, and had in fact knowingly permitted the logger to sell such logs as his own. The court found it "too well settled to admit of argument or doubt" that in this situation a bona fide purchaser for value held the logs free of any prior incumbrance.[51]

Within the frame of reference adopted by the court, *Andrews v. Jenkins* had, indeed, been an unremarkable decision, applying a familiar estoppel in favor of a good-faith buyer for value against a rival guilty of creating a misleading situation. What was striking was the total absence of reference, either in argument of counsel or in the opinion of the court, to a statute dealing explicitly with the recording of chattel mortgages on logs. General Laws, 1864, Chapter 167, section 12 (which became section 1739 of the Revised Statutes of 1878) provided that (1) a log owner might record with a public officer — the district lumber inspector — such distinctive identifying mark as the owner had adopted for his logs, (2) the lumber inspector should also record "all mortgages, liens and bills of sale, or other written instruments in any way affecting the ownership of any mark of logs in his district," and (3) "no conveyance, lien, mortgage or transfer shall be valid until the same is recorded." [52]

This declaration of legislative concern to enforce regularity of titles in logs had not gone without previous notice in the court. Indeed, the issue of the statute's relation to general public policy favoring the ready transfer of property had been squarely raised. In *McCutchin v. Platt* (1868), one who had failed to record his chattel mortgage on logs according to the 1864 act argued that this omission should not subordinate him to an attaching creditor of the mortgagor, for the statute "was unconstitutional, as interfering with the right of alienation of property." The mortgagee thus boldly anticipated doctrine which would not be firm for another twenty-five years under the as yet unratified 14th Amendment; his confidence reflected the common acceptance of marketability as a basic value which the law should protect. However, the court had shortly rejected his argument, pointing out in effect that the plea begged the question; in their own way, recording acts, too, promoted marketability, and their adoption represented a reasonable exercise of legislative discretion:

It is certainly competent for the legislature, upon any supposed ground of public policy, to provide in what manner the alienation and transfer of particular kinds of property shall be accomplished. Such laws do not interfere with the right of alienation, but only regulate the mode. The act in question is in this respect similar in principle to the general law relating to chattel mortgages, and also to the statute of frauds.[53]

If this act of 1864, dealing specifically with mortgages on logs, was for some reason not applicable, there also stood on the Wisconsin statute books, as it had existed since the beginning of the state, a general law on chattel mortgages. This, too, declared in strong terms a legislative policy favoring formal regularity of security titles. It declared invalid, except as between the parties thereto, any mortgage of personal property, unless the mortagee held possession of the mortgaged property, or unless the mortgage was recorded with the clerk of the town where the mortgagor resided, or (if the mortgagor were not a resident of the state) in the town where the mortgaged property was when the document was executed. Until superseded by the 1864 statute, this act had been routinely applied to mortgages of logs already cut at the time the mortgage was executed. But neither counsel or court showed any awareness of its possible relevance to the issue in *Andrews v. Jenkins*.[54]

The relation of legislatively declared policy to security interests asserted in the product of standing timber was finally made an explicit issue in *Cadle v. McLean* (1880). By this time the legislature had added yet a third provision possibly relevant to the competing claims of a security holder and a buyer in the market. Chapter 113 of the Laws of 1873 declared that "no contract . . . for the sale of personal property, by the terms of which the title . . . is to remain in the vendor, and the possession thereof in the vendee, until the purchase price is paid, or other conditions of sale are complied with," should be valid except between the parties, unless reduced to writing and recorded with the clerk of the town where the buyer resided, or (if the buyer were not a resident of the state) in the town where the contract was made. Plainly the legislature desired that the same recording policy applied to chattel mortgages should govern the newer device of the conditional sale; its extension of this policy emphasized its general concern for formal regularity in the definition of security interests. Nothing on the face of the statute or in the legislative record shows that this act responded to concern for the peculiar problems of the timber industry. On the other hand, the statutory language was clearly broad enough to include sales of logs already cut, and was so treated by the court.[55]

There were serious doctrinal and practical difficulties in allowing a man to pursue his security interest in standing timber into its derivative logs or lumber, against one who bought the cut product in the market. In other connections the law was accustomed to rule that things which, as growth of the soil, were real property became personal property when severed.

These analogies favored application of the recording acts regarding chattel mortgages or conditional sales, as soon as the subject matter of a secured interest was transformed from standing timber into logs. However, out of fear of fraud on other creditors, judicial doctrine was also firm that a chattel mortgage (and, by like policy, a conditional sale) might not be made to cover logs first acquired by the mortgagor after execution of the mortgage. Perhaps the legislature might, if it willed, create a valid procedure by which new property might be brought under an old security instrument. But the contrary rule was too well established to be changed by anything short of express legislative command. The ban on inclusion of after-acquired property in a mortgage thus posed a formidable legal barrier to a security holder who wished to record as a "mortgage" or a "conditional sale" an instrument dealing with title to logs not yet in existence because still standing in the tree. On the other hand, to require him to keep close trace of the progress of logging, and to record successive small lots of logs as they were cut, would require an accurate flow of information and nicety of business practice such as it might be impractical to exact in the rough-and-ready conditions of pinelands operations. Recording the original security instrument in the register of real property titles would not answer either doctrinal or practical problems; as we have seen, recorded or not, the security holder's interest was probably safe so long as it remained a real estate interest; when the cut timber became personal property in law, the business fact also was that it acquired a mobility which would make grossly unfair to buyers any requirement that to protect their titles bought in market they must search the books of the registers of deeds in all land areas from which the logs might conceivably have been cut. Yet, in the face of these difficulties of legal concept and business fact, the general chattel mortgage recording act, the specific statute for recording of log transfers, and the act for recording of conditional sales added up to an impressive legislative declaration that public policy required that security interests in personal property be cast in regular form.[56]

In *Cadle v. McLean* (1880) a contract of sale of standing timber allowed the buyer to cut the timber and manufacture it into lumber, but stipulated that, until the buyer paid the purchase price of the standing timber, he might not dispose of any of it or its product without the seller's consent, and that meanwhile the seller should hold title to the timber and its product, and be entitled to repossess it on ten days' notice, if the buyer made any default or if the seller at any time felt insecure. The seller duly recorded the contract in the proper office under either the general chattel mortgage or the conditional sales recording act, but did not record it with the lumber inspector as provided in the special recording statute for log transactions. The court ruled that the contract created either a chattel mortgage or a conditional sale — the opinion inclined to the view that "the contract might be sustained as a mortgage of personal property, taking effect when the timber was cut"—and that, being properly filed, it prevailed against a bona fide purchaser of the logs, who bought from the chattel "mortgagor" without the "mortgagee's" knowledge or consent. The opinion made no reference to the rule against mortgaging after-acquired property. The court shortly dismissed the special log-recording act as not relevant; "a bare reading of the statute" showed that it applied only to already-cut logs, since it stipulated that the instrument recorded "shall specify the marks placed upon said logs and when they were cut." It is clear that a "bare reading" of the statute had not previously so convinced lumbermen or their lawyers; the registers of the lumber inspector for the important Sixth District (Chippewa County), for example, in the years before 1880 abound in recorded contracts stipulating that sellers of timberland or standing timber, or financiers of logging operations, should have a security title not only in standing timber, but in the logs yet to be cut from it.[57]

A year later, *Bunn v. Valley Lumber Company* (1881) neatly presented the court with a situation the converse of that in *Cadle v. McLean*—with a seller who had recorded his contract under the special log-recording act, but not under the conditional-sales or chattel-mortgages recording statutes. Since the contract here explicitly transferred title to the buyer, and provided no reconveyance of a security title, it could hardly be treated as a conditional sale or as a chattel mortgage. For by the contract the seller sold all his interest in the pine standing on a given tract, stipulated that the buyer should put a specified mark on what he cut, and have possession of the logs, run them to market, and make due efforts to sell them, the seller meantime "reserving and having a lien in and to said pine timber and saw logs cut therefrom," until paid. The trial court held that, by virtue of the recording of the contract with the lumber inspector, the seller—being unpaid—was entitled to recover logs cut under the contract, from a purchaser from the contract vendee; no issue was made at the trial, or evidence taken, as to whether the defendant was a bona fide purchaser or took with knowledge of plaintiff's lien. In a 4 to 1 decision the Supreme Court tersely reaffirmed the earlier interpretation of the special log-recording act as applicable only to already-cut logs; ruled the conditional-sales recording act irrelevant, since no title, but only a contract lien, was here reserved by the seller; made no reference to chattel mortgage theory (probably because under established Wisconsin doctrine there must have been a transfer of title back to the seller to create such a security interest); and, declaring that "there is the strongest reason for enforcing [a contract lien] . . . as against one who purchased with actual notice," remanded the case for trial on the question whether defendant was or was not a bona fide purchaser. Agreeing that this "equitable lien" represented an "important class of contracts" in the timber business, which the courts should be careful to protect, Mr. Justice

Lyon vigorously dissented. *Cadle v. McLean,* he would agree, was rightly decided on its facts, since it involved a duly recorded conditional sale, and the broad, unqualified terms of the 1873 conditional-sales recording act might well be held impliedly to supersede the special log-recording act of 1864 and to govern all conditional sales of logs; but the interpretation of the log-recording act as applicable only to already-cut logs was hence only dictum in *Cadle v. McLean,* put undue stress on incidental language of the statute at the expense of its clear purposes, and should now be disregarded; the special log-recording act should be held to embrace all instruments affecting log titles not otherwise provided for.[58]

Whatever doubts there might be as to its reasoning, the court seemed at this point to have achieved a well-defined compromise between the policy of the recording acts and the more general legal policy favoring transferability of title, at least as to one kind of transaction: The seller of standing timber might insure his right to pursue a reserved security title into the cut product by recording the contract as a conditional sale. But in *Lillie v. Dunbar* (1885) the court completely reversed its stand. Here, one who owned standing timber by contract of purchase from the landowner undertook by contract to sell it to another, with the right to cut, but on the express condition that title remain in the contract vendor until the price was paid. Disavowing the two prior cases on this score, the court invoked the analogy of its decision that a contract to sell standing timber concerned an interest in land within the Statute of Frauds, and ruled that such a contract "was not, therefore, a conditional sale of personal property within the meaning of [the conditional sales recording section of the Revised Statutes] . . . and is not in any way affected by said section." The court made no mention of the general chattel-mortgage recording act, but in *Bent v. Hoxie* (1895) it made clear that the reasoning of *Lillie v. Dunbar* also excluded application of the chattel-mortgage act to such a contract. Though Mr. Justice Lyon was still a member of the court in 1885, he made no apparent effort to reopen the issue of applying the special log-recording act. The opinion in *Lillie v. Dunbar* conceded that the conditional-sales recording act evidenced the policy of the legislature "to prevent . . . the means of doing an injustice to a purchaser of personal property dealing with one having the actual possession thereof and the apparent ownership, without knowledge of the secret agreement that the title was to remain in the vendor, notwithstanding the possession and apparent ownership was in the vendee." The opinion made no effort to offset this consideration by any counterargument from policy, but was content to rest on the logic of legal concepts: "The difficulty of the argument of [the buyer in market] . . . is the assumption that the standing timber . . . was personal property in the hands of [the contract vendor] . . . by virtue of his contract" by which the latter had bought it from the landowner. This brought the question back to familiar common law estoppel doctrine. The holder of the reserved security title should prevail against the buyer in market, unless the buyer was a bona fide purchaser and the security title holder had negligently allowed a misleading situation to arise to the buyer's detriment. By familiar standards of judge-made doctrine, one who conferred on another merely the contract rights to cut and hold possession, and explicitly withheld any right of the cutter to sell, was not guilty of creating a misleading impression. Though the opinion had in effect conceded that the recording acts expressed a sharply different legislative estimation of the equities of the situation, the judges put this aside without comment. Thus, though the deal was in movable items, the buyer in market was held chargeable with notice of the recorded state of the real estate title to the land from which the movables derived.[59]

From this point on, the court adhered firmly to the doctrine that the holder of a contract security interest in standing timber and its product might pursue it into the derivative logs or lumber against a later buyer in market, unless (1) the later buyer were a bona fide purchaser for value without notice of the security interest, and (2) the security holder had given the contract vendee possession and the right to sell, or had knowingly acquiesced in prior sales by the contract vendee, or knew or should have known that the contract vendee was about to make an unauthorized sale and, knowing so, did nothing to stop him. The legislative record shows no attempt to override the judges' treatment of the problem. This is some evidence, of course, that the court correctly interpreted legislative policy. But this conclusion seems doubtful in the face of the breadth and strength of the three statutes covering the recording of instruments of title in personal property, and particularly in the face of the acts dealing specifically with title in logs. The court's eventual doctrine certainly gave timberland owners all the protection they could justly claim. Lumber inspectors' records show that contracts reserving to the seller a security title in standing timber and its cut product continued to be recorded through the years after 1880; so far as this was not done out of inertia or abundance of caution, it was probably done less out of felt need to protect vendors than to protect the contract vendee against inconsistent dealings in the timber by the contract vendor. The favor here shown the landowner against the buyer in market stood in contrast to the avowed concern for the bona fide purchaser in contemporary decisions which restrictively interpreted the statute setting an enhanced measure of damages in timber-trespass cases. Since by the late eighties the important buyers in market tended to be the big companies, it is rather surprising that they did not seek new legislation more favorable to the buyer's position. That they did not do so may attest their practical power to look out for themselves in out-of-court maneuver, or may reflect simply a lack of deliberate policy planning, characteristic of the time.[60]

Granted the existence of doctrinal and practical difficulties, the application of some recording act offered an administratively simpler and less costly way to regularize log titles than by lawsuits, each of which must turn on evidence of the particular conduct of the parties in each particular situation. The prevailing judicial opinions offered no satisfactory discussion of the policy issues implicit in their rulings. Mr. Justice Lyon's dissent in *Bunn v. Valley Lumber Company* put a persuasive case for a liberal interpretation of the log-recording act, which no one answered:

The object of the statute manifestly is, to enable persons dealing in logs readily to ascertain if logs marked with any particular mark are incumbered in any manner; and to this end the office of the district lumber inspector, with which such dealers are presumably in frequent communication, was thought by the legislature to be the most convenient and accessible place of record for all parties. The location of the pine lands of the state, mainly in regions sparsely settled, in which organized towns are frequently of large territorial extent, and the public offices therein difficult of access; and the nature of the lumber business, which cannot regard town lines, but must be carried on where the timber grows, and where there are rivers upon which to run the logs or lumber to a market,—rendered it necessary that some more convenient place than the office of a town clerk, or even a register of deeds, should be designated for the recording of log contracts . . . Every reason for the enactment applies with as much force to contracts affecting logs, made before the logs are cut, as to those made after they have been cut. In either case the objects to be accomplished are, to protect persons not in possession who have an interest in logs marked with a particular mark, and to enable persons proposing to purchase such logs to ascertain the true state of the title thereto.[61]

The court's unsympathetic handling of the special log-recording act did not end with its refusal to apply the act to transactions in logs yet uncut. The 1861 and 1864 acts were awkwardly written, declaring it the duty of the lumber inspector

to record all mortgages, liens and bills of sale, or other written instruments in any way affecting the ownership of any mark of logs in his district . . . *provided,* that said instrument shall specify the marks placed upon said logs, and when they were cut, and shall be recorded in the office of the inspector in which the said marks are recorded . . .

But the statute did go on then firmly and without qualification to declare,

. . . and no conveyance, lien, mortgage, or transfer shall be valid until the same is so recorded . . .

In common-sense fashion, *McCutchin v. Platt* (1868) interpreted the statute to provide for recording instruments affecting title to "logs of any particular mark," and gave effect to the requirement of recording:

The statute, it is true, speaks of conveyances "affecting the ownership of any mark of logs," etc., but that phrase was evidently used as the equivalent for "logs of any particular mark" . . . It seems clear . . . that it was the intent of this statute to require owners of logs to record their marks, and to provide that no conveyances of the logs should be valid unless such conveyance was recorded or filed in compliance with its provisions.[62]

However, despite this strong beginning, the court went on to later rulings which first severely limited and then destroyed the required character of recording of chattel mortgages on logs. In *Morrow v. Reed* (1872), without explicit analysis of the market policies at stake, the court in effect interpreted the recording acts as qualified by the common law doctrine that a mortgagee would acquire a complete security interest against the world if the mortgagor made delivery to him of the mortgaged chattels. Because the property was bulky and incapable of manual delivery, the court held that there had been an effective delivery within this doctrine when the mortgagor accompanied the mortgagee to the bank of the stream where a large mass of logs lay, pointed out to the mortgagee the logs in question, and declared that he transferred possession of them to the mortgagee. Given the unqualified language of the special log-recording act, the court might well have ruled that statutory policy recognized that there were special aspects to trade in logs. Obviously the legislation showed a desire to regularize log titles for the protection of third parties. In this context the court might well have held that such a token delivery as satisfied the common law would not overcome the unqualified command of the statute that every instrument creating security in logs be recorded. That the judges did not even acknowledge the existence of a possible issue on the point underlined their readiness to protect the security holder.[63]

But the court gave the recording act still rougher and more summary handling. Without indicating intent other than to smooth out the awkward language defining the coverage of the act, the Revisers of 1878 rewrote the recording provision, as section 1739, to state:

All mortgages, liens, bills of sale or other written instruments in any way affecting the ownership of any marked logs in any lumber district which shall specify the marks placed upon said logs and when they were cut shall be recorded in the office of the lumber inspector in which said marks are recorded; and no such conveyance, lien, mortgage or transfer shall be valid, except as to the parties thereto, until the same is so recorded . . .

Without explanation, and disregarding the established rule that revisers' language should not be interpreted to effect material substantive changes in law without clear evidence of intent to do so, the court ruled in *Cook v. Van Horne* (1890) that

. . . it is competent for parties dealing in marked logs so to frame their contracts in respect thereto as to bring them within sec. 1739, or, at their option, to frame them differently, so they will remain subject to the common law and statutory rules which would govern them had sec. 1739 not been enacted.

Accordingly, it held that a bill of sale of logs which did not specify the marks placed on the logs nor when they were cut was not required to be recorded, nor was it subject to record, under section 1739, and that the buyer under the unrecorded bill of sale might prevail against attaching creditors of the seller. The court thus deprived the statute of practically all meaning as a protection to third parties; such effect as the statute still might have would be only to protect a suspicious buyer who did not at once take possession, by allowing him to press for a recording which would bar any later sale of the logs by the contract vendor. We can only infer the court's policy from its behavior, since it did not explain itself; evidently the judges disliked special, statutory formalities which limited the freedom of contract accorded by general law. Had the *Cook* decision come twenty years earlier, when shoestring finance probably bulked relatively larger in the industry than in the nineties, it might have stirred legislative reaction. As it was, the legislature showed no immediate interest in the ruling, and nine years later (1899) the Senate summarily rejected a bill which in effect would have overruled the decision.[64]

Adoption and Recording of Log Marks

The registration of log marks was the one respect in which the recording act successfully provided a formal pattern for dealing in the timber industry. Though nothing at common law required a man to place an identifying symbol on his logs, probably most operators did, in common-sense self-protection. The industry was rather tardy in writing this practice into law. However, Chapter 83, Laws of 1861 (soon broadened in application by Chapter 167, Laws of 1864) finally recognized the log mark as a basic device to prevent title disputes. These statutes (1) authorized every log owner to use as his own any identifying mark not before used and recorded by another in the district, and put indirect pressure on him to adopt a distinctive mark by providing that all unmarked logs should be "prize logs," to be divided among the owners in each "subdistrict" according to the number of logs each owned; (2) required the owner to record his mark with a public official, the district lumber inspector; (3) forbade anyone else to use a prior, duly recorded mark (unless properly assigned to him by its owner), on pain of a $10 forfeit, part at least of which should be paid to the aggrieved owner; (4) made the register of marks a public record, required that lumber inspectors' scale bills describe the mark on the logs measured, and—until the strange ruling in *Cook v. Van Horne* in 1890—apparently required that all instruments affecting title to logs specify the marks thereon. The 1861 and 1864 provisions on log marks were not well drawn, nor were they materially improved as they were carried into sections 1738–1740 of the Revised Statutes of 1878. In total effect, the fair reading was that they required the adoption as well as the recording of log marks, but they said so largely by awkward indirection. Their clear implication was that a recorded log mark should be at least presumptive evidence of ownership of the marked logs by the recorded owner of the mark, but the legislature nowhere said so in its general statutes, though it declared this rule expressly in an 1889 act specially requiring adoption and recording of marks on logs rafted or floated on the Wisconsin River system. Basic and sensible as they were, the log-mark record provisions were thus drawn with a lack of craftsmanship characteristic of nineteenth-century state legislation.[65]

This general pattern of policy regarding log marks and their recording remained stable throughout the history of Wisconsin lumbering. The only important qualification concerned the geographical areas of its application. The 1861 statute applied only to the northwestern lumber regions of the state; an 1863 act brought in the Green Bay timber district, but expressly excepted the area tributary to Lake Winnebago and the Oshkosh mills; acts of 1863 and 1876, applying standard procedures for measuring lumber by the office of lumber inspector in the Oshkosh area, at first omitted reference to the log-mark record system and then expressly declared that none of it should apply in that region; the 1878 Revisers finally brought the Oshkosh territory within the log-mark record scheme. Beginning with an 1862 statute, the legislature agreed to special provisions for mark records in District Number Four (the St. Croix section), under which recording (and, apparently, adoption) of a mark was required under compulsion of a penal sanction; one who cut and manufactured logs and lumber in that district without filing a mark was guilty of a misdemeanor, punishable by a $25 to $100 fine. An 1866 act declaring navigable the West Twin River and its tributaries in Brown and Manitowoc counties incidentally required the owner of any logs on such waters to put his own distinctive mark on them, or be liable to their confiscation for the benefit of all other owners of logs on these streams; oddly, however, the requirement to mark was here unaccompanied by any requirement to record the mark. In 1889 the legislature singled out the Wisconsin River and streams flowing into or tributary to it as another area in which everyone rafting or floating logs must adopt, use, and record a distinctive mark; yet, though putting this much emphasis on recording, the statute did not borrow the specific penal sanction applied in the St. Croix district statute, but resorted to the indirect sanctions of the general mark-recording laws. Save for the total exception of the Oshkosh area in the years before 1878, none of these special aspects of log-mark record legislation departed from the basic policy of the central statutes. Their presence on the books is notable mainly as another reminder of the legislature's tendency to respond to local pressures, without responsibly fitting particular proposals into the state's general policy.[66]

The record books of the lumber inspectors, and the evidence of business practice preserved in the testimony of many log-contract lawsuits, show that the use of recorded marks was the commonplace practice. The complete absence of any Supreme Court litigation over the validity and application of the mark-records system, and its unquestioned continuance and expansion on the statute books, tend to confirm the appraisal of an historian of the industry that "such marks . . . were as effective in determining ownership as a real estate deed." Among the scores of contract and conversion cases involving logs or lumber before the Wisconsin Supreme Court over the years in which the log-mark record laws stood on the books, only a trivial number presented issues of fact on owner identification, and there was no case in which an issue arose out of failure of the mark-recording procedure.[67]

Provision of Public Facilities for Log and Lumber Measurement

In addition to procedures to regularize evidence of title, Wisconsin statute law created another important device to standardize transactions. This was an administrative machinery to provide authoritative, impartial measurements of the quantity of logs or lumber involved in sales contracts. Legislation divided the pinelands region into "lumber inspection districts," authorized the Governor to appoint for a two-year term a "lumber inspector" for each district, empowered each inspector to appoint subordinates, and made it the prime duty of the inspector and his deputies (under bond, and for stipulated fees) to scale (measure) logs or lumber of private owners and furnish scaling bills stating the measurements so made. That lumbermen took the official scaler's role seriously was indicated when, in 1885, in the full tide of the industry, the legislature was moved to define heavy criminal penalties for fraudulent measurements by these officials, or the procuring of fraud in measure; the penalties imposed might run as high as a $5000 fine or five years in the state penitentiary. The statutes emphasized the official scaler's impartial position by declaring that his certificate should not be received in evidence of the correctness of a measurement made by him on logs in which he had an interest. On the other hand, the statutes underlined the fact that the scaler's services were made available because of the social interest in the growth of lumber marketing — rather than out of any more abstract concern for public order — by allowing an action on the inspector's bond not only at suit of the public, but also of any person aggrieved by the inspector's misconduct.[68]

Determination of quantity was always the main purpose of this "lumber inspection" procedure. The legislature made no general venture into definition or application of quality standards in timber marketing. The statute did require the lumber inspector to "make such allowance for hollow, rotten or crooked logs as would make them equal to good, sound, straight, merchantable logs," but this was only incident to the measurement job. Apparently trade practice and custom satisfied the industry's need for minimum quality standards, for there is no evidence that there was ever any serious interest in promoting legislation in the matter.[69]

To prevent disputes over quantity, the statutes did undertake to define standards of measurement as well as to provide a procedure of measurement. In each aspect of the legislation, the basic question was whether the statutes would require that all transactions be governed by their provisions, or would merely furnish definitions and procedures which the parties might use or not, at their election.

The general "lumber inspection" acts went part way toward laying down a required standard of measurement. The basic 1861 and 1864 acts stamped with public favor a rule of measurement well recognized in the trade, by declaring,

> The "Scribner Rule" shall be the standard rule for scaling or surveying logs in the said districts, but other rules may be used at the request of the owner or owners of logs, but in all such cases the bill of the [official scaler] . . . shall state by what rule the logs were surveyed or scaled.

The preferred position of the Scribner rule continued until a 1901 act set out a measuring schedule in detail in the statute book — still, however, with an option in the log owner to use another rule, to be acknowledged in the scaling bill. The option tendered by the statutes was not, it would appear, a complete freedom for all parties. The declared right to contract out of the rule preferred by statute lay only in the log owner; if he chose to abide by the statutory rule, the buyer had no standing in law to object. There were substantial dollar stakes in the issue of how far the law should go in preferring or even requiring a given scaling rule. The industry in the Lake States knew over forty different rules, and these varied greatly in effect, some preferring the seller, some the buyer of logs. Between the two most used rules, the Scribner rule and the Doyle rule, there might be variations of from 5 to 20 per cent in measuring large logs. The Scribner rule gave smaller readings for the larger logs, but yielded larger totals for the smaller logs; the preference expressed for it in the 1861 and 1864 acts may have reflected the influence of the large buyers, but continuation of the statutory preference into a later generation — when the prime trees had been cut, and the bulk of logs marketed were smaller — worked, so far as it went, to favor the sellers. Confronting these divisions of interest within the trade, the legislature made plain its general preference for freedom of contract; four proposals to enact a required standard for measurement met summary defeat on the floor or in committee.

The general "lumber inspection" laws likewise left it optional with the log owner whether he made use of the services of the official scaler; the statute limited itself simply to declaring that it was the official's duty

> at the request of any owner of logs, timber or lumber, after a scalement or measurement, to make out a bill, stating therein the number of logs, the num-

ber of feet, board measure, contained in such logs and lumber, and the number of feet, cubic, running or board measure, contained in said timber, and at whose request the same were scaled or measured, and by whom scaled or measured . . .

However, apart from the intrinsic attractiveness of the proffered service, the statutes offered some special inducement to use of the official scaler by declaring that a duly certified bill stating a measurement made by an officer under the statute "shall be received in all courts of this state as *prima facie* evidence of the truth and correctness thereof." [70]

Though the general frame of the scaling laws was clear and remained stable, policy was clouded somewhat by the same legislative yielding to localism as we saw in the regulation of log marks. There was a cumbersome amount of amendment and revision in the establishment and redefinition of lumber inspection districts and offices under the general statute, involving fifteen separate pieces of legislation. It was sensible to conform district lines and office locations to the geographic growth and shifts of focus of the industry. It would have been more efficient lawmaking to delegate these adjustments on suitable terms to the Governor or other executive agency. But the Wisconsin executive establishment of the 1870's and 1880's was small, and there was yet no tradition, and little experience, in delegation of rule-making powers. With all its clumsiness, the process of piecemeal legislative adjustment was characteristic of the time. It was also characteristic that the legislature seems always to have been content to leave the creation and redefinition of districts to its lumber-area members, acting either on such standing committees as specialized in timber matters, or through select committees of members from the immediately affected districts.[71]

However, this mode of establishing inspection districts does not explain, and certainly does not justify, important statutory variations in substantive policy according to locality. If an official measuring service was a useful contribution to the industry's growth, it was sensible that it be available wherever the industry operated. However, the pioneer measurement act (1848) applied only in the St. Croix valley. The next statute (1854), without qualification, declared that the Scribner rule "shall hereafter be the uniform and legal rule for scaling logs on Black River," but provided no official measuring service. The first broad act, in 1861, serviced only northwestern Wisconsin. In 1862 the legislature declared Winnebago County (Oshkosh), in the northeast, a survey district, under a "surveyor general" appointed not by the Governor but by the county board, with many variations from the pattern of the 1861 statute. In 1863 the 1861 act scheme was applied to the other major northeastern lumbering area, that about Green Bay, but the special provisions for the Oshkosh territory were expressly continued, as they were also in the general revision of the system under the act of 1864. In 1876 the area tributary to Oshkosh was finally brought under the general scaling system, though even now with a separate schedule of scaling fees.[72]

The legislature also made an important local variation from the general policy that use of the official scaling service be at the option of the log owner. Chapter 388, General Laws of 1862, without qualification, made it the "duty" of the lumber inspector for the Fourth (St. Croix) District

> to enter upon and scale all logs and lumber cut in said district; and when the said logs, lumber and marks shall have been so surveyed, the owner or owners of the same shall pay, upon demand therefor, as the fees of said [officer] . . . the sums now provided by statute therefor.

Despite its firm command, the 1862 act imposed no penalty if a private owner refused to allow the official scaler to measure his logs. In 1875 the legislature continued this special requirement for the Fourth District, and added stiff, if indirect, sanctions; without qualification the new act declared invalid all sales or contracts of sale of logs or lumber cut in that district, unless scaled by the public officer with a certified bill of scale delivered to the owner, and bulwarked this sanction by further declaring that no action might be maintained on any such sale or contract, unless the required scaling had been done. Moreover, the lumber inspector was now given a lien for his fees, a remedy he did not enjoy under the general inspection statute. The Revised Statutes, 1878, continued special provisions for the Fourth District, in section 1743. However, the revisers indicated some lack of enthusiasm for this situation:

> The requirement is peculiar to the 4th district and seems singular, but is retained as enacted heretofore.

Despite this statement, the revision in fact substantially changed the law. It materially weakened the requirement of scaling; sales and contracts of sale of logs and lumber in the Fourth District were now declared "void, except as to the parties," unless officially scaled, and the new statute eliminated the provision banning all suits on deals involving logs or lumber not officially scaled; in other words, use of the official scaler was optional, so far as concerned all dealings immediately between buyers and sellers. In 1885 the legislature tinkered again with Fourth District provisions, but not in a fashion to require that the official scaler enter into all private bargainings; the emphasis now shifted to obtaining a general report of all cutting in the district; all lumbermen therein must report either on the basis of official scaling, or make sworn statements of their production based on their own scaling arrangements, except that logs driven down the St. Croix River must be officially scaled (a measure apparently intended to prevent their scaling in Minnesota). In this form a special provision for the Fourth District remained on the Wisconsin statute books until the whole system was revised and consolidated on a four-district basis in 1919, in reflection of the diminished importance of the industry.[73]

There were other variations from the general policy of optional measurement provisions. From 1862 to 1876 the legislature apparently imposed the Scribner rule as the required standard for measurement in the Oshkosh area, while log owners elsewhere had the right to elect to use other measuring formulas. An 1870 act required that log owners operating on the Black River cause their logs to be scaled "by some competent person," and record the scale bill with the District Two lumber inspector, under pain of suffering a $10 to $50 fine for each violation; some indirect pressure was applied to induce the log owner to use the services of the official scaler, by requiring that any private scaler so employed must take the same oath and post the same bond as the public officer, if his scale bills were to be accepted as complying with the act. This provision peculiar to the Black River area was carried into the Revised Statutes of 1878, and was not repealed until 1919. The limited terms in which these statutes moved toward required official scaling in the Second and Fourth Districts indicate that there was a substantial issue at stake in the matter. This implication was underlined by the defeat in 1885 of two bills which would have imposed limited requirements of official scaling also in the Sixth and Thirteenth Districts. Still another local variation was provided, however, by an 1882 act, which required that a special scaler's mark be put on every log scaled by the lumber inspector in the Ninth District. Though the bill went through without sign of serious opposition, it undoubtedly grew out of local tension over the operations of the great Mississippi River Logging Company ("the pool"), which declined to use the services of the official scaler; apparently the bill was designed to advertise which logs were those of "independent" owners, who regularly used the lumber inspector's service. This anomaly, too, was carried forward into the compiled statutes of the state, and lasted as long as there was any statutory lumber inspection scheme. Finally, a variety of special laws incorporating or regulating particular boom companies, or companies combining boom and sawmill operations, required the use of the services of the appropriate official scaler to measure all logs, or at least "prize logs," handled by these concerns.[74]

Wisconsin common law struck a nice balance of policy in handling parties' private contract arrangements for log measurement. If the parties agreed upon a scaler and a procedure of scaling, the Supreme Court in effect welcomed the stipulations as useful preventives of dispute, and of its own initiative construed determinations made under such contract procedures as entitled to great weight. Thus, where a written record was made according to such a contract procedure, it was the best evidence of the quantity measured, and the court enthusiastically applied familiar evidence doctrine to exclude other evidence of quantity, unless good reason was shown for not producing the scale book. Where, pursuant to contract stipulation, a scaler proved competent was named by mutual agreement, his scale might not be overturned, especially on matters admittedly requiring exercise of judgment, unless impartial tests or strong evidence showed that he was prejudiced, or careless, or acted under fundamental mistake of fact, or in gross violation of his contract authority.[75] On the other hand, this was a court always keenly conscious of the public policy which favored freedom of contract. The judges were pointedly careful, therefore, to hold the private scaling procedure within the limits set by the parties' agreement. The authority of a private scaler must be clearly established under the contract; thus, where the contract said that it was "mutually agreed . . . that the scaling . . . shall be under the supervision of a competent person," this must be taken to require mutual assent to appointment of any supervisor; an agreement on scaling to determine quantity would not be construed to authorize the scaler to pass on the quality of the tendered logs, save as this was implicit in measuring quantity; the private scale would bind only on the question on which the contract made it determinative, so that provision for scaling of logs delivered on the mill deck of a sawmill did not preclude proof of a scale made in the woods, to help determine the separate issue of the quantity delivered at the sawmill boom.[76]

The court's treatment of the statutory measuring procedure completely paralleled its common law attitudes. Unlike other instances, however, this seems a case in which the result was properly to implement the legislation. The lumber inspector was an early kind of state administrative officer, and it is tempting to think that in the court's treatment of his handiwork one might find clues to the judges' initial attitudes toward the administrative process. But the opinions give no indication that the court saw much that was distinctive in the inspector's official status, save as the judges emphasized the ideal of impartiality. The decisions are more consistent with the view that the legislature had given formal standing to a procedure familiar in private contract, to provide an auxiliary service helpful to market operations. The doctrine of the cases on the scope of judicial review of the inspector's work so followed the common law approach regarding analogous private contract arrangements as to suggest that it was attitudes generated out of contract law, rather than policy especially related to review of administrative action, that guided the court.

The court acknowledged that the statute evidenced a public policy favoring methods to promote impartiality and regularity in measurement of this important commodity, and in effect recognized that this policy originated not with the legislature but with the industry: "The scaling and measuring of logs and lumber has been deemed of sufficient importance to those engaged in the business of lumbering to call for legislative regulation . . ."

The court would give full weight to the official measurements. Where there was evidence that a log owner had used the services of the official scaler, the general presumption of the regularity of official action would support the application of the official measuring system; thus the court would presume that the scaler had, according to statute, delivered to the log owner a duly certified scale bill. Where a scale bill had been duly made by the lumber inspector or his deputy, certified and recorded and delivered to the log owner, it was the best evidence of the quantity of logs involved, and might not be impeached, or the quantity proved by parol evidence, unless sufficient cause were shown why the official scale bill was not produced. Any other ruling "would be most dangerous"—presumably to the market regularity which the statute sought to promote. Where the official scaler acted pursuant to stipulation in the parties' contract, his measurement would prevail, unless clear showing were made of fraud, negligence, mistake, or violation of his statutory duty. In that particular, the court gave practical force to the official scaler's effort by ruling that his measurement might not be rebutted merely by evidence of opposing estimates; the contesting party must go to the trouble and expense of procuring independent, detailed measurements. On the other hand, the court was careful to hold the official system to its limits. Use of the official scaler was recognized to be optional, under the general statute; hence no effect would be given a determination of grade or quality, rather than quantity, made by the official scaler at the request only of one party to the contract. So, too, the statute meant what it said when it authorized the official scaler to make a "scalement or measurement," and the presumption favoring the validity of his scale bill was held overthrown by evidence that he had in fact measured only 784 out of 2145 logs, for this "was at best a mere estimate, and the section does not authorize the making of an official scale bill out of an estimate." On the other hand, the court held that the lumber inspector's bonded obligation to discharge his duties "to the best of his knowledge, judgment and ability" did not make him a guarantor of the accuracy of his measurements, or those of his deputies, but made him liable only for bad faith or negligence. Conforming to general doctrine on public officers' liability, this interpretation also was reasonable in view of the optional nature of the official measurement service; had the intent been to make the lumber inspector a warrantor of his measurements, presumably the legislature would have made the system compulsory in application and would have backed the inspector's work with a more solid guaranty than exaction simply of a $5000 bond.[77]

The official measurement procedure was useful and was used. So much appears from the variety of attention given to the system—from its expansion to cover all Wisconsin lumbering regions, from the repeated redefinitions of district lines and office locations to meet demands for more convenient service, from the effort exerted to obtain special procedures or fee schedules for particular areas, and from the evidence of contracts stipulating its use. But we do not have the data by which to make any close estimate of the relative extent of its use in the industry as a whole. A priori, one might expect that the larger companies buying logs as middlemen or as mill operators would enforce their own notions of scaling, out of their superior bargaining power. Of course, companies would have no call on the official service so far as concerned their own integrated logging operations. Likewise, we would expect that use of the official service would decline in relative importance in proportion as large firms came to dominate purchase and production. Lumber inspectors annually reported the total amount of logs and lumber that they scaled. Their reports contain matter that bears out the propositions stated a priori. Thus the inspector for the Tenth District (Marathon County, on the Wisconsin River), in the fifth year of the district's existence, warned that the figures which he reported for the year ending December 31, 1875—1,251,844 feet of logs scaled, and 49,394 feet of sawed lumber measured, for a total of 1,301,238 feet—

do not approximate to a representation of the annual product of logs and lumber in this district, for the reason that the inspector of the district, as the law now stands, is not called upon by loggers and lumbermen to perform any official duties, except now and then upon small outside lots.

His estimate was that in the reported period 87,100,000 feet of logs had been cut, and 53,500,000 feet of lumber manufactured in his district. In contrast, however, District Two (Black River) reported for the same season 188,344,640 feet of logs scaled, and 23,500,000 feet of lumber measured. In 1877, when the Tenth District inspector repeated his observation (6,883,417 feet of logs scaled, 5,375,140 feet of lumber measured, compared with estimated total district production of 90,000,000 feet of logs cut, and 60,000,000 feet of lumber manufactured), District Seven (Eau Claire) reported scaling 41,323,229 feet of logs cut and made into lumber in that district; the Seventh District report noted, also, that several million feet of logs were cut and run out of the district, to be manufactured and scaled in the adjoining area. The extent to which both integrated operations and large-scale middleman purchasing took scaling business away from the official scaler was early and repeatedly noted by the disgruntled inspector for the Ninth District (lower Chippewa Valley), who commented, for example, in 1875 that he had scaled 29,477,456 feet of logs, and that

aside from this amount, 104,000,000 feet was scaled by employees of the Mississippi River Loging Company who have not had any logs scaled for three years by the regular inspector. Not from any dissatisfaction with the appointee. I have asked a statement from them, but have so far been unable to get one.

Data from the lumber inspectors' reports, and general data on timber production from other sources such as the trade press, unfortunately cannot be sufficiently correlated as to time and place of origin to permit charting the relation between total output and the amount of logs or lumber scaled or measured under the official procedure.[78]

Correspondence preserved in the state's executive files shows that sometimes, in some places, the office of lumber inspector was regarded as important enough to warrant energetic campaigns of letter writing and personal interview to obtain from the Governor the appointment of one candidate rather than another. So far as the preserved materials show, most of this activity centered on the Chippewa and Black rivers; there is no evidence of sustained interest in the appointments most of the time in other districts. The contemporary evidence tends to play down the worth of the office, so far as its fees are concerned. A candidate urging himself for the Chippewa Valley appointment in 1862 thought "the position . . . a good thing," warning that "while our people are asleep in the matter some *democrat* will apply . . ." In 1868 a leading lumberman of LaCrosse, Angus Cameron, described the post in the Second (Black River) District as "a lucrative one — worth at least $5000 a year." In contrast, about the same time the inspector in the Green Bay area estimated that office to be worth at most $500 a year in fees, and in 1884 a detached observer described the post in District Eleven (Ashland) as "of little pecuniary benefit" to the holder. The general indication is that most of the inspectors handled a modest to substantial amount of business, yielding relatively small income.[79]

When it was time to make a new appointment, the Governor might receive a fair number of charges of inefficiency, negligence, or even corruption, against the incumbent inspector. We noted that most of the controversies of which there remains official evidence centered in the Chippewa or Black River areas. Charges against the effectiveness or honesty of the system were in no instance backed with evidence; in the one suit which reached the Supreme Court, on an inspector's bond, for alleged negligent performance of duty, the complainant lost for failure of proof. The Governor's correspondence files make plain that the office was often valued highly as a political prize or symbol, despite its modest financial return; as many correspondents wrote the Governor to praise or condemn the politics of an incumbent or an applicant as wrote to appraise his technical fitness or honesty; in an 1874 controversy a prominent operator reported to Governor Taylor that Chippewa Falls lumbermen "look at the appointment with considerable interest and make it a party issue." This partisan aspect colored the history of the lumber inspection system, and makes it difficult if not impossible at this distance to judge how valid may have been the charges of inefficient or even dishonest administration sometimes leveled at the official scalers. The available evidence does show, on the other hand, that there were real and material conflicts of interest on trade grounds within the industry; clearest was the alignment of loggers as sellers against millmen as buyers, but there is indication, too, that the smaller loggers sometimes distrusted nominees of larger loggers who operated partly as middlemen, buying logs from lesser woods operators. Writing in 1895 of a sharp contest for an appointment in the Sixteenth District, a Merrill lumberman advised Governor Upham in terms which caution us that there was some trade dissatisfaction with the system, and at the same time show the confusing crisscross of political and industry factors coloring contemporary judgment:

> You know [he assured the Governor] as much about lumber scalers as any of us, what they will do, and what they will try to do, and how they will be influenced, and it seems to be the prevailing idea among most of the lumbermen in this Valley that [First Applicant] . . . would do most anything that he was paid for doing. He has that reputation among the lumbermen . . . I am satisfied that Clifford [another lumberman] signed the petition for what good [First Applicant] . . . done him on a scale of logs, and the result of which is a lawsuit between Clifford and the party of whom he bought the logs . . . All of the mill-men here signed [the petition for Second Applicant] . . . except the Merrill Lumber Co. [which, he says he is informed, signed First's petition in ignorance that there was another choice], and I think that we have some rights to guard in the appointment of a lumber inspector. You take the mills of Merrill, eight of them, six at Wausau. We have to derive our supply of timber and logs from the 16th district to a very large extent, practically all of it, and of course we are interested in having a good man, that is all right, and one of our own political faith, appointed as lumber inspector.[80]

After due allowance for politics, and despite the fact that important amounts of logs and lumber were not handled through the public scalers, the net impression from sifting the pleas and counterpleas which came to the Governors over lumber-inspector appointments is that the industry felt that the office had substantial importance to it, and put high value on the goal of impartial measurement. The atmosphere of controversy surrounding the marketing operations carried on at Beef Slough by "the pool" (through the Beef Slough Manufacturing, Booming, Log-Driving & Transportation Company) highlights this conclusion in one instance. Though this company and the big firms associated with it were content to scale their own logs, they did not regard it as a matter of no concern who the district lumber inspector might be, and exerted their influence — sometimes without success — to place in the post a man of their appointment. A communication to Governor Jeremiah Rusk from J. G. Thorp, for example, announced acceptance by "the pool" of the continuance in office of a Ninth District inspector not of their choosing, in terms whose peremptory quality hardly fits the dignity of the Governor's office, but which also show the lively attention these men gave to the appointment:

> Confidential: I have seen and talked that matter over with Weyerheauser. He agrees with me that it is best to make no change at present and you can say to McKay [the incumbent inspector] that you will not. It is not of such importance

that a change should be made now. You, of course, cannot do it consistently without good reasons for it.

Log owners not in "the pool" valued highly an independent log inspector, to bulwark them in their dealings with the larger firms. A local newspaper editor urged that the Governor recognize that it was the independents "for whom the scaler works," and a Prairie du Chien lumberman wrote earnestly in favor of reappointment of the incumbent inspector, because "by his strait forward way of business he has secured us our rights in geting Logs due us from Beef Slough what we think never would have got by any other men . . ." One of the Governor's correspondents made clear the importance of an honest and independent official scaler, in the tug of interests between logger-sellers and millmen-buyers, when he replied to Governor Rusk's observations on the strong support given a particular candidate:

You speak of the petition as being a very strong one. It is indeed a strong one in one sense, viz: being signed by the entire manufacturing interests of the district. But therein lies its weakness. A lumber inspector should be a man that would deal fairly between the log sellers and buyers, one whom to be useful should have the confidence and respect of both parties. To allow either party to choose or dictate if you please the appointment would not be fair to the other. In this case, in the recommendation of Mr. Young exclusively by the log buyers, his appointment would be very unsatisfactory to the log sellers who generally get the worst of the deal in any event.

At other times, in other districts, independent lumber firms wrote to the Governor in terms which emphasized the weight they assigned to the ideal of impartiality and neutrality, as contributions which the office of lumber inspector could make to the trade. The matter was bound to be of acute interest because, as one lumberman emphasized, "the Inspector holds the pocket books of operators . . ." [81]

Around the statutory lumber-inspector system, the recording acts, and the reactions of the court and different segments of the industry, swirled currents of diverse policy ambition. The story of the recording acts is in the familiar nineteenth-century pattern of pull and haul of creditor and debtor and of secured as against unsecured creditor. It is characteristic of the contemporary emphasis on growth, that recording-act issues arose not from efforts to assure definiteness in titles simply to safeguard vested interests, but out of efforts to make maximum use of scarce working capital, to press the pace of production. Corollary to this trend, the court's decisions reflected the sharply felt capital scarcity of the period, when the judges subordinated recording-act protection of title regularity and of general creditors to legal encouragement of the venture of capital under security of mortgage or lien. Of course, to promote confidence in the security of market transactions was also a basic policy for economic growth; the court dramatized this value when it made timber-trespass legislation yield before the claims of the bona fide purchaser. Yet, even that darling of the market, the purchaser for value without notice of competing claims, was not allowed to prevail against unrecorded security interests in logs or lumber unless he could prove the further element of misleading conduct by the mortgagee. On the other hand, the creation, expansion, and substantial use of the lumber-inspector system bore witness how real was the demand that law provide a framework of regularity within which the market might grow. The particular crosscurrent here was produced by the development of large firms and concentration of assets; demand for the legal framework of regularity, it seems, was a phenomenon of a market in which many units operated, most of small or modest size; the larger a few operators became, the more they wished to standardize transactions on their own terms. In this, of course, the experience under the lumber-inspection procedure was part of a major policy issue unfolding in the latter nineteenth century as the market showed the capacity to develop forces that might destroy it.

BOOTSTRAP FINANCE: LOG LABOR LIENS

Labor accounted for a substantial part of the cost of getting timber out of the woods, through the mills, and to market. The labor was that of men who usually had little capital except their own skill and muscle. In the first half of the life span of Wisconsin lumber, most of the men who paid for the labor had little working capital with which to meet payrolls; logging and driving contractors must find credit enough to equip and feed their crews and teams through the winter and early spring, and the woods and river crews thus received part of their pay in kind, in food and rude shelter, as they worked. But the typical independent contractor could only meet his full wage bill out of the proceeds of his contracts and sales when these were completed by delivery of logs in the spring; his men must wait till then for their pay, and the contractor himself must trust to the credit of his contract buyer. In later years, insofar as large log buyers began to assure their supply by becoming timberland owners and loggers for their own account, their credit was generally sufficient in itself to assure payment of wages, or fulfillment of the percentage of contracts that they continued to make with independent loggers. But the small or medium-scale operator continued to figure in the industry throughout its span. His typically uncertain financial condition continuously posed serious problems, both of justice (how should men have reasonable assurance of pay at the end of months of hard and dangerous work?) and of economic function (where cash was scarce, how persuade men to lend their time, effort, and skill — literally to lend their "working capital" — to the industry, on promises of deferred payment?). The one stable, tangible resource in this pattern of shoestring finance was the logs themselves; logs were not paper promises which could become empty symbols when a June account book showed an excess of debits over credits; logs could not pack a carpetbag and fly the state overnight.

The common law had provided a limited answer to analogous problems, before the lumber industry presented its peculiar difficulties. Men's freedom of contract included scope to agree upon security interests in goods on behalf of those who contracted to do work on the goods; Wisconsin common law, thus, had no difficulty in recognizing and enforcing between the parties an agreement that one who cut and drove another's logs should be entitled to hold or take possession of and sell such part of the logs as necessary to pay for his services. If he asserted such a contract right against a third party, more difficult questions would arise; if he relinquished possession and then sought to retake logs, he would surely lose, as against an intervening bona fide purchaser for value without notice.

Where no contract was made to secure the laborer his pay out of the goods he worked on, the common law by its own force granted the laborer a limited security; by performing labor or service upon goods in his possession, he acquired the right to hold possession until he was paid, or upon suitable petition in a court of equity to have a decree for the sale of the goods and for payment of his dues out of the sale proceeds. The laborer's right so to hold and realize on the goods was a common law "lien." So far as it went, it offered him good security. But with characteristic care to preserve and secure the marketability of goods, the law limited its lien still more strictly than it did liens created by contract; the common law lien lasted only so long as the claimant continued in possession of the goods on which he had worked; if he voluntarily surrendered possession, his lien was gone, not only as against third parties but also as against the person who owed for the labor done on the goods.

There is no reported Wisconsin case enforcing a common law lien for work on logs, in the first generation of the industry. As we shall see, from as early a time as lumbering was a substantial factor in the economy the legislature provided the log laborer with statutory liens more generous in scope than that at common law; this in itself may account for the early absence of common law cases. When attention was finally given to the matter, Wisconsin policy treated the common law lien as having always existed and as continuing to exist parallel with statutory protections for the timber worker. Thus the Revisers took special pains to add to the Revised Statutes in 1878 a new section 3347, which stamped with broad legislative approval the whole array of such common law security devices, providing that "every person having a lien . . . existing in favor of any bailee for hire . . . or otherwise, by the common law" might have a summary remedy by public auction to enforce his claim, if the affected property were not worth more than $100; if its value were over that figure, "then such lien may be enforced against the same by action in any court having jurisdiction." Later special log-lien statutes made no reference to this sweeping general expression of legislative favor toward the common law lien. The Wisconsin Supreme Court was thus the readier to invoke the canon of construction that common law should not be held changed by mere implication from statutes setting up rights or remedies other than provided by common law; hence it ruled that one who performed labor or service upon logs in his possession enjoyed concurrent common law and statutory liens to secure his payment. The court clearly recognized that the common law lien existed to secure payment for rafting, driving, or sawing; perhaps because of doubt that the typical woods worker could be deemed ever in possession, the court indicated at one point that the lien would not lie for labor in cutting the logs, but later statements apparently rejected this view and treated the common law lien as embracing the whole range of logging operations.[82]

On forty-nine occasions, from 1849 to 1935, the Wisconsin legislature declared policy concerning the establishment or enforcement of liens for labor or supplies furnished in lumbering — from the cutting in the woods through the sawing at the mill. In some aspects the qualifications, limitations, amendments, repeals, supersessions, and plain mistakes in this flow of legislation converted essentially simple policy into as technical and involved a body of law as any outside the rarified doctrine of future interests in real estate. Only a part of this complexity derived from conflict over basic values; most of it came from the excessively localistic bias of the nineteenth-century legislature and from its characteristic lack of even elementary machinery for developing broad and systematic definition of what it wanted to do. The legislative history of log liens provides an outstanding instance of the drag of the law's own institutional inertia upon its response to legitimate, and mainly uncontested, needs of the economy. In another aspect, this tangled story shows the Wisconsin Supreme Court in a relatively more favorable light. An appellate court has its energies focused upon specific problems, posed in the context of records of evidence and briefs of arguments assembled in the perspective of general doctrines of public policy, under procedures designed to define and weigh issues, and presented with some opportunity for deliberation. An appellate court thus enjoys a creative opportunity to add coherence and direction to the products of the bustle and pull and haul of legislative sessions. In this instance the Wisconsin court realized its opportunity in a measure that added much to the workable quality of an unwieldy body of statute law. Early sensing and defining the main, unifying trend of legislative policy, the judges steadily thereafter made good the principal theme in detailed application.[83]

With remarkable steadfastness and consistency, an otherwise tangled series of statutes expressed, from 1849 on, the key value judgment that it was in the public interest to confer a broader lien than the common law provided for one who invested his "labor or service" in work upon logs or timber.

The unchanged core of this policy through all the years was to grant a lien, though the lien claimant lacked possession of the liened property. With some minor backtracking, and much needlessly confusing detail, the statutes built out from this center a lien the scope of which was consistently expanded in favor of these claimants. The objective pattern of the successive statutes early and amply justified the parallel doctrine of the Supreme Court: This was law which "was enacted in the interests of labor, and a sound public policy requires that it be liberally construed." These "are deemed remedial statutes, and are liberally construed, in order to bring within their operation all cases intended by the legislature to be within the intended benefits." At first appearance a tally of the decisions in seventy-two statutory log-lien cases carried to the Supreme Court does not yield as clear-cut a pattern as the court's announced doctrine; in thirty-five of these cases the party opposing a lien claim finally prevailed. But a mechanical count of appellate court results is misleading. The borderline issues of doctrine, and the more hardly fought cases, naturally came to the higher court. The thirty-five instances in which the contestant finally won, moreover, included six cases turning primarily on the evidence, and eleven on routine or narrowly technical points of procedure; of the eighteen cases involving primarily questions of the substantive scope of the lien, no more than six presented points of central significance. More indicative of how the general weight of the lien statutes was felt is the fact that fifty of the seventy-two appeals were by lien contestants, twenty-two by lien claimants. And when one turns from simple counting of decisions, to summarize the substantive lien doctrine represented in the legislation and the cases, the prevailing note of liberality toward the claimant becomes clear.[84]

The dominant pattern of expansion of the "labor and services" lien emerged through the interplay of statute and decisions defining the activity creating the lien, the claimants, the persons subject to the lien, and the procedures for enforcement. If the legislature's steps were halting, they were yet in the same direction.

Activity Creating the Lien

Definition evolved here with reference to the kind of work, and the place where the work was done. The first statute (Revised Statutes, 1849, Chapter 120, section 12, continued by Revised Statutes, 1858, Chapter 153, section 12) gave a lien simply for "manual labor" on land, timber, or lumber, without more specification. But General Acts, 1860, Chapter 215, which fixed the model for all subsequent, broader log-lien legislation, granted a lien for "any labor or services." *Hogan v. Cushing* (1880) realized the broad import of this recognition of "services" when it held that a lien existed in favor of a contractor who himself did little of the manual work of getting logs into the Chippewa River but relied mainly on the work of teams and servants under his supervision. Moreover, "labor" should be construed with broad realism, in the light both of the legislative favor toward workers and the way in which the labor force was ordinarily recruited; hence a farmer who in the slack winter season hired out to a logger the labor of himself with his team might have his lien for the whole sum due, and would not be limited to a claim for the value of his personal work alone. The plain indication of the statute, however, was that there must be some personal contribution by the lien claimant; thus there seems nothing unsympathetic to the legislative policy in the court's rulings that no lien existed for the mere rent of an ox or horse, involving no labor of the animal's owner. The pattern statute — Chapter 215, General Acts of 1860 — specified that the labor or service creating a lien should be rendered "in cutting, falling, hauling, driving, running, rafting, booming, cribbing or towing any logs or timber." So much detail might ordinarily be taken to imply an intent that the legislation should not be extended by implication to any other than the named activities. But with characteristic liberality the court inferred an underlying intention to give a lien for any work closely incident to such logging operations, and so ruled that the work of a contract scaler or of a camp cook, and "clean up" operations at the end of a drive, were covered by the reasonable implication of the law. However, even the most flexible judicial approach must acknowledge ultimate limits set by the legislature's words; plainly the legislative formula as fixed in 1860 did not cover sawmill operations, and thus when the lien was extended to cover labor or service in converting logs into lumber, it was by action of the legislature, in 1867. Likewise it was reasonable, in light of the coexistence of the general mechanics-lien acts, and of the focus of the log-lien statutes on activity in the flow of production, to rule that the log-lien act did not cover work in the construction or out-of-season repair of permanent installations like a mill or railroad, though it would cover a millwright's labor in repairs incident to current sawing operations.[85]

As with its log-mark and measurement statutes, the legislature developed the "labor or services" lien out of a series of laws limited in application to particular districts of the pinelands. True, section 12 of Chapter 120, Revised Statutes, 1849 (and its continuation in Revised Statutes, 1858, Chapter 153, section 12, and in section 3341, Revised Statutes, 1878) applied anywhere in the state; but it was limited to manual labor and to cases where some privity of contract might be traced between claimant and log owner. The broader lien later embodied in section 3329 was first created by General Acts, 1860, Chapter 215, for a group of northeastern counties; in 1861 it was applied in Chippewa County, and in 1862 in a group of northwestern counties; 1865 and 1866 acts added more counties, mostly in the north-central pinelands; newly created northwestern counties were brought into the scheme in 1876, 1877, and 1879, with some older counties included in 1878 and 1879. Not until Laws, 1881, Chapter 330, did the legislature finally make the basic lien apply regarding labor or services rendered "in any of the counties of this state." Even as it accomplished this simplification of the basic lien, the legislature introduced a new geographical complexity by launching an additional series of area-limited liens for work done on specified secondary forest products (e.g., cordwood, bark), which it continued to elaborate until it reduced them, likewise, to a simple "all counties" coverage in section 3329 of Statutes, 1898. There was much less doctrinal variation among the basic lien statutes in the years of limited-area coverage than occurred in the growth of the log-mark and measurement laws. But the very fact that the legislature was so content with the model set by General Acts, 1860, Chapter 215, underlined the needless complexity and costs of extending it piecemeal. This fragmentary approach also hampered the court's constructive service in adding to the life and direction of the central policy of the liens; the sharply marked area limits of the statutes could not be interpreted away, and the necessary obedience to them inevitably introduced into the law of lumber liens an element of rigid technicality counter to the prevailing liberal temper of the substance of the legislation and decisions. We shall hereafter note that geographical limits on creation of suppliers' liens seem to reflect genuine conflict of interests over extension of substantive doctrine. But the record shows no evidence that opposition to the substance of the basic labor lien accounted for the area-by-area approach to its extension. Thus, after the court in 1876 felt compelled to rule that the lien existing by statute in Chippewa County was inoperative in territory shifted from Chippewa to the new Taylor County, the next succeeding session of the legislature without fuss enacted the lien for the new county. Again, we seem to confront simply the force of institutional inertia; the nineteenth-century Wisconsin legislature was not accustomed or equipped to operate other than by specific response to specific demands in most fields of policy. The court's work made vivid the paradoxical character of this growth of log-lien legislation. With good sense the court pierced the apparent diversity of area-limited lien acts to recognize the common values which they represented; it construed them as statutes *in pari materia,* freely borrowing its interpretations of one act to apply under another; it held that, despite their geographical limits, the particular statutes were not "local bills" within the meaning of the restrictions which the Wisconsin constitution laid upon such measures, but were rather "general laws" of the state.[86]

The Claimants

Log-lien claimants were likely to be men pressed for cash and anxious quickly to realize their claims. Their anxiety would be the greater if there were serious contest of their liens; the average worker or small contractor feared the mysteries and the time and money cost of lawsuits, even though the legislature adopted procedural devices to aid the small claimant. On all counts achievement of the practical benefits of the liens would be helped if lien claims were readily assignable, so that the worker might sell to a financial middleman, or might consolidate his claim with others for a lawsuit.

This was a matter in which the legislature definitely took the creative initiative. Common law tradition was not hospitable to the ready transfer of lien rights. The general rule was that a lien was a personal right, the beneficial interest in which might not be transferred at all. Especially where a lien was entirely the creation of statute, the courts required clear provision by statute to make it assignable. The legislature itself moved cautiously in this field, probably because this was an aspect of liens on which lien debtors could plausibly press their own claims for fair treatment. Beginning with an amendment to the general mechanics-lien act in 1859 (applicable to enforcement of manual labor liens under section 12, Chapter 153, Revised Statutes, 1858), and the broader model log-lien statute of 1860 (Chapter 215), the legislature allowed one of several lien claimants to receive assignments of other lien claims against the same property, and thereupon to enforce all of such claims as his own. Then Revised Statutes, 1878, section 3316, declared that all lien claims under Chapter 143 — the Chapter which included log liens — should be generally assignable. The Revisers explained,

> It is often a hardship on a mechanic that he cannot raise any money but by the result of an action, and there is no reason why such demand should not be assignable like others.

By continuing the older provision allowing assignments to a lien claimant of other claims on the same property, section 3336 might have cast some shadow of doubt whether the sweeping terms of section 3316 did, after all, include log liens. But no contest was ever made on the point in the Supreme Court. And when the effect of a complex new statute in 1891 was, perhaps inadvertently, to restrict the general assignability of log liens, Statutes, 1898, section 3333, promptly settled the point beyond doubt, declaring that "any claim" for a lien under the general log-lien provision (section 3329) "shall be assignable" without specified limit.[87]

Persons Subject to Log Liens

Under any of the various log-lien acts it was elementary that the owner of the beneficial interest in lienable property was subject to the lien on his goods where he had himself contracted for the labor or services which created the claim. It was likewise elementary that the person who contracted for the work was personally liable for the money due the worker, apart from the matter of the worker's security in the lienable goods.[88]

Because laborers or small contractors so often had their direct contract relations with men whose credit was undependable, the practical benefit of a log lien was largely determined by the extent to which it might be asserted against persons with whom the claimant had no contract. Thus the earliest statutory log lien, as established in the Revised Statutes of 1849 and 1858, was relatively narrow, granted only when the manual labor had been performed "for or on account of the owner, agent, or assignee" of the lienable goods. We have seen that section 3341 of the Revised Statutes, 1878, provided a residual, or catch-all, lien for manual labor on logs, timber, or lumber not within the coverage of the more specifically defined lien under section 3329. This residual lien was limited by language similar to that under the 1849 and 1858 provisions, and stayed so until Statutes, 1898, section 3329, amalgamated all the log liens under a formula which did not require a showing that the work was done "for or on account of" the owner or his representative. The court recognized that these earlier limitations were not accidental, but were intended "to protect the owner by restricting the lien." "The statute seems to imply a privity or some contract relation between the laborer and the owner of the property, where the lien is given . . ." Thus, if the owner of goods otherwise subject to lien made an agreement with an independent contractor for operations upon the goods, the contractor's employees might not establish a lien for their manual labor under these provisions; the intervention of the independent contractor insulated them from the property owner, so that they could not be deemed working "for" him or "on [his] account." This outcome betokened no lack of judicial sympathy toward these provisions, which, we shall see, in other respects received a liberal construction in favor of lien claimants; the court merely applied familiar privity doctrine implied by the terms which the legislature had chosen.[89]

Against the background of the limited lien under the 1849 and 1858 Revised Statutes we see more sharply how far the legislature advanced in liberality to claimants by the new model statute which it passed in 1860. Chapter 215 of the General Acts, 1860, granted a lien based simply on performance of labor or service on logs or timber. With full readiness to take the statute's unqualified language at its face value, the court's rulings made clear that the legislature had now favored the lien claimant with a "real" security, a right *in rem,* so far at least as concerned his capacity now to obtain what was due him out of the lienable goods themselves. Thus *Munger v. Lenroot* (1873) held that under such a statute claimants might assert a lien for their labor against the owner of logs who had agreed with an independent contractor for their cutting, though the laborers had no contract with the log owner, and though the statute had no provision for giving notice of the prosecution of the claim to anyone other than the one personally liable for the debt. Moreover, since the lien rested simply on the fact of the work done, it stood against the log owner, though the contractor had not yet performed by completing delivery to the owner. That the new, broader form of statute made the logs themselves "debtors" for the work done on them, though there was no contract between the laborer and the owner, led to other decisions favorable to the claimant. His suit to establish the whole amount due from two defendants as a lien on the logs was not demurrable for misjoinder of causes of action, though the defendants held different interests in the lienable goods. The work done created the lien, regardless of defendants' diverse interests; the lien action, though it "is *sui generis* . . . bears some analogy to a libel *in rem,* and proceedings thereon in admiralty"; in any case, to allow joinder of all interested defendants in the one lien proceeding promoted the remedial purposes of the statute, for the result "certainly saves a multiplicity of suits," and "further saves the common property from being consumed by costs in several suits, which seem unnecessary for the protection of any right." Seizure under the lien was held so far a proceeding *in rem* as to be the basis of subsequent service by publication upon a nonresident. And because it was the work done that created the lien, if the work was done in the life of the one contracting for it the lien might be prosecuted against his administrator, though the petition for the lien was not filed in his lifetime. Likewise, the independent "liability" of the logs meant that the log owner might not complain against a partial assignment of a lien claim, where consented to by the one personally liable for the work debt; by his work done, the claimant acquired a kind of title in the logs, of which he might dispose. Again, any portion of a lot of logs might be seized under the lien for the whole sum due for any work upon the lot, without need to show that the particular logs seized were those whereon the claimant had done his labor. The basis of the lien was the whole job done:

> If the statutes giving a lien for labor on logs and timber were so strictly construed as to limit the lien upon each part of the property to the work done on such part, it would defeat the entire scheme, designed to protect a favored class of laborers, for reasons sufficient to satisfy the legislature of its wisdom. No such construction was intended, as is abundantly evidenced by the history of legislation on the subject, all tending in the one direction of securing, so far as practicable, to laborers on logs and timber security against any probability of losing their wages.

So, too, the claimant was entitled to pursue his lien without first exhausting his remedy by personal judgment and execution against the one who contracted for the work; it would not consist with the legislative policy to treat the liened property simply as a secondary security, though if the property owner discharged the lien he would have a surety's rights for reimbursement against the one personally liable for the work debt.[90]

The shape of the log-lien cases brought to the Wisconsin Supreme Court graphically demonstrated the practical importance of the security in the tangible property. In not one of seventy-two appeals involving statutory log liens was the burden of battle borne by one whose sole interest was that of contractor for labor, services, or supplies. This cannot be explained on the ground that the contractor had no possible stake in a lien proceeding; save in two instances where it proved impossible to obtain personal jurisdiction of the contractor, these log-lien suits all involved entry of a judgment *in personam* against the one who contracted for the labor or supplies, as well as a judgment fixing a lien upon the property. More specifically, in each of the twenty-seven cases in which labor or supplies were contracted for by one who had no ownership interest in the liened goods, and against whom personal jurisdiction was obtained, the only contest on appeal was by others whose interest was solely that of owners of the liened logs or lumber. Even in the fifteen instances in which the defendant owned the liened goods and had also himself contracted for the labor or supplies, in but four cases did the issue on appeal focus on the basis of personal liability. And in twenty-eight of the seventy cases where personal jurisdiction was obtained against the one who contracted the debt, the substantial contest on appeal was, nonetheless, by a transferee of the liened goods.

Because logs so often were the security on which men obtained advances to finance woods operations, and because they were so freely bought and sold, and indeed passed often as a substitute for cash in the pinelands, the practical worth of the log-labor lien depended much on the extent to which the claimant might assert priority over persons who acquired interests as secured or unsecured creditors, or as purchasers before or after the lien accrued. For example, as has just been noted, in twenty-eight of the seventy-two statutory log lien cases brought to the Wisconsin Supreme Court, the real contest was between the lien claimant and a purchaser or other transferee of the liened goods. In a fashion similar to the development of doctrine on security interests in standing timber, the court felt moved in the log-lien cases to strike a balance between the financing needs of the business and the demand for security of market transactions. But, in striking contrast to its silent acquiescence in the judicial handiwork in the broader field, the legislature dramatized its continuing favor to the "labor or services" lien claimants by overruling the court and assigning these lien holders an unqualified preference, even over that favorite of the market place, the bona fide purchaser for value without notice.

The first log-labor lien, under the Revised Statutes of 1849 and 1858, was within the general mechanics-lien law. As such it was declared by the statute to be a claim "before any other lien which originated subsequent to" the doing of the work which created it. The court early extended the scope of this preference, to give the log-labor lien priority also over security interests existing before the labor lien came into being. *Paine v. Gill* (1861) thus held that a labor lien under section 12, Revised Statutes, 1858, Chapter 153, prevailed over a lien created previously by contract between the log owner and the contract vendee of the logs, to secure the latter for advances by him to finance the logging operation. The court had "no doubt" that this was the legislature's intent; the statutory lien simply extended and made more effective the equitable policy behind the common law's grant of a lien to one whose labor enhanced the value of property. A year later, reaffirming this principle, the court explained in more detail. The question was

whether, where a party has contracted with another for the delivery of a lot of logs at a particular place, and for a prior lien on those logs as a security for a debt, such lien, so contracted for, cuts off the lien which the law gives the workmen who perform the very work on the logs necessary to a fulfillment of the first contract.

The court was no less convinced than it was on the first occasion that the legislature intended the log labor lien to prevail:

. . . it was the intention of the statute to give such workmen an absolute lien, where they were employed to do the work by anyone having competent authority, as against everybody, upon the principle that their labor enhanced the value for the benefit of everybody who had any interest in the property.

The broader lien created by General Acts, 1860, Chapter 215, for any "labor or services" on logs or timber was by that act declared to "take precedence of all other claims" on the liened property. Under this type of declaration the court found it too clear to require discussion that a general creditor who attached logs for a debt not growing out of labor on them was subject to a lien thereafter filed for labor done on the attached goods. Likewise, the log owner's trustee in bankruptcy might not set aside assignments of log-labor liens as preferences, because the laborer's interest under his lien was "an interest *in praesenti*" in the liened property, "acquired by performance of the labor"—that is, for value—and hence no violation of the bankruptcy act.[91]

But these late-nineteenth-century judges—working in a society organized largely about the market, and in a legal order in which no body of doctrine was more central than that of contract—placed a high value on maintaining the ready marketability of titles. In *Smith v. Shell Lake Lumber Co.* (1887), over the vigorous dissent of Mr. Justice Taylor, the court held that one who bought shingles as a bona fide purchaser after their manufacture, before expiration of the allowed statutory time for filing a claim for a labor lien on them, but also before the claim had in fact been filed, should take them free of the lien. The majority found ambiguity, where the dissent found only plain meaning, in the statute's stipulations that the laborer's claim should "remain a lien" unless it were not filed within proper time, and that it should "take precedence of all other claims or liens." Passing other arguments as convincing, but secondary, the majority rested its ruling firmly on a presumption in favor of maintaining titles obtained by good faith purchases in the market. The statutory terms might be interpreted to prefer the laborer's claim. Yet:

If this peculiar language of our statute can have force without violating the great principle and clear public policy of the law that protects *bona fide* purchasers in the usual course of trade without notice of the lien, then such should be its construction.

"The gravity of such an innovation" as would subordinate the bona fide purchaser must be weighed in the light of the market's key importance in this society:

The paramount importance and incalculable value of personal property in these modern times makes its ready and easy transfer from hand to hand, and the protection of *bona fide* purchasers thereof, absolutely essential to our modern systems of trade and commerce. Secret trusts, liens, and incumbrances, and unknown and concealed claims and interests, in and upon personal property, and especially that kind of personal property that enters so largely into the general commerce of a country, would, if enforced by law, work the greatest injustice and be utterly destructive of the greatest financial interest that any country can have . . . Logs, timber, lumber, including boards, shingle, and lath, constitute most valuable and important articles of our trade and commerce, and are readily and necessarily and almost constantly being transferred and sold in wholesale and retail in open market, and carried and scattered over vast distances by land and water. This interest and trade are too vast and important to be clogged, impeded, and incumbered by secret liens, following them into all the distant markets of the land, to be enforced in violation of such a cardinal principle to facilitate and protect the sale of personal property . . . The laborer's lien should, of course, be protected as far as possible without inflicting too much loss and damage upon interests paramount by reason of their magnitude and their general bearing upon the trade and commerce of the country.

The majority decision stands out the more sharply in its affirmation of market values because the dissent pressed with vigorous competence the argument from the court's own well-established rule of construing the lien statute so as to promote its equitable purpose. The lien act itself expressed deep values of the legal order — against unjust enrichment, and "to secure the laborer pay for his wages out of the property which his labor has created or enhanced in value, which . . . is in accordance with natural justice." The majority made no answer to Justice Taylor's warning that its decision facilitated evasion of the lien, without extending regular market processes any protection that they really needed:

. . . it is especially likely to defeat the object of the statute in all cases where the laborer most needs its protection, viz., in those cases where his employer is impecunious, dishonest, or neglectful of his business. Such employers are the ones most likely to make sales of the materials produced by the labor of their employees, without regard to the question of whether their wages are paid or not. When the laborer has an honest or responsible employer, there is generally little need of the protection of the statute.[92]

The legislature lost little time in declaring in favor of the dissenting Justice's appraisal of the competing values. In 1889, at its next session after *Shell Lake Lumber Co.,* it provided that any person who intended to file a labor lien claim might file a preliminary notice of intent with the clerk of the circuit court at any time after he contracted to do the work, which "notice when so filed shall be in all things a notice to all persons . . . who shall become purchasers of any such logs, timber, lumber . . ." or other derivative products, and that once the notice of intention was filed,

> no person purchasing such logs or timber or otherwise acquiring any interest therein, before the time for filing such petition or statement [for a lien] has expired, shall be considered a bona fide purchaser as against the rights of any such lien claimant . . .

Laws, 1891, Chapter 139, repealed these portions of the 1889 legislation, and declared the priority of the labor lien claimant in absolute terms, free of the provision for filing a notice of intention. Until the time for filing the petition for the lien expired, the labor lien should

> take precedence of all other claims, liens or incumbrances . . . [on the liened property] or sales thereof, whether such claims, liens, incumbrances or sales are made, created or accrue, before or after the time of doing such work, labor and services.

This language was too plain to be denied, and the court applied it without further objection. However, the judges continued to show the importance they assigned to the ready marketability of goods, by insisting on strict enforcement of the statutory time limits for filing lien claims.[93]

Log-Labor Lien Procedure

Given the limited means of the typical beneficiaries, a direct and simple procedure was as important as liberal substance, if the objectives of the log-labor lien were to be achieved. So far as its deliberate efforts went, the legislature quite consistently recognized and sought to act according to this proposition, and in general the log-lien procedure expressed the high favor accorded the claimants. The significant qualification upon this pattern was created not by legislative intent, but by the legislative ineptness which needlessly complicated the lien law through crude localism, haphazard variation of detail, and ill-fitted amendments, re-enactments, and repeals.

In main outline the statutory lien procedure adhered to canons of simplicity and economy. Save for the short-lived 1889–1891 provision for filing a preliminary notice of intent against intervening purchasers, there was no formal step needed to create the lien, which came into being simply when work was done. True, a claim must eventually be filed if the lien were to "remain." But formality was at a minimum here—"Such statement or petition shall briefly set forth and state the nature of such claim or demand, the amount due, and a description of the logs or timber upon or against which the lien is claimed"—and there was generous leeway of time to file for the bulk of claims originating in woods and driving operations, which might be put on record anytime up to June 1 following the winter-spring work season. Under the first log-lien act, suit might be brought only in county or circuit court. But this remedy "was not found to be very efficacious or complete in all cases." Hence, General Acts, 1860, Chapter 215, set a new pattern, "to give a more adequate and prompt remedy," by allowing suit before justices of the peace for claims within their ordinary jurisdictional amount. Though over the years the legislature somewhat tightened requirements as to parties, it took general care not to impose burdensome forms upon untutored plaintiffs. With a short-lived exception, the one personally liable for the work debt was the only person who was required to be made defendant. Again subject to this short-lived exception, there was not even provision for notice of the proceeding to a log owner who had not contracted for the work but whose property was subject to the lien. The most favor the legislature showed toward lien contestants was to grant them permission to intervene: first, only to one asserting general ownership, if he applied before entry of final judgment; later, to anyone asserting any beneficial interest, at any time before or after judgment and before an actual sale on execution. The model 1860 act relieved the lien claimant of the unvarying requirement under the general attachment statute for posting a bond with surety for costs and damages; the lien contestant might obtain a court order requiring such bond only on a special affidavit showing a good and valid defense. A provision first enacted only for the Chippewa River valley, but later generalized, permitted constructive levy by the sheriff in place of actual seizure. Though claimants must, of course, plead and prove that they fell within the statute's benefits, the court was not disposed to hold them to very formal requirements. Various claimants might attach the same logs; the court would determine their relative priorities, according to time of filing. Only the elements plainly required by the log-lien acts themselves must be pleaded thereunder. A claimant need not prove that the particular logs he caused to be attached were those on which he had labored, if he showed that they derived from the general operation of which he was a part; a more strict requirement of proof would in practice defeat most woods workers' claims.[94]

Only where procedure bore close relation to maintenance or impairment of the ready marketing of forest products did the court show disposition to go counter to the legislative policy of flexible and liberal procedure. The otherwise generous model act of 1860 declared straitly that one who did labor or service on "any logs or timber" should have his lien "thereon." The earlier Chapter 153, section 12, of the Revised Statutes of 1858 had allowed its more limited lien to be pursued against "any timber or lumber," and in view of the generally liberal character of Chapter 215, General Acts, 1860, it seems likely that if the letter of its remedy was less generous, this was the result of poor draftsmanship rather than deliberation. Nevertheless, on frank-spoken grounds of market policy, the court chose to apply the legislation within its literal limits, and to hold that a lien for labor on logs might not be pursued by attachment of the manufactured product of the logs. *Babka v. Eldred* (1879) thus held that the similarly narrow language of Laws, 1862, Chapter 154, did not allow enforcement of a lien on lath or lumber:

While the property remains in the form of logs or timber, it can easily be traced, described and identified by reference to location and marks; but after it is cut or sawed into lumber, it becomes more portable, more liable to be scattered, and more difficult to describe or identify. It also then becomes more peculiarly an article of commerce, and more liable to pass into the hands of innocent purchasers. It may well be the legislature were of the opinion that while the property remained in the form of logs or timber, comparatively little injustice would be done the owner by the severe remedy of the special act of 1862; but that after the property became *lumber,* the owner should not be subjected to a law so severe in its operation; hence the omission of the word lumber from the body of the act of 1862.

Inept drafting surely explained why the legislation of 1879 extending the labor-and-services lien to cover "sawing or manufacturing [logs or timber] . . . into lumber" yet declared the lien simply "upon such logs or timber." *Gross v. Eiden* (1881) acknowledged that this language was "strangely confined," and that "of course it is an absurdity to say that the laborer shall have a lien upon logs and timber for work done in manufacturing them into lumber." Nonetheless, the court there again refused to find a basis in the underlying policy of the statute to permit enforcing the lien against the manufactured product: "We cannot by construction cure what appears to be an omission or defect in the statute." As it had done in the matter of the bona fide purchaser, the legislature—finally spurred to clear speaking—overruled the judges' preference for marketability; Laws, 1882, Chapter 319, declared that the statutory lien might in all instances be enforced against "lumber." Laws, 1889, Chapter 413, made the point still more explicit when it declared that the labor lien should survive any change in the liened property by manufacture, and that the claimant should "have his right of a lien upon the manufactured product the same as though the labor had been performed directly upon the manufactured article." [95]

Save for their differences over the relative favor to be shown lien claimants as against purchasers in the market, legislature and court both thus developed the main aspects of the "labor or services" log lien without serious conflict, and with favor toward lien holders. However, there was one issue of substantive policy on which legislation moved with such marked and abrupt changes of course as to evidence real value conflicts. This matter was the provision of a comparable lien for persons who furnished supplies used in logging and driving operations. In 1867 the legislature granted a lien to anyone who furnished "any supplies" used in such operations in several northwestern and extreme northeastern counties, with preference along with labor liens over all other claims on the liened property; in 1870 it narrowed this same lien to cover only "feed used for teams, and the food necessarily used in camp to support the men, and no other thing," at the same time extending its coverage to Waupaca County; in 1872 it removed the Wolf River area from the scope of this amended supplies lien, while in 1874 it brought Outagamie County within its coverage; in 1874 it created a new broad supplies lien for the Wolf River area and the counties of Shawano, Waupaca, Outagamie, Winnebago, and Fond du Lac (thus withdrawing the first three named counties from inclusion in the previous narrow lien), embracing not only "all . . . supplies, rafting materials or other materials, or articles . . . usually used" in woods and river work, but also any "clothing, groceries, provisions and ordinary articles used by a laboring man or his family, and which may be furnished by an employer to an employe, or to employe's family" while the employee was so engaged in lumbering, where furnished to pay for labor or services, not exceeding the amount due for wages — but not to include "horses, cattle, teams, bobs, or bob-sleds, sleighs, sleds, wagons, harnesses, beds, bedding or bed clothing" — declaring further that this supplies lien should be subordinate to any lien for labor or services, but prior to other claims; in 1876 the legislature brought Taylor and Lincoln counties under the earlier, narrow supplies lien; also in 1876 it gave a lien in Chippewa County "for hauling supplies for men or teams engaged in" lumbering, but said nothing of a supplies lien; in 1877 it declared the Chippewa County lien applicable also in Taylor and Lincoln counties, without indicating whether this was in addition to, or repealed, the 1876 act, which had put these counties under the first type of supply lien.

This absurdly complicated melange was all woven into sections 3329 and 3330 of the Revised Statutes of 1878, with some minor changes. Soon, however, the history unrolled into further complication. In 1879 the legislature extended the broader supplies lien to Lincoln, Marathon, Portage, and Wood counties, and in another act apparently extended the peculiarly limited provision for Chippewa County to the newly-created Price County; in 1881 in a drastic reversal of direction it swept the board clean, repealing all existing supplies liens, broad or narrow; in 1885 it created a lien for "any person furnishing any supplies" in woods, river, or sawmill operations on logs or timber, or in woods operations on an enumerated list of secondary forest products, such as railroad ties, cordwood, and telegraph poles, in a number of northeastern and north-central counties, but omitting the whole northwest and the area tributary to Lake Winnebago; in 1887 it extended the new

question whether wide use of the device was not incompatible with maintaining a healthy, democratic dispersion of power in the community. This debate focused on banks, and to a less degree on insurance companies and internal improvements ventures. Fear of the banks' power over the lives of ordinary people provoked and fanned the battles over the corporation provisions in the Wisconsin constitution. Neither in the years 1846–1848, nor in later years when the Wisconsin timber industry flourished, did anyone raise any substantial issue over use of the corporate form in any other than a business which common opinion recognized as of public-utility character. The only aspects of lumbering which contemporary opinion so treated were the transport and power use of streams, and, later, the extension of the railroad network. Significantly, these were the only areas of the timber industry in which there was any dispute over legal franchises or forms of operation. Significantly, too, even here dispute was over undue lobby pressures, or omission of regulation, rather than over the grant of corporate power as such. This was so even when, as with the Mississippi River Logging Company, men were finally stirred to uneasiness over the extent of market control achieved through certain corporate combinations; such opposition as appeared on this score — never much, in total range or effect — rested not on the grant of corporate status, but on charges of excess or abuse of particular economic power.[99]

The unquestioning acceptance of the corporation in the maturity of Wisconsin lumbering attested that the constitution makers of 1848 correctly gauged the community's readiness for the device when they gave the legislature broad authority to provide for non-banking charters. In 1858 Governor Randall made the pattern of values more explicit when he explained the only veto ever exercised against a charter for a general lumbering business. The governor disapproved a special charter for the Oconto Lumbering Company, because there was no showing that the objects of the enterprise could not as well be achieved by a charter under the general incorporation procedure provided by statute. Though he noted that the bill before him "creates a powerful corporation," he took pains to emphasize:

> There is no objection to the character of business intended to be done under the charter, or to the extent of the business which the corporators may desire or design to do, or to the gentlemen who are named as corporators.

Rather, he indicated, sound policy should support the kind of venture proposed, since "the business, while it is of a private character, may be considered incidentally a public benefit." [100]

Wisconsin policy was thus hospitable to the use of the corporation by lumbermen from the earliest years when their industry took on substance. Within the flexible range allowed by the Wisconsin constitution this was policy given shape almost entirely by the interplay of business demand and legislative response. A relatively small number of questions came to the state Supreme Court under charters of ordinary lumbering firms. These matters were invariably disposed of routinely within the frame of general corporation law doctrine, without definition of policy peculiar to use of the corporation in the timber industry.

Revised Statutes, 1849, Chapter 51, made a general incorporation procedure available to "any three or more persons desirous of forming a corporation for the purpose of carrying on any kind of manufacturing . . . lumbering . . . [or] mechanical . . . business." Still more broadly, Chapter 73 of the Revised Statutes of 1858 provided a general incorporation procedure for any three or more persons "for the purpose of engaging in and carrying on any kind of manufacturing, mechanical, mining, or quarrying business, or any other lawful business . . ." Under the broad terms of these laws men organized not only corporations for logging and milling timber, but also for the waterway improvements and activities necessary to move logs to mill and market. Moreover, starting with Acts, 1850, Chapter 257 (incorporating the Upper Wisconsin Navigation Improvement Company), and until the practice was ended by constitutional ban in 1872, the legislature repeatedly showed its willingness to issue special charters for lumber-industry ventures, and to do so, as the legislative journals indicate, generally upon request and substantially on the terms proposed by the petitioners.[101]

However, the law's readiness did not of itself produce a considerable use of the corporation. The chronology suggests how much the stage of economic development and the character of enterprisers' plans and ambitions determined the actual role of the corporation. Of eighty special charters substantially involving the industry, enacted in the years from 1848 to 1872, thirty-four fell within the fifteen years from 1850–1864, and forty-six in the seven years from 1865 through 1871. Only eight corporations were organized for clearly defined lumber-industry operations under the general incorporation acts available in this same period; a few of the forty-two manufacturing companies formed under the general laws in this same time declared the manufacture of lumber to be one among various enumerated purposes of their operation. On the other hand, in 1896, of the 168 Wisconsin lumber firms of sufficient consequence to be included in the first organizational data compiled for the industry by the state Bureau of Labor and Industrial Statistics, one hundred were corporations; though comparative asset figures are lacking, the roster of contemporary firms organized as corporations indicates that by the 1890's the great bulk of investment in the industry was in corporate form.[102]

Further evidence that the initiative lay in the course of business (rather than legal) development, appears in the markedly different timing in resort to the corporate form for transport ventures on the one hand, and general lumbering or milling enterprises on the other. Thus, of fifty-two special charters enacted from 1850 to 1871 for river improvement, boom, or log-driving companies, twenty-five were passed in the years 1850–1864, and twenty-seven in the span 1865–1871; in contrast, of twenty-eight lumbering firm charters specially enacted over the same general period, only nine were passed in the years 1850–1864, with nineteen put on the books from 1865 to 1871. Throughout the industry's history transport operations on any sizable scale were almost invariably cast in corporate form. The most striking change in the industry's organizational history was the transition of established lumbering-milling-marketing firms of medium or large size from partnership origins to ultimate incorporation. This shift became practically a standard phase of development for long-lived companies originating from about the 1850's on. However, the trend did not show itself until after 1865. Thus, in 1866 a partnership in which Joseph G. Thorp was a key figure translated itself into the Eau Claire Lumber Company, with all the shares of the new corporation held by the former partners. The Union Lumbering Company of Chippewa Falls was chartered in 1869, in part out of a former partnership in which Thaddeus C. and Albert E. Pound were prominent. In 1874 the partnership known as Daniel Shaw & Company took on the corporate form, as the Daniel Shaw Lumber Company. In 1878, after a famous growth that began in 1846, the partnership known as Knapp, Stout and Company incorporated as The Knapp, Stout & Co. Lumber Company.[103]

The contrast in timing between use of the corporation for water transport ventures and for logging, milling, and marketing firms points toward a variety of reasons for employing the corporate device. Evidently as a logging-milling-marketing concern moved from small to medium or large-scale operations, the entrepreneurs felt strong pressures of business expediency which enforced a shift into the corporate form. Business ambition, competition, and technological advance drove lumber firms into larger and larger capital commitments; men recurrently strove to "corner" timberland, or at least secure enough control to measurably affect prices; as demand pressed harder on supply, security, too, required larger investments in timberland to provide an assured basis for sustained operation; woods and mill equipment became more elaborate and costly, and payrolls rose. Though the lumbermen have left us little direct testimony of their calculations, their circumstances fairly indicate their reasons for incorporation.

The larger the capital commitment, and the more extensive the venture, the more practical benefit the corporate form offered. Under it liability might be limited to the stockholder's investment, continuity of firm life be assured, and a constitution provided for resolute central direction of pooled interests. Where the venture was new and untried, or where it was very large, with a considerable diversity of participants, it seems likely that the prime inducements were the opportunity afforded for limited commitments both of capital and of attention — attractive to distant investors, or to men busy with other concerns — combined with the means of centralized decision making. For an untried venture, in addition, the corporate form allowed promoters to offer well-defined investment units or shares, which could be flexibly adapted in size and number to the needs and prospects of the undertaking, and which could be made payable in installments or by periodic assessment, under the discipline of provision for forfeiture of shares or sums paid in, in case the subscriber defaulted. For a large-scale program, corporations might themselves be used as building blocks; as the timber industry gained in organizational sophistication, at the end of the century, the holding company appeared. However, both the speculative venture and the corporate empire seem to have been exceptional, though Moses Strong's unsuccessful Nekoosa Lumbering Company (1856) offers an example of the first, and the Weyerhaeuser alliances a dramatically successful example of the second use of the corporation.

The typical Wisconsin lumbering corporation, of whatever size, was a relatively tight-knit concern with few stockholders, and these usually the insiders who ran the business. Plainly this was so of the important group of corporations organized out of well-established partnerships. That the situation continued thus appears from the 1896 data of the Bureau of Labor and Industrial Statistics, in which 100 lumber industry corporations reported a total of only 541 stockholders.

For closely-held corporations the promise of continuity, and ready means to maintain unquestioned control in familiar hands, were probably the prime inducements for adopting the corporate form. Thus when the partnership of Daniel Shaw & Company became the incorporated Daniel Shaw Lumber Company, $430,000 of its $500,000 of capital stock was owned by the senior Mr. Shaw and his brother-in-law. For even so well established a firm of the middle size as was the Daniel Shaw Lumber Company, the limited liability of stockholders for debts of the business was likely to prove more a matter of form than of substance; in tight times the chief stockholders typically must use their personal credit, by endorsement of the firm's notes, to obtain badly needed working capital.

The river-improvement, boom, and log-driving companies present a somewhat different background of likely motives for incorporation. Significantly, of fifty-two special charters enacted for such ventures from 1850 to 1871, twenty-one set some fixed term of years for the life of the corporation (eleven for fifteen years or less, eight for twenty years, two for twenty-five years), whereas not one of the twenty-eight special lumbering-company charters of that period set any limit on the life of such a firm. Still more notable was the fact that thirty-five of these fifty-two waterways charters set a relatively low dollar maximum on the capital stock of the enterprise (twenty-one of $25,000 or less — eight of these of $20,000 and five of $25,000 — one of $40,000, eight of $50,000, four of $150,000, and one of $250,000). Seventeen of these charters said nothing about capital stock; in the context of most of them the omission indicated that the organizers assumed an undertaking so limited in fact as to raise no issue. In contrast, only one of the twenty-eight lumbering-corporation special charters of the period was silent on the amount of capital stock, and the other twenty-seven fixed maxima predominantly of a high order (one of $100,000, one of $200,000, four of $250,-000, one of $350,000, thirteen of $500,000, and seven of higher amounts, of which three were of $1,000,000, one of $1,500,000, and one of $3,500,000). In many cases the higher capital-stock maxima should be taken as symbols of ambition or vainglory, rather than as sober evidence of actual investment. But the figures chosen in the two types of corporation contrast so sharply as to evidence different concepts of the nature of the two kinds of enterprise. Most of the waterways ventures were apparently regarded by their participants as limited commitments, many of them apparently as more of the nature of profit-sharing cooperatives than profit-making businesses. It fits this pattern that, whereas forty-six of these fifty-two waterways charters were silent on what should be the consideration paid for capital stock, thirteen of the twenty-eight lumbering-firm charters expressly, and another five by implication, authorized receipt of property for stock; these men contemplated putting their major existing business assets into their lumbering corporations (as by transition from a partnership), but expected in a waterways corporation to join with others in limited investments of cash. The corporate form was thus particularly useful for men who sought a means to make strictly limited commitments. The larger river-improvement or boom operations, moreover, inherently required firm, continuing traffic management, which could be most readily legitimated by adopting the corporate form of organization. These considerations make clear why incorporation of waterways undertakings had an earlier start than incorporation of lumbering firms.[104]

Governor Randall's 1858 veto of the bill to incorporate the Oconto Lumbering Company points to a more limited issue in the evolution of nineteenth-century policy toward the corporation. This concerned the extent to which incorporation should be by special act of the legislature in each case, or under general incorporation statutes. Through the 1830's special charters were the norm throughout the United States. But by the time Wisconsin became a state, Jacksonian opinion had prevailed to the extent that legislatures were commonly providing general procedures under which charters might be obtained, thus making incorporation less objectionable as special privilege, and providing some standards of form and content for charters. At mid-century, policy was still ambiguous, however, since legislatures continued wholesale enactment of special charters. This ambiguity was carried into the Wisconsin constitution (1848), when it declared:

Corporations without banking powers or privileges may be formed under general laws, but shall not be created by special act, except for municipal purposes, and in cases where, in the judgment of the legislature, the objects of the corporation cannot be attained under general laws.

The constitution makers thus stated a presumption against special charters, but left its enforcement to legislative discretion. The statute books and legislative journals show that the legislature used its discretion so freely as in effect to cast the practicing presumption in favor of special charters. From 1848 through 1871, 1130 business corporations were organized in the state by special act. Only 143 corporations were organized in this time under the broad range of general incorporation acts available; some of these firms were later reincorporated under special laws, and in other instances the legislature amended general-act corporations by special legislation, as if the charters themselves had been special. The journals show occasional refusals of bills for charters, and occasional instances of conflict or reshaping of terms in course of adopting special charters. But these occasions are relatively so few, scattered, and uneven in character as to furnish little evidence of other than haphazard concern over the special status of such incorporations. Several Wisconsin governors used their veto power freely in vain efforts to stem the tide of special charters. However, the main theme of their vetoes was the excessive burden which the volume of special charters put on doing general legislative business; rarely did a veto indicate concern lest excessive or undesirable power be given a private group in this guise. There was no end to the predominance of special charters until in 1872 a constitutional amendment deprived the legislature of power to grant them.

Incorporation of timber-industry companies completely fitted this general pattern from 1848 to 1871. There at least an 8 to 1 ratio in favor of special charters prevailed, as in the case of all business charters; eight corporations were formed under general laws in that period for the clearly declared, primary purpose of lumbering operations (one boom company, seven logging or sawmill firms), but sixty-three corporations were organized primarily for such purposes by special act. The legislative journals show a little more than ordinary attention to inclusion of regulatory stipulations in charters affecting navigable waters. But this pattern only highlights the legislature's general readiness to approve, substantially as introduced, charter terms which did not have this public-utility aspect. Lumber-industry corporations figured in only four of the numerous vetoes with which governors recurrently sought to dam the flow of special charters. Three of these vetoes involved a public-utility type of issue, all arising from contests for control of the Chippewa River, between Eau Claire and Chippewa Falls interests; this line of controversy in no way turned on policy toward incorporation as a business device. The fourth veto was that of the charter of the Oconto Lumbering Company. Governor Randall did there warn in general terms of the balance-of-power problems that special charters posed; but in context—especially in view of a veiled reference to the railroad land-grant scandal of 1856 or the Chippewa River fight — his fears plainly ran to the imbalance of power in the public-utility type of situation:

> Extraordinary powers ought never to be conferred, by special act, upon a private corporation except in case of some great necessity, and when the public are to receive a benefit commensurate with the powers conferred upon the corporation, as an equivolent for the use of the powers which the public authorizes private corporations to enjoy. There is immense power, now held, in this State by private corporators under special acts, and it has already been demonstrated to the mortification and disgrace of the people of this State, that the power and influence and wealth of a corporation, created by special act, may be prostituted to dangerous purposes.

As we noted, the veto of the Oconto Lumbering Company's charter expressly disavowed any objection of substance to the enterprise there proposed, and the Governor finally rested his action on his repeated objection to the interference which the whole flow of special bills presented to the conduct of general policy making.[105]

Why the lumbermen shared the general businessmen's preference for special charters over incorporation under general act, so long as the choice was open, is another question to which they have left us no direct answer. There is ground for inference that the general incorporation acts carried limitations and regulations which the lumbermen preferred to avoid, if they could. On the other hand, the record offers little evidence for melodrama. Characteristically, it is only in the public-utility type of situation — notably, again, the 1856 railroad aid-lands scandal, and the "Dells bill" fight between Eau Claire and Chippewa Falls — that special corporate franchises bear unmistakable taint of devious maneuver and satisfaction of private ambition at public cost. Close comparison of texts shows that the "special" character of all the statutory charters for logging-milling-marketing firms lay in differences from charter terms imposed under the general incorporation statutes, and not in significant variations among themselves. These special charters early fell into established forms of their own. Their marked lack of individuality bears witness that businessmen sought a special charter for this kind of activity not to gain advantage peculiar to their particular enterprise, but simply to obtain corporate status on terms somewhat different from those under the general acts. There was relatively more variation in the terms of waterways charters, but significant differences here almost invariably concerned the "public utility" aspects of such ventures — the scope of control over improvement or traffic management on a stream, standards of service and of tolls. In those parts which provided the frame of corporate organization, the waterways charters fell as much into a few standard patterns as did charters for lumbering firms.

To read significance into many of the textual differences between the general incorporation statutes and typical special charters would be to proceed on sheer conjecture. However, two types of difference stand out so clearly as to indicate features of the general acts which businessmen undoubtedly preferred to avoid so long as they could. In some measure, undefinable from the face of the statutes, these features touched general public interest or the interest of stockholders; but, consistent with the prevailing emphasis of contemporary policy on the protection of market processes, they seem primarily directed to the protection of creditors.

In the first place, the general statutes imposed affirmative duties of disclosure of corporate affairs to the public and to stockholders, to a degree which — though limited and ineffective by standards of twentieth-century regulation — was yet materially greater than obligations imposed under special charters. Thus, before beginning business, a corporation organized under Chapter 73 of the Revised Statutes of 1858 must cause its articles of association to be published in full in two newspapers in or adjoining the county of the company's location, and must file with the secretary of state and the appropriate town clerk a certified statement of its purposes, the amount of its capital stock and the amount actually paid in, and the names of the shareholders with the number of shares owned by each; additional certificates must be filed upon the payment of any installment on the capital stock called for by the directors, or upon any increase in the capital stock. Moreover, once during each year of operation, a Chapter 73 corporation must file with the town clerk a certificate "containing the amount of their capital actually paid in, the amount of their debts and credits at the time of the making of such certificate, as nearly as the same can be ascertained, with the name of each stockholder, and the number of shares held by him at the date of such certificate." Privacy in the transfer of shares was further limited by the stipulation that no transfer would be valid against any creditor of the stockholder until a certificate of the transfer was put on record with the town clerk. In addition, Chapter 73 guaranteed stockholders a legal right "at all reasonable times" to inspect "the books of every [Chapter 73] corporation, containing their accounts," as well as the right "as often as once in each year" to receive from the directors "a statement of the accounts of such corporation." In contrast, of fifty-two waterways charters granted by the legislature from 1850 to 1871, only ten stipulated for stockholder access to company records, and only two for making any regular report to stockholders; of twenty-eight lumbering-corporation special charters in the same period, only three stipulated for stockholder access to records, and none contained any provision for a regular report to stockholders; of the total of eighty special charters none provided for any public filing of financial data or the making of public reports of operations. That the question of public and stockholder access to corporate books or information was a matter of lively interest appeared from three successive statutes, 1866–1868, which undertook to impose disclosure duties upon all corporations, specially chartered or not. The 1866 act directed that a record book be kept of all the formal actions of the stockholders and directors, to be open at all times to the inspection of the stockholders or legislative committees, and to be subject to the process of the courts in judicial proceedings; in 1868 the legislature added some sanction to this requirement by declaring that no resolution or proceeding of the directors or stockholders should be of any effect until recorded in this book. Most revealing, however, was an amendment in 1867 which added "all the other books" of the corporation to the record of formal proceedings, as material open to stockholders, legislature, or court, and the repeal of this addition in 1868; the 1868 act did impose the requirement that, each January, "a balance sheet and exhibit of the transactions of such corporation" be made out and filed in its office, but declared no procedure by which either stockholders or any public officer might obtain access to this statement.[106]

In the absence of contrary provisions in charter or statute, incorporation conferred on stockholders freedom from general liability for debts of the firm, other than for sums unpaid on stock subscriptions, and apart from the risk of loss of what they had paid for their shares. Likewise, unless there was provision to the contrary, corporation directors were not liable to firm creditors, even though the directors caused dividends to be paid while the corporation was insolvent, or which caused it to become insolvent. By their silence, forty-five waterways charters and twenty-seven lumbering-corporation charters specially enacted, 1850–1871, conferred this pattern of immunity on stockholders; one lumbering-firm charter expressly denied any liability of stockholders to firm creditors; seven waterways charters declared stockholders liable for debts of the ventures — two without limit, three to double the par value of stock held, one to the amount of any unpaid subscription, and one to the value of stock held. The twenty-eight lumbering-corporation special charters contained no provisions of any kind imposing any debt liability on directors, nor was there anything said on the matter in forty-nine of the waterways charters; three of the waterways charters forbade payment of dividends from capital, but said nothing of any penalty for violation. There were certain somewhat ambiguous provisions in the 1858 Revised Statutes that may have created a new, separate liability of directors of any corporation to creditors as well as to stockholders to the extent of corporate property misappropriated, lost, or wasted by breach of the directors' duty. But, restricted as this liability was to the value of the misapplied assets, at most it fell far short of a general liability for debts.

Under a Chapter 73 charter, if any of the capital stock was withdrawn and refunded to stockholders before payment of all debts of the corporation "for which said stock would have been liable," a creditor might recover his claim from any of the stockholders to the extent of the sum which the stockholder had received in the distribution. In putting the stockholder under this liability, Chapter 73 probably did no more than declare doctrine which Wisconsin common law would apply in any case to any corporation. But for corporations organized under its provisions, Chapter 73 created a liability unknown to the common law, and potentially of heavy burden, when it said that stockholders should be jointly and severally liable for all debts due employees of the corporation for services performed within six months preceding demand for payment. Likewise, Chapter 73 laid novel liabilities on management. Directors or officers who intentionally refused or neglected to file the required public statements on behalf of the corporation were jointly and severally liable for all debts of the firm contracted during the period of such a breach of duty. Directors who knowingly declared and paid a dividend when the corporation was insolvent, or the payment of which rendered it insolvent, were jointly and severally liable for all debts due from the corporation at the time of the dividend; and directors who ordered or assented to any violation of any provision of Chapter 73 which caused the company to become insolvent were jointly and severally liable for all debts of the firm contracted after the violation.[107]

The terms of the special charters suggest another less tangible consideration — one rather of convenience or reassurance than of sharply felt business need — which perhaps induced men to prefer this avenue to incorporation. Both statutory and judge-made doctrines as to corporate powers were in lively growth in the years when Wisconsin lumbering developed stature. In important respects it was not yet clear toward what outcomes this growth moved. The general law still left much to implication. One could find in judicial decisions and opinions encouragement to construe corporate powers broadly and flexibly, to the end of best serving business purpose; one could also find decisions and opinions which in particular matters laid stricter limits on what corporations might do. Without seeking any hidden or impolitic gain, businessmen might understandably like to write their own charter in language of their own choice, which on key points would give the benefit of doubt to their freedom of action. Moreover, though the Wisconsin constitution followed the pattern set throughout the country in reaction to the *Dartmouth College Case,* by reserving broad power to the legislature to amend or repeal any corporate charter, whether under general or special act, one might calculate that the legislature was less likely to tamper, unbidden, with the framework of an established special-charter firm than with the structure of companies originally created under a general statute which from the outset expressed a fair measure of public regulation.

Four aspects of the lumber-industry special charters stand out as probably illustrating this total pattern of cautionary thinking. We have noted that all of the lumbering-corporation special charters, and thirty-three of fifty-two waterways charters, either explicitly or by silence conferred perpetual succession. In contrast, the course of general legislative policy showed some wavering on this matter; Chapter 51 of the Revised Statutes of 1849 had limited incorporation thereunder to thirty years; Chapter 73 of the Revised Statutes of 1858 set no time limit, and hence implied perpetuity; yet, again, in creating a general act for incorporation of mining or manufacturing companies, by Chapter 166, General Acts of 1864, the legislature returned to a thirty-year limitation. General policy had shifted, too, on the definition of the allowed capitalization of companies; the 1849 act set no limit; Chapter 73, in 1858, however, limited capital stock to a maximum of $300,000, raised by Chapter 77 of the General Acts of 1860 to $500,000. As we have seen, this $500,000 figure was that most favored by draftsmen of special charters for lumbering firms, and capitalization was typically set quite low in waterways charters. Even so, it is not insignificant that most lumbering-firm special charters chose capital maxima higher than that which the legislature had once shown itself to favor, and that seven charters provided maxima ranging from $600,000 to $3,500,000.

Another point of peculiar interest to men who wished to transform an existing partnership into a corporation was assurance that stock might be issued in exchange for property (the assets of the partnership). Chapter 51 of the 1849 Revised Statutes had declared without qualification that "nothing but money shall be considered as payment of the capital stock" of any corporation organized thereunder. Chapter 73 of the 1858 Revised Statutes was silent on this important point. In 1860 the Wisconsin Supreme Court held that a railroad's special charter, in authorizing the directors to make contracts as the company's convenience and interests might require, and to determine the "manner" in which stockholders should pay on their stock subscriptions, authorized payments for stock in notes secured by real estate mortgages. The opinion spoke in favor of liberal construction which would arm a corporation to fulfill its purposes. This was all helpful to interpretation of Chapter 73 charters as implying a power to accept property in payment for stock; but businessmen to whom this was a critical point in the shift from a partnership to a corporation might well see no reason to leave the matter to argument. The limited commitment implied in the typical waterways venture would not occasion this problem; hence it is not surprising that forty-six of fifty-two waterways special charters said nothing of it. But, in contrast, eighteen of twenty-eight lumbering firm charters dealt more or less explicitly with the question. Thirteen of these expressly authorized payment for stock other than in cash; as close as any to a standard provision was that in the charter of the Eau Claire Lumber Company (1866) stating that the board of directors might "accept payment" for capital stock "in money or in property at a valuation to be fixed by them, which shall be as near the true value as may be." Five special acts were content with language borrowed from the railroad charter which the court had liberally construed in 1860 to authorize acceptance of property for stock; in these, as for example in the charter of the St. Croix Manufacturing and Improvement Company (1857), the directors were empowered "to prescribe by their by-laws or by resolution, the time, manner and proportions in which the stockholders shall pay the amount of their respective subscriptions."

Finally, men who expected to deal largely in timberlands might wish to leave no question of their corporation's full discretion to do so. From old English mortmain statutes our legal tradition inherited distrust of the accumulation of land titles in corporations of perpetual succession. Like other aspects of policy concerning corporations, this general attitude never came to any pointed conflict over the general run of business corporations, even when, as in the timber industry, the massing of land holdings proved a prominent feature of operations. Yet there were aspects of the general statutes which might concern prudent men. Chapter 51 of the 1849 Revised Statutes had expressly limited holdings of real estate of corporations organized under it to a maximum of "forty acres to each stockholder." Even within this limit, Chapter 51 had simply authorized the purchase and holding of "any real and personal estate whatever, which may be necessary" to the firm's declared purposes. Chapter 73 of the 1858 Revised Statutes dropped the acreage limit, but it continued simply to authorize such property holdings, including real estate, "as shall be necessary" for the corporation's purposes. In comparison, sixteen of twenty-eight lumbering firm special charters conferred on the companies power typified by the grant to "receive, buy and hold such lands, real estate and other property, and such lumbering and mining rights and rights of way as may be deemed necessary by them to the successful prosecution of their business and the execution of the powers herein granted"; seven charters conferred authority to buy and hold any land "for the use" or "for the use and benefit" of the firm; four conferred authority simply to buy and hold land, without further words; one was silent; none of these charters merely defined the authority as one to acquire land "necessary" for the firm. Consistent with the more limited scope of the typical waterways venture, fifteen of fifty-two special charters authorized landholdings such as might be "necessary" or "necessary and proper" to the undertaking, and nine charters were silent. Three set dollar limits to the value of real estate that might be held ($10,000, $20,000, and $25,000 respectively); since the last two of these also set dollar limits ($40,000 and $50,000) on the amount of personal property that might be held, they reflected concern to define the general limits of the undertakings, rather than policy aimed especially at landholding. On the other hand, even among the waterways ventures, twenty-eight of fifty-two charters authorized landholdings in terms not limited to what was "necessary." [108]

The end of the period of special charters, and the enactment of Laws, 1872, Chapter 144 (carried into the Revised Statutes of 1878 as Chapters

85 and 86), creating a new general procedure for incorporation of business firms, substantially ended such distinctive role as the timber industry played in the development of Wisconsin corporation law. From this point on, the industry's use of the corporate device merges indistinguishably with general business usage, save as the statutes continue and somewhat enlarge controls peculiar to the limited public-utility aspects of lumbering in its use of streams and railroads.[109] Of course, this does not mean that the availability of incorporation had less practical meaning to the industry; indeed, such distinctive attention to the forms of incorporation as appeared in the years of special charters was of a character which pointed up businessmen's appreciation of the general and continuing advantages of incorporation to the good order of an enlarged scale of operations. Men of affairs showed by their conduct that they rated the availability of this instrument as one of the law's prime contributions to expansion and consolidation of market activity.

PART III

THE POLICE POWER

INTRODUCTION

THE ADJUSTMENT OF CONTINUING RELATIONS

"THE POLICE POWER," said Mr. Chief Justice Shaw, is "the power vested in the legislature by the Constitution, to make, ordain, and establish all manner of wholesome and reasonable laws, statutes, and ordinances, whether with penalties or without, not repugnant to the Constitution, as they shall judge to be for the good and welfare of the Commonwealth and of [its] subjects." It is, he conceded, "much easier to perceive and realize the existence and sources of this power, than to mark its boundaries, or prescribe limits to its exercise." (For, indeed, said Mr. Chief Justice Taney on another occasion, "the police powers of a State . . . are nothing more or less than the powers of government inherent in every sovereignty to the extent of its dominions.") Nonetheless, Shaw reminds us, this power of police enters into and qualifies all rights of property and contract, because "it is a settled principle, growing out of the nature of well-ordered civil society," that it should be so. Thus:

> Rights of property, like all other social and conventional rights, are subject to such reasonable limitations in their enjoyment, as shall prevent them from being injurious, and to such reasonable restraints and regulations established by law, as the legislature, under the governing and controlling power vested in them by the Constitution, may think necessary and expedient.[1]

To cite so broad a concept asserts rather than demonstrates the legitimacy of any particular use of law. Yet, taken in the full meaning which Shaw gave the term, police power expresses a valid general criterion of legitimacy. A working society means relationships which must be kept in some minimum balance, and requires fulfillment of some minimum functions of reciprocity and mutual restraint. The order which law helps create and keep is not a static condition. It is a moving equilibrium, product of continuous adjustment to diverse pressures and to the constant press of changed circumstance. The police power refers to the role of law in effecting these adjustments.

The law of property provides a kind of constitution for allocating power over the economy and, by derivation, over many other aspects of social life.

PRIVATE AND SOCIAL COSTS OF LUMBER INDUSTRY OPERATIONS

Whatever ultimate judgments be passed, certain facts of lumber industry operation stand out sharply concerning losses in physical productivity. Woods operations showed two persisting sources of large loss in real yield. Through the principal years of Wisconsin lumbering loggers cut with no regard to preserving young or immature growth. No care was taken to avoid loss to this unready timber; all that mattered was the expeditious delivery to stream or railhead of what was deemed marketable. Apart from direct loss or damage in course of logging, new growth suffered heavily from fire—by far the most serious source of loss, because severe fire destroyed soil conditions favorable to succession of first-rate timber. Indeed, some opinion among professional foresters at a later day was that fire damage to new growth and to the soil conditions favorable to new growth was so basic that adequate fire prevention and fire fighting measures would have served, without more, to preserve the Lake States forest as a major producer into the twentieth century, however much other waste attended current logging. Like wind, fire was partly a hazard of nature. But logging practice greatly increased the risk and destructiveness of fire, for large quantities of slash (topping and limbs cut from the trees felled and removed) were left strewn among the standing timber. The slash might have been carted off, or piled and burned under supervision, but instead it was left to provide ready fuel for untamed fire.

In any case, whatever the causes of fire, neither government nor industry devoted any substantial investment to preventing it or fighting it. On top of these sources of loss of real production through logging practice, nineteenth-century accounting was crude, so that practical men of business deceived themselves that they were cutting profitable timber when by more sophisticated reckoning they were taking out of the forest considerable amounts of trees the yield of which did not meet their costs. Losses in stream transport were a heavy price paid for a tempo of exploitation that would not or could not wait upon the extension of a rail net. Sawmill machinery and organization improved in physical efficiency with great strides in the decades of Wisconsin lumbering. However, the very scale of improvement emphasized the amount of material lost in manufacturing lumber with inefficient equipment, as production pressed on with imperious disregard of the state of technique. The general condition of the lumber market provided less tangible but no less costly sources of loss in real productivity. The impatient press of midwestern growth demanded lumber and wood for its most obvious and immediate needs—for buildings, railroad ties, telegraph poles, fence posts, and fuel. This demand allowed no time to develop a comprehensive market pattern which might afford profitable outlet for a wide range of qualities and types of forest products. One result was that much of the forest's yield was destroyed for want of buyers. Another was the readiness to supply wood to any market that would pay something above out-of-pocket cost. Thus prime timber went into uses which might have been served by inferior product, had there been time to develop a more selective pattern of demand. Finally, there were profound implications for real costs in the fact that we committed substantially all decision making for the development of the industry to market processes. Market decisions and market discipline worked simply in terms of money calculus. This was a stern reckoning, calling men to account at relatively short intervals in terms which could be satisfied only with liquid assets. This discipline meant a constant urgency to subordinate real-productivity considerations to considerations of achieving at least minimum liquidity. Insistent enough under prosperous conditions, this market drive for liquidity became most painfully urgent when business turned bad. Interest must be paid, mortgages must be discharged, current bills must be met; these pressures of the market stood always in the background to deny decisive weight to physical efficiency as the criterion of woods operations.[2]

The crude official data and the fragmentary private business records which come to us from nineteenth-century Wisconsin permit no close assessment of the costs in real productivity which flowed from these aspects of lumber industry growth and operations. The best opinion is that contemporary real costs were high. Probably two thirds of the timber cut was wasted. Probably not over 40 per cent of the timber available in the northern Wisconsin forest reached the sawmills. Fire loss most likely exceeded the cut in some areas; over-all, sober estimates put destruction by fire at 7 to 10 per cent of the standing timber. Inefficient saws, it has been calculated, cost a billion board feet of pine in Wisconsin mills from 1872 to 1905. Comparison with twentieth-century experience provides some check on the probable extent of losses in real productivity in nineteenth-century operations. Comparison with the twentieth-century record serves, also, to set some realistic limits upon judgment of the efficiency that might have been exacted of earlier generations. Except for developments in paper and chemical production, the forest-products industries as a whole were backward in improving productive efficiency. By mid-twentieth century they still showed the smallest increase of productivity per worker of any major industrial group in the United States. Thus it chastens our judgments on the nineteenth-century industry to learn that in the 1940's about 30 per cent of the total annual drain on the country's forest resources was preventable waste, that half of the timber cut for lumber was wasted in the process, and that losses by fire, disease, pests, and storms accounted for about 11 per cent of the annual decrease in stock. On the other hand, there was conservative measure of the real losses from nineteenth-century inefficiency and inattention in the history of some twentieth-century improvements. Fire loss was reduced 62 per cent in the United States from 1910 to 1950. Increased efficiency in logging and milling since 1910 achieved economies equal perhaps to 10 per cent of the wood being cut at mid-century.[3]

Big as they were, the real costs of current operations were far less important than the real costs represented by long-term depletion of the Lake States forest. A rough kind of selective cutting went on in the Wisconsin forest in the second half of the nineteenth century. Men logged off what seemed the most valuable timber at a given time. Despite their heedlessness of future growth, the forest was so rich as to leave standing a great deal of timber which took on value in later years in tightening markets, so that loggers returned to take what they had earlier disdained. Toward the end of the period, demand finally gave value to the hardwoods, and extension of the railroad network made it economic to bring these to market. Thus logging proceeded by stages in the same general areas. But one stage followed hard on the other, with no time taken to convert this curiously accidental selective cutting into even an elementary kind of planned sustained-yield operation. The Lake States forest area was relatively level country, well laced with streams for water carriage, offering no great natural obstacles to rapid construction of railroads, and favorable to the large-scale, clear-cut kind of operation made possible by development of the donkey engine, the logging railroad, and the high-capacity sawmill. The fifty years after the Civil War saw the greatest timber cutting in the country's history, concentrated with the most complete preference for present over future yield in the Lake States. The great forest of Wisconsin, Michigan, and Minnesota was clear-cut in this period in degree beyond what was practiced in any other major forest region of the United States. Fire prevention and protection did not exist. By 1915 — when, symbolically, Wisconsin witnessed its great constitutional lawsuit over creation of a state forest reserve system — the white pine and related species of prime conifers were almost gone, and commercial forest production was fast becoming concentrated in limited hardwood stands. Experience showed that to grow sawlog timber by deliberate planting and cultivation was so costly that the undertaking would probably always be the exception; natural growth, aided by fire protection and other quite limited care, must be the main reliance. Yet, experience also soon showed that first-rate pine was slow to reseed itself in land clear-cut and subjected to such fire damage as was the Wisconsin forest. Thus inferior conifers and low-value broadleaf species proved the sole replacement of the original pine. In most places in Wisconsin of the first half of the twentieth century there was too little saw timber left to allow profitable integration of production of saw timber and pulpwood. Thus with little exception the great lumber industry of the last quarter of the nineteenth century had wholly lost its production base. Save as slow-paced growth promised limited future stands of saw timber, the Wisconsin forest-products economy had dwindled to the growth of inferior species for the paper industry, and such marginal production as that of Christmas trees. Despite twenty-five years of rapid increase in logging, northern Wisconsin in 1875 had contained some 130 billion board feet of white and Norway pine and some 16 billion board feet of merchantable hardwood timber. By 1925 this resource had been reduced to hardly a billion feet of white and Norway pine, and some 7 billion feet of hardwoods. Of the 18 million acres of the original northern forest, by 1925 five sixths stood under only inferior growths such as jack pine, popple, and scrub oak, a great deal of which was so understocked or so handicapped by growth conditions that it would never provide merchantable forest under natural conditions.[4]

Depletion of the Lake States forest involved several kinds of long-term real and money costs which both the national and the regional economies felt as a continuing burden from the nineteenth-century lumber boom. These costs did not, it is true, include the national timber famine which the first generation of ardent conservationists had predicted. Keener cost consciousness, both in private and in public accounting, produced more careful cutting practice after 1910; new growth rates improved. Government developed more effective programs for preventing and fighting forest fires. Processors learned to saw with less waste and to find profitable by-products in materials formerly thrown away. By 1950 it appeared that the third quarter of the century would see annual growth and drain in balance in the nation even for saw logs, which were the slowest division of production to mount toward this mean. This was all to the good. We never suffered absolute physical deprivation in the general lumber market. But this outcome did not mean that we did not pay a continuing price for the nineteenth century's urgent preference for present over future yield.[5]

To the national economy, exhaustion of the Lake States forest meant adding to social overhead cost by contributing to the large quantity of unused or under-used land which stood on the national books. "It is useless to decry the generous use which American industry has made of our forests," the Chief Forester of the United States observed in 1921. "The forestry problem does not result from the liberal use of our forests, but from our failure to use our forest-growing land." In 1950 the United States contained over 88 million acres of natural forest land partly idle and another 75 million acres entirely unemployed for forest production. The depleted northern Wisconsin forest alone contributed almost 10 per cent of these totals of unproductive and under-productive timberlands.[6]

Clear-cutting also spelled long-run higher money and real costs for consumers of lumber. In part these costs were experienced over the long term in lowered quality of products as well as in higher prices reflecting the greater investment of time, energy, and capital needed to process lower-quality wood and to produce less desirable and more costly substitutes for prime wood. The nineteenth century's preference for present yield also added to the direct overhead costs to be borne by future demand. Rapid clear-cutting increased depreciation charges for timber stands, much of which were cut before they reached their point of maximum economic return; to clear-cut the forest as it stood meant at best to gain what the fortunes of the current market allowed for the yield that could be had at that moment, foregoing the gains both of quantity and quality that a managed, growing forest might yield over time. It increased the costs of basic camp, transport, and mill investment, much of which became obsolete by exhaustion of timber long before the facilities had worn out. It increased the interest cost of capital, by putting the industry under recurrent, necessitous pressure to return to capital markets. The pattern of rapid clear-cutting meant that the industry's financial needs were never limited to borrowing current working capital at simple interest, but that it must borrow at high-risk, short-term rates reflecting a boom and bust career. Lumber prices moved in long-range cycles, and there was no simple correlation of them to total timber supply; indeed, as we have noted, the country never faced absolute timber scarcity. But, if no precise calculation of added costs could be made, it was nonetheless clear that the pattern of sheer depletion of major forest-production areas added substantial elements of cost in lumber supply. One item on which estimates could be made was the loss of site value, as exhaustion of nearer sources forced buyers to look further abroad for supplies and hence to pay increased freight. Between 30 and 50 per cent of the price paid by the consumer of lumber in the mid-twentieth-century United States went for freight, reflecting the poor distribution of timber supply sources in relation to major markets. Since the country's over-all freight bill for lumber ran to as much as $500 million, the continuing costs of lost site value were high. From 1890 to 1920, the Chief Forester of the United States noted, the Chicago area imported over two billion board feet of lumber per year; in 1920, 60 per cent of the imports went into local construction and manufacturing. In 1900 the average freight paid was less than $3 per thousand board feet; by 1921 the average freight paid was $12 per thousand board feet, based on increased haul and not on freight rate advances. At this point nearly 23 billion timberland acres stood idle in the nearby Lake States; the Chief Forester estimated that the annual extra freight bill being paid in the Chicago lumber market would meet the costs of planting each year 1½ million acres to new growth.[7]

These elements apart, exhaustion of the Lake States forest represented a long-term cost to the national economy in consequent loss to flexibility in decision making, not only in economic but also in political terms. A continuously productive forest meant possession of a range of productive potential, to stand as a buffer against changes in demand; given appropriate arrangements as to land tenure, finance, and taxation, such a forest factory could be operated without impossible cost to curtail output in a falling market and to expand yield in response to rising demand. In contrast, the nineteenth-century pattern of clear-cutting made for considerable inelasticity of supply. Lumber price cycles reflected the fact that within this framework we were unable readily to adjust to shifts in demand that accompanied changes in population and in the location and scale of industrial and agricultural activity. Moreover, the exhaustion of northeastern and Lake States timber sources meant greater demands upon manpower, capital, and transportation facilities to bring forest products to market from the far South and West. We learned in World War II that this added up to a serious loss in capacity to respond to emergency; it was not lack of timber, but rather scarcity of men and of transport means, which caused wartime shortages of lumber, pulp, and paper.[8]

These long-term costs made themselves felt primarily on the demand side, and in the national or the Mississippi Valley economy. They had their counterparts on the supply side, but with much more concentrated effect — in long-term costs to the general economy of Wisconsin. The Mississippi Valley market could find its supplies elsewhere, and though the added costs were large, they were nonetheless marginal to the users' economy. But the forest was the single major economic resource of most of northern Wisconsin, and bulked up in the state's over-all nineteenth-century economy second only to agriculture. Its exhaustion meant much more sharply focused additions to social overhead cost for Wisconsin than to the out-of-state economy, measured in abandoned or under-used villages, mills, roads, and railroads. The overshadowing capital loss, of course, was in the timberland itself. At the end of the nineteenth-century lumber boom a scant 10 per cent of the original northern Wisconsin forest remained capable of producing merchantable saw logs. By the late 1920's it was plain that the bulk of the cut-over lands was wholly unsuitable to farming; after fifty years' effort, less than 7 per cent of the cut-over had been "improved" for farming, and much of this land was not capable of supporting its owners. The area offered the basis for an expanding recreation industry, but this business was not capable of providing a self-sustaining economy; indeed, the development of the resort business underlined the fact that the north country should have found its firm economic base in a multiple-use forest. Disappearance of the forest as a production factor also meant, of course, disappearance of the wage earnings and the investment opportunities and processing and trade profits that went with an active forest-products industry. At the peak of the Wisconsin industry it produced about four billion board feet of lumber per year which it sold for about $10 to $12 per thousand, on which it netted a little short of 10 per cent on a capital investment of about $45 million and paid wages of about $7 million a year.[9]

The greatest long-term loss to the Wisconsin economy from depletion of the forest was the opportunity cost which this depletion represented. The market value of the raw material supplied by the forest was a relatively small part of the ultimate market worth of the goods and services produced with the aid of these raw materials; so measured, the forest product was much less important relatively than the investments of labor and mobile capital (including investments in technique and organization) that went along with its use. The basic significance of the forest yield to the state's economy was not monetary but real, measured not in dollars but in the fact of physical availability which made possible a wider range of alternatives for economic growth than otherwise existed. Depletion of the forest did not merely cost the state the loss of its existing forest industry. Forest depletion cost Wisconsin all the possibilities of expansion and diversification held out by the original site value of this raw-materials stock, located close to a Central States market which by 1950 included about a quarter of the country's population, to which the state's forest industries had privileged access guaranteed by favorable freight differentials. But this loss was not the whole of the matter. In addition, Wisconsin must henceforth pay the cost of bringing into the state from distant, freight-costly sources the bulk of the lumber still needed for its own industry and consumers; one estimate in the 1920's put the price of this adverse balance of lumber trade at an amount about equal to the whole budget of the state government; in the late 1930's it was estimated that 45 per cent of the lumber used by Wisconsin planing mills, box factories, furniture makers, and panel and veneer mills was imported.

Another cost to the flexibility of the state's economy was the loss of tax base represented by exhaustion of the timberlands. Most obvious evidences of this loss were the mounting total of tax-delinquent lands in northern counties and the lands there withdrawn from the tax rolls by public purchase for rehabilitation. Large-scale delinquency began even before the turn of the century, and ranged from 20 to 50 per cent of all land in the cut-over counties in the full ebb tide of the depressed 1930's. In 1936 in six northern counties, lands withdrawn from the tax base ranged from 29 to 44 per cent of the area of the counties. This was the negative side of the story. The positive side was the potential productive contribution which the timberlands might have made to the state's economy, had the forest remained in operation. If a sustained-yield forest had remained on even one half of the original 18 million acres of the northern forest, its fair assessed value for taxation in mid-twentieth century would probably have been about $500 million. This wealth would have represented a net addition to the taxable value of the depleted north country as it actually existed seventy-five years after the industry's peak. This figure would have added 5 per cent to the existing total assessed value of real property in the state. Of course, even under conservative assumptions such estimates are open to wide error. But some cross check was afforded in the late 1950's when impending transfer of the Menominee Indian Reservation from federal trusteeship to tribal or private ownership required formal estimate of the taxable value of 220,000 acres of timberland constituting the main worth of the reservation area. The Wisconsin Department of Taxation then valued this prime timber stand at about $12 million for tax purposes, or about $54 per acre. This figure assumed sale of the timber at mean contemporary market prices on a clear-cut basis or without restriction against clear-cutting. Were the timber sold to be maintained as a sustained-yield operation, presumably the present cash value would be less, but still substantial. These estimates deal only with the contribution that continuingly productive timberland, as such, might have made to the tax base. To them would be added the taxable value of primary and secondary processing plants and related income from processing, trade, and transport activities. Wisconsin production of sawed lumber topped four billion board feet per year at the height of the nineteenth-century boom. If the state and local tax yield would run to $1.50 per thousand feet of lumber produced, from taxes on mills and other equipment and on inventory and income, at only one half of the industry's boom production rate the tax flow from processing would have amounted to an additional $3 million per year, exclusive of taxes on timber holdings.[10]

With the wisdom of hindsight, we know that such were current and long-term costs involved in depleting the North Wisconsin forest. How far did contemporaries see what was at stake? Within what limits of knowledge, imagination, and will did the legal process operate in assessing the implications of the stakes for public policy? What scope did nineteenth-century constitutional doctrine allow policy makers, to come to terms with the choices posed by the facts? Hindsight does not provide a fair measure for moral judgment. But it offers proper help in defining what issues were implicit in the facts, and what facts affected the way in which lawmakers handled the issues.

THE AVAILABLE POLICE POWER

When we ask what authority nineteenth-century lawmakers had to regulate how private owners harvested their timber, we must deal largely with hypothesis. Through the boom years which exhausted the Lake States forest there was little done by statute or judge-made law to occasion answers to the question. Nonetheless we should try to answer it. We can hardly appraise the meaning or quality of what the nineteenth-century policy makers did, without some idea of the freedom of maneuver their constitutional powers afforded.

Since the United States held original power and responsibility, as first owner of the forest, there is logic in beginning with its potential powers as conservator. So long as it held the ultimate disposal of the land, the federal government might by contract, lease, or condition have fixed the terms on which those taking under it should use the land's timber wealth. Consistently, from years contemporary with the great Wisconsin lumber industry, the United States Supreme Court ruled that the property clause of the federal Constitution gave Congress unqualified discretion to name and enforce terms of use and disposal of the public domain. Neither in the nineteenth century or thereafter did the Court recognize any substantial limitation upon this authority, derived from the 5th or 10th Amendments, or from any other part of the Constitution. So, for example, Congress might effectively stipulate that land claimed under the federal homestead law should not become liable to the satisfaction of any debt contracted prior to the issuing of a patent, and though the provision of remedies to enforce local private debts would normally be within the exclusive power of the state, the state's process must yield in this instance to the command of the federal statute.

As an authority over federal "property," however, such rights of control came to an end once the United States conferred a fee simple absolute title upon its grantee. Given the sweeping character of the power enjoyed under the property clause, it is clear that one decisive point was passed when Congress allowed the fee of the forest to pass to private hands. Once lost, this position of leverage could hardly be regained. By the Weeks Law of 1911, Congress provided for the purchase of land deemed most suitable for return to the condition of public timber reserve. But procedures of repurchase were slow and costly. Federal purchases in the Lake States area by 1950 were substantial, but far from enough to make the United States a major influence for development of sustained-yield forestry; in Wisconsin the United States by 1950 had acquired about 1,300,000 acres for national forest purposes.

For economic reasons there was much more involved in nineteenth-century federal timberland policy than relinquishment of the United States' powers as landowner. The wholesale transfer of federal land put into private ownership vast quantities of virgin timber at very low prices and without restriction as to manner of use. The lasting result well into the twentieth century was severely to limit the practical freedom of private entrepreneurs to adopt, or of the states to impose, forest-use regulations, in the absence of unified regulation by all major producing areas. So long as vast inventories of cheap virgin timber hung over the market, available for clear-cutting without regulation, private operators faced substantial deterrents to pursuing more costly sustained-yield procedures in havesting timber, and state legislators faced difficult choices in subjecting home industry to regulatory burdens from which out-of-state competitors were free.[11]

Conceivably the Congress might have made good under its fiscal and commerce powers the care of national timber resources that it had so casually and precipitately relinquished as property owner. By the last quarter of the nineteenth century, constitutional doctrine was firm that Congress might act to provide and preserve conditions reasonably deemed necessary to a sound and healthy national economy, including a national bank or national banking system, a protective tariff, federally chartered and subsidized interstate roads, canals, and railroads, and a federal monopoly of the currency. Within this doctrinal framework it would be dogmatic in the extreme to assert that the late-nineteenth-century Congress might not have found regulatory power to protect the Lake States forest as a national asset, had it willed to do so. Quite plainly what stood in the way was not reasoned conviction of want of authority, but want of conception that there was cause to invoke authority. Congress followed the common opinion of the time, which accepted as a self-evident proposition the social utility of maximum extension of the private fee simple. This policy included uncritical extension of the private fee to natural resources like timber, though their exploitation posed problems of cost which market processes were incapable of handling. Again we recall the question which Wisconsin's Timothy Howe put to the United States Senate in 1876, confident that reasonable men could see but one answer:

> If a full-grown man with unimpaired intellect really thinks he needs to cut down and cut up a pine tree, and is willing to pay a fair price for it to the owner of the tree, why should he not have that privilege?

Moreover, the definition and regulation of "property" was ordinarily the familiar domain of state law. Thus the transfer of fee simple title from the United States to other ownership promoted uncritical acceptance of the notion that all the incidents of use thereupon fell within the care of state policy.[12]

In one aspect late nineteenth-century policy showed such marked inconsistency as to make it clear that Congress was far from achieving a coherent pattern of ideas regarding national timber wealth. By the "timber culture" acts of March 13, 1873, and June 14, 1878, Congress made special offers of homesteads to settlers who would plant certain numbers of prairie acres to trees, and cultivate and care for them for sufficient years to see them established. The statutes produced little result, save to provide another opportunity for fraudulent claims. This legislation stood in remarkable contrast to the contemporary policy which allowed ready grant or easy sale of prime federal timberland, without provision to protect its continued productivity. The fact did not pass without notice. In 1876 the Commissioner of the General Land Office criticized the early timber culture legislation:

> It is an anomalous fact that the Government is giving away the rich alluvial soil of Iowa, Nebraska, Kansas, and Minnesota to any citizen who will plant a

general interest in economic development. Later generations may question the pace and scale at which forest use was thus pressed. This question does not lessen the constitutional significance of the tax exemptions. In their impact on other property than that exempted, these statutes were a kind of economic planning regulation. They enforced extra contributions from other property — and most notably from the generality of private timber holdings in the north country — to give what was accepted as desirable direction to realizing Wisconsin timber wealth. Of like import was the great volume of special charters for river-improvement and boom companies. Again, in the interest of what was deemed socially efficient development of the north country, the legislature in effect delegated powers of taxation to dam and boom owners to lay tolls upon owners of forest products seeking market. Thus, contemporary policy boldly asserted the state's authority to regulate framework conditions for realizing upon its resources. On this record it would be a wholly unproved case to assert that the state did not do more or differently in regulating the efficiency of forest use because the legislature lacked power, or thought it lacked power. As with the Congress, the record's import is that there was want of will or imagination to adopt other regulations for forest efficiency, not want of authority, nor want of confidence in the scope of authority.[15]

USE OF THE POLICE POWER

INVESTIGATION AND DATA COLLECTION

There was one kind of legislative activity so necessary to rational government that there was never serious question of its legitimacy. This was the authority of the Congress and the state legislature to cause investigation of facts that might reasonably be deemed of public interest. The legislative branch could not properly exercise any of its powers — or properly determine that there was no occasion to exercise power — without some minimum knowledge of what was going on. Presumably the legislators brought with them some knowledge of the community, learned simply by being part of it. But the existence of legal order attests the need for processes of decision that reach beyond men's immediate concerns and everyday experience. The extent to which the legislature or its delegates use this power offers valid measures of the maturity of the legal process.[16]

By the measure of resort to the power of legislative investigation, federal and state policy making affecting the Wisconsin forest was conducted by crude makeshift. Through the years that committed the Lake States timberlands, Congress made no effort to acquire reliable data on the extent or nature of the forest or the conduct of the lumber industry and its relation to the general economy. Most damaging to the national interest in the long-term productivity of this resource was Congress' complete indifference to investigating the facts of the fire hazard and the practical possibilities of limiting the extent of almost irreparable fire damage. Congress did stipulate that in the survey of federal lands note should be taken of such special features as the presence of valuable timber stands. But neither in Washington nor in the field was the General Land Office ever adequately staffed for its responsibilities, and the surveyors' field notes were notoriously inaccurate. Moreover, the survey system was oriented entirely to marketing land, and not at all to appraising alternative modes of realizing the land's productivity.

The power of legislative investigation has no greater utility than to help overcome the normal inertia of society; here, with minimum initial commitment, the community has means to combat mindless drift and default. It is a mark of the force of social inertia, and of the easy acceptance of the familiar, that it has proved most difficult even to set in motion the process of inquiry. Time and again original initiative lies only with interested individuals — in or out of government — who procure the introduction of bills which at first come to nothing, but over the years bring questions to sufficient definition to obtain investigation, debate, and perhaps eventually new policy. So it was here.

From about 1870 a trickle of bills introduced in Congress marked beginning concern over inefficient realization of national forest wealth. In 1873 the American Association for the Advancement of Science memorialized the President to cause study of the country's timber assets. In 1876 Congress appropriated funds for appointment of one man in the Department of Agriculture, to collect statistics and distribute information about forestry. The appointment of Franklin B. Hough to this post led by 1881 to creation of a small Forestry Division in that Department. In 1888 the American Forestry Association presented a bill for establishing national forest reserves. Nothing was more characteristic of the back-door approach to policy making in this area than the next step. In 1891 Congress authorized the President to create national forest reserves. It accomplished this important advance by a terse provision inserted at executive initiative without legislative inquiry or debate, as a last-minute conference committee amendment to the statute of March 3, 1891, which repealed the ineffectual Timber Culture Act.

The first major investigation of the grounds of public policy toward national timber wealth took advantage of the fact that when Congress incorporated the National Academy of Sciences it charged that body to investigate and report upon any subject of science when requested by any department of the national government. Maneuvers by a few interested private citizens procured a formal request early in 1896 from President Cleveland's Secretary of the Interior, Hoke Smith, to the National Academy to investigate the need for national forest reserves, for general forest conservation measures in those parts of the country where public timberlands were mainly situated, and for legislation to remedy evils threatening national timber wealth. In response the National Academy created its Forest Commission. Though its final report had little effect, the Commission's interim report of January 29, 1897 — compiled largely out of the field work of Gifford Pinchot — provided President Cleveland with the data on which he could proclaim over 21 million acres of new national forest reserves in the Far West. From this point on, mainly through the executive branch, Congress grudgingly provided increasing means to learn facts and define the questions to be put in shaping a national forest policy. The chief relevance of this recital to our purpose is to emphasize that federal policy-making processes moved so slowly against the inertia of the situation that there was no significant investigation or debate focused on the Lake States forest while logging there was mounting to its peak.[17]

The state legislature could not command as much money or experience as Congress. Even so, conceivably it might have done better than Congress, at an earlier time, in collecting data about the uses of the Wisconsin forest and in bringing to definition the issues implicit in the data. Northern Wisconsin was a long way from Washington, and was only a small part of a continental nation. But to the state this northern resource was a massive, present fact. This fact projected itself into state policy in scores of statutes concerning state lands, log measurement, river-improvement charters, and the incorporation of lumber companies, as well as in hundreds of lawsuits over dealings in timberland, logs and lumber, and use of streams. When state policy thus reacted in hundreds of specific ways to the specific felt presence of this resource and its industry, one might think that the sheer weight of business might push the legislature into more general and more self-conscious exploration of the framework of policy. Yet, events did not work out so; up to the turn of the century the state made no more aggressive or creative exploration of the circumstances and choices involved in forest use than did the federal government. The fact deserves emphasis for its relevance to issues broader than those of the present study. There is no substitute for deliberate focus of will and intelligence upon general relations. The push and bustle of "practical" affairs and immediate operations do not of themselves generate attention to the cumulative direction and force of events. Indeed, the urgent detail of current business creates much of the inertia of the total social situation, which it is a prime function of law to combat.

That the Wisconsin legislature never stirred itself to make a broad examination of policy toward use of the northern forests stands out the more because the legislature was not wholly indifferent to the need to inform itself. As early as 1857 the legislature declared it the duty of town and city assessors, at the time of the annual assessment of property for taxation, to collect statistics "of all the agricultural, mineral, and manufacturing interests carried on in their respective towns and cities," and to return the data to the clerk of the county board of supervisors, who in turn should compile the information and return it to the Secretary of State for general tabulation and report to the legislature. Though the lumber industry had then been growing to considerable size for a decade, its activities did not yet enough impress the legislature to procure its inclusion in this general economic census. However, in 1861 an amendment added sawed lumber, various finished wood products, and "capital invested in manufactures" to the list of items to be returned. With the short-sighted economy typical of this capital-scarce community, the 1862 legislature changed this economic census from an annual to a decennial affair, which it thereafter remained through the great years of lumbering. So far as this procedure was concerned, the items affecting the lumber industry remained the same throughout. But state legislation characteristically moved in jerky, disjointed fashion. In 1874 the legislature created a separate duty of town assessors to return annually to the county clerk for compilation and report to the Secretary of State "the whole number of acres of each kind of farm products" then growing in the town, specifying grain and grass crops. In 1877 this list was expanded to include "the whole number of acres of . . . timber" growing in each town. Carried into the Revised Statutes of 1878, this separate annual report on standing timber likewise continued through the great lumber years. These decennial and annual economic censuses were wholly decentralized in operation, relying without check on the dubious zeal, conscientiousness, and intelligence of local officials. The Secretary of State early complained of inadequate reporting, and contemporary evidence attests the crude and uneven operation of local government, particularly in the thinly populated timber counties. Moreover, neither of these general census procedures collected any but the most simple, gross data on the lumber industry. In 1880, after reporting need to tighten discipline over reporting by local officers, the Secretary of State recommended that the economic census be much more specifically fashioned to reflect the importance and complexity of the lumber as well as the mining industries:

The present law wisely provides for the gathering of full returns of the products of the farm, but when it is remembered that stretching across the northern half of the State is the second largest forest in the United States, annually producing millions of dollars worth of lumber, and that the state also contains mines of great richness, which are being advantageously worked, concerning neither of which can any exact information now be had, it should seem that the law ought to go farther and be made comprehensive enough to exhibit the full strength of these other resources of the commonwealth.

This pointed recommendation may have inspired the introduction in the next (1881) legislative session of a bill ambitiously designed to create a state bureau of statistics. But pennywise policy again prevailed; this bill was summarily rejected on recommendation of the Assembly committee concerned with state finances, and the only measure enacted was one which, significantly, aimed to tighten the procedure for obtaining more complete data on tax assessment and yield.[18]

Though the legislature made no adequate provision to survey the growing lumber industry through a general economic census, it might have recognized the distinctive importance of the industry by taking special measures to gather information about it. Here again we find enough false starts and incomplete action to highlight the poor craftsmanship of the legislative process. In 1855 the Wisconsin Natural History Association, led by Wisconsin's first scientist, Increase Lapham, memorialized the legislature to create a zoological and botanical survey, in part to study "our noble forest trees . . . [which] are useful in various ways, besides the more obvious ones of affording a supply of lumber and fuel. The nature of the soil, and the general character of a country may be known by the forest trees." But this recommendation produced no result helpful to the study of the northern forest. In 1853 the legislature had created the first of several geological surveys; these ran at different intervals through the latter nineteenth century, attended by a good deal of acrimony, both because the surveys became involved in political patronage and because, insofar as their published results ran cautiously close to their evidence, they disappointed the speculative hopes of mining entrepreneurs. Reports of the various geological surveys made passing mention of the wealth of north-country timber, but their focus was never sufficiently on the forest to yield any data significant for policy making. It is of interest, however, that the legislature was thus induced to make recurrent provision for rather elaborate factual surveys of minerals resources and mining industry needs throughout decades in which it paid no comparable attention to studying the more important lumber industry. Mining was a sick industry, and timber was not; the miners and mining companies themselves desired more information, because they hoped that it would lead them to new productive veins; the lumber industry saw its opportunities so obviously close at hand and all about it as to have no like cause to press the legislature for this kind of help. The legislature thus attended to gathering facts about the mines, not out of any marked foresight or breadth of plan, but primarily in response to a special interest.[19]

A like limitation of interest robbed of its potential effect the one scientific and technical study of timber wealth that the Wisconsin legislature provided when there was yet time to affect use of the northern forest. At the urging of Increase Lapham, whose scientific activity and publications in Wisconsin then already embraced work in archaeology, botany, and geography, the 1867 legislature authorized the state agricultural and horticultural societies to create a commission to report to the next legislative session "whether the destruction of the forests of this state, now going on so rapidly, is likely to prove as disastrous to the future inhabitants of the state as is claimed by many." The statute gave the commission a broad mandate — to inquire whether the state should by law limit cutting or encourage cultivation of trees, whether scientific experiments or investigations should be made "to ascertain the best methods of growing and managing forest trees," whether wood substitutes might be found, and "to report generally such facts, in a condensed form, as will be of the most practical use to persons desirous of increasing the growth and preservation of forest and other trees." This measure passed without delay or apparent controversy, supported by a warm endorsing report of an Assembly select committee. However, the committee's emphasis foreshadowed the limited scope of what finally came of this venture. The committee made relatively brief reference to "the northern woods," which it noted "are fast passing from us — faster than those who have given but little thought to the subject believe." It warned that "the expense of procuring timber from those distant woods is already so great as to put it almost beyond use for fencing purposes," and that "the price will continue to rise more and more as the distances and difficulties increase." But the bulk of the report dealt with excessive destruction of trees in the southern, agricultural part of Wisconsin, and the committee reserved its specific criticisms and recommendations for the uses of timber stands in connection with farms. It attempted no analysis of the problem of maintaining the pinery as a continuing productive timber resource. In total context, the committee quite plainly regarded this study bill as one related primarily to the interests of agriculture.[20]

Increase Lapham headed the three commissioners appointed under the 1867 statute. Within the year they filed a monograph-size report, "On the Disastrous Effects of the Destruction of Forest Trees Now Going on so Rapidly in the State of Wisconsin." In range of description and analysis the report went considerably beyond the Assembly's select committee. In suggested program, however, it stayed within similar limitations.

Lapham's report has particular relevance to our immediate concern with state action to obtain the facts and definition of issues necessary for making policy. The commissioners astutely noted that the police power of the state was not limited to regulation of private behavior. As the state might regulate the taking of fish or game, so surely it might regulate the cutting of timber, to preserve continuity of supply. But this was not all it might do. It was entitled either "to regulate the cutting of timber where it would obviously entail a public calamity, or to encourage its production where it is so much needed for the public good." One measure to encourage production might be to provide for gathering and publishing more information on forestry and its relation to the general economy. This was especially a proper exercise of police power because it concerned conditions which were important to the general framework of the economy, but for which market processes would probably not provide:

> Private enterprise cannot be expected to perform the herculean labor of making the experiments necessary to procure all [the knowledge] that is needed; and in-as-much as when obtained it must result more to the public good, than to that of the individual, so no one man or society of men ought to be expected to perform it, without public aid.[21]

The Lapham report presented the first attempt at a broad estimate of the quality and size of the northern forest. Moreover, it made brief but highly suggestive reference to points which experience would prove to be critical to an adequate general forest policy. Thus the commissioners indicated concern not simply over the rising current scale of lumber consumption, but over the implications for loss of the forest as a base for long-term, sustained industrial development. They expressed worry over the loss of potential tax base and consequent increased burdens on other parts of the economy that would go with forest depletion. They at least introduced the subject of greater efficiency in modes of logging. They looked in the direction of broad-scale zoning: "Lands least suitable for ordinary culture should remain, or be planted in wood."

Despite these provocative leads and insights the Lapham report was disappointing in total impact so far as concerned the northern forest. Its references to problems in the use of the great forest were relatively brief. It attempted no bill of particulars for regulations to implement its suggestions for zoning land or setting minimum standards for efficiency in cutting. These limitations take on the more meaning in light of the dominant emphasis of the document. This emphasis, like that of the Assembly select committee of 1867, was upon the relation of timber assets to agriculture. The main concern of the report was with the effect of wholesale cutting of trees upon climate, water supply, and the blasts of wind and cold upon fields and orchards. The commissioners' most concrete and elaborate recommendation was for legislation to encourage the planting and maintenance of narrow tree belts to protect the land for farming. Underlying this pattern of emphasis was the assumption that agriculture was the normal economic destiny of most of the land. The tree belts were important because they would protect the "permanent occupants" of the land after the absentee owners of timberland had reaped their profit from the forest and gone their way. It was not surprising, but entirely in accord with the practical force of the Lapham report, that the only evident result of it was the enactment in 1868 of a statute to promote the growth and maintenance of agricultural tree belts by grant of tax exemptions and bounties in the form of tax credits. There is no evidence that the tree-belt law had significant effect, even within its limited objectives.[22]

The general economic census did not develop into a procedure adequate to inform general policy making concerning the forest. More specialized scientific and technical studies, represented in the geological survey and the Lapham commission, did not gather sufficient momentum to meet the need. There remained the possibility that more specialized official procedures for collecting marketing data might develop into assembly of broad-ranging information and help generate interest and debate on the general course of the lumber industry.

Both the promise and the difficulties of this approach through market data were suggested early, when in 1858 a resolution of the Wisconsin Senate created a committee "to collect statistics in relation to the lumbering and fishing interests in the State" and report to the Senate. The committee produced a brief table showing the number of sawmills, the amount and value of manufactures of lumber, shingles, lath, square timber, and logs, and the number of men employed in this production in 1857 in the northern forest area. Conceding that its data were often based only on "partial returns," the committee thought that the material was "sufficient to show the amount and importance of those industrial interests in our State." It had "met with considerable difficulty and delay in procuring accurate statistics from remote parts of the State, where a great portion of such business is carried on," and hence was forced to some extent to rely on estimates made from the committee's own knowledge. Moreover, "from the counties bordering on Lake Superior, no statistics have been received, and having no data upon which to form an estimate of those branches in that vicinity, your committee do not feel justified in making any estimates including that portion of the State." The committee recommended that the legislature provide for future collection of fuller information on these industries.[23]

In 1861 the legislature not only added sawed lumber and wood products to the items included in the general economic census, but also enacted the first of a long series of measures creating "lumber inspection" districts in the north country. For each district the governor would appoint a surveyor general or inspector to measure logs, timber, and lumber at the request of their owners and render official bills stating the measurements. The statute made it the duty of each of these officials to report each year to the legislature the amount of logs, timber, or lumber measured by him in his district the previous year. Despite its obvious limitations, here was a procedure which might have developed to present a comprehensive moving picture of the course of the lumber industry and of the timber inventory.[24]

In the outcome, the limitations triumphed over the potentialities. The system was in general voluntary; the lumber inspectors measured only such forest products as the owners chose to tender for measurement. Only in the Fourth (St. Croix) District, bordering Minnesota, did the legislature attempt to require measurement of all logs and lumber produced. The procedure was less decentralized in organization than that for the general economic census; the lumber inspectors were state appointees. But no central officer supervised their work, and there was not even provision for the inspectors to come together to plan uniform methods, let alone to consider enlarging or improving the system. Local politicians fought over lumber-inspector appointments as worthwhile patronage. There was a good deal of charge and countercharge concerning the competence and honesty of some appointees.[25]

The lumber inspectors themselves complained that the data they collected had only limited meaning, because private owners chose not to submit large quantities of their output to the official scaling. In 1877 Governor Ludington told the legislature:

> It would afford statistics of considerable value, if a law were passed compelling persons cutting logs within this State to report to the lumber inspector for the district the number of feet cut, and requiring inspectors to consolidate such reports and forward them to the proper authorities of the State.

The Assembly Committee on Lumber and Manufactures responded by introducing such a bill. But it was promptly tabled on recommendation of the watchers of the purse, the Joint Committee on Claims. True to a familiar pattern, a legislature which did not balk at spending state timberlands lavishly, and which was indifferent to the efficiency or inefficiency of forest use, was alert to the short-range, small economies achieved by not expanding a state service which did not promise obvious pocketbook returns. Again, in 1881 Governor Smith made the modest suggestion that the lumber inspectors be required to report to the Secretary of State as well as to the legislature, in order to facilitate consolidation of the data they collected. The legislature showed no interest even in so short a step toward more centralization of the procedure. The lack of growth energy behind the system stood most sharply revealed in the failure to enforce the requirement that all production in the Fourth District be reported through the office of the lumber inspector. This rule, the district officer complained in 1891,

> is practically a dead letter for the reason that no penalties are imposed for a non-compliance with the law, and no person engaged in cutting logs in this district has ever made any report of any kind. The great bulk of the logging and lumbering in this district is done by non-residents, who do not pretend to pay any attention to the requirements of the laws of Wisconsin that can be avoided.

Again, the legislature was indifferent.[26]

The sum of this history is that the lumber-inspector system originated as a convenience in the immediate conduct of transactions in market, and never grew beyond that character. Hence it began and continued, on the whole, as a purely optional and limited service. The Senate Committee on State Affairs accurately reflected the situation at the start of the system. When the first report was received from a lumber inspector, it was referred to the committee. A day later that body reported rather helplessly that, having had the document under consideration, "and having no knowledge of the subject referred to in the report, [the committee] return the same to the Senate without recommendation." In the long run the market itself would have been better served, had the lumber-inspector system grown into a procedure for collecting complete data on forest and mill operations — including a broader range of information on quality, costs, and prices, as well as on quantity cut and processed — and providing bases for better defining the issues of social as well as private income and outgo involved in use of the forest. The record shows that nineteenth-century market processes were not capable of generating demands on government of this sweep and sophistication. Likewise, it is plain, the legal process lacked the resource to take the initiative itself, even when market pressure had provided a starting point for growth.[27]

When at last some substantial beginning was made toward use of the state's powers and means of inquiry to lay an adequate base of fact and discussion for a broad forest program, the initiative came neither from within government nor from the market, but from private associations interested in science and in an emerging concern for social income and cost accounting. In December 1893 the Wisconsin Academy of Sciences, Arts, and Letters created a committee to press the legislature to set up a broader geological and natural history survey on a permanent and professional basis. A bill to this effect failed in 1895. But in 1897 the legislature constituted the body which was to become the permanent survey of the state, charged among other things to conduct "a study of the plants of the state, and especially of the forests, with reference to their cultivation and preservation." Likewise in 1893, efforts of a few laymen, with the sympathy of the Wisconsin State Horticultural Society and the general encouragement of the American Forestry Association, produced the organization of a Wisconsin Forestry Association. In 1897 this group successfully urged the legislature to create a temporary state Forestry Commission, to draw a plan to protect and use the forest resources of Wisconsin.

If general public opinion was still relatively indifferent, it was at least not actively hostile; these Wisconsin study measures were taken, significantly, at almost the same time that a small group of men interested in the national forest resources situation moved through the National Academy of Sciences to obtain the first broad appraisal of federal forest policy. Both the Wisconsin study bills went through the legislature with no record opposition on the merits; both proposals were enacted with the definitions of objective and organization as laid down in the bills as introduced. It was consistent with the past, and symbolic of the difficulty of creating a generous-minded enthusiasm for careful planning in the legislature, that the only obstacles raised concerned the appropriation of the modest funds sought to sustain the two projects. In each instance the legislature's influential financial watchdog, the Joint Committee on Claims, raised objection on this score; the committee was finally persuaded to yield in the matter of the temporary forest commission, but it had its way in cutting by two thirds the money and the working time authorization asked for the new geological and natural history survey. However, the important point was that these measures finally committed the state to at least some minimum responsibility for long-term programming of policy toward its forest wealth. The change came too late to affect the fate of the original forest, and thus the work so begun relates rather to the last part of this history, dealing with the aftermath of the great lumber industry, than to our present concern.[28]

DIRECT REGULATION OF INDUSTRY OPERATIONS

In nineteenth-century legislatures, whose instincts were to operate by rule of thumb, indifference to finding and analyzing facts did not necessarily mean that nothing would be done. The Wisconsin legislature churned out scores of statutes disposing of state lands, chartering companies, and providing systems of liens and security-title recording, for example, without ever an inquiry or hearing on the general policies involved. On the other hand, given a want of special pressures for action, the want of regular official procedures for collecting and analyzing information relevant to general policy reinforced the inertia of the situation. It is not surprising, then, to find that the thin record of official inquiry is matched by a thin record of legislation relevant to the efficiency of forest yield.

The times put a high premium on rising productivity. So far as government could affirmatively assist present production, Congress and the state legislature were generally prepared to act. Chronic scarcity of fluid capital made the state, at least, wary of action which called for cash outlay. We have seen, however, that both the national and the state governments acted on the almost unquestioned principle that it served public interest to spend public lands generously in aid of private production. It was consistent with this pattern that without controversy the Wisconsin legislature answered the recommendation of the Lapham commission in 1868 by tendering tax exemptions and credits to induce planting of tree belts, as a few years later Congress offered homesteads to those who would cultivate trees in the prairie. More efficient forestry might also be promoted by publicly supported research and education. The 1867 Wisconsin statute authorizing the Lapham commission asked the commissioners to determine whether a want of information was contributing to undue destruction of the forest, and whether scientific investigations should be made "to ascertain best methods of growing and managing forest trees." And the legislature instructed the commissioners "to report generally such facts, in a condensed form, as will be of the most

Costs of nineteenth-century woods operations: The aftermath of clear cutting.

The aftermath of fire.

Labor in the woods: A north woods camp, 1885.

Cook shack, 1907.

practical use to persons desirous of increasing the growth and preservation of forest and other trees." The bulk of the commissioners' report was descriptive or concerned the general relation of tree cover to the physical and economic condition of the state. However, the commissioners did make modest suggestions for more efficient harvesting of timber. Subsequently the legislators provided for publication and distribution of two thousand copies of the Lapham report. It was characteristic of this cash-scarce community that a year later, on grounds of economy, a legislative committee recommended against a second edition.[29]

To teach men or to offer them bounties to practice more efficient forestry — as also to encourage them to enter production in the first instance by cheap sale or grant of public timberlands or by public subsidies to improve or build transport facilities — was in a sense to regulate their conduct. But these measures were more obviously also forms of public subsidy to private production. Prevailing nineteenth-century opinion believed in such state intervention. Wherever the law could mobilize resources and use them with multiplier effect to increase production and to release the creative energy of individual will and shrewdness, men were willing to employ it so. They held back only where a procedure would call for spending much cash, for cash was scarce.

So long as government sought to shape situations by subsidy, public policy and private desire were likely to have enough in common so that conflict was at a minimum, save as men competed for the prize. Of course, where law undertook to fix standards by teaching them or by offering bounties for observing them, it undertook to regulate men's wills; there would be no occasion for instruction or bounty if operators would behave the same without them. But there was little overt coercion here; those immediately affected did not pay the bills, and the tax collector operated some steps removed from the legislature's appropriations. The times were too preoccupied with the heady bustle of market growth, to give attention to close points of political analysis. Thus regulation by subsidy did not raise in men's minds sharp issues either of principle or practice concerning the relative spheres of public and private decision making. If little was done to affect efficiency of forest use by teaching or bounty, it is plain on the record that this omission was due to indifference and not to formed opposition.

To fix minimum standards of efficient forest use by law, and to implement them not by persuasion but by penalty, presented the competition of public and private decision making in sharp and unmistakable outline. Passing for the moment the problem of fire, the facts of forest-use control may be quickly summarized, for they are few. The federal government did nothing, either as landowner or as sovereign, to impose any minimum requirements for efficient realization of the Lake States forest wealth. The nearest it came to doing anything was by some measures to improve and police the navigation of principal rivers; most of this effort — discussed elsewhere in this book — represented the familiar pattern of subsidy rather than direct regulation. The state likewise enacted no direct regulations to exact minimum efficiency in logging or milling. It, too, paid attention to stream traffic, but where it did not act by subsidy to improve carrying capacity its regulatory concern was primarily for traffic control. First the state, and later the federal government, attempted some regulation of rates charged loggers and lumber dealers by water and rail transport facilities; this is discussed later in this Part. In a sense this effort involved some regulation in the interest of socially efficient return. Producers and consumers of lumber should not have to suffer exactions of strategic middlemen, unrelated to functional costs of serving the lumber market. However, shippers of logs and lumber, and not consumers, were the prime movers in such regulation as was attempted. The record shows little to indicate that their concern was other than for the incidence of costs and the distribution of rewards within the framework of the industry as given by circumstances. In any case, transport rate regulation had no traceable effect on efficiency of woods and mill operations, whatever its impact on distribution costs.[30]

Through all the years of lumber boom the Wisconsin legislature enacted only two regulations of industry costs in the woods, apart from fire control. As early as the Revised Statutes of 1849 the district overseer of highways was given an action for treble damages for injury done by drawing logs or timber on the surface of any road or bridge. It was, of course, a harm that came close home to local taxpayers. Beginning with the Revised Statutes of 1858 it became a misdemeanor to fell any tree so that it blocked a navigable stream.

Wisconsin law did not concern itself with direct production efficiency of sawmills. However, two types of regulation developed in recognition of two important types of indirect social costs of milling operations. One of these types of regulation was the imposition of penalties for dumping into streams such mill refuse as slabs and sawdust. This kind of law began as early as 1856, when a statute made it a misdemeanor "for any owner of a saw mill or saw mills on the Wisconsin river, or any of its tributaries, or any other person to throw or permit or suffer to be thrown [therein] . . . any slab or slabs or other substances which may obstruct or hinder the navigation of said river, from or near said mill or mills." This style of regulation in part served a general interest of the lumber industry itself. Beginning with an act of 1860, however, the legislature developed another line of prohibitions on dumping wood-processing refuse into streams, to the damage of the fisheries.

Logging, driving, and sawmill operation also involved serious costs in injuries to workers. Nineteenth-century legislation paid no attention to the human costs of woods or river hazards, a fact not unrelated to the absence of any successful labor organization in that sector of the industry before 1900. In the absence of legislation, the only provision for administering the private and social costs of industrial accidents in the industry was by lawsuits, in which the injured employee faced the formidable obstacles of the employer's defenses under the fellow-servant rule and the doctrines of contributory negligence and assumption of risk. Labor organization first showed some force and effect in the mills and woodworking plants. The first statute requiring guards to be placed about dangerous mill machinery was passed in 1887, at a time when the passing success of the Knights of Labor had focused attention upon labor's claims.

Both in the matter of stream obstruction and pollution and in the case of industrial accident, there were immediate savings to mill operators in not spending to comply with statutory standards. Thus these were types of regulation the practical significance of which depended almost entirely on the vigor of enforcement. Wisconsin's twentieth-century experience was that specialized administrative procedures were essential to obtaining any decrease in latter-day pollution of streams from industrial waste. Since the nineteenth-century statutes left enforcement wholly decentralized and at the dubious initiative of local prosecutors, it is a fair inference that they had little effect. No suit arising under these statutes reached the state Supreme Court, and no suit under them appeared in the circuit court of a great lumbering county, Chippewa, whose business we know in detail. The case was better regarding the factory safety laws. In 1883 the legislature created the office of labor commissioner. At first the commissioner did little save receive and compile data. But the 1887 law which required guarding of dangerous machinery also gave the commissioner an increased staff of inspectors, and the official reports from then on show considerable activity in field work to check on observance of the statutory standards. Evidence indicates that by the mid-1890's there had been a substantial decrease in the more serious mill accidents.[31]

However, the scattered legislation that dealt with secondary social costs of logging and milling is all collateral to the central concern of this chapter, which is the relation of law to the real productive efficiency of the forest itself. On this front there was no legislation at all. The point is sharpened by the summary defeat in 1872 of the one proposal that was made for some measure of state care for maintenance of timber stock. Bill 265A would have created a state superintendent of forestry, charged to acquire "seeds and seedlings of timber and forest trees" out of moneys there appropriated, to distribute these free on request to "owners of occupied lands," to collect data and circulate information on timber culture, and regularly to inspect in the field the progress made in planting and cultivation of timber stands. Like the Lapham commission and the tree-belt statute of 1867–1868, this 1872 proposal had more apparent pertinence to the agricultural south of

Wisconsin than to the pineries. Yet, like the lumber-inspection system, it might have given the necessary first momentum to a wider program related to the great forest. Without opinion, the Assembly Committee on State Affairs recommended that bill 265A be indefinitely postponed; this was a committee much concerned with state finances. The bill's sponsor succeeded in keeping it alive by recommittal to the Committee on Agriculture—which, presumably, might be more attentive to its substance than to its draft upon the state budget. However, the Committee on Agriculture likewise reported adversely, and the Assembly put aside the proposal. The second committee spelled out its reasons. It was not opposed to the substantive objectives of the proposal, but simply to the likely cost in direct cash outlay:

> . . . the state would be greatly benefited, made more beautiful, healthy and valuable by adopting some cheap and efficient system to encourage the growing of forest trees in lots along the highways, and by the homes of its inhabitants. The exemption of timber plots from taxation for a term of years might be a sufficient incentive, or small premiums might be preferable. [But] the bill or the system proposed to be adopted is too expensive, in the opinion of the committee . . .[32]

The indifference of Congress and the state legislature to levels of efficiency in forest production would have had only limited long-term effects, however, had it not included throughout the nineteenth century a want of any adequate attention to loss by fire. There are no cumulative data on total fire losses in Wisconsin during the years of the great forest. Informed estimates for years in which spectacular fires called special attention to this destruction show the scale on which loss could occur. Thus in 1886 it was estimated that 500 million board feet of pine burned in the Menominee River district alone. In 1908 more than 1,200,000 acres were burned over; loss in new growth alone was estimated at over $6 million. Though fire caused substantial losses in mature timber, the more serious cost was in loss of new growth and—after repeated burnings in the same area—in the loss of soil conditions favorable to new growth. Some professional foresters of the twentieth century estimated that had there been only modestly effective fire-prevention and fire-fighting provisions in the late nineteenth century, this without other use of the police power would have preserved the Lake States forest as a major saw-log producer for an indefinite future.[33]

Congress paid no attention to the fire hazard to national timber resources until nearly the turn of the century. In 1897 it took its first step—the most limited measure open to it—by enacting criminal penalties for willful or negligent setting of fire on the public domain. Meaningful federal effort to do something about forest fire prevention dates in effect from the 1905 publication of the new Forest Service's handbook, "The Use of the National Forest Reserves." As in all other major respects, so here the United States did not begin to act in time for its action to have important effect upon the handling of Lake States timber.[34]

In 1838 and 1841 the legislature of Wisconsin Territory declared it a misdemeanor maliciously, intentionally, or negligently to set fire to woodlands to another's damage. According to the particular provision violated, the sole penalties provided ranged from minima of $5 or $10 to maxima of $50 or $500; no jail sentence was provided, though the offender might be held until he paid the fine and costs. The Revised Statutes of 1849 incorporated the substance of these provisions. No more attention was paid the matter until an 1873 act fixed a term of 30 to 90 days in the county jail or a fine from $20 to $500 for anyone who should set on fire any woods not belonging to him during any August, September, October, or November. Taken literally, the 1873 act required no showing of intent or negligence, but merely the fact of causing the fire. The Revisers of 1878 merged all these provisions and markedly narrowed their scope. One was liable who fired another's woodland "wilfully, maliciously or wantonly." But whereas the earlier law had rendered liable one who "intentionally or by neglect" allowed fire to escape from his own land to another's, the 1878 version limited this liability to a landowner who allowed fire to escape him "intentionally, or by gross negligence." Moreover, the Revisers limited the apparent stringency of the specific dry-months provision by declaring explicitly that the state must show that the accused intentionally set the fire. Again several years passed with no action. In 1887 the legislature authorized any town board of supervisors to make a finding that the setting of fires or burning of logs or brush in the town would be a public danger because of dry weather, and thereupon to publish and post a fire ban for such period as the board might deem necessary, to "terminate when the cause or reason . . . shall cease to exist." For violation the act stipulated a maximum fine of $50 or a maximum term of 30 days in the county jail, or both. This was the closest the legislature came to any regulation of the fire hazard created by slash. This limited measure only set bounds on initiative taken by the property owner or logger, imposing on him no affirmative duty to dispose of forest waste in a way to reduce the danger of fire.

In 1883 an Assembly resolution created a select committee "to inquire into and investigate the causes of the destruction of so much valuable growing timber by fire, and to consider the best action to be taken by the legislature for the protection of the forests of the state from fire." But this promising declaration proved to be too far ahead of its time for action, and nothing came of it. Finally, after some unusually severe losses in 1895 the legislature enacted the first statute to deal with the forest fire problem in any elaboration. This act designated as state forest warden the chief clerk of the commissioners of public lands, and as town fire wardens the town supervisors and road superintendents under the chairman of the town board.

It charged these local officers "to take precautions to prevent the improper setting or progress of fires," to summon and direct men in efforts to extinguish fires, to limit and license fire setting in the dry months, and to post fire warning signs in places frequented by sportsmen. In vague terms it charged the state forest warden to supervise the local officials, to draw and publish regulations to govern their activities, and to keep records of fires reported and complaints of violation of the act. The statute tightened the duties laid on the general public. One who set a fire on land not his own or under his control must extinguish the fire completely before leaving it. One who "wilfully or negligently" set a fire on another's land or allowed fire to escape from his own land to the injury of another was guilty of a misdemeanor; there was no longer a separate provision against malicious firing, nor was it any longer necessary that the state prove gross rather than ordinary negligence. New detailed duties were put upon railroads, which must at least once a year cut grass and weeds on their rights of way, equip all locomotives with spark arrestors, forbid dumping of fire or live coal or ashes outside of yard limits, and instruct employees to report all observed fires and to act to help put out fires. In general, penalties were set more modestly, however, than under earlier statutes; for most offenses the act set a maximum fine of $100 and a maximum jail term of one month, though it did authorize the court to impose both sanctions. In each session after 1895 the legislature tinkered with details of this system without altering its substance, save to press a little toward more centralization of authority by giving the state forest warden authority to appoint local officers and by enlisting as fire wardens ex officio the newly created state fish and game wardens. The next substantial change in state policy came with Chapter 450 of the Laws of 1903, establishing a system of state forests and creating a state forestry office. But the 1903 statute seems functionally a part of the history of the policy developments dealing with the aftermath of the great lumber industry, and will be noted later in this context.[35]

The only praise that can be spoken of this legislative history is that the policy makers early acknowledged that there was a social interest in limiting forest fire damage. Otherwise the record is a classic of ineptitude, characteristic of nineteenth-century statute law making. The legislature wavered unreasonably in its attitude toward the relative importance of the social interest in timberlands productivity, compared with immediate freedom of individual action. Penalties changed toward leniency; from 1878 to 1895 the careless fire setter went free, unless the state could prove gross negligence. Until the modest act of 1887 authorized town boards to prohibit fire setting in dry months, the law relied wholly upon penalties after the event; not until the 1895 statute was there even formal declaration that preventive action was at least as important a general form of regulation as was punishment for damage done. It was characteristic that a legislature which did nothing about fire prevention was yet ready to make humanitarian grants of money and seed to local governments and people who had been burned out. Policy could encompass after-the-event sentiment. Until 1895 enforcement was wholly decentralized. Before 1887 responsibility for the public interest was committed solely to local constables, sheriffs, and public prosecutors; the dry-months preventive act in 1887 brought town boards into the picture, but the initiative was theirs alone, with no central supervision. When the 1895 act first recognized that the interests of the state economy required some positive direction from the state government, it did so only in grudging fashion, committing the state's role to an officer of clerical rank and of clerical functions, rather than to an experienced field officer. Since the 1895 statute represented at least a formal break with the past and a move toward emphasis on centralized preventive action, its grudging financial provisions were especially revealing of the difficulty legislators had in taking a new policy line of substance. The designation of the chief clerk of the state lands office as state forest warden carefully stipulated that he should have no extra pay for the added job; the short-sighted economies of such ex officio appointments had already cost Wisconsin dearly in the handling of its public lands. The 1895 act's further restrictions on the pay of the local fire wardens cumulated into what might seem a parody of economy in government. Local wardens and those whom they employed to help in putting out fires should be paid by the towns not more than $1.50 per day; no warden should be paid in any one year for more than ten days' service in preventing or putting out fires, or for more than five days' service in posting fire notices, nor in total for more than fifteen days' service altogether; no one employed to help put out fires should be paid for more than five days' service per year; no town should spend more than $100 of public money under the act in any one year; no payment should be made to any warden or person employed by him until the claimant should have rendered a sworn, signed, itemized account for his services; for the state's role in the system the act appropriated an annual sum not to exceed $300. The best that can be said for the limits the legislature put on field fire-prevention and fire-fighting expenditures is that they may reflect a realistic judgment that local officials would use more money only as a means of political hand-outs or patronage. In proportion as this was so, however, it underlined the need for stronger central action, for which the legislature also so signally failed to provide.[36]

DELEGATED REGULATION OF INDUSTRY OPERATIONS

If the record of direct public action to conserve timberland productivity is so thin, perhaps this means only that we should look in another direction. Public policy favored delegating to private hands decisions over the objects and manner of using resources — through the law of property and contract, with supplementary regulation from the law of tort. Conceivably this is where we might find our conservation law. In fact, however, if we look here we find only confirmation of the limits of will and imagination that mark the public record.

Of course, landowner-logger and logger-buyer contracts typically stipulated as to the size and condition of timber to be cut or furnished. But this was done simply incident to immediate transactions for the sale of labor or goods. There is no evidence that nineteenth-century dealers made any material use of contract terms to protect the productivity of the forest.

Closer to the point, at first appearance, are the rather numerous reported cases which define as "waste" certain wrongful cutting of timber and give redress therefor, whether directly by judgments for damages or injunction, or indirectly, incident to lawsuits to try title or to enforce security interests in land. Again, however, the time point of reference proves to be quite other than protection of the real productivity of timberland. These lawsuits sought to redress or prevent cutting which was condemned not so much for impairing real productivity as for impairing the land's present money worth; most of the cases concerned protection of the money value of land as security for unpaid balances on the price, represented in a vendor's lien or in a mortgage. The measure of the stakes by a money rather than by a real calculus was especially plain in rulings that an injunction would lie against a mortgagor's logging only on proof that the mortgagor was insolvent or that the value of the land and timber was scant coverage for the debt.

Nineteenth-century policy also relied on private initiative to set in motion the processes of defining and legitimating social values through tort litigation. The law of trespass dealt with invasion of a timberland owner's interest in deciding for himself when he would bring timber to market. Theoretically the landowner might seek damages for being deprived of the opportunity to exercise a long-term future preference. In practice both statute and case law in Wisconsin assumed that damages would be measured in loss of present market value; the significant indicator of prevailing opinion is the fact that the value of future opportunity costs was not discussed, nor was loss of immature timber pressed as a special item of damages, in any reported Wisconsin case. Conversion apart, the only civil wrong affecting timberland values with which the courts found occasion to deal was for damages caused by wrongful handling of fire. Doctrine developed here along familiar lines of liability based on fault. The Wisconsin Reports were liberally studded with such lawsuits. Significantly, few cases reached the Wisconsin Supreme Court — and no case appears in a detailed study of the business of the circuit court of a busy lumber county, Chippewa — in which the primary damage claimed lay in destruction of standing timber or destruction of the productive capacity of a forest tract. Almost invariably the suits were for present losses of immediately marketable goods — logs, cordwood, lumber — or of such production facilities as woods equipment, camps, or sawmills. It seems more than coincidence that the five cases in which loss of productive forest was a prime item of damages claimed came at the turn of the century — when cutting had proceeded far enough that men could at last foresee the end of the great forest, and hence began to value the forest as a productive facility and not merely as a storehouse of an inexhaustible inventory.[37]

COMMUNITY VALUES AFFECTING USE OF THE POLICE POWER

The record offers plenty of evidence to help explain why nineteenth-century federal or state law gave so little attention to levels of efficiency in harvesting the forest. Point for point, the reasons are substantially the same as those which induced easy disposal of public timberlands. There is no need again to rehearse them. However, certain points take on special emphasis in the context of law's relation to private owners. Uncritical popular and official acceptance of the fee simple as the norm of title promoted ready disposal of public timberland. The same attitude even more naturally deflected attention from serious attempt to calculate social income and outgo consequent upon the uses that private owners made of the initiative which the law granted them. Equally powerful and much more explicit were two assumptions of fact by which men estimated the common sense of their situation. They assumed that the forest was "inexhaustible for ages," "practically without limit," "endless." They assumed that agriculture was by nature inevitably Wisconsin's fortune, and that the normal use-history of any given tract of land in northern Wisconsin would be for a farm to follow the forest when the pine should "fail," as they assumed it must finally do in any particular area subjected to logging. Common opinion held to the first assumption into the eighties, and to the second into the early twentieth century. All of these attitudes gained momentum and were the less subject to scrutiny because they operated within the field of the deep popular faith in production. A sharply rising over-all material productivity was the peculiar dynamic by which we would make a new kind of society, capable of realizing a fuller life for more people. This we experienced as the central challenge of an unopened continent, in times of an expanding market and an unfolding science and technique. Spurred by the challenge, we were the more impatient with the frustrations of the gross imbalance in the factors of production available to us, when land seemed so obviously cheap and fluid capital was so obviously expensive. Thus the situation bred an overriding present preference for employing the physical wealth of land to obtain the highest immediate yield from cash and labor. The policy of ready and cheap transfer of public lands to private ownership not only expressed this point of view, but — once well launched — tended to confirm it as the common sense of the situation, by emphasizing the easy availability of raw materials.[38]

Conflict is markedly absent from this record. The lumbermen were politically powerful, thanks to the wealth and position which public policy helped them gain. They were powerfully represented in Congress and in the state legislature. Had serious efforts been made to impose legal restraints upon their manner of operation, they would undoubtedly have opposed these, as Minnesota operators did oppose belated efforts there to regulate disposal of slash. Probably in such a contest the lumbermen would have had their way. At least up to the nineties, had events required so, probably they could have used their political power to insure that some embarrassing proposals were not even presented to the state legislature. This is all supposition, however, for there was almost no occasion for the industry to muster its forces against outside intervention. In Congress the proposal in 1876 to reserve southern federal timberlands in order to control use of this resource presented the closest analogy that arose for any test of force relevant to handling the Lake States forest. This suggestion was turned back with little difficulty, and with little effective counterattack by its proponents. In Wisconsin the only substantial conflicts at law concerning development of the north country did not occur between the lumber industry and proponents of other interests. The battles fought were more like family rows, in which factions representing different segments of forest industry or competitors of like general character fought for market advantage or for subsidies. Investors in standing timber had an interest in more stringent laws on trespass that was not shared by millmen and distributors. Mill operators at Chippewa Falls fought bitterly with those at Eau Claire for control of the Chippewa River. Rival lumbermen were among those who competed through various railroad promotion schemes to lay hands on the richly timbered railroad-aid lands of the "St. Croix grant." But none of this controversy posed issues of conservation policy. So far as there was controversy in Wisconsin over modest proposals affecting forest conservation, it took the form of successful opposition to increasing state cash outlays, whether for subsidy, for data collection, or for regulation. In a broad sense this alert regard for holding down the state cash budget undoubtedly expressed the point of view of the lumber industry as well as that of business generally. But it was a viewpoint which expressed itself in other fields, in a common nineteenth-century acceptance of the notion that most public business would be done through local government and that there was need only of minimum state apparatus. There is no evidence in the Wisconsin record that the budget paring done on the few forest conservation proposals that came up expressed any special reaction to this particular subject matter.[39]

The pattern would be simpler, if we could leave it thus — a pattern formed not much of overt conflict, but mainly of uncritical acceptance of certain very generally held symbols of value (the fee simple title, the virtue of productivity), fashioned upon the warp and woof of circumstance (the imbalance of production factors, the pace of population growth and technical development). The pattern is not so simple, however, because it does include some suggestions of very different design. When the Wisconsin legislature authorized the Lapham commission in 1867, its statute in effect posed quite sharply some basic questions of social cost accounting related to timber wealth: (1) to what extent present use was competitive with future use; (2) whether the market calculus governing timber use was consistent with adequate attention to natural balance and to the economy as a whole; and (3), most pointedly,

whether, owing to the want of information in individuals and the shortness of their lives, it is the duty of the state to interfere [interpose] its authority to prevent an undue destruction of forest trees where they now exist, and to encourage their cultivation where they are deficient.

The Assembly select committee which urged this enactment spoke likewise in terms that reminded the legislature that even in a society which dispersed power broadly through the market, care must be taken for factors and programs that pertained to the strength of the commonwealth; only with the support of measures which they took in common could men enjoy the freedom otherwise to go their separate ways. So the need of wood

is a want felt by every man in the entire community; and being a common want to the whole people, we are of opinion that it is the duty of the government to take steps to preserve what remains, and increase its quantity where it is scarce.

Some measures needed to promote the general economy could be taken only by the broad concert of action for which law must take the lead. So it was, the committee argued, of the proposal for public subsidy for the growth and maintenance of tree belts:

. . . to be successful, more than one individual must engage in the work. Single groves and belts would be prostrated by winds which have swept uninterruptedly from the summits of the Rocky mountains, and yet be entirely broken up by the groves and belts of a town or county.[40]

Some of the arguments with which Wisconsin policy makers first urged Congress to grant federal lands in aid of an agricultural college, and later pressed for speedy action under the Morrill Act grant, acutely presented the problem of valuing present over future yield, and the need to recognize that the "cheapness" of land might weigh too heavily in resources use, if the market alone governed decision. "[P]artly the cheapness of our lands, but more largely . . . the lack of scientific knowledge" explained why "we have been characterized the most thriftless class of farmers in the civilized world," the Committee on Agriculture and Manufactures explained to the Wisconsin Senate in 1858. In 1863 the same committee urged acceptance of the federal grant, because

as an agricultural State of immense resources, and yet already suffering the evils of that unsystematic and exhaustive practice which is the bane of American agriculture, Wisconsin cannot, consistently with a true political economy neglect to adopt every measure that may reasonably be taken to foster and build up this primary interest of the State . . .[41]

It is significant that these were arguments for public intervention by subsidy and not by command. Though contemporaries did not see that the analysis was as relevant to the forest as to the farm, when men did begin to apply such arguments to forest conservation the more prudently practical of them did so likewise with emphasis on persuasion and education rather than on mandates. In 1885 Governor Rusk transmitted to the Wisconsin legislature, without recommendation or comment of his own, a memorial of the American Forestry Congress bearing the supporting endorsement of the United States Commissioner of Agriculture and his Chief Forester. Recommending government action to develop and to spread knowledge of better forestry practice, the Forestry Congress was cautious in venturing further: there should be study to recommend desirable legislation, as the country's wants in this matter were better understood. For "the differences that exist between the American states and countries of Europe, as well with respect to the tenure of the land as the structure of the laws . . . would prevent any one of these European codes of forestry from being applied in America . . ." Enlightened cooperation must be government's main theme here:

Since in the states and territories of the United States, as well as in the Provinces of Canada, most of the settled portions of the country belong in fee simple to private owners, who are usually the actual occupants; and since the entire care and cost of management of the woodlands upon these estates must devolve upon these owners, it is evident that there can be no more effectual means devised for promoting this object than by the diffusion of correct ideas among the owners of these lands with reference to the forest-interests of the country.[42]

The cast of such contemporary comment teaches us caution in speculating what different directions nineteenth-century policy might have taken. Had law given more leverage to innovators of conservation policy than it did, its force would more likely have been exerted along lines of subsidy than of command. But the legislature, we must recall, also made some small start toward requiring that the lumber industry account for indirect social costs of its operations. It declared loggers liable for damage to public roads. It declared millmen liable for damage to fisheries and to the navigable capacity of streams, and required them to reduce the human costs of production by guarding dangerous machinery. It is hardly surprising that the seventies and eighties did not produce regulations in the pattern of 1950. Nonetheless, the nineteenth-century record does show the stirring of a new kind of legal initiative. That these beginnings did not develop more rapidly or on a wider front is not explained, on the record, by controversy and identified opposition. Undoubtedly the opposition would have mobilized effectively and controversy would quickly have become overt, had the initiative been pressed. Two facts which do make a pattern on this record are (1) the limited and episodic expression of early interest in adding a calculus of social income and outgo to the accounting of the market, and (2) indifference and lack of conviction toward developing the fact basis of public policy. The issue, we have seen, was no lack of constitutional authority; our tradition was pragmatic and flexible in using law to promote community development. Not constitutional limitations, but limitations of will, imagination, and energy determined the working scope of the police power. Conflict cannot enter history until scarce values, and interests relevant to them, are brought to some definition. At once the most radical and the most subtle use of police power is that which employs resources mobilized by law to bring issues implicit in circumstances from below the level of conscious interest into the arena of choice.

CHAPTER VII

REGULATING USE OF PRIVATE MOBILE CAPITAL: THE COSTS OF MONEY AND LABOR

THE analogies that exist between ideas and action embodied in economic analysis and decision and in law reflect the reality that developments in both fields respond largely to facts of scarcity. Scarcity tends to make markets and to invite the making of law. Both market and law offer processes to resolve tensions created when wants and means are not in ready adjustment. Abundance may remove the need of markets, or at least make them very simple. Abundance, likewise, blunts the edge of demands that might otherwise produce greater or less response in public policy. We were slow and inept in defining and responding to what was at stake in the use of timber wealth, because for a long time we thought of it as practically a free good. On the other hand, the limited and irregular concern shown in law about the supply and administration of mobile capital essential to the lumber industry's growth demonstrated that scarcity alone does not necessarily produce legal response. By 1900 the larger companies in Wisconsin had so prospered that they had no difficulty in supplying their own working capital needs; indeed, some of them shortly began to seek profitable investment opportunities out of state to occupy their surplus funds. But through the first generation of Wisconsin's lumber boom all operators, and thereafter all but the biggest ones, experienced chronic scarcity of working capital. The men who came to work in the woods, on the river, or in the mills had their own experiences of capital scarcity, too. For want of current cash they must borrow at high interest from their employer, though the employer owed them deferred wages. Lacking a broad market for their labor in the typically isolated situation of the north Wisconsin camps and mills, they chafed at ties to a particular job market and a dominant employer: "Work is uncertain unless you do just as employers wish you to do." A man might fear to commit himself to home ownership and permanent roots, because: "The town is too small. There is only one company here, so that a man for almost any cause might be obliged to go elsewhere to seek work, and so abandon his home at a loss." The pressures of scarcity were real. That they evoked quite limited response in law poses a question the exploration of which shows further aspects of law's interplay with circumstance and men's values.[1]

The immediate reason is plain. The presumption of nineteenth-century public policy was that problems of scarce economic means in relation to economic and social wants should normally be resolved in the market. It would put an unrealistic gloss upon our practice, however, to read it as expressing doctrinal insistence on laissez-faire principle. True, like the popular prestige of the private fee simple title to land, popular acceptance of contract dealings in market (as the normal procedure to adjust scarcities) in some degree conformed to our faith in the virtues of a society in which decision making was widely dispersed. So we would check abuse of power and encourage the creative energies of the greatest number of minds and wills. By the same token, however, acceptance of the market was in substantial part the product of custom and drift, following in the wake of familiar but vaguely apprehended symbols which enjoyed a prestige little analyzed in official or popular discussion of Wisconsin's booming lumber industry. Moreover, always in the background was a deeply felt principle of popular policy broader than reliance on the market. This was the belief that all public policy, including the favor shown market processes, was in the end only an instrument to serve individual life. We believed that life's meaning lay in growth and enlargement of individual power and experience. Our circumstances in a rich and relatively unoccupied land taught us that the primary means of growth were rising material productivity and expanding chances for men to exercise their wit and energy upon the economy. This was our belief in social process at a deeper level than our faith in the market. If we saw positive or regulatory uses of law which might be useful to this more profound level of social process, we showed little hesitation in using the law so. Especially where we could believe that legal intervention might multiply opportunities for releasing individual energy, we did not find it inconsistent with our belief in dispersing power to use law for such multiplier effects. Our use of the public lands made this clear. Thus, the balance we struck between relying on the market or on the law to adjust to problems of scarcity seems typically to have turned rather on our judgments of what was the most practical means to a given end, and upon the extent to which our imagination grasped all the operational factors in a situation, than upon a lively regard for general principle.[2]

Experience taught us eventually to recognize two kinds of limitation on market processes which might make them inadequate to render socially satisfactory solutions of problems of scarcity. One of these limitations might derive from circumstances outside the market. The market assumed that on the whole those who entered it were equal enough in means so that one or a few could not by their own manipulation coerce the rest. Where some dealers in market showed persistent relative weakness so great as regularly to subordinate them to the manipulation of more powerful dealers, the weaker interest was likely over the long run to try to better its bargaining position by law. Another kind of limitation was functional, inherent in market processes themselves. The market's social utility was to allow wide dispersion of decision among uncounted centers of energy, and yet to channel these energies into effective total results. To be effective, this process required that projects not demand resources on a scale practically beyond the reach of numerous decision makers. Yet, the market could work only within a sturdy social context. Just because of its dispersion, it required the support of framework institutions outside it — as, for example, the law of property and contract. A good deal of the history of the police power is the history of change in official and popular concepts of what was needed to maintain this sturdy social context, and hence by definition what must be supplied from outside the market, by law. Another functional limitation of the market inhered in its reliance on individuals or small groups to initiate and sustain action within its procedures. Again, this was an aspect of the market which contained much of its strength; this was its function as a setting for release of creative energy. It was nonetheless also a limitation, for it meant that only those items of income and outgo which impinged on the affairs of individual or small group dealers with sufficient precision to be counted and weighed by them would figure in market decisions. Gains and costs, however real, would not enter the market calculus, if their incidence was too general to be defined in private bookkeeping or to move private interest. Thus another main current in development of police power was to write into law critical heads of accounting for social income and social costs.[3]

In various reactions to capital scarcity in Wisconsin's lumber industry, we see men grope for workable definitions of a sound relation of law and market as they experience market imperfections. What we find in result is no well-ordered or comprehensive pattern of public policy. The point is, rather, to see the difficulty of bringing policy choices to definition. As in development of forest conservation policy, so here the primary problem was not to decide among actively competing interests, though explicit conflict figured more in our present subject than it did in analogous phases of conservation policy. Men cannot press interests, let alone fight over them, until they acquire some idea of interest, of the relation of interest to situation, and of means for achieving what they want or think they want. In this light the history of police power in any field begins with men's perception of their situation, with the effect of this perception (be it more or less adequate to realities outside the observers) upon the demands men make on law, and with the role of law in bringing perception into focus or blurring it. Because law offers regular procedures and holds out legitimated rewards for bringing interests to definition and lending them force, one of its prime functions may be to help overcome the inertia of circumstance and the mindless drift set in slow current by mere cumulation of events. This stage prior to handling overt conflict may lack the drama of legal battle. But it involves law at a still deeper level of cause and effect. How effectively or ineffectively law contributes to enlarging the action of mind in events thus becomes a significant part of the history of law's social functions.

THE TERMS OF SUPPLY OF MONEY CAPITAL

In four respects law reflected concern to regulate terms of capital investment during the sixty years of Wisconsin's lumber boom. These legal activities had to do with the cost of borrowed money, protection of corporate creditors and stockholders, scrutiny of private combinations to stabilize product prices, and use of federal tax power to close the market against Canadian competition. We speak here of regulations affecting the flow, safety, and returns of invested capital as such; where capital was invested in public utilities offering water or rail transport facilities for the lumber industry, special issues of regulation were presented, which are discussed in Chapter IX. Too, we speak here of regulation in its strict sense, of the law's intervention to impose duties or disabilities without regard to the consent of those so burdened. In Chapter V we examined the terms on which Wisconsin law made available the corporate instrument as a device to help mobilize capital and put it under private discipline. There, also, we examined the more grudging terms on which the legislature provided suppliers' liens on logs and lumber, to encourage extension of credit to the industry's shoestring operators. These were uses of law relevant to the supply of money capital. But they show another function of law, its promotional or "subsidy" role, in contrast to its role of imposed adjustment which is our present concern. Here we consider only aspects of the state's business laws that imposed the state's will on the parties, rather than tendering legal facilities to assist the parties' wills.

Statutory limits upon the maximum interest which lenders of money might fix and enforce against their debtors were a familiar type of legal redress of unequal power in the market even before Wisconsin became a state. Justification for such regulation sometimes ran in terms of preventing oppression offensive to the popular sense of justice. On the whole, however, more weight seems to have been given an argument of economic function: that usury laws preserve socially desirable productive energy from such undue burdens as might discourage its expenditure. But a community keenly conscious of capital scarcity would weigh carefully the counterconsideration — that the law should not be invoked in a fashion to discourage potential lenders from coming forward with badly needed funds. The argument carried the more weight because capital scarcity was likely to characterize a rude new economy, where risks ran high. The pull and haul of these contentions was plain on the Wisconsin legal record, though they never came to special focus upon the lumber industry in distinction from other fields of investment.[4]

Borrowing from older states, the Statutes of Wisconsin Territory in 1839 fixed 12 per cent as the "legal and valid" maximum rate of interest "which persons may agree upon," and 7 per cent as the rate to prevail under instruments "wherein the rate of interest is not otherwise specified." But the new state's Revised Statutes of 1849 declared that "any rate of interest agreed upon by parties in contract, specifying the same in writing, shall be legal and valid," though in the absence of specification 7 per cent should be the legal rate. However, in 1851 the legislature fixed the maximum contract rate of interest at 12 per cent per annum; in 1860 it put the lawful maximum at 10 per cent; in 1862 it lowered it to 7 per cent; in 1866 it returned to a 10 per cent maximum, and this remained thereafter the lawful ceiling.[5]

The period of lively concern and change was thus the time when the establishment of farms and small towns was fixing the social scheme of Wisconsin south of the great forest. The defense of usury came into play before the Wisconsin Supreme Court, then and later, almost always in connection with dealings in farmlands or town lots or the affairs of merchants or general manufacturers.[6] Logging contractors and millmen, including some whose operations were substantial, commonly paid the maximum lawful interest and more than that, to get working capital. When business was bad generally, and the lumber market sagged and customers were slow to pay their bills, loggers and millmen paid what they had to, to get the funds that spelled survival. But there is no evidence that lumbermen invoked the law's help to improve their bargaining position in obtaining credit. So far at least as the more substantial operators were concerned, the usury laws were too negative a legal measure to play a significant role in the financial problems of heavy industry.[7]

The timber industry needed positive provision of reliable, long-term credit more than negative defenses against high charges for short-term loans. The practices we followed in disposing of public timberland in effect subsidized the industry to meet its fixed capital investment. Had this *de facto* policy of subsidy been examined and given explicit definition, we might have seen logic in supplementing it with public aid to make working capital more readily available for harvesting the forest. For the constant pressure for liquidity to carry short-term, high-cost borrowings was a large factor in focusing the industry upon present, at the expense of future, yield.

Moreover, for want of substantial, assured working capital at moderate and stable rates, the industry was very vulnerable to the downswings of the business cycle. This was another source of importunate pressure for liquidity, which gave further impetus to preferring present yield to the exclusion even of planning for long-term forest harvest. Both the chronic lack of an assured flow of long-term capital and this special exposure to hazards of business downturn narrowed the industry's choices in using its fixed capital. In addition, these factors worked to raise the industry's financing costs and pressures by insuring that lenders would view it as a high-risk borrower, to be kept on short rein and premium payments.

The lumber industry needed ample long-term capital at a time when the financial market was prepared only to offer short-term capital, and not even an easy supply of that. This was not surprising. During the decades which set the pattern of use of the Wisconsin forest the country's growth was running ahead of the development of its financial institutions. It was all that the market could do to manage the financing of the railroads, with generous subsidies of public land and public money to increase the equity available to secure private lending. It took the impetus of railroad promotion and the Civil War money needs of the national government, to create an investment banking industry with the knowledge and means to assist the growth of heavy industry, as earlier and more limited banking had assisted the growth of commerce. The Jacksonian attack on the Second Bank of the United States had ended the nineteenth-century chance to develop a central bankers' bank capable of disciplining credit and braking the financial impact of the ups and downs of general business. The lumber industry thus confronted financial problems which pointed to need of institutions and framework facilities beyond the capacity of the contemporary market. The lumbermen's eager response to the opportunities that public lands policy afforded them showed that they were ready to seek service and subsidy from government where they saw practical forms and uses of service or subsidy. Why then was there not further pressure for using law to mobilize fluid capital?

For one thing, the public lands policy represented public service and subsidy in kind and not in cash. The nineteenth century did not welcome programs that spelled taxes or any other influence toward withdrawing cash from private circulation; recurrent controversies over hard and soft money, the tariff, and tax exemption of railroads remind us of this attitude. Businessmen lived in this climate of opinion, and many shared it; if they did not share it, they could not escape its teachings concerning the practical bounds of policy. A second element to be considered was inertia. The lumbermen fell into the advantages of public-lands subsidy of their fixed capital needs; this was policy fixed by the drift of practice, never by explicit deliberation. Land subsidy for railroads was debated and made a matter of explicit decision. But railroads so obviously concerned the planned framework for economic growth that government's role regarding them could not emerge out of inattention. The special and explicit concern directed at railroad subsidy, attended sometimes by scandal and always by sharp competition for public favors, encouraged men to think of such issues as peculiar, to a degree that did not suggest similar handling of more "ordinary" business. Such general influences on attitudes reinforced customary reliance on the market and on private contract as the normal means by which to come to terms with scarcity. True, this presumption in favor of market processes was in practice open to rebuttal. But the presumption itself carried a force of custom which weighed against resort to positive action in law. Unless government itself provided some initiative—through legislative investigation, for example, or through the momentum which specialized, sustained executive attention might give to exploring issues and framing questions for debate and decision—the prevailing set of the situation was not such as to promote perception of the problem as one ripe for positive government action, even by subsidy.

As their enterprises became established and their investment grew larger, from the late 1860's entrepreneurs tended to cast their business organizations into corporate form. Chapter V pointed out that, despite early political controversy over incorporating banks, by the time the lumbermen sought the privilege of incorporation state policy was prepared to offer it without serious question, as a facility in aid of the market. This was so much the dominant attitude that the functional place of the corporation in the history of Wisconsin's lumber industry plainly is under the head of contract rather than of police. Likewise, it seems clear that the practical significance of the corporation lay in its convenience to businessmen, rather than in any opportunity it afforded to impose regulation for ends of primary concern to the community at large. The principal operative meaning of incorporation for Wisconsin forest industry was the assurance it gave of continuity and closely held control in firms which usually had a few stockholders. Such regulations as the statutes did set upon chartering were aimed mainly at protecting stockholders or corporation creditors from fraud. That is, they were regulations to support the integrity of market processes. This was the meaning of the minimum stipulations made for publicizing corporate capital and net worth, for keeping records and allowing stockholders or public officers some access to corporate books, and for preserving the corporation's capital and its solvency against withdrawals by stockholders or grossly improvident dividends by directors. The legislature provided no executive agency regularly to check compliance with these safeguards of the market; any policing that was done would be done, it was clear, by lawsuit. The statutes provided no other type of regulation of corporate finance, and in particular no regulation of the procedures of floating loans or issuing stock. These limitations of subject matter, of course, fitted the typically close-held character of most firms and the rudimentary state of the capital markets available to Wisconsin lumbermen in the second half of the nineteenth century. Likewise characteristic of the times was reliance on the interest of aggrieved suitors or the uncertain will and energy of local prosecutors to enforce statutory standards by court action; it fitted the contemporary emphasis on loosely dispersed power arrangements, not to provide for sustained executive surveillance. But these congruities of law and market attitudes simply point up the absence of initiative to use law affirmatively outside the market framework to mobilize fluid capital, as it was used in effect through the public lands to give the industry ready access to its raw material.[8]

A few companies, like Knapp, Stout and Company, out of superior leadership and strategic location, achieved considerable security from financial need and business depression after the seventies by plowing back earnings into large-scale timberland investment and improvement and enlargement of plant, as well as by integrating and diversifying operations. Size alone could not create power or security, however. This had dramatic demonstration in the failure in 1873 of the great Union Lumbering Company of Chippewa Falls, defaulting bonds of some $650,000, and other debts to about a quarter of a million dollars.[9] There was selective strategy behind successful growth. Bigger, more efficient mills provided economies of large-scale production. But the bulk and weight of the product made transport costs very important to the lumber industry. This factor imposed sharp limits on the gains to be had from increasing the size of processing operations in a given locality; the greater the local mill capacity, the sooner the exhaustion of nearby timber stands and the more certain that the cost of hauling logs from a distance would eat up the profits of the mills. Cost structure was more effective, thus, than any legal control that the late nineteenth century was likely to create, in curbing the scale of growth in particular lumber manufacturing enterprises and in holding open opportunities for processors of modest size, if they could find economical sources of raw material.

Favorable large-scale investment in timberland was the critical determinant of successful growth for individual enterprises or for enterprises closely linked by common ownership of stock, interlocking directorates, or pooled investment in land or transport improvements. This was the basis of the profitable expansion of the Weyerhaeuser lumber interest, and of the interests joined in the great Mississippi River Logging Company. To what extent entrepreneurs could have surmounted their financial insecurities by pursuing shrewd investment in timberland out of their own unaided resources, we cannot know. They did not have to try. Without ever a clear-cut definition or explicit decision in the matter, government disposed of public timberland on terms that greatly reduced the industry's need to put scarce funds into acquiring secure sources of raw material. This *de facto* capital subsidy, within the legitimacy conferred by the respected forms of private fee simple title and contract, made it practically possible for the more vigorous and able entrepreneurs to win superior profit and security by building enlarged, integrated operations on the base of relatively low-cost raw materials.[10]

Even when public subsidy relieved much of the potential pressure, integration still called for assembling considerable investment and working capital in close-order drill. This requirement limited the scale within which production and distribution could be planned or manipulated against the swings of general business activity. In the disastrous mid-seventies, lumber firms began to try to control output and price competition by treaty among themselves. They sought this end by national organization in 1874 and 1876, and by combinations of Lake States producers in 1884 and 1885 and again in 1891. Within the Wisconsin industry itself local organizations took shape, in the Chippewa Valley (1875), the Oshkosh area (1879), and the Wisconsin River valley (1895). Until the nineties, and except for limited local co-operation, these national and regional efforts were failures. Individual participants either would not pledge to submit to group decisions, or, if they did, were quick to break their word under pressure of ambition for profit or fear for survival. The early agreements lacked ready means to measure compliance or sanctions to enforce it. More successful was the Mississippi Valley Lumbermen's Association, organized in 1891, which by 1905 enlisted 90 per cent of the white pine production in the upper Mississippi Valley and sought directly to fix prices at levels which could command the assent of 80 to 85 per cent of its membership.[11]

Large integrated lumber enterprises, and compacts to regulate output or price, implicitly challenged the market as an institution for resolving issues of scarcity by unmanaged cumulation of particular decisions. Where the people confronted great private economic power fulfilling functions which obviously set the framework conditions for the existence of the market — notably in banking and transportation — they had already begun to demand some legal control of the terms on which such power was exercised. Apart from the crude beginnings of "public utilities" law, however, there was little doctrine relevant to preserving the balance of power in the market. Contract and property doctrine, of course, furnished concepts and procedures for orderly administration of private dealings. Tort law recognized rights of aggrieved private parties to recover for harm caused by competition conducted by other means than exchange — as by violence or fraud. There was some common law doctrine condemning conduct "in restraint of trade." But through the seventies and eighties this doctrine dealt mainly with the special problem of defining the allowable scope of covenants not to compete, given by the seller of a business. Such emerging common law precedent as went beyond this point, to deal with interference with the exchange process, was scant and ill-defined. Before Congress put the Sherman Act on the books in 1890 there was no major legislative declaration of public policy assigning government a positive responsibility to preserve the market against private subversion.[12]

Through the years in which large Wisconsin lumber firms began to reach for a measure of market control, the law of restraint of trade remained unimportant to the practical affairs of the Wisconsin industry, and the industry's affairs were unimportant to the general development of antitrust law. There was grumbling among small-scale logging operators over the influence that the Mississippi River Logging Company exerted on the price of saw logs and on the availability of ready river transport when traffic ran heavy.[13] This fear and dislike produced legal challenges to the Company's conduct affecting water transport, which we shall note in Chapter IX. But there was no effort by lawsuit or legislation to challenge the Company's power in the commodity market. In 1892 a federal grand jury invoked the new Sherman Act to indict leaders of the Mississippi Valley Lumbermen's Association for a conspiracy in restraint of interstate trade in lumber by combining to raise prices in the Mississippi Valley area. The district court sustained defendants' demurrer to the indictment. In part the court grounded its ruling on the failure of the indictment sufficiently to charge that the members of the alleged combination enjoyed any practical power to affect the market. In part, however, the court acted on an interpretation of the statute which in effect denied it any operation to penalize market combinations of dangerous tendency: "Unless the agreement involves an absorption of the entire traffic in lumber, and is entered into for the purpose of obtaining the entire control of it with the object of extortion, it is not objectionable to the statute . . ."[14] No more was heard of this experiment in legal control. Wisconsin enacted its first counterpart of the Sherman Act in 1897. In 1907 the legislature created a joint committee to investigate lumber industry agreements in restraint of trade, but the committee published no report, nor did any action under Wisconsin law follow from its establishment or otherwise in the years which saw the rounding out of the history of the lumber industry.[15]

The slight effect which the Wisconsin industry and the developing field of antitrust law had on each other is not surprising. The law was in an elementary stage of development through most of the Wisconsin lumber boom. The industry's efforts at manipulating the market were inconclusive enough in effect to make its activity of relatively less interest or concern than more massive efforts of much more powerful industries in other sectors of the economy. The practice and attitude of the Wisconsin lumber industry did show the pervasive effect of the shortcomings of the capital market and the hazards of the uncontrolled business cycle. Plainly these entrepreneurs wanted to escape free market processes and to install informal controls in their place. If the law had little to say about the capital market or the business cycle, this omission did not mean that there would be no response to these problems, but only that there would be response not subjected to the test of public interest.

The protective tariff is part of the history of the lumbermen's concern for the supply and safety of money investment. By law they might thus limit access to the market. By eliminating Canadian competition they could reduce the capital they must commit to maintain their growth, and cut the risks of competitive price movements which added to the hazards of the business cycle. Until the seventies the Wisconsin industry evidently felt secure behind the carrying-cost advantage which it had because of the natural waterways linking it to the midwest lumber market. It was eastern lumbermen who procured the termination in 1866 of an early treaty of reciprocity which since 1855 had allowed timber and certain timber products to move free of duty between the United States and Canada. The spread of the railroad network after the Civil War, however, exposed a broader reach of markets to Canadian competition, as it also enlarged the areas which Wisconsin sellers might hope to enter. The change from comparative indifference to lively concern for lumber tariff advantages was marked by the memorial which the Wisconsin legislature addressed to Congress on February 6, 1872, opposing remission of duties on building materials imported to aid the reconstruction of Chicago after its great fire. The Wisconsin legislators sympathized with their neighbors' trouble, but so long as there was profit in it, saw no reason why the profit should go to foreigners to the detriment of home business:

> [In] . . . Michigan, Wisconsin and Minnesota . . . immense amounts of capital are invested, and thousands of persons are engaged in the manufacture of lumber, brick and other building materials, whose interests would be seriously crippled by the introduction of foreign competition with free duties, and . . . while individual capitalists of Illinois, and especially of the city of Chicago, would reap all the benefit, the mercantile and manufacturing population of the neighboring states would sustain all the damage, and an incredible derangement of their business . . .[16]

From this point on Wisconsin loggers and millmen regularly appeared in the industry's sallies upon Washington, whether in defense of outstanding tariffs as in the seventies and eighties, or on the offensive to restore and increase tariff protection, as in the late nineties after the Wilson-Gorman Act of 1894 put rough lumber on the free list. Two cross currents in these controversies revealed the extent to which attitudes related directly to immediate investment interests. The tariff was, of course, a classic issue between Republican and Democrat, as it had earlier been between Whig and Jacksonian. For a time party loyalty brought Democrats among the Wisconsin lumbermen to the support of free trade, and timberlands investor William F. Vilas stood by this position throughout his political career in the state. But over the long pull, commitment to business proved stronger than that to party on the tariff question, and by the nineties most Wisconsin Democrats among the lumbermen were actively supporting the Republican position. The other cross current of interest made manifest by the tariff battles revealed a cleavage according to business function, which was also partly a cleavage on sectional lines. Lumber wholesalers and retailers, notably in the central Chicago distribution point, turned up in the ranks of those arguing to reduce or eliminate the tariff on Canadian forest products. After the turn of the century lumber manufacture moved toward relative decline in Wisconsin's economy; there emerged into more prominence the interest of consumers and of manufacturers for whom forest products were a raw material for further processing. In this phase sentiment for free admission of Canadian lumber showed itself once more in the state.[17]

The growth of the Wisconsin lumber industry's active interest in the tariff indicates, again, that the demands men made on government were shaped less by general attitudes than by specific perception of interest or specific blocks to perception set by custom, habit, and preoccupation with the day at hand. These men did not press government to provide public credit or public market management to help their search for investment capital and to protect their investment against the swings of the business cycle. But the energy with which they sought legal regulation to exclude Canadian competition, like their own groping efforts to control output and price at home, shows that they were ready to manage the market so far as they conceived practical means of doing so, with or without the positive help of law. The lumbermen argued that their private combinations and their tariff program were also in the public interest. Though in no other respect did they raise the problem of preference for future over present forest yield, they urged that their combinations and tariff demands would work to conserve timber supply, by reducing pressure for liquidity and allowing more orderly cutting than could be afforded in the face of the selective competition of cheap rough lumber from outside. Thus the lumbermen in effect conceded that they sought such power over investment as to require that they submit their conduct to scrutiny in a public forum. Issue was never squarely joined over the industry's desire for a controlled market, however, because industry thinking was not yet ready to draft a comprehensive code for its self-regulation; NRA was still a generation away.

THE TERMS OF SUPPLY OF LABOR

Like cash, manpower was a scarce and expensive item in the working capital needs of Wisconsin's expanding lumber industry. Before 1870 the prime difficulty was to muster into the rude north country the skill and muscle to serve the ambitions of logging contractors and millmen. Thereafter the price and terms of labor offered more pressing issues than absolute scarcity. Yet, scarcity was still much at the heart of the matter. Costs were the higher because woods labor turnover was great; the percentage of transients was large. This reflected the slow development in the north of the balanced economy and settled communities which could have provided a stable body of workers. Moreover, from the nineties on the industry felt the competition of other bidders for labor, both from the general expansion of the economy and from new forest-production areas in the South and West.

The supply of labor, like that of money, affected the realizing of economic values. The terms of labor supply in forest industry influenced the pace and manner of exploiting a basic natural resource, and hence affected general interests in economic growth as well as the special interests of the lumber trade. But labor supply directly concerned social and political values that were only indirectly involved in what we did about credit and the supply of investment money. Labor supply drew upon men's own vital store; the terms of labor supply bore directly on what men got out of life, compared with what they spent to live. Popular philosophy was crude and ill-defined in many ways, but it held a clear central belief by which it sought to rationalize society and the sacrifices that social living called for. We believed that all forms of social order, and hence all forms of social power, must be legitimate and not arbitrarily imposed. Order and power were legitimate, if in the long run and on the whole men felt that their institutions promised them, or at least their children, lives of richer, more satisfying experience. Thus, though it might take time, demands for legal regulation were likely to be heard sooner where action laid demands directly upon men's life opportunities than where the first impact of action involved impersonal institutional ends and operations. So the aspects of public lands policy which alone evoked much popular attention were those which touched the farmer's access to a homestead or his chance to move his produce to market. For similar reasons there eventually was more activity relating to legal regulation of labor supply than ever showed itself concerning the implications of public timberlands policy for general economic growth, or the promotion of the supply of investment capital. Corollary to this trend was the fact that controversy specially marked the growth of policy in this area where interests of personality were felt to be at stake; in this context men were likely sooner to bring issues to definition because the issues were close to their experience. In contrast, though high controversy eventually emerged over a more impersonal value like conservation of natural resources, for a long time policy in that matter grew by drift rather than by debate and conflict. Indeed, Theodore Roosevelt and Gifford Pinchot made conservation a popular issue in the first instance by dramatizing it as a battle for the "little man" against the "interests."

Growth of policy on institutional matters that could be translated in terms of personal involvement was likely to be affected directly by power politics. The weight of the balance of power was felt much less in the evolution of policy that concerned impersonal institutional functions, because there, typically, for long periods no one challenged prevailing currents of action in such a way as to evoke counterorganization. Lumber entrepreneurs showed their power in the legislature more sharply over questions of legal regulation of the terms of labor supply than in any other broad area of the industry's concerns. This was because eventually they faced more defined challenge to their wishes in that area than in others. It is part of this pattern that there was little legislative activity related to lumber industry labor until the late eighties, after northern Wisconsin witnessed the first serious efforts at trade union organization.

Before the late eighties, legislation creating liens on logs and lumber for those who worked on them was the only broad policy specially related to lumber labor. In function, these statutory liens seem to have served more as a device to promote production than as a regulation imposed to redress market inequality; hence they were examined in detail in Chapter V. The pattern-setting lien act of 1860 came long before there was any effective organization to press labor interests on the legislature. Over the years the legislature expanded the application of the labor lien and showed marked generosity to the lien claimant, preferring his claim over that of general creditors and subsequent bona fide purchasers, allowing him to proceed for his whole claim against any of the logs or lumber on which he had worked, providing simple and inexpensive procedures of enforcement. This was done with the support of legislative committees on which leading lumber entrepreneurs were commonly key members. This early, consistent favor toward a labor interest, in years in which labor wielded no independent lobby power, probably resulted from the fact that the lien likewise served the interest of the more important firms. The lien might be invoked against a solid firm as a spite proceeding incident to some other dispute, but there was no practical need for labor to rely on the lien to collect wages from established companies. Labor had grievances against such employers, but these complaints ran to the terms of payment and not to the employers' inability to pay their debts. The practical significance of the lien was to protect employees of shoestring operators, and at the same time to enable such operators to persuade men to work for them for wages deferred until after the spring drive or the milling of the logs, where their credit was of itself too shaky to induce the workers' reliance. For many years larger millmen and wholesalers counted on the output of small operators, particularly small contract loggers, to provide them with profitable volume; the statutory labor lien was an important underpinning of their own commodity supply. Had the question of the labor lien first come up in the nineties, when the big firms had become more self-sufficient, perhaps their representatives in the legislature would not have troubled to support the lien. It may be significant that the legislature was much more grudging in creating suppliers' liens; there is some testimony that the larger firms opposed the supply lien, because it allowed smaller suppliers the margin of safety they needed to compete with the larger concerns in furnishing independent loggers their winter's camp needs. In another aspect the legislative favor toward the log-labor lien was not perhaps as surprising as it first appears. The lien was a device to enforce contract claims for money fairly due, against the hazard of an insolvent or fraudulent employer. The dominant attitudes of the community strongly supported the regime of contract; the lien simply bulwarked that regime in a most elementary way, hardly open to argument. The same may be said of the liability which, at least after 1878, applied to stockholders of any non-railroad corporation, specially chartered or not, to stand good to the value of their stock for up to six months' unpaid wages owed by the corporation to its laborers.[18]

There were more direct ways by which the state might have promoted a more adequate and stable labor supply for the timber country. Given the industry's demonstrated readiness to take public assistance where it could see that it might be had and could be useful, the fact that there was not more promotional activity was traceable to no principle of self-restraint, but to the limits of state machinery and of men's current notions of what they could do with it. The direct way to ease the labor situation was to encourage movement of people into the north country. During the years of the lumber industry the Wisconsin legislature created six executive agencies (1852–1855, 1867–1871, 1871–1876, 1879–1887, 1895–1901, 1907–1915) to encourage immigration into the state from other states and from abroad. The emphasis throughout these efforts, however, was always on settlement for agriculture; the lumber industry figured in the picture only because official pamphlets pointed to it as a source of supplementary employment and as a market for produce, to assure prospective settlers that they would find ready means to earn the purchase price for farms in the north. Through the relevant period, the immigration agencies were small and rudimentary organizations. To service the lumber industry's labor needs would have taken a more aggressive and effectively organized office than any for which the simple executive branch of the nineteenth century offered precedent. The transient character of its labor cost the industry a good deal and was the subject of complaint. Yet, the lumbermen showed no imagination concerning the benefits they might gain if government collected reliable labor market data or maintained public employment offices. The state's nineteenth-century machinery for collecting economic data was limited, and it was especially inadequate for the north country. The legislature first created state free employment offices in 1901, and enlarged the system in 1903. Prime impetus for this step seems to have been the concern to prevent frauds on workers by private employment agencies, more than to provide for efficient allocation of labor supply. Though one of the first offices was located at Superior, where it served the lumber area, the impetus for creating the system came from organized labor concentrated in southeastern Wisconsin. However, like the log-labor lien statutes, effective state agencies to help build a reliable labor market in northern Wisconsin would not have been a sharp departure from previous nineteenth-century policy, but would simply have supported and improved market processes. That so little was done seems to have been due mainly to inertia. True, the immigration agencies suffered ups and downs in legislative favor. In 1887 the system was ended for a time in response to fear of labor radicalism, connected with the rise of the Knights of Labor and the Haymarket Riot. But recurrent disfavor for the immigration agency idea rested mostly on narrowly practical grounds — on criticisms that the agencies were instruments of political patronage or were ineffective, and on demands for public economy.[19]

Apart from the log-labor lien statutes, for the first generation of the lumber industry men left the terms of the wage contract to be fixed in market. There was one exception, but of form rather than of substance. Beginning in 1867 the legislature forbade employment of women or of children under eighteen in manufacturing plants for more than eight hours a day. In 1877 it added a flat prohibition of employment of any child under twelve years during the school year at any manual labor in any factory employing three or more persons. Until the beginnings of a state factory inspector system in 1885, there was no provision to enforce these regulations save the uncertain zeal of local prosecutors. In substance, thus, the legislation set little practical limit on employers' discretion. There was no vigorous enforcement by state officials until the late nineties. In any case, the matter was not important to the lumber industry, which never employed women or children in large numbers. No attempt was made to except lumber from the terms of these statutes, though the industry exerted itself to secure exemption from other labor regulations. The contrast itself attests that lumbermen did not see in these limits on employment of women and children even a formal issue of much meaning to their business.[20]

The manner of paying wages, on the other hand, made a real pocketbook difference to lumber industry employers. Here they were alert and largely successful in avoiding legal curbs on their bargaining power. Three points came to issue: time of payment, medium of payment, and disciplinary forfeitures of pay.

Employers' concern with time and medium of wage payment originated in the shortage of working capital, which was chronic with small and medium-sized firms, and which troubled larger companies through the first generation of the industry. Not at least until after the spring drive, and perhaps not until after logs were turned into lumber at the mill and sold in the Chicago or Mississippi River wholesale markets, could employers find means to settle the balances owed their labor beyond the food and shelter furnished during the working season. Moreover, hard-pressed employers might wish to pay off their men in notes or time checks, to win extra time to realize cash out of the season's output. Beyond these reasons of necessity there was profit to be had by requiring labor to help finance the business. Thus even well-established firms followed the tradition of deferred wage payments, reducing the amount of cash tied up in operations during the winter and spring. If the worker himself needed cash or supplies during the months while he waited for his pay, the employer was glad to take on the role of small loans company, discounting the worker's time checks at a high rate of interest, whether for small cash advances or for purchases made at a company store. The company store itself offered an additional chance for the employer to profit by selling clothing, tobacco, and other items at markups as high as the traffic would bear. By geographical remoteness, the company store often had the advantages of a monopoly, and some firms reinforced this position by requiring as a condition of employment that the worker agree to buy there all he needed of the articles the store had to sell.

The employer interest scored its first defensive success in 1878. Supported by Milwaukee and Racine workingmen, bill 7A of 1878 would have forbidden the issue of any written instrument in payment of wages, unless made redeemable by the issuer in lawful money of the United States. In the course of legislative maneuvers, the sponsor offered an amendment, undoubtedly to ward off north-country opposition, declaring that nothing in the measure should be taken to forbid employers "to deduct from the wages of employees any amount due them for rent or necessaries of life furnished to such employees, when such necessaries of life are produced or regularly kept for sale by such employers." With this and other qualifying amendments, the bill passed the Assembly, only to meet decisive defeat in the Senate following an unfavorable report from a Senate committee.[21]

No more was heard of this kind of regulation for a decade. Then Laws, 1889, Chapter 474, required that all wages for labor be paid weekly or biweekly in cash, unless the parties stipulated otherwise by written contract. However, Chapter 474 excepted from its command farm laborers, traveling salesmen, traveling employees of railroad or express companies, "persons employed in logging camps, or 'driving', running or manufacturing of logs or lumber," as well as any other "whose occupation is such as to render him inaccessible on the regular pay day." This act bore plain testimony to the defensive alertness of lumber employers; the logging and driving exception was put in the original draft of the bill despite the presence of the broad contracting-out clause, and the exception was broadened to cover lumber manufacture by an amendment offered by the bill's sponsor — again, it seems plain, to ward off opposition from the north. Like the 1875 bill, the successful 1889 measure was promoted by Milwaukee labor, and did not represent a prime concern with lumber industry problems, though in 1887 the commissioner of the new Bureau of Labor and Industrial Statistics reported that procedures of wage payment were a sore point in the lumber areas. Exception of sawmills from the statutory ban on deferred wage payment seems particularly strong testimony to the political power of the lumbermen; the arguments of isolation of work area and necessary seasonal postponement in bringing product to market, which gave woods operations their special character, did not typically apply to the manufacturing stage of the business.[22] A revision of the statute in 1915 dropped the exception of lumber "manufacturing" for the more ambiguous term "lumbering"; employees "engaged in lumbering and logging operations" were now the excepted class. Even if this phrasing somewhat narrowed the exception, it was striking that the exception was still continued at all, since the same revision drastically tightened the regulation by dropping the authorization to contract out of the statute's command. In 1933 the exception occasioned a belated flurry of controversy, the outcome of which showed that the lumber interest, though much diminished, was still an effective lobby. Bill 395S of 1933 would have brought the whole lumber industry under the statute requiring cash wage payments on at least a biweekly basis. But an amendment in committee restored the exception of "employees engaged in . . . logging operations." Thus — as enacted — Laws, 1933, Chapter 303, merely corrected what had been always an anomalous situation, by bringing under the general wage payment provision men who worked in sawmills or other primary forest-products processing plants.[23]

There was closer pull and haul of forces over legal regulation of the medium of payment of deferred wages. No question of legislation was raised for many years after the lumbermen obtained their exemption from the general cash payment bill which failed in 1878. But in Laws, 1891, Chapter 430, the legislature required, under criminal penalty, that any employer "engaged in lumbering . . . or in the manufacture of lumber" in Wisconsin who "shall defer payment of wages" should give his employees on demand "written evidence of indebtedness, which shall be negotiable for all amounts due such employees." Enactment of this rule showed the depth of dissatisfaction in the lumber country over procedures of wage payment, which the state's labor commissioner had noted four years before; the special force of labor's victory here is the more notable because in 1893 the legislature rejected a strong drive to enact an unqualified cash payment law applicable to all industry.[24] The revised Statutes of 1898 evidenced continued close contention over the lumber regulation; the revision enlarged the rule by applying it where "all or any part" of the wages were deferred, but eased its rigor by stipulating that the writing be "assignable" rather than negotiable, and removing the stigma of criminal sanction by providing instead for a "forfeiture" upon violation.[25] Labor won back lost ground when Laws, 1901, Chapter 47, restored the requirement that the employer give "negotiable" evidence of his wage debts. The last word on this subject was for the employer, for in 1915 the legislature repealed the requirement of negotiability and left standing simply a requirement that all "time checks" be assignable. However, by this time the industry had passed its peak and the matter was no longer important to a large class of lumber workers.[26]

The statutes requiring payment of deferred wages in negotiable instruments, and to a less extent the requirement that payment be tendered at least in assignable paper, limited employer insistence that workers take pay in drafts on a company store. But in 1895 the legislature summarily rejected bills which would have flatly prohibited employers from conducting "stores known as company stores or general supply stores." In 1913 the Attorney General ruled that no law of the state forbade a pulp and paper company to pay for logs and pulp only in orders for goods drawn on a general merchandising company of which it was part owner, where the company would pay cash on demand, though only upon presentation of the order at the store. In 1939 the legislature finally banned conduct of company stores by any firm "engaged in any enterprise" in Wisconsin, except for sale of meals, candy, and tobacco, or of specialized appliances or tools needed on the job or for the employee's safety. In view of these exceptions, there was, again, tribute to the lobby of a now much diminished lumber trade in the fact that the statute also excepted altogether "lumber producers and dealers." But this phrasing perhaps carried some ambiguity, and so an amendment in the same session added the exception of "loggers." If the company store became less important in the balance of advantage between lumber employer and employee, this change was not due to direct intervention of law, but to a tighter labor market and the widening of trade areas that came with the automobile and improved roads.[27]

Production organized on a crew basis in the woods and on the river, and on a factory basis in the mill, implied a system of order and discipline. In this market-oriented economy, contract provided the first, and usually the last, means of defining and enforcing the necessary order. In its most elementary form this discipline meant that the boss told the men what to do, and anyone who did not do it could quit or be fired. Especially in the woods, in a showdown the camp boss's fists enforced the terms of the employment contract. But the industry's practice of deferring wages put another, more formal sanction in the employer's hands by giving him an easy means to impose forfeits by subtracting from wages due and yet unpaid. Larger firms used this device particularly to cut the costs of transient labor,

Hauling logs by oxen, about 1890.

Hauling logs by steam sled, 1907.

TURN OUT!

RALLY EVERY MAN!

Court House To-Night!

CITIZENS:---It having been represented to the Governor that the sentiment of the people here is opposed to giving lands to North Wisconsin Railway Co. for road already built, turn out to the

Court House To-Night,

and say whether or not the representation is true.

A. J. Goss,	H. A. Taylor,	P. Q. Boyden,
J. W. Bashford,	J. R. Ismon,	H. L. Humphrey.
J. O. Henning,	H. C. Baker,	O. F. Jones,
T. Dwight Hall,	C. R. Coon,	T. E. Williams,
W. S. Evans,	A. E. Jefferson,	S. C. Johnson,
J. A. Bunker,	D. C. Fulton,	Frank Harding.
		John Comstock.

And 100 Others.

Hudson, Saturday, Dec. 5th, 1874.

Local politics and economic development: Poster for political meeting on railroad-aid lands, Hudson, Wisconsin, 1874.

by stipulating that the employee stay on the job for defined periods, on pain of forfeiting a substantial percentage (perhaps 20 per cent) of pay already earned if he quit before his contract period ran out, or if he were discharged for insubordination. Such provisions responded to real problems of organizing efficient production. Obviously, also, they invited abuse of superior market power when there was no effective trade union organization to bargain on over-all terms or to enforce fair grievance procedures. Reporting in 1888, a state factory inspector found "the dockage system" to be a "bad feature" of employment conditions in the timber country, and observed that it was harshly enforced:

> Peter Larson got killed in the woods while in the company's employ. His wages were docked 20 per cent. for not working his time out. His widow returned to Norway. I don't blame her.[28]

Such beginnings of union organization as marked the years of lumber boom were too much concerned with the basic issues of the right to organize, and the fixing of rates and hours and manner of wage payment, to press questions of due process in handling individual labor contract enforcement. There was no effort to set statutory limits upon the employer's disciplinary procedures. Conceivably the courts might have hammered out standards to prevent overreaching, but judges can make law only as litigants give them opportunity. The log-lien statutes provided a surer and simpler means of enforcing wage claims than did an ordinary action for debt, where the question was simply that of collecting money. Where the issue was more complicated, involving construction of the wage contract or determination of the employer's right to withhold wages as a disciplinary measure, the circumstances typically would discourage lawsuits. The bulk of lumber industry workers were unskilled hands, with little means or education. Though the money at stake might loom large to an aggrieved worker, it was unlikely to be enough to warrant the costs and risks of suit, and even less to warrant appealing a case to the Wisconsin Supreme Court, whence only would issue any declared contribution to common law doctrine. Even on a liberal classification of matters in controversy, in all the years of Wisconsin's lumber industry the state Supreme Court confronted common law questions under woods-labor contracts in but a score of cases, and dealt with only another sixteen suits involving mill-labor contracts. Almost all these cases presented only disputed questions of fact, or, at most, general issues of contract law, such as the hiring authority of the employer's supervisory agents, or defenses under negotiable instruments. Hiring terms and disciplinary issues peculiar to lumber industry practice were central in only two woods-labor contract lawsuits (1874 and 1904) and two mill-labor actions (1883 and 1893). The Supreme Court held for the employee in all of these. In line with the court's prevailing reluctance to raise questions of the legality of contracts, all these decisions rested on interpretations of the agreements favorable to the employees. Thus the court did not need to decide whether the terms argued for by the employers would be enforceable consistent with sound standards of public policy. An agreement to defer wage payment until end of the working period, said the court, should be construed nonetheless to allow immediate recovery of what was due, where the employee was wrongly discharged before his time was up. In three cases where the employer claimed that, by the alleged custom of the firm, the employee's wages might be docked for quitting the job before the end of the season, the court ruled that the employer had the burden of proving that the parties intended the alleged general practice to apply in the specific employment contract in suit, and that in these instances the employer failed to sustain his burden.[29]

Policy making in this instance proceeded in a way that suggested an important functional division in the separation of powers. Legitimate management needs were being pressed through contract in the context of inequalities of bargaining power too great for the market to adjust. The weaker interest had not yet found the leadership or morale to make effective group demands for statutory redress of the balance. Individuals of the weaker interest could expect no help from the legislature, where an effective organized counterinterest was already in the field. But litigation allowed a few individuals, outweighed in the market, to bring their weaker interest to sharp definition in a public forum the processes of which could be set in motion without group action. Confronted thus with their own responsibility for contributing to the order of the community, the judges did not so much indicate a favor for the labor interest as a disinclination to lend their processes to aggravate the market imbalance which these lawsuits dramatized. Apart from the immediate tensions of these cases, this was a style of judicial administration calculated to force more open expression of conflicting interests in market bargaining, and to foster a climate of legal doctrine and political opinion in which legislative action would eventually find acceptance.

As matters stood in mid-twentieth century the most important developments in the law affecting labor were generally related to collective bargaining and the hazards of industrial accident and unemployment. These developments responded to problems which market processes had proved unable to handle efficiently. In one aspect they were problems of the organization of power. Since it is the law's role to assume the legitimate monopoly of violence in society, law cannot long remain indifferent to force exerted on some men's wills by the private organizations of others. Where power rested primarily on command of economic means, law delegated to the market the initial responsibility for maintaining checks and balances. However, there was in any case a public interest in preventing malicious damage to productive capacity. Even where the market developed private forces of sufficient equality to exert some check on each other, there might be need to impose legal limits to their competition. Where the market signally failed to develop balancing private forces, law might have to intervene lest persistent frustration goad the inferior interest into violence or sink it into degraded apathy. In another aspect the problems to which labor law responded were problems of economic function. They involved the incidence of social costs of labor supply which the market either did not acknowledge, or for which it did not make socially efficient provision from the standpoint of earning a net social return or maintaining productive facilities in condition to continue productive.

Two notable bursts of creative political energy impelled law into these areas of fresh policy development, both in Wisconsin and in the nation. One of these periods fell in the years 1905–1915, in delayed reaction to the urban-industrial growth of the last quarter of the nineteenth century. The other period was that of the 1930's, when measures earlier urged with little effect by the political advance guard were adopted by groups in power under the spur of the great depression. A key feature of legislation enacted in both periods was the emphasis on creating new administrative agencies to give living substance to general policy declared in statute.

The decade 1905–1915 saw Wisconsin's lumber industry pass its peak and begin to fall fast in relative importance in the state's economy. By the 1930's lumber was a minor factor in total productive activity in Wisconsin. Facts thus dictated that lumber should not figure centrally in what proved to be the most important labor law developments in the state. The twentieth-century changes have significance to this history by pointing up the extent of the problems which the nineteenth century left to the market, and again revealing the limited capacity of nineteenth-century legal processes to bring serious emerging issues to reasonably quick definition and decision.

There is historic logic in giving first place to the law affecting collective bargaining, for labor's effectiveness in private group action determined its effectiveness in the legislature.

Given the social pattern of the lumber country — the slow growth of settled communities, the large preponderance of men over women in the population, the transient character of much of the unskilled labor, the rough and dangerous nature of the work, the boldness bred by regularly meeting emergencies — violent self-help was more likely than in other settings. It was a remarkable tribute to the hold that market processes had on popular attitudes, that there was, nevertheless, little violence in lumber-labor disputes in Wisconsin. Violent crimes against person and property bulk up far larger in the business of the circuit court for Chippewa County during the lumber decades than later, but the court records do not indicate that any significant number of such cases grew out of labor troubles. The one major challenge to order occurred in Eau Claire in 1881, when mill workers striking for a ten-hour day forcibly ejected "scabs" from the mills and stopped their operation with some damage to machinery. After a request from local officials the Governor sent in militia to police the city. A serious armed clash was narrowly avoided, but the mills resumed operation. However, they soon went on an eleven-hour day.[30]

There was but one line of statutes reflecting a serious threat to order in lumber-industry labor relations. The danger of sabotage was enough to produce Chapter 154, General Laws of 1867, declaring guilty of a felony punishable by one to three years in state prison anyone who should drive any nail, spike, or other metallic thing into any log banked for marketing or milling; a spike in a log would badly damage the mill saw which struck it, possibly to the danger of the mill workers. This statute remained a subject of legislative interest, in ways which reflected employers' lively concern with it, as well as legislators' perception that rigid severity might be self-defeating. In 1878 the Revisers of the statutes sharply cut the penalty, reducing the crime to a misdemeanor carrying a maximum of six months in county jail or a fine of $100. But in 1880 the legislature broadened the offense, to include driving any rock or stone, as well as any spike, into a log, and authorizing penalties up to six months in county jail or five years in state prison or up to a $500 fine. Despite its increase in penalties, the 1880 act as it was shaped by the Assembly Judiciary Committee represented a more lenient scale than that pressed by the sponsors of the measure; the original bill would have set a mandatory minimum sentence of two years in state prison, with fines from a minimum of $500 up to $1000. In 1882 the Assembly passed a lumbermen's bill still more drastic than the original proposal of 1880, fixing mandatory minimum penalties of three years in state prison or $1000 fine, with no top limits stated. However, the Senate rejected this effort, following the recommendation of its Judiciary Committee. Thereafter the law remained as declared by the act of 1880. The lawyers in the legislature thus curbed the demand of a strong economic interest to write an unrealistic, rigid severity into the statutes.[31]

The sabotage statute was the only item of legislation dealing with permissible limits of labor conflict which grew out of or was specifically framed in terms of the lumber industry. From the mid sixties, episodes of strike and mass picketing by workers, and of lock-outs and blacklisting by employers, made relevant to the industry such general development as there was in nineteenth-century laws affecting bargaining. This was not much, however; in lumber as in other branches of industry, organized self-help was the main reliance of both parties.

In 1885 the legislature charged the newly created bureau of labor statistics "to examine into the relations between labor and capital," including "the causes of strikes and lockouts," and in 1895 it created a procedure for voluntary submission of labor disputes to a state board of arbitration and conciliation. During the years of large-scale lumber activity these were the only affirmative measures taken in law to acknowledge the existence of the social costs and the general interest in labor disputes. The state board of conciliation intervened in a number of lumber industry controversies, as much as its small authority allowed, but with no evidence that it affected their outcome. The reports of the Bureau of Labor and Industrial Statistics preserve the most reliable evidence of labor tension in the lumber country from the mid eighties on, but do no more; it was too early for a fledgling administrative office to show that data collection, skillfully handled, might exert positive influence on trends of legislation.[32]

It takes less invention and initiative to confront new problems with prohibitions than with positive action. Thus the main legal developments about the labor bargain in late nineteenth-century Wisconsin looked merely to limits on organized effort. During the state's first generation its statutes laid penalties upon twelve or more persons with arms, or thirty or more unarmed, "unlawfully, riotously or tumultuously assembled." The Revised Statutes of 1878 marked the new importance of industrial labor conflict by adding to the offenses of unlawful assembly and riot. The new provisions penalized any three or more persons assembled "in a violent or tumultuous manner to do an unlawful act," or attempting "a lawful or unlawful act, in a violent, unlawful or tumultuous manner, to the terror or disturbance of others," or committing such acts when so assembled. Invoked to prosecute Milwaukee "Eight Hour Day" demonstrators in 1886, these provisions were strengthened by laws passed in 1887 when alarm over labor "radicals" was at its height. Another provision of the 1878 revision had limited the crime of conspiracy as it stood at common law, by requiring proof of an overt act done to effect the object of the conspiracy. Now Chapter 287, Laws of 1887, penalized any combination, agreement, or concert of two or more persons willfully or maliciously to injure another in his trade, business, or profession, or otherwise to coerce his lawful will. The state need show no overt act to make out this kind of conspiracy. And Chapter 427 (1887) penalized any person who by threats or force should hinder another from engaging in lawful work. The 1887 legislature made a concession to labor in penalizing blacklisting combinations formed by employers to bar employment or to procure discharge of employed persons. However, in the absence of any provision for a special enforcement agency, such a provision had little practical meaning for the weaker interest it purported to protect. The unions, of course, opposed the statutes limiting workers' organized effort. But the Wisconsin Supreme Court indicated that, were the acts not on the books, the judges would announce substantially similar doctrine as common law, and in the face of such consensus of legislative and judicial attitude the unions did not press hard for repeal. Moreover, save for the exceptional disturbances of

1886, there was no wholesale resort to law to fight the trade unions. From the nineties on the lumber country felt increasing competition for labor from other employments in and out of Wisconsin, and though wages tended to hold within a rather limited range through the remaining years of the large-scale lumber industry, there was a marked tendency toward the reduction of working hours for which labor had unavailingly spent its earlier organized effort.[33]

Nothing could more plainly show how important is organized effort, if law is to respond to values and costs reaching beyond market processes, than what happened about the human and economic costs of industrial accident and sickness in the lumber industry. Work was dangerous in the woods, on the river, and in the mills. Woods camps were rude affairs built and maintained with no special concern for sanitation. Tens of miles of rough country usually separated loggers from the nearest doctor, and not until the seventies were hospitals available in some of the mill towns.

After romantic reminiscence of the lumberjack's life has been filtered out, contemporary testimony tells plainly that his work subjected him to great hardship. Until the turn of the century, however, there are no reliable data by which to measure the extent of accident and illness growing out of logging and milling operations. From the mid-eighties the new Bureau of Labor and Industrial Statistics produced some field reports of industry conditions, and finally some accident reports; so long as lumber was a major industry there never was any connected reporting of health conditions in the woods or mills. In 1901 a smallpox epidemic threatened to spread through the north country, borne by men moved out of lumber camps into surrounding communities; the danger was seen as great enough to call for mobilizing medical aid under direction of state officials. Nor was this an isolated episode. In 1913 the Industrial Commission reported that incomplete records from seven cities showed that within three years local health officers traced to labor camps, mainly lumber camps, 159 cases of typhoid, 12 of smallpox, 6 of diphtheria, and 28 of other communicable diseases. Incomplete accident reports showed that the industry contributed substantially to the total industrial accident count in Wisconsin. In the woods the main cause of injury was falling trees; an early report of the Industrial Commission noted that in less than two years it had record of 138 accidents from this source, 18 of them fatal. In the mills, saws produced the bulk of injuries, due in earlier years to lack of safety guards and later in large part to workers' carelessness or impatience in not using guards provided. Saw accidents loomed large in general accident totals. Of 2326 accidents reported to the state board of health for the twelve months ending October 1, 1906, arising out of use of machinery, 433 occurred in the use of band and circular saws and 202 in the use of other wood-processing devices.[34]

Until 1887 the only legal regulations protecting workers in the lumber industry against the hazards of accident or illness growing out of the job were those provided by judge-made law. At common law the employer owed the employee the duty to furnish him a place and conditions of work reasonably safe by the standard of what men of ordinary prudence would provide for their own well-being. But the common law furnished the employer with three defenses, broad in total scope and of such nature that at least one of them could be invoked plausibly in almost any situation. The employee might not recover anything if his own negligence contributed in any degree to his injury, if he were injured by the negligence of a fellow servant, or if his injury proceeded from a risk so ordinary to the job that he might be deemed to have assumed the risk by agreeing to take the job. The Wisconsin Supreme Court committed itself to these doctrines before it confronted a lumber industry case, but it adopted them in lawsuits growing out of another heavy industry, the railroad, which presented hazards comparable to those of lumbering. The court first encountered the issues of industrial accident with enough misgiving so that it originally refused to adopt the fellow servant rule from the prevailing doctrine of other states. But shortly it reversed direction and declared that the values in observing judicial precedent outweighed the judges' worry whether the common law made a fair or efficient allocation of accident costs. For fifty years following, the court administered the common law rules in their severity against plaintiffs' claims in nearly seven-score cases growing out of lumber industry accidents, mainly in the mills. In the later years of this span plaintiffs' rights were in a measure expanded by legislation, but until the enactment of the workmen's compensation act in 1911 the dominant doctrine, however qualified by statute, followed the pattern of liability based on some fault of the employer.[35]

The practical result of common law doctrine was to cast on the employee the whole first incidence of injury or illness growing out of the job. To administer these costs wholly by lawsuit meant, in the first place, to give advantage to the longer purse, which was usually the employer's. An injured man of little education or means would hesitate to incur the costs and risks of a lawsuit. If he did sue, the uncertainty involved in establishing that the employer was at fault in not providing safe conditions of work, and the strength of the common law defenses, gave the defendant great advantage in procuring delay and in bargaining for an out-of-court settlement on favorable terms. There is evidence how potent were the hazards and obstacles of the common law to discourage suit, in the fact that through all the great lumbering years there were brought to a circuit court sitting in one of the state's busiest lumbering sectors (the Chippewa Valley) only four or five tort actions growing out of accidents on the job. If this state of law meant immediate hardship and injustice to aggrieved individuals, it meant also real, if less directly felt, costs to the community. There was avoidable loss to the general stock of labor when law put employers under no pressure to manage operations to cut risks, and made no adequate provision that an injured worker have access to care that might restore him quickly and as fully as possible as a productive member. Industry operations were in effect subsidized out of general taxes, insofar as sick or injured men fell on the care of municipal welfare agencies.

There is a grim record of death and mutilation in the 137 forest-industry accident cases that reached Wisconsin's Supreme Court up to 1915, when the workmen's compensation and safe place statutes put the issues in a new frame of reference. But we need not go to the appealed cases to learn of these costs. The distinctive value for us in these 137 appeals is in what they suggest of the limitations of making public policy by lawsuit. (1) First is the time lag between trouble and legal response. Work in the woods or mill was always dangerous, and accidents were always frequent and costly to the injured. Yet, it was not until the eighties, when the industry neared its peak, that the Wisconsin Supreme Court had first to consider the common law of industrial accident in relation to this great sector of the state's economy. The first appeal came in 1882. Before 1890 the court had handled only four such cases. After 1890 the appeals came in substantial flow — 42 in the 1890's, 37 from 1900 to 1909, 54 from 1910 to 1915. Plainly, litigation through to appeal was a phenomenon only of the full growth of the industry. Probably the development of relatively more settled mill communities in the north affected this trend. (2) The second limiting aspect of policy making by lawsuit was the high selectivity of the process. Of the 137 appealed cases under the common law and the old employers' liability statutes, only 14 grew out of woods accidents; only two of these were for injury in felling trees — the rest all arose from dangers of land transport of logs; no case ever came to the court out of the hazards of the river drives. Thus, of 137 appealed cases 123 related to accidents at the mill stage of lumber operations; all but about a dozen of these were for injuries inflicted by saws or other moving machinery. The want of cases growing out of the woods or streams stages was not because there was no serious problem of death and injury there. Probably the difference reflected the greater isolation of logging operations, the greater likelihood that the workers there were transients, the greater ease of finding witnesses and preserving objective evidence and obtaining timely legal advice where accidents occurred in the mill and in a settled area. (3) The 137 appealed cases reflected the practical need of a fair and economical procedure to translate into rules the vague standards of care which the common law and the early employers' liability statutes laid upon the employer, to furnish a safe place and conditions of work. In 9 of the 14 woods-accident appeals, and in 80 of the 123 mill-accident cases, the central issue on appeal was the definition of the employer's responsibility; these totals should perhaps be increased by adding the 14 cases in which the prime issue was the closely related matter of the employee's alleged assumption of risk of the prevailing conditions of the job. Contributory negligence was obviously a powerful defensive weapon for the employer — in the instances in which it provided the main issue on appeal, the employer prevailed in 26 — but, relatively, it was far less a source of the costs of repeated and extended litigation than the primary question of defining socially acceptable conditions of labor. The contrast suggests the practical impetus of the facts toward the creation of workable administrative rule-making procedures. With all the uncertainties of suit, the ultimate odds were against the injured man, though not by so clear a preponderance as wholly to discourage his attempting the lottery. Out of 14 woods-stage cases, the employee was the winner at the appellate level in 9 instances; in 123 mill-accident suits the employer prevailed on appeal in 76 cases, the employee in 47.[36]

The Wisconsin court might have mitigated the hardships of litigation, had it pursued its first inclination to declare a body of common law doctrine more favorable to plaintiffs. But at best there were close limits to what the judges might have done. In the first place, the court owed deference to common law doctrine already well-established before Wisconsin law first encountered the accident risks of heavy industry. True, the law of civil wrongs does not present the same reasons for maintaining established expectations as enjoin regard for precedent in property law. But our constitutional tradition assigns primarily to the legislature the job of striking balances in complex conflicts of interest. Industrial accident and illness posed hard problems in fair and rational allocation of costs. Even if the Wisconsin court had rejected the dubious defenses of fellow servant and assumption of risk — both of rather recent origin when first presented in Wisconsin cases — still it would undoubtedly have felt itself bound by the older and very effective defense of contributory negligence, which was so logically part of doctrine that made fault the critical issue. Certainly compelling precedent held the court to the test of fault as the normal common law measure of liability for unintended harm. Moreover, the fault test fitted into the central policy presumptions of the middle nineteenth century. The times favored creative action, and found it natural to require the damaged party to show cause why law should redress loss, rather than requiring the actor to justify his initiative. Large-scale use of machinery was new. It promised great gains for communities acutely conscious of capital scarcity. If law laid upon entrepreneurs liabilities of unpredictable consequence, who could tell, in the sixties and seventies, what crushing financial weight might be laid upon these exciting economic developments, to the detriment of everyone's hope for a better standard of living?

Yet, the fault criterion was irrelevant to other great social interests that pressed their claims on men's hearts and good sense, as the costs of accident and sickness grew with the growth of heavy industry. The community had concern for the human dignity of the worker and his dependents. The general economy had concern to keep and repair its manpower. The polity had concern for realistic social cost accounting; if legal doctrine were in effect to subsidize industry or its consumers by relieving them of predictable accident and illness costs of production, good public order required that at least the subsidy be acknowledged, measured, and judged for its public worth. All these considerations presented a crisscross of questions of fact, interest, and judgment which could hardly be handled comprehensively within the limits of private lawsuits. However unready it was in will or experience to come to grips with such complexities, the legislative branch had the basic responsibility here, imposed on it by the character of the problem.[37]

So far as woods operations were concerned, workers' rights were defined solely by common law until the first workmen's compensation law of 1911. Such limited legislative activity as concerned lumber industry accidents before 1911 dealt only with safety in sawmills and woodworking plants. The difference fits the fact that the only substantial trade union organization in the industry took place in the mill towns. However, the lumber workers never won enough organized strength of their own to exert significant pressure for legislation. Employer liability laws developed only as organized labor developed, but the lumber workers owed such added protection as the statutes gave them chiefly to the success of organized labor in the railroads and in Milwaukee and other factory cities of southeastern Wisconsin. The first legislative episodes revealed the lines of force. In 1872 a Milwaukee Assemblyman introduced a bill for better security of life in steam mills and factories. P. Sawyer and Son, C. N. Paine and Company, and other sawmill operators of Oshkosh memorialized the legislature against the bill, which died in committee. In contrast, the fledgling railroad unions managed to get on the books in 1875 a short-lived statute abolishing the fellow servant rule in application to railroads. Even this proved a premature show of strength, for the roads procured the act's repeal in 1880. Not until 1889 did the unions muster enough force to put on the books further legislation, much less satisfactory than their 1875 victory, dealing with railroad job hazards. By then other unions had begun to build strength in Wisconsin factory industry. Moreover, in 1883 the legislature created a Bureau of Labor and Industrial Statistics. With a scant budget, small field staff, and leadership cautious to the point of timidity, the Bureau for some years made little contribution to the growth of public policy. Its detachment seemed often to be born of resignation. Addressing himself specifically to what he had observed in the lumber country, the Bureau's field inspector commented in 1888,

> Some of the features of the work [in the lumber industry] are rather unpleasant to the workmen, namely, the shortness of the season; the manner of payment of wages; the mode of living; the utter dependence of the men upon circum-

stances, and the danger to life and limb. However, these features seem to be inseparable from the work, and it remains an open question whether the conditions could be improved at the present time, in any other way than by more regular payment of wages in cash.

Within the limits of its means and attitudes the new Bureau did, nonetheless, begin to publish field observations and some data on industrial accidents such as would tend to build attitudes receptive to legislation.[38]

Legislation regarding fire hazards in factories began in 1878, but the year 1887 marked a much more vigorous response to problems of safety in the factory. This year witnessed lively concern in the state over labor radicalism, following the fast, though short-term, rise to power of the Knights of Labor, and the Chicago Haymarket bomb outrage and the Milwaukee eight-hour-day demonstration of 1886. In this year the Wisconsin legislature tightened the law against conspiracy to coerce men in their business or employment. However, the 1887 legislature in effect also acknowledged that order was not maintained simply by negatives, and that it was the business of law to relieve tension by removing sources of grievance. Chapter 549, Laws of 1887, first recognized that factory sanitation was a matter of public concern by forbidding employment in any work place of "more persons than the laws of health will warrant, as shall be determined by the state board of health." The same act made the first positive move by law to foster reduction of the human costs of work with dangerous mill machinery. Of particular relevance to the lumber mills was the following new statutory prescription:

All belting, shafting, gearing, hoists, fly-wheels, elevators and drums of manufacturing establishments so located as to be dangerous to employees when engaged in their ordinary duties shall be securely guarded or fenced so as to be safe to persons employed in any such place of employment.

Conspicuously absent from this or any later statute was direct prescription of tighter safety standards regarding saws, which occasioned the single greatest number of lumber industry accidents. In 1905 the legislature removed the defense of assumption of risk in any case where injury was caused by the employer's "negligent omission . . . to guard or protect" the kinds of machinery specified in the 1887 act. In view of this strengthened provision, the continued omission of regulations regarding saws stands out the more. In 1909 the legislature required safety devices upon "any machine for sawing wood," but the Attorney General promptly ruled that, as a statute carrying a criminal penalty, this act must be strictly construed, and hence it should be limited to saws used for cutting wood for fuel. Another 1909 act pointed up the limited extent to which the legislature was ready to deal with mill hazards, by banning employment of persons under sixteen years of age to operate any saw. The statute reflected dismay at harsh decisions in the courts under existing law. But such piecemeal response only emphasized the failure of policy to come to terms with the true reach of problems. Until 1911 Wisconsin law thus left the greatest sawmill hazard to the always controversial, after-the-event remedies of the common law requirement that the employer furnish a safe place to work.[39]

Provision of particularized rules and enforcement to prevent factory accidents, and provision of a reasonably assured basis and procedure for money payments and medical care to help restore the injured worker as a self-sustaining member of the community, began in 1911 with enactment of the safe place and workmen's compensation acts under administration of the new Industrial Commission. As usual it was labor organized in southeastern Wisconsin that provided the main pressure for the new laws. But to the climate of opinion in which the new legislation was adopted, the mill workers of the lumber country at least made a contribution; the need of change was dramatized by their hard experience of uncompensated injury and death, as recorded in the reports of the old Bureau of Labor and Industrial Statistics and in numerous decisions before the Wisconsin Supreme Court. Moved by dislike of the rules which it felt bound to apply in some harsh mill cases, the court itself added to pressure for the new laws by calling in its opinions for statutory reform of basic doctrine, as well as by quiet behind-scenes work by Mr. Justice Marshall. The lumbermen scored considerable success in exempting their industry from statutes regulating terms of wage payment, but in this matter they could appeal to familiar reliance upon customary freedom in market bargaining. It had always been artificial to treat as just another point for contract bargaining the incidence of the costs of accident on the job. By 1911 it was so plain that there should be a wholly new basis for adjusting industrial accident losses that employers did not attempt to fight on the basic issue. From the start the legislature's disposition was for broad coverage of industry under safe place requirements, spelled out by administrative rule and compensation provided without regard to the fault issue and under administrative award. Thus the new statutes contained no exceptions of the lumber industry. Quite to the contrary, from the outset no matter was dealt with in more detail by the general safety orders of the Industrial Commission than the proper guarding of all types of saws.[40]

Theoretically, the common law obligation of the employer to furnish a safe place to work might include some duty regarding sanitary conditions in labor camps and regarding care of men who fell ill as a result of working conditions. There is no recorded case before the Wisconsin Supreme Court to show that anyone attempted to realize these common law obligations, nor did any such suit appear in the Chippewa County circuit court, in a busy lumber area. In 1899 and 1901 concern over the spread of certain contagious diseases led the legislature to appropriate substantial sums to the state board of health for preventive measures. In the years 1901–1903 particularly, lumber camps showed up as focal points from which there was alarming spread of smallpox. The state board of health and the governor sent state medical officers into lumber areas to aid in diagnosis and quarantine and decontamination measures. The episode revealed complete lack of adequate medical provision in the isolated logging sectors, irresponsibility of camp managements in unloading their sick workers on unprepared local communities, and an absence of strong local government measures to identify and confine disease. The reaction to this smallpox scare threw into relief certain elements of attitude which would figure in the development of twentieth-century administrative regulation. There was a thrust of facts toward more action by the state government in reaction to the weakness and inertia of localities, realization of a new interlock of living conditions which no longer permitted the treatment of public health as a matter solely of narrow local concern, and an impulsion to action born of more knowledge of the conditions under which disease flourished and spread. Though these developments were partly realized in reaction to the lumber camp smallpox scandal, their real fruit was for the future. Bills for regulating labor-camp sanitation failed in 1905 and 1911 despite favorable committee reports, as a result of floor maneuvers which suggest successful lobby action, probably by the lumber and railroad interests which were then mainly affected. Two acts were passed in 1905, in effect conceding the existence of a problem which called for broader response; these statutes, however, merely strengthened the hand of local governments in controlling contagion, and did nothing to promote basic preventive measures. Over ten years after the first smallpox fright, the Industrial Commission could still report gloomily that there was not even a minimum legal standard for sanitation in labor camps. "By spreading disease," the Commission bluntly noted, "these men pay back the state for its neglect to compel sanitary conditions in the camps." But an invigorated state board of health, and the Industrial Commission, were new sources of initiative to focus opinion and generate energy against inertia. In 1913 the legislature finally directed that deputy state health officers should inspect labor camps and enforce therein regulations laid down by the board of health.[41]

Of course, the safe place and workmen's compensation measures that began in 1911 and 1913 came too late to affect handling of the social costs of labor supply while lumber was still a major Wisconsin industry. The new laws and the further broadening developments under them provided important framework conditions for the minor lumber industry that continued into mid-twentieth century. Their principal significance for this history, however, is that by contrast they reveal the limitations of earlier public policy. By comparison with this twentieth-century legislation we can see more clearly how far-reaching and important were issues involving the social context of the industry, of which nineteenth-century law took tardy and scant account.

The wage contract offered explicit drama of conflict among readily identified actors. Considered as a problem in cost allocation, the mill accident had an impersonal aspect toward which the community needed time to develop attitudes and ideas fruitful for action. But the accident had obvious, poignant human meaning, and — however wide of the mark they were otherwise — the common law doctrines of fault and contributory negligence and assumption of risk provided a framework within which men could bring to definition issues of individual and social responsibility. In contrast, unemployment was a risk which derived from such broad causes, rooted in general patterns of production, marketing, and finance, not only in one industry but in the whole economy, that men first endured it as they did bad weather, and, when their frustrations drove them to despairing anger, took refuge from their confusion in such empty symbols as free coinage of silver.

Unemployment was a large cost of the job to men who worked in the lumber industry. Operations were sharply seasonal throughout the boom years in Wisconsin. In the first twenty years, when the streams afforded the only practical transport, there was hard technological reason for this. The building of a trunk-line railroad network to mill and market, and the development of practical take-up-and-put-down logging railroads for the woods, eventually made it possible to log and mill in more steady order. But it took time to extend the use of the railroads in a capital-scarce country where sparse settlement offered little general traffic to help bear the costs. By the time the railroad became the transport framework for the industry, production and marketing procedures on a seasonal basis were the familiar custom of lumbermen. It took a generation to conform operations more to the full flexibility which the railroad allowed. In any case, technology was never more than part of the problem. The lumber industry was highly vulnerable to general declines in economic activity. To this weakness the drift and default of public policy made material contribution; the capital subsidy tendered *de facto* by fast and cheap disposal of public timberlands encouraged chronic overproduction, at the same time as want of low-cost, long-term working capital put the industry under regular pressure to convert timber to cash. But, of course, the troubles of the lumber industry were only a minor part of the great swings in the economy which spelled boom and depression. Public policy did not begin to exert material control upon the range of cyclical movements in business until creation of the Federal Reserve system in 1913, and did not mobilize major public resources to this end until the 1930's. Through the years when lumber was a major Wisconsin industry, thus, national and state policy had nothing significant to do with the costs of seasonal or cyclical unemployment, save for the modest beginnings of free employment offices at the turn of the century.

When Wisconsin gave attention to unemployment compensation, the lumber industry—though much diminished in relative importance in the general economy of the state—scored remarkable, consistent success in keeping itself out of the new program. This outcome can be traced mainly to the weight cast in the legislature by the new paper industry.

A broad measure introduced in 1921 failed after considerable legislative maneuver. One explicit objection was that the proposal did not limit unemployment compensation in seasonal industries, such as logging and sawmill operation, by reference to their ordinary periods of work. A bill introduced in 1923 took account of this objection, measuring compensation rights against the longest seasonal periods of work which the best practice of a given industry permitted. But it, too, failed of enactment.[42]

The depression of the 1930's created the impetus to enact a state unemployment compensation program, four years before Congress made national provision therefor. The first bills introduced in 1931 were of broad scope, and again encountered the lumbermen's objection that such a program was not well adapted to their seasonal operations, and especially not when the industry was in a depressed condition. Given the large problems connected with enacting so novel an experiment, the sponsors of unemployment compensation treated the lumbermen's opposition as a marginal factor, to which they could give ground without serious loss to their main objectives and so eliminate one block to legislation. Thus Chapter 20 of the Laws passed by the Special Session of 1932 excepted from the new unemployment compensation system "employment in logging operations." In 1935 the legislature provided that an employer otherwise excepted from the act might, with the approval of the Industrial Commission, elect to bring his workers within the system. In a separate enactment in 1935 it declared explicitly that such an option was open to any employer engaged in logging operations. Beginning with the 1949 session, every Wisconsin legislature through 1957 saw the introduction under trade union sponsorship of bills to repeal the logging exclusion; in each session the industry successfully opposed the repeal, generally managing to delay the proposals until they were lost in the clean-up bustle at the end of the session. This was a striking victory for a rather limited lobby to achieve, especially in view of the explicit inclusion in the system of a conspicuously important seasonal industry of Wisconsin, the commercial canning business. No like exclusion was contained in the unemployment compensation act of any other state. The industry valued its exclusion from the state law no less for the fact that after 1936 it had to contribute to the federal unemployment compensation system, without benefit of the offsetting credit it might have had for money paid into the state fund, and without any benefit to its logging employees for the funds their employers thus paid to the United States. The paper makers particularly argued that an unfavorable experience in the logging phase of their operations would, under Wisconsin's experience-rating plan, subject their whole payroll to the higher rates of unemployment insurance tax. The unemployment compensation division of the Industrial Commission denied that the industry's experience would likely subject it to increased rates. Under the election clause added to the logging exclusion in 1935 there was a way to resolve the matter by actual trial, but generally the industry showed no disposition even to bargain its way under the system through such voluntary contributions. The stubborn success with which it thus held itself out of a major new public policy in mid-twentieth century is suggestive of the likely strength that the lumbermen would have shown, had more issues gone to conflict and fewer by default in the making of nineteenth-century policy.[43]

CHAPTER VIII

THE OVERHEAD COSTS OF LEGAL ORDER

THE existence of formal legal order was one of the basic conditions on which men did business in nineteenth-century Wisconsin. They took for granted that there was a law of property, contract, tort, and crimes, and that there existed public agencies to give some force to this law. They assumed, too, that the political community commanded economic resources (the public lands, tax-derived moneys) and might sanction economically valuable privileges (franchises for transport improvements), for the disposition of which there existed a legislative, executive, and judicial apparatus. Exchange of promises, commitment of labor and funds over time, peaceable adjustment of disputes and disappointments in doing business, effective concentration of means and enlistment of will for work — all went on and had effect in Wisconsin's lumber industry, as in the rest of its economy, in part because men assumed a sustaining framework of law. So viewed, legal order was an important economic asset, part of the capital on which the lumber industry was conducted.

Unlike the forest, legal order was not a gift of nature to the community's capital resources. It had to be made by investment of men's time and effort. To sustain it required continuing investment of manpower and funds put into buildings and equipment. Somebody must decide what investments in legal order were to be made, and who should make them. Thus public fiscal policy, and most obviously the law of taxation, were part of the legal history of economic life. This was the supply side. Part of this history, too, concerned the demand side — policy about the kind and scale of legal apparatus which the community should support. Logically this might seem the prior issue. In practice, fiscal policy got earlier and more sustained attention, and hence we begin with it.

The provision of legal order affected both the money and the real income and costs of the lumber industry. The special feature of this part of industry accounting was that these were income and costs not primarily determined by market processes. The creation and existence of distinct institutions of law meant the operation of inducements and compulsions distinctive to this separate apparatus — the operation of the ballot and the efforts to command ballots, the drives and contests over patronage, the jousting of local and sectional voting interests in the legislature and in county boards, the efficiency or inefficiency of officials, the momentum of established budgets and action patterns of government units. Out of its own internal house-keeping, the law impinged on the lumber industry in ways and with weight most peculiarly its own.

Taxation is the only aspect of public finance which calls for detailed discussion at this point. True, government can affect private economic activity by borrowing on public credit and spending the proceeds so as to influence the total pattern of investment. Conceivably public borrowing might have been used to supply long-term, low-interest credit to extractive industries like the Lake States lumber industry, to promote preference for future over present forest yield. But the Wisconsin constitution makers of 1848 had strictly forbidden the incurring of state debt (as well as the lending of state credit to private enterprise or state participation in works of internal improvement); the Congress showed no conception of this possibility of fiscal maneuver; and a cheap public lands policy helped lull nineteenth-century lumbermen into contentment with short-term production programs, so that they pressed for no alternative. Government can also affect private economic activity by regulating the supply of money and credit. But through the decades that determined use of the Lake States forest, national policy lived with the legacy of Jackson's defeat of the Second Bank of the United States. Not only did we lack the discipline of a strong central bank; we had also foregone the possibility of learning by continuous fiscal management something of how public policy might confine the range and violence of general shifts in business activity. Thus, in this area, the defaults of public fiscal policy left the lumber industry the more exposed to pressures for liquidity. This factor, again, encouraged short-run rather than long-run patterns for operation. Government can also affect private economic activity by the ways in which government disposes of its own property. Of the fiscal techniques available, other than taxation, this was the one which both the United States and the state used with massive, if little calculated, effect upon Wisconsin's lumber industry. The public lands policies of both sovereigns in substance offered the industry a subsidy in fixed capital, minimized its working capital needs, held down operating costs by foregoing any attempt to impose minimum standards of efficient forest use, and relieved the industry of any obligation to create depletion reserves against the costs of a substitute economy for the cut-over region. These matters were explored earlier in this volume; it suffices here to recall that they must be reckoned in a full inventory of the relations of public fiscal policy to the use of the forest.[1]

TAXATION

Ideally the history of tax policy relative to economic growth or the course of a specific industry should examine a broad range of facts and values. It should explore purposes (both those formally legitimated and those defined simply by practice), effects (calculated or unforeseen), and administration (formal and informal). Since taxation is purely instrumental and neither in form nor substance an end in itself, its history cannot avoid passing beyond relation of facts to judgments of value — of justice, and of efficiency in achieving formally or informally defined purposes and in striking a balance of calculated and uncalculated effects. This is a tall order. To fill it, the legal historian needs more help from economic theory and economic history than these fields have yet provided. This is especially so regarding the incidence of taxes and their relative weight in business decision making. Where, as in the case of Wisconsin's lumber industry, the general property tax dominates the story in a framework of highly decentralized administration, the difficulties are compounded for want of adequate studies of nineteenth-century local government in operation. Thus what can be presented here must fall far short of a desirably complete telling. At least, however, we can mark out the range of tax issues that became explicit in the lumber industry's life, note some large questions implicit in the facts but ill-defined or wholly ignored in policy-making processes, and strike a rough judgment on the likely total effect of the tax system on the course of forest use.

Of prime economic relevance in the history of tax policy are the questions whether and how far taxation was used to pay the out-of-pocket costs of the legal establishment, and how far it was used to regulate behavior the focus of which was outside the operations of government. This is a question of the intended objects of taxation, as distinct from its effects; of course, by compelling one use of means which might have been spent otherwise, all taxation in effect in some degree regulates private behavior.

By the protective tariff against Canadian competition, regulatory taxation played a small part in the life of the Lake States lumber industry. But lumber was a relatively minor item in the whole movement of federal tariff policy, and Wisconsin lumbermen rather belatedly enlisted in the struggle for tariff protection. Apart from their lobbying, there is no convincing evidence that the fact of tariff or no tariff materially affected the conduct of their business. Even with enlargement of the railroad net, the industry enjoyed a built-in protection of its prime Mississippi Valley market, created by the high transport cost in the sale of its heavy and bulky products. Moreover, cheap and ready disposal of federal and state timberlands so lowered the industry's capital costs as to cushion it against foreign competition. In part because of overproduction fostered by easy access to public timber, in part because of scarcity of working capital and the fact that it dealt in commodities peculiarly sensitive to general business decline, the industry had more reason to worry over its own tax bills than those of its foreign competitors.[2]

So long as lumber was a major Wisconsin industry the state made no use of taxation to regulate its development. Wisconsin lumbermen lived with a tax system which formally, and on the whole in substance, was designed simply to produce government revenue. From 1868 to 1905 there stood on the Wisconsin statute books a grant of tax exemption, and for a time of bounties in the form of tax credits, designed to encourage landowners to plant and maintain tree belts. There is no evidence that the system had important effect. In any case it was conceived solely in aid of agriculture, and has relevance to our concern only as it shows that regulatory use of the tax power was an idea not beyond the imagination of the times. Not until the forest was gone, in the second quarter of the twentieth century, did the state legislature adopt a tax program designed to influence the use made of timber-productive land. But the Forest Crop Lands tax law of 1927 belongs to the story of the effort to rehabilitate the north country, and not to our present concern with the relation of tax law to the primary exploitation of the great forest.[3]

Not only was there no effort by taxation to regulate the use of the forest. There was no effort even to adjust the administration of taxes for revenue to the peculiar facts of the northern forest. And yet, this was a part of the state's economic base so distinctive in location and in economic character as to invite tax treatment specially adapted to it.

In most of the forested north the soil was economically best suited to timber production. The untouched forest contained a high proportion of mature timber which was the free gift of nature; it made productive sense to harvest this wealth as fast as there was use for it, not only to speed the growth of the midwest farm belt, but for the advantage of future forest growth. The facts invited a nonrecurring capital levy on such mature forest product to capture for necessary government operations as much free capital of this type as could effectively serve public interest. This unearned increment apart, if the forest were kept as a continuing productive asset, men must expect slowly maturing yields and must commit large areas to programs which would not furnish annual income from every unit. The forest was a resource which should be taxed mainly according to yield, and by levies timed to coincide with yield. To levy annual toll, measured as a percentage of the capitalized value of timber on small standard units of land, introduced into forest operations an overhead cost item which was almost wholly arbitrary by economic criteria. So far as it affected business decisions, this kind of cost encouraged men to liquidate forest capital rather than to maintain it. So far as it was rationally calculated, the "value" on which the general property tax was assessed was exchange value, or the present discounted worth of future net income. Under an annual general property tax, the longer the asset was held the more often a future unit of net income was taxed before it was realized; thus the tax tended to encourage the property owner to realize what he could sooner rather than later, and so added to the other pressures for present over future preference in obtaining yield from the forest.

Another aspect of forest economy related to the desirable location of tax policy making and administration. The most practical means of keeping the forest in existence as a productive resource was some minimum control of fire. The original stand was so rich and various that if men could limit the loss which fire caused to young growth and to soil fertility, natural growth would suffice to replenish the forest without elaborate forestry practice or accounting controls such as would exceed the knowledge or the working capital of the times. But fire did not respect township boundaries, or suit itself to the means or efficiency of town government. To police against it and to mobilize the means to fight it called for a quality of management and a command over sizable areas such as only centralized organization could supply. Since central organization against fire would underwrite the productive life of the whole forest, a case could be made for a state tax to pay this underwriting cost, with appropriate distributions from the centrally administered tax to meet the general needs of local governments in the timbered area. Of course, this puts the facts with advantages of hindsight. Yet we may properly use hindsight at least to define more sharply what was done and what it meant in effect.

The only fiscal policy the state followed through the years of the lumber industry was to apply the general property tax to timberland and to forest products, almost entirely under local administration with no effective central supervision. At no point in the record was there a focused decision to this effect; the familiar tax became the central item of fiscal policy in the days of Wisconsin Territory, and was carried into statehood before lumbering loomed as a major factor in the economy. Thus the Statutes of Wisconsin Territory, 1839, provided simply that there be a tax on the unimproved value of all real property as well as on "merchandize," without special reference to timber. In 1848 the Wisconsin constitution makers authorized the state legislature to prescribe the property which should be taxed, stipulating only that nonresident proprietors be not taxed higher than residents, and that "the rule of taxation shall be uniform." The Revised Statutes of 1849 imposed a tax on all real and personal property within the state, and in particular specified taxation of land improvements, mines, minerals, quarries, and "fossils, in and under" the land, as well as of "all goods and chattels" and intangibles, but said nothing specifically of standing timber or saw logs. So matters stood until, in 1866, the legislature provided for appointment of commissioners to revise the tax laws. By then the lumber industry had shown the potentiality of its growth, and the strivings of timberland speculators and millmen attested the emergence of a major new factor in the state's economy. This, it might seem, was the point when a deliberate survey of the whole tax system would yield some special adaptations to the peculiar conditions of realizing forest wealth. But the simple processes of the mid-nineteenth-century legislature, and the resources of its characteristic instrument for major policy making — an *ad hoc* commission — again showed their want of force to develop searching decisions. The 1866 statute directed the commissioners simply to "revise and codify" the tax laws. They should recommend such improvements only as would "render said laws consistent throughout and make their construction clear . . . and . . . secure greater uniformity, and prevent frauds and evasions . . ." The commissioners' report of 1867 showed implicitly that they focused on the farm and town economy of southern Wisconsin, and assumed that the norm of policy was to rely on the general property tax, locally administered. Such was the pattern continued by Chapter 130 of the General Laws, 1868, which was the commissioners' product, and which in substance remained the frame of tax law under which the lumber industry ran its course.[4]

Though it taxed timberland and forest products as it did all other property save that expressly exempted, the 1868 statute made the first explicit references to timber wealth in connection with state tax programs. Real property should be assessed "at the full value which could ordinarily be obtained therefor at a private sale," and

> in determining the value the assessors shall consider as to each piece, its advantage or disadvantage of location, quality of soil, quantity and quality of standing timber, water privileges, mines, minerals, quarries, or other valuable deposits known to be available therein, and all buildings, fixed machinery and improvements of every description thereon, and their value.

Also, for the first time "saw-logs, timber and lumber, either upon land or afloat," were listed among items to be treated as taxable personal property — along with such other items as "toll-bridges . . . ships . . . all debts due from solvent debtors . . . ; and all goods, wares, merchandise . . . having any real or marketable value . . ." That the policy makers felt no reason to make special provision for taxing forest wealth was implicit in the character of these first direct acknowledgments of its presence in the tax base. The references were auxiliary only to the general assessment process. The attitude was pointedly implied in the fact that the one provision by which the 1868 statute recognized a tax problem peculiar to the lumber industry concerned a point of administration. In specifying the tax situs of various kinds of personal property, the policy makers found it desirable to make particular reference to "saw-logs, timber and lumber and any article of personal property in transit" — which should "be assessed in the town or ward where the owner resides; but if such owner be a non-resident of this state then such property may be assessed wherever such owner's agent or place of business may be located." In these various contexts forest-derived wealth was wholly assimilated to general categories of assets that yielded regular recurring income in market operations.[5]

The lack of will, ideas, and procedure to bring legislative policy to bear on issues not on the surface of affairs was dramatized by events of the 1867 and 1868 legislative sessions. In 1867 as the commissioners pursued their assignment to revise the tax system, the legislature — prodded by the state's first scientist, Increase Lapham — created a commission to report on all matters of concern to "persons desirous of increasing the growth and preservation of forest and other trees." In 1868, only two days apart, the governor signed bills on the one hand offering tax exemptions and credits to promote growing of tree belts, and on the other continuing the general property tax in application to forest-derived wealth. The two statute-making procedures went on independent of each other, without cross reference either in matters of policy or detail. True, the Lapham commission and the legislation which it sponsored paid only passing reference to the pinery, and concentrated on the uses of timber stands to protect farm soil and homesteads. Nonetheless, the existence and work of the Lapham commission, and its legislative proposal, did briefly single out, in the policy-making process, questions of the distinctive nature and worth of timber assets. Moreover, they put these questions in the immediate context of using the taxing power to effect economic ends related to the utility of timber stands. That, with this much impetus, there was no broad consideration of the wider issues implicit, shows the limits which prevailing attitudes set on lawmaking, and the force which the prevailing drift obtains when there is no agency equipped to give sustained attention to exploring policy.[6]

In the generation after 1868, legislation repeatedly acknowledged in effect that there were special problems in applying the general property tax to forest wealth. Yet, the attention given to these questions served only to emphasize the inertia of central policy. Both local tax assessors and lumbermen pressed their particular difficulties on the legislature. Legislators responded only within the frame of reference given by particular complaints or desires. The legislature showed no capacity to generalize from the immediate pressures put on it; successive governors ignored the bearing of tax policy on the state's second industry; there was as yet no central tax commission to direct sustained attention to the broad issues of the field. We have noted in other parts of this story how the press of detailed business blocked out consideration of long-term policy. So it was here. Men were so engrossed with the "practical" problems of the urgent present that they showed themselves quite impractical about the future.[7]

Chapter 130 of the General Laws, 1868, contained directions regarding assessment of timberland by which a bold central executive might have reshaped the substance of tax policy. The 1868 act directed assessors to "consider as to each piece" of land "its . . . quantity and quality of standing timber . . ." This might have been treated as authorization to classify timberland for taxation according to conditions of growth and variations of species, in a way to reduce the weight of current taxes on land given over to growing stock, and to put the greater load on land as its crop came to maturity.

However, Chapter 130 created a decentralized tax administration which provided no means to develop such policy initiative. The state senate and secretary of state, it is true, were constituted "a state board of assessors" — but only to determine the relative total value of all taxable property in each county, in order to apportion among the counties the small share which the state took (as "the state tax") out of the whole levies made on taxable property in Wisconsin. Otherwise all action under the very general standards of valuation set by the statute was committed to town and city assessors, subject to some powers in county boards to equalize total assessments among cities and towns. If any particular local officers were to take a constructively inventive line, there were no means to generalize their initiative.

In any case, events showed how little likely were local officials to develop any but the most blunt procedures of valuation. In 1873 assessors in Barron County adopted and applied rules by which they valued timberland. Their scale had three crude categories: "First. Pine on first class driving streams . . . at $2 per M. within the limits of two miles hauling. Second. Pine on such streams of more than two miles hauling, at $1.50 per M. Third. Pine on second class driving streams, as Moose Ear and other streams mentioned, at $1.50 per M. within two miles, and $1.00 per M. beyond." In *Hersey v. The Board of Supervisors of Barron County* (1875) — the only case in which it was called on to declare basic principles of timberland taxation — the Wisconsin Supreme Court sustained an injunction issued at the suit of a taxpayer against enforcement of levies made under these rules. The 1868 act commanded that "real property shall be valued by the assessor from actual view" — a requirement somewhat relaxed, before the assessment here under challenge, by an 1873 amendment allowing the assessor to attest that he had valued each parcel of land by actual view "as far as practicable." The court would be sensible about the matter: "It might not be practicable for the assessor to go over every foot of ground" in such wild country. "But there should be an attempt to substantially comply with the law." Plainly the formula followed by the Barron County officers was not even an "attempt" to value by view. More fundamental was their neglect of the legislative policy on the substance of valuation. Here, said the court, was no mere mistake in applying valid rules of assessment; the statutory standards "were utterly ignored and disregarded." To value pine according to the "class" of driving stream and the location of trees on one side or the other of a two-mile line meant that

the real estate is valued solely with reference to the quantity of pine timber standing upon it, without taking into account the quality of the pine, its location up or down the stream, or the character of the soil, or those other elements which determine the value of land, and which the statute says the assessor shall consider . . . True, it is alleged in the answer that it was only practicable to make a fair and equitable valuation of the real estate under these rules, which is equivalent to saying that the law upon the subject cannot be complied with. But if no valuation can be made as the statute requires, we fail to see upon what ground the tax can be sustained. A valuation is essential to lay the foundation for the tax; and if no legal valuation was practicable, it follows that any tax based upon a wholly unauthorized valuation would be illegal and void. The allegation is a *felo de se*.[8]

As *Hersey* read the statute, the law gave the taxpayer great leverage to move the assessment process toward an exacting sophistication in valuing timberland. The direction that assessors consider the "quality" of standing timber might especially require that, out of this awkwardly decentralized tax administration with its natural bent to rule of thumb, there be developed somehow a classification of timberland that would fit significant criteria of forest economics. There had already been indication that local politicians feared that the 1868 statute set stricter standards of administration than they wished to meet; we have noted that in 1873 the legislature amended the terms of the affidavit which the assessor must attach to the assessment roll, so that he need make oath only that he had valued all land from actual view "as far as practicable." Now, after there had been time for the implications of *Hersey* to sink in, there was sharp legislative reaction toward less exact standards for the local assessment process. In successive steps the legislature whittled away the requirement of "actual view." In 1877 it said that valuation must be done by actual view in towns containing 108 square miles or less, but that in towns of greater size "real property shall be assessed either from actual view or the best information that the assessor can practically obtain, except improved lands, which shall be assessed from actual view." That this change responded particularly to demand of local officials in the forest area was made plain when the Assembly rejected an amendment which explicitly would have continued the requirement of actual view in assessing "timber" as well as improved lands. The session laws of 1878 again amended the view requirement to limit it to towns of not more than 48 square miles in area. Then in short order the general revision of the statutes in 1878 made assessment by view altogether optional. Section 1052, Revised Statutes, 1878, stated, "Real property shall be valued by the assessor either from actual view or from the best information that the assessor can practicably obtain . . ." The 1878 session brought a second change, this time in substantive definition, to complete the retirement of the *Hersey* decision. The legislature struck from the valuation formula the stipulation that assessors consider the "quality" of standing timber; assessment simply according to "quantity" would now fit the statutory standard.[9]

Antedating and interweaving with this chain of events was another thread of tax procedure legislation which further reflects the bias of the times for full decentralization of practical tax policy, as well as the odds against contrivance of broad, original patterns. North-country politicians were well aware of both the rich tax base which the forest afforded, and the practical difficulties of exploiting it. They sought authority to pursue timberland tax revenues by specialized informational and enforcement procedures, so long as these grants carried no strict standards of rationality imposed by central authority but simply enlarged the opportunities of local government. Thus in 1871 the legislature authorized (but did not require) the county boards of twelve named northern counties to appoint and pay "land inspectors" to compile by direct view data relevant to assessing land, with emphasis on the presence or absence of timber and the quality of soil. Any county board electing this procedure was authorized (but not required) to publish the resulting data and distribute it to town assessors, who must examine and preserve these lists, and who were authorized (but not required) to value lands on the basis of this data without personal examination of the listed tracts. Authorization of such "land inspector" procedures was carried into the Revised Statutes of 1878, where it was broadened to empower "the county board of any county in this state" to adopt the device; despite the sweep of the language, the Revisers' Notes make plain that the significance of the change was seen as making the procedure available to all the northern forest counties. Apart from its utility for increasing local revenues, the "land inspector" system seems likely to have had attractions as an added source of patronage; Oneida County, it was said, laid out between $12,000 and $15,000 in one year to pay expenses of the system. The state supreme court upheld the procedure against a taxpayer's challenge: the statute was a reasonable exercise of legislative judgment to cope with problems of assessment in large, sparsely settled areas; the court "must assume" that a county board was motivated by honest purposes of tax administration and not merely by jobbery, in providing for "land inspection"; the county board of review might properly base its findings on the inspectors' reports as at least prima-facie evidence of facts relevant to land values.[10]

One feature of this "land inspector" legislation seems especially revealing of the incongruities in contemporary policy making. The 1871 act declared:

> Every assessor who shall know or shall be reliably informed that the pine timber has been cut off of any tract of land in his town, shall, in assessing such land, affix a nominal value thereto.

Moreover, all lands which "land inspectors" listed "as having the timber cut off"

> shall be assessed at a nominal value, unless the assessor of his own knowledge or from reliable information shall know or learn to the contrary.

In 1875 the legislature repealed the first-quoted provision, which in terms had imposed a flat requirement on the local officers.[11] So again the pattern of legislation limited centrally defined standards and stressed the discretion won by the local politicians. On the other hand, these same provisions paid implicit tribute to the distinctive importance of timber values in north-country economy. Thus in effect they pointed to the good sense of a special, centrally defined tax program for that area. But current policy processes proved incapable of bringing the matter to definition for debate, let alone for decision.

The same incongruity — implicit acknowledgment that taxing forest wealth presented special problems, but explicit attention only to the secondary and most narrowly instrumental aspects of those problems — appeared in other legislation on tax enforcement procedure. One development was that of special statutory liens for taxes on timberland. From 1859 unpaid taxes were made a lien on the land taxed, in all cases. But in 1872 the legislature added a lien for real property taxes on all logs or timber cut on the taxed lands after assessment, where the landowner was not a resident of the town in which the land lay. In 1878 the Revisers extended this lien to "wood." In 1901 the lien was made generally applicable, regardless of the landowner's residence.[12] A more limited special lien dated from an act of 1868, which authorized the county treasurer to issue a warrant for seizure and sale of logs or timber found within the county, cut from land on which the county owned a tax certificate at the time of cutting. In 1897 a like lien was given to any tax certificate holder other than the county.[13]

Definition of the tax situs of logs, lumber, and other forest products was a third subject of considerable legislation, the quantity of which reflected the dollar stake, both for the tax base of particular localities and for the tax bills of loggers, processors, and dealers. The situs laws left only a scanty and unrevealing legislative history, and at this distance it is not plain what cross currents of interest moved behind particular changes. One concern of the legislature clearly was to define situs with sufficient precision to reduce opportunities for tax evasion, particularly by nonresident owners. Precise definition inherently served another legitimate objective, in protecting property owners against double taxation or the harassment of double claims for tax.[14] The greatest ambiguity lies in the jockeying among alternative places for assessment — whether assessment should be at residence of owner or agent-in-charge, place of actual location on assessment day, or millsite to which logs were destined for processing. Choice of situs meant increase of the tax base for one type of industry locale and loss of potential tax base for another — logging and woods storage areas, as against milling areas, as against investor areas. Taxpayers who wielded business and political influence might calculate that they could exercise more control over local tax officials in one type of locale than in another; but this seems a judgment so peculiar to the conditions of particular firms and localities as to baffle generalization at this distance in time.[15] Whatever the background, two trends stood out in various shifts in the situs statutes: preference tended to fasten on the districts of actual location of the property at assessment day, or on millsite districts to which logs were destined for processing. Emphasis on location of the property itself was the more marked because this was the regular statutory formula fixing the situs of dealers' stocks of goods and merchandise. The Wisconsin Supreme Court interpreted this general provision to cover all forest products held for wholesale or retail sale, though other specific statutory formulas fixed the residence of the owner or his agent as the situs for taxing forest products which were not considered as manufacturers' stock by their owners. That the legislature left this arguable interpretation undisturbed suggests that the test of actual location tended to win out because it was relatively simple to administer and best reflected the operations of the industry.[16]

What has been said so far sums up the formal tax policy of the state regarding timber wealth, as the legislature fashioned policy. We have noted that the governors played no apparent role in the business, and that not until the lumber industry had run its course was there created a specialized central tax commission to provide a new source of potential executive leadership in programming.

This leaves the Wisconsin Supreme Court as a possible policy maker at the center. On the whole, it is not surprising that the court played a minor role in this field. By history and constitution, in our legal system taxation is the classic area in which policy initiative is with the legislature. Especially in this field the court could declare important policy only as statutes created the occasion. But the legislature took no significant initiative to adapt state tax policy to the distinctive problems of forest wealth, and this course of affairs effectively blocked any chance for the judges to make a major contribution. When the court in the *Hersey* case vigorously applied those terms of the 1868 statute which might have supported a distinctive forest tax program, the legislature quickly retreated to more conventional generality. Thereafter, in some fifty appeals involving validity of forest industry taxes, the court dealt typically with the sufficiency of evidence to support an assessment, or with such fine points of procedure as notoriously abound in the law of tax administration and the issue of tax deeds. More than in any other field touching the lumber industry, the court's role here was to superintend administration of a narrowly technical body of procedure. The sum of these cases added up to business which had most dubious claim on the time of a top appellate court; if the court's tax decisions had little perceptible effect on the course of the industry, the industry probably had more effect on the court by thus cluttering its docket. Taxes interwoven with the conduct of substantial business inevitably generate issues of technical detail, most of them significant only to the immediate controversy. An inventory of the forest-industry tax cases in Wisconsin's highest court suggests that such business would better have been committed to a specialized tribunal.[17]

Of course, thus far we have spoken only of formal tax policy. The substance of policy may take quite different meaning when measured by informal practice. We cannot be sure we have told the whole story, unless we consider the possibility that practice defined some distinctive forest tax policy of which the statutes give no clue. And perhaps in this realm we might find some important contribution from the court, whose particular business is to give law content in application.

Letter files of timberland investors and lumbermen, squabbles preserved in the faded pages of north-country newspapers, and some litigation bear witness to chronic dissatisfaction of taxpayers and of local officials with the operation of the general property tax in the lumber region. It marks the limits within which men conceived their problems, that no significant controversy centered about the basic question whether the general property tax made sense as applied to forest wealth. Men argued about charges of discrimination within the accepted framework of that tax. Small property owners alleged that assessors were in the pocket of the big resident lumber firms. Public officers complained that timberland owners deliberately let their taxes go in arrears, so that they might bargain for advantageous settlements with financially hard-pressed local governments, or that loggers timed their cutting and movement of logs to evade levies. The big lumber companies justified close dealings on taxes on the ground that they were regarded as fair game by local politicians and small property owners. Nonresident landowners complained that local politicians and resident landowners, large and small, loaded a disproportionate share of the costs of government on absentee owners.[18]

Most of what we know of discriminatory tax administration in the lumber country we know only by hearsay, and that almost always from witnesses spurred by self-interest of one kind or another. Reliable evidence indicates a good deal of over-assessment of nonresident timberland owners, at least of those who were not Wisconsin millmen. This was part of the current pattern of prejudice against the "speculators" who held land without fitting their holdings into a production program.[19] Passing by for now the question of sheer jobbery in creating unnecessary units of local government, the evidence is less satisfactory concerning the relative importance of assessment manipulation by resident lumbermen, as against unfair assessment of lumbermen to the interest of homesteaders and local politicians. Plainly there was contest among such interests. Weighing the net effects is another matter.[20]

One thing is clear, however. The continual skirmishing over tax assessment in the forest country rarely mounted to the level of battle. This was not a field of campaigning, but of guerrilla warfare among partisan bands which formed no broad alliances and concerted no general strategies. It is consistent with this relatively formless and noncumulative character of pull and haul over tax administration on the local scene that most of it apparently consisted of contention among local officers and taxpayers and their lawyers. There was some litigation — more at the trial court level, it is fair to conclude, than reached the supreme court — but such general evidence as we have indicates that lawsuits over tax administration were exceptional. The detailed inventory taken of business in the circuit court for Chippewa County shows no direct challenges to taxes there at all from 1855 to 1894, and then an isolated clump of eleven suits, 1895–1904, with none thereafter during the remaining years of the industry. Informal tax practice developed no pattern of deep significance — no effort, for example, to employ discriminatory assessment for a calculated capital levy upon the unearned increment of timber value in the mature virgin forest. The logic of the situation was against the possibility that tax practice would develop broad policy *de facto,* in contravention of the policy set forth in the statutes. A sensible alternative to the formally declared policy would be one which took account of the forest's significance for the total Wisconsin economy. But if the formal agencies of state policy making could not evolve such a program, one could hardly expect it to grow out of the competing interests and emotions of local tax assessors and politicians, homesteaders and businessmen, whose worlds revolved about their immediate enterprises. Comprehensive policy can emerge out of administrative and lay practice only where that practice operates in a theater of broad action. The highly local and episodic character of tax discrimination not only denied it broad effect in shaping policy, but meant further that it had only marginal effect upon timberland profits and the course of the lumber industry. At least for those able to hold for appreciation, the inefficiency of local officials and their lack of information meant that assessed values, discriminatory or not, lagged substantially behind the increasing market value of good timberland.[21]

Since informal practice developed no alternative pattern of policy to that of the statutory general property tax, our story may properly return to inquire whether there was authority, model, or means by which formal policy making might have developed an alternative.

There was no want of constitutional authority in Wisconsin's legislature to shape a tax program special to the forest economy — a program which might include a capital levy on the unearned value of original mature stands, assessment and collection adjusted at least not to discourage operations for long-term production, and central definition and scrutiny of policy to treat the northern forest as the economic and social entity that it was. The Wisconsin constitution empowered the legislature in sweeping terms: ". . . taxes shall be levied upon such property as the legislature shall prescribe." True, it also declared, "The rule of taxation shall be uniform." But this was constitutional language. In the absence of qualifying terms — and here there were none — familiar principle enjoined that this language be construed to create a broad standard of fairness and not a rule of mechanical and purely formal equality. Nothing in the records of the Wisconsin constitutional conventions persuades to another reading.[22]

By Chapter 74 of the General Acts, 1854, the legislature made practical construction of its powers as entitling it to create a special tax structure with a central administration, adjusted to the circumstances of a distinctive sector of the economy. The 1854 statute imposed on all railroads and plank roads organized and operating within the state an annual state tax of 1 per cent of gross earnings, in lieu of all other taxes upon the companies, their property, or their stock; the act forbade any other assessment or tax upon the roads or upon the interests of their stockholders.[23] Railroad promoters had obtained this legislation. Partly, no doubt, they expected to pay less tax thereunder. Partly, however, they could make plausible arguments that the system would favor general economic growth. Railroads were operations of framework significance to the general economy. They were new, risky, and expensive undertakings. Typically their operations fell within the territory of many local taxing units, but grew out of and had meaning for much wider areas. An earnings tax, even one on gross earnings, would weigh less on fledgling enterprises than a tax on the theoretical exchange value of invested capital. And central administration would protect such far-flung business against the heavy and perhaps crushing burdens that might grow out of the inefficiency, honest variations, or prejudices of local assessors.[24] The case to be made for special-sector treatment and central administration in railroad taxes, if it won its way into state policy, offered suggestive analogy for special tax treatment of the forest.

In an unreported decision in 1855, the Wisconsin Supreme Court upheld the railroad tax act as against challenge under the uniformity clause. But in 1859 the court made an abrupt about-face. By a 2 to 1 vote it then held in violation of the uniformity clause a local act which provided that farm land might be taxed at a different level from land used for other purposes. The court seemed to say that, under its constitutional authority to "prescribe" what property should be taxed, the legislature had broad discretion to grant total exemption from taxation, but that the uniformity clause denied it any power to classify taxable property so as to apply different rules of assessment and administration, however reasonable the grounds of classi-ation.[25]

So rigid a reading of the constitutional standard could perhaps be explained, but it could not be justified, and it did not stand long. Explanation lay probably in timing. In 1859 it was familiar doctrine that the court had power to pass on the constitutionality of legislation. But this was doctrine which, so far, the Wisconsin court had had relatively little experience in applying. Uneasily aware of the extent of power they wielded, the judges seemed anxious to disclaim the full scope of their responsibility, by finding a rigid command in the language used by the constitution makers. Moreover, in 1859 Wisconsin's farmers were in full cry against alleged abuses in financing railroads, in aid of whose construction thousands of them had given their notes and mortgaged their farms to buy stock. The court already confronted the unhappy prospect of enforcing these notes and mortgages in suits by bona fide purchasers against whom farmers' pleas of railroad fraud and abuse must be held unavailing. In upsetting the farm land classification statute, the court's 1859 opinion indicated intent to overrule the unreported 1854 decision in favor of the special railroad tax system; indeed, there is convincing evidence that the 1859 lawsuit was a test case, brought by a partisan of the farm mortgagors to reopen the doctrine underlying the railroad tax system. Thus the pressures of the situation invited the court to interpret the tax uniformity clause in a way which would dramatize even-handed justice; while they did their hard duty in enforcing the farm mortgages, the judges would also deal a blow to the railroads by overturning their special tax plan. In 1860 — again in a test case, and again by a 2 to 1 vote — the court confirmed the implications of its 1859 opinion and held that Chapter 74, General Acts, 1854, violated the uniformity requirement.[26]

Had the state's constitutional doctrine long remained in this posture, there would be explanation enough, in simple want of legislative authority, why no one considered applying to forest wealth some special tax program in distinction to the general property tax. But in 1862 the court once more reversed course, repudiated the decision of 1860, reinstated the ruling of 1855, and again — and this time, it proved, finally — held that the special railroad tax system did not violate the uniformity clause. True, the court left the state of doctrine not altogether clear. It now upheld the 1854 statute not on the judges' reading of the constitutional language, but rather because the judges now saw in the statute a legislative construction of the uniformity clause, entitled to great weight. For this practical construction had entered so fully into the pattern of state fiscal policy and tax administration and into private business and title dealings conducted in reliance thereon, that it could not be upset without too costly disturbance to public and private affairs.[27] Meanwhile, however, the legislature had further demonstrated in practice its belief that the uniformity clause did not bar it from making tax classifications. The court had repeatedly conceded the legislature's broad discretion to grant tax exemptions. In 1860 the legislature built on these dicta by passing laws that were in form distinct, one of which exempted railroads from all taxation, the other of which exacted from them what it termed a fee for a "license" to operate, measured as 1 per cent of annual gross earnings. This pattern of railroad tax laws stood thereafter unchallenged for a generation; when it was changed about the turn of the century, the change arose not out of constitutional doubts but out of new legislative concepts of desirable sharing of tax burdens.[28] The period of real doubt over constitutional authority for tax classification was thus a short one: 1859–1862. The legislature's bold and subsequently unchallenged maneuver of 1860, and the court's deference to legislative construction in 1862, indicate that want of will or imagination rather than any strong denial of constitutional power must account for the failure to bring forth a distinctive program of forest taxes.

Constitutional questions apart, the times were not wholly wanting in models of tax programs cut to the contours of distinctive sectors in the economy. We have seen that, continuously from 1854, Wisconsin's statutes provided one form or other of special tax system for railroads. A special set of tax exemptions for railroad-aid lands offered another analogy — especially pertinent, since so much of the lands' value was in standing timber. Though its purpose was regulatory rather than simply to adjust revenue collection to the character of peculiar property, the tree-belt tax exemption statute from 1868 to 1905 implied the suggestion that a tax program might be specifically adapted to timber wealth. Statutory procedures special to the relation of the general property tax to forest assets — the land inspection system, the liens for taxes on forest produce — implicitly acknowledged timber as a distinctive factor in the tax base.

Particularly interesting are two short-lived models of special tax treatment for another extractive industry. Generally, from territorial days to 1915, mineral land, like timberland, was taxed like any other realty; in the general pattern the only notice of mineral wealth was by direction that assessors include it in striking the over-all value of the property. But from 1864 to 1872 the legislature provided that a mining corporation chartered under Wisconsin law should pay an annual state tax based upon actual output, which should be "in lieu of all state taxes to be paid by such corporation." This was a good deal less generous than the special railroad tax, which relieved the taxpayer and its stockholders of all other taxes whatsoever, state or local. Nonetheless, it did make limited provision for a yield tax as the only state tax upon processing concerns in an extractive industry. Again, in 1877 the legislature directed assessors to value land containing iron ore by appraising the land without the ore deposits and then adding to this figure the "net value" of ore mined in the preceding year (after subtracting from the "gross value" of the ore its cost of production). This provision, which offered obvious analogy for taxing timberland, lasted but a year until it was repealed in the Revised Statutes, 1878. The Revisers saw the only substantial issue as one of administration, rather than of substantive policy. Indeed, they said that they rejected the 1877 formula in part because, if it stood, in principle it should be applied to a wide range of resources, including timber — a result which obviously they thought so unsound as to require no analysis for its rejection:

> The principle of [the 1877 law] . . . is wholly at variance with the principle of assessment in this state, and equally applicable to timber lands, quarries, etc. . . . [W]hile there may be some evil in the assessment of mineral lands, it results from the ignorance or viciousness of assessors in particular cases, rather than the defects of the law, and the revisors recommend no peculiar relaxation of principle to cure disorders in peculiar cases, not due to vice in principle but in practice only.[29]

Both the ore-lands tax provision and the Revisers' explanation for striking it reveal implicitly why there was no consideration of a tax program distinctive to the forest. Wisconsin's people were busy with private business, from which public affairs were an annoying distraction. Their instinct was to borrow their public policy ready-made from older states. So Wisconsin had uncritically adopted the general property tax as its "principle of assessment." The generation of 1878 was, if anything, yet more engrossed in its private affairs than was its predecessor, and even more disposed to economize its investment in public policy making. Thus, entrenched forty years in Wisconsin law, the borrowed "principle of assessment" effectively limited the imagination of the Revisers appointed *ad hoc* for the 1878 recasting of the statutes, as well as the imagination of the contemporary legislatures which accepted the Revisers' product.[30] A powerful and dissatisfied interest group might have made a difference. Given the prominence of lumbermen in local politics and in key legislative committees, the absence of industry effort to obtain a special tax system is strong circumstantial evidence that, despite complaints of over-assessment and discrimination, the general property tax did not impose on lumbermen burdens which materially affected the way in which they were content to do business.[31] Of course from a twentieth-century retrospect this is far from stating all the factors in the situation. A central state tax based on yield would have been a proper part of a rational and modest program in the interest of the general economy, to help keep the forest in being as a continuing productive asset. In the absence of interest group pressure, however, such initiative must depend upon legislative and executive apparatus geared to produce comprehensive and long-term policy planning. The act of 1877 — applying only to iron ore lands a policy that made sense for the whole range of extractive industry —

symbolized the episodic, limited-term activity to which alone the state's nineteenth-century policy processes were adjusted.[32]

LOCAL GOVERNMENT

The costs of state government or state policies never were big enough during the lumber years to become of active concern to lumbermen, save in disposition of state lands, where lumbermen were too much beneficiaries to complain of costs. The nineteenth-century state government apparatus remained small and simple. The nineteenth-century Wisconsin statutes set up no regulatory schemes, whether for conservation, marketing, or any other aspect of forest use, to add materially to the costs of business. Like the disposal of public timberlands, the grant of transport franchises was in substance a promotional more than a regulatory application of law, the benefits of which so far outweighed the burdens that the industry was concerned primarily for preferred or at least equal access to these privileges, caring less about the fees or regulations that went with them. The lumbermen had no serious complaint that the state was subsidizing other interests at their expense; subsidies to agriculture out of state lands were completed before lumber took on stature; subsidies to railroads in the north stirred local or factional rivalry, but the rivalry was to capture the prize, not to object to the giving of it, for a rail net was essential to realizing the full wealth of the forest. Federal policy was still more remote from laying any material burdens on the Lake States lumber industry. By its public lands policies the United States made the industry a capital subsidy, without using its original ownership to impose any terms of forest use which might increase present costs of production. No federal tax bore on the industry; to the contrary, the federal tariff offered bounty. Thus, so far as the expense of public machinery and substantive programs became an issue between law and the industry, it was at the level of local government.

Here was a crisscross of interests, in part familiar to the prairies, in part peculiar to the forest. In their desire to economize on public services, their skepticism of local budgets, their conviction that they were targets of demagogues and political pirates, timberland owners and millmen reacted to local government like big taxpayers anywhere in their time. For their part, in these close-to-home matters of public service and patronage, homesteaders, small merchants, and county-seat politicos reacted to the big lumber companies much as their counterparts in the wheat and corn country did to the symbols of "the railroad" and "the money power." [33] But in addition the attitudes of both sides were specially, though differently, affected by their shared conviction that the forest would pass, perhaps to be succeeded by farms. This conviction did not produce an articulated philosophy, but it offered implicit support to certain working attitudes. If the forest were only a passing source of wealth, this fact pointed up the conclusion that the forest represented largely a gift of nature, not earned assets; why then should not local settlers seize on it so far as they could, to furnish themselves with the roads, bridges, schools, and court houses that settled communities needed and were otherwise hard-pressed to supply in this difficult country? Urges of cupidity and community alike could find rationalization here. In particular, why not provide local services and well-paying public jobs for local residents, if this could be done at the expense of nonresident landowners or millmen, who would only take off the forest wealth and invest its proceeds elsewhere? Looked at from the owning and production side — why not use every means to beat down assessments, defer payment, oppose settlement and the extension of public services, when the industry would live but a brief decade or so in any given locality? Why not import gangs of fraudulent voters to carry the day, when the problem was simply to hold things steady for the limited time it would take to cut and get out, especially when the other side was plainly moved by a like short-run calculus of its own interests?[34] Such attitudes grew largely out of the ideas and definitions of the situation which were expressed in the public lands policies of state and nation. They were not attitudes in the first instance generated by law. But the law — above all, the subtly educating example of the public lands policy itself — became part of the interplay of experience which taught men their "common sense"; so viewed, the short-run, limited-reckoning views with which men handled local government in the forest country must be counted as part of the intricate pattern of costs of our policy toward the public domain.

The political chronicles of the north country are long and involved. Men fought over the creation of new counties out of old ones, over the division of towns, and over contracts for building schools, court houses, jails, and county poor houses, as well as roads and bridges.[35] It would be unprofitable to try to set out all these disputes. There is no evidence that in net result they — any more than burden or discrimination under the general property tax — had distinctive effect upon the main course of the lumber industry. This excessive busy-ness about local government has relevance for our story only in its general, not in its particular, impact. These local controversies bear the stamp of a good deal of nineteenth-century legal administration: too much time invested in matters of narrow significance, too much repetition in handling questions which should have been settled by vigorous general policy. In this fashion they added to the overhead cost of production, and also — and more significantly — tended to divert attention from larger questions and to increase the pressures which worked in favor of present over future preference in the use of forest wealth.

We may usefully venture far enough into one avenue of this maze of local government business to get a more immediate sense of its density. Consider, for example, the attention which the legislature gave to drawing and redrawing the lines of one county in a great lumber section. As it stood defined in its first revised form under statehood in 1849, Chippewa County in northwestern Wisconsin had an area of about seven thousand square miles. This expanse embraced nearly all the basin of the Chippewa River and of some smaller rivers, forming one of the great natural log transport systems around which the lumber industry grew. From this beginning the lumber decades saw ten more statutes redefine the area of Chippewa County down to about one seventh of its original size:

General Acts, 1854, Chapter 1, transferred land from Chippewa to Buffalo County, an existing county.

General Acts, 1854, Chapter 7, transferred land from Chippewa and other counties to Dunn County, here created.

General Laws, 1854, Chapter 100, transferred land from Chippewa to Clark County, an existing county.

General Acts, 1856, Chapter 114, transferred land from Chippewa to Eau Claire County, here created.

General Laws, 1860, Chapter 235, transferred land from Chippewa to Dallas (later renamed Barron) County, an existing county.

General Laws, 1864, Chapter 462, transferred land from Chippewa to Burnett County, an existing county.

Laws, 1875, Chapter 178, transferred land from Chippewa and other counties to Taylor County, here created.

Laws, 1879, Chapter 103, transferred land from Chippewa and Lincoln counties to Price County, here created.

Laws, 1883, Chapter 47, transferred land from Chippewa and Ashland counties to Sawyer County, here created.

Laws, 1901, Chapter 469, transferred land from Chippewa to Gates (later renamed Rusk) County, here created.[36]

The bills that passed were only part of this legislative story, however; time and energy went also into rejected proposals to redraw the lines of Chippewa County. Thus bill 438A of 1871 would have organized "Pine County" out of Chippewa. Timberland owners and millmen more often opposed than supported proposals to create new units of local government, but the Pine County bill had the backing of some large landowners who apparently hoped that they could persuade a new set of local officials to lower taxes. However, bill 438A did not get past the Assembly, which rejected it after it was reported without recommendation by the Committee on Town and County Organization.[37] Proponents of a "Pine County" tried again, with bill 175A of 1872, in behalf of which the legislature was memorialized by persons "representing," they said, tens of thousands of acres of pinelands in the affected territory. On recommendation of its committee the Assembly tabled the bill.[38] In 1874 bill 297A presented the first effort to create Taylor County, in part from territory to be detached from Chippewa; after an adverse report from its Committee on Town and County Organization the Assembly gave the bill polite burial by referring it to a select committee composed of its sponsor. A second try succeeded in 1875, when the legislature created Taylor County out of areas drawn from the counties of Chippewa, Clark, Lincoln, and Marathon. This measure went through the Assembly with little apparent trouble. It passed the Senate only after considerable maneuver, which resulted in the last-minute defeat on the floor by a 15 to 14 vote of a competing bill which would have set up a new county out of territory withdrawn simply from Clark, Lincoln, and Marathon.[39] In 1876 and again in 1877 the Assembly summarily rejected proposals to carve "Ludington County" out of Chippewa and Lincoln. A parallel bill was buried in the Senate in 1877. On recommendation of its Committee on Town and County Organization, each house in 1878 tabled parallel proposals to erect a new county (to be called Flambeau) out of Chippewa and Lincoln. It was only after these maneuvers of 1876–1878 that a bill could be drawn satisfactory to a sufficient balance of forces to pass into Laws, 1879, Chapter 103, creating Price County out of areas withdrawn from Chippewa and Lincoln.[40]

Committee reports on the Senate bill of 1877 made articulate some of the competing interests and values behind such a pull and haul over organizing a new county. Arguments over carving a new county (here "Ludington") out of an old one (here Chippewa or Lincoln) fell mostly under two heads. One might take the proposal at face value, and argue whether it made sense for the public service and in terms of the available tax base. Or one might look behind such legitimate criteria, to test how far the proposal was the product of special or private advantage only.

Dividing 2 to 1 as it reported unfavorably to the Ludington County bill in 1877, the Senate Committee on Town and County Affairs gave equal attention to both heads of argument. The majority thought that the proposal would not serve public interest. ". . . the territory from which it is proposed to form the county of Ludington is, as yet, a wilderness, in which there are few actual *bona fide* residents, not exceeding three hundred, and those residing in two or three small settlements along the line of the Wisconsin Central Railway . . ." The dissenter, who favored the bill, took issue on a basic question of fact, asserting from his own knowledge that the proposed county contained several thriving villages with an actual population over five hundred. To him the presence of the new railroad was more than incidental, for it sustained the promise of rapid increase in population by opening the area to readier settlement. A town organization already existed for the territory of the proposed county, and in the rude stage of its development the majority thought that this much organization was "amply sufficient for all . . . present wants of the residents, and under which they can make all the local improvements that just regard for the rights of the property owners should permit." To the dissenter this analysis ignored the facts of legitimate local interest which, especially while so much of the land was wild and unsettled, gave people a proper reason to have all their agencies of local government close to home: ". . . I believe there is no identity of interest in the section of Chippewa county lying near Chippewa Falls and that lying along the Central railroad, because of the intervening wilderness, almost one hundred miles in extent . . ."

The majority report mixed diplomatic innuendo and blunt misgivings, in appraising the private interests which the majority obviously believed were supplying the main drive for the bill. The proposed county had "few actual *bona fide* residents." (A similar report, recommending rejection of the 1878 bill for "Flambeau" County, similarly stressed as an undesirable feature that the area's "population is too small; what there is being mostly transient . . .") The committee majority held stoutly to the idea that a responsible electorate was one that had tangible stakes in the community life. Yet, of the small population of the proposed new county "very few pay any taxes whatever, and altogether not one per cent of the taxes assessed . . ." In the absence of good reason to the contrary, preponderant weight should be given to the interests of the substantial taxpayers, and these the majority thought were clear-spoken against the bill:

> . . . the actual tax-payers are almost unanimous in their opposition to this measure, assigning reasons which appear sufficient to your committee, — the principal one being that it will greatly increase the already large taxation to which they are now subjected, without any corresponding benefit . . . [T]hese lands have recently been burdened with the expense of buildings in Chippewa and Lincoln counties for court houses and jails; also for the construction of roads and bridges; and it would be extremely onerous and unjust at this time to impose additional burdens for similar purposes.

Finally, given the narrow base of "bona fide" settlers and the temptations inherent in the spending that would go with setting up a whole new county machinery, the majority voiced full distrust of the likely quality of government the new county would have. ". . . the multiplicity of offices which it will create and the irresponsible class of men who will attain them, in that sparsely settled region, will tend to increase the burden of taxation."

The dissenting committeeman felt that his colleagues presented an incomplete, if not warped, estimate of the pattern of private pressures behind the Ludington County bill. The majority, he thought, spoke without necessary evidence; no evidence had been submitted by which the committee could determine the percentage of taxes paid by the bill's proponents, or by which to prove that the big taxpayers opposed the measure; the dissenter's own knowledge led him to believe that property owners near the railroad would prefer that their taxes go to benefit that area, rather than the distant southern part of Chippewa County. The last point led the dissenter to a counter-thrust on the question of motive. What was so pure about the opposition to the bill? ". . . no reason can possibly exist for opposition in that [southern] part of the county, except the loss to them of taxable property; for the friends of this bill have offered to submit to limitations that would reduce taxation for county purposes to one-half what it now is, and thus afford an ample protection to all property owners." The opposition, he thought, was seeking simply to preserve a rich tax base provided by an over-large county, for continued exploitation to match a shameful past: ". . . the facts presented to the committee justify me in saying that this legislation is actually necessary as a measure of protection from the wholesale robbery in the shape of taxation in Chippewa county." Amendments which the dissenter would propose to the bill (attaching the new county to Taylor County for judicial purposes, and stipulating that the new county should consist of but one town) would limit taxes for local improvements, "and thereby the bill have been made a measure of protection against the villainous system of taxation that seems so popular in the southern portion of Chippewa county . . ." [41]

Separated by eighty years from the contest, we can hardly pass confident judgment on the quality of the particular legislative decision which rejected the "Ludington" county bill. The questions of value presented, and the competition of interests, legitimate or illegitimate, were such as clearly to be proper only for legislative disposition. Adjudicating such issues over north-country local government provided a substantial volume of legislative business in Wisconsin throughout the lumber decades.

The legislature rejected quite a number of the proposals made to it to expand and shuffle local government organization in the forest area; certainly its record of rejections, over-all, is sizable enough to show that legislative control in the matter was by no means insubstantial. If we look not at particular instances, but at the whole record, there is objective basis for saying that the legislature did not use its veto strongly enough. Twentieth-century Wisconsin inherited an overexpanded apparatus of local government in the north. This inheritance charged heavy and unwarrantable overhead costs upon the economic and social growth of the state as a whole, as well as upon its northern region. The dollar load resulted mainly from an excess number of school districts and towns; policy-making efficiency was burdened particularly by the arbitrary pattern of county government. In 1920, when the great lumbering era had finally closed, Wisconsin had seventy-one counties, more than half of which lay in the region that had belonged to the forest economy. In 1920, of forty northern counties six had populations of no more than 10,000, and sixteen more fell in the range between 10,000 and 25,000; fourteen counted between 25,000 and 50,000 people, and four had populations over 50,000. Experience showed that this distribution was close to a stable equilibrium. In 1950, of these forty counties six still had no more than 10,000 people each, seventeen had between 10,000 and 25,000, twelve between 25,000 and 50,000, and five over 50,000. With about a third of the state's population, these 40 counties contained 782, or about 61 per cent, of the state's 1280 towns, and 3747, or a little over 50 per cent, of the state's 7400-odd school districts. By mid-twentieth century the record showed fourteen "cut-over" counties whose history proved them entirely unsuited to a farm and town economy. In 1920 these fourteen counties mustered 192,927 of the state's 2,632,067 inhabitants; a little over 7 per cent of the state's population lived in about 20 per cent of its counties. The situation had reached apparent equilibrium here, also. In 1950 these fourteen cut-over counties held 194,045 of the state's 3,421,316 people, or a little under 6 per cent. In 1935 they included some 236 towns (about 18 per cent of the state total), and some 679 school districts (about 9 per cent of the state total). They included the state's least populous counties: Florence (1920 population, 3602; 1950, 3756), Forest (1920, 9850; 1950, 9437), and Vilas (1920, 5649; 1950, 9363). They included half a dozen towns with less than 100 people. In 1935 an average Wisconsin town had an equalized value for tax purposes of $1,259,460, and an average county an equalized value of $60,054,160. One cut-over town had an equalized value as low as $64,000. Of the forty forest-economy counties, only five had equalized valuations equal to or above the average; this number, of course, included none of the hard-core cut-over fourteen counties, ten of which ranked among the thirteen counties of lowest equalized valuation in the state. The number and location of northern local government units helped form interests and traditions which by 1950 had long contributed to overrepresentation of the north in the legislature, at the expense of southern Wisconsin. With about 6 or 7 per cent of the state's people, the fourteen cut-over counties through 1950 held 9 per cent of the Assembly seats and 10 per cent of those in the Senate. The forty-counties area held about 45 per cent of the seats in each house of the legislature, with about a third of the population.[42]

Obviously there were pressures behind the multiplication of northern counties which no mere change in legislative procedure would have removed. However, more specialized and formal procedures for framing issues, investigating facts, and rendering decisions might have raised higher the barriers to excessive creation of local government units. The nineteenth-century Wisconsin legislature had its standing committees on local government. But these had no staff nor any regular procedure or formal requirements to press the proponents of a given change into making a substantial record for their contentions. Indicative of the looseness of the prevailing procedure are the striking differences between majority and dissenting reports from the Senate's Committee on Town and County Affairs on the "Ludington" County proposal of 1877. Whether the population of the proposed new unit did not exceed three hundred or was over five hundred, how many "thriving" villages the area already contained, how many tax dollars were paid by how many actual taxpayers, what proportion of the population had other than transient connection with the section, and how many taxpayers, and of what consequence, favored or opposed the proposition — all these matters, on which the majority and minority reports took flat issue, presented questions of fact and not of judgment. Yet, so little approach was there to insistence on a factual record that the majority report rested on appeal to opinions of unnamed and unnumbered residents, and the minority member invoked his own knowledge alone.[43]

There was one striking and characteristically limited nineteenth-century acknowledgment of the need to regularize legislative handling of local government problems. This was the proposal and ratification in 1870 and 1871 of an amendment to the Wisconsin constitution creating sections 31 and 32 of Article IV to limit special and local legislation. The amendment prohibited the legislature from enacting any special or private laws:

> 2d. For laying out, opening or altering highways, except in cases of state roads extending into more than one county, and military roads to aid in the construction of which, lands may be granted by congress.
>
> . . .
>
> 5th. For locating or changing any county seat.
>
> 6th. For assessment or collection of taxes or for extending the time for the collection thereof.
>
> 7th. For granting corporate powers or privileges, except to cities.
>
> 8th. For authorizing the apportionment of any part of the school fund.
>
> 9th. For incorporating any town or village, or to amend the charter thereof.

So far as it went, the amendment attested a felt need to channel within general procedures the handling of a variety of matters affecting local government, its creation, policies, and administration. It was typical of the times that the response to what was a general problem of orderly legal process was made by a collection of *ad hoc* propositions addressed to particular pressures, and then mainly by negative fiat with only a vague affirmative statement of policy. The enumeration makes the more notable the absence of a ban on local legislation creating counties or defining the territorial limits of counties, towns, villages, or cities. The Wisconsin Supreme Court emphasized the narrowness of the amendment by ruling that the ban on special laws conferring corporate status on local units of government contained nothing to limit legislation creating or changing local units by definition of their territory and by conferring legal capacity short of corporate quality.[44] In any case, the 1871 constitutional amendment seems to have grown out of general concern with the weight of special and local business pressing upon the legislature, without particular relation to problems of north-country local government.[45] It is not surprising then that the amendment or the attitudes behind it had limited effect upon the continued flow of legislative business concerned with lumber-area counties and towns. As has occurred so often in this history, the relevance of a particular policy measure — here the adoption of the 1871 amendment — figures less in its direct effect upon our subject than in pointing up the fact that while certain issues and responses were not wholly outside current thinking, there was a characteristic lack of will or imagination to generalize them.

With its ban on special and private laws the 1871 constitutional amendment coupled the direction that "the legislature shall provide general laws for the transaction of any business" within the enumerated categories. The admonition to generality implied more wisdom than the times knew how to practice. Events were in tumultuous movement. After 1865 the north-country economy grew faster than men's conception of it. Nature inclined the region to scattered and isolated settlement, and lumber-trade expediency made these settlements transient and shifting. These facts made difficult, if not irrelevant, the application in the north of the pattern of rather small units of government firmly tied to particular localities, familiar in the farm and town country of southern Wisconsin. The facts implied the practical sense of providing in the north a few large formal units of local government which might use circuit-riding officers to bring the necessities of government close to the people. So government might be adapted flexibly to area growth, with a minimum of vested interest created in local offices, fixed institutions, and county-seat land speculation. But opposing factors of custom and interest were so strong that such arguments for the efficiency of more generalized handling of the local government problem never entered discussion. Political tradition, difficulties of travel and communication, the restless search for place and fortune bred deep in a striving and mobile people, the sharp sense of capital scarcity which fostered demand to spend tax dollars close to home — such elements contributed to a powerful bias for localism in government. On but one occasion in the lumber years was the legislature faced with the question of constructing a master plan for the political organization of the north country as a whole. Significantly, the abortive proposal was framed solely in terms of increasing the number of fixed local units, and the two main criteria suggested were the "convenience" of the residents and "securing to them the expenditures within their bounds, of all taxes levied for local purposes on the property of such counties . . ." No master plan was debated, let alone adopted. The legislature continued to respond to the immediate pressures of growth and ambition by creating many formal units of government *ad hoc*. There was no leadership nor any special stress of circumstance to bring Wisconsin's legislature to ways of action radically different from those generally prevailing; like most state legislatures most of the time, it operated as a meeting of local ambassadors rather than of state representatives. If realism requires that we recognize the environmental factors against any very different handling of the local government problem, it is no less realistic to recognize the consequences of what was done, as these consequences stand defined in light of rational alternatives. By multiplying narrowly bounded formal units of local government the legislature added enormously to the inertia which in any formal power organization works against efficient adaptation to change. In contrast, skillful use of breadth and generality in framing organization would have made it more practicable to match the uncertain flux of affairs with more capacity for maneuver.[46]

Second best to a tighter constitutional or legislative control over multiplication of local government units in the north would have been effective central definition and enforcement of standards of local government policy and administration. Given the simple conditions of the forest country, it would have sufficed if there had been effective central superintendence of taxing and spending.

Once again, just enough happened of a positive character to show that although the nineteenth-century state government was not wholly incapable of glimpsing relevant general policy, its processes were usually incapable of translating particular insights into firm and general form. One such incident which promised central leadership, but came to little, was Governor Harvey's veto of bill 62S of 1862, which would have donated to the counties in which they lay all the swamplands that the state might receive under the federal grant of 1850. Harvey put his veto squarely on the responsibility of the state government to preserve against parochial interest resources significant for the total state economy. The vetoed bill had the defect of "gross partiality . . . since the most cursory examination cannot fail to detect the unequal relation which the expectants of the enormous benefits attempted to be conferred by this bill hold to the population and tax-paying interests of the State." It was a point of view which in many later connections might have been invoked to warrant positive state action to preserve the forest economy as of framework significance to the economy of all Wisconsin. The Governor foresaw that, as might have been argued of the forest, the swamplands would pose problems of engineering and finance exceeding the capacities of particular local governments. Beyond these considerations were deeper problems of controlling substantive policy in the use of the lands. These swamplands were given to the state in the first instance, and taken by it, as a trust — partly for their reclamation, so far as their proceeds were needed for that purpose (a trust declared in the Congressional grant), the remainder for education (under declaration of the state constitution). An economic resource which had this meaning for the general economy should be cared for by an authority of comparable generality; fractionalized policy making and administration could not be expected to preserve the general interest. Probably some members of county boards would be qualified to share the trust.

But I think one hundred and fifty citizens, divided into and acting as perhaps fifty different and distinct tribunals, and acting in each case independently and finally, without restraint of any other official power or department upon their determinations, are far less likely to deliberate cautiously and administer a great pecuniary trust justly and fairly, than the Legislature of the State, composed as it is of two separate organizations, each acting independently of and as a check upon the other, and both liable to be restrained to the careful reconsideration of all measures determined upon, and to great unanimity in their final enactment, by the power over legislation lodged with the Executive, in the exercise of which he is expected to regard, in the broadest sense, the interests of the whole State.[47]

Moreover, Governor Harvey argued, the problems were not only those of general compared with local concepts of public interest. The law must always be alert to guard public interest against invasion by private calculation. The swamplands were a rich prize. The disposition of the great railroad land grant of 1856 had brought shocking corruption into the operation of the state government itself. How much less could county governments be expected to withstand the pressures of private interest, to which they stood so much closer in day-to-day relations than did legislature and governor? The counties, he predicted, would come under pressure from their towns for further division of the lands, and the towns under pressure from road districts, and the road districts from individuals. The danger was aggravated because so much of the swamplands lay in the newer, thinly settled counties, where it was peculiarly hard for local officials to withstand private importunities and peculiarly difficult to staff the public offices with men of probity and detachment from personal fortune.

. . . in those counties where the population is newest and sparsest, where the mass of the people are least acquainted with each other, and where, as a consequence, unfit or designing men find the fewest obstacles to success in their schemes after places of trust and responsibility, or to enrich themselves at the expense of the public, there the greatest portion of these lands lie — and there the duties connected with the administration of this trust are most delicate and important . . .

Experience of affairs in the newer counties did not breed confidence. "The past financial history and experience of some, at least, of the new counties, is not encouraging to those who would entrust them with the grave interests involved in this grant." The Governor received "almost daily" letters complaining of irregularities in the discharge of the limited duties already resting on these local governments. "The enormous taxation in some of these counties, without any corresponding benefits, manifestly inuring chiefly to the private gain of a few, ought to admonish to great care, lest we transmit this fund for disposal by stewards deeply practiced in enriching themselves out of public trust." Nor did the danger lie only in the greed of men who made a career of county-seat politics. The lumber interest loomed large in these new counties of the north.

Again, there are in the State some counties, the inhabitants of which are almost exclusively engaged in the manufacture of lumber. Of course many of the swamp lands in these counties are chiefly valuable for the pine timber thereon. Each mill owner has in his employment many voters who have no pecuniary interest in the county — who are solely desirous that the business which gives them employment may prosper. A combination of such proprietors so engaged and so interested, is at any time sufficient to control the politics of the county. Is it wise to place within reach of men so situated, the sole control of thousands of acres of the most valuable timbered lands of the State? The past history of these counties shows that men so employed have not hesitated to appropriate large quantities of such timber without making any compensation therefor to the owners, and in plain violation of law. Is it too much to expect them to avail themselves of lawful means to attain similar ends? [48]

Governor Harvey thus made powerful affirmation of central responsibility for public programs that underwrote the general economy. The veto message of April 5, 1862, stands as one of the handful of first-rate state papers to come out of the generally weak office of the nineteenth-century Wisconsin governor. The Senate, which had passed the offending bill by the narrow margin of 16 to 14, now voted 17 to 11 to sustain the veto. But in the aftermath no one followed the lead which Harvey had given toward a strong central policy for the north country. The legislature never openly challenged the values affirmed in the 1862 veto. But subsequently it deprived them of much substance, by piecemeal grants of swamplands to particular local governments without effective supervision. Of still broader significance was the failure to develop an adequate system of classifying public lands for their unique timber resources — a failure protested unavailingly by Governor Harvey himself, again in 1862. This want of a rational land classification was fundamental not only to the course of policy regarding the public lands; it had radical meaning, too, for the development of a sharper consciousness that the northern forest constituted a natural social and economic as well as political entity. To see this region so was practically prerequisite to making firm denial of parochial pressures in the area and asserting positive central superintendence of local government there. But no such conception emerged as the legislature plodded about its business through the last quarter of the century. The legislators continued to deal with highly specific bits of local business and local ambition as they arose out of north-country affairs, without ever exploring a master plan for local government in the forest region.[49]

More typical of state policy toward the special problems of local government in northern Wisconsin is a story which stands in revealing contrast to the bold generalization in Governor Harvey's veto message. A part of the total general property taxes in the state was levied by the legislature for state purposes, to be collected by the counties and remitted to the state. The legislature made general provision that county treasurers pay over the state tax in February of each year. Apparently this requirement proved to be a hardship on lumber counties, whose economies were not geared to providing cash in midwinter; it was after the spring drives and the bringing of logs to the mills by summer that collections could best be pursued. There may have been an equity for special treatment of the forest region as a whole here. Characteristically, the legislature did not respond to the problem save by accretion of local instances. In 1859 it remitted a penalty due from Douglas and LaPointe counties for delinquency in paying the state tax for 1858, and provided that thereafter the state tax should not be due from these counties until July. In 1875 the legislature amended this exception to apply to Douglas, Bayfield (the former LaPointe), Polk, and Ashland counties. In 1879 it extended the exception to include Burnett and Marathon counties. In 1895 it brought Oneida County into the special category. It repealed the whole exception in 1899, by which time the industry had reached a maturity which the legislature evidently felt removed such reason as had warranted the favor.[50]

The want of broad, firm, central supervision of local government finance and operations was the more marked because the Wisconsin constitution from the outset gave the legislature a mandate that it should "establish but one system of town and county government, which shall be as nearly uniform as practicable." The state supreme court indicated that it would have found no barrier in this uniformity requirement to reasonable classification of local governments according to objective differences in the character of the areas to be administered.[51] The original constitution also declared it "the duty of the legislature . . . to provide for the organization of cities and incorporated villages, and to restrict their power of taxation, assessment, borrowing money, contracting debts, and loaning their credit, so as to prevent abuses in assessments and taxation and in contracting debts by such municipal corporations." In omitting reference to towns and counties, this command could not be taken to warrant fiscal license by these local units which, as unincorporated administrative subdivisions of the state, were in any case under the plenary power of the legislature.[52] Despite the policy indication and warrant that might have been drawn from these constitutional provisions for a strong central leadership, the legislature did little to fulfill its charge or its opportunities. Whatever the constitution makers might formally declare, the considerable isolation of settlements and difficulties of travel furnished practical reasons for reinforcing an inherited tradition of highly decentralized local government, attested as early as the provision for townships under the federal survey ordinance of 1785. By the time the north country felt the full surge of forest production, the people of Wisconsin already had behind them another generation's usage in extremely decentralized local government.[53] Experience brought some disquiet with this tradition. A constitutional amendment of 1874 forbade any county, city, town, village, school district, or other municipal corporation "to become indebted in any manner or for any purpose, to any amount including existing indebtedness in the aggregate exceeding five per centum on the value of the taxable property therein" as measured by the last previous assessment for state and county taxes. The 1874 amendment was unusual in passing beyond mere prohibition to affirmative regulation from the center. For it added the command that any such local government incurring permitted debt "shall, before or at the time of doing so, provide for the collection of a direct annual tax sufficient to pay the interest on such debt, as it falls due, and also to pay and discharge the principal thereof, within twenty years from the time of contracting the same." The 1874 provision primarily reflected problems of north-country growth, seeking to curb the impetuous zeal with which that region pursued the subsidy of railroad construction. But, again, policy was not generalized. The 1874 amendment was relatively limited. It defined a central policy only for term indebtedness of local government and by its own terms made no new provision to superintend the standards of current budgeting and administration.[54] Thus the responsibility continued primarily in the legislature.

Legislative initiative for policing or invigorating local government was rare, episodic, and marked by that relative indifference to adequate implementation that stamped most nineteenth-century statute law. This was true especially in the critical area of fiscal policy. Complaints were regular of excessive cost and waste in tax delinquency procedures, notably in the generous provisions for advertising the lists of delinquent properties, which furnished lucrative subsidies to local newspapers. Despite occasional official acknowledgments that simpler and less expensive tax administration procedures might be contrived, nothing substantial was done during the lumber decades when the great tax delinquency problem lay in the forest region.[55]

The legislature did, from 1852 on, make varying provisions for a state board of tax equalization. But such a body had only the limited function of curbing competitive undervaluation among local taxing units and did not otherwise touch their fiscal programs. Moreover, through the nineteenth-century the boards of equalization were the familiar ex officio kind of agency, made up of busy officers on whom this business was simply another load. Even within the narrow range of tax equalization effort the state officials experienced continuing difficulty in requiring adequate information from local governments, and waited upon the legislature until the late seventies for the first provision of some sufficient means to enforce reporting discipline. The decisive step toward effective central leadership in tax administration standards did not come until the latter years of the lumber industry. In 1897 the legislature at last abandoned the ex officio style of agency and set up the first state tax commission; in 1901 the new commission was given the job of equalization.[56]

This measure of itself did not touch the general efficiency of local tax administration or local government finance. Through the nineteenth century there was no effective supervision of local assessment or budgeting; occasional admonitions and complaints from the Secretary of State stood as the sole central contribution to standards. A flurry of particular regulations of local government finance witnessed grudging legislative reaction in 1877–1879 to mounting complaints and scandals. Towns must not issue tax anticipation warrants without prior referendums approving taxes to retire the warrants; town officers must not deal in town warrants or contracts; fair adjustments should be made of accounts between old and new counties. None of this activity could be said to add up to a program.[57] In 1901 the legislature provided for county supervisors of assessment. In 1905 and 1911 the legislature imposed on a reorganized state tax commission the duty generally to supervise tax administration, further to require the keeping of forms of local accounts which would facilitate comparative study of local government finances, and to make available on a voluntary basis advice and help in improving the quality of local fiscal administration.[58] This new pattern of central leadership emerged only after the lumber industry had passed its peak. Its relevance to this story is, therefore, to throw into sharper relief the limited outline of state action during the lumber years.

CHAPTER IX

THE USE OF COMMON TRANSPORT FACILITIES

AMONG THE legal adjustments of relations which expressed the police power there were few in which the lumber industry appeared as direct beneficiary of the state's action. More often, as was true for business generally, the actual or potential impress of the police power was to put limits upon the industry's action; this was the relation normal in a legal system that broadly delegated initiative of decision to private persons. There is a sense, however, in which this generalization is more nearly true to appearance than to fact. Like all other private dealers, timberland owners, loggers, millmen, and woods, river, and mill workers depended ultimately on the law of property, contract, and tort to support the framework of reasonable expectations necessary to market operations. Thus, measured by social functions, these staples of the law which armed private will also expressed the police power insofar as they used legal sanctions to support or adjust private power relations deemed socially productive. However, there are also functional reasons why property, contract, and tort law are not customarily analyzed as exercises of police power, and why in various aspects they have been separately treated in this volume. In usual course the policies represented in these bodies of doctrine acquire content and operate only by the initiative of private persons. Though property, contract, and tort are major heads of law because they represent public concerns, they express values that ordinarily come to a focus in the immediate interests of private persons. Where affected interests are more diffuse, or at least less sharply differentiated, different problems arise both in bringing interests to definition and in generating action on those interests. The slow and relatively ineffective growth of public policy concerning forest conservation offers a telling example of both aspects of a typical police power area. Where circumstances and the alignment of interests cast a greater burden of initiative and sustained energy upon the legal process itself, there is functional logic in treating the development of policy as a subject somewhat apart from policy growth in areas marked by more focused and persistent private zeal.

In fact — quite apart from their status in law — waterways and railroads were common facilities for transporting logs and lumber. By nature, unimproved navigable water was useful to all operators within its watershed. By the logic of overhead cost, the considerable capital investment in improved streams and the great investment in railroads invited the traffic of all who could efficiently reach markets by such means. Partly by nature and primarily again by overhead cost accounting, railroads and improved streams were of necessity common facilities for all shippers within given areas, for such facilities could not readily be duplicated or closely paralleled.

Where functional imperatives thus bring men into mutual dependence, they also generate social interests in the good order, the fairness, and the efficiency of the relationships which express this dependence. Wisconsin public policy dealt with these relationships in two ways. First, expressing a key organizational faith and value of the nineteenth-century United States, the law delegated to private decision makers broad initiative of decision in forming up such relations. To the quite limited extent to which the issue was of the use of waters in their natural (unimproved) state, the law gave form to this delegation by the doctrines of property, contract, and tort, which favored and protected private initiative in decision; these familiar rules cast on a complainant the burden of establishing that a given use of private title, or a given private agreement, should be deemed illegal as against public policy, or that detriment caused by private action was the product of the actor's fault. Most transport, however, proceeded by means which changed the natural situation, whether by improvement of the navigable capacity of streams or by construction of railroads. The law made more formal expression of its policy of delegation in these cases — the bulk of important transport situations — by granting titles to act, including titles to make impositions upon the wills of other persons. This was the law of franchises, which we considered in Part One as a type of property.

The second way in which Wisconsin law dealt with men's relations to common transport facilities was by regulating or adjusting the consequences of these delegations of power, as these consequences unfolded in action. Concern with continuing, evolving relations among types or patterns of action is the most distinctive expression of the police power; the police power consists in the law's attention to the integrity and productivity of relations in behavior.

Police power activity affecting common transport facilities dealt with two types of relation, consensual and nonconsensual. Law developed some doctrine and rules to deal with problems created simply by the fact that different persons sought to use a common facility, with no other basis of relation between them — as when a raftsman complained of delay or damage alleged due to the improper condition of another's private (non-toll) dam. Because all persons had some rights guaranteed by constitution, statute, and common law upon navigable waters, whether improved or not, nonconsensual claims for access to stream use might arise against the holder of a stream-use franchise as well as against one who used the stream without a franchise. Because a railroad was wholly an artificial contrivance, problems between the road and its users as to terms of access to its benefits could arise only out of relations originally consented to, or for which suitable consent was sought; so far as the utility of a stream was created by artifice, claims of access to these benefits likewise could arise only out of an existing or desired consensual relation. For both improved streams and railroads, claims based on the manner or results of operations might have either a consensual or nonconsensual basis.

Thus our consideration of the police power in this context involves two categories of policy, concerning (1) relations derived from shared or interdependent use of a common transport facility, apart from consent, and (2) relations in use of a common facility based on some consent. The first category is most analogous to the law of tort, the second represents at least a rudimentary development of the law of common carriers or public utilities. The most elaborate, because the most functionally complicated and important, developments were in the growth of public utility doctrine.

The charges and service standards of private persons providing bulk transport facilities for industry presented problems that produced distinctive police power response, though with strong overtones of private-law-style activity. How bulk transport facilities were provided was a major influence on the content and directions of the whole economy. This matter thus presented values and tensions which reached so wide and deep into the general life that by familiar paradox their importance made it difficult to bring them to adequate expression. On the other hand, so far as these matters touched business enterprisers they were assured a sharper scrutiny and a stronger application of private will than other economic issues which did not have such pointed relevance to market operations. As related to industrial or merchant shippers, thus, common carrier regulation was marked by a combination of appeal to the general mediating and planning roles of law characteristic of the police power, and of private prosecuting energies characteristic of the staple fields of private law. This was so of the lumber industry.

Because of the relative inaccessibility of its raw materials sources and the unwieldy size and weight of its products, the lumber industry was peculiarly dependent upon bulk transport facilities. The availability of waterways and of railroads, the level and equality of tolls or charges, and the adequacy and equality of service were facts which could determine for a given enterpriser whether he could enter or stay in the business and with what measure of success. Most timberland owners and mill firms commanded capital or bargaining power of modest scale. But more critical for larger as well as for smaller exploiters of forest wealth were the hard facts of operating relationships. Those who controlled bulk transport facilities occupied key positions in the pattern of productive relations, and could dictate their own terms simply because of position, unless some larger pattern of relations could be imposed on them. Unlike the eastern oil industry, the Wisconsin lumber industry never developed the generalship or the means to impose its will upon the opposing power of common carriers; where the will of the lumbermen controlled, it was not by superior maneuver but by gift of circumstance. Developments thus took a different course as to water and as to rail transport. In both fields the lumber entrepreneurs exerted positive efforts to obtain the transport terms they wanted. In both fields there was enough movement toward the growth of crude forms of public utility regulation to testify to pressures of position which produced rudimentary moves toward public regulation.

REQUIREMENTS OF MUTUAL ACCOMMODATION

COMMON LAW

Judge-made (common) law provided three main doctrines to regulate relations among those who used the streams. These were the definition of rights of riparian ownership, the doctrine of nuisance, and the doctrine of reasonable use of waters.

These doctrines served functions in the light of which all three might be considered part of the law of property (title), and might enter into the earlier chapter dealing with stream-use franchises. To hold title to property means to enjoy legitimated, protected initiative of decision over a range of uses of resources; all three of these common law doctrines in part served to define who might exercise protected initiative of decision in using streams. However, one who holds a title or franchise does not thereby enjoy legal immunity for all consequences of exercising his title. The concern for the good order of social relations which leads to creating titles leads also to legal action to review and adjust some of the consequences that follow when one titleholder's initiative impinges on that of another, or on the affairs of broader groups or of the general community. This attention to the interplay of areas of initiative of will (for the sake of the individual and group interests that depend upon decent and efficient social organization) is what we mean by the police power. As they achieved definition in particular situations, the common law doctrines of riparian right and nuisance sometimes had effects akin rather to police regulation than to definition of title. The doctrine of reasonable use was in function primarily an expression of the police power.

By the doctrine of nuisance the common law distinguished between legitimate and illegitimate continuing uses of streams; this was the title analogy in the doctrine, since it marked out sanctioned spheres of initiative. Common law was modest in its claims in this area. To grant special rights of use in such a commonwealth resource as a stream was recognized by the courts to be on the whole the prerogative of the legislature.[1] However, to obstruct a channel regularly navigable was to create a public nuisance, which might be proceeded against at common law as a crime or enjoined at suit of the public prosecutor, or enjoined or made the occasion of a judgment for money damages at the suit of a private person who showed that he suffered special damage from the obstruction — unless the legislature had positively authorized the obstructing works.[2] The combination of obstruction plus want of statutory warrant made a clear case of nuisance at common law.[3] If the legislature positively required its license, the court would make the common law remedies available at suit of one who held a statutory franchise against an unlicensed structure simply on the basis of the want of license (without proof of obstruction in fact), even though the legislature had provided no sanction itself.[4] Thus the common law declared that there was a public franchise to use streams without impediment, unless the legislature had conferred some contrary special right. Common law qualified this public franchise by certain rights which it recognized as elements of the ownership of land which abutted a stream. So long as he did not flow another's land, a riparian owner might dam a nonnavigable stream.[5] After earlier indicating that one created a nuisance who built a dam without statutory warrant on any stream navigable at any time for logs, the Wisconsin court drew back; it now construed the legislature's general requirement of license not to forbid a riparian owner on his own initiative to dam a stream navigable only for logs at spring freshet time, if he did not wholly obstruct the passing of logs at that limited period.[6] Even on regularly navigable waters a riparian owner might without statutory license build a pier or boom to give him access to the channel, to secure or harbor logs or rafts for his mill.[7] Nonetheless, the public franchise to navigate sharply qualified these riparian rights; the dam which did obstruct logs at freshet time, or the boom which extended into the navigable channel itself, was a nuisance against which one specially aggrieved might have an injunction or judgment for damages.[8] At this point nuisance doctrine takes on as much of a regulatory as a title character. Where the legislature required its warrant, the matter was one of clear-cut title; with a license, the presence of works was lawful, without a license unlawful. Where the legislature had not spoken, or where it was deemed to tolerate marginal activity which rendered riparian land more useful, the court must measure the particular extent and practical consequences of private action by the general standard which said that there must be no material obstruction to navigation; in effect this assessment amounted to regulating the way in which riparian owners put their ownership to use, rather than defining their title to act at all.[9]

The common law doctrine of reasonable use said that in any particular activity one who employed a stream for his own purposes must do so for a rationally productive end, and with such care as made his employment consistent with use of the common resource by other persons.[10] Except perhaps in adjusting claims of competing dam owners upon the flow (where doctrine seemed shaped rather by the analogy of property than of tort), the standard of reasonable use was applied with favor for action; one who sought law's help to shift from his shoulders a loss caused by another's activity must ordinarily persuade the court that the actor was at fault.[11] This allocation of the burden of persuasion on underlying value issues is the title analogy in the doctrine of reasonable use, tending as it does to legitimate continuing initiative in the will to action. Otherwise, the standard of reasonable use seems wholly in the domain of police power, adjusting relations to experience.[12] Though in some applications the law of nuisance and that of reasonable use closely resembled each other, they operated in different situations; the law of nuisance concerned continuing patterns of behavior; the reasonable use doctrine measured liability for episodes of behavior.[13] All three doctrines — riparian right, nuisance, and reasonable use — had in common that they expressed a strong presumption in favor of multiple productive employment of streams, admitting as many users and uses as possible, denying as few as possible.[14] They realized this common value by different forms of encouragement to the exercise of will. The doctrines of riparian rights and nuisance invited enterprise on the streams by helping create predictable protection for stable patterns of behavior; under their protection a man might determine to build a mill or commit himself to a timberland investment dependent upon an assured waterway to market. The doctrine of reasonable use invited enterprise by reducing the risks of the unpredictable incident; the actor could know that generally he did not act at his peril, but that he was held only to the requirement of using reasonable care toward others as he coped with the shifting moment.[15] The different range of remedies reflected the difference in situations dealt with. Because the law of nuisance (including the qualifications made by the law of riparian rights) defined the legitimacy of continuing patterns of behavior, and demarcated types of sanctioned activity, it bore the more drastic sanctions of the criminal law and the injunction as well as the judgment for money damages. Because the doctrine of reasonable use was designed to adjust dispute and loss arising out of particular incidents of action — usually unintended in the aspect later complained of — there was no occasion to apply the injunction, and it would usually be unfair to invoke the criminal law; thus the reasonable use standard was implemented only by the judgment for damages.[16]

The common law doctrines of riparian right, nuisance, and reasonable use were well established in English and American precedent before Wisconsin judges put them to service; only in its bold extension of the log-transport definition of navigable waters (which affected the application of all three doctrines) did the Wisconsin court make a marked addition to the inheritance.[17] This existing body of common law provided definitions and arrangements of values which colored the role of legislation in regulating stream use. Indeed, from 1853 on the legislature declared that so far as it did not provide the law of stream use, "the . . . rights of individuals, the state, and its citizens, in respect to all such [riparian] lands and waters, shall be determined in conformity to the common law, so far as applicable . . ." [18] When the legislature in 1858 provided that remedies theretofore available only in equity should be available in an action at law against a private nuisance, clearly it assumed that the wrongs which the new remedy should redress included those which the common law defined as nuisances.[19] In suits arising under statutes regulating stream use, the court assumed that where the legislature had not clearly provided otherwise, and where the result would not be plainly inconsistent with the statutory policy, the content of duties should be construed to embody the common law standard of reasonable use, common law remedies should be deemed available to enforce statutory standards or rules where the legislature had not itself created a remedy, common law defenses should be available, and the common law allocation of burden of proof should prevail.[20] Thus the existence of the three common law doctrines helped give momentum to the development of Wisconsin policy.

However, having paid our regards to the common law element in Wisconsin stream-use policy, we must take care not to exaggerate its extent. The range and content of police power regulation of streams was primarily determined by legislation, just as was the body of franchise policy. Indeed, the primacy of the legislative element in stream-use regulation derived largely from the legislature's use of its exclusive prerogative to create special franchises. Most regulatory issues arose out of the enlarged use of streams made possible by franchises to alter the natural conditions of the waters. Since franchise legislation generated the new problems, men tended to come to the legislature to provide bases for adjusting them.

The primacy of the legislative element in stream-use regulation is, of course, first evidenced in the statute books. Without break from 1841 on the legislature pre-empted the definition of nuisances on regularly navigable waters by requiring its license for construction of works.[21] The largest single category of stream-use regulations was the body of navigation guaranties which conditioned the bulk of the 802 special franchises granted for dams, booms, and channel and flow improvements.[22] If the Wisconsin Supreme Court used the common law to supplement the terms of stream-use legislation, with no less emphasis the judges turned whenever they could to statutes to define the particular content of the rights and duties which they enforced through common law remedies. Given the legislature's broad assertion of its licensing power, the court ruled that the simple existence of an unlicensed structure on a regularly navigable channel constituted a public nuisance.[23] To prove that the operator of a boom or dam had violated a navigation guaranty stipulated in his statutory franchise, to the detriment of a navigator, was sufficient to establish a wrong.[24] So, too, a dam built under color of statutory warrant, but to a height exceeding that limited by the statute, was thereby a nuisance.[25] In the absence of other provision, the common law would redress the flowage of another's land caused by a power or flooding dam, but almost every flowage case that came to court was decided under some statutory procedure; of forty-four suits over permanent flowage that reached the Wisconsin Supreme Court from the forest region, only three (about 7 per cent) were decided on a common law basis.[26]

The Wisconsin Reports bear witness that common law doctrine alone governed only a minor proportion of the problems of stream-use regulation. Of 158 cases brought to the Wisconsin Supreme Court over forest-area uses of inland streams for navigation or power, only 23 (about 15 per cent) turned on rights or duties defined simply by common law; in 85 per cent of these lawsuits the parties' positions were established within a framework of statute law.[27] Though (as was characteristic of nineteenth-century legislation) the statutes regulating stream use fell far short of constituting a comprehensive code, they were nonetheless the principal body of public policy.

STATUTES

Statutory regulation of operations on streams reached large totals — because of the hundreds of use conditions imposed in special franchises — but dealt on the whole with only a narrow range of situations and interests. The basic limit upon the pattern of regulation was the counterpart of that which marked the range of interests considered in creating franchises. The legislature showed concern to regulate only those aspects of operations which came to a focus upon the immediate increase of production of tangible goods for market. Given the emphasis which contemporary policy put on dispersed decision making through contract and property, this prime concern with production for market usually implied exclusive concern with interests which were sharply defined by the ambitions of particular private enterprisers, and which were capable of translation into the money values of trade. There was lively concern with the public interest insofar as this was served by fostering movement of larger quantities of forest products to market. Thus, both for public and private advantage the legislature attended chiefly to protecting the continued free flow of transport, as this was affected either by the operation of permanent works (dams and booms) or by the manner of driving logs or navigating rafts. However, just as decisions affecting works construction, investment, and timing in the creation of franchises paid little heed to more diffused or long-term interests, so statutory regulation of operations looked little beyond the closest problems of transport.[28] Thus statutory regulation attended to the safety of goods in transit, but had little to say of safety of navigators.[29] Flow must be managed with care for efficient movement of commodities, but the legislature was content to leave to the common law such redress as downstream riparian owners or towns might have for ill-timed floods.[30] The legislature penalized those who dumped mill refuse into waters to an extent which impeded navigation, but until the end of the lumber era showed little worry over damage which the industry's pollution might do to fisheries, or to domestic or municipal water supply.[31] As with franchise terms, some problems of social capital involved in stream use might perhaps fairly be left without regulation in the conditions of the wild, thinly settled north country, in order to advance immediate productivity. The disturbing feature is not that more diffuse or long-term interests were subordinated to more focused and immediate concerns. The root objection is that this pattern of regulatory attention was set rather by default than by decision, by indifference and not by distinction. Within the values this legal order attached to rational, responsible use of power, such a record cannot be ruled irrelevant or trivial in appraising law's performance of its social functions.

The most common statutory regulation required that those who built or operated permanent works to develop water power or enlarge the carrying capacity of a stream must manage their affairs so as not materially to hinder general navigation. Basic to such regulations was the statute which from 1841 on required legislative warrant for works affecting regularly navigable waters.[32] The keystone of regulations in behalf of transport was this insistence that all enterprises to alter the natural condition of streams must meet certain statutory criteria, or at least survive the opportunity for contest which the legislative process afforded. Bulwarked by common law doctrine and remedies against public nuisances, this general requirement of legislative license made franchise a factor in the lumber industry's use of the streams, by generating obvious concern among lumbermen to achieve legitimacy for their access to waters.[33] The regulatory impetus of the general licensing requirement was to promote navigation guaranties as conditions of franchises, under both general and special laws. However, this stimulus exhausted the regulatory influence of the general license requirement; it insisted that an improver show title to alter a stream, but once he showed his warrant any further complaint against him must rest on proof that he had violated the terms of the specific title.[34]

The operations regulation most often imposed on improvers was the requirement that they not only build but maintain in good condition physical facilities to assure safe and ready passage of logs or rafts. For a dam, this meant a slide or chute; for a boom, a regular point of exit.[35] Most often the requirement was phrased as a standard with minimum specifications; the franchise holder must provide and maintain a "suitable" slide or exit point, or facilities which "shall not obstruct or hinder the safe navigation of said river for the running of logs or lumber," or "a good chute or sluiceway

. . . of sufficient size and capacity for the running of logs and timber with safety, and without unnecessary delay," or a boom "so constructed as to admit the free passage of logs." [36] In a minority of instances the legislature added construction specifications, most notably in legislation governing all power dams built on the Wisconsin or Black rivers, which in its final form required "a slide . . . which shall be forty feet in width, constructed of hewn timber or hewn timber and planks, made tight and sunk at the upper end, together with that part of the dam beneath it, two feet lower than the remainder of such dam; such slide . . . shall be built of sufficient length to graduate the descent from its top to the mean level of the water below, at the rate of five feet in each fifty feet of length . . ." [37] Most of these passage guaranties were contained in special franchises, but aside from the general provision for the Wisconsin and Black rivers the legislature in hit-or-miss fashion imposed briefly stated slide or boom-exit requirements for all works built on some of the smaller streams which it declared to have navigable status.[38] As the lumber industry moved into its short maturity, from the late seventies on, some special franchises laid more exacting operating requirements on grantees. To the simple requirement of a slide, a minority of dam franchises now added a positive obligation to provide a suitable flow to pass traffic safely and expeditiously; flood gates or passage facilities "shall be kept open at all times when the said river is at a driving stage and there are logs and timber to run over said dam." [39] Some boom franchises added to the conventional requirement of exit facilities an obligation of the franchise holder to provide labor at his expense to move others' logs out of the boom.[40] A few later dam franchises imposed an analogous requirement, that the owner bear the labor costs of moving others' logs safely and speedily over the dam or past any accumulations of logs gathered in the dam's backwater.[41] These more elaborate, more affirmative obligations underlined the creative possibilities in legislation, compared with the inherent limitations of court decrees, which (apart from providing for total or partial removal of an offending structure) were traditionally restricted to giving damages.[42] On the other hand, the more elaborate the regulatory patterns became, the less sensible it was to produce them by the clumsy process of legislation. Unwittingly the legislative trend built toward the shift to administrative legislation which came after the lumber industry passed its peak of demands upon law.[43]

There was substantial burden in these statutory regulations which put improvers under duty to maintain safe and ready passage for logs and rafts. It was inherent in the situation that these regulations be felt solely as restrictions by those on whom they fell. These navigation guaranties protected persons who used the stream without obligation to pay toll to the improver for such benefits as his works conferred on transport. Where the relation was one of toll required for service rendered, the law also laid duties on the improver to facilitate transport. But there the improver's duties had a different source and function, enforcing fair performance for a price; there we confront a nascent law of public utilities, explored in the next division of this chapter; the relation was presumably of mutual benefit, whether or not it was entered by full consent of each party. In contrast were the situations where an improver pursued simply his own purpose (a dam to power his mill or float his own logs, a boom for his own harborage), or where he was franchised to offer service, but only to those who freely elected to use it. Here the improver must be required to restrict his private pursuit so as to accommodate the distinct interests of others, or to manage his services so as to preserve a free election in navigators to use them or not. Public utility regulations (in the toll-for-service relation) sought to promote productivity by implementing mutuality of interest and interplay of roles in cooperative use of common stream resources. The statutory expressions of the reasonable use principle (as in the navigation guaranties which conditioned franchises for private or optional-service uses of streams) sought to promote productivity by protecting scope for separate interest and distinct roles in shared use of common waters.[44] Though the ultimate objective was positive — to enlarge shared use — to the franchise holder the operation of the navigation guaranties was inherently and solely negative or burdensome.

Apart from defining an investment minimum required for legitimate stream improvements (a feature we noted earlier as a regulatory incident of franchise), the navigation guaranties increased the operating responsibilities of improvers in two ways.[45] The common law imposed a general obligation not to obstruct navigation. But the particular content of this standard in a given situation could be determined only by lawsuit. The statutory navigation guaranties translated this obligation into more concrete definition; there must be a slide, or a defined boom exit; the gates must be open when the stream was at a driving stage and there were logs that sought passage; the boom owner must supply labor and facilities to sort out with reasonable despatch his logs from those of other owners desiring to move downstream. The statutory regulations also used the language of standards (a "suitable" slide, provision to sort and pass others' logs "without unnecessary delay"), but within the narrower frame set by their specifications. The immediately affected parties treated this legislative crystallization of standards as a matter of practical importance. An occasional franchise bill, as introduced, contained no navigation guaranties; almost invariably the omission was corrected by amendment in course of passage.[46] Improvers did not treat lightly the extent of obligation so laid on them; where they could, they inserted qualifying language, or obtained amendments reducing a requirement to provide sure passage for rafts and logs to one of passage simply for logs.[47]

The second way in which the navigation guaranties increased improvers' operating responsibilities was by fixing duties in terms of the changed situation brought about by the improvements. Implicitly, the slide, exit-point, free-channel, or flow-management stipulations of the franchises meant that the improver must make such continuing arrangements as would reasonably safeguard navigation in the altered condition of the stream. Such new obligations, said the Wisconsin court, should be defined in terms of reasonably foreseeable, ordinary conditions of using the improved waters; merely because he improved, the improver was not to be made an insurer against extraordinary floods, or required to provide for prompt passage of unprecedented volumes of traffic; public policy favored free navigation, but it also favored improvements, which must not be discouraged by legal standards which reached far beyond normal business experience.[48] On the other hand, a "suitable" slide did not mean only such a slide as the franchise holder thought suitable; the requirement subjected the grantee's operations ultimately to the review of a court and jury.[49] Moreover, the law measured the grantee's operating obligation by the reasonable possibilities for transport created by the improvement. To a navigator's suit for delay and physical damage to his rafts, allegedly caused by the improper condition of defendant's dam, it was no answer that rafts like the plaintiff's could not have navigated the stream at all in its natural condition with the water at the same stage as when these rafts were run. The dam owner must be held liable if by improperly maintaining his dam he "occasions injury to rafts properly and skilfully constructed and managed, with reference to the river as it then is . . ." The public policy which warranted the grant of an improvement franchise would be stultified, if the law did not hold the improver to a standard of care in operations measured by the enlarged carrying capacity he had brought about. "It would be wanton, arbitrary and unreasonable to say that, in such a case, those navigating the river had not a right to avail themselves of the benefit of any improvement in the navigation occasioned by the dam. The legislature expressly required those maintaining the dam, to keep 'a good and sufficient slide, that will admit the passage of all such rafts, crafts and boats as may navigate the said river.' And there can be no doubt that this act had reference to the navigable condition of the river after the erection of the dam, and intended to require it to be kept in such condition as to admit the passage of rafts properly constructed in view of that condition." [50]

Apart from the navigation guaranties which conditioned franchises, the most important regulatory legislation dealt broadly with obstructions to transport, beyond such as were inherent in given works or operations on streams. Throughout the lumber era the provision which eventually became Revised Statutes, section 1598, laid a money forfeit upon "every person who shall obstruct any navigable stream in any manner so as to impair the free navigation thereof, or place in such stream or any tributary thereof any substance whatever, so that the same may float in or into, and obstruct any such stream or impede its free navigation . . ." [51] This sweeping statute obviously might be invoked against a continuing condition which constituted a public nuisance, but its terms applied also to temporary blockage.[52] It might be invoked against intentional or careless interference with navigation, but in terms it required no showing of intent or negligence; the fact of obstruction apparently made out a violation, and since the statute created a civil forfeiture rather than a criminal penalty, it was the more likely to be applied in its full sweep, without requirement of proof of fault or wrongful intent.[53] The statute's history claimed for it an extensive rather than a restrictive interpretation, as amendments extended its reach and increased its forfeit.[54] On the other hand, section 1598 spoke with generality akin to the common law standards governing nuisance or reasonable use; where the legislature had created a more specific regulation to protect navigation — as in the conditions of a special franchise — the court ruled that section 1598 would not apply.[55]

In forty-four acts which declared named streams or portions of streams to have legal status as navigable waters, as well as in the general statute which required legislative warrant for any works built on "all rivers and streams . . . meandered and returned as navigable by the surveyors employed by the government of the United States," the legislature used its power to bring a common law standard to more specific definition — with effect analogous to the sharper definition which franchise navigation guaranties gave to the common law standard of reasonable use.[56] These declarations of navigable status did not add substantive content to the common law of waterways; their operation was, rather, to make plain beyond argument that the substantive doctrines and the sanctions with which common law protected navigation applied to the named waters.[57]

Two regulatory statutes empowered log drivers to keep traffic in efficient motion. Ordinary regard for the privacy of the fee simple yielded to the costs of driving, in a statute which went on the books in 1849; continued through the lumber era as Revised Statutes, section 1600, this act authorized the owner of stranded logs to remove them from another's property any time within three years, upon paying or tendering "the reasonable damages caused by reason of such removal and nonremoval." [58] First in 1863 for the Wolf River, and after 1881 for "any river in this state or any of its tributaries," the legislature provided (Revised Statutes, section 3337) that whenever logs of two or more owners were so intermixed that they could not be conveniently separated for driving, "and either owner shall refuse or neglect to make the necessary provision or to furnish the necessary labor and materials" for driving them, any other owner might drive the whole and receive "reasonable compensation" from the neglectful owner, with a lien to enforce the compensation.[59] The court looked favorably on this act. It summarily dismissed an objection that the law infringed constitutionally guaranteed rights of navigation; the statute dealt simply with cases where one owner drove the whole mixture "by necessity," to keep traffic moving, and his compensation rested on "an equitable principle" of payment for "the benefit of another's service."[60] Moreover, the statute was "remedial in its nature and should have a liberal construction."[61] Hence it was not limited to instances where mingling occurred before or at the beginning of a drive, but applied where mingling occurred at any time during driving and some logs were inadequately cared for.[62] If an owner had not sufficiently provided for handling his logs, he was subject to the statute, though the compensation claimant had caused the mingling.[63] The statute's authorization ran simply to any "owner" who took charge of the drive. But in view of the underlying purpose to assure uninterrupted movement, a bailee in charge of another's logs must be deemed an "owner" qualified to act under section 3337.[64]

The record offers little basis for appraising the practical impact of these various statutory regulations of operations on streams. The Wisconsin Supreme Court disposed of 135 forest-region stream-use cases which presented issues under legislation. Of these, only 12 (about 9 per cent) turned on alleged breach of statutory navigation guaranties or traffic regulations. Three times as many cases — 37 — came to the court from the forest area presenting within a statutory framework the more static title (franchise) issue of compensation for continued flowage of land.[65] An inventory of business in the circuit court for Chippewa County shows a rather small but steady flow of suits over water transport through the lumber years.[66] On the other hand, fragmentary business records indicate that, though lumbermen might file suits as tactical maneuvers in adjusting complaints over delay of drives or damage to rafts, they found litigation too slow, uncertain, and expensive to offer a ready reliance in regulating operations on the streams.[67] There were no nineteenth-century state administrators to police the observance of statute regulations. Neither the Attorney General nor any local public prosecutor brought a case to the supreme court to enforce any statutory rule on stream operations. Thus the legislative record stands as the main evidence that the statutory regulations were treated as having practical consequence. Of course, legislative declarations may be empty forms. However, this record carries persuasion that the statutory rules mattered. They specified, where the common law was vague.[68] The trend of new or amended legislation was to enlarge, rather than to restrict, the scope of the navigation guaranties, of the permission to recover stranded logs, and of the authority of the diligent driver to take charge of the logs of the neglectful.[69] The legislative journals reflect attentive pressure to insert navigation guaranties where they were lacking.[70] The limited range of the regulatory legislation — the fact that no more in this area than in any other was the nineteenth-century legislature inclined to enact comprehensive codes — itself attests the presence of specific, felt pressures to act. If men did not choose to litigate much under such statutes, nonetheless they wanted some minimum of authoritative statement to set criteria and bounds for private adjustment of differences.[71]

THE RELATION OF PUBLIC UTILITY AND CUSTOMER

Where log or lumber transport was assisted or managed for a fee under a statutory franchise, experience generated pressures for legal regulations different from those governing persons who dealt with each other only at arm's length or only by private bargain. The franchise-fee situations involved elements of superiority and dependence, of shared and competing purposes, and of voluntary and enforced coordination of activity, which worked to produce a rudimentary law of public utilities. Water transport presented experience earlier and more distinctive to the problems of the lumber industry than movement by rail; thus we begin with the waterways, and trace on them legal developments peculiarly linked to the fortunes of the lumbermen.

WATERWAYS

In its familiar, narrowly pragmatic fashion, the nineteenth-century legislative process built into toll franchises for stream use a body of rudimentary public utility obligations with minimum statement of principle. The statutory regulations which constituted this trend of policy included such variations in detail as defy significant classification. These were, however, variations within a pattern marked out by legislative practice. The pattern showed attention to three objects of concern: rates, content and quality of service, and obligation to serve. Logically, perhaps, we should first consider what the statutory regulations had to say about the franchise holder's obligation to serve. However, this was the aspect of regulation slowest to emerge and most ambiguous in statement. We are more faithful to events if we first examine regulations of rates and services, on which, it is plain, the franchise makers earlier and more consistently felt pressure to declare themselves.

Rates and services attracted attention first because they mingled the familiar with the unfamiliar. Service for a fee presented a relationship akin to contract, or at least to the common law quasi-contractual obligation to make fair payment for benefit received. In contrast, to impose a legal obligation to render service was to step outside contract analogy and squarely into the domain of police power; in market-oriented, mid-nineteenth-century Wisconsin this was a realm in which men walked more hesitantly.[72]

If definitions of rates and services presented questions akin to those which men otherwise settled by contract, nonetheless it was clear at an early point that, where they arose in the context of toll franchises for stream use, these matters could not be dealt with simply by contract. The holders of such franchises were specially privileged, as against the general community which otherwise had constitutional, statutory, and common law rights to use the

Public utilities on the streams: Dells log reservoir near Eau Claire, Wisconsin, about 1900.

Flood dam on Clam River, Wisconsin, about 1900.

The Spokesmen in the Forestry Decision. Top: Rouget De Lisle Marshall, Associate Justice of the Wisconsin Supreme Court, 1895–1918. Bottom: John Bradley Winslow, Associate Justice of the Wisconsin Supreme Court, 1891–1907, Chief Justice, 1907–1920.

streams without paying to do so. Having created special privileges, presumably for some public benefit, the legislature was under pressure to see that fair return was made for the privileges, by criteria other than the grantees' own good will.[73] There was another element, born usually of the mingled impact of law and fact. Those who had logs or lumber to move by water were dependent upon the care and fairness of those who provided improved facilities for transport. This dependence might exist because in fact there could be but one dam, one boom, or one set of coordinated works or driving operations to control flow, manage traffic, or provide harborage at a given location or over a given stretch of water. Dependence might exist, or be heightened, by law's intervention. Expressly or by clear implication, flow-management, driving, and boom franchises conferred on grantees the right to control their immediate operations, even where the franchised facilities were available only at the free choice of those served. Where the legislature authorized facilities which all transport must use and pay for (as in a franchise to conduct all general drives on the river, or to boom all logs that came downstream), dependence of transporter upon improver inhered first in the law, then in the facts as the law shaped them. So far as dependence was created by facts, pressure to protect the transporter derived from considerations both of fairness (possession of a key point should not allow the holder to play robber baron against other men's productive activity) and efficiency (forest production, recognized as important to the state's immediate economic growth, should not be hampered by excessive costs levied in a bottleneck phase of handling). So far as dependence was created by the law's grant of control, pressure to protect the transporter derived from insistence that all power, and especially law-derived power, be used responsibly; this was the value called constitutionalism, which could work with material effect upon action even in this unphilosophic Wisconsin community when a situation was sharp enough in outline to show men its application.[74]

Rates of toll for using private improvements or driving facilities on streams must be "reasonable." This insistence borrowed strength from the common law of common carriers.[75] But, common law apart, the requirement of reasonable rates had the force of a constitutional command, as applied to Wisconsin's navigable waters. For without this acknowledgment of public interest the legislature lacked authority to grant anyone the right to charge fees for legally privileged operations on waters which the Wisconsin constitution declared should be "common highways, and forever free . . . without any tax, impost, or duty therefor." [76]

Perhaps because it was from the outset conscious that it acted in this matter under a special constitutional limit, but more likely because it was unpracticed in delegating power to executive or administrative officers, the legislature itself early undertook to fix the precise limits within which stream franchise tolls might be deemed reasonable.[77] The first forest-region toll franchise granted by the state legislature, Acts, 1849, Chapter 46, provided that the grantees might charge those served by the boom there authorized on the Chippewa River

> for every twenty days or any shorter time any property shall remain within or be protected by said boom, and at the same rate for a longer period of time, viz:

Shingles per thousand, and the same for shingle timber or bolts, ...	$00	12½
For building timber, per 100 feet, ...		25
For every log twelve feet long, and in the same proportion, when longer or shorter, ...		10
For square ton timber per ton, ...		15
For round ton timber per ton, ...		20
Sawed lumber per thousand feet		20.

The first toll franchise granted for improving a channel by dams and slides, Acts, 1850, Chapter 257, declared that the Upper Wisconsin Navigation Improvement Company might charge "on all lumber and on all hewed or squared tim[b]er at the rate of one cent. per thousand feet of board measure, for each and every foot of elevation in any dam made by [grantees] . . . and over which such lumber or timber shall be run; on all logs or timber which cannot be conveniently reduced to board measure thirty cents, for each raft, for each and every foot of elevation in any dam made by [grantees] . . . and over which such raft shall be run," with similarly phrased charges of five mills per thousand on shingles or lath, and five mills per ton burden on boats, scows, barges, or other vessels.

Following this pattern, though often in less detail, the most common style of toll grant continued to state a flat sum allowed for a described service, as Laws, 1874, Chapter 153, authorized the grantees of a franchise for a dam on the north fork of the Clam River in Barron County "to receive and collect from the owners of all logs, timber and lumber passing over such slides or driven by the aid of such dam, as a compensation for keeping and maintaining such dam, the sum of ten cents per thousand feet board measure, the amount to be ascertained by scale on the landings in the woods . . ." [78] Where the franchise contemplated improvement of a stretch of water by several works, the grant often authorized a flat sum for the use of each of several described divisions of the improved stream. So, having stipulated that the total improvement there authorized to be made by the Northeastern Improvement Company on the Peshtigo River "shall be divided into three divisions" which the act specified by section lines, Private & Local Laws, 1870, Chapter 93, authorized the Company to charge as tolls "for all logs and lumber run through said improvement . . . Through said first division or any part thereof, fifteen cents per thousand feet; through said second division or any part thereof, fifteen cents per thousand feet, and through said third division or any part thereof, ten cents per thousand feet; the number of feet to be determined by scale measurement, unless the parties can agree upon an estimate of the same, and such measurement to be made in a boom or booms provided by said company for that purpose, unless the parties can agree to make such measurement elsewhere."[79] Taken together, the single-location and divisional fixed-sum toll grants made up 61 per cent of all toll boom franchises and 78 per cent of all toll flow-improvement franchises, or 75 per cent of all toll franchises. Though the statutes did not specify, and the court never had occasion squarely to decide the point, probably these flat-sum authorizations were intended only to set legal maximum charges, so that the grantees might charge less.[80] Next to the flat-sum authorizations, the most common styles of toll grant authorized "reasonable" fees (15 per cent of toll boom grants, 8 per cent of flow-improvement grants, or about 10 per cent over-all), or fees "not exceeding" specified sums (19 per cent of toll boom grants, 5 per cent of flow-improvement grants, or about 8 per cent over-all).[81] Under a stated-maximum grant it was clear that the grantee might charge less than the allowed ceiling.[82] These three modes of grant (flat-sum, reasonable, or specified maximum) account for all but a handful of toll grants. There were some marked variants. One franchise authorized the grantee "to establish a uniform rate of tariff for transportation, and to alter the same at pleasure," subject to a requirement of posting or publication and adherence to the published schedule; a driving grant scheduled rates according to difficulties of handling the logs; one dam franchise adjusted rates downward as the total volume of logs passed through in a season was the greater.[83] However, these variations were so few, as well as so strikingly different, as only to emphasize the scant range of invention shown in the bulk of toll provisions.

The basic policy implicit in all these statutory stipulations was that there must always be ultimate legislative control of rates. The legislature might lawfully delegate initial rate-making discretion to the private enterprisers who improved the stream, as it did most openly when it authorized "reasonable" tolls, or tolls "not exceeding" a specified sum.[84] But the court was clear (if the legislature was less explicit) that the grantees did not act at their sole discretion within such formulas; whatever rate the franchise holder set was subject to judicial review by the criterion of reasonableness, if challenged.[85] Moreover, parties might not by private agreement exceed the statutory maximum; such a contract was against public policy, and the court would deny it enforcement.[86] Transporters and improvers alike showed their awareness that the ceiling on tolls was set effectively in the pull and haul of the legislative process. Many toll provisions were amended in course of enactment of franchises; the changes usually made substantial reductions (25 to 50 per cent) in the charges set out in the introduced bills.[87] There were not many amendments after enactment, but enough to show that the affected interests treated the statutory maxima as meaningful; there were about as many reductions as increases in authorized rates.[88]

The nineteenth-century legislative process exhausted its creative force, however, in the simple operation of setting fixed schedules or maxima for tolls. Within these limits the law's criterion was that particular charges be "reasonable." Legislation made scant contribution to defining the content of this standard. The only specifications which achieved some pattern out of the statutory franchises were those which indicated that "reasonable" rate schedules might take account of the raw or semiprocessed state of the commodities transported (logs as compared with rafted lumber, notably), the distances over which improvements were used (divisional rate schedules), and the extent to which the log or raft owner supplied necessary labor and equipment to handle his goods.[89]

Basic to the idea of reasonable rates were definitions of the investment on which the improver was entitled to a return, and of a fair rate of return. Legislation said nothing about rate of return, save as some judgment might be implied as a necessary corollary from rate ceilings in relation to the nature of authorized works and expected volumes of traffic.[90] The statutory franchises evolved no formulation of a standard of prudent investment. With utmost vagueness, franchises in effect required a minimum capital investment as a base for tolls when they conditioned the toll right, as they always did, upon completion of works or provision of facilities calculated to achieve the declared objectives of aiding transport.[91] Out of 235 toll improvement franchises, 21 (about 9 per cent) conditioned toll upon investment of stated dollar minima, ranging mostly from $1,000 to $15,000.[92] Franchises set no ceilings upon the investment base on which tolls might be earned.[93] Where the improver was a corporation, its charter usually limited the authorized capital stock. In context, these provisions seem designed only as part of the internal government of the corporations; so far as they expressed a general public concern, they pertained to vague distrust of the monopoly power of concentrated wealth; nothing in their terms suggests that legislators thought of these capital stock limits as protections to customers against inflated rate bases.[94] Since the franchises simply linked tolls to the volume of commodities serviced (subject to flat-sum or ceiling designations as to unit charges), improvers perhaps had no selfish interest to serve by inflating their investment figures beyond whatever minima the legislature imposed. Had the tolls weighed more heavily on transporters, so as to produce more controversy, the investment ceiling would undoubtedly have become more important as a factor in the legitimacy of charges.[95]

In only three cases did the Wisconsin Supreme Court confront issues which allowed it to add to definition of the standard of reasonableness governing rates. The cases came too late in the course of the lumber industry (1890, 1898, and 1912) to exert much influence on courses of dealing already set by fifty years of industry and legislative practice. With so few opportunities for action it is not surprising that the court did little but enumerate factors relevant to a rational rate structure, without providing a formula to bring them into a weighted total. The genius of litigious law-making is to build tough-fibered doctrine out of inclusion and exclusion of instances grouped about a few organizing ideas; without business in some volume the courts ordinarily can make little significant contribution to policy.[96] The want of rate litigation carried through to appeal does not mean that toll rights as such were unimportant; toll clauses attracted more attention than any other topic in amendment during and after enactment of franchises; of the twelve customer-relations cases that came to the Wisconsin Supreme Court under improvement franchises, ten centered on collection of tolls (including the three cases specifically focused on reasonableness of rates). There were reasons offsetting the intrinsic importance of tolls to explain why there were not many appealed lawsuits (and probably few suits at the trial court level) over rates. Under the many franchises which set flat sums as allowed charges, probably the franchise holders simply levied the statutory rate, and this ended the matter. So far as franchises left leeway for specific determination of rates, several factors militated against litigation. Ordinarily the amounts of money at stake in particular instances were not large. The transporter was under severe business urgency to move his logs or rafts through to market without delay. Most stream situations did not involve many operators; relations tended to be close and personal, and often the associated improvers or the stockholders of an improvement corporation constituted most of the customers of the improvement; in such contexts law-suits over rates were less likely to occur than in more impersonal relationships.

Official action on rates thus centered almost wholly in the legislative process. This was a procedure inherently too cumbersome to allow fine-drawn elaboration in rate making. A richer doctrinal development waited upon growth of the administrative process. By 1890 the Wisconsin court was ready to acknowledge "the difficulty in fixing such standard of [reasonable] compensation from evidence" in view of the variety of factors which might enter rational judgment. The court cited Mr. Justice Cooley on the embarrassments which courts must encounter in such matters, and quoted with approval his suggestion of "the importance of some legislation for fixing the charges, either by statute directly, or by some board or local authority empowered for the purpose." [97] But so far as it prescribed delegation, the diagnosis called for more sophistication than went into shaping nineteenth-century stream franchise policy.

The legislature could more readily specify toll rates than it could the elements that should enter into giving good service. Service needs varied materially according to the peculiarities of particular geographic and business situations; the court acknowledged this stubborn reality by insisting that statutory service obligations must be read as exacting only what was reasonable in the circumstances.[98] So far as a franchise aimed at the creation of physical works — a dam, a boom, a cleared channel — the best the draftsman could ordinarily do was to stipulate that the facilities be adequate to their declared purpose, and that they be kept in a good state of repair.[99] If the franchise were simply for the provision of a navigable flow, the operations duties of the grantees could only be stated with like simplicity and generality, for what would constitute good management of flow in the particular instance would be definable only upon a particular conjunction of time, place, and circumstance.[100] Thus, for example, Laws, 1880, Chapter 84, followed one standard formula when it authorized Charles W. Hanson "and his assigns" to build and maintain a dam across Christmas Creek, Chippewa County, on a specified tract "and to otherwise improve said creek by clearing and straightening its channel, so as to facilitate the driving of logs down and out of said creek from said described land." So much was permissive. To this title the franchise added obligations concerning construction and operation: "The said Charles W. Hanson shall build suitable slides and gates in said dam for the purpose of sluicing logs and timber through said dam, and shall keep the same in good repair, and operate the same for the purpose of facilitating the driving of logs down and out of said creek from above said dam." Finally, the franchise set a general service standard, restating its purpose and making fulfillment of that purpose a condition of the toll right conferred: "When said dam shall have been constructed, and said creek improved as before provided so as to render the floating of logs down and out of said creek from above said dam reasonably certain, the said Charles W. Hanson, in consideration of the building of said dam and the making of said improvements, shall be entitled to charge, sue for and collect tolls on all logs cut and banked on said creek to be driven down and out of the same, ten cents per thousand feet board measure for logs banked above said dam; and five cents per thousand feet board measure for logs banked below said dam." The other, most common statement of operating duties under a flooding dam franchise was that contained, for example, in Laws, 1875, Chapter 70, which stipulated that the grantees "shall build suitable slides in said dam for running logs, timber and lumber over the same, and keep the same in repair; the same shall be kept open at all times when the river is in a driving stage, and there are logs, timber or lumber to run over said dam, and when it is not necessary to hold the water back for the purpose of driving or flooding logs, timber or lumber below said dam, for which purpose flood gates shall be kept in repair and built in such manner as to be shut or opened, as the case may require, to flood the said logs, lumber or timber."[101]

Franchises for booming or driving services might be as simply stated as those for the regulation of flow, but they were often more elaborate in defining service obligations because they concerned more active and more extended operations by the grantees. Booming or driving franchises were likely to include duties of safekeeping, of diligent handling, of prompt delivery or delivery at agreed times, of delivery at named termini, and of provision of suitable labor and perhaps of specialized attendants such as a boom superintendent or drive foreman.[102] The conduct of a general drive or the operation of a large sorting boom inherently demanded some cooperation from the customers. Thus the more elaborate booming or driving franchises defined the service obligations of the grantees as much by conditions imposed on those served as by the positive duties laid on the franchise holders. The customer must properly prepare his banked logs for entry into the drive; he must give reasonable notice, or notice by some specified number of days in advance, of his desire to have his logs driven or held and sorted in boom; he must provide the labor to drive his logs into the boom, and duly mark his logs to facilitate sorting; he must be prepared to take over his logs at the terminus upon due notice.[103] Not every booming or driving franchise included all these specifications, and some such grants were as spare in statement as the simplest flooding-dam license. Over-all, however, there was sufficient detailing of obligations of improver and customer to show that all parties affected realized that they stood here in no ordinary relations of private bargain and agreement.

A classic obligation of a common carrier is to serve equally all who properly ask for service and tender due payment.[104] In substance, the trend of toll franchise policy was to establish this obligation to serve. In form, the statements of the obligation lacked the simple directness with which the legislature asserted its basic control of rates and content of services.

Occasionally legislation spoke bluntly to the point. Private & Local Laws, 1864, Chapter 302, directed that the Willow River Dam Company "shall permit any owner . . . of logs [or] . . . his . . . agent driving the same down said Willow river above either of said dams, to use either of said dams and the water therein, for sluicing his . . . logs . . . through either of said dams," on three days' advance notice. Laws, 1873, Chapter 275, declared that the licensees of a dam on Moose River in Douglas County "shall have at all times the sole and full control of said dam" there authorized, "provided, that at all reasonable times, the said [grantees] . . . shall, when requested, use the said dam for the purposes" of sluicing through or running over the dam logs brought into the river above it. Having authorized a dam and channel improvements by William A. Rust, "his associates and assigns," on the Eau Claire River in Eau Claire County, Laws, 1878, Chapter 337, declared that the grantees "shall operate said dam and other improvements at all times during the season of driving logs, in an impartial manner, and in a way to afford the greatest facility to all parties driving logs down said river . . ."

Laws, 1882, Chapter 278, not only "authorized [grantees] to receive and take possession of all logs . . . coming down" the Bad or White rivers, but further ruled that grantees "shall also, at the request of the owner . . . take charge of the same and drive all such logs . . . to [the] . . . mouth" of the stream.[105]

More often the improver's obligation to serve was declared as a plain implication — but, still, only as an implication — of franchise terms primarily addressed to some other object. There were three main variants of this style of statement. (1) The bulk of franchises for particular flooding dams implied the grantee's obligation to serve all comers, incident to defining the service. Slides and flood gates "shall be kept open at all times when the river is in a driving stage, and there are logs . . . to run over said dam, and when it is not necessary to hold the water back for the purpose of driving or flooding logs . . . below said dam . . ."[106] (2) Where rights or duties of the franchise holder were defined by reference to "desires" or "requests" of customers, context usually indicated that the customer's (reasonable, or properly presented) desire or request created an obligation to respond. Where a toll boom franchise required that the boom "be so constructed that all logs . . . not owned by [grantees] . . . or other parties desiring to use said boom . . . may be assorted and passed by or through said boom . . . without unreasonable delay," there was an inference that transporters had legal claim upon the boom's services; the inference was strengthened when the same franchise conferred toll rights "whenever any logs . . . shall be secured and boomed in pursuance with the written order or request of the owner or owners thereof," or "whenever any logs . . . shall be secured and boomed (except such as the owners do not wish to have so secured and boomed)."[107] There was analogous, but implicit, reference to the navigator's desires as a criterion of the improver's duty, where franchises granted a compulsory toll (by authorizing toll for channel or flow improvements which all who would navigate the stream must in fact use), yet coupled the grant with a guaranty of clear passage — as when Laws, 1883, Chapter 355, laid toll on all who ran past the licensed flood dams on St. German Creek in Lincoln County, and also declared as a condition of the license to build the dams that they "be so constructed as not seriously to delay or prevent the transit of logs down and through said creek . . . but in aid thereof . . ."[108] (3) When a franchise imposed conditions on the customer, it implied that if he satisfied the conditions he established a claim to service. Its charter explicitly gave the Little Wolf River Log-Driving Company "the right to drive down said river all logs . . . belonging to said company or any other person . . . which now are or may hereafter be placed in said river, from upper sources to the boom in Winnebago county"; the charter did not balance this right of the company with a similarly explicit right in any log owner to obtain the company's services, yet the charter also stipulated "that the owner of any logs . . . shall have no right to demand of said company that the same shall be driven by said company, unless such owner shall place his logs . . . in the main stream of the said river, or on the banks thereof, in a workmanlike manner." [109] The charter of the Wausau Boom Company did not directly declare the company's duty to serve, but it provided that the company "shall be under no obligation to receive, retain or store any logs or timber the owners of which have not complied with" the requirement that "all log owners . . . desiring to have logs or timber retained or stored in such booms, shall on or before [the] . . . first day of April in each year, file with the secretary of said company a statement in writing . . . setting forth that he desires his logs or timber retained or stored in said booms, and specifying as near as may be the number and quantity, and . . . the particular marks or brands thereon . . ." Further, this franchise cautioned that the company "shall not be obliged to receive or take charge of any more logs or timber at any one time than the capacity of their said booms will admit . . ." [110]

By their disclaimers some franchises indicated concern to avoid or to qualify the common carrier's duty to serve all comers. A few franchises declared that the corporate grantee's stockholders might have preferential claim upon its facilities. Though it had laid upon the grantee no explicit duty to serve, the statutory license of the Suamico Boom Company took pains to provide that "the said company shall have the exclusive right to the use of said boom, upon so determining at any meeting of the stockholders, and upon publishing a notice to that effect for at least ninety days, in one newspaper printed in the county of Brown." [111] The franchise of the White River Dam, Log Driving and Boom Company sharpened the issue because of its contrasting treatment of the company's driving and booming functions. The franchise not only "authorized" but also "required" the grantee "to receive and take possession of all logs . . . hauled, driven or coming into White river . . . and to drive all such logs . . . into said corporation's booms." The franchise implied a right in log owners to demand boom service, when it stipulated that "all logs . . . the owners of which shall not file with the secretary of said company a request to have the same retained in said booms, with a description of the marks thereon, shall with reasonable diligence be passed through the boom into the river below." However, it qualified this right — and in so doing by implication confirmed its existence — by declaring that "the stockholders in said company shall have preference over any and all other persons in the storage of their logs . . . in said booms . . . [in] equal proportions to the amount of stock owned by each of them in said company . . ." [112] There is ambiguity here. The stockholder preference may have been inserted primarily to assure the stockholders' claims on the prior attention of the corporation's management; on the other hand, it inhered in the preference that it gave stockholders and management alike a defense to complaint by an outsider that the company was not fulfilling its common carrier duties. The importance of the stockholder preference was brought into particular prominence when the preference was first included and then by amendment removed from the franchise of the Wausau Boom Company.[113] Subsequently, a dictum of the Wisconsin Supreme Court found that, with the stockholder preference included, "it would be difficult to consider the [grantee] . . . other than a private corporation, for private use . . ." so that—for want of a justifying public purpose—the Company might not be given a monopoly of booming operations to the exclusion of the rights which riparian owners in the area would otherwise enjoy to maintain their private booms to service their own mills. But when the preference was removed, the court found that the franchise (which, we have noted, imposed obligations of service only as these were indicated by conditions laid on the customers) "gives an equal right in the use of the works to all the world . . ." [114]

The dominant pattern of legislation imposed on these toll improvers the common carrier obligation to serve equally all who properly asked service. Yet this important feature of regulation was most often stated by indirection or implication. Blurred statement was familiar in the growth of nineteenth-century legislative principle, which tended to unfold in a fashion more akin to the narrowly pragmatic style in common law development than according to the distinctive legislative ideal of generalization. The concepts of public utility status and obligations were crude and new in the principal years of lumber industry growth; the competing interests that met in the legislative process were groping to identify relatively unfamiliar relationships. However, apart from the novelty of the situation, there were aspects of water transport which made it unlikely that equal access to service would become a fighting issue on a scale calculated to produce sharply defined reactions in law. The legal record here is markedly free of the contests of shippers against carriers which filled the annals of the railroads. Stream improvements were functionally critical to forest production. But they did not require or invite such massive concentrations of capital, and hence of independent power, as did the growth of railroads. Improvers and transporters were not only more evenly matched in economic force, but they were also more closely identified in common ventures than was true of the railroads; those who improved the streams usually did so because they had their own logs and lumber to move to market: often the bulk of traffic serviced by toll improvements was provided by production of the improvers themselves. In such circumstances it is no small tribute to the importance of legally assured access to improvement services that the matter received the attention it did; by the same token it is less surprising that the pressures were not severe enough to produce more forthright statements of obligation.[115]

The same considerations offer explanation of the want of provisions to implement the public utility status which stream franchise legislation recognized. There were no regulations or limitations upon services which the improvers rendered to themselves and their own commodities — naturally enough, since so often the prime impetus to make the improvements was to move their own goods.[116] There were no limitations or regulations of the consolidation of improvement enterprises; to the contrary, franchises granted to individuals commonly ran also to their "associates" and "assigns," and statutes gave broad authorizations to corporate grantees to consolidate with other companies — naturally enough, where there were not sharp-felt divisions of power between improvers and customers.[117] Sanctions to enforce public utility obligations were rudimentary. The courts would ground a common law action upon a showing of negligent nonfulfillment of statutory duty, but dissatisfied customers were unlikely to find satisfaction in the delays and costs of litigation; only two such cases appeared in the Wisconsin Supreme Court.[118] A few franchises made gestures toward publicity of rates and financial statements as curbs upon excessive charges or discriminatory practices.[119] Boom franchises sometimes required the grantee to keep accurate books, to substantiate charges.[120] Occasional franchises declared the threat of forfeiture for violations, a penalty meaningless because too drastic for ordinary breaches.[121] Above all, what was lacking was any provision of ready, inexpensive, flexible administrative or executive process to back up regulatory laws.[122] Again we are brought back to the fact that the organization of water transport was not such as to create imperative pressures of widely felt disproportions of power between the utility and the customer. Where such pressures existed (as they did concerning the railroads), after much fumbling, legislation moved toward concentration upon sanctions, and especially upon administrative process.

RAILROADS

Water transport bore most forest products to market before 1870. Not until the mid-seventies had substantial rail construction reached into the northern forest, and — duplicating the course of events earlier in the southern half of the state — men's first concern was to extend the rails rather than to worry about the terms of using them. The record offers no evidence that there was substantial difference of attitude, in the promotional period of railroad growth, between railroad entrepreneurs and the general community.

Charters for the first railroads in an area consisted almost wholly of authorizations to act, and contained little to limit the discretion of the promoters to conduct their business as they saw fit. Thus, early charters authorized the roads to carry passengers and goods but declared no standards of service; they authorized connecting arrangements with other railroads, but imposed no requirements to set up through rates; they might include broad powers to consolidate with other roads, but without stipulation of terms to curb the power of the consolidated companies; typically they authorized the railroad managers to make such charges for carriage "as they shall from time to time deem a reasonable compensation therefor." [123] To read these provisions as the product of wily maneuver and sinister lobbies would be to indulge in a presumption for which there is little supporting evidence. The uniformity of pattern in these special charters of the railroads' promotional period implicitly denies that there was something "special" about the charters, in the sense of something imposed upon a gullible public or upon unsophisticated legislators. When great trunk-line empires became established, the roads indeed earned their later reputation for ruthless exploitation of power and unscrupulous pressures on public officers. But when the first roads were built into forested areas that could not develop without them but could promise only speculative returns for the ventures, the commonly accepted value put upon economic growth insured that the legislature without substantial question would give the promoters broad grants of power.[124]

If there are morals to be drawn from the legislative responses to the pioneering enthusiasm for railroads, they do not relate so much to curbing men's greed as to cultivating their foresight. In connection with land grants and tax exemptions, contemporary policy making rested on recognition of the peculiar significance of railroads as setting basic terms for general economic growth.[125] Nevertheless, with uncritical enthusiasm for growth as a self-evident good, men failed to anticipate the powers of life and death which the railroads would hold over particular enterprise. Borrowing the wisdom others had drawn from the consequences of the *Dartmouth College* case, Wisconsin's constitution makers had stipulated that all general or special laws for organizing corporations might be altered or repealed by the legislature at any time after their passage.[126] Judicial doctrine might limit the extent to which this reserved power would warrant redefining rights protected otherwise than by the contract clause of the federal Constitution (rights protected by the due process clause, for example). Still, in form the reserved power offered considerable scope for the legislature to build fresh regulation. Experience showed, however, that there was illusion in resting on this potential power as a basis for easy terms in fostering new centers of power. Power derived from key position in a network of relationships could build fast and to such effect that it became very difficult to dislodge. The railroads early developed aggressive lobbies. Their pressures were the more easily effective because in this new state, still close to frontier conditions, the legislature was a relatively crude institution, lacking in tradition, experience, or staff. Thus, although some general statutory regulation was superimposed upon the permissive pattern of the special charters in the first twenty years of railroad development, the regulation was limited in scope and wholly lacking in effective implementation.[127]

Even after a considerable rail net had reached into northern Wisconsin—by the end of the seventies—concern over rates and standards of service was limited by divergent interests among the ranks of those seeking to realize forest wealth. Investors primarily interested in profits from the sale of timberland (notably Cornell University, for which Ezra Cornell had located nearly 500,000 acres in the forest region) felt no reason to press for general railroad regulation; their attention focused on the terms on which railroad-aid lands might be sold, and on the grant of aid-lands tax exemptions which would increase tax burdens on nonexempt landowners.[128] Such pressure as came from forest interests for rate and service regulation came from the millmen. However, lines of interest were blurred here. Prominent lumber manufacturers were usually also speculators in timberland, and sometimes (like Philetus Sawyer) were substantial investors in railroads and participants in top management.[129] Thus the record showed no strong, consistent lumber industry activity for railroad regulation. Not only in the first promotional period—in 1864, 1867, and 1873—but as late as 1891, legislative committees which probably spoke the opinions of forest industry leaders recommended against strong legislative controls of railroad practices, on the ground that this might discourage needed railroad investment in the north country.[130] The lumber industry does not appear to have played a substantial or distinctive role in the Granger agitation which produced the Potter Law of 1874, though there was support for the act among legislators from lumber districts.[131] In 1903, at the height of his battle for more effective regulation, Governor LaFollette spoke his regret that lumbermen were actively supporting the railroad lobby in the legislature, and attributed their position to misunderstanding of their own interest and to threats of economic duress by the roads.[132] Millmen, large and small, felt grievances against the roads, as they made plain by complaints which they brought before federal and state regulatory agencies when these were created. The incongruity between this fact and the want of a well-defined, sustained industry demand for regulatory legislation is without good explanation on the record. The incongruity does suggest the function of legal process in overcoming the inertia of a given pattern of interest. When law offers play to wider forces, these may create a new frame of reference within which specific interests theretofore silent (if not content) may find reason for fresh expression. Business-oriented politics in late nineteenth-century Wisconsin was a very conservative politics, content with strong rule by inner circles of the Republican party which could be counted on not to press for regulatory experiments. Despite their specific grievances as entrepreneurs, the millmen were probably constrained by general political allegiance not to join those who would obtain new definitions of program and principle through the legislative branch. Ideological and party commitments apart, there is plausibility also in Governor LaFollette's belief that even substantial lumber firms might hesitate to stand in broad policy opposition to interests so much larger than they, on whose day-by-day operating practices they were as dependent as they were on the conduct of the roads. On the other hand, when public policy had taken a definite new direction, manifest in the creation of specialized administrative agencies to police the railroads, there was less overt challenge to ideology or party loyalty in testing out the benefits which the new agencies might give. Moreover, in the context of specific controversies focused in particular administrative records there might seem less daunting challenge to the total power of the railroads than in meeting them in legislative battle. The ambiguities in the lumber industry reaction to the railroad problem may thus add up not to mere confusion, but to evidence of the creative tensions that lie in the interaction of legislative and administrative lawmaking.

Millmen had their complaints about railroad service and operating rules. Sometimes they could not get cars when they needed them. The railroad would set a minimum charge per car for a stated number of board feet, and in woods operations this posed difficulties of proper calculation to avoid underloading cars and so paying uneconomically high rates. The roads imposed weight limits on the car loads they would carry; again, in the rough conditions of logging operations close calculations were hard to make, and a shipper might find his tender of goods unacceptable.[133] On the whole, however, service difficulties were much the lesser issue. Rates were the critical problem.

The central rate issue concerned access to markets. There were materials for other, familiar rate issues: the roads departed from published tariffs without formality; there were debates over the comparative fairness of short-haul and long-haul charges.[134] But through the years of Wisconsin's lumber boom (up to about 1910), lumbermen were so concerned to obtain basic transport facilities that they made relatively little effort to use law to control details of the roads' relations with particular shippers, or the impact of rates upon intrastate traffic. They were eager to use the rails to obtain access to wider sources of raw materials and to broader markets. Here the roads and their customers reached mutual accommodation through the negotiation of milling-in-transit privileges; a mill would agree to ship all or a large part of its output to market over a given railroad which in return would grant a low carrying rate to bring logs from the woods to the mill. Milling-in-transit privileges were important to the railroads, to make them competitive over wider areas with water transport; the privileges were important to the mills to help create more assured supplies of raw material.[135] Far less satisfactory to the mills were their dealings with the roads which affected the reach of selling markets open to the lumbermen.

Through peak years of Wisconsin production in the late nineteenth-century intermittent conflict flared between the mills and the railroads over rates for shipments to trans-Mississippi markets. The controversy was a legacy from the distribution patterns first set by water transport. So long as all or most lumber moved by water, the Wisconsin mills depended upon Chicago and Mississippi River wholesalers to sell their products west of the river. As the rails spread westward, some railroad companies without Wisconsin connections developed profitable traffic in lumber moving out of Chicago or Mississippi or Missouri River wholesale centers. A large consideration in the enthusiasm that Wisconsin lumbermen had for railroad building in the state was, originally, the promise they saw for freeing themselves from the Chicago and river port middlemen and moving their lumber more directly to inland dealers, to their own greater profit. They had enough success in realizing this dream to keep them fighting for it over long years. In 1878 Wisconsin and Minnesota lumbermen initiated negotiations with railroads in Wisconsin and across the Mississippi for advantageous through rates to prairie markets. In 1879 lumbermen obtained an agreement of the Omaha road with three trans-Mississippi carriers to haul Wisconsin lumber at rates which would enable Wisconsin lumbermen to compete with Chicago and river distribution centers. The agreement was expanded by inclusion of the Wisconsin Central in 1881, and through the eighties the reach of the arrangement within Wisconsin was extended as additional new construction was brought within it. This expansion of direct marketing by Wisconsin lumbermen into the trans-Mississippi territory grew by the mid-eighties to proportions which stirred retaliation from both the old distribution centers and the railroads which did not operate in Wisconsin. In 1884 major railroads reaching Missouri River markets agreed upon temporary differentials favoring Chicago over inland Wisconsin shipping points. They submitted to private arbitration the definition of permanent schedules. The arbitration produced a pattern under which lower Mississippi River shipping points paid lumber rates from one to six and a half cents per hundred pounds below the rates from Chicago, while upper Mississippi River points (notably Winona in Minnesota and LaCrosse in Wisconsin) paid one cent above the Chicago rates, and inland Wisconsin millmen (at Eau Claire, Chippewa Falls, Wausau, Stevens Point, Fond du Lac, and Menominee, for example) paid from six to eight and a half cents above the Chicago rate. Inland Wisconsin millmen complained bitterly of the discrimination and to some extent resumed water shipments of lumber. Through the late eighties the mills pressed their objections in private negotiation, but obtained only temporary and limited concessions which were soon withdrawn. As Wisconsin railroads expanded their reach, moreover, they also found reasons in their own interest to favor traffic from the river ports and Chicago, so that the Wisconsin lumbermen lost a measure of the support they had once had from the home roads.[136]

In 1887 Congress had at last put federal law into the picture, with enactment of the Interstate Commerce Act. By 1890 Eau Claire and Chippewa

Falls lumbermen had lost hope of bettering their situation by private negotiation with the roads. They pooled funds to bring a proceeding before the Interstate Commerce Commission, asking that the rate differential to Missouri River points be reduced to half a cent over the Chicago rate. Not only the railroads, but also intervening lumbermen from LaCrosse and Winona, opposed the petition. The taking of testimony and the Commission's consideration of the case dragged on until, in June 1892, the Commission ordered rates more favorable to the Chippewa Valley. However, only the Chicago, Milwaukee and St. Paul railroad complied with the order. After a few months, when it became evident that their competitors would not obey, the Milwaukee road ceased compliance. The Commission did nothing about the railroads' defiance until June 1897, when it began belated investigation of the matter. Even this tardy action came to nothing, for in that year the United States Supreme Court decided that the Interstate Commerce Act gave the Commission no authority to fix rates.[137]

The Wisconsin millmen's troubles dramatized the fact that railroad rate patterns inherently constituted major economic planning decisions affecting whole regions of the country. Private negotiation could arrange treaties in matters of this scale only among parties of substantially equal power. The lumbermen's ill success bears witness that, although the railroads were a critical influence on the lumber industry, to the railroads the lumber industry was only one more of many substantial but never dominating shipping interests. The inland Wisconsin millmen pressed persuasive claims that railroad rates should not negate their advantages of economic location. Chicago and the river port interests had their own claims based upon established and valuable distribution relations well established in time. The railroads had some legitimate considerations of traffic balance to support their positions. With due allowance for the selfish motives involved, dogmatic judgment cannot be passed that these varied elements did not present real issues of public interest. Plainly there must be some legal intervention in adjustment of relations so complex.[138]

The ineffectiveness of resort to the Interstate Commerce Commission, during years when the Commission's authority and energies were conspicuously weak and uncertain, for all practical result closed the story of public regulation of railroads as it affected the great years of the lumber industry. After bitter battles through two previous legislatures, Governor LaFollette in 1905 obtained enactment of a Wisconsin Railroad Commission act, creating a permanent regulatory body to police the intrastate operations of the roads.[139] The existence of this new agency promptly produced scores of applications for relief from various alleged misdeeds of the roads, from millmen, fabricators of wood products, and lumber retailers. The overwhelming bulk of matters brought to the state agency concerned rates; complaints over service (the furnishing of cars or the building of industrial spurs, notably) were few enough to underline that, to businesses dependent upon market calculations, the money charges for carriage were the essential point. The names of the dominant lumber concerns do not appear prominently in the state commission's flow of business; at least at this stage of the industry, the new state procedures seemed to attract the hopes rather of the smaller firms — a result which tallied with the fact that the demonstrated main concern of the bigger companies was with access to interstate markets, and hence with appeal to federal regulation. The rate issues presented challenges both to the reasonableness of rate levels and the fairness or discrimination of rate patterns; the discrimination issue predominated in the complaints, however, to a degree which suggests that the railroads' impact upon the area development of markets was the most sharply felt and recognized significance of railroad power. The new state commission recognized this as a key issue in an early ruling relied on as the framework doctrine for a lumber rate case. A railroad, said the commission, might not refuse to establish a joint rate because the rate might divert business from dealers located on the railroad company's lines. "If we had a system of state owned railways, the patriarchal idea, that the state might say who should do business in a given locality and who should not, and might establish zones where certain interests could monopolize business to the exclusion of others, and direct commerce from its natural channels to artificial ones, is an idea that would, we think, be abhorrent to the American sense of justice and fair play." [140]

Active federal and state regulation of railroad rates and practices became a reality so late relative to the life course of the Wisconsin lumber business as to have no opportunity to play a material role in shaping the character of the industry. Thus this history may properly end at the point to which we have now brought it. For the timing and manner of use of forest wealth, the promotional history of Wisconsin railroads had more significance than the regulatory story. Both this fact, and the fact that, nonetheless, regulatory issues emerged sharply in the late years of the lumber industry, attested the close relation between the growth of public policy and the growth of patterns of social interdependence.

PART IV

POLITICAL ECONOMY

INTRODUCTION

POLICY FOR A REGION

BY 1915 lumber production was no longer a commanding element in Wisconsin's economy; by 1940, after years of continuing decline, it began to stabilize at a level which left the lumber industry a substantial but relatively subordinate contributor to the state's general output. Wisconsin lumber production was estimated at 3,389,000,000 board feet in 1899, at 1,391,000,000 in 1914, at 436,000,000 in 1940; 1940 output was about 12 per cent of that of 1899. Wisconsin ranked first among the states in lumber production in 1899, and still held that place in 1904; in 1914 it stood tenth; in 1931 it was in fourteenth place. Lumber and timber basic products, which ranked first among Wisconsin industries in value added by manufacture in 1899, ranked twelfth in 1939. Over these same years of transition in forest industry, growth in production of paper and allied products was an important offset to the decline of lumber production; in 1939, when lumber production stood only twelfth among Wisconsin industrial groups in order of value added by manufacture, paper ranked fourth. But in forest economics, technology, and industrial organization, and in the relations it bore to the general community of its time, the paper industry represented a distinctive development to an extent that separates its history from that of the nineteenth-century lumberman.[1]

By its own decline, and because it no longer bulked large in a more diversified economy moved by new energies, the lumber industry ceased after 1915 to be a focus of large public policy making. Its affairs no longer provided impetus to develop broad patterns of legal action. At most they simply occasioned special adjustments in policy aimed at broader objectives. It was symbolic that the new Industrial Commission provided in its general safety orders for the guarding of sawmill machinery, while over the years it developed special safety codes for other industries.[2]

Three trends in events after 1915 combined to the end that lumber production no longer had a distinct legal history. (1) By changes in technology or in business organization, or by mere decrease in activity, some problems disappeared or no longer had significant impact. Lumbermen lost interest in navigation improvement franchises and in the services of state lumber inspectors, and legal records no longer reflect lively concern with timber trespass on private lands.[3] (2) Some problems merged into broader issues of a more interdependent and administered economy. Regulations of conditions of camp and mill labor became items to be adjusted in patterns of general labor law. Fair access to rail service for shipping logs and lumber presented only variations upon general themes in public utility administration.[4] One striking new regulatory issue emerged, peculiar to mid-twentieth-century forest products industry. This concerned pollution of streams by paper mill wastes. But this, too, was treated as a problem of a broad, new, industrial and social complex. The law's response developed in the context not merely of the affairs of the paper industry, but as an aspect of programs which dealt with the total water resources of the state and with diverse problems of pollution from human and industrial wastes.[5] This merger of lumbering problems into broader patterns of policy graduates into the third trend which ended the separate legal history of the lumber industry. (3) Wisconsin public policy was much concerned in mid-twentieth century with growth and harvesting of forest crops: for the long course, to build back a greater saw log production; more immediately, to provide pulpwood for the paper makers. But the dynamics of such policy derived only in secondary measure from the affairs of forest products industries; the principal impetus sprang from a new concern for the over-all growth of northern Wisconsin and its relation to the economic health of the whole state. In mid-twentieth century the focus of legal operations was no longer forest products industry. Attention now centered on restoring the economic vitality of a region, with lively consideration for agriculture, mining, and the recreation business, as well as for forest crops.[6] To an extent unmatched in the nineteenth-century record, conscious planning now turned to tax structure, closer calculation of public subsidies of private action, and more efficient relation of public services to area development. Whether we measure by objectives or by allocations of public resources, the legal history of forest production yields place to the legal history of regional rehabilitation. This is an important story, but it is a different story from that which is the subject of this book.[7]

The purpose of this Part, therefore, is not to relate in detail 20th-century efforts in law to guide and promote the economic and social rebirth of northern Wisconsin. The purpose, rather, is to note salient features of this new regional-focused policy, for the light which they cast on the law's handling of issues created by the nineteenth-century lumber industry. To this end we first examine some evidence of the painful emergence of new values, and then summarize certain structural features of the new pattern of policy that developed toward the old forest region.

CHAPTER X

REASSESSMENTS OF LAW'S ROLE IN THE ECONOMY OF NORTHERN WISCONSIN

MEN USED law as a major factor in developing the nineteenth-century lumber industry in Wisconsin. So, too, between 1925 and 1940 they used law as a major factor in creating new sources of economic vitality for the cutover country which the lumbermen had left. In both instances they employed law as a positive instrument to direct action toward the dominant objective of the time — to speed and increase lumber production at first, later to rehabilitate the old forest-region economy. In neither phase were they content to treat law solely as an instrument to keep the peace or enforce contracts. True, they used law to regulate as well as to promote action. But so far as they contrived new types of regulation, their concern usually was to support a wider range of private action; so the legislature created log-labor liens to encourage men to work in the woods, and included navigation guaranties in stream franchises to enlarge the volume of traffic.

Though both periods were ready in using law to shape the situation, the middle twentieth century used law with a markedly different style from that of the 19th century. The prime difference was in the immediate objects of legal subsidies or supporting regulations. Nineteenth-century law sought chiefly to encourage maximum range and intensity of private decision making: by gifts or low-priced sales of public lands, by bestowing stream-use franchises, and by its most important types of regulation (offering private timberland investors special remedies against trespass, or supporting private contracts), law focused its help almost exclusively upon men active in forest production. In mid-twentieth century the law's subsidies and creative regulations ran no longer simply to one industry, but to the region — focused no longer simply upon the supply of investment and working capital of private entrepreneurs, but upon the social capital and the overhead costs of the north-country economy as a whole.

The next chapter sketches the emergence of the pattern of action which represented this changed style in political economy. Idea mingled with action to make policy. Men applied law with new emphasis and in new directions because they came to perceive and evaluate their situation differently. As is so common in this unphilosophic society, often they made little record of their new thinking, save as their ideas were implicit in what they did; the next chapter takes note of such implicit testimony. However, not the least of the functions of legal process is to stimulate and help men to make more explicit definition of their situation and to take more explicit deliberation and decision upon it. Legal process worked so, here. Since a new assessment of the situation was in progress, it is not surprising that it proceeded unevenly. There was evident pain at the old that was left behind. There was evident confusion over the new that was emerging. The pain and confusion, but also the successful expression of fresh awareness and fresh choices, are as much part of this history as the action recited in the next chapter. We can fairly represent the emergence of new perception and new decisions by comparing (1) the Forestry Case in the Wisconsin Supreme Court in 1915, and (2) the report in 1929 of the legislature's Interim Committee on Forestry and Public Lands.[1]

Under a series of statutes enacted between 1903 and 1911, bulwarked by what appeared to be a duly adopted constitutional amendment of 1910 authorizing appropriations for "acquiring, preserving, and developing the . . . forests of the state," Wisconsin created an ambitious state forest reserve program. Largely from its own lands withheld from sale, and from lands bought for the purpose with proceeds of sales of state lands and with general funds, by 1912 Wisconsin had assembled a forest reserve exceeding 400,000 acres, under plans contemplating a total of 1,500,000 acres of reservation.[2] To increase the acreage obtainable with current appropriations, the State Forester contracted to buy lands on time.[3] Questioning the validity of such a contract or the statute which purported to authorize it, the Secretary of State refused to audit a claim for the last payment due a private seller under an agreement which provided for payments over three years. On behalf of the state the Attorney General petitioned the Wisconsin Supreme Court in its original jurisdiction to mandamus the Secretary to audit the claim. The Secretary demurred. So that "the great interests of the state could be properly conserved," the Supreme Court "exercise[d] in a somewhat extraordinary, but proper way, its power to prevent and redress wrongs, particularly where vital interests of the whole people were imperilled," by declining to decide the matter upon the Secretary's demurrer, and inviting the Secretary to file an answer designed to put in issue all matters of defense. To this answer the Attorney General demurred. Having concerned itself in "supervising the formation of issues," the court, not surprisingly, disposed of the case with an array of alternative grounds of decision and dicta which took it a good deal beyond what was necessary to resolve the case.[4]

The majority of the court chose to speak through Mr. Justice Roujet Marshall. For years before he went on the bench Marshall was leading counsel to great Chippewa Valley lumber interests headed by Frederick Weyerhaeuser. His lumber industry practice had given Marshall a personal fortune, a heady sense of personal accomplishment and influential participation in affairs of large moment to his region, and reverence for the managing genius of Mr. Weyerhaeuser. Marshall was both vain and able. His vanity assured him that he was a self-made man, and left him no room to question how far his advancement (with that of others who led in the lumber industry's affairs) had rested on appropriating social income and avoiding social costs involved in the rapid exploitation of the forest. Too acute to miss the signs of new times — he played a role from the bench in promoting the creation of a workmen's compensation system, for example — Marshall's self-regard moved him also to justify the past which had made his career; so, when public policy turned to deal positively with the aftermath of the lumber boom, Marshall was impelled to mingle grudging recognition of change with tributes to a mythical, simpler time when strong men had made their way without the law's help. Vain and able, Marshall, moreover, sought on the bench to match his success at the bar with a role appropriate to his judicial position; he saw himself called to be a great lawgiver.[5] It came easily to him to take 106 pages in the Reports to range so widely and expound so freely on the limits of legislative power that Chief Justice Winslow was moved to file a separate concurrence, because he thought that the opinion of the court "so limits and circumscribes the powers of the state with regard to afforestation and reforestation that it leaves little more than a shell behind." [6]

The case called for no ruling save on the legality of using general funds to meet the last payment due on the land-purchase contract in suit. But the court ruled that the legislature lacked authority for any major feature of its state forest reserve program. Three important aspects of the decision have no necessary relation to the substantive problems of the north Wisconsin economy which are our concern; hence they need only brief note here. First, the court held that inclusion in the forest reserve of state-owned lands withheld from sale or of lands purchased with the proceeds of sales of state lands constituted a breach of trusts created by the state constitution for the support of public education. The decision made the repairing of this breach of trust a prior charge against all lands in the forest reserve which could be traced to this wrongdoing of the legislature.[7] Second, the court held that the legislature had no authority, by the purported constitutional amendment of 1910, to appropriate general funds to buy lands for the state forest reserve, because in consequence of procedural irregularities in its passage the amendment had never been duly adopted.[8] Third, the court held that contracts to buy land for the reserve on payments stretching beyond a current budget year violated the ban which the Wisconsin Constitution laid on contracting state "debt." [9]

Three aspects of this forestry decision of 1915 bore directly on future policy concerning the general economic and social condition of north Wisconsin. Marshall's opinion for the court held that (in the absence of any legal effect in the purported amendment of 1910) to spend general funds to buy land for a state forest reserve violated section 10 of Article VIII of the Wisconsin Constitution, which said that "the state shall never . . . be a party in carrying on . . . works . . . of internal improvement." [10] Second, Marshall's opinion for the court declared that the legislature lacked authority to tax and spend to create a state-owned and state-operated forest reserve. The opinion was fuzzy in its grounds under this head. Apparently it meant that such taxes or expenditures were invalid because a reserves program was not for a public purpose, and because state money aids given the affected local governments to make good their loss of tax revenue from the reserved lands were not expenditures of state money for purposes of state-wide concern.[11] Third, in a variety of ways which constituted expressions of attitude more than rulings of law, the opinion of the court indicated a mixture of bafflement, hostility, and grudging acceptance, as the judges confronted a more closely calculated use of law to affect the economy than had marked nineteenth-century experience.[12]

We shall return to these matters. First, however, we should note legislative action taken between twelve and fourteen years after the Forestry Cases, offering comparisons to illuminate the further development of ideas and attitudes which entered into a new style of lawmaking for the north country.

With painstaking care for the procedural proprieties, and with relative smoothness, the legislature saw through to popular ratification a constitutional amendment in 1924 which authorized the state to spend money on forestry. Between 1925 and 1927 the legislature and the voters adopted another constitutional amendment, which authorized special classification of timberlands for taxation.[13] In 1925 the legislature created an Interim Committee on Administration and Taxation, which, when it reported in 1927, put its attention largely on the alarming fiscal problems of the cutover country. In response, the 1927 legislature created a special pattern of deferred real property taxes on timberland to encourage the growth of trees, and authorized counties to establish county forests.[14] All of this stir signaled that the legislature was ready for a more comprehensive look at north-country policy than any it had taken since the 1915 decision scuttled the state forest reserve program. It created an Interim Committee on Forestry and Public Lands, which reported in 1929.[15] This 1929 analysis showed that men had moved a considerable distance from their mixed attitudes of 1915, toward quite firm convictions as to what should be done and what values should guide the doing. It is this 1929 Report which provides us with useful bench marks by which to measure trends beyond the lines indicated in the Forestry Decision.

SPHERES OF PUBLIC AND PRIVATE WILL

When the legislature allocated state property (state lands or the proceeds of their sale) or spent tax-derived money to enable a state agency to create a physical condition (to establish a state forest reserve), there could be little argument but that the state so became "a party in carrying on" the works created.[16] A harder question was whether a state forest reserve should be deemed "works . . . of internal improvement" within the meaning of the ban in Article VIII, section 10, of the Wisconsin Constitution.

The framers left little helpful gloss upon their language. They worried lest Wisconsin suffer such financial difficulties as other states had incurred by too great enthusiasm to subsidize construction of bulk-transport facilities, especially canals. But they left no clear evidence as to how much broader ran their conception of the "works" to which the state must not become "party." [17] It fell thus to the Wisconsin Supreme Court to give fuller content to Article VIII, section 10.

In 1902, in *State ex rel. Jones v. Froelich,* the court ruled invalid legislative appropriations of general funds to build levees on the Wisconsin River to protect life and property from floods.[18] The levees were found to be "works . . . of internal improvement" within the constitutional ban. To rule so the court was impelled to construe Article VIII, section 10, as in effect a constitutional embodiment of laissez-faire principle. The court conceded that the words of section 10 could not be taken at their full possible extent, lest they destroy the capacity of the state government to function. Taken in full literal sweep, the words "are capable of including substantially every act within the scope of governmental activity which changes or modifies physical conditions within the limits of the commonwealth." Taken so, the words would forbid building public offices or jails. Thus it must be understood that section 10 did not forbid the state to invest in works "which primarily and preponderantly merely facilitate the essential functions of government." On the other hand, merely because given works would serve a public purpose they were not therefore outside the constitutional ban; if section 10 merely forbade public expenditures for private purposes, "the prohibition would have been needless," since other familiar constitutional doctrine already required the legislature to act only for a public purpose. Out of its search for a formula which would give sensible vigor to the constitutional language, the court emerged with the conclusion that the ban "included those things which ordinarily might, in human experience, be expected to be undertaken for profit or benefit to the property interests of private promoters." Since the dominant purpose of the levees involved "enough of pecuniary benefit to warrant belief in the possibility, at least, that they may be undertaken by private enterprise or local associations," the levees fell within the constitutional limitation on state activity.[19]

Froelich in effect held that Article VIII, section 10, created an exception to the general rule that every statute should be presumed constitutional until a challenger clearly established its invalidity. Where the legislature appropriated public property or funds so that a state agency might bring about a change of physical condition, the appropriation must be held to exceed legislative power if the objectives were such as "ordinarily might . . . be expected to be undertaken," or such as would "warrant belief in the possibility at least, that they may be undertaken by private enterprise . . ." Wherever possible, this principle preferred private over public decision making in procuring physical change. Under this formula — contrary to the general presumption of constitutionality — doubt must be resolved against the legislature's determination, not in favor of it.[20] This ordering of values was designed not only to uphold the exercise of private will, but to protect the "possibility" of its exercise. The decision in *Froelich* underlined the point, for "in human experience" there could be deemed no more than a "possibility, at least" that private enterprise would furnish flood-control levees along a rampageous stream.[21]

Marshall's opinion for the court in the Forestry Case seized upon the *Froelich* formula with an evident satisfaction that went beyond simple obedience to precedent. The limitation set by Article VIII, section 10, must be given generous extension. In *Froelich "ex industria,* the court attempted to give the broadest meaning reasonable to the particular term" in the constitutional limitation. "[T]he principle of exclusive individual and local communal instead of state enterprise was the thing the framers of the constitution had in mind and not any particular class of internal improvements." As Marshall read the historic record, "it was the principle at stake, — whether the new government should be confined to the ordinary civil affairs or be extended, at general expense, to those things of a public nature which were not really governmental in character, which was the subject of the convention labor."[22] Marshall conceded that to reserve state-owned lands for a favorable market, or to spend state funds to protect state-owned lands against fire and trespass, while they awaited sale, would not violate the internal improvements ban. But the sweep of his argument barred such indefinite reservation of state lands as would allow the state to practice forest management or administer the lands to effectuate a program of over-all resource development for the region.[23] Thus he found that important parts of "the all-pervading idea of the whole forestry scheme" were contrary to the constitutional limitation. In particular he singled out for condemnation the "handling of such lands . . . in a proprietary way, gainful use by business operations suitable to such an enterprise of forest reserve lands, with incidental gainful use of water-power opportunities in such areas, devotion of such area to the acquired use, and annual extension of the business, so far as practicable, by use to that end of all the results of such business and an additional annual appropriation out of the general revenues of the state . . ." "That some features, at least, of the scheme as pictured spell 'carrying on works of internal improvement' under the constitutional meaning of the term, there can be no manner of doubt; such as pledge of the revenues of the forestry business to augmentation of the forest area, especially, so far as relates to conservation of natural waters and regulation of the flow thereof for climatic, drainage, or waterpower purposes or creating a game-preserve territory and the carrying on of lumber operations, and permanent removal of a vast property acquired at public expense from the field of opportunity for any return to the taxpayers of the present or even of the future . . ."[24]

Though he was not disposed to challenge the *Froelich* formula, Mr. Chief Justice Winslow disagreed sharply with the extended application of it in Marshall's opinion for the court. The Chief Justice agreed that the state forest reserve program must fall, on other grounds, but not because it constituted "works . . . of internal improvement." To him the court's application of *Froelich* indicated favor for private over public will to a point beyond reason. "I affirm that it is not to be expected in the light of human experience in this land at least, that the establishment and conservation of great forest areas for the public good should be undertaken by private enterprise." By the test — also part of the *Froelich* formula — of what physical changes "primarily and preponderantly merely facilitate the essential functions of government," the Chief Justice thought that a state forest reserve program should stand, and no less so because the establishment of the policy represented fresh adjustment to a new situation. "[T]he establishment and conservation of great forest areas for the public good . . . is pre-eminently a public work and hence one of the essential functions of government. It has not been recognized as such until recently perhaps, but that is merely because the conditions which make it such have only recently arisen and become acute."[25] In its sweeping application of *Froelich* the majority made bad law — so Winslow implied — because its doctrine denied necessary capacity to make reasonable adjustment to changed circumstances which concerned the commonwealth.

Apart from the presumption which it raised, the *Froelich* opinion did as good a job as perhaps could be done, in substituting a manageable judge-made formula for the ill-defined admonition left by the constitution makers. Stated as a presumption against legislative discretion, however, and applied with the rigor of the rulings made against the levees and the forest reserve, the *Froelich* formula rested more upon the judges' obvious worry over new demands upon the state than it did upon any demonstrated intent of the framers. Mid-nineteenth-century lawmakers worried over the solvency of state government in a community plagued by want of fluid capital. But in their practice they raised no presumption against using law positively to foster conditions favorable to economic growth. Their presumption, in practice, was Hamiltonian; the community should boldly use the capacity with which its legal organization endowed it, to arrange relations or allocate resources in ways reasonably calculated to multiply total productive power. Restriction upon this inventive capacity of law was, in nineteenth-century context, the exception, rather than the rule which *Froelich* and the Forestry Case made it.[26]

The decisions of 1902 and 1915 raised barriers against creative legislation for the north country too high to be surmounted other than by constitutional amendment. There was convincing testimony to the prevailing current of attitude in the decisive margins by which between 1921 and 1924 a constitutional amendment moved through to final passage, to provide that "the state may appropriate moneys for the purpose of acquiring, preserving and developing forests of the state." Adopted in its first legislative session by votes of 22 to 5 in the Senate and 58 to 11 in the Assembly, and in the required second round of legislative consideration by votes of 26 to 1 in the Senate and 57 to 4 in the Assembly, the amendment was ratified at the polls, 336,360 to 173,563.[27] The amendment showed more continuity with 1848 policy than had the court decisions. The legislature and the voters were ready to sanction broad discretion for public intervention to invigorate the forest economy. Concern ran, not to preference of private over public will in such matters, but to the impact on state finances; the amendment limited appropriations under its authority in any one year to an amount not to exceed two tenths of one mill of the taxable property in the state as determined by the last preceding state assessment.[28]

With no problem remaining under the internal improvements clause of the Wisconsin constitution, the 1929 Report of the Interim Committee on Forestry and Public Lands could weigh the promotion of public forests simply as a question of policy. Gone was the 1915 appeal to basic political or social principle to extend or protect the sphere of private will. The 1929 Report fastened firmly upon a fact and a purpose. The fact was that the cutover area was in serious economic trouble. The purpose was to help build new sources of productive strength for the region. Within this context, the Report made pragmatic appraisal of the various means of action that promised the desired result. Its approach was operational, in contrast to the groping for abstract laissez-faire principle which marked Marshall's 1915 opinion. The committee wanted the state to economize its own direct action by promoting sound forest management among private landowners — not from preference of private action as such, but to invest public means where they would have maximum influence. "We are convinced that the state may well engage, on a reasonable scale, in developing state-owned forests, but we are likewise convinced that the best results can be obtained through private ownership, with state aid and encouragement." [29] For such state "encouragement" the committee had none of Marshall's doubts over the soundness of substantial state spending to improve the production base; the state should spend boldly to create adequate fire protection, not limited to preserving state timberland held for sale; likewise the state should underwrite the costs of extended research in good forest practice as generously as it supported research in agriculture, for "forestry is in greater need of a research program, because timber is a long time crop and mistakes cannot be corrected the next year as with farm crops." [30] If the committee did not favor large investment of the state's limited funds in state-owned forests, this attitude spelled no basic hostility to public forests as such, where these offered the most economical way to advance timber growth. The counties held increasing quantities of tax-delinquent cutover lands, and realism told that the only productive future for most of these acres lay in their incorporation into well-managed county forests. These delinquent lands "compose a tremendous potential area of public forests. Because, under our laws, these lands go to the county, state owned forests are less important in Wisconsin than in some other states. Whether we would wish it or not, the fact remains that the counties own or can take title to an enormous acreage, and will acquire still more within the next few years . . . [T]he only productive use for most of this land within any reasonable length of time is as county forests . . ." Whether they "would wish it or not," the committeemen showed no hesitation in their affirmative recommendations; the state should in effect subsidize enlargement of county forests by relieving the counties of all obligation to return to the state any moneys in lieu of taxes which the lands would otherwise bear, and by supplying technical aid through the Conservation Commission and the state university in establishment of such forests. "The Committee is impressed with the fact that the counties, because tax delinquent lands revert to them, are the chief governmental unit around which the program of public forestry in Wisconsin must be built." [31] Even the echo of the Forestry Decision was gone. Where public forests would best promote north-country rehabilitation, public forests should be advanced by the state's full-hearted effort.

PUBLIC INVESTMENT IN SOCIAL OVERHEAD CAPITAL

Marshall's opinion for the court in the Forestry Case warned that the constitutional problems of a state forest reserve program ran deeper than the ban set by the internal improvements clause. Were that difficulty removed, the creation of a large, permanent public forest reserve would still fail, because it could not be deemed a reasonable means to achieve a public purpose. Thus appropriations for such an effort would violate the basic principle that the legislature might tax and spend only for public benefit.[32]

Marshall conceded that discussion of this public purpose issue was unnecessary to dispose of the case before the court. But the issue was of such importance that "it does not seem best to pass it without at least pointing out its dangers and general limitations, even if a definite principle be not announced which will include the particular matter." [33] Thus at one stroke Marshall blandly claimed the court's right to expound vague limits upon legislative authority and disclaimed its responsibility to justify the specific implications of what it said. The court's practical objective was apparently to discourage proponents of state forest reserves from pursuing their objective merely by further legislation based on an amendment of the internal improvements clause. If creation of a state forest reserve could not be deemed within the range of reasonable legislative determination as a proper means to a public purpose, such a program might be adopted only by the cumbersome process of a specific validating constitutional amendment. Marshall made the point with characteristic grace. "I am not entirely satisfied to say, and certainly the court is not, that the power [to create state forest reserves] reasonably exercised and free from the internal improvement feature, if that be possible, does not exist at all. It is sufficient for now to point out the gravity of the subject. I am permitted to do that. It may lead to the precise question never having to be decidedly answered with detail limitations, without a change in the fundamental law. An aroused sense of legislative responsibility to vindicate it ['the fundamental law,' presumably], doubtless, will result in reasonable doubts being resolved in that field in its favor [i.e., apparently, in favor of maintaining the assumed limits of the constitution upon legislative authority]." [34] Marshall's opinion hinted, moreover, that a state forest reserve program might confront a barrier which even a specific amendment of the state constitution could not surmount. The federal Constitution (perhaps by the substantive force of the due process clause) might bar taxing and spending for an objective which could not be deemed a public purpose. "As put by the federal supreme court in *Loan Asso. v. Topeka* [20 Wallace 655. 1875] . . . an object which is not within the purposes for which governments are established, falls outside the limitation upon the right to take the property of the citizen, by way of taxation, for public use." [35]

Marshall was too shrewd a lawyer to take a position which left no room for argumentative maneuver. He would not claim that state action to promote forestry could never fulfill a public purpose. "Where reasonably necessary the conservation of forests and conservative reproduction of them has been time out of mind, one of the well recognized ways of promoting the public welfare," and — "where there is need for it" — "manifestly within the field of police power." [36]

But the court must appraise the authority of the Wisconsin legislature within the frame of the Wisconsin constitution of 1848. "Placing ourselves where the framers of the constitution were placed and the people who ratified it," and recognizing that the matter "must be looked at, not from the high-

Before and after planned reforestation: County Trunk M and Northern Highland State Forest near Vandercook Lake turn-off road, Vilas County, Wisconsin, 1911.

The same area in 1957.

power and justification for it, is that, it involves an exchange of equivalents . . . It is often difficult to appreciate just where the consideration comes in, but without it, in fact, the taking of one's property, though under the form of taxation, and, so, in form, by due process of law, would be confiscation, — an appropriation of private property for a nonpublic use, even though, in the particular case, the public might be interested in the enterprise promoted and be largely, incidentally, benefited." By this measure, a state forest reserve program exceeded legislative power. "[H]ow to square [this principle] . . . with the imposition of large public burdens upon the people of the present without any hope of return to them, in addition to the benefits they enjoy as their equivalent for ordinary taxation — burdens imposed with the avowed purpose of accumulating benefits for generations yet unborn, is somewhat puzzling . . . There must be some present benefit. It is not sufficient that the forced contribution will be a boon to some future generation. The state has no right to take the property of individuals presently and afford them no possible return, merely because the storehouse, being filled, will be opened some time, depending upon Providence and the majority as to when, for the enrichment or comfort of the people then in being in which the taxpayer has no special interest which reasonably demands any such sacrifice." [42]

As if to underline the court's zeal to forestall any further legislative maneuvers in this forbidden field of economic planning, Marshall's opinion went on to rule also invalid the provision under the state forest reserve program of payments from the general fund to the affected local governments, in lieu of the taxes which they would no longer receive from the reserved lands. He disposed of this matter in a staccato drumfire of doctrine. Such state aids violated a constitutional principle that state funds might be spent only for state purposes; they violated the constitutional requirement that taxes be uniform; they violated the constitutional requirement that there be a uniform system of local government; they violated the constitutional ban on local legislation concerning the assessment and collection of taxes. Mere recital sufficed on these counts, for "the various infractions of the fundamental law are so clear that it seems unnecessary to cite authority in respect thereto." [43]

This was an amazing sequence of constitutional interpretation, the more so since none of it was necessary to dispose of the case before the court. It was this sequence apparently which brought Mr. Chief Justice Winslow to the reluctant conclusion that he must file a concurring opinion, because the opinion of the court "so limits and circumscribes the powers of the state with regard to afforestation and reforestation that it leaves little more than a shell behind." Winslow was not certain of the full range of meaning in Marshall's turgid analysis. But "at least this is the way the opinion impresses me and the way I think it will be generally understood." [44] Obviously the Chief Justice was not happy that his colleagues should by one sweep of dicta undertake to determine the future permissible range of legislative fiscal policy.

Marshall's analysis of state fiscal power cannot be justified; it can only be explained. It has relevance to our story chiefly because the extreme positions there adopted bear witness to the conflicts of values and attitudes out of which a new style of lawmaking was born.

To identify the fiscal authority of the Wisconsin legislature with the particular policy outlook of 1848 violated the cardinal rule of constitutional interpretation. John Marshall had cautioned, "[W]e must never forget, that it is a constitution we are expounding." [45] In the absence of clear limitations, the prime function of a constitution is to legitimate authority to make rational adjustments to the unfolding future. As an essay in history, Roujet Marshall's analysis correctly stated a mid-nineteenth-century attitude made plain in legislative decisions. Acutely conscious of the scarcity of fluid capital, the nineteenth-century legislature begrudged committing mobile resources in ways which did not promise quick, multiplier returns.[46] But the point was too particular to the times to be relevant to constitutional interpretation, especially since it found no expression in specific constitutional limitations. To support his dogma Marshall could cite no specific limitation declared or plainly apparent in the Wisconsin constitution. Save for the internal improvements clause — which by its terms was limited to the one aspect of fiscal policy there stated — every provision which Marshall invoked to support his drastic limitation upon the scope of allowable public purpose was a provision governing not the objectives but the procedures of the legislature's fiscal decisions.[47] That he derived a substantive limitation by sheer implication from this web of irrelevancy ran counter to "a well settled political principle" — long acknowledged by the Wisconsin court — "that the constitution of the state is to be regarded not as a grant of power, but rather as a limitation upon the powers of the legislature, and that it is competent for the legislature to exercise all legislative power not forbidden by the constitution or delegated to the general government, or prohibited by the constitution of the United States." [48] Marshall's prime argument baffled the Chief Justice. "I confess my inability to understand the reasoning which finds [such a limitation on fiscal power] . . . in that clause of the constitution which commands the legislature to levy an annual tax to defray the estimated expenses of the state." The error here was especially grievous because it touched so central an element of government. "The power of taxation is one of the necessary attributes of sovereignty. To say that because the constitution makers thought best to make a specific provision that taxes should be levied for certain purposes they intended thereby to interdict taxation for all other public purposes is to my mind unthinkable." [49]

In its broader reach Marshall's opinion imposed an unreal and unworkable limitation upon public fiscal policy. His analysis was unworkable so far as it would invalidate commitment of public resources to investment which would yield only postponed or indirect returns. No society in our pattern could endure without public contribution to the social overhead capital required for operating a division-of-labor economy.[50] The analysis was unreal, for it ran counter to the continued practice of United States and Wisconsin public policy since the early nineteenth century. Well aware that they could raise little capital by internal taxes in a cash-scarce society, Congress and the state legislature had nonetheless made bold and generous use of the real public assets they commanded — the public lands — to subsidize bulk transport and create a broad base of agricultural and forest production.[51] So far as Marshall's opinion showed concern, as it repeatedly did, over the "vast" scale of the proposed state forestry investment, the court so clearly intruded upon the proper sphere of legislative judgment that Marshall avoided specifying the relevance of his comments.[52] Experience taught, moreover, that — for all its failings — legal process served no more distinctive function than to lend leverage to the more farsighted elements in the community, readier than the crowd to postpone present satisfactions for the gains of future growth and stability. So far as it erected impatience into constitutional principle, Marshall's opinion was subversive, not conservative, of social order. With all that it implied for rebuilding productive capacity in the cut over region, reforestation offered a classic example of such investment as provided the necessary foundations of "commonwealth." [53] So Mr. Chief Justice Winslow felt that the court's opinion, as applied to public forestry, simply gave rein to judicial value preferences in an area within the proper scope of legislative choice. To Winslow, "the acquisition, preservation, and scientific care of forests and forest areas by the state, as well as the sale of timber therefrom for gain in accordance with the well understood canons of forest culture, is pre-eminently a public purpose. It would be mere affectation of learning to dwell upon the value to a state of great forest areas." [54] Given an objective clearly within the state's competence, and absent any plainly relevant constitutional barrier, it likewise fell within proper legislative discretion to supply state aids, to make smoother the adjustment of state policy to the financial needs of local government.[55]

Had there been substance to the argument of Marshall's opinion, the 1924 amendment to the state constitution, authorizing appropriations for public forestry purposes, would not have set all question to rest. For, as the Marshall opinion itself intimated, the issue of public purpose there raised ran deep enough to present a problem not only under the state constitution, but also a problem of due process of law under the 14th Amendment. We can measure the main directions of policy, and the transition character of the arguments raised in the Forestry Case, by the fact that as the legislature developed its new program for the north country in the years after 1925 no issue was made under the due process clause.[56]

In 1929 the interim Committee on Forestry and Public Lands defined the state's responsibility to press affirmative public forestry programs with a confidence which showed that the public purpose argument had been put to rest. Indeed, the committee was critical of past doubt and indecision. "The importance of conserving the forest resources of the state was realized as early as 1867, and from time to time spasmodic efforts were made to cope with the problem of the rapidly disappearing forests." Nothing short of a sustained, positive approach would do. "Spasmodic efforts spell waste, and above all else, Wisconsin should define its policy and then the development should go forward year by year without interruption." [57] Whether the state should promote reforestation mainly through public forests or by providing incentives and aids to private action was not now treated as a matter of constitutional power but of choice of the more efficient means. "We are convinced that the state may well engage, on a reasonable scale, in developing state-owned forests, but we are likewise convinced that the best results can be obtained through private ownership, with state aid and encouragement." [58] The committee recognized that the relative roles of public and private effort would likely vary according to the length of time it took to get a given type of timber crop; private enterprisers could grow pulpwood fast enough so that this was a promising field in which to rely on their effort, but "the situation with respect to lumber companies is not so favorable, since it takes longer to produce saw logs than to grow pulpwood, especially when hardwoods are grown." [59] The committee favored county forests as the prime type of direct public operation in this field. Again, it analyzed the choice simply as a judgment of what course promised efficient results. [60] Underlying its confidence, however, was the committee's clear conviction that public forestry would serve public interest. "Unemployed acres, like unemployed men, become public charges and much of this [tax-delinquent] land will find its best use in public forests. No other form of land use is so well suited to public ownership and management as are forests. This is due largely to the fact that the public derives indirect benefits from forests in addition to the value of the timber produced." [61]

PLAN AND SKILL IN PUBLIC POLICY

Doctrinal analysis does not reveal the whole dynamics of the 1915 Forestry Case. Marshall's arguments of law are shot through with characterizations of the state's forestry program which reflect deep emotional reaction against the style of legal action that the program represented. These feelings grew naturally out of the confrontation between men bred in the buoyant opportunism of nineteenth-century action and an emerging twentieth-century insistence on closer, more professional rationalization of economic and social processes. Lawmaking works upon and through men's feelings as well as their ideas. We omit an important dimension of the problems brought to law, if we do not take account of the impulse or restraint which emotion lends to judgment or will.

Marshall, and others who had, like him, played parts in the rush and growth of late-nineteenth-century operations in market, did not fear or distrust large organized activity, where the command rested in private hands. In such maneuver they were at home. Through such operations they had achieved their own sense of status and accomplishment. Marshall properly praised the imagination and skill with which Mr. Weyerhaeuser brought into harmonious treaty the fruitlessly warring lumber interests of the Chippewa Valley; Marshall took pride in his own role in assisting into being that private venture in regional economic planning.[62]

But regional planning by government was large, organized activity of quite different import from operations in market. It spelled power in different hands, in men who worked by political rather than by entrepreneurial skill. Such men might threaten the status or self-esteem of those who had won influence or measured their accomplishment by market processes. Vain and able, Marshall was also insecure; he sought constant reassurance of his own worth by incessant effort to enlarge his role in affairs. As practitioner, he identified his chance of influence with manipulations of property, contract, and franchise. As appellate judge, he sought influence by magisterial exposition of common law or constitutional principle, ranging far beyond the job at hand; significantly he spent himself little upon statutory interpretation, reserving his major efforts for areas of judge-made law, and for constitutional doctrine limiting legislative power. To men bred and constituted like Marshall, there was threat — to their familiar world and to their sense of place — in the emergence of strong legislative and administrative leadership in affairs. The issues of the Forestry Case carried broader implications of idea and feeling because they arose within the wider frame of controversy that swirled about Robert M. LaFollette in Wisconsin and Theodore Roosevelt on the national scene. Out of involvement of self, Marshall colored his opinion for the court with emotion which had its counterpart many times over in the polemic with which inheritors of the late-nineteenth-century power structure encountered Wilson's New Freedom and Franklin Roosevelt's New Deal.

Marshall's opinion for the court reacted with scarce-concealed distaste against the central fact that the state forest reserve program represented a new range and breadth of positive action by the legislative and administrative arms of government. The temper of the times, Marshall warned, called for particular heed to constitutional limitations. "[M]ore study of the fundamental law and a better appreciation of what it means and of fidelity thereto . . . is as necessary now, if not more so, to the general welfare . . . as it ever was. Those who decry it may secure, efficiently, for a time, the public ear, at least have efficient influence with the people's representatives;

but the principle so assaulted, notwithstanding the 'black menaces' which sometimes seemingly surround it, is as secure as 'a star in the maw of the clouds.' "[63] The court felt warranted in going beyond points necessary to dispose of the case before it, because of the "dangers" implicit in the trend of affairs which generated the lawsuits; perhaps if the court now spoke its warnings, "an aroused sense of legislative responsibility to vindicate [constitutional limitations] . . . doubtless, will result in reasonable doubts being resolved in that field in . . . favor" of restraint of legislative innovation.[64] Nonetheless, the court's opinion breathed foreboding. "It is strenuously urged that there is a tendency to drift away from the old position, keenly appreciated when the people met to originate a form of government and that we should desist if not retreat."[65] The state forest reserve program might claim some basis in the police power. Yet, this authority was of such pretensions that it must be closely watched. "In its possibilities, whether the form of government be based on theory of individual rights or of mere privileges — lurks the menace of socialistic, autocratic, bureaucratic, and paternalistic despotism. Though in its nature paternalistic and beneficent and, when exercised within the field of reason, capable of inestimably promoting the general good, the suggested dangers of excessive exercise were so well appreciated when our system was created that the people set up for themselves definite standards of disability and a judiciary on the watch tower clothed with the power and the duty to save the good by the constitutional test and condemn the bad, always resolving all reasonable doubts in favor of legislative wisdom."[66]

As Marshall's last-quoted words show, he had not lost awareness of familiar canons enjoining judicial self-restraint in the face of legislative innovation. So, too, he felt impelled to pay tribute to those who worked for creative adjustment to changed circumstances. Concluding that the legislature had violated the constitutional trusts for education in appropriating school lands and their proceeds to the purposes of the forest reserve, the court's opinion yet cautioned: "Nothing in the foregoing should be taken as purposed to cast discredit on the investigator, experimenter, and promoter in the economic field. Progress would move slow without such. A monument could not be erected so high as to exaggerate the debt of mankind to them. To examine, to analyze, to think deliberately, practically, and conservatively, and to restrain reasonably and fundamentally, utilizes the good and rejects the useless and harmful. It is at this point that extremes meet and then settle to an equilibrium."[67]

However, these disclaimers ran counter to the decisive rulings in the Forestry Case. The court in effect indulged a presumption against, not for, the validity of public will compared with private will under the internal improvements clause. By implication alone, it drew from the constitutional limits set on fiscal procedure what amounted to a weighty presumption

by the statute under consideration to confer upon the state forester the extraordinary authority to sign the corporate name of the state to contracts of the nature of those under consideration . . . One would expect to see authority of the large nature suggested conferred, if at all, even upon constitutional officers of the state, clothed in plain unmistakable language. *A fortiori,* when conferred upon a legislative creation the selection of which was practically surrendered to an outside agency and a nonofficial class, not answerable to the people and not under bonds for fidelity to the trust." [76]

To pass from the emotion-laden innuendoes of the court's opinion in the Forestry Case of 1915 to the 1929 report of the Interim Committee on Forestry and Public Lands is to move from an atmosphere of bafflement, fear, and distrust into one of matter-of-fact and assured operations. The men of 1929 were, of course, still struggling for substantive answers. But they knew how they wanted to proceed to attack their problems. Confident that their procedures were legitimate, they felt no need to justify their lawmaking method.

Bold, decisive, and sustained legislative action must provide the framework; past efforts had failed for want of sufficient determination to implement policy. A "most excellent" report by a special legislative committee in 1915 had, for example, come to nothing for want of continued attention — "again indicating that spasmodic efforts on the part of the state will never develop a constructive program of forestry." [77] The promising legislative measures were not seen, however, as attempts by fiat to resolve problems which must be worked at over time and with sustained and skilled devotion. Rather, the legislature should provide for state administrative promotion and support of private and local government efforts to plan more efficient land use in the north. "The encouragement of private interests; a reasonable taxation policy; aggressive leadership on the part of state authorities, and educational work, to point out the importance and the necessity of forest development, will enable Wisconsin to solve one of its most pressing economic problems." [78] The committee accepted as self-evident wisdom the goal of planning which would seek to treat the economy of the north country as a whole. "Unless properly directed, the forest industry may develop along lines which are wholly undesirable; for this reason the Committee is firm in its belief that proper consideration should be given to forestry in all its aspects." The Report showed no hesitation in recognizing that its central prescription called for integrating knowledge and developing attention over a broad front of concerns. "This, of necessity, calls for consideration of our water supply; our recreation facilities, fish and game, industrial development — all of which are more or less dependent upon the forest." [79] Involved in this approach also was readiness to recognize the need of expertness and professional handling of difficult matters. Thus the state must be prepared to sponsor research in north-country land-use problems. "Scientific study of forestry problems will supply the information needed, not only by private interests whom the state wishes to encourage in the growing of timber, but also for the guidance of state activities in forestry. The investment which the state is making through the Forest Crop Law [deferring the bulk of real property taxes to foster timber growth] should be safeguarded by scientific knowledge."[80] So, too, there should be bold use of state administrative apparatus. Where central supervision was essential, as in fire fighting, the state's administrators should have "entire responsibility." "Closer supervision and inspection, systematized control, trained personnel, a highly specialized force of district rangers and field wardens are absolutely essential to provide a fire control system which will approximate a perfect foundation for the structure now under way."[81] With university help, county administrators must be pressed to make adequate land-use surveys as a basis for positive action to bring land use into more efficient relation to social overhead costs and regional potential. The emphasis was on welcoming positive public effort. "The Committee is impressed with a plan which begins with action by the county board, since the information secured is more apt to be applied in planning and guiding the development of the county."[82] Throughout, the state Conservation Commission and the university should lend technical aid to counties in the selection, care, and improvement of county forests.[83] The legislature should arm the counties to deal with their land-use problems as a whole, moreover, by conferring on county boards authority to zone the uses of rural land. Again, the Committee accepted with confidence the legitimacy of public planning. "Both the orderly development of northern Wisconsin and the need for reducing expenditures because of tax delinquency, require that counties be given the authority to control development."[84]

The forecast of a new style of public policy toward the north country had existed even amid the forebodings and restrictive analysis of the court's opinion in the Forestry Case. Even there, Marshall had felt uneasily compelled to make abstract concessions to the proper claims of legislative discretion, of rationally planned effort, and of the competence residing in professional knowledge or specialized administrative experience.[85] The 1929 Report showed — more by the approach it took for granted than by any felt need to argue the legitimacy of approach — that under the continued pressure of events men's ideas and feelings had attained more clarity as to objectives and methods. We need now regard this history from the standpoint of the full pattern of action which expressed the new style of policy making.

CHAPTER XI

REGIONAL PLANNING

IN SUCCESSIVE phases of enthusiasm, dispute, bewilderment, and calculation, between 1897 and 1940 Wisconsin policy makers charted three different courses for the future of the north country. From 1898 to 1915 emphasis was laid on creating a large state-owned forest reserve. Brought down by mounting political opposition and an adverse decision of the Wisconsin Supreme Court, the forest-reserve program was succeeded from 1915 through the early 1920's by an optimistic effort to encourage farm settlement. By 1925 this endeavor was clearly a failure. Both state and local leadership groped to restate objectives and methods, while the cutover region showed alarming tax delinquencies and dwindling output. On the eve of the great depression, in 1927 and 1929, the legislature took new measures. The new effort put private and public reforestation at the center of a program which, however, now looked toward a multi-factored calculus of the region's future. Rather than thwarting this fresh energy, the depression spurred it on. The shocks of the 1930's brought into stark relief the need for a positive public policy to build sources of vitality for the area. Out of fear and challenge the times cultivated attitudes receptive to new ventures. By 1940 it appeared that men had finally achieved a relatively stable pattern of objectives and procedures within which law might help northern Wisconsin toward a more productive future.

THE CONTENT OF POLICY

The cutover acres finally bulked large enough to force the future of the north country on men's attention about the turn of the century. The first appraisal translated into legal action was to combine forest reserves with more adequate fire protection. So might a substantial, continuous-yield forest endure as the base of the regional economy. This was the recommendation of the Interim Commission on Forestry of 1897. Reserve lands would protect some mature and maturing timber for proper harvesting, but their prime function would be to allow proper management of new growth. This undertaking was, the Commission felt, far too time-costly to be of interest to private enterprise. Thus a reserves program, like an improved fire protection program, called for bold public action. The thinking of the Interim Commission reflected the new interest in natural resource conservation manifest on the national scene. The response in Wisconsin legislative action over the next fifteen years gained impetus from the crusading enthusiasm aroused by Theodore Roosevelt and Gifford Pinchot. Creation of public forest reserves met significant early opposition from those who feared that forest reserves would prevent farm settlement and from those who objected to increased public expenditures and the creation of new administrative offices. The opposition balked any legislation in 1899, and prevented enactment of a reserves law in 1901. But the 1901 legislature did withdraw all remaining state lands from sale, pending reappraisal. And in 1903, with the support of Governor LaFollette, the legislature established a State Forest Commission, authorized the Commission to appoint a superintendent of forests to create a more effective organization to handle the fire hazard and to manage state timberlands, and withdrew from sale indefinitely state lands found suitable primarily for forests. The 1903 statute was ambiguous in defining the extent of the reserved lands. Amendments in 1905, 1907, 1909, and 1911 more clearly defined the reservations and empowered a reconstituted Board of Forestry to buy lands to add to the state reserves out of proceeds of timber sales and sales of tracts within the reserves found more suitable for other uses, as well as from increasingly generous appropriations from the general fund. The reserves program gained momentum under the direction of Edward M. Griffith, one of Pinchot's men in the Forestry Division of the United States Department of Agriculture, who was named Wisconsin's superintendent of forests under the 1904 law, and continued as state forester under the revised program launched in 1905. Griffith brought to his job not only the professional attitude, but also moral fervor and driving energy. By 1912 state reservations and purchases, plus a 20,000-acre grant from the United States, had created a state forest reserve of over 400,0000 acres, with further acquisitions planned to bring the total to as much as 1,500,000 acres.[1]

After opposition was overcome at the first proposal of a state forest reserve program in 1899 and 1901, the new policy grew without substantial controversy through its intermediate stages to 1910. The Wisconsin effort was riding the crest of the Progressive wave in the state and the Roosevelt liberalism in the nation; proponents of the local program strove with the vigor of full moral confidence; some press comment took pride that Wisconsin was assuming leadership in the national conservation movement. However, the absence of controversy derived in part from the fact that the reserves program was still of relatively modest scale. When conservationists put to the voters in 1910 a constitutional amendment, to resolve doubts about the legislature's authority to pursue the forest reserve policy, it was against the background of proposals that the state contract $2 million of debt to allow more rapid purchase of lands for the reserve. The proposed amendment secured a favorable vote, but against a substantial opposition:

62,406 to 45,847. It became clear in the 1911 legislative session that the greatly expanded reach of the program was now alarming those who took more conservative views of state spending and enlargement of state administrative effort. Griffith proposed that the legislature use the authority apparently conferred by the recent constitutional amendment to levy a special tax which would yield as much as $600,000 a year for twenty years to buy lands for the reserve. Modified in the Senate to reduce the program first to $250,000 and then to $200,000 a year for ten years, the proposal was cut drastically in the Assembly to yield $50,000 for five years, and was adopted in that form.[2]

From late 1911 more focused opposition developed rapidly in the north country. Agriculture had regularly followed the cutting of timber in southern Wisconsin. Throughout the nineteenth century, and now into the early twentieth, there persisted a strong body of opinion that this was the natural sequence of economic growth.[3] Measured by this faith, a large forest reserve program could only obstruct the proper prosperous course of the area. Local lumber firms which wanted to sell their cutover acres to farmers, other investors in real property, local bankers, businessmen, and newspaper editors joined in an increasingly impassioned attack upon Griffith's ambitions and the expansion of the state's reserves. The rapid spread and consolidation of this dynamic opposition made clear now that the forest reserve program suffered a structural weakness, not in technical devising but in power base. The program had grown chiefly out of efforts at the state capitol and out of the devotion of a relatively few state leaders; confronted with aggressive and not too scrupulous opposition based on sharp-felt and particularized interests, the program lacked broad-based support, for there had been no substantial effort to educate or involve regional interests in the purposes or execution of the state forest reserve plans.[4] Profiting by the rise of conservative sentiment which had so drastically reduced the scope of the program in the 1911 legislature, the hard-driving northern opposition now scored a major victory. The 1913 legislature decreed that there be no more purchases for the reserve until July 1, 1915, pending investigation by an interim committee there created to determine "what areas of land now held as forest reserve and those parts proposed to be included within such reserve within the counties of Forest, Iron, Oneida, Price and Vilas are better adapted to agricultural purposes than to forestry purposes, and whether the best interests of all persons concerned, and especially the taxpayers of the state, will not be better promoted by devoting said lands to other than reforestation purposes."[5] However, to the discomfort of the opposition, the interim committee made a report which approved the general public worth of state promotion of forestry, and concluded that the program so far developed was sound. The committee found that those administering the forest reserve program had been taking due care to exclude from the reserve lands more suitable for agriculture, and doubted that much of the cutover would ever prove more valuable for farms than for new timber growth.[6] The northerners, it appeared, must regroup for further battle, if they would halt the expansion of state forest reserves. But at this point the Wisconsin Supreme Court ruled that the state reserve program had been enacted without constitutional authority, and that moneys had been spent on it in breach of constitutional trusts for education. The decision destroyed the momentum of the existing program. For about a decade the field now belonged to those who would use law to promote farm settlement as the future of northern Wisconsin.

Legislative efforts to promote farm settlement in northern Wisconsin had overlapped the interest in forest reserves. A new board of immigration was set up in 1895, continued to 1901, revived in 1907, reorganized in 1909 to emphasize still more the effort to attract farm settlers, and merged into the operations of a new state Department of Agriculture created in 1915.[7] In 1895 the legislature directed the dean of the university College of Agriculture "to prepare a bulletin or handbook describing the agricultural resources of Wisconsin, especially the newer and more thinly settled districts, with reference to giving practical, helpful information to the homeseeker. This book . . . shall set forth the advantages of the newer portions of this state for those seeking homes on lands in the effort to draw to Wisconsin a desirable class of farmers."[8] The College produced a handbook which spoke proper cautions on selecting good land and warned of hard work, but on the whole expressed confidence that a prosperous agriculture could be built in the north. With legislative support the College of Agriculture invested much more specific effort in the next twenty-five years to assist northern Wisconsin farmers. It established branch stations, set up demonstration farms, helped counties organize county agent programs, made soils studies, tested seeds and fertilizers adapted to cutover lands, and experimented with stump-pullers and uses of explosives to make land clearing easier.[9] When the attack mounted against Griffith's expansionist program, Dean Russell of the College of Agriculture (an ex-officio member of the state Board of Forestry) lent aid to the opposition. In 1913 and 1914 Russell publicly attested his belief that the extent of Griffith's forest reserve plans encroached upon the agricultural development of the region. The dean pressed the farm promotion activities of his College in the cutover area through the early 1920's with characteristic vigor of conviction, but in 1928 in effect he conceded an excess of zeal: "Now we know there is time to grow one or more crops of pulp or lumber before these undeveloped acres that are suitable for cropping will be needed for farm use."[10]

Incident to creation of the state forest reserves out of state lands, the legislature in 1903 had sharply restricted sales on credit, and in 1905 had decreed flatly that all sales of state lands should be for cash only, to be paid upon sale. However, in 1911 it authorized sale of "all state lands valuable for agriculture" for 15 per cent down, with twenty years to pay the balance at 7 per cent, subject to full cash payment for the value of any standing timber. In 1917 the legislature continued these more liberal terms, now declared applicable to all state lands offered for sale, and this remained the law.[11] In 1911 the legislature authorized county boards to issue special improvement bonds from the proceeds of which they might pay costs of clearing lands on petition of private landowners, "enabling settlers to reclaim cutover lands." This provision was expanded in 1913 and at the same time was paralleled by a statute which authorized the state commissioners of public lands to make loans from the educational trust funds "on the security of agricultural lands," and "only for the purpose of assisting the borrower to erect necessary dwelling houses and farm buildings, to build silos and to clear his lands of stumps, trees, brush and fallen timber." In 1917 the legislature created alternative county loan funds. The county improvement bond procedure stayed on the books until 1955, though the state trust funds loan authorization was repealed in 1917 and the alternative county loan funds were cut off in 1943.[12] The state provided other direct subsidies to cutover area development by bearing part of the cost of roads for opening up unsettled sections (1919), and providing several-year tax exemptions and separate assessment categories for cleared lands (1921, 1923).[13] The legislature favored private development action by allowing incorporation of colonization ventures and land mortgage associations (1913, 1919),[14] approving favored railroad rates to land promoters (1909–1923),[15] and creating liens in favor of those who cleared land (1921).[16] Almost all these ventures in policy originated before 1925; many stood repealed or of no evident practical effect by the late 1920's.

Overhead costs — the costs that must be met, regardless of current income, if an operation is to exist — exert stern, if postponed, discipline over activity which spends limited resources. In public as well as in private enterprises, the pressure of overhead costs will at some point enforce review of basic objectives and procedures, on pain of collapse. In the middle 1920's the tax base of local government in northern Wisconsin was steadily contracting to an alarming extent, under the impact of general and local agricultural depression and rising tax delinquency. Hopes for sizable farm settlement were stubborn and died hard. But by 1940, after more than a generation of promise and effort, only 6 per cent of the total acreage in seventeen northern counties was in cultivated farm crops. Before 1927 more than four and a quarter million acres of land in these counties had been tax delinquent one or more times; two and a half million acres were offered at tax sales in 1927 alone.[17] Where market difficulties alone did not suffice to enforce sufficiently broad reconsideration of policy, the threatened collapse of social structure did. Between 1925 and 1929 the legislature sponsored two investigations, and enacted three statutes, which provided a promising new framework of policy. The new approach took more realistic account of the distinctive assets of the north country. It sought the strength to be had from applying these assets not in isolation but as parts of a regional economy.

The 1925 legislature created an interim committee on taxation and administration, which in natural course found that the facts directed its recommendations in large measure to the problems of northern Wisconsin. When the committee reported in 1927, it made a persuasive case for a special tax system designed to promote efficient reforestation. It coupled this emphasis with recommendations for extending public forests and developing more effective fire protection such as was necessary to safeguard public or private investment in the long-term enterprise of growing timber.[18] The 1927 legislature created an interim committee on Wisconsin's forestry and public lands problems. The committee served as a focal point for gathering and reviewing local opinion and bringing together the planning of state leaders, including the governor. When this second committee reported in 1929, a key recommendation — advanced within a context of deliberation which assured its favorable reception — was for a novel departure in state law, to establish rural land-use control in the former forest area.[19] The interim committees named in 1925 and 1927 did not of themselves invent elements of public policy, but they served admirably the catalytic function of legislative investigation. Their efforts brought together creative individual research and planning, and worked to develop such a consensus as permitted the relatively rapid enactment of a new statutory framework for regional rehabilitation.

The new pattern of policy had three elements of a common character:

(1) First attention went to tax policy. Initiated in 1925, and confirmed by a second legislative vote and by popular referendum in 1927, a constitutional amendment authorized the legislature to make separate provision for taxing forested lands.[20] Following recommendations of the 1925 interim committee, the legislature promptly used its new power to enact a forest crop tax law. This first act of 1927 provided that landowners (private or public) might contract with the state to use land for growing timber, rather than for other purposes, and to manage it properly as forest land for as long as fifty years; in return the state agreed that the landowner need pay only 10 cents an acre tax annually until he cut the trees, plus a severance tax of 10 per cent of the stumpage value of timber cut, payable upon harvesting; the state also undertook to pay the affected towns 10 cents per acre annually on private lands entered, until the trees were cut, so that on private lands the town would collect 20 cents per acre annually during the growth period.[21] In 1929 the legislature authorized counties to put their own timber holdings under the forest crop law free of the 10 cents per year acreage tax due from private entrants; in 1931 it further encouraged entry of county-owned tim-

berlands by obligating the state to pay to the towns on such lands the same 10 cents per acre state aid that was paid on private entries, plus an added 10 cents per acre subsidy toward the costs of county forest management.[22] To mid-century, county entries by far predominated; private entries fluctuated from 1928 to 1951 up and down within the range of about 150,000 to 300,000 acres; over that time all county entries increased steadily, if at varying rates; in 1951 when all private entries stood at 226,450 acres, the entries made by twenty-three north and central Wisconsin counties totalled 2,113,368 acres.[23]

(2) The second step in building a new policy pattern for the north country was to authorize the creation of public forests. The Forestry Decision of 1915 and the distractions of the war years had turned interest away from a state forest reserve program. Over the span of 1921–1924, leadership emerged for a constitutional amendment to authorize state appropriations for "acquiring, preserving and developing forests of the state." The amendment passed two legislatures without substantial votes in opposition and was approved at the polls, 336,360 to 173,563.[24] However, no pressure emerged to use the new power to revive a state reserve program on the pre-1915 Griffith model.[25] North-country opinion was now coming alive to the need of fresh, bold planning to overcome the decline of the regional economy; the College of Agriculture and county officials alike sponsored numerous local meetings to take stock and explore programs, and the 1927 interim committee held hearings throughout the cutover area which further stimulated local discussion. The home base of support which had been lacking for the pre-1915 state reserves program was now available for fresh effort. It was natural that the new ventures emphasized local government action, within a framework of values and procedures defined by the state legislature. In 1927 the legislature authorized counties to buy land or acquire it by tax deed to create county forests, to engage in forest management, and to enter into cooperative agreements with the state Conservation Commission particularly for fire protection and for research and the development of growing stock.[26] As we have seen, in 1929 and 1931 the legislature gave major encouragement to the creation of county forests by financial aids to facilitate entry of county-owned lands under the forest crop law.[27] By 1955 twenty-seven counties owned and had registered under the forest crop law 2,175,000 acres in county forests, most acquired on tax deeds.[28]

(3) The attention given tax reform and public forestry turned men's thinking to the total productive potential of the north country, and hence to appraisal of the range and distribution of uses of land. Governor Kohler spoke with the support of leading opinion in northern Wisconsin when he advised the 1929 legislature to empower counties to zone land to direct its use into the most efficient patterns. "County zoning serves as a deterrent against ill-advised location of farms. It permits the relocation of farms poorly situated, and the consolidation of school districts, and avoids the

construction of unnecessary highways. It will assist materially in the solution of tax delinquency, which must have our attention for it is a problem of the entire state and not only of the area immediately affected."[29] Again, we can see the pressure of social overhead costs stirring legal process free of inertia and outmoded values into fresh creation.[30] The 1929 legislature thus authorized any county board by ordinance to "determine the areas within which agriculture, forestry and recreation may be conducted, [and] the location of roads [and] schools," and to exchange lands acquired by tax deeds for other lands in the county "for the purpose of promoting the regulation and restriction of agricultural lands." Except regarding lands owned by the county, a county zoning ordinance might become operative in any town of the county, however, only upon approval of the town board.[31] By mid-century twenty-seven cutover counties enacted zoning ordinances. Uneven in effect and in enforcement, the rural zoning programs nonetheless appear to have had material impact, particularly in separating agricultural from forest-recreational uses of lands, cutting fire hazards, and reducing the social overhead costs of sparse or isolated settlement.[32]

A reformed tax system for forest lands, and county board authority to establish county forests and rural zoning programs, thus provided the foundation on which local initiative might create new economic vitality for northern Wisconsin. The state legislative process was basic to forming and adopting these new policies. Research sponsored by the governor and the legislature, and legislative committee hearings, brought both state and local opinion to a focus for action. Key legislators and the governors supplied indispensable, sustained impetus to press action through to decisions. With its control of the process of constitutional amendment and of the public purse and the structure of local government, only the legislature could effect the kinds of legal change required. Moreover, to realize the potentials of the new areas for local initiative, the continuing assistance of state appropriations and state administrative assistance was essential, as in the state aids to local government under the forest crop law, and in the helps which the state Conservation Commission and the university, for example, gave as counties deliberated, planned, and sought to operate county forests and zoning programs. Nonetheless, for all the importance of the state's roles, these three elements (tax reform, county forests, rural zoning) were realized through local education, local opinion, and local decisions. This close-to-home basis of policy contrasted with the centralized direction which marked the state forest reserve program from 1897 to 1911.[33]

There was a fourth element of policy which stood apart in character from these three. It stood apart because its realization required, not only the programming leadership, but also the operating predominance of state authority. This fourth element was the achievement of effective protection against forest fires. It was an objective critical to the success of the other items of a program to re-create the economic vitality of the north country.[34]

Before 1895 Wisconsin law dealt with forest fire hazards only by conventional criminal penalties for fires set willfully or maliciously on another's land, or negligently allowed to spread from one's own.[35] In 1895 the legislature made the first preventive (posting and licensing) provisions against forest fires, and set up the first fire fighting organization, by committing responsibility to town fire wardens.[36] With many variations, but with unfailing weakness, from 1895 to 1923 the state continued to put prime reliance on this extremely decentralized, ex-officio, unprofessional organization. At one time or another the legislature conferred modest appointment and supervisory authority over local fire wardens upon the state land commissioners, the Board of Forestry (under the state forest reserve program), and the Conservation Commission (which succeeded to the Board's responsibilities after the Forestry Decision of 1915). From 1911 on the legislature also put some positive authority, first in the State Forester, later in the Conservation Commission, to mobilize and concentrate local fire wardens in emergency, with authorization that the state relieve the counties of one-half the costs of such emergency action.[37]

State control for forest fire prevention and fire fighting became established as the fourth element of a regional rehabilitation policy by acts of 1923, 1927, and 1931. There was testimony to the urgency of the problem in the fact that action on the fire problem anticipated creation of the other items which entered into the regional program.[38] In 1923 the legislature provided for creation of fire protection districts by rule of the Conservation Commission, each district of not more than one million acres, each under charge of an experienced forest ranger appointed, paid, and supervised by the state through the Conservation Commission. The state undertook to provide the capital equipment of an adequate fire detection and fire fighting operation, including lookout towers, telephone lines, fire lanes, and tools.[39] In 1924 the voters ratified the constitutional amendment which permitted state appropriations for forestry.[40] In 1927, substantially concurrent with its authorizations of forest crop tax reform and county forests, the legislature revised the new state fire protection district system. It now firmly supplanted local control with full authority in the Conservation Commission "in all matters relating to the prevention, detection and suppression of forest fires" outside the limits of incorporated cities and villages. The Commission was authorized to enter cooperative arrangements with local government units, but it held the authority to control the terms of issue of burning permits and the organization and operation of preventive and fire fighting forces. Fire fighting expenses were still shared between the state and the counties, but the legislature put more spending authority in the Commission.[41] In 1931 it put the state effort on a new plane of expenditure with an appropriation of $600,000, most of which was allocated for fire protection.[42] By mid-century an efficient state organization, centered on ten intensive protection districts embracing over 16 million acres in north and central Wisconsin, had brought the fire hazard within bounds which assured the economic rationality of long-term public and private forestry.[43] This assurance, in turn, provided the organizing base for the total economic pattern of the region.

THE MAKING OF POLICY

Policies to re-create the north Wisconsin economy grew out of a new orientation of attitude and a new style of action by legal agencies. Details belong to a different history from that of the nineteenth-century lumber industry. However, it is relevant to note aspects of decision making in this twentieth-century context which by contrast more sharply define the character of nineteenth-century process.

The key factor after the mid-1920's was less the entry of any specific new attitude or value than a new balance of attitudes and values, expressing more closely calculated uses of law in economic growth. Nineteenth-century policy had involved strong, affirmative uses of law to promote forest production, as by public lands subsidies and grants of transport franchises; a crude laissez faire was never the guiding doctrine.[44] But the nineteenth-century legislature had given public support to private action uncritically and with scant limitations. Twentieth-century tax reform, and provision for county forests under the encouragement of forest crop law payments, were accompanied by conditions on the state's bounty, requiring that the public or private lands which benefited from these laws be managed according to the canons of good forestry technique.[45] State policy continued to favor private decision making to develop north-country resources. But now the legislature also authorized rural land-use zoning, and through the facilities of state administrative apparatus (especially the university and the Conservation Commission) aided and encouraged counties to use their new powers.[46] With uncritical faith in the productive virtues of the settlement that — it was assumed — would follow the forest, and confident that wisdom lay in speeding exploitation of timber by a liberal land policy and by promoting improvement of streams and construction of railroads, the nineteenth century had pursued an ill-defined objective of regional development for the north. But it did so with exaggerated focus upon one element — maximum forest production, achieved as rapidly as possible. Twentieth-century policy looked toward building the productive strength of the region out of a necessarily slower, more balanced, multi-factor approach. This approach made reforestation the core element. But it took account also of recreation, agriculture, and mining, and sought a more efficient relation between regional development and social overhead costs.[47]

Thus the shift from nineteenth- to twentieth-century public policy was not from laissez faire to state intervention, but from a relatively simplistic to a relatively complex calculation of the relations between public and private decision making. A good deal of policy was still rough-hewn in making and execution. There was pull and haul between more general and more local interests, between the professionalism of state administrators and the narrow pragmatism of legislators, and among all who, so humanly, defined their horizons by their immediate concerns — sportsmen, resort keepers, small farmers and small cutters of pulpwood, county seat bankers and realtors, papermill operators, real estate taxpayers.[48] No age of reason suddenly dawned, either over the state capitol or over northern Wisconsin. Yet, there was so different a style in using law as to produce significantly different effects, and significant difference in effectiveness.

The components of this distinctive style were (1) a value calculus which embraced more factors than usually figured in nineteenth-century policy making, and (2) a more balanced division of labor among legal agencies responsible for making and effecting policy, marked by firmer use of legislation and by new continuity of working relations between state and local government administration.

(1) This policy pattern widened the range of values weighed to determine the law's proper interplay with private activity. The broader calculus emphasized elements of social income and social costs which did not readily find expression through the money reckoning of the market.[49] Thus the state underwrote the long-term risk costs of private or public forestry investment by creating for the first time an effective system of fire protection. It met the costs of rationalizing general conditions of forest production by scientific research and by professional techniques of forestry and pest and game control. It sought to reduce social overhead costs by promoting withdrawal of population from areas relatively inefficient for other uses than forest crops or recreation. It fostered the recreational utilities of timber land and waterways, encouraging reforestation, preserving the wilderness character of streams, checking pollution of waters by human and industrial wastes. By deferred private tax payments on forest land, offset by state advances to local governments, and by state subsidies paid on account of county forests, the state in effect drew on the general tax base of the state to finance the carrying costs of long-term investment in reforestation. Such varied measures had a common character. They used law to install social accounting alongside the private books of account which the nineteenth century had thought sufficient for calculating the input and output of the economy.[50]

Emphasis upon fiscal technique marked this social accounting approach as distinctively of the twentieth century. (a) Raising taxes was never easy. But the productivity of twentieth-century Wisconsin allowed its government such a command of money as the nineteenth-century Wisconsin legislature had never known. The incomparably greater money resources of twentieth-century government gave its operations a flexibility akin to that which once only the market had enjoyed.[51] Thus state money subsidies to local government figured as critical elements in tax reform to promote reforestation. State appropriations were the life blood of programs of fire protection and pest and game control. The nineteenth-century legislature was liberal in subsidizing economic growth, so long as it could pay subsidies in kind — out of public lands. This practical limitation upon its generosity inherently limited the purposes it could promote by subsidy. The legislature could thus manipulate events on a large scale only by subsidies of fixed capital (land, or land-derived resources) to private enterprisers, who might thereby extend their scarce fluid capital (money and labor) to enlarge their operations in market. This was the principal avenue of maneuver, because at that time only private enterprisers could mobilize substantial fluid capital. The greater availability of tax-derived moneys in the twentieth century meant that public subsidies could be applied to promote activity which market incentives did not at all support (notably, large-scale long-term investments in saw-timber growth), and to underwrite operating as well as carrying costs of long-term commitments of capital (notably, in fire protection to support long-term timber-growth investments).[52] (b) Apart from the limited, if important, objectives which it could achieve by subsidies in kind out of public lands, when nineteenth-century law undertook to channel men's wills it usually found no instrument other than a regulatory rule backed by conventional penal sanctions.[53] With its greater fiscal flexibility, twentieth-century policy relied more upon guiding private action to desired ends by indirection. Such indirect pressure might be extended by tendering benefits or privileges, or by organizing a total situation to make one course of action more profitable or easy than another. Reforestation (which implied continuous-yield management) was the key element in rehabilitating northern Wisconsin. To this end the legislature did enact new regulations against fire, and provided tighter regulatory enforcement. But the legislature did not pursue the key objective of good forest management by enacting a regulatory code to govern timber growing and harvesting. Instead, it held out inducements of deferred taxes and state subsidies for county forests, and exerted pressure for sound forestry practice by the stipulations for good timber-crop management which it attached to these aids.[54] The legislature did authorize counties to zone land, to restrict to timber crops lands not suitable for farming. It authorized enforcement of such zoning regulations by conventional sanctions of fine, imprisonment, or injunction. However, conventional enforcement was subject to all the pressures against action which nearby constituents can bring on local government; especially in areas fringing resort towns a good deal of violation was winked at. Again, it was indirect pressure created by using state fiscal resources which accomplished the principal results achieved in barring uneconomic agriculture from lands best suited to growing trees; county ownership, encouraged by state subsidies, was the main factor in

no force large enough to revive so centralized a program, though the constitutional amendment of 1924 endowed the legislature with the necessary authority.[68] The regional rehabilitation program set up after 1925 derived largely from a sharp change in north-country attitudes, amid disillusionment with farm settlement and concern over mounting tax delinquency. True, the new program would not have been enacted, and would not have taken the shape it did, without knowledge and leadership at the capitol. The significant new style of approach, however, was the close interplay of action between the center and the cutover country. Legislative hearings conducted throughout the affected area gave local opinion full expression, built local understanding, and mustered local support for the new frame of policy.[69] Because effective fire protection called for uniform and firm regulation, and required the means and authority to concentrate men and equipment, the legislature asserted full state control.[70] On the other hand, to bring private or public lands under the forest crop law, to foster the development of county forests, and to procure use of rural zoning powers required primarily persuasion, adaptation to local circumstance, and care for local government budgets, rather than sheer executive drive. So, having legitimated new land-use arrangements, the legislature committed their realization to private and local government decisions.[71] Yet, the delegation was not complete. By the stipulations with which it conditioned tax deferments and state subsidies, and by advice, persuasion, and technical knowledge supplied through university and administrative staff, the state made itself an active partner in building from the statutory blueprints.[72] This creative interaction of center and field depended upon wisely calculated delegations of power, upon continuous attention to unfolding experience both at the center and in the field, and upon opportunity for professional knowledge and practical accommodation to play upon one another. So far as the experiment promised success, it pointed especially to the fruitful contributions which the administrative process could make to the business of legislation. The experiment risked failure most from the tendency of short-run local interest to defeat the more general objectives set by central policy; in proportion as the new arrangements developed new wealth in the north country (as in more profitable recreation business, or in maturing county-owned timber stands), so parochial interests were stimulated to seek illicit privileges or quick returns which would imperil long-term growth or cheat the state's general taxpayers out of proper returns from their investment. So much had been conceded to the localities that, if basic change proved necessary, the most likely need would be for stronger assertion of state interests through the legislature. Between 1925 and 1940 the statutes had created a new framework of policy for the development of northern Wisconsin. But it inhered in growth, that it should outgrow its earlier impetus. Not only the threats of parochial interest, but also the challenges of success, would bring fresh problems of the north country to the legislative process.

RETROSPECT

IT ACCORDS with the character of law in United States society that the legal history of an industry involves experience at many levels and with different ranges of cause and effect. We can regard the legal history of the Wisconsin lumber industry both in terms of a given flow of action in a given time and place, and in terms of what the story suggests regarding more general issues of social structure and process affecting and affected by law.

This was a society keenly conscious of its opportunities to manipulate its environment. Impatient and activist, the nineteenth-century Wisconsin community was ready and quick to use law as an instrument for practical ends. Within this context of attitudes and ideas, men often defined their purposes and their opportunities more narrowly than they needed to. But in any case the insistently practical bent of their interest in law brought legal action intimately into the conduct of particular economic operations. It fits reality to inquire into the functional relations of law to the course of the Wisconsin lumber industry; law and the industry were much involved with each other.

A good deal of the approach to law in the nineteenth-century United States was narrowly practical. But it was also true that the people made broader demands upon law and made broader use of legal process than can be encompassed by the history of any specific operations. Our tradition included the constitutional ideal — that all public and private power must be exercised responsibly and, in the end, in ways which served individual life. This standard meant that the uses of power must be consistent both with some measure of rationality and with purposes legitimated by disciplined emotion. Corollary to the constitutional ideal, we emphasized procedural values in law. Law offered ordered procedures for exploring facts, mustering evidence, defining values and the range of value choices, and setting decent bounds and terms for controversy. Thus the legal order held promise of enlarging the application of reason and responsible feeling in adjusting men's relations to material circumstances and to each other. In the background of concern for the constitutional ideal and for procedural regularity was, also, the sobering realization that the law held the ultimate legitimate force of the community; to use law was serious business.

In addition to these considerations — which tend to be cautionary or restrictive in emphasis — were more positive aspects of our working concepts of legal order. Tradition and experience fostered a high regard among us for individual life and will; in particular we had profound faith in the beneficent, creative possibilities that resided in broad dispersion of initiative of decision concerning the use of material resources and social opportunities. However, other potent factors taught us that individuals obtained leverage upon life by mobilizing their resources as members of a society. Our constitutional tradition looked to combined action through an elected representative assembly as the prime means of expressing common interests. Entry upon a rich, undeveloped continent taught successive generations confidence that they could materially affect the content of their lives by cooperative manipulation of their environment. The surge of scientific and technical advance gave new energy to our readiness to organize effort and muster resources on increasing scales of size and complexity. Such elements of experience found legal expression in the uses we made of law to affect the organization and allocation of economic resources. Significant aspects of property, contract, and tort law fostered the release of private energies of mind and will. By creating franchises we encouraged more effective group action through the corporation, and conferred authority and privileges which enabled entrepreneurs to multiply the power and transport uses of streams and to build a railroad network. Hard pressed for mobile capital, we disposed of the real wealth of the public domain in ways which enhanced the immediate productivity of scarce cash and manpower, and channeled economic action in directions which promised quick and substantial multiplier effects on total output.

It is thus inherent in the demands and expectations which this society addressed to law that general issues of social structure and process are woven into the history of particular uses of law. Conversely, apart from its own distinctive character the legal history of Wisconsin's lumber industry exemplifies themes of general significance for the roles of law in United States history. That the particular and the general should so intertwine creates special challenge in studying the social history of law. Because law was deeply involved in practical objectives and operations, generalization about the law's roles must stand firmly on fact. For this legal order it was indeed true that "the life of the law has not been logic: it has been experience." [1] But the practical uses of law here went on in the context of values attached to constitutionalism, procedural regularity, and the enlarging of men's command of circumstances. Hence significant generalization about law in the growth of the United States must also relate law to the whole pattern of social organization and processes. The character of such a legal system invited men to undertake "the large survey of causes"; to the perceptive observer the history of law in this society was "no less . . . than that of the moral life of his race"; to grasp the meaning of law in its social context was to accept a challenge "to see so far as one may, and to feel, the great forces that are behind every detail — for that makes all the difference between philosophy and gossip, between great action and small." [2] Such an approach sets goals perhaps impossible to reach. But the sweep of the subject — the life of the law in the life of the society — admits no less a measure of its challenge. We can learn even from failures in the search, so long as we put to ourselves significant questions.

SOURCES

NOTES

INDEX

PRINCIPAL SOURCES CITED

OFFICIAL DOCUMENTS: UNITED STATES

(Research monographs published under the auspices of government agencies are listed hereafter under Books and Monographs)

Statutes at Large of the United States, 1789–1873, 17 vols. (Boston, 1845–1873); vol. 18 *et seq.* (Washington, 1875–).

Documents Illustrative of the Formation of the Union of the American States, 69th Cong., 1st Sess., House Doc. no. 238 (Washington, D.C., 1927).

American Charters, Constitutions, and Organic Laws, ed. Francis Newton Thorpe, 59th Cong., 2d Sess., House Doc. no. 357, 7 vols. (Washington, D.C., 1909).

United States Reports.

Federal Cases (to 1880).

Federal Reporter (since 1880).

Agriculture, Department of, Weekly News Letter, July 27, 1921 (Washington, D.C., 1921).

Commerce, Department of, Bureau of the Census, *Census of Manufactures, 1900* (Washington, D.C., 1902), vol. II.

——— Bureau of Corporations, *The Lumber Industry,* 2 vols. (Washington, D.C., 1913, 1914).

Interior, Department of, General Land Office, *Reports of the Commissioner of the General Land Office, 1836–1932* (Washington, D.C.).

National Academy of Sciences, *Report of the Commission upon a Forest Policy for the Forested Lands of the United States* (Washington, D.C., 1898).

OFFICIAL DOCUMENTS: TERRITORY OF WISCONSIN

Session Laws of the Territory of Wisconsin, 1836–1848: 16 vols., bearing a variety of title-page designations, apparently according to the taste of different printers.

Statutes of the Territory of Wisconsin, 1839 (Packard, Van Benthuysen & Co., Albany, N.Y., 1839).

Journals of the Territorial Legislature: of the Council, and of the House of Representatives, 1836–1848.

Reports of Cases Argued and Determined in the Supreme Court of the Territory of Wisconsin, 1839–1848, by S. U. Pinney, 2 vols. (Callaghan & Co., Chicago, 1872, 1874), cited as 1 and 2 Pinney.

OFFICIAL DOCUMENTS: STATE OF WISCONSIN

Constitution of the State of Wisconsin, in *The Revised Statutes of the State of Wisconsin, 1849* (C. Latham Sholes, Southport, 1849).

Session Laws of the State of Wisconsin: published as a single volume per session, 1848–1852, in separate volumes of General and of Private & Local Laws, 1853–1872, and thereafter as a single volume per session; before the middle 1860's individual volume titles vary erratically in designating their contents as Acts or as Laws. The legislature met annually through 1883, after which (save for occasional special sessions) meetings were biennial.

Revised Statutes of the State of Wisconsin were published in 1849, 1858, 1872, 1878, and 1898, with supplements in 1852, 1868, 1883, and 1889. The revision of 1898 bore the title of Wisconsin Statutes, and the same title was used on the continuously updated compilation of the general laws of the state provided by the Revisor of Statutes, whose office was created in 1909.

Wisconsin Statutes: see last preceding reference.

Journals of the Senate and of the Assembly of the Wisconsin legislature were published annually until 1883 and biennially thereafter.

Public Documents, volumes containing messages of the governors, reports of other major executive officers, and legislative committee reports, were published beginning in 1853; appendices to legislative journals contain like material in prior years.

Reports of Cases Argued and Determined in the Supreme Court of the State of Wisconsin, 1849–1853, by S. U. Pinney, 2 vols. (Callaghan & Co., Chicago, 1874, 1876, cited as 2 and 3 Pinney.

Wisconsin Reports, vol. 1 *et seq.*, 1853– , being reports of cases determined in the Supreme Court of Wisconsin.

Attorney General of Wisconsin, *Opinions* (Madison), through the nineteenth century published episodically in the text or appendices of the legislative journals, first collected into a separate volume as an appendix to the Attorney General's report of 1900–1902, recognized as a numbered serial beginning in 1912–1913.

Bureau of Labor and Industrial Statistics, *Reports,* 1883–1912 (Madison).

Commissioners appointed under General Laws, 1867, Chapter 36 [Increase A. Lapham, J. G. Knapp, and H. Crocker], *Report on the Disastrous Effects of the Destruction of Forest Trees Now Going on so Rapidly in the State of Wisconsin* (Madison, 1867).

Commissioners of School and University Lands (entitled Commissioners of Public Lands after 1878), *Reports,* 1850–1912 (Madison).

The Natural Resources Committee of State Agencies, *The Natural Resources of Wisconsin* (Madison, 1956).

The State-Wide Highway Planning Survey, *A History of Wisconsin Highway Development, 1835–1945* (Madison, 1947).

Wisconsin Industrial Commission, Bulletins, vol. 2, no. 9, *The Wisconsin Free Employment Offices,* and no. 12, *Accidents Caused by Objects Striking Workmen* (Madison, 1913).

Wisconsin Industrial Commission, *Labor Camps in Wisconsin* (Madison, 1913?).

Wisconsin Legislative Reference Library, *The Wisconsin Blue Book, 1925* (Madison, 1925).

Wisconsin Regional Planning Committee, *A Study of Wisconsin* (Madison, 1934).

Wisconsin Special Legislative Committee on Water Powers, Forestry and Drainage, *Reports,* 1910 (Madison, 1911).

Wisconsin Special Legislative Committee on Forestry, of the Senate and Assembly, *Report,* Pursuant to Chapter 670 of the Laws of 1913 (Madison, 1915).

Wisconsin Special Legislative Committee on Land Use and Forestry, *Forest Land Use in Wisconsin* (Madison, 1932).

Wisconsin State Forester, *Reports,* 1906–1912 (published biennially, Madison).

Wisconsin State Planning Board, *Local Government Study in Wisconsin, 1927–1936,* 4 vols. (Madison, 1941–1943).

MANUSCRIPTS

(in the collections of the State Historical Society of Wisconsin)

Holt Lumber Company, of Oconto, Wisconsin, and Chicago, Illinois: Record Books, 1839–1943.

Ingram Papers: of Orrin H. Ingram of Eau Claire, 1857–1904.

Knapp, John Holley: Diary, 1848–1881.

Knapp Papers, 1889–1932.

Lincoln Lumber Company, of Merrill, Wisconsin: Articles of Incorporation and Minutes of Meetings of the Board of Directors, 1881–1884.

Washburn & Woodman Letter Book, 1844–1855.

Wisconsin Archives: Attorney General of Wisconsin, Letter Books, 16 vols. (1859–1862, and 1896–1907).

——— Executive Records: Lumber Inspector, 1861–1905.

——— Executive Records: Timber Agents' Reports, 1858–1890.

DISSERTATIONS

Canuteson, Richard Lewis, "The Railway Development of Northern Wisconsin" (M.A., University of Wisconsin, 1930).

Dykstra, Daniel J., "Law and the Lumber Industry, 1861–1881" (S.J.D., University of Wisconsin Law School, 1950).

Goodman, Archie Bristol, "Assessment Administration in Wisconsin" (Ph.D., University of Wisconsin, 1942).

Greiner, Gordon Oswald, "Wisconsin National Railroad Land Grants" (M.A., University of Wisconsin, 1935).

Haygood, Tyler Francis, "State Control of Local Finance" (Ph.D., University of Wisconsin, 1935).

Helgeson, Arlan Clayton, "The Promotion of Agricultural Settlement in Northern Wisconsin, 1880–1925" (Ph.D., University of Wisconsin, 1951).

Kleven, Bernhard, "The Wisconsin Lumber Industry" (Ph.D., University of Minnesota, 1941).

Lake, James A., Sr., "Buried Wealth and Unwanted Water" (S.J.D., University of Wisconsin Law School, 1952).

Nybroten, Alfred Norman, "Tax Assessment and Equalization in Specific Reference to the Wisconsin System" (Ph.D., University of Wisconsin, 1941).

Schmidt, Gertrude P., "History of Labor Legislation in Wisconsin" (Ph.D.. University of Wisconsin, 1933).

Van Brocklin, Ralph M., "The Movement for the Conservation of Natural Resources in the United States Before 1901" (Ph.D., University of Michigan, 1952).

Young, William Henry, "The Present State of Research in Wisconsin Local Government" (Ph.D., University of Wisconsin, 1941).

UNPUBLISHED TEXTS

Kanneberg, Adolph, "Log Driving and the Rafting of Lumber in Wisconsin" (Wisconsin Public Service Commission, mimeographed, Madison, 1944).

Mermin, Samuel, "Law and the Promotion of Enterprise" (manuscript on deposit in University of Wisconsin Law Library, Madison, 1959).

BOOKS AND MONOGRAPHS

Adams, Brooks, *The Degradation of the Democratic Dogma* (The Macmillan Company, New York, 1920).

Andersen, Theodore A., *A Century of Banking in Wisconsin* (The State Historical Society of Wisconsin, Madison, 1954).

Anderson, William H., *Taxation and the American Economy* (Prentice-Hall, Inc., New York, 1951).

Angell, Joseph K., and Samuel Ames, *The Law of Private Corporations* (7th ed., Little, Brown, & Company, Boston, 1861).

Auerbach, Carl A., Lloyd K. Garrison, Willard Hurst, and Samuel Mermin, *The Legal Process: An Introduction to Decision-making by Judicial, Legislative, Executive, and Administrative Agencies* (Chandler Publishing Co., San Francisco, 1961).

Bailey, W. F., *Digest of Decisions of the Supreme Court of Wisconsin,* 3 vols. (Callaghan & Company, Chicago, 1902).

Barnett, James D., *The State Administration of Taxation in Wisconsin,* Transactions, Wisconsin Academy of Sciences, Arts, and Letters, vol. XV, part I (Madison, 1905).

Beuscher, J. H., *Law and the Farmer* (Springer Publishing Co., Inc., New York, 1956).

Bishop, Joel Prentiss, *Commentaries on the Criminal Law,* 2 vols. (6th ed., Little, Brown, & Company, Boston, 1877).

Black, Henry Campbell, *A Treatise on the Law of Tax Titles* (2nd ed., West Publishing Co., St. Paul, 1893).

Blaisdell, J. J., *Forest and Tree Culture in Wisconsin* (Wisconsin Horticultural Society, Madison, 1893).

Bogue, Allan G., *Money at Interest* (Cornell University Press, Ithaca, 1955).

Bowman, Francis F., Jr., *Industrial Wisconsin* (Wisconsin Department of Commerce, Madison, 1939).

Brown, Ray A., *The Law of Personal Property* (2nd ed., Callaghan & Co., Chicago, 1955).

Buttrick, P. L., *Forest Economics and Finance* (John Wiley & Sons, Inc., New York, 1943).
Cameron, Jenks, *The Development of Governmental Forest Control in the United States* (The Johns Hopkins Press, Baltimore, 1928).
Carstensen, Vernon, *Farms or Forests: Evolution of a State Land Policy for Northern Wisconsin, 1850–1932* (University of Wisconsin College of Agriculture, Madison, 1958).
Chamberlain, Lawrence H., *The President, Congress, and Legislation* (Columbia University Press, New York, 1946).
Ciriacy-Wantrup, Siegfried von, *Resource Conservation* (University of California Press, Berkeley and Los Angeles, 1952).
Clark, James I., *Cutover Problems: Colonization, Depression, Reforestation* The State Historical Society of Wisconsin, Madison, 1956).
——— *Farming the Cutover: The Settlement of Northern Wisconsin* (The State Historical Society of Wisconsin, Madison, 1956).
Clark, John Maurice, *Social Control of Business* (University of Chicago Press, Chicago, 1926).
——— *Economic Institutions and Human Welfare* (Alfred A. Knopf, New York, 1957).
Clawson, Marion, *Uncle Sam's Acres* (Dodd, Mead & Co., New York, 1951).
Clawson, Marion, and Burnell Held, *The Federal Lands: Their Use and Management* (The Johns Hopkins Press, Baltimore, 1957).
Cochran, Thomas C., *Railroad Leaders, 1845–1890: The Business Mind in Action* (Harvard University Press, Cambridge, 1953).
——— *The American Business System* (Harvard University Press, Cambridge, 1957).
Cochran, Thomas C., and William Miller, *The Age of Enterprise* (The Macmillan Company, New York, 1943).
Commager, Henry Steele, *The American Mind* (Yale University Press, New Haven, 1950).
Commons, John R., *Myself* (The Macmillan Company, New York, 1934).
Conover, Milton, *The General Land Office* (The Johns Hopkins Press, Baltimore, 1923).
Crevecoeur, Hector St. John, *Letters from an American Farmer,* ed. W. Barton Blake (J. M. Dent & Sons, Ltd., London, 1945).
Cummings, Homer, and Carl McFarland, *Federal Justice* (The Macmillan Company, New York, 1937).
Cunningham, R. N., and Forest Survey Staff, *Forest Resources of the Lake States Region,* U.S. Department of Agriculture, Forest Resources Report no. 1 (Washington, D.C., 1950).
Current, Richard Nelson, *Pine Logs and Politics: A Life of Philetus Sawyer, 1861–1900* (The State Historical Society of Wisconsin, Madison, 1950).
Curti, Merle, *The Growth of American Thought* (Harper & Brothers, New York, 1943).
Dahl, Robert A., and Charles E. Lindblom, *Politics, Economics, and Welfare* (Harper & Brothers, New York, 1953).

Dorfman, Joseph, *The Economic Mind in American Civilization,* 5 vols. (The Viking Press, New York, 1946, 1949, 1959).

Duckett, Kenneth W., *Frontiersman of Fortune: Moses M. Strong of Mineral Point* (The State Historical Society of Wisconsin, Madison, 1955).

Dunham, Harold Hathaway, *Government Handout: A Study in the Administration of the Public Lands, 1875–1891* (Ph.D. thesis, published by the author, Columbia University, New York, 1941).

Dupree, A. Hunter, *Science in the Federal Government* (The Belknap Press of the Harvard University Press, Cambridge, 1957).

Fries, Robert F., *Empire in Pine: The Story of Lumbering in Wisconsin, 1830–1900* (The State Historical Society of Wisconsin, Madison, 1951).

Galloway, George B., *The Legislative Process in Congress* (Thomas Y. Crowell Co., New York, 1953).

Gara, Larry, *Westernized Yankee: The Story of Cyrus Woodman* (The State Historical Society of Wisconsin, Madison, 1956).

Gates, Paul Wallace, *The Wisconsin Pine Lands of Cornell University* (Cornell University Press, Ithaca, 1943).

——— *Fifty Million Acres: Conflicts Over Kansas Land Policy, 1854–1890* (Cornell University Press, Ithaca, 1954).

Glover, W. H., *Farm and College* (University of Wisconsin Press, Madison, 1952).

Greeley, William B., *Forests and Men* (Doubleday, New York, 1951).

Gross, Bertram M., *The Legislative Struggle* (McGraw-Hill Book Co., Inc., New York, 1953).

Groves, Harold M., *Financing Government* (rev. ed., Henry Holt & Co., New York, 1945).

Gulick, Luther Halsey, *American Forest Policy* (Duell, Sloan & Pearce, New York, 1951).

Hacker, Louis M., *The Triumph of American Capitalism* (Simon & Schuster, New York, 1940).

Haferbecker, Gordon M., *Wisconsin Labor Laws* (University of Wisconsin Press, Madison, 1958).

Hagood, Margaret Jarman, and Emmit F. Sharp, *Rural-Urban Migration in Wisconsin, 1940–1950,* University of Wisconsin Research Bulletin no. 176 (Madison, 1951).

Handlin, Oscar, and Mary Flug Handlin, *Commonwealth: A Study of the Role of Government in the American Economy: Massachusetts, 1774–1861* (New York University Press, New York, 1947).

Harris, Marshall, *Origin of the Land Tenure System in the United States* (Iowa State College Press, Ames, 1957).

Hibbard, Benjamin H., *A History of the Public Land Policies* (The Macmillan Company, New York, 1924; reissue by Peter Smith, New York, 1939).

Holmes, Oliver Wendell, Jr., *The Common Law* (Little, Brown, & Company, Boston, 1881).

Hotchkiss, George W., *History of the Lumber and Forest Industry of the Northwest* (G. W. Hotchkiss & Co., Chicago, 1898).

Hunt, Freeman, *Lives of American Merchants,* 2 vols. (Derby & Jackson, New York, 1856, 1858).
Hunt, Robert S., *Law and Locomotives* (The State Historical Society of Wisconsin, Madison, 1958).
Hurst, James Willard, *The Growth of American Law: The Lawmakers* (Little, Brown, & Company, Boston, 1950).
——— *Law and the Conditions of Freedom in the Nineteenth-Century United States* (University of Wisconsin Press, Madison, 1956).
——— *Law and Social Process in United States History* (The University of Michigan Law School, Ann Arbor, 1960).
——— ed., *A Digest of Regional Sources for the Study of the Economic and Political History of the Law* (University of Wisconsin Law School, Madison, 1941).
Ise, John, *The United States Forest Policy* (Yale University Press, New Haven, 1920).
Jacobs, Clyde Edward, *Law Writers and the Courts* (University of California Press, Berkeley, 1954).
Jensen, Vernon H., *Lumber and Labor* (Farrar & Rinehart, Inc., New York, 1945).
Johnson, E. A. J., and Herman E. Krooss, *The Origins and Development of the American Economy* (Prentice-Hall, Inc., New York, 1953).
Jones, Chester Lloyd, *Statute Law Making in the United States* (F. W. Faxon Co., Boston, 1923).
Kapp, K. William, *The Social Costs of Private Enterprise* (Harvard University Press, Cambridge, 1950).
Kent, James, *Commentaries on American Law,* 4 vols. (Halsted, New York, 1826–1830).
Kimball, Spencer L., *Insurance and Public Policy* (University of Wisconsin Press, Madison, 1960).
Kittredge, Joseph, Jr., *Forest Planting in the Lake States,* U.S. Department of Agriculture, Bulletin no. 1497 (Washington, D.C., 1929).
Knight, Frank H., *On the History and Method of Economics* (University of Chicago Press, Chicago, 1956).
Krooss, Herman E., *American Economic Development* (Prentice-Hall, Inc., Englewood Cliffs, 1955).
Kuehnl, George H., *The Wisconsin Business Corporation* (University of Wisconsin Press, Madison, 1959).
LaFollette, Belle Case, and Fola LaFollette, *Robert M. LaFollette,* 2 vols. (The Macmillan Company, New York, 1953).
Lake, James A., *Law and Mineral Wealth: The Legal Profile of the Wisconsin Mining Industry* (University of Wisconsin Press, Madison, 1962).
Lapham, Increase A.: see Official Documents: State of Wisconsin, Commissioners appointed under General Laws, 1867, Chapter 36.
Larson, Agnes M., *History of the White Pine Industry in Minnesota* (University of Minnesota Press, Minneapolis, 1949).
Laurent, Francis W., *The Business of a Trial Court: 100 Years of Cases: A census of the actions and special proceedings in the Circuit Court for Chip-*

pewa County, Wisconsin, 1855–1954 (University of Wisconsin Press, 1959).

Levy, Leonard W., *The Law of the Commonwealth and Chief Justice Shaw* (Harvard University Press, Cambridge, 1957).

Lewis, Martin D., *Lumberman from Flint: The Michigan Career of Henry H. Crapo, 1855–1869* (Wayne State University Press, Detroit, 1958).

Lillard, Richard G., *The Great Forest* (Alfred A. Knopf, New York, 1947).

Livermore, Shaw, *Early American Land Companies* (The Commonwealth Fund, New York, 1939).

Loehr, Rodney C., ed., *Forests for the Future: Diaries and Papers of David Townsend Mason* (Forest Products History Foundation, Minnesota Historical Society, St. Paul, 1952).

Machlup, Fritz, *The Political Economy of Monopoly* (The Johns Hopkins Press, Baltimore, 1952).

Marquis, Ralph W., *Economics of Private Forestry* (McGraw-Hill Book Co., Inc., New York, 1939).

Marshall, Roujet D., *Autobiography,* 2 vols. (privately printed, Madison, 1923, 1931).

Martin, Lawrence, *The Physical Geography of Wisconsin,* Wisconsin Geological and Natural History Survey, Bulletin no. XXXVI, Educational Series, no. 4 (2d ed., Madison, 1932).

Maxwell, Robert S., *LaFollette and the Rise of the Progressives in Wisconsin* (State Historical Society of Wisconsin, Madison, 1956).

Merk, Frederick, *Economic History of Wisconsin during the Civil War Decade* (State Historical Society of Wisconsin, Madison, 1916).

Merrill, Horace Samuel, *William Freeman Vilas, Doctrinaire Democrat* (State Historical Society of Wisconsin, Madison, 1954).

Morison, Elting E., *Turmoil and Tradition: A Study of the Life and Times of Henry L. Stimson* (Houghton Mifflin Co., Boston, 1960).

Morton, Walter A., *Housing Taxation* (University of Wisconsin Press, Madison, 1955).

Murphy, Earl F., *Water Purity* (University of Wisconsin Press, Madison, 1961).

Myrdal, Gunnar, *An International Economy* (Harper & Brothers, New York, 1956).

Nichols, Philip, *The Law of Eminent Domain,* 2 vols. (2d ed., Matthew Bender & Co., Inc., Albany, 1917).

Norton, Matthew G., *The Mississippi River Logging Company: An Historical Sketch* (published by author, n.p., 1912).

Palmer, Edgar Z., *The Prewar Industrial Pattern of Wisconsin,* Bulletin of the University of Wisconsin: Wisconsin Commerce Studies, vol. I, no. 1 (Madison, 1947).

Paul, Arnold M., *Conservative Crisis and the Rule of Law* (Cornell University Press, Ithaca, 1960).

Peffer, E. Louise, *The Closing of the Public Domain* (Stanford University Press, Stanford, 1951).

Phelan, Raymond Vincent, *The Financial History of Wisconsin,* Bulletin of the University of Wisconsin, no. 193, Economic and Political Science Series, vol. 2, no. 2 (Madison, 1908).

Pinchot, Gifford, *Breaking New Ground* (Harcourt, Brace & Co., New York, 1947).
Polanyi, Karl, *The Great Transformation* (Farrar & Rinehart, Inc., New York, 1944).
Quaife, Milo M., ed., *The Movement for Statehood* (The State Historical Society of Wisconsin, Madison, 1918).
——— ed., *The Convention of 1846* (The State Historical Society of Wisconsin, Madison, 1919).
——— ed., *The Struggle over Ratification* (The State Historical Society of Wisconsin, Madison, 1920).
——— ed., *The Attainment of Statehood* (The State Historical Society of Wisconsin, Madison, 1928).
Randall, Thomas E., *History of the Chippewa Valley* (Free Press Print, Eau Claire, 1875).
Raney, William Francis, *Wisconsin: A Story of Progress* (Prentice-Hall, Inc., New York, 1940).
Rector, William Gerald, *Log Transportation in the Lake States Lumber Industry, 1840–1918* (A. H. Clark Co., Glendale, 1953).
Reimann, Lewis C., *When Pine Was King* (Northwoods Publishers, Ann Arbor, 1952).
Reynolds, A. R., *The Daniel Shaw Lumber Company: A Case Study of the Wisconsin Lumbering Frontier* (New York University Press, New York, 1957).
Robbins, Roy Marvin, *Our Landed Heritage: The Public Domain, 1776–1936* (Princeton University Press, Princeton, 1942).
Rostow, W. W., *The Process of Economic Growth* (W. W. Norton & Co., New York, 1952).
Roth, Filibert, *On the Forestry Conditions of Northern Wisconsin,* Wisconsin Geological and Natural History Survey, Bulletin no. 1, Economic Series no. 1 (Madison, 1898).
——— *Forestry Conditions and Interests of Wisconsin,* U.S. Department of Agriculture, Division of Forestry, Bulletin no. 16 (Washington, D.C., 1898).
Schafer, Joseph, *A History of Agriculture in Wisconsin* (State Historical Society of Wisconsin, Madison, 1922).
——— *Four Wisconsin Counties: Prairie and Forest* (State Historical Society of Wisconsin, Madison, 1927).
——— *The Wisconsin Lead Region* (State Historical Society of Wisconsin, Madison, 1932).
——— *The Winnebago-Horicon Basin* (State Historical Society of Wisconsin, Madison, 1937).
Schlesinger, Arthur M., *New Viewpoints in American History* (The Macmillan Company, New York, 1923).
Schumpeter, Joseph A., *History of Economic Analysis,* ed. Elizabeth Boody Schumpeter (Oxford University Press, New York, 1954).
Sedgwick, Theodore, *The Measure of Damages,* 4 vols. (9th ed., Baker, Voorhis & Co., New York, 1912).
Smith, Henry Nash, *Virgin Land* (Harvard University Press, Cambridge, 1950).

Snider, Guy Edward, *The Taxation of the Gross Receipts of Railways in Wisconsin,* Publications, American Economic Association, 3rd series, vol. VII, no. 4 (The Macmillan Company, New York, 1906).

Solberg, Erling D., *New Laws for New Forests* (University of Wisconsin Press, Madison, 1961).

Soule, George, *Economic Forces in American History* (Sloane, New York, 1952).

Sparhawk, William N., and Warren D. Brush, *The Economic Aspects of Forest Destruction in Northern Michigan,* U.S. Department of Agriculture, Technical Bulletin no. 92 (Washington, D.C., 1929).

Stephenson, Isaac, *Recollections of a Long Life* (privately printed, Chicago, 1915).

Stone, Robert N., and Harry W. Thorne, *Wisconsin's Forest Resources,* Lake States Forest Experiment Station, Station Paper no. 90, August 1961 (St. Paul, 1961).

Story, Joseph, *Commentaries on the Constitution of the United States,* 3 vols. (Hilliard, Gray & Co., Boston, 1833).

Tocqueville, Alexis de, *Democracy in America,* ed. Phillips Bradley, 2 vols. (Alfred A. Knopf, New York, 1948).

Ward, Dudley, ed., *Goals of Economic Life* (Harper & Brothers, New York, 1953).

Wharton, Francis, *Criminal Law,* 3 vols. (Lawyers Co-operative Publishing Co., Rochester, 1932).

White, Leonard, D., *The Jacksonians* (The Macmillan Company, New York, 1954).

——— *The Republican Era, 1869–1901* (The Macmillan Company, New York, 1958).

Williams, Robin M., Jr., *American Society* (Alfred A. Knopf, New York, 1951).

Williston, Samuel, *Life and Law* (Little, Brown, & Company, Boston, 1941).

Wisconsin Commercial Forestry Conference, *Forestry in Wisconsin: A New Outlook* (published by the Conference, Milwaukee, 1928).

Wood, John A., *A History of Lumbering in Maine, 1820–1861* (University of Maine Press, Orono, 1935).

Woytinsky, W. S. and E. S., *World Population and Production* (Twentieth Century Fund, New York, 1953).

Wright, Benjamin F., Jr., *The Contract Clause of the Constitution* (Harvard University Press, Cambridge, 1938).

Zon, Raphael, *Timber Growing and Logging Practice in the Lake States,* U.S. Department of Agriculture, Bulletin no. 1496 (Washington, D.C., 1928).

Zon, Raphael, and R. D. Garver, *Selective Logging in the Northern Hardwoods of the Lake States,* U.S. Department of Agriculture, Technical Bulletin no. 164 (Washington, D.C., 1930).

ARTICLES

Abramovitz, Moses, "Introduction," in National Bureau of Economic Research, *Capital Formation and Economic Growth* (Princeton University Press, Princeton, 1955).

Adler, Edward A., "Business Jurisprudence," *Harvard Law Review,* 28:135

(1914), reprinted in Association of American Law Schools, *Selected Essays on Constitutional Law,* II, 436 (Foundation Press, Chicago, 1938).

Aubrey, Henry G., "Investment Decisions in Underdeveloped Countries," in National Bureau of Economic Research, *Capital Formation and Economic Growth* (Princeton University Press, Princeton, 1955).

Bancroft, George, "The Office of the People," in *Literary and Historical Miscellanies* (Harper & Brothers, New York, 1855).

Brown, Betty R., and Willard Hurst, "The Perils of the Test Case," *Wisconsin Law Review,* 1949:26.

Brown, Ray A., "The Making of the Wisconsin Constitution," *Wisconsin Law Review,* 1949:648, 1952:23.

Chandler, Alfred D., Jr., "Henry Varnum Poor, Philosopher of Management, 1812–1905," in William Miller, ed., *Men in Business* (Harvard University Press, Cambridge, 1952).

Clark, John Maurice, "Aims of Economic Life as Seen by Economists," in Dudley Ward, ed., *Goals of Economic Life* (Harper & Brothers, New York, 1953).

Cochran, Thomas C., "The Entrepreneur in American Capital Formation," in National Bureau of Economic Research, *Capital Formation and Economic Growth* (Princeton University Press, Princeton, 1955).

Conover, Milton, "The Abandonment of the 'Tidewater' Concept of Admiralty Jurisdiction in the United States," *Oregon Law Review,* 38:34 (1958).

Dykstra, Daniel J., "Corporations in the Day of the Special Charter," *Wisconsin Law Review,* 1949:310.

——— "Legislation and Change," *Wisconsin Law Review,* 1950:523.

——— "Legislative Efforts to Prevent Timber Conversion: The History of a Failure," *Wisconsin Law Review,* 1952:461.

——— "Federal Government, State Governments, and Natural Resources," *Minnesota Law Review,* 37:569 (1953).

Espeseth, Edmund C., "Early Vilas County — Cradle of an Industry," *Wisconsin Magazine of History,* 37:27 (1953).

Fegtly, Samuel M., "Historical Development of Land Surveys," *Illinois Law Review,* 38:270 (1944).

Gates, Paul Wallace, "The Homestead Law in an Incongruous Land System," *American Historical Review,* 41:652 (1936).

Goldsmith, Raymond W., "Financial Structure and Economic Growth in Advanced Countries," in National Bureau of Economic Research, *Capital Formation and Economic Growth* (Princeton University Press, Princeton, 1955).

Greeley, William B., "Forest Management on Federal Lands," *Journal of Forestry,* 23:223 (1925).

Harvard Law Review, 48:485 (1935): Note, "Damages for Exploratory Trespass to Lands."

Helgeson, Arlan Clayton, "Nineteenth Century Land Colonization in Northern Wisconsin," *Wisconsin Magazine of History,* 36:115 (1952).

Hibbard, Benjamin H., "Land in Relation to Forestry," in Wisconsin Commercial Forestry Conference, *Forestry in Wisconsin* (published by the Conference, Milwaukee, 1928).

Hostak, Kenneth F., "Wisconsin Ground Water Law — A New Era," *Wisconsin Law Review,* 1957:309.

Innes, Harold A., "Economics of Conservation," *Geographical Review,* 28:137 (1938).

Jewkes, John, "The Economist and Economic Change," in Arthur Smithies and others, *Economics and Public Policy* (Brookings Institution, Washington, D.C., 1955).

Kane, Lucile, "Federal Protection of Public Timber in the Upper Great Lakes States," *Agricultural History,* 23:135 (1949), reprinted in Vernon Carstensen, ed., *The Public Lands* (University of Wisconsin Press, Madison, 1963), 439.

Kneipp, L. F., "Federal Activities in Wisconsin Forestry," Wisconsin Commercial Forestry Conference, *Forestry in Wisconsin* (published by the Conference, Milwaukee, 1928).

Kull, George F., "What Forests Mean to Manufacturers in Wisconsin," in Wisconsin Commercial Forestry Conference, *Forestry in Wisconsin* (published by the Conference, Milwaukee, 1928).

Lake, James A., Sr., "Legal Profile of the Mining Industry," *Wisconsin Law Review,* 1955:399, 566.

Landis, James M., "Constitutional Limits on Congressional Power of Investigation," *Harvard Law Review,* 40:153 (1926).

Levy, Marion J., Jr., "Some Social Obstacles to 'Capital Formation' in 'Underdeveloped Areas,'" in National Bureau of Economic Research, *Capital Formation and Economic Growth* (Princeton University Press, Princeton, 1955).

Lundsted, James E., "Log Marks, Forgotten Lore of the Logging Era," *Wisconsin Magazine of History,* 39:44 (1955).

Mallare, Frank L., "Wisconsin's Internal Improvements Prohibition," *Wisconsin Law Review,* 1961:294.

Marshall, Marjorie Loomis, "History and Analysis of the Wisconsin Securities Law," *Wisconsin Law Review,* 1942:552.

McCullough, M. P., "What Forests Mean to the Lumber Industry," in Wisconsin Commercial Forestry Conference, *Forestry in Wisconsin* (published by the Conference, Milwaukee, 1928).

Mills, Lewis R., "Government Fiscal Aid to Private Enterprise in Wisconsin: A Quantitative Approach," *Wisconsin Law Review,* 1956:110.

——— "The Public Purpose Doctrine in Wisconsin," *Wisconsin Law Review,* 1957:40, 282.

Page, Barbara, "Statutes in Derogation of Common Law," *Wisconsin Law Review,* 1956:78.

Quaife, Milo M., "Increase Allen Lapham, Father of Forest Conservation," *Wisconsin Magazine of History,* 5:106 (1921).

Rae, John B., "Commissioner Sparks and the Railroad Land Grants," *Mississippi Valley Historical Review,* 25:211 (1938).

Rector, William Gerald, "The Birth of the St. Croix Octopus," *Wisconsin Magazine of History,* 40:171 (1957).

Reis, Alvin C., "What Forests Mean to the Public," in Wisconsin Commercial Forestry Conference, *Forestry in Wisconsin* (published by the Conference, Milwaukee, 1928).

Remington, Frank J., "Liability Without Fault Criminal Statutes," *Wisconsin Law Review,* 1956:625.

Robbins, Lionel R., "Freedom and Order," in Arthur Smithies and others, *Economics and Public Policy* (Brookings Institution, Washington, D.C., 1955).

Robinson, Richard R., and Willis J. Zick, "Liability Without Fault Criminal Statutes," *Wisconsin Law Review,* 1956:625.

Russell, H. L., "Wisconsin's Idle Plant," in Wisconsin Commercial Forestry Conference, *Forestry in Wisconsin* (published by the Conference, Milwaukee, 1928).

Scott, Anthony, "Conservation Policy and Capital Theory," *Canadian Journal of Economics and Political Science,* 20:504 (1954).

Smithies, Arthur, "Economic Welfare and Policy," in Smithies and others, *Economics and Public Policy* (Brookings Institution, Washington, D.C., 1955).

Spengler, Joseph J., "Capital Requirements and Population Growth in Undeveloped Countries: Their Interrelations," *Economic Development and Cultural Change,* 4:305 (1956).

Van Alstyne, W. Scott, Jr., "Land Transfer and Recording in Wisconsin," *Wisconsin Law Review,* 1955:44, 223.

Vickery, William, "Goals of Economic Life: An Exchange of Questions between Economics and Philosophy," in Dudley Ward, ed., *Goals of Economic Life* (Harper & Brothers, New York, 1953).

Waite, C. Graham, "Public Rights to Use and Have Access to Navigable Waters," *Wisconsin Law Review,* 1958:335.

Williamson, Harold F., Comment on essay by Thomas C. Cochran, in National Bureau of Economic Research, *Capital Formation and Economic Growth* (Princeton University Press, Princeton, 1955).

Wilson, F. G., "Zoning for Forestry and Recreation: Wisconsin's Pioneer Role," *Wisconsin Magazine of History,* 41:102 (1957).

Young, George H., "Some Comments on the New Wisconsin Business Corporation Law," *Wisconsin Law Review,* 1952:5.

Zon, Raphael, "Forest Conditions in Wisconsin," in Wisconsin Commercial Forestry Conference, *Forestry in Wisconsin* (published by the Conference, Milwaukee, 1928).

NOTES

ABBREVIATIONS

AJ	Assembly Journal	PD	Public Documents
HRJ	House of Representatives Journal	SJ	Senate Journal

In addition, the standard abbreviations for references to Reports, Statutes, etc.

Introduction: Law and the Commonwealth

1. For data on the original condition and rate of exploitation of the northern Wisconsin forest, see Robert F. Fries, *Empire in Pine: The Story of Lumbering in Wisconsin, 1830–1900* (151), 17–18, 103, 206; Frederick Merk, *Economic History of Wisconsin during the Civil War Decade* (1916), chap. ii; Bernhard Kleven, "The Wisconsin Lumber Industry" (unpublished Ph.D. thesis, 1941), 1–2, 10–11, 75–90, 123–126; Filibert Roth, *Forestry Conditions and Interests of Wisconsin* (1898), 22, 25, 29, 47–48; Wisconsin Commercial Forestry Conference, *Forestry in Wisconsin: A New Outlook* (1928), 10, 18, 26, 32. The 1950 estimate of Wisconsin's timber situation will be found in R. N. Cunningham and Forest Survey Staff, *Forest Resources of the Lake States Region* (1950), 6, 7, 14, 15.

PART I: PROPERTY

Introduction: The Constitution of Power

1. The chain of public title to the land area of Wisconsin is set forth in Marion Clawson, *Uncle Sam's Acres* (1951), 19; Benjamin H. Hibbard, *A History of the Public Land Policies* (1924), 7–13; and Roy Marvin Robbins, *Our Landed Heritage: The Public Domain, 1776–1936* (1942), 6; see In re Veeder v. Guppy, 3 Wis. 502, 520 (1854). Clawson, p. 71, notes the extent of federal land grants to Wisconsin in comparison with those to other states (e.g., 16.4 million acres to Minnesota, 12.1 million acres to Michigan, 8.1 million acres to Iowa, compared with 10.2 million acres to Wisconsin). Congress imposed the condition of no interference with the primary disposal of the soil by the United States, in the Act of Aug. 6, 1846, sect. 7(5), authorizing the formation of a state government in Wisconsin Territory (9 Stat. 56). This conformed to the stipulation laid down in the same terms in "Article 4th" of the "articles of compact between the original States and the people and States in the said territory," in the (Northwest) Ordinance of 1787. See *American Charters, Constitutions, and Organic Laws,* ed. Francis Newton Thorpe (1909), II, 961.

2. Fries, 13, 100–106; A. R. Reynolds, *The Daniel Shaw Lumber Company* (1957), 129–139.

3. It is suggestive of the potential importance of the leverage opportunity afforded by possession of the public domain to enable relatively untried governments to adopt a stronger line of forest policy, to note that before the Revolution, out of its stronger tradition of power as sovereign and proprietor, the Crown launched the first forest reserve policy on this continent, symbolized by the "broad arrow" marking trees reserved for naval masts. See Jenks Cameron, *The Development of Governmental Forest Control in the United States* (1928), 20, 22.

Chapter I. General Public Lands Policy

1. Cf. Luther Halsey Gulick, *American Forest Policy* (1951), 19; Cameron, 99; Henry Nash Smith, *Virgin Land* (1950), 124, 126, 128.

2. The overshadowing importance of federal public lands policy was given vivid contemporary expression as Governor Farwell observed in his general message to the legislature in 1853 that it was "a question of momentous importance to us, what disposition ought to be made, by the Congress of the United States, of this two-thirds portion of our State" then yet held by the central government: SJ, 12 (1853). That the years in which Wisconsin enjoyed status as a separate Territory gave its territorial government no scope to shape public lands policy was often cited by its officials. See Reports of Committee on Schools, HRJ, 85 (1837), HRJ, 76, 80 (1841); Report of Committee on Territorial Affairs, Council Journal, 150 (1845); Report of Select Joint Committee on State Government, Council Journal, App., 336 (1846); General Message of Governor Doty, Council Journal, 34 (1841).

3. Sensitivity to the competing pressure of federal land policy is reflected, for example, in Governor Harvey's General Message, SJ, 17 (1862); cf. Report of Senate Committee on Agriculture and Manufactures, SJ, 360 (1863). Typical hindsight lamentations, which at the same time reflect the general past emphasis on speedy disposal in the state's land policy, are the statements in the Annual Reports of the University Regents, 1874 (PD, I, 4), and 1875 (PD, I, 7).

4. See, e.g., John Ise, *The United States Forest Policy* (1920), 40–45, 50, 55, 58; Gifford Pinchot, *Breaking New Ground* (1947), 27, 84–90, 107.

5. Merk, 49, notes that by 1860 the choice public lands in the southern half of the state were practically all transferred to private ownership.

6. Details of federal land policy as it stood by the 1860's may be found in Hibbard, *Public Land Policies,* esp. chaps. vi, ix, xiii, xiv, xvi, and xvii; and in Robbins, *Our Landed Heritage,* esp. pp. 7–12, 39, 217.

7. Typical characterizations of the state lands as a "trust," or a contract, carrying obligations on the state to employ them to produce stipulated works, or a money fund for objects designated by Congress or the state constitution, are contained in the General Messages of Governor Farwell, SJ, 16 (1852), and of Governor Randall, SJ, 48 (1858); compare the implications of the strenuous efforts to justify strict construction of the state's obligations under the swampland act, in order to preserve maximum freedom of action for the state legislature, in Report of Select Committee, SJ, App., 6, 7 (1855). Federal grants for designated objectives were not self-executing, in the sense that breach by the state would result in automatic reversion of title to the United States; Congress must take positive action to reclaim the land — which Congress never did as to any of the bounty it extended Wisconsin. Cf. Schulenberg v. Harriman, 21 Wallace 44, 60, 63–64 (U.S. 1874). Town of La Pointe v. Town of Ashland, 47 Wis. 251, 2 N.W. 306 (1879) declared that the trust imposed by the federal swamplands act was obligatory only on the state and would not fasten upon the title passed by the state to its grantee, and ruled that no such trust attached to proceeds of such lands in the hands of the state's transferee. However, the court upheld the authority of the legislature to conduct investigations necessary to enforce the state's trust obligations owed the United States: In re Falvey, 7 Wis. 630 (1859). And it did not hesitate to set aside legislative or executive action of the state which it found contrary to obligations owed the United States under land grants, where the impact of the court's ruling would be directly on the state rather than upon the state's grantees, or would not divest title. State ex rel. Douglas v. Hastings, 11 Wis. 448 (1861) (state treasurer may not withhold swampland proceeds moneys due county under state disposing statute, to pay unrelated county debt to state); Denniston v. Unknown Owners, 29 Wis. 351 (1872) (realty taxes on Fox–Wisconsin canal aid lands invalid, because levied while land was still held by state for accomplishment of federal purpose); State ex rel. Sturgeon Bay and Lake Michigan Ship Canal and Harbor Co. v. Commissioners of School and University Lands, 34 Wis. 162 (1874) (where state disposing act is ambiguous, it must be given the construction best calculated to secure the purpose of the federal granting statute); Chippewa Valley & Superior Railway Co. v. Chicago, St. Paul, Minneapolis & Omaha Railway Co., 75 Wis. 224, 44 N.W. 17 (1889) (contract not to compete in bids for railroad-aid grant unenforceable, as opposed to fulfillment of federal trust implicit in the grant); State ex rel. Sweet v. Cunningham, 88 Wis. 81, 57 N.W. 1119 (1894) (legislative reservation of school lands for state park invalid, as violating state

constitution's devotion of such lands to education); State ex rel. Owen v. Donald, 160 Wis. 21, 151 N.W. 331 (1915) (legislative reservation of state lands for forest reserve invalid, as violating trusts under federal grants and state constitution, devoting lands to education); cf. Sloan, Stevens & Morris v. State, 51 Wis. 623, 8 N.W. 393 (1881) (state constitution ban on state debts for works of internal improvement bars contracting against general funds for protection of state lands, though state owes trust obligations to attend to their protection). Though governors used their veto power against diversion of "trust" lands only intermittently, yet their action attested that the trust concept was not altogether lacking in the pull and haul of ideas that contributed to the state's public lands program. See Governor Farwell's Veto Message, SJ, 591 (1852) (Fox-Wisconsin grant); Governor Harvey's several Veto Messages regarding swamplands diversion, including those in SJ, 641 (1862), 809 (1862), and AJ, 406 (1862), 481 (1862); cf. Veto Message of Governor Fairchild, AJ, 899 (1871) (normal-school fund); Veto Message of Governor Rusk, SJ, 747 (1887) (lieu lands); Veto Message of Governor Hoard, AJ, 754 (1889) (drainage funds of towns). Reflection of the withholding of action by Congress or the federal executive as a means of bringing pressure on the state may be seen, for example, in dealings for a settlement of the Rock River canal fiasco, as in Governor Farwell's Veto Message, SJ, 809 (1853); Governor Barstow's General Message, SJ, App., 8 (1855), and his Special Message, AJ, 284 (1855); Governor Randall's General Message, SJ, 20 (1859); and Governor Fairchild's General Message, SJ, 21 (1866). Compare the concern regarding renewal of the lapsed portion of the St. Croix area railroad-aid grant, in Governor Washburn's General Message, SJ, 30 (1873); and the summary of various counterclaims of the state and the United States, in Governor Rusk's Special Message, SJ, 55–63 (1882).

8. Even by 1860 state lands legislation had ramified into a large body of technical detail, rendered the more difficult to trace in particulars because of the careless and ambiguous legislative practices of the times. It is sufficient for our present purpose to note merely the central policies, which remained fairly stable throughout shifts in detail. See, thus, Laws, 1849, c. 212 (general sales procedure regarding school and university lands), and Revised Statutes, 1849, c. 24 (incorporating the substance of c. 212); General Acts, 1853, c. 98 (disposition of Fox-Wisconsin improvement aid lands shifted to private company); General Acts, 1855, c. 21 (oath of purchase for actual settlement, and within 320-acre limit); General Acts, 1855, c. 84 (swamplands pre-emption rights); General Acts, 1856, c. 112 (trust of Fox-Wisconsin improvement aid lands); General Acts, 1856, c. 122 (grant of federal aid lands to LaCrosse & Milwaukee Railroad Co.); General Acts, 1856, c. 125 (general sales and pre-emption provisions for swamplands); Revised Statutes, 1858, c. 28 (sale of school and university lands), and 29 (swamplands). For abortive suggestions of a possible homestead-grant policy regarding state lands, see Report, Joint Committee of Investigation regarding the state departments, SJ, App., 14 (1858); proceedings regarding Resolution 45A, AJ, 147, 159 (1860); Memorial 178A, and Resolution 106A, AJ, 570 (1860).

9. The predominance of disposition under the standard combination of public offerings and private sales is reflected in the Annual Report of the State Superintendent of Public Instruction, PD, 670–672 (1863). That public offering became a routine preliminary to the effective practice of private sales is indicated by the Annual Report of the State Treasurer, PD, 9 (1868); the only occasion on which a public sale was set aside on a finding of collusive bidding seems to be that noted in Biennial Report, Commissioners of Public Lands, PD, II, 6 (1884). Early concern with the extent to which state land sales were producing large accumulations of acreage in a few private hands is expressed in Report of the State Superintendent of Public Instruction, SJ, App. F, 12 (1853); in Report of Select Committee under Resolution of Inquiry regarding the Commissioners of School and University Lands, SJ, App., 4–7 (1855); and in Report, Joint Committee of Investigation Regarding the State Departments, SJ, App., 14 (1858). See also Governor Barstow's General Message, SJ, App., 9–10 (1855).

10. James Willard Hurst, *Law and the Conditions of Freedom in the Nineteenth-Century United States* (1956), chap. ii.

11. General Message of Governor Dewey, SJ, 21 (1851); General Message of Governor Farwell, SJ, 12, 13 (1853); General Laws, 1867, Memorial No. 4. On recognition of the positive power and responsibility represented by government landholding, see Clawson, *Uncle Sam's Acres* (1951), 1–2, 32, 42; Herman E. Krooss, *American Economic Develop-*

ment (1955), 102, 116. The special leverage importance of the public lands for transport development is discussed in U.S. Bureau of Corporations, *The Lumber Industry* (1913), vol. I, part I, p. 248, and in Thomas C. Cochran, "The Entrepreneur in American Capital Formation," in National Bureau of Economic Research, *Capital Formation and Economic Growth* (1955), 347, 353–354.

12. Milo M. Quaife, *The Movement for Statehood* (1918), 215, 225, 357, 364, 374, 486 shows the contemporary eagerness for federal grants as a prize of statehood. The limitation and exception regarding internal improvements are in Wisconsin Constitution, Art. VIII, sect. 10; cf. Quaife, *The Attainment of Statehood* (1928), 417, 449, 581–584. The characterization of the importance of the "school and swamp land bureau" is in Report of the Joint Committee of Investigation regarding State Departments, SJ, App., 11 (1858). The "trust" concept of the state lands is discussed in note 7 above. Governor Harvey's Veto Message is in SJ, 809 (1862); the quotation is from p. 816.

The clearest expression of the opinion holding the state responsible as architect of community growth, through the power given by its landholdings, was General Acts, 1855, c. 21, requiring that everyone who applied to buy school or university lands should file his affidavit that he bought for his own use by settlement, and that his purchase would not bring his total holdings of such lands to exceed 320 acres. The background of this act in the demand to curb speculative holdings which would distort community development may be seen in Governor Barstow's General Message, SJ, App., 9–10 (1855), and in the Report of Select Committee under a resolution for inquiry into the business of the land commissioners, SJ, 6–7 (1855). The same attitudes found logical expression in distaste for any hint of policy which would have the government itself hold onto its land for an indefinite period. See, e.g., regarding swamplands disposition, the Barstow Message cited above, at pp. 21, 22, and Report of Select Committee regarding the Governor's message concerning swamplands, SJ, App., 6–8 (1855).

The state's plenary power (and, inferentially, responsibility) in disposal of its lands was asserted by the court in Smith v. Mariner, 5 Wis. 551 (1856) (constitution leaves legislature free to devise other terms of secured sale than by mortgage); State ex rel. McIndoe v. Commissioners of School and University Lands, 6 Wis. 334 (1858) (320-acre limitation imposed by General Acts, 1855, c. 21, is within legislature's dispositive authority, and may apply to prior applications to purchase where title has not yet passed); Wisconsin Central Railroad Co. v. Taylor County, 52 Wis. 37, 8 N.W. 833 (1881) (legislature enjoys power to create tax exemptions reasonably related to achievement of objectives of a "trust" of public lands); State ex rel. Sweet v. Cunningham, 88 Wis. 81, 57 N.W. 1119 (1894) (Laws, 1885, c. 222, directing that state lands re-offered after having been withdrawn from sale be first offered at public sale, is a remedial act, entitled to a liberal construction in the state's interest, and hence should be construed to apply to lands withdrawn before as well as after its passage). See, also, Zemlock v. United States, 73 Wis. 363, 41 N.W. 445 (1889) (state might in grant of its lands reserve right to take them back later for a public purpose without compensation).

The state court was equally forthright in acknowledgment of the plenary power of disposition enjoyed by the United States over its lands, so far as it did not transfer them without condition. Woodward v. McReynolds, 2 Pinn. 268 (Wis. 1849), and Knight v. Leary, 54 Wis. 459, 11 N.W. 600 (1882) (no adverse possession against the United States); Paige v. Peters, 70 Wis. 178, 35 N.W. 328 (1888) (United States may exempt homestead from liability to satisfy homesteader's debt contracted before patent); see Farrington v. Wilson, 29 Wis. 383 (1872) (United States may impose restraints on alienation in disposition of its property, but presumption in construction favors intent to maintain alienability); Quinney v. Town of Stockbridge, 33 Wis. 505 (1873); cf. Sitzman v. Pacquette, 13 Wis. 291 (1860). Questions of public land titles apart, it should be noted also that the court recognized that the general ("police") power of the legislature extended to making reasonable regulations in the public interest defining the incidents of private property in land, and this nonetheless because the constitution declared that all lands in the state should be "allodial"; the impact of the latter provision was held to be simply to reinforce the prohibition of feudal titles: Barker v. Dayton, 28 Wis. 367 (1871).

Acceptance without argument of the legitimacy of using the public domain as the lever for economic and social development is implicit in the debate over seeking federal aid lands for railroads as against channeling all public lands at minimum prices to farmers. See contrasting General Messages of Governor Farwell, SJ, 17, 25 (1852), and of Governor Barstow, SJ, 22, 24 (1854). Report of Committee on Internal Improvements, regarding memorials for federal railroad-aid grants, AJ, 699–703 (1852), presented the most extended discussion of public land use policy in the Wisconsin record, accepting fully the planning role of government in use of the public domain. Cf. Quaife, *The Struggle Over Ratification* (1920), 650–655; Quaife, *The Attainment of Statehood,* 553–559. Of like import regarding use of the swamplands is the Report of a Select Committee regarding the governor's message on swamplands, SJ, App., 6–8 (1855).

13. Clawson, *Uncle Sam's Acres* 43–47, 58, 61, 62, 64; Gulick, 19, 155; Robbins, *Our Landed Heritage* 8, 9, 10–20, 37, 50, 89, 91, 107, 109, 277, 278. The relation of the settlement emphasis of federal land policy to general humanitarian and political reform notions of the mid-nineteenth century is noted in Joseph Dorfman, *The Economic Mind in American Civilization* (1946), II, 684. The quoted observations of Jefferson are placed well in context with the American agrarian myth in Smith, 128, 203. The tenacious grip of the abstract ideal of favor to the settler was nowhere more clearly shown than in the difficulty in obtaining repeal of pre-emption provisions which had clearly outlived their function and become primarily instruments of fraud. See Harold Hathaway Dunham, *Government Handout: A Study in the Administration of the Public Lands, 1875–1891* (1941), 154, 210, 269, 336. Typical of strong official statements of the pro-settler bias of formal federal land policy are the Reports of the Commissioner of the General Land Office, 1868, p. 96 ("The title to the soil thus acquired it has ever been the policy of the United States to transmute into individual ownership in the shortest possible space of time"); 1869, p. 20 (Pre-emption Act of 1841 recognized unsoundness of policies restrictive of settlement); 1870, p. 7 ("It would be unwise financial economy to administer this trust in such a manner as to raise a million or so more of dollars, if this involved the necessity of impeding the settlement of the public domain by raising the price, or by withdrawing any of the present facilities for their appropriation which the law holds out to the humblest industry"); 1871, p. 30 (frauds on the pre-emption system criticized solely as they constitute an impediment to bona fide settlement).

14. On the general pattern of the Wisconsin laws referred to, see note 8 above. Agencies to promote immigration had a tangled history of creation, repeal, and re-creation over the years, but the difficulties arose from doubts of their effectiveness rather than of the public value of the objective. See, e.g., General Acts, 1853, c. 53, and Report of the Committee on State Affairs regarding the Emigrant Agency, SJ, App., 4 (1853) ("Of the policy of encouraging immigration, there ought not to be a doubt, since the present prosperous condition of the United States affords so striking and forcible an illustration of its practical results, and the rapid growth and prosperity of our State is no less demonstrative of the same fact"); Report of Emigrant Commissioner, SJ, App., C, 10 (1853) ("Every unprejudiced citizen and observer of our State must confess and acknowledge that the continued prosperity of its inhabitants, is in a great degree dependent upon further accessions to its population from abroad"). Characteristic statements of freeholder values as the basis for proposed federal land policy may be seen in Laws, 1849, Joint Resolution No. 7 ("That the toleration and protection by the government of the monopoly of the soil in this country, is greatly to be deplored, and we believe it inexpedient and unjust for our national government to dispose of the public domain to any persons except actual settlers"); General Acts, 1853, Joint Resolution No. 2 (urging that Congress pass a homestead bill, because "the rapid settlement of our government lands, and the consequent cultivation of the soil, would add to the wealth, happiness and social elevation of our people"); and General Acts, 1854, Memorial No. 31 (further urging that Congress pass a homestead bill, and also grant aids for transport improvement, "that the vast number of our fellow citizens who are now compelled to surrender their labor to capital for a mere subsistence, may speedily find a home on the great domain of our nation"). See also Report, Committee on Public Lands, regarding recommendations in the Governor's message on public lands sales, SJ, App., 819 (1849). One should note, further, that the consistent, practical tolerance shown both by federal and state authorities toward squatters was a significant reflection of the depth of

feeling in support of settlement as an end value. So viewed, the pre-emption statutes expressed government's yielding to a demand for family farmland use so massive that it could not be resisted. See Robbins, *Our Landed Heritage,* 54, 99; cf. General Acts, 1851, c. 35, amending General Acts, 1850, c. 236 (to limit Wisconsin's pre-emption law to settlers, explicitly excluding those who improved but did not settle); Report of Committee on Internal Improvements, AJ, 720–721 (1852) (recommending a memorial to Congress to extend allowed time within which pre-emptor may perform requisite improvements). With explicit confidence in the righteousness of the request, the state legislature memorialized Congress to grant pre-emptive rights to "thousands" who had settled on federal lands in advance of survey or otherwise without legal permission. See e.g., Acts, 1850, Memorial No. 3; Acts, 1852, Memorial No. 18. Another sign of the acceptance of the rightness of settlement, as such, was the provisions carried forward from territorial into state law, allowing possessory actions to any one who had settled on and improved federal lands, without requirement that he show rightful possession as against the government (Revised Statutes, 1849, c. 118; cf. Territorial Laws, 1837–1838, p. 309).

15. General Messages of Governor Dewey, SJ, 21 (1850), and Governor Bashford, SJ, 23 (1857). The argument for state agencies to promote immigration to Wisconsin ran typically in terms of increasing the State's productive capacity. See note 14 above. On the whole, emphasis on family-farm-type settlement as important for increasing the productive power of the community tended to dominate public lands policy expression through the 1850's. To this effect, for example, see Acts, 1852, Memorial No. 11 (urging removal of Chippewa Indians east of St. Croix River, lest development of resources be retarded); Petition of citizens re Fox-Wisconsin aid lands pre-emption rights, SJ, 250–251 (1852); Report, Committee on Internal Improvements, regarding memorializing Congress to liberalize pre-emption laws, AJ, 720–721 (1852); General Message of Governor Farwell, SJ, 17, 23, 25 (1852) (multiply producers, increase tax base); General Acts, 1854, Memorial No. 31 (plea for railroad aid grants coupled with stipulation for settlers' pre-emptions, all "that agriculture, commerce, and the arts may be rapidly advanced," and "that our vast forests of pine, immense plains of prairies, and inexhaustible mines may be approached and made to yield by man's labor, to his wants"); Report, State Superintendent of Public Instruction, SJ, App., 12 (1852) (large speculative holdings bad, because keep land out of production awaiting a rise in values, "instead of becoming at an early day the homes of industrious and valuable citizens"); General Message of Governor Barstow, SJ, App., 21, 22 (1855), and Report, Select Committee on the Governor's message, regarding swamplands, SJ, App., 7, 8 (1855) (so dispose of swamplands as to increase production, broaden tax base); General Laws, 1858, Memorial No. 39 (urging United States to buy Oneida Indian land, to promote bringing it into production, and enlarging tax base). Typical linkage of transport improvement aid grants, and increase of productivity by stimulating settlement, may be seen in Acts, 1850, Memorial No. 11; General Acts, 1854, Memorial No. 31.

16. A strong statement both of the practice and of the sustaining sense of righteousness behind the transfer of the fee simple absolute as the standard title passed by the United States to its grantees was made in the Report of the Commissioner of the General Land Office, 1870, pp. 28, 29: "The individual title derived from the Government involves the entire transfer of the ownership of the soil. It is purely allodial, with all the incidents pertaining to that title as substantial as in the infancy of Teutonic civilization . . . The liberal principles embodied in our public-land policy have reconstructed to a great extent the legal basis of our social order by liberalizing the idea of land ownership." Graphic relation is made of this "fee simple empire" to the contemporary ideal of the republic of yeoman farmers as the prime objective of social development in this country, in Smith, 133, 170, 172. The attractiveness and tenacity of the ideal of private fee simple ownership as the norm of public lands policy is attested by the continued invocation of the same symbols in twentieth-century debate on conservation and the disposition of the public domain in the Far West. See the resolutions of a western states' governors' conference on conservation, in June 1913, quoted in Robbins, *Our Landed Heritage,* 382: "The best and most economical development of this western territory was accomplished under those methods in vogue when the states of the Middle West were occupied and settled. In our opinion, those methods have never been improved upon,

and we advocate a return to these first principles of vested ownership with joint interest and with widely scattered individual responsibility." Compare Daniel J. Dykstra, "Federal Government, State Governments, and Natural Resources," *Minnesota Law Review,* 37:569, (1953), n. 1 (federal ownership of natural resources as a campaign issue in 1952).

Typical of Wisconsin legislation expressing the legislature's concern to promote the marketability of government-derived land titles are Acts, 1850, c. 4 (recording of United States patents); General Acts, 1856, c. 30 (admissibility in evidence of certified copies of U.S. Land Office records); General Laws, 1866, c. 19, General Laws, 1867, cc. 3, 11, and 29 (availability of public records to facilitate transfers). The right of the state's defaulted vendee to any excess realized on foreclosure was declared in Laws, 1849, c. 212, sect. 18. The presumption in favor of construing government grants to pass a fee simple absolute may be seen in Sitzman v. Pacquette, 13 Wis. 291, 313–314 (1860) (U.S.); Farrington v. Wilson, 29 Wis. 383, 397 (1872) (U.S.); Stephenson v. Wilson, 37 Wis. 482, 490–491 (1875) (U.S.); Knight v. Leary, 54 Wis. 459, 469, 11 N.W. 600, 604 (1882) (U.S.); Town of La Pointe v. Town of Ashland, 47 Wis. 251, 2 N.W. 306 (1879) (state); Green Bay & Mississippi Canal Co. v. Hewitt, 62 Wis. 316, 336, 21 N.W. 216, 222 (1885) (state). Cf. Conklin v. Hawthorn, 29 Wis. 476 (1872), and Burrows v. Rutledge, 76 Wis. 22, 44 N.W. 847 (1890) (grants construed to pass accrued causes of action); Knapp v. Alexander-Edgar Lumber Co., 145 Wis. 528, 130 N.W. 504 (1911) (similar, obiter, as to perfected homestead). Court rulings expressing the policy in favor of recognizing the existence of disposable title in a public grantee as soon as he had met substantial requirements of ownership include Bracken v. Preston, 1 Pinney 584 (Wis. 1845) (ejectment lies on Land Office receiver's certificate, though no patent issued); Dillingham v. Fisher, 5 Wis. 475 (1856) (one who has completed requirements for pre-emption has transferable interest before patent); Lefferts v. Board of Supervisors of Calumet County, 21 Wis. 688 (1867) (opinion rendered though action discontinued: holder of school land certificate has sufficient title to contest taxes); Knight v. Leary, 54 Wis. 459, 11 N.W. 600 (1882) (patent presumed valid, if there was any law under which it might validly have issued); Cornelius v. Kessel, 58 Wis. 237, 16 N.W. 550 (1883) (entrant having completed all substantial requirements has equitable title effective against later patent wrongfully issued).

17. Cameron, 13, 20, 21, 25, 65, 76; Robbins, *Our Landed Heritage,* 7–12. Though pressure for political and social equality was probably the main element in the origins of this exaltation of the fee simple, it seems realistic to include in the pattern of popular values from an early time the high status accorded the venturer or entrepreneur, the individual assuming responsibility for the planning and administration of the use of resources, grounded as much in faith in the beneficent increase in production to be expected from the flowering of entrepreneurial energy, as in regard for an ideal of individual dignity. Cf. St. John Crevecoeur, *Letters from an American Farmer,* ed. W. Barton Blake (1945), 40, 56–60. A society whose constitution depended as much as ours on the institution of private property naturally bred respect for, and acceptance of the legitimacy of the power wielded by the entrepreneur; and, given the inertia of social attitudes, this respect and legitimacy could easily develop to sanction the big operator as well as the modest one. Thus the growth in the social status of the entrepreneur lent added force, but of potentially different emphasis, to the social and political ideas which sanctioned the delegation of decision making initiative represented in the fee simple absolute. Cf. Thomas C. Cochran, "The Entrepreneur in American Capital Formation," 342, 347, and comment there by Harold F. Williamson, 381, 383. The quoted provision of the Wisconsin Constitution is from Art. I, sect. 14.

18. The history of the lead lands leasing experience in the Illinois-Missouri-Wisconsin lead region is set forth in James A. Lake, *Law and Mineral Wealth: The Legal Profile of the Wisconsin Mining Industry* (1962), chap. ii. See also Hibbard, *Public Land Policies,* chap. xxv; and Joseph Schafer, *The Wisconsin Lead Region* (1922), chaps. iii and ix. A contemporary review of this experience was set out in a statement by the Commissioner of the General Land Office, in 1850, Senate Exec. Doc. No. 2, pp. 19–20, 31 Cong., 2 Sess., recommending sale of mineral lands in California, in light of the failure of a leasing policy in the Mississippi Valley: (1) such lessees had no stable interest in the locality, but rather were possessed with "a spirit of wild, speculating hazard"; (2) they were

disposed to evade the laws, because "the government was regarded as a rich landed monopolist, that had no feeling in unison with the interest of the tenant, but was rather trying to strip him of part of the earnings of his toil and privations"; (3) "The leasing system was also the cause of much irritation, jealousy and complaint on the part of the States within whose jurisdiction it was practiced. They justly complained that they were crippled in their resources by being restricted in the power of taxation — that the general government, instead of selling the land, retained the fee, and stood in the attitude of a powerful and opulent landlord, holding a large portion of this population in an abject and servile state of tenancy, destroying all inducements to cultivation, or the making of permanent improvements or settlements upon the land . . ." See, accord, statement of the Commissioner of the General Land Office in 1852, Senate Exec. Doc. No. 1, p. 77, 32 Cong., 2 Sess.

Pressure to speed the federal survey and offering of land may be seen in General Message of Governor Dewey, SJ, 10 (1849) (Indian title); Acts, 1850, Memorial No. 3 (Indian title); Acts, 1851, Memorials, c. 1 (survey; Indian title); Acts, 1852, Memorials, c. 17 (additional land offices); General Acts, 1854, Memorial No. 10 (survey; Indian title); General Acts, 1856, Joint Resolutions No. 2, 4, 5, 6 (survey; Indian title); General Laws, 1858, Memorial No. 37 (Indian title); General Laws, 1864, Memorial No. 3 (survey); General Message of Governor Lewis, SJ, 21 (1865) (remove Indians). The federal survey was completed in Wisconsin by 1866. See Report, Commissioner, General Land Office, 1866, pp. 45, 78. In retrospect it does not appear that the survey moved much slower than facts warranted, and in rare moments of patience local opinion sometimes recognized this. See Report, Committee on the Governor's message as to mining lands, AJ, 263 (1852). If eastern industrial or speculative interests felt cause to use influence earlier against speeding the opening of "western" lands, the belief in the greater gains to be had from expansion had overcome this sentiment by the 1860's, as the Homestead Act symbolized. See Robbins, *Our Landed Heritage,* 107, 109. The state's own press-to-market policy is particularly reflected in Laws, 1849, c. 212, sect. 8 (water-power sites); General Acts, 1856, c. 122, sect. 7, and c. 137, sect. 12; Private and Local Acts, 1857, c. 230, sect. 2; P & LL, 1866, c. 314, sect. 11, c. 362, sect. 12, and c. 365, sect. 8 (railroad grantees to expedite sales); General Laws, 1868, c. 105, sect. 4 (in consideration of grant of aid lands, Sturgeon Bay & Lake Michigan Ship Canal & Harbor Co. "shall use all due diligence in disposing of said lands at a fair and equitable price, and they shall not be held by the said company for speculation"). Compare the reservation to the legislature of future disposal of waterpower sites in connection with the Fox-Wisconsin Rivers improvement so long as the state was proprietor thereof, and the omission of any public reservation of such sites when the aid lands for the project were transferred to a private company, reflected in Laws, 1848, p. 58, sect. 16; Laws, 1850, c. 275 (unimproved waterpower sites within prior, valid pre-emption claims shall pass to claimants); and General Acts, 1853, c. 98. The possibility of an indefinite landlord's role for the state was regularly rejected when it was mentioned at all. See General Message of Governor Dewey, SJ, 25, 30, 31 (1849); Report of Committee on Public Lands, on the Governor's message, SJ, App., 819 (1849); and Laws, 1849, Joint Resolution No. 7; Report, Committee on State Affairs regarding "the capitol lands," AJ, 169 (1857); Report, Commissioners of School and University Lands, PD, Doc. D, p. 38 (1860); General Message of Governor Harvey, SJ, 17 (1862).

19. See Alexis de Tocqueville, *Democracy in America,* ed. Phillips Bradley (1948), II, 157. The general absence of any effort to tie men to the land is pointed up by the unique character of Laws, 1848, p. 58, sect. 38, barring from pre-emption rights in Fox-Wisconsin improvement aid lands any person "who quits or abandons his residence or his own land in this state to reside on the said lands . . ." A characteristic statement of the distinction which contemporary opinion drew, though not always with desirable clarity, between profiting by growing up with the country, and profiting simply by the growth of the country as shaped by others' effort, may be seen in Governor Dewey's General Messages in SJ, 31 (1849), and 21 (1851) ("every purchaser of the soil from the common trustee, should increase its aggregate wealth, by bestowing similar labors upon his own"); Acts, 1852, Memorials, c. 19; Report, Committee on Public Lands, SJ, 444 (1852) (speculator's command of cash allows him to garner the increased value which settlers' toil should reap); Report, Joint Committee on

Investigation of State Departments, SJ, App., 14 (1858) (suggested, state might properly forego principal, limit itself to asking interest, for land sold actual settlers). Governor Farwell put the issue on railroad-aid grants with sharpness of definition unusual in current statement of policy questions, in his General Message, SJ, 23–25 (1852); the contemporary lengthy report on the same subject by the Committee on Internal Improvements, AJ, 699–703 (1852) is interesting because of an ambivalence which makes almost painfully clear men's desire to promote railroads and their fear that the necessary aid lands grants would throw out of market immense quantities of lands, "placing them in the hands of our non-resident owners, and thus, according to our imperfect system of assessment . . . increasing largely the burthens of taxation, upon a people who are already outraged almost to the point of endurance, by being compelled to sustain the burthens of thousands of capitalists who refuse to share in the expenses whilst they reap all the benefits of government" (p. 703). Compare the state's railroad aid grant legislation of 1856 and 1857, cited, note 18 above. The terms of contemporary pleas and arguments for using the public domain to promote railroad development made plain that the assumption, and indeed the rationale, of the process was that the railroad itself would open up the country and enhance the value of the lands. See, e.g., General Acts, 1854, Memorial No. 31; proceedings regarding Resolution 57A, AJ, 296–297 (1866); General Laws, 1867, Memorial No. 27. That in practice the sale of railroad-aid lands typically lagged behind the pace of rail construction, see Krooss, 429.

20. Laws of Wisconsin Territory, 1846, Memorial, p. 225; Report, Select Committee on bill and petition of citizens of Rock County regarding land limitation, SJ, 399 (1848).

21. Acts, 1852, Memorials, c. 1. Typical statements of concern over long-term federal landholding in the state, because of the block to productive use and the limiting of the tax base may be seen in Laws, 1849, Joint Resolution No. 7, and the related passages in General Message of Governor Dewey, SJ, 30, 31 (1849) (Governor argues that federal land grants to states are just, "as some equivalent" for the inequality of position between old states unencumbered by federal landholding, and the new states: "New States are admitted into the Union upon an equal footing with the original States, but no such equality can exist until the new States possess as well the ultimate right of soil as of jurisdiction, and the right also to tax the whole domain wherein her limits [sic]"); Acts, 1852, Memorials, c. 22 (asking federal land grant in aid of state institutions for the blind and deaf, observing that the expense of such provisions "falls heavily upon a state, which like all new states has no means of revenue but direct taxation, and that revenue derived from one-half of the real estate within our borders, the other half being by the laws of the United States exempt from taxation"); General Acts, 1854, Memorial No. 14 (tax status of Fox-Wisconsin project aid lands); Report, Select Committee on the Governor's message regarding swamplands, SJ, App., 8 (1855) (object of public policy in fixing burdens that go with swampland disposal should be "to reduce as greatly as possible the necessary expense which is borne by the citizen, and to add to his means of payment"); General Message of Governor Barstow, SJ, 27–28 (1855) ("The general government still owns nearly or quite one third of the domain embraced within the boundaries of the state, while it does not afford any assistance to our treasury, or lighten, in the smallest degree, the burdens of taxation"); General Laws, 1858, Memorial No. 39 (asking that United States buy up and market a strip of Indian land, since other growing areas "are cut off from proper thoroughfares with the interior, by this strip of Indian land exempt from taxation, and of course not subject to the usual burthens of constructing roads and other improvements necessary to the opening of a new country"); General Laws, 1859, Joint Resolution No. 2, and Memorial No. 2 (protesting offset by United States against payments due Wisconsin under the law granting states 5 per cent of proceeds of federal land sales within their borders, of unrelated fiscal claims against state; 5 per cent law viewed as acknowledgment of equity owing state because federal landholdings deprive it of tax base), with which compare Opinion of Attorney General on claim against state by Milwaukee & Rock River Canal Co., SJ, 141–142 (1862). Criticism of the barrier to expansion of the production and tax base of the economy posed by long withholding from market or by tax exemption of state-owned or state-derived lands, including the lands allotted in aid of railroad building, may be seen in the General Message of Governor Lewis, SJ, 12 (1865) (mingled criticism of

Governor Dewey, SJ, 21 (1851), and in Report, State Superintendent of Public Instruction, SJ, App., 11 (1853); see also Memorial No. 75A, AJ, 151 (1877) (opposition to tax exemptions for privileged groups expressed by "the undersigned farmers . . . having left the eastern states and foreign countries many years ago with all our moneys and effects, and in addition have [sic] laid out from twenty to thirty years of hard labor, thus using up the better part of our physical ability for the aggrandizement of our state, and for the benefit of future generations"). Cf. Gile v. Hallock, 33 Wis. 523, 527 (1873), and Paige v. Peters, 70 Wis. 178, 35 N.W. 328 (1888) (upholding plenary authority of United States to exempt its homestead grants from execution for debts contracted before patent). The practical preference expressed in our production methods and law for man-hour over per-acre measures of satisfactory productivity is discussed in Johnson and Krooss, 243, 247, 257; cf. Gunnar Myrdal, *An International Economy* (1956), 100, 161, 163.

27. The consistent, unquestioning discussion of agriculture as just one of the branches of "industry" in the state, albeit the most important single branch, tells much of how investment-minded, managerial- (entrepreneurial-) bent was the general faith and strategy of this agrarian community. See General Messages of Governor Farwell, SJ, 22 (1852); Governor Randall, SJ, 13–14 (1860); Governor Harvey, SJ, 27–28 (1862); Governor Fairchild, SJ, 30 (1866), and 18 (1867); Governor Taylor, AJ, App., 23–24 (1874); Governor Smith, PD, I, 16 (1878). To like effect, see, e.g., Report, Committee on Agriculture and Commerce, AJ, App., 1126 (1851); Report, Committee on Agriculture and Manufactures, SJ, 233–235 (1858); Report, Committee on Incorporations, SJ, 135–137 (1866). The middle-class, small-businessman values underlying the flamboyance of agrarian polemic is noted in their general relation to our economic and legal history in Cochran, "The Entrepreneur in American Capital Formation," 342, 347, and in Hurst, *Law and the Conditions of Freedom*, 74, 83, 95.

28. The degree to which issues of regulation of currency, banks, and the manner and timing of enforcement of creditors' rights engrossed political attention in the nineteenth-century United States is both evidence of this tendency to monetize policy and in itself an influence working cumulatively to distract men's attention from the reckoning of "real" costs and income. Compare, generally, Myrdal, 75, 154, and see, as to the extent of preoccupation with these kinds of issues in mid-century Wisconsin, Theodore A. Andersen, *A Century of Banking in Wisconsin* (1954), chaps. i, ii and iii; Merk, chaps. vii and ix. The development of more wide-ranging markets, involving greater distance and impersonality of dealing, exerted a strong pressure toward habits of money calculation. Cf, Siegfried von Ciriacy-Wantrup, *Resource Conservation* (1952), 13. With its focus almost wholly on analysis of exchange equilibrium, contemporary economic thinking contributed nothing to offset the practical bias of the times toward monetizing all issues; growth of a theory of "real" social costs and returns as a part of a complete economic calculus lagged until after the turn of the century. See K. William Kapp, *The Social Costs of Private Enterprise* (1950), pp. 5, 9, and chap. iii.

29. The "funded" treatment of the public lands in public discussion and official proceedings connected with the transition to statehood in Wisconsin may be seen throughout the materials collected and edited by Quaife, thus: *The Movement for Statehood,* 62–64, 215, 224, 225, 277–280, 317, 357, 364, 375, 486; *The Convention of 1846,* 92, 135, 193, and 637 (Milwaukee and Rock River Canal aid lands discussed chiefly in money terms), 189, 196, 307, 683–684 (unquestioning acceptance of propriety of using public lands proceeds to pay current expenses of convention), and 298–300, 335, 337, 407–408, 485–486, 754 (500,000-acre grant from the United States discussed simply as source of funds); *The Struggle Over Ratification,* 653 includes the passage quoted in the text from a letter of Feb. 29, 1847, from "Old Crawford Forever" to the Prairie du Chien *Patriot; The Attainment of Statehood,* 3, 115, 135–140, 144, 158, 444–445, 448, 451, 589, 835. This tendency to view public lands programs mainly in money terms was already well developed in Wisconsin's territorial years. See, e.g., Report of Commissioners on the Milwaukee and Rock River Canal Project, Council Journal, App., Doc. No. 8, p. 227 (1839), and Report, Committee on Internal Improvements, HRJ, 85–87 (1839); Report, Committee on Schools, HRJ, 93 (1842); General Message of Governor Doty, Council Journal, 15–17, 28–29 (1841). The quoted provisions of the Wisconsin Constitution are from Art. X, sects. 7 and 8. The quoted counsel of Governor Dewey is from his General Message, SJ, 25 (1849). Compare the tone of

Governor Fairchild's General Message, SJ, 22 (1866), explaining that pressure to secure full payment for state lands was related to the state's fiscal needs growing out of the War. Compare, of like emphasis, the recommendation in the Report of the Secretary of State, PD, 27 (1868), that proceeds of state land sales be invested primarily in obligations of the state of Wisconsin, rather than bonds of other sovereigns, for "all the money and capital arising from the sale of lands in charge of the state, can find ample sources of use and investment at home." Typical statements of the argument for enlarging the tax base are cited in note 15 above. The nearest approach in Wisconsin to official consideration of the possibility of money grants to subsidize settlers' land purchases seems to have been in an 1849 Report of the Committee on Public Lands, rejecting the general idea (SJ, App., 823). Cf. Report, Committee on Railroads, AJ, 320–321 (1855) (state constitutional ban on state participation in works of internal improvement must also be reckoned with). Note 14 above includes references reflecting the typical sentiment back of the creation of state offices to promote immigration. Official comments on this policy sometimes made quite explicit the emphasis on accumulation of mobile capital, as such. See, e.g., Report, Commissioners of Immigration, AJ, 60 (1880) ("We need their capital, their labor, their energy and enterprise here"); General Message of Governor Smith, AJ, 32 (1881) (" . . . their strong arms and willing hands will assist very materially in developing our dormant resources. It is probable also, that these immigrants brought with them and added to the cash capital of the State not less than a million and a half of dollars in ready money"), and Report, Board of Immigration, AJ, 111 (1881) (estimating on same data that previous year's immigrants brought to state at least $1,272,-540: "If to this be added their ability to work, their skill, energy and enterprise, it must be plain that the benefit to be derived from this great influx of people is of very great moment to the growth and prosperity of the state"); General Message of Governor Rusk, AJ, 27 (1882) ("the immigrants have been of the best agricultural and industrial classes, and have made material additions to our capital and wealth"). The quoted explanation and praise of state grants of rights of way to transport companies is from Report, Committee on Railroads, AJ, 321 (1855). Compare Governor Rusk's Veto Message, AJ, 973 (1887), rejecting a bill to allow certain towns to tax state lands in aid of a reclamation project, with the fact that he was willing to let another bill granting state lands to the towns to be sold for the project to become a law without his signature, as noted in his Special Message, AJ, 1183 (1887). Typical of the scores of acts granting state lands to be sold in aid of local transportation improvements, or granting rights of way to transport companies, are General Laws, 1859, cc. 92, 94, 132, 177 and 183; P & LL, 1860, c. 78, 1861, c. 26, 1866, c. 71. The federal government's use of its lands to pay veterans' bonuses is noted in Clawson, *Uncle Sam's Acres,* 63; Dunham, 4. Clawson, 64 comments on Congressional pennypinching in the survey; see also Hibbard, *Public Land Policies,* 488, 490, 494–496. The weight with which the felt need to economize on public expenditures bore down to limit the state's field work regarding its landholdings is reflected in Laws, 1849, c. 68, as recommended in General Message of Governor Dewey, SJ, 25 (1849) (suspension of appraisals); Special Message of Governor Farwell, SJ, App., 5 (1853) (independent survey of overflowed lands too costly); cf. Governor Dewey's General Message, SJ, 29 (1849) (geological survey would be valuable, but state can't now afford it), and Lake, *Law and Mineral Wealth,* chap. vi (on the subsequent tortuous history of state geological surveys and appropriations therefor). The story of the action and the subsequent regrets and recriminations concerning the state's decision to economize by relying on the federal survey field notes as the basis for selecting lands under the federal swampland grant is too involved to be documented here. A typical contemporary view is put in Governor Randall's General Message, SJ, 23 (1860). See, generally, James A. Lake, "Buried Wealth and Unwanted Water" (unpublished S.J.D. thesis, 1952), part II, chap. i.

30. Acts, 1850, Memorials, c. 11; Acts, 1852, Memorials, c. 22.

31. The quoted language on "application" of a railroad-aid grant is from General Laws, 1866, Memorial No. 8 (Winnebago & Superior Railroad Co.). Precise emphasis was put on the importance of regulating the disposition of lands granted in aid of a privately-directed transport improvement so as to assure a regular flow of money proceeds for construction, in State ex rel. Sturgeon Bay & Lake Michigan Ship Canal & Harbor Co. v. Commissioners of School and University Lands, 34 Wis. 162

(1874). This decision construed General Laws, 1868, c. 105 (that as quarters of canal construction are done, state might pass to company "the proportion of said lands" the company should be entitled to on the basis of work completed), to authorize the Governor to certify the land by money value rather than by acreage, since this interpretation would best serve the grant's objective of procuring a finished canal; this would not unduly hamper the company, since within these limits "those selections may be made [by the company] which are most to be preferred on account of their marketable situation and present cash value" (34 Wis. at p. 168). See the recital of the history of this canal project aid grant, in Annual Report, Secretary of State, PD, I, 47 (1876) ("The directors and incorporators of the new company seem to have been from the outset impressed with the impracticability of constructing a semi-public work of such magnitude as this, with exclusively private capital"), and 51 (company mortgaged entire land grant under trust deed to secure $350,000 of 20-year 7 per cent bonds, as apparently the best way to realize on the aid lands). Cf. General Laws, 1864, c. 439, sect. 2 (authority to mortgage state-granted aid lands, conferred on grantee counties, to finance reclamation). Statutes stipulating prompt sale of aid lands by assisted railroads are cited in note 18 above. Authorization was given assisted railroads to mortgage aid lands in such acts as General Laws, 1864, c. 324 (note, especially, tax exemption unaffected by mortgage, though terminated upon the making of a sale, contract of sale, or lease of earned lands); P & LL, 1866, cc. 365, 366 (general authorization to mortgage); P & LL, 1869, c. 51 (mortgage will not end tax exempt status); P & LL, 1870, cc. 94, 242, 251, and 326 (general authorization to mortgage); P & LL, 1871, cc. 119, 160, 299, 334, 391, 457, and 487. After the end of the special charter period, general authorization to mortgage railroad aid lands was conferred by Laws, 1874, c. 303, Revised Statutes, 1878, c. 87, sects. 1858 and 1859. See also Laws, 1879, c. 22. Realization that the aid lands' utility to the assisted railroads was primarily as a base for mortgage credit was made dramatically clear in instances where the legislature amended statutes terminating tax exemptions upon mortgage as well as upon sale of aid lands, to provide that mortgaging the lands should not end the tax exemption. See P & LL, 1863, c. 243, as amended, 1866, c. 323; General Laws, 1864, c. 420, as amended, 1865, c. 228. See also the authorization to mortgage lands given in aid of the Sturgeon Bay canal project, in Laws, 1873, c. 89. The quoted committee report on tax exemption of railroad-aid lands is that of the Committee on Railroads, AJ, 125 (1874); the contemporary railroad promoters' statement quoted in the text is that of the North Wisconsin Railway Company, Memorial 11S, SJ, 79, 81 (1874). In support of similar policy see also Report, Committee on Assessment and Collection of Taxes, AJ, 182 (1877) ("Its term of exemption from taxation upon these lands has already expired, and as the annual taxes are more than they could possibly receive from the sales of land in any one year, unless this bill passes they will lose the main consideration for which they have invested so much money in this state"); Report, Committee on Railroads, AJ, 248 (1879) ("The honest execution of the trust upon the part of the state requires that the railway company shall be permitted to utilize these lands to the fullest extent for the purpose of completing and equipping the road, unembarrassed meanwhile by state and local charges"); Report, Committee on Railroads, AJ, 429 (1879) ("Unless sufficient exemption is granted to enable the company to sell lands enough to afford large aid in construction, the object and intent of the land grant are defeated, and unless relief is granted the company is deprived of all the advantages offered in the grant"); compare the retrospective comments of railroad managers on the linked problems of tax exemption and mortgageability of aid lands, in the report of a conference between railroad representatives and the Joint Committee on Assessment and Collection of Taxes, SJ, 335–336 (1903) ("You wanted the capitalists from outside of the state to come here, bring their money here and you promised them a low rate of taxation if they would come here and furnish the capital and build up the railroads"), 338 ("ask the state . . . to adhere to the promise that they held out when the road was built and when the money was invested there, that they would favor these small roads and favor the building of new ones"), 342 ("I remember that members of the legislature voted against the proposition on the general principle that the railroad should pay their share of taxation, that we should not make it burdensome upon the people up there, and the next day the members of the legislature were satisfied they did wrong and they re-considered that vote and exempted those lands from taxation, enabling the Wisconsin Central railroad to borrow money and open up that country to Ashland, since which time that country has improved"). As the last quotation indicates, there were sometimes — usually belatedly — sharp legislative contests over tax exemption of railroad-aid lands, but in the event the exemptions were granted. Note especially in this aspect Laws, 1877, c. 245, 1878, c. 261, 1879, c. 22, and the legislative record on these acts. Contributing to the monetized context in which public lands played their role as the underpinning of railroad finance are the statutory authorizations of school fund loans to local governments, to enable payment of maturing local government railroad aid bonds. See Laws, 1880, cc. 85, 136, 189, 280; Laws, 1885, cc. 77 and 143; compare the involved sequel to Laws, 1880, c. 280, in Laws, 1881, c. 314, 1883, c. 260, and 1885, c. 331. The relation between this monetized evaluation of the aid lands, for their utility as a credit base, and the insistence on increasing productivity, is underlined in the General Message of Governor Randall, SJ, 22 (1859), as he insists that railroad-aid lands may properly be mortgaged only to borrow for new construction, not to pay off past debts. Johnson and Krooss, 247, in effect show that this Wisconsin emphasis was typical of the general importance the country gave to using land as the basis for investment. In practical acknowledgment that the low price of public land in effect also subsidized private lending on the security of farmland bought from the government, see General Messages of Governor Randall, SJ, 26 (1859) ("in newly settled states, where immigration is active and lands cheap, numerous cases will occur, where high rates may [be] paid on small sums, for short loans, with a large margin for profit in the rapid rise in the value of property. Such in very numerous instances has been the case in Wisconsin. But now that our lands are largely taken up and occupied, such cases more rarely occur. There is no branch of business in this State, except money lending, sufficiently profitable to warrant the payment of twelve per cent"), and SJ, 14 (1860) ("The entry of large quantities of the best agricultural lands belonging to that [school] fund by speculators, to be put into the market at speculating prices, has retarded the settlement and cultivation of the lands, and financial reverses have returned them by thousands of acres, upon the fund"). Compare letter from the Commissioners of School and University Lands to the State Superintendent of Public Instruction, quoted in the Superintendent's report, PD, 30, 34 (1860).

32. Materials reflecting the tendency to value "settlement" for its implications for rising production are cited in notes 21 and 23 above; on the tendency to offset concern for frauds on the settlement laws with comfort taken in the extent of settlement in fact going on, see material cited in note 24 above; on acceptance of the squatter's equity, see material in notes 13 and 18 above, and Robbins, *Our Landed Heritage,* 9–12, 26, 54, 99; since timber trespass provided the clear-cut revelation of the monetized approach to handling depredations on the public domain, material on this topic is postponed to a later chapter. In addition to the statutory provisions regarding reservation of a security title in the state for the unpaid price of land (e.g., Acts, 1849, c. 212, sects. 15–25), the preoccupation simply with the fiscal interests of the state in its land is suggested by the readiness with which a majority of the court, in Smith v. Mariner, 5 Wis. 551 (1856) (Smith, J., dissenting), construed Art. X, sect. 8 of the Wisconsin Constitution — despite its explicit references to the "sale" of state land and the taking of "security by mortgage" for any unpaid price — to leave the legislature discretion to enact a drastic system of reserved security title carrying no equity of redemption; this broad construction of the legislature's powers, the majority thought, met the constitution's "main, its primary, its only object, the protection of the school fund" (5 Wis. at p. 582). See Gough v. Dorsey, 27 Wis. 119 (1870), and Harrington v. Smith, 28 Wis. 43 (1871) (will enforce careful adherence to statutory payment procedures); Weber v. Zeimet, 30 Wis. 283 (1872) (strong enforcement of state's security interest). Cf. Eldred v. Sexton, 30 Wis. 193 (1872) (requirement in federal act of prior offerings at public auction should be liberally construed in favor of protecting government's maximum realization). One may also read the temper of the times in the fact that, while no official agency paid attention to any aspect of protecting the physical-productive ("real") potential of the land, striking initiative and assumption of responsibility were shown in steps taken by the authorities administering school fund loans to individuals to tighten security requirements by administrative regulation after the legislature had failed to act. See note 23 above. Compare the Attorney General's rather impressive recital of the school fund commissioners' care and ingenuity

in exacting on their own initiative more adequate security for school fund loans to local school districts. Communication of Attorney General in response to Resolution 40A, AJ, 461 (1870); see Report, Committee on State Affairs, AJ, 457–458 (1872). Characteristic criticisms of the state's disposition policy solely in terms of the too low money realization may be seen in Report, Commissioners of School and University Lands, PD, Doc. D, pp. 37–38, 58 (1860); Report, Select Committee on the University Appropriation, AJ, 499 (1870); General Messages of Governor Fairchild, SJ, 6 (1871), and of Governor Washburn, SJ, App., 18 (1872). The monetized concept of the "trust" of the state lands will be found throughout the materials cited in note 7 above. Nothing more strikingly testifies to the centrality and tenacity of this way of looking at the public lands program than the extent to which monetized values dominated the discussion of the meaning and appraisal of the school lands policy in the elaborate opinion written for the court by Marshall, J., in the lawsuit that brought a round-up accounting of the "trust" and defined the nature of the legislature's continuing powers in its fulfillment. State ex rel. Owen v. Donald, 160 Wis. 21, 104, 115, 151 N.W. 331, 358, 362 (1915).

33. On the use of public lands proceeds to pay current operating costs of government, see Quaife, *The Convention of 1846,* 189, 196, 307, 683–684 (convention expenses); Quaife, *The Attainment of Statehood,* 3, 144 (5 per cent fund). The 1842 report is that of the Committee on Schools, HRJ, 96 (1842); the citizens' petition resulted from a meeting of settlers of Port Hope, Feb. 10, 1852, printed in SJ, 251. Implications of like attitudes based on the urgency of presently felt hardship may be read in Acts, 1850, Memorials, c. 3 (urging Congress to extend pre-emption rights, consistent with "the fostering care which the general government has so liberally and so promptly extended to all those who have led the way into the unsettled portions of our western country"); General Message of Governor Dewey, SJ, 21 (1851) (public land policy should not be such "that capital should reap the enhanced value added to it by pioneer labor"); Memorial No. 75A, AJ, 151 (1877) (protesting tax exemptions as unjust to farmers who "have laid out from twenty to thirty years of hard labor, thus using up the better part of our physical ability for the aggrandizement of our state, and for the benefit of future generations").

34. The culture context which made this an entrepreneurial, investment-minded society is characterized in Cochran, "The Entrepreneur in American Capital Formation," 342, 347, and in the same author's, *The American Business System* (1957), 2–6, 28–29. The importance allowed the impersonal, money-calculus market as an organizing institution, and the premium placed on individual and firm mobility, both for egalitarian and market goals, contributed to a context of social values encouraging ready manipulation of landed resources, unhampered by the reverence for the land as the base of social order that was felt in more status-bound societies. See Ciriacy-Wantrup, 13; Hurst, *Law and the Conditions of Freedom,* 13, 35, 38; W.W. Rostow, *The Process of Economic Growth* (1952), 75, 104. This distinctive bias of the nineteenth-century United States emerges the more sharply in comparison with the mid-twentieth-century problems of savings and capital accumulation in countries that faced the problem of industrialization without the benefit of a supporting middle-class value tradition, under the political urgency of meeting the minimum subsistence needs of overcrowded populations learning to make more insistent demands on government. See Myrdal, 161, 163, 206 (especially, at p. 206, the quoted observation of India's Prime Minister Nehru: "A poor country, poor in resources, has not got large resources for investment, for building up tomorrow. And if you want a surplus, well you have to be strict with yourself in the present generation: and democracy does not like stinting the present . . . [T]hat is a tremendous advantage, from that limited point of view, which an authoritarian government has, which can build for tomorrow, not paying too much attention to things of today . . . We cannot do it"); Joseph J. Spengler, "Capital Requirements and Population Growth in Undeveloped Countries: Their Interrelations," *Economic Development and Cultural Change,* 4:305, 306, 314–316 (1956).

35. The emphasis and belief in the social desirability of a speedy, exploitative push to open up the land, and the constant pressure of settlers' and entrepreneurs' impatience, as dominant in the shaping of federal public domain policy, are noted in Clawson, *Uncle Sam's Acres,* 61–66, 81–86; Gulick, 19, 20–25; Ise, 19, 22, 26, 35, 41, 50; Robbins, *Our Landed Heritage,* 10–12, 50, 89, 91. The objection to public or private landlordism on grounds of delay to "improvement," as well as on egalitarian grounds, will be found to run

through materials cited in notes 15 and 18–24 above. The quoted 1842 committee discussion is contained in Report, Committee on Schools, HRJ, 97 (1842); the 1846 editorial was from the Mineral Point *Democrat*, and is quoted as it appears in Quaife, *The Movement for Statehood*, 364; comparable contemporary statements may be found on pp. 223, 357, 362, 363, and in Quaife's *The Convention of 1846*, 327–328, as well as in the same editor's *The Struggle Over Ratification*, 586, 653–654, and *The Attainment of Statehood*, 136, 146, 149, 197, 594, 616, 741, 835, 862.

36. General Message of Governor Barstow, SJ, App., 20 (1855); General Laws, 1871, Memorial No. 1. Other typical statements of the felt need to use the public lands to promote development of "overhead" or framework capital facilities like railroads, to speed general economic growth, may be seen in Quaife, *The Attainment of Statehood*, 136, 417, 581; Acts, 1850, Memorials, c. 11, Acts, 1851, Memorials, cc. 3, 5, and 14. Though he did not win his point, in arguing that the federal government's 500,000-acre gift to the new state be devoted to transportation improvements rather than schools, "Old Crawford Forever," writing to the Prairie du Chien *Patriot*, Feb. 29, 1847, expressed with particular clarity the booster criterion which was invoked to justify so much of contemporary policy toward the public domain. "Again, suppose," he argued, "by applying these funds to their originally intended use [for internal improvements] our dividend of school money should be a little less than it otherwise would be. Yet, if by having good roads our farmers, miners, and lumbermen can save twenty-five or fifty per cent on the expense of transporting their surplus produce to market, would they not be the gainers by paying the little addition to their school bills? Would they not have more means to do it with? Would they not have more neighbors to help them pay it? They certainly would" (reprinted in Quaife, *The Struggle Over Ratification*, 654). In his Special Message of Apr. 12, 1905, AJ, 1051–1052, urging greater care in defining public interest protective terms in dam franchises than he thought had been taken in earlier state promotion of other framework investments, Governor Robert M. LaFollette noted that the past popularity of the booster theory had fostered its uncritical application: "In the early life of states and municipalities franchises are freely granted for the building of ferries and bridges, turnpikes, railroads, and street railways. Liberal donations of moneys and lands are frequently bestowed upon those receiving the franchises. Eager to secure rapid development, little thought is taken for the future, and no consideration given to the proper restrictions or limitations to be imposed upon those who are the beneficiaries of these valuable public grants."

37. The cheapness of land on the market, combined with the experience of large returns from labor and cash investment, together with the high community prestige of capital management as a constructive type of endeavor (qualified, but only qualified, by fears of bigness) made for a heady, optimistic popular acceptance of the desirability of pushing growth as fast as possible, and, as a corollary, using cheap land in generous mixture with scarcer factors of production. See note 34 above. The inadequacy of standard nineteenth-century economic theory to deal with evaluation of growth processes and costs is discussed by Arthur Smithies, "Economic Welfare and Policy," in Smithies and others, *Economics and Public Policy* (1955), 6, 7, 13, and by John Jewkes, "The Economist and Economic Change," in the same work, 90, 91; cf. Joseph A. Schumpeter, *History of Economic Analysis* (1954), 964, 1160. Later students of economic growth would concede that the booster theory might represent wisdom for particular circumstances — but with a more sophisticated attempt to weigh costs against returns than ever entered nineteenth-century enthusiasm for the notion. See Ciriacy-Wantrup, 19, 20 (costs of future preferences); Marion J. Levy, Jr., "Some Social Obstacles to 'Capital Formation' in 'Underdeveloped Areas' ", in National Bureau of Economic Research, *Capital Formation and Economic Growth*, 491, 492 (what is "waste" of resources can be determined only according to the context of circumstances; otherwise "wasteful" expenditure may be justified as "a sort of bunched stimulus to development of effective productive capital"); Rostow, 42, 43 (current investment is comparatively marginal, even in societies where there is relatively high proportion of investment of current income); Anthony Scott, "Conservation Policy and Capital Theory," *Canadian Journal of Economics and Political Science*, 20:507, 508 (1954) (costs of future preferences); Spengler, 311 (key importance of availability of raw materials for general growth). Economic historians, moreover, agree that application of landed resources on the booster theory did in fact contribute material impetus to the pace of growth in the nineteenth-century United

States, and some of them, at least, would hold that the process represented a socially efficient mixture of the factors of production. Varying estimates may be found in Ciriacy-Wantrup, 10, 16; Clawson, *Uncle Sam's Acres,* 4–5, 8, 81–88; Gulick, 19; Johnson and Krooss, 243, 247; Krooss, 107–108, 115, 116; Spengler, 311, 324. Particularly graphic is the characterization of our nineteenth-century public-domain policy as one of "real" deficit financing, in Krooss, 115.

38. John Quincy Adams' vision of the public domain as the source of a long-term, if not permanent, income for national learning and science is set out in Brooks Adams, "The Heritage of Henry Adams," in his edited volume of the later essays of Henry Adams, *The Degradation of the Democratic Dogma* (1920), 25, 27, 31; see also Krooss, 102, 116, and materials cited in note 13 above. The consistent fiscal treatment of the significance of the state lands as the basis of a permanent school fund may be seen in the volumes edited by Quaife on the development of the Wisconsin constitution: *The Movement for Statehood,* 62–64, 224, 277–280, 364 (the "immediate disposal" reference noted in the text), 374–375; *The Convention of 1846,* 298–300, 335, 337, 407–408, 485–486, 754; *The Struggle Over Ratification,* 41, 86, 343, 490, 509, 652–653, 666–667; *The Attainment of Statehood,* 444, 448, 451, 481, 589. The remarks of "A Farmer of Grant" are quoted in *The Movement for Statehood*, 317. The materials on the concept of the state lands as a "trust," in note 7 above show that popular and official opinion included the notion of a long-term policy connected with the state lands, but only in a fiscal sense. In other respects, moreover, state policy was capable of the long view, so long as the issue was seen as that of preserving the monetary integrity of a school "fund." An example is the unusual energy shown by the executive branch in leading the way to safer investment of these moneys. See note 23 above. So, when latter day critics complained that the education funds had been wronged by the policy of rapid disposal at cheap prices to promote settlement, the legislature was willing to take some tangible remedial steps, always, however, by way of fiscal redress. See General Laws, 1866, c. 56, General Laws, 1867, c. 140, and Revised Statutes, 1878, c. 15, sect. 190 (expenses in care and management of the various funds derived from state lands and of the state lands themselves to be borne by the general fund); General Laws, 1872, c. 100 (proceeds of a special tax appropriated to compensate university fund for losses from cheap land sales policy); the materials cited in note 32 above offer a fair sample of the large amount of hindsight criticism of the earlier low-price-sales policy toward the state lands, always in fiscal terms.

39. "Rough Hewer's" letter is quoted in Quaife, *The Movement for Statehood,* 464; the Burlington resolution of Feb. 14, 1848, and the comment of the Southport *Telegraph,* Feb. 18, 1848, are quoted in Quaife, *The Attainment of Statehood,* 70, 72; the last-cited volume at pp. 553–556 reports the discussion initiated by Sanders; see also pp. 480–482, 557–558. In discussing the advantages of the land grants to be expected from the United States, as an incident of achieving statehood, the Report of the Select Joint Committee on State Government, HRJ, App., 336 (1846), observed that when Wisconsin became a state its legislature would have power "to fix upon some settled policy in relation to our school lands; and as soon as permanent leases or conveyances can be made, many of them will doubtless soon yield a revenue which will greatly aid the cause of education." It made like reference to the legislature's authority "to sell or lease" university lands. No more was said, and in context the references to leasing seem of no emphasis.

40. To complete the record, we should note some other scattering suggestions of a policy to realize continuing, long-term returns from the state land. Like the incidents noted in the text, the relevance of these further matters is chiefly to point up the fact that the times were not wholly lacking in the presentation of issues in other than a "funded" fashion, but that there was so little drive or action toward any other way of realizing on the state lands as to emphasize the unquestioned acceptance of monetized ways of stating the program. The General Message of Governor Bashford, SJ, 929 (1855), thus, suggested that in consideration of railroad-aid grants, the state should receive some percentage on the road's gross earnings annually for an indefinite term. Though the legislature subsequently from time to time provided for gross receipts taxes on railroads, these were enacted as another form of aid, being in lieu of all other taxes and a safeguard against parochial tax administration. See Betty R. Brown and Willard Hurst, "The Perils of the Test Case," *Wisconsin Law Review,* 1949: 26,

46, 57, 59–60. However, Laws, 1879, c. 22, sect. 2, in analogy to federal stipulations on federal railroad-aid grants, exacted as part consideration for tax exemption of aid lands granted the North Wisconsin Railway the railroad's undertaking for thirty years to transport free all troops and property of the state. We should note also an abortive suggestion in Report, Joint Committee of Investigation of the State Departments, SJ, App., 14 (1858), that the state sell lands to settlers without exacting any principal sum, but only annual interest — for a period unspecified. Finally, we should note that the statutes recognized the possibility of leasing state land but authorized only short-term leases, obviously intended to allow the state to obtain interim earnings pending sale. Acts, 1849, c. 212, sect. 53; Revised Statutes, 1858, c. 28, sect. 70; Revised Statutes, 1878, c. 15, sect. 194.

41. See Report, Committee on Schools, HRJ, 93 (1842); Quaife, *The Struggle Over Ratification,* 652, 653, and the same editor's, *The Attainment of Statehood,* 446, 481, 558. The episode of the university lands policy may be traced through Annual Report, Board of Regents of the University, AJ, App., 601, 602 (1850); Acts, 1850, c. 176; General Message of Governor Dewey, SJ, 10–11 (1851); Annual Report, Board of Regents of the University, AJ, App., 943 (1851), and Annual Report, Board of Commissioners of School and University Lands, AJ, App., 999–1001 (1851); Annual Report, Board of Commissioners of School and University Lands, AJ, App., 202 (1852); Acts, 1852, c. 493, carried into Revised Statutes, 1858, c. 28, sects. 15, 20 and 32, reaffirmed in General Laws, 1864, c. 455, carried into Revised Statutes, 1878, c. 15, sect. 202; General Laws, 1863, c. 287, carried into Revised Statutes, 1878, c. 15, sect. 204 (graduated price law, allowing reduction up to one third on unsold university lands); cf. General Laws, 1866, c. 121 (Morrill Act lands at $1.25). General Laws, 1872, c. 100, in its preamble recited that the most valuable university lands had by that time been sold, as it provided a special tax and appropriated the proceeds thereof to compensate the university fund for the deficiencies of the low-price policy of the past.

42. See pp. 10–11, 34–40, and especially notes 25, 26, and 29 above. General judgments on the significance of the imbalance in the factors of production for the currents of political, social, and economic development in the United States through the nineteenth century are expressed in Clawson, *Uncle Sam's Acres,* 87, 88; Johnson and Krooss, 243, 247–249; Arthur M. Schlesinger, *New Viewpoints in American History* (1923), 59, 64. The Wisconsin wheat story is told with particular reference to its relation to public lands policy in Joseph Schafer, *A History of Agriculture in Wisconsin* (1922), chap. v, especially at pp. 84–85, 91, 95. Land as a security base for mortgage lending is discussed in materials in note 31 above; cf. Allan G. Bogue, *Money at Interest* (1955), 5, 265; Hacker, 120. The buoyant, activist optimism to which the general economic situation contributed received classical characterization in the contemporary accounts of Crevecoeur, 57–62, and Tocqueville, II, 156–157; a good twentieth-century retrospect is presented in Henry Steele Commager, *The American Mind* (1950), 5–13.

43. The dynamic importance of leadership, in interplay with the pattern of the factors of production, is put forcefully in Moses Abramovitz, "Introduction," to National Bureau of Economic Research, *Capital Formation and Economic Growth,* 11. The general significance of middle-class ways of thinking for their impact on legal development in this respect is suggested in Hurst, *Law and the Conditions of Freedom,* 7, 8, 35–36, 38, 73, 75, 94–95. The quoted observation of the Wisconsin Supreme Court is from Woodward v. McReynolds, 2 Pinney 268, 273 (Wisconsin 1849); that of Governor Farwell is from his General Message, SJ, 17 (1852).

44. See Tocqueville, II, 156–157; cf. Hurst, *Law and the Conditions of Freedom,* 29, 40–41. The undistinguished career of Wisconsin's Philetus Sawyer, in the federal House and Senate, devoted as it was to legislation and executive affairs essentially local or special in character, exemplifies in tangible detail the typical, limited character of mid-century dealings in public affairs. See Richard Nelson Current, *Pine Logs and Politics* (1950), 61, 79, 82, 98–101, 196, 198–200.

45. The limitations of the market as an institution of social order are analyzed in Robert A. Dahl and Charles E. Lindblom, *Politics, Economics and Welfare* (1953), 385–393; Kapp, 11, 15–18; Frank H. Knight, *On the History and Method of Economics* (1956), 270–272; Karl Polanyi, *The Great Transformation* (1944), chaps. xi–xv. The soil costs of the Wisconsin wheat boom are noted in Merk, 20, 22, 43–47, 276; William Francis Raney, *Wisconsin: A Story of Prog-*

ress (1940), 218, 224; Schafer, *A History of Agriculture in Wisconsin,* 105, 110. The limited horizons of contemporary economic theory are characterized in materials cited in note 37 above.

46. There is graphic description of the difficulties of financing the opening of agricultural land in Wisconsin of the generation before the major growth of the lumber industry, in the majority opinion in Pratt v. Ayer, 3 Pinney 236, 258, 259 (Wisconsin 1851) ("Money was scarce, land was cheap. Settlers and capitalists made such arrangements as suited their respective circumstances and interests; and the result was unquestionably a mere division of profits . . . Shall we say that the amount of the commission in this case is evidence of a design to cover usury, when we can see clearly, and the parties themselves believed, that the amount agreed upon was an equitable division of profits, in proportion to their respective undertakings and risks? . . . Those who operated in the purchase of lands at an early day may have avoided or evaded the usury law of the territory. But was that the prominent design of the parties? . . . It seems to me gratuitous and unjust to draw such a conclusion when other and abundantly good reasons existed for the very arrangement that was made"). Cf. Fisher v. Otis, 3 Pinney 78, 95 (Wisconsin 1850). The limitations of nineteenth-century financial machinery, as they bore on land development and the possibility of long-range planning, are reflected in general outline in Bogue, 265–267; Alfred D. Chandler, Jr., "Henry Varnum Poor, Philosopher of Management, 1812–1905," in William Miller, ed., *Men in Business* (1952), 257, 283–285; Thomas C. Cochran and William Miller, *The Age of Enterprise* (1943), 52, 76, 149–150, and Cochran, "The Entrepreneur in American Capital Formation," 361, 367; Paul Wallace Gates, *Fifty Million Acres: Conflicts over Kansas Land Policy, 1854–1890* (1954), 94, 95, 99–101; George Soule, *Economic Forces in American History* (1952), 213–216. The quoted remarks of Governor Farwell, on farming as the principal use of Wisconsin's land, are from his General Message, SJ, 22 (1852). Compare the confidence in the general utility of the American land for "settlement" (meaning, in the context of the times, use for farming) in the recitals by which General Laws, 1864, Memorial No. 15 urged the good sense of federal aid to transport improvements which would more quickly realize the country's potential: "Of unsold public lands we have over one million and a half square miles. These lands, if settled as densely as Massachusetts, would contain over two hundred and fifty millions of inhabitants, and presenting every variety of climate, soil and production, under the operations of the homestead law, are now being most rapidly peopled by both native and foreign settlers . . ." See, generally, Smith, 123–124, 155, 159, 170, 172, 174, 181, 195–200, exploring the varied ways in which the "myth of the garden," the image of the central United States as by destiny a republic of family farms, tenaciously fastened onto our policy-making toward the land and economic growth as a whole.

47. The classic statement of the equality of new and old states is quoted from the ordinance of July 13, 1787 (the Northwest Ordinance), Art. 5th, *American Charter, Constitutions, and Organic Laws,* II, 957, 962. Governor Barstow's remarks are quoted from his General Message, SJ, 27–28 (1855). The legislature's opposition to the agricultural colleges aid bill was expressed in General Laws, 1862, Joint Resolution No. XIV; see proceedings of June 10, 1862, SJ, 899. Compare Report, Committee on Agriculture and Manufactures, SJ, 360 (1863), complaining that the grant when made was "less than our fair and equal share of the entire amount of land donated to the several States for a like purpose," but, in fashion characteristic of the time, expressing gratification at receipt of "a munificent grant" and urging quick acceptance so as not to lose out to other states or to homesteaders in selection of desirable tracts. There is hot expression of the suspicion and resentment of a "new" state against efforts of interests in an "old" state to procure forfeit of unrealized railroad-aid grants to the newcomer, in the Address of the Superior Board of Trade to the Wisconsin legislature, AJ, 330–337 (1870), especially at p. 336: "It is reserved for a member from Pennsylvania, a state running over with abounding wealth, to arise in Congress and ask the government of the United States to put its hand upon the young, struggling state of Wisconsin, and push down its springing activities in its own proper field, avowedly on the ground that to give them scope and afford her pioneers, who for half a generation have endured the privations of backwoods life, an opportunity (constantly hitherto had in prospect) of bettering their condition, under the ameliorations that follow the construction of railroads, might operate competitively against

the interests of men of Pennsylvania—referred to as men of immense wealth . . . Verily, wherewithal have these pampered capitalists been wont to be indulged, that to make them richer, the humble people of Northwestern Wisconsin must be made poorer!" Cf. General Laws, 1872, Joint Resolution No. 10, on the same subject. That such attitudes sprang easily out of the experience of opening new country is suggested by the raising of like issues between the frontier portions and the older-settled areas within Wisconsin itself. See Veto Message of Governor Harvey, SJ, 809–819 (1862), especially p. 811 (disapproving bill 62S, to donate the state's swamplands grant to the counties wherein the lands lie, because the bill would show "gross partiality" to the newer regions of the state, because of "the unequal relation which the expectants of the enormous benefits attempted to be conferred by this bill, hold to the population and tax-paying interests of the State"); the veto was sustained in the Senate, 11 to 17, June 6, 1862, SJ, 881. Cf. General Laws, 1871, Memorial No. 13 (Congress should relieve farmers who mortgaged lands to aid railroad construction, because this helped growth of state as a whole, and hence sale of federal lands therein as a whole). See also the argument of a Report, Finance Committee, SJ, 619–624 (1864), that taxes on railroads should not be such as to discourage capital investment, because the undeveloped parts of the state are entitled to as good a chance to attract capital as the older portions enjoyed in their development. The persistence of the "new" state claims to enjoy similar growth opportunities as the old, as public domain issues came to focus on the south and far west, may be seen in Ise, 13, 50, 166, 254, and in Robbins, *Our Landed Heritage,* 373, 382. On the significance of federalism for expanding the scope for play of supply and demand and movement of money and labor, see Hurst, *Law and the Conditions of Freedom,* 44–52; cf. Wabash, St. Louis & Peoria Railway Co. v. Illinois, 118 U.S. 557 (1886). Suggestive comments on the impact of broad, fluid capital markets, both for reach and weight of influence and for the quality of impersonal, tradition-free impetus, may be found in Ciriacy-Wantrup, 13, and Myrdal, 18, 60, 73. The impetus given Wisconsin development by outside markets for lead and wheat is described in Schafer, *A History of Agriculture in Wisconsin,* 82–83, 92–93, and *The Wisconsin Lead Region,* 2–5, 10, 31–32. If federalism made the individual state more vulnerable to sectional or nationwide forces, in fairness it must be noted, on the other hand, that able students credit the balance of sectional interests in the broad national forum with bringing about as broad a family-farm pattern of growth as we achieved, against the possibility of great private landlordism. See Clawson, *Uncle Sam's Acres,* 8, 81–86; Gulick, 19, 155; Pinchot, 161, 197, 200.

48. See the introductory part of the present chapter, and especially materials cited in notes 1–3 and 5–7 above on federal land policy as a frame for state programs. Cf. Spaulding v. Martin, 11 Wis. 262, 273 (1860) (general United States land policies are "well understood"); Harrington v. Smith, 28 Wis. 43, 66 (1871) (Wisconsin public lands act properly construed according to terms of analogous federal and state laws and the practice thereunder, because of "the points of similarity between them and the law we are construing, showing that the latter is based upon and intended as an enactment or adoption of very nearly the same system"). The Wisconsin court's remarks on the federal government's established tolerance of squatters are from Pratt v. Ayer, 3 Pinney 236, 256 (Wisconsin 1851). Cf. Revised Statutes, 1849, c. 118 (possessory actions given to "any person settled upon" United States land), originating perhaps in Session Laws of Wisconsin Territory, 1837–1838, p. 504. See Hurst, *Law and the Conditions of Freedom,* 3–6, on the values explicit and implicit in the organization and conduct of squatters' "claims associations." Governor Barstow's counsel on obtaining for the state whatever the federal government was willing to give, regardless of the unwisdom of its policy, is quoted from his General Message, SJ, 24 (1854). Compare (Majority) Report, Committee on the Judiciary, SJ, 276 (1869), praising land grants in aid of railroads, urging relaxation of the state constitution's ban on creation of state debt in aid of more construction, but nonetheless analyzing the need of careful limitations where money is given, with no reference to comparable caution in land grants.

49. The Wisconsin court made a fair characterization of the condition of federal land law, in its opinion on rehearing in McCord v. Hill, 111 Wis. 499, 523, 87 N.W. 481, 482 (1903): ". . . in 1820 Congress for the first time attempted something like general legislation which should affect and regulate

the management and disposal of the then considerable public domain . . . Since then legislation has been almost as frequent as the sessions of Congress; some of it scientific and deliberate, but much of it seemingly accidental and aimed at special details. Many of these acts, if construed according to their exact words, would have been subversive of certain phases of the obvious general policy of Congress with reference to this important subject; and in this field, perhaps more than almost any other of congressional legislation, the construction placed upon the land laws from time to time enacted both by the department and by the courts has been that they were intended merely to be added to and become a part of a consistent system of legislation, and as enacted one by one were, so far as possible, to fall under and be controlled by those general provisions and regulations evidently intended to cover the whole field." Cf., accord, Hibbard, *Public Land Policies,* 547, 548, 550, 554, 557. The bias toward *ad hoc* legislation in the states generally is characterized in Chester Lloyd Jones, *Statute Law Making in the United States* (1923), 32–42. Bancroft is quoted from his essay, "The Office of the People," in his *Literary and Historical Miscellanies* (1855), 408, 410. The roots of our mistaken conception of the "practical" and our depreciation of thoughtful deliberation of policy are discussed in Merle Curti, *The Growth of American Thought* (1943), 142, 268; Robin M. Williams, Jr., *American Society* (1951), 402–404.

50. Report, Select Joint Committee on State Government, HRJ, App., 335 (1846); Report, Joint Select Committee on charges of legislative corruption, SJ, 766 (1867). Typical of later criticisms of the inadequate selection of lands for the state under federal grants are Governor Randall's Special Message, AJ, App., 1–61 (1860), and his General Message, SJ, 26 (1866), on swamplands selection, and Annual Report of the University Regents, PD, 180 (1872), on the university lands. The strongest forebodings on the capacity of the state government to withstand the improper pressures that might be generated by availability of rich prizes in railroad-aid lands were expressed in the General Message of Governor Farwell, SJ, 23–25 (1852). See, in substantial accord, Report, Committee on Internal Improvements, AJ, 701–703 (1852); cf. General Message of Governor Bashford, SJ, 928 (1855) (desirable to distribute aid lands to create competing railroads, for balance of power). The confirmation of Farwell's fears in the 1856 scandal is related in Robert S. Hunt, *Law and Locomotives* (1958), chap. i. See also the rueful contemporary review of this situation in Report, Committee on Railroads, AJ, 843–844 (1859). Other aspects of the handling of state lands and their trust fund yields occasioned many expressions of contemporary lack of faith in the competence of government or its capacity to withstand improper pressures. See, e.g., (1) Report, Select Committee on Memorial of Milwaukee & Waukesha Railroad Co., AJ, 105 (1850), expressing fear of the patronage dangers and the likely subordination of trust fund interests to those of embarrassed debtor groups, in conducting a large-scale small-loan business out of the school land funds; (2) crisscrossing arguments for or against state or local government control of public land assets, according to differing judgments of which level of government would be most vulnerable to corrupt or unduly narrow pressures, as in Minority Report, Judiciary Committee, on bill 289A, AJ, 654 (1857) (counties' closer interest would make them better guardians of school fund than state legislature), and on the other hand in the Veto Message of Governor Harvey, SJ, 809–819 (1862) (disapproving donation of swamplands to counties wherein lie), and in an address by the President of the Board of Normal School Regents, appended to the Annual Report, State Superintendent of Public Instruction, PD, 201 (1868) ("raids" on swamplands "under the cover of benefits to various enterprises and localities"); (3) criticisms of elementary inefficiency, and occasional charges and a handful of apparently proved instances of dishonest dealings for their own interest by members of the staff in the land commissioners' office, as in General Message of Governor Randall, SJ, 42–43 (1858); Report, Committee on School and University Lands, SJ, 503 (1859); Report of Joint Committee on Investigation of the State Departments, SJ, 620 (1859), and a general report of the same committee in 1859, PD, 14, 15; various proceedings critical of the commissioners' fees and the quality and integrity of the work done for the fees, in AJ, 96, 115, 329–330 (1870) (resolution asking information on fees and replies of commissioners, Jan. 27, 28 and Feb. 12), SJ, 311–312 (1870) (Memorial, Feb. 14, of late school land commissioners protesting innuendos against their integrity); (4) criticisms of the efficiency and impartiality of adminis-

tration of the school lands fund loan system, as stated in materials cited in note 23 above, and in Special Report of Commissioners of School and University Lands (in response to Resolution 85A), AJ, 567–585 (1861); General Messages of Governor Harvey, SJ, 17 (1862), and Governor Fairchild, SJ, 22 (1866); Annual Report, Secretary of State, PD, 28–29 (1868). Though little expressed, distrust of legislative yielding to special interest importunities seems clearly implicit in the insistence with which the Wisconsin Supreme Court construed tax exemptions of aid lands strictly against the grantees. See, e.g., West Wisconsin Railway Co. v. Board of Supervisors of Trempealeau County, 35 Wis. 257 (1874), affirmed, 93 U.S. 595 (1876); Weston v. Supervisors of Shawano County, 44 Wis. 242 (1878); Wisconsin Central Railroad Co. v. Comstock, 71 Wis. 88, 36 N.W. 843 (1888); State ex rel. Bell v. Harshaw, 76 Wis. 230, 45 N.W. 308 (1890); Grunert v. Spalding, 104 Wis. 193, 80 N.W. 589 (1899); cf. Chicago, St. Paul, Minneapolis & Omaha Railway Co. v. Douglas County, 134 Wis. 197, 114 N.W. 511 (1908) (state not estopped to tax because made wrongful claim of ownership to lands now sought to be taxed). As the United States Supreme Court's affirmance of the first case cited above reminds us, this doctrine of strict construction conformed to well established doctrine of that court, concerning both state and federal grants of privilege to private persons or groups. The basic declaration of principle was, of course, that by Taney, C. J., in Proprietors of the Charles River Bridge v. Proprietors of the Warren Bridge, 11 Pet. 420, 547–548, 552–553 (U.S. 1837). The same approach as taken toward public lands questions may be seen, for example, in Rice v. Minnesota & Northwestern Railroad Co., 1 Black. 358 (U.S. 1862), and Leavenworth, Lawrence & Galveston Railroad Co. v. United States, 2 Otto 733 (U.S. 1876). On reactions of distrust toward the capacity of the nineteenth-century federal legislative and executive to withstand pressure, see material quoted and cited in Ise, 59, 64–66, 70, 87, 92, 133, 141.

51. The property clause of the United States Constitution is in Art. III, sect. 3. The breadth of construction given the grant, in accordance with the sweep of its terms, may be seen in the opinions in United States v. Gratiot, 14 Pet. 526, 538 (U.S. 1840) ("this power is vested in congress without limitation"; under it Congress may lease as well as sell, in its discretion, and a state may not complain of leasing as an encroachment upon state rights by creation of a numerous tenantry); Gibson v. Chouteau, 13 Wall. 92, 99 (U.S. 1871) ("Congress has the absolute right to prescribe the times, the conditions, and the mode of transferring this property, or any part of it, and to designate the persons to whom the transfer shall be made. No State legislation can interfere with this right or embarrass its exercise . . ." and hence a state statute of limitations may not begin to run against a federal patentee before the patent issues). Cf. Schulenberg v. Harriman, 21 Wall. 44, 64 (U.S. 1874) (Congress has full discretion to determine the mode of asserting or resuming a forfeited grant of public lands of the United States); General Message of Governor Washburn, SJ, App., 30 (1873). The relevant provisions of the Wisconsin constitution are Art. X, sects. 2, 7, and 8. That the state constitution is a limitation and not a grant of legislative power, and that hence the legislature enjoys all general policy-making authority not denied it, was early declared in Bushnell v. Beloit, 10 Wis. 195, 225 (1860). In the years of major state lands disposition, the Wisconsin Supreme Court never delivered an adequate exposition of the over-all constitutional relations between the legislature and the commissioners of school and university lands. Some of its decisions and opinions enforced the exclusive constitutional authority of the commissioners, as against legislative encroachment, in language vigorous enough to suggest the court's sympathy for protecting the commissioners in a wide range of policy control of the lands; but, on their facts, these cases dealt only with control of the detailed administration of the land system. State ex rel. Byrne v. Harvey, 11 Wis. 33 (1860) (held, mandamus does not lie against secretary of state to issue patent, partly because only the commissioners as a body may patent state land); McCabe v. Mazzuchelli, 13 Wis. 478 (1861) (held, alternatively, patent executed by governor and secretary of state is ineffective, since only commissioners may execute patent); Harrington v. Smith, 28 Wis. 43, 66 (1871) (obiter, acknowledging force of the prior decisions, noting legislative obedience to them). The only exception to this focus simply on administrative authority was the ruling in State ex rel. Sweet v. Cunningham, 88 Wis. 81, 57 N.W. 1119 (1894), that the commissioners enjoyed a constitutionally exclusive power to

determine the timing of offerings of state lands, under the grant by Art. X, sect. 8, to them of authority "to withhold from sale any portion of such lands when they shall deem it expedient . . ." The Sweet case disapproved the analysis, without finding it necessary formally to overrule the doctrine laid down previously in State ex rel. Holston v. Commissioners of Public Lands, 61 Wis. 274, 20 N.W. 915 (1884), that all state lands, once publicly offered, must thence remain on the market for private sale. In a majority opinion tinged with strong prejudice against concession of any authority in the legislature to use state lands for economic planning, Marshall, J., in a later generation spoke in language indicating that the commissioners had full substantive authority, exclusively theirs by constitutional mandate, to determine the uses of state lands as well as the timing of their sale; his intimation was that legislative authority was in effect limited to caretaker functions, such as protecting the land against trespass or fire. See State ex rel. Owen v. Donald, 160 Wis. 21, 94, 95, 143, 151 N.W. 331, 354, 372 (1915). Marshall invoked the Sweet case as authority, but, though that decision did enforce the constitutional trust of certain lands as school lands, invalidating a legislative effort to devote them to state park purposes, this aspect of the decision did not amount to a broad declaration in favor of the commissioners' authority, as such, over the legislature's; and on this particular matter of the division of power between legislature and commissioners, the Sweet opinion took pains to emphasize very broadly that though the commissioners might decide when lands should be offered or withheld, "there is no question of the power of the legislature to direct the manner in which the public lands shall be sold" (88 Wis. 81, 85, 57 N.W. 1119, 1121). The court strongly enforced legislative limitations on executive procedure in state lands disposal. State ex rel. Delaplaine v. Commissioners of School Lands, 2 Wis. 423 (1853); State ex rel. Mariner v. Gray, 4 Wis. 380 (1855); State ex rel. McIndoe v. Commissioners of School and University Lands, 6 Wis. 334 (1858); Krebs v. Dodge, 9 Wis. 1 (1859); State ex rel. Crampton v. Commissioners of School and University Lands, 14 Wis. 345 (1861); Gough v. Dorsey, 27 Wis. 119 (1870). With reference to the school trust lands themselves, the court in Smith v. Mariner, 5 Wis. 551 (1856) upheld the legislature's authority to choose a reserved title as security for the price as against a purchase money mortgage, though the latter was explicitly referred to as a mode of security to be taken by the commissioners. See criticism of this ruling in the dissent of Smith, J., 5 Wis. 594, conceding legislative discretion to dispose other than by "sale"; and later criticism by Paine, J., in State v. Weston, 17 Wis. 107, 108 (1863), emphasizing the force of the original decision. In the McIndoe opinion cited above, 6 Wis. 334, 336, the court observed that the legislature's authority to set acreage limits upon the amount of state land purchased by any one buyer "will be undoubtedly admitted." The Sweet case opinion (88 Wis. 81, 85, 57 N.W. 1119, 1121) carefully noted that "as to lands other than school lands, there is no constitutional direction" limiting the legislature's control over disposition. Consistent with this observation was the wide discretion which other decisions conceded the legislature in determining policy for disposition of lands that had not come under the school trust (subject, of course, to obedience to obligations the state assumed in accepting lands granted it for designated purposes). State ex rel. Douglas v. Hastings, 11 Wis. 448 (1861), and Town of La Pointe v. Town of Ashland, 47 Wis. 251, 2 N.W. 306 (1879) (force of legislative determination of extent to which swamplands necessary for devotion to land reclamation); State ex rel. Sturgeon Bay & Lake Michigan Ship Canal & Harbor Co. v. Commissioners of School and University Lands, 34 Wis. 162, 169 (1874) (legislature may delegate value determinations essential to administration of aid grant); Grunert v. Spalding, 104 Wis. 193, 80 N.W. 589 (1899) (recitals in patent issued by governor do not bind state, where legislature gave full dispositive authority to commissioners of its own creation); see also Chippewa Valley & Superior Railway Co. v. Chicago, St. Paul, Minneapolis & Omaha Railway Co., 75 Wis. 224, 44 N.W. 17 (1889) (court zealous to protect legislative dispositive power and trust over aid lands, against agreement of potential bidders to limit competition in seeking grant). The court also carefully curbed the activities of the land commissioners on occasions where it saw these as infringing the legislature's control of the public purse. Orton v. State, 12 Wis. 509 (1860) (commissioners may not bind state to pay for special counsel services); State ex rel. Harney v. Hastings, 12 Wis. 596 (1860) (commissioners held closely to statutory authority to bind state for advertising of land sales); Lawton v. Howe, 14 Wis. 241 (1861) (commissioners may sell only with strict compliance with statutory procedures, to protect state against claims for defective titles); cf. Denniston v. Unknown Owners, 29 Wis. 351, 361 (1872) (construe legislation to guard against premature divestment of public title before trusts accomplished). Thus, whatever the force of Mr. Justice Marshall's determination of the breadth of the commissioners' substantive authority and the limits of the legislature's, as of 1915, the record seems plain that in the years of principal disposition the court in fact treated the legislature as the dominant policy maker within the limits of constitutional trusts or contract obligations.

52. Report, Joint Committee to formulate a plan for investigation of the state offices, AJ, 647 (1855); compare comments of Governor Harvey on the difficulty of holding the legislature to sustained, conscientious surveillance of executive operations, in his General Message, SJ, 13–14 (1862). The quotation from John Quincy Adams is from Brooks Adams, "The Heritage of Henry Adams," in *The Degradation of the Democratic Dogma,* 31. Examples of the occasional legislative investigations of the Wisconsin land commissioners' office are included in materials cited in note 50 above and note 54 below. Legislative preoccupation with narrowly local and special matters in the nineteenth century is discussed in materials cited in note 49 above.

53. Limitations of Presidential leadership in the mid-nineteenth century are noted by Leonard D. White, *The Jacksonians* (1954), 20–22, 49. The two Wisconsin gubernatorial messages referred to in the text to point up, by contrast, the general lack of executive leadership are the General Message of Governor Farwell, SJ, 10–26 (1852), and the Veto Message of Governor Harvey, SJ, 809–819 (1862).

54. Lack of adequate attention to policy making at the Cabinet level in the federal government is noted by Dunham, 17, 18; compare his comments on the indifference of the political parties toward public domain policy, pp. 25, 26, 144, 167, 317. The quoted Senate report is that of the Committee on Public Lands, Senate Report No. 362, p. viii (1882), 47 Cong., 1 Sess. See Dunham's comments, p. 132. The discussion in the Wisconsin constitutional convention is reported in Quaife, *The Attainment of Statehood,* 554–557. It further reveals (p. 556) the extent to which men monetized land issues, that the only other point of discussion on the commissioners' office turned on the safekeeping of funds. See also Harrington v. Smith, 28 Wis. 43, 69 (1871) (practical construction of governing statutes by commissioners entitled especially to weight in interpreting the legislation, because one of the commissioners has always been the state's highest law officer; commissioners' entitlement to fees no basis for assuming prejudice in their construction of their powers). The quoted criticisms of the ex officio arrangement are those in Governor Barstow's General Message, SJ, 16 (1854), and in Reply of Investigating Committee to Memorial of former Secretary of State Alexander T. Gray and former Attorney General George B. Smith, AJ, App., II, 22 (1856). The characterization of the "vast" trusts deriving from the school lands was in Report, Land Commissioners, regarding the Contingent Fund, SJ, 530 (1861). The Annual Report of the Board of Regents of the University, AJ, App., 951–952 (1851), has interesting praise for "the wise economy of the early creation of a special guardianship of the University interests" in the state lands, but neither the Regents nor anyone else applied the moral to the care of the state lands in general. (See end of note 55 below.) Other contemporary comment in effect pointed to the unwisdom of overloading general officers with ex officio duties as to complex issues, but without drawing any conclusions in favor of more vigorous land policy provisions. See Special Message of Governor Randall, SJ, 418 (1859) (commissioners, not governor, should be given land survey duties); General Message of Governor Harvey, SJ, 15, 17 (1862) (recordkeeping hasn't kept pace with tempo of business; loans business cannot be handled adequately within total pressure of affairs); General Message of Governor Fairchild, SJ, 21 (1866) (with all care, commissioners could not cope with detail of adequate loan fund administration). Strong criticism of losses to the school fund from commitment of the lands to officers burdened with other "engrossing" duties was voiced in retrospect in the Report of the State Superintendent of Public Instruction, PD, I, 141 (1892); cf. General Message of Governor LaFollette, AJ, 52 (1901) and Laws, 1901, c. 432. Compare discussion of the unwisdom of overloading the secretary of state with ex officio duties as insurance commissioner, in Annual Report, Secretary of State, PD, I, 73 (1877) (put land

funds management under a separate official); General Message of Governor Smith, PD, I, 13 (1878); Report, Committee on Finance, Banks and Insurance, SJ, 266 (1878); and compare the general retrospective criticism of an undue tendency to create ex officio boards, in Report, Interim Committee on Administration and Taxation, AJ, 273–274 (1927).

55. The inadequate provision by Congress for the General Land Office, the resulting inefficiency and lack of force in that office, and the unavailing protests on policy by various Commissioners, are portrayed in Clawson, *Uncle Sam's Acres,* 64, 82, 104, 110–111; Milton Conover, *The General Land Office* (1923), 22–23, 47, 51–52; Dunham, 17, 18, 23, 51, 58–59, 132, 140, 260, 293, 294, 332; Paul Wallace Gates, "Homestead Law," 652; John B. Rae, "Commissioner Sparks and the Railroad Land Grants," *Mississippi Valley Historical Review,* 25:211 (1938). Pinchot, 80–82, 111, 161–170, vividly depicts the lack of efficiency and imagination in the General Land Office, and with less charitable appraisal than is made by other commentators of either its integrity or various Commissioners' efforts to obtain more means. No one has yet studied the operations history of the office of the Wisconsin land commissioners, but contemporary criticisms and characterizations make plain its limitations. See, e.g., Report, Joint Committee of Investigation into Affairs of the State Departments, PD, 14, 15 (1859) (naïve acceptance of failure of appraisal process, and measure of staff needs of land office wholly in terms of routine handling of sales); Report, Committee on Education and School and University Lands, AJ, 612 (1860); Report, School Land Commissioners, in Reply to Resolution 37A, AJ, 441–468 (1862), and Reply of School Land Commissioners to Resolution 22A, AJ, 329–330 (1870) (lack of records to answer elementary questions on past administration); Report, Committee on State Affairs, in Reply to Resolution 24S, SJ, 343 (1879) (school lands office employs eight clerks, one messenger); compare, generally, material cited in note 50 above. Laws, 1909, c. 527 placed the land office staff under civil service for the first time. The lack of an adequate system for classification of federal lands according to quality or special values is discussed in Clawson, 84, 89; Conover, 51–52; Hibbard, *Public Land Policies,* 488, 490–491. In Wisconsin, contemporary discouragement with the cost of pursuing an adequate field classification finds typical expression in the Report of the Commissioners of School and University Lands, PD, Doc. D, pp. 37, 38 (1860). As late as his Report of Dec. 10, 1892, the State Superintendent of Public Instruction found basis for bitter criticism of the continuing failure to provide adequate classification of state land holdings, concluding that "by this shortsighted economy the state has lost millions to save thousands" (PD, I, 141 (1893). Later commissioners found much to praise in the operation of classification provisions belatedly made under Laws, 1897, c. 367, as amended by Laws, 1899, c. 345, and under Laws, 1901, c. 458. Thus, the Report of the Commissioners of Public Lands, PD, 38 (1902), observed that "prior to 1897 no examination of state lands was ever made," and that "had provision been earlier made for the examination and appraisal of state lands, the trust funds would be larger by some millions of dollars." See also their Report, PD, 4 (1904), and General Message of Governor LaFollette, AJ, 100 (1905). During the years of major state lands disposition, the record shows no material criticism by the commissioners of the adequacy of the state's basic policies toward the lands; as commentators upon, let alone as initiators of, general programs, these ex officio commissioners made no discernible impact. The initiative taken by them to correct by their own rules and interpretations an unduly loose system of small loans from the school land fund, when the legislature remained slow to act in the face of executive warnings of losses and troubles, shows that they were not wholly incapable of creative action. See note 23 above. It is another mark of the deep, monetized bias of official conceptualization of policy that this rare example of initiative took place to the end of preserving an existing money fund. During the years of principal state land disposition, it was the governors who displayed occasional, if disconnected, attention to the more fundamental aspects of the program. See note 53 above. There is some evidence of the impetus toward policy initiative that may go with a sufficiently specialized interest, that the university board of regents complained (in retrospect, it is true) of deficiencies in state lands disposal at a time when the commissioners, with material quantities of land still in hand, showed no concern to correct past mistakes. See Annual Reports of the University Regents, PD, 33 (1870), 8 (1871), 181 (1872); cf. the 1851 report of the Regents, cited in note

54 above, on the virtue of a "special guardianship of the University interests."

Chapter II. The Disposal of the Public Timberlands in Wisconsin

1. The naval timber reserves are discussed, in the light of their significance as the sole positive federal forest program before the seventies, in Cameron, 31–33, 62–89, 99, and in Ise, 23–26. See also United States v. Briggs, 9 Howard 351 (U.S. 1849). The lack of special provisions in federal law for the classification or sale of timberland or forest products is noted in Clawson, *Uncle Sam's Acres,* 84, 89–90, 104–106; Homer Cummings and Carl McFarland, *Federal Justice* (1937), 261–262; Dunham, 2, 10, 17–19, 46–49, 53; Hibbard, *Public Land Policies,* 458–469; Ise, 55, 58, 122–128, 141, 229, 232, 237; Pinchot, 80–90; Robbins, *Our Landed Heritage,* 240–247. The absence of explicit classification provisions, coupled with the plain implication, nonetheless, of intent to limit settlement laws to land suitable for farming, may be seen in the Pre-emption Act of Sept. 4, 1841, 5 Statutes at Large 453, 455, and in the Homestead Act of May 20, 1862, 12 Statutes at Large 392. The Sundry Civil Expenses Appropriation Act of July 31, 1876, 19 Statutes at Large 102, 121, specified that no lands should be surveyed under the funds there granted the General Land Office except those falling within specified classes, the third of which was that of "timber lands bearing timber of commercial value." The history of the 1878 legislation is summarized in Hibbard, 463–469, including the acreage entered under the Timber and Stone Act in Wisconsin to June 30, 1923. The act of Aug. 4, 1892, 27 Statutes at Large 348, extended the Timber and Stone Act to all "public-land States." Cf. act of May 18, 1898, 30 Statutes at Large 418 (Timber and Stone Act extended to include offered as well as unoffered lands). The General Land Office estimated the total of unappropriated federal land in Wisconsin on July 1, 1893, as 627,774 acres, all in northern Wisconsin (Report General Land Office, 1893, p. 131). See Dunham, 336, noting the awareness shown in Congressional debate that through 1891 there was no provision for sale of federal timberland as such in areas east of the Mississippi, including Wisconsin. The outrage of conservationists at the extension, rather than the repeal, of the Timber and Stone Act is evidenced in E. Louise Peffer, *The Closing of the Public Domain* (1951), 46, 48–49, 53, 104, and in Pinchot, 246. The Timber Cutting Act likewise was not repealed, but administrative action much limited its abuse after 1898. See Ise, 240. The Weeks Law was the act of Mar. 1, 1911, 36 Statutes at Large 961. Congress first authorized the creation of national forests by Presidential reservation of unappropriated federal lands in the act of Mar. 3, 1891, 26 Statutes at Large 1095, sect. 24, the policy implications of which are discussed hereafter in this volume; the act came too late, however, to figure as a part of the pattern of public lands disposal policy by the United States in Wisconsin; indeed, first planning for federal acquisition of land suitable for timber growth in the state under the extended Weeks Law did not come until the late 1920's. See L. F. Kneipp, "Federal Activities in Wisconsin Forestry," in Wisconsin Commercial Forestry Conference, *Forestry in Wisconsin* (1928), 44–46.

2. That the bulk of federal timberlands in the Lakes States, including Wisconsin, passed by sale, see Bureau of Corporations, *The Lumber Industry,* vol. I, part I, p. 257; Gates, "Homestead Law," 652, 660, 666–667; William B. Greeley, *Forests and Men* (1951), 33. In the Report of the General Land Office, 1874, p. 6, Commissioner Burdett characterized as "entirely ineffectual" the discretion given the executive to cause timberlands to be sold at public auction to the highest bidder; the field notes were unreliable, and there was "little room for doubt" that combinations regularly held bids to "merely nominal" figures in the sale of eastern as well as western pineland. On the failure of competitive bidding in the face of inadequate classification of land and collusion of bidders in the Wisconsin pinelands, see Current, 122; Fries, 167; and Paul Wallace Gates, *The Wisconsin Pine Lands of Cornell University* (1943), 111–113. The evidence clearly contradicts the sweeping denial of existence of collusion in auctions in Wisconsin which was made by Wisconsin's Senator Howe in the debate over opening Southern lands to sale, 44 Cong., 1 Sess. *Cong. Record,* IV, 1084 (1876). The curb put on Timber and Stone Act transactions by administrative enforcement of prices reasonably reflecting values is noted in Hibbard, *Public Land Policies,* 465. Though the passage of the Homestead Act stirred up considerable unfocused sentiment in favor of ending all cash sales of United States lands (at least those suited to agriculture), direct cash sales of already surveyed land, and

sales by pre-emption, continued until both were finally ended in a series of acts passed in the years 1889–1891; thus through the years of major timberland disposition in Wisconsin and the other Lake States there remained nonfraudulent means of purchasing federal land outside the procedures of the Homestead Act. See Gates, "Homestead Law," 652, 660, 666, 678, 681; Hibbard, *Public Land Policies,* 111–112.

3. Bona fide entry of timbered land by pre-emption and as homesteads for the opening of farms is noted in Hibbard, *Public Land Policies,* 457, 459, and in Joseph Schafer, *Four Wisconsin Counties: Prairie and Forest* (1927), 107–108. Houlton v. Nichol is reported in 93 Wis. 393, 67 N.W. 715 (1896). Cf. McCord v. Hill, 111 Wis. 499, 84 N.W. 27 (1901), affirmed, 195 U.S. 395 (1904) (false statements not relevant to satisfying terms of particular land act under which claim is made do not invalidate claim). The extreme tenderness toward colorable title once vested under transfers from the United States, and the great difficulty of satisfying the court of the existence of fraud sufficient to vitiate a patent obtained by pre-emption or direct purchase, may be seen in Colorado Coal & Iron Co. v. United States, 123 U.S. 307, 316 (1887), and United States v. Budd, 144 U.S. 154, 161, 163, 167 (1892).

4. On the lack of separate pre-emption data, see Hibbard, *Public Land Policies,* 170. Homestead entries in Wisconsin were not tabulated officially in such fashion as to allow their attribution to forested areas. Raney (p. 88) thought that there was considerable resort to the Homestead Act in northern Wisconsin.

5. See Report, Committee on Schools, HRJ, 76, 80, 82 (1841), noting the Territory's dependence on Congressional sales policy, without substantial criticism of any points of substance.

6. See Chapter I, notes 29, 35, 38, and 39. The casual acceptance of the forest because of its apparently inexhaustible size, in the period of Wisconsin constitution making, may be seen in the volumes edited by Quaife, *The Movement for Statehood,* 77; *The Convention of 1846,* 561; *The Attainment of Statehood,* 3. Contemporary recognition of the lumber industry as a component of the state's economy may be seen in Quaife, *The Struggle Over Ratification* 654; cf. Merk, 60. Relatively early acceptance of the idea that the state timberlands were to be taken for granted as simply another part of the general land trust for education may be seen in State v. Weston, 17 Wis. 107 (1863) (as vendor with reserved title and cutting stipulation, state enjoys only ordinary right of such a seller to protection of the security for the unpaid price).

7. The Wisconsin Constitution, Art. VIII ("Finance"), sect. 10, recognized the possibility of receipt of land from the United States for specified works of internal improvement, when it excepted such cases from the general ban it laid upon state participation in such works. That legislative disposition of swamplands for purposes other than reclamation amounted to a legislative determination that the lands so disposed of were not needed for the primary trust to promote drainage — a determination entitled to almost conclusive deference from the courts — but that swamplands thus found not necessary for reclamation forthwith fell under the constitutional trust for education was ruled in State ex rel. Owen v. Donald, 160 Wis. 21, 93, 95, 100, 151 N.W. 331, 357, 361 (1915). Though other aspects of the Donald case should be treated with caution so far as the court's opinion claims for them the status of doctrine in fact recognized a generation earlier, this aspect of the decision seems convincing in its retrospective view, being consistent with general nineteenth-century efforts to enforce the "trust" of swamplands. See Part One, Chapter I, notes 7 and 12. Cf. State ex rel. Douglas v. Hastings, 11 Wis. 448 (1861). What the text says of the lack of special disposition provisions for timbered federal aid lands refers to affirmative program; we shall later note that some special reference was made to the timber component in the value of such lands, so far as concerned the negative problem of combatting trespass. See the comment on the absence of provision for particular realization upon the timber value of federal railroad aid grants to states, either in federal or state law, in Bureau of Corporations, *The Lumber Industry,* vol. I, part I, p. 245. The end of tax exemption of aid lands upon sale of timber was stipulated by the Wisconsin legislature in Laws, 1877, c. 21 (Wisconsin Central Railway Company), and Laws, 1879, c. 22 (North Wisconsin Railway Company); see, too, P & LL, 1867, c. 429 (military road lands). The limited scope of attention so given the matter is underlined not only by absence of like provision in earlier tax exemption provisions for the same lands (cf. P & LL, 1866, cc. 314 and 362), but by the failure to make like stipulation as to other aid lands. See General Laws, 1864, c.

324, 1866, c. 73, 1867, c. 5, 1868, c. 105; P & LL, 1869, c. 407; General Laws, 1870, c. 104. It was likewise characteristic of the episodic development of almost any phase of policy touching timberland that at first the legislature did not stipulate for termination of the tax exemption upon timber sales in earlier forms of the privilege as granted the North Wisconsin Railway Company itself. See P & LL, 1869, c. 51, and Laws, 1878, c. 261. One exception to the general neglect of positive realization upon timbered aid lands may perhaps be found in the position taken, across the board, by all three principal branches of state government, that the federal lands given in aid of construction of a ship canal connecting Green Bay and Lake Michigan — lands which in fact included valuable timbered tracts — should be disbursed in installments by equal parts of market value rather than simply by equal parts of land area. See, especially, State ex rel. Sturgeon Bay & Lake Michigan Ship Canal & Harbor Co. v. Commissioners of School and University Lands, 34 Wis. 162 (1874), and references in note 10 below.

8. 88 Wis. 81, 83, 57 N.W. 1119, 1120 (1894). An Opinion of the Attorney General to the Chief Clerk of the office of the Commissioners of Public Lands, Sept. 25, 1911, Attorney General of Wisconsin, *Opinions,* X, 89, seems correctly to apply the Sweet case in ruling invalid as to school lands the reservation by Laws, 1911, c. 452, sect. 6, of land to a chain's width, in fee, for the state along all land bordering any water navigable in fact; the reservation was made without any limitation which would show a relation to effecting the purposes of the educational trust. Like the Sweet decision, however, this ruling contains nothing adverse to a program of timber management on state reserves for the benefit of the educational trust.

9. See opinion of Marshall, J., for the court, in State ex rel. Owen v. Donald, 160 Wis. 21, 95, 115, 151 N.W. 331, 357, 362 (1915); cf. concurring opinion of Winslow, C. J., 160 Wis. 158–160, 151 N.W. 377.

10. The quoted observations from State ex rel. Sweet v. Cunningham will be found in 88 Wis. 81, 84, 85, 57 N.W. 1119, 1120, 1121 (1894); for the ruling on interpretation of the act requiring public sales of withdrawn lands when re-offered (Laws, 1885, c. 222), see 88 Wis. 81, 85, 87, 57 N.W. 1119, 1121, including the care taken to point out the "oversight or mistake," but not formally to overrule State ex rel. Holston v. Commissioners of Public Lands, 61 Wis. 274, 20 N.W. 915 (1884) (construing prior legislation as binding the commissioners to sell at private sale to any proper applicant state lands once offered at public sale and unsold, while conceding, however, the plenary power of the legislature to lay down a different rule under its constitutional authority to set terms for state lands disposal). The practical construction of its constitutional authority over terms of state lands disposition by the Wisconsin legislature is reflected in material cited in note 11 below. For Smith v. Mariner see 5 Wis. 551, 583 (1856). The liberality of the majority's interpretation of the constitutional standard is highlighted by the vigorous dissent of Smith, J., 5 Wis. 585, 587, 590, 594, against allowing the legislature to create any other kind of security for the unpaid price than the mortgage specifically referred to in the constitution; it should be noted, however, that in his own fashion Smith, J., was not disposed to make a rigid reading of the constitution, but would (5 Wis. 592) construe the constitutional authorization of "sale" to allow other transfers than an absolute passing of full title, including leases (5 Wis. 594). Though Paine, J., voiced passing criticism of the majority's interpretation of the constitution in Smith v. Mariner, in State v. Weston, 17 Wis. 107, 108 (1863), the basic decision stood as the doctrine of the court. See Smith v. Clarke, 7 Wis. 551, 562 (1858); Whitney v. State Bank, 7 Wis. 620 (1858); Gough v. Dorsey, 27 Wis. 119, 129 (1870). State v. Weston, 17 Wis. 107 (1863), contained no discussion of the statute, the validity of which it accepted without question, as the act limited timber cutting by the buyer from the state who had not yet paid in full for the land; the opinion does indicate that probably the court found it the easier to take the statutory stipulations in stride because it found them so fully analogous to familiar rights enjoyed by a private seller under his vendor's lien (17 Wis. 107, 108). The Sturgeon Bay canal aid lands grant was interpreted in State ex rel. Sturgeon Bay & Lake Michigan Ship Canal & Harbor Co. v. Commissioners of School and University Lands, 34 Wis. 162, 168 (1874). Of similar import were the rulings in Chippewa Valley & Superior Railway Co. v. Chicago, St. Paul, Minneapolis & Omaha Railway Co., 75 Wis. 224, 44 N.W. 17 (1889) (agreement not to compete before legislature for railroad aid lands grant held unenforceable because illegal, in part because the consideration was a share in the lands,

which would thus be subtracted from the yield intended to launch the grantee road), and in Grunert v. Spalding, 104 Wis. 193, 80 N.W. 589 (1899) (tax exemption of military road aid lands in the hands of the road contractor or his assigns, held not to benefit an assignee who had not bound himself to construction or costs of any part of the road). That the Sturgeon Bay canal project lands were valuable in large part for timber is reflected, for example, in Annual Report, Secretary of State, PD, I, 50–53 (1876) (report of commissioners of investigation appointed under Laws, 1876, c. 286) (proceeds from timber trespass actions noted; company is holding non-agricultural lands off market); and in Laws, 1877, Memorial No. 1 (request for additional aid, especially because the great forest fire of 1871 severely depreciated the original grant). Interpretations enlarging the scope of requirements of public offerings may be seen in Eldred v. Sexton, 30 Wis. 193 (1872) (federal land), and in State ex rel. Sweet v. Cunningham, 88 Wis. 81, 85, 57 N.W. 1119, 1121 (1894) (state land).

11. In addition to the noted features of Laws, 1849, c. 212, suggesting a general policy of adapting sales terms to exceptional assets, note Laws, 1848, p. 58, sect. 16 (reserving for further state disposition water-power sites existing or created in building Fox-Wisconsin improvement), as qualified by Laws, 1850, c. 275 (preserving rights of previous pre-emptors to existing sites).

12. Variations on the themes specially pertaining to timberland in the 1849 act may be seen in Acts, 1850, c. 236 (thirty-year credit on tracts in the 500,000-acre grant accompanying statehood); Acts, 1852, c. 340 (Fox-Wisconsin lands, ten-year credit, no timber reference); Acts, 1852, c. 493 (commissioners may waive down payment on university lands, but with added security against waste if needed); General Acts, 1855, c. 21 (full cash payment required for pinelands at time of sale); General Acts, 1855, c. 184 (cutting limit on swampland pre-emptors); General Acts, 1856, c. 125 (minimum 50 per cent down payment required on swamplands shown by field notes to bear timber); General Laws, 1860, cc. 95 and 341 (added sanctions against wrongful cutting by certificate holder) and c. 341 (valuation of state land to determine eligibility for switch from certificate to purchase-money-mortgage status shall be made without reference to standing timber thereon); General Laws, 1863, cc. 160 and 287 (unqualified discretion to commissioners to require full payment at time of sale under graduated prices for offered and unsold swampland and school or university lands); General Laws, 1864, c. 156 (restatement of cutting limits on swampland pre-emptors); General Laws, 1866, c. 121 (Morrill Act lands); General Laws, 1867, cc. 22 and 139 (Marathon county lands conveyed to state in settlement of county debt to state, and swamplands to be sold only for cash); Revised Statutes, 1878, c. 15, sect. 209 (summarizing price terms).

13. The limited, price-security interest in the pine timber represented by such special statutory provisions as were made concerning timber values was specifically and accurately reflected in State v. Weston, 17 Wis. 107 (1863), holding that the state's measure of damages in a civil action for unauthorized timber cutting by its vendee, in possession and owing an unpaid balance on the land's purchase price, was the extent of impairment of the state's security interest. The land commissioners are quoted from their report of Oct. 10, 1882, PD, I, 28. Land unsold at public offering would go to the highest bidder among competitors at private sale, but was authorized to be sold at the statutory minimum in absence of competition; e.g., Revised Statutes, 1878, c. 15, sect. 211. In terms, the statutes for some years might appear to have denied the commissioners discretion to withhold from sale at the minimum price. But, so construed, the legislation plainly would infringe the discretion committed to the commissioners by the constitution, so that they cannot be absolved of responsibility for allowing lands to go routinely at the lowest price. See State ex rel. Sweet v. Cunningham, 88 Wis. 81, 57 N.W. 1119 (1894).

14. See statutes cited in note 12 above. Another instance of episodic development of policy, earlier noted in the text, was the initial creation of a thirty-year credit term only for the 500,000-acre federal grant, its extension to all school and university lands in 1858, and the stipulation of the old ten-year terms for the agricultural college lands in 1866, while retaining the thirty-year span for other educational trust lands until 1878. The general lack of attention to full realization of special timber values in the various grants in aid of transport development is underlined by the unique care taken for at least some aspects of the matter in Laws, 1879, c. 22. There, on the one hand, the legislature in familiar fashion stipulated for reasonably early sale by the

railroad of lands suitable for farm settlement, on terms limited as to price; on the other hand, however, the statute explicitly left authority in the railroad to realize what better terms it could on all standing timber by reserving it from the sale of land, and on all land not "agricultural" by a test stated in the act to be whether at the date of sale any given 40-acre tract contained more than 100,000 merchantable feet of pine. Of course, this simply preserved the railroad's power to exercise its own bargaining will, and put no curb on such desire as the road might have to press sales for immediate advantage at the expense of greater yield at a slower pace of exploitation. The continued failure to extend the requirement of full cash payment at time of sale for hardwood lands is reflected in an informal Opinion of the Attorney General to Chief Clerk, State Land Office, Dec. 4, 1897, Attorney General's Letter Books, III, 36 (Revised Statutes, sect. 209, authorizes commissioners to require full cash payment at time of purchase "in the sale of hard wood lands, and in fact all the lands, not included in the lands designated as swamp lands and pine lands, and which are commonly known as State Park Lands").

15. By the act of Feb. 18, 1842, Laws, 1841–1842, p. 45, the territorial legislature authorized commissioners of common schools in organized counties to lease school lands within their town limits for terms not exceeding four years and "to dispose of all fallen timber on said lands" as well as "to sell any dead or decaying timber, standing on sections numbered sixteen," but restricted their authority with the declaration that "no green or living timber shall be sold, deadened, cut down, or taken off said lots, by virtue of any lease, which may be made by said commissioners." On its face, this legislation fitted the policy of the ensuing fifty years; so far as it did provide special care for timber values, obviously this was only to preserve the immediate sale value of the land, with no thought to conferring authority to direct a program of timber cropping. Cf. Revised Statutes, 1878, c. 17, sect. 257, originating in General Laws, 1865, c. 537 and General Laws, 1869, c. 151 (town boards may license cutting of hay or picking of cranberries on swampland for up to one year, but without granting any timber cutting rights), enlarged by Laws, 1907, c. 97 (but still without timber rights); Laws, 1897, c. 367 (ambiguous provision, possibly authorizing separate sale of standing timber on state park lands). The lack of general authority to sell even burned or downed timber from state lands is reflected in P & LL, 1872, c. 104, granting such permission regarding timber on the Sturgeon Bay canal aid lands damaged by the great forest fire of 1871. See AJ, 298, 315, 339 (1872) (proceedings regarding introduced Joint Resolution 31A and Bill 434A, on the same subject); AJ, 382 (1872), Report, Committee on State Lands, in response to Resolution 44A, reporting that, since the land commissioners inform the committee that no state lands were overrun by fire, no legislation is needed to authorize the commissioners to sell timber on such lands. The commissioners finally made specific request for authority to sell timber separate from land, at least in burned-over areas, in their report of Oct. 10, 1902, PD, 39. They renewed the request, now limited to fire- or wind-damaged timber, in their report of Sept. 1, 1904, PD, 29, and the legislature responded, within these limits, in Laws, 1905, c. 322. Cf. Laws, 1905, c. 264 (broad management powers regarding timber on the new forest reserve given to the State Forester and Board of Forestry); Laws, 1911, c. 238 (amending Revised Statutes, sect. 257, to authorize Board of Forestry to make hay and cranberry bog leases). See also Biennial Report, State Treasurer, PD, 4 (1910). Laws, 1925, c. 159, finally broadened the commissioners' authority to sell timber from the state lands, without regard to fire or wind damage. Possession of such authority did not necessarily spell introduction of timber crop management techniques, it should be noted. Thus, the Report of the Interim Committee on Forestry and Public Lands, AJ, 882 (1929), sharply criticized the commissioners for selling timber-cutting rights without stipulations designed to safeguard the conditions of future growth. The creation of the state park forest reservation by Laws, 1878, c. 324, was early questioned by the land commissioners in their report of Oct. 10, 1878, PD, I, 12, both on the ground of utility — the reserved tracts being relatively small and widely scattered — and of propriety, as a diversion of assets from their trust purposes. Compare the strong criticisms of a "lock-up" policy in general, in Pinchot, 70–71, 108–109. The 1878 reserves provision was held invalid, as a violation of the constitutional trust of school lands, in State ex rel. Sweet v. Cunningham, 88 Wis. 81, 57 N.W. 1119 (1894).

16. Contemporary discussion and official

debate surrounding the framing of the Wisconsin constitution give no help in understanding how ambitious an undertaking the framers may have had in mind by provision for "appraisal." Given the current tendency to accept the northern timber resources as practically "boundless," it seems likely that no one thought of the forest as presenting an appraisal or classification problem of moment. See note 6 above.

The appraisal statutes referred to in the text are Laws, 1848, p. 123 (act of Aug. 12, 1848), repealed as to areas north of the Fox and Wisconsin Rivers by Laws, 1849, c. 68 (see General Message of Governor Dewey, Jan. 11, 1849, SJ, 25), but again extended by Acts, 1849, c. 212, sect. 105 (preserving work in progress), and Acts, 1852, c. 124 (twelve northcentral and northwest counties) and c. 490 (unappraised part of 500,000-acre grant); General Acts, 1854, c. 67 (La Pointe and Douglass counties); Revised Statutes, 1858, c. 28, sects. 8, 9, and 15. See also, Laws, 1850, c. 236 (original appraisal of 500,000-acre grant). In the swamplands selection matter, Governor Dewey initially recommended that the legislature provide him with means for firsthand examination of the lands to be selected (General Message of Jan. 9, 1851, SJ, 11), but unfortunately when he confronted the concrete issue he failed to press this procedure. The history of the swamplands selection by recourse to the federal field notes, and typical criticisms of the costs to the state in lost acreage, may be seen in Governor Randall's General Message, SJ, 23 (1860) (Dewey's choice was an "unfortunate way" to select the lands), and his Special Message, AJ, App., 1–61 (1860). Cf. Special Message of Governor Farwell, SJ, App., 4, 6–8 (1853) (state has no apparatus for its own identification and selection of swamplands). Though the appraisal system for school lands was enacted with expressions of concern for its importance and emphasis on the fact that there was no authorization of sales until appraisal — see Report, Committee on Education and School Lands, AJ, App., 83 (1848), and Report, Committee on Education and School Lands, SJ, 195 (1849) — as early as Jan. 14, 1849, in a special report on the progress of appraisals, the Secretary of State indicated that the quality of the product was uneven, with "the returns . . . in almost as many various forms as there were appraisers . . ." (SJ, App., 828).

The criticisms of the appraisal system cited in the text are from Annual Report, Secretary of State, 1852, SJ, App., 21, 22 (1853); Report, Joint Committee of Investigation into Affairs of the State Departments, PD, 14 (1859) ("The whole system of appraising lands your committee regard as worse than useless. In but very few instances comparatively are the lands appraised at more than $1.25 per acre, which is the minimum price established by law. Many complaints have been made to your committee in regard to bills allowed to the appraisers, and special investigations have been persistently urged, but we have had no time if we have the authority to go behind the vouchers on file in the Secretary's office, we have no doubt however that large amounts of money have been paid to the appraisers and their assistants for labor never performed"); Report, Committee on State Affairs in response to Resolution 43A, AJ, 498 (1872); *Report and Explanatory Notes of the Revisers of the Statutes* (Madison, 1878), p. 26 (comment on sect. 196). For other reflections of the inadequacy of appraisal or classification provisions, see Annual Report, Board of Public Works, AJ, App., 559 (1850) (criticizing legislature's failure itself to make provision, or to empower the Board to make provision, to realize best values on water-power sites); Report, Board of Commissioners of School and University Lands, AJ, App., 996 (1851) (legislature's failure to name appraisers under Laws, 1850, c. 236, regarding the 500,000-acre grant is costly because delayed sales invite timber trespass); Annual Report, Board of Regents of the University, SJ, App., 5 (1853) (criticizing lack of "painstaking diligence" in original selection of university aid lands tendered by United States); Report, Committee on State Lands, AJ, 100 (1854) (absence of swamplands classification and lack of confidence in appraisal procedure reflected in recommendation that legislature define classes of such lands for sale and appoint commissioners "to examine and classify the land under such law as being . . . much more economical than the employment of appraisers, surveyors, and assistants, as has been done in other cases"); AJ, 791 (1855) (abortive resolution for summary marketing of swampland, "without surveying, appraising, or any other expense, always intended to squander one half of these valuable grants"); Reply of Joint Select Committee to Investigate the State Offices, to Memorial of George B. Smith and Alexander T. Gray, AJ, App., 23 (1856) (no evidence that some lands within appraisal procedure were ever appraised; insufficient evidence that others, for which appraisal bills were filed and paid, were in fact appraised); AJ, 253 (1857) (resolution asking Committee on Public Lands to report on expediency of swamplands survey to determine, *inter alia,* how much valuable timber is contained; no reply); Communication from Land Commissioners in response to Resolution 60A, AJ, 331 (1863) (reflection of no existing appraisal machinery for swamplands; judgment that "the benefits to be derived from an appraisal of said swamp lands would not warrant the expense"); Annual Report, University Regents, PD, 180 (1872) ("No proper selection or appraisal of the university lands was ever made"); Annual Report, Secretary of State, PD, I, 52, 53 (1876), and Report, Committee on State Affairs, AJ, 410, 416, 417 (1879) (inefficient appraisal of Sturgeon Bay canal project lands); Biennial Report, State Superintendent of Public Instruction, PD, I, 141 (1892) (state has lost over the years, for want of adequate appraisal or classification procedures); Report, Agricultural Experiment Station of the University of Wisconsin, Dec. 12, 1903, W. A. Henry, "Brief History of the Agricultural College," PD, 13 (as to Morrill Act lands: "Had the legislature acted as it should in providing for the judicious selection and business-like sale of these lands, at least ten times as large a sum of money would have been received therefor, since there were at that time many valuable pine lands still available for selection"); Report, Commissioners of Public Lands in response to Laws, 1911, Joint Resolution No. 3, AJ, 290 (1911) (commissioners cannot answer legislature's question as to amount of timber on state lands sold in past five years, "for the reason that the commissioners are not authorized to make estimates of timber upon such lands, separate from the appraisal of the land"). The first general appraisal law in modern times was Laws, 1917, c. 454.

17. Frauds on the federal pre-emption and homestead laws affecting timberland are noted in Communication of Commissioner Butterfield of the General Land Office, Nov. 28, 1849, 31 Cong. 1 Sess., Exec. Docs., 20; Report, General Land Office, 1875, p. 10, 1876, p. 8, 1877, pp. 25, 35, 1879, p. 186, 1880, p. 178, 1882, pp. 8, 12, 250, 1883, pp. 5, 7, 1884, pp. 6, 18, 1885, pp. 4, 9, 1888, pp. 78 (report of Register and Receiver, Eau Claire, July 26, 1888) and 87 (report of Register and Receiver, Wausau, Aug. 1, 1888); statement of H. H. Schwartz, chief of field service, General Land Office, in Report, National Conservation Commission (1909), 60 Cong. 2 Sess., Senate Doc. No. 676, III, 390. Judge Bunn's remarks are quoted from Vilas v. Prince, 88 Fed. 682, 685 (Circ. Ct. W.D. Wis. 1898). That the most extensive frauds on the federal land laws were, in general, committed to obtain timber, see Cameron, 81, 82; Clawson, *Uncle Sam's Acres,* 104; Pinchot, 80–82. On this problem in the Lake States particularly, see Dunham, 153, 154, 157; Robbins, *Our Landed Heritage,* 240–247; Isaac Stephenson, *Recollections of a Long Life* (1915), 120. Frauds on the federal settlement laws to obtain timber found occasional reflection in litigation. Reports reflect no strong, sustained government initiative against frauds, however; a check by paging of the reported decisions of federal courts in Wisconsin showed only two proceedings for timberland fraud against the settlement laws, both involving the inherently small-scale problem of alleged wrongful cutting by a homesteader in excess of his needs for maintenance or clearing for cultivation. United States v. Lane, 19 Fed. 910 (Circ. Ct. E.D. Wis. 1883) (if homesteader cut in good faith to clear for cultivation, he may rightfully sell excess timber); United States v. Freyberg, 32 Fed. 195 (Circ. Ct. E.D. Wis. 1886) (where, after wrongful cutting, homestead entrant exercises his right under federal statute to commute his entry by cash payment, he thereby acquires full equitable title — though a patent be not yet issued — which relates back and is defense to action for waste); see Vilas v. Prince, 88 Fed. 682, 684 (Circ. Ct. W.D. Wis. 1898) ("The instances where homesteaders have been convicted for cutting timber from the land claimed are numerous"). Cf. United States v. McCord, 72 Fed. 159 (W.D. Wis. 1895), and United States v. Stinson, 125 Fed. 907 (C.C.A. 7th. 1903) (alleged fraudulent pre-emption entries on non-timbered land; government barred in first by statute of limitations, in second by laches); Ex parte Black, 147 Fed. 832 (E.D. Wis. 1906) (indictment for conspiracy to defraud United States by procuring fraudulent entries on Oregon land under Timber and Stone Act). The lack of resources at the center to spur federal prosecuting zeal, and the difficulties of coming to terms with local apathy or favor toward timber exploitation, are reflected in Cummings and McFarland, 261, 264, 387, 389. The state courts would relieve against fraud on the federal settlement laws in appropriate instances;

naturally, the complaining party before the state tribunal would be, not the United States, which would resort to its own courts, but a private title claimant who challenged an opposing interest as invalid for fraud. However, the fact that most fraudulent entrants were content to cut and abandon is reflected in the circumstance that but one clear timberland case appeared, even under the spur of private interest. And here, in Week v. Bosworth, 61 Wis. 78, 20 N.W. 657 (1884), the issue turned in nowise on the fact that timbered land was at stake, but wholly on a fiscal interest in enforcing the statutory ban on premature assignment of rights to military warrants. It is likewise undoubtedly significant of the cut-and-run pattern of timberland dealings that the state court confronted several cases of alleged fraud on the federal settlement laws where the stakes were agricultural lands, title to which was in itself the prize. See Nichols v. Nichols, 3 Pinn. 174 (Wis. 1851); Bross v. Wiley, 6 Wis. 485 (1857); Stephenson v. Wilson, 37 Wis. 482 (1875); McCord v. Hill, 111 Wis. 499, 84 N.W. 27 (1901), s.c., 117 Wis. 306, 94 N.W. 65 (1903), affirmed, 195 U.S. 395 (1904); cf. Gates v. Winslow, 1 Wis. 650, 659 (1853); Dillingham v. Fisher, 5 Wis. 475 (1856); Knight v. Leary, 54 Wis. 459, 11 N.W. 600 (1882). Of course, even in the northern pinery, a homestead entrant might in good faith seek to clear the land of its timber in order to get on with cultivation. Such a situation seems to be reflected in Daniels v. Bailey, 43 Wis. 566 (1878).

18. On swamplands pre-emptions, see General Acts, 1855, c. 84; General Message of Governor Barstow, SJ, 23 (1856); Special Message of Governor Barstow, AJ, 289–293 (1856) (transmitting report of William Crombie, timber agent in the field for the land commissioners, warning — pp. 291–292 — that "nearly every man who has been found engaged in cutting timber upon these land[s], either is or pretends to be a pre-emptor, and in a great majority of cases, both real and pretended . . . pre-emptors are either speculators or speculator's hirelings from other States; and it is my belief that but a small number of these persons will ever pay the State for the lands whose principal value they take away, as they are under no earthly obligation, in law or otherwise, to take the lands, and their main object is to maintain possession long enough to get the timber off and sell it"); General Acts, 1856, c. 125 (repealing the swamplands pre-emption privilege for the future); General Acts, 1857, c. 57 (restoring the privilege), carried into Revised Statutes, 1858, c. 29, sect. 14, and Revised Statutes, 1878, c. 15, sect. 198.

The threat of fraudulent pre-emption filings to obtain timberland was reflected as a continuing administrative problem in the return filed by the land commissioners to a petition for a writ of mandamus to compel issuance of a patent, in State ex rel. Anderson v. Timme, 60 Wis. 344, 346, 18 N.W. 837, 837–838 (1884). The commissioners' zeal was not so great in cases involving sizable stakes as to provoke important litigation, however; the Anderson case is the only one reported before the Wisconsin Supreme Court directly involving official challenge to a claim of title on the basis of alleged fraud to obtain timber. The only case in which a private suitor relied on official action setting aside an adverse claim on these grounds is Burrows v. Rutledge, 76 Wis. 22, 44 N.W. 847 (1890). It must be said that, within the limited opportunities presented to it by these two cases, the court ruled and spoke vigorously in definition of the state's right and intent to protect itself against such impositions. The judges seemed more inclined toward leniency in refusing to find fraud where a bona fide agricultural use was in question. See Allen v. Allen, 58 Wis. 202, 16 N.W. 610 (1883). Cf. Bross v. Wiley, 6 Wis. 485 (1857) (vigorous reaction, where fraud clear).

The reality of the continuing threat of fraudulent use of the cover of a claim of title to screen the stripping of timberland was attested by the Veto Message of Governor Rusk, AJ, 842 (1887), disapproving bill 23A consenting to recovery from the state by its grantee whose state-derived title had proved faulty under attack by an adverse claim. Noting that his views had the "full concurrence" of the land commissioners, the governor argued that bill 23A unwisely exposed the state to imposition, especially where timber was the prize: "No provision is made for giving the state notice of the pendency of the proceedings [between the adverse private titleholders, leading to invalidation of the claimant's interest], nor any opportunity to prevent or defeat the recovery of collusive judgments. A party might purchase lands of the class described, which was valuable chiefly for the timber growing thereon, cut off the timber, and then, in collusion with the original owner, allow judgment to go by default, or failure to plead a statute of limitation, and thereupon recover from the state the original purchase price, with interest" AJ, 842 (1887). The Assembly sustained the veto by a record vote of 1–58 (AJ, 843).

Of course, this was a purely negative response to the danger of fraudulent title claims. The legislature took no positive steps for vigorous action against fraud on the state land laws through all the years of the lumber boom. As late as 1911 the land commissioners could plead inability to answer questions posed them by the legislature under Laws, 1911, Joint Resolution No. 3, as to the quantity of timbered public lands sold in the past five years and the amount of lands sold to "actual settlers." Tartly, the commissioners informed their legislative questioners: first, that they had no knowledge of the quantity of timber on lands sold, "for the reason that the commissioners are not authorized to make estimates of timber upon such lands, separate from the appraisal of the land"; secondly, that they had "no knowledge nor any means of ascertaining, nor are they required by law to ascertain whether the purchasers of public lands are actual settlers or otherwise"; and, finally, that the commissioners' whole staff then consisted of a chief clerk, an assistant, and a stenographer, but that, within the limits of their working force, they would do their best to answer such questions as to which their regular office records afforded material. See Report, Commissioners of Public Lands, AJ, 289–290 (1911); cf. their Biennial Report, PD, 3, 4 (1912); General Message of Governor McGovern, AJ, 16 (1912). Though the 1911 exchange does not bear directly on fraud, it is a relevant reminder of the continuing lack of provision for any close legislative or administrative surveillance of the operation of the state-lands laws. An over-all judgment on the whole Wisconsin experience with the operation of settlement laws in the pineries was passed by Benjamin H. Hibbard, "Land in Relation to Forestry," in Wisconsin Commercial Forestry Conference, *Forestry in Wisconsin* (1928), 25–26: "The preempting and homesteading [in the Wisconsin forest belt] were transparent frauds in the majority of cases since there was no expectation of making homes and farms on the land so acquired. The purpose was the acquisition of timber." And see Fries, chap. xi.

19. For the background of the first down payment provisions, see Report, Committee on Education and School Lands, AJ, App., 87 (1848); General Message of Governor Dewey, SJ, 26 (1849). The commissioners' extremely liberal use of their statutory discretion to grant up to thirty years' credit on the security of the land itself, and the quoted criticism of the 1855 legislative committee, may be found in Report, Select Committee under a resolution of inquiry — AJ, 204 (1855) — regarding the land commissioners, SJ, App., 7–9 (1855). The Secretary of State is quoted from his Annual Report, AJ, App., Doc. B, I, 83 (1857). Cf. Report, Commissioners of School and University Lands, AJ, App. Doc. D, I, 30 (1857) (recommending repeal of requirement of greater down payment for timbered land, for want of means provided to determine land's character); Report, Joint Committee of Investigation into the Affairs of the State Departments, PD, 15 (1859) ("Your committee are not aware that the wholesome provisions of [General Acts, 1855, c. 21, on pinelands, and General Acts, 1856, c. 125, on swamplands] . . . have ever been regarded in the sale of the timbered lands of the state, the appraisers in accordance with the provisions of law, have very generally in their returns specified the character of the lands and the kinds of timber thereon, but the very best timbered lands have been sold to speculators often at the minimum price, and only ten per cent. It is well known that many of these lands are being stripped of the timber, and when this is accomplished, they will be forfeited, and thereby come back to the state"). The comment in the 1859 report on return of land as timbered in the appraisals, it should be noted, seems contradicted in substance by criticism in the same document of the appraisals as worthless because almost invariably made at the statutory minimum price, and in any case the comment runs counter to the weight of all other contemporary evidence on the unreliability of the appraisals, as this is cited in note 16 above. The further picture of frauds or breaches of contract to the detriment of the state's credit-sales program for its lands may be seen in Report, Commissioners of School and University Lands, PD, Doc. D. p. 38 (1860); Report, State Superintendent of Public Instruction, Dec. 10, 1860, PD, 33 (1861); General Message of Governor Randall, SJ, 14 (1860); General Message of Governor Harvey, SJ, 17 (1862); Biennial Report, State Superintendent of Public Instruction, Dec. 10, 1892, PD, I, 142 (1893). On the general lack of change in credit-sales terms through the late nineteenth century, see also Report, Commissioners of Public Lands, PD, I, 4, 29 (1882).

20. Annual Report, Commissioners of Public Lands, PD, I, 28 (1882). See, also, material cited in note 2 above. On but one occasion did the land commissioners set aside a sale of state lands (probably timbered) on the ground of collusive bidding, in a situation important enough to cause it to be noted in their reports. See Biennial Report, Commissioners of Public Lands, Oct. 10, 1884, PD, II, 6 (1885) (11,514.22 acres of swampland).

21. On the accepted character of the practice of holding land only for the duration of timber investment and lumbering operations, see Fries, 177; Hibbard, "Land in Relation to Forestry," 26; Reynolds, 23; cf. Agnes M. Larson, *History of the White Pine Industry in Minnesota* (1949), 402. The combination of tax delinquency and sale at very low prices to companies speculating on agricultural settlement is pictured in Arlan Clayton Helgeson, "Nineteenth Century Land Colonization in Northern Wisconsin," *Wisconsin Magazine of History,* 36:115, 119 (1952); Edmund C. Espeseth, "Early Vilas County — Cradle of an Industry," *Wisconsin Magazine of History,* 37:27, 34 (1953).

22. 21 Wis. 423 (1867). See letter of the Attorney General to J. W. Hammond, Sept. 12, 1901, Attorney General's Letter Books, VIII, 409 (correspondent states "that at an early day it was not unusual for persons buying state lands to pay for them in full, but take out no patent," and asks for a bill to patent such lands to persons who have since bought them on tax sales; Attorney General's "informal opinion" is that such a bill is either unnecessary because of protection given by running of statute of limitations, or unconstitutional where it would cut off vested rights).

23. See Maxon v. Gates, 136 Wis. 270, 292–293, 116 N.W. 758, 766 (1908); compare similar reflections of this pattern, in Allen v. Clark, 36 Wis. 101, 106–107 (1874) (levy on judgment debtor's interest "in what is called 'stump land,' having little value, and which had for several years been sold for taxes"); brief of defendant-appellant, pp. 10–13, in Lombard v. Antioch College, 60 Wis. 459, 19 N.W. 367 (1884) (cut-and-default-for-taxes practice criticized for social burden); Swift v. State Lumber Co., 71 Wis. 476, 479, 482, 37 N.W. 441, 442, 443 (1888) (court notes certain forties to be "apparently worthless," a comment to be read in light of the Report's summary of allegation of plaintiff's bill, that these were "tracts in Price county from which the timber had been cut, and the lands abandoned"). General concern over the extent of tax delinquency in areas in which land speculation was active may be seen in Annual Reports of the State Superintendent of Public Instruction, PD, Doc. K, p. 1362 (1862), and PD, Doc. L, p. 654 (1863). The thread of fraud crisscrosses that of formally proper sale followed by tax delinquency after the timber is cut, in the plaint of a Report of a Select Committee regarding Bill 114S, SJ, 306, 307 (1867). Portraying the evils brought by excessive speculation in Marathon county timberlands, the committee noted that even if the speculator were honest in intent at the outset, he typically overreached himself by investing in more land than he could carry, and then, when tax delinquency occurred, making off with the timber: "Hundreds of thousands of acres of land were thus forfeited and still the taxes keep accumulating against them. The burden which thousands of speculators either could not or would not bear, was thus transferred by operation of law to the unwilling shoulders of the few hardy, industrious settlers of the county . . . If any portion of the lands [whereon taxes fall into default] should have valuable pine timber on them, the owners, with scarcely an exception, skin such lands of the timber, between the time of sale and when the redemption will expire, and thus leave the tax certificates held by the county worthless, so far as raising money is concerned. If the lands are good for agricultural purposes it is little better, for there is no sale for lands in that country except they have a present marketable value for the pine timber on them . . ." This committee put the blame for a large social loss squarely on the inadequacy of state lands sales programs: "The state is responsible for the laws under which Marathon county today is the involuntary owner of over one hundred thousand acres of land, with tax liens thereon amounting to at least one hundred thousand dollars"; see SJ, 307–308 (1867). But this was an isolated note of realistic complaint in a prevailing silence of indifference, which was not broken until twentieth-century Wisconsin confronted the problems of a submarginal, cutover region. Low-priced sales of cutover land for farms, directly by lumber companies, or by speculators who bought in sizable quantities from them, are discussed in James I. Clark, *Farming the Cutover: The Settlement of Northern Wisconsin* (1956), 5, 11, and his *Cutover Problems: Colonization, Depression,*

Reforestation (1956), 7; Fries, 176–178; Arlan Clayton Helgeson, "The Promotion of Agricultural Settlement in Northern Wisconsin, 1880–1925" (unpublished Ph.D. thesis, 1951), 17, 19, 22, 24, 34, 123–128, 146, 156; Reynolds, 34–35. Cf. Report, State Forester, PD, 85 (1908) ("Land companies who induce settlers to purchase poor, sandy, cutover pine lands for farms, thereby often make timber thieves"). That an inheritance from the nineteenth century was a major problem of northern land unable to yield current income even to meet its share of basic public obligations, see Report, Interim Committee on Forestry and Public Lands, AJ, 866–869, 879 (1929). Cf. Hibbard, "Land in Relation to Forestry," 26 (failure of agriculture to occupy the bulk of the cutover country).

24. That government enjoys the common law remedies available to any property owner, see Cotton v. United States, 11 How. 230 (U.S. 1850); United States v. Cook, 19 Wall, 591 (U.S. 1874) (Wisconsin land); and Conklin v. Hawthorn, 29 Wis. 476 (1872) — all timber trespass cases. Cf. Sloan, Stevens & Morris v. State, 51 Wis. 623, 8 N.W. 393 (1879) (state has trustee's obligation to protect federally derived railroad aid lands from trespass, but not in sovereign capacity). Miller, J., is quoted from E. E. Bolles Wooden-Ware Co. v. United States, 106 U.S. 432, 436–437 (1882). But cf. Wright v. E. E. Bolles Wooden Ware Co., 50 Wis. 167, 6 N.W. 508 (1880) (no recovery under state statute, against bona fide purchaser from the wrongful cutter, of the value added by trespasser's labor before sale to this purchaser; as a "penal" statute, the act should be strictly construed and not applied to detriment of regular market processes; case is discussed in the later chapter dealing with private trespass, since the values stressed by opinion are put in this context). That the state's successor in interest might claim the enhanced damages under Revised Statutes, sect. 4269, was ruled in Smith v. Morgan, 68 Wis. 358, 32 N.W. 135 (1887); the approach taken is the more striking because in a further stage of the litigation, 73 Wis. 375, 41 N.W. 532 (1889), the court held that the statute, being penal in character, must be strictly construed to hold the recovery carefully within its terms, and hence the successful plaintiff might not have interest upon the enhanced damages. See, accord, Everett v. Gores, 92 Wis. 527, 66 N.W. 616 (1896); cf. Coleman v. The Peshtigo Co., 47 Wis. 180, 2 N.W. 111 (1879) (state's patentee subject to same self-denying statute of limitations binding state in such actions); Attorney General's Opinion to Chief Clerk, Land Office, July 18, 1899, Attorney General's Letter Books, V, 175 (statute of limitations begins to run against state from time of cutting, not from time of knowledge of trespass). No case brought to the Wisconsin Supreme Court presented the issue whether the state might directly sue for the enhanced damages provided by Revised Statutes, sect. 4269. Cases which recognized the cause of action in a plaintiff who had acquired the land from the state after the trespass referred to plaintiff's right as derived from a pre-existing right of the state, though the transfer statute (General Acts, 1865, c. 520) from its terms might be viewed as creating a new cause of action in the state's grantee. See Conklin v. Hawthorn, 29 Wis. 476, 480 (1872); Coleman v. The Peshtigo Co., 47 Wis. 180, 183, 2 N.W. 111, 112 (1879); Smith v. Morgan, 68 Wis. 358, 361, 32 N.W. 135, 136 (1887). The acknowledged "penal" character of the enhanced damages statute might, however, have led to a decision that, in the absence of an explicit provision that the state should enjoy this right, the act should not be construed by implication to give it. Cf. United States v. Cooper Corporation, 312 U.S. 600 (1941) (right of action for treble damages granted by Sherman Act to "any person" injured in his business by conduct unlawful thereunder, held not to apply to the United States as plaintiff); see, however, Black, J., dissenting, *id.,* 619, invoking Cotton v. United States, cited above. The burden of proving applicability of the statutory privileges to cut timber for the uses of a settler-cultivator was put on the cutter in Smith v. Morgan, cited above; see Conklin v. Hawthorn, above, 29 Wis. at p. 481. Compare the court's recognition that the burden of proof lay on defendant under Laws, 1882, c. 239, mitigating damages in cases of good-faith mistake of title. Webber v. Quaw, 46 Wis. 118, 49 N.W. 830 (1879); Fleming v. Sherry, 72 Wis. 503, 40 N.W. 375 (1888); Everett v. Gores, cited above; McNaughton v. Borth, 136 Wis. 543, 117 N.W. 1031 (1908); Fehrman v. Bissell Lumber Co., 188 Wis. 82, 204 N.W. 582 (1925).

25. The backhanded development of federal law on timber trespass as a crime was sketched at the beginning of this section. See note 1 above. The 1831 act was carried forward into United States Revised Statutes,

1878, sect. 2461. The Wisconsin development may be seen in its territorial Laws, 1836, No. 29, and Laws, 1843–1844, p. 29; after statehood, in Laws, 1848, p. 156; Revised Statutes, 1849, c. 134, sect. 47; General Acts, 1855, c. 84 (swampland), 1856, c. 125 (swampland), 1857, c. 54 (swampland); Revised Statutes, 1858, c. 165, sect. 53; General Acts, 1864, c. 261 (lands taken on mortgage foreclosure for debts owed state trust funds), 1865, c. 377 (general revision, in protection of all important types of state land); Revised Statutes 1878, c. 182, sect. 4442; Laws, 1893, c. 64, 1903, c. 450, and 1905, c. 264 (protection of the short-lived, original state forest reserves); Laws, 1949, c. 252. Another feature of the development of the substantive statutory law of timber trespass in the state, of a character familiar in over-all nineteenth-century legislative practice, was the legislature's indifference or unwillingness to give time at an early stage to providing a basic framework of sufficient generality to allow for the growth of the situation. Thus quite a few separate acts were passed to declare, in effect, that the protection of the general penal provisions should extend to types of holdings newly added to the state's landed assets. Examples of this are apparent in the legislation already cited. See also Acts, 1849, c. 74 (Fox-Wisconsin lands); General Acts, 1854, c. 55 (lands held "in trust"); General Acts, 1866, c. 121 (Morrill Act lands); General Acts, 1865, c. 537, and General Laws, 1869, c. 151 (swamplands); General Laws, 1869, c. 46 (St. Croix railroad lands); General Laws, 1870, c. 92, Laws, 1875, c. 320, and Laws, 1876, c. 224 (Sturgeon Bay canal project aid lands); Laws, 1873, c. 55 (lands taken on eminent domain). The ambiguity in definition of the mental element in the general history of Wisconsin criminal legislation, and the tendency to resolve ambiguity in favor of an interpretation requiring a showing of a general criminal intention, in the sense at least of an "objective" fault (action taken in circumstances in which a prudent man would have recognized its illegality), are portrayed and analyzed by Frank J. Remington, and by Richard R. Robinson and Willis J. Zick, in "Liability Without Fault Criminal Statutes," *Wisconsin Law Review*, 1956:625. The context of Laws, 1865, c. 377, indicates design to make timber trespass on public lands an offense of "general" rather than "special" intent; the dropping of "willfully" in the definition of the basic offense (sect. 1) was thrown into sharper relief because sect. 6 penalized one "who shall knowingly resist, hinder or obstruct" the enforcement activities of a state timber agent, or "who shall willfully" take or interfere with any timber seized by an agent. On the emphasis on the criterion of a guilty mind after 1878 in Wisconsin executive policy, see Attorney General's Opinion to Chief Clerk, Land Office, June 9, 1897, Attorney General's Letter Books, II, 273; cf. Report, Attorney General, for biennium ending June 30, 1918, Attorney General, *Opinions*, VII, xxxvii (civil remedy against trespasser who claims adverse title); Attorney General's Opinion to District Attorney, Hurley, Wis., Feb. 19, 1935, *id.*, xxiv, 98 (distinguishing, by required showing of wrongful intent, situation where criminal as well as civil sanction applicable). For the background of the broadening of the definition of the crime, in Laws, 1893, c. 64, see General Message of Governor Peck, AJ, 15–16 (1893). The legislature responded to Peck's recommendation quickly and without controversy. See AJ, 406, 422, 463 (1893); SJ, 351, 369 (1893). The interest of governor and legislature in the subject may have been spurred by the lengthy historical review of school fund losses in the Biennial Report of the State Superintendent of Public Instruction, Dec. 10, 1892, PD, I, 141–154 (1893). On interpretation of the federal statute as requiring a showing of a guilty mind, see United States v. Lane, 19 Fed. 910 (Circ. Ct. E.D. Wis. 1883). Cf. Shiver v. United States, 159 U.S. 491, 497 (1895); Union Naval Stores Co. v. United States, 240 U.S. 284, 289–290 (1916); English v. United States, 116 Fed. 625, 626 (C.C.A. 9th. 1902); H. D. Williams Cooperage Co. v. United States, 221 Fed. 234, 236 (C.C.A. 8th. 1915).

26. See Reports of the General Land Office, 1874, pp. 7, 41 (Commissioner Burdett), and 1877, p. 16 (Commissioner Williamson on "vast" trespasses); Dunham, 48–49 (1855–1877 collections); Ise, 244 (National Academy of Sciences). Other reflections of the large scale of timber trespass and the lack of effectual federal action against it may be seen in Reports of the General Land Office, 1872, p. 26, 1883, p. 205, and 1884, p. 16; Cameron, 21, 22, 76, 78, 80, 86–87, 97, 138, 144–146, 148, 152, 154, 166, 170, 172; Clawson, *Uncle Sam's Acres*, 81–86; Cummings and McFarland, 264, 387, 389; Robbins, *Our Landed Heritage*, 26, 54, 99; Lucile Kane, "Federal Protection of Public Timber

in the Upper Great Lakes States," *Agricultural History,* 23:135 (1949), reprinted in Vernon Carstensen, ed., *The Public Lands* (1963), 439. General Samuel Harriman's report to Governor Fairchild is in his letter of Apr. 7, 1869, Wisconsin Executive Records: Timber Agent's Reports, Box 1 (Wisconsin Archives, State Historical Society, Madison). On the money settlement policy, see citations above from Dunham and Ise; contrasting estimates of this policy may be seen in General Land Office Reports, 1865, p. 26, and 1883, p. 205. For what proved to be a transitory phase of complacent assurance that seizure of logs and money settlements were solving the timber trespass problem, see Reports, General Land Office, 1864, p. 22, and 1865, p. 26 (Edmunds), and 1866, p. 33, and 1867, p. 95 (Wilson); cf. Dunham, 48–49, 50, 52. The theft of timber from federal lands in Wisconsin is discussed in Reports of the Surveyor General, Dubuque, Nov. 11, 1850 ("a swarm of laborers"), in Report of the General Land Office, 1850, 31 Cong., 2 Sess., Senate Exec. Doc., No. 2, App. H, p. 43; Oct. 21, 1855, 34 Cong., 1 and 2 Sess., Senate Exec. Doc., I, 197; Oct. 10, 1858, 35 Cong., 2 Sess., Senate Exec. Docs., 188; Oct. 18, 1859, 36 Cong., 1 Sess., Senate Exec. Docs., I, 242; Oct. 10, 1860, in Report, General Land Office, 1860, p. 81; compare with the 1859 Surveyor General's report the comments in the Report of the Commissioner of the General Land Office (S. A. Smith), 36 Cong., 1 Sess., Senate Exec. Docs., I, 193. Timber trespass on federal lands is reflected as a serious problem in observations of officials of Wisconsin Territory. See Reports of the Committee on Schools, HRJ, 85–86 (1837), and 76, 80 (1841); General Message of Governor Dodge, Council Journal, 15 (1846). Comments on the serious extent of continuing timber theft in the Wisconsin pinery will be found in Reports of the Commissioner of the General Land Office, 1865, p. 26, 1872, p. 26 ("great magnitude"), 1873, p. 13, 1877, p. 16 ("vast"), 1878, p. 124 ("to a very large extent"), 1879, p. 186 ("extensively . . . for a number of years"), 1880, p. 170 ("of vast magnitude"), 1881, pp. 376–377 (quoted at length in text); see also Cameron, 152; Ise, 81 (commenting on statement of Wisconsin Congressman Wells in the House of Representatives, Dec. 7, 1894, emphasizing, perhaps with some partisan color, the large extent of past timber trespass in that state); Miller, J., in E. E. Bolles Wooden-Ware Co. v. United States, 106 U.S. 432, 436–437 (1882), quoted in text at note 24 above.

27. See General Message of Governor Dewey, SJ, App., No. 2, p. 13 (1848), and Report, Committee on Education and School Land, AJ, 87 (1848) (threat seen); Communication from the Secretary of State, SJ, 121 (1849) (field reports, regarding the 16th sections, that "persons are stripping the lands of the most valuable of the timber"); Report, Commissioners of School and University Lands, AJ, App., 996 (1851) ("the depredations constantly being committed . . . upon the school and university lands"); Report, State Superintendent of Public Instruction, Dec. 31, 1853, SJ, App., 12–13 (1854) ("There is no doubt but that the school lands in many sections of the state are suffering from the depredations of trespassers in cutting down and carrying away valuable timber"); General Message of Governor Barstow, SJ, App., 21 (1855) (since no protecting law against timber theft from swamplands "could be rendered sufficiently efficacious to prevent their being thus despoiled," state should sell quickly); Report, Commissioners of School and University Lands, SJ, App., 27 (1855) (field reports of "daily" trespasses to swampland timber); Annual Report, Board of Regents of the University, SJ, App., 10 (1855) (trespass threat); General Message of Governor Barstow, SJ, 23 (1856), and his Special Messages, AJ, 289–293, 344, 479 (1856) (field reports of "extensive" and "rapid" depredations on swamplands, especially on "almost all" such lands near the Mississippi and its tributaries); Annual Report, Secretary of State, App. G, AJ, App. B, 101–102 (1857) (report of field agent as of Dec. 31, 1856), and Report, Commissioners of School and University Lands, AJ, App. D, 29–30 (1856) (unless swamplands are speedily sold, "no other result can be expected than . . . that of waste and destruction being speedily committed" on them; "no caution can guard against such results, except the encouragement of a speedy sale . . . on such terms as our industrious population can afford to embrace"); General Message of Governor Randall, SJ, 43–44 (1858) ("Trespasses are constantly committed" on school and swamp lands); Communication of Timber Agent William Crombie to the President of the Senate, Feb. 22, 1858, responding to Joint Resolution No. 23S, SJ, 286, 308, 470–472 (field report of trespass); General Message of Gov-

ernor Randall, SJ, 14 (1860) ("The school lands still in the hands of the State unsold, are every year deteriorating in value for want of adequate legal means to prevent trespasses"), and his Special Message, SJ, 245, 247 (1860) ("wholesale trespasses . . . with . . . great losses constantly occurring for want of necessary means of protecting these lands"; attached report of land commissioners notes the danger; attached letter of the District Attorney for Oconto County, Dec. 28, 1859, reports that "extensive" illegal cutting has been there a "custom [which] has prevailed so long that the memory of man runneth not to the contrary," and that "the present season they are making an awful descent upon the tall timber, upon both the school and swamp lands"); Report, Commissioners of School and University Lands, PD, 61 (1860) (quoted in text); General Message of Governor Harvey, SJ, 17 (1862) ("More vigorous measures should be enacted . . ."), and his Veto Message, SJ, 813–814 (1862) (vetoing bill 62S, to donate swampland to counties wherein the lands lie, in part on ground that this would imperil safety of these trust lands in counties where the lumber interest predominates, for "the past history of these counties shows that men so employed have not hesitated to appropriate large quantities of such timber without making any compensation therefor to the owners, and in plain violation of law"); Annual Reports, Commissioners of School and University Lands, PD, Doc. K, p. 510 (1863) (enforcement improving, but still too limited), PD, 195–196 (1864) (quoted in text; past losses "incalculable," commissioners' means to combat trespass have been "limited," but buying rather than stealing is thought now "more in fashion than in earlier times"), and PD, 343 (1865) (quoted in text; though commissioners are optimistic, they see the need yet to build public opinion to support enforcement); General Message of Governor Fairchild, SJ, 25 (1866) ("Trespassing upon State lands, and stealing timber therefrom, has long been an existing evil, in spite of efforts made to prevent it. If there be sufficient power in the state, this crime should be stopped"); proceedings leading to Senate rejection of proposed Joint Resolution 35A, (reciting "a great extent" of timber trespass in the past and continuing on state-owned lands in Marathon County), AJ, 323, 340 (1872), SJ, 252, 270, 299 (1872).

28. A more optimistic tone appeared in some of the land commissioners' comments on the curbing of timber theft in the sixties after the legislature had provided modest additions to the enforcement force. See their reports of Oct. 10, 1861, PD, 5–6; Oct. 1, 1865, PD, 343; Oct. 1, 1866, PD, 773. In the middle seventies their estimates of the situation become still more assured. See their reports of Oct. 1, 1872, PD, 3 ("Trespassing . . . is rapidly diminishing"); Oct. 1, 1873, PD, No. 4, p. 3 (trespasses in past year "have been very limited . . . due in a measure to the efficiency of the agents"); Oct. 1, 1874, PD, I, 3 (same statement, verbatim); Oct. 1, 1875, PD, I, 3 (same). However, in his General Message of Jan. 13, 1876, PD, I, 13, Governor Ludington, himself a lumberman familiar with the problems of the north country, advised the legislature: "It has become a serious matter to provide for the protection against trespassers of the school and university lands, and the tracts [of railroad-aid lands] known as the St. Croix grant." That the problem continued to be a grave one is likewise the indication of the strong Minority Report of the Judiciary Committee on Bill 258A, AJ, 596–599 (1876). Reports of state timber agents to the governor during the eighties continue to reflect serious, continuing trespass. See Daniel J. Dykstra, "Legislative Efforts to Prevent Timber Conversion: The History of a Failure," *Wisconsin Law Review*, 1952:461, 473, 475, 477.

The turn of the century brought renewal of optimistic estimates of the success of enforcement, which seem intrinsically more plausible, both in view of the stricter law of 1893 and because by this time the heavy capital investments of large lumber companies in their own holdings inclined them to support measures against competition of illicit goods. See Report, Chief Inspector of Department of Public Lands, PD, II, 5, 12 (1898) (emphasizing changed view of established lumbermen); Biennial Report, Commissioners of Public Lands, PD, 4 (1906) ("Trespassing upon state lands . . . is being reduced to the minimum . . . Many trespasses are unintentional, from ignorance of division lines and such trespassers pay up promptly"), and First Annual Report, State Forester, PD, 19–20 (1906) (strong retrospective criticism of weak state policy); Biennial Report, Commissioners of Public Lands, PD, 4 (1908) ("state lands are receiving better protection than ever before"), and Second Annual Report, State Forester, PD, 84

(1908). Optimism as to the growth of sentiment against timber trespass in proportion as private holdings grew was expressed early in the land commissioners' report of Oct. 1, 1865, PD, 343.

Contemporary concern over timber theft from lands held by the state in aid of transport developments found particular expression regarding the "St. Croix lands" for railroad construction in the northwest part of the state. It is further indication of the generally weak and indecisive official approach to the trespass threat, however, that, though the first grant to the state came in 1856, the legislature did not provide specifically for protection of the lands until General Acts, 1864, c. 277, and that the first expression of active official concern for effectiveness of enforcement with respect to those lands did not come until, on Feb. 15, 1868, the Assembly by Resolution 56A (AJ, 416 and 443), asked the grantee railroad to report on its activity under the protection powers delegated to it. The company's report conformed to the prevailing tone of dealings with this problem, noting great difficulties of action and scant return: see AJ, 49 (1869). Trespass continued to be a serious threat to the integrity of the St. Croix grant through the seventies, though when the state resumed enforcement and provided an unusually energetic agent, Samuel Harriman, under General Laws, 1869, c. 46, it obtained the most vigorous enforcement effort of its public lands program. See General Message of Governor Fairchild, SJ, App., 20–21 (1870) (Harriman's early success), and Address of the Superior Board of Trade, asking the state legislature to press for renewal of the St. Croix grant, AJ, 334 (1870) (alleging, as "what is notorious in the St. Croix Valley," that "certain large lumber operators in Minnesota, and a few of less note and capital in Wisconsin" have committed large past depredations on these lands, of which they now falsely accuse the railroad companies, hoping to persuade Congress not to renew the grant, "and so expose the pine lands thereof to the same reckless plundering, that has devastated so much of the adjoining lands of the government . . ." or, alternatively, to secure for the larger operators a monopoly by purchase); General Message of Governor Fairchild, SJ, App., 25 (1871) (Harriman has collected to date for past trespasses $67,786); General Message of Governor Washburn, SJ, App., 24 (1872) (suits pending for additional $75,000); General and Special Messages of Governor Washburn, SJ, 10, 24–26, and App., 30 (1873) (successful progress of test cases, request by agent for authority to settle further matters), and proceedings denying authority to agent to effect compromises, AJ, 669, 738 (1873) (Bill 148S); Memorial 114A of citizens of Polk County, praying fresh grant of the St. Croix lands to North Wisconsin Railway Co., AJ, 193–194 (1873) (reciting "that said grant has for many years been the prey of trespassers, and has been greatly diminished in value"); Memorial of North Wisconsin Railway Co., on the same subject, SJ, 75–82 (1874); General Messages of Governor Ludington, Jan. 13, 1876, PD, I, 13 (trespass on St. Croix grant "has become a serious matter"), and Jan. 11, 1877, PD, I, 20 (no serious trespass in past year on St. Croix grant); General Message of Governor Rusk, AJ, 14 (1885) (St. Croix trespass account may shortly be wound up, and moneys paid to companies which have completed their roads and are now protecting their own lands). See also Schulenberg v. Harriman, 21 Wall. 44 (U.S. 1874), and Sloan, Stevens & Morris v. State, 51 Wis. 623, 8 N.W. 393 (1881). The tenfold-damages act in further protection of the railroad-aid lands was Laws, 1876, c. 308, responding to Governor Ludington's expressed concern in his message of that year.

The care of other railroad aid grants, or of the Fox-Wisconsin rivers improvement aid lands, produced no comparable record of activity on timber trespass, though the session laws indicate that the threat was experienced with regard to these lands also. See, e.g., Laws, 1849, c. 74 (Fox-Wisconsin project); Private & Local Laws, 1867, c. 334 (Lemonweir Improvement Co.); Laws, 1876, c. 379 (conveyance by state of aid lands to any railroad, on its fulfillment of terms, shall be deemed also to convey all causes of action for previous trespass not currently in litigation). The aid lands granted to assist construction of a ship canal between Green Bay and Lake Michigan had been subjected to heavy timber trespass before they were made available for the project, and continuing threat of trespass was seen as a problem after the initial grant; accordingly, protection of these lands, and recovery for past depredations, received a degree of attention matched only by that given the problems of the St. Croix grant. See proceedings leading to the prompt enactment of General Laws, 1870, c. 92, AJ, 433 and 633 (1870), and SJ, 400, 464 (1870); also, Preamble to in Wisconsin Statutes, 1898, c. 16, sect. 238; Laws, 1901, cc. 408 (state game wardens to aid against timber trespass), and 432 (striking authorization to employ agents, puts protective function in "the chief clerk" of land commissioners' office); Laws, 1917, cc. 14 and 282 (field activity concentrated in Conservation Commission); Wisconsin Statutes, 1959, c. 26, sects. 26.06 and 26.10. The only comprehensive official review of the budget history of state efforts against timber trespass is that for the years 1850–1866 in Communication of the Secretary of State in response to resolution 34A, Feb. 4, 1867, AJ, 181–182, showing a lumped total for clerk hire and land protection of $131,934.01, from 1855 to 1866, with annual expenditures varying rather sharply from a low of $6,500 to a high of over $21,000. Before the legislature provided the commissioners with a more or less regular field staff, they and the governor on their own responsibility had appointed a few agents for limited service. See Special Message of Governor Barstow, Feb. 12, 1856, AJ, 289–293 (transmitting report of William Crombie), and AJ, 506, Mar. 6, 1856, Report, Committee on Swamplands, and Mar. 17, AJ, 636 (controversy over Crombie's pay); General Acts, 1856, Joint Resolution No. 3 (authorizing Crombie); AJ, 555 (1857) (unsuccessful resolution to withdraw Crombie's authority), and AJ, App. B, Annual Report, Secretary of State, Jan. 5, 1857, App. G (report, dated Dec. 31, 1856, by Crombie); General Acts, 1857, c. 54 (governor may cause seizure of timber stolen from swamplands; no direct authorization to appoint agents therefor); General Message of Governor Randall, SJ, 43–44 (1858) (legislature should provide "some efficient means" to arm the land commissioners against trespass; questions honesty of "some of the agents" hitherto appointed), SJ, 286 (1858) (final passage of Joint Resolution 16A, rescinding Crombie's appointment), SJ, 470–472 (report by Crombie, dated LaCrosse, Feb. 22, 1858); General Laws, 1859, c. 182 (final settlement with Crombie), and Report, Select Committee on Crombie's claim, SJ, 620–622 (1859) (including reference to governor's appointment of another agent under General Acts, 1857, c. 54, above; Special Message of Governor Randall, SJ, 244–249 (1860) (responding to resolution 37S, reporting on three agents named by him, and three by land commissioners, and including commissioners' report; commissioners state that "as it was thought desirable to act promptly, the Commissioners did not await further action of the legislature"). The ending of the land commissioners' field force in 1901 responded to economy recommendations in the Special Message of Governor LaFollette, AJ, 954 (1901).

The issue of a revolving fund for federal law enforcement against timber trespass was presented squarely by a recommendation of Secretary of the Interior Schurz in 1880, which Congress did not adopt. See Dunham, 56, and Secretary of the Interior, Report, 1880, p. xix. Compare Pinchot, 257, 258, for his stress on the importance of a limited-term grant by Congress in 1905 of the proceeds of sales of forest products and forest reserve lands for protection and development programs. At most, however, such a fund was probably useful only to finance protection. Gulick, 112, indicates that a revolving fund from government's forest management activities was of too limited size, once the peak of forest exploitation was past, to support an affirmative, large-scale, forest-maintenance program.

Wisconsin Constitution, Art. X, sect. 2, afforded the possible basis for a state revolving fund, out of the implications of its grant of the "clear" proceeds of penal fines to the school fund. Cf. Lynch v. The Steamer "Economy," 27 Wis. 69 (1870) (statute giving one-half of fine to prosecuting informer, tacitly assumed constitutional within Art. X, sect. 2); Dutton v. Fowler, 27 Wis. 427, 431 (1871) (*Lynch* did not decide the point; statute here giving full penalty to informer is invalid as violating constitutional trust); State ex rel. Guenther v. Miles, 52 Wis. 488. 9 N.W. 403 (1881) (legislature may allow county 2 per cent withholding of fines collected, as legal fees); State v. De Lano, 80 Wis. 259, 49 N.W. 808 (1891) (explicitly ruling on point left open in Dutton v. Fowler, and holding valid legislation allowing two thirds of fine to informer); State ex rel. Johnson v. Maurer, 159 Wis. 653, 150 N.W. 966 (1915) (upholding law granting county one third of fine under fish and game laws for deposit with county treasurer in fund to reimburse county enforcement efforts; one third to informer); State ex rel. Owen v. Donald, 162 Wis. 609, 644, 664, 157 N.W. 794, 804, 810 (1916) (special referee's report, approved by the court, concerning correction of illegal diversions from educational trust fund; since constitution devotes "proceeds" of state lands to the trust, none of such moneys may be credited to the general fund to reimburse it for any administrative or protective expense, distinguishing the narrower intent of the constitution's grant of "clear" proceeds only of fines thereto). The last-cited reference recites the history of the policy set in the 1860's, and generally adhered to, of providing out of the general fund for administrative and protective activity affecting the educational trust lands. Conversely, the legislature early set the practice of not returning any part of the proceeds of fines to a fund for enforcement against timber trespass. See Acts, 1849, c. 212, sect. 104, carried into Revised Statutes, 1849, c. 24, sect. 107, and Revised Statutes, 1858, c. 28, sect. 127 (school, university, "or any other lands" of state); General Acts, 1857, c. 54, carried into Revised Statutes, 1858, c. 29, sect. 10 (swamplands); cf. General Laws, 1866, c. 56 (for a time, and undoubtedly unconstitutionally, altered by General Laws, 1867, c. 138, to put all sums received as penalty and damages for trespass to the lands into the general fund, all sums specifically for logs taken by trespass into appropriate trust fund), carried into Revised Statutes, 1878, c. 15, sect. 190 (all penalty moneys for trespass on state lands to go to appropriate fund, all costs of caring for the land to be paid from general fund). Provision to pay protective costs out of proceeds of prosecutions for trespass on the "St. Croix" railroad aid lands was made by General Laws, 1869, c. 46, and 1871, c. 75; like provision was made regarding protection of the aid lands for the Sturgeon Bay canal project, in Private & Local Laws, 1872, c. 104, and Laws, 1875, c. 336. Since the Wisconsin Constitution, in Art. VIII, sect. 10, authorizing the state to accept grants of land especially dedicated by the grant to particular works of internal improvement, declared that the state "shall devote thereto the avails of such grants," it might have seemed that a similar barrier existed here as under Art. X, sect. 2 (devoting "the proceeds" of school lands to the education trust) to paying protective expenses out of any moneys derived from lands thus held in trust. However, without discussion or apparent difficulty, and without reference to the school lands analogy, the Wisconsin Supreme Court held that proceeds of trespass prosecutions concerning the railroad-aid lands might be used to pay the costs of such protection: Sloan, Stevens & Morris v. State, 51 Wis. 623, 8 N.W. 393 (1881), s.c., 55 Wis. 271, 9 N.W. 795 (1882). The case may be distinguished, perhaps, on the basis that, since the court there also ruled that the constitutional ban on state debts or participation in works of internal improvement (outside the specific authorization of the exception for earmarked grants) barred paying protection costs from general funds, there was no other way in which the state could fulfill its obligation as trustee to care for the lands. See 51 Wis. 623, 628, 8 N.W. 393, 394 (1881), and Schulenberg v. Harriman, 21 Wall. 44, 64 (U.S. 1874). Likewise, it might have been felt a distinguishing fact that these lands were designed to pass ultimately to private corporations, and — in a sense — to be consumed for those corporations' purposes. Cf. Chippewa Valley & Superior Railway Co. v. Chicago, St. Paul, Minneapolis & Omaha Railway Co., 75 Wis. 224, 249–250, 44 N.W. 17, 23 (1889).

31. The lack of effort to create a body of public forest guards with a professional approach to their job, either in federal or state forest administration in the nineteenth century, is depicted graphically for the federal scene in Pinchot, 161–167, and for the state, in Dykstra, "Legislative Efforts to Prevent Timber Conversion," 461, 473, 475–477.

32. Material cited in note 26 above portrays the *de facto* "delegation of enforcement against timber theft by transfers to private ownership. The policy became more explicit in the Wisconsin legislation conferring enforcement authority on representatives of aided railroad and waterways companies as to their aid lands (see note 28 above), and in the legislation transferring to the state's patentees all the uncollected state rights of action for trespasses prior to patent. See General Acts, 1865, c. 520 (patentees of educational trust lands); Laws, 1876, c. 379 (patentees of railroad aid lands); cf. General Laws, 1869, c. 46, sect. 7; Laws, 1873, c. 176; Laws, 1882, c. 198 (money held by state for aid lands trespass to be paid to railroad after patent). Of like import were statutes giving possessory actions to preemptors and certificate holders of swamplands, such as General Acts, 1855, c. 84, sect. 13; General Acts, 1857, c. 57, sect. 2; Revised Statutes, 1858, c. 29, sect. 17; General Acts, 1864, c. 193; Revised Statutes, 1878, c. 15, sect. 199. Likewise, possessory actions were given pre-emptors or certificate holders of school and university lands, by Acts, 1849, c. 212, sect. 21; Revised Statutes, 1849, c.

B. C. Eastman were from his speech of July 22, 1852, 32 Cong. 1 Sess., *Cong. Globe*, App., xxi, 851. A good and vivid account of the short-lived special-agent effort of the early 1850's, including the activities particularly relevant to Eastman's protest, will be found in Cameron, 138–154; on the general invocation of "settlement" to justify illegal cutting, see Clawson, *Uncle Sam's Acres*, 81–86. The official actions in the 1856–1857 handling of swamplands pre-emptions in Wisconsin are cited in note 18 above; see also SJ, 745 (1856) (citizens' petition remonstrating against repeal of swamplands pre-emption privilege); Report, Commissioners of School and University Lands, AJ, App. D, 29–30 (1856). Governor Harvey's criticism of misapplication of a long-term-credit sales program by its use concerning other than "the sale of really *farming* lands in limited quantities, and in good faith, for immediate settlement and improvement," was made in his General Message of Jan. 10, 1862, SJ, 17. For the quoted estimate of the federal surveyor general (Dubuque office) on the potentialities of the northwestern Wisconsin river valleys, see Report, General Land Office, 1863, p. 60 (quoting the surveyor general's report of Oct. 1, 1863). The Commissioner's 1870 comments are from Report, General Land Office, 1870, pp. 7, 73. Optimistic statements on settlement prospects in northern Wisconsin regularly mingle with acknowledgment of the area's distinctive value as a pinery, without attempt to blend the two elements into a consistent pattern of policy. See, e.g., General Land Office Reports, 1860, p. 18, and 1870, p. 181; cf. General Land Office Reports, 1874, p. 6, and 1875, p. 10.

37. Cameron, 17, quotes and comments on the observation of Dr. Schoep; on the unpopularity of the Crown claims symbolized by the "broad arrow" in colonial days, see Cameron, 13, 20–21, 25, 27, 65, 76, and Richard G. Lillard, *The Great Forest* (1947), 123, 127, 156, 158. P. L. Buttrick, *Forest Economics and Finance* (1943), 418, 449, casts the popular tradition into a fresh projection by his observation that we never allowed any substantial development in this country of legal rights of common cutting or pasturage as in Europe — unless one counts as a kind of bastard acknowledgment of "common" right (without limitation on waste) our tolerance of timber trespass on the public domain. On the federal policies toward timber reserves for naval purposes, timber trespass, and mineral-land leases, see materials cited in notes 1, 26, and 36 above, and in Chapter I, notes 6 and 18. The increased hazard of timber theft was noted as an argument for ending the federal government's lead-lands leasing policy in the General Message of Governor Dodge, Council Journal, 15 (1846); cf. Robbins, *Our Landed Heritage*, 140–152.

38. See Report, General Land Office, 1875, pp. 10, 12; Commissioner Burdett made a similar analysis and recommendations of policy on the then pending question of the future disposition of federal timberlands in the southern states, Report, General Land Office, 1875, p. 18. On the limited scope remaining by the mid-seventies for creative federal land policy in the Lake States in view of the disposition by then of most of the United States holdings, see also Burdett's comments in Report, General Land Office, 1874, pp. 6, 7. The practicality of long-term retention of timberlands title by the United States, and the wisdom of a strong program of sales for cash, but only at prices reflecting good timber values, were urged by Commissioner Williamson, in Report, General Land Office, 1876, p. 8, and 1877, p. 25; Williamson does not seem to have had much basic faith in this analysis, however, in view of his equally strong expressions of hope that transfer to private ownership would provide more effective guardianship against trespass (Report, General Land Office, 1876, p. 20, and 1880, p. 177). Williamson observed, with uncommon realism, that continued application of the pre-emption and homestead laws in forested sections represented in effect a "license" system anyway, rather than real transfer of the fee (Report, General Land Office, 1876, p. 9). Cf. Dunham, 44 (de facto "leases" by successive incompleted entries under Desert Land Act).

39. The Congressional treatment of the southern pinelands, the general outlines of the subsequent developments leading to creation of a federal forest reserve system, and the continued hostility to this program on the part of lumber and grazing interests and political leaders in the far western states are discussed in Dykstra, "Federal Government, State Governments, and Natural Resources," 569; Ise, 50–66, 79–81, 88–89, 114, 141, 229, 232, 237; Robbins, *Our Landed Heritage*, 25, 287, 314, 339, 341, 373, 382. Relevant also to appraisal of contemporary attitudes and the scant likelihood of earlier development

of a strong program for retention of timberlands title and the sale of regulated cutting licenses is Pinchot's criticism of the timidity or inertia of the General Land Office in failing to experiment with use of its general administrative and rule-making powers to develop a procedure of regulated cutting (see his *Breaking New Ground,* 111).

40. See Report, Select Committee regarding Bill 191A (growth of forest trees), AJ, 283, 284, 285 (1867). A report of a Select Joint Committee on State Government, Council Journal, App., 336 (1846), outlining the fruitful uses to which the future state might direct lands granted it by the federal government, mentioned briefly that the legislature might provide for "permanent leases or conveyances" of section 16 school-aid lands, and decide whether "to sell or lease" the university aid lands. No more was said. In context, it seems plain that the committee was thinking of agricultural use of land.

41. On the complete commitment of the state to fee sales, and the provisions regarding lands in aid of transport improvement, see notes 7 and 13 above.

42. Bill 412A, 1878, which became Laws, 1878, c. 324, was passed in both chambers over the adverse recommendations of committees concerned primarily with state finance, at a time when legislative journal references reflect a general sense of fiscal stringency arising out of depressed business conditions. This fiscal aspect aside, however, there is no indication that the bill aroused interest as a forestry bill. It was not a recommendation of the governor. See, especially, AJ, Mar. 6, 1878, p. 554 (adverse report by Joint Committee on Claims), and SJ, Mar. 19, p. 594 (adverse report by Senate Committee on State Affairs). The bill passed both houses without a record vote; some indication that it enjoyed a comfortable margin is given by a record vote in the Senate rejecting a motion for indefinite postponement, 7–14 (SJ, Mar. 20, p. 609). Criticisms of the economic rationality of the "state park" reservation, as well as questions of its validity under the constitutional trust of school lands, were raised in Annual Reports, Commissioners of School and University Lands, PD, I, 12 (1878), and PD, I, 23 (1880). The official explanation of the "state park" as a measure by which Wisconsin "recogniz[ed] . . . the necessity for protecting the forests surrounding the sources of her rivers" in order to curb floods and preserve navigability was made in Laws, 1882, Memorial No. 8, praying that Congress add to the reservation federally owned lands in the "park" area. Abortive repeal bills were bill 81S of 1883 (indefinitely postponed without a record vote, on recommendation of Senate Committee on Public Lands, SJ, 299 and 312); bill 783A, of 1887 (see Report, Committee on Ways and Means, AJ, 822, for the appraisal on grounds of economic unsuitability, referred to in the text; the committee's substitute bill repealing the state park provision passed the Assembly without a record vote, AJ, 973, received a favorable recommendation from the Senate Committee on State Affairs, but was, nonetheless, indefinitely postponed without a record vote in the second house, SJ, 664, 680, 750); and bills 811A and 346S of 1889 (both indefinitely postponed in Senate without record votes, according to recommendations of the Committee on State Affairs, SJ, 402, 573, 607, 854, 873). Laws, 1878, c. 324, was ruled invalid as a diversion of trust funds, in State ex rel. Sweet v. Cunningham, 88 Wis. 81, 57 N.W. 1119 (1894). Appraisal and sale of the former "state park" lands was provided for by Laws, 1897, c. 367, and 1899, c. 345. The remarks of Chief Inspector E. G. Mullen of the Department of Public Lands will be found in his report of Dec. 28, 1898, PD, II, 13. There was some dispute whether the timber on the former state park lands was not more valuable in place, to help regulate water flow for water power — a subject which was beginning to be of more current interest than forest yields — but this opposition proved ineffective to block the 1897 legislation: see SJ, 1016, 1022 (1897); AJ, 1295, 1304. The beginnings of a defined interest in managed state forest reserves may be seen in Laws, 1897, c. 229 (governor authorized to appoint three-man commission to draw plans); Laws, 1899, c. 345, and 1901, c. 458 (discretion to withhold or sell, according to state's best interests); Biennial Report, Commissioners of Public Lands, Oct. 10, 1902, PD, 38, 39 (1903) (permanent forest reserves recommended of land unsuitable for agriculture, as well as authority to commissioners to sell timber separate from land where timber is land's main value); General Message of Governor LaFollette, AJ, 107 (1903) (quoted in text); Laws, 1903, c. 450 (establishing a system of state forests); Laws, 1905, c. 264 (creating a forestry board). The upset of the new system by the decision in State ex rel. Owen v. Donald, 160 Wis. 21,

26A, to repeal the West Wisconsin Railway exemption; repeal lost on third reading in the Assembly by a record vote of 39–50, AJ, Feb. 10, pp. 216–217; bill 100S, for the same object, was eventually buried in committee). The legislative journals for the late seventies continue to be full of protests and battles over railroad-aid-lands tax exemptions, but there is no relevance to our story in detailed reference to these, because typically there is no significant reference to timberlands values or problems in all these maneuvers. For unusual, explicit references to timberland values as an important part of the stakes in the holding of aid lands, including preservation of the tax-base value of the land, see Address of Superior Board of Trade, AJ, 334 (1870) (others charge railroads with delaying construction to speculate on rise in timberland values; Board of Trade thinks lumbermen the true wrongdoers, by trespass); General Laws, 1871, Memorial No. 1 (urging renewal by Congress of the northwestern Wisconsin railroad-aid land grants, "because if said grant should not be renewed, the pine lands embraced within it would soon fall into the hands of large lumbering companies, or be taken under pre-emption or homestead laws only to be despoiled of timber and abandoned," rather than "bring[ing] a large region into actual cultivation and settlement"); Laws, 1876, c. 286, and Annual Report, Secretary of State, PD, I, 50–53 (1876) (reflections of popular discontent over alleged withholding of aid lands for speculation, by promoters of Sturgeon Bay & Lake Michigan Ship Canal & Harbor Co.; investigating commissioners' report of Oct. 7, 1876, shows company is holding timbered aid lands while seeking sale of aid lands primarily agricultural); proceedings of Jan. 30, 1877, SJ, 106–108 (focusing on stipulation ending tax exemption on sale of timber or land, under bill 4S, which became Laws, 1877, c. 21, regarding extension of the Wisconsin Central Railway Co.); Memorial of Marathon County Board, opposing extension of the Wisconsin Central Railway aid-lands tax exemption, Jan. 30, 1877, AJ, 128 ("it has been the practice of said company in the sale of its lands to reserve the valuable timber thereon for a term of years, which is a matter of no concern except to the parties contracting, *in the absence of any exemption;* but in the event of the proposed exemption being granted, it will enable the company to strip off the timber during the term of exemption; and thus deprive the state of an immense volume of taxable property by running it out of the limits of the state").

47. The land commissioners are quoted from their report of Oct. 10, 1902, PD, 38 (1903). Without explicit reference to timberland, Governor Fairchild in his General Message of Jan. 12, 1871, SJ, 6, had recommended immediate withdrawal of all educational trust lands from market for reappraisal with an eye to sale at better prices, on the ground that any past "excuse" for cheap sales in aid of settlement was no longer valid: the lands "are being purchased mainly by speculators, and the actual settlers, when they buy them, will have to pay to the dealer a large profit which the [trust] funds ought to realize." The legislature did nothing. An essentially similar, and similarly limited, judgment was passed on the history of the "considerable portion" of university lands lying within the timber-valuable northwestern railroad-aid grant area, in the Annual Report, University Regents, 1874, PD, I, 4. Noting that representatives of the affected counties had always offered "sturdy opposition" to suggestions of withdrawing these lands from market on the ground that such withholding would retard settlement, the regents observed tartly: "This objection would have force, if sale was made only to actual settlers; but it is notorious that the greater portion of sales since the land grants were made, have been to speculators, who hold the lands for the increased value, which, in simple justice, ought to inure to the University."

48. The shift from political and social to economic (productivity) arguments as the center of interest in public-lands policy in general was discussed in Chapter I, notes 15, 19–24, and the accompanying text. The productivity emphasis in discussion of federal land policy as this bore on public forests is reflected in Clawson, *Uncle Sam's Acres,* 81–86; Gulick, 19, 22–24, 79–81, 84–86, 155; Ise, 50–52 (1876 southern lands debates noted in text); Robbins, *Our Landed Heritage,* 314, 339–341, 373, 382. No one portrays more vividly the late nineteenth-century dominance of productivity values (or, at least, what a substantial and influential opinion took to be such) over the older egalitarian objectives of the faith in private property and a free market than Gifford Pinchot. Indeed, such otherwise disparate leaders as Pinchot and Theodore Roosevelt, Robert M. LaFollette, and Woodrow Wilson all worked to bring political and social, more than

economic, elements to the center of national public policy debate and decision. See Pinchot, 161–172, 197–202, 505–509. Cf. Hurst, *Law and the Conditions of Freedom,* 83. The high moral confidence with which men put their faith in a rising production curve as the measure and justification of decision at the peak of Wisconsin lumbering is nowhere better reflected than in the memoirs of a lawyer who played an important role in the creation of the Weyerhaeuser interests in the northwest part of the state. See Roujet D. Marshall, *Autobiography* (1923), I, 272–303. See also Address of Superior Board of Trade, AJ, 334 (1870), where renewal of the "St. Croix" railroad-aid land grant is urged, partly on the ground that otherwise the lands will fall into the hands of great lumber companies, whereas a railroad as grantee would find it to its interest to dispose of the timbered tracts "so . . . as to secure the largest way business for its road, through the permanent settlement along the line thereof, of the largest number of independent producers; whereby the improvement, instead of the desolation, of the region would be insured." It is significant of the dominant temper of the times that these petitioners explained their concern for "the mass of poor, hard working lumbermen" in terms wholly of access to production opportunities and benefits to the over-all productivity of the economy, rather than in terms primarily political or social.

49. See Kenneth W. Duckett, *Frontiersman of Fortune: Moses M. Strong of Mineral Point* (1955), 144–152, 156, 161–162; Fries, 108, 123, and Marshall, *Autobiography,* I, 272, 323 (both on Union Lumbering Company); Reynolds, 11, 14, 19, 129–139, 146–147. Fries, 100–108, discusses the general problem of scarce fluid capital in the growth of the Wisconsin lumber industry. On the relative importance of favorable timberland investments in determining extent of gain from the business, see Bureau of Corporations, *The Lumber Industry,* vol. I, part I, pp. 4, 25, 33, 35, 37; Current, 23, 112, 114; Larson, 70, 138–139, 141–142; Marshall, *Autobiography,* I, 266, 275, 353.

50. Settlers' cutting rights under federal and state public land laws are referred to in note 35 above. Congressman Eastman's statement is cited in note 36 above; on colonial woodsmen's disdain of the "broad arrow," and the continuing claims of men on our successive westerly frontiers rightfully to use or destroy timber to their needs, see note 37 above. In support of railroad construction which would enlarge the opportunities to employ men and capital in all phases of the lumber industry in Wisconsin, see Report, Select Committee on Bill 106A, AJ, 706, 707, 709 (1873). Compare, to like effect, Laws, 1874, Memorial No. 1 (praying Congress to extend the aid grant for the Wisconsin Central Railroad, which represents that it "is filling uninhabited forests with towns, mills and factories"). Representative of many citations that could be made to official promotion of immigration through the argument of woods-labor opportunities is Annual Report, Commissioner of Immigration, 1873, PD, No. 15, pp. 17, 29, 31, 53, 55, 81, 82, 92. As it became apparent that much of the woods-labor force was of transient character, much of it made up of men not steadied by ambitions to earn farms, dissenting voices were heard, complaining of its radicalism and urging restrictions of immigration. See Biennial Report, Bureau of Labor and Industrial Statistics, PD, II, xxvi, 417, 419 (1886).

51. On attitudes toward "speculation," see materials cited in note 43 above. Larry Gara, *Westernized Yankee: The Story of Cyrus Woodman* (1956), 31–32, 38, 44–45, and Gates, *Wisconsin Pine Lands,* 68, 73–74, 86–88, 113, 140–141, 143–162, 174, present material reflecting the hostility of both popular and lumber-industry opinion to those who bought timberland merely to hold it for the rise in value, without intention to enter the lumber business. Cf. Bureau of Corporations, *The Lumber Industry,* vol. I, part I, pp. 2, 24–25 (passive holders of virgin timber add no value to the land). The praises and exhortations in favor of state action to foster entrepreneurial energy in the Wisconsin forest, quoted in the text, will be found in Governor Harvey's General Message of Jan. 10, 1862, SJ, 27–28; General Messages of Governor Fairchild, Jan. 10, 1867, SJ, 18, and Jan. 14, 1869, SJ, 22; Wright v. The E. E. Bolles Wooden Ware Co., 50 Wis. 167, 170, 171, 6 N.W. 508, 510 (1880).

For other characteristic expressions of social favor for the entrepreneur and for public assistance of his efforts in the specific context of lumber industry development, see Majority Report, Committee on Railroads, Mar. 22, 1867, SJ, 644–645 (why single out railroads for specially onerous regulation, when, for example, "princely fortunes have been accumulated all around us in the lumber manufacture and trade"; yet, this comparison is cited only to show the injustice of singling out the railroads, for the committee majority would eschew all price regulation, believing that "corporate and individual rights and interests, as regards the use of capital within prescribed and legitimate spheres are alike to be respected and protected"); Laws, 1874, Memorial No. 1 (Congress should extend Wisconsin Central Railroad aid grant, on road's representation that it "is filling uninhabited forests with towns, mills and factories, planted there in full confidence that this road will speedily be completed; and lumbermen, farmers and miners have bought lands from the state and the United States along the line of this road . . . upon this understanding"); Report, Railroad Commissioners, 1874, PD, II, 42 (new northern Wisconsin railroads are all "financially speaking . . . constructions in advance of the paying point. On the other hand, it cannot be urged that they were built out of time. They were a necessity to the sections of the state accommodated by them . . . ensuring their early settlement, and making their extensive resources available for the more rapid advancement of the whole community"—citing as an example the Wisconsin Valley road's development of the traffic for "the lumbering region tributary to Wausau"); Report, Committee on Taxes, on bill 4S, Feb. 1, 1877, AJ, 181–182 (urging tax exemption for Wisconsin Central aid lands, because the road's "speedy completion is essential to secure the benefits" of the trade of "the iron and copper regions, and pineries of our northern borders," and so "to open up and develop a most valuable portion of Wisconsin hitherto unapproachable"); Report, Railroad Commissioner, Jan. 15, 1878, PD, II, 8 (Wisconsin Central's completion "opens up a vast country for the lumberman, the miner, and the settler, hitherto undeveloped"); Minority Report, Committee on Banks, on bill 281A, Feb. 20, 1880, AJ, 341 (urging a substitute bill to repeal usury laws, to encourage inflow of capital: in Wisconsin "the demand for money is greater than the supply. In this state are great undeveloped resources . . . unrivaled forests . . . To bring out the hidden wealth of soil, mine and forest, capital is the great requirement. We have brains, energy and enterprise. What we need is money . . . and all legislation is to be deprecated which shall discourage capital and drive it away to other states"); remarks of Lieutenant Governor Sam S. Fifield upon convening of Senate, Jan. 11, 1882, SJ, 5 ("civilization is being extended into the wilderness . . . in fact a 'new Wisconsin' is rapidly assuming strength and fair proportions, and the hum of industry and the bustle of commerce are now to be heard from the state line south to the shores of the 'great inland sea of the north' "); Report, Railroad Commissioner, 1882, PD, I, xxxiv (new northern roads "are of incalculable importance . . . as they open a new way to commerce").

Similar attitudes appear in statements emphasizing the reciprocal benefits of northern railroad development by which the roads will open opportunities for lumber enterprise, which will in turn provide sustaining traffic for the roads. See Report, Select Committee on a constitutional amendment to authorize state money grants in aid of railroad construction, Feb. 27, 1868, SJ, 438 (thirty-four counties now lacking railroads "possess superior natural resources and advantages" and would grow fast in wealth, with transport facilities), 440 ("on the well ascertained facts that railroads generally create business equal to their capacities, aid by the state proposed may with safety be given"), 444 ("Whenever these roads shall reach the forests . . . labor and capital will be employed in the conversion of the forests into lumber . . . and our farmers and manufacturers will there find new markets"), 445 (northern counties contain "extensive and valuble" forests, and "their product of lumber alone, when added to that produced by other sections of the state, will equal that of the state of Michigan for the present year"); Address of Superior Board of Trade, Feb. 15, 1870, AJ, 334 (timbered aid lands represent more fruitful potential for state's economy in hands of railroads than of lumber monopolists who would quickly devastate them, for the railroad's "interest would be . . . so to dispose of them as to secure the largest way business for its road, through the permanent settlement along the line thereof, of the largest number of independent producers; whereby the improvement, instead of the desolation, of the region would be insured"); General Message of Governor Taylor, Jan. 14, 1875, PD, I, 8 (state taxes should be light on railroads in their beginnings in the sparsely settled northern country, "in view of the practical benefits of railroad construction, by way of developing our material resources and in rapidly appreciating the aggregate of taxable property, and especially in view of the rapidly increasing revenue thus ensured from the roads themselves"), 15 (favors completion of Wisconsin Central Railroad because it will "furnish a new and better market for the hard and soft lumber of our northern forests. The timber lands belonging to the Central Company would thus be largely enhanced in value both to the company and to actual settlers; the development of the vast mineral wealth of the north would be encouraged and more amply rewarded, and a new impulse given to the prosperity of the prairie districts of Wisconsin, whose progress in both manufacturing and agricultural industry is now largely impeded for the want of cheaper building and manufacturing materials, cheaper fuel and diminished cost of transporting their surplus products to distant markets"). Twentieth-century echoes of these promotional attitudes were heard when Laws, 1933, Joint Resolution No. 103, protested the proposed abandonment of certain north Wisconsin railroad trackage by a road whose "predecessors . . . received a land grant of 1,800,000 acres for the construction of this line, nearly one million acres of which were in Wisconsin, and the company itself for many years derived very large revenues from this line, which passed through one of the finest stands of pine timber in this state." In effect, another form of nineteenth-century evaluation of the importance of cheap timber and lumber for maximizing entrepreneurial energy was the recognition of the manifold industrial uses to which wood might be put, provided it were kept at low cost. Cf. Greeley, *Forests and Men,* 41, 42; Lillard, 138–139, 145–417; Rodney C. Loehr, ed., *Forests for the Future* (1952), 4–6; Pinchot, 23, 24, 28, 70, 109, 185, 201, 205.

52. On the importance which industry, official, and popular opinion attached to public land grants as aids to railroad finance, directly and as a credit base, see Thomas C. Cochran, "The Entrepreneur in American Capital Formation," 347, 353, 354, and his *Railroad Leaders, 1845–1890: The Business Mind in Action* (1953), 44, 48, 52, 97; Hacker, 232; Johnson and Krooss, 247. With specific reference to Wisconsin, see Hunt, *Law and Locomotives,* chaps. i, iii, and v; Merk, 275, 279–286; Raney, 182, 194. Economic historians differ in their estimates of how far the land grants in fact proved as essential to railroad finance as promoters' predictions would have them. Cf. Cochran and Miller, *Age of Enterprise,* 133; Krooss, 428–429; Merk, 280; Raney, 195. But, there is no doubt of the effective persuasion which the arguments for such aid grants had upon contemporary opinion in the years in which government yet had a vast domain at its disposal. See the summary statement in Cochran and Miller, 109. On the proposition that there were legitimate reasons for "special legislation" in such matters, see Hurst, *Law and the Conditions of Freedom,* 15–17.

53. For grants of right-of-way fees or easements in state land by Wisconsin as contributions in kind to railroad construction finance, see, e.g., Revised Statutes, 1858, c. 79, sect. 27 (general); P&LL, 1861, c. 26 (Peshtigo Co.); P&LL, 1869, c. 51, sect. 19 (Northern Wisconsin Railway Co.); P&LL, 1870, c. 513, sect. 10 (Chippewa Valley & Lake Superior Railroad Co.); General Laws, 1872, c. 119, sect. 27 (general); Revised Statutes, 1878, c. 87, sect. 1857 (general). Compare like grants in kind of state timberland in aid of navigation improvements, on log-driving streams, as in P&LL, 1866, c. 214, sect. 3 (Black Creek, Outagamie County), and P&LL, 1867, c. 334, sect. 6 (Lemonweir Improvement Co.); and the grant of flowage rights in state timberland in aid of a federal government project to guard headwaters in aid of logging navigation in northwest Wisconsin, under Laws, 1880, c. 316, and 1882, cc. 255 and 256, with which compare, also, Laws, 1882, Memorial No. 8. Such allocations of timberland as aid in kind to financing improvements to provide a frame for business growth take on sharpened definition by comparison with Laws, 1882, c. 276, granting an easement over state lands in aid of improvement of the Bad and White Rivers, provided that this grant "shall only apply to swamp and overflowed lands, and to lands not valuable as timber lands . . ."

The language quoted in the text from the federal grants is from the acts of Congress of June 3, 1856, 11 Statutes at Large 20, and May 5, 1864, 13 Statutes at Large 66. The other quoted material will be found in Report of President A. P. Berthard of the St. Croix & Lake Superior Railroad Co., Feb. 27, 1864, SJ, 249; Communication of H. H. Weakley, Land Commissioner, West Wisconsin Railway Co., Jan. 24, 1876, included in Communication of Railroad Commissioners, SJ, 131 (1876); Report, Committee on Railroads, Feb. 3, 1875, AJ, 131; Chippewa Valley & Superior Railway Co. v. Chicago, St. Paul, Minneapolis & Omaha Railway Co., 75 Wis. 224, 249, 44 N.W. 17, 23 (1889). Recognition of the value of timbered aid lands as attractive security for bond issues is implicit in much of the legislation authorizing the mortgage of aid lands enacted with reference to railroad development in the forested north of the state. See Chapter I, note

31. The same point is implicit also in the stress put on obtaining tax exempt status for the aid lands earned by the forest-area railroads, and in the statutory stipulations ending the exemption upon sale of either the land or the standing timber. See notes 7, 13 and 46 above, and Chapter I, note 31. Laws, 1877, c. 21, presents a particularly clear-cut indication of the linking of a tax exemption to the emphasis on timber values as the basis of the utility of forest-area aid lands to railroad construction finance. Recitals of the complicated history of statutory tax exemptions for railroad-aid lands, the cumulative detail of which is revealing of the emphasis put on the lands and their timber wealth as means of financial leverage, will be found in Wisconsin Central Railroad Co. v. Taylor County, 52 Wis. 37, 8 N.W. 833 (1881); Wisconsin Central Railroad Co. v. Lincoln County, 57 Wis. 137, 15 N.W. 121 (1883); Wisconsin Central Railroad Co. v. Comstock, 71 Wis. 88, 36 N.W. 843 (1888); State ex rel. Bell v. Harshaw, 76 Wis. 230, 45 N.W. 308 (1890). Similar stress on the timber value of aid lands for its relevance to finance appeared in connection with waterways and roads development projects. See, e.g., P&LL, 1867, c. 429 (tax exemption on Fort Howard military road aid lands to end upon recipient's sale of timber); Annual Report, Secretary of State, PD, I, 50–53 (1876) (containing report of special commissioners of investigation under Laws, 1876, c. 286, reflecting reliance of Sturgeon Bay & Lake Michigan Ship Canal & Harbor Co. upon mortgaging of timbered aid lands); Zemlock v. United States, 73 Wis. 363, 41 N.W. 445 (1889) (Fox-Wisconsin aid lands grant by United States was not a grant for direct use of the lands for right of way, but a grant to be sold to yield construction funds, hence carried no implied reservation for use in kind). See also Joseph Schafer, *The Winnebago-Horicon Basin* (1937), 113 (importance of timber values of aid lands to Fox-Wisconsin Project).

54. The case discussed in the text is State v. Weston, 17 Wis. 107 (1863). Compare, of like import, Paine v. White, 21 Wis. 423 (1867) (if timber trespasser has paid statutory price of the land plus penalty, transfer of title to the logs originally cut wrongfully by him is no less effective, though there is no evidence he chose to take a patent to the land); Conklin v. Hawthorn, 29 Wis. 476 (1872) (legislative policy is to pass to state's vendee all its rights regarding timber trespass committed before patent); State ex rel. Sturgeon Bay & Lake Michigan Ship Canal & Harbor Co. v. Commissioners of School and University Lands, 34 Wis. 162 (1874) (statute providing that governor determine "the proportion" of aid lands company entitled to receive according to stage of work done, construed to refer to market value rather than acreage of aid lands, much of which was valuable for timber); Sloan, Stevens & Morris v. State, 51 Wis. 623, 8 N.W. 393 (1881) (state has trust duty to protect financial yield from United-States-granted railroad-aid lands, hence may provide for trespass prosecution costs out of trust funds, though not out of general fund); Burrows v. Rutledge, 76 Wis. 22, 44 N.W. 847 (1890) (condemnation of fraud on state pre-emption law, for resultant obtaining of timber at low prices intended only to favor settlement); Lombard v. McMillan, 95 Wis. 627, 70 N.W. 673 (1897) (as grantor, state has some obligation to make title good to its grantee when challenged, hence legislature may provide for its intervention in title contest); Grunert v. Spalding, 104 Wis. 193, 80 N.W. 465 (1899) (timbered aid lands tax exemption construed not to extend to one who assumed no financial obligation toward completion of aided project). Monetized estimates of trespass, fraud, and pricing problems may be seen in materials cited in notes 8, 10, 13, 17, 19, 20, 22, 26, 29, 30, and 32 above. Particularly striking examples of the monetized calculus of federal and state public timberlands problems may be seen in Reports, General Land Office, 1863, p. 60, 1874, p. 6, and 1875, p. 12; General Message of Governor Barstow, Jan. 12, 1855, SJ, App., 21; Veto Message of Governor Harvey on Bill 62S, Apr. 5, 1862, SJ, 813–814; Biennial Report, State Superintendent of Public Instruction, Dec. 10, 1892, PD, I, 141–144 (1893).

55. The "bounty" deserved by the forest "pilots" was advocated by Representative Eastman in the House of Representatives. See note 36 above. The Commissioner is quoted from Report, General Land Office, 1865, p. 26. Timber thefts to provide steamboat fuel are noted in Report, Commissioners of School and University Lands, AJ, App. D, p. 31 (1856); cf. Communication of Timber Agent William Crombie to President of the Senate, SJ, 472 (1858). Consent was given to the buyer of land on credit from the state to cut firewood "necessary for the use of his family" by Laws, 1849, c. 212, sect. 22. See Report, Committee on Education and School Land, 1848, AJ, App., 87. The substance of this permission was

continued as a standard provision in the state's land laws. See, e.g., Revised Statutes, 1878, c. 15, sect. 220. Federal courts interpreted the federal land laws to give like permission. See United States v. Lane, 19 Fed. 910 (Circ. Ct. E.D. Wis. 1883). The wide range of wood uses in the nineteenth century is sketched graphically in Lillard, 138–155; on the waste implicit in the process, see Buttrick, 281, and Gulick, 21–22.

56. The text quotes Acts, 1851, Memorial No. 8. Compare the analogous contemporary demand to make timberlands more quickly available in market, to advance the development of agricultural production power. See materials cited in note 35 above. Another strong expression of the contemporary focus on capital development was contained in Acts, 1852, Memorial No. 26, praying that Congress create a new northwestern territory of "Superior" out of that part of Wisconsin north of the 45th degree of north latitude and the upper peninsula of Michigan, on the ground that the isolation of these areas and the diversity of their interests from those of the rest of their states "prevent in a great measure the proper and legitimate development of the immense natural resources of wealth" there. Emphasis in federal debate on the social value of opening up the public timberlands to increase the productive capital at work in the society is noted in Ise, 50–52 (southern timbered lands), 122–128 and 200 (presidential reserves); Robbins, *Our Landed Heritage,* 314, 339, 341, 373, 382 (recurrent battles over reserves program). The interplay between early conservationist "lock-up" language and Western protests against "hoarding," with its revealing emphasis on production, is reflected in Dunham, 3, 10, 19, 106–107, 115; A. Hunter Dupree, *Science in the Federal Government* (1957), 232, 241; Marquis, 5–6; Pinchot, 28, 70–71, 108–109; Robbins, 373. The reaction against "hoarding" must be interpreted in light of the longstanding, pervasive, popular and official faith in the beneficence of devoting the public domain, including its timber, to the rapid settlement and productive development of the continent (see Gulick, 19, 155). The tenacity of these general attitudes is evidenced in the timing of events; not until the eighties was there any considerable interest shown by the nation's press in conservation issues, and conservationists found themselves swimming against a tide of mass indifference or hostility until Theodore Roosevelt dramatized and popularized the conservation movement at the turn of the century. It is significant that well into the 1920's forest management could still be brought as a novel innovation to the remaining productive forest lands in the Lake States. See Dunham, 165, 166, 169; Loehr, 56, 59, 68, 249, 255; Pinchot, 23, 201, 205; Report, Interim Committee on Forestry and Public Lands, AJ, 882 (1929) ("The timber from some of the trust lands is being sold by the Land Commission, and cut without thought of future growth . . . [A]ll cutting of such timber should conform to good forestry practice, that these lands may remain productive. If the state is to encourage private timber owners to practice forestry, the state should first set a good example").

57. For Senator Howe's observations, see 44 Cong., 1 Sess., *Cong. Record,* IV, 1085; for the remarks of Senator Teller, see 60 Cong. 2 Sess., *Cong. Record,* XLIII, 3226. Compare comments on these views in Ise, 53, 271, 275. Marshall, J., is quoted from his opinion for the court in State ex rel. Owen v. Donald, 160 Wis. 21, 141, 142, 151 N.W. 331, 371 (1915).

58. It was characteristic of mid-nineteenth-century policy making not to make its hypotheses of action very explicit. It is not surprising, thus, that there is no one clear-cut, authoritative statement of the booster theory of use of the public timberlands, or use of public lands to develop the lumber industry. But the idea turns up, or finds reflection, in a sufficient range of specific instances to testify to its wide appeal. See Quaife, *The Struggle Over Ratification,* 652–654 (writer of letter to editor of Prairie du Chien *Patriot,* Feb. 28, 1847, urges advantages of using lands granted new state by the federal government to spur development of lumber industry, among others, by road construction); Report, Committee on Education and School Land, 1848, AJ, App., 87 ("already urgent" demand for timber for economic growth creates pressures to invade public timberland); Report, Select Committee under a resolution of inquiry regarding the Commissioners of School and University Lands, 1855, SJ, App., 8–9 (impatient, growing lumber industry a threat to state's timberlands in sparsely settled areas); State v. Weston, 17 Wis. 107, 109 (1863) (to avoid doctrine under which the state's purchaser of land on credit "might suffer great loss," held that in action for vendee's wrongful cutting of timber, state may recover not value of timber but only extent of impairment of the security thereby; if remaining timber offers ample security for unpaid balance, the contract vendee should be able to exploit within that

limit); Report, Select Committee on Bill 191A, Feb. 13, 1867, AJ, 283–285 (recognition of popular demand for economic growth, which has led to generous or wasteful spending of timber). The tendency to translate all policy choices regarding the public timberlands into money rather than "real" terms, discussed in the previous subdivision of this chapter, was generally manifest in a way which also evidenced the belief in the booster hypothesis.

59. See General Laws, 1868, Memorial No. 18; General Laws, 1871, Memorial No. 1; Report, Select Committee on Bill 106A (to grant "St. Croix" aid lands to the Milwaukee & St. Paul Railroad Co.), AJ, 705, 706 (1873); Report, Select Committee on Bill 178A (to grant "St. Croix" aid lands to Chicago & Northern Pacific Air Line Railroad Co.), AJ, 675 (1873); First Annual Report, Railroad Commissioners, PD, II, 42 (1874). Other expressions or reflections of the booster theory applied to railroad development in the pinery may be found in General Laws, 1867, Memorials No. 13, 27, and 33; Laws, 1874, Memorial No. 1; Laws, 1875, Memorial No. 1. See, likewise, General Message of Governor Dewey, SJ, 30–31 (1849) (for Congressional aid to connect the pinery with markets); Third Annual Report, Commissioner of Immigration, 1873, PD, No. 15, p. 29 (importance of railroads opening possibilities for lumber development); General Message of Governor Taylor, PD, I, 15 (1875) (Wisconsin Central will give "a new impulse" to growth); Report, Committee on Railroads, AJ, 131 (1875) (railroad building must precede economic growth in the northwest of Wisconsin); Report, Committee on Assessment and Collection of Taxes, on Bill 4S, AJ, 181–182 (1877) (Wisconsin Central's "speedy completion . . . essential . . . to open up and develop a most valuable portion of Wisconsin hitherto unapproachable"); Report, Railroad Commissioner, Jan. 15, 1878, PD, II, 8 (Wisconsin Central's completion to Lake Superior "opens up a vast country for the lumberman, the miner, and the settler, hitherto undeveloped"); Report, Railroad Commissioner, Jan. 6, 1882, PD, I, xxxiv (1883) (Chicago, St. Paul, Minneapolis & Omaha Railroad, and Chicago, Milwaukee & St. Paul Railroad developments of recent years are "opening up large tracts of undeveloped country, valuable for its pine forests," offering outlets "to . . . large lumbering establishments" and generally constituting extensions "of incalculable importance to the western and northern parts of our state"). This preoccupation with the strategy and tactics of giving impetus to the economic development of the forested north, and overcoming the inertia created by lack of fluid capital, was highlighted from another angle by an abortive effort in the late sixties to bring to the voters a constitutional amendment allowing state money aid to railroad construction on the ground that the aid lands were not enough. See Report, Judiciary Committee, on Joint Resolution No. 16S, SJ, 600–603 (1867) (though in 1848 other states' experience was viewed as a warning of what to avoid, "lapse of time has served to demonstrate, that the aid given in these several states, greatly accelerated the public improvements therein"); Report, Select Committee on Joint Resolution No. 12S, SJ, 440, 444 (1868) ("railroads generally create business equal to their capacities"; "emigrants, with their industry and skill; speculators, with their capital, will follow the lines of railroad" into the undeveloped north, and trade will build by cumulation, for "whenever these roads shall reach the forests of pine or other timber, or the mines of iron and copper, labor and capital will be employed in the conversion of the forests into lumber, and in the raising the minerals ready for the market, and our farmers and manufacturers will there find new markets for the sale or exchange of their productions for those required for consumption or for the purposes of manufacturers in their home market"). The proposal for this money-aid-to-railroads constitutional amendment lost by tabling in the Assembly at the required presentation of the amendment to the second successive legislature: see AJ, 970 (1868). Concern to keep northern railroad development in motion was also appealed to, unsuccessfully, by opponents of rate regulation; see, e.g., Report, Railroad Committee, SJ, 116 (1873). The faith in the social merit of spending timbered aid lands on road and waterways improvements for the impetus this would give general growth made a consistent pattern with the more important theme of such aid for the railroads. See Acts, 1850, Memorial No. 13 (upper Wisconsin River improvements are "of vast importance to this state and the whole valley of the Mississippi River"); Majority Report, Committee on State Affairs, on proposed investigation of progress of Sturgeon Bay and Lake Michigan ship canal project, AJ, 407–408, 411, 412 (1879); cf. Denniston v. Unknown Owners, 29 Wis. 351 (1872) (Fox-Wisconsin project aid-lands grant to be construed to fulfill purpose of underwriting work); Winslow v. Crowell, 32 Wis. 639 (1873) (Fort Howard military road aid lands'

importance to pressing construction, indicated by parties' dealings); State ex rel. Sturgeon Bay & Lake Michigan Ship Canal & Harbor Co. v. Commissioners of School and University Lands, 34 Wis. 162 (1874) (court should interpret grant act to fulfill policy of using lands to underwrite and push construction); Attorney General's Opinion to District Attorney, Oconto County, Feb. 26, 1902, Attorney General's Letter Books, IX, 196 (Sturgeon Bay project aid lands to be protected to fulfill their underwriting purpose). In various rulings, the Wisconsin Supreme Court recognized that booster or impetus use of public timberland to put railroad development in motion (for the further impetus that railroads would give to the economy) was an important point of legislatively established public policy and should be implemented so far as the constitution allowed. See Sloan, Stevens & Morris v. State, 51 Wis. 623, 8 N.W. 393 (1881) (state obligated to use aid-lands proceeds to maintain assets in realizable condition); Wisconsin Central Railroad Co. v. Taylor County, 52 Wis. 37, 8 N.W. 833 (1881) (legislature is authorized to exercise reasonable discretion in grant of tax exempt status to aid lands when transferred to railroad, where this may be deemed necessary to fulfill aid purpose); Chippewa Valley & Superior Railway Co. v. Chicago, St. Paul, Minneapolis & Omaha Railway Co., 75 Wis. 224, 44 N.W. 17 (1889) (contract to share aid lands held illegal and unenforceable where its effect might be to deprive aided road of intended impetus to development). Because the purpose of the grant of tax exemption to earned aid lands was so plainly to enhance the booster value of the lands by helping the grantee to maximum realization upon them, and because the law's policy did not, in general, favor restrictions upon the taxing power, the use of the device bore especially strong witness to the faith in the social values of booster-type public investment. See West Wisconsin Railway Co. v. Board of Supervisors of Trempealeau County, 35 Wis. 257 (1874), affirmed, 93 U.S. 595 (1876) (where ambiguity exists, statutory tax exemption should be construed as revocable favor rather than as contract); Wisconsin Central Railroad Co. v. Lincoln County, 57 Wis. 137, 15 N.W. 121 (1883) (tax exemption construed to maintain larger rather than narrower scope of tax power); Wisconsin Central Railroad Co. v. Comstock, 71 Wis. 88, 36 N.W. 843 (1888) (similar).

60. The characterization of the "continual striving" of American society is Tocqueville's, in his *Democracy in America,* II, 42. On the imbalance of the factors of production generally as an environmental influence, see Chapter I, above, pp. 10–11, 34–40 and especially notes 25, 26 and 29.

61. See General Land Office, Report, 1870, p. 73; Report, Select Committee on petition of Menominee Indians, AJ, 69–71 (1853). For other indications of the weight with which the fact of physical abundance pressed on men's concept of Wisconsin forest wealth, see Quaife, *The Movement for Statehood,* 77, as well as his *The Convention of 1846,* 561, and *The Attainment of Statehood,* 3; Acts, 1852, Memorial No. 26; General Acts, 1854, Memorial No. 31; Report, Select Committee on the Governor's Message regarding the Emigrant Agency, SJ, 177 (1854); Report, Committee on Railroads, AJ, 825–826 (1855); Report, Select Committee on Joint Resolution 12S, SJ, 445 (1868); Report, Select Committee on Bill 106A, AJ, 706 (1873); Third Annual Report, Commissioner of Immigration, 1873, PD, No. 15, pp. 82, 92. On the complacence toward theft of public timber, the ready destruction of trees to improve land, and the legitimacy of relatively short-term capital gains as deeply ingrained elements of popular opinion, see, e.g., Buttrick, 410; Cameron, 21, 22, 76, 78, 80, 86–87, 97; Dupree, 232, 241; Ise, 19, 22, 26, 35, 41; Lillard, 123, 127, 156, 158; Pinchot, 23, 121, 201, 205. See also notes 48 and 56 above, on popular reaction against official or private activity that locked up timberland, delaying its active exploitation.

62. That the relative scarcity and expense of mobile capital tended to assign residual value to timberland, see Buttrick, 366, and Marquis, 29, 37, 123, 166. It puts the same point another way to note that the cost of the raw material was typically a minor part of the market price of forest products, the bulk of which represented the cost of labor and money. See Buttrick, 175, and Spengler, 305, 306, 308, 311. Part of this pattern is the fact that availability of cheap public timberland accelerated present economic growth by helping overcome the limits ordinarily set to growth by the marginal scale of the rate of current new investment. Cf. Rostow, 42, 43.

63. See General Land Office, Report, 1873, p. 13; Senator Timothy O. Howe, 44 Cong. 1 Sess. *Cong. Record,* IV, 1085 (1876); Majority Report, Committee on Railroads, on Bill 29A, SJ, 644–645 (1867). The relation between the prevailing middle-class philosophy and popular acquiescence or approval toward a maximum of freedom for private will in de-

termining the uses of the forest, is discussed in Cameron, 5, 8, 13, 20, 21, 25, 65, 76, 78, 80, 86–87, 97; Clawson, *Uncle Sam's Acres,* 81–86, 90, 97; Gulick, 19, 155; Ise, 50–53, 77–81, 166, 254; Pinchot, 17, 23, 28, 70, 108, 185, 201, 205. The common counsel of mid-century, to speed transfer of timberland from public to private ownership as the answer to the trespass threat, rested as much on a kindred faith in the energy and wisdom of private will as on despair at government's ability to enlist popular support for strong enforcement or its capacity to carry out enforcement. See, e.g., Reports of Surveyor General, Dubuque, Nov. 11, 1850, in General Land Office, Report, 1850, 31 Cong., 2 Sess., Senate Exec. Doc. No. 2, App. H, p. 43, and Oct. 1, 1863, *id.,* 1863, p. 60, as well as recommendations of the Commissioner, *id.,* 1874, pp. 7, 41, 1875, pp. 12, 18, 1876, pp. 8, 20, and 1880, p. 177.

64. See General Message of Governor Harvey, SJ, 27–28 (1862); General Message of Governor Taylor, AJ, App., 23–24 (1874). The observation of the American Forestry Congress leaders was embodied in a memorial adopted by the Congress at its St. Paul convention, Aug. 9, 1883, forwarded with approval by the United States Commissioner of Agriculture, and transmitted to the Wisconsin legislature without comment by the Special Message of Governor Rusk, AJ, 62 (1885).

65. Compare the nineteenth-century attitudes summarized in Pinchot, 23, 27, 80, 86, 109, 175, 201, 205, with the recognition of the existence of a public interest in forestry reflected in the pattern of twentieth-century policy noted in Gulick, chap. ii. The limited scope of public affairs touched by men active in the lumber industry's relations to legislative policy is symbolized in the activities of a lumberman-politician of Wisconsin, Philetus Sawyer, and of a lawyer prominent in Chippewa Valley industry development, Roujet Marshall. See Current, 199–200; Marshall, *Autobiography,* I, 272–276, 282, 341, 342, and II, 462.

66. The governor of Idaho Territory is quoted in Secretary of the Interior, Report, 1879, p. 419. The not unexpected focus of market transactions upon the immediate interests of the contractors is highlighted in the rare cases where a court held a contract involving public timberland to be illegal and unenforceable because it unduly neglected or invaded broader public interests. See, e.g., Chippewa Valley & Superior Railway Co. v. Chicago, St. Paul, Minneapolis & Omaha Railway Co., 75 Wis. 224, 44 N.W. 17 (1889). Cf. Week v. Bosworth, 61 Wis. 78, 20 N.W. 657 (1884). The effect of the market's money calculus in blurring the view of all the relevant interests at stake in public forest disposal is indicated in Buttrick, 68, 98–99, 237, 263, 401, 410; Ciriacy-Wantrup, 5, 69, 70, 73, 189; Marquis, 3, 24, 25, 29, 37, 120, 123, 166. The pressure toward present, maximum realization on timberland, fostered by the sensitivity of the market's money calculus to production cost reductions made possible by such technical improvements as the donkey engine, had the freer play in the Lake States because the relatively even terrain and the abundance of waterways facilitated clear cutting in great volume. See Loehr, 6; William N. Sparhawk and Warren D. Brush, *The Economic Aspects of Forest Destruction in Northern Michigan,* U.S. Department of Agriculture, Technical Bulletin No. 92 (1929), 7.

67. The general significance of the difficulty of raising long-term investment capital, in affecting the course of the lumber industry, is indicated in National Academy of Sciences, *Report of the Commission upon a Forest Policy for the Forested Lands of the United States* (1898), 42–43; and in Buttrick, 98–99, 263–264, 410, 448. An instructive case history, from the experience of a leading Wisconsin firm of the late nineteenth century, is presented in Reynolds, chap. viii. Compare Gates, *Wisconsin Pine Lands,* chap. x. See also Abramovitz, 3; Cochran, "The Entrepreneur in American Capital Formation," 367; Goldsmith, 140; Loehr, 4, 260–262. The liquidity problems of the weaker members of the industry — reflected in price cuts — pressed on all, moreover, and this became the more felt as the scale of investment and production capacity mounted. See Greeley, *Forests and Men,* 41–42; Marquis, 96, 98, 102, 111, 182. Marquis (p. 96) suggests that a cycle of cause operated here: over-liberal policy in disposing of public timberland encouraged over-investment by private holders, putting them eventually under pressure to liquidate.

68. The assumption that most timbered land was ultimately best suited to farming was basic to federal policy in disposal of the public land of the Central States, and was the more easily applied to the northern and western forest country because it fitted so well into the sentiment attached to the symbol of the free farmer as the guardian of a sturdy and healthy politics. See Bureau of Corporations, *The Lumber Industry,* vol. I, part I,

p. 245; Clawson, *Uncle Sam's Acres,* 81–86; Gulick, 19, 155; Loehr, 5; Marquis, 3, 94; Smith, 124, 126, 159, 200; also, material cited in note 46, Chapter I. The influence of these assumptions of fact and political faiths in the handling of Lake States public timber land is indicated in Sparhawk and Brush, 14, 71, 73; Raphael Zon and R. D. Garver, *Selective Logging in the Northern Hardwoods of the Lake States,* U.S. Department of Agriculture, Technical Bulletin No. 164 (1930), 2; cf. Zon, *Timber Growing and Logging Practice in the Lake States,* U.S. Department of Agriculture, Bulletin No. 1496 (1928), 15. The official record abounds in statements which reflect the common assumption that the ultimate normal destiny of the public forest of northern Wisconsin was to be turned to agricultural use. See General Land Office, Report, 1860, p. 18; Communication of Commissioners of School and University Lands, AJ, 343 (1855); General Laws, 1860, c. 95; Report, Select Committee on Bill 211A, AJ, 639 (1862); Majority Report, Select Committee on Bill 198S, SJ, 916–917 (1862); General Laws, 1863, Joint Resolution No. 1; General Message of Governor Fairchild, SJ, 22 (1867); General Laws, 1867, Memorials No. 4, 15, 27, and 31; Memorial of Chippewa County Board, SJ, 244–246 (1867), and Memorial of Citizens of Polk County, AJ, 326–327 (1867); General Laws, 1868, Memorial No. 1, and General Laws, 1869, Memorial No. 3; General Laws, 1868, Memorials No. 16 and 18; General Message of Governor Fairchild, SJ, 24 (1869); General Laws, 1869, c. 161; P & LL, 1870, c. 225; Second Annual Report, Commissioner of Immigration, 1872, PD, 150; Report, Select Committee on Bill 301A, AJ, 726 (1873); First Annual Report, Railroad Commissioners, 1874, PD, II, 43; Annual Report, Railroad Commissioner, PD, I, 382 (1879); Laws, 1879, c. 22; Annual Report, Board of Immigration, AJ, 111–113 (1881); General Message of Governor Rusk, AJ, 19, 21 (1883); Annual Report, Board of Immigration, AJ, 44–45 (1883). One may cite the official record many times over after this point, but henceforth the references tend more and more to relate to dealing with already cutover lands, rather than reflecting attitudes which influenced public policy while large quantities of public timberland yet remained for disposal.

The reality and magnitude of the factual error in overestimation of the agricultural possibilities of the Lake States forest were attested in retrospect by the poor farm production records of such areas, and by such striking forest rehabilitation measures as repurchase of land by the United States under the Weeks Law for national forests, and provision by the Wisconsin legislature of procedures for encouraging abandonment of farms in the cutover country and return of the land to timber. See, e.g., Clawson, 133; Sparhawk and Brush, 10–11, 20, 21; Report, Interim Committee on Forestry and Public Lands, AJ, 882–885 (1929).

Events on the scale of the transformation of the north Wisconsin forest are not likely to run to a single neat and consistent pattern. Moreover, shrewd men looked at facts and were not always beguiled by catchwords. So the record also shows occasional recognition that important sections of Wisconsin pineland were best suited, if not solely suited, to bearing timber, and that the whole forest was not potential farmland. See Reports of Surveyor General, Dubuque, in General Land Office, Report, 1850, 31 Cong., 2 Sess., Senate Exec. Doc. No. 2, App. H, p. 43, and General Land Office, Report, 1863, p. 60, as well as comments of Commissioner Burdett, General Land Office, Report, 1874, p. 6, and 1875, p. 10; Report, Select Committee on Bill 114S, SJ, 306–307 (1867) (cf. General Laws, 1867, c. 22); General Laws, 1871, Memorial No. 1; Memorial of Polk County Board of Supervisors, AJ, 46 (1872); Report, Select Committee on Bill 106A, AJ, 705–707 (1873). See the retrospective recognition of past errors in estimating the utility of much northern land, in Biennial Report, Commissioners of Public Lands, Oct. 10, 1902, PD, 38, 39 (1903).

After due allowance for these notes of realism, it must be said, first, that they represent a distinctly minor theme in the general pattern of contemporary opinion. Second, they point directly into the deficiencies of technical knowledge on which the text comments. Insofar as contemporary opinion estimated the worth of particular land as lying solely in its timber, almost invariably it is plain that this timber value was seen as if it were a nonrenewable, once-reaped-forever-gone asset, with no serious thought to the possibility of managing the forest for continuing crops. See, e.g., Annual Report, State Superintendent of Public Instruction, AJ, App., 889–890 (1851); Report, Select Committee under resolution of inquiry regarding the Commissioners of School and University Lands,

1855, SJ, App., 8–9; Governor Barstow's General Message, SJ, 21 (1855), and his Special Message, AJ, 598 (1855). Cf. Paine v. White, 21 Wis. 423 (1867). On the lack of elementary technical knowledge of sustained-yield forestry in the nineteenth century, see Dupree, 239–243; Ise, 122–128; Marquis, 6, 11–12, 119–122, 176, 179; Pinchot, 23, 27, 50, 134, 137–138, 147–153. The contemporary lack of forestry knowledge and technique is underscored by the comments of later students who point out that in the original state of the forest very simple steps would have sufficed to maintain a substantial continuing productive power. Cf. Buttrick, 452; Zon, *Timber Growing and Logging Practice,* 13, 16, 26, 29. Later technical developments which ended the threat of timber "famine" serve also, by contrast, to point up the crudity of the earlier operations which gave plausibility to the gloomier prophecies. See Gulick, 80, 81, 91, 117, and Marquis, 5–6.

In proportion as heavier drains were made on the forest, the extent of technical knowledge needed to maintain efficient, continuing production became much more exacting, however, and plainly far beyond late nineteenth-century achievements (see Loehr, 68). The development of twentieth-century economic theory concerning social accounting and economic growth and change, as well as twentieth-century developments in ecological studies, underlines the limits of conceptualization within which nineteenth-century policy had to be made. The point is all the clearer because the development of theory by the mid-twentieth century still fell so far short of the knowledge needed for adequate policy planning. See Ciriacy-Wantrup, 230, 231, 239, 241, 248–249, 350; Levy, "Some Social Obstacles to 'Capital Formation,'" 491, 492; Fritz Machlup, *The Political Economy of Monopoly* (1952), 299, 300, 463; Spengler, 305, 308, 311, 314–316. On the influence of cheap public timberlands in reducing market incentives to improvement of technical and accounting knowledge, see, e.g., Gulick, 80, 87, 96, 161.

69. The extent to which the "worship of the fee" clouded definition of the real issues and choices in federal public land policy has impressed economic and social historians. See Cameron, 17, 396; Lionel R. Robbins, "Freedom and Order," in Smithies, *Economics and Public Policy,* 148; Robbins, *Our Landed Heritage,* 7–10, 382; Smith, 133, 170, 172. On the potency of the market and hence of contract as symbols of reliable, self-sufficient procedures of decision, compare Myrdal, 73. And note, as particularly relevant to our concerns in view of the influence of popular and official belief in the inexhaustible quantity of forest wealth, the observations in Rostow, 40, 42, 43, indicating that the more slowly the pressure of physical facts is felt, the more scope is allowed for the operation of cultural factors.

70. See Acts, 1851, Memorial No. 8; remarks of Representative B. C. Eastman, July 22, 1852, 32 Cong., 1 Sess., *Cong. Globe,* XXI, App. 851; General Land Office, Report, 1874, pp. 7, 41. Invocation of the idea of transfers in fee simple as the satisfactory solution to disposal of public timberlands may be seen in various other contexts. On the application of general "sale" patterns to federal and state forest land, see materials cited in notes 1–3, 11–16 above. Note the unquestioning appeal to the fee simple concept in pleas by the Wisconsin legislature to speed application of the pre-emption laws in heavily timbered areas. See Acts, 1852, Memorial No. 18; General Acts, 1854, Memorial No. 14. In various respects the treatment of timber trespass reflected the uncritical value assigned the fee simple as the symbol of a satisfactory resolution of policy. Thus, greater leniency was shown in the terms of settlements with trespassers where the wrongful cutting was done by the holder of a certificate of sale from the state, than when it was by one who had no inchoate title; it was part of the same pattern that the state allowed possessory actions to its certificate holders, and provided for transfer to its patentees of all rights of recovery for trespasses by others committed before patent. See Dykstra, "Legislative Efforts to Prevent Timber Conversion," 461, 466, 467. The despairing counsel of federal and state officials who urged the speedy sale of public timberlands as the only promising step to curb the theft of timber was another tribute to the appeal of the fee simple as the answer which avoided the truly difficult problems of public lands administration. See notes 26 and 29 above. The Wisconsin court found nothing to question in the statutory policy giving the state's patentee full rights to any uncollected trespass claims of the state accrued before patent. See Conklin v. Hawthorn, 29 Wis. 476 (1872); Burrows v. Rutledge, 76 Wis. 22, 44 N.W. 847 (1890); cf. Knapp v. Alexander-Edgar Lumber Co., 145 Wis. 528, 130 N.W. 504 (1911) (analogous federal policy).

71. On the limited scope of state interest

in the cut-and-run danger under its sales contracts and in redress of trespass, see notes 13, 21, and 22 above. The Commissioner of the General Land Office is quoted from the Report of 1876, p. 9; compare the contemporary and — more frequently — retrospective comments to similar effect on the practical operation of the state's policy of "selling" its lands, in notes 19 and 21 above. The lumbermen's interest in timber rather than timberland is given contemporary recognition in effect in Paine v. White, 21 Wis. 423 (1867); Allen v. Clark, 36 Wis. 101 (1874); brief of defendant-appellant, pp. 10–13, in Lombard v. Antioch College, 60 Wis. 459, 19 N.W. 367 (1884); and Swift v. State Lumber Co., 71 Wis. 476, 37 N.W. 441 (1888). It is acknowledged retrospectively in Letter of Attorney General to J. W. Hammond, Sept. 12, 1901, Attorney General's Letter Books VIII, 409, and in Maxon v. Gates, 136 Wis. 270, 116 N.W. 758 (1908). Though perhaps in logic this was not the only inference that could be drawn, in the context of the times the many references in contemporary decisions to the timber value of litigated tracts as their "chief" or "sole" worth imply the same matter-of-fact acceptance of the industry's interest in the present standing timber rather than in the forest as a resource for continuing production. See, e.g., Haight v. Lucia, 36 Wis. 355 (1874); Riemer v. Johnke, 37 Wis. 258 (1875); Marsh v. Bellew, 45 Wis. 36 (1878); Hyslip v. French, 52 Wis. 513, 9 N.W. 605 (1881); McInnis v. Lyman, 62 Wis. 191, 22 N.W. 405 (1885); Atkinson v. Hewett, 63 Wis. 396, 23 N.W. 889 (1885); Tyner v. Cotter, 67 Wis. 482, 30 N.W. 782 (1886); Hicks v. Smith, 77 Wis. 146, 46 N.W. 133 (1890).

72. See Schriber v. LeClair, 66 Wis. 579, 581, 583, 597, 29 N.W. 570, 571, 891 (1886). Cf. Houlton v. Nichol, 93 Wis. 393, 67 N.W. 715 (1896), as well as notes 1, 3, 16–20, and accompanying text, above. Examples of the matter-of-fact contemporary acceptance of timberland values patterns which spelled substantial capital subsidies to private operators out of public assets may be seen in the cases cited at the end of note 71 above.

73. Examples of the moral claims to growth opportunities with the help of federal timberlands, urged by newer regions on the precedent of the Wisconsin and general Mississippi Valley experience, will be found in Ise, 50, 166, 254, 271, 275; Robbins, *Our Landed Heritage,* 314, 339, 341, 373, 382; cf. Gulick, 79, 88. The decisions implicit in federal railroad-aid land grants, in favor of faster and larger exploitation of the forest, receive suggestive comment in Harold A. Innes, "Economics of Conservation," *Geographical Review,* 28:137 (1938). For typical expressions of the moral claim of the undeveloped sections to such railroad-aid grants, see General Laws, 1871, Memorials No. 1 and 5. Compare, generally, Report, Finance Committee, SJ, 619–624 (1864). The quotations in the text are from Acts, 1851, Memorial No. 8, and from Address of Superior Board of Trade, AJ, 336 (1870). See also Report, Judiciary Committee on Joint Resolution 2S, SJ, 284 (1869); General Laws, 1872, Joint Resolution No. 10. It was an old story, dating from colonial years, that local interests of timbered sections made themselves felt more effectively on public forest policy than did ideas of the utility of such public wealth to the underpinning of the more general economy. See Cameron, 13, 20, 21, 25, 65, 76, 83, 85–87, 97; Cummings and McFarland, 264, 387, 389.

74. Formal acknowledgment was paid to environing federal policy in decisions which, in order to illuminate state policy, invoked federal legislation because it represented statutes *in pari materia.* See Harrington v. Smith, 28 Wis. 43, 66, 67 (1871); cf. Houlton v. Nichol, 93 Wis. 393, 67 N.W. 715 (1896). Likewise there was formal acknowledgment of a rare instance of federal leadership concerning public timberlands disposition, in Art. VIII, sect. 10, of the Wisconsin Constitution, excepting the use of special grants in aid from the ban on state participation in works of internal improvement, as also in the Wisconsin Supreme Court decisions which in various respects enforced the trusts attaching to timberlands given the state by the United States to aid transport improvements. See, e.g., Wisconsin Central Railroad Co. v. Taylor County, 52 Wis. 37, 8 N.W. 833 (1881); Chippewa Valley & Superior Railway Co. v. Chicago, St. Paul, Minneapolis & Omaha Railway Co., 75 Wis. 224, 44 N.W. 17 (1889). The poor example set to public opinion and state action by federal inaction and feebleness in handling federal timberland is noted by Cameron, 78, 81–82, 86, 89, 99; Clawson, *Uncle Sam's Acres,* 82, 84, 89–90, 97, 110–111; Dunham, 23, 25, 28, 49, 53–54, 58, 60; Ise, 23–26, 39, 40–45, 55, 58–59, 64–66, 79–81, 89, 114; Pinchot, 24, 80–82, 86, 109, 175. Such criticisms of twentieth-century federal forest policy as those in Buttrick, 448, implicitly emphasize how totally lacking was

any leadership by the United States in the nineteenth-century decades which fixed the uses of the Lake States forest. Apart from the influence of federal railroad-aid grants in expanding the arena within which market forces might gather impetus (see note 75 below), these grants likewise contributed to emphasize speedy present realization on these assets, by fixing time limits to press construction. See Bureau of Corporations, *The Lumber Industry,* vol. I, part I, p. 248; cf. Schulenberg v. Harriman, 21 Wall. 44 (U.S. 1874), and the Chippewa Valley & Superior Railway case, cited above in this note.

75. The forests of Michigan and Wisconsin, the Commissioner of the General Land Office observed in his Report, 1881, p. 377, "have furnished much of the building material that has supplied for years the great demand from the growing Western States." His comment stands almost alone in the record in noting this fact in a critical context; he pointed to this imperious midwestern demand as encouraging what had been "the most extensive trespassing in the country" (p. 376). The official records of the territory and state reflect a lively, continuing consciousness of the implications of the expanding Mississippi Valley lumber market for Wisconsin's economy — seen always as an opportunity, never as a threat. See General Message of Governor Dodge, Council Journal, 185–186 (1838), and proceedings in Council thereon, pp. 201, 204, 248; Laws, Wisconsin Territory, 1848, Memorial of Feb. 26, p. 333; Letter of "Old Crawford Forever" to Prairie du Chien *Patriot,* Feb. 29, 1848, in Quaife, *The Struggle Over Ratification,* 654; General Message of Governor Dewey, SJ, 30–31 (1849); Acts, 1850, Memorial No. 13; Acts, 1851, Memorial No. 5; Acts, 1852, Memorial No. 26; Reports, Committee on Railroads, AJ, 320, 825–826 (1855); Governor Barstow's Special Message, AJ, 292 (1856); General Acts, 1858, Memorial No. 37; General Laws, 1867, Memorials No. 13, 27, and 33; proceedings on resolution 74A, AJ, 600–601, 618 (1867); Report, Select Committee on Joint Resolution 12S, SJ, 437, 438, 440, 444 (1868); Report, Judiciary Committee on Joint Resolution 2S, SJ, 279, 280 (1869); General Laws, 1871, Memorials No. 1 and 11; General Laws, 1872, Memorial No. 1; Report, Select Committee on Bill 178A, AJ, 674, 675 (1873), and Report, Select Committee on Bill 106A, AJ, 706, 707, 709 (1873); Report Railroad Committee, SJ, 116 (1873); Third Annual Report, Commissioner of Immigration, 1873, PD, No. 15, pp. 29, 82, 92; Laws, 1874, Memorial No. 1; First Annual Report, Railroad Commissioners, 1874, PD, II, 42, 43; Laws, 1875, Memorial No. 1, and Report, Railroad Committee, AJ, 131 (1875); General Message of Governor Taylor, PD, I, 15 (1875); Report, Committee on Assessment and Collection of Taxes, AJ, 181–182 (1877), and Report, Railroad Commissioner, PD, II, 8 (1878); Laws, 1880, c. 292; Report, Railroad Commissioner, PD, I, xxxiv, xxxix–xl (1882). Indeed, those of the foregoing cited materials which deal explicitly with grants of public timberland in aid of northern Wisconsin railroad construction argued vigorously and confidently for public action to render the pinery more quickly and more broadly responsive to the expanding consumer market.

We lack the clear-cut contemporary authority which would allow us to make confident prediction of the scope which might have been allowed bold economic planning efforts, either by the nation or the state, concerning the pace and scale of development of Lake States forest industry. Gibbons v. Ogden, 9 Wheat. 1 (U.S. 1824), and The Passenger Cases, 7 How. 283 (U.S. 1848), do not encourage the idea that the United States Supreme Court would have conceded large scope to the state's authority to control imports of labor or capital or exports of produce. The twentieth-century cases plainly bar such scope of economic planning to the states, so far as it might stem from their general legislative powers. See Oklahoma v. Kansas Natural Gas Co., 221 U.S. 229 (1911); Pennsylvania v. West Virginia, 262 U.S. 553 (1923); Foster-Fountain Packing Co. v. Haydel, 278 U.S. 1 (1928); H. P. Hood & Sons, Inc. v. DuMond, 336 U.S. 525 (1949). This doctrine seems now so well established as to render it unnecessary to consider possible due process limits upon the state's economic planning power. Cf. New State Ice Co. v. Liebmann, 285 U.S. 262, 309–311 (1932).

Because of the failure of both the United States and the states to attempt any bold use of their land ownership as the leverage for economic planning in the years that fixed the fate of the Lake States forest, there was no occasion for contemporary definition of the limits which the commerce clause, the 10th Amendment, or any of the emerging concepts that would eventually reach definition as substantive due process of law, might have set to employment of public land ownership as a means of directing economic growth. Both the "practical" and the explicit construction of their constitutional powers in this regard by Congress and state legislatures was clear enough in the readiness with which they in fact disposed of public wealth to promote the growth of towns, commercial agriculture, and improved transportation. As the text notes, the most striking peculiarity of the relation of public land policy to the development of the lumber industry was the combination of an important subsidy program (in practice) with an almost complete failure to define it, let alone argue it.

The Wisconsin Supreme Court saw no state constitutional barrier to disposal of state timberlands in promotion of present maximum economic activity, so long as the disposition was in a manner which yielded funds for the educational trust or bore reasonable relation to execution of the internal improvements in aid of which the land had been given. See the broad affirmative power conceded the state legislature for such present, promotional use of lands held in the state's disposal, in State v. Weston, 17 Wis. 107 (1863); Paine v. White, 21 Wis. 423 (1867); Conklin v. Hawthorn, 29 Wis. 476 (1872); Wisconsin Central Railroad Co. v. Taylor County, 52 Wis. 37, 8 N.W. 833 (1881). Cf. State ex rel. Sweet v. Cunningham, 88 Wis. 81, 57 N.W. 1119 (1894). In retrospect the Wisconsin court was prepared to criticize the state's past failure to use its timberland ownership to brake the pace or limit the scale of development in the interest of the long-run stability and power of the general economy. See City of Milwaukee v. State, 193 Wis. 423, 452–453, 214 N.W. 820, 831 (1927). One may well doubt whether in the climate of opinion of the years 1850–1890 the court would not have found a withholding policy contrary to its reading of the constitutional trusts. See notes 6–10 above.

The United States Supreme Court has invariably spoken of Congress' "Power [under Art. IV, sect. 3, of the federal Constitution] to dispose of and make all needful Rules and Regulations respecting the Territory or other Property belonging to the United States" as subject to no judicially enforceable limit as to the general objects of its use. The Court's only pertinent ruling contemporary with the decisive period for the Lake States forest lands would seem to indicate that Congress was free to use the leverage of the federal ownership for any economic planning it desired. This seems the force of United States v. Gratiot, 14 Pet. 526 (U.S. 1840), sustaining in the broadest terms Congress' right to lease rather than sell public mineral lands. Pollard's Lessee v. Hagan, 3 How. 212, 224 (U.S. 1844), contains a broad dictum indicating that the federal government held the lands derived by cession to it by the states of their western claims, in trust only to sell the land ("to convert the land into money") to pay the public debt from the Revolution and to erect new states over the ceded territory. The matter there at issue concerned only federal rights in soil under navigable water, however; the court in effect ruled that control of such land was so important to the governance of affairs that the federal government might not undertake to dispose of the soil after the creation of a state embracing the area in question (*id.*, 230). In no later decision did the Supreme Court extend Pollard's Lessee v. Hagan beyond its facts; to the contrary, the prevailing trend of later opinions taking note of the Hagan case was to emphasize the breadth of Congress' constitutional authority over the public lands so long as Congress had not passed title to them. See Van Brocklin v. Tennessee, 117 U.S. 151, 165 (1886); Shively v. Bowlby, 152 U.S. 1, 28 47–48, 50 (1894); Brewer-Elliott Oil & Gas Co. v. United States, 260 U.S. 77, 83 (1922). The twentieth-century decisions seem clearly to concede to Congress unchecked discretion to use public property as the means of economic planning for the development of the national or regional economies. See Light v. United States, 220 U.S. 523 (1911); Ashwander v. Tennessee Valley Authority, 297 U.S. 288 (1936); United States v. City and County of San Francisco, 310 U.S. 16 (1940).

76. On the different and possibly conflicting interests of a broad market possessing alternative supply sources and any given supply area, compare Buttrick, 301; Ciriacy-Wantrup, 252–254; Gulick, chap. iv; Scott, 509–512. Thus what may be "waste" from the standpoint of a given supply source may represent a contribution to an efficient factors mix, from the standpoint of a given consumption area (see Scott, 509). The calculation of the safe minimum of depletion of a natural producing medium above the point of irreversible loss of productive potential may be different for a small, compared with a larger, production area (see Ciriacy-Wantrup, 255). Comparison of the implications of particular resource allocations between present

and future yields may be the more difficult because the same pattern of use may spell immediate financial gain in highly concentrated form to the supplier, while its greatest present significance to the buyer may lie in the contribution made to "real" present productive development (cf. Gulick, 19, 155). The consumption area can look more complacently upon its drain of natural raw materials from outside supply sources because the cost of such unprocessed or semi-processed stuff is likely to be a minor part of the total cost of the productive activities in which the consumption area uses the raw materials; this is another way of saying that the consumption area may view the raw materials primarily for their "real" (opportunity-extending) utility to it, at the same time as the supply area is tempted to hurry its present pace of exploitation by the financial rewards which, concentrated within the supplier's economy, have a greater significance there than do the financial payments to the consuming areas (cf. Spengler, 306, 308). Had the central government been taking more account of the "real" capital increments and costs at stake, however, it might have found reasons to be concerned at the pace and scale of Lake States forest exploitation even from the standpoint of the consuming areas, which in the long run paid in loss of site value and lumber quality, and hence in restriction of future flexibility of economic maneuver. See Buttrick, 210, 236, 237, 451; Cameron, 418; Gulick, 84–86, 93, 96, 133, 160. The rapid expansion of the scope of sectional and national markets, and the inherent impersonality as well as the greater pressure thereby created, might be reckoned as intangible "costs" of successful federalism, insofar as they made it more difficult to focus concern even on the real national stakes involved. Cf. Ciriacy-Wantrup, 13; Rostow, 75, 104; Spengler, 314–316. Generally, on the social utility of an increasing scale of productive operations relative to the stage of general social development, compare Myrdal, 18, 59, 60, 65, 161, 174, 213–214, 224–225.

77. On the policy attitudes surrounding the failure to provide for timberland classification, and the despairing counsel to sell as the only answer to theft, see notes 1, 2, 16, 19, 20, and 26–32 above. Compare Clawson, *Uncle Sam's Acres,* 81–86; Dunham, 46, 49, 53–58, 60; General Land Office, Reports, 1874, pp. 7, 41, and 1875, pp. 12, 18; General Message of Governor Barstow, SJ, App., 21 (1855); General Message of Governor Randall, SJ, 43, 44 (1858). Tangential, but nonetheless revealing the prevailing distrust of government's efficiency or honesty in handling the public lands (including those of prime timber worth), was the appeal to this symbol in the argument of General Laws, 1871, Memorial No. 1, that if the United States did not renew the "St. Croix" railroad-aid grant, the only consequence would be that "the pine lands embraced within it would soon fall into the hands of large lumbering companies, or be taken under pre-emption or homestead laws only to be despoiled of timber and abandoned." Compare the criticism of the "singular oversight" by which the state failed to check application of railroad aid lands, in Railroad Commissioners, First Annual Report, 1874, PD, II, 58, 61. On the impact of the uncontrolled business cycle, see the suggestive comments in Cochran, "The Entrepreneur in American Capital Formation," 351, 352, 361; Fries, 11, 108–109, 111; Gulick, 30, 84–86, 129, 196, 223; Reynolds, 13, 14, 120, 130–132. Analyses of the handicaps from tax systems ill-adapted to timberland, from the want of means to insure against fire and wind loss, and from the mounting interest costs of holding such long-term investments portray the difficulties of sustained-yield forestry even in the more capital-rich twentieth century with its broader, more sophisticated apparatus of government. Likewise, in the twentieth century government-supplied services to prevent and extinguish forest fires, to combat forest pests, and to furnish expert forestry advice offered important sustaining facilities to the lumber industry. The nineteenth century even more plainly suffered the lack of what was felt sharply enough in the twentieth, and in addition the nineteenth-century forest industry was helped by none of the government aids later made available. These elements underline the practical significance for the nineteenth century of the want even of the conception of government financial assistance such as might have relieved the constant pressure to liquidate. Cf. Loehr, 4–6, 58–60, 260–262; Zon, *Timber Growing and Logging Practice,* 15. The *de facto* capital subsidy represented in the cheap prices of public timberland was a partly articulate recognition that government's financial help was needed, but that contemporary imagination and will were inadequate to providing help in more sophisticated ways which might keep alive the forest

as a producing facility. Cf. Krooss, 81–83, 509 (cheap public lands program as early, crude form of deficit finance). On the crudity of nineteenth-century forms of federal grants-in-aid to states, as of public lands in aid of transport improvement, compared with the complex federal-state interaction and the resultant pressures for improved standards and enforcement of good forestry practice in the twentieth century, compare Gulick, 68, 69, 72. However, there was material for a plea in mitigation of nineteenth-century shortcomings, in the failure of the twentieth century to achieve high efficiency in co-ordinated efforts of various agencies of government on all levels (see Ciriacy-Wantrup, 353, 357–359, 370–376).

78. Of suggestive value on the special importance of early diagnosis and positive action concerning natural resources subject to depletion are comments in Ciriacy-Wantrup, 274–276. Clawson, *Uncle Sam's Acres,* 364, 365, Ise, 39, and Pinchot, 23, 28, 70–71, 108–109, comment trenchantly on the difficulty of bringing all major interests into focus in the handling of forest wealth (cf. Machlup, 300). On middle-class responsibility to over-all function as well as to self, and the problems of instilling and maintaining this discipline, see Clark, *Economic Institutions and Human Welfare,* 8, 30, 45; Dorfman, II, 684; Smith, 159, 170, 200. Implications for the weight of the imbalance of the factors of production in increasing the need for over-all direction in the handling of timber wealth may be found in Buttrick, 98–99, 281, 366, 401, 410; Ciriacy-Wantrup, 274; Marquis, 29, 37, 166, 174. The mindless quality of the thrust of the factors situation, and the consequent need for and challenge to intelligence to assume leadership, is commented on by Abramovitz, 11; Myrdal, 59, 65, 73; Rostow, 40, 42–43, 230, 239, 244. On the tempo of economic development as a prime situational influence on the handling of public timberlands, including the rapid development of the threat of wholesale fraud and trespass, see Ciriacy-Wantrup, 5, 7, 13, and Dunham, 159. On the other hand, it should be noted that there was probably some reciprocity in causal impact here; the large-scale availability of cheap public land helped overcome the limitations on growth rate ordinarily set by the fact that new investment is usually quite marginal compared to already outstanding investment which will have set an existing pattern of tempo of development. Cf. Rostow, 42, 43; Spengler, 305, 306, 308, 311. The risk costs, not only of physical loss but also of product obsolescence, involved in capital investments or commitments which can be turned over only through a long time span are discussed with reference to extractive industries by John Maurice Clark, "Aims of Economic Life as Seen by Economists," in Dudley Ward, ed., *Goals of Economic Life* (1953), 47, 80, and by William Vickery, "Goals of Economic Life: An Exchange of Questions between Economics and Philosophy," *id.,* 165, 167, 168. See also John Maurice Clark, *Social Control of Business* (1926), 80; Henry G. Aubrey, "Investment Decisions in Underdeveloped Countries," National Bureau of Economic Research, *Capital Formation and Economic Growth,* 397, 407. The special time complication introduced where a natural resource is renewable only if its exploitation is not pressed beyond some point of irreversible depletion is discussed in Buttrick, 281, 401, 452; Ciriacy-Wantrup, 47, 252–259, 350. The added costs for wider acreage investment, for sustaining risks of physical loss and technical obsolescence, and for obtaining the positive knowledge and protective facilities needed for sustained-yield forestry are discussed in Buttrick, 68, 98–99, 301–305, 333, in Loehr, 4–5, 58–59, 68, 260–262, and in Zon, *Timber Growing and Logging Practice,* 13, 15–16, 26–27, 29. These references indicate reasons why sustained-yield practice, once launched and protected, might not have been impossibly difficult to maintain. But this possibility does not answer the hard problems of overcoming the initial inertia of the market situation, especially when many small or modest-sized operators of limited means were pioneering the industry's growth (cf. Gulick, 202, 225, and Loehr, 254, 255). The effect of stocks of virgin timber available for exploitation emphasizing present yield, as a limitation on the practicability of sustained-yield programs conducted by some but not all private competitors, is noted in Gulick, 109, 159, 200–201, and Loehr, 56. Further perspective on the limits to a more highly rationalized forestry emphasizing more long-term yield, arising out of institutional gaps in the nineteenth-century society, may be seen by comparison with the elaborate apparatus of technical and scientific knowledge and agencies for its application available in the twentieth century (see Dupree, 33).

79. The scope for forest policy making under constitutional provisions is discussed in

the Introduction to Part One, note 1, in notes 6–10 of Chapter I, and in note 75 above. For over-all judgments on the lack of will and imagination as the critical deficiency in development of public land policy, and especially regarding public timberland, see Clawson, *Uncle Sam's Acres,* 82, 104, 332, and Pinchot, 23, 27, 85, 120.

80. The quoted paean to allodial title is from the Report, General Land Office, 1870, p. 28. The reliance on the cumulative effect of the market as the equivalent of over-all policy decisions took legal form in the familiar presumption of the legality of contracts for dealings affecting entry of public lands. See Houlton v. Nichol, 93 Wis. 393, 398–400, 404, 67 N.W. 715, 717, 718 (1896).

81. For details on the organizational shortcomings of the federal and state public lands programs, in application to timber holdings, see materials in notes 1, 5–10, 15–20, 26–32; compare the discussion in Chapter I, pages 56–61, on the comparable organizational defects affecting public lands policy as a whole. The defaults of Congress are especially obvious, (1) because under the public property clause of the Constitution and the ready recognition by the Supreme Court of the unqualified discretion thereby committed to Congress, it possessed the full responsibility of initiative — see United States v. Gratiot, 14 Pet. 526 (U.S. 1840), and Schulenberg v. Harriman, 21 Wall. 44 (U.S. 1874); (2) because it held the purse, and yet did not spend to arm the executive branch with the staff and procedures to develop public opinion and program out of the experience of positive efforts to come to grips with the evolving use of forest wealth — see, e.g., the summary of fruitless complaints of the General Land Office over lack of means, in Cameron, 78, 79, 90, 91, and in Dunham, 157; (3) because it had the power to create specialized, Cabinet-level executive agents to provide leadership more adequate to the scale of problems involved in the public domain, and did not do so in the face of tardy recommendations for such action (see Dunham, 132). In terms harsh but not unfair, the Wisconsin court singled out the sheer lack of craftsmanship in the federal land laws for criticism, as it looked back over the record from the vantage point of the turn of the century. Since 1820, the court noted, federal legislation on public lands "has been almost as frequent as the sessions of Congress; some of it scientific and deliberate, but much of it seemingly accidental and aimed at special details . . . [which] if construed according to their exact words, would [many times] have been subversive of certain phases of the obvious general policy of Congress with reference to this important subject"; see McCord v. Hill, 111 Wis. 499, 523, 87 N.W. 481, 482 (1901). That the state legislature did at least designate committees specially charged to consider timber industry matters, see proceedings of Jan. 10, 11, and 14, 1867, AJ, 10, 38, 42; Jan. 19, 1881, AJ, 52; Jan. 31, 1883, AJ, 103; Jan. 19, 1887, AJ, 42. All blame for want of broad programming cannot be put on the legislative branch, however, if only because there were conspicuous examples in the record of how much could be accomplished by strong executive officers, using the leverage afforded them either by the relative indifference of the legislature or by sharpened appeal to affected interests outside the immediately affected area. The notable federal example is the inclusion of authority to the President to create federal forest reserves in the 1891 statute repealing the Timber Culture Act, a decisively important measure brought about on the initiative of Secretary of the Interior Noble and a special agent of the General Land Office with the sanction of a conference committee. See Lillard, 264, and Pinchot, 85. The notable state example was the leadership of Governor Harvey in finally defeating the proposal to turn over disposition of all the swamplands given the state by the United States to the decision of the counties wherein the lands lay. See his Veto Message of Apr. 5, 1862, SJ, 809–819. In this instance, it must also be noted, the appeal to broader interests, when well enough defined, evoked response from the legislature on its own momentum; bills to donate the swamplands to the counties had lost out in the 1861 session as well. See, on bill 98S, Reports of the Committee on Public Lands, Mar. 15, and of a Select Committee, Apr. 8, 1861, SJ, 406, 687, 691–694, and proceedings of Apr. 9, pp. 698–700, as well as the defeat of Joint Resolution 14S (for a constitutional amendment to abolish the office of land commissioners, so that all school lands might be placed at the disposal of the counties), SJ, 115–116 (1861). Resolution 33A, instructing the Assembly Committee on Swamplands to report like legislation to give such lands to the counties, was summarily defeated in that house: see AJ, 96, 105 (1861). With labored humor, a Report of the Committee on Swamplands, AJ, 810–812 (1863), on Bill 426A (entitled "A

bill to prevent lawyers from swamping the State") also paid tribute to the need of keeping the charge of the broad interests of the state in the legislature. The committee's observations of the current session, it said, had "revived many doubts as to the propriety or safety of committing the interests of the State to the hands of men engaged in barter or trade, unchecked and unrestrained by a large infusion of farmers and lawyers — for the force of habit might, unrestrained, induce them to barter away the interests of the state at large in exchange for mere petty swamp land or other local interests . . . [W]e are of the opinion that lawyers generally who are elected to the legislature, frequently at great pecuniary sacrifice to themselves, talk no more than is necessary to protect the interests of the state from the trading propensities of some who cannot, even in the legislature, overcome the force of habit, and exchange the interests of the state for the interests of a county." Hence, the committee recommended that the title of Bill 426A be amended "as to read: 'A bill to prevent lumbermen from selling the interests of the state.' And that the body of the bill be amended so as to correspond with the title." As the last incident suggests, though the various proceedings regarding decentralized administration of the swamplands did not touch the state's timbered lands as such, yet timber was a very important element in swampland value, so that this history has some relevance to our concern. To return to the executive initiative shown in the 1891 federal legislation and in Governor Harvey's veto of the 1862 swamplands decentralization bill — the prime relevance of these unusual actions is, by contrast, to highlight the otherwise almost total lack of any such cheering energy or boldness from the first or second levels of executive action. See, e.g., the criticisms of the inertia and timidity of the General Land Office in not exploring the possibilities of its rule-making powers, in Pinchot, 111, 161, 197, 200. Compare the criticisms of the past lack of focused direction by the state land commissioners to the handling of the state's natural resources wealth, in Biennial Report, State Superintendent of Public Instruction, Dec. 10, 1892, PD, I, 141–144 (1893). Finally, in the catalogue of those who failed to give policy leadership as to public timber wealth, we must include the political parties, which paid no attention to the matter until long after the Lake States forest had been committed (see Dunham, 25, 102, 144, 167, 317). On the significance of the lack of Civil Service or of a body of professional foresters, see Gulick, 58, 74; Ise, 59, 126; Pinchot, 161–170. A particularly interesting contrast of judgments is afforded by comparison of the often-quoted condemnation of past waste of the public domain under "defective and improvident laws" and "astonishing laxity of public administration . . . if not culpable recklessness," by Commissioner Sparks, in General Land Office, Report, 1885, p. 3, and the acquittal of the Land Office of substantial malfeasance and the focusing of prime responsibility on Congress in John B. Rae, "Commissioner Sparks and the Railroad Land Grants," *Mississippi Valley Historical Review,* 25:211, 212, 230 (1938).

82. On the general desire to see the public domain used to speed the growth of markets and settlement, and the implicit link between this prevailing opinion and the diverse and significant utility of cheap forest products for all manner of economic development, see Lillard, 138–155, and Hibbard, *Public Land Policies,* chap. viii. The positive force and also the limited activities of the lumbermen as politicos are reflected in Current, chaps. ii, iv, and v; Fries, chap. xiii; Gates, *Wisconsin Pine Lands,* 72, 93–94, 118, 137–176, 181–207; Horace Samuel Merrill, *William Freeman Vilas* (1954), 27–29, 139–150.

83. The Commissioner of the General Land Office is quoted from his Report, 1876, p. 9. Compare Hibbard, *Public Land Policies,* chap. xix, on the Timber Culture Act. On the gaps and internal inconsistencies and crudity of federal and state timberland policy, see notes 1, 13–17 above. A particularly striking retrospective judgment upon the poor craftsmanship represented by earlier federal land policy is implicit in the twentieth-century purchase by the United States of northern Wisconsin land to become national forest reservations (see Clawson, *Uncle Sam's Acres,* 133). As is observed elsewhere in this discussion, the defects of workmanship manifest in public timberlands policy had dual impact; not only were these defects damaging in themselves, but by blurring issues, bolstering complacency with the drift of events, and failing to press men to closer examination and definition of interests, costs, and gains, the legal process failed to contribute to the education of the community and hence failed to contribute to the desirable dynamics of public decision making. Cf. Gulick, 80, 109, 159, 200–201; Ise, 159; Robbins, *Our Landed Heritage,* 26, 54, 99. Likewise, the failure to develop a public tim-

berlands policy of more closely rationalized economic character in some respects probably had the effect of accentuating the defects in rationality of the processes of private (market) decision, as when cheap public land fostered over-investment by private operators and hence increased their need to liquidate under stress (see Marquis, 96, 98, 182).

Chapter III. Franchises to Develop Bulk Transport and Water Power

1. See, generally, Lawrence Martin, *The Physical Geography of Wisconsin* (1932), 18–21. The particular significance of particular stream systems to the growth of the Wisconsin lumber industry is sketched in Adolph Kanneberg, "Log Driving and the Rafting of Lumber in Wisconsin" (mimeographed, 1944), 12–20.

2. William Gerald Rector, *Log Transportation in the Lake States Lumber Industry, 1840–1918* (1953), 17–24; see also Fries, 75–76, 99; Reynolds, 69.

3. The importance which access to navigable water had for the value of timberland is indicated, for example, in Report, Committee on Internal Improvements, for indefinite postponement of bill 352A (to repeal Private & Local Acts, 1864, c. 84, incorporating the Black River Improvement Co.), AJ, 813 (1867) (repeal "would cause a depreciation of property on said stream to a large amount . . . at least one million of dollars on the value of pine land alone"); also in Majority Report, Committee on Lumber & Manufactures on bill 302A, 1868 (Beef Slough boom franchise), AJ, 609 (1868) (in the absence of improvements, local milling monopoly would depress pineland values); in Memorial 103A, 1874, asking repeal of Laws, 1873, c. 12 (granting a toll improvement franchise to any persons improving the Yellow River), AJ, 216–217 (1874) ("said law, if enforced, will interfere with the vested rights of those who own pine lands on said river"); and in sect. 3 of Laws, 1879, c. 22 (nothing in this statute, governing disposition by North Wisconsin Railway Co. of lands granted in aid of its construction, "shall be construed as requiring the said company . . . to sell at any price . . . any land not exceeding in extent forty acres at any one place adjacent to logging streams and suitable and proper for log landings and for the construction of dams for logging purposes, which, if owned by any individual or individuals, might be used to render less available and accessible any unsold timber belonging to said company"). Contemporary recognition of the key relation of streams to logging and milling may be seen in Memorial No. 4, 1838, HRJ, 125–127; Minority Report, Committee on Railroads (by Assemblyman T. C. Pound, prominent Chippewa Falls lumberman), on bill 95A (which becomes Private & Local Acts, 1864, c. 207, authorizing a Mississippi River bridge for the Milwaukee & Prairie du Chien Railway Co.), AJ, 387, 388 (1864); Memorial 23S, 1867, by proponents of franchise for dam and boom at Eau Claire, SJ, 190, 191 (1867); General Laws, 1868, Memorial No. 4 (asking federal aid for improving the Chippewa River, as "the only highway for the export of [Chippewa valley] . . . agricultural products, and the immense quantities of lumber"); Laws, 1880, c. 292, enlarging the authority of the Wisconsin River Improvement Co. "for the purpose of facilitating and cheapening the driving and floating of logs and timber and lumber in the Wisconsin river . . ." That the terms of access to navigable streams might determine the patterns of distribution and processing in the forest products industry was the fact on which rested the sharp rivalry between Chippewa Falls and Eau Claire mill interests and firms interested in moving logs to Mississippi River wholesale points. The pattern-setting importance of improvement franchises is especially emphasized in Report, Committee on Lumber & Manufactures, favoring bill 295A, 1870 (to incorporate the Chippewa River Improvement & Booming Co.), AJ, 778, 779 (1870) ("in this pine timber is wealth untold to that portion of the state bordering on said river, and to the state at large, provided reservoirs and booms are constructed for the safe assorting, handling and holding of logs in and along said river . . . [A]lthough much is now doing upon [the Chippewa River] . . . the lumbering business is comparatively yet in its infancy . . . The state at large is interested in it, by the accumulating wealth which helps her revenues; and where facilities are furnished this branch of business, all know that large towns of necessity spring up, furnishing a better home market to the farmer and mechanic, than any other branch of business . . . While these improvements aid in the assorting and driving of logs below them, they at the same time afford such facilities for manufacturing, that a large proportion of the logs . . . will be sawed above and below them, thereby bringing back not only the proceeds of the lumber to that valley, but also

gives to that section of the state the great benefit of the cost of manufacture, which is not less than $4 on each 1,000 feet of logs so sawed into lumber, and which would be lost to our state in case the lumber is run out of it in the log to other states"). See also Majority Report, Committee on Lumber & Manufactures, on bill 366A, 1875 (to amend the Eau Claire dam franchise given by P&LL, 1872, c. 16), AJ, 477–478, 493 (1875). Even in the late years of the industry, when the rails afforded new access to markets, there was continuing acknowledgment of the practical importance of the streams. See J. S. Keator Lumber Co. v. St. Croix Boom Corporation, 72 Wis. 62, 80, 38 N.W. 529, 535 (1888). See, also, Veto Messages of Governor Scofield (on bill 448A, 1897), AJ, 1280 (1897), and of Governor LaFollette (on bill 155S, 1903), SJ, 1084 (1903), disapproving private dam franchises for want of sufficient log navigation guaranties. Cf. Special Message of Governor LaFollette, AJ, 1050 (1905). The varying appraisals of the relative significance of promoting the new rail facilities and continuing promotion of waterways, as curbs on transport costs, are exemplified in Report, Select Committee on Bill 106A, 1873 (disposition of St. Croix railroad-aid land grant), AJ, 707 (1873) (rail economies will mean switch from stream to land transport of logs, lumber), and General Message of Governor Taylor, PD, 24 (1875) (Fox-Wisconsin improvement will mean lumber freight savings). The framework importance of the waterways to the timing and directions of lumber industry growth is discussed generally in Fries, 34, 65, 84–85, 91, 99, 225; Merk, 87–88; Raney, 199–209; Rector, *Log Transportation,* 15–34; Reynolds, 9, 13, 62.

4. Note the dates of material cited in note 3 above. Compare J. S. Keator Lumber Co. v. The St. Croix Boom Corporation, 72 Wis. 62, 80, 38 N.W. 529, 535 (1888): "The use of rivers and smaller streams for the floatage of logs is essential to the continued prosperity of the immense lumber and industrial interests of northern Wisconsin. The regulation and preservation of such use, in connection with and as facilitating navigation by reasonable and proper booms and other structures, has long been the legislative policy of this state, as frequently sanctioned by this court."

5. See General Acts, 1864, Memorial No. 18 (for federal aid land grant for Sturgeon Bay canal, as "a channel for the great fleet of lumber vessels that annually load on the western shore of Green Bay"), and Report, Select Committee on progress of the Sturgeon Bay canal project, AJ, 281 (1874) (state may depend on energy of franchise grantees because "they, as owners of a large portion of the pine forests on the western shore of Green Bay, will be directly benefited by the completion of the work"). On Mississippi River improvements for raft navigation, see Laws, 1874, Joint Resolution No. 8; Laws, 1882, Memorial No. 3. That river and harbor improvements were one focus of lumber interest in Congress, see Current, 45–46, 50, 51, 60–61, 64–66, 70, 75, 79, 85, 93, 94; Fries, 225.

6. On harbor improvement and regulation authority, see, e.g., P&LL, 1857, c. 367 (Port Washington); P&LL, 1858, c. 128 (Two Rivers); General Acts, 1865, c. 407, and P&LL, 1868, c. 178 (Racine); P&LL, 1868, c. 335, 1870, c. 275, and Laws, 1879, c. 186 (Manitowoc); P&LL, 1868, c. 254, and 1869, c. 126 (Sheboygan). Harbor control provisions first included in charters of lake cities were copied in two Wisconsin River city charters. See P&LL, 1869, c. 247 (Grand Rapids); P&LL, 1872, c. 114 (Wausau). A few lumber company charters authorized lake harbor improvements. See P&LL, 1857, c. 326 (Wisconsin Lumber & Manufacturing Co.); P&LL, 1859, c. 133 (Peshtigo Lumber & Manufacturing Co.).

7. Of 117 reported cases before federal courts sitting in Wisconsin involving movement of vessels or rafts on waters touching or inside Wisconsin, 12 can be identified from the reports as involving log or lumber transport; seven of these 12 arose on Lake Michigan (including five cases regarding collisions, one suit for towing charges, and one suit to collect stevedore charges), two collision cases and one personal injury suit grew out of Mississippi River traffic, and there were two cases over enforcement of federal statutory regulations on the Oconto River. In another case the federal court held that it lacked jurisdiction to entertain a bill by the state of Wisconsin to enjoin an allegedly unlawful pier on Lake Superior. The only federal court case of public importance concerning the state's boundary waters and the lumber industry was United States v. St. Paul Railway Co., 5 Biss. 410 (Circ. Ct. W.D. Wis. 1873). There the court interpreted federal statutes to authorize the Secretary of War to veto a particular location set by the railroad for a Mississippi River bridge, upon a finding that construction at that point would impede navigation; the decision presented no

novel doctrine. In 11 cases the Wisconsin Supreme Court dealt with lumber-connected lawsuits arising out of the movement of vessels. Only one of these arose out of a river transaction — on a contract to buy a boat for towing logs on the Chippewa River. Of the ten lake-transport cases, five concerned performance of contracts of carriage, two contracts to repair vessels, one a contract to buy a vessel, one a mortgage for the purchase price of a vessel, and one a contract of charter. There appears no significance in the timing of the appearance of either the federal or state cases; as one might expect, they cluster in the years of most active lumber industry development, with eight of the 12 federal court vessel cases falling between 1875 and 1895, and six of the ten state court cases between 1873 and 1898. The cases were located by a team of student assistants who paged the Wisconsin Reports and the relevant federal court reports; the federal court tabulation is unpublished; that for the Wisconsin Supreme Court appears in Hurst, ed., *A Digest of Regional Sources for the Study of the Economic and Political History of the Law* (1941).

8. Pound v. Turck, 95 U.S. 459 (1878) applied in a Wisconsin lumber transport context the doctrine of Willson v. Blackbird Creek Marsh Co., 2 Pet. 245 (U.S. 1829), and of Cooley v. Board of Wardens of Philadelphia, 12 How. 299 (U.S. 1851), that — in the absence of clear pre-emption of the field by federal statute — a state might make reasonable regulations affecting navigation of state waters. This ruling had earlier been made in Woodman v. Kilbourn Manufacturing Co., 1 Biss. 546 (Circ. Ct. D. Wis. 1867). The United States Supreme Court's doctrine was promptly acknowledged and applied in Wisconsin River Improvement Co. v. Manson, 43 Wis. 255, 265 (1878). See, generally, Falls Manufacturing Co. v. Oconto River Improvement Co., 87 Wis. 134, 150, 58 N.W. 257, 261 (1894).

The Reports show some doctrinal confusion as to whether the navigation guaranties in the Northwest Ordinance of 1787 created any distinct limit of federal law on state authority, as part of an "unalterable compact" between the original and the new states. The United States Supreme Court finally made clear, however, that the federal Constitution superseded the Ordinance, so that commerce clause doctrine governed and the state might regulate its internal waters so long as Congress had not legislated specifically to the contrary. The Montello, 11 Wall. 411 (U.S. 1870); Escanaba Co. v. Chicago, 107 U.S. 678 (1882); Sands v. Manistee River Improvement Co., 123 U.S. 288 (1887). During the years important for lumber transport the Wisconsin Supreme Court recognized these rulings as stating the full relevant doctrine, sanctioning a broad discretion in the state legislature over regulation of navigable waters of the state. See Falls Manufacturing Co. v. Oconto River Improvement Co., cited above; cf. In re Southern Wisconsin Power Co., 140 Wis. 245, 260, 122 N.W. 801, 807 (1909); Flambeau River Lumber Co. v. Railroad Commission, 204 Wis. 524, 536, 537, 236 N.W. 671, 675–676 (1931). Economy Light Co. v. United States, 256 U.S. 113 (1921) qualified prior decisions by ruling that the federal Constitution and the admission of a new state superseded the Northwest Ordinance only so far as to concede state legislative authority over waters wholly internal to the state; as to waters capable of bearing interstate commerce, the Northwest Ordinance was an act of Congress, no more subject to supersession by state action than any other valid act of Congress, and as such declared binding protections for interstate commerce. Cf. Woodman v. Kilbourn Manufacturing Co., 1 Biss. 546, 549 (Circ. Ct. D. Wis. 1867) (Northwest Ordinance was continued after the federal Constitution by 1 Statutes at Large 50). This ruling was acknowledged, though with some apparent cloudiness as to its practical import, in In re Crawford County Levee & Drainage District, 182 Wis. 404, 196 N.W. 874 (1924), and Lundberg v. University of Notre Dame, 231 Wis. 187, 282 N.W. 70 (1939). However, in Flambeau River Lumber Co. v. Railroad Commission, cited above, Rosenberry, C.J., for the court, took pains to affirm that the Economy Light Co. decision contained nothing inconsistent with, but to the contrary "confirms and strengthens the prior holding" that the Northwest Ordinance did not operate to limit the authority of a new state over navigable waters within its limits, subject to Congress' supreme authority to intervene. Cf. Muench v. Public Service Commission, 261 Wis. 492, 53 N.W. (2d) 514 (1952). Whatever doctrinal qualification the Economy Light Co. decision might make upon state authority, in any event it is plain that during the years important to the lumber industry the Wisconsin legislature operated without sense of any special limit laid on it by the Northwest Ordinance beyond the policy embodied in the commerce clause of the federal Constitution, and that the decisions of the Wisconsin Supreme Court during those years — cited above — found nothing in the Northwest Ordinance guaranty inconsistent with applying the doctrine of concurrent state powers as laid down in the Willson and Cooley decisions of the United States Supreme Court. Moreover, in those years the Wisconsin Supreme Court quite consistently treated the legal and practical significance of the navigation guaranties of the Northwest Ordinance as deriving from incorporation of substantially similar language in Wisconsin Constitution, Art. XI, sect. 1, and in related statutory provisions stemming from Laws of the Territory of Wisconsin, 1841, Act No. 9, and Wisconsin Revised Statutes, 1849, c. 34. See Wisconsin River Improvement Co. v. Lyons, 30 Wis. 61, 66 (1872); Wisconsin River Improvement Co. v. Manson, 43 Wis. 255, 261 (1877); Sweeney v. Chicago, Milwaukee & St. Paul Railway Co., 60 Wis. 60, 67, 18 N.W. 756, 758 (1884); J. S. Keator Lumber Co. v. St. Croix Boom Corporation, 72 Wis. 62, 80, 38 N.W. 529, 535 (1888). Cf. In re Crawford County Levee & Drainage District No. 1, 182 Wis. 404, 408, 409, 196 N.W. 874, 876 (1924); Angelo v. Railroad Commission, 194 Wis. 543, 551, 217 N.W. 570, 574 (1928); Nekoosa-Edwards Paper Co. v. Railroad Commission, 201 Wis. 40, 44, 45, 228 N.W. 144, 146 (1930); Flambeau River Lumber Co. v. Railroad Commission, cited above. In any case, the distinctive doctrinal content of the Northwest Ordinance navigation guaranty was regarded, during the lumber era at least, as directed against state imposition of political or tax burdens upon the use of waters navigable for interstate commerce, and not as a particular directive concerning physical obstructions, or physical works, as these might be authorized by state law; in this aspect the Ordinance guaranty would seem in substance to declare no more than was contained in Supreme Court doctrine under the commerce clause, against state imposition of direct or discriminatory burdens on interstate commerce. See Wisconsin River Improvement Co. v. Manson, 43 Wis. 255, 262, 265 (1877); J. S. Keator Lumber Co. v. St. Croix Boom Corporation, 72 Wis. 62, 86, 38 N.W. 529, 538 (1888), quoting to this effect Willamette Iron Bridge Co. v. Hatch, 125 U.S. 1, 10 (1888); Falls Manufacturing Co. v. Oconto River Improvement Co., 87 Wis. 134, 152, 58 N.W. 257, 262 (1894); In re Southern Wisconsin Power Co., 140 Wis. 245, 260, 122 N.W. 801, 807 (1909). Language of the court in Lundberg v. University of Notre Dame, 231 Wis. 187, 195, 205, 282 N.W. 70, 74, 285 N.W. 839, 840 (1939) casts doubt upon this limitation of the Ordinance guaranty to a ban on political rather than physical obstruction; on the other hand, the physical interference there complained of rested on a claim of land title asserted under the general law of the state, and this perhaps might be rated a "political" obstruction. See C. Graham Waite, "Public Rights to Use and Have Access to Navigable Waters," *Wisconsin Law Review*, 1958: 335, 364–365. Whatever the merits of this latter-day suggestion, it had no counterpart in the lumber era — in which period, indeed, the court decided the Keator Lumber Co. case, cited above, on the ground, in part, that the Ordinance guaranty as embodied in Wisconsin Constitution, Art. XI, sect. 1, had no relevance to the validity of a state statutory franchise for a physical obstruction created by a boom.

Federal public land policy might also have been deemed a source of some limits upon state law affecting navigable waters. However, Pumpelly v. Green Bay & Mississippi Canal Co., 13 Wall. 166 (U.S. 1872) held that land titles derived from the United States did not come burdened with easements of flowage to foster navigation improvement; thus the state might apply to such land the requirement that compensation be made for private property taken for a public purpose. Arimond v. Green Bay & Mississippi Canal Co., 31 Wis. 316, 336 (1872) followed this ruling, and decided also that incorporation of the substance of the Ordinance guaranty into the Wisconsin constitution did not limit the breadth of the ban in Art. I, sect. 13, of that document against taking private property for public use without compensation.

9. Pound v. Turck, 95 U.S. 459 (1878); Heerman v. Beef Slough Manufacturing, Booming, Log-Driving & Transportation Co., 1 Fed. 145, 154, 161–162 (Circ. Ct. W.D. Wis. 1878); United States v. Beef Slough Manufacturing, Booming, Log-Driving & Transportation Co., 8 Biss. 421, 424 (Circ. Ct. W.D. Wis. 1879). Cf. Pumpelly v. Green Bay & Mississippi Canal Co., 13 Wall. 166 (U.S. 1872), discussed in note 8 above.

10. The limited state of development of the executive branch of the federal government was another factor which restricted the significance of federal intervention in Wisconsin waterways affairs. The Army Corps of Engineers was a far more effective instrument in this area of policy than was the General Land Office as to the public domain. Cf. Leonard White, *The Republican Era: 1869–1901*

(1958), 153. However, the Wisconsin record on its face shows that the Engineers had insufficient staff to be employed on a broad range of Wisconsin projects, and Congress begrudged extended aid for in-state waters. Cf. Current, 60–61, 65; Gates, *Wisconsin Pine Lands,* 132. Recognizing the realities, the Wisconsin court held that state remedies for stream obstruction would obtain in the absence of a showing that the Engineers had taken specific action to preempt a given situation. See Sweeney v. Chicago, Milwaukee & St. Paul Railway Co., 60 Wis. 60, 18 N.W. 756 (1884).

11. The relative distribution of problems posed by private power-dam construction in the forest area is indicated by the fact that there is no Wisconsin Supreme Court decision which develops out of a north-country litigation important legal doctrine regarding flowage — the significant flowage doctrine all grows out of cases originating in the southern part of the state — though several important decisions deal with the interplay of competing transport and power interests in the forest section. Flowage problems are reflected, for example, in Large v. Orvis, 20 Wis. 696 (1865), and Durning v. Burkhardt, 34 Wis. 585 (1874). Compare the greater doctrinal complexity where navigation interests are prominent, as in Miller v. Sherry, 65 Wis. 129, 26 N.W. 612 (1886), or Falls Manufacturing Co. v. Oconto River Improvement Co., 87 Wis. 134, 58 N.W. 257 (1894).

For legal reflections of main phases of navigation use, see, e.g., (1) on difficulties between persons engaged simply in transport, Bearrs v. Sherman, 56 Wis. 55, 13 N.W. 869 (1882), and Hayward v. Campbell, 72 Wis. 321, 39 N.W. 540 (1888); (2) for a permanent improvements franchise, Private & Local Acts, 1864, c. 84, incorporating the Black River Improvement Co., (as amended by P&LL, 1866, c. 447, authorizing dam construction; Laws, 1880, c. 225, defining area in which eminent domain power may be exercised; Laws, 1882, c. 263, similar, and extending corporate life twenty-five years; Laws, 1885, c. 316, requiring bridge and boom so that ferries may operate); (3) on controversy arising out of temporary log-transport flooding techniques, Field v. Apple River Log Driving Co., 67 Wis. 569, 31 N.W. 108 (1887); (4) for a franchise emphasizing traffic management rights, P&LL, 1868, c. 430, conferring traffic control on Yellow River Improvement Co.; (5) as to the varieties of handling facilities centered about boom operation, P&LL, 1857, c. 101, as amended by P&LL, 1862, c. 43 (Wolf River Boom Co. may charge only safekeeping fees as to timber already rafted, but is authorized also to raft unrafted timber for fee), P&LL, 1865, c. 190, and 1866, c. 582 (same company now limited to safekeeping function, as here further defined), P&LL, 1868, c. 469 (same company now authorized to secure, sort and raft, and drive logs), and P&LL, 1870, c. 34 (same company now returned to safekeeping function, with right to let out to third parties the rafting of improperly cared-for logs, at their owners' expense).

12. Aid to lumber traffic on the Rock River is mentioned in Memorial No. 8, 1838, 2d Territorial Legislative Assembly, 1st Session, Council Journal, 134; and in Report, Chief Engineer of the Territory, on the Milwaukee-Rock River canal project, *id.,* 3d Session, Council Journal, App., Doc. No. 4, p. 200. Arguments for the Fox-Wisconsin project stressing benefits to lumber traffic may be seen in Annual Report, Board of Public Works, SJ, App., 14, 37 (1853); Laws, 1882, Memorial No. 3. But most of the pleas for this project were cast in terms of its benefits to agriculture and to general trade, with emphasis on facilitating the movement of vessels rather than of rafts. E.g., General Laws, 1867, Memorials No. 3 and 11; General Laws, 1871, Memorial No. 15; Laws, 1875, Memorial No. 9, and report thereon by Committee on Internal Improvements, SJ, 279, 281–283, 286, 287, 288 (1875) (almost exclusive emphasis on benefits to wheat traffic).

13. The full story of these projects is given in Samuel Mermin, "Law and the Promotion of Enterprise" (MS, 1959).

14. The Sturgeon Bay ship canal project was treated as of large significance to the lumber traffic from Green Bay into Lake Michigan. See, e.g., General Acts, 1864, Memorial No. 18, and General Acts, 1865, Memorial No. 10. General Acts, 1864, c. 365, incorporated the Sturgeon Bay & Lake Michigan Ship Canal & Harbor Co.

15. See General Laws, 1868, Memorial No. 4; Laws, 1876, Memorial No. 5; Laws, 1880, c. 316. Cf. Gates, *Wisconsin Pine Lands,* 132; Francis W. Laurent, *The Business of a Trial Court* (1959), 20–21 (total federal appropriations of $201,750 in twenty-five years following 1877).

16. The Northwest Ordinance will be found in *Documents Illustrative of the Formation of the Union of the American States* (1927), 47–54, and also in *Statutes of the Territory of*

Wisconsin (1839), 14–19; the guaranty of the freedom of navigation is in Art. IV of the "articles of compact between the original states and the people and states in the said territory," contained in sect. 14 of the Ordinance. The benefits of this, as of other grants of right made by the Northwest Ordinance, were extended in sect. 12 of Congress' Act of Apr. 20, 1836, 5 Statutes at Large 10, establishing the Territorial Government of Wisconsin. The language of the guaranty was carried in substance as a condition of the Act of Aug. 6, 1846, 9 Statutes at Large 56, sect. 3, authorizing the formation of a state government, and embodied in Wisconsin Constitution, Art. IX, sect. 1. That these provisions represent a continuity of principle, see Wisconsin River Improvement Co. v. Manson, 43 Wis. 255, 261 (1878). Revised Statutes, 1858, c. 41, sect. 1, declared the navigation guaranty as contained in these fundamental documents. See Wisconsin River Improvement Co. v. Lyons, 30 Wis. 61, 66 (1872). No such provision had been included in the first Revised Statutes, 1849, though its c. 34 included the stipulation requiring legislative assent to works placed in streams as cited in note 17 below. The 1858 section was dropped in the 1878 revision, "as being sufficiently sanctioned by its enactment in the constitution." See *Report of the Revisers of the Statutes,* 124.

17. Laws of the Territory of Wisconsin, 1840–1841, Act No. 9; Revised Statutes, 1849, c. 34, sect. 1; Revised Statutes, 1858, c. 41, sect. 2; Revised Statutes, 1878, sect. 1596; Statutes, 1898, sect. 1596. After Laws, 1911, c. 652, and 1913, c. 755, this provision stood in a formulation of broader reach, applying to any waters navigable in fact. See Wisconsin Statutes, 1959, sect. 30.10; State v. Sutherland, 166 Wis. 511, 519, 166 N.W. 14, 17 (1918).

18. Laws of the Territory of Wisconsin, 1839–1840, No. 48, repealed by Revised Statutes, 1849, c. 157, sect. 1, and revived by General Acts, 1857, c. 62; Revised Statutes, 1858, c. 56; Revised Statutes, 1878, c. 146; Statutes, 1898, c. 146; Wisconsin Statutes, 1959, sect. 31.31.

19. Laws, 1876, c. 399; Revised Statutes, 1878, sect. 1777; Laws, 1880, c. 279 (creating Revised Statutes, sect. 1777a, adding boom corporations); Laws, 1882, c. 318; Statutes, 1898, sects. 1777 and 1777a, repealed (under renumbered designation of sect. 180.17) by Laws, 1927, c. 534, sect. 18 ("but this repeal shall not effect any company now organized and operating under said sections").

20. Examples of acts declaring particular streams navigable are P&LL, 1858, c. 44 (Hemlock River, Wood County); Laws, 1873, c. 283 (O'Neil's creek, a Chippewa River tributary); Laws, 1883, c. 88 (Little Yellow River, Wood County). The sum of these acts was gathered together into Revised Statutes, 1878, sect. 1607, repealed by Laws, 1917, c. 335. Most important of the statutory stipulations as to passage facilities to be built into dams and booms to assure navigation were those pertaining to the Wisconsin River. Acts, 1849, c. 62; Revised Statutes, 1878, sects. 1601–1602; Laws, 1881, c. 239; Statutes, 1898, sects. 1601–1602, revised and renumbered as sect. 31.18 by Laws, 1917, c. 474, sect. 18; Wisconsin Statutes, 1959, sect. 31.18.

21. Majority Report, Committee on Water Powers, Forestry and Drainage, December 30, 1910 (1911), 246–248, presents data which show that a total of 665 special statutory franchises for building dams were granted from 1836 to 1909. See also Kanneberg, pp. 31–32 and Exhibit 2.

There is a substantial disparity between the inventory of stream franchises made independently for this study and that contained in the 1910 legislative report. The disparity is not so great as would be indicated by comparison of the gross total of 802 franchises tabulated for the present history and the gross total of 665 reported in 1910, for the 802 total includes a net of 99 boom franchises not counted in the 1910 report (which did include, as dam or river improvement franchises, five franchises here classed as boom licenses). Less the net number of boom franchises included, the present inventory shows 703 franchises of types equivalent to those counted by the 1910 tabulation. However, for our purposes the 1910 gross of 665 franchises should be reduced by 22 grants which the present study ignores as unrelated to the scope of its tabulations (e.g., three dams authorized for pisciculture, and two for cranberry marshes), by five corporate franchises which contain no area designations for exercise of the dam-building authority conferred on the corporation (see further reference below), and by five instances in which the 1910 tabulation counts as separate franchise grants what the present study finds to be in substance only transfers or amendments of previously granted franchises. With these 32 subtractions, the 1910 gross stands at 633, compared with the present study's total of 703 franchises under similar headings. The present study's total was checked against the separate items of the 1910 tabulation, each of which was accounted for in the present inventory.

There are classification differences between the 1910 and the present tabulations. Notably, the 1910 count does not attempt to distinguish between forest-area and nonforest-area franchises, does not list boom franchises as such, and contains differences from the present inventory — based apparently on different readings of statutory context — as to dams designed primarily for navigation improvement and those designed primarily for industrial power; on this last point there seems no basis, on data now available, to resolve the differences in estimating the primary emphasis of given grants. Other differences have been noted above.

Grants of dam-building authority in special corporate charters present a distinct problem. Where a corporate franchise contained authorization to build a dam at a specified locality, the present inventory counted it as a dam franchise; with some misgiving, there were included also a number of instances in which a corporate charter conferred authority to build dams for the corporate purposes within a rather broad territory; but where a corporate charter merely included a dam-building authority among the listed powers of the corporation, with no locational reference, such a charter was not here counted. In the last situation the reference to dams would seem merely part of the creation of the internal power structure of the corporation, with no fair ground to infer intent to make a franchise grant as against the public. See, as apparently in accord with this judgment, Opinion of Attorney General rendered to the Secretary of State, Apr. 9, 1906, PD, 708–713 (especially at 713). The 1910 tabulation includes five such instances which the present inventory rejects. For other details, see the note appended to the table of special franchises.

Available data does not permit any closer count of lumber-industry-connected franchises than the present inventory presents. Plainly the lumber industry alone is involved in improvement and boom franchises. There seems no reliable way by which to determine the distribution of dam uses between forest products processing and other operations; given the known lessened importance of grist mills, the want of large-scale development of other forms of industry in the forest region during the boom lumber years, and the tremendous rise in lumber industry output, it seems reasonable to believe that most forest-area dam franchises were for purposes of processing forest products or aiding log or lumber navigation. This is the conclusion reached independently of the present study in Daniel J. Dykstra, "Legislation and Change," *Wisconsin Law Review,* 1950: 523, 530.

22. The Tug May, 5 Biss. 449, and The Tug Oconto, *id.,* 460 (both D. C. E. D. Wis. 1873) arose under federal safety statutes as to lumber-tow tugs on the Oconto River. Two actions for private damages or injunction against alleged public nuisances by obstruction of navigable interstate streams may have grounded federal jurisdiction on claims under the Constitution or the federal statutes continuing the guaranty of the Northwest Ordinance. See Woodman v. Kilbourn Manufacturing Co., 1 Biss. 546 (Circ. Ct. D. Wis. 1867) (Wisconsin River); Leigh v. Holt, 5 Biss. 338 (Circ. Ct. E. D. Wis. 1873) (Oconto River). United States v. Beef Slough Manufacturing, Booming, Log-Driving & Transportation Co., 8 Biss. 421 (Circ. Ct. W. D. Wis. 1879) obviously rested on claims of the federal government under the commerce clause and Congressional acts affecting the Chippewa River. Cf. note 24 below. Contemporary lower federal court opinions are typically silent on grounds of federal jurisdiction, so that tallies are difficult to make from the Reports. See, e.g., Doty v. Lawson, 14 Fed. 892, and Horner v. Dellinger, 18 Fed. 495 (both Circ. Ct. E. D. Wis. 1883) (forest-area water-power title controversies related to developments incidental to Fox-Wisconsin improvement project).

23. See Heerman v. Beef Slough Manufacturing, Booming, Log-Driving & Transportation Co., 1 Fed. 145 (Circ. Ct. W. D. Wis. 1878). Cf. Pound v. Turck, 95 U.S. 459 (1878). See comment in note 22 above.

24. See, generally, Laurent, 20–21, 46, 49. The surest generalization one can draw from this inventory of the business of the Circuit Court for Chippewa County is a negative one: no significant quantity of log-or-lumber-navigation-connected litigation appears here, different from the types of such litigation appearing in the Wisconsin Reports. On the whole, a surprisingly small total of such litigation appears in the Chippewa Circuit Court records, and of all the forest-area waterways litigation that reached the Supreme Court only three cases came on appeal from the Chippewa court (two involving mechanics liens for work on river facilities, and one involving the validity of an improvement tolls franchise). The

experience of Mr. Laurent in taking the Chippewa Circuit Court inventory suggests that there would almost certainly be insurmountable difficulties about attempting a broad tally of trial court business by very specifically defined subject matter. Many papers are missing, and typically one can determine the subject matter of a given suit only from the language used in pleadings, which either adopts generalized stereotypes in allegation or otherwise speaks with such vagueness as to defy accurate, pointed classification. Some more particular points may, however, be made from the Chippewa Circuit Court tabulation. In the decade 1855–1864, two criminal prosecutions for obstructing highways involved logging use of streams; other cases under that head concerned railroads and roads. See Table 46, *id.*, 126. Two criminal prosecutions in the decade 1875–1884 were for willful destruction of booms. See Table 42, *id.*, 120. Otherwise there was no resort to criminal proceedings concerning use of waterways — a fact which is consistent with the general lack of resort to criminal sanctions in matters touching the lumber industry (or any other industry of its period, for that matter), and consistent, too, with the absence of any criminal proceeding among the cases in the Wisconsin Reports involving lumber industry use of waters. On the civil side, analogously, there was conspicuous absence of resort to actions founded on a nuisance theory. Cf. Table 86, *id.*, 164. There was no resort to the Chippewa Circuit Court to resolve eminent domain proceedings for flowage of land by lumbermen's dams; tabulated eminent domain cases and special proceedings involving court approval of commissioners' awards in eminent domain (see Table 88, *id.*, 166, and Table 204, *id.*, 278) involved highway and railroad construction. Such suits as were brought for flowage damages connected with lumber industry operations were brought as actions in tort on a theory of strict liability for injury to realty, rather than as actions to obtain the price of a flowage easement. See Table 86, *id.*, 163 ("Damage to realty: strict liability" cases). Tabulations in the present text merge tort and eminent domain cases in the heading of "flowage damage." Save for the one case challenging the validity of an improvements toll franchise — Sellers v. Union Lumbering Co., 39 Wis. 525 (1876) — no action appears in the Chippewa Circuit Court for damages or injunction for interference with navigation. In sum, we find that (1) the trial business of a circuit located in a great lumber producing area shows little lumber-connected waterways litigation, and (2) the waterways cases in the Chippewa Circuit Court do not present issues of any type different from those appearing in the Wisconsin Reports from the forest region as a whole. These circumstances suggest that in this field of litigious activity — perhaps more safely than in some others — we may take the appealed cases as a reliable indication of the kinds and relative quantities of matters likely to be found if we had a complete inventory of all trial court business in the forest region.

25. See Hurst, *Law and the Conditions of Freedom*, 23–29. Compare the invocation of the Charles River Bridge Case, in Janesville Bridge Co. v. Stoughton, 1 Pinn. 667 (Wis. 1846), for a dramatic symbol of this preference of dynamic over static significance in the definition of law-created title. The point is developed with specific reference to the lumber industry hereafter in that part of the text which discusses the promotion of productivity as the ground of both common law and statutory rights of stream use.

26. See, generally, Chippewa Valley & Superior Railway Co. v. Chicago, St. Paul, Minneapolis & Omaha Railway Co., 75 Wis. 224, 44 N.W. 17 (1889), and Gilmore v. Roberts, 79 Wis. 450, 453, 48 N.W. 522, 523 (1891).

27. A classic statement, pointing toward the conception of the police power as the authority of the state to attend to the good ordering of relationships, is that of Shaw, C. J., in Commonwealth v. Alger, 7 Cush. 53, 84, 85 (Mass. 1851). See Leonard W. Levy, *The Law of the Commonwealth and Chief Justice Shaw* (1957), 229, 247–254. The police power, said Mr. Chief Justice Shaw, is "the power vested in the legislature by the Constitution, to make, ordain, and establish all manner of wholesome and reasonable laws, statutes, and ordinances, either with penalties or without, not repugnant to the Constitution, as they shall judge to be for the good and welfare of the commonwealth and of the subjects of the same." In particular reference to the institution of private property, he found it "a settled principle, growing out of the nature of well-ordered civil society, that every holder of property, however absolute and unqualified may be his title, holds it under the implied liability that his use of it may be so regulated, that it shall not be injurious to the equal enjoyment of others, having an equal right to the enjoyment of their property, nor injurious to the rights of the community."

28. Compare Veto Message of Governor Rusk on Bill 100S, 1882, SJ, 488: "This [toll dam] franchise is one which can only be granted by the legislature, and is, when granted and accepted, property. It involves the expenditure, in a permanent improvement, of more or less money, and it is doubtful whether the same would not be beyond the power of legislative revocation or control . . ." Governor Rusk developed the same characterization of such franchises in his General Message of Jan. 11, 1883, AJ, 24, as he further argued his point that no such franchise should be granted individuals without inclusion of a broad reserved power to the legislature to amend or repeal the grant.

29. See Art. IV under sect. 14 of the Ordinance, in *Documents Illustrative of the Formation of the Union of the American States,* 47–54.

30. Wisconsin Constitution, IX, 1: "And the river Mississippi, and the navigable waters leading into the Mississippi and St. Lawrence, and the carrying places between the same, shall be common highways, and forever free, as well to the inhabitants of the state as to the citizens of the United States, without any tax, impost, or duty therefor." That this provision creates a judicially enforceable limitation on legislative power, see, e.g., Milwaukee Gas Light Co. v. Schooner Gamecock, 23 Wis. 144 (1868); Wisconsin River Improvement Co. v. Manson, 43 Wis. 255 (1877). Though no occasion arose which squarely required the court to acknowledge the provision as likewise a limitation on judicial lawmaking, the court treated Art. IX, sect. 1, as a declaration of policy to which the judges owed deference in interpreting a statutory bridge franchise as not intended to authorize a structure which would materially obstruct navigation on the Wisconsin River. See Sweeney v. Chicago, Milwaukee & St. Paul Railway Co., 60 Wis. 60, 18 N.W. 756 (1884). Cf. In re Crawford County Levee & Drainage District No. 1, 182 Wis. 404, 196 N.W. 874 (1924); Flambeau River Lumber Co. v. Railroad Commission, 204 Wis. 524, 236 N.W. 671 (1931). In their respective contexts, the navigation guaranties of the Northwest Ordinance and the Wisconsin constitution seem clearly intended as limitations only upon official, and not upon private, action. If remarks of the court — counting at most as dicta — in Lundberg v. University of Notre Dame, 231 Wis. 187, 199, 203, 282 N.W. 70, 76, and 285 N.W. 839 (1939), carry the innuendo that the court would recognize these navigation guaranties as grounds in themselves for private causes of action against private interferences, the innuendo is without support in prior doctrine. Perhaps the implication is no more than that courts should give weight to the indicated public policy, in defining common law rights of property or contract. In tort, familiar doctrines of nuisance and riparian right would seem to afford ample doctrinal scope for judges, without need of drawing upon novel interpretations of these navigation guaranties.

31. That these navigation guaranties bar direct taxation of the use of the waters protected is indicated as dictum incident to the holding, in Wisconsin River Improvement Co. v. Manson, 43 Wis. 255 (1877), that Wisconsin Constitution, IX, 1, does not forbid the legislature to grant a river improvement franchise carrying toll rights as compensation for benefits received by users of the improvements. See accord, Underwood Lumber Co. v. Pelican Boom Co., 76 Wis. 76, 85, 45 N.W. 18, 21 (1890). Cf. J. S. Keator Lumber Co. v. St. Croix Boom Corp., 72 Wis. 62, 81, 38 N.W. 529, 535 (1888); Sauntry v. Laird, Norton & Co., 100 Wis. 146, 152, 75 N.W. 985, 987 (1898); Wisconsin River Improvement Co. v. Pier, 137 Wis. 325, 340, 118 N.W. 857, 862 (1908). In an Opinion rendered the State Treasurer on Feb. 2, 1900, the Attorney General ruled that the Northwest Ordinance navigation guaranty only barred the state from taxing navigation, and did not forbid the state to impose a license fee of a percentage of gross receipts, for the privilege of doing business, upon a public utility boom corporation chartered and given a boom franchise by the state (Attorney General, *Opinions,* VI, 159). Created by Laws, 1891, c. 422, this franchise fee was carried forward by Laws, 1893, c. 260, and as Wisconsin Statutes, 1898, sect. 1222g, until its repeal by Wisconsin Statutes, 1915, sect. 4978. Collections were always small. See summary in Tax Commission, 4th Biennial Report, Nov. 10, 1908, PD, 97–98 (1909). Franchises of river improvement companies were subjected to the general personal property tax by Laws, 1911, c. 611, carried forward into Wisconsin Statutes, 1915, sect. 51.43, 1921, sect. 76.47, 1959, sect. 76.02(8). None of this river improvement tax legislation was brought to challenge in the Wisconsin Supreme Court under Wisconsin Constitution, IX, 1, but the authority of the state to impose property taxes upon such enterprises was "assumed" in Yellow River Improvement Co. v. Wood County, 81 Wis. 554, 51 N.W. 1004 (1892), holding

that the value of a flooding dam which was part of an integrated river improvement project might not be included by local assessors in the assessment for real property taxes of the land on which it was built.

32. The doctrinal contents of the Northwest Ordinance guaranty and of the commerce clause were not explicitly aligned in the decisions. The significant point is that, when the courts discussed the bearing of federal law upon the state's authority over navigable waters within its boundaries, analysis ran simply in terms of commerce clause doctrine. See Pound v. Turck, 95 U.S. 459 (1878); Woodman v. Kilbourn Manufacturing Co., 1 Biss. 546 (Circ. Ct. D. Wis. 1867); Heerman v. Beef Slough Manufacturing, Booming, Log-Driving & Transportation Co., 1 Fed. 145 (Circ. Ct. W.D. Wis. 1878); United States v. Beef Slough Manufacturing, Booming, Log-Driving & Transportation Co., 8 Biss. 421 (Circ. Ct. W.D. Wis. 1879); Wisconsin River Improvement Co. v. Lyons, 30 Wis. 61 (1872); Wisconsin River Improvement Co. v. Manson, 43 Wis. 255 (1877); In re Southern Wisconsin Power Co., 140 Wis. 245, 122 N.W. 801 (1909). Woodman v. Kilbourn Manufacturing Co. explicitly ruled that the federal Constitution had superseded the Northwest Ordinance as to the protection of interstate commerce, and then applied commerce clause doctrine without attempt to analyze whether policy indicated under the Ordinance made any distinctive contribution to policy formulated under the commerce clause. Allaby v. Mauston Electric Service Co., 135 Wis. 345, 348–349, 116 N.W. 4, 5 (1908) — a case arising other than in a lumber industry context — contained an argumentative dictum that the omission of navigation guaranties in a statutory dam franchise would put the franchise in violation of the Northwest Ordinance and the terms of the act admitting Wisconsin to the Union, unless the franchise be interpreted as implying a legislative determination that the stream was not navigable. This intimation, that the Northwest Ordinance provision laid a federal obligation of affirmative action on the state legislature, had no counterpart in any lumber-era authority, save as the legislative practice of including navigation guaranties in franchises might be deemed to imply it.

33. That the legislature assumed that men were entitled as of right to navigate these waters in their natural condition may be implied by the care it took — as the text later notes — to assert broadly its authority to license erection of any works in such streams, and to insert navigation guaranties into the bulk of special statutory franchises for dams and booms. It is characteristic of the nineteenth-century legislative process that this principle had little explicit statement in legislative materials and, on the whole, must be drawn by inference from what the legislature did. Recognition of the principle came closest to direct exposition in occasional committee reports which accepted as common ground that proponents of works which might impair navigation stood at least under a burden of going forward with evidence that a countervailing public interest was involved, and that other men's transport would not be wholly blocked. See Majority and Minority Reports, Committee on Lumber and Manufactures, on bill 302A, 1868, AJ, 608–609, 630–632 (Beef Slough bill); Majority and Minority Reports, Judiciary Committee, on bill 7S, 1871, SJ, 411, 414, 445, 449 (Eau Claire "Dells" dam and boom bill); Report, Judiciary Committee, on bill 202S, 1880, SJ, 394 (amendment of charter of Black River Improvement Co.). Though it is not clear how far the reliance was on constitutional rights and how far on judgments of policy, the conclusion was clear enough in Messages of the Governors that Wisconsin law should recognize men's right to make transport use of waters navigable in their natural condition, without being blocked and without requirement of license by the state or any delegate of its power. See Governor Farwell's Veto Message of June 6, 1853, SJ, 816 (disapproving scope of monopoly given by bill 310S to Chippewa River Improvement & Navigation Co.); his Veto Message of June 6, 1853, AJ, 841–842 (disapproving extent of traffic controls conferred on Black River Lumber Driving & Booming Co. by bill 42A); Governor Randall's Veto Message of Mar. 7, 1860, AJ, 578 (disapproving special charter of Eau Claire Manufacturing Co. for want of navigation guaranties affecting dam authorized); Governor Washburn's Veto Message of Mar. 18, 1873, AJ, 920, 921 (disapproving bill 391A, as not providing sufficient navigation guaranties incident to authorization of railroad bridge over Mississippi); Governor Scofield's Veto Message of Apr. 26, 1897, AJ, 1280 (disapproving bill 448A, for want of sufficient navigation guaranties incident to dam authorized on Oconto River); Governor LaFollette's Veto Message of May 12, 1903, SJ, 1084 (disapproving bill 155S, for want of any navigation guaranty

incident to dam authorized on Tomahawk River). Measured against the trend of legislation and the court decisions sustaining legislative discretion, these Governors' actions must not be interpreted to represent very broad assertion of navigation rights, but rather reactions against what the Governors viewed as gross failures of the legislature to balance the affected interests. As the text later shows, legislation and court decisions went far toward approving a liberal concept of justifiable "improvement" and delegated traffic management on logging streams, so that in practice there was not much scope for claims of rights of navigation relevant to use of waters in their natural condition.

34. There was the less occasion for the Wisconsin court explicitly to acknowledge Art. IX, sect. 1, of the state constitution as a limit upon judicial as well as upon legislative policy making, because from its earliest ruling the court accepted the general common law doctrine that riparian land titles were subject to an easement of public passage over navigable waters. See Jones v. Pettibone, 2 Wis. 308, 320 (1853); Olson v. Merrill, 42 Wis. 203, 210, 212 (1877); Falls Manufacturing Co. v. Oconto River Improvement Co., 87 Wis. 134, 148, 58 N.W. 257, 260 (1894). This doctrine removed the largest likely source of clash between judge-made policy and that declared in the constitutional navigation guaranty. This condition of the law left, as the other principal head under which the court's rulings might regulate access to navigation, the doctrines of public nuisance and of private redress for special damage from public nuisance. Inherently these nuisance questions were of narrower reach, always focused only upon particular situations and not upon general categories of right. Moreover, in a nuisance case the court was appealed to as protector of the navigation interest, so that its role, again, did not create an emphasized issue of its authority as related to Art. IX, sect. 1. The court did perhaps in a degree recognize Art. IX, sect. 1, as laying direct obligation upon it when it ruled that the constitutional navigation guaranty reflected the existence of a general public interest in navigable waters such that the state had an affirmative obligation to see to their protection, and hence that a complaint grounded on charges of obstruction by unauthorized works raised a question of sufficiently general public concern to warrant invocation of the original jurisdiction of the Wisconsin Supreme Court for their determination. See Attorney General v. Eau Claire, 37 Wis. 400, 447 (1875). This was the only controversy in which the court's original jurisdiction was successfully invoked in a waterways suit affecting the lumber industry. See State v. St. Croix Boom Corporation, 60 Wis. 565, 572–573, 19 N.W. 396, 399 (1884), declaring a much more cautiously restricted concept of the original jurisdiction in waterways cases.

35. Log transport was a lawful use of a stream, and the log driver was entitled to reasonable and unobstructed access therefor. Hall v. Kitson, 3 Pinn. 296 (Wis. 1851); Whisler v. Wilkinson, 22 Wis. 572 (1868); Enos v. Hamilton, 24 Wis. 658 (1869), s.c., 27 Wis. 256 (1870); Olson v. Merrill, 42 Wis. 203 (1877). See accord, Leigh v. Hunt, 5 Biss. 338, 342 (Circ. Ct. E.D. Wis. 1873); Heerman v. Beef Slough Manufacturing, Booming, Log-Driving & Transportation Co., 1 Fed. 145, 160 (Circ. Ct. W.D. Wis. 1878). So far as reasonably necessary to the ordinary navigation use of the stream, a log transporter might delay others' logs for needed sorting, A. C. Conn Co. v. Little Suamico Lumber & Manufacturing Co., 74 Wis. 652, 664, 43 N.W. 660, 664 (1889); he was not liable to a riparian owner for damage to the shore caused by a normal log jam, Field v. Apple River Log Driving Co., 67 Wis. 569, 31 N.W. 108 (1887); nor was he liable for trespass in using a logging stream for his drive up to ordinary high water mark, Polebitzke v. John Week Lumber Co., 163 Wis. 322, 325, 158 N.W. 62, 63 (1916).

36. It was perhaps the easier for the Wisconsin court to assume without question the application of common law nuisance doctrine in navigation issues because before any such issue came to the court it had already ruled that the state might have relief at common law against a nuisance to public health caused by a dam built on a nonnavigable stream: Luning v. State, 2 Pinn. 215 (Wis. 1849). Compare accord, Douglas v. State, 3 Wis. 820, 4 Wis. 387 (1854); Stoughton v. State, 5 Wis. 291 (1856) (health nuisance alleged on navigable stream); In re Eldred, 46 Wis. 530 (1879). Indicative of the judicial instinct to preserve common law, as against implications drawn from general legislation, is the contrasting treatment of statutes in the Luning and Stoughton cases. The dam challenged in the Luning case had been built under authority of the general territorial milldam statute, but the court ruled that "it cannot be con-

ceived that by erecting a dam in accordance with its provisions, license is thereby given to create and continue a nuisance. Such a construction of the power given in that act, to owners of mill property, would be utterly inconsistent with the right of every citizen to call in the authority of the state for protection against the unlawful encroachments of others" (2 Pinn. 215, 221). The dam challenged in the Stoughton case had been built under warrant of a special statutory franchise authorizing such a dam at the particular location; distinguishing the Luning decision, because there the details of construction were fixed by the dam builder while here the details were determined by the legislature, the court held that the special statutory franchise was a bar to the prosecution (5 Wis. 291, 297). Compare note 42 below.

37. Hall v. Kitson, 3 Pinn. 296 (Wis. 1851); Enos v. Hamilton, 27 Wis. 256 (1870); Weatherby v. Meiklejohn, 61 Wis. 67, 20 N.W. 374 (1884). Cf. Olson v. Merrill, 42 Wis. 203, 214 (1877) (special damage taken as conceded, where not challenged). Common law also sanctioned reasonable measures of self-help by one specially damaged by an obstruction to navigation, thus in effect adding an indirect type of delegated regulation of stream works (Whisler v. Wilkinson, 22 Wis. 572 [1868]). Cf. Richardson v. Emerson, 3 Wis. 319 (1854); Leihy v. Ashland Lumber Co., 49 Wis. 165, 5 N.W. 471 (1879). The common law defense of contributory negligence provided another indirect regulation of stream improvement activity; on this basis Miller v. Sherry, 65 Wis. 129, 26 N.W. 612 (1886) held that a dam owner might not recover for injury to his works by a log transporter where the dam (constructed without a statutory franchise) had been built without reasonable provision for passing logs.

38. See Attorney General v. Eau Claire, 37 Wis. 400, 447–448 (1875): "The actual navigation may be little, and the obstruction might be slight. So the affidavits tend to show. But neither the right nor the wrong is a question of degree. We cannot listen to one about to put an unlawful work in a public river, that it will not materially obstruct navigation." This was early asserted to be Wisconsin law, by Smith, J., dissenting in Stoughton v. State, 5 Wis. 291, 299 (1856), though the point was not there necessary to decision. See accord, Green Bay & Mississippi Canal Co. v. Kaukauna Water Power Co., 70 Wis. 635, 653, 35 N.W. 529, 535 (1888); Pioneer Wood Pulp Co. v. Chandos, 78 Wis. 526, 531, 47 N.W. 661, 663 (1891). Cf. Houfe v. Town of Fulton, 34 Wis. 608, 616 (1874). In Woodman v. Kilbourn Manufacturing Co., 1 Biss. 546, 553 (Circ. Ct. D. Wis. 1867), Miller, J., observed that "the paramount right of navigation being in the people, who may wish to use a navigable stream, any obstruction however inconsiderable without constitutional legislative authority is a nuisance, to be abated at the suit of an individual, or of the state." However, he rested this doctrine at least in part on the finding that "this principle is prescribed" by Wisconsin Revised Statutes, 1858, c. 41, sect. 2, requiring legislative sanction for erection of works in the principal navigable waters in the state. Summing up the law in State v. Sutherland, 166 Wis. 511, 521, 166 N.W. 14, 17 (1918), Rosenberry, J., said: "At common law . . . to abate an obstruction as a nuisance it was not necessary that the obstruction should actually have interfered with navigation; it was sufficient if it rendered it less convenient and less useful." There is often talk in the opinions of founding a nuisance action upon a showing of material obstruction in fact; see, e.g., State v. St. Croix Boom Corporation, 60 Wis. 565, 570, 19 N.W. 396, 398 (1884). Since, however, in almost all such cases there was a showing of obstruction in fact or of breach of navigation guaranties in a statutory franchise, such language does not carry great weight as implying that in every instance material obstruction must be shown, and often it is not clear but that the intended reference is to the need of a private complainant to establish his particular standing by proof of special damage, rather than to the requisites of establishing the basic element of the existence of a public nuisance. Cf. In re Eldred, 46 Wis. 530, 541, 1 N.W. 175, 177 (1879); A. C. Conn Co. v. Little Suamico Lumber Manufacturing Co., 55 Wis. 580, 586, 13 N.W. 464, 466 (1882).

39. A. C. Conn Co. v. Little Suamico Lumber Manufacturing Co., 74 Wis. 652, 43 N.W. 660 (1889); Eau Claire Dells Improvement Co. v. Eau Claire, 172 Wis. 240, 179 N.W. 2 (1920). See accord, Charnley v. Shawano Water Power & River Improvement Co., 109 Wis. 563, 569, 85 N.W. 507, 509–510 (1901). Compare Miller v. Sherry, 65 Wis. 129, 135, 26 N.W. 612, 613 (1886), where the question was perhaps raised but was found not necessary to decision. In cases

concerning power dams in the southern, nonforest area of the state it was early taken to be the law that such a dam on a nonnavigable stream was not a public nuisance per se, but that the burden lay on complainants to show that the dam caused unwarranted public or special private damage. See Douglas v. State, 4 Wis. 387, 390 (1855), approved in Allaby v. Mauston Electric Service Co., 135 Wis. 345, 351, 116 N.W. 4, 6 (1908); cf. Luning v. State, 2 Pinn. 215 (1849); Rooker v. Perkins, 14 Wis. 79 (1861). These cases all concerned dams. In suits involving booms the court ruled that a riparian owner might lawfully build a boom into a stream suitable for general navigation, in aid of the navigation uses of his land, without a statutory franchise, so long as he did not in fact obstruct navigation or violate any statute or infringe the rights of other riparian owners. Stevens Point Boom Co. v. Reilly, 44 Wis. 295 (1878); Northern Pine Land Co. v. Bigelow, 84 Wis. 157, 54 N.W. 496 (1893). See accord, Cohn v. Wausau Boom Co., 47 Wis. 314, 322, 2 N.W. 546 (1879). The opinions give no indication that the court perceived the apparent inconsistency between its insistence that material obstruction in fact always be proved, to show that a boom not built under statutory franchise was a public nuisance, and its differentiation on this score as to dams, according to whether the stream involved was generally navigable or navigable only in a limited season for logs. An intermediate state of doctrine was indicated in Leigh v. Holt, 5 Biss. 338, 340 (Circ. Ct. E.D. Wis. 1873), where the court charged the jury that a pier or boom built in a log-navigable stream without warrant of a statutory franchise was unlawful on that basis alone, but one who appeared to have joined in a local custom of building such illegal works was thereby estopped to claim that the offending work was illegal per se, and could sustain an action only by proving material obstruction in fact. Apparently *contra* on such an estoppel is Olson v. Merrill, 42 Wis. 203, 214 (1877).

40. Arpin v. Bowman, 83 Wis. 54, 53 N.W. 151 (1892). See, accord, In re Eldred, 46 Wis. 530, 541, 1 N.W. 175, 177 (1879); A. C. Conn Co. v. Little Suamico Lumber Manufacturing Co., 55 Wis. 580, 13 N.W. 464 (1882). Recognition of this doctrine seems implicit in the caution made in State ex rel. Attorney General v. Norcross, 132 Wis. 534, 547, 112 N.W. 40, 44 (1907), against pursuing the public remedy for breach of statutory franchise terms in a fashion calculated to work undue forfeiture.

41. See, e.g., Hall v. Kitson, 3 Pinn. 296 (Wis. 1851) (raftsman versus power dam); Stevens Point Boom Co. v. Reilly, 44 Wis. 295 (1878) (competing booms); Miller v. Sherry, 65 Wis. 129, 26 N.W. 612 (1886) (flooding dam operation versus sawmill operation).

42. Typical is the unquestioning fashion in which the court invokes common law remedies to implement the general statute requiring legislative authority for erection of works on a navigable stream, in Wisconsin River Improvement Co. v. Lyons, 30 Wis. 61, 67 (1872). On the other hand, where no particular statutory franchise terms were in issue, the provision for circuit court jurisdiction over nuisance actions was treated as declaratory of common law, so far as concerned the content of the actions authorized. See *Report of the Revisers of the Statutes,* 228. Typically the court did not trouble to refer to this provision, but discussed the nuisance jurisdiction simply in relation to common law authority. See, e.g., A. C. Conn Co. v. Little Suamico Lumber Manufacturing Co., 55 Wis. 580, 586, 13 N.W. 464, 466 (1882). And in any event, the court never felt impelled to expound the constitutional basis on which either the jurisdictional statute or the common law of nuisance was grounded.

43. See note 17 above. The court readily recognized the legislature's title to pre-empt the determination of what works should be lawful in navigable waters, whether by general or special law. Revised Statutes, 1849, c. 34, thus was held not to ban all dams, but only such as the legislature had not authorized; hence the showing of defendant's compliance with a special dam franchise was a good defense to an action for obstructing a raftsman (Jones v. Pettibone, 2 Wis. 308 [1853]). See, accord, Edwards v. Wausau Boom Co., 67 Wis. 463, 30 N.W. 716 (1886). On the other hand, erection of works without any statutory license, contrary to the general statute, would ground an action to abate a dam. Wisconsin River Improvement Co. v. Lyons, 30 Wis. 61 (1872). By virtue of the general statute (as now carried into Revised Statutes, 1858, c. 41), breach of navigation guaranties in a special statutory franchise became, per se, a basis of complaint against the works as a public nuisance. See In re Eldred, 46 Wis. 530, 541, 1 N.W. 175, 177 (1879); cf. Arpin v. Bowman, 83 Wis. 54,

53 N.W. 151 (1892). In the face of the general statute, such rights as a riparian owner would have at common law to build booms into the stream in aid of his own navigation use must be deemed strictly limited to works which would not in material measure hamper general navigation. See Stevens Point Boom Co. v. Reilly, 46 Wis. 237, 243, 245, 49 N.W. 978, 979 (1879). Cf. A. C. Conn Co. v. Little Suamico Lumber Manufacturing Co., 74 Wis. 652, 43 N.W. 660 (1889).

44. Improvement franchises containing stipulations to assure free passage for rafts and logs as well as vessels begin as early as (Territorial) Local Acts, 1839, No. 57, and Laws of the Territory of Wisconsin, 1839–1840, No. 7. See the retrospective comments of Governor LaFollette in his Special Message of Apr. 12, 1905, AJ, 1050–1051.

45. The assertion of legislative authority implicit in the enactment of declarations of navigable status seems most strong where the legislature successively redefined the area on a given stream to which the status should attach. See, as to the Sheboygan River, P & LL, 1855, c. 366, 1857, c. 221, 1858, c. 86, 1859, c. 141. The repeal of statutory declarations of navigable status likewise carried particular implication of the legislature's control in the matter. See Laws, 1875, c. 225, repealing P & LL, 1853, c. 61 (Baraboo River, Sauk County); Laws, 1882, c. 297, repealing P & LL, 1867, c. 503 (Little Wolf River, Waupaca County); Laws, 1878, c. 79, repealing Laws, 1875, c. 250 (west branch, Kickapoo River, Vernon County). All outstanding declarations of navigable streams, as listed at that point in Wisconsin Statutes, 1915, sect. 1607, were repealed by Laws, 1917, c. 335. The boldest exercise of legislative power in this matter came after the lumber era was substantially closed, when Laws, 1911, c. 652, broadened the general requirement of a statutory license to embrace "all rivers and streams, meandered or non-meandered, which are navigable in fact for any purpose whatsoever." No governor's message dealt explicitly with the scope of legislative authority to attribute navigable status to streams. However, there was perhaps implicit claim to share in this legislative function — at least to prevent its gross abuse — and hence implicit acknowledgment that a proper legislative function was involved, in the several vetoes which were made of statutory franchises for want of sufficient navigation guaranties. See note 33 above. The Wisconsin Supreme Court said little about the legislature's authority to define navigable status. Where the legislature used this authority in effect to mark the realm of action in which navigation improvements might be made without providing compensation to adversely affected riparian landowners, the court recognized a broad discretion in the legislators. Falls Manufacturing Co. v. Oconto River Improvement Co., 87 Wis. 134, 58 N.W. 257 (1894); Green Bay & Mississippi Canal Co. v. Kaukauna Water Power Co., 90 Wis. 370, 61 N.W. 1121 (1895). Cf. Allaby v. Mauston Electric Service Co., 135 Wis. 345, 349, 350, 116 N.W. 4, 6 (1908) (legislation should be interpreted to intend broad or limited concept of navigability, according to the particular public end to be served). In functional effect, the broad scope of legislative authority to decide the navigable status of waters was also upheld in decisions which sustained the legislature's power to choose among competing uses of the waters (see note 54 below). At a time when lumbering had far declined as a major interest affected by the law of waters, the court accepted without question that the legislature might expand the general statute requiring legislative license for stream improvements, to cover all streams "navigable in fact for any purpose whatsoever," in addition to those returned as navigable under the Federal Survey. See State v. Sutherland, 166 Wis. 511, 519, 166 N.W. 14, 17 (1918). Wood v. Hustis, 18 Wis. 417, 418 (1863), and Allen v. Weber, 80 Wis. 531, 539, 50 N.W. 514, 516 (1891) intimated that the court might upset a statutory declaration of navigability which interfered with riparian land owners' rights in a situation where no reasonable case could be made of benefit to navigation. And Bixby v. Parish, 148 Wis. 421, 423, 134 N.W. 838, 839 (1912) intimated that the legislature might lack authority to declare nonnavigable waters which in fact might fairly be deemed navigable. Compare In re Horicon Drainage District, 136 Wis. 227, 116 N.W. 12 (1908), and Flambeau River Lumber Co. v. Railroad Commission, 198 Wis. 134, 223 N.W. 417 (1929), s.c., 204 Wis. 524, 236 N.W. 671 (1931), for examples of the application of a rule of construction favoring the preservation of navigation rights. These indications of judicial limitation were not realized in any actual court decisions upsetting legislation, though the dictum in Bixby v. Parish is perhaps analogous to the governors' vetoes cited in note

33 above, disapproving franchise bills for want of adequate navigation guaranties.

A statutory declaration that a named stream, or a portion of a stream, be deemed navigable undoubtedly had the effect of a legislative command that the courts apply thereto all relevant common law doctrine. Compare General Acts, 1853, c. 72, amending Revised Statutes, 1849, c. 34, to add that "the boundaries of lands adjoining waters, and the several and respective rights of individuals, the state, and its citizens, in respect to all such lands and waters, shall be determined in conformity to the common law, so far as applicable . . ." The decisions unquestioningly accepted the common law of reasonable use of waters and riparian lands as governing, in the absence of contrary legislation. See, e.g., Cobb v. Smith, 16 Wis. 661, 662, 665 (1863); A. C. Conn Co. v. Little Suamico Lumber Manufacturing Co., 55 Wis. 580, 13 N.W. 464 (1882). This approach did not mean that a statutory declaration of navigability was without its own substantial force, even in a context of rights and duties the substance of which was left to common law. There was no presumption that a stream was navigable, and hence the presence of a statutory declaration on the matter would be of substantial benefit to a party whose rights depended on the assertion of navigable status. See Wood v. Hustis, 18 Wis. 417 (1863); Leihy v. Ashland Lumber Co., 49 Wis. 165, 5 N.W. 471 (1879).

A statutory declaration of navigability had relevance, also, in the context of other legislation. The milldam act by its terms applied only to nonnavigable streams; that a particular stream had been declared navigable by special statute took it out of the scope of the milldam act. Wood v. Hustis, cited above; cf. Allaby v. Mauston Electric Service Co., 135 Wis. 345, 349, 350, 116 N.W. 4, 6 (1908). The legislature enacted various regulations against obstruction or pollution of navigable waters; presumably a statutory declaration of navigable status would bring this body of regulation to bear on the given waters. Cf. Barnes v. City of Racine, 4 Wis. 454 (1855).

46. See, e.g., as a general regulation against obstruction or pollution, the provision which runs from Revised Statutes, 1849, c. 16, sect. 97, to Revised Statutes, 1858, c. 19, sect. 112, to Revised Statutes, 1878, sect. 1598, to Statutes, 1898, sect. 1598, to Wisconsin Statutes, 1917, sect. 31.23, to Wisconsin Statutes, 1959, sect. 30.15. A leading example of legislation limited to works on a particular river is the chain of statutes stipulating navigation guaranties regarding dams built on the Wisconsin River. See Laws of the Territory of Wisconsin, 1846, p. 113; Acts, 1849, c. 62; Private & Local Acts, 1853, c. 48, omitted from Revised Statutes, 1858, as a local law, though saved there from repeal (see *Report of the Revisers of the Statutes,* 125); Revised Statutes, 1878, sects. 1601, 1602; Laws, 1881, c. 239; Statutes, 1898, sects. 1601, 1602; Laws, 1917, c. 474, and Wisconsin Statutes, 1921, sect. 31.18; Wisconsin Statutes, 1959, sect. 31.18.

47. See note 18 above. That the milldam act was intended by the legislature to assert its jurisdiction specifically over nonnavigable waters as works locations, and not over navigable waters as to the effects of works built on the former category of stream, see Clute v. Briggs, 22 Wis. 607 (1868). Cf. Wood v. Hustis, 17 Wis. 416 (1863) (milldam act not applicable to stream specifically declared navigable by statute); Waller v. McConnell, 19 Wis. 417 (1865) (claimant under milldam act has burden to plead and prove nonnavigable waters are involved); Allaby v. Mauston Electric Service Co., 135 Wis. 345, 116 N.W. 4 (1908), and McDonald v. Apple River Power Co., 164 Wis. 450, 160 N.W. 156 (1916) (concept of navigability under milldam act should be interpreted to fulfill act's purpose of promoting power development; hence act should not be interpreted to exclude streams navigable only for logs in limited seasons). Though it was treated as a broad assertion of legislative control over nonnavigable waters, the milldam act was not construed to bar the application of common law doctrine in matters not within its coverage. Here the court applied the familiar tendency to preserve the existence of common law against mere implication from statute law. See Moore v. Coburn, 1 Pinn. 538 (Wis. 1845); Ulrich v. Hull, 17 Wis. 424 (1863); Arimond v. Green Bay & Mississippi Canal Co., 31 Wis. 316 (1872).

48. Various key statutes or types of legislative action particularly symbolize the legislature's implicit consciousness of its ultimate superintendence of the uses made of streams. (1) The legislature's superior claim over private decision making as to stream uses had most sweeping expression in the general statute requiring a legislative license prerequisite to construction of works on the principal streams, and in the general procedure which the mill-

dam act made available to foster power development on nonnavigable waters. See notes 17, 18, 43, and 47 above. Of analogous import was the more particular ban on unauthorized booms, created by General Laws, 1866, c. 123, and thereafter carried in Revised Statutes, 1878, sect. 1598 and its later analogues (see note 46 above). Similarly, the supremacy of legislative over private decision was strongly represented in the power reserved to amend or repeal corporate franchises (Wisconsin Constitution, Art. XI, sect. 1), and — especially after the veto-enforced insistence of Governor Rusk (note 64 below) — in the like reservation typically inserted in grants to individuals. Prudent regard for legislative supremacy was regularly witnessed in the striking care taken in amendments of special statutory franchises to assure correct descriptions of grantees and locations, as is noted in the next subsection of this chapter. (2) The legislature's superior position vis-a-vis other official decision makers likewise found well-defined symbols. Probably we should count the limited activity of the governors as part of the legislative process, especially since the dominant tone of the few vetoes in which the chief executives most forcibly expressed their role was one of expressing policy rather than of enforcing constitutional limitations. See notes 61–65 below. Because the nineteenth-century legislature acted directly as dispenser of franchises, and provided no administrative apparatus to police its franchises, there was no occasion for asserting the legislature's potential control of the executive branch in the detailed implementation of policy. Analogous expression of legislative supremacy came, however, in the delegation of limited waterways regulatory powers to municipal corporations, as creatures of the legislature. See note 6 above. Compare Barnes v. City of Racine, 4 Wis. 454 (1855), where the court ruled that it would not construe a general bridge building authority in the city charter to allow construction at a point where a bridge would necessarily obstruct navigation, since this would run counter to the policy evidenced in Revised Statutes, 1849, c. 34, requiring a legislative license for works built in navigable streams. By enacting its general requirements of legislative license in the face of the existence of familiar common law doctrine providing remedies against public and private nuisances created by the obstruction of streams, the legislature implicitly asserted the supremacy of statute over common law, as well as of legislative over private decision making. Compare the court's early, unquestioning acceptance that the holding of a statutory franchise was a good defense to a common law action for damages claimed for obstruction of navigation by a dam (Jones v. Pettibone, 2 Wis. 308 [1853]; see, also, note 53 below). The legislature's most explicit witness of its confidence that it held the authority to determine when and how far common law doctrine should define rights pertaining to stream use was by General Acts, 1853, c. 72, declaring that common law should govern in such matters, "so far as applicable." Carried into the Revised Statutes, 1858, c. 41, sect. 3, this provision became Revised Statutes, 1878, sect. 1597, Wisconsin Statutes, 1917, sect. 30.10 (by Laws, 1917, c. 335), then Wisconsin Statutes 30.01,(4),(c) (by Laws, 1919, c. 247), and was so carried until its repeal by Laws, 1959, c. 441.

49. Newcomb v. Smith, 2 Pinn. 131 (Wis. 1849) sustained the constitutionality of the territorial milldam act of 1840, as measured by the standard of the due process clause of the 5th Amendment and the Northwest Ordinance. The validity of the milldam-act type of law was upheld as state legislation in Fisher v. Horicon Iron Manufacturing Co., 10 Wis. 351 (1860); Babb v. Mackey, 10 Wis. 371 (1860). Cf. Pick v. Rubicon Hydraulic Co., 27 Wis. 433 (1871) (legislature may adopt varied eminent domain procedures, incident to dam franchises).

50. Stoughton v. State, 5 Wis. 291, 296 (1856). See, accord, Newell v. Smith, 15 Wis. 101, 103 (1862); Cobb v. Smith, 16 Wis. 661, 664 (1863). The court had earlier implicitly acknowledged the legislature's authority to make binding decisions on the erection of works in navigable streams when in Jones v. Pettibone, 2 Wis. 308 (1853) it held that the possession of a statutory franchise for a dam was a defense to an action for obstructing plaintiff's rafts.

51. 30 Wis. 61 (1872). In Jones v. Pettibone, 2 Wis. 308 (1853), and in Barnes v. City of Racine, 4 Wis. 454 (1855), the court had inferentially accepted the validity of Revised Statutes, 1849, c. 34 (the general licensing requirement statute). The federal court had earlier recognized the legislature's authority to strike the balance between power and rafting interests, in Woodman v. Kilbourn Manufacturing Co., 1 Biss. 546 (Circ. Ct. D. Wis. 1867). See also Leigh v. Hunt,

5 Biss. 338, 340 (Circ. Ct. E.D. Wis. 1873).

52. See 30 Wis. 61, 66 (1872) (language of Revised Statutes, 1849, c. 41, "pursu[es] the language of the ordinance of 1787 and of the constitution and articles of compact between this state and the United States"). See, accord, Wisconsin River Improvement Co. v. Manson, 3 Wis. 255, 262, 263 (1877); J. S. Keator Lumber Co. v. St. Croix Boom Corporation, 72 Wis. 62, 80, 38 N.W. 529, 535 (1888).

53. See 30 Wis. 61, 62, 64, 66 (1872). Cf. Durning v. Burkhardt, 34 Wis. 585 (1874) (statute validating dam erected without previous legislative grant bars common law remedy for abatement which would otherwise exist); Cohn v. Wausau Boom Co., 47 Wis. 314, 2 N.W. 546 (1879) (statutory boom franchise may supplant common law right of riparian owner to build boom to service his land); Black River Improvement Co. v. LaCrosse Booming & Transportation Co., 54 Wis. 659, 11 N.W. 443 (1882) (statutory river improvement franchise may supplant common law riparian rights to navigation of side channel, for betterment of main channel). See, generally, note 48 above.

54. See 30 Wis. 61, 67 (1872). See, accord, that the legislative prerogative includes the authority to decide among competing uses of streams, even at the risk of mistakes in judgment, State v. Eau Claire, 40 Wis. 533, 543 (1876); Tewksbury v. Schulenberg, 41 Wis. 584, 593 (1877); Wisconsin River Improvement Co. v. Manson, 43 Wis. 255, 264 (1877); Cohn v. Wausau Boom Co., 47 Wis. 314, 325, 2 N.W. 546, 549 (1879); Black River Improvement Co. v. LaCrosse Booming & Transportation Co., 54 Wis. 659, 677, 681–684, 11 N.W. 443, 448, 454–455 (1882); J. S. Keator Lumber Co. v. St. Croix Boom Corporation, 72 Wis. 62, 38 N.W. 529 (1888); Falls Manufacturing Co. v. Oconto River Improvement Co., 87 Wis. 134, 150, 151, 58 N.W. 257, 261 (1894); Flambeau River Lumber Co. v. Railroad Commission, 198 Wis. 134, 137, 223 N.W. 417, 418 (1929), s.c., 204 Wis. 524, 537–542, 236 N.W. 671, 677–678 (1931). Compare, also, accord, Heerman v. Beef Slough Manufacturing, Booming, Log-Driving & Transportation Co., 1 Fed. 145, 155 (Circ. Ct. W.D. Wis. 1878). This impressively unbroken line of authority cast into oblivion an early dictum intimating that "it would be a very serious question" whether the legislature might prefer one over another transport interest on a navigable stream, on grounds of public interest; see Enos v. Hamilton, 24 Wis. 658, 661 (1869). Ready judicial concession to the legislative authority to make the initial judgment in favor of a work on a navigable stream was supplemented by what appears an equitable discretion in the court to refuse to order abatement of a work, even though the statutory franchise contained navigation guaranties in strong terms, until there had been ample opportunity in practice to discover its actual effects upon navigation. See State v. Eau Claire, 40 Wis. 533, 543 (1876); Woodman v. Kilbourn Manufacturing Co., 1 Biss. 546, 555 (Circ. Ct. D. Wis. 1867).

55. The only qualification upon this appraisal, if indeed it be such, is Sellers v. Union Lumbering Co., 39 Wis. 525 (1876), holding that any river improvement franchise purported to be given by Laws, 1873, c. 12, on the Yellow River, "is void for uncertainty" for want of identification of grantees beyond the mere declaration that anyone who improved the river might claim tolls for its use. The court felt "that, if the legislature had power to do it, it would be most unwise and dangerous to create franchises in so loose a manner, for any one to assume and relinquish at pleasure; to grant tolls upon a public highway to any person who should voluntarily assume to spend his money in the improvement of it. Courts have always been jealous of such grants, so loose and so liable to abuse, and we hope will always be so" (*id.*, 529). The defect here seems, however, to go to form rather than substance. This point is underlined by the fact that in closely contemporary proceedings another bill, to amend the charter of the Yellow River Improvement Co. (on the tributary of that name of the Wisconsin River) encountered considerable difficulty because of fears that it violated the newly adopted (1871) constitutional amendment banning further special legislation on corporate charters (see Reports, Committee on Incorporations, SJ, 77, 92–93 [1873]). The record thus makes clear that the offending form of Laws, 1873, c. 12, represented an effort, not to achieve some particular result in waterways policy, but to conform to the new constitutional limits on special grants of incorporation. The legislators might be excused for their mistake, for the court encountered doctrinal difficulties before the judges finally recognized that the constitutional ban on special charters did not forbid amendments of already outstanding charters. See Stevens

Point Boom Co. v. Reilly, 44 Wis. 295 (1878); Black River Improvement Co. v. Holway, 87 Wis. 584, 59 N.W. 126 (1894); In re Southern Wisconsin Power Co., 140 Wis. 245, 112 N.W. 801 (1909). Such questions as arose after 1872 concerning the validity of special statutory grants for stream improvement to corporate grantees clearly represented no judicial intrusion upon legislative policy toward stream use as such, and thus put no qualification upon the statement in the text.

56. Though the court followed Newcomb v. Smith, 2 Pinn. 131 (Wis. 1849) as binding precedent, it observed in Fisher v. Horicon Iron Manufacturing Co., 10 Wis. 351, 353 (1860): "It does appear to us to be going a great length to say that land which is used for the purpose of overflowage by a mill dam is appropriated to a 'public use,' in the proper and just sense of those words." While adhering to Newcomb v. Smith, the court further indicated its doubts, in Newell v. Smith, 15 Wis. 101, 104 (1862); Ackerman v. Horicon Iron Manufacturing Co., 16 Wis. 150, 154 (1862); Pick v. Rubicon Hydraulic Co., 27 Wis. 433, 443 (1871); McCord v. Sylvester, 32 Wis. 451, 455 (1873) (trial court). As late as Attorney General v. Eau Claire, 37 Wis. 400, 436 (1875), the court said bluntly that it followed Newcomb v. Smith "in submission to the rule stare decisis, reluctantly, against its own views. But the court has never been disposed to extend the doctrine or authority of that case; probably never will." The court moved toward positive approval of this early-established doctrine only after legislative policy for promoting productive use of waters had stood firm for some seventy-five years. Allaby v. Mauston Electric Service Co., 135 Wis. 345, 350, 351, 116 N.W. 4, 6 (1908) noted that the "object [of the milldam act] was to enable the utilization of water power upon many of the important streams of the state. The ardent discussion, first of the constitutionality of the milldam act, and later of its policy, resulting in its repeal for a time, is significant of this fact." The upshot of this history was clear: "These statutes are grounded in the theory, confirmed by the courts, that there is a public interest in the utilization of the water powers of the state to run mills sufficient to justify the exercise of the power of eminent domain in the flooding and consequent taking of lands of individuals." By 1908 this result had such favor in the court's eyes that it ruled that the milldam act should be given an interpretation favorable to its broad application. Hence the court held that a stream navigable only by logs at the spring freshet season was not a navigable stream within the meaning in which the milldam act excluded such waters from its authorization to build power dams. So, in McDonald v. Apple River Power Co., 164 Wis. 450, 455, 160 N.W. 156, 158 (1916), the court ruled that construction of a dam to create water power for generating electricity fell within the statute's authorization of "mill" dams. The court plainly felt that a statute calculated to promote enlargement of the community's productive capacity deserved liberal interpretation as to its coverage: "The operation of this dam and the electric plant connected therewith for generating electric energy and distributing it for light, heat, and power purposes has grown to be as much of a necessity to meet the varying demands of the people as the furnishing of hydraulic power in former times for operating grist and other kinds of mills." The court also treated increase of productive capacity as the ultimate validation of this history, in Wisconsin Traction, Light, Heat & Power Co. v. Green Bay & Mississippi Canal Co., 188 Wis. 54, 65, 205 N.W. 551, 556 (1925): the grant of the eminent domain power to builders of water-power dams was a "somewhat drastic power," warranted, however, "because of the public interest in the utilization of such latent power."

The grant of eminent domain powers occasioned several governors' vetoes. These were episodic in incidence and cloudy in doctrine, in some uncertain measure apparently contradictory of the scope of legislative authority defined by the court. In any event, these scattered interventions by the governor had no apparent general effect upon the confidence with which the legislature delegated eminent domain authority in stream improvement franchises. See the veto messages of Governor Farwell, June 6, 1853, SJ, 817 (eminent domain grant to Chippewa River Improvement & Navigation Co. — the franchise of which is here vetoed primarily for conferring a scope of monopoly deemed contrary to public policy — is "a most unwarrantable and unconstitutional, violation of private rights, without the promise of any corresponding public benefit"; presumably the ground of veto is want of sufficient public purpose); Governor Rusk, Mar. 21, 1882, SJ, 489 (inclusion of delegated eminent domain power cited as another element justifying veto of dam franchise made to individuals without reserved power to amend or repeal); Governor Hoard,

Mar. 13, 1889, AJ, 643 (grant of eminent domain found unconstitutional because here given only for a private use — though the grant, for an industrial power development, seems no different from the type familiarly upheld under the milldam act). Summing up the lumber-era franchise practice, in his General Message of Apr. 12, 1905, AJ, 1050–1051, Governor LaFollette seemed to link the justification of eminent domain grants more strictly to navigation improvement than legislative practice showed: "It is the law that the structure must improve the navigation of the stream. Whenever those applying for these franchises have sought the authority, the legislature has freely conferred upon them the right to condemn and take the lands of others," under suitable compensation procedure.

57. Thien v. Voegtlander, 3 Wis. 461 (1854) (plea of special dam franchise demurrable, in the absence of averment of tender of payment of flowage compensation); Newell v. Smith, 15 Wis. 101 (1862) (eminent domain delegation under special dam franchise unconstitutional for want of statutory lien on mill to secure award); Ackerman v. Horicon Iron Manufacturing Co., 16 Wis. 150 (1862), and Cobb v. Smith, 16 Wis. 661 (1863) (following Thien v. Voegtlander and Newell v. Smith); Zweig v. Horicon Iron Manufacturing Co., 17 Wis. 362 (1863) (similar); Bevier v. Dillingham, 18 Wis. 529 (1864) (damages should be calculated on basis of original land value, not on value depreciated by flowage); Sherwood v. Vliet, 20 Wis. 441 (1866) (strict construction of scope of allowed taking); Darge v. Horicon Iron Manufacturing Co., 22 Wis. 417 (1868) (strict construction of dam owner's right of appeal from award for flowage); McCord v. Sylvester, 32 Wis. 451 (1873) (common law action lies for diversion of water, where award by commissioners under statutory procedure found fatally indefinite); Folsom v. Apple River Log Driving Co., 41 Wis. 602 (1877) ("it requires no argument" to show that, where special dam franchise has no procedure to compensate flowage damages, an action lies therefor at common law); Hackstack v. Keshena Improvement Co., 66 Wis. 439, 29 N.W. 240 (1886) (common law action for flowage damages lies, where as matter of statutory interpretation the statutory compensation procedure does not apply because the injury occurred outside the area of improvements contemplated by the statutory franchise); Carlson v. Stocking, 91 Wis. 432, 65 N.W. 58 (1895) (common law action lies for flooding farmlands by improvement dam). Cf. Ackerman v. Horicon Iron Manufacturing Co., 16 Wis. 150 (1862) (action lies to abate dam, despite statutory franchise, where defendant failed to pay flowage damages award duly made under the statutory compensation procedure); Darge v. Horicon Iron Manufacturing Co., 22 Wis. 417 (1866) (grantee of statutory dam franchise is bound to condition set therein making flowage damages award nonreviewable); Newell v. Smith, 26 Wis. 582 (1870) (abatement lies, upon nonpayment of flowage damages); Charnley v. Shawano Water Power & River Improvement Co., 109 Wis. 563, 85 N.W. 507 (1901) (statutory eminent domain procedure should be liberally construed to enlarge landowner's remedy thereunder). Of course, we must always remember that such doctrine applied only where the court found there was a "taking" of "private" property; no compensation was due for detriment caused by legislative regulation directly affecting only the natural flow of a stream, since the flow itself was a public and not a private asset. See, e.g., Cohn v. Wausau Boom Co., 47 Wis. 314, 324, 325, 2 N.W. 546, 548, 549 (1879).

58. See 5 Wis. 291, 296 (1856).

59. See Wisconsin River Improvement Co. v. Lyons, 30 Wis. 61, 65 (1872); Wisconsin River Improvement Co. v. Manson, 43 Wis. 255, 262, 263 (1877); Falls Manufacturing Co. v. Oconto River Improvement Co., 87 Wis. 134, 150, 151, 58 N.W. 257, 261 (1894); Sauntry v. Laird, Norton & Co., 100 Wis. 146, 152, 75 N.W. 985, 987 (1898). Cf. Woodman v. Kilbourn Manufacturing Co., 1 Biss. 546, 551 (Circ. Ct. D. Wis. 1867); Heerman v. Beef Slough Manufacturing, Booming, Log-Driving & Transportation Co., 1 Fed. 145, 165 (Circ. Ct. W.D. Wis. 1878).

60. On several occasions governors vetoed statutory improvement franchises for want of adequate navigation guaranties, but did so in terms which can be interpreted as expressing judgments of legislative policy rather than as enforcing a constitutional mandate. See note 33 above. When the court invalidated franchises because they made inadequate provision for flowage damages (note 57 above), and upset a grant the terms of which the court felt authorized its use solely for private advantage (note 62 below), its actions might be read as indirectly enforcing requirements that certain affirmative provisions be written into such legislation as a condition of its validity. The eminent domain cases, however, concerned the incidence of costs of public improvements upon

particular private property owners. This was a familiar use of judicial power to protect the individual in the application of general law, rather than judicial enforcement of positive provisions for public use. Cf. Brower v. Merrill, 3 Pinn, 46, 52 (Wis. 1850); Zweig v. Horicon Iron Manufacturing Co., 17 Wis. 362, 371 (1863); Arimond v. Green Bay & Mississippi Canal Co., 31 Wis. 316, 334, 337 (1872); Hackstack v. Keshena Improvement Co., 66 Wis. 439, 445, 29 N.W. 240, 242 (1886). The latter action would represent a kind of measure traditionally left to legislative and executive decision in our system; indeed, probably the typical reaction of a court asked to decree, directly or indirectly, that some affirmative measure be taken by the legislature or chief executive to promote a positive program in the general interest would be that this presented a political and nonjusticiable issue. Cf. State ex rel. Martin v. Zimmerman, 249 Wis. 101, 23 N.W. (2d) 610 (1946). The case cited in note 62 below turned on the inclusion in the franchise of provisions allowing its use merely to serve private interest, rather than any effort by the court to insist on affirmative inclusion of any particular public interest provisions. This reading of the decision is consistent with the prevailing willingness of the court to presume that the legislature's franchises for stream improvement were issued on the basis of its determinations that the public interest would be served. See, e.g., Tewksbury v. Schulenberg, 41 Wis. 584, 593 (1877): "If the acts under consideration . . . were enacted merely to promote private interests, and not to improve the facilities for transportation on the highway by water therein named, — they cannot be upheld as valid laws. But they are valid if it can be reasonably inferred from all their provisions that the improvement of navigation was their primary and principal object. No express declaration in the acts that such was the object and purpose of the legislature in enacting them, is required; nor is such a declaration usual in acts of a similar nature." Of like import are the decisions sustaining the legislature's authority to establish orders of preference among competing interests in stream use, even at risk of making mistaken judgments of wisdom. See note 54 above. The characteristic hesitancy of the court to appear to enforce positive mandates as to the preferred content of legislation is consistent also with the fact that the most forthright affirmative expression of the judges' solicitude for navigation interests took the form of interpreting statutes as presumably intended to embody the same solicitude. See, e.g., In re Horicon Drainage District, 136 Wis. 227, 116 N.W. 12 (1908); Flambeau River Lumber Co. v. Railroad Commission, 198 Wis. 134, 223 N.W. 417 (1929), s.c., 204 Wis. 524, 236 N.W. 671 (1931).

61. Curtis' Administrator v. Whipple, 24 Wis. 350 (1869); Whiting v. Sheboygan & Fond du Lac Railroad, 25 Wis. 167 (1870). It was significant of the framework importance of bulk transport for the general economy that the public purpose doctrine first came to expression in cases growing out of the promotion of transportation facilities. The cited cases, involving the first instances in which the court upset legislation on this basis, concerned the permissible terms of public aids to railroad construction. The public purpose doctrine had its first approval from the Wisconsin court earlier, in a dictum in Soens v. Racine, 10 Wis. 271, 279–280 (1860), involving a municipal harbor improvement. See, generally, Lewis R. Mills, "The Public Purpose Doctrine in Wisconsin," *Wisconsin Law Review,* 1957: 40. The public purpose doctrine was accepted without question by legislative opinion as fixing the frame within which legislation must fit, in matters of navigation improvement as in others. See Report, Judiciary Committee on bill 7S, SJ, 410 (1871); Report, Judiciary Subcommittee on dam franchise bills, SJ, 1038ff (1905).

62. Attorney General v. Eau Claire, 37 Wis. 400 (1875), invalidating Laws, 1875, c. 333. Cf. State v. Eau Claire, 40 Wis. 533 (1876), sustaining a later franchise (granted by Laws, 1876, c. 231), drawn to overcome the objection raised to the earlier grant. The court's handling of the first franchise was marked by little of the readiness to indulge in a presumption of constitutionality, or in the style of statutory interpretation designed to avoid constitutional doubt, which characterized the treatment of other waterways legislation. This circumstance seems probably related to the history of unusually tense lobbying and likely corruption which attended the enactment, though the court makes only the most guarded acknowlment of its awareness of the colorful aspects of this legislative history. The decision is discussed later in the present chapter, as it bears on an appraisal of the comparative interest adjudicating roles of legislature and court. See, generally, Daniel J. Dykstra, "Law and the Lumber Industry, 1861–1881" (S.J.D. thesis, 1950), chap. vii.

63. See note 33 above. These vetoes were all sustained. See the observations in Governor

LaFollette's Special Message of Apr. 12, 1905, AJ, 1050–1051, on the liberality with which special franchises were granted for stream improvements in the nineteenth century.

64. Vetoing a dam franchise granted individuals because it contained no reserved power in the legislature to amend or repeal, Governor Rusk in his Veto Message of Mar. 21, 1882, SJ, 489, carefully limited his objection to this failure to retain legislative surveillance over grants which he recognized as often in effect conferring monopoly control of streams. The implicit monopolies served the public interest in developing the productive use of the waterways, but — as monopolies — called for maintenance of the government's regulatory competence: "A large number of the lumbering streams in the state are being given over by this means to the control of individuals. This evasion of the constitutional amendment against special legislation may be in most instances necessary. The improvement may be for the public good, the tolls fixed may be fair and reasonable, but nearly all such measures affect the interests of many who know nothing of their pendency and are unheard as to their effect. In the nature of things it is impracticable for the legislature to make thorough and exhaustive investigation in each case, and to know the precise effect upon all interests of the measures asked for. Moreover, what may be an improvement in the situation of affairs to-day, may be very far from an improvement a few years hence; and what may be fair compensation for maintaining dams and other public improvements to-day, may not be fair or reasonable after the lapse of time. So that the public interest would seem seriously to demand that the legislature in all such grants should reserve to itself the right, should the public interest require it, to revoke the same, or to continue them upon new terms, and subject to additional restrictions." The force of the Governor's comments upon the effect of improvement monopolies upon other interests underlines the significance of his acceptance, nonetheless, of the basic justification of such grants, under proper regulation.

65. See note 44 above.

66. For want of a reserved power to amend or repeal, the Governor vetoed one franchise, and his veto was sustained. See note 64 above. Apparently convinced that this one drastic step sufficiently emphasized his point of policy, he further indicated disapproval of the omission of the reserved power in other franchises merely by allowing them to become law without his signature. See Governor Rusk's Special Messages of Mar. 30, 1882, SJ, 615–616, and Mar. 8, 1883, AJ, 435. Also, on Mar. 23, 1883 (AJ, 650), the Governor returned bill 475A, pursuant to request of a joint resolution, whereupon the bill was amended to include a reserved power clause, after which, with the Governor's signature, it became Laws, 1883, c. 317. In his General Message of Jan. 11, 1883, AJ, 24, Governor Rusk stated as broad policy the importance of including a reserved power clause in every franchise granted to individuals. The Governor deemed franchises granted corporations to be subject to the general legislative power to alter or amend, reserved by Wisconsin Constitution, Art. XI, sect. 1. In some measure he was correct, insofar as a franchise was contained in an act of incorporation. See Madison, Watertown & Milwaukee Plankroad Co. v. Reynolds, 3 Wis. 287 (1854); Attorney General v. The Railroad Companies, 35 Wis. 425, 569 (1874). Whether the constitutional reservation applied to a franchise separately granted to an existing corporation might seem open to argument; in terms the constitutional provision referred to acts which "formed" or "created" corporations. The point was not directly decided during the lumber era. However, in that period some opinion apparently regarded a franchise granted to a corporation as in law a part of the corporate franchise, for the purpose of determining whether it fell within the ban of the constitutional amendment of 1871 on enactment of special corporate franchises. See Stevens Point Boom Co. v. Reilly, 44 Wis. 295 (1878). This decision was overruled, but not until about the end of the lumber era, in In re Southern Wisconsin Power Co., 140 Wis. 245, 122 N.W. 801 (1909). See, further, notes 129, 220, and 221 below. Even after this overruling it appeared that, where the issue was of continuing public authority to regulate conditions of service of a public utility, the court would construe the intent of the legislature to be to make an operating franchise an inherent part of the grant of corporate existence, so that the constitutional reserved power would work upon the franchise. See City of Manitowoc v. Manitowoc & Northern Traction Co., 145 Wis. 13, 129 N.W. 925 (1911). However we estimate the pattern of this doctrine, it is clear that there were more complicated issues of law here than were grasped in Governor Rusk's confident analysis.

A reserved power clause was almost always inserted in statutory franchises for stream improvement after the issue was brought into focus by Rusk. The matter did not again come into controversy until the context was no longer lumber but water power for generating electricity. See Governor LaFollette's Veto Messages of Apr. 2, 1901, AJ, 723, and May 11, 1903, SJ, 1038–1039. In this later form, the issue was soon put to rest by being merged in the creation of a whole new pattern of administrative regulation.

67. As we shall note later, in discussing the conferring of corporate status incident to stream improvement enterprises, there was some veto activity enforcing the 1871 constitutional amendment banning special corporate charters. However, this point does not touch the substance of navigation policy.

68. The Propeller Genesee Chief v. Fitzhugh, 12 How. 443 (U.S. 1851), overruling The Steamboat Thomas Jefferson, 10 Wheat. 428 (U.S. 1825). See Milton Conover, "The Abandonment of the 'Tidewater' Concept of Admiralty Jurisdiction in the United States," *Oregon Law Review,* 38: 34 (1958).

69. See Walker v. Shepardson, 2 Wis. 384 (1853); Wood v. Hustis, 17 Wis. 416, 417–418 (1863); Whisler v. Wilkinson, 22 Wis. 572, 576 (1868); Falls Manufacturing Co. v. Oconto River Improvement Co., 87 Wis. 134, 150, 58 N.W. 257, 261 (1894). Cf. Leigh v. Holt, 5 Biss. 338, 340 (Circ. Ct. E.D. Wis. 1873); Heerman v. Beef Slough Manufacturing, Booming, Log-Driving & Transportation Co., 1 Fed. 145, 161 (Circ. Ct. W.D. Wis. 1878).

70. See note 44 above. That the legislature early regarded as navigable a stream which would carry logs or rafts was reflected when dam franchises stipulated for immediate construction of slides, while requiring the building of locks for vessels only when the river should be improved for the movement of such craft. See Laws of the Territory of Wisconsin, 1839–1840, No. 7 (Manitowoc River, Manitowoc County); *id.,* 1841–1842, p. 11 (same area); *id.,* 1848, p. 13 (Rock River, Rock County). A like example after statehood is Laws, 1851, c. 248 (Milwaukee River, Milwaukee County). See also Acts, 1852, c. 282 (Apple and Willow Rivers), and P&LL, 1858, c. 44 (Hemlock River, Wood County), for examples of the very common style of stipulations as to dam construction in which navigation guaranties were framed solely in terms of log or lumber transit. The same was commonly true of the scope of navigation guaranties laid down for construction of booms. See, e.g., General Acts, 1856, c. 76 (all booms on Wisconsin River). This concept of the carrying capacity which warranted navigable status is made particularly sharp in P&LL, 1858, c. 86, amending P&LL, 1855, c. 366, to reduce the specification of a slides requirement for milldams on a defined area of the Sheboygan River from the original requirement of slides suitable for "all logs, lumber, rafts or water craft, which may be floated on said stream" to a requirement merely of slides suitable to pass "all saw-logs which may be floated on said stream." The log-transport measure of navigability had recognition also in some of the statutes declaring particular streams or parts of streams navigable, when the legislature "declared [the named waters] navigable for the purpose of driving logs." See General Acts, 1863, c. 40 (Rush River, Pierce County); General Acts, 1865, c. 353 (Maple Creek, Waushara and Outagamie Counties); P&LL, 1866, c. 48 (Duck Creek, Outagamie County); P&LL, 1866, c. 99 (Red Cedar River, Dunn County); P&LL, 1869, c. 156 (Polk County streams). Legislative practice also adopted the log transport test as justification for granting toll franchises for the improvement of streams to carry logs and rafts as well as watercraft. This long chain of legislation begins with Laws of Wisconsin Territory, 1845, p. 90 (incorporating the Wisconsin River Navigation Co.), and *id.,* 1847, p. 44 (Grand Rapids dams, Wisconsin River). In practice almost all of these improvements had significance only for the movement of logs and rafts. This fact had common acknowledgment in the provision of tolls simply for the passage or handling of logs and rafts. See, e.g., Private & Local Acts, 1856, c. 481 (incorporating the Apple River Dam Co.); P&LL, 1857, c. 170 (incorporating Yellow River Improvement Co.); P&LL, 1866, c. 352 (incorporating Keshena Improvement Co.). In light of such a broad and consistent pattern of legislative practice — finding in log transport capacity the justification for treating waters as navigable in law — the phrasing of P&LL, 1859, c. 129, appears as a draftsman's aberration, when it approves a dam at a location on a meandered stream where the stream "is not actually navigable or used for navigation; Provided, that in the latter case they shall construct and maintain a suitable chute for logs . . ."

Characteristically, the legislature left its doctrine implicit in its actions and said little in exposition of its concept of navigability. The one notable exception at a time before the

Wisconsin Supreme Court had plainly adopted the log transport test of navigability was in the Majority Report, Select Committee on Bill 352A (to repeal the charter of the Black River Improvement Co.) AJ, 1106, and App., 1270 (1867). Issue had been made whether log owners had any legal right to use streams to move their logs. The majority of the committee (with no dissent from the minority in this aspect) found that log owners had such right. The committee mingled Wisconsin legislative precedent with citations of Maine and New York court decisions to make its point. But it relied in the first instance on the Wisconsin legislation originating in General Acts, 1856, c. 48 (carried into Revised Statutes, 1858, c. 42), stating the terms on which a log owner might enter riparian land to recover stranded logs. It concluded that this Wisconsin statute law accurately expressed community values as these came to definition in common practice: "We find, generally, that the right to float logs is as perfect as the right to float a raft or boat, and we find as a matter of fact, that logs are floated in all the lumbering streams in the state." The first decision by the Wisconsin Supreme Court declaring the log transport test of navigability as a matter of judge-made law came in the year after this legislative report. The court's brief opinion relied on common law precedent from other states, made no reference to the recognition of the log transport test implicit in a generation of Wisconsin legislative practice, but did emphasize the factual importance of the test to the economy ("It would be exceedingly detrimental to the public interests, especially in the pine-growing regions of the state, if this were not so") and the fact that the test accorded with long-established community behavior (this river "has been constantly used, at all proper seasons of the year, for the purpose of floating logs to mills below, for a period of over twenty years"). See Whisler v. Wilkinson, 22 Wis. 572, 576 (1868).

71. Wisconsin constitutional law followed the doctrine familiar in the states generally that the state constitution should be construed as a limitation rather than a grant of power. Thus the legislature might enact on any subject of public interest, so long as its enactment offended no limitation set by the state or federal constitutions (Bushnell v. Beloit 10 Wis. 195, 225 [1860]). In effect the Wisconsin court applied this approach in the strong presumption of constitutionality it cast about legislative determinations to prefer navigation interests over competing uses of streams for water power or of riparian lands for other than transport purposes. E.g., Wisconsin River Improvement Co. v. Lyons, 30 Wis. 61 (1872); Cohn v. Wausau Boom Co., 47 Wis. 314, 2 N.W. 546 (1879); Falls Manufacturing Co. v. Oconto River Improvement Co., 87 Wis. 134, 58 N.W. 257 (1894). See the broad terms in which, after the developments of the lumber era, the court finds the legislature's "plenary" power established to decide the order of preference among competing navigation and non-navigation uses, in Flambeau River Lumber Co. v. Railroad Commission, 198 Wis. 134, 223 N.W. 417 (1929), s.c., 204 Wis. 524, 236 N.W. 671 (1931).

72. See notes 44 and 70 above.

73. See notes 17 and 43 above. The explicit assertion of legislative authority to define what should be deemed navigable waters, to serve a public interest likewise defined by the legislature, seems nonetheless plain here though it is also true, as the court observed in Jones v. Pettibone, 2 Wis. 308, 318 (1853), that the definition did not purport to be binding for all purposes but only to declare when a statutory license must be had before building works in a stream.

74. See notes 54 and 71 above.

75. Even the extension of the general statutory licensing requirement, by Laws, 1911, c. 652, to embrace "all rivers and streams, meandered or nonmeandered, which are navigable in fact for any purpose whatsoever" seems no broader than the concepts declared by the court in Whisler v. Wilkinson, 22 Wis. 572 (1868), and Olson v. Merrill, 42 Wis. 203 (1877).

76. Whisler v. Wilkinson, 22 Wis. 572, 576 (1868).

77. Olson v. Merrill, 42 Wis. 203, 212 (1877). Plaintiff's argument, *id.*, 209, indicates that this was a point of first impression in Wisconsin, though it could rest on out-of-state authority.

78. This point is implicit in the decision of the case, read in light of the record (*id.*, 206, 207, 213). Plaintiff log driver forcefully argued the practical importance to the industry of recognizing that a stream should have the legal protection of navigable status, although considerable clearing work and driving energy must be invested to use it: ". . . the greater portion of the valuable pine timber in this state is situated along streams like Levis creek, and such of this timber as annually finds its way to market, is floated through the windings of such streams, men being stationed at difficult points to break jams, and push logs off the banks, rocks or bars; that, in the application of

legal principles to the actual condition and necessities of the great lumbering interests of this state, the right asserted by the plaintiff is one of great importance . . ."

79. *Id.,* 213–214. The court's active concern to foster log transport appears most strongly as Mr. Chief Justice Ryan observes of the evidence of the driver's use of the banks that "we gather from cases which have come before us, that the same practice prevails on some of the larger streams in this state. But the navigable character of a stream does not rest on the tortious practice, but on the capacity of the stream to be lawfully used. And we cannot hold that the right to use a public highway, by land or by water, is lost even by habitual trespass upon adjoining lands. The learned counsel of the appellant cites cases clearly showing that different rules prevail in other states. But we think that we are supported in principle and by authority, in adopting a rule in this state which appears to us to be important to its great lumbering interests." Though not concerned directly with the definition of navigable status, cases holding that log transporters were not liable to riparian landowners for detriment which was an ordinary incident of normal driving activity reflect the same policy preference as that declared in Olson v. Merrill. See note 35 above.

80. A. C. Conn Co. v. Little Suamico Lumber Manufacturing Co., 74 Wis. 652, 656, 657, 43 N.W. 660, 661 (1889). In Miller v. Sherry, 65 Wis. 129, 134–135, 26 N.W. 612, 613 (1886) the court found it unnecessary to decide the question argued by counsel "of the relative rights of navigators and riparian owners to the use of small streams which, although navigable in fact, are not meandered or specially declared navigable by law, and whether as to such streams the rules are applicable which prevail if the stream is meandered or has been so declared navigable." In Charnley v. Shawano Water Power & River Improvement Co., 109 Wis. 563, 569, 85 N.W. 507, 509 (1901) the court reviewed the distinction taken in the Conn Co. case and approved it, obiter. See also Allaby v. Mauston Electric Service Co., 135 Wis. 345, 349–350, 116 N.W. 4, 6 (1908) (it was an "apparently inconsiderate" use of language in Olson v. Merrill which indicated that a public highway status attached for all purposes to a stream navigable simply for logs); Eau Claire Dells Improvement Co. v. Eau Claire, 172 Wis. 240, 259, 179 N.W. 2, 9 (1920) (Conn Co. rule broadened: though lacking statutory warrant, modification of dam is not nuisance per se where navigability not materially affected on stream once navigable for boats). It must, of course, be emphasized that the distinction taken in the Conn Co. case had relevance only where the legislature had not exercised its superior authority to adjudicate competition between log transport and power uses; where the legislature had licensed log driving improvements, even though on a stream navigable only for logs, the court construed such a statutory franchise as giving the superior right of use over the claims of a water-power user to an even flow of water. Falls Manufacturing Co. v. Oconto River Improvement Co., 87 Wis. 134, 58 N.W. 257 (1894). Cf. Nekoosa-Edwards Paper Co. v. Railroad Commission, 201 Wis. 40, 228 N.W. 144 (1930) (log-transport test found intended to measure scope of requirement of administrative license for dams under 1911 law). Where a less direct or less important clash of interests was involved than that between power and navigation uses, the court continued to look with favor on broad application of the log-transport test of navigable status. See Willow River Club v. Wade, 100 Wis. 86, 76 N.W. 273 (1898) (such a stream is one on which exists a public right of fishing, at least from boats); Village of Bloomer v. Town of Bloomer, 128 Wis. 297, 311, 107 N.W. 974, 979 (1906) (statute imposing joint liability on town and village to build and maintain bridges over "navigable streams" should be interpreted to apply to any stream "of sufficient capacity to float saw logs during some portion of the year," because "that is the sense in which the term 'navigable' is ordinarily understood"); State ex rel. Attorney General v. Norcross, 132 Wis. 534, 544, 112 N.W. 40, 43 (1907) (trial court erred in taking judicial notice Rock River not navigable for logs or rafts; while may be proper to take judicial notice a stream is in fact navigable, it is doubtful that the contrary proposition is a proper subject of judicial notice, "considering the various kinds of navigation and the various appliances for the purposes of navigation and the differing conditions along different portions of the same river"). Whatever the measure of navigability, however, there was no presumption that a stream was navigable, and one whose legal position depended on the navigable status of waters must allege navigability and sustain the burden of proof thereon. Leihy v. Ashland Lumber Co., 49 Wis. 165, 5 N.W. 471 (1879). Cf. Wood v. Hustis, 18 Wis. 417 (1863).

81. See 74 Wis. 652, 656, 43 N.W. 660 (1889). The quoted justification, it will be noted, fits into pattern with the court's general

favor for using law to promote maximum productivity of economic resources. See notes 107–113 below. Compare especially with the rationale here advanced in the Conn Co. case the ruling in Volk v. Eldred, 23 Wis. 410, 412 (1868), note 115 below, that the rights of a log navigator should be measured by the improved rather than the limited natural capacity of the stream.

82. Circuit Court for Oconto County, in Falls Manufacturing Co. v. Oconto River Improvement Co., 87 Wis. 134, 58 N.W. 257 (1894), Printed Case on Appeal, p. 34. When the lumber era was about at its end, the Wisconsin Supreme Court explained that the general authorization to dam nonnavigable streams under the milldam act should not be construed to exclude from its license streams navigable only by the log-transport test. So to limit this power-dam statute would be to hamper its obvious policy of promoting development of energy for industrial uses, and to do so under a criterion expressing the urgent policy of another day: "this court, in deference of what was believed public welfare, has gone to great extremes in protecting such rights [of stream transport] in the public," establishing "the extreme doctrine of public rights in trifling bodies of water." See Allaby v. Mauston Electric Service Co., 135 Wis. 345, 349, 116 N.W. 4, 6 (1908).

83. See notes 5, 6, 9 and 12–15 above.

84. In various ways, apart from the implicit testimony of the hundreds of statutory franchises which in fact delegated decision making to private decision makers, the legislature indicated a halfway articulate recognition that it was asserting possession of authority to delegate, along with an obvious confidence that this was an authority to which it had clear title. This was implicit in its acknowledgement that a franchise which might adversely affect other persons must find justification in serving a public interest. See note 33 above. Authority to delegate was implicitly asserted, again, when a legislative committee justified an outstanding traffic control franchise on the ground that it had been granted only after fair hearing of major affected interests. See Majority Report, Select Committee on Bill 352A (to repeal charter of Black River Improvement Co.), Apr. 8, 1867, AJ, 1106, and App., 1269. Significantly, the strong Minority Report took no issue with the validity of delegating a public improvement to private hands, but only with what the dissenter conceived to have been a grossly unfair preference of one over another private interest in the matter (*id.*, p. 1271). Renewed controversy over the Black River Improvement Co. in later years gave occasion for legislative committees to make explicit assertion of the legislature's authority to delegate; the power was regarded as so well established, however, that no more was thought necessary than to declare it. See Report, Committee on Incorporations, on bill 202S, Feb. 17, 1880, SJ, 233 (franchise holders deserve consideration, because they "have honestly and diligently improved and handled the property of the community with reasonable dispatch, and at a cheap cost, for a great many years"); Report, Judiciary Committee, on same bill, Mar. 3, 1880, SJ, 394 (committee has considered the bill "as to the constitutional power of the legislature" and rules "that the legislature has the power to grant to the Black River Improvement Company authority to improve such river and erect dams, etc.," subject only to obligation to pay just compensation for others' property taken). Probably because the point was so much a commonplace of contemporary policy as not to require statement, the legislative records show no avowed reliance on the internal improvements clause of the Wisconsin constitution to establish the legitimacy of statutory delegation of public works to private hands. The constitutional records make plain that such delegation was assumed in framing and adopting the internal improvements clause of the Wisconsin constitution. See, e.g., Quaife, *The Struggle Over Ratification,* 47, 86, 211, 321, 343, 508, 666; Quaife, *The Attainment of Statehood,* 37–39, 417, 420, 581, 583–584. See, generally, Ray A. Brown, "The Making of the Wisconsin Constitution," *Wisconsin Law Review,* 1949: 648, 675, and 1952: 23, 47. The common sense of the times in this matter was expressed with forcible practicality in Report, Committee on Internal Improvements, HRJ, 179, 180 (1842). The committee found little hope in continued federal bounty, and it expressed sharp awareness of the limited means which the territory could supply to develop facilities of bulk transportation. Thus, it found, "The only remaining resource . . . for the accomplishment of these great works of improvement, is private enterprise. It is believed, that if the Legislature would grant to the individuals owning the land on these streams, the right to construct dams, with suitable locks and slides, wherever rapids occur thereon, their navigation would be speedily improved, and the agricultural, manufacturing, and commercial interests of

our territory, thereby greatly promoted . . . Let then the people immediately interested and who own the soil bordering on these streams, have the right — give them the privilege of bringing into operation the extensive water power which will be created, in connection with the improvement of these rivers — and your committee believe that these important channels of trade will be speedily opened, without the aid of the General Government, and without involving the faith or the responsibility of the Territorial Government."

85. In Wisconsin River Improvement Co. v. Manson, 43 Wis. 255, 262, 265 (1877) the court explicitly grounded the legislature's authority to delegate stream improvements to private operators, on the combined policy implications of Wisconsin Constitution, Art. VIII, sect. 10 (forbidding the state to be party to such works), and Art. IX, sect. 1 (declaring and protecting the public interest in navigation). Though analysis was not sharply focused on the delegation issue, in accord with the Manson case are Black River Improvement Co. v. LaCrosse Booming & Transportation Co., 54 Wis. 659, 675–677, 11 N.W. 443, 450–451 (1882) (improvement charter, in the absence of explicit contrary provision, should be construed to allow grantee to take state lands necessary for the improvement, without compensation; no issue made of breach of internal improvements clause); Falls Manufacturing Co. v. Oconto River Improvement Co., 87 Wis. 134, 151, 58 N.W. 257, 261 (1894), and trial court opinion therein, Printed Case on Appeal, p. 30 (delegation explicitly acknowledged below, and tacitly accepted incident to ruling that franchise validly determines best use of a public resource, without need of compensation to interest in fact adversely affected). Delegation was upheld without question in the matter of the milldam act, where the issue was on public purpose rather than on the designation of the agents to effect public policy. See notes 49, 56 and 57 above. No serious question was ever raised of the legislature's competence to delegate navigation improvements to local government units. See Barnes v. City of Racine, 4 Wis. 454, 466 (1855) (harbor improvements); Attorney General v. Eau Claire, 37 Wis. 400, 433 (1875) (channel improvements). The state government was a simple and limited apparatus through the lumber era. But, though no state administrative agency was created, there is no indication that the court would have found objection, had the legislature given a state office a role consistent with the limitations of the internal improvements clause. See City of Baraboo v. Railroad Commission, 195 Wis. 523, 525, 218 N.W. 819, 820 (1928).

86. Sellers v. Union Lumbering Co., 39 Wis. 525 (1876). See note 55 above. Cf. Attorney General Opinion to Secretary of State, PD, 713 (1906) (corporate articles may not confer broader improvement powers than provided under Statutes, 1898, sects. 1777 and 1777e). The defect in the grant involved in the Sellers case would seem to have been that in terms it conferred toll rights, without limit of the number of grantees, upon as many persons as might establish that they had effected some improvement in the stream. The legislature was not barred from making a general delegation of improvement rights to a class sufficiently limited in definition to preclude embarrassing conflicts of interest or multiplication of claims. Thus the court accepted without question the validity of Laws, 1876, c. 399, which became Revised Statutes, 1878, sect. 1777. But that law limited its grant of improvement traffic and toll rights to a corporation formed under the general business corporation law for stream improvement "which shall have taken prior possession" of the stream with reference to which its franchise was claimed. Sect. 1777 was accepted without question of the adequacy of designation of the enfranchised class thereunder, in Falls Manufacturing Co. v. Oconto River Improvement Co., 87 Wis. 134, 146–147, 58 N.W. 257, 260 (1894), as also in Heerman v. Beef Slough Manufacturing, Booming, Log-Driving & Transportation Co., 1 Fed. 145, 165 (Circ. Ct. W.D. Wis. 1878).

87. See note 62 above.

88. See Hurst, *Law and the Conditions of Freedom,* 39, 53, 63–66.

89. See note 39 above; cf. notes 34 and 35.

90. See note 38 above.

91. See notes 43–54 above, and especially note 53.

92. That a statutory grant was necessary to allow collecting tolls for a private improvement was established in the implications of legislative practice before it was established by the court. See P&LL, 1862, c. 98, repealing toll grant under P&LL, 1857, c. 360; P&LL, 1866, c. 447, amending Private & Local Acts, 1864, c. 84 (incorporating Black River Improvement Co.), to reduce the range of products for passage of which tolls might be charged; P&LL, 1867, c. 470, amending P&LL, 1864, c. 157, to enlarge the products for pas-

sage of which tolls might be charged; P&LL, 1868, c. 304, amending P&LL, 1856, c. 294, to add a toll right where none had been given. Some early franchises had included explicit denials of toll rights. See, e.g., Private & Local Acts, 1853, cc. 152 and 208; Private & Local Acts, 1854, c. 98; P&LL, 1856, c. 405. The requirement that a slide or chute be built into a dam commonly declared that this was to facilitate "free" passage of logs and rafts, as in Private & Local Acts, 1853, c. 152. Such references might imply that the legislature believed an improver might charge tolls, unless a statute forbade him to do so. But this would appear an unwarranted inference in view of the strong, continuing assertion of legislative authority to define toll rights and the early and continued eagerness of applicants to include explicit toll grants in their franchises. That the "free" passage stipulations almost always appeared in immediate connection with stipulations for slides or chutes suggests that the reference here was to freedom from physical rather than monetary burdens upon movement of logs. In the context of legislative practice, the disclaimer clauses appear simply as statements made out of abundance of caution. This reading is supported by the firm declarations of the Wisconsin Supreme Court, when the issue was presented to it after a generation of legislative practice had set the Wisconsin pattern of authority. There was no common law right of an improver to claim tolls, or to claim in quasi contract, for the help his flooding dam rendered in moving another's logs. Weatherby v. Meiklejohn, 56 Wis. 73, 13 N.W. 697 (1882). So, where a statutory franchise failed for want of sufficiently identified grantees, one claiming thereunder had no other basis for collecting tolls, though he had in fact improved the stream. Sellers v. Union Lumbering Co., 39 Wis. 525 (1876). A grantee who failed to satisfy conditions stated in the only statute under which he could claim a toll right, had no other basis for charging for the benefits of his improvements. Black River Flooding-Dam Association v. Ketchum, 54 Wis. 313, 11 N.W. 551 (1882); Sauntry v. Laird, Norton & Co., 100 Wis. 146, 75 N.W. 985 (1898). These decisions implicitly rejected the implication in Tewksbury v. Schulenberg, 41 Wis. 584, 588 (1877), that an improver might collect charges simply on a quasi contract basis for benefits conferred. There is no case in which the Wisconsin Supreme Court had occasion to decide whether a "free" passage clause incident to a slide or chute requirement had reference to freedom from tolls. It was clear, at least, that such a clause referred to freedom from physical obstruction. See, e.g., Volk v. Eldred, 23 Wis. 410 (1868), arising under P&LL, 1857, c. 195. That no one brought such a clause to the court under a claim that it had reference also to freedom from tolls probably reflected common acceptance that no toll right could exist without explicit grant.

Quite apart from any monopolies on charging toll or managing traffic, a monopoly of location and of immediate operating control was implicit in every dam franchise and every franchise for navigation improvements in a closely defined area. That the legislature's permission was essential for a landowner to use his riparian lands for an improvement purpose was implicit in the care early taken in defining locations and regularizing the title of particular grantees. See, e.g., Acts, 1852, c. 20, validating rights in "assigns" where the original franchise (Laws of Wisconsin Territory, 1842, p. 84) ran only to the named grantee and his associates; Private & Local Acts, 1855, c. 159, amending location description under Laws of Wisconsin Territory, 1846, p. 93. A fortiori, a legislative franchise was necessary to confer not merely the monopoly that was physically inherent in the occupation of a close area, but a right to exclusive management of traffic over an extended area, or the exclusive right to collect tolls for use of an extended stretch of improved waters. Legislative practice implied the requirement of a statutory warrant for such exclusive controls, especially in the care taken to define and redefine the areas or the operations in respect to which such rights might obtain. See, e.g., Private & Local Acts, 1854, c. 28, amending Private & Local Acts, 1853, c. 30 (Wisconsin River Improvement Co. area of improvements); Private & Local Acts, 1854, c. 321 (boom area), Private & Local Acts, 1856, c. 393 (driving and boom areas), and P&LL, 1857, c. 145 (limiting driving monopoly) — all amending Private & Local Acts, 1854, c. 62, incorporating Black River Lumber Driving & Booming Co.; P&LL, 1862, c. 43, amending P&LL, 1857, c. 101, to redefine area in which Wolf River Boom Co. shall have a booming monopoly, and the area in which it has only concurrent rights with others. In effect the legislature also enforced the principle that its warrant was necessary to validate an operating monopoly on navigable water, when General Laws, 1866, c. 134 (carried into Revised Statutes, 1878, sect. 4448) made it a misdemeanor, carrying heavy possible penalties, for a boom operator or owner knowingly to turn another's logs into the boom without the log owner's consent. General Acts, 1867, c. 161, amended this provision, to exclude the Chippewa and St. Croix rivers and their tributaries; it was perhaps significant of favor for the general principle of the basic act that this qualification was dropped in the Revised Statutes.

The general requirement of a statutory warrant for any monopoly position on a navigable stream, thus asserted in legislative practice, was matched by common law doctrine. There was no basis in common law by which a private person might successfully assert the right to block all passage on a navigable stream, save on his terms of managing the water flow or the movement of traffic. Any such private effort clearly constituted an actionable nuisance. See notes 36–41 above. The obverse of this proposition was the general requirement of a statutory license. See Green Bay & Mississippi Canal Co. v. Kaukauna Water Power Co., 70 Wis. 635, 653, 35 N.W. 529, 535 (1888).

93. On the rare occasions when legislative committees made explicit analysis of the grounds on which the legislature should adjudicate among interests competing with reference to stream use, the fact that argument focused mainly on considerations of wisdom and fairness implied acceptance of the view that the legislature enjoyed a broad authority to dispose of these matters. See, e.g., the discussions by leading lumberman-legislators of the issues involved in statutory approval of railroad bridges over the Mississippi, in Minority Report, Committee on Railroads, AJ, 385–389 (1864) (T. C. Pound), and Minority Report, Committee on Railroads, SJ, 422–426 (1864) (Angus Cameron). Of like import, see the various reports on bill 352A, 1867, to repeal the charter of the Black River Improvement Co.: Report, Committee on Internal Improvements, AJ, 812–813 (1867); Majority and Minority Reports, Select Committee, AJ, 1106 and App., 1266–1272 (1867). The Wisconsin Supreme Court in effect applied a strong presumption of constitutionality in favor of legislation determining orders of preference among interests competing as to stream use. See note 54 above.

94. See Hurst, *Law and the Conditions of Freedom,* 24–25, 83. The much more conservative attitudes which showed themselves in judge-made due process law in the generation after the turn of the century, in reaction to vigorous legislation to promote commonwealth interests, may be seen in Huber v. Merkel, 117 Wis. 355, 94 N.W. 354 (1903), and State v. Redmon, 134 Wis. 89, 114 N.W. 137 (1907).

95. See notes 49, 56 and 57 above, on the legislation and decisions which early recognized the need and the propriety of statutory delegations of the power of eminent domain for developing water power. That a private improver must have statutory warrant, with proper payment procedures, in order to use others' property for his works was of course established by the command of Wisconsin Constitution, Art. I, sect. 13, that "the property of no person shall be taken for public use without just compensation therefor." See Thien v. Voegtlander, 3 Wis. 461, 465 (1854). Cf. note 149 below. If the legislature failed to supply a compensation procedure, the common law would provide an action for damages, to enforce the constitution. See note 57 above. That the legislature might validly delegate the power of eminent domain for navigation improvements was early established in legislative practice. See, e.g., Acts, 1851, c. 325; Acts, 1852, c. 324; Private & Local Acts, 1853, cc. 24 (to individuals) and 30 (to corporation). The Report of the Judiciary Committee, SJ, 410 (1871), declared that the power of eminent domain "can only be exercised by the State itself," and found, hence, that the attempted delegation by bill 7S (the highly controversial Eau Claire dam-and-boom franchise) was invalid. This declaration stands alone and contrary to consistent legislative practice and judicial doctrine in Wisconsin; it can be estimated only as a reflection of the partisan fervor generated by the "Dells bill" fight. See also Report, Judiciary Committee, on bill 380A, AJ, 547 (1873) (recommending repeal of "unconstitutional" eminent domain part of corporate charter, without further specification of grounds).

The court found no difficulty in treating navigation improvement as a public use justifying resort to the power of eminent domain. See Jones v. United States, 48 Wis. 385, 404, 4 N.W. 519, 520 (1879), affirmed, 109 U.S. 513 (1883); Wisconsin River Improvement Co. v. Pier, 137 Wis. 325, 118 N.W. 857 (1908). However, no occasion presented itself for the court to rule on this matter in a case involving improvements primarily for log or lumber transport; the Jones case involved the Fox-Wisconsin improvement; the Pier decision came about the end of the lumber era, and though the court relied on the utility of the project for navigation improvement, it appears that it was primarily a development of water power to generate electricity. The absence of an important lumber industry case on the point probably reflects the fact that most stream improvements of primary importance for log transport were made in unsettled country where no practical cause would arise for seeking compensation. The cases cited in note 57 above, in which the court held that a common law action for damages would lie, to make good the want of a statutory compensation procedure, all involved flowage of farmlands, and were obviously of limited economic significance. The opinion declared in Report, Judiciary Subcommittee, on pending dam franchise bills, SJ, 1038ff (1905), that navigation improvement was the only justifying public purpose to warrant a delegation of eminent domain authority, and that the power might not be given for power development, may be taken as expressing a strong attitude of that time, but runs squarely contrary to the consistent legislative practice and judicial doctrine of the nineteenth century.

96. That a statutory grant was necessary to confer a right to tolls, see note 92 above. The legislature's authority to confer a toll right on private persons who provided a public service was early asserted, by the Committee on Judiciary of the territorial House of Representatives. See its Report, HRJ, 240 (1842). The committee justified such a grant not only as a reasonable means to promote works of public interest, but also as an indirect exercise of the legislature's power to raise and spend money for public purposes: "It is true, that the act allows the corporation to take tolls for crossing their bridge, but the amount is fixed by the Legislature, and the intention of the act is not to benefit the corporation but the public; and it is apparent that the effect will be greatly to promote the public accommodation and convenience. The tolls allowed by the Legislature are supposed to be a fair remuneration to the corporation for the cost of the bridge and no more. The committee can see no difference in principle between taxing the public for the cost of the bridge in the way proposed in the bill, and in taxing them for the gross amount at once, and collecting the tax in the usual mode." Of course, the adoption of the ban on state participation in works of internal improvement, in the Wisconsin constitution, underlined the practical importance of this argument. See note 84 above.

In one of the rare instances in which the legislature made explicit assertion of its constitutional authority, General Acts, 1853, c. 72 (amending the navigation guaranty copied from Wisconsin Constitution, IX, 1, into Revised Statutes, 1849, c. 34, sect. 1) declared, "That nothing herein contained, shall prevent the collection of all tolls imposed for necessary easements, constructed in the said navigable waters." Carried into Revised Statutes, 1858, c. 41, sect. 1, this declaration was dropped when the whole section was omitted "as being sufficiently sanctioned by its enactment in the constitution." See *Report of the Revisers of the Statutes,* 124.

Generally the legislature left its authority to be implied. That toll grants lay wholly in the legislature's determination, and that the legislature might properly decide when and on what terms to grant toll rights, were propositions most particularly implied when the legislature amended franchises to add a toll right not originally given, or to alter the terms of a right already conferred. See, e.g., P & LL, 1867, c. 470, amending Private & Local Acts, 1864, c. 157 (enlarging categories of services for which charges may be made); P & LL, 1868, c. 130, amending General Acts, 1865, c. 54 (to confer a toll right where none was originally given). The legislature's right and power had especially strong assertion in legislative practice also where the legislature took pains explicitly to deny that an improvement authorization carried a toll right (e.g., Private & Local Acts, 1853, cc. 152 and 208; Private & Local Acts, 1854, c. 98; P & LL, 1856, c. 405, amended, P & LL, 1866, c. 286), or distinguished explicitly between the right of a cooperative-style improvement association to apportion expenses among its members and the authority of a stock corporation succeeding it, to charge tolls to its clients (e.g., P & LL, 1862, c. 35, compared with P & LL, 1864, c. 157), or repealed the toll provision while letting the rest of a franchise stand (e.g., P & LL, 1857, c. 360, as amended by P & LL, 1862, c. 98), or set a term of years upon the toll right (e.g., P & LL, 1867, c. 334), or stipulated that a newly granted toll right should not apply in the driving season of the year in which it was enacted (e.g., Laws, 1874, c. 262; Laws, 1876, cc. 219 and 220). There was early consensus that the legislature held authority to grant toll rights. The existence of this consensus was reflected, for example, in the fact that Governor Farwell's Veto Message, AJ, 841–842 (1853), objected to the toll right given the Black River

Lumber Driving & Booming Co., not in itself, but only as a feature of what the Governor deemed the unfair monopoly position conferred on the grantees by the total pattern of the franchise. Similarly, Governor Rusk's Veto Message, SJ, 488–490 (1882), noted that the inclusion of a toll right gave the greater reason why the legislature should include in its franchise a reservation of its right to amend or repeal, but the Governor raised no objection to the toll grant as such.

The Wisconsin Supreme Court did not have occasion to rule on lumber transport tolls until a generation of legislative practice had set the Wisconsin patterns of authority. The court then fully upheld legislative control and the positive authority of the legislature to grant toll rights where it might reasonably deem that a public interest was served. The authority was grounded in the facts of economic potential and utility. Tolls were warranted as a means to obtain navigation improvements, because "of the frequent necessity of such dams in the great lumber regions of the state . . . for protecting and aiding the only kind of transportation to which such streams are adapted" (Tewksbury v. Schulenberg, 41 Wis. 584, 593 [1877]). This authority was the more needful because the state constitution disabled the state from direct investment in navigation improvements; on the other hand, so long as tolls were but a fair return for adding to the carrying capacity of waters, the toll grant could not be deemed to offend the navigation guaranty stemming from the Northwest Ordinance, the purpose of which was to foster navigation. Wisconsin River Improvement Co. v. Manson, 43 Wis. 255, 262–263 (1877); compare General Acts, 1853, c. 72, discussed above. Accord: Schneider v. Staples, 66 Wis. 167, 28 N.W. 145 (1886); Underwood Lumber Co. v. Pelican Boom Co., 76 Wis. 76, 45 N.W. 18 (1890); Sauntry v. Laird, Norton & Co., 100 Wis. 146, 75 N.W. 985 (1898); Wisconsin River Improvement Co. v. Pier, 137 Wis. 325, 118 N.W. 857 (1908). See also J. S. Keator Lumber Co. v. St. Croix Boom Corporation, 72 Wis. 62, 80–81, 38 N.W. 529, 535 (1888). Of course, the legislature's authority was only to make toll grants limited by some legally enforceable standard of the reasonableness of charges; a franchise purporting to allow the grantee unchecked discretion to fix rates would doubtless have been held ineffective. See Underwood Lumber Co. v. Pelican Boom Co., 76 Wis. 76, 85–86, 45 N.W. 18, 21 (1890), and the discussion in the later chapter on police power.

97. The legislature's most sweeping assertion of authority to confer exclusive toll privileges was by Laws, 1876, c. 399, continued in the Revised Statutes, 1878, as sect. 1777, which declared that any corporation formed for stream improvement under the general business corporations law, "which shall have taken prior possession of such stream for that purpose," should have power to improve the stream, take charge of all logs inadequately provided for, and charge reasonable tolls for its benefits, and the actual costs of its driving services. Interpreted to confer an exclusive toll right upon a corporation which met its terms, the statute was evidently accepted as constitutional, though the facts did not present the matter for decision, in Black River Flooding Dam Association v. Ketchum, 54 Wis. 313, 11 N.W. 551 (1882). An analogous exclusive toll right for booming corporations — conferred by sect. 1777 as amended by Laws, 1882, c. 318 — was held constitutional, though without particular point made of its exclusive feature, in Underwood Lumber Co. v. Pelican Boom Co., 76 Wis. 76, 45 N.W. 18 (1890). Accord: Menominee River Boom Co. v. August Spies Lumber & Cedar Co., 147 Wis. 559, 132 N.W. 1118 (1912). This kind of legislation presented the more sharply focused issue of toll monopoly. But of at least equal importance were the toll monopolies implicit in the situation wherever the legislature authorized works which occupied the whole stream or its main channels, so that after the works were built all who used the stream must necessarily take the benefits of the works. Again, years of legislative practice established the legislature's authority in the matter before an issue was made on the point before the court. See, e.g., Acts, 1850, c. 257, 1851, c. 88; Private & Local Acts, 1853, c. 30; P & LL, 1856, c. 481, 1857, c. 170. When such an implicit toll monopoly was challenged, the court found no trouble in sustaining it as a reasonable exercise of the legislature's power to promote the carrying capacity of streams. Wisconsin River Improvement Co. v. Manson, 43 Wis. 255, 265 (1877), discussed in note 98 below, cf. Sellers v. Union Lumbering Co., 39 Wis. 525, 529 (1876). What might be deemed the grant of an *ad hoc* tolls monopoly was held constitutional as a reasonable regulation to promote maximum navigational use, when Wisconsin River Log Driving Association v. D. F. Comstock Lumber

Co., 72 Wis. 464, 40 N.W. 146 (1888) sustained the validity of Revised Statutes, 1878, sect. 3337, as amended by Laws, 1881, c. 141; this statute authorized one whose logs became inextricably mingled with those of another who had not adequately provided for their handling to drive the whole and have reasonable compensation therefor.

98. Legislative prerogative was clearly asserted when by explicit declaration or by necessary implication the legislature conferred a monopoly of improvements and of traffic management over an extended stretch of stream. Legislative practice asserted this prerogative as early as the 1850's. See, e.g., Acts, 1850, c. 257 (repealed, Acts, 1851, c. 325); Acts, 1851, c. 351; P & LL, 1857, c. 101. This is a complex matter, explored in more detail hereafter in relation to the detailed content of franchises. Legislative control and authority in the grant of monopoly may also be seen dramatized in the successive qualifications of the traffic control powers given the Black River Log Driving & Boom Co. by Private & Local Acts, 1854, c. 62. See P & LL, 1856, c. 393, 1857, c. 145. Another especially pointed example in this period of formative legislative practice in the 1850's is P & LL, 1856, c. 405, defining the authority of the Chippewa River Improvement Co. with a care which makes plain that it is only to improve the utility of the main channel and not to claim any proprietorship over it. We have noted earlier (see note 33 above) that in two vetoes in 1853 Governor Farwell disapproved of particular monopoly grants, leaving it ambiguous how far his disapproval ran on policy or on constitutional grounds. Plainly, in light of the continuing broad current of legislative monopoly-style grants, these vetoes established no broad limitation on legislative discretion in this matter. Cf. note 64 above. Laws, 1880, c. 225, amending P & LL, 1864, c. 84, symbolizes the broad monopoly-granting power which prevailing opinion accepted as residing in the legislature at a time when legislative practice had come to maturity. Laws, 1880, c. 225, amended the charter of the Black River Improvement Company, to authorize the company to increase main-channel flow by closing off an otherwise navigable slough. The assertion of legislative authority here was the more striking because the 1880 act passed only after a sharp battle. Relevant to our present concern is the fact that none of the several legislative committees which considered the 1880 amendment seriously questioned the legislature's power to confer exclusive improvement rights, to promote general navigation. Some of the reports explicitly and broadly asserted the legislative prerogative. The only doubt expressed was whether compensation must be paid to riparian owners who had used the slough for navigation; the 1880 act preserved such rights as the riparians might be held to have, but the court later decided no compensation was due, because there was no private property in stream flow as such. See Report, Committee on Incorporations, on bill 202S, SJ, 233–234 (1880); Report, Judiciary Committee on same, SJ, 394 (1880); Report, Committee on Incorporations on same, AJ, 571 (1880). Cf. Black River Improvement Co. v. LaCrosse Booming & Transportation Co., 54 Wis. 659, 11 N.W. 443 (1882).

To implement its authority over common resources, the legislature might confer exclusive rights to manage the water flow or the booming of logs, to promote general navigation, though such grants precluded uses which otherwise would be lawful incidents of riparian ownership, and might confer such rights without providing compensation to the affected riparians, so long as the franchises did not entail physical encroachment beyond the banks of the stream. Cohn v. Wausau Boom Co., 47 Wis. 314, 2 N.W. 546 (1879) (exclusive booming right); Black River Improvement Co. v. LaCrosse Booming & Transportation Co., 54 Wis. 659, 11 N.W. 443 (1882) (right to close off side channels, limiting traffic to main channel); Falls Manufacturing Co. v. Oconto River Improvement Co., 87 Wis. 134, 58 N.W. 257 (1894) (right to manage flow for navigation despite limits imposed thereby on power use). Cf. Green Bay & Mississippi Canal Co. v. Kaukauna Water Power Co., 70 Wis. 635, 651, 35 N.W. 529, 534 (1888) (government may pre-empt all water-power development incident to general public navigation improvement). These decisions are not, however, conclusive of the full scope of legislative power. They represent the supremacy of reasonable commonwealth interest over the focused interest of particular property owners. Consistent with such rulings, the court might still have held that the legislature lacked authority to confer monopolies of improvement or traffic management, as against the general public or the general class of those who might wish to use the waters for transport. Such was part of the cloudy import of Governor Farwell's two vetoes of monopoly bills in 1853. See note 33 above. The issue was not a hard one, if the court determined that a particular monopoly was given only for private advantage; such a grant must fail, though not so much because of any peculiar defect in the monopoly grant as such as because of the general doctrine that public power must be used for a public purpose. This was the force of the decision which invalidated the first of the Eau Claire dam-and-boom franchises. See notes 61, 62, and 87 above. But, assuming that a monopoly grant fell within the area of reasonable legislative judgment as to existence of a public interest, might it nonetheless be invalid, either because there was some unfocused doctrine of constitutional force forbidding creation of monopolies in public resources, or because such a grant would offend the protection which Wisconsin Constitution, IX, 1, assured the principal waters as "common highways"? The court answered the latter question, at least, in Wisconsin River Improvement Co. v. Manson, 43 Wis. 255, 265 (1877). The court there not only upheld the legislative authority to confer a toll right, as consistent with Art. IX, sect. 1, but also sustained the toll right though the grantee's improvements were such that all navigating the stream must necessarily use them and hence necessarily pay the toll. The court would give both the toll right and the improvement works the benefit of a strong presumption that they fell within the legislature's reasonable judgment as measures to develop the carrying capacity of the waters; as such, they did not offend the constitutional navigation guaranty, but rather fulfilled it. Somewhat oddly — in view of the importance of the issue — this was the only occasion on which the court explicitly dealt with the validity of legislative grants of monopoly incident to stream improvements; and the court never expressly discussed the more general issue whether any broad policy of constitutional weight limited the legislature's competence to create monopolies. However, the court several times in effect sustained a reasonable discretion in the legislature to use the creation of monopoly position as an instrument to promote navigation. See Black River Flooding Dam Association v. Ketchum, 54 Wis. 313, 11 N.W. 551 (1882) (implicitly accepting, as against competing improver, validity of exclusive franchise conferred by Revised Statutes, 1878, sect. 1777); Underwood Lumber Co. v. Pelican Boom Co., 76 Wis. 76, 45 N.W. 18 (1890) (implicitly accepting, as against a navigator, validity of exclusive booming franchise under Revised Statutes, 1878, sect. 1777a); Yellow River Improvement Co. v. Wood County, 81 Wis. 554, 51 N.W. 1004 (1892) (implicitly accepting, as valid exercise of legislature's control of local government finance, ban on applying local realty tax to tract which is part of integrated improvement). The court did deal expressly with more limited traffic control monopolies, which it sustained as reasonable means for promoting maximum movement of traffic.

99. See, e.g., Acts, 1851, c. 351 (incorporating St. Croix Boom Co.); Acts, 1852, c. 323 (incorporating Little Bull Falls Boom Co.); Private & Local Acts, 1853, c. 30 (incorporating Wisconsin River Improvement Co.); Revised Statutes, 1878, sect. 1771 (general incorporation act). Men might claim corporate status in law only by grant from the legislature; to confer corporate status was a recognized legislative authority. See Sellers v. Union Lumbering Co., 39 Wis. 525, 527 (1876). A constitutional amendment of 1871 prohibited the legislature "from enacting any special or private laws for granting corporate powers or privileges, except to cities," so that from that point on no business corporation might come into existence by special statutory franchise. After some doubt, the court ruled that the constitutional limitation did not forbid amendment of special charters outstanding in 1871. See Black River Improvement Co. v. Holway, 87 Wis. 584, 588, 591, 59 N.W. 126, 127, 128 (1894).

100. See notes 62 and 87 above.

101. See Report, Committee on Lumber and Manufactures, on bill 98A, AJ, 327 (1876). It attests the depth of agreement on the use of law to promote productivity, that the vigorous opposition to the new franchise at no point challenged the basic authority of the legislature, but contested only what the opponents deemed a misuse of an undoubted public power, in the particular facts of the Eau Claire situation which they regarded as showing that the power was in fact being applied only for private advantage. See Minority Report, Committee on Incorporations, on bill 98A, SJ, 534–535 (1876); Memorial No. 185S, against bill 98A, by Thaddeus C. Pound, SJ, 426–428 (1876).

102. The continued readiness of the legislature over the years to delegate the power of eminent domain to private enterprisers bore witness that the court accurately expressed the

community philosophy of growth in the terms in which it sustained such delegation originally, in Newcomb v. Smith, 2 Pinn. 131, 135, 138, 140 (Wis. 1849): "We should commit a great error if we assumed that public use is confined to such appropriations as the government may have occasion to make for the common defense and safety, when acting by its officers and agents in cases of emergency. Such cases sometimes occur; but it far more frequently happens that the right to take private property for public use is exercised in a manner less direct for objects less general . . . Is there any good reason why water-mills should not be regarded as public improvements? Why the legislature should not favor their construction, especially in a new country, among a scattered population and where capital is limited? . . . No tyranny has been found more odious in the older states than is sometimes exhibited in the pertinacious obstinacy of one man, who pursues his common-law right of resisting the occupation of his land, perhaps of some small and insignificant but indispensable portion, for the purpose of a mill. The graceless privilege of commencing daily suits for the daily infringement of constitutional rights is often less detrimental to the enterprising individual, who may be its victim than to the community, who are common sufferers. The statute which cuts off the common-law privilege of such a man to indulge in litigious malice, and which secures to him ample compensation by a single action for his property taken for public use, is surely benign in its effects, and harmonious with the spirit if not the letter of the constitution. Such a law, also, by inviting capital into the interior of the state, by encouraging enterprise and diffusing the conveniences of social life, enhances the value of land, advances its settlement, and promotes general civilization." See the retrospective recognition of the primacy assigned to the promotion of productivity through such delegation of eminent domain powers, in McDonald v. Apple River Power Co., 164 Wis. 450, 455–456, 160 N.W. 156, 158 (1916), and in Town of Marion v. Southern Wisconsin Power Co., 189 Wis. 499, 502, 208 N.W. 592, 593 (1926).

103. Cf. Volk v. Eldred, 23 Wis. 410, 412 (1868); Tewksbury v. Schulenberg, 41 Wis. 584, 594 (1877).

104. See, e.g., Acts, 1851, c. 325 (grantees entitled to demand tolls "so soon as the said channel or channels shall be completed and ready for use"); P & LL, 1857, c. 360 (grantees may receive tolls "when [they] . . . shall have completed the said dam and slide as aforesaid").

105. See, e.g., Private & Local Acts, 1853, c. 30 (Wisconsin River Improvement Co. shall not be entitled to collect tolls "until the said Company shall have expended the full sum of five thousand dollars upon the said improvement, or some part of said improvement, and made substantial improvements affording increased facilities of navigation of said river, and no greater tolls shall be levied than are reasonable in consequence of such increased facilities of navigation"), relaxed by P & LL, 1868, c. 394 (striking reference to improvements-in-fact, leaving just the $5000 stipulation); Laws, 1874, c. 288 (Daniel Shaw and associates may not collect toll for Thorn Apple River improvement use "until after they shall have expended at least five thousand dollars in the work of improving said river"), qualified by Laws, 1881, c. 163 ($5000 requirement may be satisfied by payment of such amount "in the purchase of such improvements made by others since March 26, A.D. 1874, at a price not exceeding the actual cost thereof").

106. See, e.g., Private & Local Acts, 1853, c. 24 (franchise for canal at Grand Rapids on Wisconsin River shall be forfeited if works not completed within three years after commencement of construction, for which one year is allowed from passage of this act); P & LL, 1862, c. 100 (Apple River Boom Co. works "shall be constructed and ready for use on or before the 10th day of June, 1862" — no sanction stated), amended, Private & Local Acts, 1863, c. 311 (completion date extended one year).

107. The quoted phrases are from Laws, 1874, c. 228 (Fisher River, Chippewa County), and Laws, 1883, c. 317 (Robinson Creek, Jackson County). Note also the quoted terms of Private & Local Acts, 1853, c. 30, note 105 above.

108. Newcomb v. Smith, 2 Pinn. 131, 140 (Wis. 1849). The quoted observations will be found in fuller context in note 102 above.

109. Wisconsin River Improvement Co. v. Lyons, 30 Wis. 61, 65 (1872).

110. Tewksbury v. Schulenberg, 41 Wis. 584, 593–594 (1877).

111. Wisconsin River Improvement Co. v. Manson, 43 Wis. 255, 262 (1877).

112. Heerman v. Beef Slough Manufacturing, Booming, Log-Driving & Transporta-

tion Co., 1 Fed. 145, 165 (Circ. Ct. W.D. Wis. 1878).

113. Circuit Court for Oconto County, in Falls Manufacturing Co. v. Oconto River Improvement Co., 87 Wis. 134, 58 N.W. 257 (1894), Printed Case on Appeal, pp. 33, 35.

114. *Id.*, p. 34.

115. Volk v. Eldred, 23 Wis. 410, 412 (1868).

116. See Laws, 1876, c. 399, as carried into Revised Statutes, 1878, sect. 1777 ("Any corporation formed under this chapter for the improvement of any stream, and driving logs thereon, which shall have taken prior possession of such stream for that purpose, shall have power to improve such stream and its tributaries"), and sect. 1777a, added by Laws, 1880, c. 279 (applying to any such corporation formed also for "storing, sorting and delivering . . . saw logs, square and round timber, or other timber" on a given stream). These were the only lasting statutes which created open-ended classes of persons authorized to undertake stream improvements. Compare note 118 below.

117. Only rarely did the *habendum* clause in a stream franchise run merely to a named individual or individuals, without reference to associates, assigns, or other-described sharers or successors in interest. Internal evidence suggests that such phrasing probably represented crude drafting rather than sophisticated purpose. See Laws, 1873, c. 252 (granting a franchise for "a logging dam" on a specified location, to Aaron M. Chase "and assigns," and a second franchise for "a dam" at another specified location, simply to "said Aaron M. Chase"); Laws, 1875, c. 45 ("Dan. F. Smith is hereby authorized to maintain a toll-dam at Clam Falls, on the south fork of Clam River, in Polk county"; other clauses confer authority upon "Dan. F. Smith, or his legal representatives"); Laws, 1885, c. 282 ("Theodore Buettner is hereby authorized to construct and maintain a dam across the Embarrass river, on lands owned by said Theodore Buettner, in the county of Shawano"; a proviso stipulates "that the aforesaid Theodore Buettner, his successors or assigns, shall build suitable slides"). The typical style of grant to individuals ran, for example, like the franchise for a dam conferred by P & LL, 1870, c. 32, upon "John Linder, his associates, successors and assigns"; or, to a corporation, like P & LL, 1866, c. 579, which declared that "Albert G. Ellis, William Weston and Moses M. Strong, and their successors and associates, are hereby made, constituted, appointed and declared to be a body politic and corporate by the name and style of the Stevens Point boom company," and that "said corporation is hereby authorized and empowered to construct and maintain a boom" and charge tolls for its use. That a franchise running to living persons should not be construed as a grant only for their lives was ruled in Jones v. Pettibone, 2 Wis. 308 (1853). Compare, in substantial accord, Schneider v. Staples, 66 Wis. 167, 28 N.W. 145 (1886). Jones v. Pettibone, cited above, accepted also that a franchise which in terms ran simply to a named grantee and his associates (with no mention of assigns) effectively passed to a purchaser of the land on which the dam abutted. Of like effect seems the acceptance of transfer of a dam right in Arpin v. Bowman, 83 Wis. 54, 53 N.W. 151 (1892). In other decisions the court tacitly favored the transferability of dam franchises. See Clark v. Plummer, 31 Wis. 442 (1872), Spensley v. Valentine, 34 Wis. 154 (1874), and McDonald v. Apple River Power Co., 164 Wis. 450, 160 N.W. 156 (1916) (franchises given to designated grantees and assigns pass appurtenant to mill properties without specific reference in deeds of mills). In Underwood Lumber Co. v. Pelican Boom Co., 76 Wis. 76, 82, 45 N.W. 18, 20 (1890), the court observed that "since such rights . . . were . . . expressly granted, not only to the Browns and Anderson, but to their 'heirs and assigns,' there can be no question but what . . . [they] were thereby empowered to make such transfers and conveyances to the boom company." If there is some intimation here that express authorization to assign was necessary, it is muted, and there is no other Wisconsin authority to reinforce it; the court cited United States Supreme Court decisions which dealt simply with corporate powers. An occasional statute purporting to validate transfer of a franchise probably represented only the exercise of caution by careful counsel. See Laws, 1881, c. 255, as amended, Laws, 1883, c. 96. Compare the tacit acceptance of the exercise of a power of assignment as requiring no further validation, in Laws, 1889, c. 32, amending Private & Local Acts, 1853, c. 152 (private dam, Wisconsin River). Note, however, the refusal of the Senate in 1893 to pass bill 488A, to approve a sale by Dan F. Smith of the dam franchise he had by virtue of Laws, 1875, c. 45, cited earlier in this note. See SJ, 614 (1893) (Senate Committee on In-

corporations recommends nonconcurrence); SJ, 632 (so voted).

118. Sellers v. Union Lumbering Co., 39 Wis. 525 (1876); see notes 55 and 86 above.

119. Similar to the ill-defined grant by Laws, 1873, c. 12 (Yellow River, branch of Chippewa River), which the court held legally ineffective because of vagueness (see the Sellers case, note 118 above), and hence presumably subject to the same deficiency, had they been challenged, were Laws, 1874, cc. 250 (west branch, Wolf River, Oconto County), 262 (Nail Creek, Chippewa River tributary), 297 (Prairie River, Wisconsin River tributary), and Laws, 1876, c. 219 (North branch, Eau Claire River, Chippewa County). Probably in reflection of the Sellers decision, Laws, 1874, c. 250, was amended by Laws, 1882, c. 251, and Laws, 1876, c. 219, was amended by Laws, 1877, c. 42, to make the grants run to named grantees. The two most important general laws on stream use were drawn so as not to create the conflict of claims which the court feared in Sellers. The milldam act explicitly denied its protection to any dam which was built "to the injury of any mill lawfully existing . . . on the same stream . . . [or] to the injury of any mill site on the same stream on which a mill or mill dam shall have been lawfully erected and used, or is in the progress of erection . . ." Hence, one whose dam violated this term might be sued at common law. Moore v. Coburn, 1 Pinn. 538 (Wis. 1845); Irwin v. Richardson, 88 Wis. 429, 60 N.W. 786 (1894); see Large v. Orvis, 20 Wis. 696, 698 (1865); cf. Arimond v. Green Bay & Mississippi Canal Co., 31 Wis. 316 (1872) (similar ruling under special dam franchise incorporating milldam act compensation procedure). Compare also Timm v. Bear, 29 Wis. 254 (1871); Soenksen v. Weyerhaeuser, 32 Wis. 521 (1873) (actions between contending mill owners, over backing up of water, treated as common law actions). The general-improvements franchise acts as to river improvement and boom corporations conferred their privileges only upon a corporation "which shall have taken prior possession of such stream for such purpose." See note 116 above.

120. Legislative practice made clear that reference to stream improvement authority in a general incorporation statute related only to the internal constitution of a corporation created thereunder. Thus Laws, 1873, c. 126, granted a Wisconsin River boom franchise to the Stevens Point Boom Co., organized under the general incorporation law, for booms at locations indicated in the articles of incorporation. The distinction was made a matter of general law by Laws, 1876, c. 399, carried into Revised Statutes, 1878, sect. 1777, as amended, Laws, 1882, c. 318 — which granted improvement rights to "any corporation formed under this chapter [i.e., under sect. 1771] for the improvement of any stream and driving logs thereon, which shall have taken prior possession of such stream for that purpose . . ." That a corporation organized under the general incorporation act to do business as a booming company must nonetheless show a distinct legislative grant to warrant its constructing a boom across a navigable stream, was held in Stevens Point Boom Co. v. Reilly, 46 Wis. 237, 243, 49 N.W. 978, 979 (1879). Accord: Black River Flooding-Dam Association v. Ketchum, 54 Wis. 313, 316–317, 11 N.W. 551, 552 (1882). The grant of corporate existence under Revised Statutes, 1878, sect. 1771, was noted as an item distinct from the grant of improvement powers under sect. 1777, in Falls Manufacturing Co. v. Oconto River Improvement Co., 87 Wis. 134, 146–147, 58 N.W. 257, 260 (1894). The distinction was also recognized in effect when Heerman v. Beef Slough Manufacturing, Booming, Log-Driving & Transportation Co., 1 Fed. 145, 163 (Circ. Ct. W.D. Wis. 1878), held that the claim that a corporation was acting *ultra vires* its charter in conducting a log driving business offered no basis on which to seek to enjoin the erection and operation of its stream works as a nuisance. The existence of internal corporate authority to engage in stream improvements was discussed as a distinct question in Wisconsin River Log Driving Association v. D. F. Comstock Lumber Co., 72 Wis. 464, 466, 40 N.W. 146, 147 (1888), and Underwood Lumber Co. v. Pelican Boom Co., 76 Wis. 76, 82, 45 N.W. 18, 20 (1890). Cf. In re Southern Wisconsin Power Co., 140 Wis. 245, 258, 122 N.W. 801, 806 (1909) (grant of dam franchise is not grant of corporate franchise within ban of amended Wisconsin Constitution, IV, 31).

Legislative practice left an area of uncertainty in the field of special statutes incorporating companies under definitions of corporate powers including authority to improve streams. Especially in view of the distinction taken in the general statutes between a grant pertaining to the internal constitution of a corporation and a grant of stream improvement authority as such, it would seem fair to construe general language in a special corporate charter as intended to deal only with the internal government of the corporation. See, e.g., P&LL, 1866, cc. 258 andd 331; P&LL, 1868, c. 385; P&LL, 1870, c. 468; P&LL, 1871, c. 454. Cf. Janesville Bridge Co. v. Stoughton, 1 Pinn. 667 (Wis. 1846) (if ambiguous, corporate powers should be construed to the favor of public right, restrictively as against the grantees). On the other hand, where a special corporate charter included authority for stream improvements at a specified location or over a specified stretch of water, it was clear that the charter was intended at once as a constitution for the corporation and a grant as against the public. See, e.g., P&LL, 1855, cc. 330 and 354; P&LL, 1856, cc. 294 and 504; P&LL, 1857, cc. 119, 235, 326. The dual character of such a special statute is reflected, for example, in the discussion in Yellow River Improvement Co. v. Arnold, 46 Wis. 214, 221, 49 N.W. 971, 972 (1879). There was a third style of special corporate charter, falling between these two types, where the charter included stream improvement among the listed corporate powers and also limited the exercise of these powers to a given town, city, or county as the corporation's place of business. See, e.g., P&LL, 1866, c. 63 (incorporating Cedarburg Manufacturing Co., authorized to "purchase and hold such water-powers in said town of Cedarburg, or in the county of Ozaukee, with the adjoining lands, as they may need for their purposes"). In at least one instance, by amending to add details as to the scope of the corporation's dam building authority, the legislature indicated that it viewed the original special charter as having conferred an improvement franchise. See P&LL, 1855, c. 50 (incorporating Appleton Manufacturing & Water Power Co., with "power . . . to construct or own dams" on its land or land of consenting owners, "provided that the business of the said corporation shall be conducted at the village of Appleton"), amended, P&LL, 1859, c. 129.

121. See notes 38–42, 53, 80, 81 and 89–91 above.

122. A classic example of the extended-area style of improvement franchise is Private & Local Acts, 1853, c. 30, incorporating the Wisconsin River Improvement Company. An early example of the limited-area type of franchise is Laws, 1851, c. 325 (Little Bull Falls, Wisconsin River).

123. The quoted provisions will be found, for example, in P&LL, 1870, c. 32, and Laws, 1874, cc. 118 and 239. Representative of the minority of dam franchises which spelled out some construction detail as to navigation safeguards are Acts, 1851, c. 173 (dam on Little Wolf River shall have "a sufficient slide, not less than thirty feet wide and so constructed as not to cause a fall of more than three feet to every twelve feet of surface, to admit of the passage of rafts and timber down said stream"), and P&LL, 1857, c. 99 (dam on Waupaca River shall have a slide "sufficient to pass a raft twenty feet wide and drawing twenty inches of water").

124. Compare Revised Statutes, 1878, sect. 1777: "Any corporation formed under this chapter for the improvement of any stream, and driving logs thereon, which shall have taken prior possession of such stream for that purpose, shall have power to improve such stream and its tributaries, by clearing and straightening the channels thereof, closing sloughs, erecting sluiceways, booms of all kinds, side, rolling and flooding dams, or otherwise, if necessary; but shall in no case, in any manner, materially obstruct or impede navigation upon such stream, or erect any dam or other obstruction below the head of steamboat navigation, or obstruct any navigable slough, except with the written consent of the owners of the entire shores on both sides thereof."

125. See Black River Flooding-Dam Association v. Ketchum, 54 Wis. 313, 317–318, 11 N.W. 551, 552 (1882).

126. Black River Improvement Co. v. LaCrosse Booming & Transportation Co., 54 Wis. 659, 669–670, 11 N.W. 443, 448 (1882). Cf. Yellow River Improvement Co. v. Wood County, 81 Wis. 554, 51 N.W. 1004 (1892) (legislation construed to afford extended-area improvement company protection against local taxation of segments of its works).

127. About 5 per cent of the lumber-area special dam franchises stated a height limitation in the form of a standard. See, e.g., P&LL, 1855, c. 330 (dam proposed on Wisconsin River authorized, "but not so high as to interfere with the navigation of said river by rafts of logs or lumber"); Laws, 1880, c. 255 (dam authorized "for hydraulic purposes," on Prairie River in Lincoln County, "of sufficient height for the purpose for which it is to be used"). Where a specific height limit was set, this was done with erratic variations in phrasing and often with no clear definition of the base line from which the maximum height should be calculated. Thus as between two franchises enacted on successive days for dams on the same river, P&LL, 1867, c. 503, stipulated that the authorized dam "shall not raise the water to exceed eleven feet above the ordinary stage of water in said river," while c. 586 was con-

tent simply to declare "that such dam shall not raise the water to exceed ten feet." Arpin v. Bowman, 83 Wis. 54, 53 N.W. 151 (1892) resolved another point which the statutory draftsmen almost always left undefined: a stated maximum height should be construed to mean the height of the major part of the total dam, and not just of the dam at the place where a slide or chute was located. The same decision indicated that a height limit was to be taken seriously, as entering into the very title of the franchise holder. Violation of the height limit was sufficient to establish that a dam was *pro tanto* a nuisance, grounding an action for damages by a complaining raftsman. Cf. Arimond v. Green Bay & Mississippi Canal Co., 31 Wis. 316 (1872) (statutory franchise no defense to flowage action, where dam raises water level in lake, contrary to franchise limitation); Lake v. Loysen, 66 Wis. 424, 29 N.W. 214 (1886) (excessive height is basis for flowage action under milldam act). With familiar reluctance to decree a forfeiture, however, the court also indicated in Arpin v. Bowman, cited above, that the dam was an illegal structure only "so far as it exceeded the height allowed by ch. 424" (83 Wis. 54, 58, 53 N.W. 151, 152 [1892]).

128. Jones v. Pettibone, 2 Wis. 308 (1853); Schneider v. Staples, 66 Wis. 167, 28 N.W. 145 (1886) (repeal of original fifteen-year time limit, without more, converts franchise into one of indefinite duration, as against third parties).

129. Wisconsin Constitution, Art. XI, sect. 1. The Wisconsin Supreme Court never confronted an issue of repeal or alteration of a stream improvement franchise contained in a special corporate charter. Summarily, without satisfactory analysis of the difficulty raised in the present text, Attorney General v. Railroad Companies, 35 Wis. 425, 578 (1874) held that a toll franchise included in a special act incorporating a railroad company was subject to the constitutional reserved power. There is nothing in the Reports to suggest that the court would not have taken this line in a matter of stream franchise. The confused identification of the franchise to be a corporation, with a franchise to conduct particular operations not in themselves necessarily related to corporate being, seems foreshadowed in State v. Milwaukee Gas Light Co., 29 Wis. 454, 462 (1872). Cf. Pratt v. Brown, 3 Wis. 603, 611 (1854). (Compare the seeming confusion in nineteenth-century doctrine between a franchise to be a corporation and a franchise to use a stream in a certain fashion in derogation of otherwise existing public rights, as an issue particularly relevant to the scope of the 1871 constitutional amendment prohibiting special acts granting corporate powers or privileges. See notes 220 and 221 below. Without detailed analysis, Governor Rusk, when he vetoed bills to give stream use franchises to individuals because the bills did not stipulate a reserved power in the legislature to amend or repeal the grants, declared that such a reserved power would exist by virtue of the constitutional reservation over any such grant to an existing corporation. See notes 66 and 67 above. That the constitutionally reserved power provided basic authority to repeal a separate dam franchise granted to a corporation created by another act was accepted without argument in both the majority and minority Reports, Judiciary Committee on bill 14S (to repeal P&LL, 1866, c. 424, authorizing the Kilbourn dam on the Wisconsin River), SJ, 328–329 (1873). Legislative power to amend or repeal was specifically reserved in special charters which also contained grants of stream-use franchises, (1) as to general river improvement projects, in Private & Local Acts, 1853, c. 30 (reservation clause repealed, P&LL, 1854, c. 242), and P&LL, 1854, c. 62; (2) as to a toll boom, in P&LL, 1856, c. 252; (3) as to private (non-toll) booms, in P&LL, 1855, c. 331; Laws, 1907, c. 98; and Laws, 1921, c. 124; (4) as to a private, nonforest-area dam, in P&LL, 1857, c. 238. Reservation clauses were included in separate stream-use franchise acts making grants to corporations already existing under other legislation, (1) as to general river improvement projects, in Laws, 1889, c. 77, 1895, c. 341, 1907, c. 335, 1911, c. 640, 1913, c. 649, and 1919, c. 441; (2) as to private (non-toll) booms, in P&LL, 1855, c. 331; Laws, 1907, c. 98; and Laws, 1921, c. 124; (3) as to private, forest-area dams, in Laws, 1901, c. 366, 1903, cc. 24, 178, and 180, 1905, cc. 39 and 408, 1907, cc. 329, 405, 626, and 644.

130. The Wisconsin court early accepted as a proposition beyond question since the Dartmouth College Case "that a charter of incorporation, conferring certain franchises upon a company or individual, was in the nature of a grant" protected by the contract clause; see Pratt v. Brown, 3 Wis. 603, 611 (1854). Though the court's phrasing offers another example of nineteenth-century fuzziness in speaking of all franchises as acts of "incorporation," the remark was made in relation to the status of an individual's claim under the milldam act. See Ruehl v. Voight, 28 Wis. 153, 157 (1871) (accepted, *arguendo,* that a statu-

tory dam franchise with flowage rights creates a "contract" between dam owner and flowed landowner, but held — without explicit reference to the contract clause — that the state may by subsequent legislation regulate the remedies available to enforce the contract, as by imposing a statute of limitations). There appears to have been no other nineteenth-century occasion for the court to speak to the contract clause protection of franchises to individuals, but nothing in the Reports casts any doubt on adherence to the doctrine announced in Pratt v. Brown. Cf. Ryan, C. J., for the court, in Sellers v. Union Lumbering Co., 39 Wis. 525, 527 (1876) ("A franchise is property; an incorporeal hereditament . . . ; a certain privilege conferred by grant from government and vested in an individual"). Finally, The Water Power Cases, 148 Wis. 124, 143, 134 N.W. 330, 337–338 (1912) ruled unconstitutional a statute which purported to repeal dam franchises "granted to natural persons and complied with by the latter, reserving no right of repeal"; the attempted repeal was ineffective, because "these, by all the authorities, constitute a contract not subject to repeal at the pleasure of the state." Cf. Marshall, *Autobiography,* I, 336–338. Governor Rusk's insistence that dam franchises given to individuals should contain reserved power clauses seems to have rested on his assumption that such a franchise enjoyed the protection of the contract clause. See notes 64 and 66 above. There were instances of total or partial repeal of stream-use franchises granted to individuals, where the original included no reservation of legislative power to amend or repeal. See, e.g., P&LL, 1862, c. 98, repealing toll provisions of P&LL, 1857, c. 360; P&LL, 1863, c. 345, repealing P&LL, 1856, c. 469; Laws, 1876, c. 396, repealing P&LL, 1869, c. 156. This study found twelve such instances. Some on their face indicate amicable arrangements. See, e.g., Laws, 1882, c. 260, 1889, c. 235. Some probably reflect only inadvertence, or confused belief that the constitutional reservation applied. See notes 129 above, and 220 and 221 below.

131. See notes 66 and 67 above. After Rusk's intervention, from 1883 on, a reserved power to amend or repeal was omitted in only about 9 per cent of forest-area private (nontoll) dam franchises (16 of 182), about 17 per cent of all toll and nontoll boom franchises (3 of 17), and in 4 per cent (3 of 72) of the toll river improvement franchises granted to individuals. Of seven improvement franchises granted after 1882 to corporations already existing by virtue of other legislation, six contained a specific reservation of power to amend or repeal. It was not uncommon that, as introduced, a bill to confer a dam franchise upon an individual lacked a reservation of legislative right to amend or repeal, and that a reservation clause was added in committee or during floor consideration. One might interpret such incidents as efforts by the proponents to slip through a measure containing the special favor of omitting the reservation. On the whole record, however, this does not seem a plausible reading. As we have just observed, after Rusk's intervention relatively few individual franchises went through without a reservation clause. Whenever omission of such a reservation was raised in course of passage, the reservation was invariably inserted with complete want of recorded controversy and in summary response to the suggestion. The reaction indicates that prevailing sentiment was overwhelming to insert the reserved power. For typical incidents, see SJ, 608 and 634 (1887), as well as AJ, 985 (1887) (bill 19S, which became Laws, 1887, c. 386); AJ, 201 (1889) (bill 152A, which became Laws, 1889, c. 77), AJ, 315 (bill 326A, which became Laws, 1889, c. 252), as well as AJ, 688 and 795 (bills 511A and 512A, which became Laws, 1889, cc. 445 and 485); SJ, 486 (1895) (bill 271A, which became Laws, 1895, c. 114), SJ, 686 and 708 (bill 216A, which became Laws, 1895, c. 251), as well as SJ, 700 (bill 640A, which became Laws, 1895, c. 272).

132. Among thirty-eight river improvement franchises containing a stated term-of-years limitation, the terms set were: 25 years — 1; 20 years — 4; 15 years — 24; 10 years — 8; 5 years — 1. In fifteen toll boom franchises the terms set were: 30 years — 1; 25 years — 1; 20 years — 6; 15 years — 2; 12 years — 2; 10 years — 3. Two private boom franchises set terms of 20 and 10 years respectively. Ten private dam franchises in the lumber area had terms of: 30 years — 1; 25 years — 1; 20 years — 2; 15 years — 6.

133. Emphasis upon defining and fixing legislative responsibility for the scope of franchise powers conferred is especially marked in Governor Farwell's Veto Message of June 6, 1853, SJ, 815–817, objecting to the broad monopolies given the Chippewa River Improvement & Navigation Co.; and in the objections to committing the state to irrevocable franchise grants which will deny opportunity to adjust to a changing future, in Governor Rusk's General Message of Jan. 11, 1883, AJ, 24. Emphasis upon fixing the responsibility of

legislative decisions is of course implicit in Sellers v. Union Lumbering Co., 39 Wis. 525 (1876), holding an improvement franchise invalid for want of sufficiently defined grantees.

134. The function of the franchise in providing a framework within which third parties' claims must be adjusted to the privileges of the franchise holder is highlighted in cases which turn mainly on questions of defining the area of operations within which the franchise holder shall enjoy his special status. See Stevens Point Boom Co. v. Reilly, 46 Wis. 237, 49 N.W. 978 (1879); Black River Improvement Co. v. LaCrosse Booming & Transportation Co., 54 Wis. 659, 11 N.W. 443 (1882); Hackstack v. Keshena Improvement Co., 66 Wis. 439, 29 N.W. 240 (1886).

135. Over the years a substantial number of amendments were made to special franchises, dealing mainly with definitions of grantees and locations. The fact suggests some substantial measure of attention by franchise holders to the secure definition of the scope of their rights. That the largest single class of amendments concerned definition of location seems particularly to imply this concern. Of 119 amendments to general improvement franchises, 31 (26 per cent) concerned location; location was a prime subject in 9 (29 per cent) of 31 amendments to boom franchises and in 31 (50 per cent) of 62 amendments to lumber-area private dam franchises. Total repeals also, of course, bore directly on the definition of outstanding titles; there were 20 repeals of general improvement franchises, 5 of boom franchises, and 11 of lumber-area private dam franchises.

136. That a franchise was significant for conferring the right to occupy and use waters, free of recurring common law damages suits or proceedings to abate the works as a nuisance, is the plain meaning of the decisions which stress the presence or absence of a statutory license. See, e.g., Stevens Point Boom Co. v. Reilly, 46 Wis. 237, 49 N.W. 978 (1879); Cohn v. Wausau Boom Co., 47 Wis. 314, 2 N.W. 546 (1879). Long before lumber industry cases presented themselves, this assured right to occupy and use a given stream location free of ouster or continuing damages claims had emerged as the substantial benefit given by the milldam act. See Stephens v. Marshall, 3 Pinn. 203 (Wis. 1851); Newton v. Allis, 12 Wis. 378 (1860).

137. See the recognition of the care that franchise holders had for the regularity of succession to the franchise, reflected in Schneider v. Staples, 66 Wis. 167, 170, 28 N.W. 145, 146 (1886). For examples in legislative practice, see Laws, 1882, cc. 78 and 95, amending Laws, 1879, cc. 136 and 137; Laws, 1889, c. 179, amending Laws, 1879, c. 127; Laws, 1889, c. 32, amending Private & Local Acts, 1853, c. 152.

138. This point seems first to have been ruled explicitly only after the lumber era, though then at suit of a lumber transporter, in Flambeau River Lumber Co. v. Chippewa & Flambeau Improvement Co., 204 Wis. 602, 608, 236 N.W. 679, 682 (1931). However, the ruling seems implicit in earlier cases where without question the court entertained suits grounded on alleged violations by the grantee of terms of his franchise. See, e.g., Volk v. Eldred, 23 Wis. 410 (1868); In re Eldred, 46 Wis. 530, 1 N.W. 175 (1879); Arpin v. Bowman, 83 Wis. 54, 53 N.W. 151 (1892). Cf. Cohn v. Wausau Boom Co., 47 Wis. 314, 2 N.W. 546 (1879) (challenge to validity of statutory franchise).

139. See notes 49, 57, 59, 61, and 62 above.

140. Compare Cohn v. Wausau Boom Co., 47 Wis. 314, 326, 2 N.W. 546, 549 (1879) with A. C. Conn Co. v. Little Suamico Lumber Manufacturing Co., 55 Wis. 580, 13 N.W. 464 (1882). See also Black River Improvement Co. v. LaCrosse Booming & Transportation Co., 54 Wis. 659, 686, 11 N.W. 443, 455 (1882). There was some indication that a private suitor might have an order requiring that defendant's works be modified in limited measure to remove a specific violation of franchise terms, but it is not at all clear how far the court was willing to press even cautiously limited affirmative orders sought by a private suitor. See Heerman v. Beef Slough Manufacturing, Booming, Log-Driving & Transportation Co., 1 Fed. 145, 158, 159 (Circ. Ct. W.D. Wis. 1878).

141. Yellow River Improvement Co. v. Wood County, 81 Wis. 554, 51 N.W. 1004 (1892).

142. See notes 18, 49, 56, and 57 above.

143. See notes 19, 53, and 54 above.

144. In addition to the many franchises which incorporated by reference one of the condemnation procedures set up in a general statute, probably one of these procedures would be held to apply in the scattering of franchises which provided merely that compensation should be made for damages "when such damages shall be determined by due process of law," or which provided merely for "damages," or which said that the authorized dam "shall be subject to all the provisions of the laws of Wisconsin, relating to the flowing of

lands by the erection of dams." See, e.g., P&LL, 1872, c. 100, amending P&LL, 1868, c. 376 (substituting reference to milldam act procedure for original "due process of law" language); Laws, 1878, cc. 163 and 271, 1887, c. 177, 1899, c. 261. The court early and consistently recognized as effective the incorporation by reference in a special franchise of a condemnation procedure created by a general act. See Wood v. Hustis, 17 Wis. 416 (1863); Crosby v. Smith, 19 Wis. 449 (1865); Geise v. Greene, 49 Wis. 334, 5 N.W. 869 (1880); Charnley v. Shawano Water Power & River Improvement Co., 109 Wis. 563, 85 N.W. 507 (1901). Repeal of the general act had no effect upon special franchises which incorporated its condemnation procedure: Wood v. Hustis, cited above. On the other hand, incorporation by reference of a general act procedure was construed as intended to incorporate the procedure as it might be thereafter altered by amendments to the general act: Geise v. Greene, cited above. The legislature might at any time repeal the incorporation by reference of a general statutory procedure in a special franchise and substitute another procedure, which would then govern any proceedings thereafter brought: Pick v. Rubicon Hydraulic Co., 27 Wis. 433 (1871). So, too, the legislature might incorporate a general statutory procedure by amendment to a special franchise which previously contained no eminent domain grant, and thereafter an aggrieved landowner must seek remedy under this procedure rather than at common law: Durning v. Burkhardt, 34 Wis. 585 (1874). But a general statutory condemnation procedure incorporated by reference in a special franchise governed only situations within the scope of the franchise; suit for flowage caused beyond the limits defined in the special franchise thus lay at common law: Hackstack v. Keshena Improvement Co., 66 Wis. 439, 29 N.W. 240 (1886).

The milldam act proceeding was in circuit court with a jury; the river improvement corporation statute proceeding was in circuit court with the help of three court-appointed commissioners. The milldam act proceeding was at the initiative of the flowed landowner; under the river improvement act procedure, the initiative might be taken by the builder of the works. Cf. Revised Statutes, 1878, c. 146, and Laws, 1882, c. 318 (amending Revised Statutes, 1878, sect. 1777). There is no clear evidence in the statutes, the legislative records, or the reported decisions of the Wisconsin Supreme Court as to the comparative advantages franchise applicants may have seen in one over the other procedure. (See the discussion, later in this chapter, of the practical operation of franchise policy.)

145. For examples of grants of eminent domain authority regarding (1) private power dams' flowage, see P&LL, 1858, c. 254 (Menomonee River, Oconto County); P&LL, 1867, c. 568) (Chippewa River, Chippewa County); (2) flowage by flooding dams with toll rights, see Laws, 1882, c. 182 (Totogatic River, Bayfield County); Laws, 1893, c. 221 (Clam River, Burnett County); (3) pier or shore attachment sites for booms with toll rights, see P&LL, 1857, c. 101 (Wolf River Boom Co.); P&LL, 1871, c. 483 (White River Dam, Log Driving & Boom Co.); (4) pier or shore attachment sites for private booms, see P&LL, 1859, c. 134 (Oconto River, Oconto County); P&LL, 1867, c. 378 (Chippewa River, Eau Claire County); (5) rights of way for canals, see Acts, 1851, c. 325 (Wisconsin River at Little Bull Falls); Private & Local Acts, 1854, c. 123 (Wolf River to Lake Poygan, Winnebago County); (6) general improvement activities of extended-area river improvement companies, see Private & Local Acts, 1853, c. 30 (Wisconsin River Improvement Co.); P&LL, 1870, c. 38 (Pine River Navigation Co.). The range of objects for which eminent domain power was thus given suggests that legislative opinion saw no general limiting principle in this respect. The great bulk of grants were simply to acquire flowage rights, but there is nothing in the legislative records to indicate that the legislature was less willing to grant the power to help build booms or improve streams; in the absence of contrary evidence, the fair inference is that there were not more such grants related to booms, and other incidents of stream improvement than flowage, simply because grantees did not see the need to ask for more. On the other hand, grants of private dam franchises in the forest region perhaps showed sensitivity to the question whether eminent domain should be given to acquire a dam site, as distinguished from flowage rights. If the franchise were silent, there was no grant of eminent domain authority for any purpose; hence it is the more notable that, in more than half of the forest-area dam franchises in which there was no grant of eminent domain, the franchise in effect particularly negated such power with reference to the dam site by specifying that the dam be built on land owned by the grantees. See, e.g., P&LL, 1861, c. 52; Laws, 1874, c. 231. Comparably, about 40 per cent

of private boom franchises, not content with silence, specified that the booms should front only on the grantees' land, or that the shore works necessary to the booms should be on land owned by the grantees. See, e.g., P&LL, 1855, c. 273, 1861, c. 37. There was no comparable extent of such specifications in franchises for public-utility-type stream improvement or boom franchises, though a few of these specified that particular works must be on the grantees' own land. See, e.g., P&LL, 1871, c. 494 (stream improvement with toll right); P&LL, 1857, c. 70 (toll boom). Some franchises authorizing dams usable for private industrial power explicitly granted eminent domain power to acquire dam sites as well as flowage rights. See, e.g., P&LL, 1867, c. 408; Laws, 1891, c. 177. The eminent domain power was also defined to include dam sites in a number of public-utility-type (toll) franchises. See, e.g., Laws, 1889, c. 215, 1891, c. 313. Apparently the legislature held no general principle against granting eminent domain authority for this particular use. Probably most private dam or boom applicants founded their enterprise upon land they already owned, and the franchise terms merely reflected the fact; perhaps some of the instances where the grant specified that the basic works be on the grantees' own land reflected a judgment of equity that such a requirement was fair for a private project, to show earnest of intention. It must also be noted that many franchises were ambiguous on the point. Thus a number of franchises gave the eminent domain power "in order to build and maintain said dam and use the same"; this formula may well have been broad enough to embrace the acquisition of a dam site. See, e.g., Laws, 1893, cc. 96 and 138, 1895, c. 172. There is no decision of the Wisconsin Supreme Court which deals with this ambiguity. If we count such arguable instances, we could say that a substantial number of eminent domain grants covered dam sites.

146. The distinction between authorizing a right against the public and authorizing a compensated taking of private property is shown, for example, by Laws, 1880, c. 225, which empowered the Black River Improvement Co. to close the west channel of the river to enhance main channel flow, amending Private & Local Acts, 1864, c. 84, which had already conferred eminent domain authority upon the company. Cf. P&LL, 1866, c. 352 (Keshena Improvement Co. "shall have the same right to use the banks of said river in making said improvement, as the public have to use the banks of rivers in running saw-logs down the same, and no other"); P&LL, 1871, c. 357 (somewhat similar authority given Pine River Improvement Co.), and c. 467 (East Shioc Improvement Co. empowered in substantially same terms as in the 1866 franchise). See also P&LL, 1871, c. 483 (flooding dams authorized to White River Dam, Log Driving and Boom Co. at its discretion over a stretch of river without a compensation procedure, though an eminent domain grant is made with reference to damages to abutting lands from piers or shore attachments of booms).

147. Cohn v. Wausau Boom Co., 47 Wis. 314, 2 N.W. 546 (1879); Black River Improvement Co. v. LaCrosse Booming & Transportation Co., 54 Wis. 659, 11 N.W. 443 (1882); Falls Manufacturing Co. v. Oconto River Improvement Co., 87 Wis. 134, 58 N.W. 257 (1894).

148. The following table shows the absolute and relative number of eminent domain grants

GRANTS OF EMINENT DOMAIN POWER
IN STREAM-USE SPECIAL FRANCHISES

	Toll improvements		Toll booms		Private booms		Private dams (forest area)		Private dams (outside forest)	
Before 1848	1/3	33%	0/0	0%	0/0	0%	11/12	91%	30/43	69%
1848–1849	0/0	0%	0/1	0%	0/0	0%	0/0	0%	3/5	60%
1850–1859	10/15	66%	1/14	7%	0/16	0%	17/45	37%	23/54	42%
1860–1869	23/33	69%	5/16	45%	1/17	5%	4/19	20%	7/10	70%
1870–1879	11/66	17%	1/6	16%	1/8	12%	10/32	31%	9/11	81%
1880–1889	25/85	29%	2/6	33%	0/11	0%	40/91	44%	1/2	50%
1890–1899	14/25	56%	1/2	50%	0/2	0%	29/49	59%	1/2	50%
1900–1909	4/6	66%	0/1	0%	0/2	0%	63/82	76%	1/2	50%
1910–1919	3/3	100%	0/0	0%	0/1	0%	2/2	100%	0/0	0%
	91/236	38%	10/46	21%	2/57	3%	176/332	53%	75/129	58%

among types of special franchises. The first figure in each column shows the number of special franchises containing a grant of eminent domain powers; the second figure, the whole number of that type of franchise; the third figure, the percentage of such franchises containing a delegation of eminent domain powers.

149. See note 95 above. Without discussion, McDonald v. Apple River Power Co., 164 Wis. 450, 160 N.W. 156 (1916) held that the milldam act procedure applied of its own force to govern proceedings for flowage caused by a dam built under a special statutory franchise containing no eminent domain clause, on a stream that was navigable only by the limited (spring-freshet-time) log-transport test. Cf. Allaby v. Mauston Electric Service Co., 135 Wis. 345, 116 N.W. 4 (1908) (milldam act procedure supersedes equitable remedies for abatement of dam on such a stream, so long as no term of special franchise is clearly contrary). The McDonald ruling was consistent with the purpose of the milldam act to promote water-power development. And the decision was in substance consistent with the rule that the eminent domain power existed only where a statute gave it. However, the McDonald case came at a time when the dominant concern was with electric power development and when log transport was no longer an important use of streams. Though the point was never squarely raised, the nineteenth-century cases treated the common law and not the milldam act as controlling flowage problems from dams on streams used for log transport, where the legislature had made no specific grant of eminent domain powers that applied to the situation. See Durning v. Burkhardt, 34 Wis. 585 (1874); Folsom v. Apple River Log Driving Co., 41 Wis. 602 (1877); Hackstack v. Keshena Improvement Co., 66 Wis. 439, 29 N.W. 240 (1886). Cf. Wood v. Hustis, 17 Wis. 416 (1863) (milldam act does not apply on any stream which legislature has declared navigable); Waller v. McConnell, 19 Wis. 417 (1865) (action for flowage under milldam act must fail, where plaintiff doesn't aver or prove that the stream is nonnavigable, and where the court may not take judicial notice of the fact). The matter was, of course, quite different where the plaintiff complained, not of the occupying of his land incident to the regular operation of the dam, but of damage caused only by improper construction or management of the dam; the latter situation presented no problem of a "taking" of land, and the action clearly lay at common law, for negligence. Rich v. Keshena Improvement Co., 56 Wis. 287, 14 N.W. 191 (1881).

There was less emphasis upon requiring a clearly explicit grant of authority, where the question was of taking state rather than private land. Where the legislature granted a river improvement franchise in an area largely made up of state lands, the court construed the grant to imply the legislature's consent that the grantees occupy such state lands as the authorized improvements might require, without provision for making compensation therefor. Black River Improvement Co. v. LaCrosse Booming & Transportation Co., 54 Wis. 659, 675–677, 11 N.W. 443, 450–451 (1882). The court gave no indication that it saw possible conflict of this ruling with Wisconsin Constitution, VIII, 10, which forbids the state to participate in works of internal improvement. Since the state lands involved were probably within the federal government's "swamplands" grant to the state, this subsidy in kind perhaps might be ruled to fall under the exception in Art. VIII, sect. 10, allowing the state to use, in aid of an internal improvement, lands specifically granted to the state for that purpose; navigation improvement might be deemed within the scope of the land reclamation purposes of the federal swamplands gift. Where an eminent domain procedure was included in a statutory franchise for an improvement project which would involve state lands, the Attorney General ruled that the legislature had not intended to make a gift of such state lands as the project required. See Opinion of the Attorney General, July 21, 1897, Attorney General's Letter Books, V, 329–330.

150. See, for example, Laws, 1880, c. 225, and 1882, c. 263, amending Private & Local Acts, c. 84, to define an added area in which the Black River Improvement Co. may use its eminent domain right, and to clarify that the power extends to taking lands on river shores necessary to improve navigation. Cf. P&LL, 1872, c. 100, amending P&LL, 1868, c. 376 (redefining procedure applicable, regarding Apple River toll dam); Laws, 1887, c. 253, amending Laws, 1882, c. 247, and Laws, 1895, c. 82, amending Laws, 1893, c. 210 (adding eminent domain power to acquire flowage rights, incident to increase in authorized height of dams); Laws, 1889, c. 270, amending Laws, 1887, c. 512 (adding eminent domain grant, where none made originally).

151. The fresh sensitivity felt in twentieth-century policy toward the delegation of this sovereign power was first reflected in Governor LaFollette's Special Message of Apr. 12, 1905,

AJ, 1050–1054. The use of water power to generate electricity was the new factor which prompted a reappraisal of policy. In this Message Governor LaFollette somewhat surprisingly made no criticism of the liberal grants of eminent domain authority in the lumber era. But he urged, "It is . . . quite apparent that these waterpowers are no longer to be regarded simply as of local importance" (AJ, 1052). The implication was that, because of the limits of nineteenth-century technology and industrial organization, past liberality in delegating eminent domain authority had not posed a policy problem of wide significance. The critical attitudes taken toward past stream policy focused upon the extent of vested private rights given against the public, rather than upon the delegation of eminent domain authority as such. See, e.g., Governor LaFollette's Veto Message on bill 158S, May 18, 1903, SJ, 1173–1174, and Governor Davidson's Special Message of Feb. 18, 1909, AJ, 257–259. The general twentieth-century policy of conferring eminent domain power only upon public agencies and public utilities came to well-defined expression in the revision and consolidation of the statute law on the matter by Laws, 1919, c. 571. It was symbolic of the new temper that Laws, 1917, c. 474, sect. 37, had repealed the eminent domain provisions of the milldam law.

152. See notes 56 and 57 above.

153. Revised Statutes, 1878, sect. 1777; see note 19 above.

154. See the table of special franchises in the first section of this chapter.

155. See notes 92 and 96 above. The strictness of the court's insistence that a toll right rest on express grant of the legislature appears particularly in the fact that the court ruled so, even as to an artificial channel made by the claimant: Weatherby v. Meiklejohn, 56 Wis. 73, 13 N.W. 697 (1882). However, where a toll right was given, the court indicated that a liberal interpretation might properly be given to its scope; so toll might be claimed for aid given to moving logs on a part as well as over the whole of the improved stream. See Sauntry v. Laird, Norton & Co., 100 Wis. 146, 151, 75 N.W. 985, 986 (1898); cf. Hackstack v. Keshena Improvement Co., 66 Wis. 439, 29 N.W. 240 (1886).

156. See P&LL, 1867, c. 470, amending Private & Local Acts, 1864, c. 157. Cf. P&LL, 1866, c. 447, amending Private & Local Acts, 1864, c. 84 (reducing list of commodities served); Laws, 1882, c. 89, amending Laws, 1881, c. 161 (adding a category of service for which toll may be charged).

157. See, e.g., Private & Local Acts, 1853, c. 24; P&LL, 1857, cc. 101 and 170; 1859, c. 200.

158. The most used formula authorized the tolls "when the said [grantees] . . . shall have completed said dam or series of dams aforesaid" (see Laws, 1874, c. 176). Occasionally the *quid pro quo* was made more specific, and perhaps more exacting, as when Laws, 1874, c. 228, declared that the grantees "shall not be entitled to collect or receive any toll whatever, unless the expense of driving said logs or timber shall be reduced by the use of said dams." Cf. Acts. 1850, c. 257 (tolls graduated according to "each and every foot of elevation in any dam made by them . . . and over which such lumber or timber shall be run"); Laws, 1878, c. 291 ("as a compensation for keeping up and maintaining such dam, the sum of seventy-five dollars for each twenty-four hours that any person or company on request shall have the use of the waters from said dam for driving purposes"). Stipulations tying toll rights to stated minimum capital investment in improvements may be seen in Private & Local Acts, 1853, c. 30; P&LL, 1857, c. 170, as amended, P&LL, 1860, c. 34; P&LL, 1866, c. 447, amending Private & Local Acts, 1864, c. 84.

159. In addition to authorizing "reasonable and uniform tolls," Revised Statutes, 1878, sect. 1777 (note 153 above) empowered any river improvement corporation qualified thereunder to take possession of logs put into the stream by persons who failed to make adequate provision for driving them, so hindering the main drive, and authorized the improvement corporation to "charge and collect therefor the actual costs and expenses of so driving the same." A few special franchises authorized what amounted to cooperative improvement or driving enterprises, with provision hence for shared expenses rather than tolls. See, e.g., Private & Local Acts, 1854, c. 62 (Black River Log Driving & Booming Co.); Private & Local Acts, 1864, c. 126 (Little Wolf River Log-Driving Co.), and c. 131 (Log Driving Association of the Yellow River). The distinction was highlighted when Private & Local Acts, 1864, c. 157, amended P&LL, 1862, c. 35, to change the Eau Claire River Log Driving Co. from an open-ended association, which anyone interested in logs on that river might join by filing his acceptance and sharing expenses, to a joint stock corporation with a schedule of authorized tolls. Cf. P&LL, 1868, c. 304, amending P&LL, 1856, c. 294 (conferring toll right for boom service, where original fran-

chise merely allowed boom owner to cause unclaimed logs to be sawed, deduct the usual sawing rates, and hold the balance of sale proceeds for the owner).

160. See Reynolds, 75, 130. The plaintiff's brief (p. 54) in Sauntry v. Laird, Norton & Co., 100 Wis. 146, 75 N.W. 985 (1898) comments on the search for a financially responsible party in connection with enforcing tolls.

161. Incident to recognizing a common law lien of a millman for sawing logs into lumber — a situation clearly within established common law doctrine on specific liens — the court in Arians v. Brickley, 65 Wis. 26, 29, 26 N.W. 188, 189 (1885) announced in very broad terms the principle that a bailee whose work or service enhanced the value of goods held by him should have a lien for his charges. In particular, the court cited with approval Farrington v. Meek, 30 Mo. 578 (1860), which had admittedly broken new ground in approving a lien for money due men who had navigated a raft of lumber. The common law had generally refused a lien for services which merely added space or time utility to the value of a commodity. See Ray A. Brown, *The Law of Personal Property* (1955), 516–525. The Wisconsin court never had occasion to spell out how far it would press the common law lien on behalf of men involved in log or lumber transport.

162. The lien conferred by Revised Statutes, 1878, sect. 1777, on river improvement corporations qualifying thereunder, was stated in language substantially that of many special franchises: "And such corporation shall, for all such tolls, costs and expenses have a lien on the logs for which the same were incurred, and may seize, in whosoever possession found, and hold a sufficient amount thereof to pay the same, and make sale thereof, upon giving ten days' notice, in the manner provided for notifying sales on execution upon the judgment of justices of the peace, or may enforce such liens as other log liens are enforced, according to these statutes." As here indicated, the draftsman often shortened his lien provision by incorporating by reference the procedures provided for enforcing liens given for labor or supplies in connection with logging operations. Typical lien provisions in special franchises, incorporating by reference the general log labor lien procedures, may be seen in P&LL, 1868, cc. 265 and 469; Laws, 1874, cc. 204, 228, 250, 262, 288, 289, 297, and 337. Some franchises made merely the summary declaration that the authorized "toll shall be and remain a lien upon such lumber, timber and logs until the same shall be paid, and the said [grantee] . . . shall be entitled to the possession of said logs, lumber and timber at any and all times after they have become subject to such toll until the same shall be paid." See, e.g., P&LL, 1868, cc. 376 and 430. Some were even more to the point: "[the grantees] . . . shall have a lien upon all logs, timber and lumber run over said dam above specified until the charges aforesaid shall be paid . . ." See, e.g., Laws, 1874, cc. 153, 154, and 176. Occasionally a special franchise spelled out in detail the terms and procedures for the lien it created, as in P&LL, 1857, c. 101, and 1868, c. 430. The practical importance attached to statutory lien terms was shown also by amendments enlarging or contracting the scope of existing lien clauses. See, e.g., Laws, 1876, cc. 247 and 263, amending Laws, 1874, cc. 153 and 154; Laws, 1877, c. 247, amending Laws, 1873, c. 12; Laws, 1880, c. 317, and 1881, c. 295, amending P&LL, 1870, c. 24.

163. See Arians v. Brickley, 65 Wis. 26, 29, 26 N.W. 188, 189 (1885); Wilson v. Rudd, 70 Wis. 98, 104, 35 N.W. 321, 323 (1887).

164. For provisions apparently authorizing assertion of a lien against logs or lumber which had passed out of possession of the lien claimant, or had not been in his possession, see P&LL, 1865, c. 190, 1868, cc. 258 and 376. Authority to assert a lien against one lot of logs or lumber for charges due for service rendered regarding another lot of such commodities belonging to the same owner may be seen, for example, in P&LL, 1866, c. 550, and 1871, c. 483; Laws, 1874, c. 262, and 1875, c. 48. In this matter Laws, 1880, c. 319, used a formula which became popular, giving a lien "on all other logs . . . of the same mark, or of different mark, belonging to the same owner which may be in [grantee's] . . . possession." Cf. Laws, 1882, c. 224, 1885, c. 75, 1889, c. 49. Contrast Laws, 1874, c. 250 (lien for charges limited to "such logs or timber upon which the same so accrued"); Laws, 1877, c. 247 (grantees "having so driven logs and timber down said river, may continue in possession of all such logs and timber . . . until all charges for tolls and driving the same have been liquidated"). More or less ambiguous language, on the other hand, may be seen in such examples as Private & Local Acts, 1863, c. 35 ("shall have a lien upon all such lumber and timber for the said tolls, and shall have the right to take possession of any such lumber and timber, and retain the same until such tolls shall be paid"); P&LL, 1869, c. 452 (grantees "shall be entitled to the possession

of said logs and lumber at any and all times after the same shall have passed through either of said dams"); Laws, 1878, c. 206 ("And the same shall be and remain a lien upon all such logs, and also of all logs of the like mark, cut the preceding logging seasons wherever the same may be, until the amounts for driving due upon said logs are fully paid and satisfied; and if the said amount is not fully paid within ten days from and after a majority of the said logs shall have passed into the said St. Croix boom, then the said company may assume possession of said logs or any of the same wherever the same may be, and may demand and recover the same from any boom or other place, and may proceed to sell the same at public auction"); Laws, 1878, c. 163 (charges "shall be a lien" on logs put in improved stream; "unless paid within ten days after said logs . . . are driven down said creek . . . the said charges shall become a lien on such logs . . . and may be enforced by [grantee] . . . in the same manner as other liens for services and labor on logs, etc. are enforced, and for this purpose may hold and detain such logs . . . until such toll and charges are paid"); Laws, 1885, c. 372 (without further specification, declares, "For all charges of tolls herein provided which shall remain due and unpaid, [grantees] . . . shall have a lien which may be enforced the same as liens for labor may be enforced under the general statute").

The lien was the primary device relied on to bulwark the position of the toll claimant. However, franchises showed a scattering of other approaches to favor his chances of recovery. See, notably, P&LL, 1859, c. 149 (larceny, to remove logs from boom before paying charges due for boomage); P&LL, 1862, c. 98 (user of flood dam water must pay before passing); P&LL, 1867, c. 470 (liability for tolls enlarged to run not only against owner or his agent, but also against anyone who "may . . . control" the logs).

165. See notes 25, 26, 31, 33, and 34 above.

166. See note 35 above. Compare Laws, 1901, c. 413 (repealed, incident to the general revision of waterways legislation, by Laws, 1917, c. 474), amending Wisconsin Statutes, 1898, sect. 1598, to declare that "the navigation of [any navigable] . . . streams or tributaries with floating logs or timber or the use of temporary booms necessarily used in navigating said streams or tributaries with floating logs or timber shall not be deemed an obstruction" within the penalties of that section. Clearly this enactment merely declared long-standing industry practice and common law doctrine. See Enos v. Hamilton, 27 Wis. 256, 258–259 (1870); Olson v. Merrill, 42 Wis. 203, 214 (1877); A. C. Conn Co. v. Little Suamico Lumber Manufacturing Co., 74 Wis. 652, 656–657, 663, 666, 43 N.W. 660, 661, 663 (1889).

167. In characteristically pragmatic fashion, Wisconsin law divided treatment of traffic management monopolies between common law and legislation according to whether the "monopoly" was episodic and temporary or an incident of a recurrent and indefinitely continuing or long-continued program of operations. One who put a drive of logs or a raft into a stream, or who built a temporary flooding dam or a temporary boom as a means of conducting a particular drive at a particular time, perhaps in theory should have come to the legislature for a license to do so; within the physical limits and the operating demands of his driving or rafting activity, he necessarily monopolized control of some navigable water for the time being. In practice, neither the industry nor the Wisconsin legislature ever indicated serious concern that statutory license was needed for such episodic and temporary activity. See note 166 above. If such activity came into collision with competing public or private interests in navigation, the difficulty fell to common law litigation for adjustment. See notes 33, 34, and 35 above. The Wisconsin court never put into explicit formulation this distinction between navigation-connected activity subject simply to common law standards and navigation-connected behavior which must have the sanction of a statutory license. Compare, however, A. C. Conn Co. v. Little Suamico Lumber Manufacturing Co., 74 Wis. 652, 43 N.W. 660 (1889) (statutory license not required to build and maintain power dam on stream navigable only by logs at freshet time).

168. See notes 36–55 above. The preferred position of the navigation interest is also discussed later in this chapter, in appraising the working effects of declared public policy.

169. For typical slide or chute requirements in forest-area dam franchises, see P&LL, 1858, c. 254 (grantees "shall also construct and keep in good repair a sufficient slide or chute not less than twenty feet wide in the clear, so as to allow the free passage of logs and lumber over such dam"); P&LL, 1868, c. 216 (grantee "shall cause a slide to be put into said dam so as to enable the driving of logs through the

same, at any stage of water on which logs could be run to the mouth of the river below said dam"); Laws, 1878, c. 283 (grantees "shall build suitable slides in said dam for running logs and timber over the same, and shall keep the same in repair. The same shall be kept open at all times when the said creek is at a driving stage and there are logs and timber to run over said dam"); Laws, 1889, c. 53 ("said dam or dams shall be so constructed as not to obstruct the running of lumber, timber or logs down said river, and . . . a slide shall be constructed in said dam or dams at least two feet below the general height thereof, and not less than forty feet in width, and . . . constructed in accordance with the laws of this state"). As these examples show, there was considerable variation in phrasing, and in detail, among these slide or chute clauses, but substantial similarity in core requirement. For lumber-area private dam franchises which lack or appear to lack any such navigation guaranty, see, e.g., P&LL, 1864, c. 325 (stipulation "that said dam shall not interfere with the rights of any person heretofore acquired" seems probably intended as equivalent of the milldam act's stipulation protecting a previously erected dam); P&LL, 1871, c. 239 (no provision capable of interpretation as a navigation guaranty); Laws, 1879, c. 155 (stipulation like that in first-cited franchise); Laws, 1887, c. 177 (no provision capable of interpretation as a navigation guaranty).

170. The differing relative emphasis on navigation compared with power use, in the forested north and the nonforested south, appears most sharply in the decades of maximum interest in stream franchises. In the north this interest focused particularly on the 1870–1900 period, when forest production surged toward its peak and the railroad network was not yet complete; in these decades, and especially in the 1880's, there is emphasis upon slide requirements. In the south most dam franchises omitted a slide requirement in the years in which the bulk of franchises were given — the time of mill-town growth.

171. See, e.g., Acts, 1852, c. 323 (boom "shall be so arranged" as to permit passage of boats at all times, "and at times of running lumber a sufficient space shall be kept open in some convenient place, for the passage of rafts, and if any raft of lumber shall by neglect or accident be carried in said boom below the space so left open, said boom may be opened so as to take out said raft, if it can be done without injury to the company, otherwise it shall be taken out at the space kept open for the passage of rafts"; also, "Any person . . . wishing to run logs or timber in a 'drive' past said boom, shall give the said corporation notice of such desire, and the said corporation upon receiving such notice shall with as little delay as possible, proceed to arrange their boom in such a manner as to permit the passage of rafts or logs, without first requiring said corporation so to do"); Private & Local Acts, 1853, c. 128 ("in the erection of such boom, a good and sufficient channel for navigation shall be left unobstructed between said islands and the north shore of said stream"); P & LL, 1861, c. 43 ("Such boom shall be so constructed as to admit the free passage of boats or other water craft passing up or down the main

SPECIAL DAM FRANCHISES LACKING SLIDE REQUIREMENT

	Forest area			Nonforest area		
	No slide clause	Total dam franchises	Per Cent lacking slide clause	No slide clause	Total dam franchises	Per Cent lacking slide clause
1836–1848	2	12	17%	6	43	14%
1848–1849	0	0	0%	2	5	40%
1850–1859	7	45	15%	43	54	79%
1860–1869	8	19	42%	6	10	60%
1870–1879	9	32	28%	9	11	82%
1880–1889	4	91	4%	1	2	50%
1890–1899	11	49	22%	1	2	50%
1900–1909	21	82	25%	0	2	0%
1910–1919	2	2	100%	0	0	0%
	64	332	19%	68	129	52%

channel of the Chippewa river, and the proprietors of said boom shall not detain the logs of other persons longer than a sufficient time to sort out their own logs from the remainder of the drive"); P & LL, 1862, c. 45 ("Said boom or booms shall be so constructed as not to hinder the passage of boats, lumber or logs at any time, in the main channel of the Chippewa river"); P & LL, 1871, c. 94 ("Said boom shall be so constructed as not to interfere with the free navigation of said river"); Laws, 1873, c. 126 ("A free and easy passage for rafts, boats, barges, logs and lumber, of at least eighty feet shall be by said company maintained in the main Wisconsin river, until such time as [grantees] . . . shall make and file . . . their certificate that . . . the Bessie slough, has been so improved as to admit of the free and easy passage for rafts, boats, barges, logs and lumber, or property floating or running on the waters of said river"); Laws, 1881, c. 158 ("to be so constructed as to allow a channel for the passage of logs, timber, rafts and other water craft"); Laws, 1883, c. 112 (grantees "shall leave a suitable passage for logs, lumber, ties or timber, so as not to obstruct the navigation of said river").

These examples show that, as with the slide clauses in dam franchises, there was considerable variation in the detail of the stipulations, but a basic common core of guaranty for free navigation. There were not many boom franchises which contained no navigation guaranty clause. Only five franchises for purely private (no-toll) booms omitted any navigation guaranty. Three of these authorized booms on large lakes extending out from the grantee's riparian lands, and probably presented no problem in fact of interference with general navigation. See P & LL, 1868, c. 414; P & LL, 1869, c. 235; Laws, 1897, c. 170. Two authorized booms on small rivers: perhaps the facts made obvious that no other users would be affected; perhaps these were instances simply of loose drafting. See P & LL, 1854, c. 115, and 1855, c. 350. One private boom franchise contained a navigation clause which referred only to boats and rafts (Laws, 1880, c. 104). In P & LL, 1869, cc. 387 and 448, the free-passage clause referred only to boats and rafts, but a separate clause required that the booms be built so that logs of other persons not desiring them held therein might be sorted and passed out without unreasonable delay, the sorting to be at grantees' expense if they did not make prompt provision therefor; clearly, in substance, such grants guaranteed free passage for logs as well as rafts. The limited character of the omissions of free passage guaranties in private boom franchises points up by contrast the prevailing attention to general navigation rights where the boom was not intended as a public service operation. Where a boom franchise authorized toll, the implication seems strong that the omission of a free-passage clause for logs meant that a service-and-toll monopoly was conferred; sometimes a monopoly was explicitly given; in either case this type of grant seems more appropriately discussed later in this section, when we examine express monopolies.

Among forest-area franchises for private (nontoll) works, seventy-eight authorized the building of dams and booms in combination. In the inventory of special franchises at the beginning of this chapter, these are counted as dam franchises; the pattern of slide requirements in these franchises is part of the larger pattern already noted for all forest-area dam franchises. Sometimes these combined-works franchises contained no distinct passage clause regarding the booms as such, and sometimes the passage clause, whether in terms covering both dam and boom or only one of these, referred only to the passage of rafts and made no reference to logs. A dam slide requirement framed simply in terms of rafts would seem to carry no implication invidious to the rights of those driving logs; if the dam slide would pass a raft, ordinarily it would suffice also to pass logs. Read in light of the whole volume of such special franchises, the clauses referring only to rafts seem more likely the result of loose drafting than of calculated policy; certainly the statutes themselves offer no implication of formed policy less favorable to log than to raft transport past private works. Cf. P & LL, 1855, c. 330, 1856, c. 294, 1857, cc. 164 and 235. The history of one particularly important dam-and-boom franchise suggests that where a special point was made of the right of log passage, the legislature would make the right explicit. P & LL, 1857, c. 235 (incorporating the Chippewa Falls Lumbering Co.) authorized dams and booms "across" the river, with navigation guaranties that spoke only of boats and rafts. P & LL, 1862, c. 72, amended this franchise, stipulating that the booms must be so built that logs not owned by the grantees might be sorted and removed within a reasonable time, at the grantees' expense.

172. The most important general slides-and-exit requirement affecting all dams or booms on particular rivers was that for the Wisconsin and Black Rivers, created originally by Acts, 1849, cc. 62 and 98, and carried into Revised Statutes, 1878, sect. 1601. Cf. *Report of the Revisers of the Statutes,* 125. See also, e.g., Acts, 1852, c. 282 (Apple and Willow Rivers); P & LL, 1853, c. 205 (Pine River); P & LL, 1856, c. 238 (Eau Claire River); P & LL, 1867, c. 503 (Wolf River: repealed, Laws, 1882, c. 297); P & LL, 1871, c. 357 (Pine River). See also General Laws, 1872, c. 105, repealed, Laws, 1917, c. 474.

173. The inherent grant of traffic control was of course the more substantial where the franchise authorized building dams or booms across the whole breadth of a stream. In this aspect, franchises for works which appear primarily designed as auxiliary to particular mills present only with less emphasis the implication discussed in the next subdivision of the text, regarding franchises granted primarily for promoting transport.

174. The flood-gates clause was added to a slide requirement in 63 (about 19 per cent) of 332 forest-area private dam franchises; the new clause first became prominent in the later seventies, and most often appears in the later years of the lumber era. See, e.g., Laws, 1876, c. 250, 1878, cc. 283 and 284, 1879, cc. 13 and 21, 1883, c. 326. Laws, 1881, c. 266, exemplifies what came to be the standard formula for such a clause: grantees "shall build suitable slides in said dam for running logs, timber and lumber over the same, and shall keep the same in repair; the same shall be kept open at all times when the river is at a driving stage, and there are logs, timber or lumber to run over said dam, and when it is not necessary to hold the water back for the purpose of driving or flooding logs, timber or lumber below the said dam, for which purpose flood gates shall be kept in repair, and built in such manner as to be shut or opened, as the case may require, to flood the said logs, timber or lumber, provided that no charge shall be made for the passage over or through said dam of all logs, timber or lumber which may be run on said river." Again, there were variations in phrasing and in adding further details of obligation, all of which further underline the practical importance which the grantee's works and his operating decisions might have for the state of flow. This aspect of such franchises becomes vivid, for example, in the particularly stringent duties laid on the grantees by Laws, 1883, c. 224: "Whenever any general drive shall be hung up below said dam for any cause traceable to the existence of said dam, the owners of the logs in said drive, or their authorized agents whether corporate or otherwise, shall have a right to the use of the water held in the flooding dams above said drive, either on Pike Lake or Long Lake or elsewhere, by dams authorized by this act, for the purpose of moving said drive; and the said [grantees] . . . shall, upon such demand, promptly make such use of said water . . ." There is almost explicit recognition of the flow control inherent in a dam privilege in Laws, 1880, c. 49, which, having authorized a dam "to enable [grantees] . . . to use the water of said river for hydraulic and boomage purposes" (i.e., just for industrial water power and auxiliary harborage), imposed the usual slide requirement and then emphasized that grantees' flow management was to be only for power uses, by declaring that they "shall have full control of said dam, at all times, except for flooding and driving purposes . . ."

175. See the development of the statement of the grantee's affirmative obligations in Laws, 1882, c. 247 (usual slide and passageway clauses as to dam and booms, plus stipulation that "all such logs, timber and lumber as are not to be retained therein, shall with reasonable diligence be passed through said booms into the river below the same, and driven over the rapids at the sole cost and expenses of the grantees"), as amended by Laws, 1887, c. 253 (flood gate stipulation added to slide requirement; passageway requirement continued as to booms; new stipulation that "all logs and timber destined to points on said river below said dam, shall be taken by the owner or owners of said dam, when they reach the flowage thereof, and shall be driven by said owners free of charge, and with reasonable dispatch, through the pond and the flowage created by said dam, and over the same"; new provision for appointment of "commissioners" to arbitrate disputes regarding "the time or times of opening or closing said gates . . . for flooding purposes"), and as amended by Laws, 1893, c. 143 (dam owners shall take and drive logs destined below dam when they reach the dam flowage "or the rear of any log jam which may be caused by the stopping of logs at the upper dividing works of said booms, but not further up said river than the north

line of township thirty-seven north of range eight east"). The requirement that the franchise grantees, at their expense, handle all logs from the point of dam flowage to below the dam, was copied thereafter in a number of other grants. For further examples of this and other types of affirmative obligation clauses, see Laws, 1883, c. 224, 1885, cc. 139, 100, and 104, 1887, cc. 12, 41, and 223, 1889, c. 481, 1891, c. 177, 1903, cc. 145, 153, 154, 155, 156, and 239, 1905, cc. 398, 400, 407, 408, 409, 415, 457, 464, and 485, 1907, cc. 284, 286, 380, 383, 405, 409, and 514. A late variant was the additional requirement that grantees build their dam so that it would be "capable of permitting the free and uninterrupted passage through, or over, the same of any and all floods discharged by any flooding dam further up said river for the purpose of assisting in, and facilitating the driving of logs and other timber products below the dam . . . authorized by this act," plus the operating responsibility to "so maintain and operate [the authorized works] . . . as to permit the free passage of all such floods without substantial impairment of their effectiveness in assisting in the driving of logs and timber products down said river." See Laws, 1903, c. 145, and all the succeeding franchises from 1903 to 1907 cited above. This last, most elaborate, pattern of affirmative obligations entered private dam franchises at a time when the primary interest in the northern streams was turning to their potential for developing water power for generating electricity. The concern evidenced for navigation values in the face of this new competing interest reflects both the now well-established policy tradition of preference to transport uses, and the recognition of the practical importance to navigation of the flow management powers inherent in dam grants.

176. See, e.g., P & LL, 1861, c. 37, 1862, c. 72 (amending P & LL, 1857, c. 235), 1865, cc. 258 and 338, 1866, cc. 237 and 579, 1868, c. 461, 1869, cc. 387 and 448, 1871, c. 342; Laws, 1885, c. 423. Laws, 1878, c. 208, imposed stringent affirmative obligations on all boom owners on the Wisconsin River to provide both passageways and assisting labor; this act was never incorporated into the general statutes of the state, nor, apparently, was it ever repealed. Many private boom franchises contained a provision penalizing one who "willfully" damaged or destroyed the boom. See, e.g., P & LL, 1857, c. 41, 1871, c. 94. Familiar principles of construction might limit this provision to cases of malicious mischief; even so, the provision attested that a boom grant in some measure limited the practical freedom of other operators. No private (nontoll) boom franchise contained a provision requiring a log owner not desiring service to move out his logs by the regular exit or without damaging the boom. Cf. note 191 below.

177. The most important single example of the type of "monopoly" implicit in a franchise authorizing grantees to occupy the whole usable stream (and, also, to exact tolls of all users of waters so improved) was Private & Local Acts, 1853, c. 30, incorporating and bestowing improvement authority upon the Wisconsin River Improvement Co. In Wisconsin River Improvement Co. v. Manson, 43 Wis. 255, 265 (1877), the court recognized that the franchise involved this type of exclusive right as against a raftsman. On the other hand, later legislation in effect recognized that this style of grant, and specifically the grant to this company, did not of itself convey the much broader monopoly right of excluding all other improvers from the river. See amendments of Private & Local Acts, 1853, c. 30, by Laws, 1880, c. 292 (*inter alia,* empowering the Company "to consolidate with any other company or association created for the purpose of improving the navigation of said river, or running or driving logs thereon"), and by Laws, 1887, c. 13 (adding to the Company's toll authorization the qualification "that, in fixing the tolls for the running of logs and timber over or through any of the said company's improvements, no charge shall be made for, or on account of the flooding of said river or improvements with water furnished for flooding purposes, by means of a dam or dams erected and maintained by other parties than said company"). Cf. Laws, 1885, cc. 158 and 278; Laws, 1887, cc. 12 and 29 (making specific declarations regarding relation of newly authorized Wisconsin River works to works of the Company). Authority to manage flow was implicit where authorized dams might cross the whole stream, and no less so though the franchise conferred no toll right. See, e.g., P & LL, 1864, c. 325, 1865, c. 319, 1867, c. 563; Laws, 1875, c. 326, 1883, c. 88. There are few if any clear instances of authorization of no-toll booms intended to contribute to general navigation (as distinct from the private or toll booms which appear linked to operation of a particular sawmill, discussed in

notes 171 and 176 above). However, boom franchises which did not authorize the grantees to impose service and tolls on all who passed, by their guaranties for free passage reflected legislative consciousness that an across-channel boom, as such, implied powers to affect all traffic. See P & LL, 1864, c. 323 (amending P & LL, 1859, c. 153), 1866, c. 579, 1867, c. 328, 1871, cc. 45 and 483. Where flooding dams were authorized "across" streams, in aid of general navigation, the stipulations made in many grants regarding construction and operation of flood gates came close to explicit acknowledgment of the traffic management powers inherent in the grant. A familiar formula was that of Laws, 1878, c. 318, requiring that the grantees should "maintain and keep in repair suitable slides and gates in either or all of said dams for flooding purposes and for the purpose of sluicing and driving logs and timber over, through and beyond either or all of said dams down and into the Chippewa river, and shall keep the same in repair and shall hold the water back for the above named purpose whenever the case may require, and for that purpose said gates shall be closed." Even as this language put the grantees under a standard of reasonable service, it implicitly recognized that the situation authorized by the franchise of necessity devolved upon the grantees considerable discretion in management decisions. For like examples see Laws, 1881, c. 58, 1882, cc. 182, 183, 184, and 185, 1885, cc. 75, 371, and 372. Many across-stream dam franchises also came close to explicit acknowledgment of the general traffic management powers implicit in their grants when they declared that the grantees should "have full and complete control of said dam or dams, and superintend all driving or sluicing of logs, timber or lumber through the same." See P & LL, 1872, c. 112. Cf. Laws, 1882, cc. 182, 183, 184, and 185; Laws, 1891, cc. 110, 111, 149, 186, 229, and 396. A few franchises explicitly conferred exclusive control of flow; see note 203 below.

178. See, e.g., Acts, 1850, c. 257 (repealed, Acts, 1851, c. 325), 1852, c. 324; P & LL, 1857, c. 170, 1864, c. 84 (as amended, P & LL, 1866, c. 447), 1866, c. 352, 1867, c. 455; Laws, 1874, c. 250, as amended by Laws, 1882, c. 251 (cf. also Laws, 1873, c. 12, and 1874, cc. 262, 297, and 337); Laws, 1877, c. 247. Occasionally a statutory provision came close to explicitly acknowledging the flow management powers implicit in a stretch-of-stream improvement franchise. There seems such acknowledgment where a franchise explicitly authorized toll on logs put in between, or run on water favorably affected by, such authorized dams, as well as on logs run over any such dam. See P & LL, 1864, c. 302 (amended, P & LL, 1868, c. 366), 1872, c. 112, 1873, cc. 134 and 275. Again, the flow management implicit in a stretch-of-stream improvement grant was acknowledged when a later dam franchise was declared subject to "the rights granted" by the earlier franchise, or to "the consent of the party or parties owning, controlling and operating the improvements constructed pursuant to" the earlier grant. See Laws, 1883, c. 326, referring to Laws, 1877, c. 247.

179. See, for example, franchises cited in note 178 above, and also Laws of the Territory of Wisconsin, 1847, p. 44, as amended, Acts, 1851, c. 88; P & LL, 1856, c. 481, 1857, c. 360 (toll authorization repealed, P & LL, 1862, c. 98), 1867, c. 334, 1869, cc. 76 and 223, 1870, c. 93, 1871, c. 494, 1872, c. 112; Laws, 1878, c. 244, amending P & LL, 1866, c. 283; Laws, 1881, c. 58. There was perhaps recognition of the all-embracing compulsory character of toll authorizations in across-stream improvement franchises in the occasional reservation that the toll should not apply during the driving season current when the statute was passed. See, e.g., Laws, 1874, c. 262, 1876, c. 219; cf. P & LL, 1863, c. 193, amending *id.*, c. 35.

180. Men familiar with industry operations commented on the practical importance of such compelled tolls. See letter of C. S. Hazen, of John F. Stone & Co., Feb. 24, 1874, making bitter complaint of the improvement toll (as well as the driving charges) authorized to improvers of the Yellow River in Chippewa County by Laws, 1873, c. 12. "They claim," said Hazen, "that they have put $30,000 . . . worth of improvements on the river which [is] . . . doubted by many parties that have seen their improvements, now they go before . . . the Legislators . . . and ask to them to still let stand . . . a law that will net them yearly from 95 to $1.10 on the dollar for every dollar that they have expended in improvements on Yellow river." Hazen quoted Joseph Wapanea, a logger on the river for twenty-four years, foreman on every spring drive for the Union Lumber Co., to the effect that "the toll is a perfect steal on the loggers, and he is positive that it never costs more on the average than half of one

cent per thousand per mile to drive logs; therefore making just the toll alone more than the actual." See AJ, 606–608 (1874); see also Marshall, *Autobiography,* I, 334–335. Recognition of the fact that across-stream improvement tolls would of necessity affect all users seems indicated as one ground on which, in his Veto Message of Mar. 21, 1882, on bill 100S, Governor Rusk objected to omission of a reserved power to amend or repeal in flooding dam franchises to individuals; see SJ, 489 (1882).

181. See, e.g., Laws of the Territory of Wisconsin, 1847, p. 44; Laws, 1849, c. 46; Acts, 1852, c. 324; P & LL, 1853, c. 98, 1856, c. 405, 1859, c. 153, 1860, c. 91, 1862, c. 35 (as amended, P & LL, 1864, c. 157), 1863, c. 35.

182. See, e.g., Laws, 1882, c. 224 (toll authorization repealed, Laws, 1889, c. 215; all repealed, Laws, 1903, c. 25); Laws, 1883, c. 130.

183. The distinction between grant of a toll right against all in fact using the improved stream, and reservation of the right of others to do their own navigating, seems plain in such franchises as Laws, 1881, c. 311, or Laws, 1883, cc. 224 and 289, for example. The point is sharpened when Laws, 1883, c. 9, amends Laws, 1880, c. 303, to add to authorization of tolls binding on all users of the improved waters the stipulation that the grantees must "conform to all the conditions of chapter . . . 70 . . . of the revised statutes of 1878 [stating general slide and passageway requirements], so far as the same may be applicable." For qualified stipulations against obstruction, see, for example, Laws, 1881, c. 311, 1883, c. 88, 1887, c. 407.

184. See, e.g., P & LL, 1856, c. 405 (as amended, P & LL, 1866, cc. 286 and 509), 1866, c. 283 (amended, Laws, 1878, c. 244), 1870, c. 263; Laws, 1878, c. 291, 1887, c. 407. Cf. Private & Local Acts, 1853, c. 24; P & LL, 1854, c. 123, 1859, c. 153 (as amended, P & LL, 1864, c. 323), 1860, c. 91, 1866, c. 295; Laws, 1881, c. 64 (tolls authorized for use of artificial waterways created by grantees, coupled with stipulations for free use of associated natural channels). The limits set on toll rights under most boom franchises, as noted in the next paragraph of the text, serve further by contrast to point up the broad rights of toll given against all comers where franchises authorized tolls in connection with across-stream works.

185. See notes 195–210 below.

186. P & LL, 1871, c. 483, provides an example of the possible range of combination of implicit flow management powers (connected with authorization of flooding dams over a stretch of river), explicit toll grant with implicit monopoly effect (toll laid on all who use the improved stream, and all who use the stream must in fact use it as improved), and explicit monopoly of holding out to render boom service for fees (coupled with explicit guaranties of free passage for log owners not desiring boom services).

187. See, e.g., P & LL, 1860, c. 26 (tolls authorized; free passage guaranty for rafts, without mention of logs; but boom is approved only "partly across the Wisconsin River"); Laws, 1875, c. 48 (authority broad enough to allow boom grantees to require all logs use their services, but coupled with obligation "to secure through [grantees' works] . . . the restoration to navigation of the natural channel of said river, and the maintenance of the same hereafter free from obstructions").

188. See Fries, 52–53. Cf. Enos v. Hamilton, 24 Wis. 658, 662 (1869); Stevens Point Boom Co. v. Reilly, 44 Wis. 295, 305 (1878), s.c., 46 Wis. 237, 243, 49 N.W. 978, 979 (1879); A. C. Conn Co. v. Little Suamico Lumber Manufacturing Co., 74 Wis. 652, 664, 43 N.W. 660, 663 (1889).

189. See, e.g., P & LL, 1867, c. 328, 1870, c. 381; Laws, 1891, c. 313. Cf. Laws, 1897, c. 305, 1907, c. 417.

190. See A. C. Conn Co. v. Little Suamico Lumber Manufacturing Co., 74 Wis. 652, 664, 43 N.W. 660, 663 (1889).

191. See, e.g., Acts, 1852, c. 323; P & LL, 1857, c. 70, 1864, c. 161, 1867, cc. 503, 586, and 587, 1870, cc. 24 and 381. Not one private (nontoll) boom franchise contained such a clause, though many private boom franchises did penalize one who "willfully" damaged or destroyed the boom. See note 176 above. The inference may be that, where the boom was not a public service facility, a wider margin of freedom was conceded a log owner in the tactics of removing his logs from a boom whose service he did not want.

192. See, e.g., P & LL, 1866, cc. 237 and 579, 1870, cc. 24 and 381, 1871, c. 342.

193. Janesville Bridge Co. v. Stoughton, 1 Pinn. 667 (Wis. 1846). Draftsmen continued to show that they were not unaware of this restrictive doctrine. See, e.g., Laws, 1889, c. 41 (expressly declared area monopoly for toll bridge).

194. Jones v. Pettibone, 2 Wis. 308 (1853); Tewksbury v. Schulenberg, 41 Wis. 584 (1877); Wisconsin River Improvement Co. v. Manson, 43 Wis. 255 (1877). Cf. Wisconsin River Improvement Co. v. Lyons, 30 Wis. 61 (1872); Woodman v. Kilbourn Manufacturing Co., 1 Biss. 546 (Circ. Ct. D. Wis. 1867); Heerman v. Beef Slough Manufacturing, Booming, Log-Driving & Transportation Co., 1 Fed. 145 (Circ. Ct. W.D. Wis. 1878).

195. Revised Statutes, 1878, sect. 1777, took its origin from Laws, 1876, c. 399. The 1876 act, however, had the same ambiguity, or the like potential for breeding controversy, as Laws, 1873, c. 12, which the court held ineffective for want of sufficiently defined grantees, in Sellers v. Union Lumbering Co., 39 Wis. 525 (1876). The 1876 statute held out its privileges to any corporation, organized for the purpose under the general acts, which should improve a stream, but it did not explicitly confer an improvement monopoly, nor on the other hand provide any test to adjudicate among competing improvers. The 1878 revision changed the character of the provision sharply, and undoubtedly gave it a constitutional validity it otherwise lacked, by specifying that the granted privileges should go to the corporation which took "prior possession" of the stream to improve it. Claimants under sect. 1777 implicitly recognized the critical character of the "prior possession" element when they took pains to buy up previously outstanding special franchises. See Falls Manufacturing Co. v. Oconto River Improvement Co., 87 Wis. 134, 145, 146, 58 N.W. 257, 259, 260 (1894); cf. Underwood Lumber Co. v. Pelican Boom Co., 76 Wis. 76, 77, 79, 82, 45 N.W. 18, 19, 20 (1890) (under sect. 1777a). The later history of sect. 1777 is summarized in note 19 above.

196. See Black River Flooding-Dam Association v. Ketchum, 54 Wis. 313, 317, 11 N.W. 551, 552 (1882). Cf. Menominee River Boom Co. v. Augustus Spies Lumber & Cedar Co., 147 Wis. 559, 568, 132 N.W. 1118, 1121 (1912).

197. That the point of the general navigation guaranty in sect. 1777 was not to deny the monopoly of improvement and the traffic management and toll monopolies implicit therein, but simply to make clear that the grantees did not obtain a monopoly of moving logs or rafts on the stream, seems the clearer because the section further specifically protected other men's driving rights by limiting the grantee's driving claim to that of taking over logs inadequately provided for, as noted in point (c) in the text. The stated guaranty against obstruction might also be viewed as in part also making explicit the *quid pro quo;* the improver must truly improve, to claim the privileges which sect. 1777 allowed. Cf. Sauntry v. Laird, Norton & Co., 100 Wis. 146, 75 N.W. 985 (1898).

198. Compare the recognition of the distinction between tolls authorized for use of improvements and tolls authorized for rendering handling service, in Wisconsin River Log Driving Association v. D. F. Comstock Lumber Co., 72 Wis. 464, 40 N.W. 146 (1888). The court repeatedly recognized grants of toll rights given separately for use of improved water. See, e.g., Wisconsin River Improvement Co. v. Manson, 43 Wis. 255 (1877); Schneider v. Staples, 66 Wis. 167, 28 N.W. 145 (1886). Cf. Weatherby v. Meiklejohn, 56 Wis. 73, 13 N.W. 697 (1882).

199. The distinctions (1) between the improvement monopoly granted and the driving monopoly denied, and (2) between authorization to charge tolls and to charge expenses, was sharpened when Laws, 1882, c. 318, amended sect. 1777 to add that such a grantee corporation "shall also, at the request of the owner of any logs and timber put into said stream, take charge of the same and" drive them, and might charge the owner "reasonable charges and expenses for such services." In context, the "charges" would be for requested driving, the "expenses" levies for handling impeding logs. Cf. Wisconsin River Log Driving Association v. D. F. Comstock Lumber Co., 72 Wis. 464, 40 N.W. 146 (1888); Gibson v. Trow, 105 Wis. 288, 81 N.W. 411 (1900).

200. Both the exclusive right, and the right to establish a monopoly on a limited stretch of a stream, are recognized as features of the 1880 act (which became, first, Revised Statutes, sect. 1777a, and later sect. 1777e of the Wisconsin Statutes, 1898), in Menominee River Boom Co. v. Augustus Spies Lumber & Cedar Co., 147 Wis. 559, 568, 132 N.W. 1118, 1121 (1912).

201. That the 1880 act (note 200 above) created a distinct right of tolls for boom service was recognized in Underwood Lumber Co. v. Pelican Boom Co., 76 Wis. 76, 85, 45 N.W. 18, 20 (1890).

202. The 1880 act (note 200 above) also declared that a corporation qualifying under it

should have all the powers granted by sect. 1777 "so far as the same may be applicable, excepting in the cases wherein special provisions relating thereto are herein made." It may be that this provision gave a qualifying boom corporation the limited handling monopoly of taking charge of logs inadequately tended by their owners.

203. See P & LL, 1859, c. 133, as qualified by P & LL, 1870, c. 93, and 1871, c. 240, and restored by Laws, 1907, c. 672 (Peshtigo Co. authorized, *inter alia,* "to deepen, regulate, change and improve the channel of said Peshtigo river and its tributaries, from the mouth . . . up to such extent as said company from time to time may be disposed to do" with accompanying right to "charge . . . reasonable tolls for the use of such improvements by others"; in light of later qualifying acts, it appears that contemporaries viewed the original improvement grant as authorizing the grantee to assert exclusive improvement rights); P & LL, 1865, c. 444 (similar grant to Stiles Co., regarding Oconto and Pensaukee Rivers and tributaries); P & LL, 1864, c. 84 (granting Black River Improvement Co. authority "to improve the navigation of the Black river and lakes near the mouth of the same, in the counties of Clark, Jackson, Trempealeau and LaCrosse" with toll rights), especially in light of the clarifying amendment by Laws, 1880, c. 225, and the easy acceptance of the idea that the 1864 act conferred the right to assert exclusive improvement authority, in Black River Flooding-Dam Association v. Ketchum, 54 Wis. 313, 317, 11 N.W. 551, 552 (1882), and Black River Improvement Co. v. LaCrosse Booming & Transportation Co., 54 Wis. 659, 667–668, 11 N.W. 443, 446 (1882); P & LL, 1871, c. 357 (Pine River Improvement Co. "authorized and empowered to take and have the exclusive possession, occupancy and control of Pine river from its mouth where it empties into the Wisconsin river, to the mills in the village of Richland Center . . . for the purposes of navigation, and for that purpose only . . ." with "the exclusive right, power and authority to transport, take and carry property and persons upon said river, from its mouth to the said village of Richland Center, by any means known to water communication," and charge published tariffs therefor).

A not insubstantial number of such "monopoly" grants appear in years of the industry's greatest growth. See Laws, 1880, c. 151 (grantees given "the exclusive right and hereby authorized to keep and maintain a dam for flooding purposes [in one specified land section] . . . also to erect a dam or dams on the upper dells [in another specified land section] . . . and otherwise improve said river for driving logs . . ." with toll rights against all users); Laws, 1881, c. 221 (grantees "are hereby granted the exclusive right, and hereby authorized to build, keep and maintain dams on Silver creek, in Taylor and Price counties, at such points on said creek and its tributaries above the point where the railroad of the Wisconsin Central railroad company crosses said creek, as may be deemed by them best for the purpose of flooding and gathering heads of water for the purpose of drifting logs and timber down and out of said creek . . ." and also "it shall be lawful for said [grantees] . . . to open said gates and sluices [herein required to be provided on dams belonging to any other persons on these waters — presumably for power], and to make an opening in any dam or dams not provided with such sluices or gates, so far as such opening may be necessary to render practicable any drive, but after the drive is completed, they shall put said dam again in good condition for use"); Laws, 1881, c. 311 (grantees authorized "to enter in and upon and take possession of . . . Aminicon river, Aminicon lake and upper Aminicon lake, or any or either thereof, and improve the navigation of the same, or of any or either thereof" within specified stretches, including "the entire distance" of said waters within the specified areas, "so far and to such extent as [grantees] . . . may deem necessary in that behalf, so as thereby to aid and facilitate the driving, running, floating, storing, sorting, and delivering, or any or either thereof . . . logs" and other forest products, for tolls, provided grantees "shall not unreasonably obstruct or impede navigation upon such" waters); Laws, 1882, c. 278 (grantees "authorized and empowered to take and hold exclusive possession of Bad river, White river, and the tributaries of said rivers, and Long Lake and Pike Lake, and any or either of them, in the counties of Ashland and Bayfield for the purpose of rendering the same navigable or improving the navigation thereof for log driving purposes," with tolls, but "none of the works of improvement of any sort authorized hereby to be put in said streams shall in any manner unreasonably or materially impede or hinder the free navigation of such streams"); Laws,

1883, c. 224 (grantees "authorized and empowered to take such possession of [White] . . . river and all lakes and tributaries emptying into the same above [a specified point] . . . as shall be necessary to enable them to construct, maintain and operate therein such dams, booms, sluiceways and other works of improvement as may be necessary, and convenient to accomplish . . . the purpose of improving the navigation of" such waters, "and in aid of said purpose they may, and shall have . . . all powers and privileges conferred by section 1777 . . . upon corporations organized . . . for similar purposes," provided none of their works "shall materially obstruct or impede the free navigation thereof for log-driving purposes; and . . . that every person . . . having occasion to use said [waters] . . . for log driving purposes shall have an equal right to the use of the improvements [herein] authorized . . . under the supervision of the said [grantees] . . . without discrimination or cost" and provided further "that if any owner of pine timber situated upon said [waters] . . . which the said [grantees] . . . are authorized under the provisions hereof to take possession of and improve, shall so land the said pine upon such White river or any such tributary, that the same cannot be driven into waters navigable for log driving purposes, for want of a dam, temporary or otherwise, and the said [grantees] . . . shall, at the written request of said owners so to do, fail or neglect for the space of thirty days to erect such dam, the said owner may erect the same, but any dam so erected upon any such tributary shall be used under the supervision of said [grantees] . . . and shall be open to the free use, without cost or discrimination of any person . . . having occasion to use the same for log driving purposes"); Laws, 1885, c. 43 (grantees "authorized to build, maintain and operate dams across the Namakagon river and its tributaries above the village of Hayward for the purpose of equalizing the flow of water in said . . . river in aid of driving logs down the same to the store booms at the village of Hayward, and in aid of manufacturing logs into lumber at said village . . . by the waterpower created by the dam across said river at said village"; grantees "may maintain and operate said dams so as to regulate the flow of water in said . . . river, in such manner, as in [grantees'] . . . judgment will best subserve the purpose specified . . . provided, however, said dam shall not be so operated during at least twelve hours per day, in the day-time, as to diminish the natural flow of water in said river, when the same is needed to drive logs into the flowage of Hanscom dam, below the dam at said village of Hayward"); Laws, 1893, c. 302 (grantees authorized to build, maintain and operate dams across the Sand river in a designated stretch in Bayfield County, and "may maintain and operate said dam or dams, so as to regulate the flow of water in said Sand river in such manner as in [grantees'] . . . best judgment will best subserve the purpose . . . of equalizing the flow of water in Sand river, in aid of driving logs . . . down the same . . . and shall to that end have the right to store all of the natural flow of the water in said Sand river and to take possession of the said Sand river between the points mentioned . . . for the purpose of improving the same," subject to obligations to operate the dams, for fees here regulated, at request of persons desiring to move logs); Laws, 1895, c. 341 (authorization regarding Butternut Creek, Ashland County, in terms substantially similar to those of Laws, 1893, c. 302).

The general control of an improvement company over the flow management of a stream was emphasized occasionally when a special dam franchise stipulated that construction or operation of the particular works authorized should be subject to the satisfaction of a named general improvement company. See Laws, 1885, cc. 158 and 278, and 1887, c. 29 (slides in newly authorized dams must meet specifications of Wisconsin River Improvement Co.); Laws, 1891, c. 242 ("a dam for manufacturing" authorized to named individuals, but "the Big Eau Claire River Improvement company shall at all times have the full control of said dam and all the water controlled thereby, for the use and purpose only of running and driving logs and timber upon said river, free of any toll or charge of any kind"). But cf. Laws, 1887, c. 12 (dam authorized auxiliary to particular mills, with elaborate navigation guaranties, and with stipulation that "the Wisconsin River Improvement Company shall not in any manner hinder, obstruct or interfere with the construction, maintenance, use or enjoyment [of these authorized works] . . . so long as said dam shall be maintained under the authority conferred by this act").

204. See P&LL, 1856, c. 101 (repealed, P&LL, 1857, c. 101), 1866, cc. 303 and 550,

1867, c. 328, 1871, cc. 45 and 483. Cf. Laws, 1897, c. 305, and 1907, c. 417. From the face of the statutes one cannot draw assured sharp lines between (1) boom franchises which confer a monopoly in a given area of holding out to render boom services for fees, but without compulsion upon all log drivers to use the services, and (2) those franchises which (note 205 below) confer an "exclusive" booming right in the sense not only of excluding other boom operators but of requiring all log owners to use the grantee's works. There are at least two sources of ambiguity in listing franchises as belonging in type (1): (a) A franchise may confer "the exclusive privilege of booming" a described stretch of water, but subject to stipulation that the works "shall in no wise interfere with the free navigation of said river." The navigation guaranty might seem to make this a franchise of type (1) despite the "exclusive" language. Yet "the free navigation" might, in context, mean simply navigation by rafts or vessels, and the franchise might allow the grantee to insist on booming all loose logs. (See, e.g., P&LL, 1857, c. 101.) (b) A franchise containing navigation guaranties capable of embracing the movement of logs in certain channels might yet, in conferring an "exclusive" booming right over limited areas of water, in the practical facts of the location amount to conferring a right to require all comers to accept the boom's services. The most notable and controversial example of such a monopoly *de facto* if not *de jure* was that at Beef Slough, the one of the several mouths of the Chippewa River which provided the only practical natural harbor for logs destined for Mississippi River points. See Fries, 142–156, and P&LL, 1855, c. 331, 1866, c. 181, 1870, c. 299. Cf. Heerman v. Beef Slough Manufacturing, Booming, Log-Driving & Transportation Co., 1 Fed. 145, 156 (Circ. Ct. W.D. Wis. 1878). Similarly there are ambiguities in classifying some franchises as type (2), where a particular grant is of the right — sometimes the "exclusive" right — to build and operate booms to hold and handle "all logs" coming down a named stream, for stipulated fees, coupled with a general stipulation that the works shall not obstruct "navigation." See, e.g., Laws, 1881, c. 209, 1882, cc. 13 and 276. Compare also instances in which broadly framed boom franchises contain a guaranty of free passage for rafts or boats, without mention of logs, as in General Laws, 1860, c. 26, or P&LL, 1866, c. 303. Some of these grants, the effect of which is not clear from the face of the statutes, might be clarified, if we knew the facts of industry operations at the particular stream points or stretches concerned. But except for such a rare spotlighted controversy as that over Beef Slough, the materials are typically not available today from which to make such estimates. In this context, the best we can do is to note that two types of rather extensive boom monopoly seem to have been created, make conservative classification of instances which seem rather clearly of the one type or the other, and acknowledge a considerable area of doubt if we try to name the proper pigeonhole for every franchise. The doubt, after all, arises only over a finer point of classification; lumping the two types together, it is clear that there were a not inconsiderable number of special boom franchises which explicitly indicated an intent to confer some kind of monopoly right, whether against other would-be boom operators, or against owners of logs in transit, or against both.

205. Rights to take over all driving, for a fee or an allocated share of expenses, were conferred by P&LL, 1864, c. 126 (Little Wolf River Log-Driving Company "shall have the right to drive down said river all logs, lumber, timber, cants and shingle bolts belonging to said company or any other person, company or corporation, which now are or may hereafter be placed in said river, from upper sources to the boom in Winnebago county"; the driving monopoly here authorized is sharpened by contrast with P&LL, 1862, c. 35, which empowered the Eau Claire River Log Driving Company in language obviously borrowed by the 1864 charter but with the stipulation, omitted in the Little Wolf grant, that the Eau Claire franchise "shall not be construed so as to exclude other persons from running or driving their own logs . . . down said river free of charge or toll from said company: provided, such . . . persons keep their said logs . . . from mixing in the general drive"); P&LL, 1867, c. 455 (Outlet Log-Driving Association "shall have the right to drive down said outlets all logs, timber, cants or shingle-bolts belonging to said association, or either or any of the members of said association, or any other person, company or corporation that now are or hereafter may be placed in said outlets," and all persons placing their logs, etc., in said outlets in such manner as to mix with grantee's logs "shall be deemed to have consented that their said logs . . . may be driven down said outlets by said association, and shall be liable to pay said association the

same proportionate prices for driving the same as are paid by the members of said association"); P&LL, 1868, c. 398, amending P&LL, 1857, c. 170 (Yellow River Improvement Company "is also hereby authorized and empowered to take charge and control of all general drives of logs on Yellow river at any future time," and "all persons hereafter allowing or suffering their logs or timber to remain in said Yellow river [within a stretch here designated] . . . in such manner as to mingle or mix with the logs and timber belonging to the said company, or the individual members thereof, in any general drive of logs down said river, shall be deemed to have consented that his or their logs and timber shall be driven down the said river by said company, and shall pay all the expenses and charges for rolling in, driving, stopping, sorting and delivering the same")—perhaps qualified by P&LL, 1869, c. 186, but perhaps enlarged; P&LL, 1868, c. 430 (Apple River Log Driving Company authorized "to receive and take possession of all logs or timber coming down, or to be driven down the said Apple river or any of its tributaries, at certain points hereinafter described, and drive all such logs and timber into the St. Croix river," for prescribed fees; "all logs and timber on the said Apple river and its tributaries, shall be deemed to be, and shall to all intents and purposes be in the possession and under the control of the said driving company, whenever driven within any of the limits mentioned")—enlarged by Laws, 1878, c. 206; P&LL, 1868, c. 469, amending P&LL, 1857, c. 101, as amended by P&LL, 1865, c. 190 (to add to authority of Wolf River Boom Company "the exclusive and sole right, power and privilege of running, driving and booming logs and timber between [designated] . . . points on said river"); P&LL, 1870, c. 38 (Pine River Navigation Company "shall have full power to open by improvements, and thereafter to control for navigation purposes the said Pine river from . . . its mouth . . . as far up the stream as they may deem practicable"); P&LL, 1871, c. 357 (Pine River Improvement Company "shall have the exclusive right, power and authority to transport . . . property and persons upon said river, from its mouth to the said village of Richland Center, by any means known to water communication," for established uniform tariffs); P&LL, 1871, c. 483 (White River Dam, Log Driving and Boom Company authorized to build flooding dams and collect toll for all logs driven by aid of their waters, and to have the exclusive right to build and operate booms for fees, but subject to a free passage guaranty for all owners who do not request booming service; beyond this, the Company "is hereby authorized, empowered and required to receive and take possession of all logs, timber or lumber, hauled, driven or coming into White river from any of the tributaries thereof, and to drive all such logs, timber or lumber into said corporation's booms" and to collect stipulated fees "for driving the same," with no stipulation that this must be only at the log owner's request—in contrast to qualification upon Company's boom right).

The driving monopolies indicated by such franchises are thrown into sharper relief by comparison with the explicit declarations or reservations of the driving rights of others than the grantees, in P&LL, 1854, c. 62 (as amended by P&LL 1857, c. 145), and 1862, c. 35. The qualified monopolies given as to inadequately-attended or impeding logs in franchises cited in note 208 below make the same point. Compare, on the other hand, the ambiguity of Laws, 1882, c. 278, which says that its grantees are "authorized to receive and take possession of all logs . . . coming down or driven into or placed for driving within the said rivers, lakes or either of them, or any of their tributaries at any point below any improvement made therein, and shall also, at the request of the owner . . . of such logs . . . take charge of the same and drive all such logs . . . down said river to its mouth. . . ." Perhaps these terms both authorized the grantees to insist on driving if they chose, and authorized the log owners to insist that grantees drive, if the log owners so chose.

Rights to require that all log owners accept the grantee's boom services for a fee were conferred in varying degrees, in a striking series of variations, upon the Wolf River Boom Company. See P&LL, 1857, c. 101, as amended by P&LL, 1862, c. 43, 1865, c. 190, 1868, c. 469, and 1870, c. 34. That this operation embraced the whole log traffic of the area, see Current, 112, 116, and Fries, 49. Other outstanding boom monopolies of this type were those conferred at the main gathering area on the St. Croix River. See Acts, 1851, c. 351; Laws, 1874, c. 48, as extended, Laws, 1880, c. 226; Laws, 1889, c. 215, as amended, Laws, 1891, c. 478, and 1895, c. 352. See Rector, *Log Transportation,* 121, 122. See also P&LL, 1859, c. 153 (service monopoly ended, however, by P&LL, 1864, c. 323); P&LL, 1862,

c. 100 (repealed, General Laws, 1865, c. 290; revived, P&LL, 1866, c. 433; acknowledged, P&LL, 1868, c. 430; qualified, Laws, 1878, c. 206); Laws, 1880, c. 317, as amended, Laws, 1881, c. 295 (converting to a service monopoly what was a non-exclusive public utility franchise under P&LL, 1870, c. 24); Laws, 1881, c. 209, 1882, c. 13, 1891, c. 222. Borderline grants are P&LL, 1866, c. 550, and Laws, 1882, c. 276. For an explicit disclaimer of intent to create a service monopoly, see Laws, 1891, c. 313.

206. P&LL, 1857, c. 145, qualifying P&LL, 1854, c. 62; Laws, 1877, c. 247 (cf. Laws, 1873, c. 12); Laws, 1879, c. 191. Cf. P&LL, 1868, c. 469.

207. P&LL, 1854, c. 62, as amended, P&LL, 1857, c. 145 (termination by judicial sale reflected in P&LL, 1859, c. 198); P&LL, 1862, c. 35 (changed to stock corporation, P&LL, 1864, c. 157); P&LL, 1864, c. 126; P&LL, 1864, c. 131 (repealed, P&LL, 1868, c. 398, and operations taken over by stock company organized under P&LL, 1857, c. 170; cf. P&LL, 1869, c. 186); P&LL, 1865, c. 371; P&LL, 1870, c. 52 (repealed, supplanted by stock corporation, P&LL, 1871, c. 494). Cf. P&LL, 1867, c. 455.

208. Limited driving monopolies were given under several formulae. One formula turned on the fact that the log owner had not adequately provided for driving his logs; here the franchise authorized the grantee "to take possession of all logs or timber put into said stream, to be floated down and out of the same, the owner or agents of which shall not have made adequate provisions for driving, by furnishing all necessary tools, men, teams, equipments and provisions for breaking roll ways at the proper time and making in due time a thorough drive of such logs or timber out of such stream or river without hindering the main drives" (see, e.g., Laws, 1874, c. 250, 1877, c. 247, 1881, c. 311). Some franchises, often containing the formula just quoted, also more sharply limited such authority to take over the driving of another man's logs "when and when only they shall obstruct the drive" (see, e.g., Laws, 1878, c. 318, 1883, cc. 224 and 347). More often the franchise granted authority in the alternative; the grantee might take over the driving either of inadequately attended logs or of logs which were in fact obstructing the general drive (see, e.g., Laws, 1877, c. 42, 1880, c. 171, 1889, c. 83). That these alternatives represented distinct heads of authority is indicated by the care taken in committee to add to a bill already dealing with obstructing logs a separate provision regarding ill-attended logs; see AJ, 688 (1889), regarding bill 512A, which became Laws, 1889, c. 485. Some franchises authorized the grantee to take over driving logs of another man who allowed his logs to "mingle" or "mix" with those of the grantee; see, e.g., P&LL, 1867, cc. 145 and 455, 1868, c. 398; cf. P&LL, 1871, c. 494 (combining requirements that the other man's logs be mixed with grantee's and be inadequately attended). A special form of limited driving monopoly was the authority which some franchises conferred to recover and drive stranded logs and charge their owners therefor (see, e.g., P&LL, 1866, c. 437, 1869, c. 186). In a unique manifestation of uncertainty, the draftsman attached a separability clause specifically to a grant of authority to drive ill-attended logs, in Laws, 1881, c. 311.

209. See, e.g., Acts, 1852, c. 323; Private & Local Acts, 1853, c. 213; P&LL, 1855, c. 120, 1859, c. 149. Analogous provisions often authorized driving or booming companies to seize and sell as "prize logs" driven or boomed logs unmarked and unclaimed (see, e.g., P&LL, 1864, c. 157, 1865, cc. 190 and 371, 1867, cc. 145 and 455).

210. See Private & Local Acts, 1853, c. 30 (Wisconsin River Improvement Co.); P&LL, 1857, c. 101 (Wolf River Boom Co.); P&LL, 1864, c. 84 (Black River Improvement Co.).

211. Stevens Point Boom Co. v. Reilly, 44 Wis. 295, 302 (1878), s.c., 46 Wis. 237, 49 N.W. 978 (1879). Compare Sellers v. Union Lumbering Co., 39 Wis. 525 (1876), where the court refused to construe Laws, 1873, c. 12 (authorizing tolls to "any person . . . who shall have improved said Yellow river") as intended to confer an improvement monopoly upon the one corporation which had in fact already improved the river prior to enactment of the statute; the statutory terms of themselves embraced any person who at any time made improvements, and the court found claimant's argument to be the sole monopolist unconvincing, for "it is inconceivable why, if the legislature intended to grant the franchise to the appellant, it should not have said so in plain words, instead of hiding its intention in such inapplicable terms" (*id.*, 529). Finding this interpretation not available to save the statute, the court held the franchise ineffective for want of sufficiently designated grantees.

212. See Yellow River Improvement Co. v. Arnold, 46 Wis. 214, 49 N.W. 971 (1879) (amendment of river improvement franchise,

adding what court assumes — as alternative ground of decision — is exclusive driving right, held not to embrace subject different from that of original franchise in such sense as to offend constitutional requirement that local act embrace but one subject, to be expressed in its title); Cohn v. Wausau Boom Co., 47 Wis. 314, 2 N.W. 546 (1879) (explicit grant of exclusive right to maintain public-utility-style boom occupying designated stretch of navigable stream, held to bar exercise by riparian landowner of the common law right he would otherwise have to maintain a boom in front of his land to service his mill); Black River Flooding-Dam Association v. Ketchum, 54 Wis. 313, 11 N.W. 551 (1882) (exclusive stream-length improvement franchise given by special act will not be treated as repealed by implication by sect. 1777); Black River Improvement Co. v. LaCrosse Booming & Transportation Co., 54 Wis. 659, 11 N.W. 443 (1882) (exclusive river-length improvement franchise construed to include authority to close up navigable slough, to enhance flow in main channel); Edwards v. Wausau Boom Co., 67 Wis. 463, 30 N.W. 716 (1886) (where franchise allows construction of boom across whole channel, court must take care to give substantial leeway to grantees to effectuate their declared right to detain others' logs for time reasonably needed to sort out their own goods, and general penalties of Revised Statutes, sect. 1898, on obstruction should be construed as superseded by the special franchise right); Wisconsin River Log Driving Association v. D. F. Comstock Lumber Co., 72 Wis. 464, 40 N.W. 146 (1888) (Revised Statutes, sect. 3337, as amended, Laws, 1881, c. 141, authorizing "any other owner" of logs on river to drive another's ill-attended logs which become mixed with his, and charge reasonable compensation therefor, construed to include such initiative taken by a bailee of logs); Underwood Lumber Co. v. Pelican Boom Co., 76 Wis. 76, 45 N.W. 18 (1890) (where boom corporation qualifies for the service monopoly given by Revised Statutes, sect. 1777a, within the general rate standards indicated by the legislature the corporation's particular determination of rate schedules should not be upset, so long as court cannot rule them unreasonable); Yellow River Improvement Co. v. Wood County, 81 Wis. 554, 51 N.W. 1004 (1892) (stream-length improvement and driving monopoly implies legislative intent that company not be subject to tax levies by local assessors on items of its works forming part of integrated improvement scheme). Cf. Janesville Bridge Co. v. Stoughton, 1 Pinn. 667 (Wis. 1846); Chapin v. Crusen, 31 Wis. 209 (1872); Milwaukee Electric Railway & Light Co. v. Railroad Commission, 153 Wis. 592, 142 N.W. 491 (1913) (rule of strict construction of special statutory franchises, in favor of public right, treated as continuing Wisconsin policy). The court's generosity in implying auxiliary incidents to a clearly given monopoly right was consistent with its readiness to recognize the implicit toll and flow-control monopolies given where a franchise authorized works which would embrace the whole usable channel. See Wisconsin River Improvement Co. v. Manson, 43 Wis. 255 (1877), and notes 177–180 above. Cf. Pioneer Wood Pulp Co. v. Chandos, 78 Wis. 526, 531, 47 N.W. 661, 663 (1891) (reflection of statutory claim of Wisconsin River Improvement Co. to fix standards of slide construction in dams of other owners). The court showed itself ready, also, to imply auxiliary incidents to explicit monopoly grants of water-power development rights, as against competing power users. See Green Bay & Mississippi Canal Co. v. Kaukauna Water Power Co., 70 Wis. 635, 35 N.W. 529 (1888); Falls Manufacturing Co. v. Oconto River Improvement Co., 87 Wis. 134, 58 N.W. 257 (1894).

213. Gibson v. Trow, 105 Wis. 288, 81 N.W. 411 (1900). Cf. Enos v. Hamilton, 24 Wis. 658 (1869) (violation of navigation guaranty in boom franchise); Black River Flooding-Dam Association v. Ketchum, 54 Wis. 313, 11 N.W. 551 (1882) (enforcement of "prior possession" requirement for successful claim of improvement monopoly under sect. 1777); Sauntry v. Laird, Norton & Co., 100 Wis. 146, 75 N.W. 985 (1898) (franchise holder may not collect improvement toll where failed to build whole number of flooding dams stipulated as conditions of grant); Menominee River Boom Co. v. Augustus Spies Lumber & Cedar Co., 147 Wis. 559, 133 N.W. 605 (1912) (grantee's toll schedules invalid as exceeding bounds of reasonable discretion allowed by franchise). See Edwards v. Wausau Boom Co., 67 Wis. 463, 470, 30 N.W. 716, 718, 719 (1886) (obiter, boom grantee liable for improperly constructed or negligently operated works).

214. See note 99 above.

215. See, e.g., Acts, 1851, c. 351 (St. Croix Boom Co.); P&LL, 1862, c. 35, amended, P&LL, 1864, c. 157 (Eau Claire River Log Driving Co., first a cost-sharing joint stock company open to all who filed as interested

lumbermen, then changed to toll-earning stock corporation without such free access to membership); P&LL, 1866, cc. 303 (Rush River Boom Co.), 550 (Suamico Boom Co.), and 579 (Stevens Point Boom Co.); P&LL, 1869, c. 76 (Embarrass River Improvement Co.); P&LL, 1871, cc. 240 (Peshtigo River Improvement Co.), 357 (Pine River Improvement Co.), 467 (East Shioc Improvement Co.), 483 (White River Dam, Log Driving & Boom Co.), and 494 (Mill Creek Improvement and Log Driving Co.).

216. Revised Statutes, sects. 1777 and 1777a (or, under Wisconsin Statutes, 1898, sect. 1777e), held out their privileges to corporations organized under c. 86 (specifically, under sect. 1771), the general business incorporation act. The total of fifty-four special statutory stream-use franchises granted to corporations which were organized primarily for industrial purposes under special acts passed before 1871, or for industrial or navigation enterprises under the general business corporation law after 1871, included: toll booms — 4, private booms — 3, (toll) river improvements — 7, forest-area non-toll dams — 26, non-forest-area non-toll dams — 14.

217. These figures are for stream-use franchises in which the original named grantee was a corporation. The legislature conferred many franchises by special statute upon named individuals and their "assigns." There is no way now to discover how many such franchises were assigned to corporations. Such assignments did occur, and were ruled effective. See Underwood Lumber Co. v. Pelican Boom Co., 76 Wis. 76, 82, 45 N.W. 18, 20 (1890) (since dam and boom franchise ran to named individuals and their assigns, "there can be no question but what [the named grantees] . . . were thereby empowered to make such transfers and conveyances to the boom company"); Chippewa & Flambeau Improvement Co. v. Railroad Commission, 164 Wis. 105, 112, 159 N.W. 739, 742 (1916), and Case on Appeal, p. 44. Note the acknowledgment and tacit acceptance of an assignment to a corporation of a dam franchise granted to individuals, in Laws, 1907, c. 123, referring to Laws, 1901, c. 455.

The 115 special statutory stream-use franchises conferred upon corporations fell into the following pattern of years:

SPECIAL STATUTORY STREAM-USE FRANCHISES
CONFERRED UPON CORPORATIONS

	Booms		Navigation	Dams	
	Toll	Non-toll	improvement	Forest area	Nonforest area
Territory	0	0	1	0	3
1848–1849	0	0	0	0	0
1850–1859	9	2	10	10	6
1860–1869	8	0	23	4	2
1870–1879	3	0	10	2	2
1880–1889	0	0	1	0	0
1890–1899	0	0	1	0	0
1900–1909	0	1	1	10	1
1910–1919	0	0	3	0	0
1920–1929	0	1	0	0	0
1930–1939	0	0	1	0	0
Totals	20	4	51	26	14

218. See P&LL, 1856, cc. 238 (having defined terms of boom franchise to named individuals, "their executors, heirs and assigns," says that "the provisions of this section shall apply to all other acts of incorporation of booming companies on said river L'Eau Claire"), and 540 (act entitled "to incorporate the Wolf River Boom Company" runs, in text, solely to "Benjamin Brickley and his assigns" or "associates"); General Laws, 1860, c. 26 (incident to boom grant to "Ephraim Kingsbury, his successors and assigns," declares that "the said corporation" shall have defined rights as to unclaimed logs handled); P&LL, 1865, c. 338 (act, entitled "to incorporate the Chippewa river rafting boom company," runs in its granting section to "Seth Webb, John E. Stillman and Hamilton W. Hubbard, by the firm name of Seth Webb & Co., their associates, successors and assigns," with no provision creating any corporate rights as such); P&LL,

1869, c. 400 (after basic grant of flood dam franchise to five named individuals, "their associates, heirs and assigns," franchise stipulates that "the said incorporators, their associates and assigns" shall provide a suitable slide, and authorizes tolls and a lien to "the said incorporators, their associates and assigns," with no other reference to corporate status). There were similar references to named individual grantees as a "corporation" in some industrial power, bridge, and ferry franchises; see P&LL, 1856, c. 220, 1859, c. 77; Private & Local Acts, 1863, c. 298; P&LL, 1869, c. 220.

219. See Governor Ludington's Veto Message on bill 101A, Mar. 3, 1877, AJ, 581–582; Governor Rusk's Veto Message on bill 460A, Mar. 27, 1882, AJ, 748. The sharp focus upon the problem of special grants of "corporate" powers as the issue in the 1882 incident was underlined by the prompt, undisputed passing of a new bill (484A) granting the same dam privilege to named individuals, their heirs and assigns, with care taken by floor amendment to insert a reserved power to repeal or amend. See AJ, 769 (1882); SJ, 583 and 597 (1882); Laws, 1882, c. 297. Similar to Rusk's veto was one of the grounds of Governor Hoard's Veto Message on bill 41A (amending, by addition of eminent domain powers, a dam grant made by Laws, 1887, c. 29, to an existing corporation), Mar. 13, 1889, AJ, 642–643. In retrospect, in an analysis colored by his argument for the need of more rigorous general regulation of stream use, Governor Davidson thought that "some of our acts granting franchises to build dams come dangerously near, if they do not encroach upon, a violation of the spirit of the constitutional provision" prohibiting special grants of corporate powers or privileges (see Special Message of Feb. 18, 1909, AJ, 258).

There was quite uneven legislative response to the warning of Governor Ludington's veto that a franchise which ran to named individuals and their "successors" became thereby a grant of corporate powers. In a few instances cautious grantees obtained amendments striking "successors" from their statutory titles, substituting "heirs." See Laws, 1877, c. 267, and SJ, 494 (1877) (change by floor amendment, three days after Ludington's veto); Laws, 1882, cc. 78 and 95, amending Laws, 1879, cc. 136 and 137. In 1882 there was a flurry of episodes in which "associates" was stricken from the dam franchises which became Laws, 1882, cc. 137, 182, 183, 184, and 185. See also Laws, 1882, c. 78, amending Laws, 1879, c. 136. Cf. Laws, 1879, cc. 154 and 155; Laws, 1880, cc. 32, 33, 40, 41, 75, 76, 92, and 104; Laws, 1882, c. 78 (grants to named individuals on land described as owned by named corporation); Laws, 1882, c. 276; Laws, 1899, cc. 144 and 177 (introduced bills granted dam franchises in each case to a named "Company," but enacted bill changed to make grants to named individuals). Laws, 1881, c. 311, included in an improvement franchise to named individuals "their and each of their heirs, executors, administrators and assigns," an explicit disclaimer — that "no corporate powers or privileges are granted or intended to be granted by this act, and nothing contained in this act shall be construed or deemed to create a corporation, or grant corporate powers or privileges, either expressly, by implication, or otherwise." This device of draftsman's caution then dropped out of sight until, for no reason apparent on the official records, it was suddenly revived and became quite common in the late spurt of special dam franchises in the early 1900's under the impetus of interest in developing hydroelectric power facilities. The first revival was in Laws, 1903, c. 62; the corporate powers disclaimer was thereafter included in six other dam franchises in 1903, eight in 1905, and eleven in 1907, after which time the legislature granted no more special franchises to individuals. See In re Southern Wisconsin Power Co., 140 Wis. 245, 259, 122 N.W. 801, 806 (1909) (this device of "overcaution" should not be construed to bar acquisition by a corporation of dam franchise granted originally to individuals). On the other hand, before this disclaimer clause was used, and even after it had been revived, many special franchises for dams or booms ran to the "successors" of named individual grantees without indication of worry over the specter of Ludington's veto. See, e.g., Laws, 1883, cc. 112 and 289, 1885, cc. 254 and 283, 1887, cc. 338 and 377, 1889, c. 372, 1893, c. 191, 1897, c. 240, 1899, cc. 315, 320, and 331, 1901, c. 185, 1903, cc. 209, 353, and 364, 1905, cc. 350, 397, 400, and 409. A number of grants flaunted their ignorance or indifference toward the point of the Ludington veto, or their conviction that it was in error, by running to named corporations and their "successors." See, e.g., Laws, 1889, cc. 77 and 283, 1895, c. 341, 1901, c. 366, 1903, cc. 24, 178, and 180, 1905, cc. 39 and 408, 1907, cc. 98, 329, 361, 405, 626, and 644. Compare similarly phrased bridge franchises to corporations, as in Laws, 1891, cc. 324 and 416, or 1893, c. 13. Especially striking were two dam franchises granted to named corporations and their "successors," coupled with the explicit disclaimer of intent to confer corporate powers or privileges. See Laws, 1903, cc. 182 and 206. By the time the legislature got to the business of creating broad administrative regulation of water-power development, it made clear that it had no worries over granting franchises to corporations, or to "successors." See Laws, 1911, c. 652 (grants to first takers and successors or assigns; references to appropriator or his "or its" successors or assigns). Indeed, Laws, 1913, c. 755, stipulated that licenses for the larger-scale water-power projects would issue only to corporations organized thereunder; this insistence, however, was dropped in Laws, 1915, c. 380.

220. In Sellers v. Union Lumbering Co., 39 Wis. 525, 527, 528 (1876), Ryan, C. J., speaking for the court, observed the distinction between "the franchise to be a corporation" and "the particular franchise conferred" to improve a stream and charge tolls therefor. In Stevens Point Boom Co. v. Reilly, 44 Wis. 295, 301 (1878), however, Ryan, C. J., for the court, ruled that Laws, 1873, c. 126 — which authorized "the Stevens Point Boom Company, as now organized, under the general laws of this state [General Laws, 1872, c. 144] . . . to . . . build and maintain a system of piers and booms in, along and across the Wisconsin river . . . [between points here designated in the special statute] . . . at the points indicated in the articles of association of said company now on file in the office of the secretary of state" — was legally ineffective. Laws, 1873, c. 126, said Ryan, "is, beyond question, a legislative attempt to amend the respondent's charter under the general statute, enlarging its powers by special grant. It unquestionably grants corporate powers and privileges; corporate franchises, not before possessed by the corporation" — citing the Sellers case. This language seems to confuse, or treat as inherently one, a franchise for particular rights of stream use and a franchise to be a corporation, wherever a statute grants stream-use rights to an existing corporation. One might try to interpret Ryan's remarks as ruling that General Laws, 1872, c. 144, did not authorize the formation of a corporation to do a booming business; that statute made no direct reference to booming corporations, and its terms were less sweeping than those of the later sect. 1771 of the Revised Statutes, 1878. In this light, when it authorized the given boom, Laws, 1873, c. 126, would have to be treated as implicitly conferring corporate authority as well as giving a franchise in limitation of public rights of navigation, and Ryan's observations might then be construed as going to this implicit aspect of the special statute. But this seems too strained a reading of the 1878 opinion to carry conviction. Before 1909 there was no other Wisconsin Supreme Court opinion which expressly or implicitly dealt with this problem under Wisconsin Constitution, IV, 31. However, there appears an analogous doctrinal confusion over the question whether the constitutionally reserved power to amend or repeal acts creating corporations extended to acts conferring stream-use franchises upon corporations existing by virtue of other legislation. See note 129 above.

221. In re Southern Wisconsin Power Co., 140 Wis. 245, 258, 122 N.W. 801, 806 (1909). Noting that this ruling "does not appear to be in harmony with Stevens Point B. Co. v. Reilly," the court said that that case "must be considered as overruled in so far as it holds that a franchise cannot be granted to an existing corporation" (140 Wis. at 259, 122 N.W. at 806).

222. See also P&LL, 1855, c. 331, 1856, cc. 238 and 252, 1857, cc. 101 and 170, 1859, c. 153, 1860, c. 91, 1864, c. 131, 1866, c. 82 (repealed, P&LL, 1867, c. 426). These early years also saw some complex charters, however, such as Acts, 1852, c. 324, or Private & Local Acts, 1853, c. 30. The relative simplicity of some charters may reflect the fact that in the eyes of their founders they represented essentially cooperative, nonprofit ventures to supply limited services to a group of entrepreneurs, as with P&LL, 1856, c. 405 (Chippewa River Improvement Co., limited to enhancing main channel flow, and barred from tolls), or P&LL, 1864, c. 131 (quite informal Log Driving Association of the Yellow River). However, this explanation does not negate, but rather emphasizes, the matter-of-fact casualness with which men were beginning to treat the corporate device — as a handy tool of no peculiar significance for navigation problems as such.

223. See Rector, *Log Transportation,* 107, 135, 136, Reynolds, 70, 80; cf. Current, 112–114, Raney, 213. Particular franchises sometimes set large authorized maxima for capital stock. See, e.g., Acts, 1852, c. 324 ($150,000); Private & Local Acts, 1853, c. 30 ($150,000), amended, P&LL, 1857, c. 135 ($500,000); P&LL, 1870, c. 34 (increased from $20,000 to $50,000); Laws, 1873, c. 256 (increases authorized, successively, from $25,000 to $50,000 to $100,000).

224. Rector, *Log Transportation,* 103, puts

the cost of particular flooding dams in the range of $500 to $2000. At the peak of lumber industry investment the federal census of 1890 estimated that of the total capital committed to the industry, 36.93 per cent represented land, 33.34 per cent live assets, 13.21 per cent logging, and 16.52 per cent mills; investment in stream improvements would be a portion of the percentages assigned to logging and mills (see Fries, 103). Authorized capital stock was often set at rather modest levels. See, e.g., P&LL, 1864, c. 302 ($6,000; majority of stockholders may approve increase to $12,000); P&LL, 1866, c. 303 ($5,000; majority of stockholders may approve increase to $10,000). Though the evidence does not seem to warrant assigning a primary role to finance as a motive for incorporation in this area, it was natural that those who obtained rather complex corporate charters took pains that they should include pointed and specific, sometimes rather detailed, provisions as to financing. Such charters commonly set the par value of shares, stipulated the maximum authorized amount of shares, required a stated minimum paid-in amount on share subscriptions, defined a subscription procedure, and sometimes provided authority in the directors to assess shares, with penalties for failure to meet assessments. See, e.g., Acts, 1852, c. 324; Private & Local Acts, 1853, c. 30; P&LL, 1859, c. 200, 1871, c. 45. In none of these respects were corporate charters for stream improvement enterprises different from prevailing patterns of incorporation in other fields. See, e.g., George H. Kuehnl, *The Wisconsin Business Corporation* (1959), 124–125.

225. See P&LL, 1854, c. 62 (member's "property" liable for company debt or obligation accrued during his membership, after execution unsatisfied in proceedings against corporation); P&LL, 1862, c. 35 (member "individually" liable for company debts or obligations accrued during his membership, nothing said as to prior efforts to collect from corporation; nothing declared on the subject when P&LL, 1864, c. 157, changed organization from an open-ended membership corporation to a limited stock corporation); P&LL, 1862, c. 100 (stockholder liable for debts or demands against the company to double the par value — $100 — of shares held); P&LL, 1866, c. 550 (stockholders "individually" liable for company debts or demands "only to the amount of their respective stock in said company"). The record contains but one occasion on which any reference was made to the limited liability of stockholders as a significant factor in assessing the merits or demerits of the corporate form for navigation improvement enterprises. At one stage of the long-drawn-out controversy between Chippewa Falls and Eau Claire interests over authorization of a major boom at Eau Claire, the Assembly Committee on Lumber and Manufactures, reporting against the then pending bill (88S, of 1867), took as one ground of criticism that "for all such injuries and expenses [to navigating log owners], occurring without the actual fault of the company, no pretence of compensation is made, and in case of damages arising through the neglect of the company, the injured individual will have no means of redress except a right of action against a corporation personally irresponsible"; see AJ, 755 (1867).

226. See, e.g., P&LL, 1854, c. 62; P&LL, 1862, c. 35, as amended, P&LL, 1864, c. 157; P&LL, 1864, c. 84, as amended, Laws, 1882, c. 263; P&LL, 1867, c. 590; P&LL, 1868, c. 430, amended, Laws, 1882, c. 134; Laws, 1873, c. 159, amended, Laws, 1887, c. 226.

227. See P&LL, 1866, c. 303 (Rush River Boom Co.), or P&LL, 1870, c. 34, amending P&LL, 1869, c. 469 (Wolf River Boom Co.), for particularly vivid and detailed reflections of the provision of authoritative management within the frame of corporate organization.

228. See, e.g., P&LL, 1857, c. 170, as amended, P&LL, 1868, c. 398 (Yellow River Improvement Co.); P&LL, 1864, c. 84, amended, P&LL, 1866, c. 447 (Black River Improvement Co.).

229. See notes 181–183 above.

230. See Fries, 143–153.

231. Cf. Newcomb v. Smith, 2 Pinn. 121 (Wis. 1849); Pratt v. Brown, 3 Wis. 603 (1854); Stoughton v. State, 5 Wis. 291 (1856); Fisher v. Horicon Iron Manufacturing Co., 10 Wis. 351 (1860).

232. So, Pratt v. Brown, 3 Wis. 603 (1854) ruled that, since public benefit was the sole justification for delegating the eminent domain power under the milldam act, the legislature might not be restrained from judging when public benefit no longer existed to warrant the delegation, and hence a franchise acquired by private action under the milldam act was subject to withdrawal of the protecting eminent domain power, at least as to flowage rights for which the dam owner had not obtained a completed right in fee. Compare, generally, on Wisconsin policy concerning restraints on alienation and the extent of control which title might give a married woman, Barbara Page, "Statutes in Derogation of Common Law," *Wisconsin Law Review* 1956: 78, 87, 101.

233. The two social needs to which police power responds — to adjust the given pattern of relations at any point of time, and to adjust the pattern of relations to changes over time — are well stated in the Message of Governor Rusk, quoted in note 259 below. On the police power in relation to lumber industry use of navigable waters, see (1) concerning adjustment of existing relations, J. S. Keator Lumber Co. v. St. Croix Boom Corporation, 72 Wis. 62, 38 N.W. 529 (1888), and Heerman v. Beef Slough Manufacturing, Booming, Log-Driving & Transportation Co., 1 Fed. 145 (Circ. Ct. W.D. Wis. 1878) (vessels versus logs and rafts), as well as Wisconsin River Improvement Co. v. Lyons, 30 Wis. 61 (1872), Falls Manufacturing Co. v. Oconto River Improvement Co., 87 Wis. 134, 58 N.W. 257 (1894), and Flambeau River Lumber Co. v. Railroad Commission, 198 Wis. 134, 223 N.W. 417 (1929) (navigation improvement versus power development); (2) concerning adjustment to changes in time, Pratt v. Brown, 3 Wis. 603 (1854) (power development), and Cohn v. Wausau Boom Co., 47 Wis. 314, 2 N.W. 546 (1879), Black River Improvement Co. v. LaCrosse Booming & Transportation Co., 54 Wis. 659, 11 N.W. 443 (1882), and United States v. Mississippi & Rum River Boom Co., 3 Fed. 548 (Circ. Ct. D. Minn. 1880) (navigation improvement in light of growth of traffic).

234. See note 128 above.

235. See notes 131 and 132 above. Restricted as was the practice of writing time limits into stream-use franchises, it was yet in striking contrast to the abortive history of leases of public lands, and the almost unqualified adherence to the policy of granting fee simple titles as the norm of their disposition. See, for example, the belated concern regarding reservation of public interest rights to water-power development incident to public lands bordering streams, in Laws, 1909, c. 374 (repealed, Laws, 1913, c. 773), and Laws, 1911, cc. 452 and 640. Probably the almost complete lack of sustained effort to keep strings to the public lands upon transfer of their use to private hands, in contrast to the more continuous sensitivity to preserving at least residual public control over navigable waters, reflected different estimations of the normal ways of putting these resources to work. Devotion of individual energy to limited plots of ground was the obvious means to obtain productivity from the land; equally obvious was it that a navigable stream would make its contribution to the general economy as a facility of general traffic. Cf. Water Power Cases, 148 Wis. 124, 142, 134 N.W. 330, 337 (1912).

236. See notes 66 and 129 above. Sect. 1777a became sect. 1777e in Wisconsin Statutes, 1898.

237. See note 130 above.

238. Stone v. Mississippi, 101 U.S. 814 (1879). Cf. Chicago, Milwaukee & St. Paul Railway Co. v. Milwaukee, 97 Wis. 418, 72 N.W. 1118 (1897); Illinois Steel Co. v. Bilot, 109 Wis. 418, 426, 84 N.W. 855, 857 (1901). The concept of an inalienable core of police power was at least as old in American judicial review as Massachusetts decisions dating from 1839. See Levy, *Law of the Commonwealth,* 263–264, 271–277.

239. See Hurst, *Law and the Conditions of Freedom,* 8–12, 14, 27–28, 36, 75–77, 96–100, 104–105, and *Law and Social Process in United States History* (1960), 143–144, 160–162, 186–189.

240. Symptomatic is the analysis in Governor Rusk's General Message of Jan. 11, 1883 (AJ, 24), which rests on reserved powers of amendment or repeal the whole possible authority of the state to adjust existing franchises to new conditions, without reference to the idea of inalienable police power. See note 238 above.

241. See Clyde Edward Jacobs, *Law Writers and the Courts* (1954), chaps. ii and iii; Arnold M. Paul, *Conservative Crisis and the Rule of Law* (1960), chaps. ii and iv.

242. See notes 29–115 above.

243. See the Water Power Cases, 148 Wis. 124, 134 N.W. 330 (1912). Cf. Ruehl v. Voight, 28 Wis. 153, 157 (1871), and note 130 above.

244. Among 129 nonforest-area stream-use franchises (all for power dams) there were 4 repeals (3 per cent), and 12 amendments (9 per cent). Among the total of 802 stream-use franchises given in the forest and nonforest areas together, there were thus 49 repeals (6 per cent) and 108 amendments (14 per cent).

245. Among 111 Wisconsin Supreme Court cases involving nonforest-area stream-use franchises (all for power dams), none presented an issue turning on a repeal, 3 (less than 3 per cent) concerned amendments which might be counted disadvantageous to the franchise holder, and 12 concerned amendments which were to his advantage.

246. See note 24 above.

247. Marshall, *Autobiography,* I, 333–341,

tells of a legal and business battle over the repeal of the flood dam toll grant made in P&LL, 1869, c. 452, to "the Nimakogan and Totogatic dam company," a corporation chartered by that same statute. The 1869 act was expressly repealed by Laws, 1885, c. 43; in its turn the 1885 act conferred a private (non-toll) dam franchise upon a rival milling interest. The 1869 franchise holders obtained a preliminary injunction in trial court against interference with their operations by the grantees of the new franchise; upon giving suitable bond, the latter then obtained the trial court's order vacating the first injunction, and in turn enjoining interference with operations under the new grant. The legal battle ended with matters in this posture, when the parties negotiated a settlement under which the 1869 franchise holders withdrew their contest in consideration of $5000 paid by the holders of the new grant for a conveyance of the old dam and its auxiliary works. Marshall confesses that, as counsel for the 1885 grantees, he was not wholly sure of the validity of the repeal, in view of the substantial investment already committed to the older operation under the 1869 franchise. The fact that the 1869 grant ran to a corporation, and in the same statute as that which created the corporation, clearly gave substantial basis for arguing that the repeal was within the legislature's power, however, under current conceptions of that power. See notes 129, 219, and 220 above. The settlement agreed to by the old grantees implicitly attested the strength of this argument, though the substantial cash payment made by the new grantees likewise attested the existence of some substantial uncertainty in the doctrinal situation. Obviously Marshall's story shows that serious litigation over repeal or adverse amendment would not necessarily be brought to the state supreme court. On the other hand, there seems to be no evidence that the "Nimakogan" episode had any counterparts. Marshall's tale does, of course, point up the likelihood that in the background of repeals or amendments there might be negotiation between rival interests; here obviously the newcomers felt that the better tactic lay in initiating an attack on the old franchise and then negotiating from the position thereby created.

248. For examples of total or partial repeals or amendments which carry internal evidence suggesting that the change is for the convenience, or made with the negotiated consent, of the original grantees, see Laws, 1885, c. 317, repealing Laws, 1848, p. 145 (with recital that repeal is at request of new owners of the land and dam site); Laws, 1889, c. 53, repealing so much of P&LL, 1856, c. 294, as is inconsistent with the new grant (obviously clearing title as against the shadowy claim of the long defunct Nekoosa Lumbering Co.; see Duckett, 162); Laws, 1891, c. 177, in effect repealing Laws, 1887, c. 512 (conditioning franchise for a dam on payment of $3500 compensation to owners of earlier-authorized works which new dam will flood and render valueless); Laws, 1907, cc. 293 and 356, repealing Laws, 1891, cc. 110 and 111 (same grantees in both sets of franchises). Negotiated repeals seem likely in those instances where stipulation was made that the repeal of a flooding dam franchise should not take effect until the close of the current log driving season. See, e.g., Laws, 1895, cc. 27 (repealing Laws, 1889, c. 49) and 28 (repealing Laws, 1891, c. 149).

249. See, e.g., Laws, 1881, c. 255 (qualified, Laws, 1883, c. 96), 1887, c. 344, 1889, c. 449, 1903, c. 145, 1905, c. 464, 1907, c. 335.

250. See, e.g., Laws, 1885, cc. 158 and 278, 1887, cc. 29, 262, and 299, 1891, cc. 242 and 322, 1899, c. 261, 1903, cc. 172, 182, and 353, 1907, cc. 123, 335, and 675.

251. Black River Flooding-Dam Association v. Ketchum, 54 Wis. 313, 11 N.W. 551 (1882); Black River Improvement Co. v. La-Crosse Booming & Transportation Co., 54 Wis. 659, 11 N.W. 443 (1882). Cf. Jones v. Pettibone, 2 Wis. 308 (1853) (general statute forbidding obstruction of navigable waters, construed not to intend to affect dam authorized by previous special statutory franchise); Edwards v. Wausau Boom Co., 67 Wis. 463, 466, 30 N.W. 716, 717 (1886) (Revised Statutes, sect. 1598, penalizing unauthorized booms, has no application where boom exists under special statutory franchise). That, generally, the presumption is opposed to finding intent that a later statute repeal an earlier by implication, see Attorney General ex rel. Taylor v. Brown, 1 Wis. 513 (1853); Attorney General v. Railroad Companies, 35 Wis. 425 (1874); Gilkey v. Cook, 60 Wis. 133, 18 N.W. 639 (1884). That, in particular, a later general statute will not usually be held to repeal by implication a prior special act, especially not where the special act confers a franchise, see Wood v. Hustis, 17 Wis. 416 (1863); Crosby v. Smith, 19 Wis. 449 (1865). Cf. Chippewa & Flambeau Improvement Co. v. Railroad Commission, 164 Wis. 105, 118, 159 N.W. 739, 744 (1916).

252. See Report, Committee on Internal Improvements, re bill 352A (to repeal P&LL, 1864, c. 84, incorporating the Black River Improvement Co.), AJ, 812–813 (1867); Majority Report, Judiciary Committee, re bill 14S (to repeal Kilbourn dam franchise given by P&LL, 1866, c. 424), SJ, 328–329 (1873); Governor Rusk's Veto Message on bill 100S, SJ, 490 (1882); Governor LaFollette's Veto Message on bill 328S, SJ, 1035 (1901). All of the cited sources indicate that their tenderness for vested franchise rights rests on findings that substantial practical business commitments have been made in reliance on the franchises. On this element, see also Majority Report, Committee on Roads, Bridges and Ferries, on bills 34S and 35S, SJ, 116 (1857); Memorials opposing repeal of the Kilbourn dam franchise, SJ, 255–257 and 381–383 (1860); Report, Committee on Incorporations, on bill 202S, SJ, 233 (1880) (though compare Report, Judiciary Committee, *id.*, p. 394), and Report, Committee on Incorporations, on same bill, AJ, 571 (1880).

253. The emphasis is on fair hearing (1) before the legislature, in Governor Peck's Veto Message on bill 72S, SJ, 679 (1893); or (2) before the courts, in Governor LaFollette's Veto Message on bill 328S, SJ, 1035 (1901). Compare the careful reservation of litigation rights of riparians on a navigable slough which the Black River Improvement Co. was specifically authorized to close, by Laws, 1880, c. 225 (amending P&LL, 1864, c. 84), and the emphasis put on this reservation as a factor justifying the amendment, in Report, Committee on Incorporations, on bill 202S, AJ, 571 (1880). The reservation of all rights of litigation was exercised in the suit reported as Black River Improvement Co. v. LaCrosse Booming & Transportation Co., 54 Wis. 659, 11 N.W. 443 (1882).

254. See, e.g., (1) the stipulation in Laws, 1885, c. 72, repealing P&LL, 1868, c. 430, that nothing in this repealing act shall be construed to affect title to any real or personal property now owned by the company whose franchise is repealed, nor to affect the company's right to sell, transfer, or dispose of such property; (2) stipulations that repeal of a dam franchise should not take effect in the current driving season, as in Laws, 1885, c. 74 (forthwith repealing authorization of one of three dams, but withholding effective repeal of the other two until end of current season; note the winding up of this situation by total repeals in Laws, 1887, c. 344, and Laws, 1889, c. 40), or in Laws, 1885, c. 43; see also note 248 above; (3) the favor for proceedings in court rather than in the legislature, to redress grievances for alleged breach of franchise terms, as expressed in Majority Report, Judiciary Committee, on bill 14S, SJ, 328–329 (1873), as also in Governor LaFollette's Veto Message on bill 328S, SJ, 1035 (1901); (4) the ruling in State ex. rel. Attorney General v. Norcross, 132 Wis. 534, 112 N.W. 40 (1907) that Revised Statutes, sects. 3466 and 3475 — providing general sanctions for activities in violation or without justification of a franchise right — should not be construed to authorize a court to decree total forfeiture of a franchise for breach of condition, but only to decree that defendant "be excluded from such franchise" as he holds unlawfully; a less restrictive construction might bring the statutes into violation of Wisconsin Constitution, I, 12, that no conviction shall work forfeiture of estate.

255. See note 251 above, and Schneider v. Staples, 66 Wis. 167, 28 N.W. 145 (1886).

256. See Stoughton v. State, 5 Wis. 291 (1856) (special statutory franchise for dam, specifying location and construction, held bar to prosecution for public nuisance on grounds of alleged impairment of public health); Sheldon v. Rockwell, 9 Wis. 167, 183 (1859), Cobb v. Smith, 16 Wis. 661, 662–663 (1863), s.c., 23 Wis. 261 (1868) and 38 Wis. 21, 37 (1875), and Pioneer Wood Pulp Co. v. Chandos, 78 Wis. 526, 527, 47 N.W. 661 (1891) (doctrine of laches applied strongly in favor of claimants under dam franchises); Ruehl v. Voight, 28 Wis. 153 (1871) (milldam act construed to grant more than naked license, hence doctrine of rights acquired by prescription applies in favor of claimant thereunder); Green Bay & Mississippi Canal Co. v. Telulah Paper Co., 140 Wis. 417, 122 N.W. 1062 (1909) (though applicable statute of limitations in terms bars only claims for damages, held that lapse of the statutory period is effective to establish prescriptive right to maintain dam as against abatement). The foregoing are all simply power dam cases. However, in the trial court's treatment of Falls Manufacturing Co. v. Oconto River Improvement Co., 87 Wis. 134, 58 N.W. 257 (1894), Case on Appeal, pp. 39–40, the doctrine of laches is invoked in behalf of the holder of a river improvement franchise.

257. See notes 29–115 above.

258. See note 20, and compare notes 16–19 above. Compare also instances in which the legislature repealed a special statutory franchise which in terms reserved only the power to

"alter" or "amend." See P&LL, 1857, c. 159, repealing Acts, 1851, c. 259; P&LL, 1870, c. 48, repealing P&LL, 1855, c. 188. Of course, such episodes are ambiguous; they may represent a strikingly liberal interpretation of the scope of the power reserved by the legislature, or they may represent simply the fruits of consent or negotiation; the legislative records provide no evidence to resolve the ambiguity.

Among statutes imposing general regulations in protection of navigation, many were phrased without explicit indication that they had retroactive effect. So far as they carried criminal penalties, of course, the constitutional ban on ex post facto laws would prevent their being given retroactive operation to impose a penal sanction. This limitation apart, the Wisconsin court also followed the familiar presumption against construing a statute as intended to have retroactive effect, unless such intent was deemed "unmistakable." See Seamans v. Carter, 15 Wis.. 548, 549 (1862); Vanderpool v. LaCrosse & Milwaukee Railroad Co., 44 Wis. 652, 663 (1878); cf. Allen v. Weber, 80 Wis. 531, 50 N.W. 514 (1891). The court never had occasion to decide whether this doctrine should apply regarding nonpenal applications of police power legislation in protection of public rights of navigation, under the established exception that remedial statutes might be construed to have retroactive effect. See State ex. rel. Davis & Starr Lumber Co. v. Pors, 107 Wis. 420, 428, 83 N.W. 706, 708 (1900).

Stream-use regulations which declared that they should apply retroactively as well as prospectively, with specific retroactive effect upon existing commitments of capital, include Private and Local Acts, 1853, c. 205; P&LL, 1855, c. 366, and amendment by P&LL, 1858, c. 86 (both repealed, P&LL, 1859, c. 141); General Acts, 1856, c. 76; P&LL, 1857, c. 24, 1858, c. 44, 1859, c. 144, 1866, c. 175. These instances of express retroactivity tend to fall in the early years of lumber industry growth; explicit disclaimers of intent to repeal or restrict prior franchises appear most often in later years (see note 249 above). These aspects of timing perhaps reflect some change in shading of policy, from a bolder generalization in years of less developed economic activity to more caution for existing patterns in years of expansion and growing complexity of operations. But there seems confident assertion of legislative power in the very presence of such variations, and there is the tone of confident authority in the announcements of legislative intent concerning retroactive and prospective effect of legislative determinations.

259. See Governor Rusk's Veto Message on bill 100S, SJ, 489 (1882): "There is great danger that the practice of conferring these powers and privileges by way of grants of franchises to individuals and their assigns will lead to the creation by contract of a vast number of franchises in which the public have an interest, but over which they have no control. A large number of the lumbering streams in the state are being given over by this means to the control of individuals. This evasion of the constitutional amendment against special legislation may be in most instances necessary. The improvement may be for the public good, the tolls fixed may be fair and reasonable, but nearly all such measures affect the interests of many who know nothing of their pendency and are unheard as to their effect. In the nature of things it is impracticable for the legislature to make thorough and exhaustive investigation in each case, and to know the precise effect upon all interests of the measures asked for. Moreover, what may be an improvement in the situation of affairs to-day, may be very far from an improvement a few years hence; and what may be fair compensation for maintaining dams and other public improvements to-day, may not be fair or reasonable after the lapse of time. So that the public interest would seem seriously to demand that the legislature in all such grants should reserve to itself the right, should the public interest require it, to revoke the same, or to continue them upon new terms, and subject to additional restrictions." See also Rusk's General Message, AJ, 24 (1883). Compare Governor LaFollette's Veto Message on bill 371A, AJ, 723 (1901): "It is a very questionable public policy to grant privileges of this character [dam franchise on navigable stream] without reserving a power in the law-making body to repeal, alter, or amend the grant. Exigencies may arise calling for either repeal or amendment. The grant in this case is for a period of twenty-five years. The unqualified existence and continuance of the grant for the entire period might hamper and even stifle the growth of other great and paramount interests to the detriment of the public welfare." See, generally, notes 131 and 132 above.

260. Legislative authority to authorize channel or flow improvements, at the expense of styles of water-power development theretofore pursued by riparian owners, was upheld

by the decisions in Wisconsin River Improvement Co. v. Lyons, 30 Wis. 61 (1872); Durning v. Burkhardt, 34 Wis. 585 (1874); Falls Manufacturing Co. v. Oconto River Improvement Co., 87 Wis. 134, 58 N.W. 257 (1894); Wisconsin River Improvement Co. v. Pier, 137 Wis. 325, 118 N.W. 857 (1908). Legislative authority to authorize channel or flow improvements, at the expense of navigation rights which would exist in the natural condition of the stream, was upheld in Tewksbury v. Schulenberg, 41 Wis. 584 (1877); Wisconsin River Improvement Co. v. Manson, 43 Wis. 255 (1877); Black River Improvement Co. v. LaCrosse Booming & Transportation Co., 54 Wis. 659, 11 N.W. 443 (1882); Wisconsin River Log Driving Association v. D. F. Comstock Lumber Co., 72 Wis. 464, 40 N.W. 146 (1888). Legislative authority to authorize public-utility-style boom operations, at the expense of boom construction rights of riparian owners, was recognized by Cohn v. Wausau Boom Co., 47 Wis. 314, 2 N.W. 546 (1879). Cf. Stevens Point Boom Co. v. Reilly, 46 Wis. 237, 49 N.W. 978 (1879); Menominee River Boom Co. v. Augustus Spies Lumber & Cedar Co., 147 Wis. 559, 132 N.W. 1118 (1912). At the end of the lumber era, other decisions explicitly recognized the legislature's authority to impress statutory stream-use franchises for new economic uses upon uses familiar at common law or under prior legislation, as the development of hydroelectric power became a concern weightier than older water-power or lumber industry transport uses. See Allaby v. Mauston Electric Service Co., 135 Wis. 345, 116 N.W. 4 (1908); McDonald v. Apple River Power Co., 164 Wis. 450, 160 N.W. 156 (1916); Flambeau River Lumber Co. v. Railroad Commission, 198 Wis. 134, 223 N.W. 417 (1929), s.c., 204 Wis. 524, 236 N.W. 671 (1931).

261. See, for example, in Wisconsin River Improvement Co. v. Manson, 43 Wis. 255, 262, 263, 264 (1877), the repeated emphasis upon the social and economic wisdom of construing the navigation guaranties of the Northwest Ordinance and the Wisconsin constitution as permitting statutory improvement toll franchises where and of such scope as the legislature might reasonably believe would accommodate larger traffic, "to render [streams] . . . more useful to the public," to "increase the navigability of partially navigable waters, either in point of time or mode of navigation," to provide "increased facilities for travel and transportation," and "to render . . . the common right of passage . . . more valuable and safe." A contrary interpretation of these navigation guaranties, said the Wisconsin court — borrowing the language of a federal case — "would embarrass, if not entirely arrest, some of the great works of internal improvement in progress or projected in the west" (*id.*, 263).

262. Pratt v. Brown, 3 Wis. 603, 614, 615 (1854), overruling Stephens v. Marshall, 3 Pinn. 203 (1851), which had held that the milldam act conferred a vested right upon one who built a dam thereunder, even to the extent that he was entitled to rebuild promptly under the act's protection after the original dam had been destroyed by flood, though the milldam act had meantime been repealed. See, following Pratt v. Brown, French v. Owen, 5 Wis. 112 (1856). Cf. Pick v. Rubicon Hydraulic Co., 27 Wis. 433, 440–442 (1871) (legislature may repeal that part of special dam franchise incorporating by reference the milldam act flowage procedure, substituting another procedure, with effect to divest easement held under the milldam act terms). At the end of the lumber era, with reference to a new direction of policy concerned primarily with hydroelectric power development, the court held in the Water Power Cases, 148 Wis. 124, 134 N.W. 330 (1912) that neither under the reserved powers of repeal contained in the state constitution or in statutory franchises, nor under the police power, was the legislature justified in declaring a blanket repeal of all outstanding stream franchises. The court also held invalid those aspects of this legislation which declared that all existing works should be deemed public nuisances until the former franchise holders should obtain licenses from the Railroad Commission on terms set by the statute regulating money returns from the works and carrying provisions for eventual public recapture of such licenses for limited compensation or their disposition to municipalities or third persons under condemnation proceedings. The opinion contains a good deal of doctrinal exposition which seems at least to narrow considerably the scope acknowledged for police power in Pratt v. Brown. However, the heart of the Water Power Cases decision seems to be the court's finding of "confiscation," because the repeal of all outstanding franchises did not stipulate for or contemplate removal of the former franchised works, but their continuance and their continued utility in aid of

navigation and power development on drastically new and limited terms of enjoyment by the former holders. See 148 Wis. at 144, 134 N.W. at 338; cf. Superior Water, Light & Power Co. v. City of Superior, 263 U.S. 125 (1923), reversing 174 Wis. 257 (1921). Later rulings and opinions of the Wisconsin court show a temper to narrow the impact of the Water Power Cases and to emphasize the breadth of continuing police power, especially to allow changes in the terms of operation under stream franchises. See City of Baraboo v. Railroad Commission, 195 Wis. 523, 218 N.W. 819 (1928); Wisconsin Power & Light Co. v. Public Service Commission, 5 Wis. (2d) 167, 92 N.W. (2d) 241 (1958). Indeed, earlier the Wisconsin court pulled back so far from the Water Power Cases as to incur rebuke from the United States Supreme Court, in the City of Superior suit cited above.

263. See Governor Rusk's Veto Message on bill 100S, SJ, 489 (1882).

264. Apart from its help in perceiving more completely the range of uses of law to define or implement policy, this distinction helps us better appraise the relative amount of dynamic thrust which law imparted to affairs. Compare the issues raised in determining whether a relationship involved action by a state, so as to fall within the limits set by the due process or equal protection clauses of the 14th Amendment. See Barrows v. Jackson, 346 U.S. 249, 253–254 (1953).

265. Cf. Clark, *Economic Institutions and Human Welfare,* 44, 71, 140.

266. See, e.g., notes 36, 43, 63–67, and 83 above.

267. See, e.g., notes 69–82 above.

268. See notes 43 and 44 above.

269. See notes 95–98 and 161–166 above, and 321 below.

270. See, e.g., Revised Statutes, 1878, sects. 1598 and 1603–1606.

271. See notes 36, 38, 39, 53, and 76–82 above.

272. See notes 36 and 38 above, and Leihy v. Ashland Lumber Co., 49 Wis. 165, 5 N.W. 471 (1880).

273. See notes 123 and 169–172 above, and Jones v. Pettibone, 2 Wis. 308 (1853); Arpin v. Bowman, 83 Wis. 54, 53 N.W. 151 (1892).

274. The expression of public policy through rules of evidence tended quite naturally to be accomplished most plainly in court decisions. See note 35 above. However, the legislature in effect recognized industry practice in the definitions it made of types of activity authorized under stream improvement franchises. See notes 124–126 above; cf. Revised Statutes, 1878, sect. 1600 (entry on land to recover stranded logs).

275. Wisconsin River Improvement Co. v. Lyons, 30 Wis. 61 (1872), and notes 33, 34, and 51–54 above.

276. Flambeau River Lumber Co. v. Railroad Commission, 198 Wis. 134, 223 N.W. 417 (1929), s.c., 204 Wis. 524, 236 N.W. 671 (1931); Flambeau River Lumber Co. v. Chippewa & Flambeau Improvement Co., 204 Wis. 602, 236 N.W. 679 (1931). Cf. Hall v. Kitson, 2 Pinn. 296 (Wis. 1851); Volk v. Eldred, 23 Wis. 410 (1868); Stevens Point Boom Co. v. Reilly, 46 Wis. 237, 49 N.W. 978 (1879).

277. Pattern v. Marden, 14 Wis. 473 (1861); Lawson v. Menasha Wooden-Ware Co., 59 Wis. 393, 18 N.W. 440 (1884); Mack v. Bensley, 63 Wis. 80, 23 N.W. 97 (1885). Cf. Lawson v. Mowry, 52 Wis. 219, 9 N.W. 280 (1881); Green Bay & Mississippi Canal Co. v. Hewitt, 66 Wis. 461, 29 N.W .237 (1886); Green Bay & Mississippi Canal Co. v. Kaukauna Water Power Co., 112 Wis. 323, 87 N.W. 864 (1901).

278. Cf. Oliver Wendell Holmes, Jr., *The Common Law* (1881), 94–96.

279. See Hurst, *Law and Social Process,* 180–181.

280. Miller v. Sherry, 65 Wis. 129, 26 N.W. 612 (1886) (common law); Falls Manufacturing Co. v. Oconto River Improvement Co., 87 Wis. 134, 58 N.W. 257 (1894) (statute right).

281. See Revised Statutes, 1878, sect. 1600; cf. Cohn v. Neeves, 40 Wis. 393 (1876); Field v. Apple River Log Driving Co., 67 Wis. 569, 31 N.W. 17 (1887).

282. Ulrich v. Hull, 17 Wis. 424 (1863).

283. Miller v. Sherry, 65 Wis. 129, 26 N.W. 612 (1886). Cf. Coldwell v. Sanderson, 69 Wis. 52, 28 N.W. 232 (1887).

284. See note 35 above.

285. State v. City of Eau Claire, 40 Wis. 533 (1876); Woodman v. Kilbourn Manufacturing Co., 1 Biss. 546 (Circ. Ct. D. Wis. 1867). Cf. Whisler v. Wilkinson, 22 Wis. 572 (1868) (right of self-help against demonstrated obstruction); State ex rel. Attorney General v. Norcross, 132 Wis. 534, 112 N.W. 40 (1907) (policy against total forfeiture for partial breach of dam franchise conditions).

286. See notes 16 and 29–35 above.

287. See notes 17 and 43–48 above; cf. notes 36–42, on common law.

288. See notes 33 and 89–94 above.

289. See notes 173–176 above. Compare Luning v. State, 2 Pinn. 215 (Wis. 1849), and Stoughton v. State, 5 Wis. 291 (1856) (action under general milldam act no defense, but action under special dam franchise is defense, to public prosecution for nuisance to health).

290. See notes 195–209 above.

291. See notes 49–62 above. Note the favor toward broad discretion in legislative creation of policy, represented by the doctrines that the federal Constitution should be construed to meet an expanding future — Pensacola Telegraph Co. v. Western Union Telegraph Co., 96 U.S. 1 (1877) — and that the state constitution should be construed as a limitation rather than a grant of powers, so that the state legislature possessed all authority not denied it — Bushnell v. Beloit, 10 Wis. 195 (1860).

292. Development of public utilities law as a distinct field of public policy, regarded as unusual in its demands upon private will, symbolized the general preference for leaving private decision makers free in their choice of purposes of action. Cf. Hurst, *Law and the Conditions of Freedom,* 89–90.

293. On statutory favor for standards rather than rules, where policy favored a free field for improvisation, see notes 123–127 above. That adoption of a judge-made standard rather than a judge-made rule of conduct favored venture had some recognition occasionally in the opinions, though — in fashion characteristic of the times — never in a clear-cut exposition of value patterns. See, e.g., Miller v. Sherry, 65 Wis. 129, 133, 26 N.W. 612, 613 (1886); Field v. Apple River Log Driving Co., 67 Wis. 569, 575, 31 N.W. 17, 19–20 (1887).

294. See notes 18, 56, 57, and 95 above, and compare the emphasis on broad favor toward water-power development in Newcomb v. Smith, 3 Pinn. 131, 140 (Wis. 1849) with the strict insistence on the qualifying compensation requirement, beginning with Thien v. Voegtlander, 3 Wis. 461 (1854) and carried up into the lumber era developments in such cases as Folsom v. Apple River Log Driving Co., 41 Wis. 602 (1877) and Hackstack v. Keshena Improvement Co., 66 Wis. 439, 29 N.W. 240 (1886).

295. See, generally, notes 173–175 above. Typical stipulations requiring that flood gates be kept open between stated dates are P & LL, 1869, c. 452, 1872, cc. 112 and 117; Laws, 1875, c. 327, 1887, c. 218, 1893, c. 264. Cf. P & LL, 1866, cc. 48 and 328 (navigators must replace fences removed for passage of logs or reach agreement with farmers thereon); General Laws, 1868, c. 67 (amended, altering dates, General Laws, 1869, c. 79), and General Laws, 1870, c. 91 (no one authorized to drive logs on named rivers shall cause such rivers to overflow their banks, adjacent lands or meadows, within specified dates).

296. On the general prominence in nineteenth-century policy of the purpose to overcome scarcity of mobile capital, see Hurst, *Law and the Conditions of Freedom,* 59–70.

297. See notes 12–15 above.

298. See notes 96, 97, and 153–159 above.

299. See notes 123 and 168–172 above. There seems some focused reflection of the felt weight of the capital requirements implicit in construction specifications, in Laws, 1881, c. 239, which amended the general slides stipulation made for the Wisconsin and Black Rivers by Acts, 1849, cc. 62 and 98, as carried into Revised Statutes, 1878, sect. 1601, by excepting "dams maintained exclusively for flooding purposes." Occasionally the grant of a toll right for passing a power dam in which a slide was required was stated to be given "as a compensation for maintaining and keeping in repair such slide . . ." See, e.g., P & LL, 1857, c. 360 (toll right repealed, P & LL, 1862, c. 98); P & LL, 1867, cc. 586 and 587.

300. See notes 108, 142, and 143 above. The public policy of minimizing the cash capital requirements of dam builders was explicitly stressed as a basis of the milldam act, in Newcomb v. Smith, 3 Pinn. 131, 140 (Wis. 1849).

301. Wisconsin Constitution, VIII, 10. See Jensen v. Board of Supervisors of Polk County, 47 Wis. 298, 311, 2 N.W. 320, 329 (1879); Frank L. Mallare, "Wisconsin's Internal Improvements Prohibition," *Wisconsin Law Review,* 1961:294.

302. See notes 5, 6, 9, and 12–15 above.

303. See notes 96, 97, and 153–159 above. Cf. Underwood Lumber Co. v. Pelican Boom Co., 76 Wis. 76, 84, 45 N.W. 18, 21 (1890); Sauntry v. Laird, Norton Co., 100 Wis. 146, 152–153, 75 N.W. 985, 987 (1898).

304. See notes 174–176 and 192 above.

305. See notes 173, 182, and 190–192 above.

306. Wisconsin Constitution, I, 13. See Thien v. Voegtlander, 3 Wis. 461 (1854); Zweig v. Horicon Iron Manufacturing Co., 17 Wis. 362 (1863); Folsom v. Apple River Log Driving Co., 41 Wis. 602 (1877). The consistent legislative practice of providing procedures to compensate flowage in effect expressed the same judgment — that there was in this situation obviously a "taking." See Hackstack v. Keshena Improvement Co., 66 Wis. 439, 444, 29 N.W. 240, 242 (1886).

307. See notes 57, 95, 108, 114, 142, 143, 151, and 152 above.

308. The values represented by the constitutional stipulation against taking private property save for a public purpose and save on due compensation were regarded as so self-evident in meaning and weight as to provoke no discussion in the Wisconsin constitutional convention. See Brown, "The Making of the Wisconsin Constitution," 23, 57. Note the tenderness toward cautious observance of the full scope of the constitutional requirements, in the redrawing of the eminent domain provisions affecting river improvement corporations, in *Report of the Revisers of the Statutes*, 140 (on sect. 1777).

309. Brower v. Merrill, 3 Pinn. 46 (Wis. 1850). So, too, though a particular economic community had grown dependent on the power developed by a given dam, this fact — in the absence of any showing of laches on plaintiff's part — was held not to bar an action to abate the dam, when the dam owner failed to pay an award of flowage compensation duly made according to the procedure defined in his statutory franchise. Zweig v. Horicon Iron Manufacturing Co., 17 Wis. 362, 371 (1863) ("It is suggested that the existence of the dam is a matter of general concern, and that its destruction would materially affect the public, or at least the interests of many persons who have made improvements upon the idea that they could use and enjoy that water power. But we cannot for these reasons disregard the rights of the [flowed landowner] . . . For it is contrary to the first principles of natural justice, as well as the letter and spirit of our constitution, that private property should be taken for public use without compensation").

310. Pumpelly v. Green Bay Co., 13 Wall. 166, 182 (U.S. 1871), applied in Arimond v. Green Bay & Mississippi Canal Co., 31 Wis. 316, 337 (1872). Neither court felt it necessary to expound the grounds of the ruling; for the United States Supreme Court, Miller, J., observed curtly that "we do not think it necessary to consume time in proving" the proposition.

311. Falls Manufacturing Co. v. Oconto River Improvement Co., 87 Wis. 134, 58 N.W. 257 (1894). Compare, in accord on the central principle involved, Kaukauna Water Power Co. v. Green Bay & Mississippi Canal Co., 142 U.S. 254, 273 (1891). Earlier, and equally without satisfactory exposition of grounds, the court had ruled that there was no taking of private property where the legislature granted a public-utility boom company the exclusive right to use a stretch of navigable water wholly within the banks of the stream, though — in the absence of such a grant — an abutting riparian owner would have been entitled to erect his private boom there; Cohn v. Wausau Boom Co., 47 Wis. 314, 2 N.W. 546 (1879). So, also, in Black River Improvement Co. v. LaCrosse Booming & Transportation Co., 54 Wis. 659, 11 N.W. 443 (1882) — again, without satisfactory indication of any empirical bases of its policy — the court held that a duly franchised river improvement enterprise owed no compensation to riparian owners for closing a navigable slough abutting their lands, in order to improve the navigability of the main channel.

312. Compare Chapman v. Oshkosh & Mississippi River Railroad Co., 33 Wis. 629, 637–638 (1873): "It seems to us that it is unreasonable to deprive the plaintiffs of the business facilities which the river front affords for shipping lumber, and landing logs and boats [in connection with operating a sawmill], without compensating them for the loss of such advantages. Such riparian rights frequently constitute the chief value of river property . . . We therefore think the plaintiffs were entitled to be compensated for the loss they sustained by the destruction or breakage of their river frontage by the erection of the railway bridge and embankment."

313. Folsom v. Apple River Log Driving Co., 41 Wis. 602 (1877); Jones v. United States, 48 Wis. 385, 4 N.W. 519 (1880), affirmed, 109 U.S. 513 (1883); Hackstack v. Keshena Improvement Co., 66 Wis. 439, 29 N.W. 240 (1886). No Wisconsin case squarely raised the question whether compensation must be paid for detriment to a downstream millsite cause as a reasonable incident of alteration of natural flow by a franchised upstream power dam, though the doctrine an-

nounced in the Jones case (cited above) would seem to determine that no compensation was due. The same result would seem indicated by the broad tolerance with which Wisconsin common law viewed the infliction of such detriment by an upstream on a downstream power dam. See Timm v. Bear, 29 Wis. 254 (1871); cf. Philip Nichols, *The Law of Eminent Domain* (1917), II, 894.

314. In Falls Manufacturing Co. v. Oconto River Improvement Co., cited note 311 above, neither the trial court nor the state supreme court attempted to expound an empirical basis for the policy ruling that the upstream floodwaters developer "took" no "property" of the downstream riparian in impairing the latter's millsite. However, the brief for the Improvement Company pressed the argument suggested in the text, that compensation should not be deemed required where the burdens would be of such indefinite and large scope as to preclude socially desirable enterprises: "No improvement act, general or special, provides compensation for injury to riparian rights. No improvement company could live under a liability to pay it" (brief for Respondent, p. 18). The Wisconsin Supreme Court suggested an analogous rationale, to distinguish situations where there should be no compensation from those where compensation was constitutionally required, in Borchardt v. Wausau Boom Co., 54 Wis. 107, 110, 11 N.W. 440, 441 (1882). The court said that so far as boom works erected under statutory franchise caused reasonably predictable flowage of others' lands, there must be compensation, but held that compensation was not required for flowage caused without the grantee's fault in consequence of extraordinary floods not within the reasonable expectations of the industry: "This principle is of the utmost importance to the existence and purposes of corporations which are created to build and maintain works of internal improvement, in part for the public benefit, by the investment of private capital. All of the ordinary and natural consequences of their works may well have been contemplated and expected, and their ability to meet such consequences and compensate for such damages as would be likely to occur may be ample and constantly maintained; but one extraordinary and unforeseen event, happening from natural causes, against which no provisions or precautions are or could be made, may sweep away in a day or an hour not only all of their profits but their capital, and bankrupt and destroy the corporation itself. In view of such extraordinary risks and hazards, capital would not be likely to seek such an investment, and such enterprises, of great public importance and benefit, would be avoided."

315. Field v. Apple River Log Driving Co., 67 Wis. 569, 575, 31 N.W. 17, 20 (1887).

316. Laws, 1891, c. 422; Laws, 1893, c. 260; Wisconsin Statutes, 1898, sects. 1222g-j.

317. Yellow River Improvement Co. v. Wood County, 81 Wis. 554, 51 N.W. 1004 (1892). Compare, accord, Chicago, Milwaukee & St. Paul Railway Co. v. Milwaukee, 89 Wis. 506, 513, 62 N.W. 417, 420 (1895); Town of Washburn v. Washburn Water-Works Co., 120 Wis. 575, 585, 98 N.W. 539, 542 (1904).

318. Cf. State ex rel. Milwaukee Street Railway Co. v. Anderson, 90 Wis. 550, 565, 63 N.W. 746, 749 (1895). See Yellow River Improvement Co. v. Wood County, 81 Wis. 554, 559, 51 N.W. 1004, 1005–1006 (1892).

319. Aesthetic value entered the frame of statutory policy concerning dams by explicit declaration only with Laws, 1929, c. 523, authorizing the licensing commission to deny a dam permit upon a finding that a given dam "is contrary to the public interest, when the public right to the enjoyment of natural scenic beauty is considered . . ." Laws, 1947, c. 124, enlarged the formula to include authorization to weigh "recreational use" in evaluating an application for a dam permit, but qualified the administrators' authority in these respects by permitting their denial of a permit on such grounds to be overridden in some circumstances by vote of the county board of any affected locality (cf. Wisconsin Statutes, 1959, sect. 31.06). Muench v. Public Service Commission, 261 Wis. 492, 508, 512, 515g, 53 N.W. (2d) 514, 521, 523, 55 N.W. (2d) 40, 43 (1952) recognized that the 1929 law represented an innovation in legislatively declared policy, though the court properly observed that its earlier decisions had recognized public rights to hunt and fish as within the legal consequences that followed from a determination that a stream was navigable. See Willow River Club v. Wade, 100 Wis. 86, 76 N.W. 273 (1898); Diana Shooting Club v. Husting, 156 Wis. 261, 145 N.W. 816 (1914). Rarely in the earlier years, more commonly in the eighties and nineties, the legislature included a fishways requirement in dam franchises. See, e.g., Laws, 1874, c. 168, 1891, cc. 251, 448, and 451. But little actual

attention was given to fisheries interests on the streams. See 5th Biennial Report, Commissioners of Fisheries, PD, II, 16 (1893) (want of fishways in unauthorized Brule River dams); Governor Scofield's Veto Message on bill 151S, SJ, 644 (1897) (fishways requirement too vague); Governor LaFollette's Veto Message on bill 371A, SJ, 723 (1901) (dam franchise vetoed for want of fishway requirement); Governor LaFollette's Veto Message on bill 25S, SJ, 1038 (1903) (dam franchise vetoed because fishway requirement too vague); Report, Commissioners of Fisheries for Biennium ending Dec. 31, 1904, SJ, App., 15 ("At present very few dams in the state are provided with any sort of fishway, and of those that are provided, many are so built and regulated as to be of no use whatever"). Only one case before the Wisconsin Supreme Court presented an issue of enforcing a fishways requirement, and this came to nothing, partly for procedural reasons, partly because the point was found in fact irrelevant to the complaint; see State ex rel. Priegel v. Northern States Power Co., 242 Wis. 345, 354, 8 N.W. (2d) 350, 354 (1943).

320. Through the lumber era, the only provision for legal action to protect public interests of health or safety affected by dam construction was by suit to abate a dam as a nuisance. Such actions of course depended wholly on the initiative of local prosecutors or specially aggrieved private litigants; there is no indication that there was much such activity. Three Wisconsin Supreme Court cases early established the availability of the nuisance action to redress injury to public health or safety, but the Reports show no other evidence that this remedy had material operative significance. See Luning v. State, 2 Pinn. 215 (Wis. 1849); Douglas v. State, 4 Wis. 387 (1854); Stoughton v. State, 5 Wis. 291 (1856). It is probably symbolic that these cases all arose in the more settled south, and did not involve dams connected with lumbering operations; most of the lumber industry's dam building went on in areas in which there would be minimum occasion in fact for generating issues on such general community concerns as health and safety. The first instances of explicit statutory attention to public safety considerations in the specification of dam construction standards seem to be Laws, 1905, c. 400, and 1907, c. 35. Bills to allow recovery of damages caused by the breaking of dams failed of passage in 1883 and 1885, though in each instance there was some indication of substantial support for the proposals. See (1) regarding 1883, bill 202A, AJ, 126, 461, 611, 718, 740–741; (2) regarding 1885, bill 338A, AJ, 195, 648, 733, 896, 950. Despite the shadow of doubt cast by the constitutional ban on state participation in works of internal improvement, the legislature appropriated state funds for the construction of limited levees, notably on the Fox River near Neenah, and on the Wisconsin River near Portage. See Governor Smith's Veto Message on bill 192A, AJ, 772 (1881) (query, whether grant for Fox River levee near Neenah is a violation of the constitutional limitation; in any case, bill passed in end-of-session rush without deliberation due such a precedent); Governor Rusk's General Message, AJ, 22 (1883) (Fox River levee provided for by Laws, 1882, c. 138, out of the drainage funds, has been completed, with twenty-year maintenance agreement by the town of Lewiston); State ex rel. Jones v. Froelich, 115 Wis. 32, 91 N.W. 115 (1902) (general funds appropriation for Wisconsin River levee, by Laws, 1901, c. 282, held invalid as violating Wisconsin Constitution, VIII, 10); Laws, 1903, c. 419 (first of successive appropriations for Wisconsin River levee, from state drainage fund and hence presumably within exception stated in Wisconsin Constitution, VIII, 10). These levee provisions, however, developed in response to general seasonal flood threats on the rivers concerned, rather than as reflections of problems created by lumber industry stream use; see Report, Committee on Federal Relations on petitions concerning the Portage levee, AJ, 400 (1901). The limited legislation against forest-stream pollution in the late nineteenth century aimed to preserve the streams' carrying capacity for the industry itself, rather than to protect more general interests of public welfare; see Earl F. Murphy, *Water Purity* (1961), 49. Concern with the purity of the streams as a general community resource came to substantial expression only about the turn of the century (*id.*, 50, 53, 61, 84).

321. See notes 317 and 318 above.

322. Compare notes 100–115 above.

323. See Hurst, *Law and the Conditions of Freedom*, 60, 69, and *Law and Social Process*, 203, 244–246.

324. Dissenting from the original judgment sustaining the constitutionality of the delegation of eminent domain powers under the milldam act, Larrabee, J., in Newcomb v. Smith, 2 Pinn. 131, 165 (Wis. 1849), in effect

recognized, as he criticized, the subordination of farm use to water-power use represented by the statute, and found no warrant for the subordination in the precedent of like legislation in Massachusetts: "The example of Massachusetts is hardly the best one for us to follow. Her manufacturing power and wealth may have become so identified with her interests as a state as to actually give tone to her government, and even to the judicial decisions of her enlightened courts. But in the young and vigorous agricultural state of Wisconsin, no such influence can or ought to exist." Reviewing policy after many years, the court recognized a general preference in such statutes for industrial power development over particular alternative uses of land surface. The milldam type of legislation had as "its object . . . to enable the utilization of water power upon many of the important streams of the state . . . These statutes are grounded in the theory, confirmed by the courts, that there is a public interest in the utilization of the water powers of the state to run mills sufficient to justify the exercise of the power of eminent domain in the flooding and consequent taking of lands of individuals"; see Allaby v. Mauston Electric Service Co., 135 Wis. 345, 350, 351, 116 N.W. 4, 6 (1908). The subordination of rights derived from riparian land title to public rights of navigation was also expressed in terms which emphasized the overriding value put on transport as a general supporting function for the economy, compared with any particular land site use. See Wisconsin River Improvement Co. v. Lyons, 30 Wis. 61, 65 (1872): ". . . the title of the purchaser to the center of the stream was taken subject to the public easement or right of passage and navigation, and when the nature and extent of this easement or right are considered, it will be found for this purpose to be almost or quite immaterial whether he is regarded as holding to the center of the river or only to the margin of it. This easement or right of the public to regulate, control and direct the flow of the navigable waters, to impede or accelerate such flow, to deepen the channel or to remove obstructions found in it, or to change the direction of the current from one bank of the stream to the other, or to make an entirely new channel, and, in short, to do anything within the banks of the stream itself which may be considered for the benefit and improvement of commerce and navigation, will be found to be a most extensive and absolute one." The practical weight of this proposition was felt most in preferring transport over power developments; see notes 328 and 329 below. However, so far as occasion required, the court applied the general value preference here expressed, to subordinate other particular land surface use to the reasonable needs of transport. See Field v. Apple River Log Driving Co., 67 Wis. 569, 31 N.W. 17 (1887); Willow River Club v. Wade, 100 Wis. 86, 76 N.W. 273 (1898); In re Trempealeau Drainage District, 146 Wis. 398, 131 N.W. 838 (1911). Cf. S. S. Kresge Co. v. Railroad Commission, 204 Wis. 479, 235 N.W. 4 (1931).

325. See notes 102, 108, and 142–152 above.

326. See note 35 above.

327. See note 281 above.

328. Though, as usual, the official agencies did not make bold, explicit generalizations of their implicit policy judgments, one may find some expression of the distinctive importance attached to waterways because of their framework importance for general economic activity, in Governor Farwell's Veto Message of June 6, 1853, objecting to the charter of the Chippewa River Improvement Company as an undesirable grant of monopoly, SJ, 815–817, as also in his analogous Veto Message of the same date, AJ, 841–842, of a charter for the Black River Lumber Driving & Booming Company. See, also, Special Message of Governor Fairchild, SJ, 30–31 (1869) (general economic importance of Wisconsin River improvement); General Message of Governor Taylor, PD, I, 22–24 (1875) (similarly grounded plea for federal aid to waterways developments). The long legislative battle fought between competing millmen of Eau Claire and Chippewa Falls over the grant of a statutory franchise for general boom works at Eau Claire was argued in good measure by appeal to the overriding value of the general transport interest compared with the industrial development of any given locality. See notes 100 and 101 above, and Dykstra, "Law and the Lumber Industry," chap. vii. The Wisconsin Supreme Court most explicitly grounded policy on recognition of the framework significance of the navigation utility of waters in those decisions in which it broadly sustained legislative discretion to act to enlarge navigable capacity and regulate traffic. See the discussion of cases in J. S. Keator Lumber Co. v. St. Croix Boom Corporation, 72 Wis. 62, 38 N.W. 529 (1888), and note the value hierarchy indicated in the court's central analysis: "The use of rivers and

smaller streams for the floatage of logs is essential to the continued prosperity of the immense lumber and industrial interests of northern Wisconsin. The regulation and preservation of such use, in connection with and as facilitating navigation by reasonable and proper booms and other structures, has long been the legislative policy of this state, as frequently sanctioned by this court" (72 Wis. 62, 80, 38 N.W. 529, 535). The emphasis of analysis upon the framework character of given economic utilities came close to explicit legislative declaration in the contrast afforded by the strong preference which the milldam act showed for power development over other land surface use, regarding nonnavigable waters, and on the other hand the consistent care of the legislature to stipulate for safeguarding transport uses as against power uses in franchises on navigable waters. See notes 108, 109, and 168–172 above. The court gave expression to this pattern of legislative policy in Allaby v. Mauston Electric Service Co., 135 Wis. 345, 116 N.W. 4 (1908) (stream navigable solely by log transport test, for purposes of protecting transport interests, held not intended by legislature to fall outside the category of "nonnavigable" streams so as to preclude resort to milldam act to sanction flowage of farm land by developers of electric power). Of the subordinate position of the fisheries interest to the transport interest, the court had no doubt: ". . . undoubtedly the right of navigation is paramount to that of fishing"; see Wright v. Mulvaney, 78 Wis. 89, 93, 46 N.W. 1045, 1046 (1890) (dispute between net fishermen and log tow).

329. See notes 16, 17, 29–32, 35, 36, 43–45, 50–54, 68–81, 100–107, 109–114, and 168–172 above. For particularly emphatic expression of the primacy accorded the transport interest, see the rulings and the opinions in Volk v. Eldred, 23 Wis. 410 (1868) (dam owner's franchise obligation to provide free passage is measured by carrying capacity of stream after improved by his dam); Wisconsin River Improvement Co. v. Lyons, 30 Wis. 61 (1872) ("in interests of commerce and navigation" legislature may ban all dams built without its approval, though a particular dam might not in fact obstruct, and might even help, navigation); Olson v. Merrill, 42 Wis. 203 (1877) (irrelevant that plaintiff maintains an illegal dam downstream; he may still have damages for obstruction of his logs by defendant's upstream works); Arpin v. Bowman, 83 Wis. 54, 53 N.W. 151 (1892) (where dam owner exceeds height authorized in franchise, raftsman may have his action for damages without need of prior notice to defendant to correct the illegal condition); Falls Manufacturing Co. v. Oconto River Improvement Co., 87 Wis. 134, 58 N.W. 257 (1894) (upstream flood dam owner owes no compensation for impairing natural flow past downstream millsite); Flambeau River Lumber Co. v. Railroad Commission, 198 Wis. 134, 223 N.W. 417 (1929), and 204 Wis. 524, 236 N.W. 671 (1931), as well as Flambeau River Lumber Co. v. Lake Superior District Power Co., 200 Wis. 31, 227 N.W. 276 (1929) (court will not presume legislative intent to authorize administrators to enter orders preferring power development to exclusion of navigation uses). The shift in legislative policy toward a greater concern with hydroelectric power than with log transport is reflected in the second opinion in Flambeau River Lumber Co. v. Railroad Commission, above. Though the court ruled that, in the absence of clearer evidence of legislative intent so, it would not construe the statute to authorize the licensing administrators to subordinate transport completely to hydroelectric development, the opinion showed marked unwillingness also in the judges to intimate absolute limits on the legislature's authority to make a drastic new alignment of interests. Thus the opinion included a lengthy obiter discussion designed to show that precedent supported the "plenary" character of legislative control of the productive development of state waters. Especially significant in light of this analysis was the court's characterization of the conflict of transport and hydroelectric interests seen to be inherent in the facts (and hence, presumably, warranting the more firmly a broad exercise of legislative choice): "Inasmuch as log-driving is from the standpoint of one wishing to develop waterpower very wasteful of water and must be done at seasons when the reservoirs should be filled by withholding water, the conflict of interest seems almost irreconcilable"; 204 Wis. 524, 532, 537–542, 236 N.W. 671, 675, 676–678 (1931). Reflections of the changed emphasis toward hydroelectric over transport use as the prime object of public policy concern at the turn of the century may be seen in Laws, 1905, cc. 95 and 475, 1907, c. 641, and 1909, c. 518 (provisions for various official studies of waterpower potential); 12th Biennial Report, Bureau of Labor and Industrial Statistics, Sept. 30, 1906, PD, 397, 398, 406, 409–411 (1907) (many power sites now available, because of end of previous competing uses of streams for

log transport: "The general use and control of those northern rivers for logging purposes in the past tended to decrease the value of the water powers by withholding the water at times when most needed, but this use of rivers is now practically past. Railroad transportation has taken the place of river logging in all the leading river valleys, thus leaving the rivers free for the development of their water powers," p. 398); Fourth Biennial Report, Tax Commission, Nov. 10, 1908, chap. iv, PD, 97–98 (1909) ("But where such [lumbering] corporations have virtually ceased to exercise their original functions, the dams constructed by them and the rights to overflow lands by means of such dams, are for the most part held and maintained for water power purposes when serviceable for such uses . . . Indeed, there are cases in which the right to construct and maintain a dam and overflow lands has been obtained ostensibly for the purpose of improving the navigability of a stream, when in fact the only substantial use ever made of the dam has been the utilization of the water power for private purposes"); Laws, 1911, cc. 591 and 652, 1913, cc. 672 and 755, 1915, c. 380, and 1917, c. 474 (establishing the modern pattern of dam licensing laws explicitly emphasizing hydroelectric power).

330. See note 319 above.

331. See notes 16, 17, 19, 29–32, 43–54, 87, 96–98, and 100–115 above.

332. See notes 155, 173–192, 195–209, 211, and 213 above. The separation-of-powers policy which the Wisconsin court saw to be involved in establishment of legal preferences for improved over natural waters was given especially sharp expression in Sellers v. Union Lumbering Co., 39 Wis. 525 (1876), and Weatherby v. Meiklejohn, 56 Wis. 73, 13 N.W. 697 (1882); see also Black River Flooding-Dam Association v. Ketchum, 54 Wis. 313, 316, 11 N.W. 551, 552 (1882). In many if not most forest stream situations, especially in the earlier years, every person who would transport logs to market must use the available stream, and must perforce use it as it was improved; typically there was no alternative natural channel. In this light, such tolls as a statutory franchise authorized and such costs of labor, delay, and damage to goods as were necessarily incident to use of the waters as improved, were in effect exactions compelled by law to promote the public interest in maximizing over-all productivity. So viewed, both the tolls and the incidental costs authorized or sanctioned by improvement franchises bore so much the character of taxes that it is not surprising that the legislature so firmly asserted its basic control in creating such rights, and the court so readily conceded the legislature's claim. In fashion characteristic of the times, however, the policy thus implicit in consistent patterns of official behavior was not readily brought to explicit generalization. However, in 1842 a territorial legislative committee justified the grant of toll rights in a bridge franchise on analogy to the public right to tax and spend to provide general transport facilities. See note 96, and compare note 92 above. In two vetoes of monopoly improvement and navigation grants in 1853, Governor Farwell emphasized the compulsory tolls to be exacted of all transporters as an important reason for objecting to the scope of monopoly given. Vetoing the charter of the Black River Lumber Driving & Booming Company, he objected especially to the broad monopoly given because the franchise authorized the grantee to collect all logs adrift on the river "and [in the Governor's characterization] assess a tax upon the same," and gave "power to sell the logs and timber so collected, to pay said tax." See note 96 above. Significantly, when the Wisconsin court undertook to justify toll improvement franchises, relative to the ban which the Northwest Ordinance and the Wisconsin constitution laid upon the levy of "any tax, impost, or duty" for use of the state's principal navigable waters, it did not argue that tolls were not a tax, but rather that the limitation applied only to taxing the use of waters in their natural condition. This approach is the more striking because in the leading case the challenger of the toll right specifically objected because under his franchise the claimant had used the entire channel of the river for his improvements, so that no one could use the river without using the improved water and hence incurring toll. The court treated the all-embracing compulsory feature of the toll as a reasonable incident of the legislature's general authority to promote economic development: ". . . the constitution only relates to [naturally] navigable waters, and does not deprive the legislature of the power, through the agency of a corporation or individual, to improve unnavigable, or partially navigable waters, in streams otherwise navigable. The legislature is, primarily at least, the judge of the necessity of the improvement, and when it delegates the power to a corporation, and the state does not question that the improvement made by the corporation is in conformity

with the delegated power, it seems to us that neither the necessity nor usefulness of the improvement, nor the manner in which it is made, can be called in question by private parties"; Wisconsin River Improvement Co. v. Manson, 43 Wis. 255, 262, 265 (1877). Cf. J. S. Keator Lumber Co. v. St. Croix Boom Corporation, 72 Wis. 62, 80–82, 38 N.W. 529, 535–536 (1888); Underwood Lumber Co. v. Pelican Boom Co., 76 Wis. 76, 85, 45 N.W. 18, 21 (1890).

333. Cf. Black River Flooding-Dam Association v. Ketchum, 54 Wis. 313, 11 N.W. 551 (1882) (improvement), and Underwood Lumber Co. v. Pelican Boom Co., 76 Wis. 76, 45 N.W. 18 (1890), as well as Menominee River Boom Co. v. Augustus Spies Lumber & Cedar Co., 147 Wis. 559, 132 N.W. 1118 (1912) (booms). Such a pre-emptive right would be recognized, however, only where it was plainly given by statute: Stevens Point Boom Co. v. Reilly, 44 Wis. 295, 49 N.W. 978 (1878).

334. Cohn v. Wausau Boom Co., 47 Wis. 314, 2 N.W. 546 (1879); Edwards v. Wausau Boom Co., 67 Wis. 463, 30 N.W. 716 (1886); see also note 333 above.

335. See notes 177–192 above. The point came to sharpest focus in Wisconsin River Improvement Co. v. Manson, 43 Wis. 255, 262, 265 (1877).

336. Green Bay & Mississippi Canal Co. v. Kaukauna Water Power Co., 70 Wis. 635, 35 N.W. 529 (1888), affirmed, 142 U.S. 254 (1891).

337. Lawson v. Mowry, 52 Wis. 219, 9 N.W. 280 (1881); Fox River Flour & Paper Co. v. Kelley, 70 Wis. 287, 35 N.W. 744 (1887).

338. Yellow River Improvement Co. v. Wood County, 81 Wis. 554, 51 N.W. 1004 (1892); see notes 317 and 318 above.

339. Wisconsin River Log Driving Association v. D. F. Comstock Lumber Co., 72 Wis. 464, 40 N.W. 146 (1888).

340. See notes 35, 40 and 41 above.

341. See notes 173–180 above.

342. Legislative practice made the underlying legislative policy apparent especially in the consistent insertion of navigation guaranties in franchises for power development or navigation improvement at particular locations. See notes 169–172 above; cf. Governor Randall's Veto Message on charter of Eau Claire Manufacturing Co., AJ, 578 (1860) (want of stipulation for passage facilities). The Wisconsin Supreme Court contributed to the general favor for multiple use by rulings which required clear showing to make out legislative intent to grant a monopolistic franchise, and which required the claimant of monopoly privileges to establish beyond question that he had satisfied such prerequisite conditions as the legislature imposed to its grant. See notes 211 and 213 above.

343. That legislative authorization was needed to build a bridge on a navigable stream was early announced by the Wisconsin court in Barnes v. City of Racine, 4 Wis. 454 (1855). In the Barnes case, as later in Sweeney v. Chicago, Milwaukee & St. Paul Railway Co., 60 Wis. 60, 18 N.W. 756 (1884), the court held that it would not construe legislation as intended to authorize construction of a bridge in such fashion as materially to hinder navigation for vessels (Barnes) or logs or rafts (Sweeney). This presumption as to legislative intent rested on constitutional policy; though it was never forced to such a ruling, the Wisconsin court made clear in Milwaukee Gas Light Co. v. Schooner "Gamecock," 23 Wis. 144 (1868), as well as in the Sweeney case, that the navigation guaranty of Wisconsin Constitution, IX, 1, forbade the legislature to license a bridge which would materially obstruct navigation. In Sweeney the Wisconsin court likewise recognized that a state statute purporting to license an obstructing bridge on waters subject to federal control would be invalid under Pennsylvania v. Wheeling & Belmont Bridge Co., 18 How. 421 (U.S. 1851). On the other hand, recognizing that provision of bridges in aid of land transport was in the public interest, the court would construe a statute governing municipal responsibility for bridging "navigable" streams as intended to apply to streams navigable only by the log transport test: Village of Bloomer v. Town of Bloomer, 128 Wis. 297, 107 N.W. 974 (1906). Policy declared or implicit in the operations of the legislative process showed consistent concern for the multiple-use principle. This was made most explicit in Governor Washburn's Veto Message of Mar. 18, 1873, AJ, 919–925. The Governor objected to the particular terms of a bill authorizing a railroad bridge at La-Crosse because there was not adequate safeguard to navigation, but at the same time he recognized that the river interest must yield fair scope to land transport: "While the Mississippi is a public highway, and not to be permanently obstructed, it is not to be denied that bridges are a necessity, and to accommodate such as are necessary for the railway commerce the navigation interest must submit to

some delays and inconvenience" (*id.*, 920). The Governor was especially insistent that large framework issues for economic growth were involved, and that it was this framework character of the problems which called for insistence on mutual accommodation: "This is no local question, nor should I for an instant think of interposing an objection on any such narrow grounds. As a local question it is magnified too much, and in my judgment is utterly insignificant; but as a public question affecting the interest of vast communities for all coming time, it is a question that cannot be overestimated" (*id.*, 923). See also his Special Message of Mar. 5, 1873, on the general subject, AJ, 688. The legislature's pursuit of the multiple-use principle was implicit in numerous navigation guaranties in franchises for bridges on streams important for log and raft traffic. See, e.g., P&LL, 1856, c. 89; Private & Local Acts, 1864, c. 397; P&LL, 1870, c. 139; Laws, 1873, c. 19, 1880, c. 320, 1882, cc. 77 and 128, 1885, cc. 279 and 318, 1887, c. 322, 1895, cc. 268, 271, and 345, 1901, c. 429, 1905, c. 86, 1907, c. 345. The underlying formula was that the navigation interest might be required to yield to the land transport interest, to the extent of incurring delay and risk of relatively minor physical damage, but not to the extent of permanently barring navigation of any important type. Compare J. S. Keator Lumber Co. v. St. Croix Boom Corporation, 72 Wis. 62, 87–88, 38 N.W. 529, 538 (1888), where the court held that a Minnesota boom franchise was not unconstitutional merely because as a necessary incident of the authorized business the boom operator might for a time completely block use of waters by steamboats. The opinion found it necessary to explain its relation to the earlier Sweeney decision: "Obviously the conclusions thus reached are in conflict with some things said . . . in Sweeney v. C., M. & St. P. R. Co., 60 Wis. 67 *et seq.*, as to obstructions in navigable rivers which the legislature could not authorize should they make the attempt; but the question thus suggested was not there involved, and the absolute right of authorizing the *permanent* obstruction of such rivers is not here involved. The right to authorize booms for the interception, storage, and handling of logs in a manner to materially interfere with the navigation by steamboats and other watercraft, however, is involved, and such right is not only sanctioned by the supreme court of the United States, but by numerous adjudications of this court."

344. See notes 35, 272, and 273 above. Cf. Whisler v. Wilkinson, 22 Wis. 572 (1868); Olson v. Merrill, 42 Wis. 203 (1877); Miller v. Sherry, 65 Wis. 129, 26 N.W. 612 (1886).

345. The policy of laying the burden of proof upon the interest complaining of loss from action initiated by another under a statutory franchise was early so taken for granted in both private and public prosecutions as to be applied without substantial analysis; the court's matter-of-fact application of the principle attests how deep ran the roots of the value attitude involved. See Luning v. State, 2 Pinn. 215 (Wis. 1849) (public action); Hall v. Kitson, 3 Pinn, 296 (Wis. 1851) (private action); note 273 above. Cf. Hurst, *Law and the Conditions of Freedom,* 19.

346. The characteristically operational attitude toward law is reflected at many points in the naïvely revealing *Autobiography* of Roujet D. Marshall: see, e.g., I, 274, 282, 336–341, 343, 353. Compare, generally, Hurst, *The Growth of American Law: The Law Makers* (1950), 342–352.

347. The recitals of evidence in particular lawsuits often afford insight into the broad and dense patterns of informal activity, only a small part of which took form in legal records. See, e.g., Olson v. Merrill, 42 Wis. 203, 204–207 (1877); A. C. Conn Co. v. Little Suamico Lumber Manufacturing Co., 74 Wis. 652, 659–661, 43 N.W. 660, 661–662 (1889). On the fact and the value of the marginal operation of law, see Hurst, *Law and Social Process,* 270, 302, 326, 328.

348. See Laurent, 20–21. Note the generality of the only figures presented on capital investment in the Wisconsin lumber industry, in Fries, 103.

349. Contrast the total emphasis upon the Eau Claire and Beef Slough controversies, and the unproved assertion of their "typical" character, in Frederick Merk's generally excellent *Economic History of Wisconsin During the Civil War Decade,* chap. iii, with the matter-of-fact history of the Half Moon Lake improvement, as described in Reynolds, *The Daniel Shaw Lumber Company,* 78–81.

350. See Elting E. Morison, *Turmoil and Tradition* (1960), 6, 120–122, 124; Samuel Williston, *Life and Law* (1941), 108.

351. See notes 29–35 above, on constitutional and common law assurances of rights to navigate waters in their natural condition; on common law rights to use riparian lands for water-power development on nonnavigable streams, or for access to natural navigation op-

portunities, see Timm v. Bear, 29 Wis. 254 (1871), and Stevens Point Boom Co. v. Reilly, 44 Wis. 295 (1878). Notes 36–42 (common law) and 43–62 (legislation), above, present materials regarding the insistence on legislative permission for works built in navigable waters. On eminent domain grants for power dam flowage, see notes 56, 57, and 95 above.

352. See, e.g., P&LL, 1871, c. 239; Laws, 1879, c. 55, 1897, c. 143. The tally of twenty-seven validating acts may understate the proper total by the number of more ambiguous franchises which it omits. Four dam franchises which authorized the grantees to "maintain" a specified dam, without reference to original construction, and two franchises which authorized dams "to be built or rebuilt," may also represent validations of structures already constructed without legal license. See, e.g., Laws, 1882, c. 182, 1883, cc. 21 and 355.

353. See, e.g., P&LL, 1866, c. 175, 1867, cc. 503, 586, and 587; Laws, 1887, c. 118, amending P&LL, 1871, c. 45.

354. Sometimes the want of positive legal permission appears only implicitly in the stated facts. See, e.g., Whisler v. Wilkinson, 22 Wis. 572 (1868); Falls Manufacturing Co. v. Oconto River Improvement Co., 87 Wis. 134, 58 N.W. 257 (1894), Case on Appeal, pp. 16, 17. Sometimes counsel or court take explicit notice of the want of original legal authority for works involved in the suit. See, e.g., Durning v. Burkhardt, 34 Wis. 585 (1874); Miller v. Sherry, 65 Wis. 129, 26 N.W. 612 (1886); Pioneer Wood-Pulp Co. v. Bensley, 70 Wis. 476, 36 N.W. 321 (1888), Plaintiff's Brief, p. 3.

355. See also Laws, 1889, c. 82 ("The building of said dam in 1859 and 1860, and the maintenance thereof to the present time, by said [present grantees] . . . and those under whom they claim, is as between the state and said [grantees] . . . hereby validated and legalized"), and c. 236 (apparent validation of dam built in 1854); Laws, 1903, c. 385 (dam built in 1856 "is hereby validated and legalized. Nothing in this act shall be construed to in any way affect any pending litigation or any rights or causes of action already accrued").

356. By P&LL, 1870, c. 299, "the Beef Slough Manufacturing, Booming, Log-driving and Transportation Company are hereby authorized to maintain the piers, side, shore, sheer or glancing booms already constructed and heretofore used by them, on and along the Chippewa river" and to build and maintain additional works necessary to their authorized purposes, subject to state navigation guaranties. This episode was but preface to the significant use of Beef Slough; the grantees of the 1870 validating act shortly became unable to maintain the enterprise financially, and the sorting enterprise there became important only when it thereafter came into the control of the Weyerhaeuser interests (see Fries, 145).

The history of one improvement franchise on a stream of secondary importance — the Yellow River, a tributary of the Chippewa — shows the interplay between the aggressive initiative of the lumbermen and their concern, nonetheless, to legitimize operations of some consequence. Laws, 1873, c. 12, purported to grant broad traffic management authority with toll rights to "any person, association, company or corporation . . . who shall have improved said Yellow river." The grant stated no criterion to determine priority or adjustment of control rights among possible competing claimants; inherently the situation invited a monopoly grant, but the legislature provided no such clear-cut test to decide who should be the monopolist as it did later in Revised Statutes, 1878, sect. 1777, which gave its improvement monopoly to such person as had taken "prior possession" of the stream. Sellers v. Union Lumbering Co., 39 Wis. 525 (1876) held that Laws, 1873, c. 12, was of no legal effect hence, for want of identifiable grantees. In the argument of the case, the claimant said that it had made improvements before the statute was passed, and that "[it] only having then improved the river as required by the statute, must be held to be the sole grantee of the franchise" (39 Wis. at 528–529). The court rejected the argument, construing the statute as making an open-ended tender to any improver, but the argument, of course, revealed a case where works had been built without clear statutory warrant — though on a minor stream. Even so, it should be noted that the Union Lumbering Co. held a special corporate charter which itself authorized stream improvements in this area in terms which are ambiguous as to whether they simply define internal corporate powers or were intended also directly to grant an improvement franchise as such (see P&LL, 1869, c. 86). By amending this charter so that its improvement provisions "shall be . . . construed as prohibiting the construction . . . of any dam . . . across the Chippewa river, below the base of the falls of the said Chippewa river, at the city of Chippewa Falls," P&LL 1870, c. 297, in effect treated the charter as directly conferring improvement franchises. See notes 120, 215, and 216 above. In any event, the interested individuals subsequently showed their concern to legitimize their status as improvers, when Laws, 1877, c. 247—a declared amendment, which in substance was a full substitute for Laws, 1873, c. 12—made what amounted to the same grant, to named grantees. Of similar character was Laws, 1882, c. 251, amending Laws, 1874, c. 250 (Wolf River). There was irony, as well as additional testimony to the search for legitimacy in sizable ventures, that, as enacted, Laws, 1873, c. 12, was a generalized substitute for a particular amendment of an earlier specific charter for the Yellow River Improvement Company; the generalized substitute was adopted out of concern that that particular amendment would violate the newly enacted constitutional ban on local laws. See Report, Committee on Incorporations, SJ, 77 (1873), and Report, Judiciary Committee, SJ, 127 (1873), both on bill 8S. Cf. P&LL, 1857, c. 172, as amended by P&LL, 1868, c. 398, and P&LL, 1869, c. 186.

Dykstra, "Law and the Lumber Industry," 112, suggests that the improvement monopolies conferred by Revised Statutes, 1878, sect. 1777, and the amendment by Laws, 1880, c. 279, upon any corporation "which shall have taken prior possession of such streams," were covert ratifications of previous unauthorized actions. Sect. 1777 was presented in the *Report of the Revisers of the Statutes,* 140, simply as a "somewhat condensed" version of Laws, 1876, c. 399, which in terms seems prospective only, as it declares that corporations organized therefor under General Laws, 1872, c. 144, "are hereby authorized and empowered to improve" streams. Sect. 1777 and the 1880 amendment were not in terms retroactive, and the Wisconsin court followed the familiar rule that statutes are presumptively intended to apply only prospectively. See Seamans v. Carter, 15 Wis. 548 (1862); Vanderpool v. LaCrosse & Milwaukee Railroad Co., 44 Wis. 652 (1878). Though the Revisers do not mention the point, the addition in the 1878 revision limiting the tender of franchises to those who had taken prior possession would seem probably responsive to the decision in Sellers v. Union Lumbering Co., 39 Wis. 525 (1876) that a statute tendering an improvement toll franchise to any person who "shall have improved" the Yellow River was legally ineffective for want of a sufficiently designated grantee, because it provided no criterion to distinguish among competing claimants. No case before the Wisconsin Supreme Court presented an issue of retroactive operation of sect. 1777 or its 1880 amendment. However, in Stevens Point Boom Co. v. Reilly, 44 Wis. 295, 302 (1878), Ryan, C.J., for the court, raising a point not presented or argued by counsel, and citing no justification from the legislative history of the statute, ruled that Laws, 1876, c. 399, "must have the effect of a confirmation of acts authorized by it, giving them legal right, as against the state, from the date of its passage. What the statute authorized to be done after its passage, it legalized from thenceforth, as against the state, when done before." There is no evidence in legislative records or from any other matters before the Wisconsin Supreme Court that this ruling ever had significance beyond the case in which it was made. Ryan's opinion in the Stevens Point Boom Co. case indicates that the court felt under strong pressure of policy to find a basis to sustain the authority of a traffic management facility which it regarded as in the public interest, but whose more specific statutory warrant the judges felt reluctantly compelled to invalidate as violating the new constitutional ban on special laws conferring corporate powers. See 44 Wis. at 301, 302; cf. s.c., 46 Wis. 237, 243, 245, 49 N.W. 978, 979 (1879). See notes 220 and 221 above. In any event, by their terms the acts of 1876, 1878, and 1880 would have validated only works built since enactment of the general incorporation act of 1872, since their benefits extended only to corporations organized under the 1872 statute.

357. A. C. Conn Co. v. Little Suamico Lumber Manufacturing Co., 74 Wis. 652, 43 N.W. 660 (1889); see notes 39, 80, and 81 above. Though the only issues squarely posed to the court on the matter concerned adjustment of private rights, the tone of opinions which sustained the right of navigators to create temporary works or obstructing conditions insofar as these were reasonably incident to proper current operations indicated that public policy did not insist on prior specific license, but rather accorded navigators a general permission, for such temporary arrangements. See note 35 above, and especially the A. C. Conn Co. case, cited above, this note. Compare statutes which required construction of booms with exit points "except temporary booms used for the purpose of running logs." See P&LL, 1867, cc. 503, 586, and 587, 1868, c. 265.

358. See the acceptance of the 1889 decision in the revisers' annotation to sect. 1598, Wis-

consin Statutes, 1898. The absence of legislative reaction to the 1889 decision gathers some significance, also, from the fact that some legislative notice had been called to problems of obstruction of lesser streams, by bills 555A of 1895, and 396A of 1897, both "relating to the obstruction of creeks, rivulets and small streams by overgrowth or by neglect of owners of lands adjoining such streams, by the fall of timber or accumulation of deposit from other sources." Both bills were summarily rejected on recommendation of the Assembly Judiciary Committee (Feb. 28 and Mar. 1, 1895, AJ, 413, 438) and the Assembly Committee on Agriculture (Feb. 19 and 23, 1897, AJ, 331, 361), respectively. On the likely congruity of the 1889 decision with already prevailing practice and attitudes, note the testimony in Charnley v. Shawano Water Power & River Improvement Co., 109 Wis. 563, 567, 85 N.W. 507, 509 (1901): ". . . Shawano creek is a navigable stream, within the [log transport] . . . rule . . . The dam was built on lands owned by the parties who maintained it. It had been built more than forty years prior to the time the defendant purchased Kast's interests, in 1892. The only evidence in the case regarding the right to maintain it was given by Mr. Kast when he said that he 'maintained the dam publicly, claiming a right to do so without permission from any one.' " Note the court's reaffirmation of the 1889 decision, 109 Wis. at 569, 85 N.W. at 509. Other implicit indication of industry practice in accord with the 1889 decision may be found in State ex rel. Owen v. Wisconsin-Minnesota Light & Power Co., 165 Wis. 430, 162 N.W. 433 (1917), Case on Appeal, pp. 28, 29, where the evidence showed that lumber operators built as many as seven dams on one small stream between Long Lake and the lake above it, as part of a network to obtain enough flood water to move logs into the Red Cedar River, though the only statutory authority which they held seems to have been that for a single specified dam on Long Lake River, by Laws, 1883, c. 222. The legislature finally made what seems practically an explicit ratification of the 1889 ruling when Laws, 1913, c. 755 (after expressly leaving undisturbed all dams for which there was "authority of law" or "legislative permission heretofore granted") declared that its general ban on unauthorized dams "shall not apply to dams constructed in or across nonmeandered streams prior to July 13, 1911." Compare the Charnley opinion, cited above, this note, where the court treats nonmeandered streams as likely to include streams navigable by the log transport test (109 Wis. at 569, 85 N.W. at 509). Laws, 1915, c. 380, and 1917, c. 474, made specific provision by which an applicant might obtain a permit to operate and maintain a dam built over navigable waters without legislative permission prior to stated dates.

359. Apart from the Beef Slough matter (note 356 above), the only instances of validating laws concerned with originally unauthorized works on a principal stream involved particular, local dams on the Wisconsin River (see Laws, 1889, cc. 82, 236, and 316, 1893, c. 138). For reflections of unlicensed works on secondary (tributary) streams, see Sellers v. Union Lumbering Co., 39 Wis. 525, 528 (1876), note 356 above, and Weatherby v. Meiklejohn, 56 Wis. 73, 13 N.W. 697 (1882).

360. See the accounts — which include care to obtain prior legal license — of the creation by Philetus Sawyer of the Keshena Improvement Co. on the upper Wolf River, in Current, 113–114, and of the conduct of the Half Moon Lake enterprise in Reynolds, 78–80. Marshall, *Autobiography,* I, 334–341, tells of franchise conflicts and maneuvers on the Namekagan River, demonstrating both the strategic and tactical advantage which original promoters obtained by getting a statutory franchise, and indicating that it was an unusual business oversight which led later timberland investors to neglect initially the importance of legitimating their competing navigation interest on the river. A few validating acts show that energetic operators continued to a late date to build dams without prior license; on the other hand, it is significant that these validations followed the unlicensed construction by short intervals, in contrast to validations of earlier unlicensed construction, which usually lagged by years or even (note 355 above) by a generation. See Laws, 1889, c. 316 (dam built in 1887), 1897, c. 143 (dam built in 1896), 1903, c. 400 (dam built in 1901); see Fifth Biennial Report, Commissioners of Fisheries, PD, II, 16 (1893).

361. Symbolic of the bias for promotion of economic growth which underlay specific regulatory uses of law was the phrasing of Art. XI, sect. 1, of the draft constitution proposed for the new state of Wisconsin in 1846. Though the section forbade the state to participate in works of internal improvement, it prefaced this limitation by declaring nonetheless that "this state shall encourage internal improvements by individuals, associations, and corporations

. . ." The 1846 draft constitution lost at the polls. The 1848 draft, which was adopted, included the internal improvements limiting provision without the prefatory exhortation for encouragement. The omission was stylistic; the consistent tone of the debates in the second constitutional convention makes plain that, no less than in 1846, the delegates wanted government aid to economic growth, short of direct state financial involvement in particular enterprises. See Quaife, *The Convention of 1846*, 745, and *The Attainment of Statehood*, 416–421, 443–453, 581–585. See, generally, Lewis R. Mills, "Government Fiscal Aid to Private Enterprise in Wisconsin: A Quantitative Approach," *Wisconsin Law Review* 1956: 110, and "The Public Purpose Doctrine in Wisconsin," *id.*, 1957: 40.

362. Compare the pressure for order represented by Sellers v. Union Lumbering Co., 39 Wis. 525 (1876), holding legally ineffective for want of sufficiently designated grantees a franchise inherently monopolistic but lacking a criterion for assigning priority among claimants.

363. E.g., Revised Statutes, 1878, sect. 3335. The quoted provision was basic policy under the milldam act from its inception. See Moore v. Coburn, 1 Pinn. 538 (Wis. 1845).

364. The confidence with which legislative and judicial opinion justified legal action by appeal to the self-evident virtues of promoting economic productivity evidences values prevailing in the general community. See notes 100–115 above.

365. See notes 5, 6, and 12–15 above, and, generally, Current, 45–46, 50–51, 60–61, 64–66, 70, 75, 85, 93, 94; Fries, 225; Gates, *Wisconsin Pine Lands*, 132; Samuel Mermin, "Law and the Promotion of Enterprise" (MS, 1959), Part Two. Aside from local charter provisions authorizing harbor improvements and management (note 6 above), there were a few authorizations of direct local aid to water transport development, all of closely limited character. See Private & Local Acts, 1853, c. 301 (Marathon County may make limited subscription to stock of Little Bull Falls Improvement & Stream Navigation Co.); P&LL, 1856, c. 405 (county boards of counties through which Chippewa River passes may tax in aid of improvements of Chippewa River Improvement Co., here incorporated, but not over $1000 per county per year); General Laws, 1860, c. 150 (Shawano County may use one year's drainage fund interest to improve Wolf River in that county); General Laws, 1865, c. 476 (Waupaca, Outagamie, and Shawano counties may use swampland proceeds to improve Wolf River navigation within their boundaries); P&LL, 1866, c. 509 (town boards may use drainage funds to pay for contract work by Chippewa River Improvement Co. in draining and improving swamplands); P&LL, 1867, c. 601 (repealed, P&LL, 1868, c. 77) (specified local units in Oconto County may tax to improve mouth of Oconto River); P&LL, 1870, c. 268 (specified towns in Sauk County may subscribe to limited amounts of stock of Baraboo River Improvement Co.); Laws, 1874, c. 15 (city of Stevens Point may subscribe up to $10,000 of stock of Stevens Point Boom Co.); Laws, 1874, c. 222 (city of Wausau may subscribe up to $30,000 of stock of Wausau Boom Co.). Apart from the state's connection with the principal forest-area improvements mentioned in the text (see sources cited at the beginning of this note), a handful of statutes appropriated state lands in aid of particular local projects. See P&LL, 1857, c. 135 (Wisconsin River Improvement Co. given flowage rights on any state lands needed for reservoirs, and title to any lands given state by the United States to aid the company); P&LL, 1866, c. 214 (thirty-foot right of way over state lands given for cut-offs or canals to be built by the individual grantees of improvement franchise on Black Creek, Outagamie County); P&LL, 1867, c. 334 (swampland grant in aid of Lemonweir Improvement Co.); P&LL, 1870, c. 38 (Pine River Navigation Co. given right of way, easements of access to stream, and construction materials from state lands); Laws, 1882, c. 276 (grantees of franchise for toll boom and canal connecting Bad River with Chequamegon Bay may use therefor any state swamplands not valuable for timber, or may purchase such at $1.25 per acre). Lumberman-Congressman Thaddeus C. Pound in 1879 sponsored a plan to create reservoirs at the Mississippi headwaters to regularize flow. Though enabling legislation passed, the project was long delayed by internal jealousies and distrust within the industry. When some works were built, they failed to deliver the projected benefits, and were abandoned. See Laws, 1876, Memorial No. 5; Laws, 1882, c. 256; Fries, 57. State appropriations (ultimately, in form, from swampland funds) aided construction of an important levee on the Wisconsin River at Portage. There was some effort to rationalize the measure in terms of aid to lumber transport, but the prime ob-

jective was protection of surrounding lands. See Report, Joint Special Committee on the Portage levee proposal, Mar. 2, 1870, SJ, 327; Mallare, 301–303.

366. Taking note of the slim resources for state or territorial borrowing, and the limits of Congressional generosity in granting public lands for projects of limited local concern, a territorial legislative committee early prophesied that for development of the power and transport potentials of Wisconsin streams reliance must be put on private enterprise, fostered by favorable franchise grants. See Report, Committee on Internal Improvements, on the Governor's message pertaining to river improvements, HRJ, 177, 180 (1842). Compare Report, Committee on State Affairs, AJ, 1394 (1858). Reporting on Memorial 291A, asking state aid to improve navigation of the Wolf River, the committee "believe that the navigation of said river might be materially improved with the sum asked for in the memorial, and the resources of the State in the lumber districts more fully developed thereby. The committee believe the funds of the State insufficient to meet all the wants of the numerous interests of the State, however plausible those wants may be, and therefore report the said memorial back to the house with a recommendation that no further action be taken." Note also Report, Committee on Internal Improvements, SJ, 303 (1859); recommending (without opinion) indefinite postponement of bill 153S, to provide for improving the Wolf River; no further action was taken beyond printing the bill (SJ, 476).

367. See Chap. I above, and Hurst, *Law and Social Process,* 48–52, 71.

368. See notes 116–213 above, especially notes 136 and 137.

369. See notes 234–256 above.

370. See Report, Committee on Internal Improvements, to which was referred that part of the Governor's message concerning river improvements, HRJ, 180 (1842). Compare the finding on which the court ruled that the delegation of eminent domain powers under the milldam act might reasonably be deemed by the legislature to promote public interest: "Such a law, also, by inviting capital into the interior of the state, by encouraging enterprise and diffusing the conveniences of social life, enhances the value of land, advances its settlement, and promotes general civilization." See Newcomb v. Smith, 2 Pinn. 131, 140–141 (Wis. 1849). So, generally, when legislative and judicial opinion accepted as a justifying basis for the creation of franchises the policy of promoting a multiplier style of economic growth, there was implicit testimony to contemporary confidence in the efficacy of franchises to help generate action. See notes 101–115 above.

371. The bargain form of stream-use franchises was most apparent where the legislature conditioned toll rights upon minimum investment in improvements and upon achievement of some stated improvement in navigability, sometimes with further stipulation that the works must be completed within a specified time (see notes 104–107 above). Occasionally a franchise still more explicitly declared that its benefits were conferred on the assumption that they would induce capital investment in stream improvements. See, e.g., Acts, 1850, cc. 277 and 283; Laws, 1875, c. 48, 1878, c. 281, 1887, c. 12. For examples of amendments to the *quid pro quo* terms of special franchises see P&LL, 1854, c. 28, 1859, c. 166, 1863, c. 311, 1868, c. 258, and 1871, c. 433 (extensions of time of completion clauses); P&LL, 1860, c. 34, 1868, c. 394; Laws, 1881, c. 163, 1887, c. 154 (redefinitions of minimum capital investment stipulations).

372. See Wisconsin River Improvement Co. v. Manson, 43 Wis. 255, 262 (1877). The legislature had in effect expressed the same judgment when, incorporating into the general statutes the navigation guaranty of the Northwest Ordinance, it added, "That nothing herein contained, shall prevent the collection of all tolls imposed for necessary easements, constructed in the said navigable waters" (General Acts, 1853, c. 72). The link of toll rights to the provision of public-interest improvements was early taken as a justification for granting such a franchise, in Report, Committee on Judiciary, on the Governor's objections to creation of a Janesville toll bridge company, HRJ, 240 (1842): "It is true, that the act allows the corporation to take tolls for crossing their bridge, but the amount is fixed by the Legislature, and the intention of the act is not to benefit the corporation but the public; and it is apparent that the effect will be greatly to promote the public accommodation and convenience. The tolls allowed by the Legislature are supposed to be a fair remuneration to the corporation for the cost of the bridge and no more. The committee can see no difference in principle between taxing the public for the cost of the bridge in the way proposed in the bill, and in tax-

ing them for the gross amount at once, and collecting the tax in the usual mode."

373. See, e.g., Petition of Edward T. Hooker and others, for the Kilbourn dam franchise, SJ, 79–82 (1866), and Report, Committee on Incorporations, on bill 9S, AJ, 454–457 (1866); Report, Committee on Internal Improvements, on bill 352A (to repeal charter of Black River Improvement Co.), AJ, 812–813 (1867), and Report by Select Committee, on the same bill, AJ, 1106 and App., 1266–1272 (1867); Report, Judiciary Committee, on bill 7S (to incorporate Chippewa Improvement & Boom Co.), SJ, 414 (1871), and Minority Report, Committee on Incorporations on the same bill, SJ, 664, 665 (1871); Memorial 103A (asking repeal of Laws, 1873, c. 12, creating open-end franchise for Yellow River improvement), AJ, 216–217 (1874); Report, Committee on Lumber and Manufactures, on bill 366A (Eau Claire dam and boom franchise), AJ, 477–479 (1875); Minority Report, Committee on Incorporations, on bill 98A (similar Eau Claire project), SJ, 534–535 (1876). Contemporary acceptance of the proposition that franchises, both in their original promise and in their vested assurance, significantly contributed to improvers' will to action may be seen with special clarity in Report, Committee on Incorporations, on bill 202S (amending charter of Black River Improvement Co. to authorize it to close off a navigable slough, to protect utility of main channel), SJ, 233 (1880): "The committee further deem this a case where the Black River Improvement Company should be sustained, if they possibly can, under the laws, for the reason that they had, as they supposed, obtained all the necessary legislation, and entered upon and took possession of said river, relying upon the franchise granted them by the state, and have honestly and diligently improved and handled the property of the community with reasonable dispatch, and at a cheap cost, for a great many years."

374. See note 244 above, and notes 451 and 462–464 below. It is impossible to make a precise tally of the number of bills which passed without significant amendment, compared with the number materially amended in course of passage, because there are gaps and obvious mistakes in the available legislative records, and because we often have no means to judge whether particular verbal changes in course of passage represent minor clerical alterations or the correction of clerical errors, or changes of substance. We can make a more precise count of amendments after enactment, because in their nature these present themselves in more reliable, formal embodiment.

375. Thus, of 148 specific amendments to toll improvement franchises, 24 concerned location and 44 related to tolls (45 per cent, together), and 21 concerned the internal organization and finance of improving corporations made grantees of the franchises (60 per cent, together).

376. On the invocation of common law to supplement the milldam act, see note 363 above. The recourse to common law on riparian rights generally (by the law quoted here in the text) was first by General Acts, 1853, c. 72, amending Revised Statutes, 1849, c. 34; this provision was carried into Revised Statutes, 1858, c. 41, and then into Revised Statutes, 1878, sect. 1597.

377. During the lumber years (1836–1910) eight forest-area cases in the Wisconsin Supreme Court turned on claims of tortious interference with flow for power uses; one of these involved alleged interference by a bridge, seven involved contending millsite operators. The same years saw thirty-seven forest-area cases before the court involving contract or title issues concerning power dams; of these cases, sixteen were controversies specifically about water-power rights between persons in a common chain of title, and eleven were controversies over ownership of millsites (hence, secondarily, over water-power rights) between persons in a common chain of title. Thus, of a total of thirty-four forest-area cases in the years 1836–1910 directly or closely concerning competing claims upon water-power opportunities, twenty-seven (79 per cent) turned on criteria set initially by private agreements. (See, generally, Table 3, on private power dam cases before the Wisconsin Supreme Court, in the first division of this chapter). From the one inventory of trial court business available (for the area centering on Chippewa Falls), it does not appear that a full tally of suits which did not get beyond first-instance courts would show a much greater bulk of such litigation than that represented in the Supreme Court cases (see Laurent, 25, 163, 166).

378. See Fries, chap. iv; Larson, 86–104, 185–191; Reynolds, 12, 26, 62, 63, and chap. iv.

379. See notes 35–48, 53, 54, 68–81, and 89–94 above. See Hurst, *Law and Social Process,* 3, 45, 179, 195, 270–271.

380. See Majority Report, Select Com-

mittee on Bill 352A (to repeal Private & Local Acts, 1864, c. 84), AJ, 1106, and App., 1269 (1867). In accord were findings of the Committee on Internal Improvements, AJ, 813 (1867), which declared that the company's operations were of "vast benefit" to navigation, and that repeal of its charter "would cause a depreciation of property on said stream to a large amount, in the opinion of your committee, at least one million of dollars on the value of pine land alone, and would work wrong and injustice to many." The minority member of the Select Committee saw the problem either as one of conflict between interests of log drivers and raftsmen, or as a conflict of sectional economies, but nothing in the minority report took issue with the general propriety or need of an improvement-management franchise of some type as an inducement or condition of the most effective use of such a stream, at least for driving logs: cf. AJ, 1271, 1272 (1867). The repealer was decisively defeated, on a recorded vote, 64–15 (AJ, 1150).

381. Yellow River Improvement Co. v. Arnold, 46 Wis. 214, 231, 49 N.W. 971, 975 (1879). The management authority here conferred (under P&LL, 1857, c. 170, as amended by P&LL, 1868, c. 398, 1869, c. 186, and 1871, c. 116) first (by the 1868 amendment) was a broad management monopoly "to take charge and control of all general drives of logs on Yellow river at any future time," and then (by the 1869 amendment) was qualified to an authorization "to take charge of each and all log drives on the said Yellow river belonging to the said company or the individual members thereof, or that may be entrusted to it." Even in the later, qualified form, however, the company enjoyed considerable controlling powers over traffic. Over the members' logs or logs "entrusted" to it, the company was authorized to "take the exclusive charge and possession" for such drives. Moreover, the company was also authorized "to take into its exclusive possession" and drive logs of "all persons hereafter allowing or suffering their logs, timber or lumber to remain afloat in said Yellow river . . . in such manner as to mix or mingle with the logs, timber or lumber belonging to the said company, or the individual members thereof, or with those that said company shall have in charge in such drive of logs down said river . . ." Compare Edwards v. Wausau Boom Co., 67 Wis. 463, 472, 30 N.W. 716, 720 (1886), recognizing the functional need for (defendant) grantee's franchise to halt and control all log movement incident to necessary sorting: ". . . the river was not navigable above that point for anything but running logs in this way. To make it practicably navigable for even such purpose alone, booming works for stopping, holding, and assorting the logs so running free in the river were absolutely necessary . . . It is for the interest of the plaintiff, as well as that of all other log-owners in the drive, that these works shall obstruct and stop the passage of their logs until they can be assorted, divided, etc., and the only question is, in any case, whether the works are so constructed and so managed that the logs of the various owners are not detained longer than reasonably necessary."

382. See note 159 above. Cf. Current, 112; Rector, *Log Transportation,* 119, 122, 261–264; Reynolds, 12, 69–75. Apart from the provision for sharing expenses rather than collecting tolls, there were other features of some improvement and traffic management franchises which suggest an essentially cooperative character and also point up the prime value put upon organization as such: (1) Under Acts, 1852, c. 324 (Little Bull Falls Improvement & Stream Navigation Co.), and P&LL, 1853, c. 30 (Wisconsin River Improvement Co.), if any of the named incorporators refused to accept the privileges or exercise the powers conferred, any of the others might object to the company's use of its granted powers, provided the objection was taken before the company began operation. (2) Some franchises declared that any person "interested in lumbering on said river" might become a member by filing with the company clerk his written request to that effect. See Private & Local Acts, 1854, c. 62 (Black River Log Driving & Booming Co., terminated by a judicial sale reflected in P&LL, 1859, c. 198); P&LL, 1862, c. 35 (Eau Claire River Log Driving Co., changed to joint stock company by P&LL, 1864, c. 157); P&LL, 1864, c. 126 (Little Wolf River Log-Driving Co.); P&LL, 1865, c. 371 (Oconto Log Driving Co.). (3) Stipulations for pro rata division of the proceeds of the sale of prize (unmarked, unclaimed) logs among the members, or that these funds be spent on the common improvements, seem of a cooperative character. See P&LL, 1864, c. 126 (Little Wolf River Log-Driving Co.) and c. 131 (Log Driving Association of the Yellow River); P&LL, 1868, c. 398 (though amending the stock corporation franchise of Yellow River Improvement Co., under P&LL, 1857,

c. 170, treats prize log sharing on cooperative basis), and c. 469 (Wolf River Boom Co.); Laws, 1875, c. 48 (individual grantee of St. Croix boom). (4) Provisions for sharing or accounting for particular contributions of members to the joint enterprise point up the emphasis on achieving an organization to obtain an immediate functional result, as the parties' prime concern. See P&LL, 1856, c. 393, amending P&LL, 1854, c. 62 (member who, with written consent of board of directors, spends for company's benefit shall have reimbursement with interest); P&LL, 1866, c. 352 (profits of Keshena Improvement Co. to be divided in proportion to amount of money furnished company by members to perfect its improvements); P&LL, 1870, c. 480 (Little Wolf River Improvement Co. may use eminent domain power to acquire any existing dam at specified locations and pay therefor in company stock, save for one location at which works taken to be paid for in cash: two of named incorporators designated arbitrators of value in such proceedings); P&LL, 1871, c. 45 (Wausau Boom Co. may take existing piers, booms of named incorporators, paying appraised fair value therefor, but company shall not be liable in damages for attaching a boom to any stockholder's shore); P&LL, 1871, c. 483 (similar provisions, regarding White River Dam, Log Driving & Boom Co.); P&LL, 1871, c. 467 (profits of East Shioc Improvement Co. to be divided among incorporators in proportion to money they furnish company to perfect the authorized improvements); P&LL, 1871, c. 494 (Mill Creek Improvement & Log Driving Co. may buy improvements hitherto made by company members, and count cost thereof toward satisfaction of required minimum improvement investment of $1500).

383. See, generally, Current, 112; Rector, *Log Transportation,* 137, 138, 140, 141; Reynolds, 74–75. Occasionally an improvement work strategically located was a money maker in its own right (see Rector, 142–143). More often, so far as works yielded some return, these were likely modest enough so that the recipients regarded them rather as helpful contributions toward general overhead cost than as significant sources of income for their own sake (see Reynolds, 71–73). Testimony taken in Sauntry v. Laird, Norton & Co., 100 Wis. 146, 75 N.W. 985 (1898), Case on Appeal, pp. 34, 42, 267, reflects what seems to be an essentially cooperative-style relation among persons interested in that improvement corporation, as well as the fact that for years the franchise holder had made no effort to collect tolls.

384. See notes 43–67, and 153–213 above.

385. See the account in Marshall, *Autobiography,* I, 341–342, of an embarrassment in which the great Mississippi River Logging Company involved itself because of the hesitation of its business executives to test some of its statutory privileges in court.

386. See Marshall, *Autobiography,* I, 333–341; cf. Current, 113–114.

387. See notes 217, 223–230 above. Note the implicit emphasis upon organizational values in the court's summation of the evidence in Plumer v. Wausau Boom Co., 49 Wis. 449, 453–454, 5 N.W. 232 (1880) as to the obtaining of a charter by McIndoe, Plumer, and Clark, who were already joint owners of booms at the newly franchised location: "Desiring to operate their works to better advantage, and with enlarged powers, they procured the passage of chapter 45 of the Private & Local Laws of 1871, incorporating the company, of which they were to be the principal stockholders."

388. See notes 372, 375, and 383 above. There is especially pointed consideration of the reliance on toll franchises and the difficulties of toll collection in Reynolds, 70–71. Varied reflections of the practical, felt impact of toll provisions may be seen, for example, in P&LL, 1866, c. 447, amending (and narrowing) definition of services for which toll and lien rights were given the Black River Improvement Company under P&LL, 1864, c. 84; letter from John F. Stone & Co., opposing allegedly exorbitant toll rights conferred on improvers of the Yellow River by Laws, 1873, c. 12, AJ, 606–608 (1874); Tewksbury v. Schulenberg, 41 Wis. 584, 595–596 (1877) (terms of toll collection).

389. For example, physical control of passage was the heart of controversy over the Eau Claire boom and the Beef Slough works, both on the Chippewa River. See Fries, 142–154, as well as Attorney General v. Eau Claire, 37 Wis. 400 (1875), and Heerman v. Beef Slough Manufacturing etc. Co., 1 Fed. 145 (Circ. Ct. W.D. Wis. 1878).

390. See Whisler v. Wilkinson, 22 Wis. 572 (1868); Leihy v. Ashland Lumber Co., 49 Wis. 165, 5 N.W. 471 (1879); Fries, 143; Marshall, *Autobiography,* I, 339, 340.

391. See Cohn v. Wausau Boom Co., 47 Wis. 314, 324, 2 N.W. 546, 548 (1879); cf. Edwards v. Wausau Boom Co., 67 Wis. 463, 30 N.W. 716 (1886); Wisconsin River Log Driving Association v. D. F. Comstock Lum-

ber Co., 72 Wis. 464, 40 N.W. 146 (1888); Falls Manufacturing Co. v. Oconto River Improvement Co., 87 Wis. 134, 58 N.W. 257 (1894). On the tendency for increased reliance upon coordinated traffic management arrangements, see Larson, 186, and Reynolds, 73–74. Traffic management franchises in key, busy locations were used continuously over many years — sometimes, true, with alterations or changes in the succession. Obviously such franchises fulfilled functional needs of the industry to an extent which warrants our inferring like utility of the franchises in the general run of instances of which we lack record. See Current, 112, 116 (Wolf River Boom Co.); Fries, 49–50 (Wolf River Boom Co.), 50 (Wisconsin, Black, and Chippewa river organizations), 51 (Menominee River Boom Co.); Gates, *Wisconsin Pine Lands,* 131 (Beef Slough operations); Kleven, 145 (Black River Improvement Co.), 155–159 (Wolf River Boom Co.), 191–205 (Menominee River Manufacturing Co., as succeeded by Menominee River Boom Co.); Rector, "The Birth of the St. Croix Octopus," *Wisconsin Magazine of History,* 40:171 (1957), and *Log Transportation,* 142–143.

392. See notes 121, 122, and 351–360 above. The only instance in which a major works development went on without prior legislative license was in the early construction at Beef Slough, as validated by P&LL, 1870, c. 299; this episode did not involve what proved to be the substantial, lasting activity at that controversial location. See note 356 above. Locational battles brought to the Wisconsin Supreme Court turned on interpretation of statutory franchises already outstanding. See Stevens Point Boom Co. v. Reilly, 44 Wis. 295 (1878); Yellow River Improvement Co. v. Arnold, 46 Wis. 214, 49 N.W. 971 (1879); Cohn v. Wausau Boom Co., 47 Wis. 314, 2 N.W. 546 (1879); Black River Flooding-Dam Association v. Ketchum, 54 Wis. 313, 11 N.W. 551 (1882); Black River Improvement Co. v. LaCrosse Booming & Transportation Co., 54 Wis. 659, 11 N.W. 443 (1882); Menominee River Boom Co. v. Augustus Spies Lumber & Cedar Co., 147 Wis. 559, 132 N.W. 1118 (1912). The likelihood of locational controversies was reduced because statutory franchises usually stated location authority with some leeway, especially for stretch-of-stream improvements, but even in a measure for particular works authorizations (see note 122 above).

393. See reflections of height difficulties regarding the dam finally authorized at Eau Claire, in Laws, 1883, c. 149 (amending Laws, 1876, c. 231), 1885, c. 6, 1907, c. 35. On the other hand, in quite a few instances care was taken to amend franchises specifically to authorize height increases. See, e.g., Laws, 1880, c. 208 (amending Laws, 1879, c. 147); Laws, 1889, c. 32 (amending P&LL, 1853, c. 152); Laws, 1893, c. 190 (amending Acts, 1851, c. 129, under which authorized height had earlier been increased by P&LL, 1862, c. 32). Where a dam owner repaired his dam and, incident thereto, raised it to a height beyond that stipulated in his statutory franchise, the dam was held to be a public nuisance, subject to abatement to the extent of its unlawful height at suit of a raftsman specially injured: Arpin v. Bowman, 83 Wis. 54, 53 N.W. 151 (1892). Cf. Arimond v. Green Bay & Mississippi Canal Co., 31 Wis. 316 (1872) (flowage nuisance from illegal raising of water level); Lake v. Loysen, 66 Wis. 424, 29 N.W. 214 (1886) (excessive height may be alleged as basis for flowage action under milldam act); Hackstack v. Keshena Improvement Co., 66 Wis. 439, 29 N.W. 240 (1886) (downstream landowner may have common law action for flowage damages where statutory franchise flowage procedure by its terms is limited to flowage above authorized dam).

394. See notes 168–176 above.

395. The particularity of appraisal called for in applying requirements of due care in operations is dramatized by the discussions in Enos v. Hamilton, 24 Wis. 658 (1869), s.c., 27 Wis. 256 (1870), and Edwards v. Wausau Boom Co., 67 Wis. 463, 30 N.W. 716 (1886). Trial and error applied to the construction of an effective slide at the Kilbourn dam are reflected in: Memorial of Citizens, protesting repeal of the dam franchise, SJ, 255–257 (1860); petitions pro and con on revival of a dam franchise at that location, SJ, 79–82 and 109–110 (1866), and Reports, Committee on Incorporations, SJ, 135–137 (1866), and Committee on Incorporations, AJ, 454–457 (1866) (dividing on desirability of revival); Reports, Committee on Incorporations, SJ, 229 (1873) and Committee on Incorporations, AJ, 565 (1873) (both favoring repeal of then outstanding Kilbourn dam franchise; repeal failed on ground of protection owed vested interest).

396. See note 395 above.

397. Amendments of navigation guaranty stipulations are typically unrevealing on their face as to whether they represented the initiative of works builders or men interested

in passing along the stream. The important instances in which there was a legislative battle over feared or alleged physical obstruction of navigation were clouded by their character as total contests between competing local economies (see note 399 below); even the Kilbourn dam controversy had some of this flavor, as reflected in the favoring citizens' petition, SJ, 79 (1866). Navigators' lawsuits based on alleged breach of statutory franchise navigation guaranties included Hall v. Kitson, 3 Pinn. 296 (Wis. 1851); Volk v. Eldred, 23 Wis. 410 (1868); In re Eldred, 46 Wis. 530, 1 N.W. 175 (1879); Enos v. Hamilton, 24 Wis. 658 (1869); Arpin v. Bowman, 83 Wis. 54, 53 N.W. 151 (1892). Fries (p. 73) cites two items from the Ingram Papers to support his assertion that lumbermen "frequently" sued for raft damage, relying on the navigation guaranty clauses in franchises. There seems no evidence sufficient to support the generalization. The inventory of business in the circuit court for Chippewa County (in Laurent, 120, 157, 164) shows (1) under the category of civil actions regarding transport, none, 1855–1864, and four in each decade thereafter through 1914, save for ten in the years 1875–1884, together with (2) one nuisance case in the decade 1885–1894. Two criminal proceedings in the decade 1875–1884 involved damage to booms. For an example of the use of a statutory navigation guaranty as leverage by which aggrieved navigators successfully argued dam builders into providing passage facilities, see Reed v. Jones, 15 Wis. 40, 44 (1862) (required lock built after proceedings were begun against franchise grantee). See also Rector, *Log Transportation,* 121 (guaranty of navigation for vessels imposed on St. Croix Boom Co. worked, to extent that company paid for transshipment of freight of impeded craft).

There was implicit acknowledgment of the existence, or at least likelihood, of some sharp conflict between builders of improvement works and navigators in the enactment of statutes which made special reference to criminal penalties and civil damages, or occasionally forfeits (e.g., treble damages) for willful damage done to the improvements. Judged from the statute books, the danger of malicious mischief was not a prime concern of grantees of franchises for particular dams; exceptional in making such provisions in protection of specific dams were Acts, 1851, c. 173; P&LL, 1855, c. 251, and 1858, c. 86. On the other hand, there was less need of special provisions, because a general act applied to all dams, beginning with Statutes of the Territory of Wisconsin, 1839, p. 355. Such a protective provision was included in 15 (6 per cent) of 235 franchises authorizing improvement of a stretch of stream. The 15 instances showed no significant pattern in time: 1856–1, 1857–1, 1859–1, 1864–1, 1866–1, 1867–2, 1868–1, 1870–2, 1871–2, 1880–1, 1881–1, 1883–1. See, e.g., P&LL, 1859, c. 200 (Black River Improvement Co.); P&LL, 1871, c. 357. The point took on emphasis in two instances where such protective clauses were added by amendment. See P&LL, 1856, c. 393, amending P&LL, 1854, c. 62 (Black River Log Driving Co.); P&LL, 1866, c. 171, amending Private & Local Acts, 1853, c. 30 (Wisconsin River Improvement Co.). Of 104 special statutory franchises for (toll and private) booms, 39 (37 per cent) included a malicious mischief clause. There was a marked predominance of such protective clauses in private (non-toll) boom franchises; 23 (almost 60 per cent) of the 39 instances involved private booms. The most common type of clause provided that one who should "willfully open, break, damage or destroy said boom or booms or any part thereof, or said piers or any part thereof, shall be liable to the company in a civil action for the amount of damage done, in addition to the penalties now provided by law" (see P&LL, 1871, c. 45). Cf. Laws, 1873, c. 26, and 1874, c. 48 (treble damages). The common clause was made a general act as to booms, by Laws, 1880, c. 279; it is probably for this reason that but one of the 39 instances of inclusion of a protective provision in special boom franchises occurred after 1880. No lumber industry case reached the Wisconsin Supreme Court under any of this legislation, or for that matter under the general statutes against malicious mischief (see, e.g., Revised Statutes, 1878, sect. 4439). An inventory of all criminal cases in the circuit court for Chippewa County showed two prosecutions for willful damage to booms, in the decade 1875–1884 (see Laurent, 120). In all, this evidence does not add up to an impressive showing of broad-scale or continuing acute tension over interference with transport.

398. See notes 380–384 above.

399. The competition of local or regional economies or industry functional groups comes to particularly sharp expression in Majority and Minority Reports, Select Committee on Bill 352A (to repeal P&LL, 1864, c. 84, incorporating the Black River Improvement Co.),

AJ, 1106 and App., 1266–1272 (1867); and in Majority and Minority Reports, Committee on Lumber and Manufactures, on bill 366A (the Eau Claire dam and boom franchise), AJ, 477–479, 493–494 (1875). See, generally, Dykstra, "Law and the Lumber Industry," chap. iv.

400. See notes 173 and 177–180 above. Of course, the fact that a statutory franchise existed for a dam did not necessarily mean that it was exercised. Thus, of five dams specially chartered for power purposes on the Oconto River between 1842 and 1869, two were never built, and a river improvement company which had dam-building authority from 1869 did not use it to build its largest dam until 1872. See Falls Manufacturing Co. v. Oconto River Improvement Co., 87 Wis. 134, 58 N.W. 257 (1894), Case on Appeal, p. 25.

401. The complementary aspects of the eminent domain power, as protection to the respective interests of dam builder and flowed landowner, are brought out in a comparison of Charnley v. Shawano Water Power & River Improvement Co., 109 Wis. 563, 85 N.W. 507 (1901) (though in terms Revised Statutes, sect. 1777, as amended by Laws, 1882, c. 318, simply authorizes dam builder to bring condemnation proceedings, court finds that "by a fair construction" the statute likewise allows the flowed landowner to initiate such a proceeding), and McDonald v. Apple River Power Co., 164 Wis. 450, 160 N.W. 156 (1916) (assignee of special statutory dam franchise containing no eminent domain power may bar flowed landowner's action in equity by plea of milldam act, where dam is on stream navigable only by log-transport test). The complementary policies were early recognized. See Newcomb v. Smith, 3 Pinn. 131, 140 (Wis. 1849) (milldam act policy is to conserve dam builder's capital against exorbitant bargaining of flowed landowners); Thien v. Voegtlander, 3 Wis. 461 (1854) (defendant's plea of special dam franchise demurrable, where made without averment of tender or payment of compensation for flowage).

402. See note 148 above. The assurance-in-reserve aspect of possession of the power of eminent domain probably lay in the guaranty it gave against subjection either to actions to abate one's works or to successive, harrassing actions for damages. See Newton v. Allis, 12 Wis. 378 (1860); Ackerman v. Horicon Iron Manufacturing Co., 16 Wis. 150 (1862); Durning v. Burkhardt, 34 Wis. 585 (1874). See the quite explicit recognition of this aspect of eminent domain delegations in Allaby v. Mauston Electric Service Co., 135 Wis. 345, 351–352, 116 N.W. 4, 6 (1908), and McDonald v. Apple River Power Co., 164 Wis. 450, 456–457, 160 N.W. 156, 159 (1916).

403. See notes 306–315 above.

404. Consider the implications of the court's summary of the evidence in Chippewa & Flambeau Improvement Co. v. Railroad Commission, 164 Wis. 105, 113, 159 N.W. 739, 742 (1916): "There are many owners on the various lakes [affected by the company's dams built under Laws, 1887, c. 449, which conferred the right of eminent domain] as to whom flowage rights have not been obtained unless by prescription." Compare, in the same case, Brief for the State, pp. 3, 4: "There is not even a hint that a single dollar was put into the purchase of flowage rights; on the contrary, their special statute of limitations (chapter 285, Laws of 1899), which it is safe to assume was passed at the behest of the owners of log driving charters, shows that they appealed with success to the legislature to aid them in depriving people of their property, without compensation." The judgment here passed on the interest adjudications made by the legislature will be noted in the next subdivision of this chapter.

405. See notes 29–32, and 43–67 above.

406. See notes 215 and 216 above. Enforcement of the 1871 limitation ran to some excess of zeal for a time, while officials confused the grant of stream-use franchises with those of corporate existence. See notes 219-221 above. Instances of executive and judicial action are cited in the foregoing notes. For opinions of the judiciary committees of the legislative chambers, on the compatibility of particular bills with the constitutional ban on special corporate charters, see AJ, 469–470, 476, 497, and 515–516 (1872); SJ, 127 (1873); AJ, 210 (1875); AJ, 509 (1880). When the Senate Committee on Internal Improvements questioned the constitutionality of bill 91A, 1873, approving certain Wisconsin River booms to the Stevens Point Boom Co., as granting corporate powers, the bill was referred to the Senate Judiciary Committee, only thereafter to be recalled before the latter committee gave an opinion; enacted, after a second and favorable report from the Internal Improvements committee, the bill became Laws, 1873, c. 126, which the court thereafter held invalid as

violating the 1871 constitutional amendment. See note 220 above.

407. See notes 49, 50, 54, 56, 58–67, 82, 87, 93, 100, 101, and 104–115 above. Examples of appeal to the criteria of public interest as reinforced by relevant terms of the constitution may be seen in Report, Committee on Lumber and Manufactures, on bill 88S (the Eau Claire dam and boom), AJ, 754–755 (1867); Majority and Minority Reports, Judiciary Committee, on bill 7S (Eau Claire dam and boom), Feb. 27, 1871, SJ, 409–414 (majority) and 442–453 (minority), as well as Majority and Minority Reports, Committee on Incorporations, on the same bill, SJ, 658, 660, 663; Report, Committee on Incorporations, on bill 202S (authority to Black River Improvement Co. to close a navigable slough), SJ, 233–234 (1880), and Report, Judiciary Committee, on same bill, SJ, 394.

408. See notes 16–21, 43–48, 69–73, 352, 353, and 355 above.

409. See notes 63–67, 219, 252–254, 259, and 263 above.

410. See notes 50, 53, 54–62, 74, 75, 86, 93, 94, 108–115, 118, 125, 126, 194, and 212 above. The years of the Wisconsin lumber industry cases were early, relative to the development of a fully articulated theory of the presumption of constitutionality. Thus, even while the court in effect states and follows the presumption, its formulations are typically marked by confusions of doctrine. See, e.g., Wisconsin River Improvement Co. v. Lyons, 30 Wis. 61, 67 (1872) (unclear whether decision rests on presumption favoring legislature's competence to make reasonable generalizations of policy, or on application of equitable doctrine against premature injunction); Wisconsin River Improvement Co. v. Manson, 43 Wis. 255, 265 (1877) (unclear whether decision rests on presumption in favor of legislative findings of fact and value judgments as to ends and means in public interest, or on issue of standing to raise question on part of private, compared with public, litigant).

411. See, e.g., General Messages of Governor Doty, Council Journal, 62 (1843); Governor Dewey, SJ, App., 18 (1848); Governor Taylor, AJ, 22–24 (1875).

412. See notes 66, 219, 253, and 254 above. The catalogue of vetoes in forest-area stream franchise matters ran thus:

1853, SJ, 815 (Farwell): bill 310S, charter of Chippewa River Improvement & Navigation Co.; too broad monopoly, with inadequate public safeguards; no record of further legislative action.

1853, AJ, 841 (Farwell): bill 407A, charter of Black River Lumber Driving & Booming Co.; driving monopoly objectionable per se; no record of further legislative action.

1860, AJ, 578 (Randall): bill 120A, charter of Eau Claire Manufacturing Co.; want of vessel-passage provisions in authorized dam; veto sustained, 0–78, *ibid.*

1871, AJ, 1000 (Fairchild): bill 1A, to incorporate Chippewa River Improvement & Booming Co.; corrupt influences apparent in passage; veto sustained, 32–33, AJ, 1018.

1872, SJ, 32 (Washburn): bill 1S, for continuance of legislative committee investigating alleged corruption in passage of Eau Claire dam franchise; investigators no longer legislators, hence lack title; veto sustained, 14–16, AJ, 110.

1873, AJ, 919 (Washburn): bill 391A, to authorize Milwaukee & St. Paul Railway Co. to build bridge over Mississippi River; veto overridden, 60–28, AJ, 930, but sustained by Senate, 15–12, SJ, 617.

1877, AJ, 581 (Ludington): bill 101A, authorizing individuals to improve Tupper's Creek, Chippewa County, for log driving; grant to "successors" puts franchise in violation of 1871 ban on special corporate powers laws; veto sustained, 0–87, AJ, 594.

1881, AJ, 772 (Smith): bill 192A, general funds appropriation of $5,000 to build levee to protect Fox River valley area lands; contravenes internal improvements clause of state constitution; veto sustained, 0–9, AJ, 775.

1882, AJ, 748 (Rusk): bill 460A, authorizing a dam on Little Wolf River to Wolf River Manufacturing Co.; contravenes constitutional ban on special legislation granting corporate powers; veto sustained, 0–62, AJ, 749; similar franchise thereafter granted named individuals by Laws, 1882, c. 297.

1882, SJ, 488 (Rusk): bill 100S, granting named individuals toll dam franchise, Totogaticanse River, Bayfield County; for want of clause reserving legislative power to amend or repeal; veto sustained, 0–31, SJ, 505.

[1882, SJ, 615–616: Governor Rusk noted that he had withheld his approval to bills 101–105S, granting like franchises to same individuals on same river, and that upon deposit with Secretary of State, these became law without his approval.]

[1883, AJ, 435: Governor Rusk reported that bill 175A, amending Laws, 1880, c. 303 (a franchise for a toll dam on Black River, Wood County, containing no reserved power clause), became law without his approval, as Laws, 1883, c. 9.]

[1883, AJ, 650: Governor Rusk noted his return of bill 475A, pursuant to a joint resolution; legislature inserted a reserved power clause, and bill became Laws, 1883, c. 317, authorizing an individual to improve Robinson Creek, Jackson County.]

[1887, AJ, 1015: Governor Rusk reported that bill 261A, authorizing individuals to improve Devil's Creek, Sawyer County, became law without his approval, as Laws, 1887, c. 273; this act did contain a reserved power clause; the Governor's message does not state his basis for withholding his signature.]

1889, AJ, 642 (Hoard): bill 41A, amending Laws, 1887, c. 29 (granting a dam franchise on Wisconsin River to Centralia Pulp & Water Power Co.), by adding eminent domain powers; contravenes constitutional ban on granting corporate powers by special act, and confers eminent domain authority for private purpose; veto sustained, 1–76, AJ, 682.

1893, SJ, 679 (Peck): bill 72S, repealing dam franchise on St. German Creek, Lincoln County, given individuals by Laws, 1883, c. 355; holders had no hearing, and bill was passed under mistaken impression holders had abandoned dam; veto sustained, 0–6, SJ, 680.

1897, AJ, 1280 (Scofield): bill 448A, granting individuals a dam franchise, on Oconto River, Oconto County; inadequate navigation guaranty clause; veto sustained, 8–79, AJ, 1289.

1897, SJ, 644 (Scofield): bill 151S, requiring fishways in all dams; may be invalid delegation of powers, provides no notice or hearing to dam owners, too vague as to waters covered, unclear as to liability limits, too severe in penalty; veto sustained, 0–30, SJ, 658.

1901, AJ, 723 (LaFollette): bill 371A, granting individuals a dam franchise on Wood River, Burnett County; lacks fishways provision or reserved powers clause; veto sustained, 3–86, AJ, 908.

1901, SJ, 1035 (LaFollette): bill 328S, to repeal Laws, 1893, c. 296, authorizing individuals to improve Lost Creek, Sawyer County; should not thus take away vested right, without court trial to establish violation of franchise; veto sustained, 17–12, SJ, 1036.

1903, SJ, 1038 (LaFollette): bill 25S, granting individuals a dam franchise on Kickapoo River, Vernon County; contains no reserved power clause; veto sustained, 0–22, SJ, 1057.

1903, SJ, 1084 (LaFollette): bill 155S, granting individuals a dam franchise on Tomahawk River, Vilas County; no navigation guaranty or navigation improvement purpose declared, hence bill appears to be only for private interest; veto sustained, 0–26, SJ, 1189.

1903, SJ, 1173 (LaFollette): bill 158S, authorizing any riparian owner on Wisconsin River over forty-mile stretch to build dams; against sound policy thus to resign public control over water-power development; veto sustained, 1–26, SJ, 1221.

1905, AJ, 1237 (LaFollette): bill 619A, authorizing Eau Claire Dells Improvement Co. to raise height of dam on Chippewa River at Eau Claire; against sound policy to increase private right at expense of public uses of water; veto sustained, 0–5, AJ, 2139.

1905, SJ, 1444 (LaFollette): bill 520S, authorizing increase of height of dam granted individuals on Black River by Private & Local Acts, 1853, c. 206; no fishways clause; no provision to arbitrate price for sale or rent of power created; veto sustained, 2–27, SJ, 1564.

1907, AJ, 870 (Davidson): bill 584A, authorizing a village in Oconto County to extend its sewage system into Christy Brook; contravenes constitutional ban on special legislation and is against sound policy because makes no provision for injured property owners; veto sustained, 0–85, AJ, 898.

Aside from Governor Rusk's emphasis on the reserved power clauses, there was no assertion of broad policy leadership in matters affecting use of the streams, until Governor LaFollette made an issue of public control of the newly emerging use of the waters for hydroelectric power development. See his veto of bill 158S, 1903 (above), and his Special Message of Apr. 12, 1905, AJ, 1050–1054. The breadth of problems and the urgency of attention given them in messages of later governors dealing with hydroelectric development highlight by contrast the want of large policy leadership in the nineteenth-century governors' occasional attention to stream policy. See Governor Davidson's General Message of Jan. 10,

1907, AJ, 65–65, and his Special Message of Feb. 18, 1909, AJ, 257–259, as well as the General Messages of Governor McGovern, Jan. 12, 1911, AJ, 34–37, and Apr. 30, 1912 (Special Session), AJ, 14–15. The hydroelectric power boom overlapped the dwindling end of forest production, but was not in fact a material influence upon lumber industry affairs. Symbolic of this relation was the compromise struck by the court. Interpreting new waterways legislation as not intended to authorize complete obstruction of log navigation for the benefit of hydroelectric power development, the judges nevertheless were at obvious pains to indicate that they were not foreclosing legislative authority to establish such a preference for the new power use, if such appeared a reasonable judgment in the light of the trends of economic growth. See Flambeau River Lumber Co. v. Railroad Commission, 204 Wis. 524, 537–542, 236 N.W. 671, 676–678 (1931).

413. Cf. Hunt, *Law and Locomotives,* 104–106, 158, 173–174; Hurst, *Law and Social Process,* 38, 154, 186, 197–198.

414. The Attorney General appeared as litigant in an important lumber-industry issue over the waterways only in the Eau Claire dam controversy. See Attorney General v. City of Eau Claire, 37 Wis. 400 (1875); cf. State v. City of Eau Claire, 40 Wis. 533 (1876). A handful of informal opinions of the Attorney General, preserved in the Wisconsin State Archives (Attorney General of Wisconsin, Letter Books, one vol., 1859–1862, 15 vols., 1896–1907) shows but one opinion on a point of broad policy. This was an Opinion rendered the State Treasurer (Feb. 2, 1900, VI, 159–163), that a statutory license fee of 2 per cent of gross earnings and requirements of business reports, imposed on the Menominee River Boom Co., did not violate the commerce clause of the federal Constitution or the navigation guaranties of the Wisconsin constitution. No lumber industry matter appears in the letters preserved for the years 1859–1862. Other maters in the later years were highly particular. Thus an opinion of July 21, 1897, to an unidentified addressee, ruled that the state might recover compensation for flowage of state lands by the dam authorized on the Namakogan River by Laws, 1885, c. 75 (V, 329–330); an opinion rendered Governor Scofield, Dec. 26, 1899 (VI, 56–57) refused to pass opinion on the specific rights of landowners complaining of threatened flooding of farm lands by flood dams; an opinion rendered the state treasurer, May 10, 1900 (VII, 18–19) dealt with the liability of a particular dam owner under Laws, 1889, c. 252, to comply with the reporting and gross receipts license fee statute concerning improvement companies; a letter to a lumberman, J. P. Wells, of Brantwood, Nov. 7, 1900 (VII, 365), informed the inquirer that it was not the Attorney General's function to answer his question whether he must have a statutory license to build a flood dam on Somo River; an opinion rendered the secretary of state, Apr. 27, 1903 (II, 271–272) ruled that a local dam franchise granted to named individuals need not be published in the state paper. These items underscore the fact that the Attorney General was not playing the role of a broad policy maker. See, also, note 419 below. Governor Rusk's grounds for insisting on a reserved power clause in all special dam franchises ("In the nature of things it is impracticable for the legislature to make thorough and exhaustive investigation in each case, and to know the precise effect upon all interests of the measures asked for") in effect implied the absence of any central executive staff office to help survey what was going on. See Governor Rusk's Veto Message of Mar. 21, 1882, SJ, 489, and note 263 above. Compare the like implication in Governor LaFollette's Special Message of Apr. 12, 1905, AJ, 1050–1051, and in Governor Davidson's Special Message of Feb. 18, 1909, AJ, 258. Aside from the Eau Claire dam battle, no public nuisance prosecution concerning forest-area stream use came to the Wisconsin Supreme Court. An inventory of the business in the circuit court for Chippewa County — in as busy a lumber-producing area as any in the state — showed two proceedings for obstruction of "highways" in the decade 1855–1864, and one each in the spans 1885–1894 and 1895–1904, together with one tort action for maintaining a nuisance in the decade 1885–1894 (see Laurent, 126, 164). The pattern of the times, to delegate enforcement of public rights to private suitors, was dramatized by the sweeping terms of Laws, 1878, c. 208, which declared it "the duty of each . . . boom company, and person authorized to keep . . . a boom . . . in [the Wisconsin] . . . river" to maintain the required public passageway clear at all times, "and for the purpose thereof such company . . . or persons, may in its or their own name, commence and prosecute for that purpose any and all suits necessary at law or in equity, and shall have any and all remedies or actions, public or private therefor, that may be now given or existing at law or in equity, for such purpose or purposes, without alleging or showing any special injury resulting therefrom to it

or them, or any other person, and such company . . . or persons are also authorized and empowered to summarily remove any and all such obstructions." This statute was never generalized, or incorporated into the Revised Statutes, but it was never repealed. The statute was a clear departure from Wisconsin common law, which followed the familiar doctrine that a private complainant must show special injury to himself, if he would found an action on a public nuisance. However, the indications were that the court regarded this doctrine as important primarily to prevent entry of decrees of abatement which would be severely burdensome to defendants, at suit of plaintiffs whose interest was vague or relatively small in comparison. See Remington v. Foster, 42 Wis. 608, 609 (1877); Lohmiller v. Indian Ford Water-Power Co., 51 Wis. 683, 688, 8 N.W. 601, 602 (1881). Where the court had no doubts of the materiality of plaintiff's interest, it would make liberal reading of the record to sustain his standing to sue. See Olson v. Merrill, 42 Wis. 203, 214 (1877); A. C. Conn Co. v. Little Suamico Lumber Manufacturing Co., 55 Wis. 580, 587, 13 N.W. 464, 467 (1882); Weatherby v. Meiklejohn, 61 Wis. 67, 70, 20 N.W. 374, 376 (1884).

415. On municipal harbor controls and authorization of municipal subsidies to local river improvement enterprises, see notes 6 and 365 above. Gates, *Wisconsin Pine Lands,* 143–176, graphically portrays the limitations and abuses of county government in the forest region. Statutes in Michigan and Minnesota at one point gave broad authority to county boards to license dams or improvements (see Rector, *Log Transportation,* 101–102). The Wisconsin legislature never made such a delegation of powers concerning dams or navigation improvements, though it rather freely delegated authority to towns and counties to authorize bridges, including toll bridges. See, e.g., Acts, 1852, c. 429, sect. 1, (3), carried into Revised Statutes, 1858, c. 13, sect. 28, (9), and Revised Statutes, 1878, sect. 670, (8) (counties); Acts, 1852, c. 413, declared by General Acts, 1853, c. 72, to be an exception to the requirement of legislative license under Revised Statutes, c. 34, and carried into Revised Statutes, 1858, c. 15, sect. 2, and Revised Statutes, 1878, sect. 776, (1) (towns).

416. See Attorney General v. City of Eau Claire, 37 Wis. 400 (1875), and State v. City of Eau Claire, 40 Wis. 533 (1876); Dykstra, "Law and the Lumber Industry," 267–285.

417. See Revised Statutes, 1878, sects. 3376, 3381, and 3382, on milldams; and Laws, 1880, c. 279, and 1882, c. 318, on river improvement enterprises.

418. See P&LL, 1856, cc. 58, 176, and 511; P&LL, 1857, c. 412. Compare note 421 below.

419. Laws, 1873, c. 126, was held invalid, as contravening the new (1871) constitutional ban on special legislation conferring corporate powers, in Stevens Point Boom Co. v. Reilly, 44 Wis. 295 (1878). Commissioner control of flood-gate use was provided in Laws, 1887, cc. 12 and 253, 1889, cc. 252 and 481, 1891, c. 177; these statutes contain the language quoted in the text on selection of commissioners. Beginning with Laws, 1905, cc. 290 (amending Laws, 1903, c. 353) and 350, a provision for arbitrators to resolve disputes on reasonable rentals for power created by the licensed dam became a standard clause in a new wave of special franchises for dams to provide hydroelectric power. This innovation was made in response to a specific recommendation in the Special Message of Governor LaFollette of Apr. 12, 1905, AJ, 1053. See his veto of a franchise, for failure to include the arbitrator clause, Veto Message of June 7, 1905, SJ, 1444. That so clumsy a device should have been urged on the eve of full-scale adoption of administrative regulation in the field is revealing of the extent to which use of the administrative process departed from prior habits of thought and exceeded prior boundaries of imagination. From this viewpoint, though this development concerning hydroelectric power is not part of the lumber industry history, it casts light upon the limitations of attitude and concept within which public policy operated concerning the use of the streams in the lumber era.

Another facet of the pressure for procedural innovations attending the new interest in hydroelectric power development was resolution 26A, 1905, tendered by Milwaukee's Socialist Assemblyman Brockhausen, to bring the Attorney General more intimately into the franchise problem. By the resolution, "all bills providing for special privileges and grants to private individuals, firms or corporations, such as dams and bridges, and all other privileges intended to create profit to the parties aforementioned, shall not be acted upon until the attorney general shall give an opinion as to the constitutional right of the legislature to grant such special favors." The resolution was rejected without a record vote; see AJ, 549 and 571 (1905).

Many toll-dam franchises provided for resort to determinations by the official lumber inspector of the district to resolve disputes over the quantity of logs or lumber served, on which charges were owing. See, e.g., Laws, 1877, c. 247, 1878, cc. 318 and 337. This procedure, dealing as it does with relations between the improver and his customers, relates also — perhaps more closely — to the development of public utility regulatory policies.

420. 1895, bills 501A, and 176S. See AJ, 254, 767, and 799 (1895); SJ, 147, 671, and 682.

421. Note the implications thus of the substitution of a direct statutory grant for an original franchise conditioned on commissioners' determination of public interest, in P&LL, 1857, c. 167, amending P&LL, 1856, c. 58.

422. See notes 68–82 above.

423. See notes 194, 212, and 401 above.

424. See notes 36–42 and 53 above. The common attitude, that stream-use policy rested basically with the legislative process, was symbolized when the legislature gave its warrant for the use of common law doctrine in the field, declaring in General Acts, 1853, c. 72: "The boundaries of lands adjoining waters, and the several and respective rights of individuals, the state, and its citizens, in respect to all such lands and waters, shall be determined in conformity to the common law, so far as applicable, as evidenced by judicial determinations in other states, in which the courts in such cases have adhered to its principles." This was carried into Revised Statutes, 1858, c. 41, sect. 3, and Revised Statutes, 1878, sect. 1597.

425. See notes 17 and 18 above. The court early recognized legislative practice as the first authoritative exposition of constitutional powers over the streams, in Stoughton v. State, 5 Wis. 291, 296 (1856), and Fisher v. Horicon Iron Manufacturing Co., 10 Wis. 351, 354 (1860). Of like effect, see Tewksbury v. Schulenberg, 41 Wis. 584, 593 (1877), and the trial court opinion in Falls Manufacturing Co. v. Oconto River Improvement Co., 87 Wis. 134, 58 N.W. 257 (1894), Case on Appeal, p. 33; Black River Improvement Co. v. LaCrosse Booming & Transportation Co., 54 Wis. 659, 685, 11 N.W. 443, 455 (1882); Woodman v. Kilbourn Manufacturing Co., 1 Biss. 546, 553, 554 (Circ. Ct. D. Wis. 1867); Heerman v. Beef Slough Manufacturing etc. Co., 1 Fed. 145, 160 (Circ. Ct. W.D. Wis. 1878).

426. Thus an important early exercise of legislative authority to grant a toll-right, traffic-monopoly, river improvement franchise was by Private & Local Acts, 1853, c. 30, chartering the Wisconsin River Improvement Co., the constitutionality of which was first ruled on by the court in Wisconsin River Improvement Co. v. Lyons, 30 Wis. 61 (1872). The validity of the improvement monopoly conferred on the Black River Improvement Co. by P&LL, 1864, c. 84, was not ruled on by the court until Black River Improvement Co. v. LaCrosse Booming & Transportation Co., 54 Wis. 659, 11 N .W. 443 (1882).

427. See notes 49–67, 194, and 212 above.

428. See Laws, 1876, c. 231, responding to the decision in Attorney General v. City of Eau Claire, 37 Wis. 400 (1875); cf. State v. City of Eau Claire, 40 Wis. 533 (1876). See also Laws, 1877, c. 247, "amending" (in truth, repealing and substituting for) Laws, 1873, c. 12, to overcome the objection on which the court held the 1873 act ineffective in Sellers v. Union Lumbering Co., 39 Wis. 525 (1876).

429. See Water Power Cases, 148 Wis. 124, 134 N.W. 330 (1912).

430. Cf. Hunt, *Law and Locomotives,* chaps. ii and iv.

431. Outstanding symbols of the underlying attitudes and their unquestioned hold upon community values are the decisions in Wisconsin River Improvement Co. v. Lyons, 30 Wis. 61 (1872) (public control, basically through legislation); Wisconsin River Improvement Co. v. Manson, 43 Wis. 255 (1877) (propriety of delegation of navigation improvement to private enterprisers, with toll rights); and Falls Manufacturing Co. v. Oconto River Improvement Co., 87 Wis. 134, 58 N.W. 257 (1894) (no ownership in river flow by downstream riparian, hence no "taking" and no right to compensation for interference with natural flow reasonably incident to upstream flood dam).

432. Analogy was early drawn between the authority of the legislature to tax and its authority to grant toll franchises. See note 96 above.

433. Cf. Ryan, C. J., in Sellers v. Union Lumbering Co., 39 Wis. 525, 527 (1876), and Governor Rusk's Veto Message of Mar. 21, 1882, on bill 100S, SJ, 488, 489.

434. See, generally, Hurst, *Law and Social Process,* chaps. iii and iv.

435. See, for example, cases cited in notes 35, 57, and 311 above.

436. See, especially, notes 49–62, 80–81, 108–115, 125–126, 194, 211–213, and 323–332 above.

437. See notes 426 and 433 above.

438. The log test of navigability declared in Whisler v. Wilkinson, 22 Wis. 572, 576 (1868) was in effect recognized and applied in franchises which imposed navigation guaranties specifically for logs or rafts, without mention of vessels. See, Acts, 1851, c. 173, 1852, c. 282; Private & Local Acts, 1854, c. 82; P&LL, 1856, c. 80, 1857, cc. 91, 235, 368, and 382, 1858, c. 254, 1859, c. 129 (amending P&LL, 1855, c. 50), 1866, c. 424, 1867, c. 568, 1868, c. 216. Legislative practice even more sharply adopted a log or raft test of navigability in those instances in which a franchise stipulated that a dam must at the outset be so built as to pass logs or rafts, and that suitable facilities to pass boats should be added when the stream was made navigable for boats. See Laws of Wisconsin Territory, 1839–1840, No. 7; *id.*, 1841–1842, p. 84; *id.*, 1843–1844, p. 36; *id.*, 1848, p. 13; Acts, 1851, c. 248. See also P&LL, 1858, c. 86, amending P&LL, 1855, c. 366, to reduce a requirement of slides suitable for logs, lumber, rafts, or watercraft, to such as may be suitable for saw logs. In eleven statutes enacted between 1863 and 1868 the legislature declared specified waters navigable for logs, or logs, timber, lumber, or other named forest products, without mention of vessels. See, General Laws, 1863, c. 40, 1865, c. 353; P&LL, 1866, cc. 48, 175, 214, 328, and 563; P&LL, 1867, cc. 288, 422, and 503; General Laws, 1868, c. 84. Two earlier statutes, which declared named streams to be public highways, indicated that log movement was the practical sense of this designation, when they included stipulations that any dam or other works built thereon be provided with passage facilities for logs or rafts, omitting reference to vessels. See P&LL, 1857, c. 24, 1858, c. 44.

439. Cf. Hurst, *Law and Social Process*, 138–139.

440. See, e.g., in 1873, SJ, 77 (bill 65S), and 127 (125S); AJ, 371 (48A); in 1875, SJ, 181 (bill 88S), and 470 (bill 204A); in 1882, SJ, 229 (bill 80S), 565 (bill 216S), and AJ, 206 (bill 85A).

441. Laws, 1849, cc. 62 and 98, carried into Revised Statutes, 1878, sect. 1601. The confusion of attitudes on what matters were of local and what of general policy significance is underlined by the comment in *Report of the Revisers of the Statutes*, 125, explaining why these provisions from the 1849 session laws were included in this general compilation: "These two laws have remained in force since their passage and seem expressly [to] have been saved from repeal by the statutes of 1858, and omitted there as local. They are of importance, and are general, and hence are introduced to their proper place in the general statutes."

442. See General Laws, 1858, c. 13 (never carried into any general compilation, and never repealed) (all bridges on Wisconsin River); Laws, 1873, c. 272, carried into Revised Statutes, 1878, sect. 1837 (railroad bridges on Fox and Wisconsin Rivers); Laws, 1874, c. 165, as amended, Laws, 1877, c. 125, carried into Revised Statutes, 1878, sect. 1605 (slides and guide booms required on all Wisconsin River dams or bridges). Typical *ad hoc* navigation guaranties for logs or rafts in bridge franchises may be seen in P&LL, 1868, c. 432, 1869, c. 220, 1870, c. 481; Laws, 1883, cc. 38, 275, 315, and 324; Laws, 1895, cc. 268 and 271.

443. General Laws, 1863, c. 169, carried into Revised Statutes, 1878, 3337, and generalized by Laws, 1881, c. 141. The want of justification for limiting this provision for so many years to one river is thrown into sharper relief when, sustaining the constitutionality of the statute as expanded, the court justified it as resting upon "an equitable principle" in favor of fair payment for fair service. See Wisconsin River Log Driving Association v. D. F. Comstock Lumber Co., 72 Wis. 464, 466, 467, 40 N.W. 146, 147 (1888).

444. The two principal eminent domain procedures concerning flowage were those under Revised Statutes, 1878, c. 146 (the milldam act), and Laws, 1882, c. 318 (amending Revised Statutes, sect. 1777, as to stream improvement corporations). There were no differences in declared valuation principles between the two acts. Though it was not clear on the point, the 1882 act perhaps authorized the taking of a works site as well as flowage; the milldam act seems limited to flowage. There were two notable procedural differences. The milldam act proceedings might be initiated only by the flowed landowner; the 1882 act allowed either the works builder or the owner of property taken to initiate proceedings. The milldam act put the original determination of value in a jury, the 1882 act in *ad hoc* commissioners appointed by the court. However, there is nothing apparent either from the legislative records or from litigation to indicate how or why interested parties put value upon these or any other differences between the two eminent domain proceedings. That strong interests were engaged for one or the other seems negated by the apparently arbitrary

way in which draftsmen of special franchises chose now the milldam act procedure (e.g., Laws, 1885, cc. 70, 280, 281, and 402), now the 1882 act procedure (e.g., Laws, 1885, cc. 100, 104, 180, 231, and 236), and sometimes both (e.g., Laws, 1883, c. 213, 1889, c. 45) — in franchises enacted in the same or in closely related years. For still other variants in form, see, e.g., Laws, 1887, cc. 135, 177, and 448; Laws, 1893, c. 50; Laws, 1899, cc. 134, 195, 227, and 261.

445. The competing interests and attitudes in the Eau Claire-Chippewa Falls battle may be seen in Majority and Minority Reports on bill 7S, Judiciary Committee, Feb. 27, 1871, SJ, 409–414, and Mar. 1, SJ, 442–453, and in Majority and Minority Reports on bill 7S, Committee on Incorporations, Mar. 15, SJ, 656–665. See, generally, Dykstra, "Law and the Lumber Industry," chap. vii. The atmosphere of pressure surrounding the bribery charges may be gathered better from the summaries of testimony than from the poorly (perhaps disingenuously) stated findings and conclusions in Majority and Minority Reports and Testimony of the Joint Committee of Investigation, Feb. 15, 1872, SJ, 231, and App., 135–443. An earlier investigating committee reported testimony only. See Report, Committee of Investigation under Resolution 35A, Mar. 13, 1871, AJ, 801–820. The testimony in the 1871 hearing especially conveys the flavor of a local matter to which legislators from other parts of the state were relatively indifferent. A Milwaukee lobbyist testified, for example, that he told Assemblyman Richter (Milwaukee) "that this was a private matter not affecting his constituents or the interests of the State; that he would do me a personal favor if he would fall in with the ideas of Mr. Palmer [opposing the Eau Claire grant], as I felt under obligations to him" (AJ, 816). Palmer, also a lobbyist, testified "that we in Milwaukee had no local interest in the measure" (AJ, 817). Emphasizing that the issue presented by the taint of corruption "rises far higher in importance than the well being of any man or set of men or any locality," and that "every man in the state is interested in keeping our halls of legislation untainted by even the approach of those who dare offer a bribe," Governor Fairchild intimated that, on its merits, the Eau Claire dam proposal was a local issue. See Veto Message of Mar. 22, 1871, on bill 1A, AJ, 1002. On the public waterworks form of resolving the issue, see State v. City of Eau Claire, 40 Wis. 553 (1876). The Weyerhaeuser pact is discussed in Fries, 146, 153–156, and in Marshall, *Autobiography*, I, 272–296.

Though the Wisconsin court followed familiar doctrine in refusing to examine legislative motives as a basis for refusing legal effect to a statute, it did occasionally take note that stream franchise legislation was likely to be the product of sharply focused initiative, and hence should be restrictively construed against particular and in favor of more general interests, where the two might clash over ambiguous terms of a special statute. So the court would not supply the omission of authority under the eminent domain provision of a special dam franchise to enter a corrected damages award on a successful appeal by the dam builder, though the court observed, "This is certainly a strange anomaly in the charter, and one we should not have expected to find in a law doubtless drawn up by those interested in it": Darge v. Horicon Iron Manufacturing Co., 22 Wis. 417, 419 (1868). Where a special dam franchise stipulated that the dam should not raise the level of a lake, it conferred no authority to build a dam which raised the lake level, even though it appeared on evidence that any dam at the given location would have such effect. To the argument that this decision stultified the franchise, the court replied, in effect, that it might assume that the legislature was prepared to spend only limited attention on such particular-interest applications, and hence might respond to them with such sharply contingent grants. The grantees, "representing, as we may suppose, that a dam of the height mentioned, constructed at the point named in the act would not raise the water in the lake above its ordinary level, apply to the legislature for a grant of the privilege of building it; and the legislature, not knowing, and perhaps not caring to investigate, whether a dam so constructed will have that effect or not, grant him the privilege . . . provided such should not be the effect of the dam": Arimond v. Green Bay & Mississippi Canal Co., 31 Wis. 316, 332 (1872). Where flowage occurred below the lowest stream point referred to in a statutory franchise definition of the defendant's authorized improvement area, the court might not properly construe this lower point as covered on the theory that its flooding was a necessary incident to the operations authorized above, because "so important corporate powers cannot be implied as incident and necessary to carry into full effect the powers expressly granted. It would be giving this charter a looseness and latitude of construction never before given to any other": Hackstack v. Keshena Improvement Co., 66 Wis. 439, 443, 29 N.W. 240, 241 (1886). Compare, generally, Sauntry v. Laird, Norton & Co., 100 Wis. 146, 75 N.W. 985 (1898) (strict enforcement of franchise terms set as prerequisite to toll right). The only decision which seems to run markedly counter to this trend of construction is Schneider v. Staples, 66 Wis. 167, 28 N.W. 145 (1886), where the court found that a full revision of a special franchise, containing no counterpart of the fifteen-year limit set on duration in the original franchise, must be taken as "by necessary implication" a repeal of that time limit; the opinion makes no reference to the strict construction approach familiar in dealing with such legislation. On the borderline was the early ruling in Stoughton v. State, 5 Wis. 291 (1856) that a special dam franchise, fixing location, height, and manner of construction, must be construed to bar a prosecution for a public nuisance to health alleged created by erection of the dam precisely as authorized. On the whole, the court's general readiness to adopt a strict construction of special stream franchises, where ambiguous terms collided with more general interests, seems a practical recognition of the strain put on the legislative process by invoking it for matters toward which most of the legislators were bound to be relatively indifferent.

446. See notes 236 and 237 above.

447. The official records contain no discussion for the lumber industry comparable to the forecast of hydroelectric power problems in Governor LaFollette's Special Message of Apr. 12, 1905, AJ, 1050–1054; contrast of the Message with the policy silences of prior years helps us measure the extent of decisions that had taken place by default. Such few references as one finds to the time dimension in official records invariably emphasize reducing the time costs of moving greater volumes of forest production to market. See, e.g., Report, Committee on Internal Improvements, on bill 352A, Mar. 21, 1867, and Majority Report of Select Committee on the same bill, Apr. 8, AJ, 812–813, 1106, and App., 1269, 1270 (Black River Improvement Co. operations important because the company's services help maintain timberland values, and reduce time and stranding losses in passage); Majority Report, Committee on Lumber and Manufactures, on bill 302A (regarding Beef Slough), Feb. 25, 1868, AJ, 609 (transport improvements important to timberland values); Report, Select Committee on the St. Croix railroad aid lands grant, Mar. 7, 1873, AJ, 707 (water and rail transport compared, from standpoint of promoting maximum movement of logs to market); Report, Committee on Incorporations, on bill 202S (powers of Black River Improvement Co.), Feb. 17, 1880, SJ, 233 (public interest served by improvements increasing flow of forest production to market). See the retrospective summary on the importance of waterways to forest production in 11th Biennial Report, Bureau of Labor and Industrial Statistics, 1905, PD, 303–304 ("the importance of the log-drive and the mill stream can never be overestimated in the history of Wisconsin lumbering. Without them the industry and industrial development which depended upon it would be many years behind what they are today").

448. Cf. Hurst, *Law and Social Process,* 140–143, 159–161.

449. In re Falvey, 7 Wis. 630 (1858).

450. See notes 395, 399, and 407 above.

451. Note, for example, the obvious reflections of the pull and haul of affected parties in the particular terms of Laws, 1876, c. 285 (wing dam for power, Wisconsin River), or in the amendments made by Laws, 1887, c. 253, to the dam franchise granted by Laws, 1882, c. 247, or in the terms of Laws, 1889, c. 215 (St. Croix boom) after a process of balancing amendments, recorded in AJ, 842–844 (1889). See also the summary of action on all franchise bills passed in the 1880 session, notes 501–504 below.

452. Note the complaint on procedure in Memorial 185S, of Thaddeus C. Pound, concerning the Eau Claire dam franchise, SJ, 427–428 (1876): ". . . [the bill's] general policy and consequences should command your most careful and critical scrutiny. How is this to be done? Certainly not by simply listening to the contradictory statements of the friends and opponents of the bill. Such was our position in the other branch, and we asked the committee of that house, to receive sworn testimony. This was disallowed, the statements of the opposition discredited, and consequently a favorable report and the passage of the bill by that body. We now earnestly ask your honorable body to take such steps as will enable you to determine the facts affecting this momentous question, irrespective of the *ipse dixit* of either party, before saying aye or nay thereupon. This result may be attained in two different modes, without delay: one, by requiring each and every allegation of fact made before the

committee having the bill in charge to be made under oath; the other, the personal inspection, on the part of the entire Senate, of the site of the proposed dam and other works, as well as all other improvements and interests to be affected favorably or unfavorably, in the cities of Eau Claire and Chippewa Falls and vicinity." The Report, Committee on Incorporations, on Memorial 185S, SJ, 470, declared that the prayer for testimony under oath was granted, and a maximum of two days set for hearings; in view of the near end of the session, the committee reported without recommendation the prayer that the Senate visit the affected area.

453. See note 440 above.

454. See Report, Committee on Incorporations, on bill 9S (the Kilbourn dam), Mar. 2, 1866, AJ, 456, 457; Majority Report, Committee on Lumber and Manufactures, on bill 302A (Beef Slough), Feb. 25, 1868, AJ, 609; compare Minority Report, Committee on Lumber and Manufactures, on bill 366A (Eau Claire dam), Feb. 27, 1875, AJ, 494, with which compare Majority Report, AJ, 478, 479.

455. See Newcomb v. Smith, 3 Pinn. 131, 140 (Wis. 1849) (eminent domain); Wisconsin River Improvement Co. v. Manson, 43 Wis. 255, 262 (1877) (toll right necessary to induce improvement); Miller v. Sherry, 65 Wis. 129, 134, 26 N.W. 612, 613 (1886) (sheer boom).

456. See Tewksbury v. Schulenberg, 41 Wis. 584, 593 (1877).

457. With the court's readiness, tacitly or explicitly (notes 455 and 456 above), to take judicial notice of facts of general behavior or general economic operations, contrast the care with which the court ruled that it might not take judicial notice of facts of stream flow and volume, or navigation needs, on a particular stretch of the Wolf River. See Hackstack v. Keshena Improvement Co., 66 Wis. 439, 443, 29 N.W. 240, 241 (1886). Some analogy to this ruling lies in those decisions in which the court refused to enjoin or abate works construction in navigable streams, pending development by experience of the actual consequences to navigation of the particular works at the particular location (see note 285 above).

458. See notes 351–404 above.

459. See notes 89–99 above.

460. See notes 16–19 above, and the discussion of stream-use regulatory legislation in the later chapter of this volume on police power as applied to waterways.

461. Cf. Marshall, *Autobiography,* I, 282, 334, 336, 340, 341, II, 462. See, also, the court's recognition of the enterprisers' activity in the background of franchise legislation, in cases cited in note 445 above.

462. Thus when bill 33A, which became Laws, 1880, c. 84, was amended to change the area of authorized improvement from the "northwest quarter of the northwest quarter" to "the north half of the northwest quarter" of a specified section, the alteration seems probably to have been to correct an imperfect description, to the advantage of the grantee. Likewise to the grantee's benefit (though likewise in a matter not of prime significance, since it was probably already covered by implication in the original bill) was the amendment of bill 114S, which became Laws, 1880, c. 151, adding that grantee should have the "exclusive" right to dams in specified locations. Such examples could be multiplied several score.

463. See notes 501–504 below.

464. See note 244 above.

465. See Private & Local Acts, 1854, c. 28, amending Private & Local Acts, 1853, c. 30; Private & Local Acts, 1854, c. 321, amending *id.,* c. 62; P&LL, 1856, c. 508, amending P&LL, 1855, c. 330; Private & Local Acts, 1865, c. 475, amending *id.,* c. 127; P&LL, 1870, c. 297, amending P&LL, 1869, c. 86; Laws, 1876, c. 265 (amending 1875, c. 326), 1879, c. 27 (amending 1878, c. 318), 1880, c. 208 (amending 1879, c. 147), 1881, c. 253 (amending 1879, c. 191), 1903, c. 220 (amending 1901, c. 185), 1907, c. 359 (amending 1905, c. 415).

466. The apparently arbitrary variation of height formulas may be seen handily in different franchises enacted in the same session, in Private & Local Acts, 1854, c. 98 ("Said dam . . . shall not exceed eight feet in height above low water mark"), c. 111 ("The said dam shall not exceed eight feet in height above the surface of the river aforesaid, at high water mark"), and c. 231 ("a dam . . . to the height of ten feet from the water mark" — high, low, or ordinary?). For other formulas, see P&LL, 1857, c. 335 ("a mill dam . . . at a height not exceeding four feet seven inches above the natural work at the bottom of the floom above the mill"); P&LL, 1867, c. 503 ("said dams . . . shall not raise the water to exceed eleven feet above the ordinary stage of water in said river"); P&LL, 1867, c. 586 ("such dam shall not raise the water to exceed ten feet" — above what line?); Laws, 1897, c. 145 ("said dam shall

not raise the water to exceed ten feet above the natural channel"). Like the last-cited 1867 franchise, a considerable number of such statutes specified a maximum rise permitted in the height of the water, without defining the base line from which the permitted rise should be measured. See, e.g., P&LL, 1869, c. 400, 1870, c. 270; Laws, 1874, cc. 153 and 154, 1878, c. 283. Evidently this loosely drawn formula had, or acquired, common acceptance of meaning, but no decision of the Wisconsin Supreme Court tells us what this was. There seems particular acceptance of the formula as one of some understood meaning when, within its loose terms, there is change in the allowed number of feet, as in Laws, 1880, c. 208, amending Laws, 1879, c. 143. Where the reference was to the height of "the dam," rather than to the rise in water level, in the absence of a declared base line this presumably meant the height of the physical structure from its base, if no base line was specified. Even so, the vagueness of the formula is underlined by contrast with some franchises which were more precise, such as P&LL, 1868, c. 376 ("Such dam may be erected of the perpendicular height of twelve . . . feet"); Laws, 1895, c. 346 (specified height "from the bottom of the stream"); Laws, 1897, c. 206 ("from the bed of the stream").

467. The outstanding, because potentially the most widely applicable, instance of legislation made clumsy through incorporation by reference of other statutes was Laws, 1880, c. 279, creating Revised Statutes, sect. 1777a, on boom corporations, which declared applicable to all corporations operating thereunder sects. 12, 13, 14, 15, 16, 18, and 24 of P&LL, 1871, c. 45 (the charter of the Wausau Boom Co.). The incorporation by reference said nothing of amendments to the 1871 act made before 1880; Laws, 1873, c. 256, had significantly amended sect. 15 of the 1871 charter by striking the service preference originally given thereby to company stockholders; presumably the reference to sect. 15 in the 1880 act intended to include only the remaining portion of the original sect. 15, dealing with auctions of prize (unmarked) logs. So detailed a catalogue of provisions brought into a general law by reference merely to a special charter was an unfortunately veiled way in which to define rights and duties, some of which under the 1871 act terms were of high importance (for example, setting conditions of the company's obligations to serve customers, and providing for toll maxima, for measurement of commodities serviced, and for liens for tolls). The reference to the 1871 act created substantial ambiguity as to the scope of the lien available under the general 1880 law; sect. 1777a made applicable the lien provided by sect. 1777, save as sect. 1777a made special provision otherwise; sect. 14 of the 1871 act, incorporated by reference into sect. 1777a, gave a lien of broader scope than that under sect. 1777. For other examples of clumsy incorporations by reference see Laws, 1880, c. 77 ("which lien may be enforced the same as liens for labor may be enforced under the general statutes" — but the statutes offered different lien procedures for mechanics and for log laborers); Laws, 1881, cc. 164, 326, and 327 (lien may be enforced "as the leans [sic] of mechanics and laborers" — similar difficulty); Laws, 1882, c. 316 ("said dam shall be subject to all provisions of law relating to the flowing of lands" — i.e., to tort law for damages, but also to eminent domain procedures to allow acquisition of easements?); Laws, 1887, c. 448, and many similar grants (grantees "shall pay in full for all damages occurring by reason of the erection and maintenance of said dam, said damages to be determined as provided by law" — by tort law, or by eminent domain law, and if the latter, by the milldam act in c. 146, or by the procedure under the amended sect. 1777, the latter of which in terms refers only to corporations?); Laws, 1891, c. 175, and 1893, c. 99 (grantees shall control dam and slides, "but subject to the provisions of chapter 318 of the laws of 1882 . . . ; and the dam so erected shall be subject to all the provisions of chapter 146" — creating the question whether only the navigation guaranty provisions of the 1882 act are applicable, or also its eminent domain procedure, which differs from the eminent domain procedure of c. 146).

468. See note 466 above, and 497 below.

469. See notes 50, 52, 54–62, 71, 85–87, 93–94, 108–115, 212, 242, 275, 280–281, 285, 307, 311–314, and 335 above.

470. See State v. City of Eau Claire, 40 Wis. 553, 541 (1876). In the earlier case the court indicated that it should shun inquiry into legislative motive, and held the first Eau Claire dam franchise invalid because in the court's judgment it could only be interpreted to authorize use of the granted license for private advantage solely. See Attorney General v. City of Eau Claire, 37 Wis.

400, 438, 442 (1875). See the strong reaffirmation that judicial power does not encompass investigation of legislative motive, in Diana Shooting Club. v. Lamoreux, 114 Wis. 44, 54, 89 N.W. 880, 884 (1902); cf. Fletcher v. Peck, 6 Cranch 87 (U.S. 1810).

471. See notes 455–457 above.

472. Cf. Hurst, *Growth of American Law,* 185–189.

473. Special Message of Apr. 12, 1905, AJ, 1050–1054. The absence of explicit criticism of past waterways policy is the more striking, because the Message was critical of past laxity in grant of public utility franchises (among which, again significantly, he did not mention river improvement companies, while referring to ferries, bridges, turnpikes, railroads, and street railways). That the Governor overestimated the key role that hydroelectric power would play in the state's future economy does not detract from the implications of his analysis concerning the less wide importance of past waterways uses.

474. Special Message of Feb. 18, 1909, AJ, 258.

475. See Brief for the State, pp. 3, 4, in Chippewa & Flambeau Improvement Co. v. Railroad Commission, 164 Wis. 105, 159 N.W. 739 (1916).

476. See notes 29, 30, 43–62, and 324–332 above.

477. See notes 100–115, and 322 above. Cf. Hurst, *Law and the Conditions of Freedom,* 6–7, 55, 60, 63–64, 73, 84.

478. See notes 44, 46, 50, 54, 58–67, 91, 103–107, 156–159, 168–175, 294, 295, 299, 304, and 329 above.

479. See notes 319–322 above.

480. On the fisheries, see notes 319 and 330 above. Report, Commissioners of Fisheries for biennium ending Dec. 31, 1904, SJ, App., 15 (1905), commented, "In the northern part of the state many old log-driving dams are going out of use, and the cessation of logging in these streams has made them more desirable as fishing streams, while the rapid increase of population in this region makes a greater demand for fish from the state hatcheries. At the same time, however, that these streams are going out of use as logging waters, there has begun a general utilization of the water powers for permanent use and the dams that are being built are, almost without exception, without provision for the fish to get up stream. Most of the dams in the older settled part of the state are also open to this objection." Implicit in these observations are the reflection of changing social wants with reference to streams — of new demands for fish and recreation uses — and indication that the want of provision of fishways was not peculiar to dams built to serve the lumbermen's interests in the north. Compare the reservations expressed as to the utility of fishways by the Superintendent of Fisheries in an appendix to the Commissioners' Report for 1907–1908, PD, 29–30. The numerous bridge franchises to municipalities and private persons show that land highway interests were pressed on the legislature in relation to the streams; indeed, the continuing emphasis on navigation guaranties for logs and rafts past bridges authorized in the forest region in effect acknowledged that water traffic was put under some material risk for the benefit of land traffic, merely by allowing bridges to be built. See, e.g., Laws, 1885, cc. 270 and 318, 1887, c. 322, 1889, c. 41, 1895, cc. 268, 271, 345. Beyond such provisions it seems unreasonable to expect more attention to roads compared with waterways in the north from the state legislature, at a time when there was no real state highway policy and no state agency actively concerned with highways for the state as a whole, quite apart from the sparsely settled forest area; there was no substantial state highway policy until the creation of the State Highway Commission in 1911, and during the lumber years the Wisconsin community was content, or at least indifferent enough, to fragment highway policy among the towns, not even achieving effective attention at the level of the county. See Laurent, 22–24, 95; The State-Wide Highway Planning Survey, *A History of Wisconsin Highway Development, 1835–1945* (1947), xv–xvi, 14, 20, 24, and App. B, C, D, and E. Similar judgments seem in order with reference to the want of positive attention to public health interests in connection with the granting of forest-area stream franchises. It is part of the almost exclusive preoccupation of the times with values translatable into market-money terms, that the only stream pollution legislation pertaining to the lumber industry was designed to maintain transport use unimpaired, rather than to protect the purity of water for domestic use. See Laws, 1881, cc. 134 and 220; cf. notes 319–322 above. We cannot say that there was a positive state health policy before the creation of a rudimentary state Board of Health in 1876, and this step was not backed with substantial

support until well after the end of the lumber era (see Murphy, 72–84).

481. Cf. General Messages of Governor McGovern, Jan. 12, 1911, AJ, 34–37, and Apr. 30, 1912, AJ, 13–15.

482. See notes 351–368, 386, 387, and 392–404 above.

483. Delegation of policy implementation ultimately to the litigious initiative of affected private persons was implicit in all the detailed specifications and qualifications by which the legislature marked out the scope of title and prerogative under stream franchises. See, e.g., Enos v. Hamilton, 24 Wis. 658 (1869) (definition of scope of passage right under boom franchise); Arpin v. Bowman, 83 Wis. 54, 53 N.W. 151 (1892) (definition of height limitation). In proportion as franchises declared their terms in words of standard rather than of rule, they underlined how far the legislation passed on to the stage of execution the development of the concrete content of policy, and how wide a discretion was implicitly conceded the courts, in the absence of any provision for administrative proceedings to apply the standards. See, e.g., Laws, 1874, cc. 228 and 262 ("provided, however, [grantees] . . . shall not be entitled to collect or receive any toll whatever, unless the expense of driving said logs or timber shall be reduced by the use of said dams"); Laws, 1877, c. 247, amending Laws, 1873, c. 12 (striking original flat requirement of $10,000 investment in Yellow River improvements as prerequisite to toll right, and substituting the declaration that tolls may be collected when the authorized improvements "shall be so constructed, including such works as have already been constructed, as to provide substantial and beneficial conveniences and facilities for readily and rapidly driving logs and timber out of said river with reasonable certainty, having reasonable regard for the character of the river, and the stage of water therein at the proper seasons"); Laws, 1882, c. 260 (schedule of tolls authorized "when [grantees] . . . shall have made such improvements in said stream as to make the driving of logs down or out of it reasonably practicable and certain in an average rate of water, or when the driving of the same shall have been materially benefited by reason of improvements made or purchased by him, or them . . ." plus stipulation that grantees may in interim collect "such tolls as are reasonable and just for such improvements as shall be made, or shall have been made from time to time before the whole system of improvements are completed").

484. See notes 319–322 above. Cf. Hurst, *Law and Social Process,* 89–90, 128.

485. Some dam franchises are conditioned on provision of fishways as early as the territorial years. See Acts, 1839, No. 45 and No. 49; Laws, 1845, p. 99. Though such a clause was included in a few forest-area dam franchises (see P&LL, 1855, c. 251, 1861, cc. 36, 42, and 52; Laws, 1874, c. 168), it was a rarity until the end of the century, in striking contrast to the almost invariable inclusion of a log or raft navigation guaranty. A general fishways requirement was imposed by Laws, 1895, c. 337, only to be repealed by Laws, 1897, c. 253, after Governor Scofield vetoed proposed amendments on the ground that they were too vague and too severe in penalties (Veto Message of Mar. 29, 1897, on bill 151S, SJ, 644). Particular franchises thereafter began to include fishways requirements (see, e.g., Laws, 1899, c. 207, 1901, c. 55), and Governor LaFollette's Veto Message of Apr. 2, 1901, on bill 371A, AJ, 723, rejected a dam franchise for want of such a provision. Laws, 1911, c. 282, authorized the Commissioners of Fisheries to remove old and abandoned dams; Laws, 1917, c. 668 (transferring this authority to the new Conservation Commission) added authority to provide for maintaining dams where this would serve wildlife interests. Aside from special bridge franchise acts, there was no relation of stream improvement legislation to highway problems. There was no reflection of public health issues in connection with industrial stream uses until Laws, 1911, cc. 411 and 519, made provision for state Board of Health action on complaints of stream pollution by municipalities or industries, and Laws, 1913, c. 568, provided for a survey by the state Board of Health, with emphasis on industrial wastes. On the late and episodic attention to safety considerations, see note 320 above, and Governor McGovern's General Message of Apr. 30, 1912, AJ, 8, 13, 14–15.

486. See notes 440–447 above.

487. Cf. Hurst, *Law and Social Process,* 3, 14, 82–85, 110, 161, 163, 226–243, 253–257.

488. *Id.,* 3–5, 7, 15, 86, 108–111, 154–155, 163, 180, 194–195, 267–268, 301, 311–326.

489. See notes 476–478 above. There was also, of course, a general public interest in the

security of property. Incident to a general argument that special dam franchises were entitled to little moral weight because of their selfish background, in Chippewa & Flambeau Improvement Co. v. Railroad Commission, 154 Wis. 105, 159 N.W. 739 (1916), the Attorney General (Brief for the State, pp. 3, 4) claimed as an example of abuse of legal process for special advantage the ten-year statute of limitations set retrospectively as well as prospectively for actions for flowage by "flooding dams or other dams" for facilitating the handling of saw logs on the Chippewa, Menomonee, or Eau Claire rivers, with a six-months grace period where the ten years had already expired (see Laws, 1899, c. 285). "There is not even a hint," complained the Attorney General, "that a single dollar was put into the purchase of flowage rights; on the contrary, their special statute of limitations . . . which it is safe to assume was passed at the behest of the owners of log driving charters, shows that they appealed with success to the legislature to aid them in depriving people of their property, without compensation. To shorten the period of prescription by ten years and permit but six months where the period had expired, in which to bring suits, under the circumstances, was simply to permit robbery under the guise of law. It is a vicious instance of class legislation." The 1899 statute was on its face and without deception a local-interest act, limited as it was to certain Chippewa valley streams. The Attorney General's strictures may have been justified. But they were supported by no evidence, as the state's brief in effect concedes. A ten-year statute of limitations had existed for flowage caused by "any mill dam" since General Laws, 1864, c. 184, carried into Revised Statutes, 1878, and Wisconsin Statutes, 1898, as sect. 4221(3). The court accepted the equity of the 1864 statute without question, analogizing from it to recognize a ten-year period of prescription for flowage rights. Johnson v. Boorman, 63 Wis. 268, 22 N.W. 514 (1885); see Haag v. Delorme, 30 Wis. 591, 595–596 (1872). Indeed, the court gave the statute a liberal construction in determining what types of industrial situations involved mill dams within its meaning: Green Bay & Mississippi Canal Co. v. Telulah Paper Co., 140 Wis. 417, 122 N.W. 1062 (1909). However, the court also early ruled that the 1864 statute by its terms applied only to dams erected for the primary objective of producing water power, and hence did not apply to a dam the primary purpose of which was to improve navigation: Arimond v. Green Bay & Mississippi Canal Co., 35 Wis. 41 (1874). Most log-transport improvement enterprises were carried on in wild, sparsely settled country where there would be a minimum of practical reason for riparian owners to be concerned about flowage. This is probably the reason that there was no pressure before the act of 1899, to extend the ten-year statute of limitations to apply to flooding dams. That the 1899 act was drawn only in terms of a given locality did not represent good legislative practice, but on the other hand it was a common approach to problems in those years. See notes 441–444 above. Inclusion of a short-term cut-off date concerning situations in which the standard period had already elapsed at the enactment of a new statute of limitations is a common feature of such laws, and, indeed, a necessary element of adjustment when a new standard period is laid down; the 1864 act set a one-year cut-off term in such cases; though the six-months period of the 1899 act is abrupt, in the total context it does not bear as sinister a significance as the Attorney General would attach to it, regarded in isolation. Once more, melodrama is more easily averred than proved.

490. See, generally, Current, chap. iv; Fries, chap. xiii; Merrill, 28–29, 141–150.

491. See SJ, 87, 181 (1875); AJ, 242, 280 (1875). An incident in close juxtaposition reminds us, however, that smooth sailing was not guaranteed, and that reference of a bill to its sponsor might be a polite form of burying it. On Feb. 8, 1875 (SJ, 158) Silverthorne introduced bill 108S, to authorize named individuals to build and maintain dams on the Wisconsin River in Lincoln County, for log-driving purposes. On Feb. 25 (SJ, 365), without opinion, the Committee on Incorporations recommended that the bill be referred to the introducing Senator; the record shows no further action.

492. Compare Marshall, *Autobiography,* I, 333–341, on "The capture of the Namekagan."

493. For examples of some franchise bills passed in the form in which they were introduced, see Laws, 1880, cc. 7 (bill 9S: a dam on the Kickapoo River), 77 (bill 160S: a dam on the Black River), and 184 (bill 136S: a dam and improvement on Butternut Creek, Pierce County). For examples of smooth, routine passage through the legislative mill,

see P&LL, 1867, c. 378 (bill 248A: Chippewa River boom), and c. 470 (amending charter of the Eau Claire River Log-Driving Co., bill 386S); P&LL, 1871, c. 45 (bill 109A: to incorporate the Wausau Boom Co.), and c. 433 (bill 244A, amending charter of the Keshena Improvement Co.); Laws, 1873, c. 12 (bill 65S: improvement of Yellow River); Laws, 1878, c. 113 (bill 80A: improvement of Embarrass River); Laws, 1879, c. 201 (bill 80A: dam, Little Wolf River, Waupaca County), and c. 232 (bill 255A: dam, South Fork, Clam River, Polk County). These are the two principal types of legislative record on which one must rely if he is to infer substantial amounts of improper special-interest maneuver — bills passed in the same form as that in which they were introduced, and bills passed with no evidence of particular attention or controversy, though typically with some amendments made in committee. However, as the text notes, only a minority of bills went through without amendment, and typically amendments were of a character to reduce the advantage of the grantees. It is plain that the evidence does not support a generalization that sly manipulation was the norm. Moreover, the record offers slight support even to a generalization that improper handling marked a substantial number of franchise bills.

494. So potent a firm as Knapp, Stout & Co. did not always obtain a franchise in precisely the terms asked. Thus, while the company obtained dam franchises without navigation guaranties in Laws, 1879, cc. 154 and 155 (on Ten Mile Creek in Chippewa County and Yellow River in Barron County), their application for such a franchise on Bear Creek, Barron County, was amended in committee in the second house to include such a guaranty. See Laws, 1879, c. 96; SJ, 272, 295 (1879). The record shows similar incidents in amendment of the Knapp, Stout & Co. dam franchises given by Laws, 1880, cc. 33 (bill 126A), 40 (bill 110S), and 92 (bill 113S). See also notes 451 above, and 501–504 below.

495. See Laws, 1876, c. 247, amending Laws, 1874, c. 153, and c. 263, amending Laws, 1874, c. 154. Cf. Laws, 1882, c. 89, amending Laws, 1881, c. 161. There are other aspects of payment and lien terms in toll franchises which are hard to reconcile with the hypothesis that franchise seekers were typically shrewd and hard dealers, skillful in obtaining their best advantage. Though they were few, some toll franchises omitted any provision for a lien, not even incorporating by reference the available analogies from the general statutes, such as the provision for log laborers' liens. See P&LL, 1869, c. 363, 1870, c. 93, 1871, c. 357; Laws, 1875, c. 45, 1878, c. 291 (repealed, Laws, 1882, c. 86), 1881, c. 64, 1887, c. 434. More striking, because involving substantial numbers of franchises, was the failure of draftsmen to copy the more advantageous or broader-reach forms of lien, once these appeared on the books. Thus, many lien clauses were not well phrased to allow assertion of the lien against any logs or lumber of the same owner for charges accrued against a particular lot of his goods. See note 164 above. Though many franchises declared that a lien for tolls should take precedence of all other claims or liens on the serviced logs, except for laborers' liens (e.g., P&LL, 1866, c. 352, 1871, cc. 45 and 467), this added protection was omitted in as many grants as included it. See, e.g., P&LL, 1871, c. 483; Laws, 1874, c. 262 (the more striking because the lien here is of the broader type, in clearly allowing assertion against other logs of the same owner); Laws, 1877, c. 42. Compare Laws, 1880, c. 317, amending P&LL, 1870, c. 24, to add a preference statement (of the broadest style, not excepting laborers' liens), and Laws, 1881, c. 295, revising the lien and omitting any preference statement. Lien provisions also incorporated by reference the terms of general lien statutes with ambiguity calculated to foster controversy. See note 467 above.

496. The formulas for authorizing an improver or driving-franchise holder to take over impeding, improperly-provided-for logs were stated first for operators on the Black River, by Private & Local Acts, 1854, c. 62, and P&LL, 1857, c. 145. Despite this early precedent, no such advantageous provision was included in the next six stretch-of-stream improvement franchises. The authority was next given the Eau Claire River Log Driving Co., under P&LL, 1862, c. 35; the authority was probably continued when this cooperative-style association was converted to a joint stock company under P&LL, 1864, c. 157, but it is characteristic of the loose attention to points even of self-interest that there is some ambiguity in the matter. Formulas governing intermingled or impeding logs became well fixed in terms by P&LL, 1864, cc. 126 (Little Wolf River Log Driving Co.), and 131 (Yellow River Log Driving Association). Nonetheless, after this point, and before 1881, of

sixty-four stretch-of-stream improvement franchises granted by the state only twenty-eight (43 per cent) carried this presumably advantageous authority over mingled or impeding logs. There is no evidence in the legislative record that the authority was omitted because of positive opposition to its inclusion, and contemporary recognition of both the functional and equitable character of such a provision suggests that there was no opposition to it on principle. See Wisconsin River Log Driving Association v. D. F. Comstock Lumber Co., 72 Wis. 464, 467, 40 N.W. 146, 147 (1888) ("an equitable principle"); cf. Edwards v. Wausau Boom Co., 67 Wis. 463, 472, 30 N.W. 716, 720 (1886) (boom sorting monopoly justified as functional need of the total traffic). Laws, 1881, c. 141, amended Revised Statutes, 1878, sect. 3337, to generalize the authority over intermingled logs first given (by General Laws, 1863, c. 169) only as to the Wolf River system, to apply on all rivers, and thereafter (or at least after 1888, when the Comstock Lumber Co. decision, cited above, construed the authority there given an "owner" of logs over mingled logs to apply to a bailee), it was no longer necessary to insert such a provision in particular franchises, though draftsmen did still include it in sixteen (25 per cent) of sixty-five stretch-of-stream improvement franchises granted from 1881 on.

497. See Laws, 1879, c. 213 ("The dam so erected shall be subject to all the provisions of the revised statutes of 1878, so far as the said revised statutes may apply to dams"); Laws, 1880, c. 255 ("provided, nothing in this act shall be so construed as to authorize [grantees] . . . to flow any lands of the state or individuals, other than those named herein, without making compensation therefor"); Laws, 1882, cc. 186 and 316 ("provided that the same shall be subject to the existing laws regarding the flowage of lands"); Laws, 1887, cc. 177 and 448 (grantees "shall pay in full for all damages occurring by reason of the erection and maintenance of said dam, said damages to be determined as provided by law"); Laws, 1893, c. 50 ("This act shall not be so construed as to authorize [grantees] . . . to flow any lands of the state or individuals, by virtue of the erection or maintenance of said dam as aforesaid, without making compensation therefor"); Laws, 1899, cc. 134 and 227 ("Nothing in this act shall be so construed as to authorize [grantees] . . . to flow the lands of any individual or corporation to the injury of the owner thereof, without full and just compensation therefor, in accordance with the law governing in such cases; but the said [grantees] . . . shall pay in full all damages accruing from such flowage, when they shall be determined either by agreement of the parties or by due process of law"); Laws, 1899, cc. 195 and 261, and 1907, c. 383 ("said dam or dams, shall be subject to all the provisions of the laws of Wisconsin, relating to the flowage of lands by the erection of dams"). The apparently arbitrary variations in phrasing in these instances — productive only of ambiguity, if they were selected to accomplish some special purpose — are representative of the poor craftsmanship of contemporary technique in framing legislation; cf. note 466 above.

498. See note 491 above. See, on omission of a lien provision, and on the want of a toll right for aid rendered log movement below the authorized works, note 495 above. For other examples of uncertain description of the line of title to a franchise, see Laws, 1873, c. 252 (first dam authorized to Aaron Chase "and assigns," second only to Chase); Laws, 1881, c. 163, amending Laws, 1874, c. 288 (original grant to Daniel Shaw, his associates, successors or assigns, changed to Daniel Shaw or assigns), itself amended by Laws, 1883, c. 95 (confirming the franchise to Daniel Shaw's legal heirs, their associates and assigns); Laws, 1883, c. 132, amending Laws, 1880, c. 214 ("assigns" dropped from *habendum* clause, but continued in amended toll grant section); Laws, 1881, c. 58 (*habendum* runs only to named individual, but later sections of franchise refer to this grantee "and assigns" or "associates and assigns"). That omission of "assigns" was not unimportant is indicated by the introduction of bill 488A, 1893, which would have authorized Dan F. Smith to sell, assign, and transfer his rights under Laws, 1875, c. 45. The bill had smooth passage in the Assembly, without amendment, but the Senate — on recommendation of its Committee on Incorporations — refused its concurrence (see AJ, 264, 402, 500; SJ, 614, 632). The 1895 session rejected quite a number of new franchise applications; the refusal to amend the Smith grant to allow transfer thus should be interpreted in the context of 1895, and — in light of the sizable licensing activity of the seventies — offers no evidence that omission of "assigns" in the original franchise represented a general policy against transfer, especially since

"assigns" were included in the five other toll improvement franchises granted in 1875; the incident, thus, seems another illustration of the clumsiness of draftsmen in ensuring their own interests. For clauses hedging the grantee's obligation to supply water, see P&LL, 1864, cc. 157 and 302, 1869, c. 452, 1870, c. 164; Laws, 1878, cc. 291 and 337, 1883, c. 289.

499. See Committee on Land Use and Forestry, *Forest Land Use in Wisconsin* (Madison. 1932), Table 17, p. 120.

500. These 1880 franchises included (1) for booms, cc. 21, 104, 146, and 319 (toll); (2) for toll improvements, cc. 63, 77, 84, 97, 144, 151, 171, 177, 178, 182, 184, 201, 205, 214, 241, 294, 296, and 303; (3) for non-toll dams, cc. 7, 25, 26, 32, 33, 40, 41, 49, 75, 76, 92, 102, 103, 168, and 255.

501. Enacted as introduced in 1880 were (1) for booms, cc. 21 and 104 (both non-toll); (2) for toll improvements, cc. 77, 171, 177, 178, 182, and 184 (of which the middle four were of identical terms, all running to the same grantees); and (3) for non-toll dams, cc. 7 and 103.

502. Most clear-cut of the changes effected in the introduced bills were the reductions in toll — in Laws, 1880, c. 84 (bill 33A), from 20 cents to 5 cents for logs below the dam (leaving unchanged a 10 cent charge for logs above the dam); in c. 97 (bill 73A), striking the original 5 cent toll, leaving no toll right; in c. 144 (bill 30S), a flat 10 cent toll for all logs put in the river changed to a zone schedule of 3, 5, and 10 cents (leaving unchanged zones of 10 and 15 cents for handling improperly-provided-for logs); in c. 201 (bill 194A), from 7 cents to 5 cents; in c. 205 (bill 111S), zone charges reduced from 15 to 10 cents, and 10 cents to 5 cents, respectively; in c. 294 (bill 5A), general toll reduced from 15 to 10 cents, charge for handling uncared-for logs from 25 to 15 cents; in c. 296 (bill 107A), general zone charges reduced from 35 to 20 cents, 30 to 15 cents, 25 to 10 cents, and 15 to 5 cents, respectively, and zone charges for handling uncared-for logs reduced from $1.25 to 50 cents, $1.00 to 35 cents, and 75 to 20 cents, respectively; in c. 303 (bill 301A), in place of original flat charge of 10 cents for all logs moved over or with aid of dam, a zone schedule of 10 cents for logs put in above dam, and 5 and 2½ cents, by zone, for logs put in below dam. The boom location change was in c. 146 (bill 246S). The navigation guaranty changes were in cc. 33 (bill 126A), 40 (bill 110S), 49 (bill 131A), 92 (bill 113S), 102 (bill 342A), and 255 (bill 286A).

503. Such changes in the grantee descriptions were made in cc. 25 (bill 118A), 26 (bill 13A), 32 (bill 125A), 33 (bill 126A), 40 (bill 110S), 41 (bill 112S), 49 (bill 131A), 75 (bill 109S), 76 (bill 177S), 102 (bill 342A), and 255 (bill 286A). That such changes might have substantial bearing upon secure titles, see notes 117, 219–221, and 498 above.

504. Rejected as they were introduced in 1880 were Assembly bills 76, 103, 256, 294, and 318; and Senate bills 140, 142, 143, 215, and 255. Rejected after consideration of amendments were bills 215A, 277A, and 16S.

505. The first Kilbourn dam franchise was that conferred by Private & Local Acts, 1853, c. 270. This was followed by P&LL, 1855, c. 330 (amended, P&LL, 1856, c. 508), which gave a franchise to the Wisconsin Hydraulic Co. Both of these franchises were enacted routinely and without incident.

Lumber transport and local industrial power development clashed in 1857 in a legislative hearing on bills 34S and 35S to repeal all outstanding rights. At this point no one had built a dam, though the Wisconsin Hydraulic Co. claimed to have let a contract for construction. The issue between transport and power uses was crisscrossed with the rivalry of promoters of two separate town sites, so that it is impossible to tell on which interest clash the issue most depended. See Memorial of Wisconsin River Hydraulic Co., Feb. 2, 1857, SJ, 144–149, and Memorial of S. S. Barlow and other citizens of the villages of Newport and Delton, Feb. 16, SJ, 244–248 — both focused exclusively on the town rivalries. Two Senate committees divided in reactions. The Majority Report, Committee on Roads, Bridges, and Ferries, Jan. 29, SJ, 115–118, recommended repeal of the outstanding dam franchises, because it found convincing the opinions of lumbermen that a dam at this location would so increase labor costs and damage in transit as to constitute a serious burden on the lumber traffic, which the majority viewed as of greater public importance than any particular industrial site development. The majority estimated that a dam would cause annual costs and damages to the lumber traffic amounting to $74,787 ($21,400 extra labor and delay costs; damage to rafted logs and lumber, $43,387; damage to shingles and lath, $10,000). Apart from a concern for what he deemed an interest of the

Company, vested by its actions to date, the minority member thought that a dam would make navigation safer at the Dells stretch of the river above the dam site, that a properly built slide on a dam of such moderate height as was proposed would allow safe passage, and that hence the public interest in industrial power should warrant continuance of the dam franchise. In reports of Feb. 12 and Feb. 20, SJ, 216, 298–299, the Judiciary Committee opposed repeal of the dam franchise rights outstanding. Disclaiming a verdict between the contending town promoters, the Judiciary Committee apparently took its stand on the public interest in industrial power development. It found the testimony as to likely damage to lumber transport "very conflicting," but thought that "the weight of testimony" seemed to show that a dam here would no more harm navigation than had other dams already built on the river. The Senate rejected repeal by a vote of 16 to 7 (SJ, 302). The committee dispositions seem fair efforts to strike a difficult balance of values on conflicting evidence, largely as to hypothetical problems.

A dam was then built, with a slide. As was conceded by proponents of the dam, it proved difficult to contrive a workable slide, and despite two remodelings, the works caused serious damage to log and lumber transport. See Memorial of citizens of Kilbourn, Feb. 14, 1860, SJ, 255–257; cf. Remonstrance of John B. and Wm. Vliet, Memorial 103S, Feb. 25, SJ, 381–383 (mortgagees of dam works protest impairment of vested right, by repeal). A massive attack of memorials was made on the dam in the 1860 session. Petitions signed by 496 citizens interested in Wisconsin River navigation averred that the dam obstructed transport, "causing annually the loss and destruction of large quantities of lumber, besides endangering the lives of persons engaged in running of lumber on said stream." One affidavit claimed losses to affiants of about $16,000 of lumber in the past season, because of the dam. The franchise was repealed by P&LL, 1860, c. 70, following a recommendation for repeal by Report, Committee on Internal Improvements, Mar. 2, AJ, 534–536; the committee found "that said Wisconsin River Hydraulic Company have not constructed their dam in accordance with the provisions of the act incorporating the same." Without opinion, on Mar. 21, SJ, 691, the Committee on Incorporations joined in recommending repeal. The repeal bill passed without further sign of controversy (SJ, 964).

Apparently an unsuccessful effort to obtain a new franchise for a dam at Kilbourn was made in 1864, though the legislative journals do not allow tracing it. See recital in petition recorded in SJ, 79 (1866). However, a year later P&LL, 1866, c. 424, granted a franchise to the Kilbourn Manufacturing Co., "to complete the water-power" at the controversial location "by raising the dam a sufficient hight for that purpose, not exceeding three feet above the usual low-water mark in the Wisconsin river, and so forming the same that rafts of lumber can pass safely and conveniently without hindrance or delay." This was to prove a long-lasting grant, standing unrepealed and unamended through the principal lumber transport years, until Laws, 1893, c. 118, enacted what amounted to a substitute grant, which signalized the end of important lumber transport on the river by approving a dam to ten feet above usual low water mark, and omitting any navigation guaranty.

In 1866, however, the new franchise occasioned a stiff battle in the legislature. Conflicting petitions submitted to the legislature essentially the same arguments and differing fact judgments between the power and the lumber transport interests as were voiced in 1857. See petition recorded Jan. 16, 1866, SJ, 79, and petition of "residents of the county Juneau," Jan. 23, SJ, 109–110. The "Juneau" petitioners claimed to speak for 40,000 inhabitants of five forest-area counties, producing over $10 million in industrial pursuits annually, apart from agriculture, as against a few speculators "proposing . . . to experiment again in mill dams, at a point never intended by nature for hydraulic purposes, and thereby sacrificing a vast and legitimate interest of the state and county . . . We have not yet forgotten the efforts of a former dam at Kilbourn City. Hundreds of thousands of dollars worth of property lost, a number of our citizens perished, and the navigation of the Wisconsin river was for a considerable length of time closed against the passage of all rafts and boats; and this difficulty was only remedied by the removal of the dam, at great expense, by the inhabitants on the upper waters." Two Kilbourn-area dam bills were presented to the 1866 legislature. Bill 9S produced the only explicit record of policy. Report, Committee on Incorporations, Jan. 25, SJ, 135–137, recommended passage of bill 9S, because having "taken much pains to inform itself" the committee believed that no serious impairment would occur to navigation, and that in this posture preference should be given to industrial power development. The committee's report is unusual as the only pronouncement in the legislative record of a policy preference for power over lumber industry transport. Industry would abide, the forest would pass: "But when the timber is gone . . . the country where it stood is rather impoverished than enriched. Now the erection of mills and manufactories . . . will add greatly to the permanent wealth of the country." Summarily voting down dilatory moves by the bill's opponents, the Senate passed bill 9S, 18 to 3 (SJ, 156, Jan. 26). However, Report, Committee on Incorporations, Mar. 2, AJ, 454–457, countered with a unanimous recommendation against the bill, emphasizing the special constitutional protection given to navigation, finding in past experience confirmation that a dam at this poor location would seriously impede transport (the old dam had been "very destructive to lumber," causing $100,000 loss in one season), and arguing that the lumber interest clearly overbalanced this local industrial development in importance to the general economy ("such a course of legislation will tend to destroy one of the great sources of the wealth and industry of this state, without conferring a corresponding benefit"). Parliamentary maneuver in the Assembly was close and hard fought following this report, but finally, on March 14, AJ, 604, the Assembly refused to order bill 9S to a third reading, on a vote of 45 to 45, and on March 15, AJ, 620, by a vote of 48 to 46 tabled a motion to reconsider the vote by which the bill was refused a third reading. It should be noted that the three senators who comprised the committee which reported in favor of the power over the lumber transport interest were all from east-shore, non-lumber districts (Racine, Ozaukee, and Sheboygan), and that of the four members of the Assembly committee which reported with strong preference to lumber transport, three were from areas closely related to forest production (Crawford and Portage counties in the northwest, and Oconto-Shawano-Door in the northeast — with the fourth member from east-shore Washington County), and that one of these Assembly committeemen was a leading lumber industry figure, Isaac Stephenson.

The Kilbourn dam franchise which passed was bill 425S, introduced by the sponsor of the ill-fated bill 9S, six days after 9S lost in the Assembly. Bearing the scarcely informative title of "An Act to aid in the development of the manufacturing interests of this state," bill 425S was reported favorably, without opinion, in the form in which it was introduced, by the Senate Committee on Incorporations, Mar. 23, SJ, 659. Rejecting, 22 to 6, an effort by the opposition to postpone consideration, the Senate passed the bill, 21 to 6, Mar. 27, SJ, 707; the Senate's favorable vote on bill 9S had been 18 to 3. The Committee on Internal Improvements, Mar. 29, AJ, 844, reported bill 425S, without recommendation and without opinion. Assemblyman Thaddeus Pound, leading lumberman of Chippewa Falls, led a long battle of parliamentary maneuver, Apr. 7, AJ, 1029–1034, culminating in a vote to adjourn by the margin of 39 to 38, but when the bill finally came to its third reading, on Apr. 10, AJ, 1091, it was passed without a record vote, and was thus adopted precisely in the terms in which it was introduced. The close and hard battle made clear, at least, that the lumbermen could not have their way simply for the demanding of it.

A last, determined effort to get rid of a dam at the Kilbourn location was made in 1873, by bills in each house to repeal P&LL, 1866, c. 424. The rejection of both repeal bills seems to have rested primarily on the practical respect paid to the security of committed capital. Report, Committee on Incorporations, on bill 14S, Feb. 18, 1873, SJ, 229, found on the basis of sworn testimony that the dam "is a serious obstruction to the free navigation of said river, causing a great loss of property annually and . . . ought to be removed, unless by so doing vested rights are infringed upon." Not deeming itself competent to decide the vested rights issue, the committee recommended reference to the Senate Judiciary Committee, a majority of which on Feb. 27, SJ, 328, reported that it believed that the constitutionally reserved power to repeal applied in this case, but that it might be bad policy to wield the authority: ". . . if important interests, especially of third parties, have grown up under former grants, it surely is impendant to destroy them by *summary* repeal; and this power though existing, ought never to be exercised, unless it be in a perfectly clear case of flagrant abuse." Moreover, "there are ample remedies in the courts in case the corporate franchises have been exceeded or abused." The majority of the Senate Judiciary Committee reported back the bill without recommendation; the minority pointedly limited itself to the opinion that repeal "will not be an infringement upon vested rights within the legal signification of that term" (SJ, 329). Contest in the Senate swayed back and forth thereafter. On March 3, SJ, 386, the Senate refused, 13–14, to reject the repeal bill, but then adopted, 14–12, an amendment proffered by the floor leader of the opposition, to substitute an authorization to the governor to appoint commissioners to examine the dam and improvements then in construction on it for the better safety of lumber passage, and report to the next session. However, the Senate then voted 8–20 to refuse to order the amended bill to its third reading; apparently by so drastically altering the proposition before the chamber, the original opponents added to their own strength the votes of members disgruntled at the new version of affairs. Meanwhile the Assembly had before it a repealer proposed by bill 332A. Report, Committee on Incorporations, Feb. 27, 1873, AJ, 565, recommended repeal, finding that "the evidence presented shows conclusively . . . that these conditions [for safe passage of rafts, imposed as conditions of the franchise] have not been fulfilled, but that the dam has caused a damage to rafts of lumber to the extent of many thousands of dollars annually." However, apparently there was question in this chamber, too, of the constitutional issue posed by repeal, and the bill was sent to the Assembly Judiciary Committee which, without opinion, on Mar. 14, AJ, 824, recommended rejection of the bill. The Assembly voted its indefinite postponement accordingly, Mar. 15, AJ, 852.

506. Governor Farwell's Veto Message of June 6, 1853, AJ, 841–842, disapproved a charter for the Black River Lumber Driving and Booming Co., on the ground that it was "both impolitic and unconstitutional" to confer a toll-driving and traffic-management monopoly on a general-traffic stream. This was the first such grant made by the Wisconsin legislature; earlier stretch-of-stream improvement franchises provided only for building works or clearing channels, and charging toll for use of such improved facilities. Legislative practice and judicial decision subsequently established both the constitutional authority and the prevailing judgment of policy in favor of such operating and traffic-management monopolies; thus the 1853 veto should perhaps be counted less a resolution of interest pressures than a reaction against a somewhat striking innovation. Cf. notes 54, 63, 64, 93, 98, and 195–213 above.

Obviously with an eye to the 1853 veto, proponents of organized traffic operations on the Black River qualified the franchise they obtained without incident by Private & Local Acts, 1854, c. 62, incorporating the Black River Lumber Driving & Booming Co. The Company was given driving control over logs of its members and of other persons "suffering their logs . . . to remain in said Black River, between [named points] . . . in such manner as to mix with the logs . . . of the said company, or the individual members thereof, in any general drive of their said logs . . ." but: "This act shall not be construed so as to exclude other persons from . . . driving their own logs . . . down the said river, free of charge or toll from said company: Provided, such persons keep their logs . . . from mixing in the general drive with those belonging to said company, or to the individual members thereof" (see AJ, 135, Jan. 28, 1854, on bill 42A). In the same session c. 321 reflected some sensitivity to the company's privileges by redefining, so as to narrow, the authorized location of certain booms. P&LL, 1856, c. 393, amended the company's charter with reference to matters of internal government, repealed the boom location limit set by the 1854 amendment, and changed the company's name to the Black River Log Driving Co.

In 1857 bill 223S proposed to incorporate a new organization, the Black River Improvement Co., to improve an outlet of that river in order to facilitate movement of rafts. The legislature took no action on the bill after the majority of an Assembly select committee, Mar. 3, 1857, AJ, 748, without opinion, recommended indefinite postponement. The minority committeeman filed an opinion which revealed that opposition to the bill came from men concerned to move logs, as distinguished from rafted lumber; improvement of the channel suitable for raft movement would, it was feared, draw water from another channel used mainly for running logs, and ruin mills located on the log-running channel; the conflict thus appeared one between sharply focused local interests. Assemblyman Hastings' minority report was also revealing of the localistic atmosphere in which such matters were viewed, the legislator's concept of himself as primarily a spokesman for his immediate constituents, and the want of direct investigation or knowledge of the relevant facts: "The undersigned," said Hastings, "is the only representative upon this floor of those who are asking for this measure, and if the result will be to relieve them from a tax of some 75 or 80 cents upon every thousand feet of lumber which they run out of the river, while at the same time it injures no other interest, he feels it to be his duty to his constituents to urge the passage of the bill under consideration. The undersigned has no personal knowledge of the different channels of the Black river, but he has been assured by those who have been perfectly familiar with the entire river for a long series of years, that the fear entertained by the opponents of this bill are entirely groundless; still, to prevent the possibility of this realization of the fears spoken of, he would propose the following amendment, and urge the passage of the bill as amended [reciting an amendment to strike the word 'deepen' from the improvement authorization] . . ." Apparently unconnected with this disputed incident, the legislature enacted without any unusual maneuver P&LL, 1857, c. 145, which took away any right peculiar to the Black River Log Driving Co. to drive mingled logs, and declared that any person putting his logs into that stream might drive the logs of any others which mingled with his goods, and have reasonable pay therefor; the Company thus was put in no better position in general-drive management than any diligent firstcomer.

Returning to more general stretch-of-the-river organization, the legislature without incident enacted P&LL, 1859, c. 200, incorporating for a twenty-year term the Black River Improvement Co., with authority given it to consolidate with the Black River Log Driving Co. Whereas the older company had a cooperative style (apportioning the expenses of its drives among the participants), the new organization had rather the form of a business corporation, with the right to levy tolls. In addition to broad authority to build dams and booms and make channel improvements, the company was authorized to drive for its members, and to take control of the logs of other persons mingling with the company's drive. P&LL, 1864, c. 84, incorporated for twenty-five years a new Black River Improvement Co., apparently (though not by explicit declaration) intended to supersede the 1859 entity, and authorized simply to improve navigation rather than also to conduct drives. P&LL, 1866, c. 447, operated both to enlarge and to narrow this charter; c. 447 clarified that the company had general authority to build dams; on the other hand, it struck the authorization of tolls for use of the company's improvements for running "boards, plank, square timber, cants, shingles, and lumber," leaving a toll right only for use of the company's improvements to run "logs."

Back of the 1866 restriction of the toll right was an interest-group conflict too deep to be dispelled by this concession. Bill 352A, 1867, proposed to repeal P&LL, 1864, c. 84. Amid a swirl of memorials pro and con, two legislative committees considered and recommended rejection of the repeal. Report, Committee on Internal Improvements, Mar. 21, 1867, AJ, 812–813, appealed in part to the equities of committed capital; the committee found no reason to believe there had been any violation of the charter, that up to $30,000 had been invested in reliance on the charter, and that the stockholders' rights should not be taken away unless the state were prepared to reimburse all these expenditures. On the substantive merits, the committee in veiled terms found that the charter served a preponderant interest in timberland (logging) investment, and by implication found that aggrieved local mill interests deserved less consideration from the standpoint of the regional economy; the company's works "are indispensable and are essential to the safety of property to an amount of about two millions of dollars," and repeal "would cause a depreciation of property on said stream to a large amount . . . at least one million of dollars on the value of pine lands alone, and would work wrong and injustice to many." Maneuvering to improve position, the introducer of the repeal bill obtained its reference to a select committee, including himself. However, a majority of the select committee likewise opposed repeal, finding that the charter served the more general economic interest. The Majority and Minority Reports of the select committee, Apr. 8, AJ, 1106 and App., 1266–1272, brought closer to light the nature of the competing economic interests. To the majority the prime fact was that the company's improvements assured the safe movement of logs out of the river to the Mississippi (for milling either at the outlet, at LaCrosse, or at down-Mississippi mill points); to the minority member, the point was that the improvements that aided logs created sandbars which impeded running rafts of lumber from mills upstream on the Black (AJ, App., 1269, 1271). Two conflicts seem to crisscross in the two analyses — one, an essentially mercantilist-nonmercantilist concern as to whether Wisconsin logs shall be milled in-state or out-of-state; the other, a more narrow concern with the rivalry of LaCrosse mills and mills on the upper Black. The majority chose to view the competition as the second, narrower one, probably because this cast the more parochial light on the movement for repeal (AJ, App., 1268). The dissenter put the issue as the broader one — whether the Wisconsin economy should reap the returns not only of producing but also of processing the raw material; he found the attitudes of those opposing repeal to be inconsistent, because even some of them would admit "that it cannot be otherwise than detrimental to transfer the immense product of our pineries to other states to be manufactured, thus increasing their wealth with that which should increase the wealth of our own state. Yet we find these same men in the interest of a company whose object is, and has been, to erect booms, piers and dams for the sole purpose of expediting the running and catching of logs at the mouth of the river, to be rafted and run out of the state to be manufactured" (AJ, App. 1271). However diagnosed, these positions reduced to the sheer competition of profit-seeking interests, in a situation where all could not be equally satisfied. On Apr. 9, AJ, 1150, the Assembly decisively rejected repeal, 64–15.

Past this hurdle, the company existed undisturbed until an essentially local dispute, which it won when it obtained Laws, 1880, c. 225, clarifying its power to block off a navigable slough (the Black Snake River), to assure the efficient flow in the main channel, as against the claims of a company using the slough, whose position dated only from 1876. Two legislative committees were firm in the opinion that the Black River Improvement Company's priority in time gave it a prior equity; the earlier company "had, as they supposed, obtained all the necessary legislation, and entered upon and took possession of said river, relying upon the franchise granted them by the state, and have honestly and diligently improved and handled the property of the community with reasonable dispatch, and at a cheap cost, for a great many years." See Report, Committee on Incorporations, Feb. 17, 1880, SJ, 233–234; Report, Committee on Incorporations, Mar. 11, AJ, 571. In response to insistence of the Senate Judiciary Committee, the 1880 act included a clause saving all rights of the slough user in litigation (see SJ, 394). Subsequently, Black River Improvement Co. v. LaCrosse Booming & Transportation Co., 54 Wis. 659, 11 N.W. 443 (1882) held that no compensation was due the slough user, since the 1880 statute was not a "taking" of "private" property, but an exercise of the state's sovereign control of a public highway. The last flicker of controversy over the company during the years of main lumber traffic was the introduction of bill 652A, 1889, to repeal Laws, 1882, c. 263, which had extended the company's franchise for twenty-five years from Mar. 1, 1889; several petitions were filed in favor of the repeal, but the Assembly pigeonholed the bill without action.

507. P&LL, 1864, c. 131, incorporating the Yellow River Log Driving Association, came briefly into controversy when bill 156S, 1866, was introduced for its repeal, apparently in reprisal against the activity of certain lumber interests in opposing the Kilbourn dam. See the allegations in the petition of proponents of the Kilbourn dam, Jan. 16, 1866, SJ, 79, that "all the opposition to [the dam] . . . is well understood to have originated in, and been led on by a prominent lumber company on the Yellow river, a large tributary of the Wisconsin, entering it a few miles above the dells, near the foot of which the water power in question is situated"; that the lumber company's "pretended objections" are not its real ones, for "the real objections are, that it will create a water-power below them which may be brought into competition with their own." The petitioners found the lumbermen particularly ungracious in their attitude, because they were themselves the beneficiaries of marked special privileges: ". . . those very parties who have led the opposition are themselves greatly indebted to the liberality of the legislature, in granting them, (what we shall never ask,) a real monopoly, in the driving, booming and storing of logs, by which a considerable lumbering interest is required to pay tribute to them . . . [T]hey ought to be satisfied with the advantages they enjoy, and not endeavor to arouse the whole pineries to throw obstacles in the way of their neighbors' making improvements which will in no manner injure them, but which are of the highest importance to a large section of country" (SJ, 81). Memorial 77S, of 230 citizens of Juneau, Sauk, and Columbia counties, supported repeal of P&LL, 1864, c. 131, "on the ground that the same is a monopoly" (SJ, 258). However, without opinion, on Mar. 31, SJ, 786, the Committee on Incorporations recommended indefinite postponement of the repealer. In effect, the Senate followed this recommendation, by substituting under the same bill number a

measure on a different subject (SJ, 880). P&LL, 1866, c. 437, amended the charter to strengthen the powers of the association, by adding authority to log into the river all stranded logs from prior drives, charging to the owners the expenses of doing so, supported by a lien. However, this 1866 amendment reflected some tension over the position of the company when it qualified both the amendment and the original franchise, declaring, "The provisions of this act and of the act to which this is amendatory, shall not apply to or in any manner affect the rights of any person who is not a member of the Yellow river log-driving association." Apparently this qualification repealed the authorizations in the original charter for the company to manage the driving of logs of non-members obstructing its own drives, mingled with the members' general drive, or floating unmarked in the river. Thus the 1866 amendment, in total effect, seemed to take away more than it gave. If the association suffered a setback, it promptly overcame it. P&LL, 1867, c. 317, repealed the 1866 amendment and enacted that "chapter 131 of the private and local laws of 1864 is hereby restored and declared to be in full force." This restoration of the original situation went through the legislature with a speed and smoothness which suggests adroit political management on behalf of the association. The bill was passed under suspension of the rules on the day it was introduced in the Senate, without reference to committee, and just as it was introduced; on the same day, upon receipt in the Assembly, the rules were suspended and the bill passed as it came from the Senate (see SJ, 711, Mar. 26, 1867, AJ, 856–857, Mar. 26). P&LL, 1868, c. 398, however, repealed P&LL, 1864, c. 131, and in effect merged the operating functions of the 1864 association into the Yellow River Improvement Company, amending P&LL, 1857, c. 170, to add to the 1857 company's improvement functions the driving and traffic-management powers of the old association, in their full monopoly scope. Indeed, the 1868 law conferred still broader traffic-management authority, since it empowered the grantee to take charge of all general drives on the river, without reference to obstruction, as well as retaining the authority to take charge of any logs mingled with the company's drive. The statute books continued to evidence some pull and haul over the scope of Yellow River traffic control, however. P&LL, 1869, c. 186, again amended the charter of the Improvement Company, striking the authorization given in the 1868 act to take control of all general drives on the river, and restating the traffic authority as extending to "all log drives on the . . . river belonging to the said company or the individual members thereof, or that may be entrusted to it at any future time," plus authority to take charge of logs put in the river so as to mix with those of the company or those the company had in charge, "provided, the same are not stopped and secured in side booms by the owner or owners thereof." This left the franchise holder with such traffic-management powers as became the standard type, and from this point on the several amendments that were made to the charter concerned details which assumed acceptance of the company's basic traffic-management role.

Traffic control on a tributary of the Chippewa River, also bearing the name Yellow River, came into brief controversy after Laws, 1873, c. 12, had tendered a toll improvement and driving monopoly to "any person" who should improve the stream. The rather veiled form of the grant, running to no specified grantees, probably represented an effort to avoid supposed hazard that a specific grant would offend the new (1871) constitutional amendment banning special legislation conferring corporate powers. (See note 55 above.) However, bill 65S, which became Laws, 1873, c. 12, passed with a smoothness which indicated that it had knowing and powerful supporters. Introduced by Eau Claire lumberman-Senator J. G. Thorp, the bill was referred to a select committee composed of Thorp and the Senators from the two nearest forest-area districts; recommended by this committee for passage two days after its introduction, the bill was passed on the same day under suspension of the rules moved by Thorp; on the day of its receipt in the Assembly, under suspension of the rules, the bill was moved through two readings without reference to committee, and three days later was passed. See SJ, 77, 95, Jan. 29 and 31, 1873; AJ, 173, 195, Feb. 1 and 4. The 1874 session saw a sharp attack on the franchise. Memorial 103A, of several Chippewa Falls residents, Feb. 6, 1874, AJ, 216, 217, claimed that the 1873 law "was enacted at the request of a single person or corporation, and without the consent of, or consulting the wishes of the people of the county of Chippewa, or of said Chippewa valley . . . [S]aid law is private in its nature, and passed for the exclusive benefit of the Union Lumbering Company, and . . . said corporation is the only one who could claim the benefit of said act . . . [S]aid law was passed in the Senate, under suspension of the rules, and in the house, without reference to any standing committee, and without due consideration or examination in either house . . . It is a legislation of the rich and designing politicians against the poor — of the capitalist against the laborer," causing special injustice "to those hardy, industrious loggers of the northwest." Similar charges were made in a letter from John F. Stone & Co., filed with the Assembly Mar. 3, AJ, 606–608, averring that the act "was passed expressly for the sole use and benefit of the Union Lumber Company," and elaborating complaints that the company's alleged improvements were not such in fact, and that the proposed tolls were exorbitant. The most pointed aspect of the charges was the objection that the actual grantee of this veiled franchise was not giving value for services of substance; so viewed, the controversy is further noted in Part Three of this book, in the discussion of the public utility features of stream regulation. The attacks on Laws, 1873, c. 12, produced bill 35S, 1874, to repeal the franchise. The Senate passed the bill as it came from the Committee on Internal Improvements with amendments which changed it into a measure the principal feature of which was to make substantial reductions in the authorized tolls. Reported without recommendation by the Assembly Judiciary Committee, the bill survived a motion for indefinite postponement. The Assembly rejected several amendments further to reduce the claims of the franchise holder, and then passed the bill as it had received it, to become Laws, 1874, c. 294. Opponents shifted their battle then to the courts, where they succeeded, when Sellers v. Union Lumbering Co., 39 Wis. 525 (1876) held that Laws, 1873, c. 12, was legally ineffective for want of a sufficiently definite designation of grantees. Laws, 1877, c. 247, made good the deficiency, granting a franchise to Albert E. Pound, his associates and assigns, to make improvements and charge toll for their use, and to conduct drives of the grantees' logs or of logs entrusted to grantees, with authority to take over other logs obstructing grantees' drives. However, the takeover authority was limited with unusual care, undoubtedly reflecting the impact of the controversy over the prior franchise. Grantees might manage the offending logs "when and when only they shall obstruct the drive, the owners or agents of which logs in either case shall not have inadequate [made adequate] provisions for driving the same," with authority even then limited to driving the uncared-for logs only "to such point as the same shall be taken in charge of and driven by such owner or agent with the necessary men and tools . . ." The 1877 franchise included a specific navigation guaranty: ". . . nor shall [grantees] . . . have the right, by any such works or improvements, or otherwise, to hinder, delay, impede or obstruct the driving of any logs . . ." And, not content with this much assurance, it added a positive assurance of the driving rights of others who came on the river properly prepared, subject only to grantees' right to the initiative, and gave such third persons the same rights as grantees to manage obstructing, uncared-for logs: "whenever any owner of logs . . . shall be ready and desirous to drive his logs on said river, and [grantees] . . . shall not have begun their drive, such owner . . . may drive without delay his own logs, and shall have the same right to take possession of and drive logs . . . as shall obstruct his drive . . . as is given . . . to [grantees] . . ." With painstaking caution, however, the grant qualified the rights of any driver: "provided, that neither [grantees] . . . or any such driver of logs, shall have the right in any other case to take or drive logs, without previous agreement with the owner thereof or his agent; and in no case shall have the right to control any logs . . . after the owner . . . shall have come upon the river properly prepared . . . to drive his own logs . . ." The guarded terms of the 1877 act were the product of some legislative skirmishing. The Senate Judiciary Committee first recommended indefinite postponement of bill 94S, which was to become Laws, 1877, c. 247, and then, after recommittal on the sponsor's motion, reported it for approval with amendments, one member dissenting. So amended, the bill had little trouble; it was ordered to a third reading by a recorded vote of 21 to 5, and then passed. See SJ, Jan. 26, 1877, p. 89, Feb. 9, p. 186, Feb. 12, p. 200, Feb. 24, p. 314, Feb. 27, p. 362, and Mar. 1, p. 399. With a favorable report by the Assembly Committee on Lumber and Manufactures, the bill passed the other chamber without apparent difficulty (AJ, 584, 626, Mar. 3 and 7). Despite the charges of loose or manipulated procedure at the outset of this chain of affairs, taken altogether this Yellow River history suggests that the legislative process gave substantial play to the muster and maneuver of contending interests.

508. See note 54 above.

509. See Martin, p. 170, and fig. 53.

510. P&LL, 1866, c. 181. The bill went through as introduced by Eau Claire lumberman-Senator J. G. Thorp, supported by a favorable recommendation (without opinion) from the Senate Committee on Internal Improvements. See SJ, Feb. 19, 1866, p. 314, Feb. 28, p. 374, and Mar. 1, pp. 393–394; see Merk, 91–92. Though he presents a characteristically careful and just summary of the Beef Slough story, Merk puts it out of proper perspective to the general inventory of stream franchise matters when, without presenting evidence, he finds it typical of clashes among in-state and out-state milling interests, as well as "representative also of the standard of business ethics of the time." The underlying contest for traffic control is thinly veiled behind the rationalization advanced for the 1866 franchise and the failure to utilize it, in Minority Report, Committee on Lumber and Manufactures, on bill 302A, Feb. 27, 1868, AJ, 632: "The bill of 1866, similar in its provisions to this [302A, of 1868], named as incorporators persons representing the interest in Chippewa river, so that if this work was to be done, it should be in the hands of the friends of the interests of that valley. Subsequent investigation has convinced these parties that the scheme was impracticable, and could not be carried out without seriously impairing or ultimately destroying the navigation of the river . . ."

511. Against the background of opposing petitions (see, e.g., AJ, 395–396, Feb. 21, 1867, Memorials 98A, 99A, and 100A), bill 129A was recommended (without opinion) with amendments by the Assembly Committee on Lumber and Manufactures, Mar. 1, AJ, 450, ordered to a third reading in that form by a recorded vote of 57 to 26, Mar. 12, AJ, 595, and passed Mar. 18, AJ, 694. However, without opinion the Senate Judiciary Committee recommended indefinite postponement, Mar. 29, SJ, 790. The Senate so voted, by a recorded vote of 20 to 4, Apr. 3, SJ, 842.

512. The 1858 legislative journals are studded with petitions pro and con concerning the Beef Slough Company's application. See, e.g., AJ, 247, Feb. 5, Memorial 68A, in opposition; AJ, 309–310, Feb. 10, Memorial 89A, in support. Bill 302A, to confer the franchise, was reported favorably in Majority Report, Committee on Lumber and Manufactures, Feb. 25, AJ, 608–609. Though the report is characteristically weak or fuzzy in generalizing its principles of decision, it seems to stand for one value judgment and one rule of procedure. On the point of values, the majority in effect made glancing acknowledgment of the interest conflict between in-state millmen and those interested in selling logs to out-state processors, and ruled that the general interest was best served by promoting timberland investment value, in preference to local industrial development: ". . . at nearly every session of our legislature such privileges as the one asked for in this bill, have been granted without opposition, and we cannot see why this should be made an exception, unless it is to accommodate mill owners at Eau Claire and vicinity, to contract the price of logs at the expense of all other parties interested in pine lands in the Chippewa valley" (AJ, 609). It is notable that the committee here does not put itself on what might seem the firmer ground of constitutional principle, by emphasizing the public interest in navigation as superior to that in industrial power. (Compare the wavering record on this balance of factors in the handling of the Kilbourn dam matter, note 505 above. In that controversy navigation was stressed in 1857. Then in 1866 different committee reports emphasized, in one instance, industrial power, and in the other, lumber transport. Finally, there was a merger of the two in the 1873 discussion, where navigation was argued as important to timberland values.) The 1868 majority report on the Beef Slough matter took a procedural judgment as an alternative ground for recommending the proposed franchise. The committee had heard sharply conflicting testimony concerning the likely obstruction which the proposed works might cause to navigation. The majority found the evidence not sufficient to establish that obstruction was likely; in this posture, it was fairer to allow than to disallow the franchise, for were it denied "its friends have no remedy," whereas if it were granted, and if experience demonstrated obstruction, the opposition still had recourse to injunction, as against a public nuisance. In this aspect, the majority implicitly raised a presumption in favor of initiative to improve stream carrying capacity. The minority member of the committee estimated the testimony to show a high probability that navigation would be impaired, but his strongest observations went to enforce a mercantilist judgment in favor of keeping the exploitation of Wisconsin natural resources in the hands of Wisconsin processors: "In conclusion, let us not forget that we reside in the state of Wisconsin, and that we are members of the legislature of the state of Wisconsin, and are bound to legislate for the interests of the people . . .

of this state. Let us not inadvertantly legislate to their disadvantage and for the benefit of foreign speculators who come here to seize our raw material and run it out of this state into neighboring states, there to be manufactured and given an increased value, which should properly be done within our own border. While I fully endorse the doctrine of free trade, I do not believe in the doctrine of giving *affirmative* legislation which will be to the disadvantage of the people of our own state" (AJ, 632). Thus, like the majority, the dissent seemed to have less concern for transport as a supporting facility for the general economy than as a factor in the relative advantage of economic sectors. The Assembly voted 47–36, on Feb. 25, AJ, 695, for indefinite postponement of bill 302A.

513. See Fries, 144, and Merk, 92. There are about a dozen instances in which the legislature enacted general stream-use regulations as incidents of particular franchises. See, e.g., P&LL, 1856, c. 238 (passage clause in boom franchise declared applicable to all booms on Eau Claire River); P&LL, 1867, c. 503 (navigation guaranties in particular improvement franchise declared generally applicable on Little Wolf River); P&LL, 1871, c. 357 (incident to incorporation of Pine River Improvement Co., sets bridge and dam standards for the river at large). Such legislation, however, plainly announced its character on its face; though representing clumsy technique, it did not carry deception.

514. See Fries, 145. Another maneuver in 1867 involved the building of a dam across Beef Slough by the Chippewa millmen, claiming to act under authority of an old (non-toll) franchise for general navigation improvement on the river, given by P&LL, 1856, c. 405, to the Chippewa River Improvement Company, and continued and enlarged (though there is no evidence this was done with specific reference to the competition for control of Beef Slough) by P&LL, 1866, cc. 286 and 509. Those interested in transporting logs to Mississippi River mills obtained a trial court injunction against maintenance of the dam, and their crews, aided by local farmers, promptly acted on the injunction to remove the structure. See testimony on this incident attached to Report, Select Committee on bill 302A, 1872, AJ, 633 and App., 108, 110, 120. See Fries, 143, and Merk, 92. The 1872 testimony reports the injunction as given to protect navigation; Merk, without citing his source, says that it was obtained incident to condemnation of the land for a public highway, by local authorities friendly to the log transport interest.

515. P&LL, 1870, c. 299. As bill 311A, this measure passed smoothly following a favorable report (without opinion) with amendments by the Assembly Committee on Lumber and Manufactures, Feb. 18, 1870, AJ, 411. Cf. AJ, 626, Mar. 3; SJ, 389 and 450, Mar. 8 and 10. In the prior session the Senate had quietly buried (by substituting under the same bill number a measure on a different subject) bill 274S, to require the Beef Slough company "to remove obstructions in the Chippewa river." See SJ, Feb. 15, 1869, p. 238, Feb. 18, p. 300, and Feb. 19, p. 335; cf. Laws, 1869, c. 157. The local millmen also thought to find help in sect. 112, Laws, 1869, c. 152, which — as part of a long codification of highway and bridge laws, penalized one "who shall obstruct any meandered navigable stream . . . by placing any refuse lumber, slabs or other substance calculated to impede the free navigation thereof, either in such stream or in any tributary of said stream, so that the same shall float into such meandered stream." No action seems ever to have been taken in the Beef Slough controversy under this dubiously applicable provision (see Fries, 145).

516. The general sweep of the terms of General Laws, 1872, c. 105, was quite plainly in response to efforts to avoid hazards under the (1871) constitutional ban on special legislation conferring corporate powers, and did not represent any attempt to conceal the particular bearing of the legislation on the works at Beef Slough. See the discussion of the constitutional point in Reports, Judiciary Committee, AJ, 476, 515–516, Feb. 28 and March 4, 1872. After this consideration, and following maneuvers of opposing parties to have the bill referred to a favorable committee, the bill was sent to a select committee which the Assembly Journal, Mar. 5, p. 541, frankly described as "the select committee having in charge the Beef Slough bill, No. 302,A." Report, Select Committee, Mar. 12, AJ, 661–662, reported the bill favorably with an amendment which tightened its navigation guaranties by making them apply to the channel "usually followed" as well as to the main channel. The select committee, weighing conflicting testimony, found that the company had made the slough a useful facility, and had built and used it so as not materially to impair the navigability of the main river. The committee stated bluntly the interest conflict before it, and in effect ruled that state policy should

support an enterprise which increased the overall transport capacity of the river, as against a local industrial interest: "We believe that the rights and interests of those who market logs should be taken into consideration as well as mill owners on the river, in all questions concerning its navigation. It is a question whether they shall have the monopoly of the logs of the Chippewa and its tributaries for manufacturing the same on these rivers, or whether others should be tolerated and protected in furnishing a safe harbor for logs at reasonable cost to those men running them down those rivers and seeking a market below" (AJ 661). On the floor, the more serious further question was not of navigation guaranties (the Assembly accepted its select committee's amendment and rejected two other proposals of vague import), but of the stipulation exonerating the company with reference to past deficiencies. An effort to insert provision for what would amount to suspension of the company's franchise until it made good past defects was defeated, 34–43. On a critical recorded vote, the bill was ordered engrossed, 48–28, and subsequently passed quietly (AJ, 686–688, 724, March 15 and 18). In the Senate, an amendment to strengthen the main-channel navigation guaranty by adding a specific stipulation against diversion of water from "any main river or stream, so as materially to interfere with the navigation thereof," lost on a recorded vote of 12 to 17 (Mar. 21, SJ, 590–591). In light of the fact findings of the Assembly select committee as to the successful record of the slough operation without substantial diversion of water, it would seem that the Slough interest might have safely accepted this amendment; the fact that they so successfully opposed it perhaps measures both their strength and their distrust of the opposition. Ordered to a third reading by a recorded vote of 24 to 5, bill 302A then passed the Senate under a suspension of the rules (*ibid.*). Generally, on the assumption of control of the Beef Slough operation by the Weyerhaeuser combination, see Fries, 145–147, and Gates, *Wisconsin Pine Lands,* 130–134.

517. The local mill interests tried a flanking movement against the Beef Slough company by invoking federal patent law. The slough operation required using the fin boom, on which in 1868 Eau Claire interests obtained a patent of dubious worth, the validity of which they managed to have confirmed by a Congressional act in 1872, which the Beef Slough men unsuccessfully sought to repeal. In 1874 the patent holders brought an infringement suit against the Beef Slough operators, and eventually won a trial court judgment. However, for no reason explained on the record, they resigned this victory by striking an agreement with the Beef Slough men for use of the patented boom (see Fries, 149–150). The other litigious tack taken by the Chippewa millmen was to seek relief in federal court for alleged illegal obstruction of navigation. Under Cooley v. Board of Wardens of Philadelphia, 12 How. 299 (U.S. 1851), it was quite clear that the state legislature might regulate navigation on a local river like the Chippewa, unless Congress superseded state action (see note 8 above). Since the Beef Slough company held a statutory franchise from Wisconsin, its opponents needed to find some basis in federal legislation from which to argue that Congress had taken the Chippewa under its direct protection. They claimed to find this basis in the federal appropriation made in 1876 for a survey of that river. Months of pressure from the Chippewa valley millmen finally procured an action by the United States, seeking to enjoin maintenance and operation of the Beef Slough works as a public nuisance, claiming diversion of water to the damage of main-channel use, and clogging of the channel by the large quantities of logs handled by the Beef Slough operators. The federal court made short work of this plea, however. None of defendants' actions was shown to interfere with the federal survey of the river; the federal survey act in its own terms showed that Congress did not intend to interfere with state navigation regulations on the river; since defendants operated under a valid state franchise, and their activity thereunder conflicted with no command of Congress, there was no basis on which a court might find them guilty of a public nuisance: United States v. Beef Slough Manufacturing, Booming, Log-Driving & Transportation Co., 8 Biss. 421 (Circ. Ct. W.D. Wis. 1879). The court was aided to this result by the fact that — after initiation of the Beef Slough action — Pound v. Turck, 95 U.S. 459 (1878) had made direct application of the Cooley doctrine to sustain the validity of Wisconsin P&LL, 1857, c. 235 (granting a dam and boom franchise on the upper Chippewa to the Chippewa Falls Lumbering Company), as against complaints of a raftsman. Not yet ready to surrender, the Chippewa millmen then persuaded the operator of a line of steamboats on the Chippewa to bring his actions for damages

and for an injunction, on the claim that the Beef Slough operators were preventing the operation of steamboats on the river by clogging it with their log drives. Again, the court held that, since the Beef Slough company held a valid state franchise, not superseded by any Congressional action, the mere presence of its works and their use could not be ruled to constitute a public nuisance. So far as complaint was of the detailed character of its operations, the court recognized on common law and statutory authority that transporters of logs had a right to use the stream as much as vessels, and held that if there were ground of complaint over particular unreasonable use damaging to the steamboat man, he must seek relief by an action at law for damages; equity lacked resources to police day-to-day activity: Heerman v. Beef Slough Manufacturing, etc. Co., 1 Fed. 145 (Circ. Ct. W.D. Wis. 1880). Since the principal concern of the Heerman opinion is with the scope of judicial power in regulating economic operations, further discussion is postponed to the chapter of Part Three dealing with the police power.

518. See Fries, 153–156; Marshall, *Autobiography,* I, 275–291.

519. See notes 416, 428, 430, and 445 above.

520. See note 445 above. In the evidence taken in the first (1871) investigation, there was unequivocal testimony by two Eau Claire lobbyists that they were ready to buy the vote of one legislator, and took steps to that end. See Report, Committee of Investigation under Resolution 35A, Mar. 13, 1871, AJ, 806, 808–809, 813 (testimony of Jefferson and Gleason). Eau Claire lobbyists F. O. and J. G. Thorp and O. H. Ingraham denied any impropriety on their side; their denials were stated with conspicuous care to limit them to what they knew of their own knowledge. Assemblyman Fryer testified to a payment of $50 and a promise of "a nicer thing for you in future," from F. O. Thorp; Thorp denied any such transaction, but testified that he made Fryer a loan at Fryer's request (AJ, 802, 803). Assemblyman Hoyt testified, confirming the testimony of Eau Claire lobbyist Jefferson, that Jefferson offered him money for his vote (AJ, 804). Chippewa Falls lobbyists made bolder denials of wrongdoing (AJ, 807, 812, 814, 816, 817). But Assemblyman Richter testified that a Milwaukee saloonkeeper-politician, lobbying for Chippewa Falls, offered him money for his vote; when the Milwaukeean denied the episode, the Assemblyman repeated his statement (AJ, 805, 820). Governor Fairchild took particular exception to the fact that the franchise bill passed included as one of the incorporators Gleason, whose uncontradicted testimony stood in the record to the effect that he had tendered a bribe to a legislator to procure his vote for the bill. The Senate had voted down, 18–12, a motion to strike Gleason's name from the bill (SJ, 712, Mar. 17). The Governor quite properly found this result abhorrent: "No word of condemnation has been officially expressed by the Legislature against this breach of the privileges of your honorable body and the person confessedly guilty thereof has been invested with every privilege he sought, and to obtain which he stood ready to use, and tried to use such corrupt means. Instead of being punished he has been rewarded. Instead of being rebuked, he has been by inference commended by the granting, immediately upon the disclosure of his corrupt conduct, of a franchise, which it is claimed will result in great advantage to himself" (AJ, 1000, Mar. 22). The Assembly refused to pass bill 1A over the veto by a vote of 32 to 33 (AJ, 1018, Mar. 23).

In the second investigation the Majority Report, Joint Committee of Investigation, Feb. 15, 1872, SJ, App., 166, 169, found that five legislators were offered "money or other consideration by certain of the Lobby against the bill, to influence or obtain their votes against the bill. These gentlemen severally voted for the bill"; and that proponents of the bill offered bribes to two legislators, who testified so, and who voted against the bill; that proponents admitted offering bribes to one legislator since deceased (Hoyt); that the reported bribe tendered by Thorp to Fryer, in the majority's view, was in truth a loan; that in any case Hoyt and Fryer voted against the bill. The committee minority would not affirmatively subscribe to these detailed findings, but on the other hand did not specifically controvert them (SJ, App., 186). The majority (which was made up of friends of the proposed Eau Claire franchise) found that the bribery efforts on behalf of the bill did not represent a concerted program by the proponents, but the ill-advised individual zeal of particular individuals; the majority gave similar but more qualified absolution to the bill's opposition, hedging because the committee had been unable to obtain testimony from all witnesses it sought on that side. Thus, the majority concluded that "the testimony discloses that the lobby on both sides were clouded and dishonored by two or three intemperate and senseless individuals of their number on each side, that offered, promised, or suggested the payment of money or other considerations to several members, to influence their action and vote on the bill," but that "no attorney and no lobbyist, on either side, so far as the testimony shows, was directly or indirectly authorized to use other than fair discussion and arguments for or against the bill. In especial, no money or other valuable consideration, was authorized by either party, to be directly or indirectly offered, promised or suggested to any member, to influence his action on the bill. If any attorney or lobbyist did so, it would seem from the evidence, to have been his own wrong-doing and not authorized by either party" (SJ, App., 166). The minority report found the majority's conclusions too sweeping, and stated that all that the testimony supported was a verdict of not proven: "The committee have not been able, as yet, to prove definitely the contents of any book or paper on either side; have not been able as yet to prove definitely the amount of money raised at Eau Claire, or at Chippewa Falls; have not been able as yet to prove definitely or with any fair degree of certainty, the amount of money paid out by the friends of the bill to secure its passage, nor fully for what purpose so paid, nor to whom paid; have not been able as yet to prove definitely the amount of money paid out by those opposing the bill, for the purpose of defeating the bill, nor fully to whom paid" (SJ, App., 185). In particular the minority noted that a witness on the Eau Claire side who appeared to be treasurer for the proponents' lobby testified in a hearing at Eau Claire that his books were in Madison, and at a hearing in Madison that his books were in Eau Claire, so that on neither occasion could he testify as to moneys received and paid out by him; the majority made no comment on this episode (*ibid.*). The minority also pointed out that most witnesses testified no more than that from their own knowledge they knew of no improper conduct; what it made of this aspect of the record the minority did not say, though its pointed emphasis perhaps implied that it thought the testimony was so carefully hedged by so many witnesses as to indicate less than frank disclosure; on the other hand, in fairness to the investigation, it would seem that the 1871 legislature should not have been condemned purely on hearsay (SJ, App., 185–186). Finally, the majority report absolved the legislators of accepting any bribes tendered. Three incidents claimed to involve personal loans were, the majority felt, established to have been in fact "all simply business transactions, free from any blame and entirely undeserving any suspicion." As to proffered bribes, the majority found that "for the honor of the state, each of the members so approached, indignantly refused and spurned the proffer, and voted his own convictions and judgment on the bill" (SJ, App., 166, 170). The minority did not take square issue with this finding, but again felt that the committee was not entitled to hand down more than a verdict of not proven (SJ, App., 186). It should be observed that, just as the majority of the committee was made up of men who favored the proposed Eau Claire franchise, the minority members were opposed to the grant. The proceedings of the committee of investigation and the content of both majority and minority reports are obviously colored throughout by strong feeling on the merits of the substantive issue.

Particularly revealing of the partisan temper which guided action on both sides in the treatment of the bribery issue, to an extent which warrants skepticism of the findings on both sides on the matter, was the handling of the inclusion of Gleason (the admitted offerer of a bribe) among the incorporators. As we have seen, while the bill was under consideration in the Senate, an amendment striking Gleason's name was voted down, 18 to 12, by the bill's proponents (see SJ, 712, Mar. 21, 1871). On the same day, after the bill had passed both houses and was in the hands of the Governor, the proponents suffered a belated change of heart. A friend of the bill introduced Joint Resolution 36A, reciting that bill 1A was in the Governor's hands, that testimony before the Assembly investigating committee showed impropriety on the part of Gleason, that hence the already-adopted Joint Resolution 21, barring new business, should be suspended to allow adoption herewith of an amendment striking Gleason's name from the bill. Adopted 38–31 by the Assembly (AJ, 975–976, Mar. 21), the Joint Resolution obtained an apparently favorable vote of 17 to 14 in the Senate, but the presiding officer of the Senate (Lieutenant Governor Thaddeus Pound, of Chippewa Falls, an opponent of the Eau Claire franchise) ruled that the resolution required a two-thirds vote for adoption, and hence failed. Pound's ruling seems a proper one on the merits. The Joint Resolution was in improper form; it should have requested return of the bill from the Governor, since the legislature had no power to amend the measure once it had left its hands. Moreover, to suspend the rules for such extraordinary action would properly require a two-thirds vote. On the other hand, it is clear that the reason that the requisite vote could not be mustered was because the bill's opponents would not contribute to a favorable total—a stance understandable enough from a partisan viewpoint, but hardly consistent with their pious concern at the continued inclusion of Gleason among the beneficiaries. On the other hand, the high principles spoken by Joint Resolution 36A should have moved the bill's supporters to adopt the amendment proposed when the bill was before the Senate. See the discussions in Majority and Minority Reports, Joint Committee of Investigation, Feb. 15, 1872, SJ, App., 153, 154, 182–183.

521. Somewhat analogous, but of still worse quality, was the disposition of the Beef Slough contest in 1868 by the item of general legislation inserted in a special corporate charter apparently unrelated to the Beef Slough matter (see text at note 513). The 1868 maneuver was worse, because it was clearly a piece of deception. There was never any doubt in the minds of the legislators that the Eau Claire "waterworks" bills were designed basically to accomplish the creation of the log-handling facilities desired by the Eau Claire millmen. See, e.g., Majority and Minority Reports, Committee on Lumber and Manufactures, on bill 366A, Feb. 27, 1875, AJ, 477–479, 493–494; Report, Committee on Lumber and Manufactures, on bill 98A, Feb. 16, 1876, AJ, 326–327, and Minority Report, Committee on Incorporations, on bill 98A, Mar. 7, SJ, 534–535.

522. See Wisconsin River Improvement Co. v. Lyons, 30 Wis. 61, 67 (1872); Wisconsin River Improvement Co. v. Manson, 43 Wis. 255, 265 (1877); cf. Powell v. Pennsylvania, 127 U.S. 678, 685, 686 (1887).

523. The judgment of a shrewd contemporary, intimately involved in the fortunes of the Chippewa Valley, was that the general economic interest of both cities lost because of the bitterness and duration of their contest (see Marshall, *Autobiography,* I, 275–276). So federal Circuit Judge Bunn, wise in the history of the Wisconsin lumber industry, warned that the contest of local millmen and those who would move logs out of state was one destructive to both, and to the public welfare: "The logging business, the business of sawing lumber and running rafts, and that of navigation by boats, are all interests that have grown up together on the Chippewa, as also on some of the other lumbering streams in the north half of the state. They are all interests which the law recognizes, and they are all entitled to its protection. They are not and should not be regarded as antagonistic. On the contrary, they are more or less dependent one upon the others; and by a little mutual concession and forbearance, and a just and reasonable spirit on the part of those persons severally devoted to them, they may, undoubtedly, all flourish in themselves, and be a help to one another. Such a course in the long run will be found much more advantageous to the public, and to the parties especially concerned, than the engaging in litigation, the effect of which, if successful, would be to exterminate or cripple what some are apt to regard as a rival interest": see Heerman v. Beef Slough Manufacturing etc. Co., 1 Fed. 145, 166 (Circ. Ct. W.D. Wis. 1880).

524. Cf. West v. Kansas Natural Gas Co., 221 U.S. 229 (1911); Pennsylvania v. West Virginia, 262 U.S. 553 (1923). If a state may exercise broader control over loss of natural resources which it holds in trust as commonwealth assets, such as the water of its streams, yet it may not, consistent with the commerce clause, regulate use of such resources merely to foster intrastate at the expense of interstate economic activity. Cf. Hudson County Water Co. v. McCarter, 209 U.S. 349 (1908); Foster-Fountain Packing Co. v. Haydel, 278 U.S. 1 (1928). Though all this is commerce clause doctrine elaborated past the years of the Beef Slough controversy, it is doctrine so close to the traditional strength of the commerce clause when applied to bar parochial discrimination against interstate economic growth, that it is reasonable to believe it would have been declared as the governing law, had an issue been made out of the Chippewa valley contest. See Heerman v. Beef Slough Manufacturing etc. Co., 1 Fed. 145, 166 (Circ. Ct. W.D. Wis. 1880).

525. Cf. Hurst, *Law and Social Process,* 60–61, 134, 144, 155, 160–161, 167, 247, 282, 304.

526. Cf. Hurst, *Law and the Conditions of Freedom,* 56–58, 74, 83, 89–93, 103.

527. See Fries, 34, 62, 64, 78, 84–87, 91; Gates, *Wisconsin Pine Lands,* 184; Reynolds, 12, 13, 40, 59, 62, 104, 106, 144–145. Compare Larson, 108, 220, 238, 363; Martin D. Lewis, *Lumberman from Flint* (1958), 46–48, 64, 129, 201.

528. See Raney, 179, 185, 187, 188, 194, 195. Compare, Gordon Oswald Greiner, "Wisconsin National Railroad Land Grants" (unpublished M.A. thesis, 1935), 2, 7, 11, 51–53; Merk, 276, 279, 290–299; Merrill, 32–33, 60. Cf. Memorial 11S, 1874, of North Wisconsin Railway Co., SJ, 75–82 (Wisconsin caught up in battles of interstate rail empires).

529. See Fries, 99, 139, 146, and Raney, 179.

530. Compare Acts, 1851, Memorials, c. 14; General Laws, 1862, Memorial No. VIII; General Laws, 1866, Memorial No. 8; General Laws, 1867, Memorial No. 4; General Laws, 1868, Memorial No. 18; General Laws, 1871, Memorials No. 1, 5, and 11; Laws, 1875, Joint Resolution No. 3. See also Address of Superior Board of Trade, Feb. 15, 1870, AJ, 330–337; Richard Lewis Canuteson, "The Railway Development of Northern Wisconsin" (unpublished M.A. thesis, 1930), 2–3, 108, 114, 163, and chap. v *passim*. The growth of a sense of economic needs as communities developed in the north was mirrored, also, in petitions for extension of mail service. See General Acts, 1865, Memorial No. 6; General Laws, 1872, Memorial No. 22; Laws, 1877, Memorial No. 9; cf. General Acts, 1865, Memorial No. 7 (for new federal court district).

531. See, e.g., the prime emphasis upon the values of settlement and over-all economic development in the comments of Governor Fairchild, in his General Messages of Jan. 10, 1867, SJ, 22; Jan. 14, 1869, SJ, 24; and Jan. 12, 1871, SJ, 24–25. Compare, to like effect, Governor Washburn's General Message of Jan. 9, 1873, SJ, App., 30; Governor Rusk's General Message of Jan. 11, 1883, AJ, 19, 21, 23; Reports, Committee on Railroads, responding to the Governor's suggestions on railroad aids, AJ, 320–321, 825–828 (1855); Majority Report, Committee on Finance, on bill 131S (railroad taxes), SJ, 620, 621–622 (1864); Report, Select Committee on a railroad aids amendment to state constitution, SJ, 444, 445 (1868); Report, Committee on the Assessment and Collection of Taxes, on bill 4S (Wisconsin Central tax exemption), AJ, 181–182 (1877). In reviews of railroad development in the north, the early railroad commissioners commonly note lumber industry growth, but as an item along with others in regional expansion. See, e.g., Annual Reports, Railroad Commissioner, PD, II, 8 (1878), I, xxxiv, xxxix (1882), II, vi (1888).

532. See the telling comments upon the general regional enthusiasm for railroad development in its first stages in an unopened area, in Majority Report, Committee on Railroads, AJ, 370–371 (1876), and in the unanimous Report of the same committee, AJ, 83–85 (1877), especially at p. 85, both opposing efforts to repeal statutory tax exemptions of West Wisconsin Railroad aid lands.

533. See Fries, 84, Merk, 87, and Reynolds, 24, 67, 100, 144.

534. See Raney, 179, 180; compare Duckett, 110, 162, and Fries, 84, 85, 88, 91, 92. Of course, interest in promoting forest-area railroads ran ahead of substantial use of railroads to move logs or lumber. Rector, *Log Transportation,* 203, 205, puts major use of railroads in Wisconsin lumbering well into the eighties.

535. See Raney, 186–188, 190–191. Stephenson, 174–175, notes that lumbermen pressed the North Western to extend from Green Bay to the Menominee; he also notes (pp. 213, 214) that the railroad interest was strong as a lobby in its own right before the state legislature.

536. See Current, 106, 132. However, even in such a case as that of Sawyer the railroad interest appears as a moving force in its own capacity and aims, playing a distinct role in Sawyer's political success. See Current, 69, 86, 156, 164, 165, 167, 172, 214, 229.

537. For a detailed account of this story from the standpoint of a group of railroad promoters, see Memorial 11S (of the North Wisconsin Railway Company), in SJ, 75–82 (1874). Cf. Attorney General v. West Wisconsin Railway Co., 36 Wis. 466 (1874).

538. See Gates, *Wisconsin Pine Lands,* 181, 182, 184, 185, 187.

539. See Current, 133–143.

540. See, generally, Hunt, *Law and Locomotives,* 29–30, 78, 83; Kuehnl, 143, 144, 182–185.

541. See the last section of Chapter V, dealing with incorporation as a device for extending the scale of market operations.

542. Laws of the Territory of Wisconsin, 1840, No. 48; Pratt v. Brown, 3 Wis. 603 (1854).

543. The quoted examples are from P&LL, 1866, c. 555 ("An Act to incorporate the Chippewa Falls and Mississippi railroad company"), and P&LL, 1869, c. 51 ("An Act to incorporate the northern Wisconsin railway company"). On judicial tightening of the requirement that the railroad pay the landowner before it enter on the land to build, see

Powers v. Bears, 12 Wis. 214 (1860). That, on the other hand, the roads needed liberal corporate powers, see Clark v. Farrington, 11 Wis. 306, 321 (1860).

544. See Ryan, C. J., for the court, in Attorney General v. West Wisconsin Railway Co., 36 Wis. 466, 494 (1874). Compare, to like effect, the policy implications of the attention to route designation in connection with aid land grants. See, e.g., General Laws, 1866, Joint Resolutions 4 and 5; General Message of Governor Washburn, Jan. 9, 1873, SJ, 27–28; Special Message of Governor Taylor, Feb. 20, 1874, AJ, 429–432.

545. See, e.g., Private & Local Acts, 1854, c. 269; P&LL, 1855, c. 29. Plenary legislative control of plank road routes is implicitly recognized in Madison, Watertown & Milwaukee Plank Road Co. v. Reynolds, 3 Wis. 287, 290 (1854). Cf. Madison, Watertown & Milwaukee Plank Road Co. v. Watertown & Portland Plank Road Co., 7 Wis. 59, 79 (1858). Legislative practice implicitly treated the authorization of rail routes as a problem of distinctive importance. This emerges by contrast with legislative practice concerning water transport. Charters of ordinary lumber firms sometimes authorized them to conduct a general water carriage business along with their own operations. See, e.g., P&LL, 1866, c. 284, 1868, c. 385, 1869, c. 77. Private lumber firms also came to desire authority to build and operate their own logging railroads. But here the statutes consistently distinguished these authorizations from those given to common carrier railroad companies. See Revised Statutes, 1878, sect. 1771, as amended by Laws, 1883, c. 220, and carried into Statutes, 1898, sect. 1771, making separate provision for incorporation of companies to operate private logging railroads, apart from companies to operate public railroads. Cf. P&LL, 1872, c. 125. To avoid burdensome regulations, the lumbermen themselves grew zealous to establish that their own logging roads were not common carriers, because their creation involved no grants of public lands, no charters as railroad companies as such, no general carriage, and no privileges of eminent domain. See the correspondence of private operators in Biennial Report, Railroad Commissioner, PD, II, 25–35 (1900). Cf. Biennial Report, Railroad Commissioner, PD, I, 1 (1885), PD, I, 7 (1898), PD, 131 (1902). A distinct procedure for establishing logging railroad routes was provided by Laws, 1895, c. 277 (expanding Laws, 1889, c. 402, on temporary highways for logging area uses), as broadened by Laws, 1901, c. 23, and incorporated in Statutes, 1898, sect. 1299i. Cf. Laws, 1925, c. 328 (eminent domain). In other respects besides the grant of corporate status and authorization of routes, legislative policy made separate policies for private logging railroad operations. See Laws, 1899, c. 308, and 1919, c. 244 (information reports); Laws, 1907, cc. 12 and 92, and 1911, c. 211 (particular bridge authorizations); Laws, 1909, c. 169, and 1913, c. 386 (in effect exempting logging railroads from caboose requirements imposed on common carriers); Laws, 1925, c. 268 (inspection).

546. See United States v. Milwaukee & St. Paul Railway Co., 5 Biss. 410 (Circ. Ct. W.D. Wis. 1873); Wisconsin River Improvement Co. v. Lyons, 30 Wis. 61, 65 (1872).

547. Note the various attitudes toward route competition and route monopoly as factors to be weighed in allocating the northwest Wisconsin railroad-aid lands, in Report, Select Committee on Bill 106A (aid lands grant to Milwaukee & St. Paul Railway Co.), AJ, 707–709 (1873); and in Report, Select Committee on Bill 178A (aid lands grant to Chicago & Northern Pacific Air Line Railroad Co.), AJ, 673–676 (1873). Following Proprietors of the Charles River Bridge v. Proprietors of the Warren Bridge, 11 Pet. 420 (U.S. 1837), the Wisconsin court early and consistently ruled that no route monopoly or other monopolistic privilege would pass by implication under legislative grant of a franchise of public concern. Janesville Bridge Co. v. Stoughton, 1 Pinn. 667 (Wis. 1846); City of Racine v. Chicago & Northwestern Railway Co., 92 Wis. 118, 65 N.W. 857 (1896). Cf. Chapin v. Crusen, 31 Wis. 209 (1872).

548. Compare the justification for the grant of eminent domain powers to builders of milldams, in Newcomb v. Smith, 3 Pinn. 131, 140 (Wis. 1849).

549. See Levy, *Law of the Commonwealth,* 120, 307.

550. Pratt v. Brown, 3 Wis. 603, 610 (1854).

551. *Id.,* 612.

552. Cf. Fisher v. Otis, 3 Pinn. 78 (Wis. 1850); Williams v. Phelps, 16 Wis. 80 (1862); Duinneen v. Rich, 22 Wis. 550 (1868).

553. See "Bring the Land to Market," in Chapter I.

554. Cf. Edward A. Adler, "Business Jurisprudence," *Harvard Law Review,* 28:135 (1914), reprinted in *Selected Essays on Con-*

stitutional Law (1938), II, 436. Wisconsin common law recognized the special obligations attaching to one who held himself out as a common carrier. See Doty v. Strong, 1 Pinn. 313, 324, 326–327 (Wis. 1843); Marshall v. American Express Co., 7 Wis. 1, 26 (1858); Wood v. Crocker, 18 Wis. 345, 347–348 (1864).

555. Statutes of the Territory of Wisconsin, 1839, p. 131; Revised Statutes of the State of Wisconsin, 1849, c. 43. Cf. Oscar and Mary Flug Handlin, *Commonwealth: . . . Massachusetts, 1774–1861* (1947), 76; Levy, *Law of the Commonwealth,* 256, 307.

556. The quotation is from sect. 7 of P&LL, 1866, c. 555, "An Act to incorporate the Chippewa Falls and Mississippi railroad company." Compare Ryan, C. J., for the court, in Attorney General v. Chicago and North Western Railway Co., 35 Wis. 425, 578 (1874): "We shall not discuss the question whether the defendants have a right to take toll, as intimated by Mr. Justice Strong in the *State Freight Tax Case,* 15 Wallace, 232, without any franchise to take it, as an attribute of ownership. They certainly could not have a right to exact what they might please. But the question is not here, because these corporations accepted a franchise to take toll, and must be held to take it under the franchise." Cf. Streeter v. Chicago, Milwaukee & St. Paul Railway Co., 44 Wis. 383, 385 (1878). In arguments made against railroad rate regulatory legislation, on the ground that this would discourage railroad development in the north, there were mingled implicit concessions that the legislature should be expected to decide what was wise in the community interest on this matter of fixing tolls, and that the possession of discretion over rates was a franchise of peculiar value to the roads. See, e.g., Majority Report, Committee on Railroads, on bill 29A, SJ, 643 (1867). See, generally, Hunt, *Law and Locomotives,* chap. iv; Robert S. Maxwell, *LaFollette and the Rise of the Progressives in Wisconsin* (1956), chap. iv.

557. See "The Pattern of Disposition," and "Prevailing Conceptions of the Uses of the Public Lands," in Chapter I, concerning the emphasis on settlement and complaints to interruption of the progress of settlement. The franchise aspect of the aid grants may also be implicit in the argument of Report, Committee on Internal Improvements, AJ, 699–703 (1852), that the legislature must maintain central control of the disposition of such grants, against the pressures of special, local and monopoly interest.

558. Arguments pro and con over tax exemption of railroad-aid lands, bringing out recognition of the franchise character of these exemptions, may be found in Memorial 203A, 1872, of Dunn county board, and Report, Judiciary Committee, on bill 212A (repeal of West Wisconsin Railway Co. tax exemption), AJ, 214, 622 (1872); Report, Select Committee on Bill 301A for disposal of the St. Croix grant, AJ, 726 (1873) (all other lands have borne more taxes because the aid lands have stood tax-free); Reports, Judiciary Committee, and Committee on State Affairs, on bill 201A to repeal the West Wisconsin Railway tax exemption, AJ, 601, 670 (1873); Report, Committee on Railroads, on bill 35A (similar), AJ, 124–125 (1874); Memorial 11S, 1874, of the North Wisconsin Railway Co., in justification of tax exempt status for aid lands, SJ, 75–82; Minority Report, Committee on Railroads, opposing bill 26A to repeal the West Wisconsin Railway tax exemption, AJ, 146–148 (1875); Report, Committee on Railroads, opposing bill 64A (similar), AJ, 370–371 (1876); Report, Committee on Railroads, opposing bill 25A (similar), AJ, 83–85 (1877); Report, Committee on Assessment and Collection of Taxes, favoring bill 4S (aid lands tax exemption for Wisconsin Central Railroad Co.), AJ, 181–182 (1877), with which compare the strong protest in Memorial 73A, 1877, AJ, 127–129; Report, Committee on Railroads, favoring bill 73S (tax exemption for North Wisconsin Railway Co. aid lands), AJ, 247–249 (1879). In fashion characteristic of the legislature's tendency to bring policy to analysis only under sustained pressure, the early response to complaints about the special franchise aspect of railroad-aid-lands tax exemptions was in much simpler terms than in the seventies. See Report, Committee on Railroads, on bill 41A, SJ, 431 (1867). There were protests against the creation of a special tax system for railroads — under which they paid a gross receipts tax to the state in lieu of conventional real property taxes assessed and levied by local authorities — on the ground that this constituted the grant of a special franchise, to the prejudice of general taxpayers. Compare Guy Edward Snider, *The Taxation of the Gross Receipts of Railways in Wisconsin* (1906), 2–10, and Raymond Vincent Phelan, *The Financial History of Wisconsin* (1908), 219. On local government aid controversies,

see Gates, *Wisconsin Pine Lands,* chap. ix, especially pp. 178, 179–180, 189–193. Compare, generally, Mills, "Government Fiscal Aid," especially tables on pp. 125, 127.

559. See Report, Judiciary Committee, on Joint Resolution 16S, SJ, 600–603 (1867); Report of the same committee on Joint Resolution 2S, SJ, 276–285 (1869); Merk, 287.

560. See, e.g., the lengthy Memorial No. 73A, 1877, filed by the Marathon county board opposing tax exemption for Wisconsin Central Railroad Co. aid lands, AJ, 127–129, and Memorial 75A, 1877, AJ, 151. Cf. Gates, *Wisconsin Pine Lands,* 173, 194.

561. See Gates, *Wisconsin Pine Lands,* 181, 182, 184, 187, 192, 194, 203–206.

562. Laws, 1877, c. 21, and Report, Committee on Assessment and Collection of Taxes, on bill 4S, AJ, 181–182 (1877). Compare, as symbolic of the close of a policy, the contemporaneous rejection of bill 386A, to exempt aid lands of the North Wisconsin Railway Co., despite the legislative generosity manifest in Laws, 1877, c. 218, extending by one year the time allowed for this road to earn its aid lands. See AJ, 491, 510 (1877); cf. Laws, 1874, c. 126, sect. 11, 1876, c. 158, 1883, c. 262. Mills, "Government Fiscal Aid," in Table III, p. 125, offers some measure of the time curve of interest in local government fiscal aids to railroad construction.

563. Laws of Wisconsin Territory, 1836, No. 25; cf., *id.,* No. 40 (mining company forbidden to speculate in lands). The distinct focus upon this anti-speculative policy was shown in the fact that this stipulation was added to the railroad charter by specific amendment in the course of its passage; see HRJ, 118 (1836).

564. General Acts, 1856, c. 122, sect. 7, with which compare Governor Bashford's Veto Message of Oct. 7, 1855, SJ, 1144, disapproving a prior grant for want of stipulations limiting the time of holding and creating settler preferences; P&LL, 1866, c. 314, sect. 11; General Laws, 1869, c. 161; Laws, 1874, c. 126, sect. 14, as amended, Laws, 1876, c. 392; Laws, 1879, c. 22, and Report, Committee on Railroads, on bill 73S, AJ, 249 (1879). Compare criticisms of the withholding of land from settlement, and defensive reactions against such criticisms reflected in General Laws, 1868, c. 95, and Memorial No. 16; General Laws, 1869, Joint Resolution No. 5; P&LL, 1870, c. 225; letter from H. H. Weakley, Land Commissioner of the West Wisconsin Railway Co., to the Wisconsin Railroad Commissioners, SJ, 129–130 (1876); Memorial 73A, 1877, of the Marathon county board, AJ, 127–129. The general trend in formally declared policy, to avoid conceding that pinelands might be held for long-range appreciation in the face of the general official favor for quick marketing and settlement is underlined by the elimination from the bill which became Laws, 1874, c. 126, of a provision which excluded pinelands from a stipulation for sale of railroad-aid lands within a limited term of years; see SJ, 246–250 (1874).

565. This practical direction of policy was brought to most explicit acknowledgment in debates over aid lands tax exemption, as reflected in the materials cited in note 558 above. That prevailing policy contemplated that the railroads would to some extent hold their aid lands for speculative increase in value to help meet their need of investment capital was witnessed also in authorizations to the grantees to mortgage the lands as security for bond issues. See, e.g., P&LL, 1866, c. 366, sect. 9, 1869, c. 51, sect. 20. That a significant policy choice is involved in these provisions is emphasized by P&LL, 1866, c. 323, striking a limitation on mortgaging authority contained in P&LL, 1863, c. 243. Laws, 1874, c. 303, brought into a general statement the policies of assisting the railroad grantee to muster capital through the aid lands and yet to limit the time within which the lands might be withheld from market and to set some preference for sale to settlers.

PART II: CONTRACT

Chapter IV. Contributions of Contract to the Organization of the Market

1. This chapter discusses the general body of lumber-contract cases, governed on the whole by common law doctrine; Chapter V deals with specialized aspects of the law's relation to market processes, shaped in substantial measure by legislation.

The tabulations and the discussion in this chapter rest on examination of all the lumber-contract cases that could be located in the reported decisions of the Wisconsin Supreme Court, supplemented occasionally by insights drawn from a study of the lumber industry's appearances in the work of a trial court in an important lumbering region — the circuit court of Chippewa County. Lumbermen also litigated in the federal courts of the region. It has not been possible to follow the industry into the manuscript records of these courts, however, and since by no means all lower federal court dispositions were reflected in published opinions, nothing would be gained by tabulating those opinions which exist. However, under my direction, a student team paged the appropriate federal court reports for the period preceding the Federal Reporter system, and then paged volumes 1 to 300 of the Federal Reporter, noting all cases that seemed to grow out of timber-industry affairs in Wisconsin. This search yielded eighteen lumber-contract cases falling under various of the headings of the first table in the present chapter, plus four patent cases dealing with sawmill devices. The eighteen lumber-contract cases include three dealing with timberland sales, one joint venture dispute, two cases of logging contracts, eight involving contract phases of sawmill operation, one labor lien matter, one dispute over a sale of logs, and two over sales of lumber. Disposition of all these cases followed familiar doctrine of the common law or applicable Wisconsin statutes, and in no aspect did any conflict of policy show itself between the federal courts and the law of the state.

2. In none of the Wisconsin Supreme Court cases involving resort to industry custom did the court find any issue of inconsistency between the alleged custom and public policy. Leading cases using evidence of lumber industry custom in interpreting contracts or determining whether certain conduct constituted performance according to the stipulated terms include Lamb v. Klaus, 30 Wis. 94 (1872) (interest on vendee's advances to shingle maker); Chapman v. Ingram, 30 Wis. 290 (1872) (driving contract); Smith v. Scott, 31 Wis. 420 (1872) (manner of enforcing sum due for logging); Scott v. Whitney, 41 Wis. 504 (1877) (sale of lumber); Hewitt v. John Week Lumber Co., 77 Wis. 548, 46 N.W. 822 (1890) (sawing contract); Best v. Pike, 93 Wis. 408, 67 N.W. 697 (1896) (sawing contract); John O'Brien Lumber Co. v. Wilkinson, 117 Wis. 468, 94 N.W. 337 (1903), s.c., 123 Wis. 272, 101 N.W. 1050 (1904) (logging contract); Dieck v. Oconto Co., 173 Wis. 156, 180 N.W. 932 (1921) (logging contract). Of course, there was nothing peculiar to the lumber industry in the basic rules of evidence involved in such resort to trade custom; some of the lumber cases make plain that general doctrine on custom evidence is being applied in the context of the timber business. See Lamb v. Klaus, and John O'Brien Lumber Co. v. Wilkinson, cited above, and Meating v. Tigerton Lumber Co., 113 Wis. 379, 89 N.W. 152 (1902) (general hiring authority of lumber company president includes employment of camp cook); Vogt v. Schienebeck, 122 Wis. 491, 100 N.W. 820 (1904) (sale of lumber). Interpretation of timber-industry contracts, or determination of adequacy of performance, in the light of evidence showing the functional problems and phases gone through in the course of producing and marketing lumber, presents cases substantially analogous to those invoking trade custom. For example: Morrow v. Reed, 30 Wis. 81 (1872) (problems of delivery of bulky articles affect definition of what constitutes delivery); Cohn v. Stewart, 41 Wis. 527 (1877) (driving contract construed in light of parties' knowledge of Wisconsin River conditions); McDonald v. Gardner, 56 Wis. 35, 13 N.W. 689 (1882) (lumber sale); McDonald v. Bryant, 73 Wis. 20, 40 N.W. 632 (1888) (logging-sawing contract ruled non-apportionable, in light of facts of difficulty of realizing on the logs unless operation carried through to mill stage); Garvin v. Gates, 73 Wis. 513, 41 N.W. 621 (1889) (what is good driving water); Shores Lumber Co. v. Starke, 100 Wis. 498, 76 N.W. 36 (1898) (nature of available lumber markets affects measure of damages for failure to carry); Kerslake v. McInnis, 113 Wis. 659, 89 N.W. 895 (1902) (had lumbering season ended, when logger stopped woods operations?); Mueller v. Cook, 126 Wis. 504, 105 N.W. 1054 (1906) (how much advance notice logger fairly needs, to prepare operations). Likewise analogous to use of custom in handling lumber industry contracts was the recognition that custom might establish and define revocable licenses for loggers' use of land. See Marshfield Land & Lumber Co. v. John Week Lumber Co. (Ltd.), 108 Wis. 268, 84 N.W. 434 (1900); Bassett v. Soelle, 186 Wis. 53, 202 N.W. 164 (1925).

Sufficiently specific terms in the contract would override even clear evidence of trade custom to the contrary: Williams v. Stevens Point Lumber Co., 72 Wis. 487, 40 N.W. 154 (1888); Mowatt v. Wilkinson, 110 Wis.

176, 85 N.W. 661 (1901); John O'Brien Lumber Co. v. Wilkinson, cited above; Kosloski v. Kelly, 122 Wis. 665, 100 N.W. 1037 (1904); Krueger v. Lake Trading Co., 150 Wis. 569, 137 N.W. 776 (1912). Only the custom of the area of the making and performance of the contract might be invoked: Shores Lumber Co. v. Stitt, 102 Wis. 450, 78 N.W. 562 (1899).

The absence of invocation of trade custom in defining title interests in timberland or standing timber tallies with the fact that only few, and tangential, instances of resort to evidence of custom appear in Wisconsin cases generally, regarding definition of the content of interests in land. See, e.g., Keogh v. Daniel, 12 Wis. 163 (1860) (leasehold, fixtures); Jenkins v. Sharpf, 27 Wis. 472 (1871) (custom regarding definition of metes and bounds, among dealers in particular plat); Douglass v. Ransom, 205 Wis. 439, 237 N.W. 260 (1931) (lawyers' custom, affecting definition of marketable title). On exceptions and reservations, equitable mortgages, and the presumptive inclusion of standing timber in deeds of land, see notes 10, 11, 36, and 39 below. The impact of the law's own policy and pattern upon the shape of the parties' dealing was stated most bluntly in Martin v. Gilson, 37 Wis. 360 (1875), where with some apparent reluctance the court followed Rich v. Zeilsdorff, 22 Wis. 544 (1868), simply as a matter of *stare decisis,* distinguishing a reservation from an exception of standing timber. A qualification upon this generalization was the readiness with which statute law and judge-made law recognized the special claims of squatters who improved their illegally occupied tracts on public lands. However, legislation also showed concern to regularize these "titles" by grant of pre-emptive rights. And the Wisconsin court showed its concern, in this type of matter, not so much in the title as in the contract aspects of the squatter's dealings. See Pratt v. Ayer, 3 Pinn. 236, 256 (Wis. 1851) (federal pre-emption laws recognize improver-occupier's equity; squatter's relinquishment of his "claim" is sufficient consideration to support contract of sale and of security in the land, made as incident of regularizing the title).

3. Of the nineteen railroad cases, two involved federal and nine involved state regulation of carriers; the other eight suits turned on the common law of contract, bailment, or tort.

4. In assessing cases to determine whether the court was imposing terms, a mean was sought, and both fact statements and analysis in the opinions were weighed. The typical cases of imposed terms were of implied warranties and other "implied" terms, the application of doctrines of materiality and separability, questions of allocation of risk of loss, and of the measure of damages. Especially arguable exclusions were the omission of all cases turning on failure of title, and of cases of enforcement of standard types of security, such as mortgages; in the latter instance, it was felt that the legal devices made available to the parties in the way of title and security were so well formed and commonly known that the parties might be deemed freely to be electing them. Cases applying trade custom were not included, since they did not seem to be instances in which norms were imposed by law, save in the very general sense that recognition of custom might be said to be a facet of the law's general policy to promote reasonable expectations. Cases involving statutory liens for labor and supplies were excluded, for reasons noted in the text.

5. See, e.g., Fries, 35, 75.

6. The quoted limitation on reception of the common law is stated in Coburn v. Harvey, 18 Wis. 147, 148 (1864); cf. Page, 79–89. See Marshall, *Autobiography,* I, 348. For more extended discussion of the significance of the acceptance of standing timber values as a separate item of trade, see the discussion of conservation policies in Chapter VI of this book. The validity of transactions in standing timber is tacitly recognized in such typical cases as Lacy v. Johnson, 58 Wis. 414, 17 N.W. 246 (1883); Peshtigo Lumber Co. v. Ellis, 122 Wis. 433, 100 N.W. 834 (1904); Fogo v. Boyle, 130 Wis. 154, 109 N.W. 977 (1906); Western Lime & Cement Co. v. Copper River Land Co., 138 Wis. 404, 120 N.W. 277 (1909). For Statute of Frauds decisions regarding standing timber, see Lillie v. Dunbar, 62 Wis. 198, 22 N.W. 467 (1885); Bruley v. Garvin, 105 Wis. 625, 81 N.W. 1038 (1900). Examples of recognition of the validity of a reserved security title in standing timber are Cadle v. McLean, 48 Wis. 630, 4 N.W. 755 (1880), and Lillie v. Dunbar. Typical cases in which standing timber is recognized as the main value of particular land are Haight v. Lucia, 36 Wis. 355 (1874); Hyslip v. French, 52 Wis. 513, 9 N.W. 605 (1881); McInnis v. Lyman, 62 Wis. 191, 22 N.W. 405 (1885). Beckwith v. Philleo, 15 Wis. 223 (1862) early recog-

nized the assignability of a contract for the sale of standing timber; cf. Keystone Lumber Co. v. Kolman, 94 Wis. 465, 69 N.W. 165 (1896). There was some qualification on the basic recognition of standing timber as a marketable commodity, in decisions which refused to imply a right to cut in the vendee of timberland, or a reservation of standing timber in the vendor who had not stipulated for such an interest. See Schweitzer v. Connor, 57 Wis. 177, 14 N.W. 922 (1883); Huddleston v. Johnson, 71 Wis. 336, 37 N.W. 407 (1888). Such decisions tend to maintain the market value of timberland as such; on the other hand, they set no limitation on the parties' explicit dealings with standing timber as a separate item.

7. 67 Wis. 438, 30 N.W. 621 (1886).

8. 61 Wis. 267, 269, 20 N.W. 926, 927 (1884).

9. 66 Wis. 579, 585, 29 N.W. 570, 572 (1886). The marketable character of land locators' "minutes" and skill was further recognized in Tucker v. Grover (Second Case), 60 Wis. 240, 19 N.W. 62 (1884); Walters v. McGuigan, 72 Wis. 155, 39 N.W. 382 (1888); Gates v. Paul, 117 Wis. 170, 94 N.W. 55 (1903), s.c., 127 Wis. 628, 107 N.W. 492 (1906); Ripley v. Sage Land & Improvement Co., 138 Wis. 304, 119 N.W. 108 (1909); Keith v. Rust Land & Lumber Co., 167 Wis. 528, 167 N.W. 432 (1918); Maxcy v. Peavey, 177 Wis. 140, 187 N.W. 1020 (1922). Of analogous policy import are rulings that one holding himself out to possess land locator's special knowledge or skill may be liable to his co-venturer or client for negligent performance of his role: Noble v. Libby, 144 Wis. 632, 129 N.W. 791 (1911); Knudson v. George, 157 Wis. 520, 147 N.W. 1003 (1914). See, also, the decision in Houlton v. Nichol, 93 Wis. 393, 67 N.W. 715 (1896), finding no public policy opposed to a contract to pay a land locator's fee for advice in establishing a homestead or pre-emption claim on United States land.

10. Proof of an equitable mortgage: Flint v. Jones, 5 Wis. 424 (1856); Russell v. Fish, 149 Wis. 122, 135 N.W. 531 (1912); cf. Upton v. Johnston, 84 Wis. 8, 54 N.W. 266 (1893).

No implied reservation of standing timber: Schweitzer v. Connor, 57 Wis. 177, 14 N.W. 922 (1883); Arentsen v. Moreland, 122 Wis. 167, 99 N.W. 790 (1904); cf. Westcott v. Delano, 20 Wis. 514 (1866); Western Lime & Cement Co. v. Copper River Lumber Co., 138 Wis. 404, 120 N.W. 277 (1919); Opinion Letter to Secretary of the Board of Regents of the University of Wisconsin, Jan. 4, 1910, *5th Biennial Report of the Attorney General of Wisconsin* (Madison, 1910), 723. No implied authorization to cut under unpaid land sale contract: Huddleston v. Johnson, 71 Wis. 336, 37 N.W. 407 (1888); Duluth Log Co. v. St. Croix Land Co., 145 Wis. 286, 129 N.W. 1100 (1911); cf. Beckwith v. Philleo, 15 Wis. 223 (1862); Schufletowski v. Allen, 253 Wis. 507, 34 N.W. (2d) 819 (1948).

Strict enforcement of cutting-time limitations in standing-timber sales: Sawyer v. Hanson, 48 Wis. 611, 4 N.W. 765 (1879); Golden v. Glock, 57 Wis. 118, 15 N.W. 12 (1883); Hicks v. Smith, 77 Wis. 146, 46 N.W. 133 (1890); cf. Larson v. Cook, 85 Wis. 564, 55 N.W. 703 (1893); Everett v. Gores, 89 Wis. 421, 62 N.W. 82 (1895); Western Lime & Cement Co. v. Copper River Land Co. (cited above); F. G. Hood & Co. v. Girard Lumber Co., 137 Wis. 152, 118 N.W. 552 (1908); Jens v. Habeck, 259 Wis. 338, 48 N.W. (2d) 473 (1951) (prior doctrine highlighted by change effected when Uniform Sales Act declared standing timber should be treated as personalty). The opinion in Western Lime & Cement Co. comes close to articulating a presumption in favor of keeping title components together: 138 Wis. at p. 412, 120 N.W. at p. 280. Conflict of authority among the states, on the strict effect of a cutting-time limitation, is noted in Bretz v. R. Connor Co., 140 Wis. 269, 122 N.W. 717 (1909); see Kneeland-McLurg Lumber Co. v. Lillie, 156 Wis. 428, 145 N.W. 1093 (1914).

Limited effect of a license to cut: Tyson v. McGuineas, 25 Wis. 656 (1870); Keystone Lumber Co. v. Kolman, 94 Wis. 465, 69 N.W. 165 (1896); cf. Duinneen v. Rich, 22 Wis. 550 (1868); Lillie v. Dunbar, 62 Wis. 198, 22 N.W. 467 (1885); Bruley v. Garvin, 105 Wis. 625, 81 N.W. 1038 (1900).

Proof of adverse possession: Coleman v. Eldred, 44 Wis. 210 (1878); Haseltine v. Mosher, 51 Wis. 443, 8 N.W. 273 (1881); St. Croix Land & Lumber Co. v. Ritchie, 78 Wis. 492, 47 N.W. 657 (1891); cf. Wickes v. Lake, 25 Wis. 71 (1896); Stephenson v. Wilson, 37 Wis. 482 (1875); Pepper v. Dowd, 39 Wis. 538 (1875); Wilson v. Henry, 40 Wis. 594 (1876); Ayers v. Reidel, 84 Wis. 276, 54 N.W. 588 (1893).

Possibly relevant to the policy of holding

title components together was the decision which applied to a chattel mortgage of logs the Wisconsin court's general refusal to recognize the validity of an after-acquired property clause: Mowry v. White, 21 Wis. 417 (1867).

11. See Marshall, *Autobiography,* I, 348–349. The only exception to this implicit presumption in favor of keeping together title to the value components of timberland tracts were the cases which consistently gave broad effect to exceptions of standing timber in deeds. In Chapter V we shall note these cases as an outstanding example of the impact of the law's own tradition upon the conduct of business.

12. Peshtigo Lumber Co. v. Ellis, 122 Wis. 433, 438, 100 N.W. 834, 835 (1904). Examples of waiver, estoppel, and laches doctrine are Selleck v. Griswold, 49 Wis. 39, 5 N.W. 213 (1880), s.c., 57 Wis. 291, 15 N.W. 151 (1883); Woodford v. Marshall, 72 Wis. 129, 39 N.W. 376 (1888); Combs v. Scott, 76 Wis. 662, 45 N.W. 532 (1890); Russell v. Fish, 149 Wis. 122, 135 N.W. 531 (1912); cf. Pitman v. Hill, 117 Wis. 318, 94 N.W. 40 (1903). See the vivid reflection of an atmosphere of hard bargaining in Meldrum v. Southwick-Sellers Land Co., 157 Wis. 367, 147 N.W. 1068 (1914).

13. On grounds of rescission, see Defrance v. Hazen, 2 Pinn. 228 (Wis. 1849) (standing timber, failure of consideration); Miner v. Medbury, 6 Wis. 295 (1858), s.c., 7 Wis. 100 (1858) (timberland, fraud); Danforth v. Wharton, 41 Wis. 191 (1876) (standing timber, fraud); McKinnon v. Vollmar, 75 Wis. 82, 43 N.W. 800 (1889) (timberland, mistake); Beetle v. Anderson, 98 Wis. 5, 73 N.W. 560 (1897) (sawmill, fraud); Parish v. McPhee, 102 Wis. 241, 78 N.W. 421 (1899) (standing timber, partial failure of consideration); Hansen v. Allen, 117 Wis. 61, 93 N.W. 805 (1903) (timberland, fraud); Mannel v. Schafer, 135 Wis. 241, 115 N.W. 801 (1908) (timberland, fraud); F. G. Hood & Co. v. Girard Lumber Co., 137 Wis. 152, 118 N.W. 552 (1908) (standing timber, mistake). The requirement that a would-be rescinder offer to return what he can of the consideration received by him is declared in McDonald v. Hyde, 23 Wis. 487 (1868); Hyslip v. French, 52 Wis. 513, 9 N.W. 605 (1881). Warren v. Landry, 74 Wis. 144, 42 N.W. 247 (1889) found a binding election of remedies. There are many equitable mortgage cases. Decisions in which the court seems to impose a mortgage relation as a matter of law, given certain facts of relation between the parties, include Starks v. Redfield, 52 Wis. 349, 9 N.W. 168 (1881); Hoile v. Bailey, 58 Wis. 434, 17 N.W. 322 (1883); Schriber v. LeClair, 66 Wis. 579, 29 N.W. 570 (1886); Swift v. State Lumber Co., 71 Wis. 476, 37 N.W. 441 (1888), s.c., *sub nom.* Jenkins v. Bradley, 104 Wis. 540, 80 N.W. 1025 (1899); Jordan v. Estate of Warner, 107 Wis. 539, 83 N.W. 946 (1900); cf. Pitman v. Hill, 117 Wis. 318, 94 N.W. 40 (1903) (equitable mortgagee may acquire full title by adverse possession after payment).

14. 57 Wis. 118, 123, 115 N.W. 12, 14 (1883).

15. 77 Wis. 145, 148, 46 N.W. 133, 133–134 (1890). Compare the emphasis upon the fact that the vendee of standing timber had paid the consideration, in the court's construction of a sale of standing timber as immediately passing title to the timber so as to ground an action under the special timber-trespass statute, in Kneeland-McLurg Lumber Co. v Lillie, 156 Wis. 428, 145 N.W. 1093 (1914). In the absence of any terms to the contrary, the vendee under a contract of sale of standing timber was protected by construction of the contract as giving him a reasonable time in which to cut: Donaldson v. Buhlman, 134 Wis. 117, 113 N.W. 638 (1908). An oral contract of sale of standing timber, void under the Statute of Frauds for want of a writing, was held nevertheless to confer a revocable license to cut, which was effective so far as the "licensee" did in fact cut timber before the license was revoked: Lillie v. Dunbar, 62 Wis. 198, 22 N.W. 467 (1885); Bruley v. Garvin, 105 Wis. 625, 81 N.W. 1038 (1900).

16. 58 Wis. 414, 425, 17 N.W. 246, 251 (1883).

17. Mississippi River Logging Co. v. Blue Grass Land Co., 131 Wis. 10, 110 N.W. 796 (1907). The quoted form of the recording act is that found in Wisconsin Statutes, 1898, sect. 2241, the provision governing in the cases cited in this paragraph of the text.

18. Wisconsin River Land Co. v. Selover, 135 Wis. 594, 116 N.W. 265 (1908).

19. On public land policy, see Part One. Typical of the court's unquestioning acceptance of speculative land dealing, as having proper claim on the law's protection, are Durkee v. Stringham, 8 Wis. 1 (1859) (interest of shareholders in unincorporated joint-stock company for timberland dealing is mort-

gageable); Winslow v. Crowell, 32 Wis. 639 (1873) (Fort Howard military road lands involved in private deals); Smith v. Bouck, 33 Wis. 19 (1873) (speculation in railroad-aid lands); Sawyer v. Hanson, 48 Wis. 611, 4 N.W. 765 (1879) (long-term cutting right); Wisconsin Central R.R. Co. v. Wisconsin River Land Co., 71 Wis. 94, 36 N.W. 837 (1888) (land company operation).

20. 76 Wis. 662, 666, 668, 45 N.W. 532, 533 (1890). The opinion further demonstrates the court's concern, as we have seen it in the cutting-time-limit decisions, that one who has made outlays for the development of property shall not lose, against the claim of another who has not so contributed; stress is put on the fact that plaintiff took no care of the land, and neglected to pay taxes on the various tracts, and allowed many of them to be sold for taxes, from which they were redeemed by defendant or defendant's decedent.

21. 149 Wis. 122, 127–128, 135 N.W. 531, 533 (1912).

22. See, generally, Henry Campbell Black, *A Treatise on the Law of Tax Titles* (1893). Large-scale tax-title speculation in timberland or cut-over areas is reflected in Jourdain v. Fox, 90 Wis. 99, 62 N.W. 936 (1895); Two Rivers Manufacturing Co. v. Day, 102 Wis. 328, 78 N.W. 440 (1899); Arpin Hardwood Lumber Co. v. Carmichael, 115 Wis. 441, 91 N.W. 965 (1902); Mississippi River Logging Co. v. Blue Grass Land Co., 131 Wis. 10, 110 N.W. 796 (1907).

On a tax title as giving color of title, see Starks v. Redfield, 52 Wis. 349, 9 N.W. 168 (1881); a tax title is ruled prima facie a marketable title, in Gates v. Parmly, 93 Wis. 294, 315, 66 N.W. 253, 260 (1896). If tax deeds were on their face regular under the statute, of course the least the court could do was to acknowledge their force. See note 24 below. Generally, see Marshall, Autobiography, I, 353.

23. Gates v. Parmly, 93 Wis. 294, 303, 315, 66 N.W. 253, 258, 260 (1896). See connected cases: Gates v. Avery, 112 Wis. 271, 87 N.W. 1091 (1901); Gates v. Parmly, 113 Wis. 147, 87 N.W. 1096 (1902).

24. Chippewa River Land Co. v. J. L. Gates Land Co., 118 Wis. 345, 94 N.W. 37 (1903); Pinkerton v. J. L. Gates Land Co., 118 Wis. 514, 95 N.W. 1089 (1903); Cezikolski v. Frydrychowicz, 120 Wis. 369, 98 N.W. 211 (1904); Mitchell Iron & Land Co. v. Flambeau Land Co., 120 Wis. 545, 98 N.W. 530 (1904); cf. Menasha Wooden Ware Co. v. Harmon, 128 Wis. 177, 107 N.W. 299 (1906). Where an apparently valid tax title had stood for some time, and further transactions derived from it had been for substantial consideration, the tax title was treated with respect. See e.g., Arpin Hardwood Lumber Co. v. Carmichael, 115 Wis. 441, 91 N.W. 965 (1902); Mississippi River Logging Co. v. Blue Grass Land Co., 131 Wis. 10, 110 N.W. 796 (1907). On the status of a tax-title plaintiff under timber trespass statutes, see Fleming v. Sherry, 72 Wis. 503, 40 N.W. 375 (1888).

25. Gould v. Sullivan, 84 Wis. 659, 668, 54 N.W. 1013, 1016, (1893). Accord: Bray & Choate Land Co. v. Newman, 92 Wis. 271, 65 N.W. 494 (1896); Nelson v. Churchill, 117 Wis. 10, 93 N.W. 799 (1903); cf. Williams v. J. L. Gates Land Co., 146 Wis. 55, 130 N.W. 880 (1911); Gould v. Killen, 152 Wis. 197, 139 N.W. 758 (1913). See Donnelly v. Sampson, 135 Wis. 368, 115 N.W. 1089 (1908). Compare the analogous handling of a tax-title issue in a trespass case, in Washburn Land Co. v. White River Lumber Co., 165 Wis. 112, 161 N.W. 547 (1917).

26. Edwards v. Upham, 93 Wis. 455, 67 N.W. 728 (1896); Menasha Wooden Ware Co. v. Harmon, 128 Wis. 177, 107 N.W. 299 (1906).

27. See Carter v. Smith, 23 Wis. 497 (1868) (logs); Hopkins v. Langton, 30 Wis. 379 (1872) (lumber); Blakeslee v. Rossman, 44 Wis. 553 (1878) (lumber); Norris v. Persons, 49 Wis. 101, 5 N.W. 224 (1880) (millsite); Mihills Manufacturing Co. v. Camp, 49 Wis. 130, 5 N.W. 1 (1880) (sawmill mortgage); Haywood v. Lincoln Lumber Co., 64 Wis. 639, 26 N.W. 184 (1885) (sawmill mortgage); Conlee Lumber Co. v. Ripon Lumber & Manufacturing Co., 66 Wis. 481, 29 N.W. 285 (1886) (mill labor claims); Brickley v. Walker, 68 Wis. 563, 32 N.W. 773 (1887) (logs and mill); Shores v. Doherty, 75 Wis. 616, 44 N.W. 747 (1890) (logs); Barker v. Lynch, 75 Wis. 624, 44 N.W. 826 (1890) (logging contract); Valley Lumber Co. v. Hogan, 85 Wis. 366, 55 N.W. 415 (1893) (logs); Curtis Brothers & Co. v. Hoxie, 88 Wis. 41, 59 N.W. 581 (1894) (lumber); Bradley v. Paul, 94 Wis. 488, 69 N.W. 168 (mill); Collins v. Corwith, 94 Wis. 514, 69 N.W. 349 (1896) (mill); Franzke v. Hitchon, 105 Wis. 11, 80 N.W. 931 (1899) (logging operations); Kendall v. Beaudry, 107 Wis. 180, 83 N.W. 314 (1900) (logging contract); Scott

v. Holman, 117 Wis. 206, 94 N.W. 30 (1903) (logging contract); Dockry v. Isaacson, 187 Wis. 649, 205 N.W. 391 (1925) (mortgage of standing timber). Cf. Garlick v. McArthur, 6 Wis. 450 (1857) (fraudulent effort of partner and creditor to put whole liability for debt on other partner); Allen v. Clark, 36 Wis. 101 (1874) (sale of sawmill on execution set aside for gross inadequacy of sum realized); State v. Gross, 62 Wis. 41, 21 N.W. 802 (1884) (obtaining money under false pretenses, in logging labor contract); Wunderlich v. Palatine Fire Insurance Co., 104 Wis. 382, 80 N.W. 467 (1899) (fraudulent claim for value of sawmill, under fire insurance contract); Olson v. Sawyer-Goodman Co., 110 Wis. 149, 85 N.W. 640 (1901) (gambling debt may not be offset against log laborer's wages due); Richardson v. State, 171 Wis. 309, 177 N.W. 10 (1920) (arson, for mill fire insurance).

28. Tocqueville, II, 157.

29. Fries, 17, 103. A vivid reflection of the period of small beginnings and hard scrambling for capital will be found in the story of Moses Strong's efforts to set up mills on the Wisconsin River: see Duckett, 144, 147, 151, 157–162. Cf. Gates, *Wisconsin Pine Lands,* 122–123; Reynolds, 11, 14, 107, 120, 130–139; Stephenson, 64, 144.

30. See, e.g., Hessey v. Gund, 98 Wis. 531, 98 Wis. 531, 74 N.W. 342 (1898) (enforcing vendor's interest in standing timber, as security, on sale of timberland); Spencer v. Holman,, 113 Wis. 340, 89 N.W. 132 (1902) (guaranty of performance of logging contract; loggers' chattel mortgage as prior security for performance); Wilcox v. Scallon, 133 Wis. 521, 113 N.W. 948 (1907) (vendee pays for timberland in part by logging the tract); Hemenway v. Connor Lumber & Land Co., 175 Wis. 443, 185 N.W. 524 (1921) (customer's promise to pay labor claim against his logging contractor); Underwood Veneer Co. v. Lucia, 202 Wis. 507, 232 N.W. 853 (1930) (chattel mortgage on logging equipment, securing advances to logging contractor).

31. 15 Wis. 223 (1862). This decision also ruled that the timberland sale contract was assignable by Palfrey to defendants; in effect the opinion ruled that, since there was nothing on the face of the contract to show that the rights thereunder were a personal trust confided to Palfrey, there was a presumption in favor of treating the contract as a marketable item — a ruling consistent with the opinion's implicit favor towards active development of the land. See note 6 above.

32. 42 Wis. 85, 86–87, 95 (1877). Cf. Whitney v. Ludington, 17 Wis. 141 (1863), discussed in the instant case. Note that a Maine decision is also discussed and distinguished.

33. Haskins v. Kennedy, 79 Wis. 430, 434, 48 N.W. 515, 515–516 (1891).

34. 60 Wis. 76, 83–84, 18 N.W. 725, 728 (1884) (reversed on other grounds than those discussed in the text). Quotations and fact references are from the printed "Case" filed on the appeal, pp. 7–8 (the contract), 14 (Hennessy-Tozer), 15–16 (lien), 37–38 (Tozer on the yield from the operations), 22, 23 (assignment to plaintiff). Cf. Clapp v. Webb, 52 Wis. 638, 9 N.W. 796 (1881).

35. The quotation is from plaintiff's testimony in Tucker v. Grover (First Case), 60 Wis. 233, 235, 19 N.W. 92, 92–93 (1884). Other land–locator contract cases include Flanders v. Train, 13 Wis. 596 (1861); Schriber v. LeClair, 66 Wis. 579, 29 N.W. 570 (1886); Swift v. State Lumber Co., 71 Wis. 476, 37 N.W. 441 (1888), s. c. *sub nom.* Jenkins v. Bradley, 104 Wis. 540, 80 N.W. 1025 (1888); Maxcy v. Peavey, 177 Wis. 140, 187 N.W. 1020 (1922); cf. Smith v. Putnam, 107 Wis. 155, 82 N.W. 1077 (1900). See Gara, 142, 165 (locator's shares), and 159 (resident agent's share); Stephenson, 120, 136.

Cases involving payments in logs or lumber include DuBay v. Uline, 6 Wis. 588 (1858) (millwright's pay); Spaulding v. Wood, 8 Wis. 195 (1859) (sawmill); Flanders v. Train, 13 Wis. 596 (1861) (landowner pays for logging, in kind); Edminston v. Garrison, 18 Wis. 594 (1864) (mill rental); Graves v. Gans, 25 Wis. 41 (1869) (promissory note); Smith v. Scott, 31 Wis. 420, 437 (1872) (standing timber); In re Day, 34 Wis. 638 (1874) (shingle manufacture); Lenz v. Brown, 41 Wis. 172 (1876) (groceries); Richardson v. Single, 42 Wis. 40 (1877) (standing timber); Gaston v. Owen, 43 Wis. 103 (1877) (supplies); Allen v. Seyfried, 43 Wis. 414 (1877) (mill and land); Fletcher v. Ingram, 46 Wis. 191, 50 N.W. 424 (1879) (logs); Jones v. Foster, 67 Wis. 296, 30 N.W. 697 (1886) (mill); Gaveney v. Gates, 68 Wis. 1, 31 N.W. 223 (1887) (flour); Wilcox v. Scallon, 133 Wis. 521, 113 N.W. 948 (1907) (land); Rohrer v. Lochery, 136 Wis. 532, 117 N.W. 1060 (1908) (land). See Fries, 36–37; Gara, 142, 164; Reynolds, 21–22, 25, 138; Stephenson, 44.

On the absence of security put up by the active logger in payment-in-kind situations, see, e.g., Pike v. Vaughn, 39 Wis. 499 (1876); Fletcher v. Ingram, 46 Wis. 191, 50 N.W. 424

(1879); Schierl v. Baumel, 75 Wis. 69, 43 N.W. 724 (1889). For absence of security for money advances, see Lamb v. Klaus, 30 Wis. 94 (1872); Kelly v. Schupp, 60 Wis. 76, 18 N.W. 725 (1884); McCord v. Edward Hines Lumber Co., 124 Wis. 509, 102 N.W. 334 (1905); Vollmar & Below Co. v. Bayfield Mill Co., 146 Wis. 412, 131 N.W. 899 (1911). In Haskins v. Kennedy, 79 Wis. 430, 48 N.W. 515 (1891), though the logger owned the tract logged and hence might have given security in the land for advances which the vendee of the logs contracted to make, the latter was content with a right to deduct for advances from money due for the logging operation; cf. McDonald v. Bryant, 73 Wis. 20, 40 N.W. 665 (1888). Loggers were more apt to have some independent means of giving security, as by a chattel mortgage on equipment; e.g., Shores v. Doherty, 75 Wis. 616, 44 N.W. 747 (1890); Campbell v. Dick, 80 Wis. 42, 49 N.W. 120 (1891). Of course, a logger might finance advances on the endorsement or guaranty of his customer, the buyer of the logs: see Schierl v. Baumel, 75 Wis. 69, 43 N.W. 724 (1889); Andresen v. Upham Manufacturing Co., 120 Wis. 561, 98 N.W. 518 (1904); Hemenway v. Connor Lumber & Land Co., 175 Wis. 443, 185 N.W. 524 (1921).

Standing timber as a source of finance is recognized implicitly in the contract involved in Sniveley v. Keystone Lumber Co., 129 Wis. 54, 108 N.W. 215 (1906), where the seller of land, reserving the standing timber to himself absolutely, agreed to pay taxes on the land as long as and in proportion as he continued to let the timber stand. Cf. McMahon v. McGraw, 26 Wis. 614 (1870) (reliance on timber cutting proceeds to pay taxes); Gotham v. Wachsmuth Lumber Co., 156 Wis. 442, 146 N.W. 505 (1914) (cutter to pay taxes).

36. Hoile v. Bailey, 58 Wis. 434, 17 N.W. 322 (1883); Schierl v. Baumel, 75 Wis. 69, 43 N.W. 724 (1889); Hyde v. German National Bank of Oshkosh, 96 Wis. 406, 71 N.W. 659 (1897), s. c., 115 Wis. 170, 91 N.W. 230 (1902). See also Steele v. Schricker, 55 Wis. 134, 12 N.W. 396 (1882); cf. Smith v. Chicago & North Western Ry. Co., 18 Wis. 1 (1864).

37. Contract liens to logging contractors are illustrated in Smith v. Scott, 31 Wis. 420, 437 (1872). Cf. Morrow v. Reed, 30 Wis. 8 (1872), and Morrow v. Campbell, 30 Wis. 90 (1872) (chattel mortgages on logs, to secure advances to pay labor). Salvo v. Duncan, 49 Wis. 151, 4 N.W. 1074 (1880) shows a situation where the logger took no security. On statutory liens, see Chapter V.

38. Mortgage of timberland: Durkee v. Stringham, 8 Wis. 1 (1859) (mortgage on interest in unincorporated joint-stock land company); Washburn v. Fletcher, 42 Wis. 152 (1877) (mortgage, with option to buy).

Security transfers of standing timber interests: Clapp v. Webb, 52 Wis. 638, 9 N.W. 796 (1881); Hyde v. German National Bank of Oshkosh, 96 Wis. 406, 71 N.W. 659 (1897), s. c., 115 Wis. 170, 91 N.W. 230 (1902).

Mortgage in connection with sawmill: Smith v. Potter, 3 Wis. 432 (1854) (realty mortgage on mill); Moak v. Bourne, 13 Wis. 514 (1861) (chattel mortgage); Weber v. Illing, 66 Wis. 79, 27 N.W. 834 (1886) (chattel mortgage on portable mill).

Security title taken by financier of timberland purchase: Flint v. Jones, 5 Wis. 424 (1856); Burr v. C.C. Thompson & Walkup Co., 78 Wis. 227, 47 N.W. 277 (1890); Jourdain v. Fox, 90 Wis. 99, 62 N.W. 936 (1895). Security title taken by financier of standing-timber purchase: Andrews v. Jenkins, 39 Wis. 476 (1876); Larson v. Cook, 85 Wis. 64, 55 N.W. 703 (1893); Holm v. Colman, 89 Wis. 233, 61 N.W. 767 (1895). Similar financing by mill owner: Starks v. Redfield, 52 Wis. 349, 9 N.W. 168 (1881); Merrill v. Merrill, 53 Wis. 522, 10 N.W. 684 (1881); cf. Brittingham & Hixon Lumber Co. v. Manson, 108 Wis. 221, 84 N.W. 183 (1900); Fries, 36.

39. Purchase-money mortgage: Nicholson v. Coleman, 90 Wis. 639, 64 N.W. 297 (1895).

Reserved title in timberland or standing timber: Beckwith v. Philleo, 15 Wis. 223 (1862); Smith v. Scott, 31 Wis. 420, 437 (1872); Marsh v. Bellew, 45 Wis. 36 (1878); Hoile v. Bailey, 58 Wis. 434, 17 N.W. 322 (1883).

The observation of Lyon, J., is in his dissenting opinion in Bunn v. Valley Lumber Co., 51 Wis. 376, 384, 8 N.W. 232, 235 (1881). Liens are illustrated by that case, as well as by Kelley v. Schupp, 60 Wis. 76, 18 N.W. 725 (1884) (standing-timber vendor's right to control all sales); Warren v. Landry, 74 Wis. 144, 42 N.W. 247 (1889) (vendee might cut, to extent of down payment); Long v. Davidson, 77 Wis. 409, 46 N.W. 805 (1890) (logger may cut, haul, and bank); J. V. LeClair Co. v. Rogen-Ruger Co., 124 Wis. 44, 102 N.W. 346 (1905) (logger may sell). On the importance of keeping exploitative operations in motion, see McCord v. Edward Hines Lumber Co., 124 Wis. 509, 513, 102 N.W. 334, 335 (1905).

40. Some attention was given the validity of contract liens on logs, in Cadle v. McLean, 48 Wis. 630, 637, 4 N.W. 755, 757 (1880), and in Lillie v. Dunbar, 62 Wis. 198, 200, 22 N.W. 467, 468 (1885). Morrow v. Reed, 30 Wis. 81, 84 (1872) spoke concerning an after-acquired property clause. Wing v. Thompson, 78 Wis. 256, 266, 47 N.W. 606, 609 (1890) carried the court's bluntest speaking regarding bona fide purchasers and reserved security titles in logs.

41. Bunker v. Locke, 15 Wis. 635 (1862); Fairbank v. Cudworth, 33 Wis. 358 (1873); cf. Wright v. Roberts, 22 Wis. 161 (1867); Scott v. Webster, 44 Wis. 185 (1878), s.c., 50 Wis. 53, 6 N.W. 363 (1880), 56 Wis. 356, 14 N.W. 280 (1882); Huddleston v. Johnson, 71 Wis. 336, 37 N.W. 407 (1888); Hessey v. Gund, 98 Wis. 531, 74 N.W. 342 (1898).

42. Beckwith v. Philleo, 15 Wis. 223 (1862) (assignability; no interference until foreclosure); Marsh v. Bellew, 45 Wis. 36 (1878) (waiver of security); Hoile v. Bailey, 58 Wis. 434, 17 N.W. 322 (1883) (no waste, though mistake regarding security value); cf. Burr v. C. C. Thompson & Walkup Co., 78 Wis. 227, 47 N.W. 277 (1890); Nicholson v. Coleman, 90 Wis. 639, 64 N.W. 297 (1895).

43. Tables 8–11 do not purport to measure "the amount in controversy." There would be no well-defined concept of what that amount would be in a substantial number of instances, and any effort to fix upon the record evidence to substantiate it would cost more than the useful yield would warrant. Conceivably one might use as a measure of certain aspects of these cases the amount of recovery claimed by parties seeking relief, but claims are so likely to be inflated for tactical purposes as to mean little for any other use; in any event, preliminary data on claimed recoveries was noted in seventy-two cases in which the Reports showed neither an agreed valuation nor a verdict or judgment, but these cases yielded nothing material to add to the tabulations presented in the text.

Cases surveyed for the purpose of the present tabulations included all those for which the requisite data show in the Reports, among the cases in Table 5, except those concerning transport (where the issues were typically too tangential to lumber-industry operations to be relevant to our present purpose), and those involving labor or supply liens (which typically flowed within a particular, specialized range of values, inclusion of which would seem to distort the general picture; the lien data are noted in the separate discussion of that subject).

Since the basic interest in the data presented was in its reflection of the scale of industry operations thus brought to law, where the Report showed an agreed valuation placed by the parties on their transaction this was taken as the basis of tabulation, and verdict or judgment value was used only as the basis of tabulating cases where no agreed value appeared. Agreed valuations include prices set by contract, or the stated amount of debts secured, or stipulated or otherwise-agreed-to valuations of the transaction set by the parties in the course of litigation or found and declared by the court; the great bulk of the agreed valuations used here fall under either of the first two heads. Sums were stated in round figures for these tabulations. Since the purpose was to help measure the scale of transactions involved, rather than the formal outcome of lawsuits, verdicts or judgments given in the trial court were used for the tabulation, though the judgment below was reversed.

44. Stephenson, 261, 262. Compare the shyness of legal involvement expressed in the testimony of John Hennessy, farmer and logging crew member, in the case cited in note 34 above. Stephenson probably exaggerates the extent to which he did business without legal advice. Compare Roujet Marshall's description of his long retainer by the Weyerhaeuser interests in the Chippewa Valley (Marshall, *Autobiography,* I, 271).

45. See e.g., Blewett v. Gaynor, 77 Wis. 378, 46 N.W. 547 (1890) ($35,000 logging contract); Upham Manufacturing Co. v. Sanger, 80 Wis. 34, 49 N.W. 28 (1891) ($150,000 lumber sale); Hosmer v. McDonald, 80 Wis. 54, 49 N.W. 112 (1891) ($24,000 lumber sale); Montreal River Lumber Co. v. Mihills, 80 Wis. 540, 50 N.W. 507 (1891) (disputed balance of $42,000 on lumber sale); Best v. Pike, 93 Wis. 408, 67 N.W. 697 (1896) ($38,000 at stake in logging dispute); Brittingham & Hixon Lumber Co. v. Manson, 108 Wis. 221, 84 N.W. 183 (1900) ($50,000 damages claimed for vendor's breach of output sales contract); Illinois Trust & Savings Bank v. Alexander Stuart Lumber Co., 119 Wis. 54, 94 N.W. 777 (1903) ($49,000 lumber sale, with $10,500 chattel mortgage).

46. See Upham Manufacturing Co. v. San-

ger, 80 Wis. 34, 41–42, 49 N.W. 28, 30 (1891).

47. See Fries, 103; Gates, *Wisconsin Pine Lands,* 127.

48. On the relation of timberland investors and local agents, see Fries, 164–166, 172, 175–177; Gates, *Wisconsin Pine Lands,* chaps. iv, v; Larson, 165–171, 233, 275, 277; Marshall, *Autobiography,* I, 349–357. Cf. O'Connor v. Semple, 57 Wis. 243, 15 N.W. 136 (1883); Hooker v. Hyde, 61 Wis. 204, 21 N.W. 52 (1884); Russell v. Andrae, 79 Wis. 108, 48 N.W. 117 (1891). The relation of security devices to the financing of large-scale timberland transactions is suggested, for example, in Whorton v. Webster, 56 Wis. 356, 14 N.W. 280 (1882), and Burr v. C. C. Thompson & Walkup Co., 78 Wis. 227, 47 N.W. 277 (1890); cf. Uptom v. Johnston, 84 Wis. 8, 54 N.W. 266 (1893); Gates v. Parmly, 93 Wis. 294, 66 N.W. 253 (1896). Mortgage security for long-term timberland investment was more likely to be found as an incident of the operation of integrated concerns than in isolated transactions. See references to the foreclosure of an indenture securing the $650,000 bond issue of the Union Lumber Company, in Marshall, *Autobiography,* I, 323–325. Cf. Durkee v. Stringham, 8 Wis. I (1859) (mortgage of interest in unincorporated joint-stock company for timberland exploitation). Locator-capitalist pooling ventures are reflected in cases cited in notes 8, 9, and 35 above. Cf. Current, 23, and Merrill, 27–28, 250.

49. Estoppel of an agent to assert a tax title against his principal may be seen in McMahon v. McGraw, 26 Wis. 614 (1870); Geisinger v. Beyl, 44 Wis. 258 (1878), s. c., 71 Wis. 358, 37 N.W. 423 (1888), 80 Wis. 443, 50 N.W. 501 (1891). That there should be clear proof of an agent's authority to sell in a large transaction, see Atlee v. Bartholomew, 69 Wis. 43, 33 N.W. 110 (1887). A partition decree was shaped to the requirements of economic development of timber, in Idema v. Comstock, 131 Wis. 16, 110 N.W. 786 (1907).

50. McMillen v. Pratt, 89 Wis. 612, 623, 624, 626–627, 62 N.W. 588, 591, 592 (1895).

51. Brittingham & Hixon Lumber Co. v. Manson, 108 Wis. 221, 226–227, 84 N.W. 183, 185 (1900).

52. *Id.,* 108 Wis. 221, 224, 84 N.W. 183, 184 (1900); Dells Paper & Pulp Co. v. Willow River Lumber Co., 170 Wis. 19, 37, 173 N.W. 317, 323 (1919); cf. Goodman v. Brown Land & Lumber Co., 183 Wis. 574, 197 N.W. 730 (1924). Examples of other long-term supply contracts are Rust v. Fitzhugh, 132 Wis. 549, 112 N.W. 508 (1907) (land management); Dieck v. Oconto Co., 173 Wis. 156, 180 N.W. 932 (1921); Underwood Veneer Co. v. Lucia, 202 Wis. 507, 232 N.W. 853 (1930); cf. Jeske v. Hotz Manufacturing Co., 233 Wis. 500, 290 N.W. 208 (1940), s.c., 238 Wis. 116, 297 N.W. 357 (1941) (raw materials supply).

53. Marshall, *Autobiography,* I, 266–267, 272, 279, 285, 287, 323; cf. Current, 104–105, 113, 119, 129; Larson, 231–236; Merrill, 250.

54. See Pulpwood Co. v. Green Bay Paper & Fibre Co., 157 Wis. 604, 147 N.W. 1058 (1914), s.c., 168 Wis. 400, 170 N.W. 230 (1918), certiorari denied, 249 U.S. 610 (1919); U.S. v. Nelson, 52 Fed. 646 (D.Minn. 1892); cf. Menominee River Boom Co. v. Augustus Spies Lumber & Cedar Co., 147 Wis. 559, 132 N.W. 1118 (1912).

55. Some indication of the flow of contract administration other than in the Supreme Court is given in data cited by Fries, 36, 37; Gates, *Wisconsin Pine Lands,* chaps. v, x; Marshall, *Autobiography,* I, 263, 267, 307, 340, 342, 349.

56. Marshall, *Autobiography,* I, 342–344.

57. On the Statute of Frauds, see Smith v. Bouck, 33 Wis. 19 (1873); Daniels v. Bailey, 43 Wis. 566 (1877); Langley v. Sanborn, 135 Wis. 178, 114 N.W. 787 (1908); cf. Tucker v. Grover (Second Case), 60 Wis. 240, 19 N.W. 62 (1884); Watters v. McGuigan, 72 Wis. 155, 39 N.W. 382 (1888). Disputes over the formation of contracts for speculation in timberland or standing timber, and charges of fraud or mistake, are exemplified in Miner v. Medbury, 6 Wis. 295 (1858), s.c., 7 Wis. 100 (1858); Atlee v. Bartholomew, 69 Wis. 43, 33 N.W. 110 (1887); McKinnon v. Vollmar, 75 Wis. 82, 43 N.W. 800 (1889); Kreutzer v. Lynch, 122 Wis. 474, 100 N.W. 887 (1904). On the catch-as-catch-can atmosphere of timberland speculation and logging, see Current, 22; Fries, 36–37; Gates, *Wisconsin Pine Lands,* 88; Marshall, *Autobiography,* I, 348, 350.

58. For example, Whitney v. Ludington, 17 Wis. 140 (1863); Ford v. Smith, 27 Wis. 261 (1870); Upham v. Hewitt, 42 Wis. 85 (1877); LaFlex v. Burss, 77 Wis. 538, 46 N.W. 801 (1890); Best v. Pike, 93 Wis. 408, 67 N.W. 697 (1896); Zoesch v. Thielman, 105 Wis. 117, 80 N.W. 1107 (1899); Maxcy v. Peavey, 177 Wis. 140, 187 N.W. 1020 (1922).

59. Clinton v. Webster, 66 Wis. 322, 323, 28 N.W. 349, 349–350 (1886); cf. Woodford v. Marshall, 72 Wis. 129, 39 N.W. 376

(1888); McKinnon v. Vollmar, 75 Wis. 82, 43 N.W. 800 (1889); St. Clair v. Rutledge, 115 Wis. 583, 92 N.W. 234 (1902); Mississippi River Logging Co. v. Blue Grass Land Co., 131 Wis. 10, 110 N.W. 796 (1907); Peyton v. Minong Lumber & Lath Co., 149 Wis. 66, 135 N.W. 518 (1912). Compare, generally, Glaspie v. Keator, 56 Fed. 203, 208 (C.C.A. 8th 1893), where the court declared that in view of the rude working conditions under which log scaling must typically be done, and the fact that scaling bills were typically drawn by persons of whom great precision in language should not be expected, the courts should not impose strict rules governing the procedure to validate or interpret such bills.

60. For example, Morrow v. Campbell, 30 Wis. 90 (1872); Hill v. Palmer, 56 Wis. 123, 14 N.W. 20 (1882); O'Connor v. Semple, 57 Wis. 243, 15 N.W. 136 (1883); McInnis v. Lyman, 62 Wis. 191, 22 N.W. 405 (1885); McMillen v. Pratt, 89 Wis. 612, 62 N.W. 588 (1895).

61. Morrow v. Campbell, 30 Wis. 90, 92 (1872).

62. For example, Hinckley v. Beckwith, 13 Wis. 31 (1860), s.c., 17 Wis. 413 (1863), 23 Wis. 328 (1868); Riggs v. Weise, 24 Wis. 545 (1869); Manson v. Robinson, 37 Wis. 339 (1875); Selleck v. Griswold, 49 Wis. 39, 5 N.W. 213 (1880), s.c., 57 Wis. 291, 15 N.W. 151 (1883); Rood v. Prestley, 58 Wis. 255, 16 N.W. 546 (1883); Ruege v. Gates, 71 Wis. 634, 38 N.W. 181 (1888); Russell v. Andrae, 79 Wis. 108, 48 N.W. 117 (1891), s.c., 84 Wis. 374, 54 N.W. 792 (1893); Parcher v. Dunbar, 118 Wis. 401, 95 N.W. 370 (1903).

Chapter V. Contributions of Legal Regulation to the Organization of the Market

1. See Part Two, Introduction. The two common law cases which made the most marked adaptation to the special circumstances of timber trespass did so, significantly, by modifying the common law measure of damages in two suits arising before the effective date of Wisconsin Session Laws, 1873, c. 263 (the statutory trover act), to follow the policy there indicated by the legislature: Webster v. Moe, 35 Wis. 75 (1874); Brewster v. Carmichael, 39 Wis. 456 (1876).

2. Washburn & Woodman to Matthew Newkirk, Philadelphia, Pa., Jan. 3, 1849, in Washburn & Woodman Letter Book, December 1848–April 1849, 131–132, State Historical Society of Wisconsin. See Current, 123–124; Dykstra, "Law and the Lumber Industry," 363, 370; Fries, 201; Gates, *Wisconsin Pine Lands*, 210–212; Marshall, *Autobiography*, I, 348–357. The difficulties of enforcing the law against trespass, in the face of local sentiment tolerant of the cutter and unfriendly to the "speculator," are graphically depicted in Gara, 32, 51, 143. The private employment of timberland guards is reflected, for example, in Chicago & North Western Railway Co. v. James, 22 Wis. 194 (1867), s.c., 24 Wis. 388 (1869). Correspondence quoted in Washburn v. Fletcher, 42 Wis. 152, 160 (1877) offers an example of how far success against trespassers figured in investors' calculations of their gains. The Annual Report of the Commissioners of School and University Lands, Oct. 1, 1864, PD, 196, comments on the practical connection between the preservation of standing timber and the salable value of state timberlands, in terms which might equally apply to the situation of the private investor.

3. Wisconsin Territorial Laws, 1843–1844, p. 29, Act of Jan. 27, 1844, sect. 1, and amendments, Laws, 1848, p. 156 (Act of Aug. 9, 1848), and p. 158 (Act of Aug. 19, 1848); Revised Statutes, 1849, c. 134, sect. 50, and c. 157, sect. 1; General Acts, 1854, c. 55; Revised Statutes, 1858, c. 163, sects. 56–58; General Laws, 1859, c. 180, sect. 3; General Laws, 1866, c. 17; General Laws, 1872, c. 16, sect. 3; Revised Statutes, 1878, sect. 4441; Laws, 1881, c. 170; Laws, 1889, c. 397; Statutes, 1898, sect. 4449; Statutes, 1959, sect. 26.05. Cf. Laws, 1887, c. 99. See Golonbieski v. State, 101 Wis. 333, 77 N.W. 189 (1898); cf. Francis Wharton, *Criminal Law* (1932), 1413–1416. Compare Statutes of the Territory of Wisconsin, 1839, p. 350 (penalties for arson or willful destruction or cutting and carrying away of trees, in context limited to improved agricultural land); similar were Laws, 1849, c. 60, sect. 3; Revised Statutes, 1849, c. 40, sect. 3; Revised Statutes, 1878, sects. 3178, 4441. Statutes penalizing misappropriation of floating or stranded logs or lumber had functional relevance regarding water transport; since common law doctrine would treat theft of logs as larceny, such legislation was declaratory only, save as it singled out the facts of transportation for special attention in defining the offense or fixing the penalty, as in General Acts, 1855, c. 60.

4. See, e.g., Bates v. Campbell, 25 Wis. 613 (1870) (ejectment); Warner v. Trow, 36 Wis.

195 (1874) (trespass q.c.f.); Haight v. Lucia, 36 Wis. 355 (1874) (ejectment, with temporary injunction for each party). The authorization for recovery of damages for waste, incident to a judgment in ejectment, was by Revised Statutes, 1878, sect. 3082. That this increased the previous scope of relief is indicated by comparison of Pacquette v. Pickness, 19 Wis. 219 (1865), and Riemer v. Johnke, 37 Wis. 258 (1875), with Hiles v. Atlee, 80 Wis. 219, 49 N.W. 816 (1891), s.c., 90 Wis. 72, 62 N.W. 940 (1895); Nelson v. Churchill, 117 Wis. 10, 93 N.W. 799 (1903). On provision of double damages for a willful wrong, in an action of trespass *quare clausum fregit,* under Laws, 1905, c. 264, sect. 20, see Boneck v. Herman, 247 Wis. 592, 20 N.W.(2d) 664 (1945). The terms on which an equity court would relieve against waste were defined in Wright v. Wing, 18 Wis. 45 (1864). Wolf River Lumber Co. v. Brown, 88 Wis. 638, 60 N.W. 996 (1894) discusses the conditions on which a decree might be had to quiet title, and is especially useful in pointing up the fact that land title might also be resolved in the disposition of an action at law in replevin.

5. Typical actions by persons claiming title to the land are Paine v. Libby, 21 Wis. 425 (1867), and Single v. Schneider, 24 Wis. 299 (1869), s.c., 30 Wis. 570 (1872) (replevin), as well as Tyson v. McGuineas, 25 Wis. 656 (1870), and Martin v. Scofield, 41 Wis. 167 (1876) (trover). Examples of these actions by plaintiffs claiming simply as owners of logs or lumber as personal property are Root v. Bonnema, 22 Wis. 539 (1868) (trover); Stearns v. Raymond, 26 Wis. 74 (1870) (replevin). On the monetizing of the replevin remedy, see Revised Statutes, 1849, c. 119, and Single v. Schneider, 24 Wis. 294, 302 (1869). The distinction between actions for damage to an interest in the land by wrongful timber cutting, and actions for conversion of logs or lumber derived from plaintiff's land, is pointed up in effect by Dunbar v. Montreal River Lumber Co., 127 Wis. 130, 106 N.W. 389 (1906), and Boneck v. Herman, 247 Wis. 592, 20 N.W.(2d) 664 (1945).

6. See Gates, *Wisconsin Pine Lands,* 86, 200–201; cf. Fries, 201. As one might expect, there was a strong tendency to settle most trespass cases out of court; this was especially true where the trespasser was an established and responsible firm. Settlement might be accomplished under a contract of arbitration: see Reynolds, 31–33.

A check of the reported lower federal court decisions in the Wisconsin area showed only three instances of private actions protective of land or log titles; two of these involved actions for waste by timber trespass, one an action for conversion of logs. Actions by a public authority, or turning principally on an issue of public land title, were more numerous in federal court; there were two criminal prosecutions by the United States for trespass by cutting timber, five civil suits by the United States to vindicate its property rights in timber and one by an agent of the state, together with one private lawsuit turning mainly on an issue of public land title. Of course, these are only those proceedings in federal court which led to published opinions. (See Chapter IV, note 1, above, regarding the nature and scope of this search of federal court reports). None of the federal court cases turning mainly on private law questions showed any distinction of policy from doctrine familiar in the state court. The official proceedings against trespass are noted in Part One.

7. Data from the Chippewa County circuit court show the same general relation, in the flow of business of a trial court, between trespass and contract actions. This information, however, does not, and cannot, show the relative importance of title questions in the determination of cases there; though the pleadings would sometimes show when title questions were raised, this is not always so and in any case is not reliable evidence, since pleaders may raise more issues than they seriously press and in the bulk of cases there are no trial court opinions to show the weight assigned particular questions.

8. That plaintiff's possession is sufficient title against a mere wrongdoer, see Bates v. Campbell, 25 Wis. 613 (1870) (land); Emerson v. Thompson, 59 Wis. 619, 18 N.W. 503 (1884) (logs); cf. Hungerford v. Redford, 29 Wis. 345 (1872) (plaintiff not in possession must establish complete chain of paper title). The interest of the contract vendee of timberland was upheld in Krakow v. Wille, 125 Wis. 284, 103 N.W. 1121 (1905). Liability of the good-faith trespasser was declared in Eldred v. Oconto Co., 33 Wis. 133 (1873) (logs), and Hazelton v. Weeks, 49 Wis. 661, 6 N.W. 309 (1880) (trespass q.c.f.). The plaintiff's election between contract and tort remedies was applied in Lee v. Campbell, 77 Wis. 340, 46 N.W. 497 (1890), but cf. Lillie v. Dunbar, 62 Wis. 198, 22 N.W. 467 (1885). Defendant's claim of adverse possession was denied in Austin v. Holt, 32 Wis. 478 (1873), and Wadleigh v.

Marathon County Bank, 58 Wis. 546, 17 N.W. 314 (1883); his counterclaim for taxes paid was rejected in Davidson v. Rountree, 69 Wis. 655, 34 N.W. 906 (1887). The legislature also demonstrated special concern for the security of land titles in providing for summary removal from justice of the peace courts of actions involving title to land: Revised Statutes, 1849, c. 88, sect. 52. See Lybrand v. Carson, 2 Pinn. 33 (Wis. Terr. 1847) (trespass q.c.f.); Ames v. Meehan, 63 Wis. 408, 23 N.W. 586 (1885); Maxim v. Wedge, 69 Wis. 547, 35 N.W. 11 (1887). Legislation also early gave holders of certificates of sale or pre-emptive rights to state lands the right to sue for injury to the land. Wisconsin Territorial Laws, 1839-1840, No. 10 (Act of Jan. 4, 1840); Veto Message of Gov. James Duane Doty, Apr. 5, 1843, Council Journal, 239–240; Laws, 1849, c. 212, sect. 21; General Acts, 1855, c. 84, sect. 13, supplanted by General Acts, 1856, c. 125, sect. 9; General Acts, 1857, c. 57; Revised Statutes, 1858, c. 153; General Laws, 1864, c. 156, sect. 4, and c. 193; General Laws, 1865, c. 520; Laws, 1891, c. 207; Laws, 1935, c. 541. Compare General Laws, 1864, c. 277 (railroad to protect its aid-lands grant), and General Laws, 1869, c. 146 (railroad's police function withdrawn); Laws, 1874, c. 348 (*qui tam* actions); Laws, 1876, c. 339; P&LL, 1867, c. 334, sect. 7 (Lemonweir Improvement Co.); P&LL, 1872, c. 104, and Laws, 1875, c. 336 (Sturgeon Bay & Lake Michigan Ship Canal & Harbor Co.). Compare the care taken in some plank road and railroad charters to grant eminent domain powers explicitly regarding the cutting of standing timber necessary to clear a right of way: Wisconsin Territorial Laws, 1847, p. 23, sect. 9 (Act of Jan. 25, 1847), and 1848, p. 152, sect. 7 (Act of Mar. 10, 1848); Revised Statutes, 1849, c. 16, sect. 105; Private & Local Acts, 1853, c. 40, sect. 10, and c. 55, sect. 9; Private & Local Acts, 1854, c. 73, sect. 10; P&LL, 1869, c. 51, sect. 12, and c. 407, sect. 12; P&LL, 1870, c. 38, sect. 11; General Laws, 1872, c. 119, sect. 11 (4th); Revised Statutes, 1878, sects. 1341, 1828; Statutes, 1898, sects. 1341, 1828.

9. Some specific reference to timber values or timber-industry facts will be found in Pacquette v. Pickness, 19 Wis. 219 (1865) (waste by timber cutting may be offset against action under improvements statute by loser in prior ejectment action); Webster v. Moe, 35 Wis. 75 (1874) (statutory trover act applied by analogy at common law); Haight v. Lucia, 36 Wis. 355 (1874), and Riemer v. Johnke, 37 Wis. 258 (1875) (temporary injunction against timber cutting pending determination of title, where timber is land's main value); Brewster v. Carmichael, 39 Wis. 456 (1876) (analogical application at common law of the statutory trover act); Wisconsin Central Railroad Co. v. Cornell University, 49 Wis. 162, 5 N.W. 331 (1880) (ejectment analogy: recognition of special timber-area factors in determining reasonable scope of eminent domain power for creating right of way in rough country); Atkinson v. Hewitt, 51 Wis. 275, 8 N.W. 211 (1881) (protecting timber value of mortgaged land); Nelson v. Churchill, 117 Wis. 10, 93 N.W. 799 (1903) (waste recovery incident to ejectment, regarding timberland); Boneck v. Herman, 247 Wis. 592, 20 N.W.(2d) 664 (1945) (statutory double damages for timberland trespass); Herbert A. Nieman & Co. v. Holton & Hunkel, 248 Wis. 324, 21 N.W.(2d) 637 (1946) (special damage to timberland by soil loss); Hunter v. Neuville, 255 Wis. 423, 39 N.W.(2d) 468 (1949) (adverse possession of timberland).

10. See note 3 above. The substantive scope of the statutory offense of larceny of standing timber remained in essence the same through various re-enactments. General Acts, 1854, c. 55, extended the offense to cover not only trees "upon the private property of any individual," but also those "upon any property held in trust," a provision carried into Revised Statutes, 1858, c. 165, sect. 56, and 1878, sect. 4441. General Laws, 1866, c. 17, merged into a common section definitions of larceny as covering any produce of the soil or severance of any fixture, as well as cutting of standing timber; compare the protective provisions for agriculture in Statutes of the Territory of Wisconsin, 1839, p. 350, sects. 6, 36, 38; Revised Statutes, 1849, c. 40, sect. 3, 1878, sect, 4441. Revised Statutes, 1878, sect. 4441, was apparently designed to embrace all offenses of larceny of growing things and fixtures; Revised Statutes, 1878, sect. 4978, expressly repealed Revised Statutes, 1858, *in toto,* and General Laws, 1866, c. 17, in particular. See *Report of the Revisers of the Statutes,* 302. Sect. 4441 was so condensed as to be cryptic; this probably explains the enactment of another specific provision on larceny of standing timber, by Laws, 1881, c. 170, which in substance became the central provision that stood down through the twentieth century. See Statutes, 1959, sects. 26.04, 26.05.

11. The legislation referred to is cited in

note 3 above. An additional elaboration — in General Acts, 1854, c. 55 — was the provision of higher penalties where two successive offenses included a combination of day and night trespasses. On the 1878 change, see note 10 above; the 1878 provisions for conversion of logs or lumber in transit were by sects. 4448, 4449. The particular interest in timber trespass evidenced by Laws, 1881, c. 170, is underlined by the rejection at the same session of a more general bill, declaring larceny to include any wrongful removal of any crop, trees, or fixtures. See bill 80A, 1881, and AJ, 150, 167 (1881). When Laws, 1887, c. 99, finally again enacted that theft of anything which was in law part of realty should be larceny, this was held not an implied repeal of Laws, 1881, c. 170: Golonbieski v. State, 101 Wis. 333, 77 N.W. 189 (1898). Laws, 1881, c. 170, was amended by Session Laws, 1889, c. 397, to cover the principals of cutting agents; the addition passed routinely. The 1881 scale of penalties went unchanged through Statutes, 1898, sect. 4449a; a series of changes starting with Laws, 1907, c. 149, considerably elaborated the scale of penalties by differentiating according to the value of stolen property, without substantially altering the lower-value scale, except by restoration of authority in the court to impose fine or imprisonment as alternative or cumulative penalties. All these changes were in the general scale of penalties for larceny, and did not reflect any special attention to timber trespass until there appeared the clear-cut distinction between willful and non-willful trespass in Laws, 1949, c. 252, embodied in Statutes, 1959, sects. 26.04, 26.05, hereafter discussed in the text. The legislative journals show no tinkering with criminal penalties other than the steps which eventuated in the enactments here discussed.

12. The insistence on a showing of a guilty mind as the older norm in Wisconsin criminal law is discussed in Robinson and Zick, 625. Typical cases finding civil trespass, though defendant did not intend to steal, are Single v. Schneider, 24 Wis. 299 (1869), s.c., 30 Wis. 570 (1872); Hazelton v. Weeks, 49 Wis. 661, 6 N.W. 309 (1880); Underwood v. Smith, 109 Wis. 334, 85 N.W. 384 (1901); cf. Boneck v. Herman, 247 Wis. 592, 20 N.W.(2d) 664 (1945). Laws, 1949, c. 252, went through the legislature smoothly as introduced by the Committee on Agriculture and Conservation, at the request of the Conservation Commission; it was a general revision of the law regarding trespass on public as well as private land: see SJ, 374, 670; AJ, 1112 (approved on a "required" record vote, 89 to 0). The 1949 larceny provisions were held not unconstitutionally vague, in State v. Biller, 262 Wis. 472, 55 N.W.(2d) 414 (1952) (trespass on state land).

At two points in the course of nineteenth-century legislation there was some appearance of the creation of crimes of timber conversion not requiring proof of a guilty mind. General Acts, 1866, c. 17, supplanting General Acts, 1854, c. 55, omitted the word "wilfully" or any like modifier in declaring guilty of larceny anyone who "shall sever from the soil of another" any growing tree and convert it to his own use. However, the Wisconsin court would generally imply a legislative intent to include the requirement of proof of a guilty mind, in the absence of clear legislative indication to the contrary; here the crime was explicitly designated as "larceny" — an offense involving criminal intent by ordinary definition — and it was declared that the offense "shall be punished in the same manner as is provided . . . for larcenies . . ." Again without use of words of intent, General Acts, 1864, c. 331, made it a misdemeanor for an owner or one otherwise controlling a sawmill to manufacture logs other than his own, and refuse or neglect, within thirty days after demand, to deliver to the proper owner such share or proportion of the manufactured product as he might be entitled to by contract or custom. Since the act set a penalty as high as three years in state prison, it is hard to believe that the statute would not be construed to require a showing of bad-faith withholding by the accused. Without explicitly repealing the 1864 act, General Laws, 1866, c. 134 (amended as to area covered, by General Laws, 1867, c. 161) penalized any boom or sawmill owner who "knowingly" turned into his boom or manufactured another's marked logs without the log owner's consent; this express inclusion of the intent element in a substantially contemporary act promotes the doubt that the 1864 legislature meant any radical innovation in the scope of criminal law. Cf. P&LL, 1867, c. 334, sect. 7 (willful trespass on Lemonweir Improvement Co. aid lands, punishable as malicious mischief).

13. Laws, 1931, c. 404, created Statutes, sect. 348.386 (3), renumbered to be Statutes, sect. 134.60, by Laws, 1955, c. 696, sect. 262. See Attorney General, *Opinions,* XX, 1106, Opinion of Nov. 25, 1931, to Conservation Com-

mission (act limited to regulation of commercial cutting, but all shipments, including those of gifts, may be inspected to prevent violation) — limited and distinguished in XXXIV, 433, Opinion of Dec. 22, 1945, to District Attorney, Oshkosh (license required of all dealers, regardless of origin of trees); XXII, 103, Opinion of Feb. 24, 1933, to District Attorney, Antigo (act regulates woods operations only, does not supplant municipal licensing of peddlers). The quoted summary of the act's purpose is from the Attorney General's Opinion cited above, XXXIV, 433, 435. Bill 127S, 1879, made a gesture towards preventive law; it would have barred the defense of mistaken boundaries in civil actions for timber trespass, unless defendant showed that before cutting he had caused a competent survey to be made, on which he relied: Report, Judiciary Committee, Feb. 14, 1879, SJ, 253; SJ, Feb. 20, pp. 309–310.

14. Revised Statutes, 1878, sect. 3082 (ancillary damages in ejectment); Laws, 1880, c. 147, amending Revised Statutes, 1878, sect. 2922 (full costs, including those of survey); Laws, 1905, c. 264, sect. 20 (double damages). The double damages act was in issue in Boneck v. Herman, 247 Wis. 592, 20 N.W.(2d) 664 (1945) (failure of proof of willfulness). See also P&LL, 1867, c. 334 (treble damages in civil actions for willful trespass on aid lands of Lemonweir Improvement Co.). Bill 168A, 1903, would have amended Statutes, 1898, sect. 2922, to extend the authorization of allowance of survey costs to a survey ordered within six months before and in anticipation of suit; it was indefinitely postponed on recommendation of the Assembly Judiciary Committee: AJ, 531, 550 (1903). Sect. 2922 survey costs were limited to what the court would find reasonable, in Dunbar v. Montreal River Lumber Co., 127 Wis. 130, 106 N.W. 389 (1906). And in any case, sect. 2922 applied only where the gravamen of the action was in the nature of trover, for the unlawful cutting of timber; an action in the nature of trespass *quare clausum fregit,* where the claim was for permanent damage to the freehold, though by the cutting of trees, was held not within this special costs-recovery provision: Maxim v. Wedge, 69 Wis. 547, 35 N.W. 11 (1887); cf. Wadleigh v. Marathon County Bank, 58 Wis. 546, 17 N.W. 314 (1883).

15. Laws, 1873, c. 263, as amended by Laws, 1882, c. 239, 1889, c. 95, and 1901, c. 170, was embodied successively in Revised Statutes, 1878, sect. 4269, Annotated Statutes, 1889, sect. 4269, Statutes, 1898, sect. 4269; it became Statutes, sect. 331.18, by Laws, 1925, c. 4, and so remained until its repeal by Laws, 1949, c. 252, which provided a double damages action for timber trespass (henceforth to be Statutes, sect. 26.09) and amended Statutes, sect. 331.18, to eliminate defendant's opportunity to make a tender in mitigation of damages, for timber trespass. That the substantive offense delineated by the 1873 act's principal provision was in essence the same as common law trover or replevin was ruled in Webber v. Quaw, 46 Wis. 118, 49 N.W. 830 (1879) ("wrongful" cutting under the act means any unlawful or unauthorized cutting, as at common law); Jeske v. Hotz Manufacturing Co., 238 Wis. 116, 297 N.W. 357 (1941); cf. Hazelton v. Week, 49 Wis. 661, 6 N.W. 309 (1880). The Wisconsin common law rule on damages for non-larcenous conversion was first laid down in a case not involving wrongful cutting of timber. Weymouth v. Chicago & North Western Railway Co., 17 Wis. 550 (1863) recognized a division of authority, when it chose the rule for Wisconsin; the opinion indicated that the rule would apply where "exemplary" damages might be proper — presumably in cases of willful trespass amounting to larceny. See Marshall, J., concurring, in Pettingill v. Goulet, 137 Wis. 285, 288, 118 N.W. 845, 846 (1908). Ingram v. Rankin, 47 Wis. 406, 2 N.W. 755 (1879) — not a lumber case — noted a refinement of common law doctrine in replevin, where a successful plaintiff who was unable to regain possession of the converted goods might recover the value of the chattels at the time of trial, so far as their value had risen from market fluctuations or other causes not due to defendant's labor; however, the court refused to extend this replevin rule to allow recovery of the highest value attained by the goods at any time between conversion and trial and arising from causes other than defendant's labor, because this would allow plaintiff "speculative" damages as to commodities the prices of which fluctuated considerably in the market. Single v. Schneider, 24 Wis. 299, 302 (1869) intimated approval of this rule in replevin, in a case involving wrongful cutting of timber, but no Wisconsin Supreme Court opinion seems explicitly to have held the doctrine in a timber-trespass case. The general common law rule of the Weymouth case was applied in claims arising out of wrongful cutting of

timber done without intent to trespass, in Single v. Schneider (above), and in Hungerford v. Redford, 29 Wis. 345 (1872). However, in Single v. Schneider, 30 Wis. 570 (1872) — a later stage of the litigation reviewed in 24 Wis. 299 — the Supreme Court extended the Weymouth case rule of damages even to that part of a timber trespass proved to have been done willfully — i.e., with knowledge that a trespass was being committed — though without fraud, malice, or wanton injury such as, the court said, would still allow the grant of exemplary damages. At issue in 30 Wis. 570 was the value added by defendant's labor. The statute covered both an increase in value from a rising market, as declared in Webster v. Moe, 35 Wis. 75 (1874), and an increase in value from defendant's efforts, as in Haseltine v. Mosher, 51 Wis. 443, 8 N.W. 273 (1881).

16. The instances cited in the text, in the order there given, are Fleming v. Sherry, 72 Wis. 503, 40 N.W. 375 (1888); Haseltine v. Mosher, 51 Wis. 443, 8 N.W. 273 (1881); Warren v. Putnam, 68 Wis. 481, 32 N.W. 533 (1887); Kneeland-McLurg Lumber Co. v. Lillie, 156 Wis. 428, 145 N.W. 1093 (1914); Wright v. E. E. Bolles Wooden Ware Co., 50 Wis. 167, 6 N.W. 508 (1880). See also Millard v. McDonald Lumber Co., 64 Wis. 626, 25 N.W. 656 (1895) ($1050–$4060 claim); Cotter v. Plumer, 72 Wis. 476, 40 N.W. 379 (1888) ($1.50–$3 stumpage value per 1000 feet, compared to $6–$9 value as floated to market); Ehrmantrout v. McMahon, 78 Wis. 138, 47 N.W. 305 (1890) ($22–$209); Everett v. Gores, 89 Wis. 421, 62 N.W. 82 (1895) ($120–$400); McNaughton v. Borth, 136 Wis. 543, 117 N.W. 1031 (1908) ($268–$523); Pettingill v. Goulet, 137 Wis. 285, 118 N.W. 845 (1908) ($26–$156); Gerbig v. Bell, 143 Wis. 157, 126 N.W. 871 (1910) ($12.50–$25).

17. The characterization of a suit within Laws, 1873, c. 263 (Revised Statutes, 1878, sect. 4269) as a statutory action, is in Smith v. Briggs, 64 Wis. 497, 498, 499, 25 N.W. 558, 559 (1885). The court repeatedly stressed that the new measure of damages was a legislative and not a judicial creation. See Tuttle v. Wilson, 52 Wis. 643, 9 N.W. 822 (1881); Keystone Lumber Co. v. Kolman, 94 Wis. 465, 69 N.W. 165 (1896), s.c., 103 Wis. 300, 79 N.W. 224 (1896); cf. Arpin v. Burch, 68 Wis. 619, 32 N.W. 681 (1887) (common law right of peaceful repossession remains, along with the statutory cause of action). Maxim v. Wedge, 69 Wis. 547, 35 N.W. 11 (1887) ruled on the distinction between Laws, 1873, c. 263, and an action of trespass *quare clausum fregit.*

18. That the action under Laws, 1873, c. 263 (Revised Statutes, 1878, sect. 4269), would not apply regarding out-of-state lands was indicated in Swift v. James, 50 Wis. 540, 543, 7 N.W. 656, 657 (1880); cf. Bruheim v. Stratton, 145 Wis. 271, 129 N.W. 1092 (1911) (Supreme Court left undisturbed a trial court ruling that an action did not lie in Wisconsin under a Minnesota treble damages statute, for trespass to Minnesota timberland). That plaintiff must show that he held an interest in the land, see Kneeland-McLurg Lumber Co. v. Lillie, 156 Wis. 428, 430, 145 N.W. 1093, 1095 (1914); see Keystone Lumber Co. v. Kolman, 94 Wis. 465, 69 N.W. 165 (1896). Warren v. Putnam, 63 Wis. 410, 24 N.W. 58 (1885) distinguished the bill to quiet title and against waste, from a sect. 4269 action. See also Smith v. Briggs, 64 Wis. 497, 25 N.W. 558 (1885), and Grunert v. Brown, 119 Wis. 126, 95 N.W. 959 (1903) (no misjoinder of trover and trespass q.c.f., where a sect. 4269 action); Millard v. McDonald Lumber Co., 64 Wis. 626, 25 N.W. 656 (1885) (evidence of land title need not be pleaded); cf. Wadleigh v. Marathon County Bank, 58 Wis. 546, 17 N.W. 314 (1883).

19. Marshall, J., concurring, in Pettingill v. Goulet, 137 Wis. 285, 289, 118 N.W. 845, 847 (1908). The court often ascribed enactment of Laws, 1873, c. 263, to the 1872 decision. See Webster v. Moe, 35 Wis. 75, 78 (1874); Brewster v. Carmichael, 39 Wis. 456, 460 (1876); Webber v. Quaw, 46 Wis. 118, 121, 49 N.W. 830, 831 (1879); Ingram v. Rankin, 47 Wis. 406, 415–416, 2 N.W. 755, 762 (1879); Brown v. Bosworth, 58 Wis. 379, 386, 17 N.W. 241, 244 (1883).

20. The breadth of sect. 4269 in comparison with prior Wisconsin common law may be seen most sharply by contrast of the discussion in Webster v. Moe, 35 Wis. 75 (1874) and that in Ingram v. Rankin, 47 Wis. 406, 2 N.W. 755 (1879). The legislative background of the 1873 act is set out in Dykstra, "Law and the Lumber Industry," 371, 472.

21. See Cotter v. Plumer, 72 Wis. 476, 480, 40 N.W. 379, 380 (1888); Fleming v. Sherry, 72 Wis. 503, 510, 40 N.W. 375, 377 (1888); Knapp v. Alexander-Edgar Lumber Co., 145 Wis. 528, 536, 130 N.W. 504, 507 (1911), reversed on federal grounds, 237 U.S. 162 (1915). Cf. Schweitzer v. Connor, 57 Wis.

177, 181, 14 N.W. 922, 923 (1883); Smith v. Champagne, 72 Wis. 480, 482, 40 N.W. 398, 399 (1888); Smith v. Morgan, 73 Wis. 375, 378, 41 N.W. 532, 533 (1889).

22. Webster v. Moe, 35 Wis. 75, 79 (1874); Ingram v. Rankin, 47 Wis. 406, 415–416, 2 N.W. 755, 762 (1879); Tucker v. Cole, 54 Wis. 539, 543, 11 N.W. 703, 705 (1882). Cf. Brown v. Bosworth, 58 Wis. 379, 387, 17 N.W. 241, 244 (1883); Pettingill v. Goulet, 137 Wis. 285, 287, 118 N.W. 845, 846 (1908). Even on this analysis, the statute might do less than compensate plaintiff, who would be held to the highest value of his timber achieved within the period of the statute of limitations.

23. Marshall, J., concurring, in Pettingill v. Goulet, 137 Wis. 285, 289, 118 N.W. 845, 847 (1908).

24. See Smith v. Sherry, 54 Wis. 114, 131, 11 N.W. 465, 471 (1882); Fleming v. Sherry, 72 Wis. 503, 509, 510, 40 N.W. 375, 377 (1888); cf. St. Croix Land & Lumber Co. v. Ritchie, 73 Wis. 409, 41 N.W. 345 (1889), s.c., 78 Wis. 492, 47 N.W. 657 (1891).

25. Marshall, J., concurring, in Pettingill v. Goulet, 137 Wis. 285, 290, 118 N.W. 845, 847 (1908), comments on the effect of the 1878 revision. The abortive effort at change in the following year is reflected in Report, Judiciary Committee, Feb. 14, 1879, SJ, 253; SJ, Feb. 20, pp. 309–310. On the other hand, a floor amendment to strike the repeal of sect. 4269, and thus cause bill 127S simply to amend the mistake defense thereunder by tightening it, lost on a record vote of 13 to 16 (SJ, 309).

26. The court followed familiar general doctrine in recognizing that defendant had the burden of pleading and providing his defense under such a form of the statute. Webber v. Quaw, 46 Wis. 118, 49 N.W. 830 (1879); Everett v. Gores, 89 Wis. 421, 62 N.W. 82 (1895); Fehrman v. Bissell Lumber Co., 188 Wis. 82, 204 N.W. 582 (1925); see McNaughton v. Borth, 136 Wis. 543, 117 N.W. 1031 (1908).

27. Report, Committee on Lumber & Manufactures, on bill 322A, Feb. 17, 1876, AJ, 356; AJ, Feb. 29, p. 468; Report, Judiciary Committee, Mar. 8 and 11, SJ, 579, 651; SJ, Mar. 11, p. 670. The only other legislative attention to the mistake defense was a procedural change, favoring the defendant; Laws, 1889, c. 95, allowed defendant to file his affidavit of mistake at or before the time of his answer, in contrast to the original form of the act, which required him to file within ten days of the complaint.

28. See Brown v. Bosworth, 58 Wis. 379, 387, 17 N.W. 241, 245 (1883); Ehrmantrout v. McMahon, 78 Wis. 138, 141, 47 N.W. 305, 306 (1890). Compare the observations of Lyon, J., dissenting, on the difficulties of operating within a system of recorded title in the rude conditions of the pinelands, in Bunn v. Valley Lumber Co., 51 Wis. 376, 382–383, 8 N.W. 232, 234 (1881). The most honestly and efficiently run company was bound to fall into trespass sometimes, because of the inaccuracies of the survey: see Reynolds, 32, 33.

29. Fehrman v. Bissell Lumber Co., 188 Wis. 82, 204 N.W. 582, 205 N.W. 905 (1925) made clear that Laws, 1882, c. 239, covered mixed mistakes of law and fact, and not of fact alone. The question of the constitutionality of sect. 4269, if it were construed to bar a defense of entry under good-faith belief in title, was indicated in Smith v. Sherry, 54 Wis. 114, 131, 11 N.W. 465, 471 (1882), and disavowed in Fehrman v. Bissell Lumber Co. (above), 188 Wis., at 95, 205 N.W. at 906. The opinion in the Fehrman case, however, attributes enactment of the 1882 amendment to the decision in Smith v. Sherry: 188 Wis. at 94, 205 N.W. at 906.

30. Schweitzer v. Connor, 57 Wis. 177, 181, 182, 14 N.W. 922, 923 (1883); Fehrman v. Bissell Lumber Co., 188 Wis. 82, 95, 96, 204 N.W. 582, 205 N.W. 905, 906 (1925); see note 29 above.

31. Brewster v. Carmichael, 39 Wis. 456, 460 (1876). The court had already so ruled in Webster v. Moe, 35 Wis. 75 (1874), where recovery of a rise in the market value of the stumpage since the time of cutting was the issue; the Brewster case appears to have gone further, and to have allowed recovery at common law of an increased value of the logs probably attributable, in part at least, to defendant's labor. Despite the liberality towards extension of the impact of Laws, 1873, c. 263, the court made plain that it did not regard the legislature's policy judgment as extending beyond the special problems of the timber industry, and that the former common law rule of damages remained for conversion of other property than wrongfully cut timber. See Ingram v. Rankin, 47 Wis. 406, 416, 2 N.W. 755, 763 (1879).

32. See Underwood v. Paine Lumber Co. (Ltd.), 79 Wis. 592, 595, 48 N.W. 673, 674 (1891); Marshall, J., concurring, in Pettingill v. Goulet, 137 Wis. 285, 289, 118 N.W. 845,

847 (1908). Dykstra, "Law and the Lumber Industry," 477–478, points out that the record conveys some indication that the Supreme Court was more sympathetic to sect. 4269 than were the trial courts. Appraising twenty-seven cases under that act, Dykstra points out that in eight of them the Supreme Court reversed trial court refusals to apply sect. 4269, and reversed trial court applications of the section in only four cases; the Supreme Court sustained application of the act in eighteen of the twenty-seven cases. Dykstra sees the most restrictive Supreme Court rulings as mostly concentrated in a limited period of years (1888–1893), under certain judges (*id.,* 480).

33. On the scope of sect. 4269, see Webber v. Quaw, 46 Wis. 118, 49 N.W. 830 (1879) (mistake of fact no defense, aside from the statutory proviso); Haseltine v. Mosher, 51 Wis. 443, 8 N.W. 273 (1881) (value in last market is applicable); Schweitzer v. Connor, 57 Wis. 177, 14 N.W. 922 (1883) (mistake of law); Everett v. Gores, 89 Wis. 421, 62 N.W. 82 (1895), s.c., 92 Wis. 527, 66 N.W. 616 (1896) (cutting in breach of contract; McNaughton v. Borth, 136 Wis. 543, 117 N.W. 1031 (1908) (negligent mistake); Pettingill v. Goulet, 137 Wis. 285, 118 N.W. 845 (1908) (covers any marketable timber or its product); Jeske v. Hotz Manufacturing Co., 238 Wis. 116, 297 N.W. 357 (1941) (lack of title). See also Dykstra, "Law and the Lumber Industry," 382, 386.

34. The quoted cases are, in the order of the text, Tucker v. Cole, 54 Wis. 539, 542, 11 N.W. 703, 704 (1882) (partner); Lee v. Lord, 76 Wis. 582, 586–587, 45 N.W. 601, 602 (1890) (principal); Underwood v. Paine Lumber Co. (Ltd.), 79 Wis. 592, 594, 595, 48 N.W. 673, 674 (1891) (independent contractor); Brown v. Bosworth, 58 Wis. 379, 386, 388, 389, 17 N.W. 241, 245 (1883) (negligent mistake); Warren v. Putnam, 68 Wis. 481, 488, 32 N.W. 533, 537 (1887) (no title defense if no mistake of fact involved). See also Gerhardt v. Swaty, 57 Wis. 24, 14 N.W. 851 (1883) (partner); Fleming v. Sherry, 72 Wis. 503, 40 N.W. 375 (1888) (title defense not applicable where pure mistake of law); Cook Land, Construction & Producing Co. v. Oconto Co., 134 Wis. 426, 114 N.W. 823 (1908) (similar). Particularly strict application of the doctrine refusing to allow negligent mistake as a defense was made in McNaughton v. Borth, 136 Wis. 543, 117 N.W. 1031 (1908), ruling in effect that the cutter owed an obligation to take affirmative steps of reasonable nature to determine whether he had a right to cut a given tract. See also Kneeland-McLurg Lumber Co. v. Lillie, 156 Wis. 428, 145 N.W. 1093 (1914).

35. See notes 26 and 33 above.

36. On the mistake affidavit as a binding admission, see Ehrmantrout v. McMahon, 78 Wis. 138, 140, 47 N.W. 305, 306 (1890); Underwood v. Paine Lumber Co. (Ltd.), 79 Wis. 592, 596, 48 N.W. 673, 674 (1891). The principal's assertion of the defense of mistake, based on his agent's conduct, is dealt with in Underwood v. Paine Lumber Co. (Ltd.), above. The requirement that defendant continue in his good-faith mistake of fact or title was enforced in Warren v. Putnam, 68 Wis. 481, 32 N.W. 533 (1887); Arpin v. Burch, 68 Wis. 619, 32 N.W. 681 (1887); Fleming v. Sherry, 72 Wis. 503, 40 N.W. 375 (1888); Cook Land, Construction & Producing Co. v. Oconto Co., 134 Wis. 426, 114 N.W. 823 (1908); cf. Befay v. Wheeler, 84 Wis. 135, 53 N.W. 1121 (1893). See also the enforcement of the statutory requirement that defendant's reliance on his title be in good faith, in St. Croix Land & Lumber Co. v. Ritchie, 73 Wis. 409, 41 N.W. 345 (1889), s.c., 78 Wis. 492, 47 N.W. 657 (1891). The mistake-of-title defense might be successfully invoked, though defendant in fact had never had a good title (Befay v. Wheeler, above). But, as the court there noted, this ruling marked no liberal extension of the statutory defense, which must be so construed if it were to have any meaning (84 Wis. at 140, 53 N.W. at 1123).

37. Sometimes the court designated sect. 4269 as "penal" even while giving it an extended application, as in Haseltine v. Mosher, 51 Wis. 443, 8 N.W. 273 (1881). But the "penal" label, in the court's opinions referred to in the text (in the cases next cited in this note), usually spelled some restriction on the application of the act. This was so in decisions insisting on clear proof of title: Paige v. Kolman, 93 Wis. 435, 67 N.W. 700 (1896); Knapp v. Alexander-Edgar Lumber Co., 145 Wis. 428, 130 N.W. 504 (1911), reversed on federal grounds, 237 U.S. 162 (1915). The tax-title plaintiff not in possession must especially well prove his claim, ruled Smith v. Sherry, 54 Wis. 114, 11 N.W. 465 (1882), and Fleming v. Sherry, 72 Wis. 503, 40 N.W. 375 (1888); that his burden of proof was not impossible to meet, however, is shown by St. Croix Land & Lumber Co. v. Ritchie, 73

Wis. 409, 41 N.W. 345 (1889), s.c., 78 Wis. 492, 47 N.W. 657 (1891). More general insistence that plaintiff make clear proof of all elements of his cause of action under sect. 4269 is shown in Tuttle v. Wilson, 52 Wis. 643, 9 N.W. 822 (1881), and Smith v. Champagne, 72 Wis. 480, 40 N.W. 398 (1888). Interest on the judgment under sect. 4269 was denied in Smith v. Morgan, 73 Wis. 375, 41 N.W. 532 (1889), and Everett v. Gores, 92 Wis. 527, 66 N.W. 616 (1889). Enhanced damages were denied against an administrator, in Cotter v. Plumer, 72 Wis. 476, 479, 40 N.W. 379, 380 (1888), and allowed against the receiver of a partnership, in Everett v. Gores, 89 Wis. 421, 62 N.W. 82 (1895). In Peshtigo Lumber Co. v. Ellis, 122 Wis. 433, 100 N.W. 834 (1904), the court ruled that the mere threat of action under sect. 4269 by another claimant — though it carried the danger of enhanced damages — did not warrant a contract vendee of standing timber in staying idle while the contract time limit for cutting expired.

38. Wright v. E. E. Bolles Wooden Ware Co., 50 Wis. 167, 6 N.W. 508 (1880).

39. *Id.,* 50 Wis. 167, 6 N.W. 508, 510 (1880); Tucker v. Cole, 54 Wis. 539, 542, 11 N.W. 703, 704 (1882).

40. 50 Wis. 167, 170, 171, 6 N.W. 508, 510. The balance struck among competing values by the Wisconsin court is the more striking in view of the opposite determination made in a strong opinion for the Court by Miller, J., in Wooden-Ware Co. v. United States, 106 U.S. 432 (1882), involving the same defendant, again in its stance as bona fide purchaser for value from a willful trespasser, who this time had cut from lands held by the United States. See Chapter II, note 24.

41. Tuttle v. Wilson, 52 Wis. 643, 9 N.W. 822 (1881); Millard v. McDonald Lumber Co., 64 Wis. 626, 25 N.W. 656 (1885); Joseph Dessert Lumber Co. v. Wadleigh, 103 Wis. 318, 79 N.W. 237 (1899); cf. Tucker v. Cole, 54 Wis. 439, 11 N.W. 703 (1882) (burden of proof satisfied); Smith v. Briggs, 64 Wis. 497, 25 N.W. 558 (1885).

42. Gerhardt v. Swaty, 57 Wis. 24, 14 N.W. 851 (1883). On the effect of notice as taking defendant out of the Wright doctrine, see that case, 50 Wis. 167, 171, 6 N.W. 508, 510; Smith v. Briggs, 64 Wis. 497, 499, 25 Wis. 558, 559 (1885); Grunert v. Brown, 119 Wis. 126, 95 N.W. 959 (1903).

43. The federal land survey as a measure for standardizing trade is reflected in J. H. Beuscher, *Law and the Farmer* (1956), 103–107; Samuel M. Fegtly, "Historical Development of Land Surveys," *Illinois Law Review,* 38:270 (1944); Marshall Harris, *Origin of the Land Tenure System in the United States* (1957) 391, 399. W. Scott Van Alstyne, Jr., "Land Transfer and Recording in Wisconsin," *Wisconsin Law Review,* 1955:44, discusses the contribution of the recording system to marketability. A vigorous statement of the court's valuation of the Statute of Frauds as giving desirable regularity to transactions in timberland is made in Smith v. Bouck, 33 Wis. 19 (1873); the policy exposition is the more notable because made in a case involving the finance of speculation.

44. The Wisconsin Reports abound in cases illustrating the function of the government survey in providing a framework of definition for land trade. This function is given especially explicit recognition in Fuller v. Worth, 91 Wis. 406, 64 N.W. 995 (1895); Blatchford v. Voss, 197 Wis. 461, 269–470, 219 N.W. 100, 101 (1929); Baackes v. Blair, 223 Wis. 83, 87, 269 N.W. 650, 652 (1936). Timberland was involved in Gerhardt v. Swaty, 57 Wis. 24, 14 N.W. 851 (1883); Whitney v. Detroit Lumber Co., 78 Wis. 240, 47 N.W. 425 (1890); Lally v. Rossman, 82 Wis. 147, 51 N.W. 1132 (1892). Legislative and judicial formulas to cure defects in the survey are discussed in Jones v. Kimble, 19 Wis. 429 (1865), and Westphal v. Schultz, 48 Wis. 75, 4 N.W. 136 (1879); cf. Combs v. Scott, 76 Wis. 662, 45 N.W. 532 (1890).

45. The leading decision applying the Statute of Frauds to standing-timber transactions was Daniels v. Bailey, 43 Wis. 566 (1877). English and American precedent was laid before the court in Young v. Lego, 36 Wis. 394 (1874), where it was found unnecessary to decide the point. Strasson v. Montgomery, 32 Wis. 52 (1873) interpreted Revised Statutes, 1858, c. 120, sect. 51 (removal from court of justice of the peace of "every action where the title to lands shall in any wise come in question") to cover a claim of right in standing timber. On Wisconsin law's recognition that standing timber often constituted the principal value of land, see, e.g., Revised Statutes, 1858, c. 28, sect. 52 (certificate of sale of school or university lands does not authorize timber cutting without commissioners' written consent); Revised Statutes, 1878, sect. 4269 (enhanced damages for timber trespass on private lands); Haight v. Lucia, 36 Wis. 355 (1874); Hyslip v. French, 52 Wis. 513,

9 N.W. 605 (1881); McInnis v. Lyman, 62 Wis. 191, 22 N.W. 405 (1885).

46. Routine application of the recording acts to deeds of timberland may be seen, e.g., in McInnis v. Lyman, 62 Wis. 191, 22 N.W. 405 (1885), and Daggett v. Reas, 79 Wis. 60, 48 N.W. 127 (1891). Mississippi River Logging Co. v. Blue Grass Land Co., 131 Wis. 10, 110 N.W. 796 (1907), and Wisconsin River Land Co. v. Selover, 135 Wis. 594, 116 N.W. 265 (1908) are the two cases in which a party's status as a bona fide purchaser under the recording act was a principal issue. On the deficiencies of the grantor-grantee index, see Van Alstyne, 226–238.

There were many clashes more or less directly involving recording-act issues, between tax-title claimants and persons deriving title by original grant of public lands or by private purchase. But, by definition, these are not part of our present story of the relation of the recording acts to ordinary, private market dealings in land. Conflicts immediately between grantor and grantee over the passage of title did not involve the recording acts; an unrecorded deed or contract would not prevail against a third-party, bona fide purchaser from one who had an interest to transfer, but an unrecorded conveyance or contract was valid between the parties to it, both by inference from the recording act — which, at most, generally penalized failure to record only as against third parties — and by judicial doctrine, which recognized no obligation to record such an instrument to make it effective between those who were parties to it, unless the statute very clearly so commanded. See Cadle v. McLean, 48 Wis. 630, 636–637, 4 N.W. 755, 756 (1880) (not later disapproved in this respect, though qualified in other regards).

47. A reserved title in standing timber was valid between the parties, Lillie v. Dunbar, 62 Wis. 198, 200, 22 N.W. 467, 468 (1885), as was the relation of equitable mortgage between a financier and the person for whose benefit the financier bought realty, Starks v. Redfield, 52 Wis. 349, 9 N.W. 168 (1881). Both the number of Supreme Court cases involving both types of deal, and the number of recorded contracts reserving a security title in the product of standing timber, as found in the Register of the Lumber Inspector for the Sixth District (Chippewa County), 1864–1892 (preserved in the Wisconsin State Historical Society Library), show how prevalent and standardized were such agreements.

48. The finance of long-term speculation in timberland is discussed in Gates, *Wisconsin Pine Lands,* chaps. iv, v. The limits of title, so far as based on the mere fact of possession, are indicated in Bates v. Campbell, 25 Wis. 613 (1870) (ejectment for timber tract). The Wisconsin court indicated — in dicta, since the immediate issue in each instance concerned title to logs or lumber, and not to realty — that recording of a deed or contract reserving a security title in standing timber to the original vendor thereof was not necessary to preserve his rights in the standing timber as against a third party. See Cadle v. McLean, 48 Wis. 630, 637, 4 N.W. 755, 756 (1880) (conceding that a contract of sale of standing timber is for a sale of an interest in realty under the Statute of Frauds, yet cautioning that "it does not follow from this that it was essential that defendant [vendor] should record his conveyance in the office of register of deeds in order to preserve his rights. We have not been referred to any statute which requires that such an instrument should be recorded as a conveyance, to secure the rights of parties to it"); Lillie v. Dunbar, 62 Wis. 198, 203, 204, 22 N.W. 467, 470 (1885) ("The owner of real estate who makes a contract for the sale thereof in writing, and lets the purchaser into possession, is not required to have the contract of sale recorded in order to protect his interest in the lands so sold"). Lyon, J., dissenting in Bunn v. Valley Lumber Co., 51 Wis. 376, 384, 8 N.W. 232, 235 (1881), properly pointed out that failure to record the deed or contract might, however, be disastrous to the equitable owner (i.e., the contract vendee) as against a subsequent bona fide purchaser from the original grantor, for the latter would have something to transfer: "Neither would the recording of [the contract for sale of standing timber] . . . in the office of the register of deeds have been of any benefit to the vendor, although, while the timber remained standing, such record would have protected the purchaser. The vendor needed no protection until the timber was cut and became personal property, and a record in the register's office would then have been entirely inoperative to give such protection. Such a record would operate as constructive notice of the condition of the title to the realty alone, not of personal property severed therefrom."

49. Hyland v. Bohn Manufacturing Co., 92 Wis. 157, 163, 65 N.W. 170, 369 (1896) (reserved title); see Andrews v. Jenkins, 39

Wis. 476, 480 (1876) (equitable mortage); Jourdain v. Fox, 90 Wis. 99, 102, 62 N.W. 936, 937 (1895) (equitable mortgage); Mississippi River Logging Co. v. Miller, 109 Wis. 77, 84, 85 N.W. 193, 196 (1901) (reserved title). The dicta cited in note 48 above indicate that recording was not necessary to the efficacy of such security between the original parties.

50. Wing v. Thompson, 78 Wis. 256, 266, 47 N.W. 606, 609 (1890). See accord, Bardon v. McCall, 108 Wis. 181, 185, 84 N.W. 168, 169 (1900); Mississippi River Logging Co. v. Mills, 109 Wis. 77, 92, 93, 85 N.W. 193, 196 (1901); Stubbings v. Curtis, 109 Wis. 307, 85 N.W. 325 (1901); Oconto Land Co. v. Wallschlaeger, 155 Wis. 418, 144 N.W. 979 (1914); Jeske v. Hotz Manufacturing Co., 233 Wis. 500, 290 N.W. 208 (1940).

51. Andrews v. Jenkins, 39 Wis. 476, 480 (1876).

52. General Acts, 1864, c. 167, sect. 12, continued a like provision first established by General Acts, 1861, c. 83, sect. 12. The provision remained in the general statutes of the state as Revised Statutes, 1878, and Statutes, 1898, sect. 1739, and Statutes, 1925, sect. 108.11. It continued until the lumber-inspector system was ended, as obsolete. See Report, Interim Committee on Administration and Taxation, Jan. 26, 1927, AJ, 269, 314; Report, Special Committee on Obsolete Laws, June 30, 1927, SJ, 1709, 1710. McCutchin v. Platt, 22 Wis. 561 (1868) gave a common sense construction to the statute's awkward terms, in ruling that the requirement of recording instruments affecting the ownership "of any mark of logs" meant, in context, to refer to the marked logs and not to the marks alone. The statutory language was formally corrected to accord with this construction in Revised Statutes, 1878, sect. 1739. General Acts, 1866, c. 57, sect. 3, added to General Acts, 1864, c. 167, sect. 12, the declaration that recording of an instrument under the basic act should have the same effect as the recording of deeds and mortgages in the office of the register of deeds; if this addition did more than clarify, it perhaps added the doctrine that, among competing interests, the one first recorded would prevail.

53. McCutchin v. Platt, 22 Wis. 561, 564–565 (1868).

54. Revised Statutes, 1849, c. 76, sect. 9; Revised Statutes, 1858, c. 107, sect. 9; cf. Revised Statutes, 1878, sect. 2313, qualified as to logs by sect. 1739, the origin of which is stated in note 52 above. The chattel mortgage recording act was routinely applied regarding mortgages of already-cut logs, Mowry v. White, 21 Wis. 417 (1867), or lumber, Valley Lumber Co. v. Hogan, 85 Wis. 366, 55 N.W. 415 (1893). McCutchin v. Platt, 22 Wis. 561 (1868) recognized that the general chattel-mortage recording act was superseded as to logs by General Acts, 1861, c. 83, sect. 12, which became Revised Statutes, 1878, sect. 1739. That there was no basic inconsistency in policy between the general chattel-mortgage recording act and the application of such a recording requirement to mortgages on logs was indicated by the legislature when in General Acts, 1860, c. 213, it provided for recording chattel mortgages on logs in the lumber area tributary to Oshkosh in terms nowise different from those of the general statute. Passage of the 1860 act seems only another reflection of the lack of coordination in legislative action to provide suitable generalization of policy. Compare also General Acts, 1860, c. 340, and Laws, 1909, c. 460, including timber among the listed commodities, possession of which must be delivered to a warehouseman or common carrier to permit effective issue of warehouse receipts or bills of lading.

55. See Perkins v. Best, 94 Wis. 168, 68 N.W. 762 (1896). Cf. Cadle v. McLean, 48 Wis. 630, 4 N.W. 755 (1880).

56. On severance as determining that timber should be treated in law as personalty, see the discussion in Golden v. Glock, 57 Wis. 118, 15 N.W. 12 (1883). The ban on inclusion of after-acquired logs in a chattel mortgage was declared in Mowry v. White, 21 Wis. 417 (1867).

Special provision was sometimes made by statute for recording ownership of logs delivered to a boom company, for protection of the company in its deliveries of logs serviced by it: e.g., P&LL, 1862, c. 100, sect. 20 (Apple River Boom Company). Such provisions seem auxiliary to policies promoting smooth handling of log transport, and hence are not discussed here.

57. Cadle v. McLean, 48 Wis. 630, 636, 4 N.W. 755, 755–756 (1880). Examples of recorded contracts providing security titles in not-yet-cut timber, taken from the Miscellaneous Records, Lumber Inspector, District No. 6, in the Wisconsin State Historical Society, Series 33/0/2, include the following varieties of transaction, in contracts between the named parties: (1) title, ownership, and possession of standing timber and cut timber to remain in vendor of standing timber, though contract

allows vendee to cut: Daniel Shaw Lumber Co. and Alex. A. Nicolls, filed Oct. 1, 1878, Miscellaneous Records, 1878–1879, 7:123; David A. Kendall and C. W. Lamberton, filed Oct. 19, 1878, *id.*, 7:124; H. C. Putnam and John Redman, filed Oct. 29, 1878, *id.*, 7:126; E. P. Hastings and Gilbert & Bernier, filed Nov. 8, 1879, *id.*, 7:392; L. H. Celler and Joseph O. Hebert, filed Nov. 8, 1879, *id.*, 7:394; (2) contract lien on logs to be cut, given to vendor of timberland, who, however, is to take and hold possession of the logs under the lien: Charles Allen and Duncan Collins, filed Dec. 19, 1870, *id.*, 3:226; (3) similar contract lien of sawmill-operator-financier, on yet-to-be-cut logs, and on lumber he processes therefrom and possesses: Owen McGinty and John Barron, filed Apr. 20, 1870, *id.*, 3:116; (4) title and possession to logs yet to be cut, given as security for advances to logger: Charles Longency and Young & Felix, filed Nov. 3, 1870, *id.*, 3:197; George Tupper and Henry Hewitt, Jr., filed Nov. 3, 1879, *id.*, 7:388.

58. Bunn v. Valley Lumber Co., 51 Wis. 376, 380 (majority opinion), 382 (dissent), 8 N.W. 232, 233, 234 (1881). The majority also held that replevin was a proper remedy to enforce such a contract lien, within its proper scope.

59. Lillie v. Dunbar, 62 Wis. 198, 202, 203, 22 N.W. 467, 469, 470 (1885); Bent v. Hoxie, 90 Wis. 625, 64 N.W. 426 (1895). The contract of sale of standing timber in Lillie v. Dunbar was oral, and hence unenforceable under the Statute of Frauds, but since the contract vendee had been permitted thereunder to enter and cut timber, the opinion treated the contract as effective by performance, and ruled that plaintiff's measure of recovery was properly determined by the trial court to be the contract price with interest, rather than tort damages for the value of the cut timber or lumber. There was evidence "tending to show at least" that the buyer had been informed by the security holder of the latter's interest after he had contracted to purchase but before he took delivery; however, the opinion does not rest itself on this ground, but on the inapplicability of the recording act and an implicit ruling that the contract vendor was not guilty of misleading conduct. It should be noted that the decisions, as distinguished from the grounds of decision, in Cadle v. McLean and Bunn v. Valley Lumber Co. (notes 57 and 58 above) were consistent with the doctrine that the holder of a contract right to a security in standing-timber product might prevail over a buyer from the contract vendee. In Cadle v. McLean the security holder won, having given the contract vendee no more than possession and the right to cut and manufacture; the contract stipulated that the contract vendee should not sell without the contract vendor's approval; the sale in question was made without the contract vendor's knowledge or consent and without evidence of any past practice of acquiescence in sales by the contract vendee. In Bunn v. Valley Lumber Co., the later buyer prevailed where the contract vendor had not only given the contract vendee possession and a right to cut and drive to market, but had given the contract vendee title and an explicit right to sell

60. In addition to cases cited in note 50 above, the leading decisions reaffirming and applying the test of the misled bona fide purchaser included Long v. Davidson, 77 Wis. 509, 46 N.W. 805 (1890); Burr v. C. C. Thompson & Walkup Co., 78 Wis. 227, 47 N.W. 277 (1890) (bona fide purchaser from equitable mortgagor prevailed against equitable mortgagee, who had advanced the money to buy the timberland from which the logs in question were cut, where the parties' arrangement contemplated that the "mortgagor" would cut and sell, and where there was evidence that the remaining standing timber was sufficient security for the debt); Bent v. Hoxie, 90 Wis. 625, 64 N.W. 426 (1895); Hessey v. Gund, 98 Wis. 531, 74 N.W. 342 (1898). See Ford Motor Co. v. Maeder, 171 Wis. 263, 265, 177 N.W. 39, 40 (1920). Cf. Hoile v. Bailey, 58 Wis. 434, 17 N.W. 322 (1883); Warren v. Landry, 74 Wis. 144, 42 N.W. 247 (1889); Jourdain v. Fox, 90 Wis. 99, 62 N.W. 936 (1895). The emphasis on the question whether the rightful owner out of possession was guilty of creating a misleading situation so as to estop him against a bona fide purchaser from one possessing personal property was, of course, old in the common law of conversion. See Root v. Bonnema, 22 Wis. 539 (1868) (rightful owner may recover from boom owner who took logs not as a bona fide purchaser, though nonnegligently and without wrongful intent); Eldred v. Oconto Co., 33 Wis. 133 (1873) (owner may recover from bona fide purchaser from the converter, without necessity of prior demand, where no grounds of estoppel existed).

The Miscellaneous Records of the Lumber Inspector, District No. 6 (Chippewa), show the recording of numerous contracts reserving security titles in standing timber and its cut product after 1880: (1) after Cadle v. McLean: e.g., contracts of Bassett & Revoir and C. L. Coleman, filed Sept. 22, 1882, *id.*, 9:38; F. S. G. Varner and Gower, Johnson and Lee, filed Sept. 21, 1882, *id.*, 9:48 (sale of timberland with reserved title in standing timber and its cut product); P. M. Agnew and Anne Carroll, filed Feb. 26, 1883, *id.*, 9:354; (2) after Lillie v. Dunbar: W. F. Hinz and James Sweeney, filed Feb. 2, 1888, *id.*, 11:101; Stein & Levy and Demers & Bessette, filed Feb. 20, 1888, *id.*, 11:106 (contract lien on cut timber); Perry Hopkins and A. G. Beiler, filed Dec. 12, 1889, *id.*, 11:197 (reserved title in standing timber); H. W. Sage and Lucy Bassette, filed Jan. 18, 1890, *id.*, 11:205; Charles Revoir and Lucy Bassette, filed Jan. 18, 1890, *id.*, 11:206; Adolph Barnier and Peter Bolier, filed Dec. 10, 1892, *id.*, 11:261; N. E. Kraemer and F. G. and C. A. Stanley, filed Nov. 14, 1893, *id.*, 11:271; Allen Cameron and N. L. Camerson, filed Nov. 28, 1893, *id.*, 11:275; Mississippi River Logging Co. and George B. Best, filed Dec. 31, 1897, in Miscellaneous Records of Lumber Inspector, District No. 11, 1897–1911, in the Wisconsin State Historical Society Library, Series 33/0/7, 2:53.

The legislature did once put its approval on allowing the seller of standing timber a priority security interest for its price in the logs or lumber derived thereform. But the instance was so narrow in definition, and so short-lived, as to make no material qualification of the pattern of events set out in the text: Laws, 1876, c. 372, sect. 1, amended General Laws, 1861, c. 186, sect. 1, creating a lien for labor and services upon logs and lumber in Chippewa County, to provide that this laborer's lien should be "subject to any debt owing for the purchase-money of the timber or stumpage." This declaration is the more interesting because of its differences from an earlier form of the bill which became c. 372; bill 13A, of 1876, was reported from the Assembly Committee on Lumber and Manufactures with an amendment which would have added only "that all indebtedness that may have accrued for stumpage on said logs and timber, shall constitute and be a lien thereon, that shall take precedence of all other claims except for labor performed on said logs and timber." As enacted, the measure was less favorable to the stumpage seller, in striking the general declaration of a lien in his favor preferred against all other claims except for labor; it was more favorable to him, in preferring the stumpage claim over the laborer's claim — a rare step in the general flow of labor-lien legislation, which otherwise consistently preferred the laborer's claim. The enacted form of the bill was the product of its re-reference to the Assembly Judiciary Committee: see AJ, 20, 327, 378, 476, 594 (1876); SJ, 599, 660. However, this novel statutory favor to the stumpage seller did not last long; Laws, 1877, c. 95, omitted the stumpage-lien language in recasting the terms of the basic section.

As introduced, bill 46S, 1879, would have revived the notion of a lien to the seller for the price of standing timber; the bill included in an amendment of Revised Statutes, 1878, sect. 3329 — the basic provision giving a lien for labor or services performed on logs or timber — the stipulation that a lien of like force should be given anyone who sold or conveyed any timber or trees for the purpose of being manufactured into logs, lumber, shingles, or timber: SJ, 79 (1879). After some floor fighting had indicated opposition to some provisions of the bill, its author secured its resubmission to him, and then reported it back with amendments, one of which struck out the lien for the price of stumpage: SJ, 121, 158–159, 176–177 (1879). The sponsor's amendments were approved on a requested record vote, 22 to 5. The bill eventually passed, without any stumpage lien, to become Laws, 1879, c. 167.

61. Lyon, J., dissenting, in Bunn v. Valley Lumber Co., 51 Wis. 376, 382–383, 8 N.W. 232, 234–235 (1881).

62. McCutchin v. Platt, 22 Wis. 561, 564 (1868). Sympathetically broad application of the special log-recording act was also made in Steele v. Schricker, 55 Wis. 134, 12 N.W. 396 (1882), ruling that a recorded contract was notice of the interests of all parties thereto, though the logs bore a mark indicating ownership by but one particular party.

63. Morrow v. Reed, 30 Wis. 81 (1872).

64. Cook v. Van Horne, 76 Wis. 520, 44 N.W. 767 (1890). Bill 236S, SJ, 201 (1899), would have stricken all references to log marks from Statutes, 1898, sect. 1739, thus making the recording requirement apply without qualification to all instruments concerning log titles. The Senate Judiciary Committee (the chairman of which was, oddly, the introducer of the bill) recommended that the bill be indefinitely postponed (SJ, 730), and it was so voted without evidence of contest (SJ, 745).

The only other case in the Supreme Court under sect. 1739 had treated the section more sympathetically, though no more so than its plain meaning warranted: Steele v. Schricker, 55 Wis. 134, 12 N.W. 396 (1882), summarized in note 62 above.

Recordings of bills of sale fall off after 1890 in the Miscellaneous Records of the Lumber Inspector, District No. 6, preserved in the Wisconsin State Historical Society Library. However, their number had been declining in prior years, probably in reflection of the increasing prevalence of the large companies' integrated operations.

On the presumption against an intention materially to change the substance of law by a revisors' bill, see Mundt v. Sheboygan & Fond du Lac Railroad Co., 31 Wis. 451, 463 (1872); Van Brunt v. Joint School District of the Town and Village of Fairfax, 185 Wis. 493, 201 N.W. 755 (1925).

65. General Acts, 1864, c. 167, continued the frame of policy and procedure set by General Acts, 1861, c. 83, regarding log marks. The Wisconsin River act was Laws, 1889, c. 441, sect. 4 of which stated, "Any logs, timber or other floatables having any such recorded mark or marks impressed thereon, shall be presumed to belong to the party or parties in whose name said mark or marks shall have been recorded." It was characteristic of the loose draftsmanship of the time that the general statute did include explicit declaration that lumber inspectors' scaling bills should be presumptive evidence of the facts stated therein (Revised Statutes, 1878, sect. 1735).

Changes over the years in the basic log-mark legislation included these: (1) By General Acts, 1861, c. 83, one might record a mark not before used by another in the district in which he did business; General Acts, 1864, c. 167, broadened the reference to the "districts" in which he did busineses; Revised Statutes, 1878, sect. 1738 and subsequent versions, made the test the distinctiveness of the mark "in the same district" in which a recording was made. (2) General Acts, 1861, c. 83, said that prize logs should be divided in proportion to the number of logs owners had "on the river or its tributaries" in their district; General Acts, 1864, c. 167, apparently removed this tie to logs moved by water, making the division "in proportion to the number of logs owned by each person or company, respectively, in such subdistrict"; Revised Statutes, 1878, sect. 1740 and subsequent versions, seemed to retain this broader definition, though there were other references which seemed to continue to assume that all the logs referred to were being moved by water, especially the definitions of prize logs as those unmarked, and those bearing marks, but not recorded or claimed within one year "after any general drive." Probably the point was not of practical significance; logs moved by rail were less likely to become involved in the title disputes common in mingled driving by stream. (3) General Acts, 1864, c. 259, sect. 2, clarified General Acts, 1864, c. 167, sect. 12, by fixing the liability for the recording fee for an instrument transferring title or creating a security title, on the person for whom the document was recorded. (4) General Acts, 1866, c. 57, sect. 2, struck from General Acts, 1864, c. 167, sect. 11, words limiting the time for recording marks to the interval from Nov. 1 to Mar. 1 each year; the result was to allow — or require, according as one interpreted the basic terms of the 1864 act — the recording of a mark at any time one was adopted. (5) The recorded log mark was made the determinant, absent other evidence, of liability to liens for labor or services, under General Acts, 1862, c. 154, sect. 12, as amended by Laws, 1873, c. 139, and under a broader act, of state-wide application, Laws, 1889, c. 378 (excepted from repeal of prior lien legislation in the general revision by Laws, 1889, c. 413, sect. 18).

66. General Acts, 1861, c. 83, applied to the Wisconsin, Black, Chippewa, and St. Croix rivers, and their tributaries. General Acts, 1863, c. 208, brought into the system the area of Green Bay and its tributary streams within the state; the general revision of the lumber-inspector system by General Acts, 1864, c. 167, continued this inclusion of the Green Bay area, but likewise continued expressly to exclude the Lake Winnebago country. General Acts, 1862, c. 189, in providing for application of the lumber-scaling system in Winnebago County, made no reference to adoption or recording of log marks, and Laws, 1876, c. 254, in redefining the application of the scaling system under General Acts, 1864, c. 167, to the lumber area tributary to Oshkosh, expressly declared that none of the mark provisions of the 1864 act should apply in that area. Laws, 1876, c. 254, was repealed by Revised Statutes, 1878, sect. 4978. The requirements for recording marks in Dist. No. 4 were laid down, successively, by General Acts, 1862, c. 388, and Laws, 1875, c. 299; these provisions were "rewritten but not changed," in Revised Statutes, 1878, sect. 1742 (penalty modified

from \$25–\$100 to \$5–\$100). See *Report of the Revisers of the Statutes,* 136. The Twin Rivers act was P&LL, 1866, c. 175; its marking requirement was probably thought of only as an incident of river-traffic control; though the marking requirements were central, and not incidental, the special act regarding the Wisconsin River — Laws, 1889, c. 441, carried into Statutes, 1898, as sect. 1738a — seems also primarily concerned with log identification as a device to promote smooth functioning of the log-transport system. Compare General Laws, 1869, c. 80, adding to General Acts, 1864, c. 167, sect. 7, authorization to anyone whose logs become mingled with "prize logs" not conveniently separable, to drive the prize logs and collect suitable compensation therefor; and P&LL, 1870, c. 133, requiring that logs rafted on Black River be so fastened as to display their marks.

The considerable legislation redefining old, and creating new, lumber-inspection districts is noted hereafter, as bearing primarily on operation of the scaling system. Apparently it was the creation of Dist. No. 11 (out of part of the northern Chippewa Valley district) by Laws, 1875, c. 242, which eventually brought Laws, 1883, c. 267, requiring that men rerecord — on pain of being declared to have abandoned — all marks recorded with the lumber inspector for Dist. No. 6 before Jan. 1, 1875. The 1883 act was, in turn, repealed by Laws, 1889, c. 441, which required rerecording of all marks recorded in Dists. 1, 10, 14, or any of them before Jan. 1, 1880, lest they be held abandoned. It seems characteristic of the loose legislative practice of the times that no general provision was ever made for a regular procedure to accomplish the sensible objective of periodically eliminating the deadwood from the mark records.

With other lumber inspection district records, the log-mark records were transferred to the offices of the four new districts set up to supplant the prior organization under Laws, 1919, c. 502; the statutory provisions on log marks were otherwise untouched, until repeal of the whole lumber-inspector system as obsolete, under Laws, 1927, c. 474; see Report, Interim Committee on Administration and Taxation, AJ, 269, 314 (1927); Report, Special Committee on Obsolete Laws, SJ, 1709, 1710 (1927).

67. The use of recorded log marks as everyday industry practice is reflected in Fries, 40, 238; James E. Lundsted, "Log Marks, Forgotten Lore of the Logging Era," *Wisconsin Magazine of History,* 39:44 (1955); Larson, 143, 179, 366 — from p. 143 of which is taken the quotation in the text. A typical log-mark register is that of Dist. No. 6, Log Mark Records, 1864–1881, in the Wisconsin State Historical Society Library, Series 33/0/8, Vol. I.

68. The pattern of lumber-inspector organization was set by the first statute of reasonably broad application, General Acts, 1861, c. 83, as revised and broadened by General Acts, 1864, c. 167, carried into Revised Statutes, 1878, sects. 1730–1737, and continued in the general statutes of the state until repealed as obsolete by Laws, 1927, c. 474. The act singling out official scalers' frauds for strong criminal punishment was Laws, 1885, c. 290. Industry reliance on the official scaler as an impartial intermediary is further reflected in contract provisions consenting to his intervention, if the parties were unable to agree on a private scaler. See Gardner v. Wilber, 75 Wis. 601, 44 N.W. 628 (1890); McIntyre v. Rodgers, 92 Wis. 5, 65 N.W. 503 (1895); McCann v. Doherty, 98 Wis. 335, 73 N.W. 782 (1898) (official scaler used to check private scaler); Loree v. Webster Manufacturing Co., 134 Wis. 173, 114 N.W. 449 (1908). However, a lumber inspector was not precluded by his position from contracting to procure for a log owner a buyer for logs in another district many miles from his own, and hence might collect his commission by suit when he performed his contract: McKenzie v. Lego, 98 Wis. 364, 74 N.W. 249 (1898).

Minor changes regarding the administrative organization of the lumber-inspection system were made by General Acts, 1861, c. 167 (correcting a reference to the Scribner Rule); General Acts, 1861, c. 188 (when inspector's term begins); General Acts, 1864, c. 167 (fee changes); General Acts, 1864, c. 259 (inspector's fee payable by owner of logs scaled); General Laws, 1866, c. 57 (reports required from deputy inspectors); General Laws, 1869, c. 52 (broader authority to appoint subinspectors); Laws, 1874, c. 220, and 1875, c. 107 (special fee scale in Dist. No. 2); Laws, 1919, c. 502 (new terms of appointment and fees).

The legislature commonly authorized local governments to regulate measures, and occasionally defined standard units of dealing in local wholesale and retail trade, including transactions in lumber or other timber products. However, this legislation presents no aspect distinctive to the timber industry, and seems, rather, part of the general policy of the law to provide the minimum standardization of dealings necessary to ordinary market functioning. See, e.g., Statutes of the Territory of Wisconsin, 1839, p. 126 (governor may appoint for each organized county, as an inspection district, "an inspector of shingles, wheat and rye flour, buck-wheat meal, pork, beef, fish, butter, lard, domestic spirits, and pot and pearl ashes"; includes standards of sizes for shingles); Territorial Local Acts, 1839, No. 52, sect. 22 (amended charter, town of Green Bay: president and trustees shall have sole right "to . . . regulate the measuring of boards, plank, timber and lumber of every kind"), and *id.,* No. 53, sect. 21 (charter, town of Miles); 4th (Territorial) Legislative Assembly, HRJ, 152 (1845), presentation of petition, and report of Judiciary Committee thereon, regarding appointment of lumber inspectors; Report, Committee on Agriculture and Manufacturies, on Racine petition for lumber inspectors, SJ, 220 (1848) (municipal charter provision found adequate); Private & Local Acts, 1856, c. 535, sect. 15 (14) (charter of Borough of Fort Howard: Borough Council authorized "to regulate the place and manner of weighing hay, of measuring and selling wood, lumber, timber, stone, lime and sand, and when necessary to appoint suitable inspectors and surveyors to superintend the same"); Private & Local Acts, 1857, c. 176, IV, sect. 3 (16) (charter, town of Beloit, similar); Private & Local Acts, 1857, c. 276, sect. 15 (14) (charter, town of DePere, similar); P&LL, 1866, c. 150, IV, sect. 3 (20th) (revised charter, city of Green Bay: authority in council to appoint a lumber and shingle inspector); P&LL, 1869, c. 440, IV, sect. 3, 27th, and Laws, 1873, c. 169, V, sect. 3, 20th (original and revised charters, city of Chippewa Falls: common council may regulate measurement and inspection of lumber, shingles, timber and building materials); P&LL, 1869, c. 449, IV, sect. 3, 20th (charter, city of Oconto: common council may appoint a lumber and shingle inspector); Laws, 1875, c. 228, VI, sect. 7, XXXVII (amending Charter, city of Manitowoc — P&LL, 1870, c. 275 — to add to common council's power the authority "to regulate the measuring and inspecting of lumber, cord and fire-wood, shingles, timber, posts, staves, heading, and all building materials, and to appoint inspectors and prescribe their duties"). See also P&LL, 1868, c. 158, incorporating the Chamber of Commerce of the city of Milwaukee, with authority, *inter alia,* to establish boards of arbitration to settle disputes voluntarily submitted by members or other persons, with the effect of awards at common law, and certain special sanctions. Sect. 11 provided: "Said corporation shall have power to elect or appoint one or more persons, as it may see fit, to examine, measure, weigh, guage or inspect flour, grain, provisions, liquor, lumber or any other article or produce or traffic commonly dealt in by the members of said corporation, and the certificate of such person or inspector as to the quality and quantity of any such article, or their brand or mark upon it, or upon any package containing such article, shall be evidence between buyer and seller of the quality, grade or quantity of the same, and shall be binding upon the members of said corporation or others interested and requiring and assenting to the employment of such weighers, measurers, guagers or inspectors; nothing herein contained, however, shall compel the employment by any one of such appointed." See, accord, P&LL, 1868, c. 482, incorporating a board of trade in the city of Green Bay and the Borough of Fort Howard.

69. The quoted statutory quality reference is from General Acts, 1861, c. 83, sect. 8; Revised Statutes, 1878, sect. 1736. Many cases reflect the fact that customary trade standards of quality were taken for granted as providing workable criteria. See, e.g., Ketchum v. Wells, 19 Wis. 25 (1865); Scott v. Whitney, 41 Wis. 504 (1877); McDonald v. Gardner, 56 Wis. 35, 13 N.W. 689 (1882); Early v. Chippewa Logging Co., 68 Wis. 112, 31 N.W. 714 (1887); Butterfield v. Herren, 80 Wis. 240, 49 N.W. 826 (1891). Quantity-quality determinations were recognized as sharply distinct issues. Early v. Chippewa Logging Co. (above) held that a contract provision for scaling should not be construed as authorizing the scaler to determine the quality of logs delivered, save, presumably, as incident to the scaling operation itself. Accord: Stubbings v. McGregor, 86 Wis. 248, 56 N.W. 641 (1893); Magee v. Smith, 101 Wis. 511, 78 N.W. 167 (1899).

70. The general statutes' favor to the Scribner rule began with General Acts, 1861, c. 83, sect. 9 (as corrected, 1861, c. 167). The specification of a measurement schedule referred to in the text was by Laws, 1901, c. 451, supplanted by Laws, 1913, c. 275. The import of adoption of one scaling rule rather than another, for the competing interests of sellers and buyers of logs, is discussed in

Rector, *Log Transportation,* 81–83; that the problem was an old one, see John A. Wood, *A History of Lumbering in Maine, 1820–1861* (1935), 146–150. On the defeated bills for required standards of measurement, see (1) bill 375A, of 1891, passed in the Assembly after it was reported without recommendation by that chamber's Committee on Lumber and Mining, AJ, 210, 740, and 852 (1891), but defeated in the Senate after the Committee on Commerce and Manufactures recommended non-concurrence, SJ, 608 and 642; (2) bills 344A and 346A, of 1893, both recommended for indefinite postponement by the Assembly Committee on Lumber and Mining, and so voted, AJ, 276, 344 and 362 (1893); (3) bill 448A, of 1895, buried when the Assembly Committee on Public Lands recommended its re-reference to its sponsor, AJ, 247 and 759 (1895); (4) bill 187A, of 1905, passed in the Assembly after that chamber's Committee on Lumber and Mining had recommended indefinite postponement, and after adoption of floor amendments, AJ, 259, 1723, 1802, 1845, 1919 (1905), but indefinitely postponed without reference to committee in the Senate, SJ, 1561. Cf. Hospes v. O'Brien, 24 Fed. 145 (Circ. Ct. D. Minn. 1885) (Minnesota statute requiring official scaling held valid under due process clause, as reasonable regulation against fraud and to improve quality of goods for market, and under commerce clause as placing no undue burden on interstate commerce). The new form of statutory prescription on measurement standards, adopted by Laws, 1901, c. 451 (bill 611A), and Laws, 1913, c. 275 (bill 497A), went through smoothly as introduced. The standard statement of the optional character of the lumber inspector's services, but with the inducement of the presumptive evidentiary force of his scaling bills, will be found in Revised Statutes, 1878, sect. 1735.

71. Changes of district lines or the location of district offices were made by General Acts, 1863, c. 74 (change of office location, Dist. No. 1); General Acts, 1864, c. 167 (general redefinition of districts); General Laws, 1868, c. 129 (creating Dist. No. 7, about Eau Claire River); General Laws, 1869, c. 74 (Red Cedar River in Barron County declared Dist. No. 8); General Laws, 1870, c. 90 (lower Chippewa Valley declared Dist. No. 9); General Laws, 1871, c. 70 (Wisconsin River area in Marathon County declared Dist. No. 10); Laws, 1875, c. 242 (Ashland, Bayfield, and northernmost part of Chippewa County declared Dist. No. 11); Laws, 1876, c. 254 (declaring Wolf and Fox River areas Dist. No. 12); Laws, 1881, c. 225 (Douglas County declared Dist. No. 13); Laws, 1882, c. 193 (divides Wisconsin River area into old Dist. No. 10 and new Dist. No. 14); Laws, 1885, c. 112 (Namakogan River system in Sawyer County declared Dist. No. 15); Laws, 1885, c. 288 (district office, Marinette); Laws, 1889, c. 387 (new Dist. No. 16 defined in part of Wisconsin River Valley); Laws, 1895, c. 54, as modified by Laws, 1897, c. 74 (Dist. No. 17 laid out in part of Wisconsin River system); Laws, 1897, c. 35 (change of office location, Dist. No. 4); Laws, 1919, c. 502 (complete revision of districts, reduced to four). Probably in reflection of the diminished importance of the industry, the complete revision and reduction of districts by the 1919 act went through both houses smoothly as introduced: SJ, 776, 1085, 1205 (1919); AJ, 1620.

Typical of the legislature's readiness to commit lumber-inspection bills to the most immediately affected lumbermen was the handling of the 1864 revision. Assemblyman Thaddeus Pound, himself a leading figure in the industry, introduced bill 72A of 1864, following Memorial No. 15A presented by him on behalf of a number of Chippewa County lumbermen, asking for division of Dist. No. 3 into two districts: AJ, 88 (1864). On Pound's motion the bill and the memorial were recalled from the Committee on Towns and County Organization, to which they had been initially referred, and were sent to a select committee drawn from the northwest lumber section, composed of Pound (from the Eau Claire, Chippewa Falls, Dunn assembly district), Elwell (St. Croix and Pierce), Johnson (Jackson and Clark), Phillips (Portage) and Abrams (Brown): AJ, 100. Evidently this group decided on a somewhat broader approach. Pound now introduced bill 114A, which became General Acts, 1864, c. 167; referred to the select committee, the bill was reported for passage with an amendment: AJ, 164, 207. With minor changes, it then went through both houses without incident. Compare the routine creation of new Dist. No. 14 under Laws, 1882, c. 193: SJ, 147 (1882) (bill 124S); AJ, 610. The story of the creation of Dist. No. 15 likewise shows the typically routine character of such legislation: AJ, 196, 273, 465 (1885); SJ, 368 (bill 342A).

Occasions when lumber-inspection-district bills met opposition and defeat include no instance where there is any evidence that the matter became an issue of general concern in the legislature; lumber members' initiative generally appeared determinative. See, e.g., SJ, 834, 844 (1871) (bill 709A, to define duties of lumber inspectors in Dists. No. 3, 6, 7, 9, indefinitely postponed on motion of Senator Price of Black River Falls, himself a lumberman of prominence); SJ, 313 (1881) (bill 219S, on duties of inspector in Dist. No. 9 recommitted to its sponsor, on recommendation of Committee on Manufactures and Commerce); AJ, 859 (1885) (bill 558A, redefining duties of inspector, Dist. No. 10, indefinitely postponed, according to recommendation of Committee on Lumber and Manufactures); AJ, 166, 284, 772, 816 (1887) (bill 182A, to amend Laws, 1885, c. 266, regarding required scaling in Dist. No. 4, favorably reported by Judiciary Committee, but then recommitted to Committee on Lumber and Mining, and thereafter indefinitely postponed according to the latter's recommendation); AJ, 352, 391–392 (1887) (bill 156A, amending Laws, 1885, c. 112, to enlarge Dist. No. 15, indefinitely postponed according to recommendation of Committee on Lumber and Mining); AJ, 417, 459 (1891) (bill 583A, to secure proper returns of logs cut in Dist. No. 13, and to prevent scaling by nonresidents, indefinitely postponed, according to recommendation of Committee on Lumber and Mining); AJ, 506, 531 (1893) (bill 423A, for recording in lumber inspector's office of scale bills, indefinitely postponed according to recommendation of Committee on Labor and Manufactures); cf. SJ, 265, 271 (1893) (bill 270S, on recording of log marks, indefinitely postponed according to recommendation of Judiciary Committee), and SJ, 730, 745 (1899) (bill 236S, amending Statutes, sect. 1739, on recording title instruments to logs, indefinitely postponed on recommendation of Judiciary Committee).

72. Territorial Laws, 1848, p. 67; Private & Local Acts, 1854, c. 198; General Acts, 1861, c. 83, 1862, c. 189 (explicitly left in effect, 1864, c. 167), 1863, c. 208, 1864, c. 167; Laws, 1876, c. 254; Revised Statutes, 1878, c. 84; Statutes, 1898, c. 84; Laws, 1919, c. 502. A Report of the Committee on Agriculture and Manufacturies, SJ, 220 (1848), made what proved a faulty prediction, that legislation for a lumber-inspection system in one county might be extended to all.

Laws, 1919, c. 491, required a special official measurement system for pulpwood in carload lots. In 1923 Gov. Blaine vetoed an amendment to this regulation which would have added penal sanctions to the law; the Governor found the basic measure of no public utility, because its limitation to carload deals meant that it did not benefit the small cutter or seller who needed protection, and because the stipulation that the official scalers be selected from employees of pulpwood purchasers further impaired the impartiality of the law: Veto Message of July 13, 1923, AJ, 1902–1903. In his regular message to the next legislature, Gov. Blaine urged repeal of the system on the same grounds: Message of Jan. 15, 1925, AJ, 19, 23. The measure was repealed by Laws, 1925, c. 24.

73. The Fourth District scaling requirement was laid down in General Acts, 1862, c. 388, sect. 3; Laws, 1875, c. 299, sects. 3, 4, 5; Laws, 1876, c. 296; Revised Statutes, 1878, sect. 1743; Laws, 1885, c. 266; Laws, 1903, c. 421. Bill 182A, AJ, 105 (1887), reflected trouble in the Fourth District, by proposing to amend Laws, 1885, c. 266, to stipulate a $250 forfeit for noncompliance, half to go to the lumber inspector at his suit, half to the state. Though favorably reported by the Judiciary Committee, AJ, 166, the bill was recommitted to the Committee on Lumber and Mining, which recommended its indefinite postponement: AJ, 772. The Assembly so voted, without apparent conflict: AJ, 816. The incident suggests the legislature's tendency to let such matters be determined by the dominant lumbermen in the body. There was no further legislative initiative in the matter, though in his report dated at Hudson, Jan. 24, 1891, SJ, 114–115 (1891), the lumber inspector for the Fourth District complained of wholesale violation of Laws, 1885, c. 266, for want of adequate penalties; the bulk of the logging in that area, he reported, was by Minnesota residents, who had their logs floated out to Minnesota waters for scaling there.

Chapter 421, Laws, 1903, simply expanded the definition of logs under the Fourth District requirement, without adding new sanctions. Even so, it provoked a sharp battle in the Senate after routinely passing the Assembly, and at one critical point, after an adverse committee report, survived in the upper house by only a one-vote margin on a demand record vote: SJ, 1141 (Report, Committee on Manufactures and Labor, recommending noncurrence, two members dissenting), 1232, 1248–1249 (1903). See also AJ, 235, 417, and 459 (1891) (bill 583A, to prevent scaling by nonresidents in Dist. No. 13 — a part of

Douglas County — indefinitely postponed on recommendation of Committee on Lumber and Mining).

Various laws declared that services of the lumber inspector should be used to measure either all the logs, or at least unclaimed logs, handled by particular boom companies or boom and sawmill operators: e.g., General Acts, 1864, c. 161, sect. 3, and c. 300, sect. 2, 1865, c. 258, sect. 3, and c. 349, sect. 2; P&LL, 1866, c. 237, sect. 3, 1867, c. 328, sect. 18, 1868, c. 461, sect. 2, 1870, c. 339 (repealed, however, by Laws, 1883, c. 339); Laws, 1873, c. 12, sect. 1.

74. On the Scribner rule in Winnebago County, see General Acts, 1862, c. 189, sect. 7, superseded by Laws, 1876, c. 254. P&LL, 1867, c. 291, applied that rule in Manitowoc County. The Black River scaling requirement was by sects. 1 and 4 of P&LL, 1870, c. 133. The two defeated, limited-requirement scaling laws were introduced in the 1885 legislature: (1) Bill 558A would have forbidden any boom, dam, or log-driving company in Dist. No. 13 (Douglas County) to collect any fee for services rendered unless the logs serviced had been scaled by the lumber inspector for the district, imposed a duty on the official scaler to scale all logs cut therein, and gave him a lien for his fees. Some of the bill's terms indicated more explicitly than was usual in such legislation that the prime purpose of the regulation was to facilitate tax assessment. The bill was indefinitely postponed, according to recommendation of the Committee on Lumber and Manufactures, AJ, 233, 735, 859 (1885). (2) Bill 124S would have increased the minimum of the lumber inspector's bond in Dist. No. 6 to $50,000 (compared with the standard $5,000 in other districts), and would have required official scaling of all logs cut in Dist. No. 6 to be driven down the Chippewa River or its tributaries, unless the appropriate party in interest filed an affidavit with the lumber inspector that the logs were to be manufactured by the person who caused them to be put in the stream and were not to be sold on the market; should they thereafter be sold in market, the lumber inspector should be entitled to the fees he would have collected for scaling them. Having passed the Senate without sign of serious opposition — SJ, 136, 571, 601, 624 (1885) — the bill was reported without recommendation by the Assembly Committee on Lumber and Manufactures, AJ, 1086, and then twice refused concurrence, after some maneuvering on the floor, on the same day, AJ, 1095, 1097, 1101. Laws, 1882, c. 214, imposed the special scaler's mark requirement for Dist. No. 9; see also note 78 below. Bill 307A, which became c. 214, Laws, 1882, went through both houses smoothly. See AJ, 215 (1882); SJ, 420, 438, 509. Compare, generally, Laws, 1899, c. 282, defining a standard weight unit of hemlock bark.

75. The best-evidence rule was applied regarding a scale book made according to private contract stipulation, in Tewksbury v. Schulenberg, 48 Wis. 577, 4 N.W. 757 (1880). Resort to such evidence may also be seen in such typical cases as Vaughan v. Howe, 20 Wis. 497 (1866); Kelly v. Berry, 39 Wis. 669 (1876); Christie v. Keator, 49 Wis. 640, 6 N.W. 334 (1880); Day v. Gumaer, 80 Wis. 362, 50 N.W. 182 (1891); Kennedy v. South Shore Lumber Co., 102 Wis. 284, 78 N.W. 567 (1899). The presumption in favor of upholding findings of a private scaler operating under private contract procedure is strongly stated in Early v. Chippewa Logging Co., 68 Wis. 112, 31 N.W. 714 (1887), and Stubbings v. McGregor, 86 Wis. 248, 56 N.W. 641 (1893); reversal for fraud was treated in McCann v. Doherty, 98 Wis. 375, 73 N.W. 782 (1898).

76. Early v. Chippewa Logging Co., 68 Wis. 112, 31 N.W. 714 (1887); Magee v. Smith, 101 Wis. 511, 78 N.W. 167 (1899); Peterson v. South Shore Lumber Co., 105 Wis. 106, 81 N.W. 141 (1899). Cf. Hosmer v. McDonald, 80 Wis. 54, 49 N.W. 112 (1891) (lack of agreement on scaler); McIlquham v. Barber, 83 Wis. 500, 53 N.W. 902 (1892) (which scale was adopted); McIntyre v. Rodgers, 92 Wis. 5, 65 N.W. 503 (1895) (lack of agreement on rescaler); Kennedy v. South Shore Lumber Co., 102 Wis. 284, 78 N.W. 567 (1899) (construction of parties' rights regarding dismissal of scaler, appointment of substitute); Loree v. Webster Manufacturing Co., 134 Wis. 173, 114 N.W. 449 (1908) (construction of procedure for correcting scale).

77. The quoted recognition of the statutory policy is from Steele v. Schricker, 55 Wis. 134, 139, 12 N.W. 396, 398 (1882). Presumptions favoring the regularity of the lumber inspector's actions and determinations were applied in Fornette v. Carmichael, 41 Wis. 200 (1876) (inspector's scale may not be upset by mere counter estimates); Steele v. Schricker (above); Peterson v. South Shore Lumber Co., 105 Wis. 106, 81 N.W. 141 (1899); cf. Gardner v. Wilber, 75 Wis. 601, 44 N.W. 628 (1890). The best-evidence rule was applied to the

official scaler's bills in Fornette v. Carmichael, and Steele v. Schricker (above). On the general sympathy of judicial decisions toward full implementation of the log-lien statutes, compare Glaspie v. Keator, 56 Fed. 203, 208 (C.C.A. 8th, 1893) (given field conditions under which scaling is done, and untutored character of most men drawing scaling bills, court should not construe statute to require official seal on each bill to establish authenticity, and should liberally construe terms of scale bill to derive needed information from it). Limitations on the official scaler were acknowledged in Magee v. Smith, 101 Wis. 511, 78 N.W. 167 (1899) (no authority to determine quality as such); Hurst v. Webster Manufacturing Co., 128 Wis. 342, 346, 107 N.W. 666, 667 (1906). That the official scaler was not a guarantor of his determinations was ruled in Gates v. Young, 78 Wis. 98, 47 N.W. 275 (1890), s.c., 82 Wis. 272, 52 N.W. 178 (1892), though the opinions were somewhat ambiguous as to what standard of performance he was held to meet; some language of the court might seem to indicate that he would be liable only for bad faith, but the general tenor of the court's discussion was to indicate that negligence, also, might be the basis of an action on his bond. On the consistency of the Gates decision and the general doctrine regarding the liability of public officers, compare Lowe v. Conroy, 120 Wis. 151, 97 N.W. 942 (1904); the Gates opinions themselves invoked Wisconsin cases dealing generally with the liability of public officers. The focus of the Gates case was the ruling that negligence in scaling at the boom was not proved by showing merely a discrepancy with prior scaling on the bank at a distant point; the ruling was common sense, in view of well known industry facts regarding losses of logs in transit; cf. Day v. Gumaer, 80 Wis. 362, 50 N.W. 182 (1891). A subsidiary ruling in the Gates case was that the duty of the lumber inspector's deputies, and his liability for them under his bond, were to be defined by the same measure as the duty of the inspector.

78. SJ, 166 (1876), Report, Lumber Inspector, Dist. No. 10, for year ending Dec. 31, 1875; SJ, 14 (1876) (Dist. No. 2); AJ, 72 (1877) (Dist. No. 7, Eau Claire), 605 (Dist. No. 10). The Beef Slough Company continued to use its own scaler: see AJ, 39 (1880) (official scaled 70,042,540 feet of logs; Company in addition ran about 250,000,000 feet); AJ, 46 (1881) (77,942,270 feet of logs officially scaled, 173,119,800 feet by the Company); AJ, 29 (1883) (165,031,050 feet of logs officially scaled, 252,450,250 by Company); SJ, 118 (1887) (168,351,260 feet of logs officially scaled in 1886, 295,523,300 feet by Company); SJ, 105 (1891) (56,124,580 feet of logs officially scaled, 92,806,610 by Company). This story ends with the report of a new Ninth District lumber inspector, SJ, 26–27 (1893), who reported that in his term of office he did not scale or measure any logs or lumber, because the rafting and booming business theretofore done by the Company at Beef Slough, Wis. (for itself and others) had been transferred to West Newton, Minn. Matthew G. Norton, *The Mississippi River Logging Company: An Historical Sketch* (1912), 75–79, relates that this change of location was due to increasing navigational difficulties at Beef Slough, due to silting of the channel.

79. Material in this paragraph of the text is derived from examination of the file, Lumber Inspector, 1861–1905, in the Wisconsin State Historical Society Library, Series 1/1/2–3, Box 1. The quotation regarding the 1862 appointment in the Chippewa Valley is from a letter to Hon. S. D. Hastings from John T. Tinker, Eau Claire, Mar. 27, 1862. Cameron's estimate of the Dist. No. 2 post is contained in his letter to Gov. Lucius Fairchild, from LaCrosse, Jan. 17, 1868; cf. his letters on the same appointment, Jan. 23 and Feb. 24, 1868. The estimate of the worth of the Green Bay office is made to Gov. Fairchild in a letter of Apr. 15, 1867, from W. I. Abrams, Green Bay; cf. letter to Gov. James T. Lewis, regarding the same office, from a resigning inspector, J. V. Snydam, Green Bay, Oct. 20, 1864. On the Ashland office, see letter to Gov. Jeremiah Rusk from W. R. Durfee, United States Indian Agent, LaPointe Agency, July 3, 1884; cf. letter to Gov. W. D. Hoard from Lewis C. Thompson, Superior, May 20, 1889 (duties of Dist. No. 13, at Superior, are light, "but should be looked after by some competent person"). See a lumber inspector's comment on the limited, seasonal nature of the work, quoted in 3rd Biennial Report, Bureau of Labor and Industrial Statistics, PD, II, 91 (1888) (Edwin C. Erickson, LaCrosse).

80. The comment on the 1874 Chippewa Falls controversy is in a letter from J. B. Kehl to Gov. W. R. Taylor, Chippewa Falls, Apr. 21, 1874. (For source of this and other items in this note, see note 79 above.) The letter from Merrill is that of H. W. Wright to Gov. W. H. Upham, May 2, 1895. Examples of the many letters to Governors in which great, and

sometimes exclusive, emphasis is upon the partisan affiliations of candidates are those of C. Leymond to Gov. Lucius Fairchild, LaCrosse, Dec. 4, 1867 (Black River); Frank I. Theep to Gov. Fairchild, Onalaska, Feb. 11, 1870 (LaCrosse); William Richardson to Gov. C. C. Washburn, Chippewa Falls, Feb. 7, 1872 (Chippewa Valley); John A. McRae to Gov. C. C. Washburn, Alma, Apr. 20, 1872 (candidate for appointment signs himself "a true Republican"); H. M. Stocking to Gov. W. E. Smith, Eau Claire, Jan. 16, 1877 (opposing reappointment of incumbent because latter is "an uncompromising Democrat"); J. M. Brackett, newspaper editor in Eau Claire, to Gov. Smith, Jan. 16, 1878; George B. Shaw, secretary, Daniel Shaw Lumber Co., Eau Claire, to Gov. Smith, Jan. 17, 1878; C. C. Washburn, from Philadelphia, Pa., to Gov. Jeremiah Rusk, Dec. 12, 1881; Benjamin F. Bryant to Gov. Rusk, LaCrosse, Mar. 5, 1884; Roujet Marshall to Gov. Rusk, Chippewa Falls, Apr. 12, 1884; C. J. Jellison & Co. to Gov. W. D. Hoard, Wabasha, Minn., Feb. 7, 1889; G. W. Bishop to Gov. W. H. Upham, Rhinelander, Dec. 24, 1894.

81. The letter from J. G. Thorp to Gov. Jeremiah Rusk is in the file cited in note 79 above, as are other materials cited in this note; this letter was written by Thorp to the Governor from Minneapolis, Apr. 16, 1884; cf. his letter to Gov. Rusk, Eau Claire (no date, probably 1884). The local newspaper editor who spoke for the independents was J. W. DeGraff, writing to Gov. Rusk from Alma, Jan. 5 and Feb. 1, 1884; the Prairie du Chien operator was the firm of Stauer & Daubenberger, McGregor, Iowa, writing to Gov. W. D. Hoard, Jan. 10, 1889. Related items on this same appointment in Dist. No. 9 include letters of W. J. Bradley, president, State Lumber Co. of Milwaukee, to Gov. Hoard, Jan. 10, 1889; Gardiner, Batchelder & Welles, Lyons, Iowa, to Gov. Hoard, Jan. 11, 1889 ("The outside interests should be protected" — i.e., the independents compared with "the pool"). "The pool" chose not to contest the pressure for reappointment of the popular Ninth District incumbent. See letter of Frederick Weyerhaeuser, as president of Chippewa Logging Co., to Gov. Hoard, Chippewa Falls, Jan. 26, 1889; letter of Thomas Irvine, secretary of the Beef Slough Company, wishing success to the opposing candidate, but announcing the company's decision not to support any candidate, in letter to James Russell, Jan. 31, 1889. The quoted letter to Gov. Jeremiah Rusk was from M. H. McCord, Apr. 8, 1886. The "pocket book" quotation is from a letter of H. M. Stocking to Gov. Rusk, Eau Claire, Apr. 7, 1884; communications of like tenor include those to Rusk from Murphy & Gaynor, Chippewa Falls, Apr. 14, 1884, and T. B. Scott, Merrill, Apr. 18 and May 15, 1884, as well as a letter to Gov. W. D. Hoard, Chippewa Falls, Jan. 10, 1889.

82. That a lien for services in cutting and driving logs might be created by contract "as well as by statute," and exist concurrently with statutory remedies, was held in Smith v. Scott, 31 Wis. 420 (1872); that case likewise ruled that, at least between the parties, the contract terms would control as to when the lien attached, and whether it might be asserted though possession was not retained. That an intervening bona fide purchaser would prevail over a contract-lien claimant who did not keep continuous possession, compare Andrews v. Jenkins, 39 Wis. 476 (1876), and Smith v. Shell Lake Lumber Co., 68 Wis. 89, 31 N.W. 694 (1887).

Arians v. Brickley, 65 Wis. 26, 26 N.W. 188 (1885) held that a common law lien existed concurrently with statutory liens in favor of a sawmill operator on lumber in his posession which he had sawed for the lien defendant from the latter's logs; a dictum recognized that such a lien would be available likewise to a raftsman on lumber rafted for the owner: see 65 Wis. 26, 29, 26 N.W. 188, 189; cf. Rohrer v. Lockery, 136 Wis. 532, 117 N.W. 1060 (1908). The dictum in Arians denying that a lien would be given for labor in cutting trees (65 Wis. 26, 31, 26 N.W. 188, 190) was rested on Oakes v. Moore, 24 Me. 214 (1844); the Maine opinion gave no satisfactory rationale for the ruling. That a common law lien would be recognized for cutting services had early been indicated in Gay v. Fretwell, 9 Wis. 186, 193–194 (1859); this proposition was later apparently supported by Keystone Lumber Co. v. Kolman, 94 Wis. 465, 469, 69 N.W. 165, 167, and dissenting opinion of Cassoday, C. J., *id.*, 94 Wis. 471, 69 N.W. 168 (1896); and by Dresser v. Lemma, 122 Wis. 387, 392, 100 N.W. 844, 846 (1904), where the court's sensitivity to prevention of unjust enrichment suggested that it would have recognized a common law lien for cutting services, had the question been squarely presented. The issue of a common law lien arose in the Arians case because the lien claimant had not complied with statutory requirements for obtaining a lien under Laws, 1882, c. 319

(which extended Revised Statutes, sect. 3329, to grant a lien for labor or services in sawing or manufacturing logs into lumber).

The complete congruence between the common law liens recognized in the lumber industry and in other economic areas may be seen by comparison with the stress on the critical importance of possession in the claimant, in Sensenbrenner v. Mathews, 48 Wis. 250, 3 N.W. 599 (1879) — see also Wilson v. Rudd, 70 Wis. 98, 104, 35 N.W. 321, 323 (1887) — and in rulings on the availability of equitable relief to adjudge and enforce the lien, as in Brickley v. Walker, 68 Wis. 563, 577, 32 N.W. 773, 780 (1887). See Wilson v. Johnson, 74 Wis. 337, 43 N.W. 148 (1889) (equitable action to foreclose pledge exists concurrently with statutory action at law to enforce right of lien under Revised Statutes, 1878, sect. 3347).

83. There was no Wisconsin Territorial legislation concerning lumber industry liens. The chain of state legislation ran thus: (1) Revised Statutes, 1849, c. 120, sect. 12, made the general mechanics-lien remedy apply to anyone performing manual labor on land, timber, or lumber, for or on account of the owner, his agent, or assignee. (2) Revised Statutes, 1858, c. 153, sect. 12, continued Item 1 unchanged.

(3) General Acts, 1860, c. 215, created a lien for labor or services from the stage of cutting through those of driving and towing logs, but not including milling, in Shawano, Waupaca, Outagamie, Winnebago, and Fond du Lac counties, and Lake Winnebago. (4) General Acts, 1861, c. 186, made Item 3 applicable in Chippewa County. (5) General Acts, 1862, c. 154, made Item 3 applicable in Pierce, St. Croix, Polk, Dallas, Burnett, Douglas, LaPointe, and Ashland counties; Dallas County later became Barron County, and LaPointe County became Bayfield County. (6) General Acts, 1863, c. 169, applied the lien and procedure created by Item 3 to embrace service in mingled driving on the Wolf River or its tributaries.

(7) General Acts, 1865, c. 517, included Jackson, Clark, LaCrosse, and Manitowoc counties in the coverage of Item 5. (8) General Laws, 1866, c. 66, included Marathon, Portage, and Wood counties in the coverage of Item 5. (9) General Laws, 1867, c. 100, added Shawano, Door, Kewaunee, Oconto, and Brown counties to the coverage of Item 5, and included a lien for supplies; Shawano County had already been included in the coverage of Item 3. (10) General Laws, 1868, c. 99, made a procedural amendment to Item 5. (11) General Laws, 1869, c. 144, amended Item 3 to add procedural protections to the log owner where he was not originally a party to the lien proceeding.

(12) General Laws, 1870, c. 4, amended the procedure under the general mechanics-lien act in a fashion relevant to Item 2. (13) General Laws, 1870, c. 120, added Waupaca County to the counties named in Item 5 as amended by Item 7, and amended Item 7 to narrow the scope of the supplier's lien; Waupaca County had already been included in Item 3. (14) P&LL, 1872, c. 71, added to the coverage of Item 3 those parts of Oconto and Brown counties where logs were brought out to go down the Wolf River and its tributaries, apparently removing these areas from the coverage of Item 5, as amended by Items 9 and 13. (15) Laws, 1873, c. 139, amended Item 5 to require that the log owner, if known, be made a party defendant to the lien proceeding along with the person liable for payment of the debt. (16) Laws, 1874, c. 17, amended Item 5 as amended by Item 9, to add Outagamie County to its coverage; that county had already been included in Item 3. (17) Laws, 1874, c. 161, amended Item 4 (lien in Chippewa County) to require that the log owner, if known, be made a party defendant. (18) Laws, 1874, c. 267, amended Item 3 as amended by Item 14 to include the labor of cooks, and to grant a broadly defined supplier's lien, as well as making some procedural changes.

(19) Laws, 1876, c. 32, amended Item 5 as amended by Items 9 and 16, to add Taylor and Lincoln counties to their coverage. (20) Laws, 1876, c. 372, amended Item 4, to add cooking and hauling and labor done at the owner's request to the activities creating a lien, as well as creating a lien for the purchase price of stumpage, and broadening the terms on which log owners must be included as parties defendant. (21) Laws, 1877, c. 95, amended Item 4 as amended by Item 20 to cover Taylor and Lincoln counties; these counties had been under Item 5, by virtue of Item 19; Item 21 also repealed the stumpage-price lien created by Item 20. (22) Revised Statutes, 1878, sects. 3329–3342, included Eau Claire County in a list which covered all counties under any of the prior acts, merged into one statement with some changes in detail the outstanding definitions of labor and supplies liens in named counties, continued

Item 2 by sect. 3341 as a limited, residual-lien provision for areas not specifically named, adding bark to its scope, broadly preferred labor over supplies liens, and made various procedural changes. (23) Laws, 1879, c. 167, amended Item 22 to add Monroe County to the basic list of named lien areas, and included Lincoln, Marathon, Portage, and Wood counties in the broader of the two supplies liens carried into Item 22 from Items 9 and 18, as well as creating a lien for sawing or manufacturing logs into lumber. (24) Laws, 1879, c. 103, seems in effect to have extended the lien provisions applicable in Chippewa County to the newly created Price County.

(25) Laws, 1880, c. 62, corrected verbal errors in Item 23. (26) Laws, 1880, c. 192, amended Item 22 to change the place for filing liens in the Fourth Lumber Inspection District. (27) Laws, 1880, c. 222, created a special procedure for levies on logs subject to lien in the Chippewa River area. (28) Laws, 1881, c. 330, amended Item 22 as amended by Items 23 and 25, repealing these last two items, and thus eliminating all suppliers' liens and making the basic labor and services lien applicable in all counties, while creating a lien for named types of work on certain named secondary timber products (telegraph poles, railroad ties, etc.) in Door, Lincoln, Marathon, and Waupaca counties, and a lien for described work on cordwood or bark in Door County, as well as making various changes in procedural detail. (29) Laws, 1882, c. 273, expanded the right of one claiming an ownership interest in liened logs to intervene in the lien proceedings. (30) Laws, 1882, c. 319, amended Item 22, as amended by Item 28, to permit enforcement of the lien for labor or services on logs or timber against lumber derived therefrom, amending various procedural details to conform to this change, and repealing the special provisions of Item 28 as to secondary forest products, except for Door County bark.

(31) Laws, 1885, c. 192, amended Item 22 to change the time for filing of lien claims. (32) Laws, 1885, c. 225, fixed a special place, time, and procedure for lien proceedings in the Chippewa River area. (33) Laws, 1885, c. 469, amended Item 22 as amended by Item 30, to restore a lien for supplies in connection with getting out logs or timber in Door, Florence, Kewaunee, Marathon, Langlade, Marinette, Oconto, Portage, Shawano, Taylor, and Waupaca counties, as well as declaring a lien for both labor and supplies in connection with work on named secondary forest products (railroad ties, cordwood, etc.) in the same named counties, while continuing the general lien for labor or services on logs or timber in all counties, and changing requirements as to time of filing of lien petitions. (34) Laws, 1887, c. 530, provided that Item 33 (amending Item 22) should apply in Wood, Lincoln, and Oneida counties. (35) Laws, 1889, c. 145, amended Item 34 to strike Wood County from its declaration. (36) Laws, 1889, c. 378, in effect made generally applicable the special procedure for levy on logs created by Item 27. (37) Laws, 1889, c. 413, amended Item 22, striking all provision for supply liens, including work of cooks, and work on stave bolts and staves in the general lien for labor on logs or timber, for all counties, and bringing the lien for work on secondary forest products under the limited terms of sect. 3341 (continuing Item 2), altering the terms on which these liens should be preferred over other claims, and making some changes in detail in the statement of procedures. (38) Laws, 1889, c. 454, amended Item 22 and Item 37, strengthening and clarifying the preferred status of the lien claimant over others claiming interests in the liened property, and broadening the authorization of assignments of lienable claims.

(39) Laws, 1891, c. 139, amended Items 37 and 38 to make the lien cover the manufacturing of logs into timber, and made detailed procedural changes, especially liberalizing venue for lien suits in favor of claimants, and adding penalties against persons acting fraudulently to defeat the lien. (40) Laws, 1895, c. 72, amended Item 39 to add work on the manufacture of pulpwood or charcoal to the activities creating the basic lien. (41) Laws, 1895, c. 285, created a broad supplier's lien in Door County. (42) Statutes, 1898, sects. 3329–3342b, brought the lien for work on secondary forest products within the definition of the broad, basic lien, strengthened a lien claimant's remedies against persons acting to defeat the lien, and put a limited liability for costs on an unsuccessful lien claimant. (43) Laws, 1899, c. 351, amended Item 42, correcting verbal awkwardness in statement of required place for filing a certified copy of a writ of attachment.

(44) Laws, 1913, c. 241, amended Item 42, to add "heading" and firewood to the articles for work on which a lien might be pursued. (45) Laws, 1913, c. 773, amended Item 42 to simplify levy procedure under Item 27 as

carried into Statutes, sect. 3342b. (46) Laws, 1919, c. 484, amended Item 42 to extend the basic lien to cover work done by any person or by his beast of burden. (47) Laws, 1925, c. 25, extended the basic lien under Item 42 to cover work done by tractor or similar motor vehicle. (48) Laws, 1931, c. 15, restored a lien for supplies furnished at the request of the contractor for labor or services on forest products. (49) Laws, 1935, c. 483, made many detailed changes of a minor character in a general revision of timber industry labor liens.

84. That the heart of the distinction between the statutory and the common law liens was that the former did not turn on possession by the claimant, see Arians v. Brickley, 65 Wis. 26, 30, 26 N.W. 188, 189 (1885); Taylor, J., dissenting, in Smith v. Shell Lake Lumber Co., 68 Wis. 89, 102–103, 31 N.W. 694, 700 (1887). Compare the cases which stress that the determining fact in the creation of the statutory lien is simply that labor or service has been done — as in Munger v. Lenroot, 32 Wis. 541 (1873), and Collins v. Cowan, 52 Wis. 634, 9 N.W. 787 (1881) — regardless of the particular state of contract arrangements between the worker and his employer, or the employer and the latter's customer. The objective evidence from the face of the statutes that the heart of new policy in the statutory lien was the dropping of a requirement of possession was, of course, the detailed provisions for seizure of the logs or timber by attachment, a step unnecessary if the claimant held possession in the first instance: e.g., General Acts, 1860, c. 215, sects. 3–14; see Arians v. Brickley, 65 Wis. 26, 30, 26 N.W. 188, 189 (1885).

The quoted statements of the court's rule of liberal construction of the log-lien statutes in behalf of claims based on labor or services are from Winslow v. Urquhart, 39 Wis. 260, 268 (1875) (cook's services held covered, though not explicitly within act), s.c., 44 Wis. 197 (1878), and Lohman v. Peterson, 87 Wis. 227, 229, 58 N.W. 407 (1894) (holding, however, that separate rental for services of animal was not covered). The liberal construction rule may also be seen invoked as a ground of decision in Paine v. Gill, 13 Wis. 561, 563 (1861); Paine v. Woodworth, 15 Wis. 298, 303 (1862); Munger v. Lenroot, 32 Wis. 541, 545 (1873); Jacubek v. Hewitt, 61 Wis. 96, 102, 20 N.W. 372, 374 (1884); Johnson v. Iron Belt Mining Co., 78 Wis. 159, 161, 47 N.W. 363 (1890); DeMorris v. Wilbur Lumber Co., 98 Wis. 465, 74 N.W. 105 (1898); Engi v. Hardell, 123 Wis. 407, 100 N.W. 1046 (1905); see also Kendall v. Hynes Lumber Co., 96 Wis. 659, 661, 71 N.W. 1039, 1040 (1897); Carpenter v. McCord Lumber Co., 107 Wis. 611, 614, 83 N.W. 764, 765 (1900); Edwards v. H. B. Waite Lumber Co., 108 Wis. 164, 165, 84 N.W. 150 (1900). Compare similar, liberal treatment of log-labor liens in the ruling by a federal court in Wisconsin declaring that a like lien created under Michigan law was not invalid as a preference, nor among the types of liens dissolved by an adjudication in bankruptcy, under federal bankruptcy legislation: In re Kerby-Denis Co., 94 Fed. 818 (E.D. Wis. 1899), affirmed with emphasis on lien's origin in labor done, 95 Fed. 116 (C.C.A. 7th. 1899). Cf. Bailey v. Hull, 11 Wis. 289 (1860) (liberal interpretation of Revised Statutes, 1858, c. 153, sect. 12, regarding labor on land, to include building a fence); Paine v. Gill, cited above (implicit holding that sect. 12 gives lien for labor in running logs); see Young v. French, 35 Wis. 111, 118 (1874) (cook's labor is within sect. 12). Though the court tended to interpret suppliers' liens with more caution, in view of the obviously limited terms of the statutes, a rule of liberal construction within this framework was announced as a ground of decision in Kollock v. Parcher, 52 Wis. 393, 399, 9 N.W. 67, 69 (1881).

85. Consistent with its favor towards claimants of labor liens on logs, the court held that Revised Statutes, 1858, c. 153, sect. 12, continued to be available as a parallel, if more limited, protection to the worker after the passage of the broader acts typified by General Acts, 1860, c. 215, and that, indeed, the enforcement provisions of this latter legislation were available as more effective remedies to implement the earlier lien. Paine v. Gill, 13 Wis. 561 (1861); Paine v. Woodworth, 15 Wis. 298 (1862); Babka v. Eldred, 47 Wis. 189, 2 N.W. 102 (1879); see Shevlin v. Whelen, 41 Wis. 88, 93 (1876). Revised Statutes, 1878, sect. 4978, repealed sect. 12, but sect. 3341 continued it in substance (with the addition of bark and logs to the lienable subjects), as a residual clause giving a more limited lien for types of labor and areas of the state not within the more specifically defined lien under sect. 3329; cf. Gross v. Eiden, 53 Wis. 543 (1881). Statutes, 1898, sect. 3329, merged the lien formerly under sect. 3341 with the basic, general lien.

In interpreting the lien for "services" to embrace supervision of the operations of a logging crew, Hogan v. Cushing, 49 Wis. 169, 5 N.W. 490 (1880) relied not only on the breadth of the key word of the statute, but also on the fact that the legislature said that its lien was given to any company or corporation, as well as any person (which "must necessarily" include labor or service done by others than the claimant), as well as on common law doctrine ruling that work done by an agent might be included in the common count in assumpsit. Accord: Glover v. Hynes Lumber Co., 94 Wis. 457, 69 N.W. 62 (1896), s.c., 101 Wis. 279, 77 N.W. 1119 (1898) (mill foreman). In the Hogan case "the liberal construction properly accorded to the log lien statutes was given full effect," the court later observed, in Edwards v. H. B. Waite Lumber Co., 108 Wis. 164, 165, 84 N.W. 150 (1900). Hogan v. Cushing represented the law throughout the period of the timber industry's importance. Laws, 1919, c. 484, and Laws, 1925, c. 25, revised the basic lien, as then stated in Statutes, sect. 289.18 (the old sect. 3329), to confer the lien upon "any person who shall, by himself or by his beast of burden or by his tractor or similar motor vehicle, do or perform any labor or services . . ." The record on its face carries the inference that this statutory change was intended simply to overrule the cases which had properly interpreted the prior statutes not to grant a lien for the rental of draft animals. See Lohman, v. Peterson, 87 Wis. 227 (1894), and related cases hereafter noted. But in John v. Flanner Co., 211 Wis. 424, 248 N.W. 436 (1933), on a purely textual analysis, ignoring this sequence of decision and statute, the court held that a logging contractor might no longer claim a lien for work done by his crew. It is difficult to believe that this interpretation would have been permitted to stand, had it been given in a day when the logging contractor played the central role he did in the nineteenth century.

The general liberality in application of the terms "labor or service" may be seen in the holdings in Winslow v. Urquhart, 39 Wis. 260 (1875), s.c., 44 Wis. 197 (1878) (cook); Minton v. Underwood Lumber Co., 79 Wis. 646, 48 N.W. 857 (1891) (clean-up after drive); Kennedy v. South Shore Lumber Co., 102 Wis. 284, 78 N.W. 567 (1899) (contract scaler); Engi v. Hardell, 123 Wis. 407, 100 N.W. 1046 (1905) (repairs incident to sawmill's current operation). That a lien would obtain for the labor of the claimant with his team was not directly ruled by the Supreme Court, but appears as clear policy from comparison of Kuntz v. Kinney, 33 Wis. 510 (1873) and the opinion in Lohman v. Peterson, 87 Wis. 227 (1894) (the ox case). Cf. Edwards v. H. B. Waite Lumber Co., 108 Wis. 164 (1900) (no lien for rent of horses); St. Croix Timber Co. v. Joseph, 142 Wis. 55 (1910) (similar). The first inclusion of sawing or manufacturing logs or timber into lumber within the broad lien under sect. 3329 was created by Laws, 1879, c. 167. The considerable lag in this further expansion of the scope of the broad log-labor liens might seem to present some qualification on the general record of legislative liberality, especially in view of the failure of passage of bill 44S in 1869, bill 55S in 1870, and bill 104S in 1877; the first and third would have included the manufacture of lumber, and the second all activity enhancing the value of forest products. This appearance of the record is much tempered, however, by the fact that the manual laborer in the sawmill was covered by sect. 12, c. 120, Revised Statutes, 1849, and c. 153, Revised Statutes, 1858, and after its repeal, by Revised Statutes, 1878, sect. 3341. Cf. Paine v. Gill, 13 Wis. 561 (1861) (where the contract vendor of logs was to deliver by running the logs to a named point, the workers on the drive acted "for or on account of" the contract vendee of the logs, within sect. 12, since the vendor must be deemed agent of the vendee to employ the necessary labor); Paine v. Woodworth, 15 Wis. 298 (1862) (accord); Battis v. Hamlin, 22 Wis. 669 (1868) (sect. 12 lien to employees of shingle maker, as against vendee); Babka v. Eldred, 47 Wis. 189, 2 N.W. 102 (1879) (similar, under sect. 12, as to employees of lath maker, as against vendee); Gross v. Eiden, 53 Wis. 543, 11 N.W. 9 (1881) (shingle maker's employees within sect. 3341, as against owner of the raw material, mill, and product). In this light, prior to 1879 the interest not protected by lien was that of the sawmill operator or owner. Any lag in providing for him may be part of the policy story hereafter related concerning suppliers' liens, though once he was included the sawmill operator continued to hold his position while the supplier found his advantage a precarious one. Laws, 1876, c. 372, which applied only to Chippewa County, may have covered all sawmill activity by its blanket inclusion of "any other labor or services performed upon any logs, lumber or timber" in that county in addition to the conventional formula; however, this was shortly repealed, by Revised Statutes, 1878, sect. 4978. The inclusion of sawmill labor or service through Laws, 1879, c. 167, was continued by Laws, 1881, c. 330, 1882, c. 319, and 1885, c. 469; carried into Statutes, 1898, sect. 3329, the provision was clarified (rather than enlarged, since the prior language would seem broad enough to cover the point) by an additional stipulation that the lien should lie for manufacturing logs or timber "into timber" as well as into lumber. Cf. Kollock v. Parcher, 52 Wis. 393, 399, 9 N.W. 67, 69 (1881) ("timber" means "stems or trunks of trees when cut and shaped for use in the erection of buildings or other structures, and not to be manufactured into lumber within the ordinary meaning of the word 'lumber'"). The court was plainly right in ruling that the legislature's generosity stopped at the sawmill, so far as the special log-lien statutes were concerned, and that these statutes thus did not embrace the hauling of lumber after the milling or manufacture of further products out of the milled lumber: Bernhardt v. Rice, 98 Wis. 578, 74 N.W. 370 (1898); McGeorge v. Stanton-DeLong Lumber Co., 131 Wis. 7, 110 N.W. 788 (1907). The rulings that the log lien did not lie for work on permanent installations not proximately related to current lumbering operations include Glover v. Hynes Lumber Co., 94 Wis. 457, 69 N.W. 62 (1896), s.c., 101 Wis. 279, 77 N.W. 1119 (1898); Kendall v. Hynes Lumber Co., 96 Wis. 659, 71 N.W. 1039 (1897); Carpenter v. McCord Lumber Co., 107 Wis. 611, 83 N.W. 764 (1900).

It was in line with the legislature's combination of generosity and clumsy execution that it responded piecemeal to the growth in diversity of forest-products manufacture by adding liens of varying, but broadening, scope based on work performed on such secondary products as telegraph poles, cordwood, posts, ties, and bark. Sometimes these liens were limited to the scope of sect. 3341 (i.e., manual labor, performed "for or on account of" the owner); sometimes they were merged into the broad basic lien of the sect. 3329 type. See Revised Statutes, 1878, sect. 3341; Laws, 1881, c. 330, 1882, c. 319, 1885, c. 469, 1887, c. 530, 1889, cc. 145, 413. Statutes, 1898, sect. 3329, as amended by Laws, 1913, c. 241, finally incorporated all the recognized forest-products liens into its formula. See also 1882, bill 29A; 1883, bill 18A. Most of this verbal tinkering was probably unnecessary, insofar as the more general liens covered labor or service in the manufacture of logs or timber into lumber. See Babka v. Eldred, 47 Wis. 189, 192, 2 N.W. 102, 104 (1879); Gross v. Eiden, 53 Wis. 543, 547, 11 N.W. 9, 10 (1881).

A number of minor variations point up the needless complexity introduced by loose legislative practice. Thus P&LL, 1872, c. 71, omitted the "running" of logs from the list of lien-creating activities, in amending General Acts, 1860, c. 215 — probably by inadvertence, since the term continued in similar acts relating to other areas, and was included in the all-embracing terms of Revised Statutes, 1878, sect. 3329. Again, Laws, 1876, c. 372, amended the lien created for Chippewa County by General Acts, 1861, c. 186, by adding to the lien-creating activities the hauling of supplies for men or teams, as well as an omnibus grant of the lien for "any other labor or service performed" apart from the standard list; Laws, 1877, c. 95, extended the Chippewa County lien, so amended, to Lincoln and Taylor counties; Revised Statutes, 1878, sect. 3329, then dropped the omnibus clause, but kept the supplies-hauling clause for Chippewa and Taylor, though not for Lincoln; Laws, 1881, c. 330, then dropped the supplies-hauling clause altogether. The cook's lien, likewise, received a confusing amount of special attention. Laws, 1874, c. 267, had explicitly covered the services of any cook for men performing other lienable work under General Acts, 1860, c. 215. The court was ready to cover the cook under the general terms of Revised Statutes, 1858, c. 153, sect. 12 — see Young v. French, 35 Wis. 111, 118 (1874) — and soon held that he was included by reasonable implication as one performing labor incident to the listed lien-creating activities under General Acts, 1862, c. 154, the terms of which were the same as those of General Acts, 1860, c. 215, before its amendment in 1874: Winslow v. Urquhart, 39 Wis. 260, 268 (1875), s.c., 44 Wis. 197 (1878) ("the person who cooks the food . . . equally with those who use the axe, the saw or the team . . . are all engaged in the business of manufacturing trees into logs and timber, and transporting the same from the forest to a market; and to accomplish the common purpose the labor of each in his department is necessary"). Cf. Bradford v. Underwood Lumber Co., 80 Wis. 50, 48 N.W. 1105 (1891) (contract to furnish board not apportionable so as to allow lien for the personal services in cooking involved). This common-sense outcome by judicial construction should have sufficed, but the statutory tinkerers would not leave the situation alone. Laws, 1876, c. 372, added a lien for "cooking for laborers . . . upon logs, lumber or timber, in Chippewa county." Revised Statutes, 1878, sect. 3329, dropped any explicit reference to cooks in any area, but sects. 3329 and 3341 continued the general language under which the court had previously implied the inclusion of cooks in the Young and Winslow cases. However, Laws, 1889, c. 413 — carried into Statutes, 1898, sect. 3341 — saw fit to single out the cook's lien for special statement, granting it to "all persons doing . . . any labor . . . by cooking food for the men doing . . . any labor or services upon such logs" etc., "at the request of the person . . . employing said men, for whom such food is cooked or manufactured . . ." The language was simplified in a general revision, by Laws, 1935, c. 483, amending Statutes, sect. 289.31.

86. Details on the geographic coverage of successive "labor or services" liens will be found in note 83 above; successive acts creating liens for work on secondary forest products are cited in note 85 above. The atmosphere of arbitrary and technical limitation introduced by the area limitations may be seen in Shevlin v. Whelen, 41 Wis. 88 (1876) (Laws, 1861, c. 186, creating a log-labor lien in Chippewa County, held to confer no lien on logs part of which were cut before and part after the area of cutting was made a portion of the new Taylor County, where the new county was organized before any proceedings were taken to perfect a lien, and the logs lay all in the new county); cf. Abraham v. Agnew, 83 Wis. 246, 53 N.W. 504 (1892) (statement of geographical applicability of supplies lien under Laws, 1889, c. 413, held so confusing as to be of no legal effect). The court's effort to mitigate the arbitrariness of the geographical divisions of the statutes was expressed in Munger v. Lenroot, 32 Wis. 541, 545, 550 (1873) (Laws, 1862, c. 154, derives from Laws, 1860, c. 215, and these should be construed as acts *in pari materia*); Garland v. Hickey, 75 Wis. 178, 43 N.W. 832 (1889) (re-enactment of Laws, 1885, c. 469, in form extended to more counties, by Laws, 1887, c. 530, held effective); Cox v. North Wisconsin Lumber Co., 82 Wis. 141, 51 N.W. 1130 (1892) (embodiment of substance of Laws, 1885, c. 469, in Laws, 1889, c. 413, held a continuation and not a repeal and re-enactment of the former act, apart from changes in area application). Collins v. Cowan, 52 Wis. 634, 9 N.W. 787 (1881) held that Laws, 1862, c. 154, was not a "local bill" within Wisconsin Constitution, IV, 18 (limiting local bills to one subject, to be expressed in the title). Cf. State *ex rel.* Cothren v. Lean, 9 Wis. 279 (1859); Thompson v. City of Milwaukee, 69 Wis. 492, 34 N.W. 402 (1887).

The definition, redefinition, and repeal of geographic limits for the application of suppliers' liens added greatly to the complications which area limits introduced into log-lien law. This aspect of the matter is noticed hereafter, in the general discussion of suppliers' liens.

87. On the worker's need to realize his claims for cash, compare Fries, 207–208. Much of the wage claims which might have been pressed to liens were used up through the winter in debts incurred at the company store: *ibid.;* Larson, 183–184. The distaste of the typical small claimant for litigation or even dispute is vividly reflected in Kelly v. Schupp, 60 Wis. 76, 18 N.W. 725 (1884), as presented at note 34, Chapter IV, above.

A condition of active traffic in assignment of time checks is reflected in Bayfield County Bank v. Duluth Log Co., 141 Wis. 1, 123 N.W. 120 (1909). The first authorization of limited assignment of log-lien claims (included in a general treatment of mechanics liens), was by General Acts, 1859, c. 113. The traditional common law reluctance to extend the assignment privilege was expressed in Caldwell v. Lawrence, 10 Wis. 331 (1860) (lien created by mechanics lien act assignable only so far as specific legislative authorization runs); cf. Tewksbury v. Bronson, 48 Wis. 581, 4 N.W. 749 (1880) (claim for tolls due for use of river improvement, by statute made assignable on same terms as log-lien claims, is assignable only so far as statutory authority specifically runs, and hence, under incorporation by reference into toll act of General Acts, 1862, c. 154 — a log-lien statute — assignment held limited to assignee likewise holding claim on same property: ". . . *Caldwell* v. *Lawrence,* sustained as it is by legislative construction, has stood unchallenged too long to be now lightly overruled. If a different rule is desirable, it should be enacted by the legislature"); see Shearer v. Browne, 102 Wis. 585, 587, 78 N.W. 744 (1899).

By the date of Day v. Vinson, 78 Wis. 198, 199, 47 N.W. 269, 270 (1890), the course of legislative policy favoring assignment of contract claims generally inclined the court to a more sympathetic view of such transfers of interest, but it continued to distinguish the matter of lien assignments, though now making this a special instance: "The mechanic's claim for his labor or materials is a common credit, and is assignable; but this special remedy of lien is not assignable, and does not survive. It does not depend upon the liability being special, but upon the remedy being a special and peculiar one in such a case."

On the broader assignment authorization in 1878, see *Report of the Revisers of the Statutes,* 237. Bill 410A, 1887, would have broadened sect. 3336 to the full scope of sect. 3316; it was indefinitely postponed on committee recommendation in the Senate: SJ, 746, 763 (1887). Laws, 1891, c. 139 — which the legislature cast in the form of an independent statute and not simply an amendment or revision of the Revised Statutes — repealed the log-lien provisions of the Revised Statutes and contained in itself two very specific assignment authorizations, limited to a provision the equivalent of the old sect. 3336, and a new provision dealing only with time checks or time orders. Bank of Iron River v. Board of School Directors of the Town of Iron River, 91 Wis. 596, 599, 65 N.W. 368, 369 (1895), involving a claim under the general mechanics-lien act, indicated that so long as it was in effect, sect. 3316 should be treated as having the full effect its broad language indicated: "It is entirely certain from this that the legislature has changed the policy of the law as to the assignability of pure liens . . ." However, as to log liens Bernhardt v. Rice, 98 Wis. 578, 582, 74 N.W. 370, 372 (1898) declared, obiter, that the specific, more limited assignment authorization under Laws, 1891, c. 139, must be taken to limit and control the general terms of sect. 3316, especially since the independent character of the 1891 act made the log-lien law no longer part of c. 143 of the Revised Statutes (containing sect. 3316). This last observation by the court had very limited significance, in view of the restoration of broad assignability in Statutes, 1898, sect. 3333, but it serves to point up the significance of sect. 3333 as a reaffirmation of legislative policy to apply the more liberal assignability policy to log liens.

Where the legislature had made plain its favor toward assignments, the court showed a disposition towards its characteristic liberal treatment of claimants. So Kline v. Comstock, 67 Wis. 473, 30 N.W. 920 (1886) held, implicitly, that Revised Statutes, 1878, sect. 3336, since it did not forbid them, should be treated as allowing partial assignments, at least where the one personally liable for the work debt consented. Otherwise, so far as the broad pattern of statutory authority for assignment appeared, the court's role was limited to enforcing requirements of proof that an assignment had in fact been made, rather than a payment of the old claim, extinguishing it and creating a new cause of action against the original debtor. See Dirimple v. Dells Lumber Co., 101 Wis. 509, 78 N.W. 182 (1899); Bayfield County Bank v. Duluth Log Co., cited above. Greater judicial hospitality toward the assignable quality of log liens was reflected in this period in Dufur v. Paulson, 110 Wis. 281, 85 N.W. 965 (1901) (owner of part interest in logs in a mingled drive may take assignments of laborers' time checks and enforce liens based on their labor in the drive against other log owners), and Jackman v. Eau Claire National Bank, 125 Wis. 465, 104 N.W. 98 (1905) (assignee of laborers' log liens enjoys benefit of their status as holders of an *in praesenti* interest in the liened property, acquired by their performance of work, and hence may not be defeated by debtor's trustee in bankruptcy, as holding an unlawful preference).

88. Typical examples of enforcement of both the personal claim and the lien against a log owner who had contracted for the labor are Blackwood v. Jones, 27 Wis. 498 (1871), and Harder v. Smith, 33 Wis. 274 (1873). The distinct character of the *in personam* contract liability for work done, as compared with the remedy by lien on the goods, is emphasized by the rulings in O'Reilly v. Milwaukee & Northern Railroad Co., 68 Wis. 212, 31 N.W. 485 (1887) (failure to establish lien not prerequisite to rendition of personal judgment against person liable for the debt); Shafer v. Hogue, 70 Wis. 392, 395, 35 N.W. 928, 929 (1888) (where plaintiff has obtained jurisdiction *in personam,* he may have personal judgment for the debt, though trial court erred in giving judgment for lien, where the logs were in another county: "The proceeding under our statute serves a twofold purpose: Chiefly and primarily to enforce a lien upon logs and timber for labor done upon them, and also to obtain a personal judgment for the amount due the plaintiff"); Minton v. Underwood Lumber Co., 79 Wis. 646, 48 N.W. 857 (1891) (record books of one personally liable for the work debt not admissible to prove basis of lien against another, the log owner). See, further, the cases developing the *in rem* character of the log lien, note 90 below.

89. The quoted characterizations of the limited scope of the sect. 12 and the sect. 3341 liens are adapted from Babka v. Eldred, 47

Wis. 189, 191, 2 N.W. 102, 103 (1879), and Gross v. Eiden, 53 Wis. 543, 547, 11 N.W. 9, 10 (1881); the holdings in both cases were that the claimant had the burden of proving that his work was done "for or on account of" the owner, and that he had failed to sustain his burden. In Gross v. Eiden the court expressly disavowed a dictum in Babka v. Eldred that sect. 3341 might cover a laborer hired by a subcontractor, and perhaps implicitly disavowed a dictum in Paine v. Gill, 13 Wis. 561, 564 (1861) that if title could be construed to pass upon the making of a contract for sale of logs for future delivery by the vendor, laborers hired by the vendor to drive the logs to the delivery point would be employed by the vendor as the vendee's agent, so as to give them a sect. 12 lien against the vendee. Cf. Paine v. Woodworth, 15 Wis. 298 (1862) (title on similar facts construed as not passing until delivery).

Shortly after the Babka and Gross decisions had underlined the limited scope of sect. 3341, bill 29A, 1882, was proposed, striking from that provision the "for or on account of" limitation. At this point in time the proposal did not appeal to the legislature, for the bill was indefinitely postponed in the chamber in which it was introduced, on recommendation of the Assembly Judiciary Committee: AJ, 43, 422, 442 (1882). As was pointed out in note 85, above, the limited formula of sect. 3341 was given expanded life in some of the statutes which added specific liens for labor or service on secondary forest products (see Laws, 1889, c. 413; cf. Laws, 1895, c. 72), until these were merged under the one, broader formula of sect. 3329, Statutes, 1898.

90. The existence of the lien under the 1860-style statute, though no contract relation existed between claimant and property owner, was declared or implicitly ruled in Munger v. Lenroot, 32 Wis. 541 (1873); Winslow v. Urquhart, 39 Wis. 260 (1875), s.c., 44 Wis. 197 (1878); Collins v. Cowan, 52 Wis. 634, 9 N.W. 787 (1881); Kline v. Comstock, 67 Wis. 473, 30 N.W. 920 (1886); Shearer v. Davis & Starr Lumber Co., 78 Wis. 278, 47 N.W. 360 (1890); Minton v. Underwood Lumber Co., 79 Wis. 646, 48 N.W. 857 (1891); Trevitt v. Heinemann, 80 Wis. 1, 49 N.W. 109 (1891); see Union Lumbering Co. v. Tronson, 36 Wis. 126, 130 (1874); Babka v. Eldred, 47 Wis. 189, 191, 2 N.W. 102, 103 (1879); Hemenway v. Connor Lumber & Land Co., 175 Wis. 443, 185 N.W. 524 (1921). Munger v. Lenroot (above) ruled specifically that the lien stood, though the contractor had not performed his contract with the property owner. Collins v. Cowan (above) ruled against the misjoinder objection (see, especially, 52 Wis. 634, 637, 9 N.W. 787, 788); cf. Reindl v. Heath, 109 Wis. 470, 85 N.W. 495 (1901). Cox v. North Wisconsin Lumber Co., 82 Wis. 141, 51 N.W. 1130 (1892) ruled on availability of the lien proceeding as the basis for suit against a nonresident. Viles v. Green, 91 Wis. 217, 64 N.W. 856 (1895) bound the administrator. Kline v. Comstock (above) sustained the partial assignment; on the claimant's interest as a degree of *in praesenti* title, see Jackman v. Eau Claire National Bank, 125 Wis. 465, 104 N.W. 98 (1905). The subjection of any part of a lot of logs to a lien for the whole sum due for labor on the lot was ruled in the quoted case, DeMorris v. Wilbur Lumber Co., 98 Wis. 465, 471, 74 N.W. 105, 107 (1898), and in other cases noted hereafter in the discussion of procedure. DeMorris is the ruling particularly relevant at this point because it is cast more in terms of the substance of the lien, whereas other cases emphasized rather the problems of proof. Trevitt v. Heinemann (above) held that the claimant was not bound first to pursue his claim against the one personally liable for the work debt, before enforcing his lien.

Certain decisions qualified the position of the lien claimant as it appears in the doctrine above set out, but in ways which seem so clearly equitable and consistent with the logic of the lien as in no degree to diverge from the court's general liberality toward the lien holder. Losie v. Underwood Lumber Co., 79 Wis. 631, 48 N.W. 858 (1891) held that the claimant might not assert a lien against defendant's logs for services in driving others' logs downstream to the point where they became inextricably mingled with defendant's, though plaintiff might have his lien for driving services beyond that point, for the labor rendered in the mingled drive: accord, Minton v. Underwood Lumber Co. (above). The ruling merely bore out the elementary logic of the insistence that it was the work done that made the lien; likewise in obedience to that logic was the ruling in the Minton case (see also note 88 above) that record books of the one contracting for labor of men in a log drive were hearsay and inadmissible as a basis for asserting a lien against another who was general owner of the logs. St. Croix Timber Co. v. Joseph, 142 Wis. 55, 124 N.W. 1049 (1910) treated lien rights within a framework of familiar equity doctrine; since the log owner's property was subject to lien though he was not personally liable for the labor claims and though he may have paid off the contractor, he stood in the position of a "real" surety for the contractor, entitled to ordinary equity protection of his surety's position, and so might enjoin separate lien proceedings and have determined in one suit in equity all questions of liability among the parties, and particularly whether the lien claimants owed the contractor such sums for goods furnished them as to create equitable setoffs sufficient to extinguish the basis for liens against petitioner's logs. The court displayed characteristic sensitivity to the rights of the lien claimants, as it noted (1) that the log owner, to obtain such equitable relief, must admit the validity of the contractor's liability to the claimants, since equity would not allow him to seek the benefit of inconsistent arguments, and (2) that if any net sums were found due the lien claimants, they might have foreclosure of their liens therefor in the instant proceeding: 142 Wis. 55, 62, 63, 124 N.W. 1049, 1052, 1053.

The legislature, as we shall note further hereafter, tempered the rigor of the *in rem* operation of the log lien by enlarging provisions for notice to persons having interests in the liened goods. Such provisions, however, did not qualify the substantive reach of the lien, and the court ruled in Munger v. Lenroot (above) that due process of law did not require such notice provisions, so long as there was opportunity for interested persons by separate suit to challenge the basis of asserted liens.

91. The priority provision relevant to the first log-labor lien was contained in sect. 1, c. 120, Revised Statutes, 1849, and sect. 1, c. 153, Revised Statutes, 1858. The preferential position of this first log lien was extended by interpretation in Paine v. Gill, 13 Wis. 561, 563 (1861), and Paine v. Woodworth, 15 Wis. 298, 303 (1862). This liberality stood out in contrast to the contemporary ruling that, under the same act, a general mechanics lien was subordinate to a prior mortgage on the realty: Jessup v. Stone, 13 Wis. 466 (1861). The preference for the type of lien created by General Acts, 1860, c. 215, was shown in Halpin v. Hall, 42 Wis. 176 (1877) (attaching general creditor subordinated); DeMorris v. Wilbur Lumber Co., 98 Wis. 465, 74 N.W. 105 (1898) (log-labor lien proceeding is at law and not in equity, hence equitable offsets do not apply, and there is no basis for a judgment declaring the attached property free of lien to the extent of value added by bona fide purchaser's payment of transport charges to bring the liened goods to market); Jackman v. Eau Claire National Bank, 125 Wis. 465, 104 N.W. 98 (1905) (log lien prevails over trustee in bankruptcy). By implication from the statute, and probably also from familiar equity principle, among competing labor liens asserted against the same logs, priority was given according to the time when the respective lien claims were filed: Halpin v. Hall (above); St. Croix Timber Co. v. Joseph, 142 Wis. 55, 124 N.W. 1049 (1910).

92. The court upheld the priority of the bona fide purchaser in Smith v. Shell Lake Lumber Co., 68 Wis. 89, 93–94, 96–97, 31 N.W. 694, 696, 697, 698 (1887), and in Johnson v. Iron Belt Mining Co., 78 Wis. 159, 47 N.W. 363 (1890); see Taylor, J., dissenting in the first case, 68 Wis. 89, 101, 104, 105, 108, 31 N.W. 694, 700, 701, 703. The severity of the ruling was tempered by the majority's dictum — rather fuzzily put — that one who bought while work was actually being done on the liened goods would, by analogy from the common law lien's stress on the critical importance of the claimant's possession, be held subject to constructive notice that a claim was pending: 68 Wis. 89, 95, 31 N.W. 694, 697; Taylor, J., dissenting, thought that the record showed this to have been the situation in the instant case, and nothing in the majority opinion answers him. See 68 Wis. 89, 100, 101, 31 N.W. 694, 699; cf. Vilas v. McDonough Manufacturing Co., 91 Wis. 607, 65 N.W. 488 (1895), and Prince v. C. G. Bretting Manufacturing Co., 203 Wis. 504, 234 N.W. 699 (1931) (unfinished state of sawmill is constructive notice to mortgagees thereof, subordinating them to general mechanics lien for mill machinery thereafter furnished).

93. With typical clumsiness the legislature required two measures to state clearly its overruling of the Shell Lake Lumber Co. decision: Laws, 1889, cc. 413 and 454. The unqualified preference thereafter established by Laws, 1891, c. 139, was applied without debate in DeMorris v. Wilbur Lumber Co.,

98 Wis. 465, 74 N.W. 105 (1898), and Oconto Land Co. v. Wallschlaeger, 155 Wis. 418, 144 N.W. 979 (1914). Strict application of the filing time requirements, out of care for intervening purchasers or incumbrancers, whether with notice or not, may be seen in Fish Creek Boom & Log Driving Co. v. First National Bank of Ashland, 80 Wis. 630, 633, 634, 50 N.W. 585, 586 (1891) (must strictly enforce statutory requirement that labor be continuous, to forestall required filing time, for otherwise "there would surely be no safety in dealing with property"); Keystone Lumber Co. v. First National Bank of Ashland, 80 Wis. 634, 50 N.W. 586 (1891); McGinley v. Laycock, 94 Wis. 205, 68 N.W. 871 (1896) (successful contestant was a purchaser with notice); Kendall v. Hynes Lumber Co., 96 Wis. 659, 663, 71 N.W. 1039, 1041 (1897) (the three different allowed filing times constitute a classification which was "obviously made so as to fairly protect the laborers without unnecessarily inflicting loss and damage upon other interests — particularly the interests of those who might deal with the property in the regular course of trade and commerce. To that end, considering that work done from the 1st day of November to the 1st day of May following is ordinarily in logging operations, and is continuous, so that such work itself reasonably furnishes notice of the existence of the lien, and that during such period business transactions in regard to transfers of the property are ordinarily not active, the laborer is given till the 1st day of June following to file his claim for a lien therefor; and considering that from the 1st day of May to the 1st day of November following, which includes the manufacturing season, when the manufactured product is ordinarily moved within short periods, and disposed of to consumers in the due course of trade and commerce, thirty days from the completion of the work is limited for filing the claim; and considering that continuity of work reasonably furnished notice of the existence of a lien for work done, commencing on the 1st day of November, or a day prior thereto, and continuing to a day after the 1st day of May following, the laborer is given thirty days after the completion of such work to file his lien petition"); DeMorris v. Wilbur Lumber Co,. 98 Wis. 465, 74 N.W. 105 (1898). On the other hand, it should be noted that with commendable restraint of its zeal to protect intervening purchasers the court very properly declared, obiter, in Winslow v. Urquhart, 39 Wis. 260, 269 (1875), that the provision under the lumber-inspector acts requiring recording with that official of title instruments to logs should not be construed to require the recording with him of liens under the broad log-lien act.

The discussion hereafter of the development of the procedural incidents of the log liens points to an analogous legislative preference of the lien over the court's concern for market dealings, in the case-statute sequence regarding enforcement of the lien on the manufactured products of logs or timber. The legislature's disposition to enlarge the rights of owners to intervene in log-lien proceedings, also noted hereafter, seems no qualification upon the substantive value judgments in favor of lien claimants; the intervenors might assert no more interest than the statutory preference for the log lien allowed them. A gesture was made toward the intervening purchaser, by a provision in Laws, 1891, c. 139, declaring it a misdemeanor to sell or encumber property on which a lien existed, knowing the lien to be unsatisfied, without so informing the purchaser or incumbrancer. The 1891 act balanced this, however, by declaring liable for conversion to the extent of the claim any person who should cause property subject to lien under that statute to be carried out of state, hidden, destroyed, or changed in character, with the result of preventing the claimant from subjecting the goods to his lien. Statutes, 1898, sect. 3336, broadened this to include sale or encumbrance preventing subjection of the goods to the lien. Both forms of the provision were ambiguous as to whether a plaintiff thereunder must show in the defendant an intent to produce the forbidden result. It was perhaps indicative that the court had not taken full leave of the policy convictions expressed in the Shell Lake Lumber Co. case, that it held sect. 3336 unconstitutional as taking property without due process of law, without making any effort to save it by construing it as limited by a requirement of proof of guilty intent: Rogers-Ruger Co. v. Murray, 115 Wis. 267, 91 N.W. 657 (1902). Cf. Lohman v. Peterson, 87 Wis. 227, 58 N.W. 407 (1894) (1891 conversion provision "is penal in its nature, and requires a strict construction"; held, where general owner of railroad ties sold them to a railroad, the vendee's removal of them from the state thereafter was not action by, or caused by, the general owner so as to make him liable for a "conversion" under this act).

94. The basic procedural pattern under the log-lien legislation was set by General Acts, 1860, c. 215. The notice-of-intent provision was created by Laws, 1889, cc. 413 and 454, and dropped by Laws, 1891, c. 139.

The doing of the work brought the lien into being: Paine v. Woodworth, 15 Wis. 298 (1862) (sect. 12, c. 153, Revised Statutes, 1858); Munger v. Lenroot, 32 Wis. 541 (1873) (General Acts, 1862, c. 154, following the 1860 model act); Collins v. Cowan, 52 Wis. 634 (1881) (model act as embodied in Revised Statutes, 1878, sect. 3329); Viles v. Green, 91 Wis. 217, 64 N.W. 856 (1895) (Laws, 1889, c. 454); cf. Jackman v. Eau Claire National Bank, 125 Wis. 465, 104 N.W. 98 (1905) (performance of labor still treated as the operative fact creating lien, after statutory phrasing had been changed by Laws, 1891, c. 139, and Statutes, 1898, sect. 3330, to provide that no claim should "become" a lien until filed).

The quoted comment on the policy back of extension of jurisdiction over liens to justices of the peace is from Paine v. Gill, 13 Wis. 561, 563 (1861). Cf. Cuer v. Ross, 49 Wis. 652, 6 N.W. 331 (1880); Johnson v. Iron Belt Mining Co., 78 Wis. 159, 47 N.W. 363 (1890). The shortlived requirement that the lien claimant join as defendants (and by inference, serve notice on) the one personally liable for the work debt and the owner of the logs, if known, or — if these be unknown — the person whose mark was on the logs, was set by Laws, 1873, c. 139 (amending General Acts, 1862, c. 154 — applicable only in the early group of northwestern counties), and by Laws, 1874, c. 161 (amending General Acts, 1861, c. 186 — applicable only to Chippewa County); the latter provision was repealed by Laws, 1876, c. 372, and the former by Revised Statutes, 1878, sects. 3332 and 4078. Cf. General Laws, 1869, c. 144 (amending General Acts, 1860, c. 215 — applicable only in the first group of northeastern counties — to require notice of the filing of the lien petition to the log owner, if known, but not requiring that he be made a party to the suit, though permitting him to intervene if he chose).

The development of the permission to intervene, given to persons asserting ownership interests in liened forest products, may be seen in General Laws, 1869, c. 144 (amending General Acts, 160, c. 215 — applicable only in the first group of northeastern counties — and limited to persons claiming to be "the owner"); Laws, 1876, c. 372 (amending General Acts, 1861, c. 186 — applicable only to Chippewa County — permitting intervention by any person "owning an interest" in the liened property); Revised Statutes, 1878, sects. 3332 and 4978 (supplanting prior provisions with a general permission to intervene, given to "any person claiming to be the owner," at any time before final judgment); Laws, 1882, c. 273 (right to intervene in action under any log-lien act, any time before or after judgment and before actual sale, given to anyone making affidavit that he is "the owner of or of some interest" in the property, and believes himself to have a good defense, with the right to defend or appeal); Laws, 1889, c. 413 (continuing the 1882 provision, with addition of a limit of one year after judgment for the filing of an affidavit and taking of an appeal by such an intervenor); Laws, 1891, c. 139 (tacitly continuing the 1882, 1889 provisions); Statutes, 1898, sects. 3340a, 3340b.

On relief of the lien claimant from ordinary attachment bond requirements (General Acts, 1860, c. 215, sect. 7), see Halpin v. Hall, 42 Wis. 176, 185 (1877) (claimant is relieved only of posting bond for benefit of lien defendant; on proper demand, sheriff may still exact from him security under Revised Statutes, 1858, c. 130, sects. 5, 58, where sheriff has reason to believe he may not have authority to obey the mandate of an execution). The constructive levy provisions were by Laws, 1880, c. 222, and 1882, c. 319 (Chippewa River area); Laws, 1889, c. 378 (generalized); Statutes, 1898, sect. 3342b. See Shafer v. Hogue, 70 Wis. 392, 394, 35 N.W. 928, 929 (1888) (effect of Laws, 1880, c. 222, is to authorize sheriff "to make a valid constructive levy upon logs in his county . . . without the necessity of an actual view of the property. Ordinarily he would have to make a levy upon an actual view and taking possession").

The lien claimant must plead and prove the elements of his lien under the statute: Blackwood v. Jones, 27 Wis. 498 (1871); McKenzie v. Peck, 74 Wis. 208, 42 N.W. 247 (1889); Glover v. Hynes Lumber Co., 94 Wis. 457, 69 N.W. 62 (1896), s.c., 101 Wis. 279, 77 N.W. 1119 (1898); McGeorge v. Stanton-DeLong Lumber Co., 131 Wis. 7, 110 N.W. 788 (1907); see Waite v. Anderson, 152 Wis. 206, 139 N.W. 738 (1913); Hemenway v. Connor Lumber & Land Co., 175 Wis. 443, 185 N.W. 524 (1921). But, though auxiliary procedural provisions of the general statutes might apply, the log-lien acts created a self-contained right and remedy, and at basis a claimant need plead or prove no more than the key elements set out therein: Winslow v. Urquhart, 39 Wis. 260 (1875), s.c., 44 Wis. 197 (1878); cf. Wheeler v. McDill, 51 Wis. 356, 8 N.W. 169 (1881) (general statutory authorization to sheriff to release a levy on suitable security); O'Reilly v. Milwaukee & Northern Railroad Co., 68 Wis. 212, 31 N.W. 485 (1887) (garnishment available as ancillary remedy to lien claimant); Shearer v. Davis & Starr Lumber Co., 78 Wis. 278, 47 N.W. 360 (1890), and Cox v. North Wisconsin Lumber Co., 82 Wis. 141, 51 N.W. 1130 (1892) (service by publication procedure available to lien claimant); Rayson v. Horton, 90 Wis. 367, 63 N.W. 278 (1895), and Blonde v. Menominee Bay Shore Lumber Co., 103 Wis. 284, 79 N.W. 226 (1899), s.c., 106 Wis. 540, 82 N.W. 552 (1900) (venue). Halpin v. Hall (above) ruled that several attachments might be had against the same logs. That the claimant need not prove that the particular logs attached were ones on which he had worked was held in Jacubek v. Hewitt, 61 Wis. 96, 20 N.W. 372 (1884); Johnson v. Iron Belt Mining Co., 78 Wis. 159, 47 N.W. 363 (1890); DeMorris v. Wilbur Lumber Co., 98 Wis. 465, 74 N.W. 105 (1898); Blonde v. Menominee Bay Shore Lumber Co., 103 Wis. 284, 79 N.W. 226 (1899), s.c., 106 Wis. 540, 82 N.W. 552 (1900); Engi v. Hardell, 123 Wis. 407, 100 N.W. 1046 (1905).

95. The lien under sect. 12, c. 153, Revised Statutes, 1858, might be enforced against the manufactured product of timber: Battis v. Hamlin, 22 Wis. 669 (1868). Compare Babka v. Eldred, 47 Wis. 189, 192, 2 N.W. 102, 103 (1879); Gross v. Eiden, 53 Wis. 543, 11 N.W. 9 (1881) (Revised Statutes, 1878,, sect. 3341). The quotation from Babka v. Eldred will be found at 47 Wis. 189, 191, 2 N.W. 102, 103. The quoted language from Gross v. Eiden will be found at 53 Wis. 543, 548, 11 N.W. 9, 11; the legislation involved includes Laws, 1879, c. 167, 1880, c. 62, 1881, c. 330. Glover v. Hynes Lumber Co., 94 Wis. 457, 69 N.W. 62 (1896), s.c., 101 Wis. 279, 77 N.W. 1119 (1898) shows routine enforcement of the lien on lumber after the change begun by Laws, 1882, c. 319.

96. The successive suppliers'-lien acts will be found in their appropriate order in note 83 above. Whatever the particular interest group conflicts in the background, the court, while recognizing the ups and downs of this legislation, was ready to accept it as "just and equal" and resting on the same policy against unjust enrichment represented in the log-labor lien. See Patten v. Northwestern Lumber Co., 73 Wis. 233, 236, 238, 41 N.W. 82, 83 (1888). Indeed, so far as they set a plain framework, the court would apply to supplies-lien acts the kind of liberal construction it gave the labor-lien laws. Kollock v. Parcher, 52 Wis. 393, 399, 9 N.W. 67, 69–70 (1881); cf. Stacy v. Bryant, 73 Wis. 14, 40 N.W. 632 (1888); Patten v. Northwestern Lumber Co. (above); Johnson v. Iron Belt Mining Co., 78 Wis. 159, 47 N.W. 363 (1890). But the court's patience ran out when it confronted the condition of supplies-lien legislation left finally by Laws, 1889, c. 413, the terms of which it was provoked to declare so "confused and uncertain" as to create no enforceable right: Abraham v. Agnew, 83 Wis. 246, 249, 53 N.W. 504, 505 (1892); cf. Garland v. Hickey, 75 Wis. 178, 43 N.W. 832 (1889) (incorporation by reference given effect); Bradford v. Underwood Lumber Co., 80 Wis. 50, 48 N.W. 1105 (1891) (only meaningful interpretation of Laws, 1889, c. 413, is that it denies all supplies liens). The trial court in Garland v. Hickey (Case, 23; Brief for Appellant, 5) had held the statute there in issue to be void for uncertainty. Though not lacking in clumsy complexity, log-labor-lien legislation was never denied effect on such grounds in the Supreme Court, though the circuit judge for Chippewa County declared in exasperation that he could "not understand . . . nor reconcile . . . nor even guess at [the] . . . meaning" of the venue provisions of these laws: see Brief for Appellant, p. 6, in Shafer v. Hogue, 70 Wis. 392, 35 N.W. 928 (1888).

Partly, perhaps, because of the local origin of the diverse interests concerned with suppliers'-lien legislation, the conflicts over it have left scant record outside the statutes and cases. Two items are noteworthy, after suitable discount for their partisan origin. The Brief for Appellant, the lien contestant, in Patten v. Northwestern Lumber Co., (above), pages 16–17, observed that "these acts have been asked for by, and granted to, small dealers in the several counties, as a matter of legislative history," and argued hence that they should be construed to give a lien only to suppliers located in the area of application; without denying counsel's history, the court held that it was the place in which the supplies were delivered for use, rather than the supplier's home base, that determined application of the statute. Respondent's brief in the same case, page 6, suggested that the localism of these statutes reflected "the increased difficulties attendant upon putting in logs in such counties," and that the acts hence were passed "as an inducement thereto." Petition 183A, supporting bill 51A of 1893, advanced the analysis of big jobber versus little jobber. Signed by forty-eight persons of Oconto county (nineteen of whom identified themselves as "farmers," twenty-two as "laborers," six as "jobbers," and one as "merchant"), the petition declared, "The object of this petition is to prevent ordinary jobbers from being compelled to buy their supplies of wealthy lumber concerns at such high prices as such wealthy concerns see fit to impose. Without a lien for supplies the ordinary jobbers in Northern Wisconsin are compelled to contract with the wealthy lumber concerns and take such prices as such wealthy concerns see fit to pay, and are compelled to take supplies of wealthy lumber concerns at such prices as the wealthy concerns see fit to impose. While on the other hand with a lien for supplies the ordinary jobbers can contract wherever they see fit for supplies and sell their product where they see fit unembarrassed by iron-clad contracts made by the wealthy and forced on the poor." This, and petitions 41A and 42A, of fifty-nine residents of Bayfield, all identifying themselves as local merchants, may be found in the Wisconsin State Historical Society, Secretary of State: General Records: Wisconsin Legislative Papers: 1893, Series 2/2/2, Box 114. Bill 51A, 1893, was indefinitely postponed on recommendation of the Assembly Committee on Lumber and Mining: AJ, 63, 110, 208, 237, 325, 702, 733 (1893); cf. SJ 91, 96, 124, 130.

97. On the smooth passage of General Acts, 1860, c. 215, see AJ, 796 (1860); SJ, 693. Philetus Sawyer's committee report favoring General Acts, 1861, c. 186, will be found in AJ, 521 (1861). The over-all nature — or absence — of protective labor legislation in the timber industry is noted hereafter, in discussion of the law's concern with the general balance of interest groups in the timber industry. That the woods worker's bargaining weakness and ill-protected situation was not peculiar to the Wisconsin industry, see Vernon H. Jensen, *Lumber and Labor* (1945), 58; Lewis C. Reimann, *When Pine Was King* (1952), 65–68.

Comparison with earlier or contemporary log-lien legislation in other states indicates that, though it found the general concept an already familiar one, Wisconsin hammered out a good deal that was original in shaping the liberality of its lien laws. Hogan v. Cushing, 49 Wis. 169, 170, 5 N.W. 490, 491 (1880) thus noted that Wisconsin went much beyond the scope of Maine legislation, in providing a lien for "services" as well as personal labor. Michigan's first comparable log-lien act was Michigan Session Laws, 1873, c. 185; cf. Smith v. Shell Lake Lumber Co., 68 Wis. 89, 97, 31 N.W. 694, 697 (1887), and, in that case, Brief for Appellant, p. 9, and Brief for Respondent, p. 14.

There is at least symbolic significance in the fact that bill 571A, which became General Acts, 1860, c. 215, was introduced by Gabriel Bouck, an Oshkosh lawyer who was a Democrat of Jacksonian flavor; see Current, 40, 41. It was not out of character for Bouck's political opponent, Philetus Sawyer, to sign a report favoring extension of the 1860 lien law, however. Though he was a leader in an organization of Oshkosh lumbermen to combat a strike by the city's mill workers, on the whole, Sawyer, to his biographer, "appears to have been a well-meaning and somewhat paternalistic employer," whose genuinely unpretentious social democracy was part of his political appeal even as a United States Senator (*id.*, 28, 29, 179, 245). The quoted observation by the court on the policy of the debtor's exemption laws is from Kuntz v. Kinney, 33 Wis. 510, 514 (1873). Cf. Laws, 1883, c. 349 (contract preference for wages of laborers earned within six months before any assignment for the benefit of creditors, excepted from ban on preferences); Laws, 1885, c. 48 (preference by law under every voluntary assignment for benefit of creditors, for all wage claims owing from assignor for services performed within three months preceding assignment); Lang v. Simmons, 64 Wis. 525, 25 N.W. 650 (1885); Coblee Lumber Co. v. Ripon Lumber & Manufacturing Co., 66 Wis. 481, 29 N.W. 285 (1886). The quoted comment on the humanitarian policy of the log-lien legislation is from Carpenter v. McCord Lumber Co., 107 Wis. 611, 614, 83 N.W. 764 (1900).

The Wisconsin court drew on Chancellor Kent's statements of the grounds for laborers' liens, in Paine v. Woodworth, 15 Wis. 298, 303 (1862), and Arians v. Brickley, 65 Wis. 26, 28–29, 26 N.W. 188, 189 (1885). The quoted recognition of the high proportion of value contributed by labor to the worth of logs and lumber is from Winslow v. Urquhart, 44 Wis. 197, 200 (1878); the quoted remarks of counsel are from the Brief for Appellant, p. 19, in Haven v. Markstrum, 67 Wis. 493, 30 N.W. 720 (1881). Paging briefs of counsel in log-lien cases before the Supreme Court, one is struck by the ease with which most legal argument flows simply within the channels of the law's own concepts; few briefs discussed underlying policy. But among those that did, it was common both to claimants' and contestants' presentations to emphasize the key policy of redressing an inequality of bargaining position in order to safeguard the claimant's equity based on his enhancement of the value of the goods on which he had worked. See, e.g., Brief for Respondent, pp. 14–15, in Clapp v. Webb, 52 Wis. 638, 9 N.W. 796 (1881); Brief for Respondent, pp. 8–9, 12, in Jacubek v. Hewitt, 61 Wis. 96, 20 N.W. 372 (1884) ("Over 18 months have passed since this plaintiff and his co-laborers, by the 'sweat of their brows', earned the wages due them for labor on these logs. Despairingly they wait while Shylock Hewitt by false swearing, trickery and chicanery, seeks to cheat them out of their earnings. Hewitt has long since converted these logs into money, and has stubbornly refused to pay these poor laborers, except on a compromise of less than 50 cents on the dollar"); Brief for Appellant, p. 14, in Smith v. Shell Lake Lumber Co., 68 Wis. 89, 31 N.W. 694 (1887); Brief for Appellant, pp. 6–7, in Shafer v. Hogue, 70 Wis. 392, 35 N.W. 928 (1888); Brief for Respondent, pp. 3, 10, in Cox v. Northwestern Lumber Co., 82 Wis. 141, 51 N.W. 1130 (1892); Brief for Appellant, p. 16, in DeMorris v. Wilbur Lumber Co., 98 Wis. 465, 74 N.W. 105 (1898); cf., accord, opinion of circuit court, Marinette County, in Blonde v. Menominee Bay Shore Lumber Co., 103 Wis. 284, 79 N.W. 226 (1889), in "Case," p. 22. The special hazards to the laborer arising from the fact that his contracts were most likely to be with financially irresponsible employers receive typical emphasis in the Brief for Respondent, p. 3, in Munger v. Lenroot, 32 Wis. 541 (1873) (purpose of lien is to secure worker his pay out of the logs, "whether it should be attempted to thwart him through the medium of impecunious contractors and rascally non-resident owners, or by means of the often employed dodge of a bill of sale and a transfer"), and in the Brief for Appellant, pp. 6, 7, in Gross v. Eiden, 53 Wis. 543, 11 N.W. 9 (1881) ("The vast business of lumbering is almost entirely done by contract involving the employment by the jobber or contractor of all the men required to do the work . . ."; hence the lien must be firm on the logs, else "the labour [*sic*] would be defeated nine times in ten; he would always be at the mercy of the state of the accounts between his employer and the owner of the property"). On the difficulties which the industry from time to time experienced in securing labor, see Merk, 108; Reynolds, 51, 52, 53; cf. Fries, 210. The use of independent contractor loggers is depicted in Reynolds, 55–56, 63–66, 141.

98. Reflections of typical private contract forms of doing business in the lumbering industry may be seen in Bigelow v. Goss, 5 Wis. 421 (1856) (logging partnership); Flanders v. Train, 13 Wis. 596 (1861) (independent logging contractor); Parcher v. Dunbar, 118 Wis. 401, 95 N.W. 370 (1903) (joint venture in timberland investment). An unincorporated joint-stock company for timberland speculation was involved in Durfee v. Stringham, 8 Wis. 1, 123 (1859): "This joint stock company, which was not incorporated, as respects its liabilities, to third persons, would be generally governed by the rules and principles of common commercial partnerships . . . [Yet] it had some features peculiar and not found in commercial partnerships. No time was fixed for the continuation of the company, and a shareholder might sell and transfer his certificates and release himself from all liabilities of the association (at least as among the stockholders), and the transferee would become a member of the association, and entitled to the same rights and privileges as an original subscriber"; cf. Shaw Livermore, *Early American Land Companies* (1939), 7–8, 215, 273. See First National Bank of Green Bay v. Goff, 31 Wis. 77, 80 (1872). General Acts, 1857, c. 97, provided for the limited partnership; it remained substantially the same, in its translation into Revised Statutes, 1858, c. 64, Revised Statutes, 1878, c. 81, and Statutes, 1898, c. 81. This statutory partnership carried its own special limitations; notably, special partners' contributions must be paid in, and paid in cash, and special partners might be required to restore payments made to them by the firm which impaired the firm's capital.

99. The story of the development of Wisconsin policy toward the corporation is told in Brown, "The Making of the Wisconsin Constitution," *Wisconsin Law Review,* 1949: 648, 676–681, and 1952: 23, 48–51; Kuehnl, 6–173; see also Theodore A. Andersen, *A Century of Banking in Wisconsin* (1954), chaps. i and ii. Fries, 146–160, relates the succession of legal battles against the Mississippi River Logging Company, none of which turned on a point of corporate status.

100. SJ, 595 (1858) (Veto Message of Mar.

6). The veto was sustained by unanimous vote, Mar. 11, 1858, SJ, 657.

101. George J. Kuehnl, in his unpublished S.J.D. thesis entitled "The Wisconsin Business Corporation," in the University of Wisconsin Law Library, 1953, later published in book form, lists in Table 11, p. 495 (a tabulation not carried into the published work) no cases before the Supreme Court of Wisconsin involving timber-industry corporations, in an every-fifth-year sampling of reported decisions through the years 1839–1869; in the sample years beyond that point he lists 4 in 1874, 9 in 1879, 3 in 1884, 6 in 1889, 5 in 1894, 19 in 1899, 14 in 1904, 6 in 1909. The tabulation is simply of cases in which a corporation was a party; the points at issue generally had nothing to do with corporate status as such. My independent tabulation of cases before the Wisconsin Supreme Court specifically involving issues of corporation law, concerning lumber-industry corporations, shows one case each in the years 1879, 1880, 1884, 1885, 1886, 1893, 1894, 1897, 1898, two in 1899, one each in 1901, 1902, 1912, 1915, 1920, two in 1922, and one each in 1927, 1933, and 1939 — for a total of twenty-one cases.

Revised Statutes, 1849, c. 51, was the first general incorporation act for business purposes in the Wisconsin area; the Wisconsin territorial legislature had provided general incorporation laws only for several classes of non-profit associations, and had enacted no special charter for a timber-industry enterprise. See Kuehnl, 18, 123. Revised Statutes, 1858, c. 73, did not mention "lumbering" companies, or, indeed, ventures in the improvement or use of streams. But, it did broadly authorize incorporation for "any other lawful business," apart from those enumerated specifically. So important a firm as the Beef Slough Manufacturing, Booming, Log Driving and Transportation Company was organized under c. 73. See Heerman v. Beef Slough Manufacturing, etc. Co., 1 Fed. 145, 147 (Circ. Ct. W.D. Wis. 1880). That a lumbering firm might be incorporated under General Acts, 1853, c. 68 — which became c. 73, Revised Statutes, 1858 — was declared by Gov. Randall, as the ground of his veto of a special charter for the Oconto Lumbering Company. SJ, 595, 596 (1858) (Message of Mar. 6), and 621 (Supplementary Message of Mar. 9). Though c. 73 denominated bodies created under it "joint stock companies," these were in all respects corporations: see First National Bank of Green Bay v. Goff, 31 Wis. 77, 80 (1872). Revised Statutes, 1878, c. 86 (derived from Laws, 1872, c. 144) again expressly included "lumbering" firms in its list.

102. The tabulation of eighty special charters in the lumber industry, on which the text rests, draws upon my own examination of the session laws, checked against the independent studies by Dykstra, "Law and the Lumber Industry," Exhibits I–IV; and Kuehnl, Tables IV and VI. There is room for reasonable difference among classifiers as to whether particular charters should be counted as primarily within the timber industry; my figure of eighty special charters so related to the industry derives from a classification deliberately more expansive than that which yielded a figure of sixty-three such charters for certain discussion points in Kuehnl, Table IV. My figure on charters under c. 73, Revised Statutes, 1858, is taken from Kuehnl, 141. The 1896 data on lumber corporations will be found in the report of the Wisconsin Bureau of Labor and Industrial Statistics, 1897–1898, PD, II, 559 (1899). Fries, 103, generalizes on the tendency of the bulk of lumber-industry invested capital to come under corporate form; the text receives further support from the roster of incorporated firms, and the accounts of their status in the industry, in the various chapters of George W. Hotchkiss, *History of the Lumber and Forest Industry of the Northwest* (1898).

103. Tabulation of the eighty special charters by years shows the following distribution:

	Waterways ventures	Lumbering firms
1850	1	
1851	1	
1852	2	
1853	2	1
1854	2	
1855	3	
1856	2	2
1857	3	6
1858		
1859	2	
1860	1	
1861		
1862	2	
1863		
1864	4	
1865	1	2
1866	5	4
1867	5	2
1868	1	4
1869	3	2
1870	7	1
1871	5	4

The consistent importance of the corporate device in aiding the raising of capital for river-traffic companies is noted in Rector, *Log Transportation,* 136, 138, 140. The brief lumber firm histories in Hotchkiss furnish many examples, beyond those given in the text, of the common transition from partnership to corporation. See, e.g., examples simply from the Chippewa Valley, *id.,* 477–483, of the transformation of the Smith and Buffinton interests into the Valley Lumber Co. (1872), of Porter, Moon & Co. into the Northwestern Lumber Co. (1873), of interests derived from John Barron & Co. into Badger State Lumber Co. (1874), of Prescott, Burdett & Co. into Dells Lumber Co. (1879), and of the N. C. Foster firm into N. C. Foster Lumber Co. (1891). The Wisconsin Supreme Court occasionally took note of the partnership origins, or closely-held character, of typical successful lumber firms. See In re Klaus, 67 Wis. 401, 408, 29 N.W. 582, 584 (1886); Will of Porter, 178 Wis. 556, 559, 190 N.W. 473, 474 (1922).

P&LL, 1866, c. 284, chartered the Eau Claire Lumber Co.; see also P&LL, 1869, c. 307, and Laws, 1880, c. 22, amending this charter to enlarge its scope; on the partnership antecedents of the corporation, see Fries, 103, n. 4, and Hotchkiss, 482. The Union Lumbering Co. was incorporated by P&LL, 1869, c. 86; see Hotchkiss, 478, and Larson, 97, 135. The corporations formed by the Shaw and the Knapp, Stout groups were organized under general laws: see Fries, 124–126; Reynolds, 14–15, 146. Earnest and detailed discussion and planning attending the Knapp, Stout Company shift from partnership to corporation are reflected in brief entries for Mar. 9, 11, 13, 14, 15, and 18, 1878, in the diary of the founder, John H. Knapp, in the Wisconsin State Historical Society (e.g., Mar. 15: "All busy at the office trying to satisfy ourselves as to the proper articles of Association and By Laws provided we decide to go into this incorporated co. and do business as a corporated co. instead of a firm as heretofore"). Various special charters reflect the likely partnership origin of the new corporations in the care taken to authorize subscribers to pay for stock in property, or in provision for the new concern to adopt and perform any existing contracts or leases. See Private & Local Acts, 1857, c. 220 (Trempealeau Lumber Manufacturing Co.); P&LL, 1867, c. 9 (Barnum Lumber Co.); P&LL, 1867, c. 403, amended, 1871, c. 37 (Northwestern Furniture Co.); P&LL, 1867, c. 454 (Mechanics Union Manufacturing Co.).

104. The general parallel between resort to incorporation and growth in the scale of individual firm operations is noted in Current, 112–113; Fries, 103; Gates, *Wisconsin Pine Lands,* 119, 123, 124, 127, 130, 132–133; Larson, 360. Rector, *Log Transportation,* 136–140, portrays the problems of raising capital for waterways ventures. Data on the number of stockholders in the Wisconsin lumbering corporations in 1896 will be found in the report of the Wisconsin Bureau of Labor and Industrial Statistics, 1897–1898, PD, II, 559 (1899). The inducements of maintaining continuity of a tight-knit business group are implicit in the examples cited in note 103 above, of transitions from partnership to corporation. See, more particularly, Reynolds, 15, and note the undertones of the correspondence attendant on dissolution of the partnership of Holt & Balcon and formation of the Holt Lumber Co., in the "Chicago Letter Books" of the Holt Lumber Co., Nov. 14, 1887–Apr. 30, 1888, in the Wisconsin State Historical Society. That the limited liability feature might have more formal than substantial importance, in view of the practice of the principal stockholders in tight-knit or medium-commitment loosely-woven ventures, in pledging their personal credit for debts of the firm, see Thomas E. Randall, *History of the Chippewa Valley* (1875), 68, 73 (H. S. Allen's difficulties, despite transformation of H. S. Allen & Co. into the Chippewa Falls Lumber Co.); Rector, 139, 140; Reynolds, 281; cf. Current, 113; Peyton v. Minong Lumber & Lath Co., 149 Wis. 66, 135 N.W. 518 (1912).

Moses Strong's hopes for his Nekoosa Lumbering Company (Private & Local Acts, 1856, c. 294) dramatized the promoter's sense of the utility of the corporate form to attract distant or detached capital to limited commitments: see Duckett, 151–153. Typical provisions for forfeiture of stock, or of sums paid in, upon a stockholder's failure to pay installments or respond to assessments, may be seen in Private & Local Acts, 1857, cc. 91 (St. Croix Manufacturing & Improvement Co.), and 235 (Chippewa Falls Lumbering Co.), and in articles of the Lincoln Lumber Co., organized under Revised Statutes, 1878, c. 86 (papers in Wisconsin State Historical Society). Cf. Wisconsin River Lumber Co. v. Walker, 48 Wis. 614, 4 N.W. 803 (1879), and Wisconsin River Lumber Co. v. Plumer, 49 Wis. 666, 6 N.W. 319 (1880) (suits to collect assessments on stockholders), and Clarke v. Lincoln Lumber Co., 59 Wis. 655, 18 N.W.

492 (1884) (being *in pari delicto,* stockholder may not recover what he paid for stock issued illegally, because issued below par).

Frederick Weyerhaeuser's empire building highlighted the utility of the corporation as a constitution for association of large, somewhat diverse interests. See, particularly, Larson, 135–141; Marshall, *Autobiography,* I, 276–294. The flexibility for expansion through holding of the stock of one corporation by another, as well as through interlocking directorates, is reflected in the histories of Philetus Sawyer, (Current, 130), Orrin H. Ingram, (Fries, 138–139), and most particularly Frederick Weyerhaeuser, (Fries, 139; Marshall, I, 260, 275). The only explicit declarations of authority to hold stock in other corporations under special charters were by P&LL, 1868, c. 385 (Green Bay Lumber Co.); P&LL, 1871, c. 242, amending Private & Local Acts, 1856, c. 504 (Peshtigo Lumbering & Manufacturing Co.); cf. P&LL, 1859, c. 200 (Black River Improvement Co. authorized to consolidate by contract with any other company on the river, to achieve its objectives). The legislature wavered in its general policy declarations on holding companies during the years of the timber industry's prime growth. Revised Statutes, 1849, c. 51, sect. 10, said that it should "not be lawful for" a corporation organized thereunder "to use any of their funds in the purchase of any stock in any other corporation." Revised Statutes, 1858, contained no limitation on holding companies, and in sect. 7 of c. 73 broadly authorized any business "joint stock company" organized thereunder "to acquire and hold . . . all such property of every kind, as shall be necessary for the purposes of said corporation . . ." Revised Statutes, 1878, sect. 1775, in contrast, unqualifiedly stated that "no such corporation" formed under c. 86 "shall take or hold stock in any other corporation." This was, shortly, modified to permit holding companies for river improvement, by Laws, 1879, c. 82; cf. Laws, 1883, c. 221, and 1891, c. 127 (c. 86 corporations may acquire any outstanding "right, privilege or franchise" whose possession "would be in direct aid" of the acquiring company's business). Statutes, 1898, sects. 1775a and 1777f, continued the special river improvement and franchise provisions, and sect. 1775 granted a broad authorization to any business corporation to hold another's stock, on assent of three fourths of the stockholders of both companies involved.

105. The over-all history of the development of Wisconsin policy toward the special charter is told in Kuehnl; the data on totals of special charters and of charters under the general act will be found on pp. 143, 144. More detailed consideration of the special chartering of lumber-industry corporations during the industry's upcurve of growth will be found in Dykstra, "Law and the Lumber Industry," Part Two, much of which is summarized in the same author's "Corporations in the Day of the Special Charter," *Wisconsin Law Review,* 1949: 310. That the Wisconsin story followed the general outlines of contemporary legal development in the United States, see Hurst, *Law and the Conditions of Freedom,* 15–17.

Examples of the typically routine passage of special charters for lumbering firms may be seen in the legislative history of the bills which became the charters of the Peshtigo Lumber & Manufacturing Co., AJ, 485 (1861), SJ, 406; and of the Barnum Lumber Co., SJ, 58, 65, 91 (1867). Of particular interest is the record of passage of bill 364S, of 1866, which became the charter of the Eau Claire Lumber Company (P&LL, 1866, c. 284), because a company founder was both a leading figure in the industry and in the legislature which passed the charter, and because the charter, as much as any, became a model. On Mar. 17, 1866, bill 364S was introduced by Senator J. G. Thorp, an incorporator of the company and its first president. Instead of the routine reference to the Committee on Incorporations, the presiding officer sent the bill to a select committee, composed of Senators Thorp and Marcus A. Fulton, the latter representing the St. Croix River lumber district. On Mar. 20 the select committee reported back the bill with the recommendation that it pass, and later that day on Thorp's motion the bill was taken from the general file, and passed under suspension of the rules. On Mar. 21 the bill was received in the Assembly and referred to its Committee on Incorporations, which promptly on Mar. 22 recommended approval, which was voted with some minor amendments on Mar. 27. The Senate concurred in the amendments on Apr. 4, and the Governor reported his signature affixed to the bill, on Apr. 6: SJ, 564, 593, 597 (1866); AJ, 698, 714, 806; SJ, 733, 906.

The legislative journals are bulky and their detail tedious. It is not hard, therefore, to overlook matters. But in a close search of the history of bills introduced — as well as of those actually passed — for the chartering of lumbering companies, I found only one occasion on which such a charter bill was defeated. It is characteristic of the history of the field that the incident seems to have turned largely on an issue of access to waterways. In 1854 C. C. Washburn and Cyrus Woodman and associated capitalists, probably incident to long-standing ambitions of Woodman to secure a timberland monopoly, caused to be introduced bill 45S, to incorporate the Wisconsin Lumbering & Manufacturing Company, with an original capital authorized at $500,000, subject to increase to $1,000,000 on vote of two thirds of the stockholders (cf. Gara, 116–118). Senate amendments to the bill struck out the bracketed words, and inserted the underlined words, in the critical sect. 6, which, as it went to the Assembly, read thus: "It shall be lawful for said Company to purchase and hold pieces or parcels of land on the Mississippi River [or any other river or lake bordering upon, or running through the state] *or any of its tributaries, excepting so much of the Wisconsin River and its tributaries as lies above Point Bas* which they shall deem necessary for the uses and purposes of said Company *in amount not exceeding the capital stock of said Company. Provided that nothing in this section shall be so construed, as to permit said Company to purchase and hold pieces or parcels of land on any of the waters running into Lake Michigan, or Green Bay of said Lake.*" The Senate also struck entirely sect. 9 of the bill, which had — though clearly to no effect, in view of the general reserved power to repeal or amend under Art. XI, sect. 1, of the Wisconsin constitution — declared that the act should continue in force "without being altered, modified or repealed or in any way affected or suspended, without the consent of said Company," save for authorization of forfeiture proceedings by the Attorney General for "any willful violation" of the act's provisions: SJ, 69, 208, 258, 265, 275, 293 (1854); see text of introduced bill and amendments in the state archives, in the Wisconsin State Historical Society. With the recommended addition of a standard clause to guarantee that none of the Company's works would physically obstruct navigable waters, a majority of a select committee of the Assembly reported that the bill as it came from the Senate should be approved. True, as introduced the bill had "contained certain objectionable features and conferred certain extraordinary powers, which, had the bill been passed in its original form, might have been exercised in a manner detrimental to both public and private rights and interests," but the Senate's amendments had removed "most of such objectionable features," and in the amended form the Assembly committee's majority were "unable to discover any extraordinary corporate rights or powers conferred" that would "interfere with or injure either public or private right." The chairman of the select committee dissented vigorously, apparently on grounds that ran to his distaste for the accumulation of timberlands he foresaw in the company if successful, and with an invocation of Jacksonian principle rarely applied in the Wisconsin record to business corporation charters. The dissenter objected on the grounds "that the said bill confers a grand monopoly upon some half a dozen land speculators; that but one of the corporators is an actual resident of the country proposed to be monopolized, and that he is not engaged or interested in any manner in the manufacture of lumber; that the other corporators are two bankers and land speculators; and we all know who 'Nat Dean' is, and that the 'balance' of the corporators are not even residents of the State, but are residents of the State of Maine. That the measure is odious and offensive to the people of the north and northwestern part of the state, and particularly to all who are interested or actually employed in lumbering — that the bill confers privileges which the company could not have under the statutes of our state regulating corporations. (See Revised Statutes, p. 286). [The reference is to c. 51, Revised Statutes, 1849, which the dissenter neglected to note had been supplanted by General Laws, 1853, c. 68, which became c. 73 of the Revised Statutes, 1858; if the dissenter's particular reference was to the limitation of landholding by corporations under c. 51 to forty acres per stockholder, it should be noted that this limit was omitted in the 1853 act]. And that if the bill does confer such privileges, that it is unwise and impolitic to confer such privileges upon incorporated companies, to the great disadvantage of all who are actual residents of the country . . . [T]he measure is uncalled for and unnecessary, and . . . is a species of *class legislation,* to which the Democratic party, and all reasonable men, have heretofore been opposed." Following these reports from the select committee, the bill was indefinitely postponed: AJ, 719–721, 761 (1854). The only other instance found of a rejected proposal regarding the corporate powers of a concern that may be counted within the logging-milling-marketing phases of the lumber industry was the indefinite postponement, on recommendation of the Assembly Committee on Lumber and Mining, of an amendment to the charter of the LaCrosse Manufacturing Co., extending the geographical area within which it might build dams: AJ, 260, 697, 719 (1895).

Gov. Randall's veto of the Oconto Lumbering Company charter will be found in SJ 595, 597 (Message of Mar. 6, 1858). Apart from the policy grounds of the veto, noted in the text, the Governor cited as an alternative ground of his action that, absent a showing that the ends of this business could not be served by chartering it under the available general act, the special charter was unconstitutional as in violation of the presumption against special charters created by Art. XI, sect. 1 (SJ, 596). Compare Kuehnl, 157–164, on the general, unavailing efforts of various governors, through the veto power, to enforce Art. XI, sect. 1. Though the point seems never to have been squarely raised or decided, it appears that the contemporary court regarded Art. XI, sect. 1, as not in character "anything more than directory to the legislature, as it leaves the whole matter, after all, to its judgment." See Clark v. City of Janesville, 10 Wis. 136, 175 (1860); cf., however, State *ex rel.* City of Shawano v. Engel, 171 Wis. 299, 306, 177 N.W. 33, 35 (1920).

106. On the significant policy change, from common law, involved in conferring upon stockholders — let alone the public — a right to see the books, see State *ex rel.* Bergenthal v. Bergenthal, 72 Wis. 314, 319–320, 39 N.W. 566, 568 (1888); cf. James Kent, *Commentaries on American Law* (1827), II, 240; Joseph K. Angell and Samuel Ames, *The Law of Private Corporations* (1861), 634. The court carefully applied c. 73, Revised Statutes, 1878, as creating a liability on directors or officers for refusing or failing to file the required public statements, only where their refusal or neglect was "intentional": Seaman v. Goodnow, 20 Wis. 27 (1865). By provision applicable to all corporations, Revised Statutes, 1858, c. 78, sect. 22, empowered the attorney general (on direction of the governor), or any legislative committee (on direction of the legislature or either house), to make thorough investigation of the affairs of any corporation, including examination of its books. This act required the taking of special, official initiative, however; there is no evidence that it ever had any practical bearing on the timber industry. The particular acts referred to in the text, regarding keeping of corporate records and access to them, were General Laws, 1866, c. 22, 1867, c. 42, and 1868, c. 152. The compilation of data on the eighty special charters is discussed in notes 103 and 104 above. On books and reports requirements, or the absence of them, in contemporary Wisconsin corporation law generally, see Kuehnl, 184, 185.

107. In the absence of contrary provision by statute or articles of association, stockholders and directors enjoyed freedom from general liability for the firm debts: Kent, II, 216; Angell and Ames, 31, 309, 554, 569. Wisconsin followed familiar common law or equity doctrine, however, which recognized a non-statutory liability of stockholders to creditors, on an action brought by or on behalf of the corporation, for firm debts to the amount of sums unpaid on stock subscriptions, when execution went unsatisfied on a judgment against the corporation: Adler v. Milwaukee Patent Brick Manufacturing Co., 13 Wis. 57, 60–61 (1860); accord: Angell and Ames, 566. In the same opinion it was also indicated that Revised Statutes, 1858, c. 148, sect. 31, which on its face might appear to create a statutory cause of action of this type generally, was, in context, applicable only to banking and insurance corporations: 13 Wis. 57, 64–65. See Cleveland v. Marine Bank of Milwaukee, 17 Wis. 545, 549 (1863); Sleeper v. Goodwin, 67 Wis. 577, 588, 31 N.W. 335, 340 (1887); Whitehill v. Jacobs, 75 Wis. 474, 481–482, 44 N.W. 630, 632 (1890); Gores v. Day, 99 Wis. 276, 279, 74 N.W. 787, 788 (1898). No contemporary decision squarely held that stockholders were under a common law obligation to creditors to the extent of refunds of capital stock, but the proposition was stated as unquestioned doctrine in a dictum in the Adler case (above), 13 Wis. 57, 60. See Angell and Ames, 563–564. Cf. Gogebic Investment Co. v. Iron Chief Mining Co., 78 Wis. 427, 432, 47 N.W. 726, 727 (1891); Killen v. Barnes, 106 Wis. 546, 569, 82 N.W. 536, 544 (1900). The court repeatedly recognized that such liability as the stockholder owed a corporation employee for wages was exclusively the creation of statute. See Sleeper v. Goodwin, 67 Wis. 577, 589–590, 31 N.W. 335, 340 (1887); Day v. Vinson, 78 Wis. 198, 199, 47 N.W. 269, 270 (1890); Finney v. Guy, 106 Wis. 256, 266, 82 N.W. 595, 598 (1900). Revised Statutes, 1858, c. 148, sects. 15 and 17, gave jurisdiction to circuit courts, at the suit of creditors among certain other persons

named, over officers or directors to compel them to pay to the corporation and its creditors moneys misapplied, lost, or wasted in violation of duty. These provisions do not make plain on their face whether the legislature intended thereby to declare substantive rights as well as remedies of the creditors, or merely to invest the circuit courts with jurisdiction to enforce such rights as creditors might have at common law or by other statute. No contemporary decision clarifies the point. At a much later stage in development of corporation law, the court said that the general equity powers of the circuit court had always sufficed for jurisdiction over such suits by the corporation to redress breaches of loyalty or care owed it by its management, but that sect. 17 of the 1858 Revised Statutes, in particular, was creative in allowing direct suit on behalf of creditors as a remedy distinct from action by the corporation or, derivatively on its behalf, by stockholders, creditors, or its assignee or successor. See Gores v. Day, 99 Wis. 276, 280, 74 N.W. 787, 788 (1898); cf. Haywood v. Lincoln Lumber Co., 64 Wis. 639, 645, 26 N.W. 184, 186 (1885) (within general equity powers to appoint receiver of corporation on plea of stockholder that directors are fraudulently dissipating its property, as by mortgage to themselves); Pietsch v. Wegwart, 178 Wis. 498, 509, 190 N.W. 616, 619 (1922) (director does not, as matter of general doctrine, owe a trustee's obligations to corporation creditors, as *cestuis,* so as to suspend running of statute of limitations on basis of alleged fraud on creditors). Directors of a lumber company were found guilty of no fraud, and hence under no liability to corporation creditors, on account of their purchase of some of the corporation's stock, when it was solvent, and some months prior to the incurring of the debts to the complaining creditors, in Shoemaker v. Washburn Lumber Co., 97 Wis. 585, 73 N.W. 333 (1897). The care with which the court scrutinized pleadings, in comparison with the text of the statute, in Seaman v. Goodnow, 20 Wis. 27, 30–31 (1865), implicitly witnessed the judges' recognition of the exclusively statutory character of the liability of directors to creditors under c. 73, Revised Statutes, 1858, for failure to file required public statements.

108. A corporation enjoyed perpetual succession, if no limit was set on its life by statute or its articles: Kent, II, 216, 224. Though there is little else in the record to suggest an acute concern with the maxima set for business corporation capitalization during the special-charter era, the incident of the defeat of bill 45S, of 1854, for incorporating the Wisconsin Lumbering & Manufacturing Company contained an aspect that might warn of potential issues, in the Senate amendment which limited land acquisitions to a value not exceeding the firm's authorized capital. See note 105 above. The same incident also carried warning that a potentially explosive issue might slumber in charter terms which did not plainly grant broad discretion to deal in timberlands. Special charters enacted in territorial years commonly included rather strict limits on corporation land holding: see Kuehnl, 22, 30, 32. None of these territorial instances concerned a lumbering company, but the analogy for a policy limiting land investment by such concerns might be seen in a mining company charter which limited the firm's land ownership to 640 acres: Laws of Wisconsin Territory, 1836, Act of Dec. 8, 1836, p. 75 (Peketonica Copper Co.). So far as the Wisconsin lumbering-firm special charters were phrased with some apparent care to grant liberal discretion in landholding, they perhaps reflected more the high importance of the topic in that type of enterprise than a need to eliminate any very serious doubt on the score of ordinary corporation authority. Incorporation would almost certainly be held to confer power by implication to buy and hold any property necessary to the corporate objectives, and the English mortmain statutes were held not to have been received as part of the common law in Wisconsin. Kent, II, 224, 227–229; Dodge v. Williams, 46 Wis. 70, 1 N.W. 92, 50 N.W. 1103 (1879); cf. Page, 87. In its context, the declaration of Wisconsin Constitution, Art. I, sect. 14, that all lands within the state be allodial aimed at no other result than abolition of feudal tenures. See Brown, "The Making of the Wisconsin Constitution," *Wisconsin Law Review,* 1949: 684, note 98 (constitutional limit on extent of landholding by any one person rejected, in first convention); *id.,* 1952: 23, 59. In a provision applicable to all corporations in the state, Revised Statutes, 1858, c. 78, sect. 7, said that any corporation "may hold land to an amount authorized by law, and may convey the same," in effect thus inviting the draftsman of a special charter to write in his own terms. Liberal interpretation of a corporate charter, to authorize receipt of property for stock, was applied in Clark v. Farrington, 11 Wis. 306 (1860), and even somewhat enlarged in Western Bank of Scotland v. Tallman, 17 Wis. 530 (1863). The

frank and strong announcement by the court of its concern to construe railroad charters, in a time of promising but difficult beginnings, in a fashion which would not unduly hamper provision of important facilities to the economy was a factor which a cautious contemporary might well have taken as implicitly limiting the application of these precedents for the interpretation of other types of charters: see 11 Wis. 306, 321. This would be so especially because the Wisconsin court from the outset was accustomed to acknowledge and often apply the rule of the Charles River Bridge Case, that in important respects nothing should pass by implication under a corporation franchise. See Janesville Bridge Co. v. Stoughton, 1 Pinn. 667, 672 (Wis. Terr. 1846), and note the not infrequent inclusion in a special corporate charter of a declaration that its provisions should be "liberally construed in all courts to promote the ends sought to be attained," as in P&LL, 1866, c. 284 (Eau Claire Lumber Co.).

109. After 1872, apart from a handful of amendments to earlier special charters, the only corporation legislation containing any distinctive reference to the timber industry embraced Laws, 1874, c. 113 (general incorporation act for business associations or corporations formed by "five or more adult persons living in any county of this state"); Laws, 1881, c. 133 (adding authorization of holding-company operations in lumber industry, under Revised Statutes, 1878, sect. 1775, on consent of three fourths of the capital stock of acquiring corporation); Laws, 1883, c. 220 (adding to enumerated types of business corporations which may be organized under c. 86, Revised Statutes of 1878, private logging railroad companies); Laws, 1883, c. 221 (authorizing any business corporation organized under Wisconsin law for any of the purposes listed in c. 86 to acquire any outstanding franchise, in aid of its business); Laws, 1891, c. 283 (amending Revised Statutes, 1878, sect. 1775, as amended, to authorize acquisition by one business corporation of another's stock, on assent of three fourths of stockholders of both companies); Laws, 1891, c. 27 (extending authority of c. 86-type corporations to acquire franchises hereafter issued); Laws, 1895, c. 64 (authorizing formation of lumber retailers' mutual fire insurance companies); Laws, 1897, c. 160 (amending Annotated Statutes, 1889, sect. 1775, as to terms on which a business corporation may dispose of, or mortgage, its property); Laws, 1899, c. 192 (relieving lumbermen's mutual insurance companies of certain reporting requirements); Laws, 1905, c. 356 (amending reporting requirements of lumbermen's mutual insurance companies); Laws, 1911, c. 201 (simplifying authority and procedures for dealing in real estate by corporations generally empowered therefor).

PART III: THE POLICE POWER

Introduction: The Adjustment of Continuing Relations

1. See Shaw, C. J., in Commonwealth v. Alger, 7 Cush. 53, 84–85 (Mass. 1851); cf. Taney, C. J., in The License Cases, 5 How. 504, 583 (U.S. 1846).

Chapter VI. Regulating Use of Fixed Private Capital: The Costs of Exploiting Timberland

1. Effective interest in forest conservation developed only after the Wisconsin forest had been irrevocably committed to the practice and policy of clear-cutting without regard to current real costs or to future growth. Compare Ise, 91, 162, 200; Pinchot, 23, 27, 84–85; Woytinsky, 690, 707. This lag in timing between the industry's growth and the development of conservation interest was in effect acknowledged in the opinions in State ex rel. Owen v. Donald, 160 Wis. 21, 141–142, 151 N.W. 331, 371 (Marshall, J., for the court), and 160 Wis. 21, 159, 151 N.W. 331, 377 (Winslow, C. J., concurring) (1915), and in City of Milwaukee v. State, 193 Wis. 423, 452, 214 N.W. 820, 831 (1927).

2. See Fries, 53, 62, 105, 108, 245; Reynolds, 129–133, 149; Zon and Garver, *Selective Logging*, 2, 39; Zon, *Timber Growing and Logging Practice*, 30–31.

3. Judgments on the general extent of losses in real production in Wisconsin and in the Lake States area from inefficient logging and cutting practices and from fire may be found in J. J. Blaisdell, *Forest and Tree Culture in Wisconsin* (1893), 8, 13; Fries, 105, 245; Gates, *Wisconsin Pine Lands,* 70, 88; Kleven, 223, 224; Sparhawk and Brush, 10. Such estimates of the extent of real loss in Wisconsin lumbering tally with estimates of the extent of loss in lumbering generally in the United States. See Gulick, 20, 21, 88, and Marquis,

3, 94. On the lumber industry's twentieth-century record, see Buttrick, 400, 408–412; Gulick, 80, 81; Loehr, 260–263.

4. The unplanned "selective" cutting in the Wisconsin forest grew out of factors noted in Buttrick, 210, 236; Fries, 240; Reynolds, 64, 149. The favor which Lake States topography and the technological developments in the industry showed toward the practice of clear-cutting is discussed in Loehr, 6; Marquis, 136; Sparhawk and Brush, 7. That the extent of clear-cutting in the Lake States exceeded the practice anywhere else in the country, see Marquis, 174. Experience of the very slow natural reseeding of the white pine under the conditions created by the lumber industry in northern Wisconsin, and the practically decisive replacement of prime by inferior conifers and other less valuable species, are discussed in R. N. Cunningham, *Forest Resources of the Lake States Region,* U.S. Department of Agriculture, Forest Reserves Report No. 1 (1950), 6, 7, 14, 15; Marquis, 174; Zon, *Timber Growing and Logging Practice,* 13, 16. Cf. Ciriacy-Wantrup, 57–58. For comparison of the productive potential of the northern Wisconsin forest in 1875 and in the 1920's, see Zon, "Forest Conditions in Wisconsin," in Wisconsin Commercial Forestry Conference, *Forestry in Wisconsin* (1928), 32, 33. That natural growth rather than plantation effort must be the chief reliance for development of saw-log-quality timber, see Gulick, 112, 113; Joseph Kittredge, Jr., *Forest Planting in the Lake States,* U.S. Department of Agriculture, Bulletin No. 1497 (1929), 5–6. Cf. Robert N. Stone and Harry W. Thorne, *Wisconsin's Forest Resources,* Lake States Forest Experiment Station, Station Paper No. 90 (1961), 9, 15. By the turn of the century, Wisconsin official documents began to acknowledge the fact of decisive forest depletion. See, e.g., *Report* of the Forestry Commission of the State of Wisconsin (Madison, 1898), 13, 25; testimony of representative of Western Wisconsin Railroad Co., in Report of Conference between Joint Committee on Assessment and Collection of Taxes, and Railroad Representatives, Feb. 18, 1903, SJ, 336–337; First Biennial Report, Railroad Commission, Part II: Formal Complaints No. 68, In re North Wisconsin Farmers Association, Aug. 25, 1906, PD, 267–268; Report, Interim Committee on Forestry and Public Lands, Apr. 19, 1929, AJ, 851, 866–869, 879; Laws, 1933, Joint Resolution No. 103

5. See Chief Forester W. B. Greeley, speech to National Association of Wood Turners, in United States Department of Agriculture, Weekly News Letter, July 27, 1921, p. 14; Gulick, 79–81, 88, 93, 96; Marquis, 5–6, 32, 63, 137.

6. Chief Forester Greeley's speech, cited above, p. 14; Gulick, 22; Zon "Forest Conditions in Wisconsin," 32, 33.

7. Increased costs to consumers in terms of lowered quality, increased processing investment, higher depreciation, and higher borrowing expenses are discussed in Buttrick, 299–305; Chief Forester W. B. Greeley, "Introduction," in Zon, *Timber Growing and Logging Practice,* 3; Gulick, 21, 86; Marquis, 8, 25, 38–40, 95, 98, 115, 174. Increased transport costs as the reflection of lost site value are discussed in Bureau of Corporations, *The Lumber Industry,* vol. I, part I, pp. 84–85; Buttrick, 237, 451; Greeley, speech cited in note 5, p. 14; Gulick, 121, 93, 96.

8. See Ciriacy-Wantrup, 322–326, 334, 335, and Gulick, 84–86, 96, 133, 137.

9. See, generally, Wisconsin Commercial Forestry Conference, *Forestry in Wisconsin* (1928), and in particular the observations therein on costs of abandoned or under-used facilities, in Alvin C. Reis, "What Forests Mean to the Public," *id.,* 50–51; on depleted timberland and submarginal farm tracts, in H. L. Russell, "Wisconsin's Idle Plant," *id.,* 24; Benjamin H. Hibbard, "Land in Relation to Forestry," *id.,* 26; and Raphael Zon, "Forest Conditions in Wisconsin," *id.,* 32, 33. On the plant and wage labor investment in the lumber industry during its boom period, see Fries, 103, 206, 207; Wisconsin Regional Planning Committee, *A Study of Wisconsin* (1934), 221, 226. The failure of agriculture to follow the forest in the north country is depicted in James I. Clark, *Farming the Cutover,* and *Cutover Problems,* and in F. G. Wilson, "Zoning for Forestry and Recreation: Wisconsin's Pioneer Role," *Wisconsin Magazine of History,* 41:102 (1957).

10. On the significance of the forest due to its real, opportunity value rather than for its monetary, exchange value, compare Ciriacy-Wantrup, 71–72, 88, 118; Spengler, 308, 311. See also Clark, *Economic Institutions and Human Welfare,* 271, observing from the experience of the western industrial revolution that growth of industry requires growth in the rest of the economy. There is, in other words, an ecology of economic development, as well as of natural growth, which must be respected on pain of stunting. Cf. Hacker,

401–405: Hurst, *Law and the Conditions of Freedom,* 78–82. The loss of the midwest market, not only for primary forest products but also for semi-processed and finished goods involving substantial consumption of wood, is reflected in M. P. McCullough, "What Forests Mean to the Lumber Industry," in Wisconsin Commercial Forestry Conference, *Forestry in Wisconsin,* 53–54, and George F. Kull, "What Forests Mean to Manufacturers," *id.*, 70–72; Francis F. Bowman, Jr., *Industrial Wisconsin* (1939), 34–37; Sparhawk and Brush, 20, 24, 27–28. Tax delinquency and withdrawals of land from the tax rolls in the northern counties are discussed in Hibbard, "Land in Relation to Forestry," 30; Reis, 51; Wisconsin Regional Planning Committee, *A Study of Wisconsin,* 501; Wisconsin State Planning Board, *Local Government Study in Wisconsin 1927–1936,* vol. I, *Assessments and Levies* (1941), pp. 80, 83, 91–95, 117. On the potential positive contribution which a continuing, sustained-yield forest, half the size of the original forest of northern Wisconsin, might have made to the tax resources of the state, compare McCullough, 55; Erling D. Solberg, *New Laws for New Forests* (1961), 158–160; Sparhawk and Brush, 27–29, 34, 45, 54. The valuation of the timberlands in the area of the former Menominee reservation was summarized in Wisconsin Department of Taxation, *Menominee Indian Reservation: Recheck Report,* May 27, 1960 (Madison, 1960), 4. For Michigan, Sparhawk and Brush estimate that the aggregate value at the mill at the time of cutting of all sawed lumber cut from the Michigan forest was $2.5 billion. This sum was three times the assessed value in 1922 of all the property, including mines, in the state's forty-six northern counties, nearly eight and one-half times the value of all farms in that area, and more than the assessed value of the real property in the whole state, excluding its three metropolitan counties. Adding the value of timber products other than sawed lumber, the value taken from the Michigan forest totalled over $3 billion, or twice the value of gold taken in the California gold rush and ten times the value of the gold taken in the Alaskan gold boom (Sparhawk and Brush, 10–11).

11. On the constitutional power of Congress over the public lands, particularly with reference to the unqualified scope of the property clause (U.S. Constitution, Art. IV, sect. 3), see in this volume, Chapter II, note 75. The issuing of a patent to land by the United States marked the decisive point regarding its authority as owner to control the land's use. Before patent, the United States enjoyed complete control: Wilcox v. Jackson, 13 Pet. 498 (U.S. 1839). This control included the authority to determine the incidents of title with reference to acts or transactions antedating patent, as in Congress' authority to exempt a homestead from levy of execution for debts contracted before patent: Paige v. Peters, 70 Wis. 178, 35 N.W. 328 (1888); Ruddy v. Rossi, 248 U.S. 104 (1918). Cf. Spiess v. Neuberg, 71 Wis. 279, 37 N.W. 417 (1888). But see the strong dissent by Holmes, J., in Ruddy v. Rossi, 248 U.S. 104, 107, especially at 111 (1918). However, after patent by the United States, and with reference to subsequent transactions, the definition of permissible modes of conveyance and use among private persons was generally in the sole determination of the state wherein the land lay: Langdon v. Sherwood, 124 U.S. 74 (1888).

The Weeks Law was the act of Mar. 1, 1911, 36 Statutes at Large 961. Marion Clawson and Burnell Held, *The Federal Lands: Their Use and Management* (1957), 30, note that of 18 million acres purchased under the Weeks Law of 1911 and the Clarke-McNary Act of 1924, about half was authorized for purchase during the three years 1934–1936, and that purchases after 1942 were small. The difference between the leverage enjoyed by the United States through its original land ownership, and the limitations other than fiscal involved in any substantial effort to regain a comparable position of power, was tangibly attested in the stipulation of sect. 7 of the Weeks Law, requiring consent of the state to federal purchases within its borders. Wisconsin cautiously gave its consent for a maximum federal acquisition of 100,000 acres, in Laws, 1925, c. 411; the allowed maximum was raised to 500,000 acres by Laws, 1927, c. 461, to 1,000,000 acres by Laws, 1929, c. 457, and finally to 2,000,000 acres by Laws, 1933, c. 272. In addition, the state conditioned its consent to any particular federal acquisitions upon prior approval of national forest boundaries by the governor, affected state agencies, and the interested county boards. By the mid 1950's two national forests had been established, embracing a little short of 1,500,000 acres. See The Natural Resources Committee of State Agencies, *The Natural Resources of Wisconsin* (1956), 96. Qualified students of the twentieth-century forestry problems of the United States emphasized the decisive character of the wholesale transfer of federal timberlands to private ownership in the nineteenth century. See remarks of Chief Forester H. S. Graves in address to National Lumber Manufacturers Association, Apr. 14, 1919, quoted in Cameron, 396; see also Loehr, 83, commenting on report of Willett F. Ramsdell to the advisory committee to President Hoover's Timber Conservation Board. The existence of a vast inventory of virgin timber subject to unrestricted cutting limited the installation of sustained-yield practices or regulation by industry or individual states. See Ciriacy-Wantrup, 177, 182; Gulick, 80, 87, 96, 109, 159, 200–201; Ise, 159. Loehr, 263, quotes a bluntly spoken summary of this estimate of the situation by Elers Koch of the U.S. Forestry Service in 1927: "My guess is that continued glutting of the market with private timber is inevitable until the bulk of virgin timber is gone. We might as well have it over and be done with it and then build up on National Forest timber plus second growth. I doubt if any action anybody can take will change this course of events except to a very limited extent. It is one of the penalties the country has to pay for relinquishing title to the bulk of the timber which was once in Government ownership." A rare foreshadowing of this judgment may be implied in the 1876 comment of Commissioner Williamson of the General Land Office on the policy inconsistency involved in the Timber Culture Act, quoted at note 13 below. A twentieth-century Wisconsin judgment upon the effect of nineteenth-century federal policy in hampering development of forest conservation efforts within the framework of any one state may be found in Laws, 1923, Joint Resolution No. 13. See generally Chapter II, notes 38, 39, 76, and 78.

12. The development of nineteenth-century constitutional doctrine up to the peak years of Wisconsin lumbering followed the mainstream of Hamiltonian values; wherever ultimate choice must be made, we elected to interpret broadly the authority of Congress to promote and regulate in the interests of general economic growth. Cf. McCulloch v. Maryland, 4 Wheat. 316 (U.S. 1819) (national bank); Gibbons v. Ogden, 9 Wheat. 1 (U.S. 1824) (federal licensing power over interstate commerce); Brown v. Maryland, 12 Wheat. 419 (U.S. 1827), and Joseph Story, *Commentaries on the Constitution of the United States* (1833), vol. II, sects. 1077–1095 (tariff); Veazie Bank v. Fenno, 8 Wall. 533 (U.S. 1869), and Juilliard v. Greenman, 110 U.S. 421 (1884) (currency); Pensacola Telegraph Co. v. Western Union Telegraph Co., 96 U.S. 1 (1878), and California v. Central Pacific Railroad Co., 127 U.S. 1 (1888) (promotion and protection of interstate communications facilities). Congress in effect asserted its authority to take measures for protection of renewable natural resources forming national assets, in its Joint Resolution No. 22, of Feb. 9, 1871, 16 Statutes at Large 594, directing the President to name a commissioner of fish and fisheries, charged to investigate whether action was needed to preserve the supply of food fish in waters within the jurisdiction of the United States: see Manchester v. Massachusetts, 139 U.S. 240, 257, 265 (1891). The quoted statement from Senator Timothy Howe is cited and discussed in its context of contemporary opinion in Chapter II, note 63 and accompanying text, above.

13. The "timber culture" legislation is discussed generally in Hibbard, *Public Land Policies,* chap. xix. For the quoted comment by Commissioner Williamson, see General Land Office, Report, 1876, p. 9. In 1872 one house of the Wisconsin legislature was prepared to memorialize Congress in favor of the "timber culture" type of federal legislation, but the second chamber disagreed. See note 32 below.

14. On the state lands, see general discussion in Chapter I, and the more particular matter in Chapter II, notes 7, 10, 11, 13, 15, 16, 45, and 46, and accompanying text. Especially persuasive are the implications of Smith v. Mariner, 5 Wis. 551 (1856); State v. Weston, 17 Wis. 107 (1863); Chippewa Valley & Superior Railway Co. v. Chicago, St. Paul, Minneapolis & Omaha Railway Co., 75 Wis. 224, 44 N.W. 17 (1889). See City of Milwaukee v. State, 193 Wis. 423, 451, 214 N.W. 820, 830 (1927) (state's trusts of land should be viewed as carrying active management responsibilities).

The full acceptance of fisheries protection as a legitimate object of state police power in the early nineteenth century is reflected in Smith v. Maryland, 18 How. 71 (U.S. 1855). Cf. Lawton v. Steele, 152 U.S. 133 (1894); Geer v. Connecticut, 161 U.S. 519 (1896). See Handlin, 76, 93, 97, 222.

The *Report on the Disastrous Effects of the Destruction of Forest Trees Now Going on so Rapidly in the State of Wisconsin* (1867), issued by Commissioners appointed under General Laws, 1867, c. 36 (Increase A. Lapham, J. G. Knapp, and H. Crocker), was forthrightly confident that the legislature possessed broad power to act to protect forest resources needed for the well-being of the general economy: "A state that finds authority to regulate the times and seasons when its citizens may catch fish, or shoot game, may certainly assume such as may be needed to preserve the civilization of the present times; it would require no greater stretch of power to regulate the cutting of timber where it would obviously entail a public calamity, or to encourage its production where it is so much needed for the public good" (*id.,* 36). Winslow, J., referred to this field of public action as one which had always been unquestioned in Wisconsin law, in State v. Nergaard, 124 Wis. 414, 420, 102 N.W. 899, 901 (1905).

It is true that it took until the turn of the century to obtain the basic decisions sustaining regulations protecting stocks of nonrenewable natural resources. See Ohio Oil Co. v. Indiana (No. 1), 177 U.S. 190 (1900) (oil and gas); Lindsley v. Natural Carbonic Gas Co., 220 U.S. 61 (1911) (mineral waters); Walls v. Midland Carbon Co., 254 U.S. 300 (1920) (wasteful use of natural gas compared to alternative uses). Cf. Hudson County Water Co. v. McCarter, 209 U.S. 349, 355 (1908). But, especially in view of the strong terms in which regulatory authority was defined in these cases, this lag reflects nothing more than the lag in legislative perception of need. Of like import is the fact that it took from 1868 to 1903 for discussion of legislative authority in conservation matters to enter Cooley's treatise; when the matter finally came under examination, there was no hesitation in acknowledging the legitimacy of such regulation. Compare Thomas M. Cooley, *Constitutional Limitations* (Boston, 1868), 594–596, and *id.* (Boston, 1903), 880. Of course, we deal here in prediction, and there is a substantial margin for misjudgment.

Most damaging to the estimate made in the text is Huber v. Merkel, 117 Wis. 355, 94 N.W. 354 (1903), where the court held unconstitutional a statute forbidding a landowner to let an artesian well on his land run to waste to the detriment of his neighbors. The decision was at variance with contemporary authority on oil and gas waste, which the Wisconsin court distinguished because the artesian water situation involved a resource "constantly being renewed" — an emphasis which, it would appear, could be rebutted as to the forest. See Kenneth F. Hostak, "Wisconsin Ground Water Law — A New Era," *Wisconsin Law Review,* 1957: 309.

That the factual requisites for reasonable exercise of the police power exist where problems of social cost accounting are so complex, and the minimally effective state intervention so modest, compare Buttrick, 307, 400, 452 (large results open to modest regulation); Ciriacy-Wantrup, 117, 302 (socially satisfactory objective need be no more than maintenance of a safe minimum margin against irreversible depletion; flexibility so injected into economy reduces need for closer calculations); Chief Forester W. B. Greeley, "Forest Management on Federal Lands," *Journal of Forestry,* 23:223 (1925) (proper social concern for stability of economic areas as wholes); Marquis, 102, 111, and Myrdal, 224–225, 255 (social interest in overall capacity of economy to withstand jars of market); Spengler, 308, 314–316 (the less directly overhead capital investment is tied to immediate consumer satisfaction yields, the less likely that private investment will suffice in early growth of an economy).

15. See Marshall, J., for the Court, in State ex rel. Owen v. Donald, 160 Wis. 21, 141, 151 N.W. 331, 371 (1915). Though there was no direct ruling by the Wisconsin Supreme Court on legislative power to conserve the forest by regulating the use private owners might make of their timberland, we should note Barker v. Dayton, 28 Wis. 367 (1871). Wisconsin Constitution, Art. I, sect. 14, declares that all lands within the state are allodial, and feudal tenures are prohibited. This constitutional provision was argued to invalidate a statute forbidding a husband to mortgage or otherwise alienate the homestead without his wife's signature. Dixon, C. J., readily rejected the contention; the allodial clause showed by its context that it was intended only to assure the marketability of land titles free of feudal incidents, otherwise leaving the legislature free to regulate dealings with the land as an ordinary part of the police power, "as all our governmental experience, from the foundation of the state to the present day, fully demonstrates . . ." (*id.,* 385). The basic legislative precedent for regulation directed against woods fire hazards rests in Wisconsin Territory, Laws, 1838, p. 350, and Laws, 1840–1841, No. 6, and — upon statehood — in Revised Statutes, 1849, c. 134, sect. 52; the fire laws

are examined in more detail later in this chapter. Cf. *Wisconsin Law Review,* 4:49 (1926). Legislation granting tax exemption to earned railroad-aid lands is noted and discussed in Chapter II, notes 7, 46, 50, 51, 53, and 58, and accompanying text. Such legislation was held constitutional in Wisconsin Central Railroad Co. v. Taylor County, 52 Wis. 37, 8 N.W. 833 (1881), as within the legislature's authority to make reasonable classifications of property for taxation, to achieve purposes of public interest. The river-improvements and boom legislation is discussed in Chapter IX. See also Hurst, *Law and the Conditions of Freedom,* 63–65. On the authority assigned to legislative construction of the constitution, see Dean v. Borchsenius, 30 Wis. 236 (1872); Town of Blooming Grove v. City of Madison, 275 Wis. 342, 346, 347, 81 N.W. (2d) 721, 723 (1957).

16. See In re Falvey, 7 Wis. 630 (1858); Kilbourn v. Thompson, 103 U.S. 168 (1880); and the legislative precedent discussed in James M. Landis, "Constitutional Limits on Congressional Power of Investigation," *Harvard Law Review,* 40:153 (1926).

17. See Ise, 40–45, 50–52, 55–59, 64–66, 91; Pinchot 84–85, 88–90, 107, 121. On sources of initiative in bringing matters of general concern to the level of effective interest and discussion in Congress, compare Lawrence H. Chamberlain, *The President, Congress, and Legislation* (1946), chaps. i and xii, and especially pp. 462–463; George B. Galloway, *The Legislative Process in Congress* (1953), 38–44; Bertram M. Gross, *The Legislative Struggle* (1953), 167–170, 193–197.

18. Provision was made for the general economic census by General Acts, 1857, c. 52; Revised Statutes, 1858, c. 16, sect. 1; General Laws, 1861, c. 166 (first including sawed lumber and wood products); General Laws, 1862, c. 207 (converted to decennial economic census); General Laws, 1866, c. 69 (Secretary of State, in tabulating data, shall estimate lumber at its valuation at Chicago or St. Louis); Revised Statutes, 1878, sect. 992; Wisconsin Statutes, 1898, sect. 992. The growing-crop census legislation includes Laws, 1874, c. 38 (agriculture only); Laws, 1877, c. 224 (timber included); Laws, 1878, c. 260; Revised Statutes, 1878, sect. 1010; Laws, 1897, c. 24; Wisconsin Statutes, 1898, sect. 2020; Laws, 1899, c. 210. Complaints of incomplete and unreliable reporting by the local officials may be seen in Annual Reports, Secretary of State, Oct. 10, 1857, SJ, App., Doc. B, pp. 93–94; *id.,* Oct. 10, 1861, PD, 22; *id.,* Oct. 10, 1880, PD, I, 29. Cf. Laws, 1899, c. 210. Conditions of local government in the pineries counties are described in Gates, *Wisconsin Pine Lands,* 144–145, 151, 157. The Secretary of State is quoted from the 1880 report cited above. The summary history of bill 264A in 1881, proposing creation of a bureau of statistics, may be seen in AJ, Feb. 9, 1881, p. 199 (introduced), Feb. 18, p. 302 (recommended for indefinite postponement by Committee on State Affairs), and Feb. 21, p. 336 (so voted, without a record vote). The narrower measure, dealing only with improved tax assessment data, was bill 364A, which became Laws, 1881, c. 236. This was originally rejected on recommendation of the Committee on State Affairs: AJ, Feb. 18, 1881, p. 302; Feb. 21, p. 336. But its sponsor succeeded in obtaining reconsideration, and on a second look the immediate pocketbook appeal of the proposal apparently prevailed, for the bill went through routinely in both houses: AJ, Mar. 7, p. 473; Mar. 17, pp. 565–566; SJ, Mar. 23, p. 478; Mar. 25, p. 517. Compare generally the requirement of assessment of real property, including timberland, by actual view: General Laws, 1868, c. 130; Revised Statutes, 1878, sect. 1052; Wisconsin Statutes, 1898, sect. 1052. Compare, also, Minority Report, Committee on Charitable and Penal Institutions, Mar. 14, 1873, AJ, 813 (legislators should have right, ex officio, to "become familiar with the soil, climate, productions, manufactories, water courses, forests and mines of the entire state, and thus be prepared to act much more understandingly in their official capacity").

19. The Memorial of the Wisconsin Natural History Association will be found in AJ, 497–500 (1855); the quoted passage is at p. 499. The fluctuating fortunes of the various geological surveys are summarized in James A. Lake, Sr., "Legal Profile of the Mining Industry," *Wisconsin Law Review,* 1955: 575–582. Cross currents of the times are reflected in the majority and minority reports of a committee considering continuation of the geological survey, in AJ, 1005–1010 (1858), with the majority recommending discontinuance "in view of the stringency of the times and for other reasons," and the minority urging continuance largely as a contribution to economic growth. "Such a survey," the minority argued, "vigorously prosecuted to a conclusion, would bring into notice our

unexplored and thinly settled districts, and make known the value of our mines, soils, vast lumber interest, and the rich provisions Nature has made for man in our noble State. Through its annual reports, it would send abroad accurate and reliable information to invite population and capital, and thus advance general prosperity . . ." (AJ, 1009). No action one way or the other resulted from this phase of the geological survey history. See Lake, "Legal Profile of the Mining Industry," 578–579. Very generalized references to forest resources, noted incident to minerals surveys, may be found in Report, Commissioners of the Geological Survey, PD, 6, 10 (1859); Annual Report, Wisconsin Geological Survey, PD, II, 32–40 (1878).

20. Merk, 99, note 3, attributes to Lapham the "inspiration" for bill 191A, which became Laws, 1867, c. 36, creating the study commission. See also Milo M. Quaife, "Increase Allen Lapham, Father of Forest Conservation," *Wisconsin Magazine of History,* 5:106 (1921). There is a similarity between the terms of c. 36 and the terms of the commission report which strongly indicates common authorship. See also Report, Select Committee on Bill 191A, Feb. 13, 1867, AJ, 283–286. Though the committee expresses passing concern at the quantitative extent of cutting in the northern forest, its detailed and highly critical comments on inefficient and wasteful cutting all seem to relate to disposition of timber incident to clearing southern lands for farming or to the conduct of farming. Note especially the summary verdict passed at pp. 284–285: "Many of our inhabitants have come from parts where trees and woods were considered nuisances, to be removed. Such have never asked themselves what would be the consequences to themselves and their posterity, by this systematic destruction of the natural groves of the state. Others, who first selected for a building spot, for their home, a site sheltered by trees, have found themselves, a few years after, exposed to the full blasts of the winds, because in a thoughtless moment, they cut a field out of their protecter." Bill 191A encountered no material obstacle in the legislature; on its routine passage, see AJ, 362 (1867), and SJ, 571. However, it received oddly noncommital treatment from the Senate Committee on Agriculture. The latter reported that its members, "though deeply impressed with the importance of the subject, do not feel at liberty to make a recommendation, further than to ask the attention of the Senate to the provisions of the bill, and its bearing upon the interests of the state" (SJ, 417). The contrast with the report of the Assembly select committee implies that the latter was more closely tied to ardent advocates of the study measure outside the legislature, and suggests how little initiative for such action resided within the legislature itself.

21. Lapham *et al., Report on the Disastrous Effects,* 36, 101.

22. *Id.,* 8, 25, 26, 88–89 (loss of industrial base), 28 (loss of tax base), 40 (zoning), 45–46 (cutting efficiency), 8, 12, 15, 23 (effects of deforestation on agriculture), 36 and 40 (tree-belt recommendation). The tree-belt statute was General Laws, 1868, c. 102, amended by General Laws, 1871, c. 138 (conditions), and Laws, 1876, c. 258 (limiting benefits to those who plant trees, excluding those who reserve existing trees), continued as Revised Statutes, 1878, sects. 1038 (18), and 1469–1471, Wisconsin Statutes, 1898, sects. 1038, (18) and 1469–1471, repealed by Laws, 1905, c. 264, sect. 23 (creating a state board of forestry).

23. See Resolution 19S, SJ, 85, 90 (1858); Report, Select Committee under Resolution 19S, SJ, 1180–1183 (1858). Data was collected and identified by regions, defined as Sheboygan County, Manitowoc County, Green Bay and tributaries, the Fox and Wolf rivers and tributaries, the Rock River and tributaries, the Wisconsin River and tributaries, the Black (including the LaCrosse and Chippewa) and the St. Croix rivers and tributaries. The Committee also noted that its employment figures understated the situation, because in most instances the figures returned were only for men employed in the mills, "not including a greater number, probably, engaged in logging, making shingles and square timber, during the winter months, or rafting on the various rivers in the summer" (SJ, 1181).

24. See General Laws, 1861, c. 83, and the discussion in Chapter V, especially notes 68 and 78 and accompanying text.

25. See Chapter V, notes 70, 77, and 78 (general voluntary character of the lumber inspector system), 73 and 74 (limited requirements of use of the system), and 79 and 80 (personnel problems).

26. For lumber inspectors' reflections of the limited scope of the data they collected, see, e.g., AJ, 265, report of lumber inspector for the Third District, West Eau Clair, Jan. 27, 1869; AJ, 34, report, Sixth District, Chippewa

Falls, Jan. 1, 1871; SJ, 21, report, Ninth District, Alma, Jan. 9, 1875. See especially the recommendation of the inspector for the Tenth District, that the law be amended to require report of all production to the official scalers, in SJ, 167, 168 (1876). Gov. Ludington's recommendation will be found in his General Message of Jan. 11, 1877, PD, I, 17. The bill which responded to his suggestion was bill 456A of 1877: see AJ, 544, 548, 632 (1877). Gov. Smith made his proposal in his General Message of Jan. 13, 1881, AJ, 38. The quoted comment on the failure to enforce the requirement of full reports of production in the Fourth District is from the report of the district lumber inspector, dated Hudson, Jan. 24, 1891, SJ, 114–115. Compare the failure of other bills to tighten reporting. See, regarding bill 583A of 1891, AJ, 235, 417, 459 (1891). And see, regarding bill 423A of 1893, AJ, 225, 462, 506, 531 (1893). Note also the existence of controversy over expansion of the Fourth District procedure under Laws, 1903, c. 421, as reflected in SJ, 1141, 1232, 1248–1249 (1903).

27. The Report, Committee on State Affairs, quoted in the text is that of Mar. 13, 1862, SJ, 485. Compare, accord, Report, Committee on State Affairs, Feb. 4, 1862, AJ, 214, making like disposition of another district report. That the lumber-inspector system was a response to the immediate needs of the market, rather than a start upon a comprehensive procedure to provide the bases for broad policy planning, is implicitly recognized in the characterization of the laws in Steele v. Schricker, 55 Wis. 134, 139, 12 N.W. 396, 398 (1882). As the lumber industry mounted in importance, the legislature created standing committees to deal with measures concerning it: see AJ, 42 (1867), and AJ, 42, 54 (1887). But these dealt only with the current flow of bills dealing with the immediacies which the current operations of the industry pressed upon the legislature. In contrast, a Joint Forestry Committee oriented toward broad policy planning was created only at the turn of the century: see SJ, 56, 58, 65 (1899).

28. Laws, 1897, cc. 229 (commission on planning forest-resources use) and 297 (natural and geological history survey). The course of the forest commission bill is reflected in AJ, 203 (1897), 583 (favorably reported by Committee on State Affairs), 811 (recommended for indefinite postponement by Committee on Claims), 820, 866 (recommitted to Claims, and by it thereafter recommended for passage), 955 (passed on required vote of 86 to 0); SJ, 728 (favorably reported by Committee on Agriculture), 750 (approved by required record vote of 24 to 4). Provision of a new permanent geological survey was first proposed in bill 32A of 1895, but the measure was indefinitely postponed after some maneuvering, following an adverse report by the Committee on Claims in the Assembly. See AJ, 54 (1895), 132 (favorable report by Committee on Education), 374 (adverse report by Committee on Claims), 389, 917 (recommitted to Claims, but thereafter indefinitely postponed without record vote). Enactment of the geological survey measure (bill 390A) in 1897 is reflected in AJ, 218 (1897), 865 (favorably reported by Committee on Ways and Means), 1024 (reported for passage in limited form by Joint Committee on Claims), 1065, 1123 (after maneuvers over Claims recommendations, adopted by required record vote of 65 to 13); SJ, 846 (favorably reported by Committee on State Affairs), 889 (passed by required vote of 20 to 5). The activity of the Wisconsin Academy of Sciences, Arts, and Letters in behalf of the extended geological and natural history survey is noted in Lake, "Legal Profile of the Mining Industry," 581. See also SJ, 124, 158 (1895); AJ, 172, 535, 556, 574, 594, 609. The background activity on behalf of the forestry commission bill by the newly organized Wisconsin Forestry Association, supported by the American Forestry Association (which had stimulated the local organization), is reflected in the General Message of Gov. Upham, Jan. 10, 1895, AJ, 27. Cf. Blaisdell, *passim,* and Fries, 248, 250. The most immediate products of the two study enterprises were (1) the Forestry Commission's *Report* (1898), emphasizing new growth through a system of state forests (*id.,* 13, 25); (2) from the Geological and Natural History survey, the first really comprehensive and scientific appraisal of the northern forest, done by an expert lent to the enterprise by the United States Department of Agriculture: see Filibert Roth, *On the Forestry Conditions of Northern Wisconsin* (1898).

29. The Wisconsin tree-belt statute was General Laws, 1868, c. 102; see note 22 above, on this act, and note 13 above, on the federal "timber culture" acts. The general acceptance in our nineteenth-century policy of public subsidies to economic growth, especially where no direct cash outlay was involved, is sketched in Hurst, *Law and the*

Conditions of Freedom, 7, 17, 53–65. General Laws, 1867, c. 36, authorized the Lapham commission. There was central stress on the ignorance of the men exploiting the land, as a prime cause of waste of trees, and on the consequent promise and utility of research and the spread of better information as a mode by which government might affect the situation for the better, in the report of the legislative committee which ardently supported creation of the Lapham commission. See Report, Select Committee on Bill 191A, AJ, 283–286 (1867). Compare the modest suggestions for more efficient cutting practice, with no clear specifications of implementing regulation, in Lapham *et al., Report on the Disastrous Effects,* 45–46. The report adverse to a second printing and distribution of the Lapham report was that of the Senate Committee on Agriculture, SJ, 387 (1869).

30. The general inattention of Congress to issues of efficiency of realization upon timberlands resources is noted in Clawson, *Uncle Sam's Acres,* 97, 104–106, 110–111, 296–301, and Clawson and Held, *The Federal Lands,* 64, 65; Gulick, 20, 23–25, 68, 69; Ise, 23–26, 39, 40–45, An over-all verdict pertinent both to federal and state inaction may be found in Pinchot, 23, 27, 70–71, 85. Retrospective judgments on the absence of attention to forest use in Wisconsin public policy may be seen in the General Messages of Gov. Upham, AJ, 27 (1895); Gov. LaFollette, AJ, 60 (1903); Gov. Davidson, AJ, 47, 48 (1909). On waterways and rail franchises, see Chapter III; on rate regulation, see Chapter IX.

31. Damage to public roads by loggers was dealt with in Revised Statutes, 1849, c. 16, sect. 108; Revised Statutes 1858, c. 19, sect. 125; General Laws, 1869, c. 152, sect. 125; Revised Statutes, 1878, sect. 1327; Wisconsin Statutes, 1898, sect. 1327. Cf. General Laws, 1872, c. 54; Revised Statutes, 1878, sect. 4559; Laws, 1897, c. 374; Wisconsin Statutes, 1898, sect. 4559 (injury to telegraph or electric lines by felling timber). Statutes penalizing felling timber or dumping mill waste to impairment of navigation included General Acts, 1856, c. 73; Revised Statutes, 1858, c. 19, sect. 112; P&LL, 1859, c. 141 (Sheboygan River, Sheboygan County); General Laws, 1861, c. 285 (broader coverage, stiffer penalties); General Acts, 1864, c. 269 (Fox River at Depere); General Laws, 1869, c. 152, sect. 112 (general law restated); P&LL, 1870, c. 38 (Pine River); Laws, 1875, c. 211 (Wolf and Embarrass rivers within stated limits); Revised Statutes, 1878, sect. 1598 (foregoing general acts restated in comprehensive formula); Laws, 1880, c. 314 (Wisconsin River and tributaries); Laws, 1881, c. 134; Wisconsin Statutes, 1898, sects. 4570a and 4570c. As the foregoing citations indicate, the legislature mixed local and general legislation indiscriminately in dealing with this problem of stream obstruction by loggers or millmen. The amount of local legislation was compounded by the practice of including in local government charters authority to deal with such stream obstruction. See Private & Local Acts, 1855, cc. 221 and 308 (Watertown and Burlington); Private & Local Acts, 1857, c. 136 (Neenah); P&LL, 1858, c. 128 (Two Rivers); General Laws, 1863, cc. 124 (Appleton) and 131 (Neenah); P&LL, 1866, c. 174 (Menasha), and 240 (Manitowoc), Laws, 1876, c. 47 (Appleton). Statutes penalizing the dumping of logging and mill waste to the damage of fisheries include General Laws, 1860, c. 314; Laws, 1887, c. 490; Laws, 1893, cc. 307 and 308; Laws, 1897, c. 188; Wisconsin Statutes, 1898, sect. 4567; Laws, 1901, c. 358, 1905, c. 402, 1913, c. 571. The turn into a new sweep of regulation is signalized by provision for a survey of water pollution generally in the state, under Laws, 1913, c. 568. See creation of the interagency committee on water pollution under Laws, 1927, c. 264. Legislative records show little evidence of controversy over most of these pollution measures. Efforts to repeal Laws, 1893, c. 307, in this respect stirred a flurry of petitions and counterpetitions in 1895. See AJ, 179, 194, 195, 210–211, 226, 240, 289, 308, 363, 422, 453, 555; (1895); SJ, 306. Such controversy as appeared seems focused on efforts to obtain exemptions or exceptions, directly or by redefinition of general coverage of the regulations; by the turn of the century it appears that conservation values had come into sufficient public respectability so that it was not expedient to attack them head-on. Cf. Wisconsin Statutes, 1898, sect. 4567, with the varying amendments made to it under Laws, 1901, c. 358, and 1903, c. 325; see also AJ, 141, 504, 609 (1901); AJ, 461, 530, 564–565 (1903). Maneuvers of local interests for relief from sawdust-pollution control failed, however, in bill 20S of 1903: see SJ, 1005, 1095, and AJ, 1271, 1291. And see an opinion of the Attorney General, to the District Attorney, Richland Center, May 20, 1907, PD, 916, (1909), strictly construing an exception clause in the sawdust-pollution statute. Judicial doctrine and legislation pertaining to industrial accidents in the lumber industry are

noted in Chapter VII. On the power position of labor, especially its want of organized expression in the woods side of operations, through most of the nineteenth century, see Fries, 210–220. The machinery-guards act was Laws, 1887, c. 549; see also Wisconsin Statutes, 1898, sect. 1636j, Laws, 1905, c. 303, and 1911, c. 396. On the failure of an effort at sawmill safety legislation in 1872, see end of note 39 below. See, generally, Gertrude P. Schmidt, "History of Labor Legislation in Wisconsin" (Ph. D. thesis, 1933), especially at p. 60. The typically grudging creation and expansion of a central administrative agency to bring policy definition and enforcement to the range and force appropriate to problems important to the over-all state economy is well reflected in the course of legislation establishing the Bureau of Labor and Industrial Statistics. See Laws, 1883, c. 319, 1885, c. 247, 1887, c. 453, 1891, c. 226, 1895, c. 140, 1897, c. 355; Wisconsin Statutes, 1898, c. 46a.

32. The history of bill 265A of 1872 may be traced in AJ, 258, 351 (adverse report without opinion by Committee on State Affairs), 472 (recommittal), 536–537 (adverse report by Committee on Agriculture, quoted in text), 586 (indefinitely postponed without a record vote). In the same session the Assembly adopted Joint Resolution 23A, introduced by the sponsor of bill 265A, instructing the Committee on Revision of Statutes to consider "the propriety of requiring all railroad companies hereafter incorporated, to plant forest or timber trees upon both sides of their track, within a limited time after its completion." The Senate concurred, but no further action was taken (AJ, 172, 179; SJ, 166). The same sponsor in the same session introduced Memorial 19A, urging Congress to make provision for planting trees on the western plains. This recommendation of such "timber culture" laws as Congress did enact in 1873 and 1878 passed the Assembly without a record vote, under a favorable report from that chamber's Committee on Federal Relations, and despite an adverse report from its Committee on Agriculture: AJ, 258 (1872), 497 (Committee on Federal Relations, 536 (Committee on Agriculture), 654 (adopted without record vote). However, on recommendation of its Committee on Federal Relations, the Senate rejected the Memorial (SJ, 533, 551). The planting-subsidy and intervention-by-education approach of bill 265A of 1872 anticipate the pattern of recommendations for a beginning forestry policy for the states made in a Memorial of the American Forestry Congress of 1883, transmitted without comment by Gov. Rusk to the legislature by his Special Message of Jan. 26, 1885, AJ, 60–64; the Memorial did bear supporting endorsements from the United States Commissioner of Agriculture and his Chief Forester (see note 42 below).

33. See E. M. Griffith, Superintendent of State Forests, "Forestry for Wisconsin Farmers," app. to Annual Report, State Board of Agriculture, 1904, PD, 293, 294 (1905); Griffith, "Forestry Conditions in Wisconsin," app. to Annual Report, Wisconsin State Horticultural Society, 1907, PD, 178, 180, 181 (1907); First Annual Report, State Forester, Nov. 8, 1906, PD, 9, 10, 15–17, 30–35, 45 (1907); Report, State Forester, Dec. 1, 1908, PD, 27, 32, 50–67 (1909). See, generally, Kleven, chap. viii. A typical statement of the twentieth-century emphasis on the relative importance of fire prevention and fire fighting to satisfactory forest growth may be seen in Natural Resources Committee of State Agencies, *The Natural Resources of Wisconsin,* 92–94.

34. See Ise, 91, 162, and Pinchot, 144, 276–279.

35. See Wisconsin Territory, Laws, 1838, p. 350, and Laws, 1840–1841, No. 6; Revised Statutes, 1849, c. 134, sect. 52; Revised Statutes, 1858, c. 165, sect. 59; Laws, 1873, c. 295; Revised Statutes, 1878, sect. 4406; Laws, 1887, c. 343, 1895, c. 266, 1897, c. 362, 1899, cc. 258 and 353, 1901, cc. 408 and 432, 1903, c. 450. We should note as relevant, but tangential, to this main stream of policy, General Laws, 1860, c. 95 (stipulating manner and extent of land clearing by holder of unpaid certificate of sale of state lands, to make out defense of clearing for cultivation as against charge of wrongful timber cutting); General Laws, 1861, cc. 59, 185, and 190 (steamboat spark arrestors required, but with many exceptions reflecting local pressures); Laws, 1874, c. 27, 1875, c. 228, and 1878, c. 154 (special provisions as to fire wardens and lumber yard hazards in charters of lumber-area municipalities). Contrary to the judgment passed in the text, the *Report of the Revisers of the Statutes,* 301, asserts that sect. 4406 presented "sec. 59, ch. 165, R.S. 1858, and chapter 295, Laws, 1873, condensed, but substance retained." Sect. 4406 very likely did only make explicit the meaning of the 1873 legislature, in declaring that the offense defined in its c. 295 included a showing of intent. But the Revisers' penalty for "gross

negligence" in allowing fire to escape from the accused's own land seems clearly to narrow the scope of the 1858 version, which penalized "neglect." On the other hand, though their Notes do not acknowledge it, the Revisers probably extended a little the offense of setting fire to another's land, when to the cases of "willful" and "malicious" action penalized under the 1849 and 1858 Revisions they added a penalty for one who "wantonly" so acted. The 1883 investigating committee on causes of fire loss was created without apparent controversy under Resolution 13A: see AJ, 29, 85, 111 (1883). There is no record of further action, beyond appointment of the committee of five. On the background of the 1895 act, see Ralph M. Van Brocklin, "The Movement for the Conservation of Natural Resources in the United States Before 1901" (Ph.D. thesis, 1952), 62. Bill 519A, which became Laws, 1895, c. 266, was subjected to considerable amendment in minor detail, but the record reflects no major controversy over it.

36. The failure of nineteenth-century legislative policy to come to grips with the forest fire problem stands out the more sharply because, though confidence in the "inexhaustible" extent of the forest set the general tone, there was some contemporary recognition of fire loss as a major cost, and, indeed, as sometimes a threat to the very existence of individual business firms. Social costs of fire damage were partly acknowledged in the humanitarian responses hereafter noted. But there was also some recognition of more focused economic cost. Thus the peculiar hazards of the industry led to the organization of specialized insurance companies. See, e.g., Private & Local Acts, 1853, cc. 107 (charter of Wisconsin Lumbermen's Fire & Marine Insurance Co.) and 284 (Dodge County Fire & Marine Insurance Co., empowered to insure lumber shipments by raft or other means); P&LL, 1868, c. 507 (charter of Lumbermen's Insurance Co.). See Fries, 106; Spencer L. Kimball, *Insurance and Public Policy* (1960), 9–13. In a few particular contexts the legislature recognized the seriousness of fire loss as a burden on private effort toward economic growth. See General Laws, 1872, Memorial No. 25 (asking United States to remove roving Indian bands, in part because of fire losses caused by Indians); Laws, 1873, c. 89 (fire losses on lands granted by United States in aid of Sturgeon Bay canal project as ground for grant of greater financial flexibility in use of the lands); Laws, 1877, Memorial No. 1 (asks supplementary federal money grant for Sturgeon Bay canal project, in view of depreciation of aid lands from fire loss); AJ, 69, 85, and 111 (1883) (creation of committee under Resolution 13A to study causes of forest fires and protective action; no report recorded). Calls for humanitarian aid to burned-out areas in the north were made in the General Messages of Gov. Washburn, SJ, App., 3, 9–10 (1872), and Gov. Upham, AJ, 25 (1895). See General Laws, 1872, cc. 97 and 173, and List of Appropriations, *id.*, p. 342, Item No. 25; Laws, 1895, c. 146, 1899, c. 110. On the possibility of political abuse of local fire warden organization, as a basis for a tight rein on the fiscal provisions therefor, see the comments — at a later stage — in the First Annual Report of the State Forester, Nov. 8, 1906, PD, 10 (1907). The State Forester's observations on the grudging attitude of local officials toward spending money to put out fires which endangered new growth serve also to emphasize the need of informed central direction. See *id.*, 9, 15–16, and also Report, State Forester, Dec. 1, 1908, PD, 32 (1909) ("still there are even some town boards who object to fire wardens 'wasting time and money' in putting out fires which are burning on cut-over lands, and such men seem to think that no damage is being done unless merchantable timber is being destroyed").

37. On contract "regulation" of timber cutting, see Chapter IV, especially at notes 10, 20, 41, 42, and 50. The limited kinds of cutting regulations fixed by private contract — directed, it seems clear, at considerations pertinent to immediate transactions rather than to maintenance of the land's productive potential — may be seen in Lacy v. Johnson, 58 Wis. 414, 419, 17 N.W. 246, 248 (1883); Everett v. Gores, 89 Wis. 421, 422, 62 N.W. 82, 83 (1895); Holm v. Colman, 89 Wis. 233, 234, 61 N.W. 767 (1895); Shores Lumber Co. v. Stitt, 102 Wis. 450, 452, 78 N.W. 562, 563 (1899); Fogo v. Boyle, 130 Wis. 154, 109 N.W. 977 (1906). Most often it appears that the price and finance terms of sales of timberland or standing timber were such as to enforce pressure for quick cutting without limitation of method. See, e.g., Beckwith v. Philleo, 15 Wis. 223 (1862) (no infringement of vendor's lien by cutting, where contract obviously contemplated land should be paid for with proceeds of logging operations); O'Connor v. Semple, 57 Wis. 243, 15 N.W. 136 (1883) (terms of offer: 20 per cent or 25 per cent down on price of timberland, balance to be paid as timber cut, with 8 per cent interest per annum); Snively v. Keystone Lumber Co., 129

Wis. 54, 108 N.W. 215 (1906) (grantor reserving timber agrees to pay taxes so long as timber is uncut). Cf. Idema v. Comstock, 131 Wis. 16, 110 N.W. 786 (1907) (sale rather than partition should be decreed, where partition would leave timber tract of uneconomic size for logging). In his first Annual Report, the State Forester in effect noted that private contract was not significantly used to erect safeguards on the long-term productivity of timberland. He warned that "owners of stumpage must soon come to realize that they cannot afford to sell to lumber companies without some restrictions on the cutting, for otherwise their timberlands will be either destroyed or at least seriously injured in the lumbering": see Report of Nov. 8, 1906, PD, 55 (1907).

The typical tie of questions of "waste" to monetized questions concerning security for unpaid balances or other money-measured interests in the land may be seen in Wright v. Wing, 18 Wis. 45 (1864) (tax deed grantee); Smith v. Bouck, 33 Wis. 19 (1873) (parol contract, accounting claimed); Riemer v. Johnke, 37 Wis. 258 (1875) (ejectment); Huddleston v. Johnson, 71 Wis. 336, 37 N.W. 407 (1888) (executory contract of sale of timberland does not imply contract vendee's right to cut). Cf. State v. Weston, 17 Wis. 107 (1863) (conversion by state's vendee). See also Krakow v. Wille, 125 Wis. 284, 103 N.W. 1121 (1905) (protection of contract vendee against his vendor's post-contract cutting).

Timber trespass law, statutory and judge-made, is discussed in Chapter V. The characteristic measurement of trespass damages in terms of present market values, with no attention paid to loss of opportunity to exert future preference, is reflected in the very legislation which expanded the scope of plaintiff's recovery for timber trespass to allow him to recover "the highest market value of such logs, timber or lumber . . . between the time of such cutting and the time of the trial . . . and while it remained in the possession of the party so cutting the same or any purchaser thereof with knowledge of such wrongful cutting . . ." See Laws, 1873, c. 263, and discussion of its application to allow recovery for a rise in the market value of stumpage since the trespass, in Webster v. Moe, 35 Wis. 75, 78–79 (1874). The point is sharpened because the opinion of the court by Cole, J., acknowledged loss of future opportunity value as a real damage, but clearly saw no expansion of the idea beyond such rise in value as might occur within the relatively short period of the statute of limitations (six years): "The timber may be cut by the wrongdoer when it is low, and when the owner wishes it to stand in order to give him the benefit of a rise in the market . . . And if one chooses to invade the rights and property of another, he cannot complain if the owner waits for a rise in the market and delays bringing his action" (*id.*, 79). Typical of the limitation of litigants' and courts' interest to questions of the present market value of mature timber is Nelson v. Churchill, 117 Wis. 10, 93 N.W. 799 (1903). Standard nineteenth-century damages doctrine in general provided no clear recognition of future opportunity value as a recoverable element. See, generally, Theodore Sedgwick, *The Measure of Damages* (1912), sect. 933. Cf. Ciriacy-Wantrup, 45, 67, 70, 73–74 (concepts of competitiveness and complementarity in resource uses and use rates); *Harvard Law Review*, 48:485 (1935), note on damages allowed for exploratory trespass on oil and gas lands.

Claims for loss to standing timber by defendant's fire may be seen in suits against the Minneapolis, St. Paul & Sault Ste. Marie Railway Co. by plaintiffs Montague, 96 Wis. 633, 72 N.W. 41 (1897), Cook, 98 Wis. 624, 74 N.W. 561 (1898), and Clifford, 105 Wis. 618, 81 N.W. 143 (1900); Bratz v. Stark, 138 Wis. 599, 120 N.W. 396 (1909); Kazmierczak v. Kokot, 154 Wis. 599, 143 N.W. 661 (1913). None of these cases contains any discussion of the measure of damages; in particular, there is no discussion of loss of future opportunity value as an element of damages. Compare, however, Miller v. Neale, 137 Wis. 426, 430, 119 N.W. 94, 96 (1909), where, in an action for fire damage to a farm woodlot, the court brushes aside as "trivial" an objection to introduction of plaintiff's testimony "that the timber was good timber, also that he was keeping it for his own use . . ." It is not always clear which of the numerous tort actions for fire damage in the Wisconsin Reports pertain to timber values or the conduct of the lumber industry. With some likely margin of error, one may count between twenty-five and thirty such reported cases up to mid-twentieth century. Examples of suits where the forest-products-connected damages claimed were other than damages for loss of standing timber are Read v. Morse, 34 Wis. 315 (1874) (sawmill); Whitney v. Clifford, 46 Wis. 138, 49 N.W. 835 (1879) (lumber); Stacy v. Milwaukee, Lake Shore & Western Railway Co., 85 Wis. 225, 54 N.W. 779 (1893) (sawmill and piled lumber).

38. See the discussion of background attitudes affecting policy toward public timberland, in Chapter II. The influence of popular and official images of desirable trends of community development, framed in the symbol of the fee simple title, is contained more in implication than in declaration, and is hard to untangle from other closely related ideas.

Reliance on unqualified private ownership of property as an assurance that growth was channeled in the right directions was especially apt to appear in conjunction with expressions of the deep faith which the times had in productivity as a social dynamic and in the desirability of a fast over-all pace of growth. See, e.g., General Message of Gov. Dodge, Nov. 27, 1838, Council Journal, 9 (should give every "indulgence" to settlers, for "we would then have a population, owners of the soil, attached to the republican institutions of our country . . . [and] where the unsubdued forest now stands, we would see a free enterprising population"); General Message of Gov. Fairchild, Jan. 14, 1869, SJ, 22 (manufacturing and mining developments and "the activity displayed in our pineries, all tell of a day, not distant, when the labor of the state will be so diversified as to vastly multiply its wealth and resources, and present additional attractions to those seeking homes in the west"); Address of Lt. Gov. Fifield, Jan. 11, 1882, SJ, 5 ("The people of Wisconsin may well feel proud of the rapid advancement which is being made. Throughout the northern portions of the state, where but a few years ago solitude reigned supreme, the iron track has been advanced through our matchless forests; industries have multiplied and sprung into existence, and civilization is being extended into the wilderness. Emigration is pouring in, the rich virgin soil is being taken up and cultivated, in fact a 'new Wisconsin' is rapidly assuming strength and fair proportions, and the hum of industry and the bustle of commerce are now to be heard from the state line south to the shores of the 'great inland sea of the north' "); General Message of Gov. Rusk, Jan. 11, 1883, AJ, 19 (thanks to railroad construction, "an almost trackless region of unbroken forest is now being transformed into thriving villages and prosperous homes . . . vastly increasing [the state's] . . . taxable values").

This uncritical confidence in the application of the fee simple and of the fast-paced-productivity faith to the north country was backed until the eighties, at least, by the popular assumption that the resources of the forest could not be exhausted within any foreseeable future. See, e.g., Communication from Hon. Alfred Brunson to Gov. Doty, on travels in "the northwestern wilderness of Wisconsin," Dec. 6, 1843, HRJ, 33, 35; Ellis, "The 'Upper Wisconsin' Country," in Annual Report and Collections, Wisconsin State Historical Society, AJ, App., 444 (1857); report of Edward Daniel, appended to Report, Commissioners of the Geological Survey, PD, 10 (1859); Report, Committee on Incorporations, on bill 9S, AJ, 454–457 (1866). Compare, however, Report, Committee on Incorporations, on bill 9S, SJ, 135–137 (1866); Report, Committee on Education, AJ, 73 (1871); Report, Select Committee on the St. Croix aid lands, AJ, 706 (1873); Annual Report, Commissioner of Immigration, PD, Doc. No. 15, p. 82 (1873); Annual Report, Board of Immigration, AJ, 113 (1881).

One may be sure that, had conservation measures been seriously pressed, they would soon have encountered the argument that the existing market for forest product was not yet well enough developed to support any elaborate differentiation of uses and qualities. See, e.g., Minority Report, Select Committee on Bill 298A, AJ, 732–733 (1864). Material already cited in this note indicates, likewise, how much a part of the pattern of ideas was the assumption that agriculture would normally follow clear-cutting of the timber. A related factor in deflecting particularized attention from the future of forest industry was the common acceptance of agriculture as so unquestionably the long-term mainstay of the Wisconsin economy as to subordinate public interest in other aspects of the economy. Stress on the primacy of agriculture, and emphasis that what the law did in support of the farm was its really meaningful contribution to the general economy, were staples in public policy discussion through the nineteenth century. In comparison, references to the lumber industry were surprisingly few, casual, and secondary. Particularly striking is the unbroken continuance of this over-all pattern of reference into years of lumber boom. See, e.g., General Messages of Gov. Fairchild, SJ, 18 (1867), and SJ, 22 (1869); General Message of Gov. Taylor, AJ, App., 23–24 (1874); General Messages of Gov. Smith, PD, I, 16 (1878), and AJ, 31 (1879); General Messages of Gov. Rusk, AJ, 13 (1882), AJ, 19 (1883), AJ, 22 (1885), and AJ, 23 (1887) — and compare the fact that Gov. Rusk transmitted without comment, by his Special Message, AJ, 60–64 (1885), a strong, action-oriented memorial of the American Forestry Congress on the desirability of forest conservation efforts by state legislatures; General Message of Gov. Hoard, AJ, 23 (1889); General Message of Gov. Peck, SJ, 24 (1891). It is somewhat ironic that in his General Message, AJ, 23 (1895), Gov. Upham complained, with reference to the north country, that "the mineral and lumber interests have for so many years claimed the attention of capitalists and workingmen that the agricultural interests of a portion of our state have been somewhat neglected."

Stress on the primacy of agriculture runs through legislative records as it does in recommendations of the chief executive. See, e.g., Reports, Committee on Agriculture and Manufactures, AJ, 539 (1854), and Committee on Agriculture and Manufactures, SJ, 233, 234 (1858) (recommending establishment of an agricultural department in the state university); report of the same Senate Committee, on Morrill Act aids, SJ, 359 (1863); Report, Commissioner of Immigration, PD, 9, 13, 24, 27 (1871); Eulogy of Senator John Carey, by Senator William Nash, SJ, 96–97 (1889); Biennial Report, Bureau of Labor, Census, and Industrial Statistics, PD, II, 2, 145 (1896). See, also, Stroebe v. Fehl, 22 Wis. 337, 344 (1867) (not "waste" in tenant to cut to clear in unsettled country); Stephenson v. Wilson, 37 Wis. 482, 494 (1875) (agriculture primary, but not sole, use of land that will qualify as adverse possession).

In significant association with these stereotypes of the primacy of agriculture was the paradox that, despite long-held uncritical belief in the inexhaustible wealth of the forest, there was also unquestioned acceptance that under logging the pine would inevitably "fail" in any given timbered area. See, e.g., Communciation to Secretary of State from Ira Brunson, appraiser of school lands in Chippewa County, SJ, App., 829 (1848); Report, Committee on Incorporations, on bill 9S, SJ, 135–137 (1866); Biennial Report, Bureau of Labor, Census, and Industrial Statistics, Sept. 30, 1888, PD, II, p. xxviii (1889); Minority Report, Committee on Town and County Organization, on bill 57A, AJ, 511–512 (1893).

39. The contests that occurred over timber trespass, acquisition of railroad-aid lands, and control of the Chippewa River left ample evidence behind them. The times were not given to subtlety, and it would strain credulity much too far to imply a conspiracy of silence to keep from the record evidence of comparable controversy over conservation policy. Consider, for example, the matter of the investigation of causes of forest fires, authorized under Resolution 13A of 1883, note 35 above. The resolution was approved without a record vote two days after its introduction, in the form in which it was introduced, without evidence of dispute. On the other hand, the record shows no further action beyond appointment of a committee thereunder. It would be mere surmise, to infer that this silence indicated behind-scenes maneuvering to bring the project to nothing; in the context of the scanty over-all legislative record from 1838 to 1883, the more reasonable inference is that the project did not arouse enough interest to lead anywhere. The course of bill 390A, which became Laws, 1887, c. 343 (authorizing town boards to ban burning of logs and brush in specified dry months) offers another example. The legislative record on its face suggests the possibility of some conflict of interest behind this enactment. Routinely passed by the Assembly in the form in which it was introduced, bill 390A was first recommended for indefinite postponement by the Senate Judiciary Committee (two members dissenting without opinion), and the Senate so voted: see SJ, 383, 384, 408 (1887). But the Senate then asked and obtained return of the bill from the originating house, whereupon — all on the same day — the Senate voted to reconsider the bill, recommitted it to the Committee on Manufactures and Commerce, and subsequently adopted it, with an amendment as recommended by that committee, in which form the bill was accepted without demur by the Assembly: SJ, 585, 588, 592 (1887); AJ, 953. Despite all this maneuver, all that was changed in the bill as introduced was to add the month of August to the list of months in which town boards might prohibit setting fires. This record provides scanty circumstantial evidence to prove that lumber lobbyists were alertly active behind scenes to balk fire-control regulation.

Opposed to the verdict here passed on the absence of lobby conflict in this field of policy, however, are two comments of Wisconsin's first State Forester. In his essay, "Forestry Conditions in Wisconsin," in the Annual Report of the Wisconsin State Horticultural Society, 1907, PD, 178, E. M. Griffith asserted that "Wisconsin would have passed . . . a law [regulating slash disposal] many times if the lumbermen had not defeated such laws in the legislature — I am not blaming the lumbermen for everything, possibly business conditions demanded it to a large extent, but they often blame the state for not having enforced the rules, but the lumbermen came to Madison and to St. Paul, and they defeated those bills in the legislature . . ." The First Annual Report of the State Forester, Nov. 8, 1906, PD, 45, (1907) commented to like effect: "If, twenty years ago, the state of Wisconsin had passed a law providing that in all lumbering operations the slash must be piled and burned, there would still be millions of acres of standing timber in the northern part of the state, which is now a burned, desolate waste. Several attempts were made to pass such a law, but in every instance it was killed by the lumbermen." The State Forester there quotes Mr. F. E. Weyerhaeuser, speaking at an American Forestry Congress meeting in Washington in 1905, that lumbermen caused a Minnesota slash-control bill to be killed in committee. There may have been a similar quiet, off-stage throttling of fire-control proposals in Wisconsin. But I could find no evidence of this except the undocumented assertion of the State Forester. If the lumbermen had their way so easily, the fact would reflect the general inertia and indifference which generally characterized the situation.

A further instructive instance supporting this reading concerns the record evidence of successful blocking of an early attempt to require guarding mill machinery. Bill 213A of 1872 was first referred to the Assembly Select Committee on General Laws. Apparently distrusting its future there, the bill's sponsor obtained its withdrawal and recommitment to a special select committee of which he was named chairman. Despite these presumably more favorable auspices, no more was heard of the bill, which evidently died in committee. The record shows the filing of a memorial in opposition to the bill by seventy manufacturers of Oshkosh, impressively headed by C. N. Paine & Co., and P. Sawyer & Son: see AJ, 220, 241, 482 (1872). This is a record which offers at least circumstantial basis for an inference of effective behind-scenes lobbying; there is no comparable incident in the record affecting the problem of fire control. The debate in the U.S. Senate on the southern timberlands is noted in Ise, 50–52.

40. The quotation in the text from the legislative mandate to the Lapham commission is from General Laws, 1867, c. 36, sect. 1: III. On the competitiveness of present and future use patterns, compare Annual Report, State Agricultural Society, 1858, SJ, App., Doc. L, p. 8 ("comparatively productive as our agricultural labor no doubt is, yet in most cases it is so at the expense of the means of subsistence of the next . . . generations"). The quoted comments of the Assembly select committee on bill 191A will be found in AJ, 283, 284 (1867).

41. The arguments made in favor of public support to agricultural experiment and education will be found in Reports, Committee on Agriculture and Manufactures, SJ, 235 (1858), and SJ, 359 (1863). Cf. Annual Report, University Regents, 1867, PD, 730 (experience shows that it is difficult to sustain agricultural schools, and "this difficulty will be increased in the West where land is yet so cheap and so productive that there is less inducement to study agriculture than in older portions of the country where land is much dearer and where a more scientific cultivation of the soil is demanded").

42. For the memorial of the American Forestry Congress of Aug. 9, 1883, see Special Message of Gov. Rusk, AJ, 62 (1885). The implication of the abortive 1883 provision for investigation of the causes of forest fires in Wisconsin seems probably along similar lines of emphasis on study, persuasion, and cooperation. See end of note 35 above.

Chapter VII. Regulating Use of Private Mobile Capital: The Costs of Money and Labor

1. See Fries, 102, 207–208; Third Biennial Report, Bureau of Labor and Industrial Statistics, Sept. 30, 1888, PD, II, 24, 25 (1889).

2. See Hurst, *Law and the Conditions of Freedom*, 6–8, 12, 32.

3. See *id.*, 75–76, 103, 104–105.

4. See Regular Message of Gov. Randall, SJ, 25–27 (1859); Report, Committee on Land Limitation, AJ, App., 1106, 1107 (1851); Minority Report, Committee on Finance, on bill 12S, SJ, 56–57 (1862); Fisher v. Otis, 3 Pinn. 78, 95 (Wis. 1850); Pratt v. Ayer, *id.*, 236, 258, 259 (Wis. 1851); Riley v. Gregg, 16 Wis. 666, 672 (1863); Lee v. Peckham, 17 Wis. 383, 387 (1863); Morton v. Rutherford, 18 Wis. 298, 309 (1864); McArthur v. Schenck, 31 Wis. 673, 676 (1873). Cf. Merk, 235–237.

5. Statutes of the Territory of Wisconsin, 1839, p. 156; Revised Statutes, 1849, c. 45; General Acts, 1851, c. 172; General Laws, 1860, c. 202, 1862, c. 252, 1866, c. 120; Revised Statutes, 1878, sects. 1688 and 1689; Statutes, 1898, sects. 1688 and 1689.

6. Note 4 above. Cf. Gara, 57, 77, 82, 88. See, generally, Bogue, 263, 265, 266, 272.

7. See Fries, 14, 100, 105, and Reynolds, 130–137, 139, 146.

8. Compare Marjorie Loomis Marshall, "History and Analysis of the Wisconsin Securities

Law," *Wisconsin Law Review,* 1942: 552, and George H. Young, "Some Comments on the New Wisconsin Business Corporation Law," *id.,* 1952: 5, with Daniel J. Dykstra, "Corporations in the Day of the Special Charter," *id.,* 1949: 310, 316, 323–330.

9. See Fries, 126–127; Marshall, *Autobiography,* I, 272–274.

10. See Fries, 103, 138–140, 154, 160; Reynolds, 10, 18, 23, 37.

11. See Fries, 129–138; Larson, 241, 382, 390.

12. See Hurst, *Law and the Conditions of Freedom,* 89–94.

13. See Chapter V, note 81.

14. United States v. Nelson, 52 Fed. 646, 647 (D. Minn. 1892).

15. Laws, 1897, c. 357; Laws, 1907, Joint Resolution No. 20. Cf. Pulp Wood Co. v. Green Bay Paper & Fiber Co., 157 Wis. 604, 147 N.W. 1058 (1914), s.c., 168 Wis. 400, 170 N.W. 230 (1919), cert. den., 249 U.S. 610 (1919) (pulp and paper).

16. General Laws, 1872, Joint Resolution No. 1. Cf. Laws, 1878, Memorial No. 14; Laws, 1883, Memorial No. 11. Unavailing minority sentiment for tariff reductions affecting forest products are reflected in Minority Report, Committee on State Affairs, AJ, 983, 986 (1867); proceedings on proposed Joint Resolution 15S, AJ, 376–377 (1880). See, generally, Fries, 113–121.

17. See Laws, 1909, Joint Resolution No. 50; Laws, 1921, Joint Resolution No. 27.

18. On the policy background of the log-labor lien acts, see Chapter V, notes 83–85, 90, 93 (in general), 96 (suppliers' liens) and 97 (relation of liens to various interests involved). Stockholder liability for six months' wages owing and unpaid to corporation employees was declared, for corporations organized thereunder, by General Acts, 1853, c. 68, sect. 25, carried into Revised Statutes, 1858, c. 73, sect. 25, and Revised Statutes, 1878, sect. 1769. Incorporation of the provision into the 1878 revision was held to apply it to corporations theretofore organized under special acts, as well as to those formed under the general laws: Sleeper v. Goodwin, 67 Wis. 577, 31 N.W. 335 (1887). For other recognition of the social importance of a reliable labor supply to productive growth in the lumber country, see General Laws, 1872, Memorial No. 16 (seeking citizenship and freehold rights for Oneida Indians, praising their quality as good axmen); Laws, 1883, c. 270, and Statutes, 1898, sect. 2525 (jury duty exemption for key mill personnel).

19. The main steps in the wavering course of state immigration agencies may be traced in General Acts, 1852, c. 432, and 1853, cc. 34 and 53, repealed by General Acts, 1855, c. 3; General Laws, 1867, c. 126, supplemented by General Laws, 1868, c. 171, 1869, c. 118, and 1870, c. 50; General Laws, 1871, c. 155, terminated as of January 1876 by Laws, 1875, c. 238; Laws, 1879, c. 176, supplemented by Laws, 1880, c. 194, and 1881, cc. 222 and 223, and repealed by Laws, 1887, c. 21; Laws, 1895, cc. 235 and 311, supplemented by Laws, 1897, c. 327, and 1899, c. 279, and brought to termination by Laws, 1901, c. 291; Laws, 1907, c. 407, supplemented by Laws, 1909, c. 444, and 1913, c. 772, with transfer of functions to the state department of agriculture by Laws, 1915, c. 413.

Examples of the farm settlement focus of this effort and the absence of stress on the lumber industry may be seen in the General Message of Gov. Farwell, SJ, 26 (1852), and Report, Select Committee on Bill 356A, AJ, 534–536 (1852); General Message of Gov. Fairchild, SJ, 22 (1867); Reports, Board of Immigration, AJ, 56–61 (1880), and AJ, 112–113 (1881); General Message of Gov. Rusk, AJ, 21 (1883), and Report, Board of Immigration, AJ, 44–45 (1883); Report, Board of Immigration, Dec. 20, 1898, PD, II, 1, 2 (1899). See, generally, Helgeson, "Promotion of Agricultural Settlement," chap. iii.

The opposition that flared from time to time against the investment of state money in promoting immigration represented no firm stand on laissez-faire principle, but involved a hodge-podge of objections on grounds of political patronage, alleged inefficiency of the agencies, some nativist sentiment, and occasional concern over the social radicalism of foreign labor. Little of this had any direct reference to the lumber industry as such. See, e.g., AJ, 801 (1852) (statement on patronage); Report, Select Committee for Repeal of Emigration Agency Legislation, SJ, 66 (1855) (ineffective; nativist sentiment); Report, Committee on State Affairs, AJ, 550 (1864) (need to economize); General Message of Gov. Taylor, AJ, 10 (1874) (need to economize); Biennial Reports, Bureau of Labor and Industrial Statistics, Sept. 30, 1886, PD, II, xxv, xxvi, xxviii, and 416ff (1887) (doubts of efficiency; fear of political unrest; some general opinions of this sort quoted from lumbermen), and Sept. 30, 1888, PD, II, xxv, 6–7, 11, 12 (1889) (workingmen's opinions opposed to promotion of immigration because of urban labor market competition); Special

Message of Gov. LaFollette, AJ, 954 (1901) (board ineffective and a waste of public funds).

The lumbermen's interest in immigration promotion, favorable or unfavorable, was limited and marginal to their concerns. Except in hardwood lumber areas, where transport costs favored development of a combination of seasonal local logging and manufacturing effort and hence gave particular reason for employers to stress a stable labor supply, the lumbermen looked with some doubt on immigration, as likely to promote demands for local public services costing more tax dollars. See Helgeson, "Promotion of Agricultural Settlement," 21, 23, 108. Some lumber interest in promoting settlement came with the turn of the century, when companies sought markets for cut-over lands (*id.,* 83, 105–108). For references in immigration promotion literature to the job opportunities for future farmers in the pineries, see Report, Select Committee on Bill 106A, AJ, 706 (1873), and local statements appended to Report, Commissioner of Immigration, PD, Doc. No. 15, pp. 20, 31, 53, 55, 81, 82, 92 (1873), and in Report, Commissioners of Immigration, AJ, 56 (1880); see also Helgeson, 6, 15, 16, 73, 75, 88. That the temper of the times would not easily conceive that official promotion should go beyond general advertising (for example, to extend capital finance aid to settlers), see Helgeson, 96, 238.

Immigration promotion was not limited to state action. The legislature chartered companies to encourage immigration for settlement in the north country. See P&LL, 1864, c. 382, 1870, cc. 212, 225, and 327; cf. Helgeson, "Promotion of Agricultural Settlement," 12–14, 20. Also, from the nineties at least, various county boards on their own initiative appropriated sums to foster immigration within their borders, and by Laws, 1905, c. 458, the legislature somewhat distrustfully resolved doubts of the legality of such action by authorizing it within close limits: see Helgeson, 100, 101, 231. At least through its years of most conscious scarcity of people, the state repeatedly conveyed to Congress its official desire that federal policy foster and protect immigration. See General Laws, 1864, Memorial No. 14, 1865, Memorial No. 11, 1867, Memorial No. 14; General Message of Gov. Fairchild, SJ, App., 21, 24–31 (1871); cf. Laws, 1893, Joint Resolution No. 3.

On the rudimentary state of official "statistics" on lumber country labor in the first generation of the industry, see Report, Committee on Lumbering and Fishing Statistics, SJ, 1181, 1182 (1858); compare comments on the limitations of data-collecting facilities in Report, Bureau of Labor and Industrial Statistics, Sept. 30, 1888, PD, II, ix, xxiii (1889). On the first state public employment offices, see Laws, 1901, c. 420, and 1903, c. 434; cf. General Message of Gov. LaFollette, AJ, 100 (1903). The office in the city of Superior was first set up by local ordinance in February 1899. See *The Wisconsin Free Employment Offices,* Wisconsin Industrial Commission, Bulletins, vol. 2, no. 9 (1913), 196. On the utility of such offices to the lumber industry, with its high demand for unskilled labor, see, *id.,* 198, 200; Gordon M. Haferbecker, *Wisconsin Labor Laws* (1958), 115.

20. General Laws, 1867, c. 83; Laws, 1877, c. 289, as amended by Laws, 1878, c. 187; cf. Laws, 1883, c. 319, as amended by Laws, 1885, c. 247 (factory inspector under Bureau of Labor and Industrial Statistics). Sanctions under the 1867 act were vague, since the employer was penalized only if he "compelled" any woman or child under eighteen to work more than eight hours per day, or if he "permitted" any child under fourteen under his control to work more than ten hours per day. The legislature did not move specifically to ban employment of minors in the operation of saws until Laws, 1909, c. 338, broadened by Laws, 1913, c. 466, into a general ban on employing children in conditions inherently dangerous. On the relatively limited employment of women and children in the lumber industry, see Fries, 205–206. Compare, generally, Haferbecker, 69, 71, 72, 74, 77.

21. AJ, 37, 116, 124, 132, 305, 504 (1878) (floor amendment by sponsor, in effect approving deduction from wages of debts due at company store, adopted 80 to 7); SJ, 506, 558.

22. On the background of Laws, 1889, c. 474, see AJ, 123, 719, 796, 870 (1889) (sponsor's floor amendment, enlarging lumber industry exception), 1016; SJ, 780, 816, 821, 848. Cf. Biennial Report, Bureau of Labor and Industrial Statistics, Sept. 30, 1886, PD, II, 244–245 (1887); *id.,* Sept. 30, 1888, PD, II, xxiii (1889).

23. Laws, 1915, c. 114, repealed and recreated Statutes, 1898, sect. 1729a, which had carried forward Laws, 1889, c. 474. On bill 117S, which became Laws, 1915, c. 114, see an opinion of the Attorney General, raising but not passing judgment on the question whether the bill's exceptions might create a doubt of the bill's validity, consistent with

the equal protection clause: AJ, 715–717 (1915); cf. SJ, 405–406. On bill 395S of 1933, see SJ, 114, 1334, 1367 (1933); AJ, 1950, 2074, 2191.

24. On the background of Laws, 1891, c. 430, see note 22 above. Bill 83A, 1893, would have stricken all exceptions from Laws, 1889, c. 474, thus enforcing generally the requirement of at least biweekly cash payment of wages. The bill was adversely reported by the Assembly Committee on State Affairs, AJ, 828–829 (1893). The committee believed that cases of abuse by deferred payment were rare, and doubted its ability to draft legislation "so as not to interfere with the right of contract or mutual agreement and to meet certain contingencies in the employment of labor." Moreover, the committee argued that blanket prohibition of deferred wages would harm laboring men by imposing unduly harsh requirements on shoestring employers who provided a good deal of demand for labor: "It must also be considered that not all employers of labor are capitalists. That as a rule employers are themselves laborers, contractors and subcontractors, sometimes depending upon their ability to pay promptly upon the elements and other contingencies and all laws having a tendency to drive this class of people from the field as employers, would be rather detrimental to labor, and might have a tendency to reverse a condition which has been the pride of our country, that the laborer of today may become the employer of tomorrow." The committee report made no explicit reference to the lumber industry, and its last-quoted argument seems as applicable to the urban building trades as to woods operations. On the other hand, bill 83A was introduced by a member from the lumber and mining country (Ashland): AJ, 85 (1893). The sponsor tried to save bill 83A by obtaining its further reference to the Judiciary Committee, but there also it received an adverse report (AJ, 914), and it was then rejected by the Assembly (AJ, 939).

25. Statutes, 1898, sect. 1729a. Cf. sect. 3333 (assignability of log-lien claims).

26. Laws, 1915, c. 114.

27. Bill 247A, 1895 ("to prohibit persons . . . from engaging in the business of carrying on stores known as company stores or general supply stores") was rejected on recommendation of the Assembly Committee on State Affairs, AJ, 745, 799. Bill 30S, 1895 ("to prohibit mining and manufacturing corporations from engaging in business of carrying on stores known as company stores or general supply stores") was rejected on recommendation of the Senate Judiciary Committee, on a requested record vote of 17 to 6, SJ, 500, 526. The Attorney General's opinion of Apr. 24, 1913, to the District Attorney, Oconto Falls, ruled most specifically that the conduct of a company store on the terms stated did not constitute action in restraint of trade in violation of the state's antitrust laws (Attorney General, *Opinions,* II, 283). Cf. Olson v. Sawyer-Goodman Co., 110 Wis. 149, 85 N.W. 640 (1901) (worker may recover logging wages without deduction for company store goods furnished other persons in discharge of plaintiff's gambling debts, under arrangement made by storekeeper, which is held against public policy). Laws, 1939, cc. 129 and 490, belatedly dealt with the company store outside the lumber industry. Fries, 209, notes the conviction of several lumbermen at LaCrosse in 1875 for violation of the federal currency laws by issue of scrip payable at their stores. However, United States v. Van Auken, 96 U.S. 366 (1877) subsequently held, in a case arising in Michigan, that this practice was lawful, so far as existing federal statutes were concerned.

28. Biennial Report, Bureau of Labor and Industrial Statistics, Sept. 30, 1888, PD, II, 113 (1889); cf. *id.,* 91, 112. Examples of such disciplinary provisions may be seen in the contracts at issue in Pearson v. Kelly, and in Kosloski v. Kelly, 122 Wis. 660, 665, 100 N.W. 1064, 1037 (1904). See, generally, Fries, 209–210, and Reynolds, 52, 53.

29. See Knutson v. Knapp, 35 Wis. 86 (1874) (wrongful discharge); McElroy v. Eau Claire Lumber Co., 57 Wis. 189, 15 N.W. 132 (1883) (applicability of alleged docking practice to particular contract); Brunnell v. Hudson Saw Mill Co., 86 Wis. 587, 57 N.W. 364 (1893) (similar); Kosloski v. Kelly, 122 Wis. 665, 100 N.W. 1037 (1904) (similar). Inventory of the cases brought to the circuit court for Chippewa County showed no actions for wages there growing out of the lumber industry in years when lumber dominated the local economy; on the other hand, suits upon statutory log-labor liens, and even more the filing and docketing of log-lien petitions as steps to establish lien claims, amounted to sizable items of circuit court business: see Laurent, chaps. iii and iv.

30. See Current, 28, 29, 43, 243–244; Fries, 213–218; Kleven, 455–460; Reynolds, 51–53, 90–92. On reflection of the situation in the business of the Chippewa circuit court, see Laurent, chaps. i and ii.

31. With General Laws, 1867, c. 154, com-

pare Revised Statutes, 1878, sect. 4451; Laws, 1880, c. 94; Statutes, 1898, sect. 4451. See bill 35A, 1880, and AJ, 111, 144 (1880); SJ, 316, 336–337, 350 (1880) (passed on requested record vote of 18 to 8). See also bill 426A, 1882, and AJ, 472, 578 (1882); SJ, 529, 552 (1882).

32. Laws, 1885, c. 247, 1895, c. 364; Biennial Report, Bureau of Labor and Industrial Statistics, Sept. 30, 1886, PD, II, 238–246, 310–313, 388–389, 407, 411–412, 431–432 (1887); Report, State Board of Arbitration and Conciliation, Jan. 2, 1899, PD, I, 5, 25.

33. See Revised Statutes, 1849, c. 147, 1858, c. 168, 1878, sects. 4511 (new, enlarged offense of unlawful assembly or riot), 4513 (continuing old provision on same subject), and 4568 (overt act necessary for conspiracy); Laws, 1887, cc. 287, 349, and 427, carried into Statutes, 1898, sects. 4466a–c. Cf. State ex rel. Durner v. Huegin, 110 Wis. 189, 261, 85 N.W. 1046, 1066 (1901) (Laws, 1887, c. 287, is declaratory of common law doctrine of conspiracy, requiring no proof of overt act, and insofar in effect repealing Revised Statutes, 1878, sect. 4568). See, on this legislation, Schmidt, 13, 15, 264, 273; on the pressure felt from about the turn of the century by lumber industry employers from competing labor markets, Fries, 214–215, and Reynolds, 52, 53, 92.

34. Biennial Reports, Bureau of Labor and Industrial Statistics, Sept. 30, 1886, PD, II, 246, 495 (1887); *id.,* Sept. 30, 1888, PD, II, ix, xxiii, 245, 254, 276 (1889); *id.,* Sept. 30, 1906, PD, 76 (1907); Report, State Board of Health, Sept. 30, 1902, PD, 23–28, 47–49, 55–57, 68–69, 71, 211 (1903), and *id.,* Oct. 1, 1904, PD, 38, 39, 60 (1907); Wisconsin Industrial Commission, *Labor Camps in Wisconsin* (1913?), 15, 23, 34; *Accidents Caused by Objects Striking Workmen,* Wisconsin Industrial Commission, Bulletins, vol. 2, no. 12 (1913), 301.

35. See Carl A. Auerbach, Lloyd K. Garrison, Willard Hurst, and Samuel Mermin, *The Legal Process* (1961), 149–162, 401–490, 534–637.

36. On industrial accident litigation at the trial court level, and the general costliness and hazards of litigation as barriers to the injured man, see Laurent, chap. iii, and Marshall, *Autobiography* II, 53–83. For a tabulation of cases appealed to the state supreme court, reliance has been placed on Hurst, *Digest of Regional Sources,* cross checked against W. F. Bailey, *Digest of Decisions of the Supreme Court of Wisconsin* (1902).

Cases appealed to the Wisconsin Supreme Court involving accidents (1) in the woods phases of operations and (2) in the sawmill or analogous forest-products-processing phases of operations are analyzed in the accompanying tables.

	Employer as appellant		Employee as appellant		Primary (and secondary) doctrine determining result on appeal				
					Employer's duty	Contributory negligence	Assumption of risk	Fellow servant	Other
Years	Won	Lost	Won	Lost					
Woods accidents									
1840–1879									
1880–1889									
1890–1899	2	1	2	1	4	1(3)	1	(1)	
1900–1909		1			(1)	1			
1910–1915	2	2	3		5(1)	2(1)	(1)		
Totals	4	4	5	1	9	4	1		
Mill Accidents									
1840–1879									
1880–1889	1		2	1	2	(1)	2		
1890–1899	16	8	3	9	21	10(6)	3(5)	(3)	2
1900–1909	12	10	6	8	22(5)	8(10)	5(1)	1(5)	1
1910–1915	14	14	4	15	34(2)	8(8)	3(4)	(2)	1
Totals	43	32	15	33	79	26	13	1	4

Classification of these lumber industry industrial accident cases by individual years yields no significant inference different from the tabulations by decades; there were heavy concentrations of appealed cases, 1895–1899 and 1907–1914, but no difference in patterns of doctrine or disposition in those years. Given the prominence of the fellow servant rule as an issue in state legislation regarding railroad accidents, the slight role of this doctrine in lumber industry cases is the more striking; high preponderance of lumber industry cases involving fixed machinery suggests that the employer's safe place duty was the prominent fighting issue there because mill hazards tended to center more on the set conditions of work, whereas railroading presented dangers more likely to arise out of the specific incidents of operations.

The substantial preponderance of employers' appeals over appeals taken by employees does not give a basis for clear-cut inference. It may indicate that where the case got to the jury, there was marked advantage to the plaintiff. But the smaller number of employee appeals may reflect as well the greater difficulty of employees in sustaining the costs of extended litigation. General worker dissatisfaction with litigious handling of the industrial accident problem suggests that the second inference is the stronger.

A check by paging reported federal court decisions through the years of Wisconsin lumbering revealed only three industrial accident cases from the Wisconsin industry. Of these, Nyback v. Champagne Lumber Co. is worth citing for its extreme demonstration of the hazards of litigation in this field. The case was in the lower federal courts one way or another over a span of ten years. See 90 Fed. 774 (C.C.A. 7th. 1899), 109 Fed. 732 (C.C.A. 7th. 1901), 130 Fed. 784 (Circ. Ct. W.D. Wis. 1903), 130 Fed. 1021 (C.C.A. 7th. 1904). Compare Jahn v. Champagne Lumber Co., in which the assignee of the successful plaintiff sought to pursue assets of the dissolved defendant corporation in the hands of its stockholders: 147 Fed. 631 (Circ. Ct. W.D. Wis. 1906), 152 Fed. 669 (Circ. Ct. W.D. Wis. 1907), 157 Fed. 407 (Circ. Ct. W.D. Wis. 1908).

37. See Winslow, J., for the court, in Monte v. Wausau Paper Mills Co., 132 Wis. 205, 209–210, 111 N.W. 1114, 1115–1116 (1907); Marshall, J., for the court, in Houg v. Girard Lumber Co., 144 Wis. 337, 352, 129 N.W. 633, 639 (1911); cf. General Message of Gov. LaFollette, AJ, 102–104 (1903).

38. See bill 213A, 1872, AJ, 241, 482 (Memorial 468A in opposition). Cf. Laws, 1874, c. 165 (safety booms required, to guide rafts past dams and bridges on Wisconsin River). On the railroad accident legislation, see Auerbach, Garrison, Hurst, and Mermin, 401–490, and Schmidt, 60. The factory inspector of the Bureau of Labor and Industrial Statistics is quoted from Biennial Report, Sept. 30, 1888, PD, II, xxiii (1889). On the policy-generating possibilities of executive action to collect and analyze data, in the specific context of the growth of Wisconsin industrial accident law, see Auerbach, Garrison, Hurst, and Mermin, 584–593.

39. Laws, 1878, c. 212; Laws, 1887, c. 549, carried into Statutes, 1898, sect. 1636; Laws, 1905, c. 303; Laws, 1909, c. 139, and opinion of Attorney General thereon, to Commissioner of Labor, Sept. 14, 1909, in Report, Attorney General, 1909–1910, PD, 815 (1911) (safety devices on machines for "sawing wood"); Laws, 1909, c. 338, broadened by Laws, 1913, c. 466 (employment of minors on saws or in other inherently dangerous circumstances). That the employers liability and safe place statutes prior to the workmen's compensation act did not apply regarding saws hazards, which were still governed wholly by common law doctrine, was ruled in Schmitt v. Seefeld, 139 Wis. 459, 121 N.W. 136 (1909). This had been the plain implication of numerous earlier decisions. See, e.g., Darcy v. Farmers' Lumber Co., 87 Wis. 245, 58 N.W. 382 (1894); Kutchera v. Goodwillie, 93 Wis. 448, 67 N.W. 729 (1896); Shepherd v. Morton-Edgar Lumber Co., 115 Wis. 522, 92 N.W. 260 (1902); Yazdzewski v. Barker, 131 Wis. 494, 111 N.W. 689 (1907).

40. Laws, 1911, cc. 50 (workmen's compensation) and 485 (Industrial Commission and safe place standards); see, generally, Auerbach, Garrison, Hurst, and Mermin, 632–637. Cf. John R. Commons, *Myself* (1934), 141–143, 153–160. On the court's role, see notes 36 and 37 above. The full applicability of the new system to the lumber industry was underlined by the dramatic contrast between the careful exceptions of the industry from wage legislation and the detailed treatment of sawmill hazards in the orders and reports of the new Industrial Commission. See *General Orders on Safety,* Wisconsin Industrial Commission, Bulletins, vol. 1, no. 1 (Madison, 1912), especially p. 13; *Jointer Accidents, id.,*

vol. 2 (Madison, 1913), Shop Bulletin no. 1, p. 4, and *Accidents Caused by Objects Striking Workmen,* Bulletin No. 12, pp. 284, 301. An Industrial Commission pamphlet, *Saw Injuries in Wisconsin* (Madison, 1921?), reported that from 1915 to 1920 (both inclusive) saws operations resulted in 3319 injuries to workmen, of which 2566 were temporary, 734 produced permanent partial disability, and 18 were fatal — for a total time loss of 663,010 days. Circular saws apparently caused the bulk of accidents. Most injuries, the Commission reported, resulted from lack of machine guards, or from workers' failure to use those which were provided.

41. On liability of the employer for illness other than occupational disease growing out of employment, compare Vennen v. New Dells Lumber Co., 161 Wis. 370, 154 N.W. 640 (1915); Yellow Cab Co. v. Industrial Commission, 210 Wis. 460, 246 N.W. 689 (1933). On the 1901 smallpox scare, centering in lumber camps, see note 34 above; Laws, 1899, c. 24, and 1901, c. 200 (special fund); Laws, 1905, cc. 192 and 198 (local control); Laws, 1907, c. 113 (free vaccination); and Laws, 1913, c. 674 (state inspection of labor camps to enforce state regulations). Compare earlier unsuccessful efforts to enact regulation of camp sanitation: (1) bill 460A, 1905, AJ, 728, 859; cf. bill 192S, 1905, SJ, 396, 1183, 1218; (2) bill 406A, 1911, AJ, 495, 537, 580, 833, 896; (3) bills 798A and 818A, 1913, AJ, 404, 437, 470, 837, 858. Whatever currents of industry opposition moved in these episodes, there was present also another factor — disagreement over the extension of central at the expense of local control and the shouldering by the state of financial burdens which some felt to be the proper load of localities. This seems the background of disputes which, in 1903, led to the discontinuance of special emergency appropriations to combat contagious disease emergencies. See, on bill 43A, 1903, AJ, 286, 637, 666, 697; SJ, 725, 777, 800, 821, 907; AJ, 929, 1083, 1312. Cf. bill 441A, 1903, AJ, 395, 415. The care of the sick in the lumber camps was calculated to raise the question of relative responsibility among different government authorities with particular acuteness, because of the transient character of so much of the industry's labor. See, e.g., Opinion of the Attorney General to District Attorney, Park Falls, Jan. 14, 1927, Attorney General, *Opinions,* xvi, 13.

42. On bill 122S, 1921, see SJ, 926, 980–981, (including, on p. 981, objection regarding too broad coverage of seasonal industries), 1010, and 1020–1021. Cf. bill 53S, 1923, and Haferbecker, 122–124.

43. On a lumberman's objection to the 1931 measures, see Schmidt, 352; cf. Haferbecker, 125–127. Laws, 1935, c. 192, authorized voluntary entry under the system by employers otherwise excepted. Laws, 1935, c. 446, restated this option expressly as to lumbering. Unsuccessful efforts to repeal the exception of "logging operations" from the state unemployment compensation act include bill 302A, 1949; bill 236A, 1951; bills 202S, 327A, and 662A, 1953; bills 17S and 177A, 1955; bill 549A, 1957. See, generally, "Appearance for Bill No. 549A," at the Apr. 25, 1957, hearing of the Assembly Committee on Labor, made by the statutory advisory Committee on Unemployment Compensation.

Chapter VIII. The Overhead Costs of Legal Order

1. On the regulatory potentialities and the state constitutional limitations of public borrowing as a basis for public spending in aid of industry, see the first section of Chapter VII; Brown, "The Making of the Wisconsin Constitution," *Wisconsin Law Review,* 1949: 674–676, and 1952: 46–47; Gulick, 198, 201; Mills, "Government Fiscal Aid," 114–115. On the significance of the lack of business cycle controls, see Chapter VI, notes 2 and 12. Public lands policy as a form of subsidy was discussed in Chapters I and II; the unused potentialities of public land ownership as a basis of use regulation, and the consequent avoidance of higher current operating costs to the industry, are reflected in Chapter VI, notes 11, 15, and 30, and Chapter VII, note 10.

2. See the first section of Chapter VII; cf. Gulick, 37.

3. The absence of a regulatory tax policy in the years of the great lumber industry is made graphic in the tabulations in Mills, "Government Fiscal Aid," 111, 124–125. Cf. Laws, 1927, c. 454. The tree-belt-promotion legislation was inaugurated by General Laws, 1868, c. 102, carried into Revised Statutes, 1878, sects. 1469–1471. See Chapter VI, note 22.

4. See Statutes, Wisconsin Territory, 1839, pp. 44–45; Wisconsin Constitution, 1848, Art. II, sect. 2, and Art. VIII, sect. 1; Revised Statutes, 1849, c. 15, sects. 2 and 3 (definition of taxable realty and personalty), and sects. 17

and 28 (valuation formulas — at "full cash value," with no specifications concerning valuation problems regarding timberland or forest products or any other property); Revised Statutes, 1858, c. 18, sects. 2, 3, 17, and 31 (continuing the 1849 provisions); General Laws, 1866, c. 127 (commissioners to revise tax laws); *Report,* Commissioners under Chapter 127 (Madison, 1867), 5–8. Public documents of this first generation of territorial and state tax policy show implicitly that men did not foresee that timber wealth would pose any distinctive issues for the tax system. In such tax policy discussion as went on, there was never a clear-cut decision that timber resources should be treated the same as other property. Rather, there was a failure to conceive that there might be any reason to raise a question in the matter. See, e.g., Report, Committee on Judiciary, Council Journal, 635–650, 654–656 (1842) (exemption of improvements desirable to foster cultivation: emphasis on obtaining extractive yield); Minority Report, Select Committee on Bill 298A, AJ, 727–728 (1864) (acceptance of system which taxes "unproductive" land); Annual Report, Secretary of State, Oct. 10, 1866, PD, 42 (discrimination against nonresidents as only criticism of substance of land tax system); Report, Select Committee on Bill 114S, SJ, 305–309 (1867) (recognition of liquidity pressure and cumulative overhead cost load on holding timberland, under general property tax, but no suggestion for change of tax type). The inattention to the substance of timberland tax policy is underlined by the fact that the scattering few references to timber wealth as presenting tax issues deal solely with administration. See AJ, 128 (1853) ("remonstrance" apparently concerning assessment procedures in Portage and Marathon counties); bill 154, 1865 (collection sanctions regarding pineland), adversely reported by Committees on Town and County Organization and on Assessment and Collection of Taxes, AJ, 253, 401 (1865), and indefinitely postponed accordingly, AJ, 421; Annual Report, Secretary of State, PD, I, 34 (1874) (concern for evasion of taxes on logs). That application of the general property tax to timber values was an undebated extension of attitudes originally set in regard to other kinds of resources, see *Report,* Wisconsin State Tax Commission, 1898 (Madison, 1898), 26, 28; Harold M. Groves, *Financing Government* (1945), 19, 56, 95 (in general); Phelan, 330 (1867 commissioners' focus on agriculture).

5. General Laws, 1868, c. 130, sects. 1, 15, and 17 (all real and personal property taxable, if not expressly exempt), 16 (valuation formula for realty), 19 (specification of taxable personalty), and 20 (situs of personalty). That this act contained the first specific reference to forest-derived wealth in the state's tax base was noted in Mitchell v. Town of Plover, 53 Wis. 548, 553, 11 N.W. 27, 29 (1881). Semple v. Langlade County, 75 Wis. 354, 357, 44 N.W. 749, 750 (1890) recognized that the state's tax policy made no provision for separate assessment of standing timber. In 1861 the legislature added sawed lumber and various finished wood products to the list of items on which, since 1857, local assessors were directed to collect statistics for an economic census of the state; in 1877 the legislature made separate provision for collecting data on growing timber. These crude and poorly applied census procedures had indirect relation to tax programs, but seem also intended to furnish general economic information on the state's condition. See Chapter VI, note 18.

6. See Chapter VI, notes 20–22.

7. There is no reference whatever to tax policy as it bore on the lumber industry in any of the messages of the governors to the Wisconsin legislature before the forest was depleted. Cf. General Message of Gov. Fairchild, SJ, 18 (1867) (lumber industry lumped without distinction with mining and manufacturing as deserving state's "encouragement"); General Message of Gov. Scofield, AJ, 23 (1899) (assumption that lumber industry should fall routinely within general tax system). The want of attention to lumber tax policy on the part of the chief executive through the principal years of the industry was pointed up by the first call for a distinctive forest tax program, in the General Message of Gov. McGovern, AJ, 33 (1913). Cf. McGovern's General Message, AJ, 42 (1911). Though a specialized state tax commission, as distinguished from merely ex officio bodies, was finally created by Laws, 1899, c. 206 (following recommendations of the temporary commission created by Laws, 1897, c. 340), during the remaining years of the great lumber industry the new commission's principal assignment was viewed by the legislature as that of state-wide equalization within the framework of the existing tax system. See Laws, 1901, c. 220; cf. *id.*, c. 445. See, generally, James D. Barnett, *The State Administration of Taxation in Wisconsin,* in Transac-

tions, Wisconsin Academy of Sciences, Arts, and Letters (1905), vol. XV, part I, pp. 166–172; Archie Bristol Goodman, "Assessment Administration in Wisconsin" (Ph. D. thesis, 1942), chap. ii; Alfred Norman Nybroten, "Tax Assessment and Equalization in Specific Reference to the Wisconsin System" (Ph. D. thesis, 1941), chap. ii.

8. See Hersey v. Board of Supervisors of Barron County, 37 Wis. 75, 79, 80 (1875). Accord: Marsh v. Board of Supervisors of Clark County, 42 Wis. 502, 516, 518 (1877), and Philleo v. Hiles, *id.,* 527 (1877) (enforcing substantial effort to meet view requirement and rejecting argument from alleged total impracticability). Cf. Fifield v. Marinette County, 62 Wis. 532, 537–538, 22 N.W. 705, 707 (1885). That the Hersey decision rested on a finding that the assessors had followed "an arbitrary rule of assessment, regardless of value," see Webster-Glover Lumber & Manufacturing Co. (Ltd.) v. St. Croix County, 63 Wis. 647, 651, 24 N.W. 417, 418 (1885). The valuation formula and the "actual view" requirement applied in Hersey were those declared in General Laws, 1868, c. 130, sect. 16, amended in effect as to the view requirement by Laws, 1873, c. 166. The 1873 amendment was not explicitly referred to in Hersey; that it was taken into account in the court's remarks on practicability of view was acknowledged in Marsh (cited above), 42 Wis. at p. 515 (1877).

9. Reduction and eventual elimination of the strict mandate for actual view may be traced through Laws, 1873, c. 166, 1877, c. 250, 1878, c. 77; and Revised Statutes, 1878, sect. 1052, carried into Statutes, 1898, sect. 1052. The specific attempt and refusal to continue the actual view requirement for valuing "timber" under the 1877 act may be traced in AJ, 473, 591, 612, 626 (1877). That the relaxation of the actual view requirement was constitutional, as within the reasonable discretion of the legislature, see Plumer v. Board of Supervisors of Marathon County, 46 Wis. 163, 177, 50 N.W. 416 (1879). Cf. Marsh v. Board of Supervisors of Clark County, 42 Wis. 502, 518 (1877). Under the new dispensation, see Boorman v. Juneau County, 76 Wis. 550, 45 N.W. 675 (1890) (want of actual view no longer of itself fatal to validity of assessment, where it may be reasonably inferred that assessor had other information of land value derived from past presence of standing timber; assessor's ignorance that timber had been cut since last assessment was a mere error in application which did not vitiate assessment). Revised Statutes, 1878, sect. 1052, dropped the inclusion of timber "quality" from the valuation formula; the Revisers' Notes made no comment on the change, nor was any note taken of it in any subsequent opinion of the Wisconsin Supreme Court.

10. The large importance which the taxable wealth of timberland held in the eyes of north-country politicians, townsmen, and homesteaders is discussed in Gates, *Wisconsin Pine Lands,* chap. viii. The "land inspector" system was made available to northern counties by General Laws, 1871, c. 79, as amended by Laws, 1875, c. 115 (repealing mandate that cut-over pineland be valued nominally), and Revised Statutes, 1878, sects. 1053 and 1054 (option extended to any county); see *Report of the Revisers of the Statutes,* 92. The authorization was repealed by Laws, 1891, c. 438, which, however, allowed counties "the use and benefit" of data on file from the old system. On the patronage implications of the procedure, see allegations recited in T. B. Scott Lumber Co. v. Oneida County, 72 Wis. 158, 161–162, 39 N.W. 343, 344 (1888). The same case contains the rulings noted in the text on the legality of the system. Accord: Hixon v. Oneida County, 82 Wis. 515, 52 N.W. 445 (1892).

11. Laws, 1875, c. 115. Opinions of the Wisconsin Supreme Court in the lumber decades contain many casual, taken-for-granted acknowledgments of the extent to which the whole value of northern lands resided in the standing timber, along with equally matter-of-fact acceptance of the normal fate of such lands to be cut and allowed to default for taxes. See Chapter II, notes 23, 53, 71, and 72.

12. For the general lien situation, see General Laws, 1859, c. 22, sect. 31; General Laws, 1872, c. 179; Revised Statutes, 1878, sect. 1088 (see *Report of the Revisers of the Statutes,* 94), carried into Statutes, 1898, sect. 1088; Laws, 1901, c. 190, carried into Statutes, 1957, sect. 74.01. There is probably testimony to the countervailing power of logger-landowners in the legislature, in the failure of efforts to enact much broader bans and penalties on cutting timber before taxes were paid. Bill 154A of 1865, the title of which said that it was designed to facilitate collection of taxes on pinelands and other unoccupied and unimproved timbered lands, forbade the cutting of timber unless taxes were

paid up on the land, and to assist local officials sought to spur recording of all instruments of title regarding timbered land by denying actions for timber trespass to owners who had not recorded their interests. After adverse reports from the Assembly Committees on Town and County Organization and on Assessment and Collection of Taxes, the bill was indefinitely postponed: see AJ, 253, 275, 401, 421 (1865). By Memorial 147A, 1876, the Barron County Board petitioned the legislature "for a law prohibiting the cutting of timber from lands upon which the taxes are not paid." The memorial was referred to the Assembly Committee on Assessment and Collection of Taxes, which buried it; bill 200A, entitled to the same effect as the Barron County memorial, was never reported out of a select committee of the Ashland County delegation, to which it had been referred: see AJ, 152, 200 (1876).

13. General Laws, 1868, c. 116, carried into Revised Statutes, 1878, sect. 1195; Laws, 1897, c. 48, carried into Statutes, 1898, sect. 1195.

14. Evasion of the personal property tax in its application to saw logs was noted as a "considerable" problem, in Annual Report, Secretary of State, Oct. 10, 1874, PD, I, 34. Both prevention of evasion and an effort to avoid double taxation were stressed as objects of legislative concern under the procedures created by Laws, 1891, c. 473, in State ex rel. Holt Lumber Co. v. Bellew, 86 Wis. 189, 194, 56 N.W. 782, 783 (1893). Cf. C. N. Nelson Lumber Co. v. Town of Loraine, 22 Fed. 54, 59, 60 (Circ. Ct. W.D. Wis. 1884).

15. The most clearly defined effort reflected in the legislative record was that directed toward creating situs at the mill rather than in the forest. Clearly some interests centered in the mill towns were active here. But evidence is lacking to tell whether the prime movers were the politicians of millsite taxing districts or the larger entrepreneurs who may have felt that their millsite investments tended to give them more influence in the towns than in the forest. The first situs legislation specifically referring to forest products did not mention millsites: see General Laws, 1868, c. 130, sect. 20. General Laws, 1872, c. 148, provided that saw logs and timber to be sawed or manufactured in mills in the state owned or leased by the owners of the logs or timber should be assessed as manufacturers' stock in the towns where the mills were located. Laws, 1883, c. 354, made the millsite the assessment district, whether or not the mill was owned or leased by the log owner. Laws, 1891, c. 473, returned to a narrower allocation of millsite taxing jurisdiction. Now millsite situs depended not only on the fact that the logs were to be processed in a mill owned by the log owner (a leased mill no longer fell within the provision), but apparently this was also limited to logs cut within six months before April 1 in any given year (i.e., logs cut and stored a longer time would be taxable at the place where they were held on assessment day), and to fix situs at the millsite even within these limitations before June 15 the log owner must file with the assessor of the district of storage an affidavit of ownership and of intent to mill the logs at a defined millsite, accompanied by a verified statement of the millsite assessor that the logs had been listed with him for assessment.

Three aspects of the 1891 law point to the presence of some real interest maneuvers in its background. First, as introduced, bill 632A (which became Laws, 1891, c. 473) would have wiped out all situs tests except the simple declaration that forest products be assessed where located on assessment day; the first move to restore a millsite provision was made by an assemblyman from the important mill county of Winnebago, and though his initial effort failed, a millsite provision finally entered the measure: see AJ, 931, 962, 981, 996 (1891). Second, it is significant that despite this focused effort to restore a millsite provision, what was finally written was narrower than the terms of Laws, 1883, c. 354. Third, the requirement of the filing of affidavits of the log owner and millsite assessor not only emphasized the competition and administrative problems involved in maneuvers between storage area and millsite area, but, as the court pointed out in State ex rel. Holt Lumber Co. v. Bellew, 86 Wis. 189, 194, 56 N.W. 782, 783 (1893), in effect converted what had been a primary head of situs under the 1883 law into an exception to a general rule favoring situs according to location and putting the burden of initiative on the taxpayer who would shift situs away from the place where the property was physically present. Cf. Day v. Town of Pelican, 94 Wis. 503, 69 N.W. 368 (1896) (given general statutory emphasis on location as of April 1, where before April 1 logs are shipped out of the district where originally banked into the mill district they are assessable in the mill district without need of filing the statutory affidavits). Laws, 1893, c. 179, repealed Laws, 1891, c. 473, and reinstated the law as it stood in 1883, thus restoring the broad primary provision of millsite situs.

The 1895 session produced several efforts to eliminate millsite situs, all of which failed; the quick reaction to the events of the 1893 session, however, suggests the presence of some real, focused interests in tension over this matter. Bills 449A and 29S of 1895 would have substituted the simple test of location on assessment day for the variations then existing in the statutes; bill 28S would have repealed Laws, 1893, c. 179, and reinstated Laws, 1891, c. 473, with its narrowed concessions to millsite jurisdiction; bill 246S would have given millsite taxing jurisdiction only where the logowner owned the mill. Bill 449A was supported by several petitions, apparently from local government officials; it had short shrift from the Assembly, however, which rejected it on recommendation of the Committee on Assessment and Collection of Taxes: see AJ, 356, 399, 454, 515, 663, 701 (1895). The three Senate bills were treated with like lack of ceremony. The Senate tax committee recommended indefinite postponement of all three bills, and the Senate voted so on bills 28S and 29S: see SJ, 460, 474–475 (1895). Bill 246S was referred to the Committee on Town and County Organization, from which it never emerged (SJ, 572). Bill 459A of 1899 (like bills 449A and 29S of 1895) would have created the simple and exclusive test of location as the criterion of situs. It was indefinitely postponed, on recommendation of the Assembly Committee on Assessment and Collection of Taxes: see AJ, 874, 890 (1899).

Laws, 1913, c. 81, struck out the millsite provision. But Laws, 1913, c. 497, then partly restored it, now limited to logs "in transit" to mill; this was defined to cover logs while being transported or while temporarily stored in another district than that in which they had been cut; if they rested still in the cutting district, they should be assessed there, though they were destined for a mill in the state. This second 1913 act underlined the concern for the cutting district by requiring the log owner to file verified statements with the assessors of the districts both of log location and of millsite, in order to claim millsite situs; the legislature thus restored a procedure substantially like that created by Laws, 1891, c. 473. Laws, 1921, c. 69, simplified the language without changing the substance of the provision for situs at the mill. But Laws, 1927, c. 349, reduced the application of what might be termed constructive millsite situs to the case of logs actually being transported; logs temporarily stored were to be assessed where found on assessment day, though they might be destined for a mill in another area. The provision continued thus into Statutes, 1957, sect. 70.13(2). It seems a fair summary to say that, over-all, the trend of legislation was to prefer the simple test of actual location on assessment day as the criterion of situs. Cf. Laws, 1949, c. 123 (logs or timber removed from public lands in year preceding May 1, or in transit after such removal on May 1, to be assessed where the public lands are located).

16. The course of tax situs legislation affecting forest products may be traced through General Laws, 1868, c. 130, sect. 20 (residence of owner or agent, but merchants' stock to be assessed where located); General Laws, 1872, c. 148 (millsite situs introduced; property location made residual test, wanting others); Revised Statutes, 1878, sect. 1040; Laws, 1879, c. 244 (cordwood); Laws, 1882, c. 258 (location test, for logs owned by nonresident who has no resident agent); Laws, 1883, c. 354 ("agent" means agent-in-charge; millsite provision broadened); Laws, 1891, c. 473 (location made the criterion, save for narrowed millsite provision); Laws, 1893, c. 179 (repealing the 1891 act and restoring Revised Statutes, 1878, as amended in 1882 and 1883); Statutes, 1898, sect. 1040; Laws, 1909, c. 70 (logs of nonresidents assessable where piled for shipment); Laws, 1913, cc. 81 and 497 (first striking, then reinstating millsite provision, though in form limited to logs "in transit"; simplified location criterion as general rule); Laws, 1921, c. 69 (simplified restatement); Laws, 1927, c. 349 (generalizing the location criterion, limiting millsite provision to logs in course of transport); Statutes, 1957, sect. 70.13 (2). A reflection of the special problem of tax evasion posed by the movement of logs in the St. Croix area into Minnesota waters may be seen in Laws, 1875, c. 299 — a part of the lumber inspection system of statutes — requiring that all logs be submitted for official measurement and record in the Fourth Inspection District. Of related significance was the provision made by Laws, 1882, c. 258, for assessment of logs of nonresidents during April, in contrast to the May 1 date fixed by the statutes for the assessment of other personalty. The latter pro-

vision was held a reasonable adjustment to the situation created by movement of logs out of state, and not in violation of the equal protection clause of the 14th Amendment or the tax uniformity clause of the Wisconsin constitution, in C. N. Nelson Lumber Co. v. Town of Loraine, 22 Fed. 54 (Circ. Ct. W.D. Wis. 1884), which also ruled that the possibility that the logs might be taxed as personalty first in Wisconsin and later in Minnesota after their removal there did not offend due process of law. Compare also Laws, 1903, c. 417, adding forest products to the personalty which under Statutes, 1898, sect. 1044, might be assessed to the person in charge or possession, if the owner were not in charge. This seems simply to have declared and clarified the law as it stood after Laws, 1883, c. 354. See City of Merrill v. P. B. Champagne Lumber Co., 75 Wis. 142, 43 N.W. 653 (1889); cf. State ex rel. Edward Hines Lumber Co. v .Fisher, 129 Wis. 57, 108 N.W. 206 (1906). The Wisconsin Supreme Court gave a strong push toward the primacy of the criterion of physical location of the property on assessment day, by a consistent line of decisions construing the general language of the merchants' stock clause to apply to forest products held for wholesale or retail sale. The leading case was Mitchell v. Town of Plover, 53 Wis. 548, 11 N.W. 27 (1881). The court's warmth for this approach was particularly underlined by Sanford v. Town of Spencer, 62 Wis. 230, 22 N.W. 465 (1885), holding that lumber piled for shipment in a Wisconsin town was assessable there as merchants' stock of nonresident dealers who filled orders taken out of state by shipment from the Wisconsin storage point. See also Torrey v. Shawano County, 79 Wis. 152, 48 N.W. 246 (1891).

17. By a conservative estimate, about thirty-odd appeals came to the Wisconsin Supreme Court during the lumber industry decades involving taxes on timberland, about a score more concerning taxes on logs or lumber, and about half a dozen regarding assessment of mill property. There is no significant grouping of cases by period. This count includes only reported cases in which there is some positive evidence indicating timber industry involvement. Probably some other tax appeals from northern counties grew out of the industry's affairs, though the reports do not affirmatively show so; insofar as this is so, examination of the run of these other northern county tax cases shows only more of the same succession of narrow issues. Claims resting on tax title figure in a substantial minority of appeals over timber trespass; insofar as a tax-title issue was central, such cases enter into the totals given at the outset of this note. The count given here is based on the listings in Hurst, *Digest of Regional Sources,* 99–100, 108 (compiled by check against the records and briefs in appealed cases, where the reported decision was unrevealing), cross checked against Bailey, *Digest of Decisions.* For typical tax cases turning on procedural points, see Union Lumbering Co. v. Board of Supervisors of Chippewa County, 47 Wis. 245, 2 N.W. 281 (1879) (land); City of Merrill v. P.B. Champagne Lumber Co., 75 Wis. 142, 43 N.W. 653 (1889) (logs); Brown v. Oneida County, 103 Wis. 149, 79 N.W. 216 (1899) (mill). Paging of reported decisions of federal trial courts in the Wisconsin area showed only three cases (all in 1884) involving direct taxpayer challenges of local taxes; two of these three involved interstate commerce issues. No doctrine of new significance was at stake in any of these instances.

18. See Current, 124, 135; Fries, 175–176; Gara, 164; Gates, *Wisconsin Pine Lands,* 140, 143, 144, 146, 148, 151, 158, 163; cf. Larson, 342–344.

19. See Gates, *Wisconsin Pine Lands,* 158, 160, 163–164, 168. Cf. Annual Report, Secretary of State, Oct. 10, 1866, PD, 42 ("Nonresident landowners are as a general rule not presumed to have rights which any man is bound to respect"); Semple v. Langlade County, 75 Wis. 354, 44 N.W. 749 (1890); Boorman v. Juneau County, 76 Wis. 550, 45 N.W. 675 (1890); *Report,* Wisconsin State Tax Commission, 1898, p. 91.

20. Note the allegations of local political manipulation which the court found it unnecessary to pass upon in Hersey v. Board of Supervisors of Barron County, 37 Wis. 75, 78 (1875), and Marsh v. Board of Supervisors of Clark County, 42 Wis. 502, 504 (1877). In T. B. Scott Lumber Co. v. Oneida County, 72 Wis. 158, 39 N.W. 343 (1888) the court held that it "must assume" honesty of motive in the county board in exercising its statutory authority to appoint land inspectors to value timberlands. Cf. Brown v. Cohn, 88 Wis. 627, 60 N.W. 826 (1894) (defendant must rely on change of venue statute to protect him against alleged hazard that plaintiff, as dominant mill owner, can control any jury named in area). See also *Report,* Wisconsin State Tax Commission, 1898, pp. 90–92, 96. On local government

financial embarrassment from failure of "speculators" to pay taxes promptly, see, e.g., Annual Reports, State Superintendent of Public Instruction, Dec. 10, 1863, PD, 654, and Dec. 10, 1864, PD, 373 (Wood County report).

21. See Laurent, chap. iii, on the tax cases in the circuit court for Chippewa County. The typical reliance on bargaining and threats of suit, to secure tax adjustments by local executive action out of court, is reflected in Gara, 164, and Gates, *Wisconsin Pine Lands,* 163–164, 168, 169. On the limited economic impact of discriminatory taxation, see Gates, 175–176.

22. See Brown and Hurst, "The Perils of the Test Case," 35.

23. *Ibid.,* p. 43, notes 46 and 47; pp. 44–45; and p. 59, note 95.

24. *Ibid.,* pp. 45 and 46, and note 58 on p. 46.

25. *Ibid.,* pp. 28–34, 41–43.

26. *Ibid.,* pp. 39, 43–52.

27. *Ibid.,* pp. 53–59.

28. *Ibid.,* pp. 30, 44, 59. Of course, the legislature's "practical construction" of the constitution as allowing a special railroad tax system was made in response to a powerful railroad lobby. This does not alter the relevance of this legislative history to our concerns, however, but simply serves to highlight the lack of interest among lumbermen in securing a special tax program for themselves. Recurrent criticism of the special railroad tax program is significant, too; the issue did not go wholly by default, and, again, the analogy would have been spotlighted, had there been lumber industry interest in using it. Cf. Phelan, 219–221; Belle Case LaFollette and Fola LaFollette, *Robert M. LaFollette* (1953), I, 139.

29. See, generally, Lake, "Legal Profile of the Mining Industry," 411–413, and Phelan, 315. The processing corporation tax was created under General Acts, 1864, cc. 166 and 394, repealed by General Laws, 1872, c. 144. The legislative record is unrevealing of the background of repeal. Compare Laws, 1883, c. 203, carried into Statutes, 1898, sect. 1038 (22), and Statutes, 1947, Section 70.11 (17), until repealed by Laws, 1949, c. 63 (three-year tax exemption for Wisconsin corporation processing native zinc ores). The iron ore lands tax statute was Laws, 1877, c. 269, repealed by Revised Statutes, 1878, sect. 4978. See *Report of the Revisers of the Statutes,* 92.

30. On the uncritical borrowing of the general property tax principle, compare the comments in *Report,* Wisconsin State Tax Commission, 1898, pp. 26, 28. On the imitative borrowing characteristic of growth of nineteenth-century state public policy, see, generally, Hurst, *Law and the Conditions of Freedom,* 10.

31. Gates, *Wisconsin Pine Lands,* 175–176, estimates that the total tax cost of holding these timberlands for twenty-five and thirty years was less than one tenth of the gross income from sales. Cf. Current, 114, 135; Phelan, 319. The lumber industry's lack of interest in securing a special tax program for itself is underlined by the contrast with the vigorous and considerably successful efforts of some of its leaders in their capacity as railroad promoters to obtain tax exemption for railroad-aid lands, the value of which lay largely in their timber. See e.g., Current, 133–135; Hunt, *Law and Locomotives,* 143, 144. The point is sharpened because the idea of tax exemption for timber holdings as such was raised, unsuccessfully, by the Cornell interests as to their holdings for educational purposes, notably in 1866 and 1877: see Gates, *Wisconsin Pine Lands,* 138–141.

32. Cf. William H. Anderson, *Taxation and the American Economy* (1951), 54, 117, 155–156, 445, 448, 449; Ciriacy-Wantrup, 168, 172, 178; Groves, 367; Walter A. Morton, *Housing Taxation* (1955), 74, 101, 103, 104, 109.

33. These various attitudes are reflected in Gates, *Wisconsin Pine Lands,* chap. viii. For other reflections (a) of complaints of nonresident owners of land, see Semple v. Langlade County, 75 Wis. 354, 44 N.W. 749 (1890); (b) of local residents allegedly suffering oppressions from control of local politics by large lumber interests, see Brown v. Cohn, 88 Wis. 627, 60 N.W. 826 (1894), and Petition 431A, AJ, 783–784 (1885).

34. On the common view of the forest as a passing resource, normally to be succeeded by agriculture, see Chapter II, notes 21, 23, 35, 55, 56, and 68. Local politicos' justifications for a firm hand toward the big taxpayers may be seen in material quoted in Gates, *Wisconsin Pine Lands,* 141, 147. A few abortive demands to curb the voting power of transients in the north reflect resentment at the voting crews the lumbermen could mobilize and maneuver at need. See Petition 84A, 1875, AJ, 156 (praying a statute "to prevent lumbermen from voting in the spring where they worked during the winter"); Memorial 34S, 1877, SJ, 122 (for stricter residence requirements for voting); Petition 431A, 1885, AJ, 783–784 (alleging local residents were

"out voted by a horde of transients, employees of George Hiles and tramps run in for that purpose").

35. Most political maneuver over local government in the forest region was carried on at the home scene and by informal tactics which left little record. See Fries, 221–224; Gates, *Wisconsin Pine Lands,* chap. viii. The large volume of business which was brought to the legislature affords impressive circumstantial testimony to the time and effort poured into seeking the prizes at stake in control of local politics. Examples fill the legislative journals by the score. The location of county seats was a fruitful source of contest. See, e.g., in AJ, 982–986 (1858), Report, Select Committee on Bill 12S (removal of Juneau County seat); and AJ, 1696–1697 (1858), Report, Select Committee to investigate charges of corrupt influences in connection with this proposal. Compare the controversy over bill 167S, 1876, to legalize an election about the Taylor County courthouse location; see SJ, 252, 401, 460 (1876) (conflict of citizens' petitions and committee report and Senate action); AJ, 592, 634, 643 (1876) (bill rejected on recommendation of two Assembly committees, contrary to Senate action). Cf. bills 406A and 152S, 1877. Particularly numerous and time-consuming were bills to shift land areas among different units of local government; many of these proposals involved corollary controversies over apportionment of financial rights and liabilities among the affected local governments. See, for example, the legislative background of Laws, 1879, c. 114, creating New County, and Laws, 1880, c. 247, redefining the territory, powers, and fiscal relations of New (Langlade) County, and Shawano County — including not only bill 155S, which passed in 1880, but also bills 114A and 290A, which did not. Compare the further involved maneuvers as to the territorial definition and fiscal relations of Langlade in relation to other counties, in the legislative history of Laws, 1883, cc. 295 and 303, and of the unsuccessful bills 2A, 276A, 301A, 317A, 325A, and 411A of 1883. Ordinarily the legislative journals do not spell out the considerations behind the considerable record on such bills. For a rare revelation of some of the bargaining maneuvers in the background of the shuffling of county lines and the fiscal considerations entering into a deal ratified by statute, see Minority Report, Committee on Town and County Organization, AJ, 276–278 (1887) (Oneida and Lincoln counties). A temporary resort to ease the fiscal burden of new counties was to attach them for judicial purposes to an existing circuit already being supported by older units of local government. The device is itself revealing of the shaky financial underpinnings of the expansion of counties in the north, and also added to the flow of highly particular business that this condition entailed for the legislature. For a particularly involved legislative history see the background of Laws, 1887, c. 488 (bill 315S, 1887), creating the Fifteenth Judicial Circuit.

36. See Laurent, chap. i and Table 1.

37. AJ, 325, 852, 974 (1871). See Gates, *Wisconsin Pine Lands,* 146.

38. AJ, 152, 296–297, 595–596, 729, 838–839 (1872).

39. On bill 297A of 1874, see AJ, 199, 377, 552, 613, 654; cf. SJ, 192–193, 242. For the background of Laws, 1875, c. 178, creating Taylor County, compare the legislative history of the successful bill 86A with the blocking of bill 19S (introduced by a Senator who unsuccessfully moved for indefinite postponement of bill 86A), SJ, 22, 194, 269, 306, 342–343 (1875). This was an instance in which proposals for redrawing of local political lines produced, as they often did, a considerable number of petitions pro and con from interested individuals. See AJ, 37 (1875); SJ, 22, 35, 39, 60, 78, 85, 93, 173–174.

40. The "Ludington County" proposals included (1) bill 179A of 1876, as to which see AJ, 136, 518, 569; (2) bill 60A of 1877, as to which see AJ, 41, 62, 178, 614, 632; (3) bill 243S, 1877, as to which see SJ, 145, 177–178, 202–204, 209–210, 223–224. The "Flambeau County" proposals involved (1) bill 47A of 1878, as to which see AJ, 52, 77, 87, 125, 282, 312, 401; and (2) bill 179S, regarding which see SJ, 190, 317, 354. Bill 4S of 1879, to create "Flambeau County," became Laws, 1879, c. 103, creating Price County; see especially the reflection of active reworking of the bill in the Town and County Affairs Committee of the Senate, in SJ, 215, 277, 287, 324; AJ, 364, 387, 395. A lengthy chain of effort also lay back of the creation of another county in the north-central area. See the legislative history of bill 277A, 1885; bill 527A, 1887; bill 561A, 1891; bills 314A, 492A, 522A, and 626A, 1895; bills 411A, 517A, and 571A, 1897.

41. The majority report of the Senate Committee on Town and County Affairs on bill 243S of 1877 is found in SJ, 177–178; the minority report is given in SJ, 202–204. For other explicit analyses of considerations back of local government proposals, see Report,

Select Committee on bill 12S (to shift Juneau County seat), AJ, 982–986 (1858), especially pp. 982–983 (land speculators); Minority Report, Committee on Town and County Organization, on bill 523A of 1885 (to create Oneida County), AJ, 718–720 (local politicians); Minority Report of the same committee, on bill 57A of 1893 (to create Vilas County), AJ, 510–512 (transient population inflates apparent size of community; inadequate tax base).

42. See *The Wisconsin Blue Book*, 1931 (Madison, 1931), 591–627; *id.*, 1952, pp. 250–251, 536; Margaret Jarman Hagood and Emmit F. Sharp, *Rural-Urban Migration in Wisconsin, 1940–1950* (1951), 8–10; Wisconsin State Planning Board, *Local Government Study in Wisconsin, 1927–1936*, I, 33–41, 44, and Table 1 and Chart no. 5; William Henry Young, "The Present State of Research in Wisconsin Local Government" (Ph. D. thesis, 1941), 102–103. Rejected proposals to create additional northern counties include bill 432A of 1872; bills 105A, 124A, 242A, 6S, and 183S of 1877; bill 6A of 1878; bill 205A of 1879; bills 368A and 185S of 1880; bill 197S of 1882; bills 44A, 301A, and 411A of 1883; bills 275A, 277A, 339A, and 524A of 1885; bills 224A and 527A of 1887; bills 64A and 337A of 1889; bills 170A, 561A, and 579A of 1891; bill 310A of 1893; bills 314A, 428A, 492A, 522A, and 626A of 1895; bills 411A, 517A, and 571A of 1897; bill 395S of 1905; bill 534A of 1907. That other successful bills enacted at least some of the substance of some of these proposals does not negate the burden of the failed bills or the evidence they offer of lively interest in the subject. The few scattering and episodic references in the official records to the problems of an over-expanded county apparatus in the north serve mainly to highlight the general want of attention to the matter. See Minority Report, Committee on Town and County Organization, on bill 57A of 1893 (to organize Vilas County), AJ, 510–512; Opinion of the Attorney General to the Wisconsin Tax Commission and the Secretary of State, Feb. 10, 1919, Attorney General, *Opinions*, VIII, 69–71 (dual office holding in Florence County is nonetheless contrary to statute, though the business of the officers there is very light); recitals in abortive Joint Resolutions 31S and 42S of 1933 and 14S of 1951 (seeking studies of reduction of number of counties and towns in sparsely populated sections of state). Contrast the broad range of policy considered in the report of the Wisconsin State Planning Board, *Local Government Study in Wisconsin, 1927–1936*, vol. IV, *Summary* (1943), especially chap. i, pp. 10–14.

43. Occasional reference to, or even criticism of, the catch-as-catch-can informality of legislative consideration of local government proposals shows that there was not a total lack of contemporary recognition of the problem. But, as often in other connections in this history, recognition is so episodic and so little productive that its relevance is only to point up the fact that the machinery of official decision making was not well geared to responding to tentative initiation of general policy issues. See Majority Report, Senate Committee on Town and County Organization, on bill 174S (to organize the town of Grant in Wood County), SJ, 362–363 (1885), with which compare the majority and minority reports of the counterpart committee in the other house, AJ, 822–823 and 848–851; Minority Report, Assembly Committee on Town and County Organization, on bill 57A (to create Vilas County), AJ, 510, 511 (1893).

44. Cathcart v. Comstock, 56 Wis. 590, 14 N.W. 833 (1883); Chicago & North Western Railway Co. v. Langlade County, 56 Wis. 614, 14 N.W. 844 (1883); State ex rel. Graef v. Forest County, 74 Wis. 610, 43 N.W. 551 (1889). The limited policy making represented by the detailed enumerations of the new constitutional ban on special and private laws, as applied to creation of counties, was underlined by the doctrine that the legislature possessed authority to create counties despite the absence of express constitutional warrant therefor, since the state constitution must be deemed a limitation rather than an original grant of legislative authority: State ex rel. Hicks v. Stevens, 112 Wis. 170, 88 N.W. 48 (1901). Cf. Bittenhaus v. Johnston, 92 Wis. 588, 595, 66 N.W. 805, 806 (1896) (significant that Art. IV, sect. 31, contains only a partial, not a total ban on local legislation). These cases ruled also that for the legislature to define the territorial bounds of new or existing units of local government and, incident thereto, to adjust fiscal relations among them were not such regulations of the system of county or town government as to be subject to the requirement of statutory uniformity in that matter, under Art. IV, sect. 23. See Forest County v. Langlade County, 76 Wis. 605, 610, 45 N.W. 598, 600 (1890).

45. The prime impetus to the 1871 constitutional amendment seems to have been reaction against the burden of business created by petitioners for special legislation affecting the

rights of private parties, most notably for corporate charters and franchises for stream improvements. See, generally, Kuehnl, chaps. xii and xiii; cf. General Message of Gov. Fairchild, SJ, 205–206 (1870). In State ex rel. Keenan v. Supervisors of Milwaukee County, 25 Wis. 339, 348 (1870), the court makes a graphic statement of the burdens and ill-considered fruit of pressures for local legislation, as explaining the requirement of a uniform system of county and town government as laid down in the original constitution. State ex rel. Busacker v. Groth, 132 Wis. 283, 287, 112 N.W. 431, 432 (1907) carried this explanation into the 1871 constitutional amendment. As the terms of the 1871 provision show, this was, indeed, part of the total problem of "special or private laws." But on the whole record the local government aspect of the matter does not appear the salient one. The relative inattention to problems of local government as a source of evils in special legislation is reflected in the fact that but two governors' vetoes — both by Gov. Rusk — turned on such questions of policy. See veto of bill 313A, AJ, 749–750 (1882) (bill changing voting place in a town in Price County raises "serious doubts" of constitutionality, under the uniformity clause, and in any case offends sound policy of home rule); veto of bill 80S, SJ, 418–419 (1887) (bill authorizing Douglas County to spend money on a particular harbor improvement disapproved in part because its benefits seem too local to warrant county-wide support).

46. The very tentative suggestion of a master plan for north-country county organization in 1872, and the limited grounds of policy expressed in connection with it, are reflected in the adoption of Resolution 12S, SJ, 115, 121 (1872); and in the submission of bill 143S under a report of the Senate Committee on Town and County Organization, SJ, 225–226 (1872). Bill 143S was, however, later rejected on recommendation of the Senate Committee on the Judiciary, SJ, 273, 361, 472. That, despite the initial emphasis on generalization, this project was either a cover for a local effort, or became fatally entangled in a particular local interest, is indicated by the fact that its rejection was urged by a flow of petitions from citizens of Oconto County (SJ, 327, 341–342, 410, 456). The 1875 session saw the introduction of proposed resolution 20S, to create a joint select committee to consider the erection of counties in the general region opened to settlement by the building of the Wisconsin Central Railroad northward from Stevens Point; but neither the resolution nor any other general planning action analogous to its call emerged after the proposal was sent to the Senate Committee on State Affairs: SJ, 330 (1875).

There was recurrent interest in the classic general issue whether county government should be in the hands of a small number of commissioners or of a board constituted of town supervisors and county supervisors elected by city wards. See, e.g., proceedings under Resolution 12S of 1875, SJ, 74, 78, 87–88, 196. But this issue originated in the differing political traditions of the state's first settlers, and in its later form tended to reflect more the interest distinctions of rural and urban areas than any concern focused on the peculiar problems of organizing the north country. See Raney, 133; Wisconsin State Planning Board, *Local Government Study in Wisconsin,* IV, 18–25.

The theme of preserving flexibility through generality and wide range of organization received just enough rare comment amid the bustle of local government affairs in the legislature to remind us of the prevailing want of attention, or at best want of conviction, regarding such ways of thinking about the matter. See Report, Select Committee on Bill 12S of 1858 (change of Juneau County seat opposed, because flux of the region cautions against prematurely freezing its organization), AJ, 984 (1858); Minority Report, Committee on Town and County Organization, on bill 523A of 1885 (opposing creation of Oneida County, because in a fast-changing situation action should wait upon more definition of trends), AJ, 719–720 (1885); Minority Report of same committee on bill 57A of 1893 (creation of Vilas County opposed, on principle that assurance of adequately broad tax base should be prerequisite), AJ, 511 (1893).

The limitations of imagination within which policy was conceived emerge by implication from complaints or resigned observations on difficulties of local government in the sparsely populated north, in none of which is the conclusion drawn of need for a different approach to government organization in that region than by simple multiplication of small units. See, e.g., Report, Select Committee on a contested election in the northwestern Assembly district, AJ, 54 (1859) ("The extent of territory embraced in said district is greater than some of the States of this Union, and in parts thereof it is difficult to obtain suitable men to conduct elections in a proper manner, and many irregularities have been brought to our notice"); Report, Committee on Town and County Organization, on bill 52S (to organize Shawano County for judicial purposes), AJ, 667 (1859) ("It appearing that said county is very sparsely settled, and that the organization of said county for judicial purposes, would be attended with great expense to the few inhabitants," the committee recommended rejection, but the legislature nonetheless passed the bill into General Laws, 1859, c. 68); Report, Committee on Education, on bill 307A (to substitute county for town superintendent of schools), AJ, 643–644 (1860) (bill opposed because "at the present time . . . considering the sparseness of the population in the majority of the counties of the State, the extended geographical limits of many of these counties, and the unsettled and shifting character of the population, the committee do not think it wise to adopt the measure"); Annual Report, State Superintendent of Public Instruction, Dec. 10, 1864, PD, 344 (appended report of Manitowoc County school superintendent, noting existence of many unfit school buildings there, but continuing: "There is, however, much palliation for this fact, when it is known, that this county is heavy timbered land, and not easy of cultivation, and that many of our people are yet poor, it being only three or four years since the county itself has been independent in its resources for home consumption"); Report, Committee on Roads, AJ, 273 (1868) (opposing substitution of a town highway superintendent for district overseers: ". . . any change from the present system, by which the roads over a large tract of country would be under the control of a single person, would not be attended with any good results, for the reason that some of the roads must necessarily be remote from his residence and in which he would not, naturally, take as much interest as an overseer residing in the immediate neighborhood . . . These reasons are particularly applicable to the more northern parts of the State, where a town contains a great extent of territory, over the roads of which a superintendent living in one part of the town might never travel, and of which he would know but little, and in which he would take but little interest"); Report, Committee on Railroads, submitting a report on railroad taxes by the Railroad Commissioner, SJ, 242 (1880) (Commissioner observes that "economy and good management are not distinguishing features of their local government" in some northern counties); Report, Committee on Railroads, on need of state care for unpaid railroad construction workers in north, SJ, 114 (1882) ("It seems to be inevitable that these men must be subsisted by the state, as remote and sparsely settled counties cannot be expected to maintain them; or should subsistence be withheld from them, the inevitable result will be tumult and danger to persons and property").

Typical emphasis upon the desire to keep government close to the people in conditions of scant population and difficult transport may be seen in Resolution 12S of 1875, SJ, 74 and 78; Majority Report, Committee on Town and County Organization, on bill 174S (to create the town of Grant, Wood County), SJ, 363 (1885), and Minority Report of the counterpart Assembly committee on the same bill, AJ, 850. An unusually direct statement of the competition for tax base involved in jockeying for territory among existing or proposed units of local government will be found in Minority Report, Committee on Town and County Organization, on bill 122A of 1887 (shift of area from Oneida to Lincoln County), AJ, 277. On the tenacity of the tradition of highly localized government, see Gates, *Wisconsin Pine Lands,* 153, 167; Groves, 95; Wisconsin State Planning Board, *Local Government Study in Wisconsin,* I, 25, 33.

Under the cases cited in note 44 above, it was made plain, in years in which the forest region was still undergoing fluid growth, that the requirement of uniformity in the system of county and town government imposed by Wisconsin Constitution, Art. IV, sect. 23, did not affect the legislature's discretion in fixing the territorial bounds of local government units. Moreover, the court early indicated that it would construe Art. IV, sect. 23, to permit the legislature to classify counties to provide somewhat different frames of organization according to reasonable criteria growing out of the varying situation of different regions; what the constitution sharply forbade was simply the making of local distinctions without functional justification. See State ex rel. Peck v. Riordan, 24 Wis. 484, 489–491 (1869). Compare Cathcart v. Comstock, 56 Wis. 590, 603, 14 N.W. 833, 839 (1883), with State ex rel. Town of La Valle v. Board of Supervisors of Sauk County, 62 Wis. 376, 378, 22 N.W. 572, 573 (1885). So long as the legislature provided but one model or "system" of county or town government, the cases allowed considerable scope to the legislature in providing variations within the system, in recognition of the limited mandate of the constitution that the system be "as nearly uniform as practicable": see State ex rel. Busacker v. Groth, 132 Wis. 283, 300–301, 112 N.W. 431, 437 (1907). On the face of the constitutional language the requirement that "the legislature shall establish but one system of town and county government" appeared to deprive the legislators of discretion other than to choose one model and adhere to it. In practice the decisions showed that the court could permit considerable variation in organization of offices and definition of powers, by pursuing a liberal conception of what details touched the nature of the "system" and what touched only subordinate aspects within a system. In particular the court was liberal in validating administrative variations based on differences in population which could be deemed reasonably related to the problems of local government. See, e.g., State ex rel. Grundt v. Abert, 32 Wis. 403 (1873). State ex rel. Adams v. Radcliffe, 216 Wis. 356, 257 N.W. 171 (1934), long after the great era of lumber, held invalid — as violating the constitutional stipulation of "one system" — a statute allowing the voters of a county to elect a commission form of county government, rather than the supervisor form. Though the opinion is not clear, probably the key reason why the statute here in question was deemed to provide a second "system" was that it set up a second distinct basis of representation, by districts rather than by towns. But the legislature was conceded so broad a discretion to arrange the geographical outlines of county and town government that adherence to the town basis of representation should never have been a serious obstacle to devising a local government pattern adapted to the peculiar condition of the north country; cf. Cathcart v. Comstock (cited above). So far as the "uniformity" requirement was concerned, State ex rel. Adams v. Radcliffe cautions that the legislature might have had difficulty in responding to the special problem of the north country by the simple grant of local option, but obviously this would have posed no obstacle to firm legislative classification of whole regions, for distinctive variation of the details of "one system."

47. Message of Apr. 5, 1862, SJ, 810, 811, 812, 814. See Chapter I, note 7, and Chapter II, notes 6 and 7.

48. SJ, 812–814 (1862).

49. For the sustaining of the veto, see SJ, 881 (1862). The only other instance of use of the veto power to uphold the values of a central surveillance of local government administration in the north country was in Gov. Rusk's Veto Message of Mar. 18, 1887, SJ, 418–419. Rusk disapproved bill 80S, which would have authorized Douglas County to effect a particular harbor improvement in St. Louis Bay. He rested partly on the ground that the project bore insufficient promise of general benefit to the county to warrant burdening all taxpayers therein with the expense. The Governor commented on the general state of local government fiscal affairs in a fashion which carried implicit criticism of the lack of central discipline: "In many parts of the state the rights of property owners have been largely disregarded, and in many instances burdensome taxes have been imposed upon the property of large districts for the purpose of making purely local improvements, without any benefit to a large proportion of the property taxed. This evil, I believe, should not receive legislative sanction. Because of the rapid increase of special and local legislation, and of the usual lack of consideration given to such measures, I have felt it my duty in this connection to call your attention to the growing danger of such bills." The veto was sustained by unanimous vote (SJ, 419). Cf. General Message of Gov. Upham, AJ, 26–27 (1893) (need to rearrange judicial circuits to take more regard for population and territory); General Message of Gov. La-Follette, AJ, 29, 30 (1903) (need for more prudence in dispensing state aids to sparsely populated school districts).

50. Revised Statutes, 1849, c. 15, sect. 84, continued by General Laws, 1859, c. 14, and Revised Statutes, 1878, sect. 1121; General Laws, 1859, c. 50; Laws, 1875, c. 290; Laws, 1879, c. 202; Laws, 1895, c. 43; Statutes, 1898, sect. 1121; Laws, 1899, c. 164.

51. See note 46 above.

52. Wisconsin Constitution, Art. XI, sect. 3. That, subject to this provision, the legislature had the power and the duty (though it be a judicially unenforceable duty) to regulate local finance was declared in State ex rel. Marinette, Tomahawk & Western Railway Co. v. Common Council of the City of Tomahawk, 96 Wis. 73, 71 N.W. 86 (1897).

53. Phelan, 198. Compare the implications of the quoted remarks of Gov. Rusk in 1887, note 49 above.

54. Wisconsin Constitution, Art. XI, sect. 3, as amended under the submission by Laws, 1874, c. 37. That this amendment grew out of railroad construction experience, see Hunt,

Law and Locomotives, 85; cf. Kyes v. St. Croix County, 108 Wis. 136, 83 N.W. 637 (1900). The amendment did not limit the authority of local government units to incur current debt for current operations: Herman v. City of Oconto, 110 Wis. 660, 86 N.W. 681 (1901).

55. See Report, Select Committee on tax delinquency notices, SJ, 698 (1863); Report, Committee on the Judiciary, on tax advertisements, AJ, 1131–1132 (1866); Report, Select Committee on bill 114S, SJ, 305–309 (1867); Resolution 13S of 1868, and Report, Committee on State Officers, SJ, 82, 562 (1868). Cf. Gates, *Wisconsin Pine Lands,* 141, 148, 153–154.

56. See Barnett, 166–172; Goodman, chap. ii, and pp. 150–156, 201; Nybroten, chap. ii.

57. See, generally, Gates, *Wisconsin Pine Lands,* 170–173; cf. Laws, 1877, cc. 169 and 286; Laws, 1878, cc. 133, 290, and 308; Laws, 1879, cc. 190, 205, and 209. Just as such control legislation as was passed was spotty and limited in concept, so the legislative record shows no sustained attention to surveillance of local government administration or to devising central administrative means to accomplish effective surveillance. The episodic and unrelated character of such complaining references or actions as appear in the record helps make the point. See, e.g., Report, Committee on Railroads, SJ, 242 (1880); Report, Committee on State Affairs, opposing bill 223S of 1881 (to authorize the town of Barron to borrow money), SJ, 489 (recommendation followed, SJ, 504); Annual Report, Railroad Commissioner, Jan. 6, 1882, PD, I, xxxix–xl; Majority Report, Committee on Town and County Organization, on bill 174S of 1885, AJ, 823, with which compare Minority Report of same committee, AJ, 849, 850. See also the veto message of Gov. Rusk, quoted in note 49 above. This is the only Governor's message which reflected any active concern in that office to develop some superintendence of the quality of local administration. Conceivably bold, sustained vigor in the office of the Secretary of State might have made the Secretary the effective inspector of the honesty and efficiency of local government in some fields of action. But the action of that office was weak and ineffective: see Goodman, 201.

There was no more effective central scrutiny of local government in political than in fiscal matters, though scattering episodes in the official record attest occasional uneasiness as to what was going on. See General Message of Gov. Ludington, Jan. 11, 1877, PD, I, 18 ("The facility with which voters who have not registered can make the affidavit required by the statute defeats the object of registration and unquestionably opens the way for a great deal of illegal voting"), with which compare material cited in note 34 above on abortive suggestions to regulate voting by transient lumber crews; Joint Resolution 2A of 1889, AJ, 33, and Report, Committee on State Affairs, SJ, 197–198 (1889) (lack of complaints to governor in past regarding failure of local government to deal with bawdy houses in north shows situation not out of hand); General Message of Gov. Peck, SJ, 23 (1891) ("The testimony introduced in a recent trial of an election contest in the northern part of this state demonstrates how easily under existing laws fraud and corruption, banished from the polling place, find encouragement and refuge in official dishonesty or negligence in higher places"); *Second Biennial Report of the Attorney General,* for the biennial period ending June 30, 1904 (Madison, 1904), 20 (investigation by Attorney General on request of Gov. LaFollette, and hearings by Governor, on charges of misconduct made against sheriff and district attorney of Iron County, led to order removing the two officials).

58. Laws, 1901, c. 445, 1905, c. 380, 1911, c. 523. Cf. General Message of Gov. Philipp, AJ, 23 (1917). See, generally, Tyler Francis Haygood, "State Control of Local Finance" (Ph. D. thesis, 1935), 79, 110.

Chapter IX. The Use of Common Transport Facilities

1. See Wisconsin River Improvement Co. v. Lyons, 30 Wis. 61, 65–67 (1872); Sellers v. Union Lumbering Co., 39 Wis. 525, 527 (1876); Wisconsin River Improvement Co. v. Manson, 43 Wis. 255, 262, 265 (1877).

2. General common law precedent recognized obstruction of a navigable stream as a crime. See Joel Prentiss Bishop, *Commentaries on the Criminal Law* (1877), I, 297. Though the Wisconsin Reports show no prosecution, the court indicated the Wisconsin common law to be so, in Walker v. Shepardson, 2 Wis. 384, 396 (1853). In re Eldred, 46 Wis. 530, 541, 1 N.W. 175, 177 (1879), and State v. St. Croix Boom Corporation, 60 Wis. 565, 570, 19 N.W. 396, 398 (1884) likewise broadly refer to such obstruction as "indictable." Revised Statutes, 1878, sect. 1598, had created in broad terms a money "forfeit"

for all such obstructions, without also providing that the conduct should be deemed a misdemeanor or providing any conventional criminal sanction; by sect. 3294, in this situation the forfeiture was to be sued for in a civil action; thus the references in the Eldred and St. Croix Boom Corporation cases apparently recognize the continued separate existence of a common law offense. Before the 1878 revision the applicable statutes (likewise creating only forfeitures) were narrower, applying only to obstruction by felling trees, placing slabs or other impeding objects in the water, or erecting unauthorized booms. See Revised Statutes, 1858, c. 119, sect. 112; General Laws, 1861, c. 285, 1866, c. 123, 1869, c. 152. That on general equity precedent the state might enjoin the creation of a public nuisance by works built in a stream was held in Luning v. State, 2 Pinn. 284 (Wis. 1849) (health), and in Attorney General v. City of Eau Claire, 37 Wis. 400, 447 (1875) (obstruction). The common law and equity relief available to a private litigant showing special damage was recognized as available in Wisconsin in Walker v. Shepardson, 2 Wis. 384, 396 (1853), without reference to General Acts, 1853, c. 72, in which the legislature declared that common law should generally govern rights as to navigable waters. Barnes v. City of Racine, 4 Wis. 454, 465 (1856) in any event recognized the 1853 act as warranting the availability of the common law remedies. That a specific statutory franchise barred public or private prosecution for maintaining a nuisance was held in Jones v. Pettibone, 2 Wis. 308 (1853), and in Stoughton v. State, 5 Wis. 291 (1856). See, generally, cases cited in note 1 above. To an ill-defined extent, the common law recognized a limited right of self-help in the aggrieved party against either a temporary unreasonable use or a continuing nuisance: see Harrington v. Edwards, 17 Wis. 586 (1863) (unreasonable use); Whisler v. Wilkinson, 22 Wis. 572 (1868) (nuisance).

3. Walker v. Shepardson, 2 Wis. 384 (1853) (injunction); Whisler v. Wilkinson, 22 Wis. 572 (1868) (self-help); Olson v. Merrill, 42 Wis. 203 (1877) (damages).

4. Wisconsin River Improvement Co. v. Lyons, 30 Wis. 61, 64, 67 (1872) (injunction granted against unlicensed dam, where facts show challenged dam would not in fact obstruct navigation). So in Woodman v. Kilbourn Manufacturing Co., 1 Biss. 546, 553 (Circ. Ct. W. Wis. 1867), Miller, D. J., ruled that, by the general Wisconsin statute requiring legislative license for works in navigable waters, "this principle is prescribed," that "the paramount right of navigation being in the people, who may wish to use a navigable stream, any obstruction however inconsiderable without constitutional legislative authority is a nuisance, to be abated at the suit of an individual, or of the state." In the context of citation of the Lyons decision, and more especially of a strong emphasis upon the obligations laid on the state by the act of admission to the Union as reflected in Wisconsin Constitution, IX, 1, the court held that mere maintenance of a dam without valid statutory warrant made out a public nuisance in Attorney General v. City of Eau Claire, 37 Wis. 400, 447 (1875): "The actual navigation may be little, and the obstruction might be slight. So the affidavits tend to show. But neither the right nor the wrong is a question of degree. We cannot listen to one about to put an unlawful work in a public river, that it will not materially obstruct navigation." See, also, In re Eldred, 46 Wis. 530, 541, 1 N.W. 175, 177 (1879). Rosenberry, J., for the court in State v. Sutherland, 166 Wis. 511, 521, 166 N.W. 14, 17 (1918), spoke of the common law rule in terms almost as absolute as those announced by the nineteenth-century decisions in light of the Wisconsin licensing statute: "At common law . . . to abate an obstruction as a nuisance it was not necessary that the obstruction should actually have interfered with navigation; it was sufficient if it rendered it less convenient and less useful." However, on their facts, the nineteenth-century decisions in Wisconsin which the court treated purely from a common law standpoint (that is, without invoking the licensing statute) all involved proof of material obstruction in fact.

The nineteenth-century decisions made the common law remedies available to implement substantive policy declared by statute, though the statutes themselves created no sanctions. Revised Statutes, 1858, c. 144, explicitly authorized the entry of injunctions in private actions at law concerning private nuisances, and by its terms recognized as existing by common law the availability of money judgments for damages. Earlier decisions had recognized both remedies as available in Wisconsin on the basis of common law and equity precedent. See note 2 above. The 1858 statute merged the legal and equitable remedies, and otherwise seems to have been treated as simply declaratory. See Remington v. Foster, 42 Wis. 608 (1877) (statute abolishes former separate relief in equity); Lohmiller v. Indian Ford Water-Power Co., 51 Wis. 683, 8 N.W. 601 (1881) (statute allows uniting prayers for damages and for abatement; obiter, equitable defenses to abatement may be pleaded thereunder); St. Croix Consolidated Copper Co. v. Musser-Sauntry Land, Logging & Manufacturing Co., 145 Wis. 267, 269, 130 N.W. 102, 103 (1911) ("Without conceding that the statute is exclusive we may well say that it is substantially a declaration of the unwritten law with added features and simplification of procedure"). Until about the end of the lumber era there was never any general statute providing a remedy for the state against public nuisances. Common law remedies were made available unquestioningly at suit of the state, whether to implement substantive doctrine found in common law or in the navigation guaranty of the Wisconsin constitution or in the general statutory requirement that legislative warrant be had for works on navigable waters. See note 2 above. The invocation of the common law remedies was especially pointed in State v. St. Croix Boom Corporation, 60 Wis. 565, 570, 571, 19 N.W. 396, 398 (1884), where in contrast the court cited Revised Statutes, 1878, c. 137 (the former Revised Statutes, 1858, c. 144) as the basis for a private action against nuisance. The general reliance on common law as supplying the remedies for infringement of the public right was attested also by the history of the one statute which provided a public remedy in two limited situations. Acts, 1849, cc. 62 and 98, provided that dams built on the Wisconsin or the Black rivers without specified styles of slides should be deemed public nuisances. The 1849 act also provided redress by a summary proceeding before any justice of the peace, who should summon a jury, and upon a jury finding that a dam was so illegally maintained, should issue his writ to the sheriff or any constable "commanding him forthwith to remove and abate such nuisance . . ." Revised Statutes, 1878, sect. 1603, substituted for this procedure — to implement the slides requirements specified for dams on these two rivers — the declaration that every offending dam "is declared a nuisance, and may be abated by the circuit court at the suit of any person aggrieved thereby." *Report of the Revisers of the Statutes,* 125, explained this change in terms which impliedly emphasize their reliance upon the general doctrines of the common law to set the terms of the available remedy. "A mere glance at the summary provisions of the acts of 1849," said the Revisers, "will show them to be unsafe, leaving it to a jury impaneled by a justice to cause expensive works to be torn out upon an ill regulated examination, and without the right of appeal." The prior reliance upon the common law as the source of sanctions against public nuisances was underlined when Laws, 1911, c. 652, amended Wisconsin Statutes, sect. 1596 (the general requirement of legislative warrant for works built in navigable waters), to add the declaration that an offending structure "is hereby declared to be a public nuisance, and the construction of any such dam . . . or other obstruction may be enjoined or its maintenance abated by action at the suit of the state or any citizen thereof," as well as provision for a $50 per day forfeit to be sued for by the Attorney General, and a direction that the Railroad Commission report to the Governor any violation of the amended sect. 1596. So amended, the statute was thence continued: see, e.g., Wisconsin Statutes, 1959, sect. 30.15. Note the recognition of the change effected by the 1911 amendment in the doctrinal basis for public nuisance proceedings, in State v. Sutherland, 166 Wis. 511, 519, 166 N.W. 14, 17 (1918). See also Laws, 1905, c. 145 (general authorization of actions in circuit court to enjoin public nuisances, by Attorney General, or by a private relator on leave of court; no specific reference to streams); this act was carried into the compiled statutes, to become, for example, Wisconsin Statutes, 1959, sect. 280.02.

5. Douglass v. State, 4 Wis. 387, 390 (1856); Leihy v. Ashland Lumber Co., 49 Wis. 165, 5 N.W. 471 (1880). Cf. Ulrich v. Hull, 17 Wis. 424, 427 (1863).

6. Compare Olson v. Merrill, 42 Wis. 203 (1877) with A.C. Conn Co. v. Little Suamico Lumber Manufacturing Co., 74 Wis. 652, 43 N.W. 660 (1889). See Chapter III, notes 76–82.

7. Stevens Point Boom Co. v. Reilly, 44 Wis. 295 (1878); Northern Pine Land Co. v. Bigelow, 84 Wis. 157, 54 N.W. 496 (1893). See, accord, Cohn v. Wausau Boom Co., 47 Wis. 314, 322, 2 N.W. 546 (1879).

8. Weatherby v. Meiklejohn, 61 Wis. 67, 20 N.W. 374 (1884) (boom); A.C. Conn Co. v. Little Suamico Lumber Manufacturing Co., 74 Wis. 652, 43 N.W. 660 (1889) (dam). Compare, also, Stevens Point Boom Co. v. Reilly, 44 Wis. 295 (1878) (riparian owner may build boom to reach navigable channel), with s.c., 46 Wis. 237, 49 N.W. 978 (1879) (ruling does not mean that one who owns opposite banks may thereby throw boom across

whole stream, a step for which a statutory license is required).

9. That, where a statute did not clearly confer a franchise or plainly forbid a use, the determination of whether a nuisance existed or not took on the character of deciding upon the reasonableness of particular details of construction or maintenance in a fashion akin to the application of the doctrine of reasonable use as applied to particular incidents of stream operations, see Whisler v. Wilkinson, 22 Wis. 572 (1868) (self-help allowable where dam owner did not properly maintain passage facilities); Gates v. Northern Pacific Railroad Co., 64 Wis. 64, 24 N.W. 494 (1885) (railroad bridge may "temporarily" obstruct navigation, if railroad provides proper bridge attendants); A.C. Conn Co. v. Little Suamico Lumber Manufacturing Co., 74 Wis. 652, 657, 43 N.W. 660, 661 (1889) ("Dams, booms, mills, and bridges, even, may be constructed on some navigable streams in such a manner as not to seriously affect the navigation thereof or infringe upon common right . . . 'There may be, and there must be, allowed of that which is common to all a reasonable use . . .' ").

10. See Rich v. Keshena Improvement Co., 56 Wis. 287, 14 N.W. 191 (1882) (liability for negligent construction of dam, treated as matter outside company's statutory franchise); Miller v. Sherry, 65 Wis. 129, 62 N.W. 612 (1886) (reasonable use rule includes obligation of complaining dam owner to maintain his works in condition to deal with waters discharged from above in ordinary course of stream use); Wright v. Mulvaney, 78 Wis. 89, 46 N.W. 1045 (1890) (reasonable care owed by log tow to fishermen). Perhaps there is less elaborate exposition of the reasonable use doctrine in such forest-area cases because the Wisconsin court had earlier so firmly and repeatedly declared it to be the law of the state in cases arising outside the forest region in contests between competing power users. See, e.g., Moore v. Coburn, 1 Pinn. 538 (Wis. 1845); Mabie v. Matteson, 17 Wis. 1, 8 (1863); Timm v. Bear, 29 Wis. 254, 265 (1871).

11. The allocation of burden of persuasion on fault or no-fault is implicit rather than explicit in the cases. See Rich v. Keshena Improvement Co., 56 Wis. 287, 14 N.W. 191 (1882); Miller v. Sherry, 65 Wis. 129, 26 N.W. 612 (1886). The point came closer to exposition where the court ruled that plaintiffs failed to make out a case of negligent log handling against defendants whose logs were in the control of independent contractors at the time plaintiffs suffered detriment. See Bearrs v. Sherman, 56 Wis. 55, 13 N.W. 869 (1882); Jacobs v. Hershey Lumber Co., 124 Wis. 54, 102 N.W. 319 (1905). Timm v. Bear, 29 Wis. 254 (1871) held that plaintiff downstream mill owner established a claim for damages against defendant upstream mill owner for wrongful withholding of the ordinary flow of the stream, by showing "the fact of detention and of injury by reason thereof to the lower mill," casting on the defendant the burden "to show the necessity and reasonableness of the detention." Blandly ignoring the well-established contrary general rule in tort both elsewhere and in Wisconsin — see Brown v. Kendall, 60 Mass. 292 (1850); Richards v. Sperry, 2 Wis. 216, 223 (1853) — the court declared, "If not universal, it may at least be said to be a rule of very general application, that a sufficient *prima facie* case is made when it is shown that loss or injury has been sustained by the plaintiff, and that the same was caused by the act or default, or what may have been the default, of the defendant. A . . . rule . . . no less general in its application, is, that matters of excuse or justification must be shown by the party claiming the benefit of them. It is generally true of such matters, also, that they are peculiarly within the knowledge of the party seeking the benefit of them, and are not known or are incapable of disproof by the other party . . ." (*id.*, 263). This pronouncement seems contrary to the earlier statement of the law as between millmen competing for flow in Patten v. Marden, 14 Wis. 473, 478 (1861), to which Timm v. Bear makes no reference, and to the tone, later, of Miller v. Sherry, 65 Wis. 129, 26 N.W. 612 (1886) and Coldwell v. Sanderson, 69 Wis. 52, 28 N.W. 232 (1887). The court never thereafter cited Timm v. Bear for its burden-of-proof ruling. Perhaps the doctrine there announced may be explained as resting on an unspoken conception of the parties' rights as to be determined by analogies of suits for conversion or misuse among owners in common, rather than by principles of tort among strangers. Cf. Dahl v. Fuller, 50 Wis. 501, 7 N.W. 440 (1880); Earll v. Stumpf, 56 Wis. 50, 13 N.W. 701 (1882); Sullivan v. Sherry, 111 Wis. 476, 87 N.W. 471 (1901). Perhaps the ruling in Timm v. Bear should be viewed as an unacknowledged resort to the doctrine of *res ipsa loquitur,* in view of the court's concern that the matter in issue was specially within the

knowledge of defendant. But the mere showing of detriment from detention of water seems not to constitute an injury such as in the ordinary management of affairs would happen only by want of due care in the actor. See Cummings v. National Furnace Co., 60 Wis. 603, 612, 18 N.W. 742, 745 (1884).

12. Compare the care with which the court expounds the utility of the standard of reasonable use, to protect enterprisers against unfair or crushing blows of liability for damages caused by them to other persons incident to unforeseeable and extraordinary circumstances, in Borchardt v. Wausau Boom Co., 54 Wis. 107, 110, 11 N.W. 440, 441 (1882) (applying the common law doctrine to define the content of liability under a statutory franchise).

13. The distinction is dramatized in two lawsuits which expound different bases of liability for downstream flowage caused by the same improvement company. Compare Rich v. Keshena Improvement Co., 56 Wis. 287, 289, 14 N.W. 191, 192 (1881) ("The *gravamen* of the action is the negligence charged in respect to the construction and management of the dams which were carried away, and the consequent injury to the plaintiff's land and crops. For such an injury a common law action may be maintained; and the mill-dam act affords no remedy therefor. Whether or not under the charter of the defendant the mill-dam act furnishes the exclusive remedy for the recovery of damages for injury to land below the dams, caused by discharging sufficient water to float logs down the stream, we do not decide. No such question is presented on this appeal"), with Hackstack v. Keshena Improvement Co., 66 Wis. 439, 445, 29 N.W. 240, 242 (1886) (where the court finds that "the lands of the plaintiff, by the reason of this flowage, were rendered entirely and permanently worthless for the purposes for which they could be used, and have been abandoned by the owner," and that the defendant's charter, and more particularly the eminent domain procedure therein provided, do not authorize flowage below the licensed works, held, that there has been "a complete taking of the property for the public use" and that "the plaintiff is entitled to the common-law remedy he has sought in this action"—obviously for recovery against special damages suffered from a private nuisance).

14. See, on mutual accommodation within the exercise of riparian rights, Hazeltine v. Case, 46 Wis. 391, 1 N.W. 66 (1879) ("each riparian proprietor was entitled to the use and enjoyment of the stream in its natural flow, subject to its reasonable use by other proprietors"); on the wisdom of requiring public nuisance doctrine to yield to the riparian owner's right to dam a freshet-navigable log stream, A.C. Conn Co. v. Little Suamico Lumber Manufacturing Co., 74 Wis. 652, 656, 43 N.W. 660, 661 (1889) ("a distinction should be made in view of riparian rights. For if the right of floatage is paramount, so that no bridge or dam or other obstruction can be placed in or over the stream by the riparian owner, his use and enjoyment of his property are unnecessarily abridged and restricted . . . The rights of the riparian owner and of the public are both to be enjoyed with due regard to the existence and preservation of the other. The right of floatage of logs is not paramount in the sense that the using of the water by the riparian owner for machinery is unlawful so long as he does not materially or unreasonably interfere with the public right"); and on the favor for multiple action under the doctrine of reasonable use among navigators, Wright v. Mulvaney, 78 Wis. 89, 93, 46 N.W. 1045, 1046 (1890) ("The benefit which the navigator is entitled to claim by reason of his paramount right is . . . that when the two rights necessarily conflict the inferior must yield to the superior right. But he may not by his own negligence unnecessarily force the two rights into conflict, and then claim the benefit of the paramount right").

15. That the protection which nuisance doctrine cast about the use of the waterways was valued because it provided an assured framework for industrial investment and sustained productive effort, was indicated in Olson v. Merrill, 42 Wis. 203, 212, 214 (1877), where Ryan, C. J., justifies this doctrinal pattern as one "essential to the public interest in the pine-growing regions of the state" and "important to its great lumbering interests." Similarly, that the qualifications allowed upon nuisance doctrine through the elements conceded to riparian ownership were important to the value of land, is indicated in A.C. Conn Co. v. Little Suamico Lumber Manufacturing Co., 74 Wis. 652, 656, 43 N.W. 660, 661 (1889), quoted in note 14 above. That the doctrine of reasonable use, as a basis for adjusting detriment caused by incidents of operation on the streams, served to foster enterprise by hedging it against crushing responsibility for the unpredictable

episode, was emphasized in a decision applying the common law doctrine to supplement the definitions of a statutory franchise. Borchardt v. Wausau Boom Co., 54 Wis. 107, 110, 11 N.W. 440, 441 (1882) held that the company was not liable when at a time of extraordinary flood its works caused flowing of the land of an upstream riparian: "This principle is of the utmost importance to the existence and purposes of corporations which are created to build and maintain works of internal improvement, in part for the public benefit, by the investment of private capital. All of the ordinary and natural consequences of their works may well have been contemplated and expected, and their ability to meet such consequences and compensate for such damages as would be likely to occur may be ample and constantly maintained; but one extraordinary and unforeseen event, happening from natural causes, against which no provisions or precautions are or could be made, may sweep away in a day or an hour not only all of their profits but their capital, and bankrupt and destroy the corporation itself. In view of such extraordinary risks and hazards, capital would not be likely to seek such an investment, and such enterprises, of great public importance and benefit, would be avoided," did not the law set reasonable bounds upon their possible liability.

16. Apart from the aspects of the case which dealt with the claim of public nuisance and the defense provided against that charge by the existence of defendant's statutory franchise, Heerman v. Beef Slough Manufacturing etc. Co., 1 Fed. 145, 158–159 (Circ. Ct. W.D. Wis. 1880) also involved a prayer by the complainant steamboat operator for an injunction against defendant's log driving on the Chippewa River. In this phase the case presented a claim under the doctrine of reasonable use, rather than under that of nuisance. Circuit Judge Bunn, denying the prayer, explained that equity could not undertake to regulate the detailed incidents of common use of a waterway: "That the defendants, in common with the public, have the right to float logs down the Chippewa and other rivers and streams in the lumbering portions of the state seems to me incontestable; and if so, then equity will not grant relief against the use of the river for such purpose, although the right may be so exercised on occasions as to cause damage to individuals engaged in other departments of commerce. Equity will follow the law and not go in opposition to it. And if there happen, as there may, an abuse of the legal right, equity will not interfere to forfeit or crush out the right, but will turn the injured party over to his action at law for damages. This is the only practicable course; for the manner in which the river shall be used by these several interests cannot, from the nature of things, be regulated by injunction. All a court of chancery could do, if anything, would be just what the plaintiff asks for in this case — enjoin and prohibit the use of the river for the purpose of floating logs, the use of the improvements placed in the river to facilitate that interest, and compel them to be taken out; which would amount to a complete denial of the right, and the forfeiture of an immense property, and a vast commercial interest. It cannot by a decree regulate the use of these improvements, nor the use of the river as a common highway by the proprietors of these different interests, any better nor so well as they are now defined and regulated by the law. The Chippewa river is a public highway, and if a person should, without authority from the state, and disconnected with any purpose of improving or facilitating any branch of commerce upon the river, place obstructions in it, perhaps, in a suit brought by the proper party, equity would relieve against such obstructions to commerce by declaring them a nuisance, and enjoin their use. But the court cannot, by its decree, regulate the use of the river by the several classes of persons pursuing different lawful branches of commerce and navigation, any more than it can that of any other highway. The rights of these several classes are already well enough defined by law, and if any person or class of persons take or exercise a too exclusive or unreasonable use or possession of the river, to the injury of other persons, the law affords an adequate remedy in damages. What would be an unreasonable use of the stream for purposes of commerce, such as would give a right of action at law, must be determined from a full consideration of all the facts and circumstances of the particular case, and should be submitted to a jury." Compare Ryan, C. J., that even in nuisance cases the courts should be cautious in granting injunctions, lest that drastic remedy be made available prematurely or with severity of consequences disproportionate to any offense done: Remington v. Foster, 42 Wis. 608, 609 (1877).

17. See notes 2 and 6 above.

18. General Acts, 1853, c. 72; Revised Statutes, 1858, c. 41; Revised Statutes, 1878, and Wisconsin Statutes, 1898, sect. 1597; Wis-

consin Statutes, 1959, sect. 30.10(4)(b). As established by the 1853 statute, this adoption was of the applicable common law "as evidenced by judicial determinations in other states, in which the courts in such cases have adhered to its principles." *Report of the Revisers of the Statutes,* 124, explained that the qualifying words were stricken in shaping sect. 1597, so that the revised version should "not limit the knowledge of the common law, which our court has by the want of knowledge of other courts."

19. See note 4 above.

20. Resort to common law may be seen in filling out the content of rights and duties under statutory franchises, in Borchardt v. Wausau Boom Co., 54 Wis. 107, 11 N.W. 440 (1882), and Field v. Apple River Log Driving Co., 67 Wis. 569, 31 N.W. 17 (1887) (standard of reasonable care); in providing remedies, in Enos v. Hamilton, 24 Wis. 658 (1869), and Wisconsin River Improvement Co. v. Lyons, 30 Wis. 61 (1872) (abatement of nuisance), as well as in Field v. Apple River Log Driving Co., cited above, and Polebitzke v. John Week Lumber Co., 163 Wis. 322, 158 N.W. 62 (116) (claims for damages from alleged negligent exercise of navigating rights); in applying common law defenses, in Lohmiller v. Indian Ford Water-Power Co., 51 Wis. 683, 8 N.W. 601 (1881) (equitable defenses available under statute allowing abatement of nuisance in action at law), Arpin v. Bowman, 83 Wis. 54, 53 N.W. 151 (1892) (notice to offender is prerequisite to abatement of nuisance on private property when continued by lessee or grantee, but not of nuisance by maintaining obstruction of public highway), and Charnley v. Shawano Water Power & River Improvement Co., 109 Wis. 563, 85 N.W. 507 (1901), and McDougald v. New Richmond Roller Mills Co., 125 Wis. 121, 103 N.W. 244 (1905) (estopped).

21. See Chapter III, notes 17 and 43.

22. See *id.,* notes 44, 169–172, and 181, 182.

23. See note 4 above.

24. Hall v. Kitson, 3 Pinn. 296 (Wis. 1851), and Volk v. Eldred, 23 Wis. 410 (1868) (dams); Enos v. Hamilton, 27 Wis. 256 (1870) (boom).

25. Arpin v. Bowman, 83 Wis. 54, 53 N.W. 151 (1892).

26. Common law actions for damages for permanent flowage were maintained in Folsom v. Apple River Log Driving Co., 41 Wis. 602 (1877); Hackstack v. Keshena Improvement Co., 66 Wis. 439, 29 N.W. 240 (1886); and Carlson v. Stocking, 91 Wis. 432, 65 N.W. 58 (1895). To be distinguished, because they rested on a different showing of injury, were actions based on violations of the doctrine of reasonable use, for temporary flooding allegedly due to improper construction or handling of dams. See Rich v. Keshena Improvement Co., 56 Wis. 287, 14 N.W. 191 (1882); Miller v. Sherry, 65 Wis. 129, 26 N.W. 612 (1886). Cf. Boyington v. Squires, 71 Wis. 276, 37 N.W. 227 (1888) (negligent discharge of sawmill dam waters, to injury of downstream farm land, in a non-lumber area).

27. Of these 23 cases, in four the primary issue was under the riparian rights doctrine, in 12 under nuisance doctrine, and in seven under the doctrine of reasonable use. Of the four riparian right cases, two (note 7 above) concerned the riparian's right to maintain a private boom. Two cases concerned the riparian's claim upon the passing flow: see Grand Rapids Co. v. Bensley, 75 Wis. 399, 44 N.W. 640 (1890); West v. Fox River Paper Co., 82 Wis. 647, 52 N.W. 803 (1892). Distinguish cases turning on the construction of titles to water right, derived from statute or private agreement: see Kaukauna Water Power Co. v. Green Bay & Mississippi Canal Co., 75 Wis. 385, 44 N.W. 638 (1890). Of the 12 nuisance cases, six involved alleged injuries to navigation; see the Conn Co., Gates, and Whisler cases cited in note 9 above, the Olson case in note 6, and the Weatherby case in note 8, as well as Union Mill Co. v. Shores, 66 Wis. 476, 29 N.W. 243 (1886). One nuisance case involved damage to a public road: Town of Lawrence v. American Writing Paper Co., 144 Wis. 556, 128 N.W. 440 (1911). Three nuisance cases involved common law relief for permanent flowage of riparian land: see note 26 above. Two nuisance cases involved damage to waterpower developments: see Potter v. President and Trustees of Village of Menasha, 30 Wis. 492 (1872); Green Bay & Mississippi Canal Co. v. Telulah Paper Co., 140 Wis. 416, 417 (1909). Of the seven reasonable use cases, two involved alleged injury to navigation: see Harrington v. Edwards, 17 Wis. 586 (1863); Bearrs v. Sherman, 56 Wis. 55, 13 N.W. 869 (1882). Three involved injury to riparian land by activity in aid of navigation: see the Jacobs, Polebitzke, and Rich cases, notes 11 and 20 above. Of the other two reasonable use cases, Miller v. Sherry concerned navigation-use injury to a dam, and

Wright v. Mulvaney involved damage by log tows to fishermen's nets: see note 10 above. The dating of the 23 common law cases shows a marked clustering in the years of major initial exploitation of the forest, rather than in either the years of ascension to peak production or the plateau years of achieved major production. Only two cases date before 1870 (1863, 1868), four are in the seventies (1872, 1877 — two cases — and 1878), eight are in the eighties (three in 1882, three in 1886, and one each in 1884, and 1885), five in the nineties (all in the first part of the decade, two in 1890, and one each in 1892, 1893, and 1895), and four after the turn of the century (1905, 1909, 1911, and 1916). This particular pattern fits the larger outline of forest-area streams litigation brought to the Wisconsin Supreme Court as shown in the table in the first section of Chapter III; appellate work in volume characterizes the period of full-scale growth rather than of beginnings or of achievement.

28. Cf. Chapter III, notes 319–322.

29. Laws, 1874, c. 165, is the only statute which explicitly recognized safety of the person as an objective of a stream-use regulation. Requiring slides and guide booms on all dams or bridges over the Wisconsin River, c. 165 provided that one whose property was injured by a violation of its requirements might recover treble damages therefor, with a lien for the judgment against the offending works when these were privately owned. The statute then declared that an action should lie, with the same remedies, "in all cases where injury to person, loss of limb or life occurs, caused by the neglect or refusal to provide such piers and guide booms, or by the insufficient or defective construction thereof . . ." There were passing references to loss of life or danger to life as arguments against authorization of a dam below the Dells of the Wisconsin River, but in context these passing references were submerged in the main current of argument over property values. See petition of "residents of the county of Juneau," SJ, 109–110 (1866); Report, Committee on Incorporations, on bill 9S, AJ, 456, 457 (1866); Majority Report, Judiciary Committee, on bill 7S, SJ, 412 (1871). Cf. Fries, 71–72.

30. See Chapter III, note 320. The criticisms of want of safety regulations concerning dam construction in Gov. McGovern's General Message of Apr. 30, 1912, AJ, 13–15, throw into sharp relief the past inattention of the legislature.

31. See Chapter III, note 319, and Revised Statutes, 1878, sect. 1598; cf. Murphy, 49. The central position of lumber industry operating values was vividly reflected in the qualifications of Laws, 1887, c. 490, which declared guilty of a misdemeanor any operator of a portable sawmill who deposited sawdust or other refuse in streams into which the fish commissioners had deposited fry or in which brook trout were naturally abundant, but which also stipulated that "this act shall not apply to streams used for the purpose of driving logs or to portable mills erected on such streams previous to the passage of this act."

32. Revised Statutes, 1878, sect. 1596, originating in Laws of the Territory of Wisconsin, 1840–1841, Act No. 9, and continued into Wisconsin Statutes, 1959, sect. 30.10.

33. See Chapter III, notes 346, 351, and 356–360, and notes 1–9 above in this chapter.

34. Jones v. Pettibone, 2 Wis. 308 (1853) (plea of special dam franchise is good defense to action for obstructing rafts; special franchise supersedes ban of general licensing act). Of course, the general licensing act had regulatory impact simply as a requirement that stream users show title; lacking title, a dam builder committed a public nuisance, against which the holder of a special improvement franchise might have relief under sect. 1596: see Wisconsin River Improvement Co. v. Lyons, 30 Wis. 61 (1872). We must also read against the supporting background of the general licensing statute the court's readiness to find that a nuisance was proved on showing that an improver had violated limiting terms of his special franchise: see In re Eldred, 46 Wis. 530, 541, 1 N.W. 175, 177 (1879); cf. Arpin v. Bowman, 83 Wis. 54, 53 N.W. 151 (1892). Broad as it was, we should recall that sect. 1596 was not without limitations. It applied, by its terms, only to waters meandered and returned as navigable by the federal survey: see State v. Sutherland, 166 Wis. 511, 519, 166 N.W. 14, 17 (1918), commenting on the enlarged reach of Laws, 1911, c. 652, which applied to all waters navigable in fact. Moreover, as the court construed it, sect. 1596 had regulatory as well as franchise effect only with reference to the navigable channel. Thus the court held that the legality of booms built in shoal water abutting the channel rested on the riparian owner's common law right of access to navigation; sect. 1596 entered the picture only when the riparian man's works reached into the navigable channel: see Stevens Point Boom Co. v. Reilly, 46 Wis. 237, 243, 49

N.W. 978 (1879), clarifying s.c., 44 Wis. 295, 305 (1878).

35. Occasionally the prime guaranty to navigation took the form of a requirement that the franchise holder keep free and open a natural main channel or natural alternative channel: see Acts, 1852, c. 324; P&LL, 1854, c. 123; P&LL, 1871, c. 45; Laws, 1880, c. 279 (creating general boom franchise authorization of Revised Statutes, sect. 1777a). See also in P&LL, 1870, c. 299, the unusually specific construction stipulation imposed on the Beef Slough booming operation, that the grantee "shall pave with stone the bed of the said Beef Slough, at or near its connection with the Chippewa river, for the purpose of maintaining the present level of said bed and preventing an excess of the volume of water flowing over the same." Closely related to safe passage was the setting of height limits on dams; since this is more properly an element of the title of the dam builder than a regulation of operations, it was noted in the discussion of franchises (see Chapter III, note 127). A height limit would be strictly construed, against the grantee, as applicable to the entire structure and not satisfied simply by the height of a slide which formed part of the dam: see Arpin v. Bowman, 83 Wis. 54, 53 N.W. 151 (1892).

36. Examples of the quoted formulas may be seen in P&LL, 1856, cc. 238 and 294, 1867, c. 568, 1871, c. 326; Laws, 1874, c. 239. See also Chapter III, notes 169–172.

37. Wisconsin Statutes, 1898, sect. 1601. Sect. 1602 stated equally specific, but somewhat different, requirements affecting works on tributaries of these rivers. As first enacted, in Laws, 1849, cc. 62 and 98, these provisions applied to "every dam" built and maintained on these streams, and — in addition to the required slides — stipulated for construction of "aprons" and "floating spars or fingers" to help guide rafts off the slides. Laws, 1881, c. 239, amended the requirements with effect to lighten the construction burden on dam owners. The required width of slides was reduced from 60 to 40 feet on the main rivers and from 40 to 25 feet on the tributaries, and the gradation was allowed as five (rather than three) feet of descent per 50 feet of length on the main rivers, with a separate formula (a maximum of one inch in each foot of length) for the tributaries, which before had been governed by the principal statute. The amendment dropped the requirement of aprons and fingers. Most striking was the change which excepted from the statute "dams maintained exclusively for flooding purposes." On the other hand, the 1881 amendment added a statement of operating obligations, carefully distinguished by type; on the tributaries of the Wisconsin and Black, a dam owner must at any time, on request and without fee, open the slide to pass any raft or logs or water craft, but he should not be required otherwise to aid in putting such logs, rafts, or craft through the slide. This new detail on obligations of the works proprietor regarding operations affecting passage matches the development of analogous operating stipulations in a minority of special franchise acts in the later nineteenth century: see notes 39–41 below.

38. See, e.g., Acts, 1852, c. 282 (Apple and Willow rivers); Private & Local Acts, 1853, c. 205 (Pine River); P&LL, 1856, c. 238 (Eau Claire River); P&LL, 1858, c. 44 (Hemlock River); Laws, 1881, c. 196 (Knapp's Creek, Richland County).

39. Laws, 1878, c. 284. An especially firm, explicit obligation was stated in Laws, 1883, c. 224. Such provisions were included in enough private dam franchises to amount to about 20 per cent of the total granted through the lumber era; see Chapter III, note 174.

40. P&LL, 1861, c. 37, 1862, c. 43. More often the substance of this provision was effected by a stipulation that boom grantees might detain others' logs only a reasonable time for sorting, and that if passage were not duly facilitated by the grantees the owners might move their logs out, at the grantees' expense: P&LL, 1864, c. 161, 1865, cc. 258 and 338, 1866, cc. 237 and 579, 1868, c. 461, 1869, cc. 387 and 448, 1870, c. 24, 1871, c. 342. An uncommonly drastic statute was Laws, 1878, c. 208, applying to all boom operators on the Wisconsin River, who must supply the labor and facilities to pass out of their booms without unreasonable delay all logs and rafts moving downstream, on pain of becoming liable to the owners for the value of the detained commodities.

41. Laws, 1882, c. 247 (as amended, 1887, c. 253, and 1893, c. 143), 1883, c. 326, 1887, c. 12, 1887, cc. 41 and 223, 1891, c. 177, 1903, cc. 145, 153, 154, 155, and 156, 1905, cc. 398, 400, 407, 408, 409, and 464.

42. Cf. note 16 above.

43. There was a forecast implicit in the terms of a few franchises which specified the passage obligations of the improvers in such detail that the draftsman was impelled to create *ad hoc* administrative machinery to oversee the operation. Thus Laws, 1887, c. 12,

after stating conventional slides and boom-exit requirements for a franchise on the Wisconsin River in Lincoln County, added: ". . . all logs destined to points on said river below said dam shall be taken by the owners of said dam when they reach the flowage thereof, or the rear of any log jam which may be caused by the stopping of logs at the upper dividing works of said booms, and shall be driven by said owners, free of charge, and with reasonable dispatch, through the pond and flowage created by said dam, and over the same; and suitable gates, not less than twelve feet in depth from the top of said dam, shall be placed in said dam for the purpose of flooding said river below the same, to facilitate the running of logs down said river, and said gates shall be open whenever necessary for said purpose; but the heighth of the water in the pond created by said dam, shall not, for that purpose, be reduced below eight feet above low water mark." By March 1 in each year the dam owners and the governor of the state should each select one person, and these two a third (or, in default, the governor should select as many men as needed to create a board of three), to decide "when and how long said gates shall be opened or closed for flooding purposes; provided, that said gates shall not be opened or remain open for flooding purposes when the heighth of the water in said pond shall not exceed eight feet above low water mark." On the varieties of slight experiment with the administrative process in stream-use regulation, see Chapter III, notes 417–420.

44. The court drew the distinction between an issue of public utility law (liability for toll, for service rendered by given facilities) and an issue under the law of reasonable use (between persons whose relation, in the aspect here relevant, was simply that of sharers of the common stream), when in Wisconsin River Improvement Co. v. Manson, 43 Wis. 255, 266 (1877), it indicated that to the improver's action for tolls a raftsman might not pose a counterclaim for damages alleged due from the improver's works in an area of the stream outside the serviced area: "The matters stated in the counterclaim arise out of a different transaction, and constitute a tort." See, accord, Black River Improvement Co. v. Holway, 85 Wis. 344, 351, 55 N.W. 418, 420 (1893). The distinction between service-relationship and tort contexts was also drawn in effect by Ryan, C. J., in Attorney General v. City of Eau Claire, 37 Wis. 400, 434 (1875), when he argued that Laws, 1875, c. 333, could not be interpreted as a river improvement statute, because it included a stipulation that the waterworks dam there authorized at Eau Claire must not materially obstruct navigation. "The prohibition of material obstruction by the dam positively repels all suggestion that the improvement of the river is an object of the dam. It is little likely that the legislature should provide against obstruction by that which was to be done for the purpose of improvement." Ryan's analysis does not bring the matter to a desirably sharp focus. The basic issue was posed not by the presence or absence of navigation guaranties, but by the fact that the proposed works were primarily for the private purposes of the improver, rather than to provide a service to others; in the context of the 1875 franchise, the navigation guaranty pointed to the conclusion that the legislature saw those protected by it not as customers for the grantee's services but as navigators in an independent relation to the grantee. Nonetheless, despite its blurred statement, Ryan's comment does point to the distinction between a relation to be adjusted by contract analogies and one to be adjusted by tort analogies.

45. Cf. Chapter III, note 299.

46. *Id.*, notes 394–399, 463, 494, and 502.

47. Examples of marked qualifying language may be seen in P&LL, 1870, cc. 32 and 381; Laws, 1889, c. 215. Reductions of the scope of passage guaranties are exemplified by P&LL, 1865, c. 190, amending P&LL, 1862, c. 43; Laws, 1887, c. 253, and 1893, c. 143, amending Laws, 1882, c. 247; Laws, 1889, c. 32, amending Private & Local Acts, 1853, c. 152; Laws, 1893, c. 190, amending Acts, 1851, c. 129.

48. See Edwards v. Wausau Boom Co., 67 Wis. 463, 470–471, 30 N.W. 716, 719–720 (1886); J.S. Keator Lumber Co. v. St. Croix Boom Corporation, 72 Wis. 62, 101, 38 N.W. 529, 545 (1888); cf. Neff v. Wolf River Boom Co., 50 Wis. 585, 7 N.W. 553 (1880) (statute vesting successor company should not be construed to impose absolute liability for unknown hazards from predecessors's works); Borchardt v. Wausau Boom Co., 54 Wis. 107, 110, 11 N.W. 440, 441 (1882) (flowage alleged due to negligent works construction).

49. See Sweeney v. Chicago, Milwaukee & St. Paul Railway Co., 60 Wis. 60, 70–71, 18 N.W. 756, 760 (1884); Edwards v. Wausau Room Co., 67 Wis. 463, 471, 30 N.W. 716, 719 (1886).

50. Volk v. Eldred, 23 Wis. 410, 412

(1868). The case apparently involved the dam authorized by P&LL, 1857, c. 195. Compare the court's readiness, in effect, to presume that one who artificially increases the navigable capacity of originally navigable waters intends to dedicate the product of his artifice to public use, in the absence of clear contrary showing: see Weatherby v. Meiklejohn, 56 Wis. 73, 13 N.W. 697 (1882); Mendota Club v. Anderson, 101 Wis. 479, 78 N.W. 185 (1899).

51. Revised Statutes, 1849, c. 16, sect. 97; Revised Statutes, 1858, c. 19, sect. 112; General Laws, 1861, c. 285; General Laws, 1866, c. 123; General Laws, 1869, c. 152; Revised Statutes, 1878, sect. 1598; Wisconsin Statutes, 1898, sect. 1598; Laws, 1901, c. 413; Wisconsin Statutes, 1959, sect. 30.15.

52. See A.C. Conn Co. v. Little Suamico Lumber Manufacturing Co., 74 Wis. 652, 656–657, 43 N.W. 660 (1889).

53. Contrast the strict application of a dam height limitation to ground a nuisance action in Arpin v. Bowman, 83 Wis. 54, 53 N.W. 151 (1892), with the insistence upon proof of wrongful intent to enforce a statutory treble damages remedy for conversion of floating logs, in Cohn v. Neeves, 40 Wis. 393 (1876), as an indication of the broad approach likely to be taken to the relatively light forfeiture remedy of sect. 1598.

54. General Laws, 1861, c. 285, broadened the preventive scope of the language, and increased the forfeit from $5 to $25. General Laws, 1866, c. 123, added a ban on unauthorized booms, with a more strict forfeiture attached to the new boom clause. The breadth of terms of the statute as it had stood through the principal lumber years was underlined when Laws, 1901, c. 413, amended it to state that navigation with floating logs, or the use of temporary booms in aid of such navigation, should not be deemed obstructions within the act.

55. See Edwards v. Wausau Boom Co., 67 Wis. 463, 466, 30 N.W. 716, 717 (1886). The statutes included a miscellany of other, more limited laws concerning obstruction. See Laws, 1849, c. 62 (ban on dumping sawdust in Wisconsin or Black rivers, so as to obstruct navigation); General Laws, 1856, c. 73 (misdemeanor for sawmill operator or other person to throw slabs or other impeding substances into Wisconsin River or tributaries); Laws, 1880, c. 314 (forfeiture and civil actions against persons knowingly casting slabs or other refuse into Wisconsin River or tributaries); Laws, 1881, cc. 134 (protecting lakes against mill waste) and 220 (misdemeanor, to bank or brow logs so as materially to obstruct navigation).

56. The declarations of the navigable status in law of various described waters were gathered into Revised Statutes, 1878, sect. 1607 (repealed by Laws, 1917, c. 335). See Chapter III, notes 17 and 20, and Table 1 (general waterways legislation). Laws, 1911, c. 652 (continued into Wisconsin Statutes, 1959, sect. 30.10) broadened the general requirement of official license for works in streams, to include all streams "navigable in fact for any purpose whatsoever." See note 34 above.

57. See State v. Sutherland, 166 Wis. 511, 521, 166 N.W. 14, 17 (1918). Statutory declarations of the navigable status in law of specified streams were not construed as legislative findings that other streams were not navigable: see Bixby v. Parrish, 148 Wis. 421, 423, 134 N.W. 838 (1912). Declaration that a stream was a "public highway" was treated as conferring on it the legal status of navigable water: Wood v. Hustis, 17 Wis. 416, 418 (1863). See also Chapter III, notes 54, 71, and 7 .

58. Acts, 1849, c. 216; Revised Statutes, 1849, c. 35, sect. 1; General Acts, 1856, c. 48 (removal time extended from one to three years; more strict notice requirements laid on landowner who would take ownership of logs not reclaimed); General Laws, 1858, c. 31 (tightened notice requirements on landowner); Revised Statutes, 1858, c. 43, sect. 1; Revised Statutes, 1878, sect. 1600; Laws, 1881, c. 252 (removal time allowed log owner reduced from three years to eighteen months, but new right declared to make reasonable use of lands which must be traversed for removal; landowner's right to unreclaimed logs reduced from assertion of ownership to lien for twice removal costs and damages to his land); Wisconsin Statutes, 1898, sects. 1600 and 1600a; Laws, 1917, c. 335, sect. 16 (removal period at log owner's option reduced to six months, with satisfactory provision for damages to landowner made prerequisite; landowner's initiative to obtain or effect removal strengthened); Wisconsin Statutes, 1917, sect. 30.07. These provisions were repealed by Laws, 1959, c. 441. Where removal could be accomplished at nominal damage to the land, the common law perhaps gave as much right as the statute, according to the indication in Polebitzke v. John Week Lumber Co., 173 Wis. 509, 512, 181 N.W. 730, 732 (1921). On the

other hand, the early enactment of this provision, and the sustained attention given it by amendment over the lumber years, indicate that it dealt with a matter on which the industry desired more assurance.

59. General Acts, 1863, c. 169; Revised Statutes, 1878, sect. 3337; Laws, 1881, c. 141; Wisconsin Statutes, 1898, sect. 3337; Wisconsin Statutes, 1959, sect. 289.25. Cf. P&LL, 1857, c. 145.

60. Wisconsin River Log Driving Association v. D.F. Comstock Lumber Co., 72 Wis. 464, 466, 467, 40 N.W. 146, 147 (1888).

61. *Id.*, 467, 40 N.W. 147.

62. *Ibid.*

63. Hayward v. Campbell, 72 Wis. 321, 39 N.W. 540 (1888). But the court enforced the requirement of proof that the serviced log owner had not in fact adequately provided for handling his own logs, in Gibson v. Trow, 105 Wis. 288, 81 N.W. 411 (1900).

64. Wisconsin River Log Driving Association v. D.F. Comstock Lumber Co., 72 Wis. 464, 467, 40 N.W. 146, 147 (1888), acknowledged in effect as a liberal interpretation, in Merrill Railway & Lighting Co. v. City of Merrill, 119 Wis. 249, 254, 96 N.W. 686 (1903).

65. See Tables 3 and 4 in Chapter III. Of the twelve cases in the Wisconsin Supreme Court centering on regulatory legislation, eight seem grounded on the theory of a private action for special damage caused by a public nuisance; four of these cases came long after the principal lumber-transport era. See Hall v. Kitson, 3 Pinn. 296 (Wis. 1851); Jones v. Pettibone, 2 Wis. 308 (1853); Enos v. Hamilton, 24 Wis. 658 (1869), s.c., 27 Wis. 256 (1870); Sweeney v. Chicago, Milwaukee & St. Paul Railway Co., 60 Wis. 60, 18 N.W. 756 (1884); Flambeau River Lumber Co. v. Chippewa & Flambeau Improvement Co., 198 Wis. 134, 223 N.W. 417 (1929); Flambeau River Lumber Co. v. Lake Superior District Power Co., 200 Wis. 31, 227 N.W. 276 (1929); Flambeau River Lumber Co. v. Railroad Commission, 204 Wis. 524, 236 N.W. 671 (1931); Flambeau River Lumber Co. v. Chippewa & Flambeau Improvement Co., 204 Wis. 602, 236 N.W. 679 (1931). Cf. Arpin v. Bowman, 83 Wis. 54, 53 N.W. 151 (1892) (nuisance by violation of height limit: seems a point of franchise limits rather than regulation of operations). Four cases seem to represent actions analogous to suits in tort for episodes contrary to reasonable use standards. See Volk v. Eldred, 23 Wis. 410 (1868); Edwards v. Wausau Boom Co., 67 Wis. 463, 30 N.W. 716 (1886); Hayward v. Campbell, 72 Wis. 321, 39 N.W. 540 (1888); Wisconsin River Log Driving Association v. D.F. Comstock Lumber Co., 72 Wis. 464, 40 N.W. 146 (1888). Cf. Neff v. Wolf River Boom Co., 50 Wis. 585, 7 N.W. 553 (1880) (scope of liability for negligence imposed by succession statute on new corporation); J.S. Keator Lumber Co. v. St. Croix Boom Corporation, 72 Wis. 62, 39 N.W. 135 (1888) (action may lie for negligent nonfulfillment of handling stipulations in Minnesota boom franchise).

66. See Laurent, 46, 157. On flowage actions, see *id.*, 49, 163.

67. See Fries, 73.

68. Compare the discussion by Circuit Judge Bunn of the limits of effectiveness of equity doctrine or powers to establish detailed regulations or orders of preference among particular competing users, on a principal navigable stream, quoted in note 16 above. See also the court's insistence that administrators realize the merger of flexibility and specificity contemplated by a statute delegating rulemaking powers over the stream use, in Flambeau River Lumber Co. v. Railroad Commission, 204 Wis. 524, 543, 544–545, 236 N.W. 671, 678, 679 (1931).

69. See notes 39–41, 54, 58, and 59 above. The prevailing judicial approach to such legislation was to extend its reach. See notes 33, 49, 50, and 60–64 above.

70. See Chapter III, notes 394–399, 463, 494, and 502.

71. Rail transport, of course, also presented problems at common law and under statute concerning relations between the roads and third parties not their patrons. However, no developments distinctive to the lumber industry are represented in such matters, and therefore we do not discuss them. See, generally, Hunt, *Law and Locomotives,* chaps. iii and v.

72. The differing emphasis of the obligation to serve upon the police power and of rates and services questions upon the *quid pro quo* focus of contract law appears in comparison of Tewksbury v. Schulenberg, 41 Wis. 584, 593, 594 (1877) (validity of toll franchise depends on finding that was granted for the public purpose of "protecting and aiding" transport, and not "merely to promote private interests"), with Black River Improvement Co. v. Holway, 85 Wis. 344, 352, 353, 55 N.W. 418, 420, 421 (1893) (in action to collect tolls, counterclaim for damages to logs from improper handling is as proper as would be in an action under a

private log-running contract, for "there is really no difference between such an action for services at the price of the tolls fixed, and one for services at the price fixed by the contract or for what they are reasonably worth").

73. See the criticism of the sweep of toll privileges in the vetoed charter of the Chippewa River Improvement & Navigation Company, in Gov. Farwell's Veto Message of June 6, 1853, SJ, 815–817. Compare the pointed emphasis upon the need of finding a justifying public purpose to uphold a toll grant, in Tewksbury v. Schulenberg, 41 Wis. 584, 593, 594 (1877). See Chapter III, notes 92, 96, 110, 179, 180, and 331.

74. See the criticisms of what the Governor deemed insufficiently regulated monopoly power, in Gov. Farwell's veto of the Chippewa River Improvement & Navigation Company charter, cited in note 73 above. Compare the implications of the insistence that toll rights might be justified only on the basis of "services rendered as prescribed by law." See Sauntry v. Laird, Norton Co., 100 Wis. 146, 152, 75 N.W. 985, 987 (1898); cf. Underwood Lumber Co. v. Pelican Boom Co., 76 Wis. 76, 85, 45 N.W. 18, 21 (1890). Recognition of the transporter's dependence as created by a law-given improvement monopoly (with toll stipulated for all logs floated out of the stream, once improved), as distinguished from a free-choice bargaining relation, may be seen in the unusual terms of P&LL, 1871, c. 494, incorporating and enfranchising the Mill Creek Improvement and Log Driving Company. The franchise authorized the company "to demand and collect toll on said stream . . . For all logs or timber run or floated out of said stream, such sum per thousand feet as the board of directors shall determine, not exceeding fifteen cents per thousand feet, board measurement," with a lien therefor. On the other hand, the franchise also "authorized [the company] to carry on the business of running and driving logs and timber on said creek and upon the Wisconsin river below the mouth of said creek, in the same manner as an individual," leaving charges unmentioned and presumably to be arranged by private agreement, save that the company might take charge of others' ill-handled logs mingled with the company drive, and run them "at the cost and expense of the owners." Cf. Laws, 1897, c. 266 (fixed fee for logs aided, but grantee declared free to contract for other uses of water, subject only to requirement of scheduled charges and no discrimination). The same distinction appears, though less explicitly, in franchises which confer toll rights for use of improved waters (with a toll maximum set), and authorize management of the driving of ill-provided-for logs at cost, leaving to inference that if the grantees make driving agreements with other log owners, the charges thereunder are a matter solely for private bargain. See, e.g., Laws, 1874, c. 250, 1877, cc. 42 and 247, 1880, cc. 171 and 177, 1889, c. 83.

75. Cf. Menominee River Boom Co. v. Augustus Spies Lumber & Cedar Co., 147 Wis. 559, 569, 132 N.W. 1118, 1121 (1912) (statutory requirement of uniform rates should be construed to exact the reasonable conformity of like cases required under the common law of public-service companies).

76. Wisconsin Constitution, IX, 1, was held to require the limitation of tolls to reasonable levels, in Underwood Lumber Co. v. Pelican Boom Co., 76 Wis. 76, 85–86, 45 N.W. 18, 21 (1890). See also Tewksbury v. Schulenberg, 41 Wis. 584, 593 (1877); Wisconsin River Improvement Co. v. Manson, 43 Wis. 255, 264–265 (1877); Sauntry v. Laird, Norton Co., 100 Wis. 146, 152, 75 N.W. 985, 987 (1898).

77. Legislative control of rates authorized for a toll bridge was argued to show that a public purpose was involved and was safeguarded, in the grant of a toll bridge franchise, so as to warrant conferring the power of eminent domain on the grantee, in Report, Judiciary Committee, HRJ, 240 (1842). Flat-sum charges were authorized in different amounts according to the forest product involved, in the only toll improvement franchises granted before statehood. See Laws of Wisconsin Territory, 1845, p. 90 (act of Feb. 22, 1845, incorporating Wisconsin River Navigation Company), and *id.,* 1847, p. 44 (act of Jan. 29, 1847, authorizing Wisconsin River improvements at Grand Rapids, by Miner and Clinton).

78. For other examples of flat-sum charges authorized on account of the benefits of improvements at a single location, see P&LL, 1870, cc. 270 and 463, 1871, c. 467, 1872, c. 117; Laws, 1873, cc. 252 and 275, 1874, cc. 154, 176, 204, 228, 250, and 262.

79. For other examples of divisional tolls, see P&LL, 1864, cc. 131 and 157, 1870, c. 480, 1871, cc. 240 and 483, 1872, c. 112; Laws, 1873, cc. 12, 134, 159, and 203. An especially elaborate seven-division scheme may be seen in P&LL, 1869, c. 363. Amendments in course of passage attested the practical im-

portance attached to the specific terms in which divisional tolls were created: see, e.g., Laws, 1878, c. 281, as compared with bill 75A; Laws, 1887, c. 449, as compared with bill 125S. Alteration in divisional schedules was particularly marked in amendments of franchises after enactment. See, e.g., P&LL, 1871, c. 116, amending P&LL, 1869, c. 186; P&LL, 1869, c. 452, as amended as to division area definitions by Laws, 1876, c. 405, 1877, c. 124, 1878, c. 207, and 1885, c. 74; Laws, 1875, c. 226, amending P&LL, 1870, c. 480; Laws, 1873, c. 12, as amended, 1874, c. 294, and 1877, c. 247; Laws, 1881, c. 142, amending Laws, 1880, c. 144; Laws, 1883, c. 132, amending Laws, 1880, c. 214.

80. Cf. Ulrich v. Hull, 17 Wis. 424 (1863). Authority to charge less than the stated amount was expressed in a qualified form in an occasional grant which set a flat sum "unless otherwise agreed by the parties." See P&LL, 1857, c. 57; cf. Laws, 1882, c. 224.

81. For boom grants allowing the grantees to set "reasonable" fees, see, e.g., P&LL, 1856, c. 252; Laws, 1882, c. 276, 1897, c. 305; boom grants setting an allowed maximum included, for example, P&LL, 1857, c. 101, and 1870, c. 381; Laws, 1880, c. 319. For flow improvement franchises authorizing the grantees to charge reasonable rates, see, e.g., Acts, 1851, c. 325; P&LL, 1865, c. 444; Laws, 1882, c. 278; examples of flow improvement franchises authorizing charges within a stated maximum are P&LL, 1864, c. 84; Laws, 1874, c. 297, 1887, cc. 251, 434, and 512. Excluding scattered, miscellaneous formulas, the three main types of toll right statements totaled thus: Among 46 toll boom franchises, 28 (61 per cent) were for fixed sums, 7 (15 per cent) allowed "reasonable" tolls, and 9 (19 per cent) allowed charges "not exceeding" a stated maximum. Among 235 flow improvement franchises, 184 (78 per cent) were for fixed sums, 20 (8 per cent) allowed "reasonable" charges, and 13 (5 per cent) allowed charges "not exceeding" a stated maximum. No marked pattern emerged in the timing of the appearance of these types. The improvement and booming franchises tendered to any corporate improver who first took possession of the stream for the authorized purposes, under Revised Statutes, 1878, sect. 1777, and Laws, 1880, c. 279 (Wisconsin Statutes, 1898, sect. 1777e), empowered the grantees to charge "reasonable and uniform tolls" for the facilities made available.

82. Rates set under the statutory ceilings were involved in Underwood Lumber Co. v. Pelican Boom Co., 76 Wis. 76, 83, 45 N.W. 18, 20 (1890), and in Black River Improvement Co. v. Holway, 85 Wis. 344, 347, 348, 55 N.W. 418, 419 (1893).

83. See, respectively, P&LL, 1871, c. 357, 1864, c. 157, and 1870, c. 270. Laws, 1878, c. 291, charged according to the number of 24-hour periods in which the navigator had the help of the grantee's controlled waters. Both boom and flow improvement franchises occasionally varied charges according to the length of logs aided or handled, as in P&LL, 1856, c. 540, and 1857, c. 101 (booms), or in P&LL, 1857, c. 360, 1867, c. 586, and 1869, c. 290 (flow improvement). Analogous to the grants of authority to charge "reasonable" tolls were the provisions in eight franchises for what amounted to cooperative driving ventures, authorizing "equitable" apportionment of costs. See P&LL, 1854, c. 62, 1862, c. 35, 1864, cc. 126 and 131, 1865, c. 371, 1867, cc. 145 and 455, 1870, c. 52.

84. Delegation of rate-making initiative to the franchise holder, subject to official scrutiny or revision, was occasionally made explicit when franchises authorized such reasonable tolls as the grantees might fix, subject to change by the legislature. See Acts, 1851, c. 325, 1852, c. 324; Private & Local Acts, 1853, c. 30. That an authorization of tolls "not exceeding" a stated maximum validly conferred rate-making initiative upon the grantees was ruled implicitly in Black River Improvement Co. v. Holway, 85 Wis. 344, 55 N.W. 418 (1893), which held that the board of directors of the corporate grantee might delegate rate making to its executive committee. See also Wausau Boom Co. v. Plumer, 49 Wis. 115, 5 N.W. 26 (1880) (grantee's board of directors may effectively make rule for counting quantity of logs serviced).

85. See note 82 above, as well as Menominee River Boom Co. v. Augustus Spies Lumber & Cedar Co., 147 Wis. 559, 132 N.W. 1118 (1912); cf. note 49 above.

86. Menominee River Boom Co. v. Augustus Spies Lumber & Cedar Co., 147 Wis. 559, 132 N.W. 1118 (1912).

87. See Chapter III, notes 463, 464, 494, 495, and 501–504 (especially 502). For other examples of reductions of rates by amendments in course of enactment, see Laws, 1881, c. 77, compared with bill 62A (toll reduced from 10 to 5 cents), and Laws, 1901, cc. 262 (compare bill 603A) and 264 (compare bill 602A) (50 per cent reduction in rates stated in original bills).

88. Rate reductions may be seen in Laws, 1874, cc. 166 (amending P&LL, 1869, c. 76), and 294 (amending Laws, 1873, c. 12); Laws, 1877, cc. 257 (amending Laws, 1875, c. 169, which was, however, restored by Laws, 1878, c. 191), and 258 (amending Laws, 1873, c. 159); Laws, 1878, cc. 206 (amending P&LL, 1868, c. 430), 207 (amending P&LL, 1869, c. 452), and 236 (amending Laws, 1876, c. 298); Laws, 1880, c. 317, amending P&LL, 1870, c. 24, restored in qualified measure by Laws, 1881, c. 295; Laws, 1882, c. 154, amending Laws, 1877, c. 234. Rate increases may be seen in P&LL, 1867, c. 470, amending P&LL, 1864, c. 157; Laws, 1874, c. 69, amending Laws, 1873, c. 134; Laws, 1875, cc. 226 (amending P&LL, 1870, c. 480), and 296 (amending P&LL, 1867, c. 503); Laws, 1879, c. 165, and 1882, c. 156, amending P&LL, 1869, c. 186; Laws, 1881, cc. 188 (amending P&LL, 1867, c. 586), and 253 (amending Laws, 1879, c. 191); Laws, 1889, c. 179, amending Laws, 1879, c. 127; Laws, 1893, c. 125, amending Laws, 1891, c. 215. Occasionally amendments authorized increased rates in response to what appear increased services, or in connection at least with redefinitions of areas served. See Laws, 1881, c. 142, amending Laws, 1880, c. 144; Laws, 1883, cc. 9 (amending Laws, 1880, c. 303), and 132 (amending Laws, 1880, c. 214).

89. See notes 77–79 above. Patterning of rates according to details of services rendered may be seen at its most elaborate in P&LL, 1868, cc. 430 (Apple River Log Driving Co.), and 469, amending P&LL, 1857, c. 101 (Wolf River Boom Co.).

90. In one non-forest-area franchise, the legislature limited the tolls of the Rock River Improvement Company to such levels as would produce income calculated to keep the improvements in proper repair and pay a 7 per cent per annum dividend on the capital stock, the authorized total of which was set: see P&LL, 1866, c. 295. The idea of a varying rate according to total volume of business handled during the year was embodied in effect in P&LL, 1870, c. 270, which authorized rates ranging from six cents per thousand feet if less than 8 million feet passed the dam in a year, down to two cents if 16 million feet or more passed in a year. A rough attention to rate of return was also implicit in the occasional franchises which did not merely condition the toll right on some stated minimum capital investment, but varied the toll scale according to amount of investment. See Acts, 1850, c. 257 (toll "at the rate of one cent. per thousand feet of board measure, for each and every foot of elevation in any dam made by them"); P&LL, 1864, c. 84 (increase of stated number of cents per thousand feet authorized for every additional $5,000 of investment in improving navigation); Laws, 1878, c. 337 (stated rate if cost of improvements is at least $25,000; three fourths of stated rate authorized if improvements do not exceed $18,500; one half if do not exceed $12,500; no tolls authorized until at least $12,500 invested in improvements); Laws, 1880, c. 292 (Wisconsin River Improvement Co. may adjust its tolls to amounts invested in improvements). A crude consideration of rate of return was also implicit in authorizations of divisional tolls: see note 79 above. The legislative journals show one instance of attack leveled by disgruntled lumbermen specifically on the basis that an improvement company was being allowed to earn an unreasonably high rate of return on its investment, but no substantiating record was made by the complainants: see letter of John F. Stone & Co., seeking repeal of Laws, 1873, c. 12 (Yellow River improvement), AJ, 606, 607 (1874).

91. A typical statement authorized tolls "when [grantees] . . . shall have completed the said dam as aforesaid" (Laws, 1875, c. 70), or "whenever [grantees] . . . shall have so improved said creek . . . as to render the driving or floating of logs . . . down said creek . . . practicable and reasonably safe and certain . . ." (Laws, 1878, c. 163). Occasional franchises, though no less vague on criteria of application, were more pointed in emphasizing the consideration expected in return for the privileges conferred. Private & Local Acts, 1853, c. 30, declared that the Wisconsin River Improvement Co. should levy "no greater tolls . . . than are reasonable in consequence of such increased facilities of navigation" as the company was empowered to provide. Toll rights were withheld from the individual grantees in Laws, 1874, c. 228, and 1883, c. 317, "unless the expense of driving said logs or timber shall be reduced by the use of said dams."

92. See Chapter III, notes 105 and 158; cf. *id.*, notes 299 and 371.

93. P&LL, 1871, c. 45, incorporating and enfranchising the Wausau Boom Co., stipulated that the company must let the construction of all booms and piers to the lowest responsible bidder after publishing newspaper advertisements for bids. However, the franchise made no link between this provision and

the company's toll rights; without qualification by reference to the rate base, the company was authorized to fix tolls not exceeding 50 cents per thousand feet of logs serviced. Laws, 1887, c. 512, having authorized tolls not to exceed 5 cents per thousand feet on all logs driven with the aid of the authorized dam, qualified the grant by the stipulation that "the tolls hereby authorized . . . shall be no more than to reasonably compensate said parties for the amount invested and expended in building, maintaining and operating said dam"; this stipulation, of course, sought to limit tolls to a fair return on investment in fact made, but did not in terms limit the investment. The general inattention to the problem of rate base throughout the nineteenth-century grants of toll rights for stream improvements is highlighted by the more sophisticated treatment of the matter in a number of franchises at the end of the lumber era, reflecting rather the new concern for hydroelectric power development than for navigation uses. Under these grants the franchise holders might issue capital stock only with the prior approval of an administrative body, on the administrators' findings that the stock would be issued in consideration of money, labor, or property actually received at true money value equal to the par value of the stock; money received from capital stock should be used only for original acquisition, construction, or betterment costs of the improvements; tolls should be applied only to pay maintenance, operating costs, and a net return not over 6 per cent on net cash capital actually paid in. See Laws, 1907, c. 335, 1911, c. 640, 1913, c. 649, 1919, c. 441, 1939, c. 441.

94. See Chapter III, note 224. Authorized capital stock ceilings usually ran from $5,000 to $50,000 for stream improvement companies. See, e.g., P&LL, 1869, c. 76 ($50,000), 1870, c. 480 ($25,000), 1871, c. 483 ($5,000).

95. The charge was made — though not given substance by any tender of evidence — that a challenged improver was making inflated claims of investment, in the one instance in which a complaint on reasonableness of rates was spread on the legislative journals. See letter of John F. Stone & Co., seeking repeal of Laws, 1873, c. 12 (Yellow River improvement), AJ, 607 (1874).

96. See Underwood Lumber Co. v. Pelican Boom Co., 76 Wis. 76, 45 N.W. 18 (1890); Sauntry v. Laird, Norton Co., 100 Wis. 146, 75 N.W. 985 (1898); Menominee River Boom Co. v. Augustus Spies Lumber & Cedar Co., 147 Wis. 559, 132 N.W. 1118 (1912). Tewksbury v. Schulenberg, 48 Wis. 577, 4 N.W. 757 (1880), Wausau Boom Co. v. Plumer, 49 Wis. 115, 5 N.W. 26 (1880), another suit between the same parties, 49 Wis. 118, 5 N.W. 53 (1880), and Chandos v. Edwards, 86 Wis. 493, 56 N.W. 1098 (1893), turned on proof of facts pertinent to the earning of tolls. Delegated authority (in the corporate grantee's board of directors) to fix tolls was at stake in Black River Improvement Co. v. Holway, 85 Wis. 344, 55 N.W. 418 (1893). Enforcement of lien and garnishment claims growing out of tolls was involved, respectively, in Yellow River Improvement Co. v. Arnold, 41 Wis. 509 (1877), s.c., 46 Wis. 214, 49 N.W. 971 (1879), and Little Wolf River Improvement Co. v. Jackson, 66 Wis. 42, 27 N.W. 625 (1886). Such was the slender and spotty catalogue of Wisconsin Supreme Court cases concerning the criteria and enforcement of tolls.

The Underwood Lumber Co. opinion discussed the broadest spectrum of factors relevant to the reasonableness of rates, including (1) improved functional capacity of the stream, (2) range of operating difficulties presented by the stream, (3) range of operating difficulties presented by the customer's needs, and (4) the amount of capital invested in improvements (not limited strictly to in-stream works or labor on stream, but including necessary land investment, though excluding value attached to the natural utility of the flow). To the first of these elements the Sauntry and Spies Co. decisions added a sanction of discretion to allow charges based on stretch-of-stream improvements as a whole; charges need not be limited closely to match use of a particular area of the improved waters. The Spies Co. decision also added an element: (5) a fair uniformity was an ingredient of reasonable charges.

97. See Underwood Lumber Co. v. Pelican Boom Co., 76 Wis. 76, 84–85, 45 N.W. 18, 21 (1890).

98. Wausau Boom Co. v. Dunbar, 75 Wis. 133, 43 N.W. 739 (1889). Franchises sometimes included qualifying language to limit the operator's risks in the face of untoward circumstances, though this was done much less often than the interest of the grantees would seem to dictate. Thus P&LL, 1864, c. 157, stipulated, for the Eau Claire River Log-Driving Co., "that nothing in this act contained shall be construed to require or in any manner oblige said company to drive any logs . . . except when there is sufficient water in said river or its branches to run the same;

and if said company shall drive any logs . . . part way down said river or its branches, and shall be unable to drive the same to their destination on account of the insufficiency of water in said river and its branches, said company shall be entitled to be paid by the owner or owners thereof, *pro rata,* for the distance driven . . ." So Laws, 1879, c. 229, conditioned a toll right on provision of improvements "so as to provide substantial and beneficial conveniences and facilities for readily and rapidly driving logs and timber out of said creek over the portion of the same within the limits hereinbefore described, with reasonable certainty, having reasonable regard for the character of the creek and the stage of water therein at the proper seasons." See also Laws, 1880, c. 171, 1887, c. 386, 1889, c. 83. So by P&LL, 1868, c. 469, the Wolf River Boom Co. was obligated to "raft out each person's mark of logs and timber, so far as practicable, separately."

99. Typical requirements of sound works construction, defined by the works purposes, may be seen in P&LL, 1857, c. 70 (toll boom), 1864, c. 302 (toll flooding dam), 1870, c. 480 (stretch-of-stream toll improvement).

100. Compare Gov. Scofield's Veto Message of Apr. 26, 1897, disapproving a dam franchise provided by bill 448A, because it merely required grantees to provide facilities to allow free passage of logs or lumber "through said dam." The Governor objected because " 'free passage through said dam' might be allowed to the public as provided, and then the water be immediately shut off, and thus the passage of the lumber over the rapids below seriously impeded, if not actually prevented. In measures of this character the local conditions cannot be overlooked if the rights of the public to the use of the streams are to be protected" (AJ, 1280). This incident concerned a private (non-toll) franchise proposal, but the emphasis upon attention to the navigation requirements of particular local situations applied generally.

101. For other examples, see P&LL, 1868, c. 265 (with an additional specification that the slide be not less than 18 feet wide, and of sufficient depth to run rafts drawing 24 inches of water in an ordinary running stage); P&LL, 1870, c. 270; Laws, 1878, c. 291; Laws, 1887, c. 113 (grantees "shall, during the driving season of each year, keep a full head of water in said dam, and shall, upon demand, furnish the water accumulated therein, or so much thereof as may be necessary to afford a good driving stage of water in said river, to the owner or owners of all logs destined for points below; provided, that the person or persons demanding such water for driving purposes shall make such drive with as little delay as possible, using for that purpose as many competent men as can be profitably employed upon said drive"); Laws, 1891, c. 104.

102. The gamut of specifications for booming operations can be seen in the series of acts concerning the Wolf River Boom Co. See P&LL, 1857, c. 101, as amended, 1862, c. 43; P&LL, 1865, c. 190, 1866, c. 582, 1868, c. 469, 1870, c. 34. Other elaborate boom franchises include P&LL, 1866, c. 550, 1867, c. 328, 1868, c. 430, 1870, c. 299, 1871, c. 45; Laws, 1875, c. 48, 1880, cc. 209 and 319, 1889, c. 215, 1891, c. 215, 1897, c. 305, 1907, c. 417. Detailed service obligations laid on driving companies may be seen in Private & Local Acts, 1853, c. 30, as amended by Laws, 1878, c. 236, and 1880, c. 292; P&LL, 1859, c. 200, 1864, c. 302 (as amended, 1868, c. 366), 1868, c. 430, 1869, c. 452. The contrast in obligations laid on a grantee licensed simply to improve flow and one licensed to conduct driving is dramatized by comparison of two franchises issued for operations on the same stream. See P&LL, 1867, cc. 145 (Big Plover River Log Driving Co.), and 590 (Big Plover River Improvement Co.).

103. Franchises which elaborated the service duties of the grantees were likely also to spell out obligations of the customer. Thus the franchises cited in note 102 above provide ample illustrations of customer obligations. See also P&LL, 1863, cc. 35 and 193, 1864, c. 126, 1869, cc. 186 and 452 (and latter as amended by Laws, 1877, c. 124), 1871, c. 483; Laws, 1877, c. 234, 1882, c. 13. Of course, the toll franchises also imposed obligations on customers to pay the duly assessed charges, and created liens in aid of collection. Payment terms were often stringent, as when franchises stipulated that the franchise holder might insist on payment in full before use of his facilities (see, e.g., Laws, 1877, c. 124, 1885, c. 75, 1889, c. 49), or on partial payments at successive stages of use (Laws, 1878, c. 206, amending P&LL, 1868, c. 430), or on payment in full upon commencement of the drive (Laws, 1887, cc. 329 and 449), or within a short time after completion of service (P&LL, 1864, cc. 126 and 131, 1865, c. 371, 1866, c. 550, 1871, c. 45; Laws, 1889, c. 179, amending Laws, 1879, c. 127). Though liens were to a surprising extent defined in loose or ambiguous terms, short of promoting the improver's best interests, some were drawn

with broad and stringent force. See Chapter III, notes 161–164.

104. See Menominee River Boom Co. v. Augustus Spies Lumber & Cedar Co., 147 Wis. 559, 572–573, 132 N.W. 1118, 1123 (1912).

105. See also P&LL, 1856, cc. 481 and 504, 1857, c. 326, 1868, c. 366, 1869, c. 452 (and amendments by P&LL, 1870, c. 164, and Laws, 1877, c. 124), 1871, cc. 357 and 483; Laws, 1879, cc. 28 and 127, 1881, cc. 64 and 196, 1883, c. 289, 1887, cc. 113, 329, and 449, 1889, cc. 179 and 215 (and amendment by Laws, 1891, c. 478). Deserving of special note because it is the only instance of a general law which explicitly declared an obligation to serve, is Laws, 1882, c. 318, which added this declaration to Revised Statutes, 1878, sect. 1777, conferring improvement and driving monopolies on such improvement corporations as first took possession of their streams for such purposes. A special case of a declared obligation to serve was presented by several cooperative-styled driving associations, the charters of which provided — as in the pioneer grant to the Black River Lumber Driving and Booming Company, Private & Local Acts, 1854, c. 62 — that "any person who is an owner, or interested in lumbering on said river, or its tributaries, may become a member of this company by leaving with the clerk thereof a written request to that effect, which shall be kept on file and recorded . . ." See P&LL, 1862, c. 35, 1864, c. 126, 1867, c. 145. Contrast P&LL, 1864, cc. 131 and 157, 1867, c. 455, 1870, c. 52.

106. See Laws, 1880, c. 84, quoted earlier in the text of this subdivision. This type of provision was included in about 60 per cent of all toll dam or channel improvement franchises; those in which it was omitted were often the more elaborate franchises, which made up for the omission in other specifications. See, e.g., Laws, 1880, c. 205, 1883, c. 33, 1887, c. 218. A variant of this type of statement was that of Laws, 1878, c. 272, which said that the authorized works should "be operated so as to render the driving of logs down and out of said stream practicable and unimpeded." See, accord, Laws, 1880, cc. 77 and 184. The more common form of this operating obligation stated it both as an independent provision of the franchise and also as a condition of the toll right, as in Laws, 1880, c. 84, quoted in the text. Sometimes the performance of the described flooding service was stated only as a condition of the toll, as in Laws, 1879, cc. 53 and 229, and in Laws, 1880, c. 63. This simpler form would seem, however, to carry as much implication of legislative intention that the grantee be under obligation to render service on demand as did the more elaborate form.

107. P&LL, 1870, c. 381. Similar was P&LL, 1866, c. 237, changed to a private (non-toll) boom grant by P&LL, 1867, c. 378. For other examples of reference to transporters' "desire" or "requests," see P&LL, 1857, c. 57, 1866, c. 283 (with which compare an amendment by Laws, 1878, c. 244, of the type cited in note 108 below), 1871, c. 45; Laws, 1875, c. 333, 1878, c. 291, 1887, c. 85.

108. See Laws, 1878, c. 244, 1881, cc. 64 and 311, 1882, c. 278; cf. P&LL, 1870, c. 463. The most striking combination of a compulsory toll right with a navigation guaranty was in the general law which tendered toll improvement franchises to corporations improving Wisconsin streams. See Laws, 1876, c. 399, incorporated into Revised Statutes, 1878, sect. 1777, as a tender to corporations first taking possession for improvement.

109. P&LL, 1864, c. 126. Cf. P&LL, 1870, c. 34; Laws, 1878, c. 206, amending P&LL, 1868, c. 430.

110. P&LL, 1871, c. 45. These provisions were incorporated by reference into the general boom franchise tendered to corporations first taking possession of waters for such purposes, under Laws, 1880, c. 279, incorporated into Annotated Statutes, 1889, sect. 1777a, and Wisconsin Statutes, 1898, sect. 1777e; cf. P&LL, 1864, c. 157. Franchises which required payment of tolls in advance implied an obligation of the franchise holder to render service upon payment. See, e.g., Private & Local Acts, 1863, c. 35; Laws, 1877, c. 124, 1885, c. 75, 1889, c. 49.

111. P&LL, 1866, c. 550. Cf. P&LL, 1871, cc. 342 (boom company entitled to toll from all who use boom with grantee's consent), and 494 (company authorized to conduct drives "in the same manner as an individual," whereas regulated toll is provided as to use of improvements); Laws, 1880, c. 319 (nothing in this toll boom franchise to require grantees to handle logs other than those transported over North Wisconsin or Chicago, St. Paul & Minneapolis Railways).

112. P&LL, 1871, c. 483.

113. P&LL, 1871, c. 45, as amended by Laws, 1873, c. 256 ("The stockholders of said company shall have no preference over any other persons in the storage of their logs and timber in said boom").

114. See Cohn v. Wausau Boom Co., 47

Wis. 314, 324, 2 N.W. 546, 548 (1879). In Tewksbury v. Schulenberg, 41 Wis. 584, 593–594 (1877), the court interpreted a toll improvement franchise (of compulsory effect, since the improvements apparently occupied the whole navigable channel) as intended for "improving and increasing the capacity of the streams as public highways," and as such within the constitutional power of the legislature. The court's analysis did not speak of the grantee's obligation to serve the public, but the assumption that such an obligation was intended seems implicit in the argument. See also Yellow River Improvement Co. v. Wood County, 81 Wis. 554, 559, 51 N.W. 1004, 1006 (1892); cf. Fisher v. Horicon Iron Manufacturing Co., 10 Wis. 351, 353 (1860).

115. See, generally, Current, 113–114; Fries, 154–156; Reynolds, 69, 76–77, 79–81. Executive files show complaints, not backed by evidence, alleging unfair measurement of logs tendered for sale by independent loggers to large operators who controlled streams in the Chippewa and Black River valleys; indirectly these complaints reflected tensions resting partly on concentrated control of water transport. See Chapter V, notes 79–81.

116. The absence of provisions forbidding improvers to handle their own commodities, and indeed the characteristic situation of an improvement made primarily to handle the improvers' own logs, are highlighted in P&LL, 1859, c. 153 (incorporating the Half Moon Lake Canal Co.), and 1860, c. 91 (incorporating the Tyrone Lake Canal Co.), authorizing tolls for servicing nonmembers' logs, but making no stipulation for charges on services to members. The Half Moon Lake project was initiated by the Daniel Shaw Lumber Company, which incorporated the venture when it found it necessary to gather in the capital of other millmen of the area interested in providing themselves with storage facilities. See Reynolds, 80.

117. On the assignability of franchises, see Chapter III, notes 117, 128–132, 226, and 234–237. Broadly framed authority was given to specific improvement corporations to consolidate or merge with others to achieve the declared corporate purpose (limited, that is, to the given watershed or stream), in P&LL, 1859, c. 200 (Black River Improvement Co.); Laws, 1878, c. 244, amending P&LL, 1866, c. 283 (Menominee River Manufacturing Co.); Laws, 1880, c. 292, amending Private & Local Acts, 1853, c. 30 (Wisconsin River Improvement Co.). The absence of sharply felt pressure to weigh general public interest in such merger or consolidation powers is thrown into sharper relief by the care taken to protect stockholders in the Menominee River Manufacturing Co. act, cited above, which required affirmative votes of three fourths of the capital stock of each company involved. Cf. P&LL, 1861, c. 26 (Peshtigo Co. may consolidate with any company authorized by Michigan to improve boundary river). It should also be noted that improvement franchises granted to named individuals and their "assigns" allowed transfer to corporations, as assignees: Underwood Lumber Co. v. Pelican Boom Co., 76 Wis. 76, 82, 45 N.W. 18, 20 (1890); cf. Laws, 1883, c. 221 (general authorization of transfers of individual rights to corporations; not cited in the Underwood Lumber Co. decision). Again, there is no evidence of concern that the capacity to transfer individual franchises to corporate holders might raise issues of general public interest. The absence of any sharp-felt problem on this score during the lumber transport era is highlighted by the pronounced care taken toward such acquisitions of outstanding franchises upon the chartering of companies envisaged primarily as developers of hydroelectric power. See Laws, 1907, c. 335 (Wisconsin Valley Improvement Co.), 1911, c. 640 (Chippewa & Flambeau Improvement Co.), 1913, c. 649 (Wolf River Improvement Co.). Laws, 1877, c. 168, forbade any corporation organized under the general business incorporation statute (General Laws, 1872, c. 144) to hold stock in any corporation formed under that act; carried into Revised Statutes, 1878, sect. 1775, this ban was broadened to forbid any corporation organized under the general business incorporation act from holding stock in any other corporation; this remained the stated general policy of the state until Laws, 1891, c. 283, relaxed the ban, to permit such stockholding where it was approved by three fourths of the capital stock of both corporations. Meanwhile two general provisions of different tenor had been made regarding some stream improvement corporations. Laws, 1879, c. 82, authorized any Wisconsin corporation formed to improve any stream and authorized to drive logs therein or to maintain booms or dams thereon to buy and hold the stock or assets of, or consolidate with, any other Wisconsin corporation formed for like purposes and with like powers on the same waters. The limitation of such acquisitions to corporations operating on the same stream was the only regulation imposed by the

statute for what might be deemed a purpose of public utility regulation; it was significant of the focus of attention in the legislation that it included the more pointed requirement that the acquisition have the approval of three fourths of the capital stock of each corporation involved. The same three-fourths vote was likewise the conspicuous limitation imposed when Laws, 1881, c. 133, amended Revised Statutes, sect. 1775, to except from the general ban on holding companies any corporation previously or thereafter organized under any general or special law of Wisconsin to carry on a logging or lumbering or lumber manufacturing business, or improving any stream for log driving, or engaging in running or booming logs or other forest products. The 1881 act also sanctioned such stock acquisitions in or by any foreign corporation organized for such purposes. Unlike the 1879 act, the 1881 statute was not limited to corporations operating on the same waters. By virtue of the 1881 provision, however, the articles of a corporation organized for stream improvement purposes might not effectively confer on it the power to acquire stock in other corporations formed for other purposes; the statutory authorization to acquire stock in other stream improvement companies by implication excluded the right to deal in stocks of other types of enterprises. See Opinion of the Attorney General to the Secretary of State, Apr. 9, 1906, PD, 713 (1907).

118. Tourville v. Nemadji Boom Co., 70 Wis. 81, 35 N.W. 330 (1887); Wausau Boom Co. v. Dunbar, 75 Wis. 133, 43 N.W. 739 (1889).

119. See P&LL, 1856, c. 393 (amending P&LL, 1854, c. 62), 1857, c. 70, 1862, c. 43 (amending P&LL, 1857, c. 101, and dropped by P&LL, 1865, c. 190), 1871, cc. 357 and 494.

120. See Acts, 1852, c. 323; Private & Local Acts, 1853, c. 213; P&LL, 1855, c. 120; General Laws, 1860, c. 26; P&LL, 1870, c. 24; Laws, 1897, c. 305, 1907, c. 417.

121. See P&LL, 1862, c. 43, amending P&LL, 1857, c. 101; P&LL, 1867, c. 328; Laws, 1905, c. 290; cf. Laws, 1885, c. 316; Laws, 1891, c. 422, and substitute measure, Laws, 1893, c. 260.

122. See Chapter III, notes 414 and 417–421. The legislative record shows one inconclusive effort at legislative investigation of the reasonableness of rates and services in connection with driving and handling logs on the Chippewa River, and especially with reference to the boom at Beef Slough. Senate Resolution 13S, 1885, was passed without recorded incident, requesting the Senate Committee on Incorporations to make inquiry, supported by authority to send for persons and papers and examine witnesses under oath. There was no recorded aftermath of the resolution, however. See SJ, 401–402 (1885).

123. The quoted authorization is from sect. 17 of P&LL, 1869, c. 51, "An Act to incorporate the northern Wisconsin railway company." In its other aspects the charter illustrates the statements of the text. The fragmentary beginnings of general railroad regulations, superimposed on the broadly permissive terms of special charters with no adequate provision for enforcement, are noted in Hunt, *Law and Locomotives,* 82–84, 90, 95–96, and in Kuehnl, chap. xx.

124. Compare the observation in Freeman Hunt, *Lives of American Merchants* (1856), I, 570, that when Patrick Tracy Jackson sought from the Massachusetts legislature the first railroad charter ever there obtained, "the experiment [of building the projected road] was deemed to be so desirable, and at the same time, so hazardous, that the legislature were prepared to grant almost any terms that should be asked for." In Racine County Bank v. Ayers, 12 Wis. 512, 517 (1860), Paine, J., for the court, argued that public policy favored a liberal construction of the authority granted in railroad charters affecting the internal conduct of the business. So the court sustained the validity of stock subscriptions conditioned by stipulations as to the location of the depot in a particular town: "The charters of these companies impose such restrictions and requirements upon them as, in the opinion of the legislature, the public interest and public policy demand. And it having done so, it must be assumed that within these limits the company is left to accomplish the enterprise as best it may. And the mere fact that in some agreement it may appear that motives of private interest, may, to some extent, operate in influencing its action, ought not to be held to make such agreement void, as against public policy. It is vain to suppose that such enterprises can be accomplished without the operation of such motives. They constitute the mainspring of human action, and must inevitably operate to a greater or less extent, in the execution of all great enterprises of this character, and we think it may be safely assumed that so long as the company complies with the requirements of the charter, the struggle between conflicting pri-

vate and local interests will, from the necessity of the case, be so adjusted as best to advance the enterprise and accommodate the public generally." It should be noted that this liberal construction approach was here announced in relation to a route decision of relatively narrow scope, and on the assumption that the legislature had fixed effective stipulations as to termini. See Chapter III, notes 532, 544, and 545.

125. See, e.g., materials cited in Chapter III, notes 528, 529, and 558.

126. Wisconsin Constitution (1848), XI, 1. See Ryan, C. J., for the court, in Attorney General v. Chicago and North Western Railway Co., 35 Wis. 425, 569–577 (1874).

127. See, generally, the material cited in note 123 above. The police power limitation upon the contract clause of the federal Constitution was not explicitly invoked to decide a case in the United States Supreme Court until 1878, but it was well established in state decisions through the first half of the nineteenth century. See Benjamin F. Wright, Jr., *The Contract Clause of the Constitution* (1938), 196–197. Note, however, that in Attorney General v. Chicago and North Western Railway Co., 35 Wis. 425, 588–590 (1874), Ryan, C. J., for the court, did not believe that there was police power which might qualify an explicit charter grant of rate-fixing discretion, made by the territorial legislature before there was any reserved power to amend charters in Wisconsin.

128. See Gates, *Wisconsin Pine Lands,* chaps. viii and ix; cf. Current, 124, 135.

129. See Current, 131–143, 215, and Fries, 86.

130. See Majority Report, Select Committee on Bill 298A, AJ, 721, 723 (1864); Majority Report, Committee on Railroads (J. G. Thorp, chairman), SJ, 643–645 (1867); Report, Committee on Railroads, on proposals in the Governor's message and in resolution 9S for a board of railroad commissioners, SJ, 116 (1873); Report, Committee on Railroads, on bill 23A, AJ, 974 (1891). Cf. Majority Report, Committee on Finance, on bill 131S, SJ, 620, 621–622 (1864) (opposing tax increases on railroads, lest discourage extension of roads into unsettled areas).

131. See Fries, 94. Merk, 337, found the dominant northern Wisconsin sentiment opposed to regulation up to 1870; clearly the main impetus to the Granger legislation came from southern Wisconsin, and it seems unlikely that over so short a span of years as that from the late 1860's to 1874 the north had developed a very deep-seated demand for regulation. Cf. Raney, 247–248.

132. See Gov. LaFollette's Special Message of Apr. 29, 1903, AJ, App., 1430–1431, 1500–1501. Indexed references to bill 19S, 1889, in both the Assembly and Senate Journals show numerous lumber company petitions filed against this general railroad regulatory bill. Similar petitions appeared against bills 146A and 148A, 1895, to strengthen railroad regulation; see references in the index to Assembly Journal, 1895.

133. See Fries, 92; Lewis, 129–130, 234; Reynolds, 62–63. Some reflections of lumber-traffic service problems may be seen in correspondence between lumber shippers and the state Railroad Commissioners, in the latter's Annual Report, 1875, PD, II, 230–247; Biennial Report, Railroad Commissioner, December 1900, PD, II, 50.

134. Departures from published tariffs are noted — with approval, because they included rate concessions to help the lumbermen over periods of depressed markets — in Railroad Commissioner, Annual Report, 1881, PD, I, xxxiv. Rebates on log and lumber traffic were criticized in Railroad Commissioner, Biennial Report, 1906, PD, 20–22. Long- and short-haul rate disputes involving lumber traffic are reflected in Memorial 36S, 1875, of the Chicago & North Western and the Milwaukee & St. Paul Railroad companies, SJ, 110, 117, 129; Resolution 19S, 1876, and correspondence concerning a request for rate-distance data, between Railroad Commissioner and Milwaukee & St. Paul Railroad Co., SJ, 105, 269–273 (1876); Railroad Commissioner, Biennial Report, 1886, PD, I, ix–xiii. Cf. Minority Report, Committee on Railroads, on bill 733A (switching rates), AJ, 866–868 (1893).

135. Reynolds, 60, 61, discusses the significance of milling-in-transit rate privileges. The privilege was analyzed and validated on the basis of an Interstate Commerce Commission ruling, in Wisconsin Railroad Commission, Biennial Report, 1906, PD, 326, 333 (Formal Complaints, No. 90); cf. *id.,* 401 (Informal Complaints, No. 184).

136. See Fries, 87–89, 93–96; Reynolds, 13, 101–104, 106, 144, 145. The hopes of Wisconsin lumbermen for direct rail access to trans-Mississippi markets are mirrored, for example, in Report, Select Committee on Bill 106A (to confer the St. Croix aid-lands grant upon the Milwaukee & St. Paul Railroad Co.), AJ, 706 (1873).

137. 5 *Reports and Decisions of the Interstate Commerce Commission,* 285, 297 (1892); cf. Interstate Commerce Commission v. Cincinnati, New Orleans & Texas Pacific Railway Co., 167 U.S. 479 (1897); Fries, 96–98; Reynolds, 102–103.

138. Cf. Ryan, C. J., in Attorney General v. Chicago & North Western Railway Co., 35 Wis. 425, 530–531, 567–568, 580–583 (1874).

139. See Maxwell, 52–53, 75–76.

140. Biennial Report, Wisconsin Railroad Commission, Dec. 1, 1906, PD, II, 151, 159, Formal Complaints: No. 38 (Manitowoc Malting Co. v. Wisconsin Central Railway Co.), cited as basis for disposition of Formal Complaints: No. 36 (L.C. Whittet v. Chicago, Milwaukee & St. Paul Railway Co.,) *id.*, 150. This report contains numerous examples of lumber traffic matters presenting the types of issues noted in the text. See, e.g., among the Formal Complaints, *id.*, p. 45, No. 6 (Island Paper Co. v. Wisconsin Central Railway Co.), on alleged commodity discrimination; p. 88, No. 15 (Miner Bros. v. Chicago & North Western Railway Co.), on railroad's refusal to establish joint rates; p. 191, No. 45 (A.H. Krouskop v. Chicago, Milwaukee & St. Paul Railroad Co.), for alleged excessive rates.

Part IV: Political Economy

Introduction: Policy for a Region

1. U.S. Department of Commerce, *Census of Manufactures, 1900* (1902), II, 952, 953; Wisconsin Regional Planning Committee, *A Study of Wisconsin,* 229, 233, 234, 238; Edgar Z. Palmer, *The Prewar Industrial Pattern of Wisconsin* (1947), 107, 110; The Natural Resources Committee of State Agencies, *The Natural Resources of Wisconsin,* 88. See, generally, Francis F. Bowman, Jr., *Industrial Wisconsin* (1939).

2. See Chapter VII, note 40; cf. Haferbecker, 21–23. Compare Report, Special Committee on Obsolete Laws, SJ, 1711 (1927), recommending repeal of a special safety statute on wood-sawing machines, because the matter was now covered by the general safe place act, and because under the general statute "the industrial commission has for many years had orders prescribing what safety devices shall be used on wood-sawing machines, and this special law serves no other purpose than to confuse the owners of such machines as to the requirements they must observe." Those interested in logging (here principally the paper men, rather than producers of lumber) showed effective, sustained resistance to inclusion of their activities under the unemployment compensation system. But even here the pattern held. For this was a rear-guard action against an advancing front of policy which had momentum in no significant degree from the affairs of the loggers, but from developments in the general economy. See Chapter VII, notes 42 and 43.

3. On the decline in importance of the state's inland streams for transporting logs or lumber, see Report, Commissioners of Fisheries, for biennium ending Dec. 31, 1904, SJ, App., 15 (1905); Reports, Bureau of Labor and Industrial Statistics, for biennium ending June 30, 1904, *id.*, 303–304, and for biennium ending Sept. 30, 1906, PD, 397, 398, 406, and 409–411. Compare the reflection of the change in interest from transport to hydroelectric power uses of streams, in Gov. La-Follette's Special Message of Apr. 12, 1905, AJ, 1050–1054. Report, Special Committee on Obsolete Laws, SJ, 1710 (1927), recommended repeal of the lumber inspector statutes: "At one time this was an important law, but with the disappearance of the forests and with changes in the methods of transporting logs, the lumber inspectors have ceased to have any function." Successive twentieth-century revisions kept timber trespass laws on the books, but it is symptomatic of the changed emphasis of affairs that the only striking innovation was by laws concerning the peripheral problem of theft of Christmas trees, and the only case before the Wisconsin court concerned timber theft from state lands. See Wisconsin Statutes, 1959, sect. 134.60; State v. Biller, 262 Wis. 472, 55 N.W. (2d) 414 (1952); Chapter V, notes 12–15.

4. See Chapter VII, notes 23, 26, 27, and 40–43; Chapter IX, notes 139 and 140.

5. See Murphy, 98–105.

6. Attention to forest use as one factor in the most efficient development of the total economy had marked the Report of the Select Committee on Bill 191A, AJ, 283–286 (1867), and the report of the commissioners named accordingly, under General Laws, 1867, c. 36. See Lapham *et al., Report on the Disastrous Effects,* 9–11, 20, 21, 24. Analysis of policy in such broad terms proved, however, to be far beyond the sympathies or interests active in mid-nineteenth-century legislative process. No more was heard of this point of view in public policy making until the Report of the

Forestry Commission appointed under Laws, 1897, c. 229. That report emphasized the framework importance of forest resources for the general economy ("Important as forests are, considered as sources of revenue to the state, this consideration is greatly overshadowed in the case of Wisconsin, by their importance as the foundation of industries and the producers of certain necessaries of civilized life"), and argued that allocation of lands to growing trees, and the relative favor of public policy toward forest or farming use of acres in the north country, should be determined by a rational calculus of the pattern of uses that added up to a healthy region: see PD, II, 10, 18–19, 22 (1898). Despite foreshadowing of an approach in terms of regional planning, the Forestry Commission's focus was primarily on forestry. Thinking continued, however, to move toward programs directed at the region. See, e.g., Report, Interim Committee on Forestry and Public Lands (appointed under Laws, 1927, Joint Resolution No. 14), AJ, 855, 865, 866, 869, 885–886 (1929). The full development of this new orientation of policy may be seen by 1931. Thus in his Supplementary General Message of June 4, 1931, Gov. Philip LaFollette urged the need not simply for a forestry program, but for "a program for Northern Wisconsin." After sketching particular forestry measures, the Governor cautioned, "But this is not enough. We must restore the forest resources of the region. But to develop and reconstruct a healthy general well-being there we must undertake each separate policy, whether taxation, reforestation or roads, with our vision upon a comprehensive aim and plan . . ." (AJ, 1978, 1979). Of like tone was the Majority Report, Special Committee to Investigate the Conservation Commission, AJ, 2587–2603 (1931), especially at p. 2603: "While conservation in Wisconsin is not solely a problem of the north, the north occupies the center of the stage. While fire protection is the basic activity in any program, the problem of the north finally is one of land utilization. Fish and game propagation and preservation are as important as forests in producing tourist revenue. Game and furs provide both food and income for many thousands of residents of the state. Wisely administered forests will in the future, as in the past, supply the basic raw materials for much of Wisconsin's manufacturing. The woodworking industries must retain their prominent place in the production of wealth and in the employment of labor. All nonagricultural lands in the north should be devoted to these uses . . ." See, further, Communication of Wisconsin County Boards Assoc., SJ, 838–840 (1939); Report, Interim Committee on Land-Use and Reforestation (under Laws, 1939, Joint Resolution No. 69), AJ, 1795–1804 (1941), especially pp. 1796–1798.

7. See Vernon Carstensen, *Farms or Forests: Evolution of a State Land Policy for Northern Wisconsin, 1850–1932* (1958); Solberg, *New Laws for New Forests.*

Chapter X. Reassessments of Law's Role in the Economy of Northern Wisconsin

1. State ex rel. Owen v. Donald, 160 Wis. 21, 151 N.W. 331 (1915); Report, Interim Committee on Forestry and Public Lands, AJ, 849ff (1929). Since such frequent reference will be made to these two sources in this chapter, for simplicity they will be cited respectively as "Case" (with page references given first in the Wisconsin Reports and then in the Northwestern Reporter — as, for example, Case, 21/331) and "Report."

2. See Chapter XI, note 1, below.

3. See Report, State Forester, Dec. 1, 1908, PD, 26, 27–28, 102, 104–105 (1909).

4. Case, 52/340.

5. See, generally, Marshall, *Autobiography,* I, 5, 7, 260–305, 390–394, II, 84–88.

6. Case, 158/377.

7. *Id.,* 91–119/353–363, 154–156/376. See also the accounting and report of the referee appointed by the court to aid the execution of its mandate, 162 Wis. 609, 157 N.W. 794 (1916).

8. Case, 53–59/340–342. Compare State ex rel. Postel v. Marcus, 160 Wis. 354, 380, 393, 152 N.W. 419, 429, 433 (1915), in which the court confessed error in having announced too sweeping a formula concerning the same matters of procedure in constitutional amendment, and retreated to a narrower rule. The Postel opinion is relevant to appraising the atmosphere of the Forestry Case, since in Postel Marshall originally led the court in another excursion into sweeping analysis of constitutional doctrine, admittedly beyond what was required to dispose of the matter in court. Marshall's original opinion for the court justified the range of its analysis in terms which reflect distrust of the demands then being made for more positive action by government. "While dealing with the subject suggested might be avoided at this particular time, it does not seem best to do so; but rather to face the situation and

solve it. Much harm may come by uncertainty as to an important constitutional question being permitted to exist until affairs, public and private, shall have been adjusted to a condition apparently legitimately created by a legislative effort; — and that is most emphatically so when such condition rests on a purported but illegitimate change in the fundamental law . . . Perhaps there is no greater danger to worthwhile liberty and equality than that of undigested, undemonstrated by experience, ideas of well-meaning innovators and idealists, being, unwittingly, attached to our system of government with the power of inseparability which inheres in fundamental law . . ." (160 Wis. 354, 357, 363, 152 N.W. 419, 421, 423).

9. Case, 59–68/342–346. Since the payment in question was the last due under the three-year contract, and since admittedly funds were presently held which were available to meet the payment, a court more cautious of the proprieties of constitutional adjudication might well have ruled that all issues of the legality of the contract as creating a state "debt" had become moot, within the court's own definition of the scope of the constitutional ban on debt as a ban on commitments of funds not yet in hand.

10. *Id.*, 68–91/346–353.

11. *Id.*, 119–150/364–375.

12. See notes 62–76 below.

13. See Chapter XI, notes 20 and 24, below.

14. *Id.*, notes 18, 21, and 26.

15. See note 1 above.

16. See Mallare, 297–300.

17. *Id.*, 295–296. See also Quaife, *The Attainment of Statehood,* 417, 420, 449, 581, 582, 583–584.

18. 115 Wis. 32, 91 N.W. 115 (1902).

19. See 115 Wis. 32, 36, 38, 40, 91 N.W. 115, 116, 117 (1902).

20. The more common emphasis upon a general presumption of the constitutionality of legislation was early declared by the Wisconsin court as the norm. See Newcomb v. Smith, 2 Pinn. 131, 139 (Wis. 1849); Bushnell v. Beloit, 10 Wis. 195, 221 (1860).

21. There is evidence of the extremity to which the *Froelich* formula pushed the presumption against public action, in the sustained reaction of the legislature, which — despite the Froelich decision — continued to provide state funds to maintain the levees under direction of a state agency, by the device of appropriating general funds to "repay" sums found diverted in past from the drainage fund supposed to have been provided under the federal swampland grant, and applying these repaid drainage moneys to the purpose. See Lake, "Buried Wealth and Unwanted Water," chap. v. See, e.g., Laws, 1903, c. 419, 1905, c. 340.

22. Case, 75, 77, 81/348, 349, 350.

23. *Id.*, 90, 91/353.

24. *Id.*, 89/353.

25. *Id.*, 158–159, 160/377, 378.

26. Cf. Chapters II and III. See, generally, Hurst, *Law and the Conditions of Freedom,* 33, 34, 59. This readiness to use public power was, of course, consistent with the prevailing belief in the beneficence of releasing the creative energies of private will; the two elements were not at odds, because nineteenth-century policy sought to use the affirmative capacities of law to enlarge the opportunities for effective expression of private will (*id.*, 50–63).

27. SJ, Mar. 17, 1921, p. 482; AJ, Apr. 8, p. 786; AJ, May 11, 1925, p. 1130; SJ, June 5, p. 1117; *Wisconsin Blue Book,* 1925, p. 579. See, generally, Carstensen, *Farms or Forests,* 92.

28. That the 1924 amendment imposed no limit on legislative power "other than the expressed one of limiting the amount of tax moneys which might be devoted to forestry purposes annually," see Cutts v. Department of Public Welfare, 1 Wis. (2d) 408, 417, 84 N.W. (2d) 102, 107 (1957). Cf. State ex rel. Ekern v. Zimmerman, 187 Wis. 180, 206, 204 N.W. 803, 813 (1925).

29. Report, 852. The Report intimated that the rapid growth of the paper industry, using relatively fast-growing pulpwood, was probably the main reason that it was practicable and wise to rely on private initiative. "The Committee is impressed with the importance of state activity, but because of the conditions which prevail in our state, it never should be necessary for the state to engage as extensively in state-owned forests as is done in some other states" (*id.*, 851). After expressing hopes for the "real beginning" in good forestry management made by the paper industry, the Report observed that "the situation with respect to lumber companies is not so favorable, since it takes longer to produce saw logs than to grow pulpwood, especially when hardwoods are grown" (*id.*, 876, 877). In part, also, the committee indicated that its expressed preference for private action rested on the practical implications of extensive private ownership (presumably in contrast to states in which

the land area was still in large part public domain). "Since most of the forest land in Wisconsin is still in private ownership, the Committee feels that it is of vital importance to encourage private initiative in the rehabilitation of privately owned cut-over lands" (*id.*, 890).

30. *Id.*, 869, 878, 887. Prefacing its final, summary recommendations for state action, the committee emphasized a positive approach. "The people are keenly concerned over the future of the state's forests, and the state is now entering upon a new constructive phase of forest conservation. The situation, unless it is met immediately and squarely, threatens the welfare of the people of the state" (*id.*, 890).

31. *Id.*, 883, 891. The Report was not without an eye to the effect the increase of county forests might have on the market value of lands left in private hands. "The only real opportunity for bolstering land values lies in reducing the supply on the market . . ." (*id.*, 869). But this effect depended upon affirmative public action, including not only creation of county forests but state financial help to the counties to enable them to meet their burdens despite such formal withdrawals of land from tax rolls (*id.*, 873, 883, 891).

32. See, generally, Mills, "Public Purpose Doctrine," 40, 282.

33. Case, 120/364. Marshall's full statement of this aspect of the matter combined appeals to thrift (good argument should not go to waste, especially after the court had taken pains to arrange its presentation) and ambition (the judges should set future legislators on the straight course). "If the conclusions already arrived at [ruling invalid the 1910 constitutional amendment, and holding the reserve program to violate the internal improvements clause and the constitutional trust for education] were different we would have to face the further question next to be settled. Though the final outcome of the litigation, looking to the situation as it now appears, might render it unnecessary to even consider such question, since it is within the field covered by issues formed in response to the court's suggestion, and it appears now, as it did when such suggestion was made, to be of far-reaching public importance, and counsel for the litigants have submitted it for decision with helpful discussions and citations of authority, it does not seem best to pass it without at least pointing out its dangers and general limitations, even if a definite principle be not announced which will include the particular matter" (*id.*, 119–120/364).

34. *Id.*, 124/365.

35. *Id.*, 125/366.

36. *Id.*, 139/370–371; see also, *id.*, 141/371, 142/372.

37. *Id.*, 124/365.

38. *Id.*, 141/371.

39. *Id.*, 134–135/369.

40. *Id.*, 123–124/365. Elsewhere in the opinion Marshall conceded that it lay within the legislative power to make some expenditures upon what might prove long-term efforts in policing and conserving public timberlands against trespass or fire, and even in buying and consolidating tracts where this was incident to creating more valuable or efficient blocks of holdings. Yet these concessions, in context, simply emphasized the sharp restrictions which the opinion laid upon expenditures under the general police power. For Marshall conceded the propriety of such appropriations only so far as they went to build up the realizable value of lands held for the educational trust funds or of lands specifically granted the state by Congress or private donors for a reserve; i.e., such expenditures were proper only within the framework of specially provided, exceptional situations (*id.*, 143, 144/372, 373).

41. See note 35 above.

42. Case, 125, 126/366. Elsewhere the opinion put emphasis not only upon the length of time for which public money was here committed, but equally upon the size of the commitment. The analysis was fuzzy at this point; Marshall did not make entirely clear that the doctrine pronounced was that an expenditure might not be deemed for a public purpose simply because it was large; the want of clarity perhaps derived from an uneasy sense that to speak the meaning of the argument thus bluntly would reveal how far the judges were simply substituting their evaluation of the wisdom of given action for that of the legislature (*id.*, 132, 135, 138–139/368, 369, 370).

43. *Id.*, 150/374. Winslow, C. J., in his concurring opinion, observed that only "with considerable doubt" did he "yield assent" to this aspect of the court's decision; he felt that his "mind is not clear enough on this subject to justify me in disagreeing with the unanimous convictions of my brethren" (*id.*, 158/377).

44. *Ibid.*

45. Marshall, C. J., for the Court, in McCulloch v. Maryland, 4 Wheat. 316, 407 (U.S. 1819).

46. See, e.g., Chapter II, notes 55–59, above.

47. See note 39 above.

48. Bushnell v. Beloit, 10 Wis. 195, 225 (1860).

49. See Winslow, C. J., concurring, in Case, 159/378.

50. See Hurst, *Law and Social Process,* 7–8, 24, 26, 69, 84, 97–100, 170, 176, 183, 206, 245–246.

51. See Chapters I and II. Cf. Hurst, *Law and the Conditions of Freedom,* 35, 41, 53–55, 60, 67–69, 79.

52. See note 42 above.

53. Cf. Hurst, *Law and Social Process,* 26, 128, 233, 234, 238–240.

54. Case, 159/377.

55. Cf. State ex rel. Voight v. Hoeflinger, 31 Wis. 257, 263 (1872); Jensen v. Board of Supervisors of Polk County, 47 Wis. 298, 309–310, 312, 314, 2 N.W. 320, 328, 330, 331 (1879); Bryant v. Robbins, 70 Wis. 258, 262, 35 N.W. 545, 546 (1887); Town of Bell v. Bayfield County, 206 Wis. 297, 302, 239 N.W. 503, 505 (1931). When question was squarely made of the constitutionality of state payments to local governments to ease the burdens placed on their finances by a state program of fostering private timber growth through the Forest Crop Law, the court found no difficulty in sustaining the legislative action, emphasizing the sweep of legislative taxing and spending powers: State ex rel. Thomson v. Giessel, 265 Wis. 207, 60 N.W. (2d) 763 (1953).

56. In the only case presenting challenge to the constitutionality of fiscal aspects of the new legislative program for rehabilitating the north country, the issues were confined to doctrine under the state constitution. See State ex rel. Thomson v. Giessel, 265 Wis. 207, 60 N.W. (2d) 763 (1953) (state acreage payments to towns in token of lands entered under the forest crop law).

57. Report, 851, 890.

58. *Id.,* 852.

59. *Id.,* 876–877; see also note 29 above.

60. *Id.,* 883, 891.

61. *Id.,* 879.

62. See Marshall, *Autobiography,* I, 278, 285, 287, 291–294.

63. Case, 115/362.

64. *Id.,* 124/365.

65. *Id.,* 121/364.

66. *Id.,* 138–139/370.

67. *Id.,* 115–116/362.

68. See notes 22–25, 34, 35, 38–42, and 49 above.

69. Case, 115/362.

70. *Id.,* 116/362. Marshall reveals vividly the emotion which colored his appraisal of these years of policy transition, when he thus draws upon the metaphors of the French Revolution. The grotesque quality of this view of Wisconsin politics of the Progressive period only underlines the extent to which the Marshall opinion must be taken as a cry of feeling rather than as an intellectual exercise. Cf. Cardozo, J., dissenting, in Jones v. Securities & Exchange Commission, 298 U.S. 1, 33 (1936) ("Historians may find hyperbole in the sanguinary simile").

71. Case, 68, 70, 71, 89, 90, 115, 133/346, 347, 353, 362, 368.

72. *Id.,* 132/368; see also, *id.,* 146/373.

73. *Id.,* 115/362.

74. *Id.,* 122, 123/365. Marshall's words caricature the forest reserve purchase appropriations act, which set the frame and limits of policy with the definiteness usually required to satisfy constitutional tests of valid delegation of powers. See Laws, 1911, c. 639, fixing the annual appropriation, the total term of the purchase program (five years), the standard of expenditure ("only . . . to purchase forest reserve lands"), the general range of spending discretion committed to the administrators (unused funds of any one year "shall be available and may be used for such purpose in any subsequent year"), the contracting authority ("The state forester under the supervision of the state board of forestry is authorized to enter into contracts to purchase lands as additions to the forest reserve and to make payments on such lands from the forestry investment fund as moneys become available"), and a procedure for continuing legislative review by a legislative committee of five appointed each session ("The committee during the regular session for which they are appointed shall visit the forest reserve and report to the legislature during the same session as to the purchases of land made since the last regular session, and also prospective purchases, and so far as possible such legislative examination shall be so conducted that each legislative committee will supplement and complete the work of former legislative committees in examining both past and prospective purchases of forest reserve lands"). The forest reserve purchase statute on its face would seem clearly to satisfy the

tests of valid delegation of powers as these were declared by the court in State ex rel. Buell v. Frear, 146 Wis. 291, 131 N.W. 832 (1911). Cf. State ex rel. Wisconsin Inspection Bureau v. Whitman, 196 Wis. 472, 220 N. W. 929 (1928).

75. Laws, 1905, c. 264, sect. 2. See Pinchot, 143, 175–176.

76. Case, 63/344. The court's opinion betrayed a trace of embarrassment at its arguments, as these might be read in application to the efforts of State Forester E. M. Griffith. "Nothing in the foregoing should be taken as casting any reflection upon the incumbent of the office of state forester. Nothing which we have perceived in this case casts the slightest discredit on him personally. On the contrary, there are many evidences of most distinguished devotion to the legitimate duties he was employed to perform" (*id.*, 64–65/344).

77. Report, 852. The committee repeatedly emphasized the importance of continuity and sustained, positive effort in state forestry policy (*id.*, 851, 890).

78. *Id.*, 851, 852.

79. *Id.*, 866.

80. *Id.*, 887.

81. *Id.*, 869, 871.

82. *Id.*, 884; see also, *id.*, 883, 891.

83. *Id.*, 884, 887.

84. *Id.*, 892.

85. See notes 36, 66, 67, and 73 above.

Chapter XI. Regional Planning

1. Laws, 1897, c. 229, and Report, Forestry Commission, PD, II, Doc. No. 13 (1898); Laws, 1901, c. 458, 1903, c. 450, 1905, c. 264, 1907, cc. 96 and 491, and Joint Resolution No. 31; Laws, 1909, cc. 137 and 514, 1911, cc. 638 and 639, and Joint Resolution No. 47. Griffith's approach and the scope of his plans may best be seen in the lengthy recital of the history of the state forest reserve program given in his Report of Dec. 31, 1912, PD, 7–12, 17, 18, 20. See, generally, Carstensen, *Farms or Forests,* chap. iii.

2. See Carstensen, *Farms or Forests,* 37–43. The elements of controversy may be read between the lines in Communication from the State Forester in response to Resolution 18A, AJ, 1539–1542 (1911). The troubled progress of bill 556S, which became Laws, 1911, c. 639, may be seen in SJ, 711, 823, 890, 952–953, 1222, 1241; AJ, 1360, 1434–1435, 1570, 1606, 1610.

3. See Chapter II, notes 3, 33, 35, 48, 60, and 68.

4. See Carstensen, *Farms or Forests,* chap. iv; Clark, *Farming the Cutover,* 14–18.

5. Laws, 1913, c. 670. See, on bill 487A: AJ, 193, 666, 687, 724–725, 801, 968 (1913); SJ, 792, 869, 1038–1039, 1117–1118, 1164.

6. Wisconsin Special Legislative Committee on Forestry, of the Senate and Assembly: *Report,* Pursuant to Chapter 670 of the Laws of 913 (1915), 34, 43, 54–55.

7. Laws, 1895, c. 243, 1897, c. 327, 1899, c. 279, 1907, cc. 407 and 539, 1909, c. 444, 1911, Joint Resolution No. 2, 1913, cc. 556 and 772, 1915, cc. 413 and 604, 1919, c. 693, 1929, c. 479. Cf. Laws, 1911, c. 583 (state board of public affairs charged with similar promotional duties). County boards were authorized to assist local promotional organizations, in Laws, 1905, c. 458, 1907, c. 118, 1919, c. 695, 1927, c. 106.

8. Laws, 1895, c. 311.

9. See Laws, 1911, c. 478, 1917, c. 658, 1919, c. 116, and Joint Resolution No. 17; Laws, Special Session, 1920, c. 18; Laws, 1929, c. 98; cf. Laws, 1935, c. 221. Cf. Clark, *Farming the Cutover,* 12; W. H. Glover, *Farm and College* (1952), 281–285; Solberg, 25–27, 34–35.

10. Carstensen, *Farms or Forests,* 91; see also *id.*, 53, 61, 63, 64, 66.

11. Laws, 1903, c. 450, 1905, c. 184, 1911, c. 452, 1917, c. 454; Wisconsin Statutes, 1959, sect. 24.11.

12. Laws, 1911, c. 656, 1913, cc. 647 and 774, 1917, cc. 288, 503, and 536; Wisconsin Statutes, 1915, sect. 697.60, 1929, sect. 97.01, 1935, sect. 96.01 (by Laws, 1935, c. 550); Laws, 1943, c. 179; Wisconsin Statutes, 1953, sect. 59.92; Laws, 1955, c. 651. See also Laws, 1919, c. 596, creating Wisconsin Statutes, 1921, sect. 1458.10(1), later Wisconsin Statutes, 1925, sect. 97.16, repealed by Laws, 1927, c. 474 (creating a Wisconsin Land Settlement Board, to cooperate and contract with the United States in settling veterans on the land).

13. Laws, 1919, c. 600, carried into Wisconsin Statutes, 1921, sect. 1299g, and 1935, sect. 80.49, repealed, Laws, 1943, c. 334; Laws, 1921, c. 374, as amended, 1923, c. 349, carried into Wisconsin Statutes, 1923, sect. 1038 (48), and 1935, sect. 70.11 (30), repealed in revision by Laws, 1949, c. 63; Laws, 1923, c. 101, as amended, 1927, c. 164, and 1933, c. 423, creating Wisconsin Statutes, 1925, sect. 70.32 (3), repealed by Laws, 1955, c. 10, and restored in revised form

by Laws, 1955, c. 389; Laws, 1931, c. 427, amending Wisconsin Statute., 1929, sect. 70.32 (2).

14. Laws, 1913, c. 666, as amended, 1919, c. 629, creating Wisconsin Statutes, 1915, sect. 2024–100, and 1941, sect. 225.01, repealed by Laws, 1943, c. 399, authorized the organization and operation of private land mortgage associations, to make loans on the security of agricultural lands, wholly unimproved (1913) or limited to lands partially improved or under development by colonization companies (1919). The general incorporation laws provided the framework for land colonization companies, so far as concerned their organization. See Clark, *Cutover Problems,* 6–10, 13, and *Farming the Cutover,* 5, 9–11; Carstensen, *Farms or Forests,* 53–58, 91; Gates, *Wisconsin Pine Lands,* 239–241; Helgeson, "Promotion of Agricultural Settlement," chaps. v. and vi.

15. Laws, 1909, c. 109, as amended, 1911, c. 150, creating Wisconsin Statutes, 1915, sect. 1797–8, repealed by Laws, 1923, c. 205.

16. Laws, 1921, c. 466, carried into Wisconsin Statutes, 1923, sect. 2024–155, 1935, sect. 225.48, repealed by Laws, 1943, c. 399.

17. See Carstensen, *Farms or Forests,* 91, 94, 101, 102, and Solberg, 48.

18. See Report, Interim Committee on Administration and Taxation, AJ, 269–329 (1927).

19. See Report, Interim Committee on Forestry and Public Lands, AJ, 849, 866, 869, 885–886, 892 (1929).

20. Laws, 1925, Joint Resolution No. 61; Laws, 1927, Joint Resolution No. 13; Wisconsin Constitution, as amended, VIII, 1.

21. Laws, 1927, c. 454. See discussion of the forest crop law as a measure of tax reform rather than of tax relief, in State ex rel. Thomson v. Giessel, 265 Wis. 207, 213, 214, 60 N.W.(2d) 763, 766, 767 (1953).

22. Laws, 1929, cc. 343 and 405, 1931, cc. 39 and 455.

23. See Solberg, 430, 510.

24. Laws, 1921, Joint Resolution No. 29S; Laws, 1923, Joint Resolution No. 57; see Carstensen, *Farms or Forests,* 92.

25. Though the state did not re-embark on a program of creating its own forest reservations, the legislature consented within rather grudging limitations to acquisitions by the United States for establishing national forests in Wisconsin, and later enlarged its consent. The state's permission at mid-century allowed 2,000,000 acres to be put in national forests; up to that point the two national forests established totalled about 1,460,000 acres. Laws, 1925, c. 411, 1927, c. 461, 1929, c. 457, 1933, c. 272. See Natural Resources Committee of State Agencies, *The Natural Resources of Wisconsin,* 96. As of 1955, state forests totalled only about 277,000 acres (*ibid.*)

26. Laws, 1927, c. 57. See Report, Interim Committee on Forestry and Public Lands, AJ, 869, 883, 891 (1929).

27. See note 22 above.

28. See Natural Resources Committee of State Agencies, *The Natural Resources of Wisconsin,* 95; cf. Solberg, 214, 215.

29. See General Message, Jan. 10, 1929, AJ, 33.

30. When the county zoning measure passed, it was with the combined support especially of the populous northeastern counties, the rich farming area of south central Wisconsin, and the most depressed cutover counties; opposition appeared on the whole in a belt stretching mostly across central Wisconsin from east to west (see Carstensen, *Farms or Forests,* 106). The pattern suggests the presence of two different attitudes favorable to action: the desperation of the most hard-hit areas, and the awareness in the more prosperous areas of the drain which the submarginal north represented on the whole tax and productive base of the state. The opposition tended to center in sections which were in an intermediate position economically, and where, thus, inertia and distrust of innovation might most easily have sway. Even so, thin and scattered population, limited capital, and the prevailing small scale of operations remaining after the lumber boom might have allowed inertia and distrust of change to have their way in the cutover as in the intermediate territory, had it not been for the dynamic contributed by the broader planning horizons of the few large business firms in the north, and by the impetus given to the formulation of new elements in local opinion by the activities of the legislature's investigating committees and of the staff of the university and the Conservation Commission (see Carstensen, 94–96, 98–101, 103, 104, 107).

31. Laws, 1929, c. 356, and 1931, c. 236; Wisconsin Statutes, 1959, sect. 59.97. Detailed developments under this law are discussed in Solberg, chap. xii. Rulings in Jefferson County v. Timmel, 261 Wis. 39, 51 N.W. (2d) 518 (1952) indicated that the court was prepared to sustain the validity of the

county zoning law in its full sweep, and to give it liberal interpretation to further its objectives.

32. See Solberg, 266, 349–355, 377–378.

33. The new emphasis upon realizing a program primarily through dispersed, decentralized public and private effort within a frame of policy set by the state legislature may be seen in various aspects of the Report of the Interim Committee on Forestry and Public Lands, AJ, 851, 852, 859, 860, 878, 883, 886 (1929).

34. *Id.*, 855, 869, 871, 890.

35. See Chapter VI, notes 3 and 33–36, above.

36. Laws, 1895, c. 266.

37. See Chapter VI, notes 35 and 36, above; Solberg, 51–55, 60–63, 64.

38. Creation of a more effective fire prevention and protection system in the timbered country was recognized as "vital" to the success of any tax reform designed to foster forestry, in Report, Interim Committee on Administration and Taxation, AJ, 288 (1927). The committee was convinced that "until fires are rendered much less likely than at present, neither privately or publicly owned forests will be a success." The committee advised that not only state definition of policy, but state control of administration, was necessary to accomplish a worthwhile change. There must be a complete overhaul of the laws, "to substitute for the present unworkable system of *ex officio* town fire wardens a system of local wardens, paid for part of the year only and appointed and controlled by the conservation commission" (*ibid.*). So, too, a more efficient, state-controlled fire protection program was viewed as the necessary underpinning of county forests and county zoning, in Report, Interim Committee on Forestry and Public Lands, AJ, 869 (1929): "The key to forestry is to be found in adequate fire protection . . . It is in this field that the state must show its greatest leadership, and it must be an inspired leadership." So, also, in its formal findings the 1929 report concluded that "the key to the forestry problem lies in protecting all forest lands from fire. Without effective protection from fire, all other measures for promoting reforestation are doomed to failure" (*id.*, 890). The 1929 report was still stronger than the 1927 report in urging centralization. For a successful fire protection program, "and to provide a more determined and militant organization," the committee recommended "the immediate appointment" of a state fire warden, in whom should rest "the entire responsibility" for enforcing the fire laws, managing field forces, and conducting fire prevention campaigns. "Closer supervision and inspection, systematized control, trained personnel, a highly specialized force of district rangers and field wardens are absolutely essential to provide a fire control system which will approximate a perfect foundation for the structure now under way" (*id.*, 869, 871).

39. Laws, 1923, c. 211, as amended, 1925, c. 388; Wisconsin Statutes, 1925, sect. 26.125.

40. See note 24 above.

41. Laws, 1927, c. 29; Wisconsin Statutes, 1927, sect. 26.11. Compare Laws, 1927, c. 426, Wisconsin Statutes, 1927, sect. 23.09(7)(g), which — incident to revision of organization of the Conservation Commission — authorized the Commission "to establish and maintain an efficient fire fighting system for the protection of forests."

42. Laws, 1931, c. 67, Wisconsin Statutes, 1931, sect. 20.20. See Solberg, 72, 95, 96, 425.

43. See Solberg, 94, 100, 369, 425. Laws, 1953, c. 218, Wisconsin Statutes, 1953, sect. 26.12(1)(5), authorized the Conservation Commission to establish "extensive" as well as "intensive" forest protection districts, in which the Commission would take over the issue of burning permits and the general handling of fire hazards. After establishment of two such extensive districts, the twelve forest protection districts existing in 1956 covered 17,850,000 acres in northern, northwestern, and central Wisconsin. See Natural Resources Committee of State Agencies, *The Natural Resources of Wisconsin*, 93.

44. Readiness to make affirmative use of law to shape economic growth characterized both the general sweep of Wisconsin policy, aside from the lumber industry, and the types of resort to law in the country at large. See Hurst, *Law and the Conditions of Freedom*, 7–10, 32, 38, 51, 53–66, 82, 96, and *Law and Social Process*, 5, 116, 170–171.

45. See Report, Interim Committee on Forestry and Public Lands, AJ, 872–873, 878, 889 (1929).

46. See, e.g., Solberg, 225, 282, 376, 378, 379.

47. See Gov. LaFollette's supplementary General Message of June 4, 1931, AJ, 1978–1979.

48. See, e.g., Laws, 1939, Resolution 15A,

for an investigation of the Conservation Commission, and Majority and Minority Reports of Special Assembly Committee thereunder, AJ, 2727–2757 (1939); Report, Interim Committee (under Laws, 1939, Joint Resolution No. 69) on Land-Use and Reforestation, AJ, 1796, 1797–1798, 1802, 1803–1804 (1941); statement of Commissioner William J. P. Aberg in Senate Committee of the Whole hearing, SJ, 1375–1383, 1395 (1941); Special Message of Gov. Goodland, June 17, 1943, SJ, 1305–1307.

49. Cf. Hurst, *Law and Social Process,* 24, 26, 62, 84, 88, 110, 126–128, 144, 160–162, 210, 233, 237, 242.

50. For comment on the new directions of policy as criticisms of reliance simply upon a market calculus of income and costs, see the supplementary General Message of Gov. LaFollette, June 4, 1931, AJ, 1972, 1978, 1979, and Gov. Kohler's General Message of Jan. 11, 1951, AJ, 32–33; cf. Report, Interim Committee on Administration and Taxation, AJ, 287, 288–289 (1927). Compare, generally, Murphy, 120–121. The social accounting approach had, of course, found steady expression in the earlier state forest reserves program. See, e.g., Report, Forestry Commission, 1898, PD, II, 4, 10, 12, 22; Report, State Forester, Dec. 1, 1908, PD, 6, 8, 9, 16.

51. Cf. Hurst, *Law and Social Process,* 46, 138, 186, 239, 308, 327–329.

52. The key importance of added fiscal flexibility in the state government to the achievement of a broad economic program for northern Wisconsin was stressed in Report, Interim Committee on Administration and Taxation, AJ, 287, 288–289 (1927); and in Report, Interim Committee on Forestry and Public Lands, AJ, 859, 860, 863, 865, 873, 878, 883, 887, 891 (1929). On the nineteenth-century use of public timberlands as a substitute for money subsidies, see Chapter II, notes 49–59.

53. Cf. Hurst, *Law and the Conditions of Freedom,* 103, 106, and *Law and Social Process,* 136, 289–294, 326–328.

54. That mid-twentieth-century policy did not rely upon direct regulation of the timberland management and cutting practices of landowners, see statements by Commissioner James A. Corcoran, of the Conservation Commission, SJ, 1426, 1430 (1941), and by C. L. Harrington, State Superintendent of Forests and Parks, SJ, 1490, 1500. Indirect enforcement of good forestry practice through insistence on the stipulations made concerning entries under the forest crop law is noted in Solberg, 123, 165, 226–227, 234.

55. See Solberg, 316–318, 327, 333–334, 349–351.

56. Cf. Hurst, *Law and Social Process,* 26, 61, 97–98, 247.

57. The keystone position of state legislation in the policies represented by the forest crop law and the rural zoning law was recognized in Jefferson County v. Timmel, 261 Wis. 39, 59, 51 N.W. (2d) 518, 528 (1952), and State ex rel. Thomson v. Giessel, 265 Wis. 207, 213, 60 N.W. (2d) 763, 766 (1953).

58. For varied reflections of the new favor toward conservation values, see, e.g., resolution of American Paper and Pulp Assoc., transmitted to the governor of Wisconsin, SJ, 443–444 (1907); Report, Bureau of Labor and Industrial Statistics for biennium ending Sept. 30, 1906, PD, 360–361 (1907), and Annual Report, Wisconsin Dairymen's Assoc., of May 20, 1905, quoting a speech of former Gov. W. D. Hoard, PD, 21 (1907); Report, Commissioners of the Wisconsin Geological and Natural History Survey, PD, 8 (1908), and Report, State Forester, PD, 5, 6, 8, 9, 16, 17 (1908); Special Message of Gov. McGovern, SJ, 147–148 (1911); Address by the Chief Forester of the United States, to the Wisconsin Senate, at its invitation, SJ, 118–126 (1915); letter to the Governor from the Milwaukee Assoc. of Commerce, AJ, 299–301 (1921); speech by Gifford Pinchot before joint legislative session, AJ, 774–792 (1927); General Message of Gov. Walter J. Kohler, AJ, 41–42 (1929); General Message of Gov. Walter J. Kohler, Jr., AJ, 32–33 (1951).

59. See the summary of attitudes and concerns developed in the 1925 and 1926 hearings of the Interim Committee on Administration and Taxation, in Carstensen, *Farms or Forests,* 94–97.

60. The legislature's consciousness of its broader policy role is reflected, for example, in Report, Interim Committee on Forestry and Public Lands, AJ, 851, 857, 859–860, 862, 866, 869, 890 (1929); Majority Report, Special Committee to Investigate the Conservation Commission, AJ, 2595, 2601, 2602, 2603, (1931); Report, Interim Committee on Land-Use and Reforestation, AJ, 1797–1798, 1801–1802 (1941).

61. See the recognition paid to the creative contribution of past legislative investigations, in Report, Interim Committee on Land-Use and Reforestation, AJ, 1796, 1797 (1941).

62. Reflections of the new dependence upon

professional knowledge and skill of administrative personnel in contributing to the development of policy may be seen, for example, in Report, Forestry Commission, 1898, PD, II, 3, 4, 12–13, 15, 18–19; Reports, State Forester, PD, 5, 6, 16, 82, 100, 102–104 (1908), and PD, 8–9, 17, 18, 24, 53–57 (1912); statement of Commissioner William J. P. Aberg, SJ, 1377–1378 (1941). Compare the criticisms of the relative ineffectiveness and want of continuity in legislative investigation, matched against inquiries conducted by regular administrative agencies of the state, in Gov. McGovern's Veto Message of Aug. 6, 1913, SJ, 1286, disapproving bill 503S, providing for a legislative committee to investigate accident and health insurance; the governor cited as a disappointment the work of the interim committee on forestry of 1909. But see note 61 above. Compare the general pleas for greater continuity of policy research based on organized cooperation of the legislative and executive branches, in Gov. La-Follette's General Message, AJ, 28–29 (1931), and in Gov. Rennebohm's General Message, AJ, 28, 30–31 (1949).

63. Cf. Hurst, *Law and Social Process,* 40, 72, 154, 208–209; Kimball, 25, 47, 183, 187, 312; Murphy, 121–125, 128, 129, 140, 141.

64. See Carstensen, *Farms or Forests,* 118–122, 125, and also note 46 above.

65. Cf. Report, Interim Committee on Land-Use and Reforestation, AJ, 1796, 1801, 1802, 1803 (1941).

66. Cf. notes 35–37 above, and Chapter II, notes 35–48, and Chapter III, notes 17, 20, 44, 408, 439–445, 452–454, 460, and 486.

67. See notes 60–62 above.

68. See Report, Interim Committee on Forestry and Public Lands, AJ, 852, 859, 860, 861, 862 (1929), commenting on this history.

69. See notes 58, 59, and 61 above. Cf. Carstensen, *Farms or Forests,* 94–105, and 109–114.

70. See Report of Committee appointed under Resolution 64A, to Investigate Forest Fires in Northern Wisconsin, AJ, 1344, 1347 (1931), and Report, Special Committee to Investigate the Conservation Commission, AJ, 2588–2589, 2595, 2602, 2603 (1931); statement of Commissioner William J. P. Aberg, SJ, 1385 (1941).

71. See note 65 above, and Solberg, 272, 280–282, 378.

72. See notes 54 and 55 above. Note the attention given to central-local interplay in Report, Interim Committee on Land-Use and Reforestation, AJ, 1796, 1798, 1801, 1802, 1803, 1804 (1941).

Retrospect

1. Holmes, *The Common Law,* 1.

2. The quotations are, successively, from Holmes, "The Profession of the Law" (1886), in *The Occasional Speeches of Justice Oliver Wendell Holmes,* compiled by Mark DeWolfe Howe (Cambridge, Mass., 1962), 28, 29; "The Law" (1885), *id.,* 20, 21; "The Class of '61" (1911), *id.,* 160, 161.

INDEX